HANDBOOK OF ALUMINUM BONDING TECHNOLOGY AND DATA

J. DEAN MINFORD
Consultant
Hilton Head Island, South Carolina

Marcel Dekker, Inc. **New York • Basel • Hong Kong**

Library of Congress Cataloging-in-Publication Data

Minford, J. Dean
Handbook of aluminum bonding technology and data / J. Dean Minford.
p. cm.
Includes bibliographical references and index.
ISBN 0-8247-8817-6
1. Metal bonding--Handbooks, manuals, etc. 2. Aluminum--Handbooks, manuals, etc. 3. Joints (Engineering)--Handbooks, manuals, etc. 4. Adhesives--Handbooks, manuals, etc. I. Title.
TS718.M56 1993
673'.7225--dc20 93-3882
CIP

The publisher offers discounts on this book when ordered in bulk quantities. For more information, write to Special Sales/Professional Marketing at the address below.

This book is printed on acid-free paper.

MARCEL DEKKER, INC.
270 Madison Avenue, New York, New York 10016

Current printing (last digit):
10 9 8 7 6 5 4 3 2 1

PRINTED IN THE UNITED STATES OF AMERICA

Preface

Mankind observed many manifestations of adhesion in nature for literally thousands of years before the first surface of aluminum became available for bonding. Even cavemen must have been perplexed that different species of woods obviously had different degrees of adhesion between their component parts, a property manifested by the varying difficulties in bending and breaking them for purposes of making weapons. Further, they noted that some of the stones they hurled against their enemies either fragmented on impact or were so hard and durable they could be used repeatedly. Also, some stones appeared to be more homogeneous, such as the cooled igneous rocks that had flowed from the hot interior of the earth. Other stones obviously were formed from the joining of smaller stones through some inorganic binder that readily shattered under sharp impact. Primitive man also had opportunity to observe the adverse effects of water and weathering processes on adhering materials. Although he may not have reasoned that it was the degrading effect of water or ultraviolet on the interfacial jointures, he could see that all manner of materials that were unprotected from these exposure conditions somehow became weakened, unlike those sheltered from the same factors.

It was many thousands of years later that ideas about joining two pieces of materials that had never been together in nature began to evolve. Undoubtedly, this was accelerated by the observation that certain natural materials such as pitches and resins were sticky and could not be readily separated from a wide variety of surfaces to which they were applied. Application of such naturally tacky materials between the surfaces of materials that had no natural adherence for one another must have been one of the most important accomplishments of early mankind. From such humble beginnings, the development of adhesion science has risen to a high state of the art that would not have been envisioned even a century ago.

This present level of development has required the combined contributions of scientists and engineers working in many scientific disciplines. Their separate bits of data and information, especially in the last century, have been meticulously woven into a fabric of understanding of how the matter of our universe hangs together. Having defined the constitution and physical and chemical properties of the natural elements, it then became the job of modern adhesion scientists to synthetically create new materials from these building blocks that could serve for joining and manufacturing new products for our modern world.

In the ancient world, the most important structural materials for building were probably natural stones, baked clay (bricks), and wood. Individual pieces of these materials, in turn, could be modestly held in position and the joints sealed by applications of various naturally occurring tacky materials. Choice would be dependent on the availability in the site area. Iron, copper, and bronze were the first metals available to ancient peoples. They could not be structurally joined because of the lower strength limits of naturally occurring adhesives of the time. They could, however, be effectively sealed at jointures, indicating that satisfactory adhesion had been obtained for that purpose, but these joints could not be expected to support any appreciable structural loads. The development of synthetic adhesives capable of supporting structural loading in metal products has been a relatively recent development. It was really the pioneering research and development work of DeBruyne around 1939 in the field of high-performance synthetic wood adhesives that led to the discovery that hot curing adhesives could also adhere very well to metal oxide surfaces.

Availability of aluminum as a metal material of construction is also of relatively recent origin, as is discussed in Chapter 1 which also lists pertinent references. Early adhesive bonding of aluminum surfaces was more related to sealant functioning than to structural joining. This was a consequence of the good vapor-sealing properties afforded by aluminum even in very thin gauges. It was the need to replace wood and fabric as the structural materials for aircraft frames with lightweight aluminum sheet and extrusions in World War II that triggered and accelerated the use of adhesives and sealants for joining aluminum. As with most metals, the initial assembly procedures using aluminum depended primarily on mechanical and heat-melting procedures such as riveting and welding, with supporting assistance from various adhesive and sealant formulations. The development and extensive use of adhesives for fabricating structural aluminum honeycomb panels has led today to their use for primary bonding of aircraft structures.

The order of development for structurally bonding aluminum has been somewhat unusual. For most scientific developments, the order of development has usually proceeded from simpler to more complex requirements. In choosing the bonding of aircraft as the first major development for joining aluminum with adhesives, the immediate goal required the highest state of the art to be developed first. The adhesive strengths required to be supported by the adhesive bondline were immediately those levels approaching the inherent strength of the aluminum alloys themselves. These adhesive formulations incorporated properties to resist the highest tensile, shear, cleavage, and peeling loads demanded in a manufactured product. Finally, a structural joint was needed whose long-term service potential and overall resistance to all manner of exterior weathering conditions was the highest ever imagined.

An earlier volume *Adhesive Bonding of Aluminum Alloys* (published in 1985 by Marcel Dekker, Inc.) reviewed most of the highest state-of-the-art progress achieved in the Air Force–sponsored PABST program conducted at Douglas Aircraft. The primary movers in that development program were E. W. Thrall and R. W. Shannon, who served as the editors for the volume. Why an additional volume such as this handbook might be needed would then seem to be a pertinent question. The answer is quite simple if one asks how the specialized technical information in that book might be applied by the thousands of manufacturing engineers in general manufacturing who are restricted by economics from the high-priced processing procedures that are absolutely mandated in the construction of aircraft. Few, if any, of these manufacturers ever produce structures that require the rigid level of performance demanded in airplane construction or aerospace products.

For example, a PABST-type information book fails to consider or provide any information about the vast array of available adhesives and sealants that are the bread-and-butter lower cost and lower performance products necessary for most product manufacturing outside the aerospace industry. Thus, a handbook offering full discussion of the total range of structural and nonstructural adhesive products is still a highly desirable reference book for the general manufacturing engineer. A source is also required for citing the relative effectiveness of the lower-cost methods of treating aluminum for bonding, the lower cost commercially available adhesives, and as much information as possible on the levels of joint durability that have been established by long-term evaluations in a wide variety of different potential service conditions.

I began to develop and edit *Volume 7* of the *Treatise on Adhesion and Adhesives* series (published by Marcel Dekker, Inc.) in 1985. It was the plan to assemble in one volume separate chapters detailing the state of the art for bonding wood, elastomers, plastics, aluminum, and titanium. As I began writing the chapter on aluminum, based on 25 years of practical experience at the Alcoa Research Laboratory, I realized that the scope of information that needed to be included was too vast to consider in a book with other adherends. Thus, the volume incorporating the other adherends was published in 1991, with a separate volume on aluminum to follow. The present *Handbook of Aluminum Bonding Technology and Data* meanwhile continued to grow in scope and detail.

It became clear that detailed consideration of the bonding of aluminum for general manufacturing should include use of aluminum surfaces from mill finish to various mechanically or chemically deoxidized, conversion coated, anodized, or even primed and painted surfaces. The range of required adhesive products should include the entire spectrum of formulations available commercially. The kinds of joint designs should include all varieties. Methods of evaluating and testing should include all the procedures employed in all laboratories, including consideration of separately testing all types of structures. The theories required to explain joint strengths should include all the theories developed by a variety of branches of science and engineering. The factors in the environment that can lead to aluminum joint failures, although generally the same ones that affect bonded joints in general, need to be discussed in terms of how they might initiate the process of joint failure and continue to propagate failure to total debonding. In pursuit of presenting the full story covering the preceding aspects of aluminum bonding technology, it would seem pertinent to include a significant portion of the literature that constitutes the progress made in adhesion science in this century.

As a result of these considerations, I decided early on to expand the scope of this handbook to provide the opportunity for the reader to generally review the range and scope of investigation of the major contributors to adhesion science. It is hoped that this was achieved by fitting their various ideas and technical accomplishments into the general discussion of how other materials adhere to aluminum surfaces. In this regard, mention has been made of their contributions in the form of reference to over 4500 of their most important investigative papers and books distributed as appropriate to the development of this text. In addition, I have attempted to include the names of the contributors both in the text and cross-referenced in an authors index at the end. A subject index is also provided.

I wish to thank the Aluminum Company of America for the support of my studies on the technology for bonding aluminum over the period 1953–1983. I accept full responsibility for the interpretation of results and the conclusions expressed in this volume and presented as unpublished work of the Aluminum Company of America.

J. Dean Minford

Contents

HANDBOOK OF ALUMINUM BONDING TECHNOLOGY AND DATA

1

Introduction

Aluminum is rather astonishingly the most abundant of all metallic elements, estimated at 8.05% of the earth's crust, and a constituent of virtually all common rocks. *Alumen*, from the Latin, was a term mentioned by Pliny, although he had never seen the metal we recognize as aluminum. It remained for Sir Humphrey Davy to show a metallic base in this "alumina" in 1807. While not totally successful in chemically releasing the element from its oxide prison, he still named it *aluminum* to correspond to the name of its oxide, *alumina*. The French chemist Sainte-Claire Deville first produced marble-sized pieces of the metal in 1854 (1). Deville reasoned that it should be ideal for light but strong helmets and armor, and he persuaded the Emperor Napoleon III that he should have this wonder metal for his own military forces. However, Deville was unable to produce any quantity of the metal and Napoleon's forces were never able to employ any of the new material for war purposes. Readers with more interest in the international race to develop a viable process for extracting and obtaining the metal in volume will find fascinating the story of the simultaneous development of electrolytic processes for this purpose by the American Hall and the Frenchman Heroult in 1886. In this regard, the publications by Edwards et al. (2) in 1930, Edwards (3) in 1945, Edwards (4) in 1955, and Dunn and Sewell (5) in 1967 may be of interest.

The growth of the aluminum industry has always revolved about the unusual properties that favored the use of aluminum as compared to the already existing materials in the marketplace. Thus, it was the necessity to produce thousands of aluminum aircraft during World War II that first offered the impetus for a sevenfold increase in aluminum productivity during the early 1940s. But, the cessation of that conflict left the ability to produce quantities of aluminum far in excess of what a peacetime market could assimilate. New markets had to be developed and these applications had to be related to the excellent physical properties and outstanding resistance to weather which aluminum had previously demonstrated in its service during the war. A further significant boost in the potential use of aluminum in many new markets arose from the fact that a technology was beginning to be developed that demonstrated that aluminum could be effectively combined by adhesive bonding with many other materials in a marriage of their best qualities. A review of many of these application areas for aluminum, which came to fruition after World War II, can be further studied in the indicated references taken from a major compendium of information about aluminum assembled in the ASTM Series titled *Aluminum* (6–19).

Examples of adhesive bonding as a manufacturing process can actually be traced back as far as 3000 B.C. according to Dietz (20). Bonding was a method for joining individual components to form more complex structures when most mechanical and fusion methods of joining had yet to be devised. The late arrival of aluminum into the marketplace meant its particular history of joining has had to be a relatively recent development. The need for bonding aluminum aircraft was obviously a major accelerator of the technology. The need for better aircraft performance has made improvements in the state of the art mandatory. Nevertheless, riveting of aluminum has remained as the major procedure in aircraft fabrication, as demonstrated by the profusion of mechanical fasteners that are still seen on existing aircraft surfaces. The control and inspection of riveted structures has continued to offer support to the use of the metal fasteners but we are now encountering ever more serious problems in the sensitivity of steadily aging aircraft to fatigue failure. The possibility that bondments could be more resistant to such fatigue failures seemed to be confirmed in the middle 1970s by the completion of the U.S. Air Force research program titled Primary Adhesive Bonded Structure Technology (PABST). There has, therefore, been increasing use of adhesive bonding of both primary and secondary aircraft structures. A better understanding of the surface properties of aluminum and how those surfaces interact with adhesive polymers is most pertinent to the present and future developments in aerospace and general manufacturing technologies.

The volume of adhesive used in joining aluminum to itself and other materials has been even greater outside the aerospace industries, though these developments have generally received much less publicity and review in the literature. Probably the most important explanation for this has been the opportunity for standardization of procedures in building aircraft and the higher risk factor as concerns human life. No single entity or group of manufacturers outside the aerospace industry has had the collective financial resources to conduct the expensive and detailed search for the high state of technology that the U.S. military has conducted with the aid of the airplane manufacturers. While virtually none of these general manufacturing applications for bonding aluminum may be as demanding in terms of being fail-safe under service conditions as aircraft structures, they still all must serve satisfactorily to achieve the best performance in their specific-end purpose for the customer.

In theory, any solidifiable fluid that chemically wets the aluminum adherend surface, i.e., has a low contact angle, could be considered an adhesive. In this broad definition, welding wires and metallic solders (which can wet the aluminum in the molten state) could be considered practical metallic-type adhesives. In this handbook, we will devote our major attention only to the nonmetallic-type adhesive joining. The adhesives in this case are mainly organic in nature and generally are synthetic products of the modern organic chemical industry. There is a distinct need to bond aluminum to many other materials of construction in today's marketplace, which cannot tolerate the joining parameters necessary for welding, brazing, soldering, or sometimes even mechanical attachments.

I. BASIS OF STRENGTH DEVELOPMENT AND JOINT DURABILITY IN STRUCTURAL ALUMINUM JOINTS

Two mechanisms of adhesion are mainly responsible for developing the strength and durability potentials inherent in aluminum-bonded joints. They are (1) mechanical interlocking of the adhesive polymer into a macroscopically, or better a microscopically, rough surface oxide; and (2) various forms of chemical bonding between the adhesive and the aluminum oxide interface, as recently reviewed by Allen (21), Brockmann et al. (22), and Kozma and Olefjord (23). Many additional references will be made to this same subject as the need arises, and the subject will be the sole topic in the next chapter. The relative importance of these two mechanisms will depend greatly on the specific morphology and chemistry of the aluminum adherend surface which exists at the time of adhesive application and jointure with a second adherend surface and the rheology and chemistry of the particular adhesive selected. The actual joint strength measured by testing will also be strongly influenced by the details of the joint geometry and the mechanical strength properties of the adherend and adhesive. While the details about the effect of joint design and stress distribution within joints will be considered in detail in Chapters 6 and 7, the publications of Adams (24), Hart-Smith (25), Kinloch (26), and Wake (27) can be recommended at this time for summary descriptions about the stress distribution within bonded joint assemblies.

The observed joint durability that may be achieved with a particular surface pretreatment and adhesive selection will be highly affected by the weathering conditions and the level of stressing in the service for which the assembled product was designed to function. Finally, judgment about how well the joints will be expected to perform in service will be determined by prior testing. The choice of testing procedure and the particular manner of conducting the testing will both be highly important in determining how accurate a prediction is attained.

When the typical aluminum bonded joint is exposed to ambient laboratory weathering conditions, there is generally little decline in joint strength noted even over long periods of time. However, when hot, humid conditions exist in the testing or actual service environment, then the disruption of chemical bonds between the aluminum oxide and the adhesive may be expected, as described by Kinloch et al. (28). If the "mill finish" or "as received" aluminum oxide surface is converted by deoxidizing (29), or such an operation is followed by building up a new anodized layer of oxide (30), then the physically interlocking nature of the adhesive can become a more dominant factor, and high bond strengths and greater joint durability may be found, as discussed by Venables et al. (31), Bijlmer (32), Venables (33,34), and Packham (35). Joint failures still occur, but only after significant degradation of aluminum oxide structure or crack propagation through bulk adhesive or near the oxide/adhesive interface. Venables et al. (36) and Kinloch (28) have proposed that crack propagation during exposure to high humidity occurs as an oxide form like boehmite (for example, formed by chromic/sulfuric acid deoxidation [FPL treatment] is converted to hydroxide, which has poor adherence to the underlying aluminum. Davis et al. (37) and Ahearn and Davis (38) have stated that the original oxide may be converted to boehmite by various pretreating conditions but under high humidity conditions will undergo conversion to structurally weak bayerite oxide. Minford (39) has shown that very durable aluminum bonded joints can be formed when conditions for generating bohemite films of various thicknesses are present and immediately bonded. In addition, similarly fabricated joints were able to offer high resistance to corrosive saltwater weathering and even a combination of stressing and hot humidity soaking.

II. ADHESIVE BONDING AND THE ALUMINUM AEROSPACE INDUSTRY

Powered airflight has been credited with starting with the Wright Brothers' flight at Kitty Hawk, North Carolina, in the first decade of the twentieth century. By 1910, the English channel had been flown and interest by military authorities became serious. The year 1917 saw Zeppelin flights bombing London pioneered by the use of new German high-strength aluminium alloy extrusions. The use of wood for airframe construction early necessitated the use of adhesive bonding, which continued until the featuring of a welded steel tube fuselage in the Fokker tri-motor passenger planes. Both the Fokker plane and another successful passenger plane (the Lockheed Vega), however, still featured an all-wood cantilever wing. The water sensitivity of the casein adhesive that was employed eventually called for a change that was precipitated by a dramatic 1931 crash of a Ford tri-motor passenger plane. Within a few years, the aluminium-skinned, all-metal airplane dominated the commercial plane market. However, one of the most successful World War II aircraft in 1941 (the Mosquito) had a highly efficient birch ply skin/balsa wood core glued with a new acid-hardened urea-formaldehyde–type adhesive. Their brittle nature, however, was not sympathetic to the stresses in the cross-grain direction of the wood. High-strength metal adhesives were obviously needed and the REDUX (vinyl formal/phenolic) system was the first successful candidate developed in 1942 by DeBruyne (40).

Certain properties of aluminum such as lightness, high strength/weight ratio, general good resistance to corrosion, and attractive appearance in a variety of different forms prompted consideration of aluminum alloys for aircraft. Anthony Fokker first employed all-glued aircraft wings in Germany in the 1930s. The first truly composite wood/aluminum wing structure (REDUX bonding of alloy spar booms to wood) was found in the DH Hornet aircraft produced by Hawker-Siddley Aviation

Group in 1944. A review appears in a 1946 publication by Golf (41). The year 1945 saw metal-to-metal (stringer/skin) bonding with REDUX in the DH Dove aircraft.

In 1950, the bonded sandwich construction of aluminum skins to balsa wood cores appeared in the Chance Vought Cutlass naval fighter. This American construction used the new elastomer-modified phenolic-type adhesive. The extent of metal bonding was extended to the double-curature areas of the fuselage in the 1951 jet-powered Comet, as described by Povey (42). By 1955, the European manufacturers like Fokker had fully embraced aluminum bonding with REDUX, ending with the production of one of the most popular aircraft of all time; i.e., the Fokker F27 Friendship aircraft. In 1956, the first all-aluminum–bonded sandwich panel (aluminum skins and aluminum honeycomb core) was developed in America by Martin and Hexcel. The first test was in the Martin jet-powered flying boat (Seamaster). Convair first used a nitrile-phenolic bonded integral fuel tank wing structure in both military fighters and commercial aircraft in 1958. The higher temperature-resistant epoxy-phenolic adhesive bonded aluminum honeycomb sandwich was introduced in the Convair B-58 supersonic bomber in 1960. Concentration in 1963 by Boeing in the 727 airliner was on the use of cold-setting epoxy for pressure cabin structure using refrigerated adhesive. Variously modified epoxies were used in 1965 for extensive bonding of sandwiches for wings and fuselage. Fokker focused on improving the existing techniques for manufacturing a rivet-bonded structure in 1969 using magnetic clamping with cold-set epoxy adhesive.

The use of Alclad aluminum sheet for bonding was questioned in the early 1970s, leading to a recommendation from some manufacturers that clad sheet should not be used. Modified epoxy film adhesive was used in the Lockheed Tri-Star for bonding titanium-to-aluminum and bonding titanium crack-stopping reinforcing straps in pressure cabins in 1972. The use of weldbonding was reported in Russian aircraft in 1974. The U.S. Air Force–sponsored PABST program was initiated in 1975, leading to the recommendation that adhesives could effectively bond primary aircraft structures. Finally, in 1976, the preferred method for pretreating aluminum in American manufacturing became *phosphoric acid anodizing* (PAA).

A considerable amount of testing had to be conducted by some aeronautical laboratories with various adhesive-bonded aluminum joints and simple joined structures before the all-aluminum aircraft could become a reality. For example, it had to be shown that two bonded sheets of aluminum could be expected to behave as a solid piece of their combined thickness, as shown by Hartmann and Rondell (43) and Povey (44). Ljungstrom (45), at SAAB, contributed extensive data on bonded compression members in aluminum showing weight saving of up to 25% could be anticipated in compression-loaded bonded structures as compared to the previous riveted designs. In shear-loaded bonded structures as well, the weight savings could still be 10–15% according to this early work. Ljungstrom (45) was also instrumental in demonstrating the use of bonding in attaching reinforcements to permit a flush joint in thin skins or where fatigue considerations were critical. Schliekelmann and Cools (46) showed that bonded joints of the same overlap as a riveted one could show as much as twice the overall fatigue strength. Also of great significance was the work described by Hardrath (47) of fatigue crack propagation studies comparing riveted, bonded, and integrally machine stiffener assembled aluminum box beam constructions. Van Beek (48) described the use of many of these concepts as employed in the manufacture of the very commercially successful Netherlands Fokker F-27 "Friendship" aircraft. These aircraft have demonstrated their worthiness in long, arduous, and highly successful service in all parts of the world, often with only the minimum of maintenance facilities available. The records show that out of countless thousands of hours flown by these aircraft under the whole spectrum of possible operating and environmental conditions, structural failure could literally be counted on the fingers of one hand. Joint failures of a less critical nature were, of course, encountered but could be attributed to other failure mechanisms such as corrosion of the metal, chemical degradation of the adhesive polymer, fatigue of some material under excessive stressing, or various contaminations such as from hydraulic or de-icing fluids. The basis for this excellent performance was attributed by Fokker to careful selection of both adherend and adhesive materials, good control of tooling, and strict control of the overall bonding process through continual research and development programs and extensive structural testing. For a more detailed description of the early history and development of bonded aluminum aircraft, the author recommends *Bonded Aircraft Structures*—a collection of papers given in a 1957 Conference in Cambridge, England (49);

Symposium on Adhesives for Structural Applications—a compilation of papers delivered at Picatinny Arsenal in 1961 (50); Adhesive Bonded Metal Structures by Schliekelmann in the 1967 book on *Adhesion and Adhesives* by Houwink and Salomon (51), "Operational Experience with Adhesive Bonded Structures" delivered as part of AGARD Lecture Series No. 102 (52), "Aerospace Applications" a chapter by Albericci (53) and the introductory chapter ;by Schliekelmann in the 1985 volume *Adhesive Bonding of Aluminum Alloys* arising from the PABST technology (54).

Initially, applications of adhesive-bonded joints were limited to the solution of specific problems such as the reinforcement of basically wooden wings by means of adhesive-bonded aluminum spar caps or doubler plates. In the 1950s, however, the use of adhesive-bonded joints in aircraft structures increased considerably and the growth has been accelerating ever since. The F-27 Fokker "Friendship" aircraft mentioned above is an outstanding example showing a total of more than 4000 individual parts composed into 550 assemblies by adhesive bonding. Application of adhesive bonding seems to be moving along toward developing two kinds of bonded structures. First, there are the metal-to-metal bonded structures, which are locally reinforced by bonded doubler plates. Also in this category are components composed of bonded multiple laminations where each layer progressively increases in cross-sectional area, such as stringer and spar caps (55). The second category of adhesive-bonded structure is that which contains bonded joints between rather thin aluminum skin sheets and various low-density core materials, i.e., sandwich panel construction.

Initially, core materials were either edge grain balsa wood or plastic foams; later honeycomb was made from aluminum foil and today it is often made from an organic fiber material like duPont Nomex. Combination structures later consisted of sandwich panels to which doubler plate reinforcements and profile-type edge members were bonded. Rondell (56) showed an early 1952 comparative fatigue test result using bonded stiffened panels. Bijlmer (57) evaluated fracture toughness of multiple-layer bonded aluminum sheet in 1978. Vinyl (REDUX)- and elastomer-modified structural adhesives were employed in these early panels. It may be surprising to the reader that modern-day modifications of REDUX are still widely employed in aircraft joining, particularly in Europe. Another European developmental milestone in the use of bonding was the 1946 development of epoxy resins (58) at Swiss Ciba A.G. Aluminum-faced sandwich panels with aluminum foil honeycomb cores were reviewed by Plantema in his 1966 publication (59). In the succeeding decades, the further pursuit of data demonstrating the excellent fatigue properties of adhesive-bonded structures for aircraft manufacturing has produced military specifications for aircraft structural integrity such as MIL-STD-1530 (60) and for "damage-tolerant" structures as set forth in MIL-A-83444 (61).

The combination of adhesive bonding with other methods of joining aluminum such as welding, riveting, and mechanical joining (clinching) has been studied most recently with the possibility of manufacturing even stronger and more durable structures. These joints have been given the names weldbonding, rivbonding, and clinchbonding, respectively. The first integral airplane fuel wing tanks in 1958, for example, depended on a film-type adhesive plus riveting for assembly and effective sealing. By 1963, rivbonding was being widely used to increase the fatigue-like resistance of any structure exposed to sustained or intermittent stressing. The year 1969 brought extensive evaluation of aluminum weldbonds by Lockheed-Georgia under Air Force contract. A 1967 Russian publication titled *Glue-Weld Joints in Machine Building* was apparently responsible for triggering this interest, as weldbonding was already being employed in certain Russian manufactured products. The presence of various American investigations of aluminum weldbonding was first indicated in the journals *Iron Age* (62) in 1969 and *Material Engineering* (63). A number of quarterly progress reports were issued by Lockheed-Georgia beginning in 1971, of which the First and Sixth are cited in the references as examples (64,65). Mahon et al. (66) issued a 1973 report which was indicative of the type of investigations being undertaken at Grumman. The possibilities of using weldbonds in manufacturing trucks at White was reviewed in the journal *Metal Progress* in 1974 (67). The author issued the first of several extensive investigations at Alcoa Labs on aluminum bonding of automotive alloy sheet in 1974 (68). The same year, Hall (69) reported environmental testing of aluminum weldbonded joints. The year 1975 produced the first of a series of reports on U.S. Air Force–sponsored work on aluminum weldbonding at Northrup by Bowen and coworkers (70,71). Also, in 1975, Minford et al. (72) reported on continuing work at Alcoa Labs on the weldbonding of various automotive body sheet. The subject of how much lube contamination of aluminum surfaces could be tolerated by spot-welding and adhesive-bonding processes was reported by Minford (73) in 1981, and in the same year, the author (74) reported on the effectiveness of weldbonding achieved in the presence of varying amounts of lubricant contamination on aluminum auto body sheet surfaces.

The 1970s brought an acceleration of technological progress in structural bonding of aluminum aerospace alloys culminating in the most extensive single U.S. Air Force investigation ever carried out. As mentioned previously, this program, Primary Adhesively Bonded Structures Technology (PABST), was first reported by Douglas Aircraft in 1975 (75). Many additional reports followed which will be considered later in this book under appropriate topic headings. Some of the more prominant reports dealing with the subject can be reviewed in papers by Thrall (76) and Shannon and Thrall (77) and a review of final results offered by Thrall (78) in 1979 in the journal *Adhesive Age*. Thrall (79) discussed the overall results in the same year as they might indicate the future of metal bonding in both the automobile and aircraft industries. Finally, Thrall (80) considered the improvement in performance offered by bonding by discussing "Failures in Adhesively Bonded Structures" in a 1979 AGARD Lecture. The data developed by the PABST program offered the conclusion that aerospace primary bonded structure could indeed be manufactured with the highest confidence in their long-term service durability potential.

III. ADHESIVE BONDING DEVELOPMENTS OUTSIDE THE AEROSPACE INDUSTRY

Although mentioned first in this book, the preponderance of bonding both in terms of adhesive and aluminum metal consumed is in general manufacturing activities outside the aerospace industry. Discussions of how to bond aerospace structures have usually taken precedence in most of the publications and technical papers discussing adhesion science for several reasons. First, the problems can be much more specifically defined in terms of what aluminum alloy materials are needed, and the kinds of adhesives that can be considered are much more limited in consideration of the much higher structural strength requirements that are required. Second, more research dollars have generally been available for specific goal research and development programs, since such programs are financially supported by the military. Third, the probability that data obtained will be published is much more likely, since the funded programs fall in the area of public information.

On the other hand, the bonding programs conducted outside the aerospace industry cover all other conceivable areas of manufacturing not included in aerospace. These programs are mainly designed to answer only the individual manufacturer's concern about his or her own product line. Thus, they may deal with virtually any kind of material that can be present in all possible marketable products. While the questions that need to be answered are certainly related closely to adhesion science in a general sense, they are rarely theoretical in nature to the extent achieved by the academicians and aerospace scientists. Rather, they are quite practical in nature and involve a lot of hit or miss attempts to get something produced as fast and as economical as possible. This latter factor is not a matter of primary concern when aircraft or spacecraft are being produced. Usually, if the ordinary manufacturer does find a best solution to special bonding problems, it is rare that he or she will go to any significant lengths to publish and proclaim his or her results to the rest of the world.

There is another group of investigators, not mentioned above, who do make a considerable contribution to offering answers as to how to bond in general manufacturing. I refer here to the adhesive formulators and the engineering and equipment personnel who design the equipment to disperse and cure the adhesive in bondlines. It is generally the adhesive data sheets furnished by an adhesive supplier that ends up being the main basis for adhesive selection, application, curing, and testing.

Finally, there is the group of materials suppliers, such as Alcoa, who are seeking to expand the range of application of their product, such as aluminum. Starting in the late 1950s, while working at Alcoa Labs, the author initiated a series of investigations that continued for 25 years in an attempt to establish the relative ability of different aluminum alloys to be bonded using the whole span of adhesive families available in the marketplace. Of primary importance was the relationship between the degree of surface pretreating and the level of long-term joint durability that could be achieved. In 1967, Minford wrote an Alcoa book titled *Adhesive Bonding of Alcoa Aluminum* (81). In the same year, Minford (82) published a first paper on adhesive bonding aluminum. Also in 1967, Minford et al. (83) published a paper on the proper selection of an adhesive for this purpose. Again, in 1967, Minford (84) composed a chapter on bonding aluminum for the ASM *Aluminum* series of books. In 1972, Minford (85) produced the first of many papers dealing with the shortened service life encountered when stress is imposed simultaneously with hot humid weathering conditions. This was based on a 1967 publication by Sharpe (86).

The amount of aluminum foil and light-gauge sheet bonded to other materials in the form of laminates, packaging products, and building panels is quite astronomical. In almost all the final products and structures made with such bonded components, the aluminum is serving as a vapor barrier. In addition, the many decorative faces of aluminum have made it attractive as the outside surface of many products. For most of these applications, low unit stressing is present which permits the use of a wide range of possible adhesive products from lowest-cost vinyl or casein-neoprene latexes to pressure-sensitive–type adhesives to truly structural epoxy cements. For applications with such low unit stressing, the opportunity is presented to use a wide range of contact-type adhesives, including products like the neoprene-phenolic contact cements. Such adhesives can even by used for load-bearing panel applications where large expanses of aluminum sheet are laminated to foam insulation core materials or various wood-derivative products.

Undoubtedly, the largest poundage of adhesive is used in the bonding of aluminum surfaces in connection with packaging applications. The first packaging use of aluminum (closures on bottles) actually dates back to the early 1900s. In 1988, 365 million dollars worth of packaging adhesives were used in the United States alone. While aluminum usage does not extend across the whole spectrum of packaging applications, it still plays one of the most significant roles because of its ability to render vapor-sealing properties. For general packaging application, aluminum foil was adopted after its initial production in 1913. Thus, aluminum bonding was actually proceeding in that field before its use in bonding aircraft was initiated. Obviously, the low unit stresses and lesser strength demands for even the longest demanded service performance made the technology of bonding aluminum for packaging purposes considerably less demanding overall as compared with later aerospace structure bonding.

In general, the types of adhesives used for packaging today include dextrines, starch conversions, polyviny and acrylic resin emulsions, rubber lattices, casein, silicates, animal glues, hot-melts, isocyanates (polyurethanes), and various solvent cements. The information shown in Table 1 provides the characteristics of these adhesives. It can be shown, however, that all of these adhesives are not equally applicable for adhering to aluminum surfaces. The most widely used adhesives for aluminum from the above group appear to be (1) polyvinyl and acrylic resin emulsions, (2) rubber and synthetic lattices, (3) hot-melts of a wide variety, (4) polyurethanes, and (5) various solvent cements. It is also possible to combine more than one of these in a formulation which can be adequate but cheaper. For example, the addition of a lower-priced proteinlike casein to neoprene latex results in a commercial adhesive that is still adequate for making fairly strong laminate aluminum/plywood doors for trailer trucks.

General adhesive applications for packaging include the manufacture of corrugated board, lamination of paper and board, case and carton assembly, case and carton sealing, rigid box assembly, gummed tape and paper, envelopes, speciality boxes, cigarette manufacture, lamination of films and foils, and the labeling of various adherend surfaces. The development of better *pressure-sensitive adhesive* (PSA) formulations has more recently provided a means of durably attaching identification and product information aluminum plates to product surfaces that will experience service in virtually all forms of natural exterior weathering.

Aluminum flexible wrap applications for a wide varietey of consummable products are generally performed with adhesive products of the hot-melt, heat-seal coating, or pressure-sensitive adhesive types. Most of these applications incorporate aluminum foils of relatively high purity aluminum whose surfaces tend to be more easily wetted and bonded by adhesives. Where higher-strength wrapping is needed, the foils are rolled from some of the non–heat-treatable alloys of the aluminum-magnesium or aluminum-manganese types, which are again somewhat more easily wetted and bonded without special surface pretreatment. Especially, as compared with the higher-strength, heat-treated aluminum

Table 1 Characteristics of Important Types of Packaging Adhesives

Types	Bonding speed	Adhesion	Typical applications
Dextrine	Slow to moderate	Paper	Case cartons, rigid boxes, bag paste laminating and mounting, envelopes
Starch conversion	Moderate	Paper to glass, treated polyethylene	Bottle labeling and foil laminating
Polyvinyl and acrylic resin emulsions	Moderate to fast	Depends on formula: plastics, metal, wax, and ink-coated board	Like dextrine except where special properties are required. Heat-seal, pressure-sensitives, foil, plastic, and foam laminates
Rubber and synthetic latexes	Moderate to extremely fast	Depends on formula. May bond a variety of materials	Foil laminates, polyethylene, bag seam pastes, untreated polyethylene labeling and laminating, pressure-sensitives, surface-printed adhesives
Casein	Moderate	Paper to glass	Fully ice-proof bottle label adhesive
Animal glue	Slow to moderate	Paper, clay	Tightwrap, bookbinding
Hot-melts	Very fast	Paper, metal-coated board, plastics	Same as polyviny resin emulsions but at higher speeds and lower compression
Isocyanates (polyurethanes)	Very fast	Metal, plastics, coated board, foils	Film laminating
Solvent cement	Fast	Metal, plastic films, coated board	Heat-seal coatings, laminating, bag sealing

Source: From J. Hagquist, F. K. Myers, and S. K. M. Swanson, Adhesives Markets and Applications, Adhesives and Sealants, Engineered Materials Handbook, ASM, Materials Park, Ohio, Vol. 3, p. 45 (Table 2) (1990).

alloys used in aerospace and in some structural situations in automotive, trailer, truck, and transit car manufacturing.

Rigid aluminum containers may be made from thin-sheet, heavy-gauge foil or foil laminate materials. In these instances, the demand is for higher adhesive strength, since stiffness is demanded in the final product. Unsupported foil usually lacks the ruggedness required to withstand the abuse encountered during shipping and handling of such packaged products. For example, foil gauges of less than 0.001 in (1 mil) are generally laminated to paper or plastic films with flexible adhesives. The foil also may be coated for additional weathering or chemical resistance; in which case, the retention of adhesion under service conditions needs special evaluation. Unfortunately, the time to prove in real service conditions that a package will function for months to years can be economically unacceptable. In such instances, the use of accelerated weathering tests have been devised so that extrapolations can be made of service potential. In other situations, the aluminum may be decorated for special esthetic or individual product identification. A special evaluation of the adhesion durability will certainly be required for any adhesive that contacts any untested or physically or chemically altered adherend surface. The adhesive itself might be preapplied (as with a pressure-sensitive–

coated aluminum product) and a special release paper overlaid. This overlay must be readily detachable before the boil or laminate can be adhered to a second surface.

Foils can be laminated to a wide variety of paper materials to increase stiffness. So-called kraft (sulfate) papers are employed where high strength and extra rigidity are required at moderate or low cost. Sulfite papers are used where an added degree of surface smoothness is required, whereas low-cost ground wood papers are used for lowest-strength applications with acceptance of some slight color. Glassine, grease-proof, and parchment papers will offer a definite higher resistance of the laminate to oil and moisture but at some significantly higher materials cost.

Plastic films laminated to aluminum foil include cellophane, cellulose acetate, rubber hydrochloride, vinyls, polyvinylidene chloride, polyesters, and even nonpolar polyolefins. Adhesives must be carefully selected for chemical compatibility and resistance of the bondline to the most severe service conditions that can exist.

A wide range of water-base, organic solvent, and hot-melt adhesive products can be employed to laminate foil. Important properties that are considered are color, odor, toxicity, and resistance to heat, cold, and moisture. Of these, the water-soluble and water-emulsion adhesives are generally the lowest cost, yet can be very effective adhesives for bonding foil to fibrous material surfaces. The ready water wettability of most papers aids significantly in establishing a good bond in minimum setup times. Sodium silicate is often used as an inorganic binder adhesive between foil and paper for cigarette and soap wrappers and certain types of labels. Although casein-latex formulations and resin emulsions are somewhat more expensive than silicate, they may still be used because of their much greater flexibility. Cracking of a rigid adhesively joined foil laminate has been one of the main causes of property loss when the foil is functioning as an effective vapor barrier.

Organic solvent-based adhesives, such as the vinyl types, are chiefly employed to laminate the plastic films to foil. They are also capable of affecting the best bonding to the lower water-wettability papers like glassine and flame-proofed kraft. This type adhesive is usually applied directly to the foil and heated to remove the solvent before hot laminating. The solvents must be effectively removed before closure whether water or organic solvent. The bond strength may be inadequate for immediate handling or movement may occur whilehandling and prevent the development of optimum strength. Since the foil is a complete vapor barrier, the solvent can only escape with increasing time through the second adherend surface or through the edge of the bondline; usually a very slow process unless bondline heating is employed.

Hot-melt adhesives contain no solvent to help wetting the foil surface. Rather, they depend on a lowered viscosity level at elevated temperature to achieve this wetting. Consequently, they are mostly applied as an elevated temperature melt. Wax-based hot-melts may be employed where additional resistance to water vapor transmission in the bondline is desired. Water transmitted readily through the bondline may also degrade adhesion at the aluminum oxide interface by hydration or corrosion mechanisms.

Of the available heat-seal–type coatings, the vinyl types are considered most satisfactory for bonding foil. Copolymers of vinyl chloride and vinyl acetate are most commonly used. Other satisfactory heat-seal coatings include nitrocellulose, ethyl cellulose, methacrylates, chlorinated rubbers, and polyvinylidene chloride.

Gradually, adhesive bonding began to be employed for bonding aluminum in a wide variety of other general manufacturing industries where the standard accepted joining procedures for aluminum (welding, brazing, soldering, or mechanical fastening) would not be practical or might not be cnsidered economical. These adhesive bonding manufacturers generally lacked the resources of the typical aerospace manufacturer to adequately test and evaluate the potential service life of their products. Quite often, the practicality of using an adhesive and the entire sequence of manufacturing operations would have to be based on whatever information was readily available from an adhesive formulator. The adhesive formulator, in turn, was faced with an almost insurmountable task of trying to discover how many different adherends his or her product could be used to join. Establishing how long good adhesion could be expected to be retained with such a wide array of adherends, under an equally wide array of service conditions, was beyond the resources of most adhesive formulators. Customers of Alcoa used to typically remark to the author that a particular adhesive had been suggested for their application with little information available beyond an initial bond strength value. This situation was really the stimulus for Alcoa Labs to undertake a very broad research investigation over more than 25 yr to determine some numbers establishing the general bondability and relative joint durability of different adhesive families when used to bond aluminum alloys that might be used in all types of manufacturing.

A particular impetus was generated to study the best methods of fabricating aluminum-faced building panels in the form of the Alcoa Alply Laminated Building Panel. A number of bonding problems were posed by such a project whose solution could have wide benefit for bonding aluminum to a wide variety of building product's surfaces. Data had to be developed for evaluating the adhesion durability to wood, wood-laminates, particle and molded hardboard, gypsum boards, cement-asbestos board, paper-honeycomb, and a variety of core insulation materials for panels like polystyrene (steam expanded beadboard and extruded beadboard [Styrofoam]), polyurethane (sawed board from bun stock and foamed-in-place), vinyl foam, epoxy foam, foamglass slabstock, and lightweight inorganic aggregate slabstock.

The main question became how long such bonded composites could survive as a structure when exposed to variable environmental conditions such as water immersion, hot humidity soaking, soak/freeze/thaw cycling, and the natural atmospheric conditions (industrial or seacoast). The customer for such panels and the manufacturer need to be able to substantiate by some form of accelerated testing procedures that such products can survive and function effectively for periods of 20 yr and beyond without the need for costly special protection measures and frequent maintenance. How some of these answers were obtained by the author and his co-workers at Alcoa Labs will be reviewed at several appropriate times later in this book. (Some pertinent literature reports can be found in Refs. 87–89.)

Finding practical and economical methods for bonding aluminum to literally every other material of construction has greatly expanded the overall application of aluminum in the present economy. Literally hundreds of commercial products containing bonded aluminum exist today and that number will continue to grow in the foreseeable future. All of these applications have the advantage of being manufactured without the need to melt the aluminum locally (welding) or over the whole surface (brazing), or heat to a localized high melt temperature (soldering), or puncture through the metal creating areas of high localized stress. Even though the larger consumption of adhesives overall may exist for making various laminated, lower-stressed structures, this volume must place major emphasis on the bonding data and procedures developed for aerospace technology. It is in this area that the greatest strides have been made toward a better understanding of the total field of adhesion science. It is also in the endeavor to develop the highest state of the art of bonding for aerospace that the full range of contributions can be appreciated from all those scientific disciplines that must fuse and pool their information to understand better the basics of adhesion science. In this regard, the author has attempted in this work to call to the attention of the reader the individual contributions of the workers across the whole spectrum of modern science who have collectively contributed to the present-day understanding of adhesion science as it has been applied to the technology of bonding aluminum with adhesives. At the same time, the inclusion of so many different investigators in so many different scientific fields should offer an excellent means of referencing the overall progress in adhesion science that has occurred in this century.

2

Factors Affecting Adhesion

I. INTRODUCTION TO ADHESION THEORIES

A. General Background

We seem to generally prefer to believe that the forces of adhesion were virtually unrecognized and certainly not conjectured about very scientifically until our present time. In reality, some of our most famous scientists of earlier times had already developed a keen interest in adhesion during their times. Patrick (90) reminds us that the Egyptians had to understand the general phenomenon, as demonstrated in the veneering technology applied in their burial coffins. He also records the following words of Sir Isaac Newton some two and one-half centuries ago in his book *Optiks*. "There are agents in nature able to make the particles of joints stick together by very strong attractions and it is the business of experimental philosophy to find them out." DeBruyne (91) similarly reminded us of this fact in his paper titled "Fundamentals of Adhesion," which he delivered at one of the earliest international conferences on the bonding of aircraft in 1957. It would seem to be an appropriate subject for the managing director of CIBA whose laboratories had conceived the most practical and successful adhesive system (REDUX) for the successful bonding of the early Fokker Netherlands aircraft in 1949.

Of particular historical importance was the keen interest in adhesion of Michael Faraday (92), who wrote the following quite insightful viewpoint in 1833.

> I am prepared myself to admit (and probably others are of the same opinion), both with respect to the attraction of aggregation and of chemical affinity, that the sphere of actions of particles extends beyond those other particles with which they are immediately and evidently in union. Bodies that are wetted by fluids with which they do not combine chemically, or in which they do not dissolve, are simple and well-known instances of this kind of attraction. Still better instances of the power that I refer to . . . are air, so well known to barometer and thermometer makers for here the adhesion of attraction is exerted between a solid and gases, bodies having very different physical conditions having no power of combination with each other and each retaining during the time of action its physical state unchanged.

In effect, Faraday was saying that glue sticks because of unsatisfied molecular forces emanating from a surface, and nothing that we have found since that time has altered our belief in this supposition.

The reality of these interfacial forces can be shown in many ways but Leslie (93), in 1802, had already referred to the readily observable; i.e., the rise in temperature that occurs when a solid is immersed in a liquid. A rise in water temperature when aluminum oxide powder prepared for chromatographic analysis (with a large surface area), for example, makes contact can be as much as 1–2 °C. The heat results from the rush of water moleculars into "the sphere of action" of the aluminum surface atoms. An additional 74 erg/cm will be required to break this joint between aluminum and water, as was released in forming the joint which will be equal to the additional work performed in making a new water surface. When we calculate the force necessary to break 1 cm^2 of water/aluminum interface (divide the work amount by the distance [about 3 Å] over which the forces act), we find the stress needed is equivalent to 584,000 psi. How is it that glued (bonded) joints are so much weaker has been a continuing thrust to date? At the same time, such knowledge should be our best aid in continuing to increase our abilities to obtain greater and greater percentages of these ideal numbers in service.

Patrick (90) has provided a reminder of some of the important early literature which contributed significantly to our theoretical understanding of the science of adhesion. DeBruyne and Houwink (94) provided the first excellent text of our era in 1951. In 1950, Kline and Reinhart (95) reviewed current concepts of adhesion with Bernard and Hertz (96), discussing energy of adhesion of liquids to solids. In 1952, Brantley and Carpentier (97) reviewed the modern adhesion theories of that period. The phenomenon was becoming more important than the materials which had been emphasized in the early 1940s.

Application of modern concepts accelerated in the early 1950s with the publications of Bowden and Tabor (98) and DeBruyne (99). Papers by Harkins and co-workers (100–102), dating from 1914 to 1950, were directly applicable to the theoretical concepts of adhesion in conjunction with the even earlier classic papers of Young (103) in 1805 and Dupre (104) in 1869. Zisman's (105) first paper in 1946 was followed by many others, which will be discussed later in this chapter. Surface-energy concepts started with Harkins (102) as early as 1914, continuing with chronologically arranged contributions by Langmuir (106), Guastalla (107,108), Jenckel and Rumbach (109), Derjaguin (110), Simha et al. (111), Frisch and Simha (112), Zisman (113), Voyutskii (114), Gardon (115), and Fowkes (116).

In 1981, Baterip (117) stated no overall adhesion theory yet existed, though many models pertaining to various aspects of adhesion have been proposed over the years, as by Wake (118), DeBruyne and Houwink (119), Girifalco and Good (120), Fowkes (116), Gardon (121), Kaeble (122), Mittal (123), Huntsberger (124,125), and Good (126). A First Report of the Adhesives Research Committee in London in 1922 (127) might very well be of great historical interest concerning adhesion. A first distinction between mechanical and chemical adhesion might be attributed to McBain and Hopkins in a 1925 paper (128) followed by its mention in a Second Adhesives Research Report in 1926 (129). Vasenin (130) and Allen (131) have both pointed out the many problems of attempting to understand what theories have the most real merit owing to the insistance of authors of a particular theory that their views should be a universal panacea to the overall problem.

The ASTM adhesion committee has chosen in recent years to adopt an overall definition of adhesion as "the state in which two surfaces are held together by interfacial forces which may consist of valence forces or interlocking action, or both." We might further substitute the term *chemical* or *mechanical contributions* to adhesion as the focus for study of adhesion factors of primary importance. As early as 1970, Raevskii and Pritykin (132) had proposed more than 40 definitions of adhesion which readily demonstrated the difficulties confronting the finding of any agreement on a unified theory. In 1973, Raevskii (133) proposed a more technical definition of adhesion, stating it should be defined as a surface effect consisting in the emergence of a physical (also a chemical) attraction between the molecules (or ions) on the surface of the condensed phases of their mutual contact, which attraction leads to the union of these surfaces. He contended this definition treated adhesion as a universal physicochemical (rather than technological phenomenon), classifies adhesion as a distinct surface effect, characterizes it over the whole spectrum of attractive energies, points out the presence of molecular contact, the nature of the possible adherends (any condensed phase independent of its nature and structure), and defined the end result as formation of a real physical system; i.e., an adhesive joint.

Of the five main contributing theories most often proposed, three (mechanical, diffusion, and rheological) relate more to adhesive/adherend interlocking and increase in molecular contact area. The other two (electrostatic [electrical] and adsorption [molecular] theories) are more distinctly related to chemically inspired forces.

B. Solid Adhesion Considerations

Some important considerations about solid adhesion were summarized by Lee (134) in 1991 in an invited lecture at the 14th Annual Meeting of the Adhesion Society, especially as to the relevance of donor-acceptor interactions and hard-soft, acid-base (HSAB) principles. Studies of the former were undertaken by Gutmann (135) and Derjaguin et al. (136), whereas Jensen (137–139) has reviewed the latter. In the broadest sense, both types of interactions can be considered forms of molecular interaction but with different emphases, as pointed out by Morokuma (140,141). When considering a more covalent (or frontier orbital) type of interaction, the donor-acceptor concept seems more appropriate; however, the acid-base type interaction leans toward the more ionic (or electrostatic) form of interaction.

When considering adhesion between polymers, Fowkes (142–145), Bolger (146), and Bolger and Michaels (147) have emphasized the important role of acid-base interactions. Lee (134) further applied the role of these interactions to solid adhesion by discussing (1) molecular interactions, (2) the acid-base interaction, (3) the hard-soft acid-base principle of Pearson (148,149), and (4) the application of the density-functional theory of Parr and coworkers (150–153). In the density-functional theory, any chemical system (atom, molecule, ion, radical) may be characterized by its electronic chemical potential and its absolute hardness. The chemical potential measures the escaping tendency of an electronic cloud, whereas absolute hardness determines the resistance of the species to lose electrons. In a series of 1989 and 1990 publications, Lee (154–159) demonstrated that the extension of the HSAB principle to solid interactions is feasible in view of the electronic band structures of different solids. The HSAB principle has also been applied to the study of kinetics and equilibrium and the frontier orbital method which, in turn, has been used to illustrate the electrophilic and nucleophilic interactions in organic chemistry discussed by Fukui (160), Fleming (161), and Pearson (162). Hudson and Klopman (163), as early as 1967, had proposed an equation showing acid-base and donor-acceptor components for molecular interaction. However, molecular interactions are insignificant with interatomic distances greater than 3 Å. This may be why an acid-base interaction is not readily observed for solid-to-solid systems, although it is frequently reported.

Before considering individual theories, we might wish to consider some of the general observations made when simply bringing two bodies into physical contact. Since this writing is oriented toward the use of aluminum adherends, it would seem appropriate to consider the contacting of two metallic bodies. Such bodies are said to experience "metallic adhesion" when an observable net tensile load is subsequently required to separate them, as proposed by Keller (164). The magnitude of this adhesion has been shown to be dependent on factors such as (1) the physical and chemical properties of the metals (165–168), the nature and extent of loading (169), and (3) the characteristics of the contaminant layers on all but anatomically clean surfaces (170).

The general process involves the elastic and plastic deformation of surface asperites, deformation of the bulk adherend, rupturing, and dispersing contaminate surface films. The conclusion has been accepted that two metals could be welded with near-zero normal loads if both the surfaces are actually anatomically clean. Adsorption studies of Ehrlich (171) have also predicted this. Adhesion studies have been attempted by Gilbreath (172), including deliberate contamination from a fraction of a monolayer to exposure to ambient atmospheric conditions. Analytical problems inherent in such studies are expectedly very complex because more than one energy-transfer mechanism will likely be involved. For example, increased thermal energy could cause evaporation or dissolution of contaminant layers, or shock-wave energy from either an explosive impact or an ultrasonic source could promote contaminant dispersal along the interface. Also, since the contaminant layer could develop via alternative mechanisms such as adsorption, surface creep, or bulk diffusion to the surface procedures;

these energy inputs might even enhance the rate of contaminant film development. This explains why the most commonly employed energy input for special studies of this problem is normal compressive loading at room temperature. The "real area of contact" is the most important factor to consider and further variation will obviously occur with increasing loading time owing to creep, according to Akins and Tabor (173). Hardness measurements have also been used by Akins et al. (174).

The contacts made between two macroscopic bodies has also been discussed by Van Den Temple (175). For macroscopic bodies in contact, the energy of interaction is (to a first approximation) made up of the sum of all the interacting bodies between all the atoms in counterbodies. Hamaker (176), in 1937, developed mathematical equations for calculating the interaction between systems that contain a constant A which depends on the nature of the interacting materials. The question then becomes how to arrive at some method for calculating these Hamaker constants. This was answered by involving the van der Waals constant values as calculated from the Lifshitz (177) equations based on the electromagnetic radiation field in each body employed in a 1972 formula derived by Krupp (178).

In considering pure solid-to-solid adhesion, it is specified that no adhesive is involved in holding the solids together. This principle can be applied, however, in the case in which an adhesive is actually used. With the Lifshitz–van der Waals constants and the Hamaker constants mentioned above, it is now possible to calculate the adhesion force derived solely from van der Waals interactions. The work then performed to separate these two bodies from their equilibrium separation point out to infinity will be equal to that necessary to create two new surfaces in place of the original interface. The first equation allowing such a calculation had actually been derived by Bradley (179) as early as 1932.

Only 2 yr after the Bradley publication, Derjaguin (180) published a paper pointing out that some deformation of two solids should take place at the point of contact. The shape of such contact would obviously depend on the actual adhesive forces present. Derjaguin et al. (181) started an analysis of this matter, which they published in 1975; however, this analysis remains somewhat incomplete because it does not allow for deformation due to attractive forces close to the edge of the contact. Johnson et al. (182) also contributed an important paper about this matter of deformable solids which predates the Derjaguin work by 4 years. It relates to use of an equilibrium theory of adhesion which proposes that the process of making and breaking bonds should be reversible and, therefore, the steady state of the intact bond can be considered an equilibrium position. However, in actual practice, it is readily noted that the forming and breaking of a bond is rarely reversible. Energy can be dissipated easily during the bond making or disruption, raising the problem considered under the name of adhesive hysteresis. In a real situation with adhesive in the bondline, the contact between the two joined bodies cannot be at the molecular level as treated by theory. Instead, the adhesive thickness is always thicker than a monolayer, which makes it necessary to consider the important matter of the viscoelastic effect of the bulk of the adhesive.

Kendall (183) provided us with one such effect in 1974 which refers to what is called adhesive hysteresis. In the case of the breaking of bonds between a rubber and a solid surface, there is noted an additional considerable expenditure of energy that is beyond that necessary to simply break the surface bonds. This extra energy will be that required to deform the inelastic rubber that is some distance from the actual surface of rupture. Comparable is the common experience of pulling sticky chewing gum from a surface to which it is adhering. The breaking of the surface bonds will actually act to trigger or initiate other processes which result in expending large quantities of energy. This so-called "adhesive hysteresis" can, however, be somewhat beneficial in enhancing the overall bond strength and retarding the actual fracture. It would seem that this offers some explanation of why the actual measured joint strength of joints is generally larger than the value for the attractive forces calculated from the van der Waals attraction. However, this same adhesive hysteresis may retard the bond-forming process in the opposite situation of making a joint. Fabricating acceptably strong adhesive joints will require using certain bonding experiences that can overcome this adhesive hysteresis effect. This discussion should offer the reader some preparation for the more extensive discussion of surface energetics and fracture mechanics of joint failures that will be covered in Chapter 7.

Some words about the general nature of forces between surfaces should be pertinent before discussing specific theories of adhesion. The forces responsible for explaining adhesion have been divided into two groups; i.e., primary and secondary in nature. It is generally logical to consider secondary forces first, since they are the more omnipresent forces existing between individual molecules. The most fundamental is dispersion force involving the simultaneous excitation of the two molecules in proximity. This is the instantaneous Coulomb interaction between electrons and nuclei in the two molecules. London (184) first derived an approximate formula for this dispersion energy, E, for two spherical molecules in terms of their polarizabilities and ionization potentials. This London interaction is present in all parts of atoms or molecules. Interaction at large distances (greater than 100 Å), however, required an alteration of the London equation, which was provided by Casimir and Polder (185). This offers the proposition that as the opportunity for close contacts diminishes by 50% or more, the force of attraction between the atoms of the effective contacts decreases much faster, approaching the seventh power as a limit. Allen (186) has pointed this out in his publication on *Strength and Structures* by noting that the London molecule-to-molecule forces are most effective over a distance of 4 Å (1.6×10^{-8} in). This is approximately the diameter of a linear molecule. At lesser distances than this, the attraction is diminished because of an overlap of electron clouds, whereas at greater than 4-Å distances, the attracting forces rapidly diminishes, since it is inversely proportional to the seventh power of the distance.

Investigators such as Huntsberger (187), Orowan (188), and Tabor (189) calculated these attractive forces owing entirely to the dispersion forces between two planer bulk phases. Such calculations indicate that at a separation of 1 nm, the attractive force should be approximately 100 MPa. Obviously, this force is considerably higher than the experimental strength of most joints. The discrepancy is mainly attributed to factors like air voids, cracks, defects, or geometric features acting as stress raisers under loading. This causes much lower rupture values. Lee has presented important information for understanding the intermolecular forces between solids at a lecture presented at his organized symposium on Fundamentals of Adhesion: Theory, Practice, and Applications at New York State College at New Plantz in 1984. Lewis and Forrestal (190) and Brockmann (191) have cited that the enthalpy of the reaction of chemical bonds ranges from 250 to 600 kJ/mol, whereas the energy of van der Waals forces only range from 4 to 40 kJ/mol. This demonstrates the need to attribute considerable bonding strength to the development of chemical bonds (primary attractive forces).

Bolger and Michaels (147) and Fowkes (192) have suggested that acid-base interactions across the interface might also be considered as intrinsic adhesion forces. Bolger and Michaels (147) have made attempts to identify many of the distinguishing features of acid-base reactions in terms of the ionic character of surfaces like the hydroxylated metal oxide with the organic functional group provided by the adhesive.

Besides the basic dispersion force, there are polar interactions between molecules, which have been described by Keesom (193) as the dipole-dipole interaction, the induced dipole interaction (Debye [194]), and hydrogen bonding. Both of these investigators developed a corresponding expression for calculating the values of their respective parameters. Hydrogen bonds also can arise from dipole-dipole interactions, for which the 1960 book by Pimentel and McClellan titled *The Hydrogen Bond* (195) can be reviewed. The strength of these hydrogen bonds generally varies from 1 to 10 kcal/mol, which is about the upper limit as the dispersion force maximum. The dispersion force between two helium atoms, however, has been measured at 0.02 kcal by Hischfelder et al. in their book *Molecular Theory of Gases and Liquids*. An excellent summary of these matters can be found in Good's chapter in the *Treatise on Adhesion and Adhesive* series titled *Intermolecular and Interatomic Forces* (196).

Although it is evident that intrinsic adhesion arising from secondary bonding forces alone can create adequate high-strength joints, many adhesion scientists believe that it is the additional presence of primary bonding forces that can be of great benefit in securing environmentally stable interfaces. Unfortunately, most of the evidence presented to confirm such primary interfacial bonding in the literature has been more indirect than direct evidence, as shown in the publications of DeLollis (197,198), Rutzler (199), Vilenskii et al. (200), Koldunovich et al. (201), Trostyanskaya et al. (202), Dean (203), Ahagon and Gent (204), Andrews and Kinloch (205), Gent and Hsu (206), Lerchenthal et al. (207), and Lechenthal and Brennan (208,209). In recent years, more definitive evidence has been furnished through the use of some of the more sophisticated analytical techniques such as Koenig and Shih (210) with laser-Raman spectroscopy, Bailey and Castle (211) with XPS, Gettings and Kinloch (212,213) with SIMS, and Lewis et al. (214) with IETS.

Table 2 Bond Types and Typical Bond Energies

Type	Bond energy (kJ/mol)
Ionic	590–1050
Covalent	63–710
Metallic	113–347
Permanent dipole-dipole interactions	
Hydrogen bonds involving fluorine	<42
Hydrogen bonds excluding fluorine	10–26
Other dipole-dipole (excluding hydrogen bonds)	4–21
Dipole-induced dipole	<2
Dispersion (London) forces	0.08–42.0

Source: A. J. Kinloch, Introduction, Durability of Structural Adhesives, Applied Science Publications, London, Chap. 1, Table 3, p. 15 (1983).

A summary of some bond energy values as discussed above can be seen in Table 2. The very large contributions of ionic, covalent, and metallic bonding compared to the dispersion forces are quite evident.

II. MECHANICAL INTERLOCKING THEORY

The major source of intrinsic adhesion is here proposed to be the result of mechanical keying of the adhesive into the surface irregularities, as demonstrated schematically in Figure 1. Since good adhesion can be commonly demonstrated between relatively smooth adherent surfaces and adhesives, there must be some other significant factors which must also be operating, and this theory cannot operate as an all-encompassing explanation for adhesion. Packham (215) has discussed the surface roughness and its effect on adhesion through mechanical interlocking with the adhesive at some length.

Maxwell (216) made measurements of the effect of different degrees of surface roughness using maple wood adherends. Boroff and Wake (217) related bond strength of rubber to textiles to the degree of penetration of fiber ends into the rubber. Perrins and Pettett (218) showed the coating strengths of plastics on metals appeared to involve both a mechanical interlocking and other interactions dependent on the surface chemistry of the plastic polymer. Investigations of Packham (215)

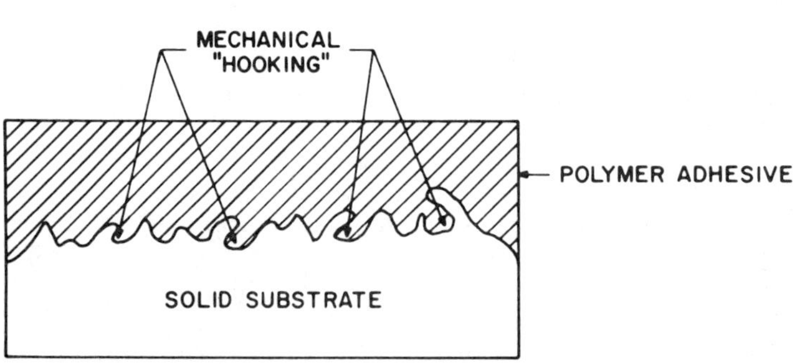

Figure 1 Adhesive bonding. (From J. D. Minford, Aluminum Adhesive Bond Performance Treatise on Adhesion and Adhesives [R. L. Patrick, ed.], Marcel Dekker, New York and Basel, Vol. 5, Chap. 3, Figure 3-1, p. 46 [1981].)

on the adhesion of polyethylene to aluminum surfaces (variously treated, including anodizing) demonstrated the importance of mechanical interlocking components to bond strength. More recently, the investigations of Venables (33,219) related the morphology of the treated aluminum oxide surfaces to affecting mechanical contributions to the joint strength. More discussion and data can be reviewed on this general subject in the publications of Bair et al. (220), Wang and Vazirani (221), Bascom et al. (222), Mulville and Vaishnov (223), Evans and Packham (224,225), and Venables et al. (226), where demonstrations have been offered of the formation of needle-shaped aluminum oxide crystals on the surface which are helpful in promoting adhesion by a mechanical keying mechanism.

The opportunity for mechanical interlocking can range from the presence of macroscopically rough surface texture to microscopically rough surface conditions, as shown by Chen et al. (227) and White et al. (228). The significant increase in joint peel strength with increased surface roughening of the adherend has been studied by Evans and Packham (225), whereas Eich et al. (229), as early as 1971, evaluated the highly significant influence of surface roughness on both wetting and adhesion in dental adhesive applications.

III. DIFFUSION THEORY

The chief supporter of mutual diffusion of polymer molecules across an interface as a major adhesion force factor has been Voyutskii (230–236). He also offered criticism of other theory advocates (237). Voyutskii proposed autohesion of two adjacent like-material adherends as a two-step process consisting of (1) initial contact and (2) final coalescence leading to a vanishing of boundary between the two surface layers. For contact between two low molecular weight liquids, coalescence occurs instantaneously owing to the high mobility of their molecules. For high molecular weight liquids (high polmers), the macromolecules are obviously not as mobile and some significant time may be required to achieve this coalescence. The emphasis in the diffusion theory, therefore, is on the second stage of autohesion.

Vasenin (130,238) also conducted extensive investigations in this area, leading to development of a series of theoretical relationships involving autohesion of a polymer to itself. Other investigators such as Skewis (239) and Forbes and McLeod (240) have also developed experimental results which are in accord with these relationships. Anand (241–244) criticized this approach in 1969–1970; Voyutskii (245) replied in 1971 with further qualitative evidence and interpretation. Although most evidence for the diffusion theory has been indirect and deductive, there has been an attempt to directly study the interfacial structure by a variety of methods, including optical and electron microscopy, radiothermoluminescence, ultraviolet (UV) luminescence, paramagnetic probe, Fourier transform, and internal reflection spectroscopy, as discussed by Sung (246). Campion (247) developed the case for explaining the autohesive properties of rubbers by citing the availability of holes in the cross section.

Titow (248) states that for mutually compatible polymers with similar solubility parameters at temperatures above their glass transition points, the prime mechanism of bonding could likely be mutual interdiffusion. Solvent bonding of two polycarbonate surfaces to one another might be one example. With the good bond that results there will be no detection of the original interface between the two adherend parts. There will, however, be a zone where part of the polymer has obviously undergone some form of structural alteration. It has been suggested that for the situation of adhesive penetrating into the fine structure of a treated aluminum oxide surface, the mechanism is one of diffusion of polymer into the interstices of the oxide.

Huntsberger (249) concluded that few adhesive system will be found for which interdiffusion is a completely satisfactory explanation for the development of structural strength. Further, the incompatibility of many dissimilar high molecular weight polymers argues against the wide application of this theory to explain high-strength jointures.

Anand and Karam (241) focused on stage one of autohesion and worked out a viscoelastic contact theory based on the shape of the adherend surfaces, the contact pressure, and temperature. Continuing to use polystyrene as the study material, Anand and Balwinski (242) considered the situation in which after contact has been established, the bonding needs to take place owing to primary or secondary forces. Since polystyrene is a nonpolar system, the intermolecular attraction must mainly be due to van der Waals forces. The strength of such secondary bonding would likely consist of an attractive component, i.e., the long-range attractive London dispersion force, and a repulsive force due to overlapping of electron clouds (250). Computations of bond strength between two flat plates of polystyrene in self-adhesion were made and good agreement was obtained with the experimentally determined value for the yield stress of polystyrene (see Refs. 251 and 252). Further experimental verification of Anand's two-stage theory of autohesion for polystyrene was furnished by Anand (253) in 1970.

More readings about relationships of interest are in publications by Flory (254), Hughes and Britt (255), and Dobry and Boyer-Kawenoki (256). Ellul and Gent (257) have published a more recent paper summarizing the overall role of molecular diffusion in the adhesion of elastomers.

IV. ELECTRONIC (ELECTROSTATIC) THEORY

Derjaguin and co-workers (258–270) have been instrumental in promoting the role of electrostatic forces in adhesion and contended that electron transfer between adhesive and adherends with different electronic bond structures can cause the formation of double layers of electrical charge at the interface. The separation of these adherends and breaking of the bond with the adhesive would be accompanied by drawing apart of the two components and the work of adhesion required could be identified with the electrical energy stored in the capacitor system.

A criticism of the theory seems to be that although some data with rather close predictions between theoretical and experimental values have been presented in the literature, the theory seems to ignore the energy which, particularly in peeling tests, is dissipated in the viscous and viscoelastic deformation of the adherends. Huntsberger (271) has critizied Derjaguin for his neglect of the dissipated energy in explaining the high work-of-peeling values that he uses as evidence for his theory. At best, the question of the relative contribution of electrostatic versus molecular forces to the strength of adhesive bonds has had to face overwhelming literature evidence that the role of the double layer in adhesion is negligibly small. Still, the Russian investigators continue to claim a fundamental role for the double electric layer in adhesion. A 1983 review of the role of the molecular and electrostatic forces in the adhesion of polymers has been published by Derjaguin and Toporov (270) which cites the present position of these authors. Their continued support for the double-layer theory is clearly indicated in the concluding paragraph of the paper, which reads, "Hence, a conclusion can be drawn that neglecting the contribution of the double electric layer to adhesion of polymers may lead to completely incorrect results and conclusions."

Skinner and coworkers (272–274) have demonstrated special situations in which these double-layer forces may be significant but they had great difficulty in measuring the charge densities and actually obtained considerably lower values than previously reported. It should be admitted that they employed direct tension test procedures instead of the peeling procedures employed by Derjaguin. Workers like Zisman (275) and Bikerman (276) have concluded that overall electrostatic forces must be considered of little importance. Other investigators like Kaelble (277), Meissner and Merrill (278), McClaren and Silar (279), Tobolsky (280), and Bueche (281) have employed rheology and the time-temperature superposition principle to quantitatively explain the rate dependence of adhesive performance, which Derjaguin contends is best explained by the use of the electrostatic force theory.

If the consideration turns from adhesion between plane surfaces to the adhesion of fine particles, then Krupp's (282–284) experiments with gold particles would furnish an example of clear evidence that the electrostatic forces seem to be significant. Mention should also be made of Comyn's (285) reference to the possible existance of electrostatic ion-pairs at the interface of aluminum/epoxy joints. Comyn suggests that this could be due to the presence of ionic oxide formed by the aluminum surface preparation and the fact that the mechanism of epoxy curing is also ionic. Comyn also refers to an electrostatic model of joint strength decay due to the presence of water in the service environment. He suggests the water may enter the adhesive and thereby increase the relative permittivity at the interface. As a consequence, electrostatic forces present will be weakened. More will be said about

postulation that the maximum adhesive fracture surface energy should be attained when the solubility parameters of adhesive and adherend can be matched. Originally based on the work of Hansen (337), this mutual solubility approach to adhesive bonding has been further justified by the results described by Harrison (338). Barton (339), in 1982, further stated an equation which related surface tension and solubility parameter whereby the surface tension equals a constant times the square of the solubility parameter times the cube root of the molar volume. Barton points out that the solubility parameter is simply one of a series of factors which can be called "cohesion parameters." The solubility parameter of Hildebrand and Scott (321) was originally intended to describe the solution properties of liquids which did not undergo any significant mutual chemical interaction; however, similar cohesion parameters may be applied to the prediction and description of interfacial properties of a wide range of materials. This review by Barton (339) attempts to outline the application of the cohesion parameters to both wetting and adhesion. It is proposed that both interfacial and adhesive properties should be closely related to those cohesion parameters of the respective component materials. Barton has defined cohesion parameters as quantities with dimensions of pressure which describe the extent of cohesion within condensed phases.

Williams (340) developed a blister test in 1969 which was used to measure the fracture surface energy between an elastic adhesive and a rigid adherend like aluminum. The fracture surface energy is calculated from a knowledge of the pressure at which the crack in the joint propagates, the geometry of the test joint (341), and the modulus of the adhesive. The test has been used in diverse applications such as determination of the fracture surface energy between rocket liners and propellents (342), and adhesion of molluscs to dental adhesives, as described by Anderson et al. (343).

Fowkes (344,345) proposed that this work of adhesion might include contributions from dispersion forces, hydrogen bonding, dipole/dipole and dipole/induced dipole forces, acid/base interactions, and possibly covalent bonding. Fowkes and coworkers (346–348), Bolger and Michaels (147), Lipatov and Sergeeva (349), and Sorenson (350) discussed the importance of acid-base interactions on adhesion of organic substances to inorganic adherends. The term *polar* has been the dominant word used to explain intermolecular forces since the work of London (250) in 1937. It was true in the dense gases studied that dipole/dipole and dipole/induced dipole interactions could explain molecular interaction between pairs of molecules. It was presumed that these forces could also account for the intermolecular interactions in liquids and solids. This belief continued even after the discovery of hydrogen bonding, which was considered at first to be a kind of dipole interaction. Even after Pimental and McClellan (195) showed the independence of the hydrogen bond strength to the magnitude of dipole moments, the term *polar* continued to be used to describe interactions involving hydrogen bonds.

In 1962, Fowkes (116,351) proposed the separation of dispersion forces, hydrogen bonds, and polar interactions as separate parts of the work of adhesion. In 1966 and 1967, Gardon (322) and Hansen (352) proposed similar separations for enthalpies of mixing and solubility parameters. In 1983, Fowkes (348) showed that hydrogen bonds are a subset of acid-base interactions and that dipole/dipole interactions are probably of negligible importance in liquid and solid interactions.

In 1983, Fowkes (348), employing the work of Drago et al. (353,354) on enthalpies of acid-base pair reactions as a base, demonstrated that the dipole interactions did not contribute measurably to the enthalpies of molecular interaction. This appeared to be in conflict, however, with the 1921 work of Keesom (193) and DeBye (194) on intermolecular forces until it was recognized that the latter were dealing with the long-distance intermolecular forces between gas molecules as compared with the much shorter-distance forces in condensed phases where molecules are close and dipole interactions are minimized by multiple nearest neighbors. Fowkes and Maruchi (355) obtained measurements on a series of polymrs of controlled acidity and basicity, permitting the development of a set of smooth curves relating acid-base contributions to the work of adhesion with the degree of acidity and basicity. Further adsorption studies of the interaction of polymers in solution in different solvents with inorganic solids confirmed a competition between the dispersion force and acid-base interactions between polymer and the inorganic solid, polymer and solvent, and solvent and inorganic solid. Fowkes and Mostafa (144) in the next year stated that the differences between the dispersion forces could cancel out leaving the acid-base competition to control the adsorption behavior. A survey of the practical use of adhesives, however, has indicated that acid-base (electron donor-acceptor) interactions have still been largely utilized in empirical, trial and error methods.

While the above discussion has involved only secondary interaction (van der Waals) forces in the form of dispersion forces, electron pair interactions, and dipole interactions, their summation is still more than adequate to provide the high bond strengths desired. Instances can be shown beyond these interactions where primary covalent bonding is probably also involved. As mentioned earlier, considerable indirect evidence has been reported (197–209) for this primary bonding, whereas more direct evidence has been obtained using laser-Raman spectroscopy (210), XPS (211), SIMS (212,213), and IETS (214). Other indirect evidence has come from investigations of Allen, Greenwood, and Wake (356) utilizing thermodynamic techniques showing that the secondary interactions are insufficient to account for the observed properties of strength and durability shown by bonded joints. Direct evidence has been provided by Wu (357) of the increase in joint strength achievable from the introduction of quite small amounts of reactive functional groups into the adhesive.

Basin and Berlin (358–360) in the Soviet Union have attempted to combine the ideas of several authors in this field to evolve what they have chosen to call the "molecular" theory of adhesion. Raevskii (133) compared this molecular theory of adhesion to the rheological theory of Bikerman (361–363) which will be discussed in detail in a later section of this chapter. Raevskii (133) further proposed that the molecular and rheological theories together seem to include the ideas that have been separately touted in the theories described in the literature as diffusion (230–236,364), electric (258–270), microrheological (365), electrorelaxation (366), chemical (367), adsorption (296,299), and mechanical (368).

When ionic, covalent, or metallic bonds operate across the interface with their high typical energies (see Table 2), we find them associated with the name *primary interactions*. As discussed earlier, the secondary interactions occur from operation of the van der Waals forces (permanent dipole/dipole interactions and dispersion forces of London), and offer significantly lower magnitudes of bond energy, as seen in Table 2. Schneberger has offered the visual conception in Figure 2 of interaction occurring between clouds of charged entities. Actual calculations (188,189,369,370) show that these secondary attractive forces between two planer bulk phases will still be considerably higher than the machine-tested strength for joints. This failure of practical joints to achieve even these lesser theoretical strength values is mostly attributed to factors of imperfection in the bondline such as air voids, omnipresent cracks, or internal "built-in" stress. We must believe that the additional presence of primary bond energies will be of additional value in increasing the measured strength of joints and of special benefit in promoting environmentally stable interfaces. Comyn (285) proposed the presence of coordinate covalent bonds between phenolic oxygens of phenolic-based adhesives (as used

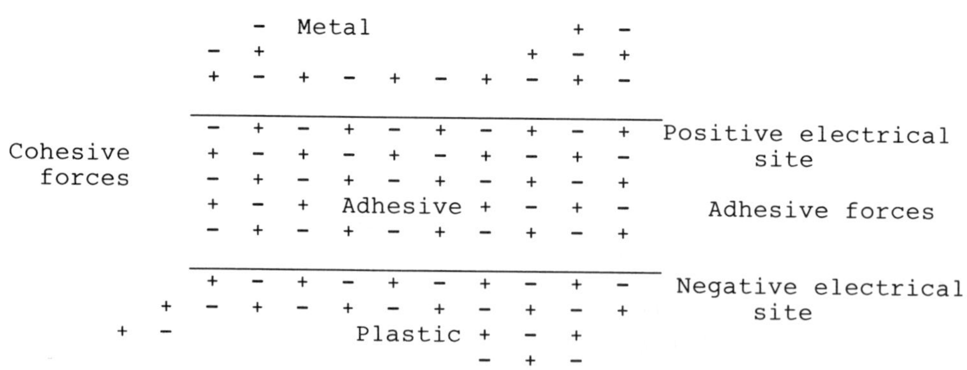

Figure 2 Visualization of interaction between clouds of charged entities. (Adapted from lecture by Schneberger on Joining by Adhesive Bonding.)

in bonding European aircraft) and aluminum ions at the bonding interface. This, in Comyn's opinion, should be a good basis for explaining the extensive demonstration of high durability in weathering of European aircraft joints. Minford (371,372) also has shown the ability of various phenolic structural film and tape adhesives (such as vinyl- and nitrile-phenolics) for durably bonding 6061-T6 alloy adherends in 2- to 8-yr exposure tests in accelerated laboratory and industrial and seacoast natural atmospheric weathering conditions. There would seem to be some basis for this complexing possibility based on demonstrations of better aluminum joint durability owing to the formation of complexes between aluminum and 8-hydroxyquinoline by Brockmann (373). Allen (374) and Kiel and Heertjes (375) also demonstrated the formation of complexes between aluminum oxides and various alizarin dyes that is pertinent to this discussion. Giles et al. (376), Giles (377), and Cummings et al. (378) have also suggested coulombic attraction as being responsible for the binding of acid dyes to anodized aluminum oxide. Lewis et al. (214) developed the IETS spectra of phenol and hydroquinone on a thin layer of aluminum oxide and showed the entities were predominately phenol or hydroquinone anions, giving support to the concept of coordination of phenolic oxygens to aluminum ions.

A. Wetting at the Interface

It would seem clear from all of the above theories that the establishment of intimate molecular contact at the interface is very highly desirable, though sometimes insufficient for developing the strongest joints if some contaminating intervening entity is present. Thus, the property for achieving strongest joints involves the adhesive being able to spread over the aluminum oxide surface displacing air and all contaminants in a process called "wetting."

One method of judging the potential wettability of a particular adherend (including aluminum oxide) has been to study the ability of a single drop of distilled water to spread on the adherend surface. The wetting or contact angle is as shown in Figure 3. If the drop is not spreading, as shown by the extreme right-hand situation in Figure 3, then the surface is considered one of "low energy" and the value of wetting angle is high. For example, the literature value for the wetting angle on low-density polyethylene is 94 according to Jeandrau (379), who also measured the wetting angle on polytetrafluoroethylene as 108. Tira (380), in searching for a gas plasma treatment to improve the bondability of polypropylene, measured an angle of 90. Such difficult to bond adherends show a low surface free energy value which agrees with their low spreading angle value. Typical surface free energy values are shown in Table 3 for a number of adherend materials. The lowest surface free energy value shown is 18.5 for polytetrafluoroethylene (Teflon), which predicts a very high wetting angle response which is found. The higher surface free energy of an epoxy surface would indicate an easier wettable surface and a higher surface reactivity potential than for adherend polymer materials which are above it in the listing. The even higher surface energies of the metal surfaces such

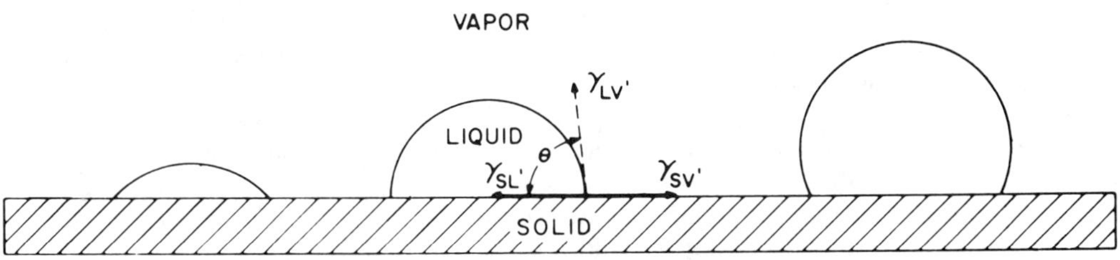

Figure 3 Wetting and contact angle. (From J. D. Minford, Adhesives, Durability of Structural Adhesives [A. J. Kenloch, ed.]. Applied Science Publications, London, Chap. 4, Figure 2, p. 142 [1983].)

Table 3 Surface Free Energies of Surfaces of Various Adherends

Adherend	Surface free energy F (dyne/cm^2)
Teflon	18.5[a]
Polyethylene	31[a]
Polystyrene	33[a]
Polyvinyl chloride	40[a]
Nylon	46
Epoxy	47
Water	73
Aluminum	~500[b]
Tin-lead solder	~500[b]
Mercury	~500
Silver	~900[b]
Copper	~1100[b]

[a]Critical surface tension for wetting.
[b]Near melting point.

as aluminum (actually the oxide rather than metal atoms) in the listing would indicate they all should be reactive with and would be effectively wet by a flowable epoxy adhesive.

The wetting angle can be highly influenced and lowered by pretreating of an adherend surface in order to produce more durable final joints. The indication of success will be to find a lower contact angle. It will be expected then that if a procedure can be found for treating a known low-energy surface which results in a low measured contact angle, effective bonding can result. Air plasma cleaning of low-energy surfaces like polyethylene has produced near-zero contact angles according to Mylonas (381). Other investigators like Barbaris (308), Boucher (382), and Cherry (383) have demonstrated the principle that the strength of a joint will increase with the recording of decreasing contact angle values as it is variously treated. Lee (384) has written extensively on the subject of surface wettability and its relationship to thermodynamic, relaxational, rheological, and mechanical properties of polymers through the glass transitional temperature. The importance was attributed to the revelation of a strong tie between the surface and bulk properties of liquidlike polymers through cohesive-energy density. At the time of this publication in 1968, Lee also admitted that further refinement would be needed with respect to allowances for hydrogen bonding. A 1983 summary of the literature on the subjects of contact angle equilibrium and the spreading of liquids on solids has been compiled by Marmur (385). In this report, the Young equation has been critically reviewed and redeveloped to apply equally to rough and heterogeneous through rigid and insoluble solid surfaces. Also critically examined are line tension and molecular interaction approaches to contact angle equilibrium. Marmur discusses three types of spreading situations, including (1) forced spreading, (2) spontaneous spreading to an equilibrium contact angle, and (3) spontaneous spreading when no equilibrium contact angle exists.

From various earlier comments, it should be quite clear that a very important factor relating to effective bonding should be the various measurements involved in the study of the wetting condition. For the purposes of this general review of the subject of wetting, it will not be possible to elaborate in detail on the many separate investigations that were conducted on wetting beginning in the early 1960s. It may be of interest for study by the reader to list some of these investigations which appear in the literature, such as those of Zisman (317–319,386–388), Fowkes (344,389), Good and Girifalco (120,390), Good (327,391–395), Good and Eibling (396), Neumann and Good (397), Good and Shoraka (398), Johnson and Dettre (399,400), Sharpe and Schonhorn (295), Owens and Wendt (316), Kaelble and Vy (328), Sheriff (401), and Schultz et al. (402).

Two 1991 investigative papers on the subject of the adsorption of polymers have been issued by Janardhan et al. (403) on the effect of polymer molecular weight in the adsorption process and Kawaguchi and Arai (404) on effect of surface geometry on adsorption and desorption of polymer chains.

When discussing the meaning of wetting potential in bonding different adherends, the earliest fundamental work of Zisman and coworkers (275,317–319,386–388) should be reviewed. Zisman established that a rectilinear relationship frequently exists between the cosine of the measured contact or wetting angle and the surface tension of the wetting liquid. He defined a critical surface tension of wetting as the value obtained as the cosine of the contact angle approached unity; i.e., as the contact angle approaches zero. Sharpe and Schonhorn (295), in 1964, supported Zisman in focusing attention on this role of spreading. They contended, however, that the adhesive must exhibit a surface tension at bonding temperature which is less than this critical surface tension measured on the adherend if a satisfactory adhesive bond is to be achieved. Sharpe has further stated that the effect of creating a spontaneously spreading situation involves the conditions that (1) the real surface area of contact must be increased and (2) the stress concentrations must be minimized. As a liquid, the adhesive (in the forms of solution or dispersed solid polymers, or molten polymer) must be able to flow more completely into the submicropores and crevices in the adherend surface, thereby displacing adsorbed gases or other contamination entities. Also, there needs to be a distinguishment between the terms *adhesion* and *practical adhesion*, which are not the same.

Adhesion refers strictly to an interfacial phenomenon, whereas practical adhesion refers to the strength with which two materials actually stick together; i.e., the force required to separate them in the form of a joint. While the processes of making and breaking adhesive joints really do not have a fundamental relationship to one another, the breaking strength will, in general, be more a consequence of how the joint was made. The time to achieve this state of strength, however, can be vastly different depending on the particular materials involved. For example, consider the combination of a low-surface free-energy solid like polyethylene and a high-surface free-energy material like a cured epoxy adhesive (an aluminum oxide surface would be another similar-acting high-surface free-energy adherend). If the high-energy epoxy material is in liquid form and the low-energy surface is solid polyethylene, then the rate of approach to an equlibrium condition in the bondline area will be intolerably slow. However, if the low surface energy polyethylene can be made fluid by melting, then its flow over the solid cured epoxy (or aluminum oxide) can approach equilibrium at a greatly accelerated rate.

The major consideration is to increase the probability for developing stronger van der Waals forces. Sharpe and Schonhorn mention that chemisorption cannot be excluded from consideration of what causes adhesion strength. However, they postulate that this mechanism is probably one that occurs after molecular contact, at which time the van der Waals forces are already in operation. Thus, the presumption is that the more important aspect of chemisorption is to increase the permanence of the adhesive joint by retarding or preventing the destruction of the interface by weathering rather than significantly influencing the actual mechanical joint strength. The reader should be reminded that many investigators have shown little or no relationship between the initial strength of a joint failing entirely cohesively and the ultimate durability of the same joint in aggressive weathering service.

DeBruyne (405) as early as 1939 postulated an adhesion rule stating that "strong bonds can never be made between polar adherends and nonpolar adhesives or between nonpolar adherends and polar adhesives." Sharpe and Schonhorn's work with molten polyethylene (as a nonpolar adhesive) and epoxy or aluminum oxide surfaces (as polar adherends) caused them to suggest that the rule may be only half true. It is true that polar adhesives will likely have higher surface free energies than nonpolar adhesives, so that wetting would be expected to be poor. However, surface energetics are actually favorable for the polar adherend and nonpolar adhesive combinations, and strong bondings were fabricated by Sharpe and Schonhorn by melting polyethylene as an adhesive over strongly polar cured epoxy or aluminum oxide adherend surfaces. Later in 1963, DeBruyne admitted that only half of his rule could be supported in view of newer technical disclosures (406). In addition to obtaining strong joints between melted polyethylene and cured epoxy or aluminum oxide, Sharpe and Schonhorn also secured good bond strengths to copper and stainless steel adherends. Obviously, the secret is to employ the nonpolar materials in the melt condition, in which case, they do have the ability to readily wet, spread, and adhere with high force to these highly polar solid surface metal adherends.

Huntsberger (407) contended that the criteria for satisfactory bonding when the adhesive exhibits a surface tension at temperature which is less than the critical surface tension of the adherend is not entirely satisfactory because application of thermodynamics has been made to a system which is not representative of a real adhesive system. Subsequently, he has stated a preference for considering wetting as the process of achieving molecular interfacial contact and the state of wetting as the number of molecular contacts between the phases comprising the interface as compared with the maximum number of interfacial contacts possible for the system at equilibrium.

Schonhorn et al. (408,409) and Cherry and Holmes (410) have studied the kinetics of spreading of polymer melts by heating them under nitrogen with an aluminum cylinder surface. They observed that the flow is composed of small drops of molten polymer which are in the shape of spherical segments. The controlling factor for flow seemed to be the rate of change of the contact angle rather than the initially measured contact angle. Dettre and Johnson (411) have especially investigated the case where the equilibrium contact angle is low and the bulk viscous forces should offer significance to the spreading. They made the observation that a foot did develop at the base of the spreading drop; a prediction that had been made earlier by Schonhorn and coworkers but never actually observed. Van Oene et al. (412) also studied the rheology of wetting with polymer melts. They viewed the spread of poly melts on clean aluminum as an example of adhesion to a high-energy surface, but they also studied the results using fluorinated polymer (Viton B), which would be a very low-energy surface.

VI. RHEOLOGICAL (VISCOELASTIC) THEORY

It would appear that research into the theory of adhesion has presented both challenges and stimulation to many branches of science. If a science of adhesion does exist, it must be regarded as one of very broad scope whose various disciplines have not been totally identified or intergrated. We have just finished a discussion of one class of theory (adsorption), which normally seems to involve various surface properties based on a thermodynamic frame of reference. A second major area of scientific development in adhesion studies more relates to the theory of how the adhesive joint itself responds. This theory attempts to interpret basic adhesion phenomena in terms of the analysis of the consequences of stress and strain in an assembly of adhesive interfaces and adjacent bulk phases. Such theories most commonly evolve from considerations of engineering mechanics and stress analysis. Their great importance stems from their abilities to provide special criteria for practically designing and making test evaluations of fabricated joints.

The more perfect understanding of the theory of adhesion would, of course, come from being able to correlate the different sets of data that arise from the use of these different theory approaches. An early example of such dual theory application can be cited from the papers of Dahlquist in 1964 and 1966 (413,414). He was able to find a correlation between the peel force of pressure-sensitive tape bonds to various adherends and the respective surface free energies measured for these materials. This appears to be the advantageous combination of a fundamental theory (adsorption) and an engineering theory (rheological). While the former theory relates well to understanding the surface and surface free energies in terms of reversible thermodynamic equilibria, only the latter theory properly considers the rate-dependent adhesive to cohesive failure transitions. The thermodynamic criteria of polymer adhesion would seem to be applicable only to the flow region of response.

That there is a close connection between the rheological and adhesive properties of polymers has been well established. The molecular properties of polymers that control such responses could broadly be separated into the three categories of (1) chemical composition (monomeric units, side groups, secondary bonding forces, and cohesive energy density), (2) molecular structure (tacticity, main chain symmetry, molecular weight distribution, chain branching, entanglements, and cross-linking, and (3) molecular free volume, degree of crystallinity, temperature, pressure, and time. A complete discussion of polymer adhesion must consider each of the properties of the bonding phenomena, the stable bond, and finally, the situation of bond fracturing.

Rheological theory is closely related to the current concepts of polymer deformation and fracture (254,280,415–420). This theory treats adhesion phenomena as a rate process, as described by

Hatfield and Rathmann (421) and Kaelble (422–426), and can be complementary to thermodynamic response data. In particular, the viscoelastic-related data lends itself to studying rate-dependent adhesive to cohesive failure transitions which are not predicted or adequately explained by the thermodynamic approach to understanding adhesion. As early as 1954, however, Bright (427) had reported that rheological processes apparently controlled both the rate- and temperature-dependent transition from interfacial to cohesive failure in an assembled joint.

It must be admitted that the adhesion properties of polymers are still primarily determined by mechanical testing of the adhesive in a bonded joint. All mechanical tests employed, in turn, are subject to the disciplines and analyses that apply to rheological experimentation. There are three classic methods of rheological or adhesion measurement: (1) static, (2) transient, or (3) dynamic with respect to the factor of time. The three classic methods of measurement above are, in turn, divided into two groups. The first imposes strain as the independent variable and measures stress as the dependent variable and offers analysis of the rheological properties in terms of the relaxation spectrum of the polymeric adhesive. The second imposes a time history of stress and measures strain as the dependent variable with final interpretation in terms of the retardation spectrum. Overall, this covers six classic methods of measurement for obtaining information concerning rheological or adhesion properties of polymers, as listed below:

Type	Imposed fixed condition with time	Measured variable condition with time
1 Static	Strain	Strain relaxation
2 Transient	Strain rate	Stress and strain
3 Dynamic	Sinusoidal strain and frequency	Sinusoidal stress
4 Static	Stress	Creep
5 Transient	Stress rate	Stress and strain
6 Dynamic	Sinusoidal stress and frequency	Sinusoidal strain

A few examples of the utilization of some of these six methods can be cited from the literature. Tobolosky (280) has reviewed the extensive contribution of method 1 (static stress relaxation) toward understanding the relations between polymer structure and rheological properties. Weidner and Crocker (428) have shown the utilization of method 2 (constant rate of strain) testing for investigating the rheological and fracture properties of adhesives in joints. Dynamic mechanical measurements by either methods 3 or 6 are essentially analogous. Ferry (416) has adequately reviewed such dynamic measurements. The data accummulated via these methods have been especially valuable in establishing basic structure-property relationships, as shown by Ferry (416) and Bueche (281). Oberst (429), Kerwin (430), and Snowden (431) have also used these methods in predicting vibration-damping properties of adhesive bonds where the joint testing can be considered a nondestructive method of evaluation. Static and transient-creep measurements, as by methods 4 and 5, have found primary application using un–cross-linked adhesives, as reported by Dalhquist and coworkers (432, 433). Finally, a 1962 discussion of expeirmental apparatus utilizing constant stress-creep by Leaderman (434) as applied to cross-linked polymers may be pertinent.

Bikerman (361,363,435–437) has had his name often associated with the rheological theory in the literature and it was so-named by him in publication. We can interpret some of the more fundamental differences between the rheological and molecular theories by considering what is required in forming an adhesive joint. There appears to be stages of such formation involving first a macroprocess of establishing molecular contact between adhesive and adherend, and the further growth of the area due to wetting and flow processes. The second stage has been described as a microprocess of attractive interactions related to sorption. While proponents of both molecular and rheological theories might agree on this general series of events, it should be pointed out that their respective needs for these two stages is quite diverse. Practically speaking, rheological theory advocates have no real need for the second stage mentioned above, since the sorption mechanism does not really affect the strength of joints according to their theory. In like manner, the molecular theory has no applicable explanation involving molecular interaction along the interface of two phases being described by volume processes of reversible and irreversible phase deformations. However, both stages do have significant effects on adhesive joint strength and need to be considered in any unifying theory.

The concept of a possible weak boundary layer being present in an improperly fabricated joint needs to be mentioned when discussing the rheological theory. Bikerman (361–363) contended in the 1960s that joint strength was independent of the specific adhesion energy but was determined by the cohesional strength of the weakest element in the cross section of the joint. He reasoned that joint failure in such improperly made joints must lie in this "weak boundary" area. Furthermore, failure should always take place either in the adhesive (cohesive-type failure) or in the adherend in properly fabricated joints. Patrick (438) first reported the fact that interfacial bonds could not be failed mechanically, with later confirmation by Bikerman (436,437). This hypothesis has been shown, however, not to be universally correct by Briggs et al. (439). The existence of such a layer in certain instances has, however, been demonstrated by the use of some of the surface-sensitive analytical tools now available.

Bikerman also introduced the concept of crack-growth into this theory, pointing out that simple physicochemical and thermodynamic arguments together with this probability of crack-growth during joint rupture might cause us to believe that the energy of interfacial interaction should always exceed the cohesive energy of a weaker phase. Thus, the probability of crack-growth along the geometrical contact boundary is immeasurably smaller than that of a rupture within a phase.

Additional proof for the weak boundary layer theory has been gathered by a number of investigators of the apparent interfacial failure (not visually cohesive) in the form of weak bonds between aluminum and polyethylene. While it was contended that no visual evidence of cohesive failure would dictate an interfacial failure, Clark et al. (440), Clark et al. (441), Briggs et al. (442), and Briggs and Brewis (443) have all contributed evidence from XPS examination of the failed surface that about one or two monolayers of polymer can be detected on the aluminum surface. Obviously, a weaker layering condition in the polyethylene near to the interface had failed cohesively. However, Crocombe and Adams (444) have stated that it is not necessary to explain the failure mode in terms of any weak boundary layer condition, since the peel test geometry used is such that stresses are going to be highest closest to the interface. The argument of Bikerman (445,446) almost 20 years earlier had been his reasonable supposition that the surface layer of polymer with its different chemical composition from the bulk could mean that poorer mechanical properties could also be present.

Berlin and Basin (358–360) and Kardashov and Vakula (447), though proponents of a molecular theory, accepted the Bikerman concept. Raevskii (133) stated that the strength of joints is apparently determined by the degree of completion of molecular contact between the adhesive and adherend more than the degree of interaction along boundaries, therefore, Bikerman and the rheological theory is the only correct theory that actually relates to the quantities responsible for the true strength of joints. It needs to be remembered that joint rupture in cohesion must be considered possible when attractive interactions are actually in place along the interface whose energy exceeds the cohesional energies in the phase volumes.

Kaelble (422–425) has measured surface energies of certain polymer solids and showed this data did not account for the contribution of three-dimensional network structure of the polymeric bulk phase to its total work of cohesion. This was in conflict with the conclusion of the absorption theory supporters that true interfacial failure was an extremely unlikely possibility (since the attractive forces are invariably greater than the cohesive strength of the adhesive). Kaelble has claimed evidence of finding true interfacial failure in polymer-to-polymer bonding. This evidence included supporting critical surface tension measurements. Further, he found that transitions from interfacial to cohesive failure were (in all his examples) associated with the coincidence of the entanglement slippage rates with the rate of adhesive deformation. Peel adhesion properties could only be correlated to the calculated works of adhesion and cohesion obtained by surface free-energy measurements during the flow (or terminal relaxation) region of adhesive rheological response.

In 1991, Chung (448) published a paper offering a unified theory on adhesion with added guidelines on adhesion.

VII. TECHNIQUES FOR MEASURING ADHESION

Mittal (449) has published an ASTM special report titled "Adhesion Measurement: Recent Progress, Unsolved Problems, and Prospects" which offers an excellent review of the measurement procedures. Mittal (450) has suggested that the material which adheres to the adherend be designated "adherate." If the adherate is liquid, then basic adhesion can be theoretically calculated or indirectly determined. Muir (451) has shown the calculation of the thermodynamic work of adhesion in terms of the wetting behavior of such a liquid adherate. Techniques for measuring practical adhesion of thin-to-thick films and coatings have been reviewed by Mittal (452,453). Direct film pull-off methods have been reviewed by Jacobsson (454). For information on practical adhesion measurements involving thick films, the reader can be referred to reports and technical publications by Hitch and Bube (455), Jacobson (456), Anjard (457), or Hitch (458).

VIII. SUMMARY

The most satisfactory statement for achieving good adhesive performance in a bondline will be the situation existing when the interfacial "molecular-contact" density is large. This must be so regardless of the specific nature of any interactions along the interface, since dispersion forces alone have been calculated to be more than sufficient for preventing failure in a properly fabricated joint.

Poor performance in a joint must be closely identified with a nonuniform contact between two material surfaces. Interference with such contact might be due to (1) molecular configuration of the two materials, (2) any inability to achieve maximum wetting or thermodynamic equilibrium at the interface, or (3) energy effects associated with the polymer morphology may be sufficiently large that the point of thermodynamic equilibrium does not correspond to the state of maximum contact. Where maximum wetting at equilibrium is achieved, however, the kind or magnitude of interfacial reaction may be relatively unimportant unless steric limitations prevent achievement of anything but a low contact density.

Complete wetting will be achievable at equilibrium for most practical systems, but real systems can exist in nonequilibrated states with corresponding interfacial discontinuities. Such conditions will obviously constitute a major factor in establishing the adhesive performance. Other important factors that determine the state of real systems will be: (1) the rate of wetting, (2) the viscosity of the adhesive, and (3) the magnitude of the interfacial reactions. Evidence has been developed to show that interfacial separation can occur in spite of its low projected probability based on certain adhesion theories.

It can still be stated with certainty that no single theory of adhesion proposed to date can explain the phenomenon of adhesion entirely satisfactorily using only the tenets of that single theory. Because certain terminology has been established in the literature relative to the variety of theories proposed, it will probably remain impossible to use any single name entity that can be all inclusive for explaining all aspects of adhesion science. However, there should be no reason why present workers in the field cannot accept appropriate parts of the several existing theories discussed in this chapter to gain the widest possible view of the overall subject.

3

Surface Characteristics of Aluminum

I. GENERAL CONSIDERATIONS

The nature and character of the surface of an adherend is very important in establishing the strength and permanence of the adhesion of an adhesive to that surface. The critical need for proper aluminum surface pretreatment for bonding in aerospace has been widely established by investigations conducted in that industry. Representative investigations at Boeing to demonstrate the critical importance of the nature of the aluminum oxide on joint durability are those publications by Benoliel (459) and Smith (460–462). Recently, Clearfield (463) reviewed various oxide conditions used most effectively for durable bonding in the 1990 *Engineered Materials Handbook* of ASM. As additional background on how surface oxides on aluminum can be studied, the author suggests a 1970 publication on *Characterization of Solid Surfaces* edited by Kane and Larabee (464), 1980s chapters by Buckley (465), Davis and Venables (466), and Danforth (467) and a recent 1990 chapter on "Surface Analysis Techniques and Applications" in ASM *Engineered Materials Handbook* by Davis (468). For background reading on the beginning of the aluminum industry in the United States, the reader may find an early book by Alcoa pioneers Edwards et al. (469) very interesting.

An amazing aspect of aluminum and other metals is that no adhesion between adhesives and the metal atoms is ever involved. Instead, the jointure is between the adhesive and a metal oxide which forms immediately when aluminum atoms are contacted by oxygen atoms in the environment. Such oxides can exist in a wide variety of forms, and it is the nature of these oxides that needs to be evaluated and controlled to make the most durable joints. The nature of these different oxides are influenced by the manufacturing process but may also be highly influenced by environmental conditions present during shipping, handling, and storing. The most widely observed oxide variations are amorphous, bayerite, and boehmite in adhesive-bonding situations. The literature is often confusing on this subject, and the reader is referred to Table 4 for nomenclature as published by Wefers and Bell and Wefers and Misra at Alcoa in "Oxides and Hydroxides of Aluminum" (470). Other discussions pertaining to nomenclature and various identifications presented in the literature for specific types of aluminum oxide can be found in references by Stumpf et al. (471), Ginsberg (472),

Table 4 Comparison of Aluminum Oxide Nomenclatures

Mineral name	Chemical composition	Alcoa (1930)
Gibbsite	Aluminum trihydroxide	Alpha alumina trihydrate
Bayerite	Aluminum trihydroxide	Beta alumina trihydrate
Nordstrandite	Aluminum trihydroxide	
Boehmite	Aluminum oxide hydroxide	Alpha alumina monohydrate
Diaspore	Aluminum oxide hydroxide	Beta alumina monohydrate
Corundum	Aluminum oxide	Alpha alumina

Source: From Ref. 470 (Table 1.1, p. 2).

Moiseev (473), Teter et al. (474), Bohm (475), Papee et al. (476), Tertian and Papee (477), and Frederickson (478).

In adhesive-bonding literature, the most common oxide designations encountered have been amorphous, bayerite, boehmite, pseudoboehmite, and hydroxide. Since the term *pseudoboehmite* does not appear in Table 4, a short description may be needed. Pseudoboehmite is a poorly crystallized form of boehmite whose x-ray pattern shows broad lines that coincide with the major reflections of well-crystallized boehmite oxide. The amorphous form is described as x-ray indifferent, whereas the next three offer some form of crystallinity. From table 4 we see that bayerite is a form of the trihydroxide, whereas boehmite and pseudoboehmite are aluminum oxide hydroxide forms. The infrared spectra differ and have been one means of identification (478). In ambient temperature water, bayerite is the form usually produced, so it is not unexpected that this form of aluminum oxide is reported on aluminum exposed to natural weathering conditions. This trihydroxide will be transformed at higher water temperatures to boehmite (the oxide hydroxide condition). Harrington and Nelson (479) reported that exposure of aluminum oxide to water temperatures above 176°F (80°C) causes a transformation to boehmite, and Tolley (480) suggested a lower transition temperature of 167°F (75°C). Bryan (481,482) and Altenpohl (483–487) described the conversion reaction as initially fast followed by a fairly quick slowing after about 4–6 hr. Altenpohl also studied the ability of boehmite oxide to provide corrosion protection to the aluminum surface under corrosive environmental conditions. Hart (488) concluded, in 1954, that immersion times from 1 to 60 min afforded virtually linear oxide development rates. This was the basis for the author's selection of 1- to 60-min boiling water conditions for producing varying thicknesses of boehmite for study of bonding strength potential and the influence on producing added durability responses (39). Murphy and Page (489) also published early observations on the formation of both bayerite and boehmite oxides under water-immersion conditions below 160°F (71°C). The boehmite form predominated above that temperature and at the boiling point. The latter observation would predict the presence of the boehmite form on anodized aluminum surfaces that have been sealed in boiling water, which was confirmed by a number of investigators. These authors also evaluated the joint strengths obtained when chromic acid deoxidized surfaces were rinsed for 10 min in various temperature waters and dried and bonded. Thelen (490) also found best bonding strengths were obtained when rinsing after aluminum deoxidation was conducted with water kept around 155°F (69°C).

Other investigators, like Hamilton and Lyerly (491) at Picatinny Arsenal, have referred more to the degree of hydration rather than to the identification of a specific-type oxide when discussing bondability to aluminum surfaces. Studying both bare and clad 2024-T3 aluminum after alkaline or chromic/sulfuric acid deoxidizing, they found the oxide after alkaline soaking was most highly hydrated, apparently brittle, and cohesively weak. A thinner and less hydrated oxide resulted from FPL (dichromate/sulfuric acid etching), whereas the oxide after the anodizing process was substantially thicker but even less hydrated. They performed limited, but similar, studies using 6061-T6 adherends and offered the same conclusions.

Calculations from thermodynamic considerations have not always agreed with the above observations as stated from conclusions of McDonald and Butler (492). These investigators studied the thermodynamics of aluminum-water systems at temperatures from 77 to 572°F (25 to 300°C) with calculations that confirmed the oxide obtained in high-temperature water was indeed boehmite. However, they also stated that boehmite should be metastable at temperatures less than 303°F (150°C), which would not predict the long-term durability shown by boehmite-surface interface joints as prepared by Minford (39) and exposed for up to 2 yr to hot humidity soaking conditions at 125°F (52°C).

The lack of agreement among many investigators about the oxide present under particular conditions is especially shown by the investigations of Weber and Johnson (493) and McCarvill and Bell (494,495). As pointed out in Chapter 1, the results obtained by many investigators such as Bowen and co-workers (70–71) seemed to confirm that the oxide produced by FPL deoxidation was boehmite. Nevertheless, both Weber and Johnson and McCarvill and Bell have contended that the oxide produced by this treatment was gamma-alumina.

It is a thin, invisible, and extremely adherent oxide layer on aluminum adherends that must be controlled to promote high strength and good durability in aluminum-bonded joints. This oxide instantaneously forms over any freshly abraded or chemically etched aluminum adherend when oxygen is present. The high chemical polarity of this oxide layer promotes the additional layering of polar water molecules, as shown in Figure 4. The presence of up to 20 molecular layers of water has been shown on so-called dry aluminum by Bolger (496).

The most comprehensive study of the nature of the natural oxide film on aluminum was initiated at the Alcoa Labs in the 1940s and reported in publications by Keller and Edwards (497,498). The complexity of the oxide formed in 10-mo outdoor weathering on a 99.3% pure aluminum surface can be seen from the work of Stroup reported by Keller and Edwards (497) in Table 5. It can be seen that the largest fraction of water and gas that is driven out of this oxide layer is liberated at temperatures up to about 329°F (165°C). However, substantial additional water and gas are not liberated until the heating temperature is up to 680°F (455°C). Additional traces are still tied up and require heating to the vicinity of 932°F (500°C). The variety of gases bound up in the natural aluminum oxide surface have been further identified as consisting of hydrogen, carbon monoxide, and dioxide, with some traces of sulfur-containing gases. It may be of interest to the reader to relive some of the early attempts in the lab to measure the speed and conditions under which the native aluminum oxide forms

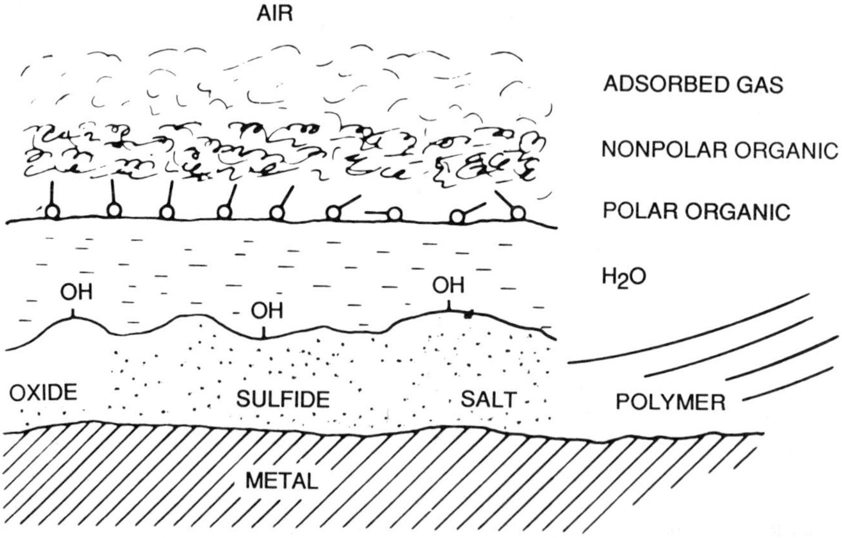

Figure 4 Hierarchy of spontaneously adsorbed layers on a metal surface. (From J. D. Minford, Adhesives, Durability of Structural Adhesives [A. J. Kinloch, ed.]. Applied Science Publications, London, Fig. 3, p. 150 [1983].)

Table 5 Water and Gaseous Liberations from Aluminum Surface[a] Under Various Heating Conditions

Preheating in dry air conditions before heating in vacuum to collect evolved gases	Evolved water vapor[b]	Evolved gas[c,d]
None	69	68
2 hr at 329°F (165°C)	24	41
2 hr at 500°F (260°C)	20	33
3 hr at 680°F (360°C)	3.5	15
4 hr at 850°F (455°C)	1.9	9.8
3 hr at 865°F (465°C)	1.0	10.6
12 min at 932°F (500°C)	1.5	7.0
26 min at 932°F (500°C)	Trace (4)	3.8

[a]Specimen was extruded 2-in rod (99.3% Al) exposed outdoors 10 mo before testing. Samples were short lengths with surface area of about 196 cm².

[b]Units represent milliliters in gas form under standard pressure and temperature (760 nm, 0°C) per 1000 cm² of aluminum surface.

[c]Gas evolved on subsequent heating in vacuum to about 1110°F (600°C). Units as above in [b].

[d]Gases identified as hydrogen, carbon dioxide, carbon monoxide, and traces of sulfur-containing gases.

Source: Data from Ref. 497 (Table II, p. 197).

from the early literature reports from 1927 to 1946 by Vernon (499), Gulbransen (500–502), and Haas (503).

Bainbridge et al. (555) conducted investigations relative to the comparative adhesion developed to oxides formed in a variety of ways, as shown in Table 6. In Table 6, they compared the peel strength of polyethylene coatings to air-formed oxide films and barrier oxide films that had been hydrated in water. Further, the coating was performed in the presence of a nitrogen or an air atmosphere. While the initial peel strengths were very similar for all conditions before hydrating, those air-formed oxide joints could be nine times stronger when coated in air compared with nitrogen. With barrier oxide joints, a difference was observed but the strength values dropped to only three times greater.

While somewhat self-limiting, natural oxide film can reach a thickness as great as 1000 Å. It is dependent on the weathering conditions, as shown in Figure 5, where the increasing thickness of the naturally forming oxide is plotted in terms of increasing surface electrical resistance. The natural oxide generally has a low structural integrity and variable chemical wetting potential with many adhesives. Such factors obviously decrease the potential for bonding if high structural strength is demanded. The native oxide can also undergo hydration to form the bayerite form of oxide in ambient water-exposure conditions. This form is undesirable for many demanding structural joints because of its readiness to fracture under stressful service conditions below the cohesive strength of many structural-type adhesives. The desired oxide should always be stronger propertywise than cured adhesive to achieve the desired cohesive-type failure in service. Thus, the oxide structure and means of converting from one type of structure to another must be carefully studied to ensure that the excellent wetting potential desired for durable bonding is achieved together with a good cohesive strength in the oxide layer itself.

Table 6 Ninety Dead-Load Peel Strength Values of Polyethylene Coatings on Aluminum with Varying Hydrated Oxides Following 14-Day Immersion in Water

Surface preparation	Initial peel strength (kN/m)	Final peel strength (kN/m)
Hydrated air-formed film (coated in nitrogen)[a]	2.68	0.21
Hydrated air-formed film (coated in air)	2.14	1.80
Hydrated barrier oxide (coated in nitrogen)[b]	2.18	0.09
Hydrated barrier oxide (coated in air)[c]	2.54	0.28

[a]Declined linearly through 225 days then reached a plateau.
[b]Declined most steeply through 125 days then reached a plateau.
[c]Declined linearly through 225 days then reached a plateau.
Source: Data from Ref. 555.

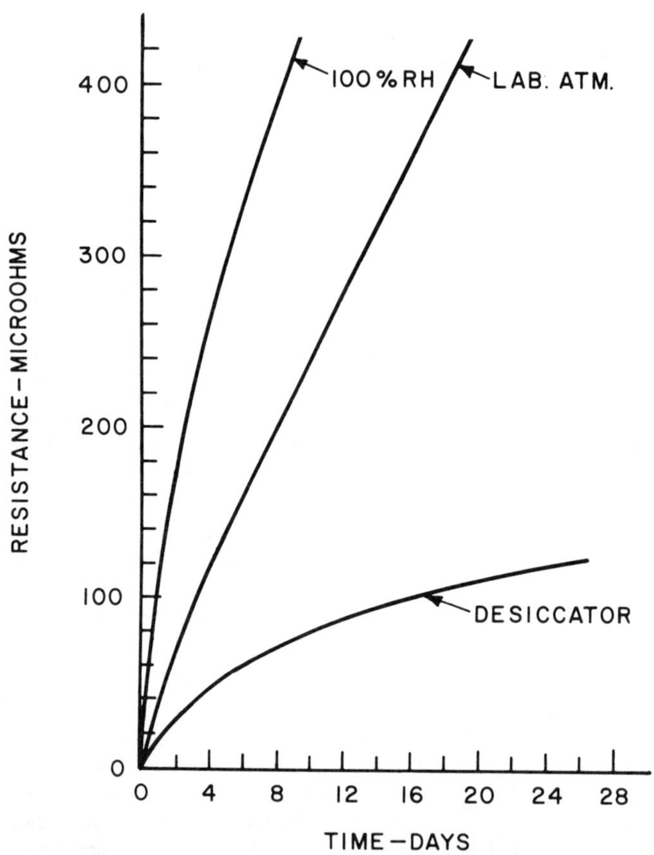

Figure 5 Effect of storage environment on 1100 alloy treated in R5 bright dip. (From an Alcoa Lab Report.)

Various interactions between surface oxide and environmental factors such as water, temperature, and atmospheric chemicals can take place and create bonding complications. Polar organic contaminants attracted by the polar character of the oxide can be a shielding between the oxide and the adhesive, as shown in Figure 4. Since the majority of structural adhesives are polar organic materials in themselves, the polar constituency of aluminum oxide is a positive factor in promoting good bondability. Bolger (496) lowered the flow of oxygen to an aluminum surface and found bonding could not proceed in a cracked oxide surface area where the polar oxide could not adequately form. Because nonpolar organic contaminants like lubes, oils, and waxes can be deposited on the aluminum surfaces, they can interfere with bonding. This is of special importance in industries like truck and automotive manufacturing where such materials are deliberately added for other purposes. There has, therefore, been an emphasis in recent years on formulating special adhesives which can effectively bond to aluminum adherends even when such agents are present. The most effective ones have been those which literally mix with the contaminate in an effort to displace it under high temperature-curing conditions. The high mobility of adhesive under such heating conditions obviously is an important factor and serves to explain why ambient temperature-curing adhesives have had less success. The main incentive to use such adhesives in nonaerospace manufacturing has obviously been related to the economies affected by not needing to specially clean and chemically pretreat the aluminum surface in the ways prescribed by aerospace manufacturing methods.

II. CHEMICAL AND PHYSICAL ASPECTS OF ADHESION BETWEEN ALUMINUM SURFACES AND ADHESIVES

Before considering the actual chemical structure and the designation of various oxides that form on aluminum surfaces, it seems pertinent to discuss the general background of what constitutes a joint involving those oxides with adhesive polymers. Brockmann (504) provided a summary of the facts which relate to the constitution of a metal joint from which the author has made interpretations with regard to aluminum adherends. Brockmann reminds us of the fact that as a first approximation there are up to five layers in the typical adhesive joint; i.e., aluminum/oxide/adhesive/oxide/aluminum. The strength and mechanical properties of the adhesive joint will, in turn, depend on the strength and mechanical properties of the polymer, mechanical properties of the oxide layer, and the adhesive forces between aluminum and its oxide and the oxide to the polymer layer.

With regard to the chemical aspects of adhesion between aluminum and adhesives, the various binding forces between atoms, ions, and molecules need to be considered with dimension in the 0.1- to 1.0-nm range. In contrast, the actual bondline dimensions are more likely to be in the 0.1- to 5.0-nm range. Further, the chemical structure of solids like aluminum and its oxide need to be described in terms of three-dimensional lattice structures of either positively charged electrons or of oppositely charged ions. By comparison, the chemical building blocks of polymers are molecules which stick together as solids only by intermolecular forces. The effects of a great number of intermolecular forces (though lower strength than metallic, ionic, or covalent bonds by 1–2 orders of magnitude) can still give rise to the high cohesion of polymers. (For some selected references that offer further data about adhesives, plastics, and polymers, see Refs. 122, 505–519.) Sharpe (520) has also offered a very helpful glossary of terms relating to adhesion science as well as an introductory article titled "Overview: Adhesives Technology" in the 1990 *Engineered Materials Handbook* (521). Zalucha (522) has contributed a review article titled "Adhesives and Adhesion" in the same book. If the oxide is stable and similar in lattice spacing to the metal, it will strongly adhere to the underlying metal, which is certainly the case with aluminum adherends. Practically speaking, bonding is present in a small boundary layer which varies in composition with increasing amounts of metal toward the surface and increasingly higher in percentage of oxide toward the top of the layer exposed for bonding. Although the chemical structure of organic polymers and oxides are fundamentally different, they do have adherence potential to one another which seems to depend strongly on the nature and activity of the oxide and structure and functional groups present in the adhesive polymer. For example, polymers with hydroxyl and amino groupings can form hydrogen bonds which seem to adsorb very well on aluminum oxide.

Only a limited number of polymers used as adhesives show good adsorption on aluminum oxide resulting in the consideration of the use of adhesion promotors and primers. Such materials must have the property of binding well to both oxide and polymer. Brockmann and coworkers (523,524) have suggested the rate of aluminum oxide hydration in the aluminum joint bondline as a primary factor in governing the joint durability performance. The means of slowing the rate of hydration should, therefore, result in enhanced bond joint durability. The most obvious procedures might involve reducing the amount of water present at the bondline, or using inhibitors which function by stabilizing the oxide against hydration, or the use of chemical agents which can bond between oxide and adhesive. Venables, Ahearn, Davis and co-workers (36,37,525–529) have been active recently in pursuing a method of adding stabilizing inhibitors using the phosphonic acid family of compounds. Specifically, they have shown that an adsorbed monolayer of an amino phosphonic acid (nitrilotris methylene phosphonic acid [NTMP]) could improve the stability of FPL-treated aluminum surfaces so that their performance in bonded "wedge-type" joints was almost comparable to similar aluminum adherends treated by the phosphoric acid anodize (PAA) procedure.

While the majority of the statements in this section have been devoted to the chemical aspects of aluminum/adhesive interfaces, there are some quite important physical aspects which relate intimately to the mechanical interlocking of adhesive into the aluminum oxide structure. Very recently, Davies and Ritchie (530) produced a related paper on this subject titled "Future Design Concepts for the Development of New Pretreatments of Aluminum Alloy/Metal Matrix Composites for Adhesive Bonding." These investigators stress the possibility that the morphology produced by the pretreatment of aluminum and any subsequent penetration into surface oxides profoundly influences the joint durability performance. Undoubtedly, the contribution of interlocking adhesive attachment between the adhesive polymers and various aluminum oxide morphologies is one of the most important factors in explaining the excellent aluminum joint durability performance published in the literature by Minford (531–536), Brewis (537), Kinloch (538), Venables and co-workers (31–34), Ahearn and Davis (539), Hardwick et al. (529), Ahearn et al. (540), and many others who will be considered in more detail in Chapter 10 on joint durability. For additional background information, Davies and Ritchie (530) refer to a 1989 publication of Davies and Kinloch (541) and a 1989 report by Davies and Moth (542) which describe a novel technique for aluminum oxide characterization within adhesively bonded aluminum alloys. They propose the concept of a "micro-composite" interphase within the joint and use to model the potential adhesive/oxide interfacial stress conditions, the cracking behavior observed, and the subsequent effect on environmental performance. Finally, they propose the optimization of future structural adhesive joints using metal matrix composites by producing "designer" oxides prior to the bonding. They found that while the locus of environmental failure in their pretreated double cantilever beam (DCB) joints appeared to involve fracture through the oxide layer, it might be described as fracture through the "micro-composite" interphase. Of many factors which may influence aluminum joint strength and durability, the pore penetration by the adhesive and corresponding stress state at the oxide/adhesive interface is apparently among the most influential.

A variety of investigators have produced papers relating to matters concerning the aluminum oxide/polymer interface and properties of alumina oxides in the period from 1984 to 1991. These include Carre and Schultz (543) on the role of the adhesive and cohesive properties of the polymer in aluminum/polymer adhesion in 1984. In 1986, Kurtz and Anderson (544) studied metal/polymer interfaces as observed with space-charge mapping techniques. In 1991, Vangani et al. (545) studied the thermodynamics of adsorption of polybutadiene on alumina and silica gel, Caper and Kirby (546) discussed the meaning of surface area measurements for microporous surface materials, and Papirer et al. (547) studied surface characteristics of aluminas in relation to their polymer absorption. In 1990, Blesa et al. (548) made a critical analysis of the thermodynamic data relating to the surface acidity of metal oxides when immersed in water. In 1991, Berteau et al. (549) studied acid-base properties of modified aluminas, Fleisher et al. (550) the electronic charge distribution and Lewis activity of surface aluminum atoms (a quantum chemical model), and Kanai et al. (551) the adhesion of chromium metallization on alumina surfaces as prepared by sol-gel techniques.

III. FORMS OF ALUMINUM OXIDE ON THE ADHEREND SURFACE

A. Natural Oxide

The first questioning about the invisible oxide on aluminum appeared in an 1890 publication by Richards (552). A main concern of the period was that this oxide was interfering with making solder joints to the metal surface. It has remained a problem in the welding and brazing of aluminum up to the present time. For the adhesive bonder, however, this oxide is a definite positive factor with the possibility for presenting the right surface chemistry for organic molecule attachment. There is, of course, some significant variation in bondability depending on the actual form of oxide present.

Oxide formed at room temperature was reported to be amorphous by Brouckere (553), Gulbransen (501), and Keller and Edwards (554). As mentioned earlier, the poor mechanical strength of the natural oxide can seriously limit the joint strength, especially where high structural strength is demanded. For low-strength joints with relatively low stress per unit area present in service, this natural oxide can still be acceptable providing it is effectively wet by the adhesive. It is for this reason that solvent-dispersed forms of adhesive are most often used when desiring to bond such surfaces in general manufacturing. The question becomes what is achievable with native oxide surfaces using the economical forms of surface preparation such as solvent cleaning or abrasion. In the former situation, it should be obvious that no physical or chemical change in the form of the oxide will result, whereas in the latter situation, some form of oxide regrowth can result depending on the particular weathering environment present. In either case, deposited soils, greases, lubricants, or corrosive-inhibiting films need to be removed before trying to bond. For many manufactured products even simple solvent cleaning has proved adequate for creating adequate service under mild or protected environmental conditions. This will usually be less than desired, however, for the more demanding structural applications, particularly those required for aerospace products.

Some investigators observed quite early that low surface energy polyethylene can bond unusually well to clean aluminum providing the polyethylene is applied in the melt condition. Typical authors would include Arbit et al. (556) in 1957 and Bockhoff et al. (557) in 1958. Bright and Malpass (558) in 1968 and Sykes and Hoar (559) in 1969 found that coatings do not adhere well when applied in an nitrogen atmosphere or with high concentrations of antioxidants present in the polyethylene. However, a special surface preparation might allow good adhesion even under nondeoxidizing conditions. For example, thick anodic films on the aluminum have led to good adhesion results even in the absence of any oxidation conditions due to mechanical interlocking of the polymer with the pores in the anodize layer according to Packham et al. (560) in 1974.

B. Regenerated Aluminum Oxides

1. *EFFECT OF ENVIRONMENTAL CONDITIONS*

The variation in aluminum joint strength with different forms of regenerated oxide using the same one-part, heat-curing, nitrile-modified adhesive can be seen in Table 7 as determined by the author. All the joints listed, with the exception of those made with vapor-degreased 6061-T6 aluminum, would fall under the classification of being regenerated forms of oxide. Even though the oxide in these regenerated forms would, in all cases, be much more uniform and reproducible than the naturally grown oxide, there is still a wide range of different strength potentials because of the varying thickness and morphology of the oxide structures. Thus, we can surmize that there are a number of different factors that can affect oxides and serve to produce diversity in their initial strengths and potentials for long-term durability service. In Table 7, it can also be seen that a very wide variation in joint strength can occur when bonding to the very thick anodic oxide (sulfuric acid [SAA]) type as compared to the thinnest anodic oxide type (phosphoric acid [PAA]). This difference is influenced by two important factors; i.e., the relative thickness and the morphology permitting more mechanical interlocking with the PAA-type oxide, as will be discussed in more detail later. Minford designed an experiment to clearly show the specific effect of oxide thickness by SAA anodizing 6061-T6

Table 7 Initial Bond Strength Values for 6061-T6 Aluminum Alloy Joints Pretreated with Various Etching and Anodizing Procedures

Surface pretreatment	Avg. shear strength (psi)	Avg. shear strength (MPa)
Vapor degreasing	4330	29.83
Deoxidine 526 etch at RT for 5 min	4970	34.24
VD + CrO_3/H_2SO_4 hot etching	5330	36.72
VD + Hcl/$K_2Cr_2O_4$ RT etching	4970	34.24
Chromic acid anodize	5513	37.98
Phosphoric acid anodize (30 V, 30 min)	5700	39.27
Phosphoric acid anodize (60 V, 18.5 min)	6030	41.55
Phosphoric acid anodize (110 V, 6 min)	6070	41.82
Phosphoric acid anodize (Boeing conditions)	6480	44.65
Sulfuric acid anodize (0.8 mil) (unsealed)	3940	27.15
Sulfuric acid anodize (0.8 mil) (boiling water seal)	3550	24.46

All joints fabricated with a one-part, nitrile-modified epoxy adhesive cured 15 min at 400°F (204°C) using 0.125-in (3.18 mm) sheet adherends. ASTM Specification D-1002 type lab specimens except individual coupons are sheared and assembled as individual lap joints rather than bonding two larger sheets and shearing as called for in the specification.
Source: Data from Refs. 533 and 534.

adherends to form different oxide thicknesses with the same anodic form of oxide. The data in Table 8 show the corresponding joints made with oxide thicknesses of 0.12, 0.30, 0.60, and 0.80 mil could range from 3550 to 5620 psi (24.46 to 38.72 MPa). Increasing joint strength being developed as the oxide thickness correspondingly decreased.

Wegman (561,562) has shown the effect on strength of regenerated oxides formed under controlled atmospheric conditions after abrading. With his best efforts to restrict the presence of oxygen, there was still sufficient oxygen present to form joints which showed cohesive failure when deliberately tested afterwards in a tensile test machine. This observation was not significantly altered by extending the conditioning time. He could only conclude that no detrimental reactions to adhesion had occurred even when the best efforts to exclude oxygen and moisture were employed. When 20% oxygen and 50% relative humidity (RH) conditions were employed, the bond strength decreased linearly to half the initial values within 30 min accompanied by increasing percentages of adhesive appearing failures. Higher joint strength values and higher percentages of cohesive-type failure followed the reduction of humidity to only 4%. The effect at the higher humidity conditions was explained by having a greater conversion of the oxide to the bayerite form (trihydroxide) and its poorer resistance to cracking failure at relatively low stress levels.

The effect of immersion in elevated-temperature water to convert bayerite to boehmite was mentioned earlier, which should lead to higher strength and more durable joint relationships. More discussion of the effect of various rinsing water conditions will be given in Chapter 4 on aluminum surface pretreating.

2. *RELATIONSHIP OF SURFACE CONTACT RESISTANCE*

The contact surface resistance has been used at Alcoa Labs for many years as one means of demonstrating the relative rate of oxide regrowth under varying environmental conditions, as shown in

Table 8 Effect of Oxide Thickness on the Strength of Aluminum Adherend Joints[a]

Electrolyte	Current amps/ft	Voltage	Time (min)	Approx. oxide thickness (mils)	Avg. joint strength (psi)	Avg. joint strength (MPa)
Phosphoric acid	—	5–10	20	0.05–0.08	6480	44.65
Phosphoric acid	—	30	30	0.10	5700	39.27
Phosphoric acid	—	60	18.5	0.10	6030	41.55
Phosphoric acid	—	110	6	0.10	6070	41.82
Chromic acid	—	40	30	0.12	5513	37.98
Sulfuric acid	11.25	—	7	0.12	5620	38.72
Sulfuric acid	11.25	—	21[b]	0.30	4960	34.17
Sulfuric acid	11.25	—	42[c]	0.60	4440	30.59
Sulfuric acid	11.25	—	60[d]	0.80–0.85	3940	27.15
Sulfuric acid	11.25	—	60[e]	0.80–0.85	3550	24.46

[a]All joints fabricated with 6061-T6 aluminum adherends with the following steps of pretreatment:

1. Clean 5 min in Diversey 914 at 150°F.
2. Etch 8 min in Alcoa A1 (5% AE16) at 120°F.
3. Desmut for 3 min in MT324 at RT + light wiping with wet cheesecloth.
4. Anodizing was followed with deionized water rinse at RT.

All joints bonded with EC-2086 one-part, nitrile-modified epoxy cured 15 min at 400°F (204°C).

[b]Same as Alcoa Alumilite 202 anodize coating.

[c]Same as Alcoa Alumilite 214 anodize coating.

[d]Same as Alcoa Alumilite 215 anodize coating but unsealed.

[e]Same as Alcoa Alumilite 215 anodize coating but sealed and desmudged by the Alcoa A482 process.

Source: Data from Ref. 534 (Table I).

Figure 6. In this instance, various commercial aluminum sheet was cleaned by vapor degreasing, deoxidized in hot chromic acid solution (Alcoa A3 treatment), and exposed to humid conditions. The contact resistance was shown to increase as the oxide regrowth occurred. We can see that the rate and pattern of oxide regrowth can be quite variable for different commercial alloys. Higher alloyed 2024, 5052, 6061, and 7075 alloys have distinctly higher regrowth rates as compared to lesser alloyed 1100, 3003, and 3004 alloys. We can interpret from Figure 6 that the higher strength (more alloyed products) while initially rising fast in their surface resistance, also continue to develop still thicker oxide layers at some significant rate. In contrast, the higher-purity alloys (1100, 3003, and 3004) show slower initial rises in surface resistance and leveling off after some relatively short incubation. With its modest addition of magnesium, 5052 alloy represents an alloy of intermediate strength and shows both sets of responses; i.e., high initial oxide growth rate with relatively rapid leveling off.

We might anticipate that the longer exposure time of any aluminum deoxidized surface would eventually lead to thicker oxides even with lesser alloyed products. This can be seen in comparing the relative oxide build-ups on architectural aluminum structures which have been exposed to the same atmospheric weathering conditions for varying periods of time. This results in an increasingly greater variability in the uniformity of the oxide layer and explains why attempting to bond weathered aluminum surfaces makes it necessary to thoroughly clean and establish a new regenerated oxide film on the surface. This may be accomplished by either mechanical abrasion or chemical deoxidizing techniques or sometimes a combination of the two. In the aerospace industry it has been mandated

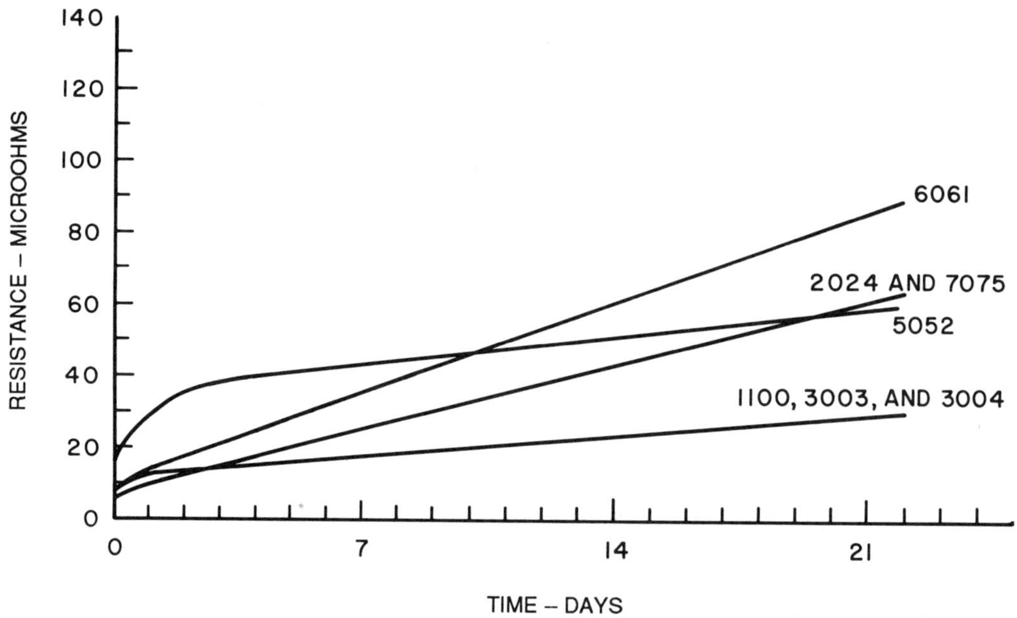

Figure 6 Contact resistance vs. time-surfaces treated in H_2SO_4-CrO_3 (5 min). (From J. D. Minford, Aluminum Adhesive Bond Performance, Treatise on Adhesion and Adhesives (R. L. Patrick, ed.). Marcel Dekker, New York, Vol. 5, Chap. 2, Fig. 3-7, p. 58 [1981].)

in manufacturing aerospace structures that the adhesive be immediately applied for joining purposes or at least a primer be applied with final bonding planned later.

3. SURFACE AND SURFACE/ADHESIVE INTERFACE ANALYTICAL PROCEDURES

Absolute characterization of the different oxide forms that develop from different kinds of surface pretreatment on aluminum has been significantly advanced in recent years with the development of various new analytical tools such as electron microscopy (EM), electron microprobe, scanning electron microscopy (SEM), ellipsometry, Auger electron spectroscopy (AES), ion-scattering spectroscopy (ISS), x-ray photoelectron spectroscopy (now referred to as XPS, but also referred to in the literature as electron spectroscopy for chemical analysis [ESCA]), secondary ion mass spectroscopy (SIMS), inelastic tunneling spectroscopy (IETS), Fourier transform infrared spectroscopy (FT-IR), x-ray emission spectroscopy (XES), high-resolution scanning electron microscopy (XSEM), energy dispersive analysis of x-ray fluorescence (EDAX), and specular reflectance infrared spectroscopy (SHIRS).

Because of the very high initial purchase price of such specilized pieces of equipment, the support of grants from the U.S. Air Force has been very important in accelerating the acquisition of such equipment for adhesion science investigations. On the other hand, this has made it quite difficult for organizations like the Alcoa Labs to compete, since such equipment had to be purchased by the corporation without the aid of government grants.

Baun et al. (563) wrote an important ASTM monograph discussing metal surface studies with SIMS, ISS, and AES. Powell (564,565) published a review of developments in the quantification of AES, XPS (ESCA), SIMS, and ISS techniques for surface analysis of different adherends. With respect to the anodized oxide on aluminum, Hanlin and Solomon (566) reported their study on anodic oxide characteristic variations due to changes in electrolyte, pH, bath temperature, anodization time, and applied potential. The AES and SEM techniques were primarily used in these investigations.

Although AES utilization has been frequently employed by many investigators in recent years, there can be a drawback due to heating effects imposed by the electron beam which can instigate some possible decomposition of the adhesive surface. The SIMS technique by itself has been shown to be a powerful tool for elemental surface characterization by Benninghoven (567) and Schubert and Tracy (568). However, uncertain or rapidly changing secondary yields due to changes in chemical bonding makes quantitative analysis virtually impossible, as shown by Schubert (569) and Werner (570). The SIMS technique can be useful when combined with other techniques such as ISS and AES which use beams of correct energy.

For the reader with special interest in investigation of specific aluminum alloy surfaces, investigations of 2024 alloy by McDevitt and Baun (571) and 6061 and 7050 alloys by McDevitt et al. (572) may be of interest. In both these investigations, the ISS, SIMS, and AES techniques were employed following pretreatment of the surfaces under various conditions. In 1975, McDevitt et al. (573) showed that 2024 alloy surface was magnesium rich, whereas Smith and Martinsen (574) reported similar observations on the 6061-T6 surface. This might not be wholly expected, since both these alloys contain other alloying elements in higher concentrations than magnesium, which is a major alloying element only in the 5000 series of aluminum alloys. The presence of significant magnesium in the 5000 alloy series has the effect of significantly elevating the tensile strength (as compared to commercial-purity 1100 alloy), yet with the maintenance of good resistance to atmospheric weathering and corrosive saltwater conditions. The alloying ingredients in the 6061 and 2024 alloys both support a higher tensile strength level in these alloys as compared to the 5000 alloy series. They also impart a much greater sensitivity to all manner of exterior weathering and corrosive saltwater exposures. This latter fact supports the aluminum joint durability evaluation results obtained by Minford. Using the same adhesives, Minford (575) found that the relative order of joint durability using hot condensing humidity or wet/dry saltwater cycling test conditions was 5052 joints > 6061-T6 joints > 2024-T3 joints.

McDevitt found distinct differences in the surface chemistry of 6061 and 7050 alloys depending on the surface pretreatments which included vapor degreasing, alkaline etching, nitro-hydrofluoric acid etching, or chromate conversion coating (Alodining). McDevitt did not conduct any long-term joint durability tests, so it is not clear what might be found as a correlation between his surface analysis results and durability test data. Based on the above background, McDevitt and Baun (576) have investigated the chemical characteristics of the composite oxide film found on 2024-T3 aluminum when an anodic barrier layer was formed after chemical pretreatments. It was apparent from these data that chemical elements which could be present in the surface layer formed by the surface cleaning could end up in the final surface layer when anodizing. This should be expected when it is recognized that the thickness of the anodized layer is generated continuously at the metal surface so that the original surface is the top surface of the anodized layer. McDevitt et al. (577) investigated the surfaces of PAA-pretreated aluminum and found the voltage and phosphoric acid concentration importantly influenced structure, morphology, and the general properties of the PAA oxide. Minford (533,534) investigated the long-term joint durability of Boeing PAA-specified treatment on 6061-T6 surfaces using one-part, heat-cured, nitrile-modified epoxy adhesive with exposure for 2 yr to room-temperature water soak, soaking in 100% RH at 125°F (52°C), intermittent corrosive saltwater spray, or 4 yr in an industrial or corrosive seacoast atmosphere. Excellent resistance to all weathering was obtained. Four different variations of voltage/time ratios were also employed and evaluated as to the initial joint strength and the ability to survive under a variety of stressing conditions in hot soaking humidity. While some modest difference in the initial joint strength was noted, variations in terms of displacement in the plotted stress-endurance curves proved negligible.

The interphase can be studied by microscopic, spectroscopic, thermodynamic, and kinetic procedures. Emphasis in this section will be on the first two analytical procedures. Baun (578) categorized 54 surface characterization methods according to which one of the six aspects of adhesion is being studied. Wightman and coworkers (579,580) have prepared two summaries of important analytical methods of surface characterization in adhesion. The most recent being via an invited lecture at the 1984 Symposium on Fundamentals of Adhesion: Theory, Practice, and Applications organized by Lee at the State University of New York at New Paltz.

It was recognized as early as 1972 at the National Aeronautics and Space Administration (NASA) that interdiscplinary research programs would be required to solve many of the problems that arise in the application of adhesive bonding to aerospace technology. Wightman and co-workers (581) issued their first report detailing their investigations in this regard using SEM, EDAX, XPS (ESCA), and SHIRS for characterizing adherend and fracture surfaces of special interest to NASA. Their final results were issued as a final report from Virginia Tech (582) and in Polymer Preprints (583) in 1975. Additional investigations of the same nature undertaken at NASA Langley Research Center were represented by publications by St. Clair and Progar (584), St. Clair and St. Clair (585), and Progar and St. Clair (586).

Neal (587) produced a detailed discussion and review of the literature dealing with the optical examination and monitoring of the surface of various adherend materials. The need to examine surfaces of any possible adherend is most important to the understanding and control of the manufacturing processes involving adhesive bonding. In addition to elucidating surface characteristics to help explain the variations in bond strength and durability found in fabricated joints, an adherend surface may be contaminated, damaged, or have an adsorbed gas layer which could cause poorer than expected performance.

Some of the most important techniques for analyzing the surface of metallic adherends and their interfaces with adhesives will be described below in some detail along with a variety of pertinent published references describing their origins and development.

a. General Background

The study of surfaces and interfaces with adhesive polymers has involved a multidisciplinary study using the fields of physics, chemistry, materials science, and a variety of engineering disciplines. The analysis of the properties of surfaces and the failures of these surfaces in bonded interfaces is absolutely necessary for any understanding of the processes of making bonded structures and predicting their service success. The fact that the properties of these surfaces and thin films on those surfaces can be quite different from the bulk materials has become increasingly recognized as where the focus of study must be if the fine-tuning aspects of adhesion science are to be understood. These differences are caused by differences between "bulk" and "surface" atomic bonds, as well as by the formation of reacted layers, segregation of alloy components, incorporation of electrolytic ions, adsorption of molecules from gases and liquids, or other phenomena. A complication for materials is that the chemical and physical properties may change dramatically from the first monolayer of atoms to the second, whereas for others, the change may occur only over hundreds of nanometers. This makes the actual thickness of a surface often difficult to ascertain and a factor which needs to be described in some fashion. Davis (588) has suggested that, unless otherwise indicated, the surface may be assumed to include the volume that can be investigated by the various surface-sensitive analytical techniques described below.

Quantification may be considered the most important chemical information that can be derived from the surface-sensitive analytical procedures in addition to the initial elemental identification which is possible. It can be achieved from XPS and Auger electron spectroscopy scanning Auger microscopy (AES/SAM) survey spectra showing the entire spectral range at low resolution. Alternatively, for XPS, it can also be achieved from high-resolution spectra of individual peaks. This quantification is usually obtained by using what has been given the name "sensitivity factors" (see Refs. 588–591 for more details). Special compilations of such factors for XPS analysis can be found in publications by Wagner et al. (592,593), Wagner (594–596), Nefedov et al. (597,598), and Evans et al. (599,600). Similar compilations using AES have been published by Davis et al. (601), McGuire (602), and Sekine et al. (603).

Another important kind of data for surface analysis concerns chemical state information. This is especially available from the use of XPS spectra, although the use of AES spectra is also possible but has been less used. Presumably this is because the AES spectra are customarily acquired with less energy resolution than the XPS spectra. This information is conveyed by finding small shifts in the binding energy of the peaks. Tabulations exist of the chemical shifts of the principal line of an element for many compounds which can assist in chemical state analysis. Data of this nature has been provided by Wagner et al. (592), Carlson (604), Wagner and Bickham (605), and Wagner (606). In many cases, the Auger peaks present in the XPS spectrum can also provide chemical state information as discussed by Wagner et al. (592,607), Wagner and Bickham (605), Wagner (606), and Wagner et al. (608). Davis et al. (601) have discussed detectable energy shifts with AES, whereas Haas et al. (609) and Davis et al. (610) have shown the line shape can serve as a fingerprint for a compound. Davis et al. (611) and Ramaker (612) have shown how line shifts can also be used to obtain a species-specific density of valence states.

Another important use of such techniques is to ascertain the distribution of elements in the near-surface region. Powell (510,511) indicated such simple distributions can include a random mixture, a surface enrichment of one species, an overlayer, a buried layer, or laterial inhomogeneities. Furguson (613) has described how features larger than the analysis area can be examined by point spectra, line scans, and maps. A depth profile on a surface is most commonly obtained by ion sputtering combined with AES/SAM, although small-area XPS may also be employed. A bean of ionized inert gases is used to erode away the surface while the AES/SAM or XPS measures the composition of the newly exposed surface. The beam allows the detection area to be representative of a single depth. Sputtering can be useful in probing depths from 2 nm to 2 μm (0.08 μin to 0.08 mil). (For some additional references of possible interest, see Refs. 614–621.) There is an alternative method of depth profiling with small probe beams (AES or SIMS) that is better suited for thicker regions (1 to 100 μm or 0.04 to 4.0 mil). Amplifications of the technique can be found in investigations of Thompson et al. (622), Walls et al. (623), Brown et al. (624), Lea and Seah (625), and Bierlein et al. (626).

Also, the use of surface behavior diagrams (SBDs) should be discussed briefly. The SBD is a graphic means for the display and analysis of quantitative XPS and AES/SAM data as discussed by Davis. They are particularly useful in the detailed investigations of surface and interfacial interactions. The SBD resembles ternary or quarternary phase diagrams, since they represent surface composition as a weighted sum of three or four basic compounds. Unlike true phase diagrams, they offer surface compositional information rather than bulk phase information. Changes in surface composition (during nonequilibrium reactions) can be traced as a function of reaction time, solution concentration, applied potential, and depth penetration of sample, as well as other parameters of possible interest. By comparing the path along which the composition evolves to those predicted by a proposed model, the mechanisms involved during a reaction on the surface or at an interface might be established. Surface behavior diagrams have been used by Davis et al. (37), Venables (34), Ahearn and Davis (38), Nefedov et al. (597), Davis (627), and Davis et al. (528) to study the evolution of the surface composition under hydration conditions. Davis (36,627), Ahearn and Davis (38), and Davis et al. (525) similarly studied the adsorption of hydration inhibitors from aqueous solution.

Finally, the importance of these analytical tools for studying the nature of bond failure should be mentioned, since failure analysis is a critically important aspect of both manufacturing and scientific investigaitons as well. Identification of the locus of failure is certainly necessary to establish both the cause of a product's failure and to establish a possible fix or remedy. It should be possible with these instruments to understand the mechanisms of crack initiation and propagation; thus identifying the weakest link in the overall joint structure. Four modes of possible bond failure are recognized in studying the failed surface of joints. Failure is classified as cohesive, adhesive, interfacial, or mixed mode. We generally think of cohesive failure as being in the adhesive cross section; however, a cohesive failure is also one that occurs within the oxide film, within a composite member of the joint, or even in the metal itself. Such a failure indicates that ultimate performance may have been reached for the particular system studied; however, improvement can obviously be made by redesigning the joint or replacing the weakest component where the cohesive failure occurred with a stronger or more durable component. An adhesive-type failure is generally recognized as being between the adhesive or a primer and the adherend surface. Obviously, the fix is to work on strengthening the interface where the failure occurred. An example of an interfacial failure would be one between the oxide film and its underlying metallic adherend. The fix is obviously on which creates a strengthening of that interface. Because the freshly exposed metallic adherend surface can reoxidize immediately, it can be very difficult to distinguish between such interfacial failures and one that is cohesive in the oxide, as shown by Clearfield et al. (628).

The mixed mode–type of joint failure is probably most often viewed when well-bonded joints fail, since there is likely to be some sort of competition in different areas of the bondline as to whether cohesive- or adhesive-type failure may occur first. Regardless, examination of the failed surfaces should be made initially with both electron microscopy and XPS or AES/SAM techniques, as discussed by Venable (33) and Goldstein and Yahowitz (629). It is important to remember that the development of SEM specimens utilizes a conductive film over the surface which prohibits effective surface chemical analysis. Thus, the investigator must employ separate, but supposedly identical, samples when using both SEM and XPS or AES/SAM techniques. Depth profiles might be needed to identify the specific locus of failure (especially in complex structures with multilayers), as discussed by Davis et al. (630). Once the locus of failure is determined, the cause of the failure can generally be visualized from the chemistry or morphology data.

Some of the causes of failure in aluminum adherend joints have been listed as (1) surface contamination prior to bonding (see Ref. 466), (2) poor surface preparation resulting in a smooth oxide surface (see Ref. 33), (3) hydration of the oxide surface (see Refs. 33,36,37,630), (4) incompatibility of surface treatment with adhesive (see Refs. 36,37), (5) disengagement (either partial or complete) of the adhesive from the porous oxide (see Refs. 36,528,631), and (6) elastoplastic facture of the adhesive (see Refs. 37,631).

(1) AUGER ELECTRON SPECTROSCOPY/SCANNING AUGER MICROSCOPY (AES/SAM). As mentioned earlier, the description and use of this technique can be reviewed in the *Handbook of Auger Electron Spectroscopy* by Davis et al. (601) and the chapter by Davis (588) in the 1990 edition of the ASM *Engineered Materials Handbook*. The AES technique involves the analysis of secondary electrons that are emitted from a surface that has been bombarded with a monochromatic electron beam. The energy distribution of these Auger electrons reflect the density of electronic states in the valence band according to Coghlan and Clausing (632) and Madden (633) and can be used as a "fingerprint" to identify different compounds. The elements present in the first three to four atom layers are revealed in the Auger electron spectrogram produced. The comparison of the Auger peak-to-peak heights (APPH) for a given peak can be significant with respect to the changes of surface composition after some pretreatment. The AES technique also has the capability of providing depth profiles and elemental maps with high lateral resolution that are difficult, if not impossible, to obtain with the XPS (ESCA) technique, which is discussed below.

By combining AES with inert ion sputtering, it is possible to measure the composition of the surface as material is milled away exposing the underlying surface layers. This procedure allows both elemental concentrations to be determined for the first few hundred nanometers of the sample and the thicknesses of thin overlayers. This sputtering rate for the removal of material does require that each spectrometer be calibrated for one standard material. Wehner (634) has published a compilation of the relative sputter yields for different materials.

In determining the thickness of layers, it is necessary to remember that sputter time will be proportional to the amount of material removed rather than the actual thickness, as would be determined with an electron microscope. Thus, if the sputtering rate has been calibrated for a dense oxide, the apparent thickness obtained by AES results will likely be a lower number. Sun et al. (635) have used these differences in measurements as indications of relative porosity of the oxide present.

While it is generally the condition with most materials and primary electron beam energies that fewer electrons are ejected from a surface than are impinging upon it, differences can arise in the testing of materials depending on whether the test sample is a conductive or insulative material. With the former situation, no extra problem develops because the electrons can flow to ground, maintaining the charge neutrality of the sample. In the latter situation, however, with an insulative material, the lack of electron flow away from the material can cause the sample to charge negatively and make it difficult to utilize the AES technique. In this case, a thin metallic overlayer can be one means of altering the situation, permitting electron microscopic studies. Any severe charge situation, however, will prevent the development of an AES spectrum. Even with mild charging, the peaks of the Auger spectrum may be shifted uniformly to higher kinetic energy. This degree of charging can sometimes be reduced or even eliminated, by reducing the electron beam current, raising or lowering the beam voltage, reducing the angle of incidence of the beam, or by light inert-ion-sputtering of the surface to provide a source of positive charge. In the presence of a very thin layer of surface contamination, however, this latter procedure cannot be employed.

Finally, in addition to charging, some surfaces (particularly polymeric surfaces) can suffer electron-beam-induced damage related to either electron excitation (dose-related effect), or heating (flux-related effect). Either mechanism can result in broken chemical bonds and decomposition of the surface. Where these variously described problems cannot be overcome, the XPS analytical technique will likely be substituted. (Other technical readings on AES can be found in Refs. 565, 604, 610, 612, 636–644.)

(2) X-RAY PHOTOELECTRON SPECTROSCOPY (XPS). This is a technique widely employed for surface and interphase analysis, which is described in *Handbook of X-Ray Photoelectron Spectroscopy* by Wagner et al. (592). It should be mentioned that this method has also been widely referred to in the literature under the name ESCA, which stands for electron spectroscopy for chemical analysis. Since the development of this technique by Siegbahn and co-workers (645) as early as 1967, there have been a number of additional books giving various details about this method (604, 637,646–650). A number of reviews have been written of the various aspects of XPS applications, including its scope and limitations compared with other analytical methods (651–660). Improvements in instrumentation to provide small spot analysis has been detailed by Wagner and Joshi (661) and Yates and West (662).

The XPS technique provides very good qualitative and quantitative elemental analysis of a surface plus a reasonably good interpretation of chemical bonding. As the name implies, the procedure involves bombarding surfaces with monoenergetic x-rays in an ultrahigh-vacuum environment. As photons travel through the material, some are absorbed and their energy is transferred to electrons which are ejected from the surface. Only photoelectrons originating near the surface have high probability of being emitted without a loss of energy and only these electrons can provide useful information in the XPS spectrum.

Kinloch and Smart (663) and Gettings et al. (664) used XPS on aluminum surface investigations, with the latter authors concluding that there was no correlation between the aluminum alloy or pretreatment and the surface levels of carbon found. Seah (665) has discussed the possible systematic errors inherently present in the XPS analysis procedure, whereas Seah and Dench (666) pointed out the sampling depth in XPS analysis of aluminum surfaces is around 5–10 nm, which means the detection of an element may only mean that it is residing within this depth of the surface rather than at the actual surface. Spectral interpretation has been discussed by Briggs and Riviere (641). (For additional readings of interest, see Refs. 526, 565, 605, 636, and 667–671.)

Major disadvantages of XPS are poor lateral resolution, a relatively weak signal, and possible nonuniqueness of its chemical shift information. Recent advances in instrumentation have led to significant improvements in the first two problems. It must be admitted that XPS will always be inherently less capable in these characteristics than AES because of the greater difficulties in building intense x-ray sources (relative to electron sources) and of focusing x-rays relative to charged particles. Davis (672) has presented an excellent table summarizing the advantages and limitations of XPS, AES, SIMS, and SNMS techniques for surface analysis.

(3) SECONDARY ION MASS SPECTROSCOPY/SECONDARY NEUTRAL MASS SPECTROMETRY (SIMS/SNMS). This has been yet another important analytical tool as it is described in "Secondary Ion Mass Spectrometry" by Heinrick and Newbury (673). This technique detects ions ejected from a surface by momentum transfer from an incident ion beam. It is capable of detecting all elements (including hydrogen) and can discriminate isotopes as well. Its very good sensitivity is due to the origination of the ejected ions from the very top layers of the impinged solid surface. According to Reuter (674), this sensitivity for some elements can be as low as a few parts per billion.

Quantitative analysis can be very difficult, as discussed by Feldman and Mayer (670). They have discussed the reduction of such matrix effects by detecting sputtered neutrals; i.e., SNMS. Because sputtered atoms are much more likely to be neutrals than charged particles, variations in the ion yield have very little effect on the number of neutrals. For detection, the neutrals may be postionized by electrons (see Ref. 675), plasma (see Refs. 676,677), or a laser (see Refs. 668–681). Elemental detection limits are relatively uniform for SNMS and average in the parts per million range according to Feldman and Mayer (670). These values can be greater or less than the corresponding SIMS values depending on the specific element and material involved.

Chemical information may also be available because ion clusters are detectable. Their interpretation can be difficult, however, because such clusters may have been ejected from the sample or formed in the vacuum above it. The SIMS technique has been widely employed for the chemical analysis of aluminum surfaces before bonding and after fracture to find degrees of adhesive or cohesive failure. Briggs (682), at ICI, has focused his attention on the use of SIMS for analyzing polymer surfaces using the spectra as sort of "fingerprints" for polyfilm identifications.

Unlike AES and XPS, SIMS is destructive by its very nature and imaging is only possible at the expense of depth resolution. A minimum number of atoms are required for analysis and longer running times cannot improve this detection limit. Low concentrations in a thin surface layer are detected by a slow-sputtering rate, i.e., a spatially large incident beam, whereas imaging analysis requires a fast sputtering rate, i.e., a dynamic mode of detection.

A method referred to as static SIMS/SNMS has been discussed in the publications of Benninghoven (683), Briggs (684), and Vickerman (685). In this technique, the ion beam current density is kept very low, so that the lifetime of a surface monolayer can be several hours to several days. The focus of the static SIMS is generally on the chemistry of the surface as revealed by ionized clusters of atoms ejected from the sample instead of detection of the depth distribution of trace impurities. Hofmann (686) has determined the surface sensitivity for static SIMS is governed by the escape depth of secondary ions (typically about 0.4 nm).

In contrast, dynamic SIMS uses a high ion beam current density for a fast sputter rate, as discussed by Sykes (687). Understandably, it is more concerned with depth distributions through thin films than with surface compositions. It has been used for the low-level detection of impurities. (Other discussions relating to SIMS and its modification procedures can be found in Refs. 465, 565, 637, and 688.).

(4) INELASTIC TUNNELING SPECTROSCOPY (IETS). Another important analytical tool in recent years is inelastic tunneling spectroscopy, as described in *Tunnelling Spectroscopy: Capabilities, Applications, and New Technology* by Hansma (689). In principle, it provides a sensitive method for measuring the vibrational spectrum of a molecular species adsorbed at the surface of a metal oxide. Another metallic film is evaporated onto the doped oxide to form a metal-insulator-metal junction. The current is recorded when a voltage is applied across the junction and a plot is made to create an IETS spectrum. Coverages lower than a monolayer can be studied with relatively simple equipment and without need for an ultrahigh vacuum. However, sample preparation is difficult and the spectra have to be taken at liquid helium temperature.

The IETS technique plays an important role in developing information about the adhesive/oxide interface, which is typically of the order of 10–100 Å thick. It is particularly sensitive for monitoring the chemical and physical state of a molecular substance adsorbed on oxide surfaces. The phenomena were discovered by Jaclevic and Lambe (690,691) in 1966. In the period 1973–1977, a number of studies of molecular adsorption were conducted by Klein et al. (692), Kirtley and Hansma (693), Skarlatos et al. (694), Diaz et al. (695), and Chu et al. (696). Hansma (697) reviewed the subject in 1977. For adaptation to adhesion studies between epoxy and aluminum oxide, the publications and presentations of White et al. (699–700) are pertinent. These studies were based on the earlier work of Lewis et al. (214), who used IETS to show the bonding of phenol to aluminum oxide surfaces. Shott and Field (701) also employed the technique in 1975. More recent uses of the method will be discussed later under the subject of mechanisms of joint failure.

(5) ION SCATTERING SPECTROSCOPY (ISS). Smith (702) suggested the use of ion scattering spectroscopy could be applied to a number of chemical problems as early as 1967. Baun (703), in the early 1980s, investigated the chemistry of aluminum adherends by the use of SIMS combined with ISS, since quantitative analysis using SIMS alone was difficult to achieve because of the rapidly changing or uncertain secondary ion yields due to changes in the chemical bonding situation. The ISS uses a beam of ions of correct energy for combination with SIMS. Bond failures classed as pure adhesive type have lately been examined by ISS/SIMS, and it could be shown that the aluminum surface had obviously not been wet by the adhesive; i.e., no trace of adhesive on the aluminum could be found. Baun (704) reported his analysis by the ISS technique of aluminum using a fractured wedge-type test specimen. The ISS technique does have the detriment, however, of being extremely surface sensitive.

(6) SURFACE REFLECTANCE SPECTROSCOPY (SRS METHODS). Some experimental techniques have developed and are collectively called surface reflectance spectroscopy methods. The resurgence of interest in this method has recently occurred with the introduction of Fourier transforms, which will be discussed later.

As mentioned, the typical approach to the characterization of polymer surfaces has been to use ion and electron spectroscopies which probe to depths of 10–100 Å depending on the technique. There have been some associated problems, however, and the subtle details of the organic bonding often could not be determined. In addition, the requirements of medium- to high-vacuum conditions and possibility of sampling beam damage restricted the ability to do certain types of "in situ" exposure studies. The use of infrared and Raman spectroscopies were well known for their use to determine the vibrational structures of organic polymers, but their application to surface analysis had been limited to internal reflectance infrared spectroscopy (IRIRS). This photon in–photon out probing procedure had an advantage over ion and electron spectroscopies of not requiring vacuum conditions. Also, the probing radiation is nondestructive to the sample surface. A drawback did exist, however, in the large probing depth into the bulk of the sample, making it difficult to separate vibrational features of the outermost surface regions from features arising from up to micrometers into the material. Tompkins (705) and Harrick (706) both discussed this physical reason for lack of "surface sensitivity" of IRIRS.

More recent works by Delhaye et al. (707) and Iwamoto et al. (708) have shown that it is possible to obtain vibrational spectra of polymers with internal reflection Raman spectroscopy (IRRS). The exciting beam, normally visible laser light, is propagated along the interreflection element, usually a transparent metal oxide. A Raman shifted radiation is generated which is observed away from the element face. Allara et al. (709) have offered a further discussion of this development in their 1984 publication in *Physicochemical Aspects of Polymer Surfaces.* They also mention the 1980 development by Rabolt et al. (710) of a closely related technique called waveguide Raman spectroscopy, which may offer a large improvement in surface sensitivity.

Another approach to improving the problem of surface sensitivity has been use of surface-enhanced Raman spectroscopy (SERS). Interesting studies and the generation of theories about the approach have been published by Burstein et al. (711), Furtak and Reyes (712), and Chang and Furtak (713). The majority of studies have been concerned with organic molecules adsorbed on silver, but enhancements have also been obtained with other metallic surfaces.

Webster (714) reported in his work at Virginia Tech on a new related procedure called specular reflectance infrared spectroscopy (SRIRS), which he employed to study ultrathin silicone films on thin polymer adherends which, in turn, were deposited on metal plates. Debe and Tushaus (715) employed reflectance IR spectra to demonstrate bulk-to-surface diffusion in aged urethane pressure-sensitive adhesives. Sung et al. (716) earlier used internal reflection techniques to show that A-1100 and A-151 silane films on aluminum oxide are polysiloxane networks with incomplete cross-linking. Boerio and Gosselin (717) have also used these techniques to study the interaction of silanes on aluminum. They were able to decide that gamma-APS silane adsorbed on aluminum in the hydrochloride form from acidic solution, with this step being followed by extensive hydrolysis of the silane with the final formation of siloxane polymer.

The Fourier transform *infrared spectroscopy* (FT-IR) provides another vibrational spectrum using optical interferomic technique. In effect, a Michelson interferometer provides a Fourier transform of the infrared spectrum transmitted, reflected, or emitted by a specimen. A minicomputer performs the inverse transfer back to the frequency spectrum. More information about the chemical bonding of both organic and inorganic films can be obtained than would be possible with XPS. Also, the use of photons for the sample eliminates the need for high vacuum and permits the examination of materials like solvents which are not compatible with vacuum examinations. A negative effect of using photons exists and makes the procedure not very surface sensitive. Measuring in the reflection mode can improve this sensitivity but a deposition of the sample on to a vaporized metal mirror will be required for deposits of less than 100 Å. The FT-IR procedure is best suited for determining the types of chemical bonds present in thin (1–3 μm) organic and inorganic layers on surfaces. (Further readings are found in Refs. 718, 719.)

The ability of increasingly fast computers in recent years has been important in renewing interest in some of the older classic methods of surface examination. The computers are able to perform previously tedious exercises hitherto not practical to develop by ordinary calculator methods. Of particular interest in this regard is the study of submonolayer and monolayer coverage by reflection absorption infrared spectroscopy (RAIR). Boerio and co-workers have made extensive use of this technique in studying the characteristics of silanes on aluminum (720), iron (721), copper (722), and titanium (720,723). (More information about reflectance spectroscopy and its employment can be found in Refs. 724–729.)

(7) LOW-ENERGY ELECTRON DIFFRACTION (LEED). Low-energy electron diffraction has been used for the characterization of adherend surface atomic structure as present on surface crystalline structures. Because this device detects the surface crystal structure, single crystals are preferred for study. Carbon contamination has been viewed by this method, as described by Buckley (465). A clean surface can readily be identified as well as the film structure on a surface that is clean.

(8) ELLIPSOMETRY. Ellipsometry has played a significant role in the optical examination and monitoring of surfaces with regard to ongoing bonding. The theory relating reflection of electromagnetic radiation was actually reported in the late 1800s by Drude (730–732) but was not used as such for more than 40 yr until Tronstad (733) used it to study corrosion on metal surfaces and Winterbottom (734) used it to study contamination of metal surfaces. For some background on the subject, the book *Ellipsometry and Polarized Light* by Azzam and Bashara (735) or publications by Pssaglia et al. (736,737) and McCrackin et al. (738) can be suggested.

Archer (739) published an early 1962 review followed in 1981 by an updated review by Aspnes (740) on procedures for measuring thickness and other properties of the oxide or some overlayer. Modern technique calls for the use of an automatic spectroscopic ellipsometer which is microprocessor controlled. Since the point of extinguishment of the plane-polarized beam is the null point specific for the thickness and other optical constants of a film, any subsequent change resulting from hydrating or other oxide growth will result in some observed signal. Consequently, an incubation time prior to hydration can be determined.

Hayfield (741–743) has been active in studies of transparent films on highly reflective aluminum using the technique and concluded that single-reflection ellipsometry has adequate sensitivity for the detection of films of monomolecular dimensions. Obviously, the properties of films and layers on the aluminum surface must be of significant importance in both adhesion and corrosion sciences where aluminum is the material of construction. Although, it might be contended in theory that films on oxide layers form some protection against some forms of surface reactivity. Westwood et al. (744) proposed that the physical and mechanical properties of such films and oxides can be quite sensitive to their chemical environments. For example, they demonstrated adsorbed water on naturally occurring aluminum oxide could reduce the elastic modulus of the surface by as much as 50%, and the microhardness by 25%. Adsorbed oxygen meanwhile could increase the near surface conductivity of aluminum oxide by 2 or 3 orders of magnitude.

(9) SCANNING ELECTRON MICROSCOPY (SEM). For the study of many adhesive bonding problems, it is necessary to supplement the chemical information obtained by AES, XPS, SIMS, and SIMS/ISS with high-resolution scanning electron microscopy (SEM). Of interest are the means of developing more morphological information. Because of the scale (~ 50 Å) of microstructure features, ultrahigh-resolution (XSEM) microscopy may be required. Such resolution is only available with scanning transmission microscopy (STEM). For more details, the reader is referred to the appropriate chapters by Goldstein in *Practical Scanning Electron Microscopy*, which are titled "Electron Optics" and "Electron Beam-Specimen Interaction" (745,746). Other references for an in-depth understanding of the technique can be found in two proceedings of the IITRI Symposia of 1968 (747) and 1969 (748) and the book of Oatley et al. (749) *Electronic and Electron Physics*. More recently, Newberg and Yahowitz (750) published a chapter on specimen preparation, special techniques, and applications of SEM, whereas Ledbury et al. (751), in 1980, compared various microscopic techniques. In Volume 3 of the *Treatise on Adhesion and Adhesives*, Patrick (752) discussed a number of special applications of SEM involving aluminum surfaces in a chapter titled "Use of Scanning Electron Microscopy." Johari and Samudra (753) also furnished a chapter "Scanning Electron Microscopy" for a book titled *Characterization of Solid Surfaces*.

Conventional micrographs by their nature are two dimensional, so that using them to interpret three-dimensional surfaces can be difficult and even misleading. Newberg and Yahowitz (750) presented a solution to this problem involving taking micrographs of the same area after tilting the sample a few degrees. Thus, a stereo pair is obtained and an observer examining these two micrographs through a stereo viewer using parallax can generate a three-dimensional view of the surface. The XPS method has been used with SEM to provide the additional factor of elemental analysis. Venables et al. (754) have provided excellent examples of these matters in their description analysis of various forms of aluminum oxide produced by deoxidizing and anodizing pretreatments on aluminum. Other important publications by Venables and co-workers that relate to this subject are Refs. 33 and 34. A recent 1990 chapter in *Adhesive Bonding* by Clearfield et al. (755) also reviews the field.

(10) COMBINATIONS OF VARIOUS TECHNIQUES. It is often found that combining different techniques is particularly helpful and supportive. Allen (756) used the combined techniques of AES and ellipsometry to measure elemental chemical composition and optiproperties of clean and contaminated aluminum surfaces under vacuum conditions. He showed a linear relationship between changes in amplitude of Auger peaks associated with both oxygen and aluminum oxide and oxide thickness measured optically. If an analysis is made of commercial aluminum samples, then the surface should be cleaned by argon ion bombardment in ultrahigh-vacuum then exposed to the environment studies. This same technique can be adopted for surface examination by XPS to produce a chemical profile analysis of surface layers. Depending on the surface energy of the beam, surface damage can occur together with a modification of optical properties. Neal (587) indicated his studies were underway in 1979 to compare the results of XPS, electron microscopy (EM), and optical studies of oxide and hydroxide growth on aluminum under varying environmental conditions for both the early stages of oxide with and without prolonged exposures to special environments.

Studies by Smith (757), Pattniak and Meakin (758,759), Bowen et al. (69,70,760–764), and Sun et al. (765) are representative of investigations of aluminum surface characteristics, oxide layer thickness, and surface reproductiveness. Crystallographic studies were conducted with high-resolution electron diffraction and surfaces were elucidated by AES. The oxide layer on "as received" or vapor degreased–only aluminum was usually identified as mainly "bayerite" with varying thicknesses from 100 to 1500 Å. This also applies after alkaline cleaning at 170°F (77°C). All mill-finish contaminants seem to be removed by this hot cleaning. When the "mill-finish surface" was deoxidized (as with hot dichromate/sulfuric acid solution [FPL etch]) (766), the oxide that regrew on the surface was identified as primarily boehmite in a thickness of only 100–400 Å. Further anodizing again creates boehmite (aluminum oxide hydroxide). Pressure to eliminate chromate discharge from manufacturing plants using FPL pretreating has led to various proposals for the use of a nonchromate hot-deoxidizing process as suggested by Russell (767) in 1976. Actually, the chemical composition of this suggestion was quite similar to a solution for deoxidizing aluminum proposed in 1971 in the *Metals Handbook* (768) for pretreating aluminum before welding. Russell and Tanner (769) also promoted the same solution for pretreating aluminum for weldbonding. As compared to the deoxidized surface oxide regrowth potential, the overall thickness of the various commercial anodized oxide layers is much greater and has a distinctly different surface morphology characteristic of the type acid used in the electrolytic bath.

Pattniak and Meakin (758,759) tried to determine the effect that alloy composition has on a treated aluminum surface. They obtained distinctly different responses from 2024, 6061, and 7075 alloy adherends. The nominal chemical percentages of the most important alloying elements for these alloys are reviewed in Table 9. The SEM and STEM techniques were primarily employed for observing any of the effects of surface preparation, whereas x-ray emission and AES were used to determine the chemical composition of the surface. The bulk oxide layer differed from the outermost surface layers in each instance. Obviously it would be these outer layers that would be intimately involved in bonding. Earlier surface pretreatments for aluminum, now generally acceptable for use in the literature, had mainly been developed in the lab by trial and error procedures. Pattniak and Meakin showed clear SEM micrographs of three different aluminum alloy surfaces mentioned above in (1) "as received condition," (2) freshly FPL etched, and (3) anodized and sealed conditions. Each alloy showed a different degree of surface roughness before and after etching. The 6061-T6 and 7075-T6 surfaces had clearly an etch-pitting present. After anodizing, the 6061-T6 surface was the

Table 9 Nominal Chemical Composition of Aluminum Alloys

	Elemental Components						
Alloy	Si	Cu	Mn	Mg	Cr	Zn	Zr
2024		4.5	0.6	1.5			
7050		2.3		2.25		6.2	0.12
7075		1.6		2.5	0.3	5.6	
6061	0.6	0.25		1.0	0.2		

Source: Data from Adhesive Bonding Alcoa Aluminum, Aluminum Company of America, Pittsburgh, 1967.

most strikingly different. If alkaline cleaning preceded etching, then 2024-T3 and 6061-T6 surfaces had broken up oxide layers that were strikingly different. This same oxide break-up effect could not be induced on the 7075-T6 surface under any conditions tried. Also, no condition was found for producing the etch-pitting condition on the 2024-T3 surface as noted above for the other two alloys. In spite of this fact, the same FPL etching pretreatment continued to be recommended and employed for pretreating 2024-T3 adherends in aerospace bonding based on its acceptable joint durability response. With FPL etching, the etch-pitting condition mentioned above for 6061 and 7075 alloy was always noted. Bijlmer (770) had proposed earlier that the presence of etch-pitting on FPL-etched Alclad 2024-T6 surfaces was necessary for obtaining the good joint durability results obtained with this adherend and the FPL etch procedure. Unfortunately, he did not report the results with the bare 2024-T3 alloy adherend, which has an entirely different surface. It must be remembered that the cladding on 2024 alloy is actually a relatively high-purity aluminum material as compared to the highly alloyed 2024-T3 core alloy in Alclad 2024-T3. We do know that bare 2024-T3 alloy has been pretreated and bonded with the same processes and adhesives in bonding aircraft as with the clad alloy. Also, we will find in our later discussions in Chapter 10 on joint durability that there is a preference by many present-day aircraft manufacturers to use the bare 2024-T3 alloy in preference to the Alclad 2024-T3 alloy for a better durability response.

The earlier results of Wegman (561,562) showed weak bonds with the oxides formed on etched 2024-T3 after subsequent soaking in 140°F (60°C) deionized water. This seems explicable from the micrographs secured and discussed by Pattniak and Meakin above (758,759). Also, the anodized layers on the different alloys were recognized as offering distinct differences. For example, the anodic layer on 6061-T6 or 7075-T6 surfaces covered the etch-pits already present on the surface from the FPL etch. The cell size on the 6061-T6 surface was about 200 Å. While cell size seemed to be independent of the etch-pit size, it did seem to be a function of voltage and the choice of electrolyte. In contrast, the bare 2024-T3 aluminum with no etch-pits initially present after etching exhibited pores on a background of cells in a geometric pattern with average size cells of about 400 Å. Even the sealing in 100 ppm aqueous chromic acid affected alloys differently. Closure of the pores inside the cells was distinctly visible for both 2024 and 7075 alloy surfaces. In contrast, the sealed 6061-T6 surfaces showed an unusual structure that appeared to be only loosely bound to the surface. This would seem quite unexpected from the results obtained by Minford (771–775) which demonstrated excellent long-term joint durability in even corrosion seacoast weathering for 8 years using variously anodized 6061-T6 joint specimens. Similarly, excellent resistance of anodizied 6061-T6 joints under stress and hot humidity conditions obtained by Minford would not be expected from this situation where an anodized oxide layer was described as being loosely bound to the surface.

Pattniak and Meakin (758,759) showed micrographic evidence with "mill-finish" surfaces that general surface roughness increased in the order 7075-T6 > 6061-T6 > 2024-T3. Jennings (776) has shown that sandblasting aluminum before acid etching produced stronger joints than acid etching alone. Whether more opportunity exists for significant mechanical interlocking of surfaces with adhesive when microetching over a macroscopically rough abraded surface (as claimed by Jennings) has not been shown by any comparison micrographs to the author's knowledge. Minford (532) clearly showed the opposite effect when comparing joint strength of silica-grit-blasted 6061 adherend surface joints with FPL etching of a vapor-degreased mill-finish surface. Not only was the joint strength of the grit-blasted surface joints lower initially, but the joint durability in various water-soaking environments was distinctly inferior with an epoxy. It must be remembered, however, that the initial joint strength can rarely, if ever, be considered a good predictor of ultimate long-term durability potential. As will be shown later when discussing joint durabilities in Chapter 10, the long-term durability response of any particular joint can also fluctuate significantly depending on the nature of the particular environmental and stressing conditions imposed in the testing situation.

Pattniak and Meakin (758,759) also studied the constitution and thickness of the aluminum oxide produced on these three different alloys after varying surface pretreatments. They found some chromium remaining on the surface after chromic acid sealing of anodized layers. The AES, in combination with argon ion sputtering was used to determine oxide thickness. In the case of the 6061-T6 surfaces, there was the additional presence of some sulfur after sealing.

With the exception of a 1760-Å oxide produced by soaking the FPL-etched 2024 alloy surface in deionized water at 140°F (60°C), the oxide thickness for all three alloys after acetone degreasing, FPL etching, and a tap water soak at 140°F (60°C) showed only a modest oxide thickness from 100 to 236 Å. Oxide diffraction studies on "as received" 2024 or 6061 alloy surfaces showed the surfaces were primarily amorphous. None of these showed a single electron diffraction pattern. Pattniak and Meakin further cited a 1974 Weber and Johnson paper (493) showing that ESCA and mass spectroscopic data showed only 1% of the oxide was mono- or trihydroxide after FPL etch and washing. This is, of course, completely counter to the observations of Bowen et al. (69,70,760–764) and many others who confirmed that oxide regrowth after FPL etching formed distinctly "boehmite" oxide.

The deionized water at 140°F (60°C) soak surface showed a textured boehmite pattern according to Pattniak and Meakin. They postulate that the anions and cations in tap water retarded the formation of this type of boehmite, and thus limited the overall growth of the oxide layer, accounting for the higher strength of such joints that had been soaked in hot tap water by Wegman (561,562). This would mean that the significantly lower-strength 2024 joints observed by Wegman with hot soaking in deionized water was due to a significantly thickened layer of textured boehmite. The author's work with 6061-T6 joints fabricated with boehmite generated in boiling deionized or tap waters was quite counter to this observation, since both high-strength and excellent joint durability potential was realized using either type of water (39).

Finally, Pattniak and Meakin stripped oxide layers from all three alloy surfaces after FPL etching using the Wilson (777) technique and showed the existence of amorphous alumina. They were also able to distinguish between a bulk and a top layer of oxide of 100–200 Å thickness. The top layer they believed decided the bonding parameters to the adhesive, whereas the strength and stability of the bulk layer would likely exert the majority of control over the overall durability. It has been clear from many more recent investigations (which will be further discussed in both Chaps. 7 on failure mechanisms and Chapter 10 on joint durability) that the failure of many joints in the top layer of oxide rather than the bulk phase demonstrates that the integrity of this top layer plays a very significant role in determining the overall durability of the bonded aluminum joint.

IV. MORPHOLOGY OF REGENERATED OXIDES

The characterization of aluminum surfaces left after specific surface treating obviously should be of the greatest importance in understanding the levels of joint strength and potential joint durability that may be achieved in adhesive bonding to these surfaces. It could offer further value in understanding the location of the potential locus of failure. In this regard, the chapter in Volume 5 of *Treatise on Adhesion and Adhesives* series by Smith and Kaelble (778) on mechanisms of adhesive failure of 2024-T3-pretreated adherend joints may offer the reader an excellent review of the various procedures that might be employed to consider each of various different pretreating situations.

Smith and Kaelble (778) characterization of the aluminum surface include a wide variety of techniques of which the reader needs to be aware. For example, thickness, index of refraction of the surface films, and oxide were measured by ellipsometry. Chemical analysis of the surface layers and the chemical profile of the films have been investigated with AES in conjunction with sputter-back etching by ion bombardment. The surface dielectric properties have been determined by surface potential difference (SPD) measurements and surface topography and morphology by optical, electron, and scanning electron microscopic techniques. The sample surfaces and the adhesives have been characterized by contact angle measurements which could distinguish the surface tension properties, permitting predictions of the wettability of the oxide.

After such characterization, joints could be fabricated and tested both immediately after curing and after aging in environmental chambers. After failing joints spontaneously or deliberately, fracture surfaces could be examined by optical and scanning electron microscopy and AES to determine the region of failure; i.e., metal/oxide interface, oxide, oxide/adhesive interface, or cohesive. Finally, an assessment of the chemical constituents at the fracture interface might support a final judgment about the actual locus of failure.

A. Forest Products Lab Dichromate/Sulfuric Acid Deoxidized (FPL)

The morphology of the FPL-etched aluminum surface was one of the first to be extensively studied because at the time it was the most commonly employed pretreatment for American aircraft alloy bonding. Extensive durability testing showed that bonded joints formed over such a surface not only showed a very high initial joint strength but also a high degree of joint strength retention following exposure to hot humidity and various laboratory testing conditions.

Bijlmer (770) and Bijlmer and Schliekelmann (779) at Fokker investigated an extensive matrix of varying concentrations and mix ratios of dichromate/sulfuric acid etchant and found only a few could produce a special micro-etch–type pitting on the surface. It is understandable that the particular concentration and ratio of dichromate/sulfuric acid already chosen at the Forest Products Lab (FPL) for treating aluminum alloys produced this special micro-etch–pitting condition on Alclad 2024-T3 alloy. Venables et al. (754) provided the first high-resolution SEM stereomicrograph of this FPL oxide morphology. The oxide consists of a network of shallow pores and protrusions, or whiskers, on top of a thin barrier layer. Proturbances seem to be spaced about 300 Å apart and rise 50 Å. This degree of microroughness provides a means for mechanical interlocking between adhesive and oxide. Venables (33) has related this surface as being critical for durability of epoxy/aluminum joints in various weathering conditions. If chromate-containing solutions are buffered, the chemical aggressiveness is controlled and the result is shallower pores without the above-mentioned "whiskers."

Davis et al. (37) have shown that this film can have varying quantities of adsorbed water present that can be removed by heating, vacuum exposure, or adsorption with certain organic hydration inhibitors. When magnesium-rich surfaces like 5000-type aluminum alloys are treated with an FPL-type treatment plus heat treatment, there can be an outdiffusion of magnesium with the formation of MgO on the surface. The surface morphology, however, is not really altered. The effect of this magnesium on joint durability was reported to be favorable by Chen et al. (780) in 1979, but Kinloch, Bishop, and Smart (781) reported the effect was deleterious in their 1982 paper. The author has conducted many comparison joint-durability evaluations over many years involving 2000, 5000, and 6000 series aluminum alloy joints with epoxy adhesives and found the relative durability was always 5000 series joints > 6000 series joints > 2000 series joints. Since 5000 series alloys have magnesium as the major alloying element, as compared to lesser overall percentage amounts in the other alloys, this might be interpreted as support for the Chen position. However, the author chooses to relate these results more directly to the fact that the overall corrosion resistance of these alloys in direct weathering conditions is in the order 5000 alloys > 6000 alloys > 2000 alloys. Thus, the effect of magnesium oxide build-up on any alloy surface that contains any magnesium alloying components can still be relatively negative as compared with a lesser concentration of the surface magnesium. When comparing such an alloy with one containing significant amounts of a more cathodic metal alloying additive like copper, this effect will be overwhelmed.

B. Anodized Oxide Configurations

The high joint durability required for aluminum aerospace applications has been achieved by anodizing after FPL deoxidation. When aluminum is made the anode in a suitable electrolyte and a current is passed through the cell, the aluminum surface is converted to an aluminum oxide coating with outstanding properties for bonding because of the porous structure of the surface oxide. A number of electrolytes have been investigated over the years, as discussed by Cochran (782) in Volume 3 of the ASM *Aluminum* books. The most utility has been found from the use of sulfuric, chromic, or phosphoric acid electrolytes. For use in sheathing architectural buildings, the best application has been with sulfuric acid anodizing because of its superior thickness and sealing properties (the Alcoa Alumilite series of anodized coatings). More recently, even thicker anodized (hard coatings) have been developed such as Alcoa Duranodic-type finishes, which have been employed on some of the world's tallest buildings such as the Sears and Roebuck and John Hancock Buildings in Chicago.

Some early investigations of great historical interest are those of Jenny (783), Lewis and Plumb (784), Hoar and Mott (785), Keller et al. (786), Taylor et al. (787), and Keller et al. (788). Solomon and Hanlin (789) published a 1980 paper titled "Characterization of Anodized Aluminum Alloy Adherends for Adhesive Bonding Application" which can serve as a good review.

The theory of the pore growth of anodized aluminum oxides has emphasized field-assisted dissolution as a key mechanism; i.e., the pore grows by preferential dissolution at the bottom of a pore where the electrical field strength is greatest. Papers by Hoar and Mott (785) and O'Sullivan and Wood (790) or Cochran's chapter on anodizing (782) can be reviewed for details. Of great interest is how the pore initiates, which is a most interesting as well as a highly controversial subject. Theories suggested in the literature include those of Wood (791), Thompson et al. (792), and Heber (793). The effect of anodic oxide characteristics due to the variables of electrolyte, pH, temperature, anodization time, and applied potential have been studied by Hanlin and Solomon (794) using AES and SEM techniques. Sun et al. (765) summarized and discussed a variety of their investigations at Martin-Marietta relative to depth composition profiles, electrochemistry measurements, oxide growth mechanisms, and crystal structure identifications of oxides on aluminum after FPL etching followed by chromic or phosphoric acid anodizing. They concluded accurate interpretation of the AES depth profiles should be made with the aid of STEM morphology data. Coburn and Kay (795) reported 4 yrs earlier on the AES depth profile analysis (measurement of sputtering time) for measuring the aluminum oxide thickness. The next year Holloway (796) mentioned a high sensitivity of Auger signals from such porous oxide surfaces to the surface micrographical condition resulted in misinterpretation of composition.

Solomon and Baun (797) published in detail on the thin film Auger analyzer technique. Many others, meanwhile, have dealt with a variety of problems relating sputtering time to the oxide depth analysis. For example, Kircher and Etzhorn (798) reviewed the complications to accurate analysis due to the presence of multilayers of different elements and chemical states. Wehner (634) showed a factor of 2 difference between sputter yield of pure aluminum as compared to aluminum oxide. Manhard and Mozelewski (799) explained the sputter rate dependency on the thickness of porous oxide as due to decreasing pore diameter with greater thickness layers. Solomon (800) arbitrarily chose the oxide/metal interface as the so-called profile point representing the minimum signal near the interface.

Solomon and Hanlin (789), in 1980, attributed poor durability of "as received" 2024-T3 alloy joints to the magnesium/aluminum oxide surface composition. After ion sputtering is completed, the magnesium presence disappears, indicating the forming of the "as received" surface had depleted bulk magnesium near the surface. The same authors found that after Oakite (alkaline cleaner) deoxidizing of the 2024-T3 surface, the magnesium disappeared; however, traces of copper and elements present in the cleaner then appeared. After additional elemental sputter of the surface, a further increase in copper occurred. From such observations, it seems that bondable surface obtained with different surface treatments will be different and could help explain the varying joint durabilities that are found in service.

Solomon and Baun (801) presented data in 1978 on an analytical comparison between the surface composition of various adherends and the bulk compositions, concluding that there was a variety of unwanted chemical species that needed to be removed and the surface topography altered to achieve strong, durable bondments. Similar information was contained in a summary a year earlier by Drzal (802) titled "Summary of the Workshop Held on the Role of the Polymer/Adherend Interphase in Structural Adhesion." This critical interphase region concept was introduced by Sharpe (803) as early as 1972. Sharpe defined interphase as all material from a point in the bulk adhesive toward and through the boundary between adhesive and adherend to a point in adherend where the local properties are the same as the bulk properties. Contaminants or unwanted materials in this region (originating from the original adherend surface, adherend bulk, and bulk adhesive) can certainly adversely affect the bonding mechanisms (wetting and adhesion) as well as bond performance (strength and durability). More will be said about specific results obtained above by Solomon and Baun with aluminum alloys 2024, 6061, 7050, and 7075 in the latter section of this chapter titled "Surface Chemistry of Aluminum Affecting Bond Durability."

In 1990 Hudak et al. (804) discussed the specific use of XPS analysis to study the interphase between an anaerobic adhesive and metal adherends, whereas Ho (805) contributed a comprehensive paper on the chemistry and adhesion of metal/polymer interfaces. Nitschke (806) has specifically conducted a microanalytical study of the aluminum/polymer interface in aluminum joints. Ondrus et al. (807) reported a study of molecular structure of polymer/metal interphases, whereas Bishop et al. (808) described the influence of surface treatment and the environment on the interphasial region of adhesively bonded aluminum joints. Finally, Achary and Ramaswamy (809) introduced the concept of the addition of an interfacial agent to enhance the adhesion between nitrile rubber and metals.

1. Sulfuric Acid Anodize Oxide (SAA)

The anodic oxide produced by the sulfuric acid electrolyte has received much greater attention from aluminum producers over the years because of the small pore size and greater potential for "sealing." This is demonstrated by the relative pore diameter and cell wall thickness data in Table 10 and numbers of cellular pores in Table 11. These data can also demonstrate why the CAA and PAA forms of anodic oxide with their larger-diameter cells might permit more inflow and trapping of high molecular weight adhesive polymers.

Thus, SAA surfaces have proved to be ideal for fabricating exterior panels for monumental buildings. Since the bond strength in laminates of aluminum facings to, e.g., foam plastic insulations, woods, cement boards, need only exceed the cohesive strength of such building materials, the bonds made to SAA anodized surfaces can be more than sufficient using contact elastomer-phenolic adhesives. Since it might be desirable to create more truly structural bonds to such surfaces, Minford conducted comparative joint durability studies using 6061-T6 aluminum adherends. The commercial Alcoa Alumilite SAA finished sheet was used for such studies generally with an oxide thickness of about 0.8 mil, which is widely used on monumental buildings. The long-term durability of such joints will be discussed later in Chapter 10.

Rather poor bonding experience has been obtained in aerospace with SAA surfaces, since bonding was attempted mainly with thicker oxides present, which should provide best interfacial protection against corrosive weathering. The thick oxide characteristics that offer the highest surface corrosion resistance works negatively when bonding aircraft. Such thick layer oxide carries with it the greater statistical probability of finding physical discontinuities (weak boundary conditions) which fracture fail under high stress and peel. Reinhart (810) described some of the aerospace failures involving

Table 10 Pore Diameter and Cell-Wall Thickness Values of Oxide Coating

Electrolyte	Pore diameter (Å)	Wall thickness (Å/volt)
15% sulfuric acid, 50°F	120	8.0
2% oxalic acid, 75°F	170	9.7
3% chromic acid, 100°F	240	10.9
4% phosphoric acid, 75°F	3300	10.0

Source: From Ref. 782 (Table 3, p. 647).

Table 11 Number of Cells or Pores in Oxide Coating

Electrolyte	Volt	No./in.^{2} × 10^9
15% sulfuric acid, 50°F (10°C)	15	498
	20	334
	30	179
2% oxalic acid, 75°F (23.9°C)	20	230
	40	75
	60	37
3% chromic acid, 100°F (37.8°C)	20	140
	40	52
	60	27
4% phosphoric acid, 75°F (24°C)	20	121
	40	50
	60	27

Source: Data from Refs. 855 and 782 (Table 8, p. 667).

SAA oxide surfaces where the joint failure occurred between the metal and oxide, whereas Bijlmer (811) claimed failures had occurred in the oxide itself. Early reference to SAA procedures can be found in the 1947 publication of Mason and Slunder (812).

2. *Chromic Acid Anodize Acid Anodize Oxide (CAA)*

The first chromic acid anodizing was patented by Bengough and Stuart in 1923 at the National Bureau of Standards. The requirement for rather complex voltage control was a disadvantage. Buzzard (813) and Buzzard and Wilson (814) offered a constant voltage procedure, making the process more commercially feasible. The oxide has been used as a protective coating against corrosion and as a base under paint coatings. However, chemical conversion coatings have proved to be more convenient and economical for these applications in recent years. As a base for aerospace bonding (particularly in Europe), this type of anodizing has been prominantly used in bonding aircraft since World War II.

Bijlmer (815) has investigated the use of this anodized layer for aerospace bonding in Europe. In Table 12, Bijlmer has provided data on the weight of the anodic film, energy of formation, and the energy efficiency employing the CAA procedure on both bare and Alclad 2024-T3 alloy. The energy efficiency calculated from the coating weight and energy of formation indicates directly the amount of anodically formed oxide based on energy consumed. This efficiency is obviously lower for the nonclad alloy, so the presence of extra alloy ingredients on that surface must be exerting a negative effect. Also, at lower voltages and lower temperatures, the energy consumption can be an order of magnitude lower than for higher voltage and temperature. In Table 13, Bijlmer has recorded current density, coating thickness, energy of formation, and the corresponding lab-joint strengths using voltage and time as variables with Alclad 2024-T3 alloy. Both variables are obviously important. The CAA aluminum pretreatment continues to be the preferred pretreatment in European aircraft over either the optimized FPL or PAA procedures.

There have been differences of opinion as to whether any posttreating (sealing) of the CAA is required to offer the desired degree of joint durability under aggressive weathering and stress conditions. Bijlmer (816) at Netherlands Fokker advised against any hot rinsing after CAA anodizing, whereas Rogers (817) at Bell Helicopter supported sealing of that surface to improve both adhesion and corrosion resistance. Rogers reported in 1950 that Bell had experienced erratic results with unsealed CAA-treated joints, and accordingly had replaced the processing with the FPL etch for bonding main rotor helicopter blades. As might be anticipated, edge corrosion problems then began to surface in coastal service areas and the Bell Process Specification No. FW 4352 (U.S. Patent No. 3,414,489) for sealing CAA surfaces resulted. The confirmation of good corrosion resistance for these kind

Table 12 Weight of the Anodic Film (W), Energy of Formation (E), and Energy Efficiency (G) for 2024-T3 Aluminum Adherends Anodized in Chromic Acid[a]

Anodizing voltage	Variable	Solution temperature (°C [°F]) 30 (86) Alclad	30 (86) Nonclad	40 (104) Alclad	40 (104) Nonclad	50 (122) Alclad	50 (122) Nonclad
10	W	22	26	39	39	44	32
	E	3	6	7	12	11	22
	G	7.3	4.3	5.5	3.2	4.0	1.5
25	W	42	34	78	60	110	67
	E	13	18	26	38	53	78
	G	3.2	1.9	3.0	1.6	2.1	0.0
40	W	51	26	104	62	155	32
	E	27	22	54	64	108	125
	G	1.9	1.2	1.9	1.0	1.4	0.3

[a]Processing time 45 min; W in mg/dm^2; E in kJ/dm^2; G = W/E in mg/kJ.
Source: From Ref. 815 (Table 2, p. 34).

Table 13 Results From Chromic Acid Anodizing on Alclad 2024-T3 Aluminum Adherends

Volts	Time (min)	Variable[a]	Solution temperature [°C (°F)] 20 (68)	40 (104)	60 (140)
20	5	i	0.07	0.18	0.48
		E	0.35	0.96	2.66
		W	4.4	5.9	14.0
		P	47	48	49
20	10	i	0.05	0.19	0.52
		E	0.60	2.13	5.88
		W	5.6	11.8	26.0
		P	47	35	47
20	15	i	0.06	0.21	0.53
		E	0.96	3.52	0.27
		W	7.9	16.5	32.0
		P	44	46	37
40	5	i	0.06	0.26	0.99
		E	0.43	2.34	10.05
		W	4.9	9.5	27.0
		P	44	16	50
40	120	i	0.05	0.26	1.05
		E	0.77	5.81	23.67
		W	6.2	19.0	50.0
		P	46	21	57
40	15	i	0.04	0.26	1.10
		E	1.11	8.86	36.49
		W	8.6	28.0	67.0
		P	48	21	47

[a]Variables code: i, current density (A/dm^2); E, energy of formation (kJ/dm^2); W, coating thickness (mg/dm^2); P, lap-joint strength (kgf/mm^2).
Source: Data from Ref. 815 (Table 1, p. 30).

of joints was shown by special testing procedures developed by Akers (818) in 1966 and Rogers (819) in 1970.

Direct comparison between American and European joint successes has been difficult to make because different types of structural adhesives have been historically employed for bonding the respective aircraft. REDUX-type adhesive formulations have been widely employed in Europe since World War II, whereas a variety of modified-epoxy adhesive formulations have been the mainstay of American aircraft structural bonding. Rogers (820) reviewed the evidence for good performance with sealed CAA conditions with reference to earlier publications, citing durability exposure testing that he conducted at Bell Helicopter (821,822).

The inherent CAA oxide (i.e., without prior FPL etching or other pretreatment) is a dense structure of tall columns according to Venables et al. (745). The outer surface actually can be quite smooth, with fine pores running through most of the oxide layer at the junction of the column walls. Total thickness can be 1–2 μm (0.04–0.08 mil). The barrier layer at the bottom of the column is relatively thick because of the fairly high anodization voltage employed. Penitz and Sharp (823), in 1984, studied the effect of different aluminum alloy surface compositions on barrier anodic films. Singh et al. (824) and Csanady et al. (825) have measured these barrier layers as being about 40 nm thick at the bottom of the pores. Brockmann et al. (523), Franz (826), Kollek (827), and Abd Rabbo et al. (828) identified the upper portion as being amorphous, whereas the lower part appears to be crystalline. Franz (826) and Yaniv et al. (829) could not identify the presence of any significant amount of chromium in CAA oxide even though phosphate is definitely present after PAA processing according to Franz (826), Davis et al. (525), Kinloch and Smart (663), Solomon and Hanlin (789), Poole and Watts (830), Kinloch et al. (781), and Abd Rabbo et al. (828). Brockmann et al. (524) showed that primers can penetrate the CAA pores, which can account for some of the greater advantage for mechanical interlocking possible with PAA oxide morphology. It appears that the morphology of the outer surface of the CAA oxide can be further enhanced by choosing special types of pretreatment before anodizing. For example, FPL-type etch pretreatment before CAA processing results in a surface with a typical FPL cell-and-whisker morphology. A PAA pretreatment produces a deeper cellular structure much like the PAA morphology. The morphology can also be altered by varying the processing conditions. Poole and Watts (830) and Arrowsmith and Clifford (831) investigated higher-temperature anodizing solutions and postanodizing phosphoric acid etches. Brockmann and co-workers (523,524) indicated that adherends treated with the dual-voltage CAA processing forms more durable bonds than those from single-voltage processing. Brockmann et al. (524) and Allen and Stevens (832) stated a possible advantage of the CAA oxide over the PAA oxide could be its lesser friability.

Cotter (851) furnishes data in Table 14, comparing the retained joint strength of aluminum/epoxide joints with chromic acid etch, CAA, or PAA pretreatments after 1000 hr immersion in hot water. The relative order of durability appeared to be PAA joints > CAA joints > chromic acid etched joints, which agrees with the published data of Minford (533,534).

Table 14 Effect of Pretreatment on the Durability of Aluminum Single Lab-Joints Bonded with a Modified Epoxide and Immersed in Deionized Water at 122°F (50°C)

	Retained joint strength (%)		
Surface pretreatment	500 hr	750 hr	1000 hr
Phosphoric acid anodize	94	91	87
Chromic acid anodize	80	77	75
Chromic acid etched	76	69	65

Source: Data from Ref. 851.

3. *Phosphoric Acid Anodize Oxide (PAA)*

American airframe manufacturers originally used the FPL (later the "modified FPL" etching pretreatment) for aluminum aircraft before PAA. While optimizing the FPL, Boeing continued to search for a superior procedure and appeared to find it in the early 1970s in the form of PAA. McMillan (833), at Boeing, has summarized much of this development, describing what was unsatisfactory about FPL and "optimized FPL."

The discovery and evaluation of PAA was investigated at Alcoa Labs by Hunter et al. (834) in the 1950s. However, the relatively thin (by comparison with SAA and CAA oxide varieties) and unsealable large-pore structure was not conducive to use as a commercially acceptable protective oxide for exterior weathering. It is interesting to note that while the freely exposed surface of a PAA-treated sheet will readily show pitting corrosion in natural weathering, these same large pores offer excellent mechanical interlocking potential with an adhesive. Also important is the fact that the same unsealed PAA surface which freely pits in direct weathering exposure does present exceptional resistance to the undercutting of the bondline even in the most corrosive weathering conditions. In contrast, the same bondlines with the optimized FPL pretreatment can suffer undercutting corrosion under exposure to corrosive seacoast weathering. For this reason, there was a need for the development of special anticorrosion primers. In 1986, papers concerning interface and interphase corrosion inhibition were published by Matienzo et al. (835) and Lorenz and Mansfield (836).

Boeing recommended and used the PAA pretreatment in 1974 after studying the various variables shown in Table 15. They also showed it was possible to use polarized light for quality control. Venables et al. (31,33,34) found PAA oxide had longer "whiskers" than available with CAA surfaces, thus allowing greater mechanical interlocking. This translated into the higher joint strengths shown in Table 7 from the data of Minford. Minford also found PAA joints capable of higher tolerances for resisting stress and high humidity environmental conditions than possible with similar CAA, SAA, or FPL-type pretreated aluminum joints.

Table 15 Variables Related to Phosphoric Acid Anodizing

Preprocess variables	Process variables	Postprocess variables
Materials	Preanodize	Prepriming
Alloys	Degreasing	Handling
Forging	Alkaline cleaning	Airborne contaminants
Extrusions	Preanodize etch	Time before priming
Castings	Rinsing	Environment
Sheet		
Plate	Anodize	Storage
	Time	Surface—primer interactions not considered critical after priming
Surface conditions	Voltage (part to solution)	
Heat-treatment scale	Acid concentration	
Machined surfaces	Solution age	
Protective oils	Current density	
Corrosion products	Racking—single, multiple	
Chemical milled surfaces	Electrical contacts to parts	
Ink markings	Part orientation	
Fingerprints	Alloy	
Stretch-forming lubes		
Protective maskants		
Shot peen		

Source: From Ref. 843 (Table 4, p. 68).

The long-term durability of aluminum/polymer bonds is determined to a great extent by the environmental stability of oxide involved in making the bond and promoting the initial level of structural strength. Water intrusion into an aluminum joint can cause conversion of the oxide present to a less stable hydrated form which may adhere poorly to the underlying surface and readily fracture. Venables (33) postulated from analytical studies that oxide conversion may end as the trihydroxide with the above unfavorable characteristics. It had been known from the 1950s that PAA oxide is significantly more resistant to hydration than CAA and SAA types (35,837). Thus, it might be expected that PAA-treated joints would be more durable in natural weathering where water was the main chemical deterioration factor. Noland (838) also confirmed this greater stability in humidity of PAA oxide by XPS analysis.

Sun et al. (635) investigated the PAA process and determined that it is a two-stage process consisting of a fast linear growth of the pore cell structure followed by an additional growth of fine oxide "whiskers" on top of the pore cells. They have further stated that the oxides from FPL, CAA, and PAA appear to be amorphous but that the "boehmite" oxide structure will be generated after the oxide is treated in hot water. This is to be compared with the findings of Remmel (839) at Northrup, who concluded that the FPL and PAA oxide surfaces show diffuse ring patterns not corresponding to any known oxide or hydroxide structure. While mainly amorphous, the CAA oxide showed a higher degree of crystallinity. Vedder and Vermilyea (840) had earlier described this hydrated oxide as a "pseudoboehmite." All agreed that a boehmite crystalline structure followed exposure to hot water.

Sun et al. (635) showed that when 2024 aluminum is FPL etched, there is a definite build-up of copper at the surface. (Note: copper is the main alloying ingredient.) When PAA anodized, copper is largely removed in the first 2 min of anodization. This should aid in promoting superior durability to similar FPL-etched joints.

The fact that you cannot hydrate (seal) PAA surface pores is a positive factor for bonding, since the pores remain available for penetration and good mechanical interlocking. Sun et al. (636) showed phosphate remains on the surface after PAA processing with AES analysis, whereas Kinloch and Smart (663) confirmed this with XPS. A paper by McDevitt and Solomon (841) can be studied for establishing the relationship between alloy choice and thickness of the oxide.

It has been suggested that the high resistance of PAA oxide to hydration may be due to the presence of phosphate ions which are physically near the surface. Thompson et al. (842) obtained supporting data for the theory that the phosphate ions are probably attached by chemisorption to the whiskers, ridges of the cells, and the inner walls of the pores. The wide range of factors that exert an influence on the quality of PAA pretreatments have been itemized by Marceau, as shown in Table 15. His recent chapter in *Adhesive Bonding of Aluminum Alloys* discussing the use of PAA in bonding aircraft is also recommended for reader review (843).

The PAA anodize oxide surfaces obviously have a greater degree of microroughness as compared to FPL-, SAA-, and CAA-treated surfaces, as shown by Venables et al. (754), and reviewed by Venables (844) in *Adhesive Joints: Formation, Characteristics, and Testing* and by Clearfield et al. (463) in the 1990 *Engineered Materials Handbook* of ASM. The oxide shows a well-developed network of pores on top of a barrier layer with whiskers protruding from the pores. The total oxide thickness is approximately 400 nm. As higher and higher purity aluminum alloys are processed, the whiskers disappear from the morphology. Venables (33), Kabayashi and Donnelley (30), and Harwick et al. (529) have all used the superior mechanical interlocking achieved by most adhesives with this microroughness to explain the superior joint strength and durability present in these kinds of pretreat-aluminum joints. However, it should also be clear that this can only be true if the primer and/or adhesive are capable of physically penetrating into the structure. Venables (33) has shown the cross-sectional SEM views of the desirable entry of aerospace epoxy primers into the structure. Franz (826) has demonstrated that the alloying constituents of the base alloy will generally be absent from the "as-anodized" oxide. The author would point out that this should be a good reason for anticipating that the durability of these bonded interfaces would not be unfavorably affected by the fact that a high-strength, aircraft-type alloy like 2024-T3 was being anodized along with its attendant high concentration of cathodic copper in the base alloy. Normally, the presence of surface copper on aluminum adherends would be expected to decrease the overall resistance of the joint to all forms of weathering involving water exposure.

4. General Comparison Between the Different Anodic Oxide Surfaces

It needs to be remembered that there is an inverse relationship between choosing an anodic surface morphology for its better bondability and its ability to resist corrosion on its surface in a free exposure to natural weathering conditions. For example, the best anodized surface for creating the highest strength and most durable joints has been indicated as the PAA oxide, yet the large, open pore structure and relatively thin cross section of such an oxide makes it totally unacceptable for exposure to natural weathering. Because of the poor performance potential for resisting weather, the Alcoa Labs chose in the 1950s to reject PAA anodic oxide as a practical surface condition for architectural application facings. A considerable time gap existed between this Alcoa developmental work and the adoption of the PAA oxide for adhesive bonding by the Boeing Labs.

On the other hand, the significantly thicker and better pore-sealing characteristics of both CAA and SAA oxides caused them to be extensively investigated for usage in long-term weathering applications. The CAA oxide, in addition, found extensive use up to the present time in bonding European aircraft. The even higher resistance to atmospheric weathering of the thicker and sealable SAA anodic oxides made them most acceptable for virtually any aluminum structure that must stand unprotected in all types of natural weathering conditions. The need for optimum corrosion resistance in aircraft joints also prompted the testing and some application of SAA oxide joints in aircraft; however, certain problems of bond failure showed up under conditions of high peel and stressing in the joints. It has been theorized that a major factor in these failures may have been the higher probability of incorporating structural variables in oxide regrowth situations as they are allowed to grow to greater thicknesses. Minford (845) has offered some evidence of the degree to which this increased oxide thickness can lower the initial joint strength of SAA oxide joints which ranged from 0.1–0.8 mil in thickness. It is clear from the data of Minford shown in Table 8 that a significant lowering of joint strength did occur as the oxide thickness advanced. It is worthy of note that a similar joint strength was obtained with a 0.1-mil oxide produced by either SAA or CAA processing.

5. Other Effects Noted

Peterson and Arnold (846) produced a comprehensive report which included a number of investigations regarding the structure and chemistry of the PAA and FPL oxides on a variety of aluminum alloys, including clad and bare 2024-T3, bare 2024-T81, and bare 7075-T6. They found the measured lap-shear strength of a 350°F (177°C) curing aerospace adhesive was greater between FPL and PAA pretreatments depending on the test adherend thickness. With the higher stress gradients in thinner 0.0625 in (1.59 mm) aluminum test adherends, the FPL-pretreated joints tested relatively stronger. On 0.125 in (3.18 mm) thicker test adherends, the PAA joints now tested higher. Stress durability test results generally showed superiority with the PAA-treated surfaces. In Appendix A to their report, these authors described analytical methods for analyzing the oxide structures, primer/oxide interactions, and failure modes using secondary electron imaging in SEM and STEM, transmitted electron imaging in TEM and STEM, and surface analysis by AES and XPS. In an Appendix B, they described oxide structure and chemistries arising from the use of procedures described in Appendix A. Finally, in Appendix C, methods and results are described for chemical characterization of six adhesive primers using infrared, reverse-phase liquid chromatography, and gel permeation chromatography. Their analyses determined that failure modes of joints were relatively independent of the surface preparation but did depend on the intrinsic properties of the adhesive and primer.

V. EFFECT OF WATER-RINSING CONDITIONS AFTER DEOXIDATION

After deoxidation of FPL processing, it has been shown that the nature of the oxide film is strongly influenced by the water-rinsing conditions used after the etching and the environmental storage conditions. Comparative testing of aluminum joints made after hot deionized water rinsing as compared to hot tap water rinsing was conducted by Wegman et al. (561,562). It was shown that significantly lower-strength joints were apparently formed by the deionized water rinsing, which Wegman ascribed to generation of a relatively weak oxide layer. By adding multivalent ions to the deionized water (in essence producing a sort of tap water), Wegman produced much higher-strength joints with the capability of showing 100% cohesive-type joint failures.

Minford found this difference in joint strength was not present after either FPL (39) or hot caustic (847) deoxidizing when the postrinsing was in boiling deionized or tap waters. Further, Minford (39) demonstrated that the relative durability of deionized- or tap water–treated joints was quite similar based on years of testing under accelerated weathering conditions. It should be pointed out that the Minford's results were obtained with 6061-T6 adherends instead of 2024-T3 adherends. The author believes the most important difference was the boiling water treatment conditions after deoxidizing as compared with exposure in only 140°F (60°C) water by Wegman.

Nevertheless, it seems logical that some procedure needs to be employed where the oxide regrowth layer after rinsing and drying needs to be one of outstanding stability and strength for effective bonding of structural joints. Also, it might be expected that bonding immediately after such pretreating would offer the best chance for best performance and repeatability. Wegman and Dickerson (848) found that even when FPL-etched and rinsed surfaces were placed in a 50% RH environment for 72 hr before bonding, they could still be bonded with about the same strength. However, since no long-term durability testing was conducted with these joints, it cannot be confirmed that such a practice would not affect ultimate service life. The efforts of Minford (849) may offer more direct evidence, since long-term durability testing for several years was conducted under related circumstances. In these tests, 6061-T6 adherends were vapor degreased and deoxidized in hot chromic/sulfuric acid (Alcoa A3 deoxidizing procedure) and the surfaces were tap water rinsed and dried in a clear dry air blast. These surfaces were stored for various times up to 30 days in ambient lab conditions (~50% RH at 75°F [23°C]). Triplicate joints were fabricated with surface aging up to 30 days with a heat-curing, one-part, nitrile-modified epoxy paste adhesive. No joint durability penalty could be observed when these joints were exposed for 2 yr in room temperature water or 125°F (52°C) soak in 100% RH conditions. The only precaution taken while storing was to wrap the pretreated surfaces with many folds of cheesecloth to restrict surface contamination from air fallout.

Sun et al. (850) determined that during the etching process the original oxide is removed; i.e., the aluminum surface is dissolved and a new layer is simultaneously formed. Others argued that new oxide formed only during a rinsing step. The AES investigations of Sun et al. indicated that thickening of the oxide layer was the main action performed during actual rinsing. Earlier authors cited investigations that delineated the oxide after FPL etching as "boehmite" in character, but Cotter (851) offered a summary of literature reports which argue about the nature of this oxide. Investigations of the author have supported the contention that a boehmite oxide layer is present. There is an obvious need to believe that a relatively stable form of aluminum oxide is present, since the apparent stability of the FPL-treated aluminum oxide in various water and condensing humidity exposures has proven to be very acceptable. A more detailed discussion of the rinsing effects will be offered later in the next chapter which details the specifics of various surface pretreatments used on aluminum.

McCarvill and Bell (852) found some evidence to support the observation of Wegman that weaker joints are present when deionized water rinse is used (with water temperatures of 140°F (60°C) or below. They contended tap water rinsing should always be employed immediately after deoxidation. Scardino and Marceau (853) investigated the effect on joint durability of a number of processing factors, including a 3-min rinse delay. This proved to be a more negative factor than lowering the etching solution temperature by 5.5°C.

Ahearn and Davis (38) demonstrated with both XPS and AES techniques that the treatment of FPL-etched aluminum with a hydration inhibitor like nitrilotris methylene phosphonic acid (NTMP) involves the displacement of water and the formation of P – O – Al bonds with a saturation coverage of approximately one monolayer. In the presence of water in the service environment, they proposed a three-step process: (1) physisorption of water by FPL created oxide, (2) hydration of this oxide form to boehmite, and (3) at longer exposure times, further hydration with the formation of bayerite (the trihydroxide). Since they found no phosphorus in or below the hydrated oxide, it was presumed that with continued water exposure the inhibitor-Al complex undergoes a slow dissolution process followed by a rapid rise in hydration of the exposed oxide.

VI. CONVERSION OXIDES

Many chemical treatments and finishes have been devised for changing aluminum surfaces over the years. One early reference to such attempts has been provided in a summary of publications by Harris (854). More recent investigators have been Wernick and Pinner (855) and George et al. (856). Using the deoxidized surface (either acidic or caustic deoxidation), selected chemical treatments can be employed to interact with the aluminum atoms and produce "chemical conversion coatings." They are amorphous oxide coatings which include a portion of the base metal as one of their components. Thus, an inert film is produced which has been an excellent bonding base for organic paints and adhesives. At the same time, these coatings show excellent corrosion resistance underneath a paint or adhesive. This resistance to corrosion in a bonded joint is especially important where the joint is exposed to very aggressive weathering conditions such as the seacoast or marine environments. More recent publications by Newhard (857) and Spring and Woods (858) are recommended for further information.

Conversion coatings are actually gel-type coatings which initially contain water throughout their cross section. As this water dissipates under storage or deliberate drying under heating, the properties can gradually change from being distinctly water wettable to a surface that is virtually water repellent. Obviously, trying to bond to such a surface with any water-solvent- or latex-dispersed contact adhesive would be courting disaster. However, Minford has shown that it is quite practical to drive out this water while the adhesive is being cured at high temperature with the result that a high-strength, durable joint can still be formed. This is readily demonstrated from results shown in a Minford chapter (859) summarizing the long-term exposure of 6061-T6 joints with an Alodine 1200 conversion surface pretreatment, as shown in Table 16. It should be noted, however, that the performance with an epoxy adhesive cured at room temperature was not nearly as good as when the elevated temperature-cured epoxy was used. It seems that some ongoing process is occurring which not only permits good wetting by the adhesive but that the conversion oxide surface becomes more stable to long-term penetration by water into the bondline. Though not shown in Table 16, Minford was able to demonstrate that the surface of an already heated Alodine 1200 conversion oxide could not be adequately wetted by the same epoxy adhesives, thus demonstrating the necessity to have a water-wettable surface when the adhesive is applied. The author can cite one commercial situation in which a paint had been applied only to one side of an aluminum facing at the customer's request with the facing to be later laminated to plywood using a casein–neoprene latex adhesive. In service as the decking of a subway car, the floor panels began to show early delamination along the panel

Table 16 Effect of Better Surface Wetting on Bond Permanence

Surface treatment	Adhesive	Type cure	Range of bond survival times in seacoast (days)
Vapor-degreased	Two-part epoxy	RT	70–80
Vapor-degreased	One-part epoxy	Heat	71–270
Alodine 1200[a]	Two-part epoxy	RT	1158–1440
Alodine 1200	One-part epoxy	Heat	2920
Chromic-sulfuric	Two-part epoxy	RT	270–760
Chromic-sulfuric	One-part epoxy	Heat	760–1440

[a]Commercial proprietary aluminum conversion coating treatment of Amchem.
Source: Data from Ref. 859 (Table 3-1, p. 55).

edges and an explanation was sought. It appeared that some of the Alodine 1200 conversion coat solution was whipped around to the back edge of the aluminum sheet while being painted on a high-speed coil-coating line. At the high temperature required to bake the paint on the front side, the Alodine coating on the backside converted to a non-water wettable surface. The application of a water-latex adhesive to this edge surface readily provided the weak edge bonding situation to the plywood which showed up under later service conditions involving water contact.

The reader can be refered to several publications by Minford (533–535) which consider the fabrication and durability evaluation of structural aluminum joints made with Alodine 1200 conversion coated surfaces.

VII. HIGH-TEMPERATURE THERMAL/WATER OXIDE COATINGS

Coatings of a very complex chemical nature can be built up on aluminum surfaces when the aluminum alloy is exposed to boiler condensate water at high temperatures in an autoclave. The original purpose behind such investigations at the Alcoa Labs was to improve the overall surface resistance to various extremely corrosive service conditions. To the author's knowledge, the only work on attempting to bond such surfaces has been the author's investigations at the Alcoa Labs (860). It might be expected that these very thick oxide surfaces would be extremely sensitive to fracture within the oxide and it will be recalled that this was demonstrated by the author earlier with regard to thick SAA oxide surfaces. However, the combination of very high temperature and pressure in the autoclave seemed to create a very dense, highly compacted oxide which offered modestly high joint strength potential with epoxy adhesives. The author also conducted long-term durability testing of these joints with the result that they showed quite acceptable joint durability responses. However, the cost of creating such oxides as a preferable surface to CAA or PAA anodized oxides for bonding would not be practical or economical, nor was the overall joint durability performance as satisfactory as those obtained using the same adhesives over CAA or PAA oxide adherends.

VIII. BOEHMITE OXIDE AND ALUMINUM JOINT DURABILITY

The fact that many investigators have found "boehmite" oxide present on FPL and anodized aluminum adherend surfaces when durably bonding aluminum adherends would seem to confirm the importance of such an oxide in promoting the outstanding durability of structurally bonded aluminum joints. It was established very early in the study of oxide surfaces that the most stable form of oxide in boiling water was the "boehmite" form. An obvious way to support a correlation between the presence of boehmite oxide and excellent joint durability would be to deliberately generate a known boehmite oxide on an aluminum surface under conditions known to generate that form of oxide. This could be accomplished by subjecting a freshly deoxidized aluminum surface to boiling water conditions for varying time periods and compare the relative durability of joints made with an epoxy adhesive in a wide variety of accelerated lab and natural atmospheric weathering conditions. Minford (39) used such a procedure on 6061-T6 deoxidized aluminum. Surfaces were vapor degreased and deoxidized in hot chromic/sulfuric acid (Alcoa A3) or caustic (Alcoa A1). The surfaces were rinsed immediately in running tap water and immersed for varying times in boiling deionized or tap water. The method of achieving deoxidization did not seem to be critical. Joint exposures were conducted for periods up to 2 yr in room-temperature water, 100% RH at 125°F (52°C), salt fog cycling, or simultaneous exposure to various stress levels in 100% RH at 125°F (52°C). Depending on the length of time in the boiling water, the oxide thickness could be varied and initial joint strengths showed correlation with oxide thickness. For example, the 60-min immersion gave the thickest oxide films and the lowest joint strength of 2970 psi (20.46 MPa). The shortest immersion time conditions afforded the thinnest oxide regrowth and corresponding highest initial joint strength of 5730 psi (39.48 MPa), as shown in Table 17. The relative durability performances of these joints will be discussed in detail in Chapter 10 on joint durability.

Table 17 Initial Joint Strength Comparison of Aluminum Joints[a] with Varying Boehmite Oxide Thicknesses on Surface

Type boiling water treatment	Exposure time (min)	Average shear strength psi	Average shear strength MPa	Average type failure adhesive (%)	Cohesive (%)	% of mill-finish joint strength
None[b]	0	5120	35.28	50	50	100
None[c]	0	5020	34.59	60	40	98
None[d]	0	6180	42.58		100	121
Deionized[e]	1	5730	39.48		100	112
Tap[e]	1	4930	33.97	15	85	96
Deionized[e]	5	4750	32.73		100	93
Tap[e]	5	5930	40.86	5	95	116
Deionized[e]	10	4090	28.18	40	60	80
Tap[e]	10	5000	34.45	10	90	98
Deionized[e]	60	2970	20.46	100[f]		58
Tap[e]	60	3575	24.63	100[f]		70

[a]6061-T6 aluminum alloy joints bonded with 3M's EC-2086 nitrile-modified epoxy cured for 15 min at 400°F (204°C).
[b]Mill-finish surface condition.
[c]Vapor-degreased-only surface condition.
[d]Vapor degreased plus hot caustic etch plus HNO desmudge plus rinse under tap water and dry.
[e]Same pretreatment as [d] except transferred to boiling water of the indicated kind for the indicated time after the tap water rinsing step.
[f]Probably is a totally cohesive failure but in an extremely thin invisible layer of oxide that is invisible to the naked eye.
Source: Data from Ref. 39 (Table 1, p. 507).

IX. SURFACE CHEMISTRY OF ALUMINUM AFFECTING BOND DURABILITY

Excellent resistance of aluminum to weathering is directly related to the thin, invisible, tenacious oxide that develops naturally and instantaneously over the exposed aluminum atoms in the presence of oxygen. There can be surface heterogeneities which can cause variable resistance to corrosion over any particular surface which can be shown. The evidence of surface oxide variability can be easily viewed in several situations. For example, if a visually uniform "as received" or recently deoxidized surface is immersed in a dilute (1%) hydrochloric acid solution, the evidence of a few sites of immediate attack with hydrogen gas evolution will indicate areas of thinnest oxide protection. Within a few minutes, a multitude of additional sites of gas evolution become evident until virtually all the surface is being attacked and rapidly liberating gas, which is evidence that all oxide has been effectively removed.

It can be further observed that the speed with which the first evidence and subsequent attack develops can be distinctly related to the overall purity of bulk metal. Thus, the acid attack on 99.99% purity aluminum sheet will be slowest observed among a series of increasingly alloyed aluminum surfaces. As more and more alloying is employed, the rate of initial and overall attack per unit of time will increase until it is most rapid when exposing the high-strength alloys used to fabricate aircraft. Further, the attack rate will be most rapid with those alloys incorporating significant percentages of some highly cathodic alloying element like copper. A relationship of relative chemical resistance of different alloys to direct weathering with the overall relative durability of joints made with those alloys when exposed to the same weathering conditions would seem logical. As such, it has been verified by the author through the durability testing of literally thousands of specially fabricated test joints over many years at the Alcoa Labs.

A similar pattern of relative sensitivity to surface attack by water in the weathering environment on these same alloys might be anticipated and again has been verified by studies conducted by Minford at the Alcoa Labs. The evidence of aluminum surface corrosion under water contact conditions often takes the form of what the layman refers to as "water staining" in milder conditions or "pitting" in the more aggressive situations. Under magnified viewing, it will be readily apparent that the surface has undergone some variable chemical attack with the subsequent development of variable surface roughness. Where this roughness is relatively mild, it may only lead to some reduction in the reflection of incident light which makes the surface appear to be gray or sometimes even black to the viewer. This is the "water staining" condition that often occurs in stacks of flat or coiled sheet because moisture has invaded the space between sheets and condensed. Though the actual amount of aluminum lost from the surface may be relatively slight (based on weight loss measurements), it can still be sufficient to degrade or destroy the bond interface and eventually lead to total adhesion failure. When the surface is acid deoxidized and the oxide is regrown on such a surface, it can have a much more uniform cross section and offer some morphological characteristics which create reinforcement of the interface owing to better wetting and mechanical interlocking with the adhesive polymer.

In direct water immersion testing of a deoxidized aluminum surface, the surface may even appear to develop water staining faster than with a "mill-finish" surface or vapor-degreased surface, which is a reflection fo the fact that the overall regrown oxide is much thinner and therefore, offers less protection than thicker oxide on a mill-finish surface. It is likely that some support might be forthcoming for the application of an adhesive as soon as possible after the deoxidizing pretreatment based on the above considerations. However, as mentioned earlier, both Wegman and Minford have published data that such immediate bonding might not be necessary (at least in situations not involving the high criticality of aircraft structures), in which case, the more important argument involved would be to eliminate the possibility for subsequent contamination of the surface from the surrounding environmental conditions in the manufacturing plant.

Another situation of interest in discussing factors affecting joint durability is the heterogeneity of an aluminum surface with regard to its potential for electrochemical reactivity. If two aluminum facings of the same alloy are immersed in a saltwater solution with zones of differing concentration, they can show varying pitting patterns (at least as to the specific location of different individual pits). This is due to the fact that no two aluminum sheet surfaces have uniformly thick overall oxide. Rather, there will be individual patterns of thinner and thicker oxide growth conditions. Where the oxide is very thin, a crack may develop in the presence of chloride ion, permitting that area to become an anode with respect to the rest of the surface. The potential difference resulting will be indicated by a corrosion current flow between the areas and the development of a pit at the anodic site. An oxygen concentration cell can also develop if different quantities of oxygen are dissolved in different portions of the liquid in surface contact. This is really the condition that can exist when aluminum surfaces are protected from ready accessibility to oxygen for localized oxide repair. Water entering the area between the adherend surfaces can produce a conducting liquid. Thus, aluminum in a bonded joint can be related to this situation when the cured adhesive is acting as an oxygen-screening agent and water is encroaching into the bondline, promoting the possibility for the leaching of certain undesirable contaminants from the adhesive. That this situation is not always so threatening to joint durability is shown by the excellent resistance of aluminum joints with surfaces pretreated by procedures such as FPL etching, alodizing, or CAA, PAA, and SAA anodizing. Thompson (861) recently offered a summary of current thinking about the morphology, composition, and structure of these films on aluminum surfaces.

With the procedure for originating active pits in salt water as just described, there will be increasing numbers of new pits generated by the chloride ions until extensive pitting occurs. There are many different forms of pitting, ranging from what can be described as "general etching" to concentrated sites of attack leading to actual sheet perforation. Obviously, where this kind of intensive attack occurs, the localized adhesion may be destroyed long before perforation may be noted on the outside of the joint. When aluminum foils are laminated, for instance, the author has observed sufficiently corrosive conditions to exist that the foil was literally converted to the appearance of "lace." It will usually be found in these situations that a condition of unfavorable pH has been trapped or develops in incoming water at the interface. The opportunity for chloride ion or unfavorable pH to directly attack an aluminum surface oxide will likely be diminished by using a primer, paint coating, or adhesive over the oxide. In such a case, it may still be possible for destructive corrosion to be generated along an unprotected edge of a bondline and later encroach into the joint interior owing to a process of (1) generation of corrosion product, (2) washout of corrosion product, and (3) availability of a new adherend surface for the cycle to be repeated as often as necessary to produce debonding. The total disruption of the bondline and replacement with a relatively continuous layer of corrosion product over the debonded surface can occur within an amazingly short time frame when two mill-finish or even vapor-degreased–only 6061-T6 surfaces are bonded into a lap-joint using a two-part, room-temperature–curing epoxy adhesive and exposed to wet/dry cycling in saltwater conditions. The above process can proceed to total joint failure within about 7 days. If any significant stress is on this joint simultaneously with exposure to hot humidity conditions, then the same adhesive bondline might fail in a matter of minutes, as will be shown later in Chapter 10. Boerio and Hong (862) studied the cathodic delamination of rubber-to-metal bonds, whereas Watts (863) published a commentary on the cathodic delamination of organic coatings from metal surfaces.

Based on the above comments, it is readily apparent why there is a direct relationship between the type of oxide that is produced on an aluminum surface by pretreating and endurance of aluminum joints in all types of weathering and stressing situations. To further understand and develop ways for promoting oxides of outstanding stability, it has been necessary to analyze the surface both before and after bond failure with the most sophisticated analytical procedures as described earlier.

Solomon and Baun (797), in 1975, carried out analyses which bear further on the matter of enhancing joint durability through understanding and controlling the oxide in the interfacial area. They undertook to list principal points at which contaminants or unwanted chemical species might be introduced into aluminum alloys with relationship to manufacturing an adhesive joint, as shown in Table 18. Obviously, many impurities can be introduced into aluminum surfaces during the manufacturing and processing steps before the actual bonding is undertaken. After aluminum receipt by the bonder, it is the introduction of undesirable contaminates through the pretreatment operations that must be recognized and controlled. Such operations are deliberately designed to mechanically or chemically alter the "as received" surface and further interact with the surface. Of particular importance can be the way that the various chemical solutions may subject the alloy constituents present to segregation, selective chemical attack, or what is called "smutting." Finally, there is the environmental and handling effects on the treated surface before and during the actual bonding. These can suddenly introduce contaminants or even physically alter the surface right at the critical time of closing the bondline. In 1989, Treverton and Thomas (864) studied and published their results on the problem

Table 18 Sources of Contaminants or Species Which Could Affect Adhesive Bonding and Bond Properties

1. Raw Materials Processing
Adherend
Adhesive
2. Prebonding Treatments
Chemical solution contributions
Alloy constituents
3. Environment
Storage and handling
Bonding

Source: From Ref. 797 (Table I, p. 610).

of surface contaminants that can be present on annealed aluminum foil and the subsequent effects on adhesion.

The four aluminum alloy surfaces investigated by Solomon and Baun (801) included 2024, 6061, 7050, and 7075. Using AES, SIMS and ISS instrumentation, it was shown that the surface compositions of the "as received" alloys were totally different from the bulk alloy compositions. All, however, showed some magnesium-rich surface oxides which were believed to be due to the heat-treatment step during the final processing by the aluminum producer. For example, the oxide layer on "as recieved" 6061-T6 alloy was approximately 1000 Å thick with magnesium contant decreasing with increasing oxide thickness. In addition, segregated particles rich in the alloying constituents were shown to be present in the bulk as a result of quenching alloys after heat treating. Copper, iron, and manganese enrichments in segregated particles in the 2024 bulk could be shown in electron micrographs. Such inclusions could obviously contribute to the site of debonding, especially in exposures to high humidity capable of initiating corrosion. Also, these inclusions can be shown to have an inhibitive effect on the anodizing pretreatments.

Since it is recognized that the direct bonding of the above "as received" surfaces is generally unacceptable for long-term service under exterior weathering, some additional surface preparation needs to be undertaken. Solomon and Baun (801) undertook to investigate the resultant surface chemical alterations after seven different treating conditions, including (1) degreasing, (2) alkaline etch (NaOH at RT), (3) condition (2) + immersion in RT solution of HNO_3/HF, (4) condition (2) + FPL etching, (5) FPL alone, (6) NH_2 HF at RT, and (7) an Alodine commercial conversion coating process. From their results, a matrix of surface elemental information of the type shown in Table 19 was developed for 7075 alloy aluminum. Although these investigators did not attempt to correlate actual aluminum joint durability results with the various surface elemental detections, the author speculates that such data would likely correlate with the long-term durability results such as obtained by the author with 6061-T6 alloy joints over many years. For example, it would appear to be more than coincidence that the three surface treatments which leave no detectable copper or magnesium on the treated surface, i.e., alkaline + FPL etch, FPL etch alone, and Alodine conversion coating, are also pretreatments which afford excellent joint durability responses for the 6061-T6-treated joints evaluated for periods up to 2 yr in various water-soaking conditions by the author. Strohmeier (865) issued a 1989 report characterizing the surface of aluminum foil treated by annealing in the presence of ammonium fluoroborate which might be of interest.

As an example of a deleterious environmental effect on aluminum joint durability, Solomon and Baun have mentioned the lowered survival time in a wedge test (to be described in Chapter 9 on testing) found when a treated surface came in contact with something as apparently innocuous as kraft paper before applying the adhesive. In contrast, the significantly longer surviving control joint surfaces were only handled with clean cotton gloves.

Table 19 Major Elements Detected on a 7075-T6 Aluminum Adherend Surface After Different Surface Pretreatments Using ISS, SIMS, and AES Analytical Techniques

	Elements identified by indicated techniques						
Surface treatment	Al	Cu	Mg	Zn	Cr	P	F
Degreasing	I,S,A		I,S,A				
Alkaline soak	I,S,A	I,S,A	I,S,A	A			
Alkaline soak + HNO/HF	I,S,A	I,S,A					I,S,A
Alkaline + FPL etch	I,S,A				S,A	A	
FPL etch	I,S,A				S,A	A	
NH HF	I,S,A		I,S,A				I,S,A
Alodine treatment	I,S,A				I,S,A	A	

Source: Data from Ref. No. 797 (diagram on p. 620).

Kovacich and Lichtman (866), in 1985, published their qualitative and quantitative study of the oxides of aluminum and silicon using AES and XPS techniques, whereas Bariremow et al. (867), in the same year, published an XPS investigation of dichromate and molybdate inhibitors on aluminum. In 1989, Grant (868) offered his updated summary of the methods of making quantitative analyses on various adherend surfaces using both XPS and AES techniques, whereas Strohmeier (869) the following year suggested an ESCA (XPS) method for determining oxide thicknesses on aluminum. Cotter and Hockney (871) and McNally and Ronan (872) each offered quite comprehensive reviews of the state of the art for bonding metals as it was known up to the 1970s and still can be recommended as having some general and specific interest for the reader.

X. INTERACTION BETWEEN THE ADHESIVE AND ALUMINUM OXIDE

While there are various degrees of importance to be attached to different kinds of aluminum surfaces discussed above, it is necessary to involve the adhesive and try to understand its interaction with these surfaces in explaining all aspects of polymer/metal adhesion. In regard to the adhesive itself, it may not be sufficient to consider only bulk properties of adhesive as if it is the best representation of the unique surface involved. There also needs to be some consideration of various modifications of these bulk properties somewhere in the vicinity of the actual interfacial area. Sharpe (803,873) was the first to propose this concept of a three-dimensional "interphase" to describe a zone between the bulk adhesive and the bulk adherend. Although the use of this terminology has been slow in acceptance into the literature, it has had many more references in recent years in considering how that area adjacent to the adherend may be more specifically studied. For example, Bishop et al. (808), in the 1990 Cambridge Meetings, studied the influence of the surface pretreatment and environment on the interphasial region of bonded aluminum joints. The interphase zone referred to has been defined by Sharpe (803) as that zone extending from a point in any adherend where the local properties begin to change from the adherend bulk properties, through the interface, and into the adhesive where the local properties now approach those of the bulk adhesive. Depending on the particular adhesive/adherend combination present, this interphase region might vary from a few nanometers to a few thousand nanometers. Obviously, such an interphase must play a major role in the load transfer process between the bulk adhesive and adherend phases.

The number of investigators of interfacial relationships in polymer/metal adhesion has been so numerous that only a few can be selected here for reference. The list will continue to be augmented throughout this book as various specific subject areas are discussed which bear on that subject (e.g., see Refs. 22, 450, 452, 559, 754, 806, 807, and 874–891).

For specific investigations on aluminum oxide, the works of Venables et al. (33,754), and Packham (885) can be suggested. Hansen and Schonhorn (892) discussed the possible existence of weak boundary layer conditions, Ward et al. (893) considered internal stresses in the adhesive and toughening mechanisms, Knollman (894) the gradient in material properties in the interphase, and Kwei (895) and Theocaris (896) the extent of polymer/filler interphase. Howard and Shanks (897) studied polymer response near the glass transition temperature to investigate how the interphase polymer properties might be altered as compared to the bulk; however, Peyser (898) questioned the ability of any high-modulus surface to influence macromolecules at any significant distance from the interface. Nevertheless, mathematical modeling of the interphase to allow for increases or decreases in the glass transition temperature of a filled composite has still been introduced by Theocaris and Spathis (899).

Relatively recently, Ko et al. (900) conducted research on the influence of surface topography on the viscoelastic properties of thin thermoplastic polysulfone films (PSF) on aluminum foil that had either been treated by vapor degreasing only or PAA anodizing. The interactions were characterized using XPS and SEM to study the aluminum surface topography. They concluded (1) XPS results showed the same interfacial chemistry on degreased and anodized aluminum surfaces, (2) the PSF uniformly coated the degreased adherent but migrated into the porous PAA oxide, (3) the glass transition temperature for PSF was highest for films on porous aluminum where the penetration had occurred, and (4) the Arrhenius activation energy was observed to decrease with increasing film thickness and approach that of the neat film at about 1.4 μm for the film on the degreased surface. For the film on the PAA surface, the activation energy at 2.0 μm was actually higher than that of the neat film.

4

Surface Treatment of Aluminum Adherends

I. INTRODUCTION

The method of aluminum surface preparation ultimately selected for a particular commercial bonding application can vary significantly. Without question, surface preparation can be considered as one of the most vital factors in assuring adequate joint strength and long-service potential. In Table 18, Baun reviewed the various sources of contamination that could adversely affect the adhesive bond strength and the joint durability in manufacturing. He has further identified in Table 19 the presence of a variety of elements which he found present on a 7075–T6 aluminum surface even after pretreating in various fashions. The desired goal of any aluminum surface-cleaning and oxide-regeneration procedure is to produce a dry, firm, stable, and chemically wettable aluminum adherend surface with known and reproducible surface characteristics. The degree of the chemical stability of the treated aluminum surface necessary to achieve the desired goal of long-service life may vary depending on the strength of bond desired to maintain the structural integrity of any particular manufactured product and the severity of the service environment. Listed in order of increasing effectiveness for producing this chemical stability, the aluminum surface pretreatments can be separated into four general categories as follows:

1. Operations designed only to remove surface soil contaminants
2. Operations necessary to remove the above soiling plus some mechanical abrasion with subsequent operations to remove the residual surface contamination produced by the abrading step
3. Operations to remove surface soils plus a special chemical processing designed to create a new stable surface oxide followed by final rinsing to remove any chemical contamination or residues left by the chemical processing, followed by final air drying of the surface
4. Operations combining the steps of (3) as a pretreating before an overcoating with a special primer designed to create maximum chemical compatibility with the selected adhesive and temporary protection from environmental contamination until final bonding operations

A 1984 discussion of the need to prepare adherend surfaces before adhesive application has been offered by Adams and Wake (901). More recently, Clearfield et al. (902) developed a chapter on "Surface Treatment of Metals" with the major emphasis on treating aluminum adherends (902), including observations on morphology and relative joint durability. It should be clear that it is absolutely mandatory to have the best possible conditions for adhesion on any aluminum surfaces for best service in long-term service applications under severe weathering conditions. It may often be satisfactory for less critical applications to simply confirm whether reasonable wetting can be achieved by the adhesive with removal of gross surface contamination. The simplest test is to observe whether the flow of water over the surface to be bonded is uniformly free flowing. Those areas where the water tends to bead up must be considered suspicious at best. This latter condition is certainly not acceptable for 100% solids structural adhesive pastes, films, or tapes. For a more quantitative measure of wettability, the procedure of making accurate measurements on the spread of a drop of water of constant volume needs to be viewed through a transparent gauge placed over the drop area. As the angle of contact approaches zero, the degree of anticipated wetting is correspondingly more confirmed. Bijlmer and Schliekelmann (779), as mentioned in Chapter 4, have particularly studied the relationship of the surface condition of aluminum adherends after pretreatment for bonding Fokker commercial aircraft which have achieved good service results for many years. Bijlmer (903) has described the use of the Fokker Contamination Tester for evaluation of adherend surface cleanliness (including aluminum) for bonding. An oscillating probe is used to measure the electron emission energy of the surface in this device. (More references about the theoretical basis for operation and field success of this instrumentation will be given in Chapter 9 on testing.) The instrument reading can vary greatly depending on the degree of surface contamination and offers a testing device for some relative prediction of bonding success. It has been shown to be sensitive even to residues from an alkaline cleaning operation on aluminum where that method of surface preparation has been employed in manufacturing.

It needs to be emphasized from the beginning that no existing method has been found which is totally satisfactory for establishing the best conditions for bondability under all known treating circumstances. Best assurance of acceptability remains in exercizing the most careful control of the actual treating processes used for preparing the surface. Information provided by Steel (904) in a report of the *Internal Military Vehicles and Eng. Establishment* in Dorset, England, might be of special interest in this regard. Brockmann (905) provided a report on the effect of surface pretreatment on the properties of metal/adhesive-bonded joints. Thrall and Shannon (906) provided details on PABST Surface Treatments and Adhesive Selection," whereas Carrillo (907) published on "Alternative Cleaning Methods for Aluminum Honeycomb Core." Danforth and Sunderland (908) discussed the effects of contamination of adhesive bonded surface treatments, and Rogers (909) reported on surface preparation of metals for bonding.

Another important fact, not always recognized, is that the actual level of initial joint strength measured by lap-joint testing (widely used because of simplicity of test joint fabrication and economies of testing) is not really an accurate method for predicting the presence of bondline defects that can significantly affect long-term joint durability. Wang et al. (910) demonstrated this as early as 1971 by creating a disbond area in the center of the overlap of aluminum/epoxy structural joints by inserting a polypropylene disk in that location. Even though this large defect area was present, the joint strengths measured by lap-shear testing were largely unchanged from similarly bonded test joints without this implanted defect. Other discussions of various aspects of cleaning aluminum surfaces that relate to better bonding would include early papers by Snogren (911,912) on selecting surface preparations for bonding adherends, Spring and Peale (913) on mechanisms of metal cleaning, and Dumond (914) and Vance (915) on metal-cleaning materials and procedures. The range of potential cleaners for metals has been reviewed by Lesser (916) and Liddiard (917), with the latter focusing on metal cleaning in aqueous media. Morgan and Lankler (918) also have outlined methods for evaluating the efficiency of metal-cleaning compounds, whereas Harris et al. (919–921) have considered the use of radioactive isotopic cleaning evaluation methods.

A complete metal-cleaning bibliography covering the period from 1842–1951 has appeared in a series of ASTM Spec. Tech. Publs. (922). Spring (923) produced a 1963 book on metal cleaning;

Zelley (924) a special Alcoa Tech. Service Report on cleaning aluminum in 1965; Chessin and Curran (925) an extensive listing of preparations on aluminum for bonding in 1966; and Schneberger's (926) preparation for bonding article of 1974. Spencer (927) and Linford (928) have also considered the preparation and evaluation of various metal surfaces, whereas McKee (929) has offered an Alcoa Labs discussion on the specific cleaning of aluminum surfaces. Westrey (930) produced one of the more recent surveys of metal cleaning for bonding in 1983. In spite of repeated emphasis above on having very clean aluminum surfaces for bonding, Burrows and Crowe (931) offered data as early as 1962 challanging the principle that optimum bond strengths demand surfaces completely free of foreign matter.

II. SOIL REMOVAL-ONLY PROCEDURES

Most commonly employed aluminum surfaces for bonding applications exist in the form of various thicknesses of aluminum foil, flat sheet, and extruded shapes of varying geometries. After processing foil and flat sheet forms of commercial aluminum, the surfaces will be contaminated with some form of rolling oil applied by the producer for reducing the ingot to thinner gauges of foil and sheet. This "mill-finish" surface condition will be further contaminated in many situations with selected corrosion-preventative coatings which are applied primarily to protect the aluminum under shipping and storage conditions. Extruded aluminum sections, in turn, can have some unusual surface conditions in the form of special surface topographies along with certain forms of imbedded surface contamination resulting from the flowing of the aluminum under high pressures in contact with extrusion die surfaces. All forms of aluminum need to be handled by specific procedures during packing and shipping operations at the primary metal production facility and subsequent handling and storing at the customer's plant. Finally, all incoming aluminum for eventual bonding is initially covered with a natural or specially processed oxide which needs to be identified as to how best bonding can be achieved for final manufacturing of bonded products of the required strength and service durability potential.

A. Solvent Wiping

Solvent wiping by hand is the simplest procedure usually encountered in general manufacturing to remove poorly adhering particulate soiling and the fabricating oils and lubricants that can be present on "as received" aluminum surfaces. It is generally the least costly procedure for small manufacturing facilities even though it requires individual worker attention on each workpiece. However, special care needs to be taken to avoid leaving residual contamination on the workpiece surface through transfer from repeated use of a solvent-laden wipe rag. Unfortunately, such a procedure can become a sort of going-through-the-motions operation with too little attention paid to the fact that surface contamination is simply being moved around on the surface rather than being effectively removed. This fact is readily observed by viewing the surface after the solvent has been vaporized. The appearance of hazy residues of lube or oil with the same wiping pattern as employed for cleaning can be clearly identified. The author has been present at manufacturing facilities involved in cleaning aluminum for subsequent bonding where the aluminum sheet started out being so heavily oiled (often to reduce corrosion hazard in storage) that a signiture could be traced on the surface with a finger. In spite of observing this heavy contamination, the line operators would employ the same solvent-soaked cleaning rags repeatedly, so that the above cited situation of simply moving the contamination around on the surface was clearly in evidence. At the same time, the adhesive was being applied as though effective cleaning were being performed by the line operator.

Traces of soil residues (of any nature) left on an aluminum surface will be particularly critical when employing any form of 100% solids adhesives. The worst scenario is obviously trying to achieve any kind of effective overall wetting of the aluminum using a two-part, room temperature-curing epoxy paste which ordinarily does not absorb or disperse the contamination. Heat-curing 100% solids adhesives offers a better wetting potential situation because of the lowered viscosity (increased flowability) of adhesive under high temperature–curing conditions. Contact adhesives with significant amounts of solvent present in the "as applied" adhesive will offer a better chance for obtaining some degree of reasonable bonding with such soiling still present. Kircher (932) has provided a review of what every user should know about solvent degreasing in ASTM Bull. No. 2129. A later publication titled *Handbook of Vapor Degreasing* was published as ASTM Spec. Tech. Publ. No. 310 in 1962 (933).

A production line engineer should be assured by actual durability testing of joints that the methods employed to remove any contamination by solvent wash have been adequate for achieving the level of joint service life demanded for his particular product. Thus, it may be demonstrated that simple soil removal procedures by wiping with solvent may suffice for many of the less-demanding service products with interior weathering and low stress service conditions present.

The bonding of Alcoa Alply commercial building panels using only furnace-cleaned, oil-free aluminum facing sheets was demonstrated by Minford and co-workers to be satisfactory for even long-term atmospheric exposure conditions. However, unit-stresses were low and effective wetting was achieved using special-quality neoprene-phenolic contact cement adhesives (87–89). Later, Minford (934) showed that equivalent long-term durability could be achieved in similar panels by substituting a well-monitored vapor degreasing of the mill-finish facing sheets for the more expensive furnace-cleaned products using the same good-quality neoprene-phenolic contact cement. The reference to "special good-quality" neoprene-phenolic contact cement is predicated on the fact that there existed a small number of commercial products of this family which offered distinctly higher probability for long-term satisfactory lamination under the relatively severe service involving exterior weathering exposures. More will be said about how these unique qualities were verifiable by special testing in Chapter 10 on durability of bonded joints.

B. Vapor Degreasing

Where heavy oil, grease, or loosely adhering particle soiling is present on foil or sheet, vapor degreasing should be substituted for the hand-wiping type of solvent degreasing just discussed. This procedure has the double advantage of removing soil uniformly from the surface with a recycling flushing action plus achieving more solvent-dispersing action owing to the higher chemical action achieved by using the solvents at their boiling point. It should be remembered that the nature of some particulate soils (such as metal fines or surface corrosion products) may not be readily removed by either solvent wiping or vapor degreasing. As will be discussed later, these more adherent soils can only be removed by some sort of vigorous mechanical abrading or chemical deoxidation of the surface.

It may be important to point out that while vapor degreasing may be highly recommended as a first step of surface preparation for any mill-finish aluminum adherend, there are significant alloy difference factors which are interacting as regards ultimate joint durability performance. In Table 20, relevent data can be seen that is designed to show some examples of this different alloy effect (935). It can be observed that the level of initial joint strength was relatively undiscriminating when a one-part, heat-curing epoxy adhesive was used to bond adherend surfaces of three different alloys in the mill-finish, vapor-degreased, or chromate-phosphate conversion–coated surface conditions. When duplicate joints were placed in only 85% relative humidity (RH) conditions for 6 mo, the effect of both surface pretreatment and interaction with different alloy surfaces could be observed. As mentioned in Chapter 3, the 2000 series joints (with highest copper present for alloying strength) lost 79 and 85%, respectively, with mill-finish or vapor-degreased surfaces for bonding. This alloy would be the one with least resistance on its surface to general corrosive conditions. The 6000 series alloys would have better general resistance to surface corrosion than the 2000 series, and this is reflected in lower joint strength losses of 47 and 43% found under the same exposure conditions. Finally, the 5000 series aluminum alloy adherends would be the most resistant to general surface corrosion, which is reflected in a loss of only 17% for both mill-finish and vapor-degreased surface joints. The importance of further surface treating with a conversion coating (in this instance, Alodine 1200) is shown

Table 20 Effect of Surface Treatments[a] on Different Alloy Aluminum Joints Using a One-Part Heat-Curing Epoxy Adhesive

Exposure condition		Alloy 2036–T4			Alloy 6151–T4			Alloy X5085–H111		
		A	B	C	A	B	C	A	B	C
Int.	(psi)	1850	1930	2130	2550	2690	2530	2250	2270	2110
strength	(MPa)	(12.75)	(13.30)	(14.68)	(17.57)	(18.53)	(17.43)	(15.50)	(15.64)	(14.54)
85% RH[b]		380	313	2200	1350	1550	2410	1860	1890	1910
75 °F (24 °C)		(2.62)	(2.16)	(15.10)	(9.30)	(10.68)	(16.60)	(12.82)	(13.02)	(13.16)
100% RH[b]		20	0	690	920	1050	1100	1260	1520	1230
125 °F (52 °C)		(0.14)	0	(4.79)	(6.34)	(7.23)	(7.58)	(8.68)	(10.47)	(8.48)
5% salt spray[c]		0	0	1110	80	150	2090	690	530	1210
95 °F (35 °C)		0	0	(7.65)	(0.55)	(1.03)	(14.40)	(4.75)	(3.65)	(8.33)

[a]Code for surface treatments: A—mill finish; B—vapor degrease; C—Alodine 1200 (commerical chromate-phosphate conversion coating solution).
[b]Exposure time was 6 months.
[c]Exposure time was three weeks.
Source: Data from Ref. 935 (Table 3-2, p. 82).

by achieving virtually full retention of strength regardless of the alloy tested. With elevated temperature 100% RH continuously condensing exposure conditions, the same pattern of alloy preference was discernible; i.e., descending joint durability in the series X5085–H111 > 6151–T4 > 2036–T4. The more severe exposure conditions, however, will create a situation of total joint failure for the 2036 alloy joints, whereas the 6151 and X5085 alloy joints sustain declines of 56 and 42% respectively. The final exposure was in 5% salt spray at 95 °F (35 °C). The same order of relative joint durability continues to be demonstrated. Most corrosion-sensitive 2036–T4 alloy joints show total failure within a 3-wk period when no initial surface treatment was employed or vapor degrease was used. When Alodine 1200 conversion coating was used as the pretreatment, duplicate joints retained 52% of initial values. Meanwhile, 6151–T4 alloy joints with greater inherent resistance to the corrosive saltwater did not fail; however, joint strength retention was only at the 3 and 6% levels with mill-finish or vapor-degreased-only surfaces, respectively. The most corrosion-resistant (major alloying element, magnesium) X5085–H111 alloy joints managed to retain as much as 29 and 23% with mill-finish or vapor-degreased conditions. When the conversion coating was used on these latter alloys, the 6151–T4 joints unexpectedly retained as much as 83% of their initial joint strength compared to only 57% for the X5085–H111 joints with no explanation available. It does indicate, however, that the surface interactions to form such new conversion coatings can be more complex than anticipated. However, the basic response of all the above test results seems definitely to fit a relatively predictable pattern of behavior.

Vapor degreasing has been generally employed as the first processing step for further pretreating of the aluminum surface by chemical deoxidizing, conversion coating, and anodizing. There are interesting interactions with vapor-degreased-only aluminum adherends that might be mentioned. For example, the heat curing of an adhesive that is ordinarily cured at room temperature, i.e., two-part, room temperature–curing epoxies, may often provide significantly higher initial strength values on these vapor-degreased surfaces. This is clearly shown in the data provided by Minford in Table 21, which included both types of curing of the same epoxy (EA-9320) paste over vapor-degreased 6061–T6 aluminum surfaces. A 33% boost in initial strength found with heating must be considered clearly significant. It is also noted that starting with an abraded or HCl-deoxidized surface showed only 7 and 3% joint strength elevations. This is explained by the fact that the enhanced wetting already present owing to these pretreatments makes the increased flowability of the heated epoxy paste less of a factor in contributing to the overall level of initial joint strength.

Table 21 Comparison of Vapor Degreasing, Abrading, and RT Acid Etching Surface Pretreatments and Variable Temperature Adhesive Curing on Initial Aluminum Adherend Joint Strengths[a]

Surface pretreatment	Adhesive	Cure conditions	Lap-shear (psi)	strength (MPa)
Vapor degrease	EA-9309[b]	60 min [180 °F(82 °C)]	3400	23.43
Aloxite 60 abrade	EA-9309	60 min [180 °F(82 °C)]	5000	34.45
HCl etching with paste for 2 min at RT	EA-9309	60 min [180 °F(82 °C)]	5100	35.14
Vapor degrease	EA-9320[b]	60 min [180 °F(82 °C)]	3090	21.29
Aloxite 60 abrade	EA-9320	60 min [180 °F(82 °C)]	4380	30.18
HCl etching	EA-9320	60 min [180 °F(82 °C)]	5800	39.96
Vapor degrease	EA-9320	7 da at R.T.	2320	15.98
Aloxite 60 abrade	EA-9320	7 da at R.T.	4100	28.25
HCl etching	EA-9320	7 da at R.T.	5620	38.72

[a]All joints fabricated with 0.125 in (3.18 mm) 6061–T6 aluminum.
[b]Both adhesives are products of Hysol and are two-part epoxies formulated for high-strength applications demanding flexibility without the need for special heat-curing. EA-9309 has a T-peel of 42 pli (7355.3 N/m), whereas EA-9320 has a floating drum peel of 55 pli (9632 N/m)
Source: Data from Minford unpublished work at Aluminum Co. of America.

While the joint durability for structures which will primarily experience only interior weathering service may be adequate when employing vapor-degreased–only aluminum adherends, confirmation still is required by undertaking some testing of the final joint under the most severe conditions that might exist in the anticipated service. This often will take too long to adequately demonstrate, which is why laboratory weathering conditions such as condensing humidity at 125 °F (52 °C) has been employed by the author. The extrapolation of durability test results using such an accelerated weathering condition is still very difficult (if even possible) to translate to some measure of time for a practical warranty for a bonded product. A good performance in such an accelerated laboratory weathering test environment does provide a degree of confidence in the product, however, that cannot be obtained by any theoretical assessment.

Although vapor degreasing will usually be preferred over a simple solvent wiping for reasons already discussed, it must still be properly controlled to achieve best effectiveness. The author has encountered manufacturing operations rather frequently where sudden poor bonding results, or even early field-service failure, occurred as a result of failure to maintain proper scheduled cleaning of the vapor-degreasing equipment and recycling solvent. Although manufacturers of such equipment always provide for some form of automatic trapping and screening of major contaminant particulates, it is still possible to build significant contaminate levels of soluble organic materials, rendering the procedure as ineffective as hand wiping adherends with grossly contaminated pads. Thus, a part of maintaining effective vapor degreasing is to continuously filter and periodically regenerate the solvent by distillation or drain and refill with fresh solvent. Most manufacturers choose to use either trichlorothylene or perchloroethylene as solvents for vapor degreasing with appropriate chemical stabilizers added to prolong the service life of the degreasing charge. Pollack and Westphal (936) published a book titled *An Introduction to Metal Degreasing and Cleaning* which would likely still have relevancy though written in 1963.

C. Other Degreasing Techniques

Other degreasing techniques have been proposed for manufacturing that are more sophisticated and costly and do show excellent results. A procedure known as electrolytic degreasing was described as early as 1949 (937). Ultrasonic cleaning and pickling was discussed by Spring (938) in 1964. A test for evaluating effective degreasing called the Atomizer Test was proposed by Linford and Saubestre (939) in 1953. These specialized types of degreasing have been especially useful and effective where large numbers of smaller aluminum parts must be cleaned which have significant deposits of oil or lube on the surface. They also fine use for cleaning moving aluminum foils at relatively high speeds in a variety of laminating operations.

D. Nonetching-Type Solvent Cleaners

For most metals, including aluminum, a frequently used commercial method for removing soiling has been the use of so-called nonetching emulsion–type cleaners, as discussed by Eyles (940) and Lowe's (941) paper on precleaning with solvent emusions. Vos' (942) paper reported on detergent cleaning for removal of soils, whereas Spring (943), in 1963, discussed how the type and degree of metal surface soiling can affect the cleaner selection. Lesser (944) discussed the use of cleaners specifically for aluminum, whereas Bass (945) focused attention on special cleaning procedures for use on aluminum sheet surfaces prior to spot-welding. This has pertinence because the same cleaning that could promote better spot-welding could also promote better adhesive joining. If long-term service durability is sought, the author suggests some testing program to prove applicability, since no alteration of oxides or change in morphology of the surface will be achieved through use of nonetching cleaners. Use of nonetching cleaners does have advantages, however, in the fact that either a spray or soaking immersion procedure might be considered. In addition, the temperature and/or contact times can be conveniently altered to handle different types and degrees of soiling present.

E. Controlled-Etching (Inhibited) Alkaline Cleaners

It should be recognized that a nonetching-type cleaner (usually employed at elevated temperature to better emulsify hard to remove soil) will not offer the aggressiveness necessary to remove ingrained soiling, as discussed by Mohler (946). This includes soils caused by mild water staining or induced by more aggressive high humidity or saltwater mild pitting corrosion on the aluminum. While these kinds of soils can appear to be firmly adhering to the underlying oxide, they will not act as a good primer for subsequent bonding. They may actually constitute a weak boundary layer condition under long-term weathering exposure. The most typical situation is use of vapor degreasing to remove excessive initial oil, lube, or anticorrosion protective coatings. This is followed by cleaning in hot inhibited alkaline cleaner for many foil or sheet laminating applications. Mohler (946) also discussed what alkaline solutions can be used for effective cleaning of a variety of metals, including aluminum. Heath and Tolley (947) reported some pertinent considerations for study in a paper titled "Dissolution of Aluminum in Alkaline Solutions," while Muller and Low (948) and Streicher (949) considered dissolution of aluminum in caustic (sodium hydroxide) solutions. A related paper would be that of Prescott et al. (950) on the use of various sequestering agents on etching rates in alkaline cleaning solutions. Such reports usually relate to producing a controlled etching on aluminum, which will be considered later in this chapter under the subject of etching pretreatments. Minford (951) conducted an extensive investigation relating chemical composition of aluminum alloys with their dissolution rates in caustic solutions as part of studies related to the inhibited caustic cleaning of aluminum beer barrels. He showed an increasing rate of etched aluminum loss as the purity of 99.99% aluminum was being gradually replaced with increasing concentrations of various other alloying elements. This was the basis for employing relatively high-purity interior cladding for aluminum barrels that needed to be repeatedly cleaned, such as in the brewing industry.

Inhibited alkaline cleaning of aluminum surfaces for bonding is a very versatile method, since either spray or immersion procedures can be used. Also, different degrees of soil concentration and soil adherence can be accommodated by proper choice of solution temperature, cleaner, concentration level, and solution contacting time. For aluminum cleaning, as compared to some other metals, the caustic content of the cleaner used tends to be relatively lower. Chromates or silicates have been extensively added to such baths to control or even inhibit the relative aggressiveness, as discussed by Heath and Trolley (947). As in all chemical cleaning procedures, the practical question arises as to balancing the cleaning efficiency of the solution as compared to compositional changes, as discussed by Zelley (924), Foulke (952), George et al. (856), and Mazia (953). Examples of the chemical composition of inhibited and noninhibited cleaners are shown in Table 22 (856). Concentration of aluminum and other alloying ingredients that can be permitted to accumulate in the cleaning bath must be studied for each separate manufacturing situation. Barber (954) wrote a Boeing specification for alkaline cleaning of aluminum of possible reader interest. Rinsing of the aluminum surface after alkaline cleaning can also be of considerable importance, as discussed by Kushner (955,956) in 1949 and 1955 publications.

A common procedure used to determine if any of these degreasing procedures has adequately prepared the surface for bonding is the so-called water-break rinse test. If water is sprayed or flowed over supposedly clean aluminum and the water is shed uniformly in a continuous sheeting action, then the surface can be considered water wettable. This surface would be appropriate for most general bonding situations where further treatment like etching, conversion coating, or anodizing is not planned or economically feasible. If discrete separate beading of water is observed over any part of the sheet, then the cleaning will not be acceptable. Sometimes as different degrees of soiling might occur between different lots of material (and failure to pass a water-break test suddenly appears), it may be necessary to alter the cleaning procedure until success is attained. Even where a perfect water-break has not been obtained, it may sometimes still be acceptable for bonding with an organic solvent-rich, contact-type adhesive containing strong organic solvents like methyl ethyl ketone and toluene. This needs to be established by some form of accelerated weathering test procedure that predicts good adhesion durability for the service conditions anticipated for the laminated product. A borderline water-break test result would never be considered satisfactory when using a water-dispersed contact-type adhesive. Trevoy and Johnson (957) discussed this matter in some detail from a physical chemistry viewpoint.

III. SOIL REMOVAL–ABRADE–RECLEAN PROCEDURES

Abrading aluminum adherends would be expected to further improve the resistance of bonded joints to exterior weathering conditions as compared to mill-finish, vapor-degreased-only, or emulsion-cleaned aluminum adherends. This seems to be verified by the 6061–T6 joint durability results shown in Table 23 (774), where exposure was up to 8 yr in the natural industrial atmosphere near Pittsburgh, Pennsylvania. However, when similar joints were exposed to a more aggressive accelerated weathering test in a cycling exposure to hot-water soaking/freezing/hot-air thawing for 2 yr, as shown

Table 22 Examples of Alkaline Cleaners for Aluminum

Type	Composition	Temperature	Time
Noninhibited, etching	Na_2CO_3, 22.5 g/l $Na_3PO_4 \cdot 12\ H_2O$, 22.0 g/l	160 to 180	As required
Inhibited, nonetching	Na_2CO_3, 22.5 g/l $Na_3PO_4 \cdot 12\ H_2O$, 22.5 g/l $Na_2SiO_3 \cdot 9\ H_2O$, 15.0 g/l Kerylbenzene sulfonate (40%), 2.5 g/l	160 to 180	As required

Source: from Ref. 856 (Table 2, p. 598).

Table 23 6061–T6 Aluminum-Bonded Joints Exposed to 8 Years in the Industrial Atmosphere with Various Abraded, Etched, Conversion-Coated, Paint-Primed, and Anodized Surface Pretreatments

Surface pretreatment	Adhesive type[a]	Avg. joint stength after indicated exposure time (MPa)				% Strength retention after longest exposure
		1 yr	2 yr	4 yr	8 yr	
Vapor degrease	A	12.20	10.54	9.09	6.25	43
Vapor degrease	B	22.05	21.36	19.29	3.10	10
Belt-sanded	B	29.63	28.46	25.98	23.43	70
Silica gritblast	A	13.78	14.26	13.99	13.99	100
Chromic acid etch	A	18.60	20.67	20.23	18.71	96
Chromic acid etch	B	37.21	36.52	35.83	32.59	89
Alodine 1200 (conversion coat)	A	8.75	8.27	8.96	7.79	100
Alodine 1200	B	18.60	17.91	18.60	19.09	100
Epoxy-phenolic (paint-primer)	A	–	19.50	20.77	19.70	98
Epoxy-phenolic	B	–	33.97	30.56	33.35	91
Alumilite 215 (H_2SO_4 anodize at 0.8 mil thick)	A	16.05	14.68	12.88	10.92	63
Alumilite 215	B	19.29	19.09	19.50	18.09	80

[a]Adhesive-type A is a two-part, polyamide-cured epoxy; adhesive-type B is a one-part, nitrile-modified epoxy.
Source: Test results taken from Ref. 774 (Table V, p. 496).

in Table 24, the employment of belt-sanding as compared to only vapor degreasing was not especially enhancing. This could be due to the fact that the gross macroscopic roughening produced by the sanding operation produced an interface where the bonding was more sensitive to the alternating stresses of wet/freeze/thaw cycling. As shown, the bonding to the microscopically roughened surfaces of FPL etched or anodized aluminum surfaces could be much more resistant to such cycling exposures.

A. Grit-Particle Papers and Rotating-Wheel Grit-Paper Treating

In situations in which degreasing procedures alone are inadequate, it would seem beneficial to follow with mechanical abrasion. While it may not always be necessary to solvent wipe or vapor degrease before abrading, most manufacturers will find it helpful to do so in order to prolong the useful life of bonds to an abraded material surface. It would be evident that any solvent removable soil will immediately contaminate any abrading medium surface. Further, such contamination will continue to build-up, producing poor bonding conditions. Also, the contaminated abrading surface actually has the opportunity to imbed contamination in the mechanically roughened surface. It is for this reason that any abrasion processing must be followed by a strong solvent flushing (or vapor degreasing might be better) to dissolve soluble organic contaminants and flush loose oxide and various surface residues. Failure to remove such residues and soluble contaminates would undoubtedly create non–water-wettable areas and probably even weak boundary layer conditions.

Poor reproducibility as a surface preparation usually results from either manual or power-driven abrasion procedures, since no two line operators will have the same technique or dedication to duty.

Table 24 6061–T6 Aluminum-Bonded Joints Exposed to a Hot-Water Soak/Freeze/Hot-Air Dry Cyclic Exposure with Various Abraded, Etched, Conversion-Coated, and Anodized Surface Conditions

Surface pretreatment	Adhesive type[a]	Avg. joint stength after indicated exposure time (MPa)				% Strength retention after longest exposure
		3 mo	6 mo	12 mo	24 mo	
Vapor degreased	A	0.83	0	0	0	0
Vapor degreased	B	13.09	11.51	7.30	3.45	12
Belt-sanded	B	12.4	11.23	9.65	0	0
Silica gritblast	A	8.82	8.75	6.44	0	0
Chromic acid etch	A	17.71	13.71	13.57	14.47	74
Chromic acid etch	B	11.71	10.54	9.16	0	0
Alodine 1200 (conversion coat)	A	5.72	5.55	5.27	0.83	11
Alodine 1200	B	17.08	12.94	16.54	0	0
Alumilite 215 (H_2SO_4 anodize at 0.8 mil thick)	A	13.57	12.88	10.54	8.41	48
Alumilite 215	B	15.64	13.99	15.16	6.41	28

[a]Adhesive-type A is a two-part, polyamide-cured epoxy; adhesive-type B is a one-part, nitrile-modified epoxy.
Source: Test results taken from Ref. 774 (Table IV, p. 493).

As with manual solvent wiping, the question arises as to when an abrasive grit paper or rotating-wheel assembly is performing adequately. Lower performances might result from the build-up of contaminates on actual abrasive particles and the transfer of more contamination to the newly abraded workpiece surface. With power-tool abrading, these negative effects might be further accelerated and exaggerated by the fact that the abrading rate is much faster with a resulting faster pickup of contamination and consequent wear-down of the sharp particle edges. In addition, the higher cost of both equipment and higher labor skill required for such power equipment might make the manufacturer more reluctant to close down operations to clean up the equipment on the production line. Thus, a decision and commitment needs to be made early in the development of any manufacturing program about how much contamination (if any) will be tolerated. Hopefully, this decision would be based on tested joint strength levels achieved and the results of joint durability evaluations conducted after using a variety of levels of repeat use of the same abrading equipment.

B. Manual Wire Brushing and Power-Driven Wire-Abrading Assemblies

The same problem mentioned when using grit particles bonded to paper or cloth backings also occurs when manual or power-driven wire-abrading devices are used for pretreating. The amount and speed of contamination build-up on such devices usually has some direct relationship to the closeness of packing of the wire bristle grouping. There is some obvious relationship between the number of bristles and density of abrasion cuts; i.e., relative uniformity of the abraded surface. The individual operator must be given some method of judging what pattern and length of action time his or her tool is capable of operating without unduly contaminating the surface that is being prepared for effective bonding.

In addition, there can be a surface treatment variation due to the differences in the depth and pattern of abrasion under the direct control of each separate operator. This will result from how much pressure is exerted by the tool on the workpiece or, conversely, by the pressure of the workpiece where it is pressed against the abrading device. Probably, a robotized abrasion operation offers the best opportunity to create a more reproducible abraded surface, since the orientation and pressure of the abrasion would be precisely controlled without operator fatigue entering the picture.

Another important concern when abrading aluminum is to prohibit the use of any wire material whose residues in the abrasions could later produce surface corrosive action in the presence of water in service. Bhatt and Ramakrishnan (958) reported studies on 2014 aluminum alloy surfaces using various machining processes such as turning, milling, and shaping. The influence of tool feed rate, surface roughness, and adhesive film thickness on joint strength was considered along with the surface profile records. Ramakrishnan and De (959) also investigated the effect of adhesive film thickness on the joint shear strengths obtained with shaping, grinding, milling, and turning procedures. Their report does not, however, include details about the aluminum adherend surface roughness after such machining.

C. Shot- and Gritblasting Treatments

The other widely used abrading procedure for aluminum is shot- or gritblasting. Steel shot is commonly used for pretreating steel and other metals for bonding but should never be used for treating aluminum. This is because of the significant electrochemical potential difference the two metals can have in a conducting solution. Each steel particle embedded in the aluminum surface will act as a strong cathode site surrounded by a more anodic aluminum surface. In effect, the aluminum acts as a sacrificial anode to protect steel from corroding in the presence of water which may have invaded the interfacial area between the two surfaces. At the same time, adhesive in the bondline acts as a barrier to ready diffusion of oxygen into the area to offer oxide repair. On the other hand, inorganic grit particles like silica do not have any negative electrochemical effect and, therefore, are much more acceptable for blasting aluminum surfaces for bonding.

Minford (532,774) has reported an unexpected occurrence using silica gritblasted 6061-T6 aluminum joints that might be mentioned. When a silica gritblasted surface was bonded with a two-part, room temperature-curing, polyamide-catalyzed epoxy adhesive and the joints exposed up to 8 yr in the corrosive seacoast atmosphere, the joint durability proved to be significantly better than the same kind of joints with chromic/sulfuric acid (Alcoa A3)-deoxidized surfaces. While the silica gritblasted joints retained good strength after 8 yr of unprotected exposure, the corresponding Alcoa A3-deoxidized surface joints completely delaminated via a surface corrosion mechanism after only a 2-yr exposure. Minford explained this unexpected result as being due to the ability of a thin deposited film of silicate functioning as a surface corrosion inhibitor against chloride ion attack. Moloney (960) has shown that degreasing and gritblasting of aluminum adherends may suffice for many lower-severity aluminum-joining applications in manufacturing.

Finally, Rogers (961) has published data on surface preparation of aluminum using the mechanical abrasion qualities of both glass beads and quartz grits. He compared the bonding properties of bare 2024-T3 surfaces so treated for bonding to both epoxy or rubber-phenolic adhesives.

D. Special Abrasive Procedures

Considerable abrading of aluminum has been accomplished with 3M's special Scotch-Brite products in recent years. The technology using this special type of abrasive material has advanced to the state where Pocius and Claus (962) have compared these treated surfaces directly with aluminum that had been FPL deoxidized. They have presented data demonstrating that this three-dimensional (Scotch-Brited) surface conditioning of aluminum adherends using a special Scotch-Brite abrasive could effectively replace the FPL etch as a pretreatment step in aircraft bonding before finishing the surface with phosphoric acid anodizing (PAA) for final bonding.

IV. SOIL REMOVAL PLUS CHEMICAL DEOXIDIZING PROCEDURES

Because a form of corrosion can begin to develop on an aluminum surface after the removal of contaminates like oils, greases, smut, and metal fines, the relative oxidative resistance and general chemical stability of the aluminum oxide surface formed after a pretreatment is very important for assuring best bondability and long-service potential. Since it is still the native (naturally occurring) oxide that remains after the removal of various forms of organic surface contamination such as, e.g., the lubes and protective oils, the overall build-up of additional oxide thickness will be relatively slow compared to the much more rapid regrowth of oxide on a thoroughly deoxidized aluminum surface. The uniformity of this new oxide, however, will usually be much more effectively wet by the adhesive, permitting higher joint strength potential and better overall joint durability performance under exterior weathering conditions.

A. General Procedures

The ASTM Subcommittee D-14 on adhesive bonding has described six methods of pretreating aluminum alloys for bonding in ASTM Method D-2651-79 (963). All of these treatments fall into the general procedure of soil removal followed by some form of chemical deoxidation.

Method A is claimed as the most successful and accepted method of the group consisting of (1) degreasing, (2) FPL (hot sulfuric acid/sodium dichromate etching, (3) water rinsing, and (4) hot-air drying.

Method B is similar to method A but utilizes a milder acid/surfactant solution in the deoxidizing bath.

Method C substitutes in method A the possibility of using a commercial proprietary alkaline cleaner instead of, or in addition to, the degreasing step. Also, a higher concentration of sulfuric acid/dichromate solution can be employed in the deoxidizing step.

Method D follows the processing procedure described above in method A except a sulfuric acid-anodizing (SAA) step follows. The SAA built-up oxide layer is not sealed (Note: All the procedures used in ASTM standard practice D-2651 were suggested prior to the development of the PAA-anodizing pretreatment at Boeing.)

Method E suggest using an unspecified member of the class of proprietary commercially available metal surface preparations. The aluminum joint strength and durability performance obtained with phosphoric acid/butyl alcohol proprietary mixtures such as the Deoxidines from Amchem are representative of this general procedure. Their use and durability testing has been reported by Minford (531-534) and Eickner (964-966).

Method F furnishes a procedure for preparing a chromic-sulfuric acid paste to etch the aluminum where an immersion treating process may not be practical (such as treating surfaces in the field for initial or repair bonding in place).

While these recommended procedures (with the exception of method D) are those that could be best afforded in general manufacturing of bonded products, the author would suggest that either chromic or phosphoric anodizing (CAA or PAA) after method A would lead to the production of the most reliable and successful aluminum joints for long service under stressful and very aggressive weathering conditions.

B. Proprietary Commercial Deoxidizers

1. Deoxidines

The simplest type of chemical pretreatment that has shown a potential for developing acceptable bond strength along with relatively weather-resistant aluminum joint properties has been the use of various mixtures of phosphoric acid with butyl alcohol as represented by the Amchem Deoxidine proprietary deoxidizing mixtures. As mentioned above, both Minford (531-534) and Eickner (964-966) have provided joint durability data using 6061-T6 and Alclad 2024-T3 adherends, respectively. While Eickner focused largely on the resistance of such joints to various natural atmospheric weathering conditions at Forest Products Lab exposure sites, Minford added investigations using various accelerated laboratory weathering conditions for periods up to 2 yr. These conditions included continuous immersion in room-temperature (RT) water or 100% RH at 125 °F (52 °C) or cycling every

24 hr between soaking in 165 °F (74 °C) water, freezer-chest conditions, and drying out in 175 °F (80 °C) circulating air cabinets.

The treating procedure generally consists of vapor degreasing followed by immersion at RT in a 25% concentration (1:3 water dilution of Deoxidine) for 10–15 min. When RT-treating conditions are employed, it can represent a considerable economy as compared with heating in elaborate and expensively controlled temperature baths as required for FPL deoxidation, as reported in ASTM D-2651. At RT, as short a treatment as 5–10 min can often suffice, depending on the amount of soiling present and its adherence to the surface. If the solution temperature is raised (and consequently, the chemical etching rate advanced), the contact time to achieve deoxidation can be as short as 1–2 min.

This has been shown to be an effective treatment on 6061-T6 aluminum surfaces in spite of the considerable thickness of heat-treat oxide on these surfaces after vapor degreasing. The treated surface should be thoroughly rinsed in deionized water to remove any acidic residuals on the surface and then air dried.

Minford (967) has also experimented successfully with a surface-clinging slurry made by mixing Cab–O–Sil inorganic filler with concentrated Deoxidine solution to give a consistency that will not slide off the surface even on a vertical surface. This has been a preferred treating procedure where aluminum joints need to be made in the field over externally weathered surfaces or where a bonded joint that has failed needs to be repaired.

Minford (968) has also employed a Deoxidine-type pretreatment to prepare long aluminum extrusions for bonding to aluminum cast bases to produce aluminum highway light standards. Since the aluminum parts were so large and manufacturing costs had to be minimized, a process had to be developed for shop assembly operations where hot deoxidizing tanks were not available. The surfaces were first manually wiped with fresh solvent rags under the limitation of surface recontamination discussed earlier under solvent cleaning. A treating bath of appropriate dimensions to handle the deoxidizing of the large extrusions was economically fashioned by lining a wooden box with a thick vinyl liner. The 25% Deoxidine solution was employed in a 10- to 15-min immersion step followed by thorough overflow tap-water rinsing and hot air blowing to dry. Subsequent bonding was accomplished by pouring a commercial liquid two-part, room temperature–curing epoxy into the deep cavity between the extruded pole and the cast light standard base. A tape was used to seal off and retain the epoxy in the bondline for a full 7-day cure under ambient plant conditions. Aluminum flake powder was then mixed with some of the liquid epoxy to produce a spreadable paste sealant that could be used to fill the top gap. A bright metallic appearance was produced at the jointure intersection which resembled a brushed weldbead in appearance.

2. *Alum-Surf-Prep Deoxidation*

Another example of the many proprietary room temperature–etching solutions recommended for aluminum in the marketplace is Alum-Surf-Prep from Arcal Chemicals. This material is one of many aggressive acidic-type cleaners that have been promoted by their formulators for cleaning and brightening unevenly oxidized surfaces. These surface alterations can occur during handling, storage, or atmospheric oxidizing in aluminum manufactured products in service. Minford and Peters (969) attempted to demonstrate the possible usefulness of this relatively low-cost deoxidizer-cleaner chemical solution for preparing 6061-T6 alloy surfaces (with their heavy heat-treat oxide in place) for bonding. After an initial vapor degreasing to remove any heavy organic contaminant layers, the aluminum surfaces were immersed for only 1 min at room temperature as recommended by the chemical supplier. After thorough water rinsing and drying, these treated surfaces were fabricated into lap-shear joints with room temperature–curing epoxy adhesive. The joints offered sufficient long-term resistance to hot condensing humidity or natural atmospheric weathering conditions to be considered for use in many manufactured products. This conclusion was reached after making comparisons with 6061-T6 joints pretreated with the Alcoa A3 version of FPL deoxidizing. It was shown that comparable initial joint strengths could be achieved with either a 0.5-min immersion in a 25% dilution of bottle-strength Alum-Surf-Prep or a 1-min immersion in a 10% concentration solution.

C. Alkaline Etchant Pretreating

Smith (970) has reported his efforts at Rockwell to produce a simple, inexpensive, nonacid, nonchromate surface pretreatment for aluminum that would be capable of producing strong, durable joints in water-soak test environments. A number of investigations using an alkaline solution approach involved developing what Smith has chosen to call STAB, STAB 1, STAB 2, and STAB 3 procedures, as described in his publications of 1977, 1978, and 1981 (971–973). Joints produced by STAB 1 and STAB 2 procedures consisted of a hot-water soak or commercial detergent soak, respectively, but both seemed to lack consistent reproducibility. STAB 3 involved a soak in concentrated sodium hydroxide, which seemed to considerably enhance reproducibility. Further, Smith has investigated the surface properties affected by these procedures and sought to identify the mechanisms whereby the joints showed endurance failure.

D. Chromic Acid/Sulfuric Acid and Forest Products Laboratory Deoxidizers

The most reliable earlier aluminum bonding resulted from variations of a general pretreating scheme which included combinations of the steps as listed in Table 25. The most commonly employed pretreatment for aluminum structural adhesive bonding in the United States for many years before the development of PAA anodizing consisted of exposure to some time, temperature, and concentration of sulfuric acid/sodium dichromate. This was based on the development of this pretreatment for aluminum at the Forest Products Laboratory in the late 1940s. The credibility of this pretreatment was established by many investigations conducted by Eickner (964–966) and Eickner and Schowalter (974,975). Soon after, this treatment assumed the name FPL-type treatment for aluminum bonding. It continued to be referred to as such until some additional changes in procedure caused it to be replaced by the designation modified or optimized FPL.

Aluminum manufacturers like Alcoa had been looking at a wide range of possible deoxidizers of aluminum as a means of preparation prior to additional processes such as conversion coating, anodizing, and painting. The Alcoa process capable of producing an equivalent type of surface for bonding was designated Alcoa A3 and used a chromic acid/sulfuric acid mixture at some designated set of time and temperature conditions. The optimum conditions for the Alcoa A3 process seemed to require a higher bath temperature but for a shorter soaking time. In 1972, Minford (976) published on the first long-term durability testing using this Alcoa A3 procedure on 6061-T6 aluminum adherends and confirmed equally acceptable results to those secured with the FPL-designated procedure (977–979).

It should be remembered from Chapter 3 that the preferred oxide form on aluminum for maximum initial joint strength and long-term joint durability had been identified by many investigators as

Table 25 General Sequence of Steps for Chemical Pretreating of Aluminum Adherends

Step No.	Treatment description	Step No.	Treatment Description
1	Vapor degrease	6b	Alternative to (6) would be immersion in an anodizing electrolyte
2	Alkaline cleaning		a. Water rinse and dry if no sealing of coating
3	Water rinsing		b. If sealing desired, then it could be accomplished by boiling in any one of several bath compositions
4	Chromic acid etching		c. Final rinse and dry
5	Water rinse		
6a	Immersion in chemical conversion bath a. Water rinsing b. Air dry		

being the boehmite form of oxide. This is the oxide generally believed to result from either the FPL- or Alcoa A3-deoxidizing conditions as compared to the bayerite form of oxide which regrows naturally on aluminum surfaces in a moist atmosphere at room temperature. It has further been demonstrated repeatedly that this bayerite oxide is much less stable when water infiltrates into the bondline, being converted ultimately to the hydroxide, which is very weakly bonded to the underlying aluminum.

The FPL etch produces a thin coating or layer of boehmite oxide which may range from 100–400 Å in thickness according to Bowen and co-workers (70,71,760–764). Its morphology has been described by Venables (33) and its chemistry by Davis et al. (37). While the resistance of FPL- or Alcoa A3–pretreated aluminum joints to water-soaking exposures has been confirmed by many investigators, i.e., high resistance of the oxide to hydration, some significant sensitivity of these same joints to direct exposure to marine or seacoast conditions has also been demonstrated. At first, the full effect of this sensitivity to marine conditions in aircraft was somewhat masked by intensive application of sealants with attendant burying of the joint edges. When the design of new aircraft with more opportunity for exposure to long-term corrosive seacoast conditions occurred at the time of the Vietnam conflict, this sensitivity was more recognizable and the need for a surface pretreatment with more protective potential such as PAA anodizing was required.

McMillan (833), at Boeing, has shown that even with a corrosion-inhibiting primer present over the FPL-etched aluminum surface, the overall joint performance would be considered unacceptable in seacoast or marine service. As mentioned above, it was rather unfortunate that this fact was not truly appreciated until some near disasterous failures in the bonded segments of Vietnam vintage aircraft occurred. The evidence that this was the case, however, was probably a significant stimulus to accelerating the search for more durable performing surface pretreatments such as PAA.

Bijlmer (770) and Bijlmer and Schliekelmann (779) were among the first investigators to study the type of surface produced by the FPL procedure and attempted to explain the good adhesion and joint durability potential that existed in water-exposure conditions. They studied the surfaces produced by a wide variety of combinations of the sulfuric acid/dichromate mixture using electron micrographs which clearly showed that only certain specific combinations were even capable of producing the desired type of surface roughness, which they termed etch pitting. Most recently, in 1989, Pietsch et al. (980) have described their continuous surface roughness characterization using angle-resolved electron and photon scattering.

Miller (981), at the Alcoa Labs, as early as 1954 had begun to investigate the superior joint strength afforded by the use of the related Alcoa A3 etchant using Alclad 2024-T3 adherends. Miller also compared the strength results of Alclad 2024-T3 joints which had been pretreated by (1) solvent cleaning only, (2) alkaline cleaning with metasilicate cleaner plus a hot chromic acid dip to remove the smut, (3) phosphoric acid/alcohol etching (current Deoxidine product), or (4) a sealed sulfuric acid anodize (SAA) pretreatment. His continuing work eliminated the ALROK (Alcoa trademark) conversion coating from further consideration as a bonding pretreatment even though the prior deoxidation for the conversion coating had been the Alcoa A3. (The reader is reminded that considerable detail relating to the matter of using a conversion coating under adhesives has been covered in Chapter 3 under that subject heading. More discussion on the durability of such pretreated joints will be offered in Chapter 10 on that subject.)

Attempts were made by McGowen et al. (982–984) in the early 1960s to establish whether the FPL-etched surface was effective for bonding because of a greater cleaning efficiency or a specific chemical action on the surface. These investigators could find no relationship between the degree of surface cleanliness produced and the bond strength obtained using a variety of different treatments, including the FPL etch. By using a radioactive isotope detection procedure, they did show a direct relationship between the level of peel strength achieved and the amount of chromium adsorbed on the aluminum surface. Many investigators have reported finding insignificant amounts of chromium residual on the FPL-deoxidized surface after etching and rinsing. It really remained for Bijlmer (770) 13 yr later to offer the best evidence of what was actually taking place with his concept of significant additional mechanical interlocking due to special micro-etch pitting on the surface. However, we must remember that only the initial joint strength (not overall joint durability) was being used as the sole criterion for what would constitute a significant change. Other investigators such as Sun et al. (985), Chen et al. (986,987), and Venables et al. (31) have looked at the effect of such factors as chemical and thermal treatments of 2024-T3 aluminum, fluoride contamination on the microstructures and bondability, alloy components on durability, and the characterization of treated aluminum surfaces that had been prepared for bonding. It is unfortunate that so many investigations of the effect of various factors on bonding of aluminum in the 1960s and 1970s were based only on comparative initial joint strengths rather than on joint durability performances.

Weltman (988) studied the effect of some 60 variations and chemical substitutions in the FPL composition using standard treating procedures of (1) MEK solvent wiping, (2) vapor degreasing, (3) FPL-etchant variations, (4) tap- and distilled-water rinsing, and (5) air drying. He did not obtain any enhanced peel strength to that already observed using the already-established standard FPL-etching procedure. There was some evidence that some chromium ions on the surface were necessary for producing the highest peel-strength joints. Sodium hydroxide etching prior to FPL etching did not show any bond strength increases. Weltman claimed comparable peel-strength results with an etching pretreatment in a mixed solution of sodium hydroxide, potassium chromate, and Versene (a complexing agent), but the author is not aware of any other investigations which carried this observation further. Once again, it is pointed out that no durability evaluations were performed. The author will point out in Chapter 10 on joint durability that aluminum joints of apparently equal initial strengths can vary widely in durability test results.

For a literature report of an opposite view to the generally favorable comments published for aluminum etching pretreatments, the investigations of Severus-Laubenfeld (989) can be mentioned. He used his measurements of the critical surface tension of aluminum surfaces after treating as an indication of good bondability potential. He found that improved aluminum surface tension values were present after sandblasting, degreasing, or green chromating of aluminum surfaces. Lesser improvement resulted from his alkali degreasing or acidic pickling pretreatments. Many hundreds of aluminum joint evaluations comparing degreasing and silica gritblasting of 6061-T6 aluminum adherends by Minford have always confirmed both higher initial strength values and better overall joint durability performances when using alkaline- or acidic-deoxidized surface preparations in hot-water or condensing hot-humidity exposures.

Evans (990) also reported a conflicting viewpoint in 1970 to those reporting the presence of boehmite oxide after FPL pretreating. He contended that the bayerite form of aluminum oxide was the replacement oxide for an amorphous form actually produced by FPL etching. Further, he claimed that boehmite on aluminum adherends would actually offer a poor surface for adhering to; however, no joint durability test data was furnished to support this proposal. Meanwhile, many investigators (including the author) have continued to support the contention that excellent and durable adhesion can be expected when boehmite has been established as the regenerated oxide for bonding.

Wegman (991) has compared the bond strengths, reproducibility, and ease of application for six surface preparations on aluminum, including (1) sandblasting, (2) FPL etching, (3) sodium hydroxide etching followed by a nitric/hydrofluoric acid desmut, (4) nitric and hydrofluoric acid mix etching, (5) cold-dip etching in hydrochloric acid/potassium dichromate solution, and (6) a brush-applied hydrochloric acid/potassium dichromate etch treatment. Without the benefit of any comparative weathering durability testing, Wegman picked out treatments (1) and (6) as yielding the best overall strength, reproducibility of surface, and ease of application. This early literature disclosure of Wegman was the basis for the author undertaking a very extensive evaluation of pretreating aluminum joints with hydrochloric acid in the middle 1970s. More details will follow later in this chapter. Relative to the claim for picking sandblasting as a more acceptable surface pretreatment as compared to FPL deoxidizing, Minford (156,388,389) has conducted long-term durability test comparisons between gritblasted and FPL-pretreated aluminum joints and found the grit-abraded joints to be distinctly inferior in durability in all water-soaking weathering situations. The exception of this conclusion, as mentioned earlier in Chapter 3, was a superior response of the silica gritblasted joints in the seacoast exposure. This was attributed to the type of grit used (silica) rather than the nature of the mechanical abrasion itself. The author believes that this would not necessarily be the end result with other grit materials which might not function as corrosion-inhibitor films.

Ross et al. (992–994) have focused attention on the effect of aging time of Alclad 2024-T3 surfaces pretreated by either vapor degreasing, sandblasting, or FPL deoxidizing. Regardless of the surface exposure time after treating (SET), the FPL-etched surfaces gave significantly better bonding results. However, after 30 days SET, even the FPL-etched surface joints exhibited a 37% lower peel strength after hot-humidity exposures as compared to joints with only 1–4 hr SET. The same investigators also looked at the SET effect on FPL-pretreated 5052-H34 aluminum alloy joints. They observed the improvement produced by the FPL etching on 5052-H34 surface joints was definitely less than for similary prepared 6061-T6 joints. The author might explain this as what might be anticipated when comparing the effect of deoxidation of the heavy heat-treat oxide film on 6061-T6 surfaces as compared with a lesser alloyed and non–heat-treat oxide film on an alloy like 5052-H34. Wegman and Dickerson (995) concluded from their tests with hot chromic acid–deoxidized aluminum and a two-part, polyamide-cured epoxy adhesive that bonding should take place within 72 hr after pretreatment. If a chromic acid paste-type deoxidizer procedure was used, then bonding should take place within only 4 hr according to these investigators. Mention has been made earlier of the author's investigation of bonding and durability evaluating Alcoa A3–deoxidized 6061-T6 adherends for periods up to 30 days after treatment with no curtailment in joint durability after 2 yr in accelerated lab weathering or 8 yr in natural atmospheric conditions. The author believes these varying test conclusions from those of Ross et al. is reflecting some of the greater criticality of peel-test evaluations for delineating differences in surface pretreatment as compared to the lap tensile-shear testing used by the author for discrimination.

Brockmann (373) has used etching pretreatments on aluminum in association with chemicals which can produce coordinate covalent complexes with the treated surface. A further enhanced resistance of these joints to water-soak weathering was claimed.

It should be obvious that most commercial manufacturing operations using the FPL or other alkaline- or acidic-deoxidizing procedures use some sort of immersion-type processing. Some readers may have an interest in a publication by Herrigel (996), because it proposes an automated spray-deoxidizing procedure for faster manufacturing processing. It may be of interest also to compare the English standard for chromic acid pickling (997) with the earlier-mentioned ASTM (FPL-type) procedure (963). Michaelson (998) has written the FPL-type specification for Boeing pretreatment of aluminum alloys (BAC-5765 Rev. G), as of 1968.

E. FPL Procedure Modification

Recently published data showing the differences between the FPL procedure and similar procedures are shown in Table 26 (999). Further improvement in the FPL-type etching pretreatment was sought

Table 26 Some of the Most Commonly Used Chromic-Sulfuric Acid Etching Procedures

Process	Etch bath comp. (wt%)			Temp.		Time (min)
	$Na_2Cr_2O_7.2H_2O$	H_2SO_4	H_2O	°C	°F	
FPL etch	2.5	24.3	73.2	68	155	15–30
UK 03-2/1 (1970)	6.4	23.3	70.3	60–65	140–150	30
DIN 53 281	7.5	27.5	65.0	60	140	30
Alcoa A3	(35 g chromic acid + 100 ml sulfuric acid diluted to 1 L with water)			82	180	10
FPL optimized	5.0	26.7	68.3	65	150	10
FPL-RT	6.4	23.4	70.2	22	72	240
P2	15.0 (FeSo)	37.0	48.0	60–70	140–160	8–15

Source: Data from Ref. 999 (Table 2, p. 260).

by the aerospace industry when certain deficiencies in overall joint durabilities were uncovered. This improvement involved the rather simple addition of some aluminum and copper ions to each new bath of chemicals before its use for pretreating, as shown in Table 27. It had been recognized in the aerospace labs that the joint durability results were considerably improved when using an old (used) treating bath solution as compared to a newly made up composition bath. Bethune (1000), at Boeing, found that this increased durability could be achieved in a new bath composition by predissolving some ordinary 2024-T3 aluminum alloy. Alnar (1001), at 3M, demonstrated further that the joint longevity was distinctly improved not only for regular hot-humidity conditions but also when such joints were under sustained loading situations.

Of the very large number of investigators that have studied the FPL etch on aluminum adherends (31,32,531–536,760–764,770,771–775,977,1002,1003), only Smith (1004), Russell and Garnish (1005), Almer (1001), and Pocius (1006) evaluated the effect of this predissolved 2024 alloy. Smith and Almer agree that dissolved copper is responsible for the enhanced durability owing to the generation of cupric ion in the etchant. Smith theorized that the treated aluminum surface is being "self-anodized"; however, Pocius (1006) investigated more details about the effect with the benefit of better analytical tools available to him in the 1980s as compared to Smith earlier. No major effect of bath composition was found on the electrode kinetics by Pocius, but he did establish that both rate and morphology of oxide regrowth was definitely altered by the predissolved copper making the "optimized surface" more wettable by the adhesive. However, there needs to be chemical control of the bath composition to prevent higher levels of dissolved aluminum and copper in the etchant bath, since Bijlmer (1007,1008) demonstrated unsatisfactory bonding results with what he referred to as excess aluminum and copper present in the early 1970s. Chen et al. (227) demonstrated that chloride ion can have a negative effect on bondability in etching baths. They also, along with McMillan (833), have demonstrated a negative effect of fluoride ion in the treating bath.

There have also been other variations of the FPL aluminum surface pretreatment, as described by Eichner and Schowalter (29) in 1950. Another variation of the chromate-containing solutions has employed $NaHSO_3$ as a replacement for the sulfuric acid. This procedure only appears to have applicability to relatively low-stress applications.

A more recent development has been to avoid chromates altogether, because of their toxic nature and the cost of effectively meeting waste removal standards. A complex oxide surface morphology has still been produced using a so-called P2 etchant in which ferrous sulfate has been used as an oxidizer in place of the sodium dichromate. Details have been provided by Rogers (1009). Desai et al. (1010) demonstrated that the P2 process can produce an oxide morphology very similar to that found on chromic-sulfuric etch surfaces over a broad range of time-temperature solution concentration conditions. While mechanical testing confirms that P2-prepared aluminum joints are equivalent to

Table 27 Optimized Forest Products Laboratory Etching Procedure

Steps	Materials
1. Solvent degrease	1,1,1-Trichloroethane or other vapor degreasing solvent
2. Alkaline deoxidize	Turco 4215S
3. Rinse, 5 min. (R.T.)	Water
4. Etch, 150 F (65 C)	FPL bath solution $Na_2Cr_2O_7.2H_2O$ ([60 g/L 8 oz/gal]) 96% H_2SO_4 (173 g/L [23 oz/gal]) 2024-T3 aluminum (1.9 g/L [0.25 oz/gal])
5. Rinse	Deionized water
6. Oven dry	

Source: Data from Ref. 999 (Table 1, pg. 260).

FPL-prepared joints, the author is not aware of extensive long-term durability investigations which confirm this.

F. Effects of Rinsing After Deoxidizing

The final procedural step of rinsing after FPL-type deoxidizing has also been studied rather extensively and because of some significant effects noted must be controlled, as described by Reinhart (1011). Sun et al. (850) have proved the regrowth oxide layer actually begins to form during the actual etching process; however, the thickening of this oxide largely occurs during the final rinsing step. The author agrees with Bowen (761–764) and Reinhart (1011) that the type of oxide is beta aluminum oxide hydroxide (boehmite), although arguments have developed about this in the literature, as discussed by Cotter (1012).

Wegman et al. (561,1013,1014), using room temperature–curing epoxy adhesive, showed rinsing the FPL-etched adherends in hot deionized water led to very low-strength joints and adhesive-type joint failures. Tap-water rinsing or adding multivalent ions to the deionized water resulted in much higher-strength joints with cohesive-type failures when tested for initial strength values. Unfortunately, no durability test results in accelerated weathering conditions were provided in these reports. The author was prompted to conduct some related evaluations (with 6061-T6 alloy instead of Alclad 2024-T3) where Alcoa A3 pretreating was performed (Note: There is no reason to believe that the Alcoa A3 [chromic acid/sulfuric acid composition] produces any different bonding potential than FPL.) Both deionized and tap waters were employed after the deoxidation, as had been used by Wegman; however, the water temperatures were confined to the boiling point where boehmite oxide should be the stable form. Under these circumstances, it made little difference whether the treating bath was deionized or tap water, since both produced very strong joints with very high durability potential (39).

G. Joint Durability and FPL Etching Variables

The Boeing-developed wedge-test specimen (see ASTM D-3762) has been extensively employed to distinguish the relative acceptability of different types of aluminum surface pretreatments. It has especially been used to follow the progress of developments relative to FPL deoxidizing or CAA- or PAA-anodizing refinements (1015–1017) Scardino and Marceau (853) have employed this specimen to distinguish the effect of different processing variables in FPL etching as a pretreatment for durable aluminum bonding. Their data showed the negative effects on durability of using (1) a new FPL-etching solution, (2) etching at 5.5°C below recommended bath temperature, (3) delaying rinsing for 3 min, or (4) having fluoride bath contamination in the etching bath as shown in Table 28. Sun et al. (850) have demonstrated poorer durability performances in some aluminum alloys due to

Table 28 Effect of Variables in Chromic Acid Etching on Durability as Measured by the Boeing Wedge Test

	Crack length (mm)				
Variable description	15 min	30 min	1 hr	2 hr	3 hr
Correct processing	25	25	25	25	25
New FPL etchant solution	38	50	62	65	65
Etch solution 5.5°C below minimum temperature	55	72	83	85	85
3-min rinse lag	57	85	100	110	110
Etch solution contaminated with 200 ppm fluoride ion	65	125 (25 min)			

Source: Data from Ref. 853.

diffusion of copper to the surface during FPL etching. Chen et al. (227) have confirmed the negative effect on joint durability of having fluoride contaminants in the treating bath.

Baun et al. (1018), McDevitt and Baun (1019), McDevitt et al. (1020), and Solomon and Baun (1021) studied the surface chemical compositions of 2024, 6061, 7050, and 7075 adherends in the "as received" condition and after seven different surface pretreatments, including FPL deoxidation. They found significant changes in surface copper concentration with varying etching times and rinse times. High-strength aluminum alloys, as used in aerospace bonding, often have significant amounts of alloying copper and, as a result, we can observe an electrochemical dissolution of the aluminum adherend surface. To what degree this inherent characteristic can be overcome to create high joint durability must be considered by the structural designer. The above investigators traced the presence of most surface contaminants to manufacturing processes, prebond treatments, and the overall environmental conditions.

H. Miscellaneous Etchants

A wide variety of other deoxidizing etchant procedures have been investigated over many years on aluminum alloy 6061-T6 surfaces by Minford (1022–1025) for preparation of structural aluminum joints. These joints were all tested routinely for at least 2 yr in various accelerated laboratory weathering conditions as well as up to 8 yr in the natural industrial atmosphere and seacoast atmospheric conditions. The relative performance of the joints produced by the various pretreatments could be directly compared because the same proprietary one- and two-part epoxy adhesives were employed to fabricate all the joints, thus ruling out the effect of adhesive as a variable. Comparisons were made by comparing the strength decay curves produced by exposing the joints to the various weathering conditions outlined above. More complete test results will be presented in a later section of this chapter when discussing the individual etching compositions employed and in Chapter 10 on joint durability. All adherends pretreated in the 1960s in aerospace had been vapor degreased as a first step, so Minford employed this step before each chemical etching, so that it was a constant factor.

A partial listing of the treatments used by Minford included (1) 5-min exposure at 180°F (82°C) to a nitric acid/sulfuric acid/water solution mixture with a ratio of 5/10/85 parts by weight, (2) 5-min exposure at 170°F (77°C) to phosphoric acid/sulfuric acid/water solution with a ratio of 2/10/88 parts by weight, (3) 1-min exposure at room temperature to 25% water dilution of Alum-Surf-Prep proprietary aluminum cleaner, (4) 5-min immersion at room temperature in a 25% solution of Deoxidine 526 diluted with deionized water, (5) 3-min immersion in Ridoline 53 alkaline cleaner solution diluted to the supplier's recommendation at 180°F (82°C), (6) 20 min immersion at room temperature in Alcoa A3 (FPL-type etchant) consisting of a mix of 35 g of chromic acid with 100 ml of concentrated sulfuric acid with eventual dilution to 1 L by slowly and carefully adding distilled water, (7) 3-min immersion in Ridoline 53 (procedure [5]) above followed by water rinse and subsequent immersion in Alcoa A3 at 180°F (82°C) for 5 min, (8) 2-min soaking in Alcoa A1 caustic deoxidizer at 150°F (66°C) followed by a dip in 40% nitric acid to desmut and finally a water rinse and dry, (9) above procedure, (8) followed by deoxidizing in Alcoa A3 solution for 5 min at 180°F (82°C), and (10) Wyandotte Pondax nonchromate aluminum deoxidizer used as prescribed by the supplier.

The number (6) treatment above proved to be rather unique in view of the extraordinary degree of joint durability that was found. For example, Minford (1026) used both one- and two-part epoxy adhesives to compare this so-called "cold" Alcoa A3 treatment with the hot (180°F [82°C]) FPL-type pretreating. The important difference in joint durability noticed was in exposure to the corrosive seacoast conditions. With the two-part epoxy adhesive, the vapor-degreased–only joints can fail within 71 days, whereas the hot Alcoa A3-treated joints had all failed before 2 yr of exposure. In contrast, those 6061-T6 aluminum joints that had been treated at ambient temperature with the same Alcoa A3 etchant solution now survived for 8 yr in the same seacoast exposure with good strength retention. It was more unexpected that this kind of performance could be achieved in spite of the fact that the usual bright etched appearance of a hot Alcoa A3-treated surface had been replaced by a relatively dark and variegated appearance that was not much different from the original mill-finish appearance. Unfortunately, this performance was not anticipated and after the 8-yr testing results became available, other matters prevented an analytical study of the surface to determine why such extraordinary performance had been achieved. Also, no publication of the information was made and it remained for a later English publication to confirm that a room-temperature treatment with the FPL etchant had also offered an unusual joint durability response. Obviously, it was only after the Alcoa joints had survived 8 yr at the seacoast with only an 11% lowering of initial joint strength that such a treatment might even be suggested. Whether varying results might be found with different degrees of heat-treat film on the aluminum alloy surface has not been evaluated to the author's knowledge.

I. Hydrochloric Acid Deoxidizing

The author has chosen to describe the use of hydrochloric acid deoxidizing as a separate section from the miscellaneous listing above because of the considerable experimental work performed at the Alcoa Labs with such a procedure and the relatively scarce information found in the technical literature. Wegman (991) made an early 1961 disclosure about the use of a hydrochloric acid etching on aluminum comparing the initial strengths at −65°F (−54°C), 73°F (23°C), and 160°F (71°C) using both a two-part, room temperature–curing epoxy and an epoxidized-novolac adhesive. Using a MIL-STD-304 type of special conditioning, he further compared the relative durability of this dichromate/hydrochloric acid–brushing pretreatment with an acid paste from Hughson that also might be considered for field bonding of aluminum structures. Starting from a 43% higher initial strength, the HCl-treated joints retained 53% of that value after exposure as compared to only 23% retention for the commercial acidic-deoxidizing paste-treated joints. Since no additional reports or joint durability data appeared with other alloys or variations of the treatment in subsequent years, the author (1026–1028) attempted to further investigate. The immediate attempts were to try and optimize the treating conditions and establish the maximum level of joint durability that could be obtained using a room temperature–deoxodizing process as compared to the various heated acidic-deoxidizing conditions (FPL and Alcoa A3) previously tested.

The surfaces produced in even a very short exposure (less than 1 min at room temperature) using a diluted HCl solution appeared to significantly roughen a 6061-T6 "as received" surface under low magnification viewing. This clearly demonstrated the extremely aggressive chemical action of this treatment as compared to the almost passive nature of the hot Alcoa A3 or FPL processes. It must be remembered that a very important contribution of the chromate in these acidic pretreatments is to control the overall aggressiveness of the hot acid on the alloy surface. It appeared to the author that Wegman's original choice to use an HCl/potassium dichromate mixture was based on a substitution of the HCl for the sulfuric acid in the FPL-type bath formulation. Minford (1029) had earlier conducted an extensive evaluation of the rate of chemical attack by a dilute 1% hydrochloric acid on virtually all aluminum alloys currently produced showing a direct relationship existed between the rate of etching and the relative purity of the aluminum alloy; i.e., highest-purity 99.99% aluminum was the most slowly attacked surface, whereas the rate continued to significantly increase as increasing concentrations of other alloying elements replaced the aluminum in the high-purity product.

The above relationship between aluminum alloy purity and rate of deoxidation in dilute HCl solutions, as anticipated, was a much more influencing factor than the addition of the dichromate in the mixture originally proposed by Wegman. Since there is also a recognized important relationship between the opportunity of an adhesive to develop additional mechanical interlocking with the adherend surface and joint strength and durability, the selection of alloy for HCl deoxidation could be an additional significant factor in obtaining more or less change in the morphology of the surface. The most important observation arising from the earliest Minford investigations using 6061-T6 aluminum alloy adherends with the HCl and dichromate mixture was an immediate recognition of the high strength and high resistance to weathering of such bondments, particularly in the corrosive conditions of the seacoast. For example, unlike the FPL-pretreated joints which fail in less than 2 yr in the natural seacoast exposure, the HCl-treated joints retained high strength after 8 yr exposures. This has proven to be a performance that can only be matched by FPL-treated surface joints that have been further pretreated by some form of anodizing. The possible exception would be the recently mentioned "cold" FPL-type procedure which also shows outstanding joint resistance to this very corrosive type of weathering. The treating time of 1 min for the original HCl-etching investigations was quite short compared to the 20 min or longer employed in the "cold" FPL etching procedure. It is also quite evident that the overall surface roughness produced by the 1-min etch time is significantly greater than produced by 20 min in the relatively passivated FPL etching solution.

Having rejected the need for the dichromate in the original Wegman investigation, the author next sought to find another additive that might even more significantly control and reduce the aggressiveness of the HCl solution. This was found by substituting the butyl alcohol which has been mentioned as a key ingredient in the Deoxidine-type phosphoric acid aluminum deoxidizers. This substitution of butyl alcohol was finally finalized in an additional investigation (1030) as an 50/50 volume mix of concentrated HCl and butyl alcohol. Fine tuning was performed by studying the relative effects of variable etching times, open time after etching before bonding, and a possible post-treatment after etching at 400°F (204°C) in a recirculating air oven. The joint stress-endurance curves (discussed in more detail under Testing in Chapter 9 and Joint Durability in Chapter 10) obtained by soaking in 100% RH at 125°F (52°C) were used to show that overall performance increased when etching times increased over the range from 30 to 120 sec. While most testing was conducted with joints bonded within 4 hr of etching, the greatest increase in joint stress-endurance occurred after an induction time of 72 hr before bonding using a 30-sec etching time. Postheating of the 30-sec etching time surface within 4 hr of treatment did not significantly alter the durability response.

It was immediately observed when etching an aluminum alloy surface in an HCl solution that a dark (sometimes almost black, depending on the alloy choice) layer of residual smut remained on the surface even when immediately rinsed in running water. Furthermore, this layer had little adhesion to the surface, since it could be removed by simply wiping with a finger over the surface. Believing that this apparently poorly adhering layer could function as a weak boundary layer condition in a bonded joint, the author initially choose to remove this smut layer by simple immersion at room temperature in a 40% nitric acid rinse tank followed by thorough water rinsing and air blow drying. In an attempt to determine the extent of liability of not removing the smut layer, it was amazingly found that there was no difference in joint strength between 6061-T6 joints that had been desmutted or not desmutted after curing for 7 days at room temperature with a two-part epoxy adhesive in the bondline. Of even greater significance was the fact that there was no penalty in joint durability performance in any accelerated weathering testing or resistance to corrosive pitting in seacoast exposures. It is still not clear what happens to permit the initially poorly adhering smut layer to develop sufficient adherence after bonding to permit 100% cohesive-type failures of strong structural epoxy adhesive in the bondline and, further, to develop joints as strong as those pretreated with the hot FPL-deoxidizing procedure with even greater durability in the most corrosive weathering conditions.

In the latest refinement worked out by the author, a thick viscous paste slurry is made by mixing bottle-strength HCl with Cab-O-Sil inorganic filler. This slurry can be readily brushed over the solvent- or vapor-degreased surface and allowed to totally react; i.e., the end of any gaseous evolution of hydrogen. The reaction residues are then force flushed from the surface with water and the surface is dried immediately in a stream of compressed, filtered air. The dark surface arising after the etching will significantly lighten in color when the water is driven out. This signals the surface is ready for bonding. It should be pointed out, however, that the appearance of the etched area is always rather nonuniform to the eye and, therefore, cannot be used as any criterion for acceptable treatment, as is the case with other acidic etching procedures which leave a fairly bright uniform-appearing treated surface.

Because the degree and speed of etching and irritant nature of the evolving fumes can be so variable on different alloys, the person setting up a treating line for manufacturing purposes needs to consider the conditions based on alloy choice, size of the area to be treated, and plant equipment and personnel available. Higher-purity alloys may even require more than a single etching cycle, because of a significantly diminished reactive rate. The author has directly evaluated some of the nonaerospace alloys like 5052 in this regard and would suggest the diminished reaction rate even here might necessitate double-treatment conditions if the slurry procedure is employed.

Before retirement, the author had the opportunity to see this type of procedure successfully utilized in the preparation of very large satellite antenna structures which were to furnish long-term service in many remote locations with severe weathering present.

J. Aluminum Conversion Coating Pretreatments

In the most practical sense, the largest volume of pretreating aluminum for obtaining long-term durable adhesion of organic polymers has been the generation of various forms of aluminum conversion coatings to pretreat for subsequent painting, including both ambient- and heat-cured paint coatings. One of the main goals for developing acceptable aluminum siding over many years was to create painted aluminum sheet products that could resist exterior atmospheric weathering for 20 or more years. This conversion-coating technology has also been employed in the aerospace industry for many years as the means of enhancing both bonding and priming of aluminum honeycomb structures. The specific chromated conversion coatings developed have also permitted high resistance to the corrosive service conditions that had caused many earlier joint failures.

An extensive literature exists on the general subject of conversion coating aluminum (see Refs. 1031–1036). The specific chemical details developed by individual chemical companies or aluminum honeycomb manufacturers are not usually available. It is the purpose of such treatments, in general, to convert the discontinuous native oxide film to an inert film imparting simultaneous excellent adhesion development properties and resistance to undercutting corrosion in chloride ion exposures or high-humidity conditions.

The main coating types are (1) amorphous chromate, (2) amorphous phosphate, (3) wash primers, (4) acid clean-coats, (5) ALROK (Alcoa trademark), and (6) crystalline phosphate. The compositions, treating conditions, and appearance of these coatings can be reviewed in Table 29 (856).

For the Alcoa Labs' investigations dealing with adhesive bonding, it seemed appropriate to concentrate on the amorphous chromate, since it is commercially the most popular. Also, there has been

Table 29 Chemical Conversion Coating Processes on Aluminum

Type of coating	Solution composition	Treating conditions	Color
Alkaline oxide (0.04–0.1 mil)	3% Na_2CO_3 1% Na_2CrO_4	Boiling 5 min	Gray
	0.5% $NaAlO_2$ 0.25% Na salicylate	200°F 5 min	Colorless
Crystalline phosphate (100–500 mg/ft)	0.7% Zn^{++} 1.0% PO_4^{---} 2.0% NO_3^- 1.0% BF_4^-	130–135°F 2–5 min	Colorless
Amorphous phosphate (10–400 mg/ft)	7% H_3PO_4 0.2% KHF_2 0.4% CrO_3	100–130°F 0.5–5.0 min	Green
Amorphous chromate (10–80 mg/ft)	0.75% $Na_2Cr_2O_7$ 0.5% $K_3Fe(CN)_6$ 0.1% NaF 0.3% HNO_3 pH = 1.5	70–90°F 0.5–5.0 min	Yellow to tan
Boehmite (2000–20,000 Å)	Distilled or deionized water	Boiling 15 min–5 hr	Colorless

Source: Data from Ref. 856 (Table 5, p. 611).

considerable good manufacturing experience with this type of coating and evaluation of its inherent excellent resistance to corrosive weathering conditions. As part of this program at Alcoa, Peters (1037) studied the effect of varying coating thicknesses on bond strength using structural epoxy adhesives and confirmed the expected increasing joint response with decreasing overall oxide thickness. It was also demonstrated that the weakest structural part of such conversion coatings resided in the upper layers closest to the interface. The fracture site could be deepened by using adhesives with increasing mobility, i.e., lowered viscosity, indicating some actual penetration into the coating cross section involving some mechanical interlocking contributions. These observations would also indicate the significance of the terminology "interphase" in the explanation of how bonds fail, since the properties of the oxide near the interface with the adhesive were obviously quite different from those of the bulk oxide.

Since high heat-curing adhesives will naturally proceed through some lowered viscosity state enroute to cure (as compared with room temperature–curing paste adhesives), they might be preferred for fabricating higher-strength joints with the thicker conversion-coated aluminum adherends. There is, however, an addition factor to be considered as regards the superiority of heat-curing adhesives which is not related to the above observation. This concerns the fact that these conversion coatings during the building up process will retain considerable water within their cross section. If this freshly generated coating is permitted to dry out, i.e., age under room temperature conditions or be heated, this water will slowly dissipate with the development of significantly higher cohesive strength potentials in the oxide. With this greater densification, however, there is a penalty of being less water wettable. This is especially recognizable when attempting to use water solvent–type adhesives on ambient temperature–aged or freshly developed coatings which may have been subjected to high heat. It can be demonstrated, for example, that if a freshly generated Alodine 1200 (proprietary formulation of Amchem) oxide on 6061-T6 alloy is heated to the same temperature as would be required to cure a one-part epoxy adhesive, the surface then cannot be effectively bonded with a water-dispersed adhesive product like a casein-neoprene latex adhesive. At the same time, however, that same freshly generated oxide can be coated with a one-part epoxy and heat cured with generation of a very high-strength bond with long-term joint potential even in the corrosive seacoast environment (see Table 14).

Minford (1038) published an investigation in 1982 showing the interesting interacting effects of conversion-coating thickness, relative rigidity of room temperature–curing epoxy adhesive, and the adherend thickness of 6061-T6 lap-shear joints.

In general, we can observe that increasing the adherend thickness will increase the level of joint strength measured with a fixed oxide thickness and an adhesive of a particular rigidity in the bondline. Using a more rigid adhesive will have the reverse effect of decreasing the apparent joint strength with the same conversion coating. (More discussion about the interacting effect of these factors on lap-shear strength will be considered in Chapter 6 on adhesive joint design.)

Minford (1039) has also evaluated the relative durability of aluminum joints prepared with hot Alcoa A3–treated surfaces and two commonly employed proprietary conversion-coating treatments (Alodine 1200 and Iridite 14). With a room temperature–curing epoxy adhesive, the acidic-deoxidized adherends afforded the best durability performance in various water-soaking and hot-humidity conditions. There was an exception to this generalization in the seacoast atmospheric conditions, as can be seen in Table 14, where the hot Alcoa A3-treated joints generally survived less than 1 yr, whereas the corresponding Alodine 1200–treated joints had a survival range of 1158–1440 days. However, with a 400°F (204°C) curing one-part epoxy, the greater joint durability for all types of exposure conditions proved to reside in the conversion-coated pretreated joints.

It should be noted that conversion coatings on aluminum have been produced using both solution-brushing and tank-immersion processes. Oliva (1040) has written the Boeing Process Specification for brush alodizing. Note that using the term *alodizing* has almost become generic, because of the broad use of the Alodine-type chemical formulations from Amchem. Michaelson (1041) has written the Boeing Process Specification BAC-5719 Rev. E which deals with regular alodizing as conducted in 1969. Scalise (1042), in 1977, contributed a paper to SME relating experience in practicing energy conversion in the conversion-coating aluminum finishing processes. Tuttle (1043) has provided a description of conversion-coating use in 1979 making it a more recent contribution to the literature.

K. Sulfuric Acid Anodizing

Because of the excellent resistance to severe corrosion conditions produced by sulfuric acid anodizing (SAA), it would be desirable to use this surface preparation for aluminum structural bonding applications that must offer maximum resistance to corrosive pitting service environments. Unfortunately, as was discovered early in the bonding of aircraft, the thickness of this oxide, as applied for maximum corrosion protection, can lead to fracturing in the oxide under high peel-stressing and low-temperature service conditions. There is obviously a direct correlation between the increasing thickness of built-up anodized oxide layers and increasing corrosion resistance of that exposed surface which creates an inverse relationship with bonded joint strength and resistance to fracturing failure.

Sulfuric acid anodizing has been the most widely used anodizing procedure over many years as employed by the aluminum industry to develop thick, porous-type coatings with a wide range of desirable properties for aluminum architectural products which also could be effectively sealed to further increase the protective quality of the surface. The sulfuric acid electrolyte is also quite economical to use. Two general types of SAA coatings have been produced; i.e., the conventional- and hard-coat varieties. The conventional type is most extensively used to produce decorative and protective coatings ranging in thickness from about 0.1 to 1.2 mil. These coatings were first produced in the United States under the Alcoa trademark Alumilite. The typical conditions for producing these coatings can be reviewed in the early work of Alcoa scientists Mason and Slunder (812) and Mason and Fowle (1044) and of Cochran (1045). The conventional form of these coatings is mainly colorless and transparent on aluminum, although coloring effects have been achieved for decorative purposes where desired.

The most commonly employed thickness for providing acceptable corrosion protection for long-term architectural applications has been in the range of 0.7–0.8 mil. Aerospace builders would certainly prefer to use such a thickness for maximum resistance to corrosion of both the area under the bondline and the surrounding surface areas of the structure. However, when early aircraft joints were fabricated using the SAA procedures at protective ranges of oxide, their poor peel performances were noted, especially at high altitudes with the accompanying low service temperatures. Bondline fracture failures could literally be induced through the cross section of the anodic oxide. While this kind of performance may not be acceptable for bonded aircraft, the possibility still exists that it might be acceptable for applications where the stressing levels were low and the service temperature moderate.

Minford (532–534,772–774) has attempted to provide data that would offer assurance to manufacturers that might have a place in their product lines for aluminum that would be as weather resistant as the SAA-treated surfaces. In testing conducted over many years, the same Alcoa Alumilite 215-treated 6061-T6 surface ahderends were used with the same one- and two-part epoxy commercial products. The performance of these joints always demonstrated a high degree of durability with tensile-shear testing of lap-joints. Satisfactory joint strength retention was maintained in exposures up to 2 yr in continuous-water or hot-humidity soaking or a soak/freeze/thaw cycling. Similar good performance was afforded during 8 yr in industrial atmospheric weathering or seacoast atmospheric exposures. The thicker the oxide form of SAA employed, however, the greater is the need also to provide some testing under simultaneous weathering and stressful conditions. For many bonded products that are only infrequently exposed to exterior weathering, it is likely that thin SAA coatings might be employed which would have commensurate strength with CAA and some thicker variations of PAA and would likely perform adequately. It would have to be recognized that such thinner SAA oxides would not show the resistance to weathering that has come to be recognized for the thicker versions.

L. Chromic Acid Anodizing

The preferred surface pretreatment for aluminum adherends in European aerospace manufacturing for many years has been chromic acid anodizing (CAA), as discussed by Brockmann and Hennemann (1046), Brockmann et al. (524), and Bijlmer (815). Rogers (820) has written a corresponding chapter on its use in the United States. Good corrosion protection and joint bondability may be provided with coating weights as light as 200 mg/ft² (20 mg/dm²). This is roughly equivalent to a coating thickness of about 0.0003 in (0.3 mil). Martin-Marietta scientists have described the configuration of the oxide as looking like "cornflakes," which can be of the order of 2000 Å protruding out of the generally 15,000-Å thicker background section. Publications have been forthcoming from most of the aerospace industry.

The chromic acid anodizing of aluminum alloys is by no means a new process, however, since Buzzard (813) and Buzzard and Wilson (814) obtained early patents in 1936 at the National Bureau of Standards and the Bengough-Stuart process originated in England in 1925. Bijlmer (815) refers to this latter process as the basis for CAA anodizing as it is still performed in Europe. Unlike the American MIL-SPEC and Bell Helicopter processes which increase voltage initially to 40 V, the Bengough-Stuart process consisted of anodizing with voltage increasing to 50 V. Bijlmer (815) admits the reason for this additional increase of 10 V is still not clear and a basis for an improvement in bondability has never been established, although some improvement in overall corrosion resistance of the higher voltage-generated layer would be expected. Actually, the first stage of the anodizing process should be the most important for bonding, since the oxides formed at this stage are those found in the top layer of the coating near the adhesive.

As early as 1940, the CAA treatment was required for all aircraft in the U.S. Navy. Such coatings were used both alone for corrosion protection and as a base for organic finishes such as zinc chromate primer and enamels. During World War II, the CAA process found increased popularity and it is now being used for the treatment of various aircraft parts and aircraft assemblies. Because the CAA electrolyte does not attack the basic aluminum alloys, it poses no problem if it should happen to become entrapped in joints, seams, or pores. In Europe, the CAA-treated aluminum parts for bonding are used without postsealing treatments, as discussed by Bijlmer (815); however, Rogers (820) composed his chapter around the subject of "sealed chromic acid anodize" when discussing its use in American aircraft. Bijlmer (815) discussed the anodizing solution variables, processing variables, and the influence of variable aluminum alloy materials in considerable detail. He particularly mentions the interfering effects of chloride ions in the rinse water and Exacto's Fokker report (1047) which showed corrosion resistance in CAA oxides was decreased by the presence of sulfates which might be contaminates from previous acid pickling processes. Bijlmer (816) also warned against extended use of hot deionized water rinsing designed to promote quick drying because of some sealing action on the oxides. Reference can also be made to Exalto's Fokker report (1048) on determination of optimum CAA conditions for bonding. Russell and Westerphal (1049) have studied the characteristics of the surface produced on 2024-T3 alloy by the chromate anodize process. Also, Herfert (1050) conducted a fundamental investigation of anodic films on aluminum alloys as a surface preparation for bonding. McDevitt and Baun (1051) made a final reporting of their observations on the relation between chemical surface pretreatment and the growth of anodic barrier layer films to AFML in 1975.

A number of different procedures have been used to produce the CAA surface on aluminum. Rogers (1052) was issued a pertinent patent (No. 3,414,489) in the name of Bell Helicopter. Michaelson (1053) has written the Boeing Specification BAC-5019 Rev. D to cover their CAA procedure. Schliekelmann (1054) composed an early 1973 summary of the use of CAA on the very durable Fokker-VFW F-28 Fellowship model aircraft, which was distributed by the U.S. Department of Commerce. Elrod (1055) wrote the Boeing Specification BAC-5000 Rev. C relating to their procedure for sealing a CAA-treated aluminum surface when that is deemed appropriate. As reviewed in Table 30, the American MIL-A-8625 (1056) specifies the conditions for applying CAA pretreatment, whereas the most current Bell specification (1057) is Process Specification 4352. Other pertinent reports with reference to CAA technology can be found in a 1966 Netherlands Aerospace Lab report by Hartman (1058) and Wernick and Pinner's book on surface treatment of aluminum (1059), whereas more general information on anodizing can be found in publications by Jenny (783), Keller et al. (786), Keller (1060), Best et al. (1061), Diggle et al. (1062), and Diggle (1063).

It needs to be emphasized that the formed layers from any anodizing process are generated underneath the already-existing oxide; i.e., generally the regenerated oxide from acid deoxidizing as a pretreatment for the anodizing. Where ammonium tartrate has been used as an anodizing electrolyte,

Table 30 Processing Steps for CAA Pretreating Aluminum

Step pretreatment	Specification identification: MIL-A-8625C[a] FPL	Bell Helicopter[b] FPL	Fokker DIN[c] Chromic-sulfuric acid etch
Anodization	40–50 g/L (5.3–6.6 oz/gal) of CrO_3, 90–100°F (32–38°C), 8 V/min to 40 V, hold for 55 min	60–100 g/L (8–13 oz/gal of CrO_3, 91–99°F (33–37°C), 3–5 V/min to 40 V, hold 30–35 min	50 g/L (6.6 oz/gal) of CrO_3, 100–108°F (38–42°C), 5 V/min to 40 V and hold 20 min, raise to 30 V and hold for 10 min
Process controls			
Amps/ft² (Amps/m²)	0.9 (10)	0.9 (10)	----
Weight oz/ft² (g/m²)	0.007 (2.2)	0.10 (3.2 <0.8:free acid 4 oz/gal (30 g/L)	Coating weight
pH	0.4–0.6		Free acid 4–6.6 oz/gal (30–50 g/L)

[a]Ref. 988.
[b]Ref. 989.
[c]Ref. 990.
Source: Data from Clearfield Ref. 999 (Table 3, p. 260).

the lack of defined pores makes it actually possible for the etched surface morphology to be essentially duplicated on the anodized surface. Similarly, a fine enough porous structure also arises from the use of a sulfuric acid electrolyte to show evidence of the replicating effect. This is not true, however, for either the CAA or the PAA surface oxides that develop quite different surface morphologies that then offer the greater mechanical interlocking properties with adhesives, which explains the higher strength and probably more durable properties of such joints.

It has been mentioned that Bijlmer (816) studied the influence of hot-water rinsing on both etched and anodized surfaces, whereas Kawakernaak (1064) studied rinsing and drying effects on the surface quality of chromic/sulfuric-pickled Alclad 2024-T3 surfaces. This caused him to caution against any use of hot deionized water rinsing. Bijlmer and Schliekelmann (779) also believed that a second rinse after normal rinsing and drying might spoil the surface bondability owing to additional water adsorption on the surface. They used surface potential measurements to show such a surface obviously takes a long time to thoroughlly dry out. It was their personal bonding experience that the adhesion of either paints or adhesives to an anodized surface that has been repeatedly rinsed was poor. This contention that the CAA coating should be used in the as-formed state for maximum adhesion has been the continuing policy at Fokker (1047,1057,1065).

In the most recent book surveying the bonding of aluminum aircraft (1066), Rogers (820) describes the process used to seal CAA coatings in the American aircraft industry. Rogers states that significant improvements in adhesive formulations in recent years has made it necessary to modify existing CAA processes to attain maximum corrosion-resistant joints. Rogers also discusses the fact that CAA processes are normally restricted to those aluminum alloys that contain less than 5.0% copper or 7.5% total alloying elements, since difficulties arise in producing suitable coatings when such parameters are exceeded. Aluminum alloys containing high percentages of magnesium or silicon might also cause problems according to Rogers. Evidence that sealed (postheating in high-temperature water) CAA coatings have been successfully used for a number of years in the production of adhesive-bonded aluminum assemblies can be found in the literature. However, preparation-adhesive combinations still should be tested in weathering conditions designed to simulate service conditions to determine the final suitability according to Rogers (821).

The posttreatment after CAA proposed by Rogers consists of (1) rinse in clean water within 5 min after the voltage source is turned off; (2) clean deionized water or solutions of chromic acid, sodium dichromate, or nickel acetate can be used to obtain the best corrosion protection/adhesion compromise, but do not seal above 190°F (88°C); (3) dry in ambient air or no higher than 160°F (71°C); and (4) exercise special caution in handling to protect the surface from contamination by fumes, dust, and so forth. All parts must be handled using clean racks or clean gloves, since prepared surfaces are very susceptible to damage by wiping. Avoid any wiping or rubbing of the surface.

Visual inspections cannot be relied upon to predict bondability even though a characteristic CAA color will be produced. Each different sealing process will also provide a distinct visual signature. The floating roller peel test has been highly discriminating as a means of evaluating the relative bondability of anodic-coated aluminum surfaces in general. Rogers confirms some selectivity noted with different types of adhesive polymers. High peel strength, film-type adhesives like the nitrile-epoxies, nylon-epoxies, or elastomer-phenolics should only be used with a modified sealing process involving still lower temperatures than would be necessary to affect full sealing. Paste-type adhesives and lower peel strength film adhesives can be bonded to a full seal as produced by dichromate or boiling-water solutions. Using well-sealed coatings for bonding still does not obviate the need for some corrosion-inhibiting primer in the area of the bondline for critical areas like aircraft bondments.

It might be pertinent to mention in regard to the importance of CAA oxides for aircraft bonding that two of the three anodizing systems investigated in the military-sponsored PABST (primary adhesive bonding structure technology) program at McDonnell-Douglas were chromic acid anodizing specifications (1067,1068).

M. Phosphoric Acid-Anodized Aluminum Adherends

The preferred aluminum surface treatment for bonding American aircraft in recent years has turned from the use of the optimized FPL etch to anodizing in phosphoric acid electrolyte.

Hunter et al. (834) performed phosphoric acid anodizing (PAA) at the Alcoa Labs and published the results in the 1950s. It became clear very soon in the investigation that this form of anodized coating could not be effectively sealed as with the CAA and SAA oxide layers. As a result, the excellent resistance of this oxide form to corrosion was entirely lacking, making it unacceptable for commercial use on aluminum surfaces to protect against exterior weathering conditions as demanded for most architectural applications. For adhesive bonding, however, it was Boeing who first showed that any open, porous, columnar oxide structure as produced by the anodizing procedures on aluminum could be considered for high-strength and durable structural bonding.

The most recent summary of facts concerning the process has been assembled and published by Marceau (1069) in 1985. One of the most comprehensive early studies on factors affecting adhesion to anodized aluminum was concerned with aircraft paint systems as conducted by Popley et al. (1070) at the Atomic Weapons Research Establishment in the United Kingdom. This investigation showed cohesional failures in the anodic oxide could occur, especially where the thicker more corrosion-resistant layering had been built up. In this respect, the quite thin PAA anodize layer should be a positive factor. A 1975 study of the morphology of sealed PAA coatings issued by Thompson et al. (1071) concluded that a large open-pore structure did exist which could only be effectively sealed after prolonged 72-hr sealing processing. This should, however, be a good confirmation that production of high-strength and durable aluminum joints should be achievable. (For other pertinent investigations, see Refs. 569, 833, 1000, 1072, and 1073.)

The large open pores of the PAA oxide should enhance adhesion by offering the greatest possibility among the developed anodizing procedures for securing deep mechanical keying of the primer or main structural adhesive in the absence of a primer, as shown by Venables (33), Venables et al. (36,754), Marceau (1069), and Bethune (1974). The commercialization of the process for making high-strength and durable aluminum joints has been described by a Kabayaski and Donnelly Boeing report (30), Boeing Process Document BAC 6614, and a Boeing patent issued to Marceau et al. (1075). Sutliff (1076) provided a method of controlling the electrolyzing solution. The details of a typical process has been provided in Table 31 (463).

Table 31 Processing Steps for PAA Pretreatment of Aluminum

Step	PAA
Pretreatment	FPL deoxidation
Anodization	100 g/L (13 oz/gal) H_3PO_4
	68–77°F (20–25°C)
	10 V for 25 min
Rinse	Deionized water at 68°F (20°C)
Air dry	

Source: From Ref. 999 (Table 3, p. 260).

As noted above, because these PAA oxide layers will not hydrate or "seal," they do not provide good corrosion protection over such a treated surface. This was, of course, the reason that Alcoa had much earlier discontinued their study of the procedure. However, the combination of the oxide with an adhesive impreganted into it does provide a very high degree of resistance to any undercutting of the bondline by corrosive weathering conditions. Albeit, the unprotected PAA-anodized surface in the periphery around the joints will rapidly show pitting corrosion in a saltwater exposure unless adequately protected by some sort of polymer overcoating material. The overall durability in both water-soaking, hot-humidity, and corrosive saltwater exposures of PAA structural bonded joints has been widely confirmed in the literature (533,534,1078). (More references to PAA joint durability data will be supplied in Chapter 10 on that subject.)

The pretreatment of aluminum surfaces to receive PAA processing has also been important and Marceau's summary of the work at Boeing should be reviewed (1069). Any concern about potential sealing of the coating during rinsing operations, as previously mentioned about the CAA oxide, is obviously of no practical concern. However, certain rinse-water–quality guidelines have been developed for production purposes and the time between turning off the anodizing current and beginning the rinsing procedures must not exceed 2 min according to Marceau.

Mention should also be made that Locke et al. (1079) have reported the development of a nontank PAA technique which has been a viable process for repair surface treatment for aluminum structures. These investigators were concerned primarily, however, with the repair of aircraft in the field. A reference related to improved bond durability with PAA surfaces has been issued at Boeing by McMillan et al. (1080). For pretreatment prior to this nontank PAA-repair procedure, observations have been made using various abrasion and acid etching procedures by Horton et al. (1081). Comparison of results were made with the hand-cleaning methods in earlier use at military repair facilities. The etching procedure preferred in this investigation ended up as an application of either a Pasa-Jell 105 or an FPL ingredient paste formulation. (For other publications dealing with the standardized repair handbook, see Refs. 1082 and 1083.)

Smith (1084) described a study of the relationship between nondestructive inspection (NDI) for contaminants on PAA Alclad 7075-T6 aluminum and bond strength and degradation that results from the contamination. Contamination broadly included surface preparation process errors (e.g., incorrect anodize time or voltage), organic contamination from humans, and smog and handling damage in general. It was demonstrated that the ellipsometric technique can adequately reveal contamination at levels below that capable of significantly degrading the bond integrity. The details of such a tester have also been included.

Hennemann (1085), in discussing the adhesion between metals and polymers as a three-dimensional system, has offered the following picture of the bonding to PAA aluminum–treated surfaces. After a pretreatment such as PAA, there is first a dense barrier layer of aluminum oxide formed. Further developments generate a main oxide layer with a split top area. This whole oxide layer is penetrated by polymer, primer, or other adhesive. The oxide layer thus becomes a composite, a reinforced oxide. The reactivity of the resin in the aluminum material will be different from that of the bulk material. The interaction of the resin with the oxide can, therefore, lead to a weak boundary layer above the oxide. All the different zones of aluminum/adhesive bonds are of different elasticity and thickness—all together about 60 μm. It is a zone of different material discontinuities and all together they constitute adhesion in a technical sense. In other words, technical adhesion is the integral property of many single interfaces.

N. Duranodic Anodize-Processed Aluminum Surfaces for Bonding

Duranodic is an Alcoa trademark for quite thick anodic coatings used in recent years in many architectural applications such as the sheathing of monumental buildings. For example, the sheathing on the highest building in the world, i.e., the Sears Building in Chicago, features a special black Duranodic anodize finish. These coatings offer several differences from the earlier SAA-type architectural building finishes, including an intrinsic coloring (based on special alloying elements) as compared with the earlier dyed Alumilite finishes. The colors can range from various shades of bronze and gray to black depending on the alloy selection. An organic acid electrolyte is employed in the bath as compared to the inorganic sulfuric, chromic, and phosphoric acid electrolytes discussed earlier. These coatings are significantly thicker, harder, and more dense oxides which can be as much as 15–40 times thicker than the CAA and PAA coatings most used for bonding preparation.

Because no information was available about structural bonding to these kinds of anodized layers, Minford (1086) conducted some investigations to determine how these much thicker oxides might respond to bonding with the same epoxy adhesives previously used to investigate the SAA-, CAA-, and PAA-pretreated 6061-T6 adherends. The commercially produced Duranodic aluminum was used with only a simple degreasing operation before applying adhesive to assure no soluble contamination was present. This was followed by fabrication of standard D-1002 ASTM lap-joints using both a one- and two-part epoxy adhesive which would permit direct comparisons involving long-term weathering exposures.

It might be anticipated that oxides that are more than twice as thick as the previously employed SAA-anodized surfaces and as much as 20 times the thickness of CAA- and PAA-pretreated oxide surfaces would form relatively low-strength structural joints. However, Minford failed to find any significant decline in initial joint strength as compared to the much thinner SAA joints earlier evaluated. The author has speculated that the higher compaction and greater density of these Duranodic oxide layers must have affected a trade-off situation in which the weakest point in the oxide cross section was no weaker than that present in the much thinner SAA-treated layers. (More will be said later about the relative durability results in Chapter 10.)

V. PRIMERS AS PART OF THE PRETREATMENT PROCEDURE FOR BONDING

A great many adhesive-bonding applications employ some type of coating applied to the treated adherend surface, which usually can be described as primers or coupling agents. The former have the function of serving as adhesion promoters and protective coatings, whereas the latter serve to create a chemical bonding between the adhesive and the adherend. This distinction is not always as clear as described, however, and situations arise where primers end up acting more like a coupling agents, and conversely, coupling agents often are serving distinctly as primers in the overall jointure.

Primers are important for a number of reasons. First, they protect a freshly prepared adherend surface from contamination or changes which might otherwise occur as a result of contact of the surface with its environment. This is the case whether the adherend surface has only been cleaned to remove loose oxide or organic contamination or more specifically prepared to create a special surface condition like the "regenerated oxides" described in the previous sections. The primer is especially useful when it can penetrate to create mechanical interlocking with such specially treated aluminum surfaces. A related function of protection against the corrosive effects of the service environment is provided when special corrosion-inhibiting primers are used. A second function is to penetrate any surface roughness or porosity (microscopic with metals and macroscopic in adherends like woods). Paste-type adhesives and even the structural film adhesives at elevated curing temperatures can have relatively high viscosities which finds the penetration of the surface more or less difficult. Primers meanwhile are often 10% solutions of adhesives or a few percent solution of coupling agents which can wet out the adherend surface with good probability of success. A recent 1990 review of the subject by Bascom (1087) can be recommended.

Primers have always been considered an intimate part of the preparation of aluminum surfaces for structural aerospace bonding for all of the above reasons. However, primers can be equally important for many other manufacturing situations in which aluminum must be bonded to some other material where the other material is porous like wood or the surface has a higher degree of adhesion to some primer other than the adhesive which might best serve to bond the aluminum. Examples of the latter might be elastomers and plastics where specific primers might be recommended for special adhesion to those surfaces with a different adhesive for bonding to the aluminum. There must, of course, be a high affinity between the special primer and the adhesive for aluminum bonding.

The low viscosity of most primers, as compared with the primary adhesive formulation in the bondline, permits ready application by different methods such as spraying, dipping, brushing, or roll coating. This is applied as early as possible to the aluminum surface, for example, which has just been pretreated by some combination of solvent cleaning, abrading, deoxidizing, conversion coating, or anodizing. Falcon and Miller (1088) considered the benefit of special primers for making repairs on the main rotors in helicopters in 1977. A basic formulation for a corrosion-inhibiting primer developed at the Naval Air Development Center can be seen in Table 32. It is easily seen from observing the 12 different basic ingredients present in this primer formulation that there can be a considerable range of different formulations that are possible. Although strontium chromate has been used in this particular formulation, corrosion inhibitors such as zinc and other inorganic chromate salts have been employed.

In addition, there can be enhanced joint durability where better wetting of the adherend surface by the primer is achieved than would result from the use of the adhesive alone. Not only the adhesive but the primer solution as well may contain special wetting agents, flow-control agents, elastomeric toughening additives, and corrosion-inhibiting additives. The low molecular weight of most primers obviously is able to create the distinct opportunity to flow, wet, and develop more intimate and complete surface contact with the varying geometries of the treated aluminum surface or on

Table 32 Basic Formulation for a Corrosion-Inhibited Epoxy Primer for Aluminum Adherends

Pigmented epoxy (Component I)	% by weight
Strontium chromate	12.0
Titanium dioxide	2.3
Magnesium silicate	5.5
Diatomaceous silica	2.9
Epoxy resin (75% in toluene)	18.2
Methyl isobutyl ketone	12.6
Toluene	6.5
Polyamide (Component II)	
Polyamide (70% in xylene)	10.7
Toluene	12.9
Isopropyl alcohol	8.2
Butyl alcohol	8.2

Source: from Naval Air Development Center, Aero Materials Department, Report No. NADC-MA-7164 on Corrosion Inhibiting Primers For Aluminum Alloys, Appendix C (Dec. 31, 1971).

both aluminum and other adherends in dissimilar joining. There always needs to be some testing of any primer with a suggested adhesive where this information is not already available to make sure the primer adequately wets the adhesive. Herczeg et al. (1089) have pointed out that if epoxy primers are not properly formulated, the surface energy level of the cured epoxy primer coat can be reduced so the adhesive will not wet out the primer. For those situations where a primer is employed to more effectively join aluminum to another adherend like wood, the best primer often may be a diluted version of the adhesive. The advantage on the wood may be more related to filling the porosity of the wood surface, whereas the same dilution has the different function of increasing the degree of surface wetting on the aluminum, which does not have a macroscopic porosity similar to the wood surface.

Bishop et al. (1090) has recently offered an interesting examination of the bonding between aluminum surfaces with an epoxy with and without a primer. They sectioned across the bondline with an ultramicrotome and viewed the sections using transmission electron microscopy (TEM). In the absence of a primer, it was possible to observe trapped air in the adherend surface roughness created both by light abrasion or FPL deoxidizing. Much porosity was apparently eliminated when the aluminum adherends were primed. Reinhart (1011) has reviewed a number of concepts developed by the Air Force for priming aluminum and titanium aerospace structures. These included the development of water-soluble polymers for water-base primers, electrodeposited primers, and more effective corrosion-inhibited primers.

Ecological concerns have made it more desirable to consider the switch of water-solvent primers for the more widely employed organic solvent-based primers of the past. The driving off of solvent to activate the primer for subsequent bonding is not always acceptable, especially where large surfaces may be involved, as in aerospace and automotive manufacturing. Apart from the concern for the workers, recovery and recycling of these solvents simply is not cost effective. Accordingly, water-base primers are being developed that are mainly aqueous emulsions of an adhesive with some of the other ingredients shown above in Table 32 for formulating primers. It is anticipated that the emulsifying agent might be a nonionic surfactant which remains behind in the dried primer film. Use of cationic or anionic surfactants would be expected to attract water when left in the film because of their hygroscopic nature, leading to later degradation of the bond strength.

The solubility of some of the better corrosive-inhibiting materials like zinc chromate and strontium chromate may make them ineffective in water-base primers. U.S. Air Force studies have attempted to develop other materials like dichromate salts of organic nitrogen compounds; however, they have tended to be inherently unstable and decomposable by oxidation in air. Bishop et al. (1090), however, have reported guanidine dichromate acted as effectively as zinc chromate.

The electrodeposition of primers is a well-established technology that has found wide past use in the automobile industry prior to painting. Some modification would be expected to make such primers suitable for adhesives. The process involves an aqueous suspension of an electrophoretic (charged) organic polymer that will be attracted to the charged metal adherend. It offers the advantage over spraying and dip-coating of being more adaptable to automation, more pollution free, more uniform film distribution, better film thickness control, and more rapid film application. Because of the studies in the automotive industries for use on body steel, the development of electrodeposited technology on steel has been predominant over that for aluminum. Foister et al. (1091) have presented recent information on investigaitons of such primers on steel for structural adhesive bonding. A complex chemistry is undoubtedly involved and testing of different catalysts in adhesive formulations has even shown one agent could degrade the primer, whereas no problem was present with a shift to another catalyst.

It must be confessed that it could be very cost effective in a manufacturing process if no primer was required. This is certainly possible if certain requirements can be met. For example, the adhesive needs to be applied before any contamination takes place, which would ordinarily be prevented by use of a primer. Next, the surface treatment must produce an adhesive/adherend moisture-resistant boundary. Conceivably, this might best be achieved in manufacturing in a high production rate process with automation and a corrosion inhibitor present in the adhesive. The author has conducted thousands of long-term durability tests over many years using 6061-T6 variously treated adherends and the same one- and two-part epoxy adhesive formulations. The durability results have generally been directly predictable based on the ability of the treatment to regenerate stable, corrosion-resistant forms of aluminum oxide with the advantages of being highly wettable by the adhesive. In addition, the expected advantages due to the ability of the adhesive to form strong mechanical interlocking with the best of these oxides is quite evident whether a primer has been employed or not. While the aerospace industry has maintained their processes using selected primers, other industries, like the automotive industry, have sought to obtain acceptable bonding even in the absence of the surface-cleaning steps ordinarily considered indispensible. Minford (68,72,73) has published joint durability test results with deliberate contamination of the aluminum surface and found reasonable bond durability could be obtained. Wegman (1092) has discussed the bonding directly to oily steel surfaces in the automotive industry. The surface, of course, must be free of corrosion products and thick layers of oil (if present) must be reduced by some solvent wiping. Curing the adhesive at temperatures of 250°F (121°C) or above is also beneficial. In fact, the author has never been able to achieve satisfactory bonding at room temperature with any standard paste epoxies to deliberately oiled surfaces, although this has been claimed by some German investigators whose work will be discussed later.

The bonding of sheet molding compound (SMC) parts to either steel or aluminum autobody sheet has been achieved commercially using a two-part polyurethane adhesive. The parts need to be wiped clean with some solvent such as methylene chloride and a two-part, room temperature–curing epoxy primer is required to obtain necessary resistance to exterior weathering. Shah (1093) has obtained a 1989 U.S. patent on a two-part epoxy composition containing amine-terminated butadiene-acrylonitrile which claims to bond SMC-metal or SMC-SMC with good performance without the need for even this solvent wipe.

Marceau (1094) has stated that while most 350°F (171°C) curing primers are compatible with the PAA-anodized aluminum surface, a few rubber-containing primers have exhibited suboptimal behavior. This behavior has been characterized by reduced lap-shear strength of as much as 30% and accompanied by a failure mode that appears to be at the primer/oxide interface when examined visually. These failures have been described as being adhesive appearing or slick appearing. Further examination of their features revealed a thin rubberlike layer approximately 100 nm thick on top of the oxide, as viewed by SEM examination by Marceau (1094).

When considering the effect of primers on aluminum, it must be remembered that the overall joint is a linked system consisting of the adherend jointure to the primer plus the primer jointure to the adhesive. Each such jointure is an entirely separate interfacial situation. All of this, in turn, is only half of the total joint, which may or may not be symmetrical in design; i.e., a dissimilar materials joint. The old so-called chain-link theory certainly applies where the overall strength of a joint can be no stronger than its weakest link. The selection of a primer must be predicated on the basis that its properties do not form a weak link in this chain. In this regard, the primer may actually be a weak link in that its cohesive strength is lower than the bulk adhesive in the overall joint. Its value then becomes how much the actual service life is extended by its greater ability to wet the adherend and resist deterioration due to water or corrosive saltwater as compared to the adhesive/adherend bond that would have existed in its absence.

A. Corrosion-Inhibiting Adhesive Primers

Probably the most important event in the history of primer technology for bonding aluminum surfaces was the introduction of corrosion-inhibiting adhesive primers (CIAPs). While primers may be selected based on the compatibility with various adhesive families that are conventionally used in bonding aluminum to itself or other adherend materials, the primer may constitute a factor for improving joint durability only because it has a greater ability to wet and spread over the aluminum oxide. In a corrosive weathering situation, this advantage may not be sufficient. What is needed is the additional capacity to further protect in such service conditions as salt spray, seacoast, or marine conditions. Thus, CIAP primers are compatible with the adhesive used in the bondline but also contain dispersions of finely ground chromates (normally strontium, barium, or zinc salts). When the primer is applied, the chromate-filled polymer dispersion provides an even film of protection. Provided the aluminum adherend surface has been properly prepared, the primer should be able to protect the surface in both warm, moist environments and in corrosive chloride ion environments as well.

Rogers (1095) stated that CAA-sealed aluminum adherends can be used effectively with both CIAP and non–corrosion-inhibiting primers. He further states that the resistance of bonded joints to aggressive weathering conditions in order of increasing durability potential are (1) anodize only surfaces, (2) anodize plus noncuring CIAP primers, and (3) anodize plus curing CIAP primers. Finally, a combination of sealed CAA surfaces with CIAP primers can greatly improve the resistance of Alclad 2024-T3 and Alclad 7075-T6 alloy joints as well as the nonclad versions of these alloys. As will be discussed later, there has existed some controversy in the aerospace industry in recent years as to whether clad or nonclad versions of these aluminum alloys should be used because of the deliberately planned sacrificial nature of the Alclad surface alloy to prevent penetrating pitting corrosion of such surfaces when directly exposed to severe weathering. (This will be discussed in more detail when considering joint durabilities in Chapter 10.)

It has also been shown by Bethune (1000) that CIAP primers can positively influence joint durability under stress and hot-humidity conditions. The CIAP in the bondline was exposed to 172°F (78°C) and 100% RH conditions at a stress of 300 psi (2.07 MPa) and the failure times were extended from 28 days (without primer present) to 250 days with primer present. At 900 psi (6.19) stress conditions, the extension time was still significant from 7 to 40 days. In either case, the mode of failure was judged to be interfacial.

Early reference to the use and function of corrosion-inhibiting primers was discussed by Krieger (1096), whereas some investigations were conducted by Lorenz and Mansfield (1097) and Matienzo et al. (835). The latter dealt with the durability of aluminum/polymer systems protected with organic corrosion inhibitors.

B. Comments on Joint Durability With Primers

Sell (1098) has shown that the aluminum joint durability tends to increase with increasing primer thickness up to the point where the thickness makes the primer begin to act as a separate layer in the joint. Kinloch (1099) has recorded a situation in which the primer layer was clearly the weakest link in the bond cross section. Scardino and Marceau (853) showed "aluminum wedge" test results in salt spray weathering where the use of a primer in the bondline significantly reduced the crack growth rates for both FPL pretreating or FPL deoxidizing followed by CAA or PAA pretreatment. It has been observed that very low crack growth rates can be secured with PAA-treated adherends whether a primer is present or not. However, some investigators, like Rogers (1095), claim to have substantiating data for proving superior performance with CIAP promers used over CAA surfaces.

For more than 20 years of testing at the Alcoa Labs, the author deliberately chose to conduct all 6061-T6 aluminum joint durability testing without the use of any special primers. The purpose of such investigations was, of course, to offer durability data to those general manufacturers (outside aerospace) who might not be able to afford the additional cost of special priming. Minford (533, 534) has demonstrated that 6061-T6 alloy surfaces that had been only Alcoa A3 deoxidized, or better conversion coated or anodized after Alcoa A3 pretreatment, could offer very acceptable long-term durability responses with unsophisticated one- or two-part commercial epoxy paste adhesives when exposed for long times to water or hot-humidity soaking (wet/freeze/thaw) cycling or a natural industrial atmosphere. Even in a corrosive seacoast exposure, the performance could be very acceptable when the conversion-coated or anodized surfaces were employed with a one-part, heat-cured epoxy paste. Under simultaneous stressing conditions, the performance of two-part, room-temperature–curing epoxy pastes was relatively suspect; however, with the one-part, heat-curing epoxy pastes and conversion-coated or anodized surfaces, the survival times were remarkably long.

Reinhart (1011) issued one of the most extensive summaries of the use of primers in aluminum bonding based primarily on actual experience from aerospace investigations. He has also considered problems of toxicity and flammability which can be introduced by use of primers applied from organic solvents. The normal procedure after surface treatment has been to apply a corrosion-inhibiting primer by spray technique; however, the need to handle large volumes of organic solvent (normally methyl ethyl ketone (MEK]) has been a major drawback. Beckwith and Pollard (1100) have presented an alternative of electrodeposited water-solubilized primer particles for use with 250°F (121°C) curing epoxies. Greer (1101), at Rohr, has been working on a similar sort of electrodeposited primer that would be compatible with those epoxies that require curing at 350°F (177°C). Pocius and Wilson (1102) released their information on a sprayable water-based epoxy primer in 1987, whereas Pike and Lamm (1103,1104) presented yet another alternative in the form of inorganic primers like aluminum alkoxide. Recently, Pike and Golden (1105) introduced data on a water-soluble, hydrated polymeric aluminum oxide which had almost as good wedge-crack specimen performance as the previously mentioned aluminum alkoxide.

While most of the above comments have had special importance for improving the performance of high-strength aluminum-to-aluminum joints as employed in aerospace manufacturing, there are some other areas of aluminum joining to nonmetallic materials of construction where priming can also be important. Sometimes with dissimilar materials, the same adhesive may not be the best choice for wetting both adherends. In such cases, it may be appropriate, and even necessary, to prime one of the surfaces to enhance the joinability to the poorer-wetting adherend. A case in point would be the joining of nonporous, vapor-barrier, aluminum surfaces to wood and wood-related surface materials. If a rather high-viscosity paste epoxy adhesive is known to bond effectively to an aluminum surface, it is still unlikely that it might be able to bond to a wood surface with the desired fiber-tearing bond strength. Thus, it might be appropriate to prime the porous wood surface with some low-solids, contact-type, elastomer-phenolic contact cement that can both wet and penetrate with mechanical interlocking. The primer bond to the epoxy may then offer the opportunity for fiber-tearing bond failures as the weakest linkage in the overall bondline.

The author has conducted literally thousands of separate durability test evaluations of aluminum-to-wood dissimilar-adherend joints because of the multitude of different wood-related products that exist for use in architectural applications. As will be discussed in Chapter 10, the optimum bond for such dissimilar bondments was often only obtainable by using various priming procedures using various contact-type adhesives. Such priming materials also offer the opportunity for later reactivation to effectively wet some other adhesive for making a final joint or lamination.

VI. COUPLING AGENTS AS SPECIALIZED PRIMERS

A. General Considerations

There has been a general agreement among adhesion scientists that water in the service environment can be among the most destructive factors affecting joint failure. (For reviews of this subject, see, e.g., Refs. 33, 1099, and 1106–1108.) In an attempt to improve adhesion durability, a number of investigators have sought to use various low molecular weight coupling agents which have a potential to form chemical bonds across the polymer/metal oxide interface (1109–1117). Recently, Schmidt et al. (1118) and Schmidt and Bell (1119) have studied polymer coupling to iron oxide in steel/epoxy joints which could serve as an amenable technique for other metallic oxides as well. In this latter report, it was demonstrated that coupling agent thickness produced a pronounced effect on the bond strength as compared with the change in protection against corrosion under hydrolyzing conditions.

The mechanical and chemical conformation of these films and their thickness, cross-link density, and reactivity appear to be highly dependent on their application conditions. The most critical factors appear to be solution concentration, solution pH, drying times, and the drying temperature. Boerio (1117) has also pointed out that the acid-base character of the adherend surface can influence the polysiloxane film structure for silane coupling.

Bascom (1111) suggested very early that an adsorbed polysiloxane film might be an open network which could permit interdiffusion of adhesive molecules to form an interpenetrating network. Boerio (1117), in reviewing the state of the art in 1989, has provided both circumstantial and direct evidence which supports this hypothesis. The apparent success of silanes to promote good adhesion between high molecular weight thermoplastics and a variety of different adherends, however, presents somewhat of a challenge for this concept. Sung et al. (1120) have demonstrated the possibility for interpenetration of polyethylene into polysiloxane films deposited on aluminum oxide (sapphire) and a relationship to the joint strength. Both the concentration of the silane and the drying time before bonding seemed highly significant. Higher drying temperatures seemed to create a more highly cross-linked network into which the polyethylene could not penetrate.

B. Silane Coupling With Metal Oxides

Silanes generally have the structure R′-Si-R, where R is a functional group that is easily hydrolyzed and R′ is a functional group that is specifically chosen for compatibility with a particular polymer. The R group converts in water to a trisilanol which can then adsorb on aluminum oxide (or other metal oxide) by hydrogen bonding between these groups and the surface hydroxy groups to form siloxane polymers. The remaining R′ end of the molecule can react with its compatible adhesive polymer forming an interpenetrating polymer network by a condensation mechanism. Commercial silane coupling was being researched as early as the 1960s by Plueddemann (1121–1123). Much of the early work was actually concerned with the use of these organosilane coupling agents as pretreatments for glass fibers to reduce the moisture sensitivity in fiber-reinforced composites as reviewed in the book *Interfacial Polymer Matrix Composites* (1124). A simplified schematic of the relationship between the silane, adhesive, and aluminum oxide adherend can be viewed in Figure 7.

Ishida (1125) has offered a 1984 review of progress in studies of coupling agents and the molecular and microstructure of interfaces and interphases in composites, coatings, and adhesive joints. He placed particular emphasis on the progress involving the aminosilanes. While chemical bonding does exist between oxides on aluminum and the silane-type materials, the structure of partially cured hydrolyzate can be quite complicated. Half of the amine groups in aminosilanes may even be combined with carbon dioxide in air to form amine bicarbonate salts. This seems particularly true when the aminosilane is adsorbed from a neutral pH solution and dried in room temperature air. Surface characteristics such as the acidity topography and homogeneity of the aluminum oxide surface can significantly influence the structure of this coupling with a silane. It appears that there is a gradient in various properties within the coupling agent interphase and that both physisorbed and chemisorbed silanes may be present. Very recently, Boerio (1117) has written a quite detailed summary chapter for the *Treatise on Adhesion and Adhesives* series. This is especially pertinent for those interested in the spectroscopic studies of how silanes function.

1. *HYDROLYSIS IN WATER*

Hydrolysis of silanes in water is usually complete within 1 hr, however, the time necessary to completely hydrolyze the silane will mainly depend on the organofunctionality, temperature, and pH

```
                                   ORGANIC RESIN
------------------------R ----------------------------------------
                        |
                    -O-Si-O-                          INTERFACIAL
                        |                             REGION
                        O
                        |          INORGANIC SURFACE
__________________________________________________________________
```

Figure 7 The bridging molecular relationship of a silane molecule between the organic resin of the adhesive and the inorganic oxide surface of the aluminum adherend. From J. D. Minford, Aluminum Adhesive Bond Performance, Treatise on Adhesion and Adhesives, Vol. 5 (R. L. Patrick, ed.). Marcel Dekker, New York, p. 57 (1981).

(1123,1124,1126). Dilute concentrations of silanes (0.01–2.0%) are usually used to ensure predominately monomeric silane, because oligomeric silanes were shown by Clark and Plueddemann (1127) with higher concentrations. The silanetriol content can be monitored uniquely by laser Raman spectroscopy (1126,1128).

Erickson and Plueddemann (1124) demonstrated that aminosilanes yield oligomers by hydrolysis even at dilute concentrations and Ishida et al. (1129) have accordingly studied a wide range of gamma-aminopropyltriethoxysilane concentrations in water. With appropriate hydrolysis conditions, they found no evidence of any unhydrolyzed alkoxy groups with concentrations below 30% by weight. By contrast, Plueddemann and Erickson (1124) had earlier showed a vinyl functional silane at 10% concentration still showed a predominant monomer condition under hydrolyzing conditions where the aminosilane showed major siloxane formation. The structure of aminosilanes was studied in the early 1970s by Bascom (1111), Kaas and Kardos (1130), and Nichols et al. (1131) and reviewed by Ishida and Koenig (1132).

2. THEORY AND REACTIVITY OF SILANES WITH METAL OXIDES

While several theories have been proposed to explain silane's ability to promote hydrothermal stability at interfaces, the most frequently employed has been the "chemical bonding" theory of Plueddemann (1109,1110). Some evidence has been furnished for the presence of polysiloxane coatings on various metallic adherends (212,664,1134,1135). Recently, Plueddemann has considered an additional mechanism which includes the promotion of stress release due to a network of reversible structures.

The employment of silanes on various metal adherends to improve the water resistance of lap- and butt-joints has been widely investigated. The characteristics of silane films on metal surfaces has been the focus of wide investigations in an attempt to better understand the nature of the silanes as adhesion promotors (1136–1139). (Additional reference data for consultation about effects noted on silane films on different metal adherends can be found in Refs. 721, 723, 891, and 1140.) Sung et al. (1141) and Sung and Paik Sung (1142) have studied polysiloxane coatings on sapphire, which is a special form of aluminum oxide surface. In these investigations, Boerio and his co-workers have involved the use of reflection-absorption infrared spectroscopy (RAIR) to determine the various spectrum identifications of APS silane on aluminum, iron, or copper. As might be anticipated, the reactions were distinctly different for the different adherends, making it necessary to evaluate them separately any time the silane to be studied is changed.

The "as-formed" silane films were shown to be quite stable in a dry environment but when exposed to a humid environment or dried in an oven at 140°F (60°C) for 20 min, they showed band changes that indicated additional polymerization had occurred. The bands could also manifest significant changes when the silane adsorption was performed at varying pH levels. After any pH variations, however, the resulting spectrum was still indicative of additional polymerization if the deposited films were heated at 140°F (60°C) for 20 min. Dillingham (1143), in his thesis work under Boerio, also investigated the effect of drying on as-formed for epoxy adhesion development. There was a significant loss of reactivity potential with this adhesive after drying for 20 min at 212°F (100°C).

The study of gamma-APS films on aluminum by Boerio et al. (721) has also shown some complications due to the etching pretreatment of 2024-T3 alloy adherends at modestly alkaline pH 10.4. There was insignificant etching observable in a 1-min exposure at this pH, but a 10-min contact time definitely created an etching condition with resultant low molecular weight polymer formation by the silane on the oxide surface. When the pH was lowered to a very slightly alkaline pH of 8, similar film results were noted on both iron or 2024 aluminum. Extensions of the adsorption times failed to show any negative effect at the lower pH value.

Ondrus and Boerio (1144) have looked at thin gamma-APS films on polished aluminum, copper, iron, and titanium using aqueous solutions at pH 10.4. After 1 min, the RAIR spectra were similar for all these metals; however, even in this short time there began to be evidence that there was less polymerization on the 2024 aluminum and the copper adherends. Drying the films at 230°F (110°C) for 1 hr showed definite additional polymerization on the steel and titanium surfaces, but the change on the 2024 and copper surfaces seemed to be involved in the formation of new imines. This similar performance of aluminum and copper surfaces probably is related to the catalysis of the oxidative reaction that can be ascribed to copper. Since this element is the main alloying ingredient of 2024 aircraft alloy with some copper on the surface, this might seem reasonable to conclude. To further confirm this supposition, these investigators tested 1100 alloy (commercial-purity aluminum) surfaces which has inconsequential copper present in the bulk or surface. The resulting spectra now reverted to being more similar to those obtained with the iron and titanium adherends. Thus, the choice of aluminum alloy to be treated with a silane can be influencing. When the iron, titanium, and 2024 aluminum surfaces were reacted with silane and the films heated to 302°F (150°C), a significant difference was shown among the three. Very little retention of epoxy adhesion could be found under these conditions with the 2024 alloy, whereas there was a highly significant retention of adhesion on both iron and titanium.

Boerio and Greivenkamp (1145), using RAIR (as initially developed by Francis and Ellison [1146] and Greenler [1147]), clearly established as early as 1977 that a structural gradient seemed to exist in the silane layer which was an apparent function of silane layer thickness. They also found an intensity band which they proposed showed the coordination of the amine group to the silicon in an iron adherend. The complexity was additionally shown by noting that this band could disappear when the silane/iron specimen was simply washed in water. This band was then replaced, however, by a different band, which they attributed to the formation of a positively charged amino zwitterion. Diaz et al. (1148), at about the same time, found a similar feature with the same silane on aluminum using inelastic electron tunneling spectroscopy (IETS). Ishida (1125) proposed, however, that it may be more likely that the amine group of the silane was being hydrogen bonded to the surface hydroxyl groups of either metal oxide. Lee (1149), Plueddemann (1150), and Weetall and Hersh (1151) had all reported the possibility of hydrogen bonding with the adherend surfaces earlier. Moses et al. (1152) somewhat later postulated an intramolecularly interacting amine with the silanol groups either in a protonated or an unprotonated form. Even more recently, Chiang et al. (1153) have hypothesized an intramolecular hydrogen structure together with hydrogen bonding with surface silanols. The hydrogen-bonded structure has also been supported by the 1980 work of Boerio et al. (1154).

Thiedman et al. (1155) has looked at the performance of eight organofunctional silane-coupling agents as adhesion promoters for the bonding of aluminum with 250°F (121°C) and 350°F (177°C) curing aerospace film adhesives. Both initial strength and final durability responses using wedge-type specimens with FPL-pretreated and two nonchemical pretreated-type surfaces were positively influenced.

Minford (1156) used Dow Surface Conditioner A, a silane proposed for bonding aluminum with silicone adhesives, for bonding aluminum/epoxy joints. The use of this silane-type adhesion promoter proved to be essentially worthless as a means of enhancing the durability of these joints in aggressive water-soaking conditions. In yet another study, Minford (1157) similarly investigated 17 proprietary silane-type adhesion promoters recommended by their manufacturers for use on metallic adherend surfaces. Only one of these products had the ability to significantly improve the durability of 6061-T6 aluminum/epoxy joints in hot-water soaking conditions even though the aluminum had been Alcoa A3 deoxidized. Another Minford (1158) investigation tested the ability of a silane primer that had been recommended by an aluminum automobile radiator manufacturer for making epoxy bond repairs to such units. It was shown that the performance of specially prepared joints in hot antifreeze test solutions was actually less effective in maintaining durability than similar repairs made with several room temperature–curing epoxies applied over abraded 6061-T6 surfaces.

For a general survey discussion of the use of silanes with a variety of adherends, the author can suggest Arkle's technical paper (1159) on that subject. Kinloch et al. (1160) have reported enhanced durability of aluminum/epoxide joints through the use of silane adhesion promoters.

VII. PREPAINTED ALUMINUM SURFACES AS PRIMED ALUMINUM ADHERENDS

There is a large area of technology involving the application of organic coatings in the form of paints to aluminum surfaces that have the dual purpose of being both decorative and protective of the underlying surface. The aluminum manufacturer wishing to offer a painted aluminum sheet for architectural applications orginally faced the same situation when preparing a particular alloy for painting as if he were directly going to adhesive bond. It should be very obvious that the chemical families representating the range of paint finishes on aluminum are closely related chemically to some of the adhesives used in general manufacturing; i.e., vinyls, polyesters, acrylics, silicones, epoxies, among others. It is the prime purpose of the aluminum manufacturer to create an interfacial bondline integrity between the aluminum oxide and these paints that is capable of surviving in direct weathering for periods of 20 yr and beyond with a special ability to resist undercutting corrosion in even the most corrosive service conditions. The above description seemed to the author to constitute a possibly highly durable priming system for aluminum that could be secondarily bonded to itself or any other material of construction with an adhesive at some later date. Since information was not available in the literature as to the overall durability of bonds to such surfaces, the author undertook a number of investigations in which lap-joints made by bonding commercially painted aluminum sheet with a variety of structural adhesives were conducted in a variety of accelerated and natural atmospheric weathering conditions. The possibility existed, of course, deliberately to select the chemistry of the primer to match the chemistry of the adhesive.

A main difference between such paint primers and the structural adhesives used to fabricate aluminum structures is that the paints need not be designed to resist high stresses, since the paint is simply furnishing good adhesion without being called upon to develop a bonding interface with a second material as in a jointure. This means these paints can be formulated from considerably lower molecular weight polymers with corresponding lower viscosity numbers. This lower viscosity property directly translates, however, to achieving a greater ability to flow and wet the aluminum oxide surface. Further, paints are mostly formulated with considerable solvent present, which adds further to their ability to wet the oxide surface. (For a variety of readings on this subject, see Refs. 856, 1045, and 1161.)

It does seem clear that a distinct analogy can be made between the use of paint to offer surface protection for aluminum against direct weathering and the adhesive priming of aluminum by the aerospace industry to protect an aluminum surface for later adhesive bonding. The question becomes whether the properties of these paints, as formulated for highest degree of surface oxide wetting and subsequent long-term exposure to natural weathering, have adequate properties to offer high-strength structural joints which remain durable in all types of exposure conditions. For example, the ability of many formulated paints for aluminum to directly resist hot condensing humidity conditions, such as those used to evaluate structural bonded aluminum joints, can be suspect, since various degrees of paint degradation may be observed with long exposure conditions. However, we should keep in mind that the PAA-treated aluminum surfaces which cannot protect against simple corrosive exposures can become quite resistant to such conditions when overlaid with a good structural adhesive. Similarly, an adhesive over a paint-primed surface may protect against the degradation that severe humidity conditions would produce with the paint surface fully exposed.

There is a possible danger that needs mentioning which relates to the fact that adhesive or solvents present with the adhesive could chemically degrade the structural integrity of underlying paint film and adhesion to its underlying oxide. This has proven to be less a problem in recent years because of the increased use of heat-cured, highly thermoset paints which have much higher resistance to polymers and solvents. The development of such coatings for aluminum was a natural consequence of the continual search for higher and higher weathering resistance and better abrasion resistance.

The author has discussed both the generation and properties of special conversion coatings on aluminum which could contribute an interface with structural adhesives that could produce joints which were virtually unaffected by the corrosive seacoast atmosphere after 8-yr exposures. These same amorphous chromate conversion coatings have been the basic pretreatment for preparing aluminum alloy surfaces for painting in recent years and, hence, we should expect the adhesion of the paint to the underlying oxide to have distinct potential for long-term durability. Thus, we should anticipate that the paint/oxide interface would not constitute any weak boundary layer in the overall joint formed with the paint used as a surface primer.

For general bonding applications, the use of prepainted aluminum sheet for joining to itself or other adherend surfaces should circumvent the need to specially prepare the aluminum surface for adhesive application, since this has already been accomplished by the aluminum supplier. As mentioned many times earlier, however, there is still a distinct need to actually service test any combination of primer or adhesive for which direct experimental data on its durability potential is not available.

Another problem observed by the author in the real world of manufacturing has been the lack of recognition of poor wetting conditions on the paint surface. This can especially occur where 100% solids adhesives are used, or the adhesive selected has been dispersed in water. A water-wettability test (observing the condition of flowing water over the surface) needs to be conducted by any manufacturer with any new shipment of painted sheet received. This is especially of concern because the paint manufacturer may make his modifications based on achieving improvement in some other property such as the paint's ability to undergo forming operations without cracking or crazing. In this manner, organic flexibilizing agents (plasticizers) may have been added without the knowledge of the aluminum manufacturer or his customer. As long as an existing specification is being passed by the paint formulator, he may not deem it necessary to advise of new chemical additive changes which further improve performance in that area. As a result, the adhesive bonder may continue to use the same bonding procedure with the result that actual field delaminations have to occur before any question arises, with subsequent testing showing some change must have been made in the paint formulation.

Minford (774) has published some data that relate both to the joint strength potential of epoxy-bonded prepainted aluminum sheet joints and their long-term durability potential in accelerated weathering and natural atmospheric conditions. One such study employed epoxy-urea or epoxy-phenolic heat-curing, thermoset paints over 6061-T6 surfaces where the paint was directly sprayed over vapor-degreased–only sheet. Even without the conversion-coating preparation usually employed under commercially painted aluminum sheet products, the durability response was outstanding. (More data will be presented when aluminum joint durabilities are discussed in Chapter 10.)

VIII. SPECIALIZED TREATMENTS OF ALUMINUM FOIL

The use of aluminum foil has become firmly established as one of the major materials for use in establishing a vapor barrier characteristic to various structures. The relative effectiveness of this sealing is, of course, a function of the probability that no significant pinholes will exist in the sheathing surface of foil. Technically, aluminum foil sheet is considered to be less than 6 mil in thickness, although it can be rolled as thin as 0.17 mil under highly controlled conditions. An important characteristic of aluminum foil is its high covering area per unit of weight. For example, 0.35-mil thick foil has a covering area of 29,300 in^2 (203 ft^2) per pound. At a price of 84 cents/lb, a foil coverage in this gauge could cost as little as 3 cents/1000 in^2.

As mentioned above, the ability of foil to furnish a very low permeability to water vapor and gases is contingent on its fabricated thickness. Extremely small pinholes are really unavoidable in thicknesses less than 1 mil; however, even when a gauge of 0.35 mil has been used in packaging applications, the water vapor transmission rates (WVTR) are negligible when employed with coatings or laminated structures. Foils in the 4-mil and above range of thicknesses may usually be considered to be virtually pinhole free. It was for this reason that the author once became involved in the bonding of 4- to 8-mil aluminum foil as the weathering surface over plywood for roofing on low-cost housing units. Demonstration was made that 6- to 8-mil foil could be prepainted and laminated directly to exterior-grade plywood with a contact cement.

A. Cleaning Procedures

Through the choice of a specific chemical and/or mechanical treatment, the surface of aluminum foil can be rendered amenable to a wide variety of bonding and laminating conditions. Obviously, some form of naturally occurring oxide film will have formed on foil as it is being rolled down to gauge during fabrication. This oxide constitution will be some function of the humidity and rolling temperature conditions and would afford adequate protection of the metal for many applications (especially for inside weathering service). Since foils are generally higher-purity aluminum compositions than the hot-rolled and heat-treated alloys used for structural applications, there is less need to remove the oxide film, as fabricated, and replace it with a more compatible form of specialized form of oxide.

A commonly employed foil is 1100 alloy (commercial purity grade of aluminum), which is approximately 99% aluminum. The oxide on this surface as fabricated can be shown to be a stable film that is amphoteric in nature and protective to the underlying metal over a pH range from about 4.5 to 8.5. It is also observed, however, that at the higher levels of acidity or alkalinity, pH alone may not be a reliable criterion of the corrosivity of various chemicals. For example, glacial acetic acid, concentrated nitric acid, sodium disilicate, and ammonium hydroxide all have relatively little effect on commercially pure aluminum, and nitric acid and sodium disilicate can even have valuable film-forming characteristics. (For more details about the cleaning and pretreating of aluminum, see Ref. 856.)

1. SOLVENT CLEANING AND DEGREASING

The remarks about solvent cleaning and degreasing of aluminum alloy surfaces reviewed earlier in this chapter equally apply to foil surfaces. The presence of rolling oil residuals on foil surfaces will obviously be of greater criticality where a high degree of surface water wettability is demanded by the adhesive of choice. One of the largest volume markets for foil continues to be the various laminated products produced by adhering foil to various papers and other wood-reconstituted material surfaces. The economics often dictates the use of low-cost, water-solvent–type adhesives whose ability to wet any contaminated surface is highly suspect. Thus, if only a solvent wipe or degreasing operation is to be considered with a water-solvent–type adhesive, it is mandatory to conduct water-wettability examinations (especially with new shipments) to check on the efficiency of the cleaning process.

2. EMULSION CLEANING

Continuous feeding of webs of foil through emulsion cleaning tanks followed by thorough water rinsing and drying is widely used as an effective yet lower-cost cleaning procedure. Once again, the use of an effective monitoring system for assuring good water wettability of the adhesive on the foil surface must be installed. The removal of soils which are resistant to removal by simple solvent actions can often be achieved by such relatively economical cleaning procedures.

3. ALKALINE CLEANING

Sometimes the degree and constitution of soiling present on "as received" foil will be resistant even to good emulsion cleaners and then necessitates the use of a hot alkaline cleaning procedure. It is often critical in laminating foil for special products to impart some form of controlled etching of the aluminum surface. Inhibited alkaline cleaners can be one form of chemical treatment that falls into this category.

4. ACID CLEANING

Actually, acid cleaners are not commonly used where the primary requirement is soil removal, as they are less effective for this purpose than most emulsion, alkaline, or even strong organic solvent systems. Rather, it is the necessity of removing oxides and smutting that largely influences one to choose acidic treatments. From earlier descriptions of the special surface morphology produced by the Alcoa A3 and FPL acidic deoxidizers on aluminum surfaces, we might suppose that acidic cleaners would not need to be used to achieve sufficient bond strengths where aluminum foil adherends are being used. Even with the bonding of 4- to 8-mil foil to plywood mentioned earlier in this section, the need to use an acidic etchant for bonding with an elastomer-phenolic contact cement would not be justified.

5. ULTRASONIC CLEANING

When sound vibrations above 20,000 cycles per second are transmitted through certain liquids, microscopic bubbles rapidly form and collapse, thus creating a scrubbing action. This ultrasonic energy has been used successfully in both solvent and alkaline cleaning baths to increase the efficiency and reduce the cleaning time. There is some significant cost penalty owing to the need for an ultrasonic generator and transducer combination to affect the noted response. Thus, the use of ultrasonic energy is normally restricted to situations in which it eliminates manual operations or where the cleanliness standards are very high.

6. ELECTROSTATIC TREATMENTS

Stillwater and Butcher (1162) have described a commercial electrostatic treatment for aluminum foil which can improve the adhesion of a variety of bonding adhesives. There does not seem to be any evidence that such a treatment actually alters the naturally occurring oxide film or the amount or composition of the residual lubricant on the surface of "as rolled" foil. Rather, it is postulated that the treatment does have a slight oxidative effect on organic matter present. This is somewhat confirmed by the fact that the treatment diminishes rapidly with time, indicating a shallow penetration of the treatment. Such a polar layer may be rapidly absorbed into the mobile underlyfing layer of residual lubricant. Obviously, if the adhesive of choice works effectively with such a process, the above variety of treatments would be unnecessary.

7. CORONA DISCHARGE TREATMENT

Sonkin (1163) has written a technical article attempting to describe the situations in which corona discharge might be employed as a surface pretreatment on a variety of surfaces. The use of this treatment has similarly been employed for processing foil for subsequent lamination. Cram and Bibee (1164) have provided a general description of CDT equipment and processes. The primary effect of an air-type corona (which is most commonly used) is to introduce chemically bound oxygen into the adherend surface (1165–1167). For such pretreating of polyethylene, Kadash and Seefried (1168) found a direct relationship between the amount of surface oxygen introduced and the wettability of the surface as measured by the contact angle method.

IX. MODIFICATION OF OTHER ADHERENDS FOR SUBSEQUENT BONDING TO ALUMINUM

The concept that the solid surfaces of polymers needed to be modified to regulate their adhesive properties was discussed in the early 1950s by a variety of Russian investigators like Voyutskii (1169–1171). More recently, Korolev (1173) has discussed the relative polarities of polymers in contact as regards enhancing the bondability of specially hard to bond polymers like polyethylene, polypropylene, silicone rubbers, Teflon, and Kel-F. Having enhanced the bondability of such low polarity surfaces, Korolev points out that it is then possible to combine them with aluminum adherends.

A number of investigations in the 1980s have recognized the importance of surface modification of polymers to enhance their opportunity to develop specific chemical adhesion to another material. For example, Schreiber et al. (1174) have described the microwave plasma treatment of a polymer-filled surface to modify the acid-base balance of the surface. Samal et al. (1175) have introduced reactive groups like isocyanate, acyl chloride, and azo configurations onto a polymer surface to modify its adhesion properties. Keller et al. (1176) have provided a means of introducing reactive amine sites onto a Kelvar surface, whereas Yasuda (1177) has attempted to reduce the overall sensitivity of polymer surfaces to water by means of plasma polymerization. The development of methods to characterize these surface alterations has been the focus of discussion by Dilks and VanLaeken (1178). Oldfield and Symes (1179) have presented an excellent and comprehensive summary of the various means of surface modification of elastomers for bonding in 1983. The same authors in 1991 presented the most comprehensive summary to date of the state of the art for bonding all classes of elastomers and rubbers (1180). In 1990, Kowalski (1181) published his investigation of ion bombardment as a means of modifying the surface morphology of solids.

A. Plasma Oxidation of Polymers

The oxidative degradation of polymers effected by means of electrical discharges or plasmas (radio-frequency, microwave, corona) has been widely studied because of its importance in a number of technologies, including promotion of improved adhesive bonding. Selective control of the plasma operating parameters allows beneficial modification of the polymer surface leaving the bulk properties unchanged (1182–1183). The improvement of adhesive bonding by such techniques has been reviewed by Hudis (1185) and Blythe et al. (1184). Other investigators in the field have included Nowlin and Foss-Smith (1186), Hiraoka (1187) and Taylor et al. (1188). The rapid etching of polymers in oxygen plasmas operated at high-power levels combined with the much lower etching rates on inorganic oxide surfaces has been the basis for photoresist stripping applications. The last two investigators listed above have used the differences in oxygen plasma etching rates for different polymers or polymer mixtures for accomplishing dry development of photoresist. Such plasma modification of surfaces has major advantages in being clean, selective, and solvent-free processes. Clark and Dilks have studied both oxygen-containing plasmas (1182) and inert gas plasmas (1183,1189), demonstrating the change is mainly affected in the outermost few monolayers of the polymer. Further, it has been demonstrated that an appropriate analytical technique for studying such changes is XPS (ESCA). Dilks and VanLaeken (1189) reported such an investigation in a 1981 symposium with later publication in 1983.

The surface oxidation by an oxygen plasma is more easily understood even though the plasma is highly complex. Bell (1191) has postulated the presence of both positive and negative ions, atoms, ozone, metastables of atomic and molecular oxygen, electrons, and a broad electromagnetic spectrum. However, as far as the polymer surface regions are concerned, the greatest reactivity is shown by the oxygen atoms resulting in a wide range of free radicals that can react with the molecular oxygen to produce a variety of oxygen-containing species at the polymer surface (1184). The low activation energies involved with atomic oxygen leads to a uniform overall rate of reaction that considerably simplifies the analysis of data.

B. Surface Treatment of Polyolefins

Sharpe et al. (1192) looked at the means of developing strong bonds between polyethylene and epoxy adherends in the early 1960s. Devine et al. (1193) also conducted an early investigation of the comparative abilities of six different pretreatments for bonding polyethylene (PE) using epoxy, polyester, and nitrile-phenolic bonding candidates. They concluded the effectiveness of the pretreatment was in descending qualtitative order as follows: flame treat > acid treatment followed by acetone dry > acid treatment followed by air dry > acid treatment followed by oven dry at 194°F (90°C) > sanding. Bikerman (1194) examined the effects of various impurities on the PE with regard to adhesion to PE, whereas Schonhorn and Hansen (1195) agreed with Bikerman that various low molecular weight fractions needed to be removed from the PE surface to obtain effective bonding. Schonhorn and Hansen were not, however, able to isolate and identify these weak boundary layers.

Hansen et al. (1196) were able to develop a treatment for PE, polytetrafluoroethylene (PTFE), nylon, and polyvinyl fluoride (PVF), which permitted effective bonding to these adherends by a process which caused cross-linking on the adherend surface by use of an activated species of inert gas. This process was given the name CASING by the investigators. Additional reports on the subject of eradicating these weak boundary layer conditions on these normally difficult to bond polyolefin adherends were issued by Schonhorn and Ryan (1197,1198). The bondability of polypropylene did not seem to be enhanced by the rare gases used in the CASING process; however, Schonhorn et al. (1199) later showed good polypropylene joint strength with aluminum where the gases employed were oxygen or nitrous oxide. The cross-linked surface seemed to extend to about 300 Å in a layer whose thickness seemed relatively independent of exposure time. This theory seemed to ignore the fact that other polymers like polysulfide and RTV silicones with up to 5% plasticizers and/or low molecular weight polymer fractions can bond very well as shown by Black et al. (1200) and DeLollis and Montoya (1201).

The use of plasma pretreatment on polyolefins to make them bondable has been of particular interest to aluminum fabricators who may wish to manufacture aluminum/polyolefin laminates or joints. Methods regarding such plasma treating can be reviewed in publications by Wu (357), Hudis (1185), and Boenig (1202). Wu (357), Hsieh and Chen (1203) and Yasuda et al. (1204) have pointed out that such treatment loses its effectiveness with time, indicating that there is a need for more knowledge about polymer surfaces and interface dynamics (see also Ref. 1205). Morra et al. (1206) have attempted to undertake such a study using both PE and PP surfaces. This was predicted, however, by McGinniss' (1207) report about the different mechanistic responses of these two materials to a high-energy density media; PE undergoing cross-linking, whereas PP showed instead a chain scission response to the plasma stimulation.

Oxygen corona pretreatment of polypropylene as studied by Kim and Goring (1208) produced a great enhancement of adhesion which was apparently based on the development of carbonyl functionality on the surface. They also postulated a mechanism for enhancement of autoadhesion for corona-treated polyethylene (1209). Owens (1210), studying the same subject, postulated a mechanism based on hydrogen bonding. Peroxides formed on the surface decompose to provide ketones which can later enolize. Hydrogen bonds can then be formed between keto groups on one side of the autoadhesive bond and enol groups on the other side. XPS analysis combined with surface chemical derivatization was utilized by Briggs and co-workers (1184,1211–1213) in their investigations.

C. Pretreatment of Other Adherends for Bonding to Aluminum

While it is not the intent of this book to consider the specifics of pretreating other adherend materials than aluminum, the author believes some reference to some general sources on the treating of some other adherends which could be involved in dissimilar joining to aluminum might be of interest. In this regard, Goldman et al. (1214) have reviewed how to use fluoroplastics as actual bonding agents. DeLollis and Montoya (1215) have considered the bondability of RTV silicone rubbers. The best overall considerations of the pretreatment of other important adherends are those found in the 1991 *Treatise on Adhesion and Adhesives* (1216). River et al. (1217) have reviewed the experience of the U.S. Forest Products Laboratory records for best treatments for bonding all woods and wood-derivative materials as adherends. Symes and Oldfield (1180) have produced a similar work involving all prominant treatments of elastomers and rubbers from their vantage point at the Australian Defense Science and Technology Materials Research Laboratory. The surface treating of the huge assembly of plastic adherends available in today's marketplace has been detailed by Pocius et al. (1218) at the 3M's Research Laboratory of the Adhesives and Sealers Division. While Shaffer et al. (1219) have summarized the available literature information and their background experience on titanium at the Martin Marietta Research Laboratories in Baltimore, Maryland.

X. ADDENDUM OF ADDITIONAL REFERENCES

A. On Matters of Adhesion to Polymer Surfaces

Acid-base interactions in polymer adhesion (348)
Role of molecular and electrostatic forces in the adhesion of polymers (270)
Adsorption and contact angle studies on polypropylene and polycarbonate (1220)
Adsorption on modified silicone surfaces (1221)
Physicochemical surface properties of fiber-forming polymers (1222)
Surface characterization of polymers by inverse gas chromatography (1223)
Autoadhesion of etched polyethylene films (1224)
Mild direct fluorination of polymers studied by XPS (1225)
Chemical characterization of surface-activated polymer films using ESCA (1226)
Modification, degradation, and synthesis of polymer surfaces studied by ESCA (1227)
Localized Auger states in polyethylene (1228)
Wettability of polymer surfaces (1229)

Novel methods of studying polymer surfaces by employing contact angle gonioimetry (1230)
Surface thermodynamics of liquid polymers (1231)
Surface free energy of plasma-deposited thin polymer films (1232)
Interfacial interactions between polymers and other materials and their effect on bond durability (1233)
Chemical interactions at polymer-metal interfaces: XPA studies (1234)
Analysis of metal-polymer boundaries using ultrasonic interface waves (1235)
Adhesion of encapsulants (1236)
Alumina ceramic surface cleaning for optimum bonding of resin encapsulant (1237)

5

Selection of an Adhesive

I. THEORETICAL CONSIDERATIONS

A. Molecular Contact at the Interface

When selecting an adhesive, we must remember that maximum adhesive joint service potential is usually more important than the absolute strength. It was pointed out in the discussion on adhesion theory that this initial strength relates to the degree of physical and chemical interaction between adhesive and adherend at the interface. Maximum adhesive performance, however, is only secured when a number of additional factors are also working in concert with good adhesion. Some factors are gross joint geometry, topography of the interface, chemical nature of the materials present in the total joint, mechanical responses of the solids present, viscoelastic phases involved, strain rates, strain geometry, and finally the temperature and environmental exposure conditions in service. Obviously, the properties of the selected adhesive can have relevance to most or all of these variables.

We can refer to that ideal concept of true intimate intermolecular contact between the adherend and adhesive with a maximum effect of London or dispersion forces acting. However, in practice, adhesion will always be due to a limited interfacial contact. This assures that the practical joint strength and long-term serviceability of any joint will be less than expected based on any mathematical calculations. Accordingly, we find the calculations of Kemball (1238), Taylor and Rutzler (1239), Fowkes (1240), and Girifalco and Good (120) all showed higher adhesion force potential than experimentally measured. With any adhesive that we might choose, we are attempting to secure an increase in the number of effective molecular contacts with the adherend surface, as compared with other adhesives that we might have alternatively chosen for the same application.

B. Wetting at the Interface

An important aspect of achieving more effective contacts is the degree to which the adhesive can effectively wet the adherend. We have referred earlier to speculation about this matter as developed by investigators like Zisman (275,319,386–388), Fowkes (344–348), Good and Girifalco (120,390), Good (126,391–395), Dettre and Johnson (399,400), and Sharpe and Schonhorn (295).

Schneberger visualized the process as sort of "electron cloud mixing" between adhesive and adherend (see Fig. 2). Physically, we view the process of wetting as a uniform undisturbed flow of adhesive over the adherend surface. If water will flow over an aluminum surface without a break at any specific location, we can anticipate some measure of acceptable bonding can be achieved providing we have made a wise choice in adhesive selection. Thus, we should select a surface pretreatment to create a condition where the aluminum is "water wettable" and bondable. In effect, what we are trying to create is an aluminum surface which has a high affinity of reaction with polar chemical entities, of which water is the most common example. At the same time, we must acknowledge that if this affinity is high, then there must also be a competition between the polar sites on any adhesive and water in the surface environment. Because aluminum oxide is one of high polarity, we must anticipate that any adhesive we select with a corresponding high polar chemical structure will generate bonds of desirable high initial strength. Also, this combination should be quite resistant to service failure. In contrast, adhesive combinations which have low polarity should be less ideal adhesive candidates for adhering to aluminum surfaces unless special situations are developed in the manufacturing process to get around this problem. It will be shown later that bonded aluminum joints have been fabricated with some low-polarity organic polymers like polyethylene; however, this occurs through special fabrication procedures rather than natural affinity between polar sites. Similarly, it has been possible to affect high-strength and moderately durable joints between organic polymers and oil-coated aluminum surfaces. Once again, however, special adhesive compounding and heat-curing of the adhesive is involved to circumvent the deleterious effect on wetting normally predicted by the presence of such surface contaminants.

From a thermodynamic point of view, the condition for wetting will only occur when the free-energy change for producing the liquid/solid interface is negative, as compared to the free-energy changes for loss of the solid/air and liquid/air interfaces. To achieve this, we must be willing to spend significant time, effort, and expense in cleaning; and otherwise modify the aluminum surface so better wetting can occur.

II. ADHESIVE MATERIALS AND PROPERTIES

A. General Considerations

In preceding sections, we have given specific consideration to various methods of specially preparing aluminum surfaces to receive an adhesive, and now we need to devote similar attention to understanding how the adhesive relates. Discussions about chemical and physical property values in adhesives which relate to obtaining good adhesion to adherend surfaces can be found in publications by Kaelble (122), Zisman (1241), and Lee (1242). For structural adhesives that must bond to aluminum and other metals, there are important application, curing, dynamic flow, mechanical, and environmental resistance properties that should be present for best bonding results.

Kreibich and Batzer (1243) showed considerable variation in glass transition temperature (Tg) with varying chemical structures. Further, shift of the Tg to higher temperatures as the degree of reaction and cross-linking density increases has been reported in (1244). Lin (1245), in 1990, discussed the effect of entanglement and molecular weight on polymer glass transition temperature. Moreno-Villa Lobos et al. (1246) discussed properties of adhesives as relating to their application in 1986 and Wallach (1247) the general subject of bonding metals with adhesives in the same year. When the exotherm for a curing adhesive is measured as a function of time, in the isothermal mode, the extent of the cure can be calculated. Daniels (1248) has shown the general improvement in both mechanical joint and environmental performance that can derive from higher degrees of curing.

Another quite important structural adhesive property involved in creating good bonding is the rheological behavior of a typical thermoset adhesive. Macosko (1249) has shown how the viscosity of the adhesive can be obtained at any temperature or time at a specific heating rate before gelation occurs. A graph of the viscosity at constant temperature with time is one measure of the stability of the adhesive. When adhesives are used to bond primary metal structures, the impact resistance is important for good performance; because while structural adhesives like the epoxies can be thermally rather durable, they can be quite brittle unless specially modified. The papers of Forster (1250),

Kreibich et al. (1251), Kinloch (1252), Yee and Pearson (1253,1254), and Janardhan et al. (1255) are all pertinent to this subject in the period from 1984 to 1990.

B. Classification of Adhesive Materials

The task of classifying adhesives is complicated by the fact that most proprietary formulations consist of more than one chemical type of material. Claret (1256) reminds us that when several adhesive-like substances are combined into a heterogeneous mixture, the combination must first have stable properties to be a successful adhesive. McGuire (1257) has used the broadest possible classification for adhesive materials as being either of a natural or synthetically constituted origin. Each individual basic material will generally have both advantages and limitations as an adhesive material which need to be recognized by the formulator.

Historically, the earliest known adhesive materials in the B.C. period were lime, natural resins, gums, and bitumen (asphaltic materials). For the next 1000 yr into the A.D. period, only animal protein and starch-based and cellulosic gums were added to the list.

The main adhesives from the natural animal proteins were derived from blood, fish skin, animal hides, bones, connective tissue, and casein from milk. In more recent times, adhesives have also been derived from the natural plant protein in soybeans. All are forms of polyamino acids, so there is some polarity present which theoretically might enhance bondability to polar oxide. They have, however, lacked the level of cohesive strength necessary to structurally join rigid adherends like aluminum. They also have a low degree of flexibility, which is quite a negative factor when bonding rigid adherends. Accordingly, their major use has been for the joining of certain wool-related products; but even here, more modern synthetically constituted materials may now be used. The major use for starch-based adhesives has been the binding of wood-related products, especially papers and paper boards.

The limited strength properties of the animal-origin materials usually precludes their use for general bonding of aluminum unless they are modified with the more modern polymers of synthetic chemistry. An outstanding example that gets involved in some bonding of aluminum is the large-volume family of proprietary casein-neoprene–type cements. The synthetic elastomers and thermoplastic synthetic resins mixed with a variety of thermosetting synthetic resins have made the development of truly structural adhesives possible for use in modern manufacturing. The polar chemical sites in many of these formulations has, in turn, made them amenable for bonding aluminum oxide surfaces. The most common members of these groups are neoprene and acrylonitrile elastomers; acrylic, nylon, or vinyl thermoplastic resins; and epoxy, phenolic, polyurethane, and silicone thermosets. (Refs. 1258–1265 provide further reading about the variety of commercial adhesives in todays marketplace.)

It will be the purpose of this chapter to discuss the full range of adhesive types that can be used in manufacturing aluminum-bonded structures. Included are natural rubber and its derivatives (chlorinated, cyclized, and rubber hydrochloride), GRS (SBR) rubber, reclaimed rubber, neoprene (chloroprene) rubber, nitrile rubber, polysulfide rubber, silicone rubber, polyvinyl chloride, acrylic and modified acrylic resins, phenolic resins, vinyl-phenolic resin combinations, nitrile-phenolic resin combinations, epoxy resins, elastomer modified epoxies, nylon-epoxy combinations, other modified epoxy combinations, cyanoacrylate adhesives, polyurethane adhesives, anaerobic adhesives, hot-melts, and pressure-sensitive adhesives. The relative properties of these various adhesive classes has been summarized in Table 33.

Recent 1990 reviews on various adhesive classifications are offered in the ASM *Engineered Materials Handbook by Gauthier* (1266–1288).

C. Adsorption Properties of Polymers and Adhesives

For a complete understanding of the interrelationships between polymers and adhesives, we need to study the ultimate-strength properties of solids in general and polymers in particular. Although obviously beyond the scope of this work, a list of references on the subject can be suggested (1289–1298).

Table 33 Relative Properties of Different Adhesive Families

Adhesive type	Shear strength[a]	Peel strength	Flexibility	Outdoor resistance	Resistance to temperatures: High	Resistance to temperatures: Low
Epoxy (high strength, flexible)	A	A–B	C	B	C	A
Epoxy (high strength, semirigid)	A	D	D	A	A	A–B
Epoxy (multicomponent)	B	C–D	D	B–C	C	B
Phenolic (thermoplastic-modified)	A	B–C	C	A	B	A
Phenolic (elastomeric-modified)	B	B	B	B	A	B
Natural rubber	D	C	A	C	C	B
Chlorinated rubber	D	C	B	C	B	B
Cyclized rubber	D	C	B	C	D	C
Rubber hydrochloride	D	C	A	C	C	C
GRS (SBR) rubber	D	C	B	B	C	C
Neoprene rubber	C	B	B	A	C	B
Nitrile rubber	C	B	B	A	B	B
Butyl rubber	D	C	B	B	D	C
Polysulfide rubber	D	C	B	A	D	B
Silicone rubber	C	A	A	A	A	A
Reclaimed rubber	C	C	B	C	C	C
Polyvinyl acetate	B	C	C	D	C	C
Polyvinyl chloride	C	B	B	C	D	C
Acrylic	C	C	C	C	C	B
Hot-melt	C–D	B–D	B–D	B–D	A–D	C–D

	Creep at high temperature	General chemical characteristics: Water	Oil	Salt spray	Fuels	Solvents
Epoxy (high strength, flexible)	A	B	A	B	A	A
Epoxy (high strength, semirigid)	A	A	A	A	A	A
Epoxy (multicomponent)	A	B	A	B	B	A
Phenolic (thermoplastic-modified)	A	A	A	A	A	A
Phenolic (elastomeric-modified)	B	B	A	B	B	B
Natural rubber	D	B	D	C	D	D
Chlorinated rubber	C	B	C	C	C	D
Cyclized rubber	D	B	D	C	D	D
Rubber hydrochloride	C	B	C	C	C	C
GRS (SBR) rubber	C	B	D	C	D	D
Neoprene rubber	C	A	B	B	B	B
Nitrile rubber	C	A	A	B	A	B
Butyl rubber	D	B	D	C	B	D
Polysulfide rubber	D	A	B	B	A	B
Silicone rubber	—	A	C	A	C	C
Reclaimed rubber	C	B	D	C	D	D
Polyvinyl acetate	—	C	B	C	D	C
Polyvinyl chloride	D	C	C	C	C	C
Acrylic	C	C	B	C	C	D
Hot-melt	B–D	B–C	C–D	D	D	C–D

[a]A—very good, B—good, C—fair, D—poor.

Source: Table assembled by Minford for *Adhesive Bonding Alcoa Aluminum*. Copyright 1967 and published by Aluminum Company of America.

We also could benefit from studying the considerable body of knowledge developed since the early 1950s on the nature of the adsorption of polymer molecules on adherend surfaces. Stromberg (1297) summarized much of this information in the *Treatise on Adhesion and Adhesives* series. It seems that the degree of adsorption on aluminum adherends of polymers must be intimately related to the durability of aluminum joints. Equal importance must be attached to the number and strength of polymer attachments and availability of other portions of the polymer chain for interaction with nonattached molecules. Just as no single theory explains all the observed aspects of adhesion, so no single theory of polymer adsorption has been in agreement with all observations. Typically, each of the different theories about adsorption is in agreement with some of the experimental data, but in disagreement with other apparently valid observations.

Three different approaches have been set forth. In the investigations of Frisch and Simha (1299–1301), Simha et al. (1302,1303), and Frisch (1304), a quantitative equilibrium theory of polymer adsorption was developed. This idea of adsorption has been referred to as the diffusion-equation approach. Gilliland and Gutoff (1305) treated the adsorption of polymers from dilute solution at the theta temperature by equating partial free energies of adsorbed and dissolved molecules, and this approach has consequently been referred to as the thermodynamic approach. In attempting to overcome what was perceived as being a fundamental weakness in the Frisch et al. diffusion equation approach, Silberberg (1306,1307) developed what has become widely known as direct combinatorial evaluation of the partition-function theory. The weakness he recognized was in determining the shape of the molecule by assuming zero attractive energy.

A useful tool recently introduced for studying molecular adhesion to metal oxides has been inelastic electron tunneling spectroscopy (IETS) (214). For example, it has been possible to develop IETS spectra of the multimolecular components of an epoxy adhesive system adsorbed on aluminum oxide.

D. Rheological Aspects of Polymer Adhesion

Kaelble (1308) published an extensive review on the subject in I of the *Treatise on Adhesion and Adhesives* series. The molecular properties of polymers that control the rheological and adsorption responses can be listed as (1) chemical composition (monomer units, side groups, secondary bonding forces, and cohesive energy density, (2) molecular structure (tacticity, main-chain symmetry, molecular weight distribution, chain branching, entanglements, and cross-linking, (3) molecular free volume (degree of crystallity, temperature, pressure and time). Any complete discussion of polymer adhesion must also consider the phenomena of bonding properties of the stable bond and characteristics of bond fracture. It does not seem unexpected then to find that reported data for adhesive joints does not conform; and is not exactly predicted even when we know a lot about the properties of the bulk adhesive. Dwight et al. (1309) provided literature examples. However, some examples exist in the literature where good correlation has been reported between in situ and bulk properties; as in the report of Dolev and Ishai (1310) using FM-73 adhesive and aluminum adherends.

A large number of recent investigations have been undertaken to determine the range of property changes that can be induced in polymers alone and with adherends. Ishida (1311), in 1987, studied the quantitative analysis of polymer surfaces with Fourier transform intrared spectroscopy (FT-IR). Rakhmatkariev et al. (1312) reviewed infrared spectroscopic investigations of adsorption of polyacrylamide polymer on oxides in 1989. Brockmann (1313) reviewed his concept of adhesion of polymers to metal oxides in 1987. Kovacevic et al. (1314) discussed property changes in polymers induced by aging conditions in 1990, whereas Shmorhun et al. (1315) used fluorescence spectroscopy and mechanical stress relaxation data to study free volume changes in epoxies affected by aging. Zosel (1316) studied factors influencing polymer failures based on deformation behavior.

Tang et al. (1317) studied the effects of cure pressure on polymer flow, void formation, and mechanical properties. In 1988, Sancaktar and Jozavi (1318) studied various linking cure processes' effect on adhesive bulk strength using differential thermal analysis (DTA). Ennis (1319) commented on this work resulting in a reply by these authors (1320). In 1989, Sancaktar and Jozavi (1321) continued their study on cure effects by measuring changes in the relaxation behavior of thermosetting adhesives, whereas Mijovic and Lee (1322) attempted to make comparisons between various chemorheological models for explaining thermoset curing in the same year. Finally, Pang and Gillham (1323), in 1990, examined the competition between curing rate and the rate of thermal degradation that can occur in high glass transition temperature epoxy systems.

Another subject of recent study has been the effect of various fillers and the property changes they can affect. In 1984, Bertenev et al. (1324) made determinations of volume fraction of polymer bond on filler based on thermophysical data. Ponce et al. (1325), in 1987, used variously pigmented epoxy polymers to evaluate component interaction and property changes. Kulik and Davies (1326) produced a paper evaluating the mechanical properties of adhesives filled with metal powders, and Kulik et al. (1327) measured thermal conductivity of adhesives with fillers, both in 1989. Cantwell et al. (1328) in 1990 reported on the processes of both deformation and fracture using silica-filled epoxy polymer. Somewhat related is the paper by Prassoamakin (1329) correlating the mechanical and acoustical property changes produced by the addition of plasticizing additives to epoxy polymers.

Thickness of adhesive in a bondline is also a very important contributor to how the joint behaves. Trying to relate the adhesive fracture energies and joint time-to-failure values can be reviewed in the papers of Hyland and Sidiwell (1330) or Rose (1331). In this regard, it has been possible with aluminum/epoxy joints to find a range of adhesive bondline thicknesses (0.02–0.07 cm) where Mode I fracture energies could be optimized providing certain testing rates and temperatures were also controlled (1332–1334). It might also be worthy of mention to note that there can be a significant modification of polymer properties at the "interphase" of the polymer induced by the presence of a filler. In addition, polymer properties can be altered at the interphase by the presence of an adherend surface (1335). Peyser (1336), however, has offered his observations which question the ability of any high-modulus surface (of which aluminum oxide would be one example) to change macromolecules at any significant distance from the interface. Thus, the outcome of a linear viscoelastic test scheme on a bonded joint assembly is not easily anticipated even where the properties of the polymeric adhesive are well documented.

III. NATURAL AND SYNTHETIC ELASTOMERS IN ADHESIVES

A. Natural Rubber and Derivatives

The first natural elastomer was natural rubber whose properties of excellent initial tack, tack retention, and outstanding flexibility were readily recognized. These long-term, tack-retention properties still make it an ideal candidate material for some pressure-sensitive adhesive formulations. We have stated that all commercial adhesives are really mixtures; and even the commercial natural rubber-based adhesives must have their own particular combinations of tackifiers, plasticizers, fillers, chemical catalysts, or accelerators plus some dispersing liquid (organic solvent or water), as shown in Table 34. The variety of rubber types used in most commercial formulations is shown in Table 35.

Table 34 Typical Components of Rubber-Based Adhesives

Component	Purpose
1. Rubber resin	1. Base resin
2. Tackifiers	2. Provide tack and increase autohesion
3. Plasticizers	3. Make adhesive more permanently soft
4. Pigments and fillers	4. Control viscosity, reduce cost, and add color
5. Solvents	5. Reduce viscosity modify green strength
6. Curing systems	6. Build heat resistance, increase shear strength
7. Metal oxides	7. Participate in cure, act as acid acceptor (neoprene adhesives)
8. Antioxidants	8. Stability in UV and environmental exposure
9. Sequestering agents	9. Sequester Cu and Mn which decrease oxidation resistance

Source: From Pocius Lecture 4 on Chemistry of Adhesives, ACS Audio-Teleconference Course on Adhesion Science.

Table 35 Types of Rubber Used in Rubber-Based Adhesives

1. Natural rubber
2. Reclaimed tire rubber
3. Polychloroprene (neoprene)
4. Butyl rubber
5. Polyisoprene
6. Acrylonitrile-butadiene rubber (nitrile rubber)
7. Styrene-butadiene rubber (SBR or GRS)

Source: from Pocius Lecture on Chemistry of Adhesives, ACS Audio-Teleconference Course on Adhesion Science.

In some aluminum-bonding applications requiring relatively low bond strength per unit area and nonstressing service conditions, a natural rubber pressure-sensitive or contact cement–type adhesive could conceivably be used. With aluminum such adhesives find their widest application in attachment bonding of nonmetallic adherends such as leather, fabrics, papers, and other rubber products.

1. CHLORINATED RUBBER

Further processing of natural rubber led to additional adhesive-formulating materials such as chlorinated natural rubber (1337). Dispersed in organic solvent, this rubber showed high resistance to acidic or alkaline service conditions. It could be used to bond a rubber-to-aluminum laminate by serving as a dried tacky primer over the aluminum surface. Another adherend rubber is pressed against the tacky-coated aluminum. Ultimate bond strength and best durability of bond will require some heat cure. A caution is in order, however, about the rather poor resistance of this bond to organic solvents in the service environment. A 1990 publication by Hace et al. (1338) reports on various structural and morphological changes that can transpire after the chlorination of rubber surfaces.

2. CYCLIZED OR ISOMERIZED RUBBER

Cyclized or isomerized rubber is produced by treating natural rubber with acids or metal salts. This creates a density increase due to 65% conversion of linear polymer to a cyclic polymer. This material, as an adhesive, can also be used to prime an aluminum surface. The prime-coated aluminum is brought into contact with a similarly primed rubber surface for applications such as preformed tank linings. When the assembly is heated, the interface can be vulcanized into a sealed joint. A relatively low service temperature of about 140°F (60°C) is, however, a distinct limitation for use of such rubbers.

3. RUBBER HYDROCHLORIDE

Rubber hydrochloride is also made from natural rubber; and like the cyclized rubber just discussed, it can be used in a two-surface prime and vulcanize bonding procedure for making rubber-coated aluminum products. This rubber derivative adhesive is virtually insoluble in organic solvents like the ethers and esters and has good water resistance as well. However, it is sensitive to aromatic and chlorinated hydrocarbons in service.

B. Synthetic Organic Rubbers for Adhesive Formulation

The other rubbers of main commercial interest for adhesives have all resulted from synthetic organic processes.

1. GRS (SBR) TYPE

A very large-volume rubber that has been produced synthetically for a number of market areas has been labeled GRS or SBR. It was actually developed first for tire manufacturing during World War II. A copolymer of butadience and styrene, this rubber proved to be superior to natural rubber in its resistance to important environmental agents such as sunlight and many organic chemicals and oils. However, a lower natural tack in GRS rubber has made it necessary to compound it with special tackifying and plasticizing chemicals for use in adhesives. Ways have been discovered to increase a number of desirable properties such as strength, heat resistance, and heat flexibility. Both water-latex emulsion and organic solvent solution forms of GRS rubber adhesives can be obtained commercially. Sometimes it is important in aluminum-to-plastic laminating to use a contact-type GRS rubber adhesive formulation that can be diluted with a milder organic solvent like naphtha or hexane. Such milder solvents (as compared to ketones or toluene) might not degrade certain plastic interfaces in aluminum/plastic joints, which often occurs with contact with the stronger organic solvents.

A practical commercial example is found in laminating aluminum to polystyrene insulation beadboard where stronger solvents would collapse and chemically degrade the polystyrene. A contact cement dispersed in naphtha or hexane, however, may be applied directly to such a foam surface. This affords the opportunity to make a contact bond directly to the aluminum surface which has been precoated with the same adhesive. The modest bond strength of the GRS-type contact cements has, however, limited their use for aluminum-to-aluminum bonds, which often demand higher-strength properties. Nevertheless, the strength of these adhesives is often greater than the cohesive strength of a second adherend; for example, a low-density polystyrene or polyurethane insulation board. The strength may also be adequate in large area structures where the stress per unit area requirements are relatively low. Where aluminum is added as a vapor barrier material, it may only be necessary for the adhesive to sustain the weight of the thin aluminum foil vapor barrier.

2. RECLAIMED RUBBER

An even cheaper form of rubber material for formulating adhesives has been made available by employment of a special reclamation process in which original rubbers, after being used for a first time, are shredded and reprocessed to form a second rubbery mass which is surprisingly similar in strength to the GRS (SBR)-type rubber cements. A general joining procedure using contact bond adhesives is offered in Table 36.

The usual method of bonding such adhesives is to apply adhesive in a thin syrup or heavy mastic consistency by whichever of spraying, roller-coating, curtain-coating, brushing, or mechanical spreader procedures may be appropriate and economical for the manufacturer's operations. Adhesive should be applied to both surfaces to be joined, although only one surface really needs to have active task. If the tackiness has been sufficient for the active surface to wet the second adherend surface (could be primed earlier), more or less continuous stringers of adhesive will be observed when separation is attempted after contact. Pressure from simple contact to as much as 100 psi (0.69 MPa) might need to be used to close the bondline. This pressure might be selected based on the amount of open time the adhesive has experienced between its application to the surface and its closing. The ambient

Table 36 Joining Procedure for Contact Bond Adhesives

1. Adhesive is applied to both adherends by brush, roll bar, spraying, or curtain coater
2. Some solvent is flashed off to produce a tacky surface (open time)
3. Adherends are mated using some pressure as from hand pressing, rotary, or platen press (''knitting''—autohesion occurs)
4. Usually bonds strong enough to handle immediately, however, additional storage time may significantly increase bond strength prior to testing or placing in service (green strength)

Source: from Pocius Lecture 4 on Chemistry of Adhesives, ACS Audio-Teleconference Course on Adhesion Science.

temperature in the assembly area can also be a factor. This relates to the rate and amount of solvent lost from the bondline owing to higher temperatures in assembly. Of course, if tack is missing from either coated adherend, then generation of new tack by spray misting with adhesive or heating the bondline can be used just before laminating the two pieces.

Bondments between two vapor barriers like aluminum to itself or another me;tal can present special problems because the solvent initially present in the bondline at closing cannot be readily dissipated. Under such conditions, the maximum strength that the adhesive can theoretically achieve cannot be realized until after some uneconomically long time. Solvent cements tend to be less used, therefore, for bonding metals-to-metals. If wood is the second adherend in an aluminum laminate, however, then solvent containing contact cements can be effectively used; since the porosity of the wood structure offers a vent to solvent trapped in the bondline.

A general method also exists for developing strength faster by accelerating the removal of solvent from precoated adherends by either blowing a stream of heated air over the surface or simply heating the air over the open bondlines as in a hot recirculating air tunnel. The timing of removing all but the last traces of solvent before closing the bondling will be the most critical aspect of such processing. The loss of too much solvent will be indicated by touching the finger to the adhesive-coated surface and failing to raise any stringers of adhesive from the surface. It is usually most practical to concentrate attention on the amount of tack registered on the surface, which will lose solvent most slowly. This will work because it is only necessary to develop tackiness on one adherend surface in order to bond effectively to a second precoated surface using the same contact cement product.

Still another procedure involves the use of sufficient overall heating that all the solvent has been effectively driven from both surfaces. At this moment, there is a thermally developed tack that still exists in both surface so that lamination can be consumated. An additional benefit that may be gained by creating a hot-bonded interface in that higher initial strengths may be achieved as soon as the bondline is cooled to ambient temperature. It is obvious that using a heat-induced tack bondment can greatly accelerate the manufacturing rate, since the bonded product returned to ambient temperature can immediately be handled, stacked, and even shipped where the final field service conditions are not too stringent.

The final general procedure that might be employed with these contact-type rubber cements is to preapply adhesive to both adherends and dry out the solvent by any convenient means and set aside for later lamination. It is characteristic of these contact cements that light misting with an appropriate solvent will renew the tacky condition that was present when the last solvent was disappearing from the adhesive as originally applied. By light misting only one of the adherend surfaces, it is possible to use the tacky condition produced to wet the dried adhesive on the second adherend surface. The use of some external pressure (as from use of hand- or hydraulically-driven pinch rolls) can then create the bondline desired. Application of this pressure would be of significant benefit regardless of whether the tack had been restored to one surface by solvent misting or produced in both of two predried adhesive-coated surfaces by hot-air or hot-pressing action.

3. *NEOPRENE RUBBER ADHESIVES*

The most important rubber base for formulating contact adhesives that can be considered for structural bonding is chloroprene (designated neoprene by the inventor, DuPont). Synthetically produced, this rubber is very closely related chemically to natural rubber. However, its higher strength, temperature resistance, and superior aging characteristics (as compared to natural, GRS, or reclaimed rubbers) makes it highly desirable for manufacturing a variety of dissimilar material structures. Obviously, in the instances of bonding metal vapor barriers, a special procedure must be developed to effect bonding, since neoprene in most of its formulations is solvent dispersed.

A negative effect for joining aluminum-to-plastic surfaces is the fact that quite strong organic solvents are usually required to disperse neoprene rubber in adhesive formulations which can seriously chemically degrade many plastic surfaces. While the interfacial area is most immediately affected, such solvents can also cause later embrittlement and special sensitivity to crack failure in the bulk plastic. It should be remembered, however, that many plastics are effectively joined to themselves or other plastics through use of adhesives made by dissolving the same or similar plastic solids; i.e., the so-called plastic dope cements.

The general procedure for bonding the contact-type cements has been discussed earlier in connection with bonding procedures used with GRS or reclaim rubber products. The presence of a thermosetting phenolic resin in neoprene contact cements does not reduce the tackiness that marks all rubber-containing formulations. Like other rubber adhesives already discussed, neoprene-phenolic cements are available in both water- or organic-solvent dispersions. A typical formula for a neoprene-phenolic contact cement is provided in Table 37.

Pletcher and Yaroch (1339) discussed the formulation and use of water-latex adhesive products in *Adhesives in Manufacturing* (1340). There is also a large market for use of neoprene rubber in a latex formulation with the natural protein casein. While the largest customers for such adhesives have been manufacturers of wood adherend assemblies, there has also been considerable use for laminating aluminum sheet to wood in applications like rear door panels for over-the-road trailers, sidewall structures for various architectural applications, and floor panels for public transit cars.

a. Choosing Between Water or Organic-Solvent Cements

It may be interesting to cite specific times for bonding aluminum surfaces. The most important consideration is whether the surface is water wettable. Bare, painted, or anodized aluminum adherends can be bonded with latex adhesives providing the surface to be bonded permits overall nonbreak flow of water over that surface. Most times an "as received" surface will not have this quality owing to deposition of various soilants due to handling, shipping, and storage conditions. The cost of additional processing to remove these surface contaminants can be high and these additional costs are often offset by economies affected by use of a water-latex adhesive formulation as compared to using an organic-solvent–dispersed cement. Sometimes the manufacturer will be most influenced by the more expensive operations of removing and disposing of organic solvents as compared to removal of water from the bondline.

An additional problem encountered by the author has been the non–water wettability of certain prepainted aluminum sheet caused by deliberate addition of plasticizers. Such plasticizers are often added by a paint supplier as a means of passing certain sheet bending specifications that have nothing to do with bonding. As a consequence, the author found it necessary at Alcoa to check the water wettability of the backside paint on each separate shipment. It will be remembered that it is generally the backside of prepainted aluminum sheet that is going to be bonded to some other material of construction. The need to concentrate some attention on the backside surface qualities was generally not considered by the paint suppliers nor by the paints applicators, whose concentration was on the chemical resistance of the frontside paint to long-term weathering and the formability properties of the paint. Assuring good water-wetting potential is particularly important where final bonding will be with a water-dispersed cement.

With bare aluminum, it should only be necessary to observe that good unrestricted flow of water is possible with the current lot of mill-finish material. Often, vapor degreasing will be necessary as an

Table 37 Typical Formula for a Neoprene-Contact Adhesive

Components	Parts
Neoprene	100
Antioxidant	1
ZnO	5
Heat reactive tertiary butyl resin adhesive	45
Solvents	600

Source: Data from Pocius Lecture 4 on Chemistry of Adhesives, ACS Audio-Teleconference Course on Adhesion Science.

additional minimum surface preparation for bonding water-dispersed cements. Sometimes emulsion cleaning will affect a favorable response when vapor degreasing is inadequate. It should be pointed out that organic-solvent-dispersed rubber cements can often overcome the problems of marginally clean sheet by absorbing or dispersing the surface contaminants. Whether this is a viable process must be proved by direct durability testing related to the field service intended.

Interior weathering service bonds can survive for long periods of service using organic-solvent-dispersed neoprene-phenolic adhesives even without the best possible surface precleaning. However, it has been the author's experience in the real world that a high degree of water wettability needs to be demonstrated for any bonds exposed to intermittent or continuous exterior weathering conditions. The author cites experience from manufacturing Alcoa Alply Building Panels using selected neoprene-phenolic formulations that demonstrated the ability of aluminum facings to bond with assurance of long-term life in exterior weathering applications where only solvent cleaning of bare, painted, or anodized facing sheets was used as a pretreatment. Initially, the extra expense of a special furnace cleaning of bare sheet was employed to remove all organic surface contamination. However, curtain-coater application of 25% solids (MEK/toluene) neoprene-phenolic adhesive gave indistinguishable durability performance to earlier panels manufactured with furnace-cleaned sheet.

b. Use of Contact Cements for Bonding Dissimilar Adherend Surfaces

The situation in which aluminum is being bonded to another adherend calls for special attention to be paid to the second adherend surface. Questions that arise include (1) whether the second surface is also a vapor barrier, (2) whether the second surface is also water wettable, (3) what is the porosity condition of the second surface when not a vapor barrier, and (4) what are the chemical characteristics of the second surface as compared with the relative inert chemical character of the aluminum oxide. If the second surface is also a vapor barrier, then the use of any contact cement (water or organic solvent dispersed) will result in an unsatisfactory curing situation as long as any solvent remains trapped in the bondline. The situation is more critical with a water-solvent-type cement, because the volatility of trapped water is much lower than for most organic solvents. In either case, clear preference must be to use a procedure which begins by removing the solvent before closing the bondline. Removing water from a thickly deposited layer of a water-contact cement can be both tricky and difficult. If the heating rate is too high, then the water will cause erupting conditions, whereas a heating rate below this point will greatly extend the time before the joint can be closed. The reactivation of either surface, when using an organic-solvent-dispersed cement, involving the same solvent must be carefully controlled if optimum bonding with immediate handling potential is to be achieved.

Unfortunately, the method previously given for determining the water wettability of aluminum adherends is not always amenable to other adherend materials to which aluminum might be bonded. For example, if water is poured over any surface with significant porosity or roughness, the pattern and continuity of water flow will be difficult to determine. However, the porosity factor can be a distinctly positive one where a low-viscosity contact cement is used. The flow of the adhesive into this permeable structure will offer opportunity for significant mechanical interlocking, even in the absence of overall surface wetting. With respect to combining aluminum-to-wood, the pores in wood also permit much more rapid loss of solvent from the bondline, accelerating cure. Even so, the desired procedure may be to prime the aluminum with the solvent-rich contact cement and drive out the solvent before combining with the wood. The ability of water or solvent to escape into the pore structure of wood is readily shown by the way a thick coating of casein-neoprene latex cement can be applied to wood and immediately closed against an aluminum facing. The facing must, however, be stacked in place and held overnight under stacking pressure. Usually, the stack can be disassembled the following day with handling strength already developed. If the aluminum surface was water wettable, the bond failure should be in the wood fibers when the aluminum facing is deliberately peeled from the wood surface.

The methods by which vapor barrier aluminum surfaces can be bonded to wood adherends with organic-solvent-type mastic contact cements are considerably more varied than for the water-solvent-type cements. Both adherends will be wetted best in the presence of significant solvent, so application of such solvent-mastics to both surfaces is desirable in order to achieve highest strength and most durable joining. Usually, if the organic or water solvent is not removed from the coated aluminum before closing the bondline, the time for significant development of strength will be too long for economical manufacturing. While the procedure of bonding aluminum-to-wood with a casein-neoprene water-solvent mastic can be achieved overnight, as mentioned earlier, the commercial lamination of these adherends with organic-solvent mastics is usually varied to achieve immediate handling strength without the use of clamping or fixturing. Removal of virtually all the organic solvent before laminating can permit substantial strength to develop as soon as the bond is made. Thus, the manufacturer of aluminum/wood laminates needs to always heat the layer of deposited mastic as soon as practical. If highest joint strength and ultimate heat strength is to be achieved, then all solvent must be absent and tack, which initiates the bonding, must be developed by heat and pressure. It is precisely at this time that the contact must be made with the second adherend regardless of what it might be.

It was mentioned earlier that aluminum/wood bonds made with a water-latex adhesive might be improved by prepriming of the aluminum adherend sheet. The opposite may be true with the organic-solvent cements. The highest initial joint strength and deepest wood fiber–tearing bonds are achieved with wood that has been primed with same adhesive. If the wood is not primed, then the adhesive which is present can rapidly bleed into the porous wood structure, leaving the bondline relatively adhesive starved. In contrast, adhesive used as a primer will tend to prefill the porous wood structure and make that surface one which can be more readily wetted by a second adhesive applied to the aluminum.

An alternative procedure used by some manufacturers is to place the aluminum and wood parts in tandem on a conveyer and pass them under a spray or curtain-coat applicator to deposit adhesive in equal thickness on each surface. The coated surfaces are then passed through a drying tunnel and pressed together after exit with final passage through a rotary press. Some problems can arise in this situation if a differential rate of drying of the same thickness of adhesive can occur on the two different surfaces. However, the faster heat-up of the metal adherend facing can be offset by a faster loss of adhesive into the porous structure of the wood surface. A better procedure may be to preprime the wood and remove the solvent by fast heating. It can then be stored for later laminating to the aluminum. In this situation, it is only necessary to consider the degree of tack on the aluminum as it exits the drying tunnel. The exit temperature should be sufficiently high to create maximum heat resistance in the cured adhesive bondline.

While neoprene elastomer has inherently higher heat resistance than natural, GRS, or reclaim rubbers, it gains significantly higher heat resistance in adhesive formulations containing phenolic resins. While sufficient joint strength for handling may immediately develop when the bondline is cooled to ambient temperature after hot-press bonding, there is an additional aging factor for gaining ultimate heat resistance. This aging can be attained faster with heat aging but this may not always be practical when dealing with large area structural building panels. Generally, 24–72 hr of ambient aging can affect this heat-resistance development providing the adhesive has previously undergone some aging before use. The author found out this fact by chance when an emergency shipment of neoprene-phenolic adhesive was forwarded to Alcoa and failed to develop adequate heat resistance in prescribed Alcoa tests. It was shown that some period of aging time was necessary for a properly formulated neoprene-phenolic cement before application and use. This fact had not become known to the formulators of such adhesives because a previous manufacturing situation had not developed where batch mixing and immediate shipment occurred, accompanied by application of a special testing procedure to delineate the lack of such heat resistance.

Aluminum bonding to solid or foam plastics can be achieved in some cases with neoprene-phenolic formulations but assurance must be given by separate chemical compatibility evaluations. Further, restricted loss of solvent from the bondline with an aluminum surface can make development of joint strength slower and less economical for manufacturing. Minford has demonstrated with a variety of plastic solvent dope cements that acceptable bonds of aluminum to some solid plastic adherend surface was not achievable. Best success (though not wholly acceptable) was achieved by applying the dope cement to the aluminum as a primer. A coat of cement to the plastic surface was then allowed to dry until only a trace of tackiness remained, whereupon lamination was affected. Such joints were especially low in shear strength with virtually no cleavage strength owing to the brittleness of the bondline. Adhesion to the aluminum was very poor when exposed to any significant weathering.

In spite of the fact that strong organic solvents (used to disperse the neoprene) can dissolve a foamed polystyrene plastic surface, these adhesives have been successfully used to laminate aluminum facings to foamed polystyrene cores (see Fig. 8 for a flowsheet of the production sequence). The success of the method is totally dependent on making the bond with no solvent permitted in the bondline. This was readily accomplished by curtain-coating the aluminum and removing all solvent while conveying through an air-drying tunnel. Tackiness induced by the heating alone served to bond the aluminum to the uncoated polystyrene cores with simultaneous pressing in a heated platen press followed by passage through a rotary press.

Foamed polyurethane slabstock served as an even more thermally efficient insulator for aluminum-faced panel production with the same neoprene-phenolic adhesive. In contrast to polystyrene foam, however, the thermoset polyurethane foam board has significant chemical resistance to these same organic solvents. In addition, it was evident that the sawed surface of polyurethane was quite friable and a precoating with adhesive was necessary to strengthen the surface to prevent a weak boundary layer effect in the bondline. A wide variety of other core materials fall in this same category of requiring application of separate adhesive to form a durable bond with aluminum facings. Included are isocyanurate board, mineral fiberboard, polyvinyl foamboard, epoxy foamboard, ceramic foamboard, polyethylene foamboard, balsa wood, plywood, hardboards, wood particleboard and paper-honeycomb. A production sequence used by the author to combine all these materials with aluminum sheet can be seen in Figure 9.

4. HIGH SOLIDS, MASTIC-TYPE RUBBER CONTACT CEMENTS

While most contact cements discussed above have been commercially formulated in the range of 10–40% solids dispersed in some solvent, there are also a number of mastic-type contact cements available in the 70–85% solids range which are used extensively in the lamination of corrugated bare or prepainted aluminum facings to various native wood or reconstituted wood materials for architectural sidewalls. These panels constitute the sidewall construction of many factory-built low-cost home units. For lower-performance applications, the lower cost GRS and reclaim rubber mastics have been employed, but for better weathering and long-term heat resistance, the neoprene-base mastics should be used. However, 15–30% organic solvent can still be left in the bondline, so that such mastics have little application for bonding aluminum to itself. Even this solvent presents some

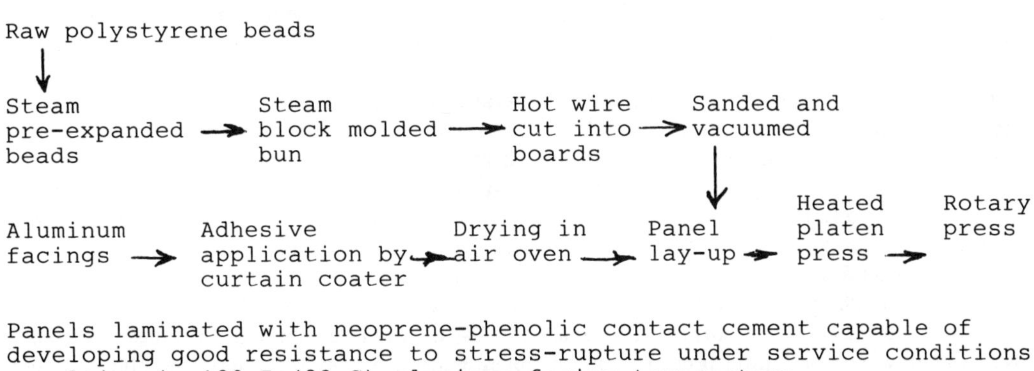

Figure 8 Flowsheet for manufacture of Alcoa Alply panels. Panels laminated with neoprene-phenolic contact cement capable of developing good resistance to stress-rupture under service conditions resulting in 180°F (82°C) aluminum-facing temperature. (From Ref. 87.)

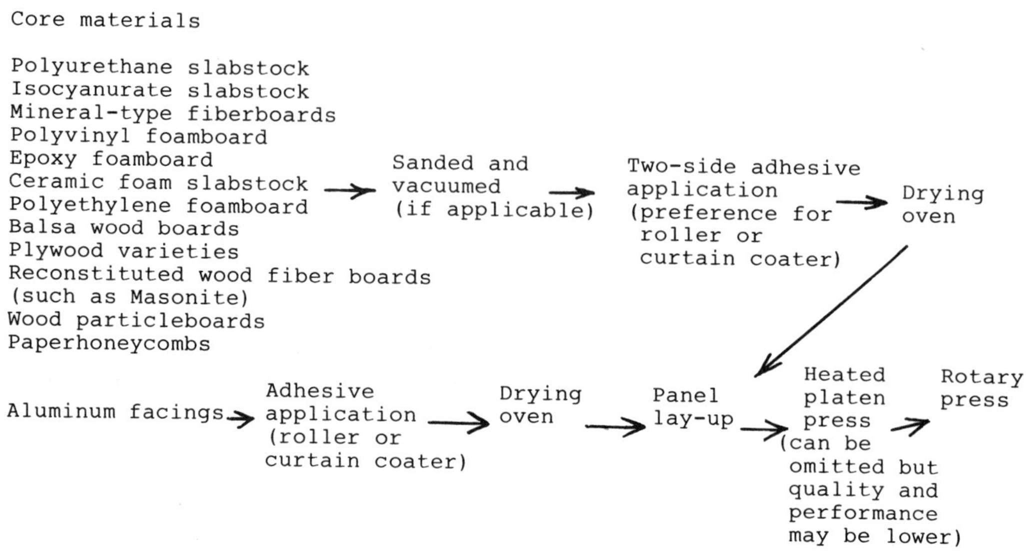

Figure 9 Fabrication of aluminum-faced sandwich panels with various insulation core materials. (From Ref. 87.)

difficulties for bonding large aluminum sheet-to-wood or wood-product panels, because of their large size and attendant difficulties associated with getting significant bondline heating. Typical production procedures have included the application of one or more heavy beads of mastic on the flat areas of the aluminum-corrugated sheet, followed by positioning of the heavy wood backup sections over the adhesive-coated flats. It is presumed that the adhesive will be squeezed uniformly by the dead load pressure of the wood backup. Sufficient volume of adhesive is applied to be sure that a fillet is formed by squeeze-out. The panels are then stacked in multiple layers to further increase the likelihood of uniform adhesive distribution. Depending upon the plant ambient temperature and humidity, the laminated panels are kept stacked until the solvent has dissipated into the wood and handling joint strength has developed. To accelerate the time to shipping and reduce the stress of the facing weight in handling, some laminators use periodically spaced nailing in the flat areas.

At one time, there was some commercial consideration of using special aluminum extrusions as a substitution for wooden structurals in buildings. A large experimental program at Alcoa demonstrated that these neoprene-phenolic high-solids mastics were quite adequate in strength to be considered for bonding the large aluminum structurals to a variety of floor- and wall-sheaving materials. One important consideration was how fast would the bond strength develop under varying ambient temperature conditions? The data obtained by Minford (1341) using aluminum extrusions and fir plywood adherends is shown in Table 38. As solvent slowly dissipated from the bondline, the tensile-peel-strength of the joints increased. At curing temperatures below 75°F (24°C), joint strength development rate was unacceptably slow. Even at 75°F (24°C), the joint strength continued to increase over a 56-day test evaluation time. In another investigation, the author determined the lap-shear and tensile-peel values for 11 different commercial neoprene high-solids mastic cements for joining aluminum extrusions to fir plywood, as shown in Table 39. Results showed as much as a 500% difference in lap-shear and tensile-peel strength values, yet all were equally recommended by their manufacturers' data sheets. The danger in selecting adhesives without specific test information utilizing

Table 38 Rate of Aluminum-to-Wood Joint Strength Build-up Under Various Temperature Conditions Using an 85% Solids Mastic Adhesive

	Tensile-peel strength after indicated days of curing at the indicated temperature (psi)		
Time (days)	0°F (−17.8°C)	25°F (−3.9°C)	75°F (24°C)
1	0	6	12
2	9	8	22
4	9	9	40
14	10	12	66
28	11	14	93
56	—	—	120

All joints made with B. F. Goodrich PL-400 high solids rubber mastic and 6063-T6 aluminum extrusion and fir plywood.
Source: Data from Ref. 1341.

a customer's specific adherends and bonding conditions is clearly indicated. In Table 40, we also note the wide variation obtained with these 11 commercial mastic adhesives in the type of joint failures encountered. Cohesive failures could range from 10 to 100%, indicating a wide range of adhesion potential to the aluminum surface. Only two of the mastics showed adhesion to the wood capable of pulling wood fibers from that adherend surface.

5. *NITRILE RUBBER-BASED ADHESIVES*

Another popular family of contact cements are formulated from synthetic nitrile (Buna N) rubber and phenolic resins. They may be used for bonding aluminum to a variety of adherends such as plastics, wood, other rubbers, paper products, fabrics, and glass. Nitrile rubber adhesive products, in

Table 39 Comparison of Tensile-Peel and Lap-Shear Joint Values for Aluminum/Plywood Laminate Bonds Fabricated with 11 Different Commercial Mastic Cements Recommended for Structurally Joining These Two Adherends

Product no.	Mfg.	Lap-shear (psi)	Tensile-shear (psi)	Differences in values	Lap-shear/ tensile-shear
1	A	303	68	235	4.46
2	A	233	63	170	3.70
3	B	165	33	132	5.00
4	B	188	42	146	4.48
5	B	60	31	29	1.94
6	C	128	15	113	8.53
7	C	133	31	102	4.29
8	D	238	74	164	3.22
9	D	220	52	168	4.23
10	D	115	38	77	3.03
11	E	248	45	203	5.51

Testing was not initiated until all joints had cured under ambient laboratory conditions (75°F [24°C], approximately 50% RH) for 28 days. Earlier testing of joints with increasing curing times had shown that maximum strength was developed after about 28 days. These mastic cements are generally about 85% solids and time must be allowed for the dissipation of the 15% of solvent into the wood structure.
Source: Data from Minford Ref. 1341.

Table 40 Comparison of Type of Joint Failure of Aluminum Extrusion to Plywood Bondments with Different Commercial Mastic Adhesives and Lap-Shear vs Tensile-Peel Testing Procedures

		Description of bond failure					
		Lap-shear specimen (% failure)			Tensile-peel specimen (% failure)		
Product no.	Mfg.	Adhesive	Cohesive	Plywood	Adhesive	Cohesive	Plywood
1	A	15	35	50	40	10	50
2	A	10	90		30	65	5
3	B	75	25		55	45	
4	B	90	10		93	7	
5	B		100			100	
6	C	25	75		75	25	
7	C	50	50		65	35	
8	D	10	40	50	50	50	
9	D	30	70		5	85	10
10	D	10	90		8	80	12
11	E	10	90		25	75	
		Avg. 33	Avg. 61		Avg. 50	Avg. 52	

Source: Data from Ref. 1341.

general, have superior resistance to oil contamination as compared to any of the other rubber adhesives. Durable adhesion to aluminum surfaces in water-service exposures has always been a plus for nitrile-phenolic contact cements. However, the greater range of adhesive properties that are present in the neoprene-phenolic formulations has enabled them to have a wider service application range. The maximum service temperature of these adhesives on aluminum is about 250–300°F (121–149°C) providing some heat-curing of the bondline is employed, as specified by the adhesive formulator. This higher level of heat resistance is not possible with the neoprene-phenolics nor is the maximum initial joint strength potential as high.

The solvent dispersed–type of nitrile-phenolic contact cement is the variety found in many hardware stores and is used primarily for moderate-strength ambient temperature contact bonding, where neither high stressing nor very elevated temperature–service conditions are present. It does offer more than sufficient strength, however, for bonding aluminum to other adherends where large surface areas may be involved. The nitrile rubber can be blended with a phenolic-novolac resin to form structural tape and film adhesives, which have had a remarkable record of excellent durability bonding to aluminum adherends in the aerospace industry. In the automobile industry, they have been used to bond brake shoes and clutch disk assemblies. These 100% solids tape and film adhesives usually must be cured under conditions such as 30 min at 300°F (149°C) and 50 psi (0.34 MPa) pressure. Of special interest for aluminum bonding was the discovery that incorporation of small amounts of carboxyl groups gave enhanced adhesion to metals (1342–1345). Minford (1346,1347) has shown the high-strength-retention after exposure to extended salt spray, water immersion, and natural atmospheric weathering conditions of chromic acid–deoxidized 6061-T6 aluminum joints when bonded with nitrile-phenolic aerospace-type film and tape adhesives. It is probably true that no other adhesive type exceeds the nitrile-phenolic's ability to maintain bond strength on steel or aluminum after extended exposure to water, salts, or other corrosive media and especially the prevention of undercutting corrosion of the aluminum bondline. This fact coupled with their relative low cost has been a trade-off against some disadvantages like their need for higher pressure curing, only modest tensile strength, and lower temperature resistance as compared to the newer modified-epoxy aerospace adhesives.

Nitrile rubbers are also extensively used to formulate sealants which have good serviceability in sealing aluminum structures.

6. POLYSULFIDE RUBBER ADHESIVES

While polysulfide rubber itself has several characteristic deficiencies that make it unacceptable as a major raw material for formulating strong adhesives, it has been used both as a modifier for some epoxy resins and as a major component of various adhesive sealant formulations. The good adhesion of polysulfide sealants to aluminum surfaces has been shown in a number of long-term atmospheric weathering studies by Minford (1348,1349).

Polysulfide-modified epoxy formulations have also been evaluated by Minford (1350) for low-strength structural applications involving both flexibility and sealing. Application examples include sealing and bonding of aluminum trailer sheet structures and bonding and sealing aluminum canoes. In the later situation, the final joining was through a rivet-bond joint. An important advantage of these adhesives has been their ability to bond durably under adverse weathering conditions without the need for elaborate or costly aluminum surface deoxidizing pretreatments. Although these adhesives have epoxy resin as their main strengthening ingredient, and cure at room temperature, they are much more flexible (though lower strength) than the typical two-part, room temperature–curing epoxies found in the hardware store. In this respect, these adhesives can be considered to act more like structural sealants than structural adhesives.

7. SILICONE RUBBER ADHESIVES

Silicone rubbers have a chemistry quite different from the other rubbers discussed earlier, since they are neither exclusively organic or inorganic in nature but a combination of the two. Pressure-sensitive silicone adhesive formulations are found on a variety of commercial tape carriers. They can exist as an adhesive on the back of aluminum foils or sheets for direct bonding over some other adherend or as transfer films which may be used as a bonding film between two aluminum surfaces or aluminum to another adherend. Nonstructural bonds between aluminum and glass, paper, plastics, and silicone rubber itself can all be made conveniently with silicone-based adhesives. In paste form, some silicones can be vulcanized to clean aluminum surfaces, and such joints can show respectible high peel strengths.

A considerable science has developed around technology which can produce room temperature–curing (RTV) silicone adhesive sealants (1340,1351). They can be formulated as either liquid or paste, one- or two-part systems that cure at room temperature by several different mechanisms. All RTV silicone adhesive-sealants are formulated using the same three basic ingredients (1352). These include a silicone polymer, various fillers, and an organoreactive silane cross-linker/catalyst system. The function of the fillers to develop a desired property profile can be demonstrated by comparing the 50- to 75-psi (350- to 525-kPa) tensile strength of a cured, unfilled silicone polymer with the 400- to 800-psi (2.80- to 5.60-MPa) range of tensile strengths obtained after addition of a reinforcing fiber.

If aluminum is to be bonded to itself or another vapor barrier material, then relatively fast and full curing can only be obtained with a two-part system. The one-part RTV adhesives usually have a curing mechanism which is initiated by moisture incursion into the bondline. For most rapid acceleration of bondline curing, assembled and fixtured joints can be actually incubated in a high-humidity condition, which is usually a situation to be avoided with most other polymer/aluminum joints while curing. Care must be taken, however, to achieve thorough mixing and precise parts metering with the two component systems, since the ratio of curing agent can be a relatively small fractional part compared to the main ingredient part.

General Electric in their *Technical Data Book S-35* (1353) has pointed out that one-component RTV silicone sealants generally bond well to a variety of adherends, including aluminum. However, they also mention that each particular cure system must be separately evaluated on each different adherend or each different surface treatment on an adherend. Further, it should never be assumed that one-component products having the same cure system, but made by different manufacturers or bearing different product numbers, will necessarily perform similarly on the same adherend surface.

Adhesion of these one-part systems may be enhanced by the presence of surface hydroxyl groups, which are usually present in abundance on glass and to a lesser extent on aluminum. One of the attractive features of such sealants on glass, aluminum, and steel is that it is only necessary to remove loose particles or corrosion from the surface to bond effectively.

One-part RTV systems cure by egress of moisture, which means that full curing will occur at edge areas first. This necessitates testing of joint strength under each different set of plant curing conditions to assure full-strength joints before exposing to service conditions. The time to secure full strength will be further slowed by the slower rate of moisture diffusion that results through the first polymerized edge areas. The curing problem is also made more difficult when two or even one vapor barrier adherend is present to restrict moisture entry primarily to the edge areas. The fact that this interior curing will eventually take place is probably good evidence that while silicone adhesive sealants are true sealants against liquid water entry, they lack the ability to keep moisture out of the bondline. Minford (1354) has also demonstrated this fact by obtaining complete protection of aluminum/epoxy-bonded joints from a corrosive seacoast water exposure for up to 8 yr. The same sealant, however, utterly failed to protect identically fabricated aluminum/epoxy joints in the jungles of Surinam for even a few weeks (1355,1356).

Minford (1357) has conducted some relatively long-time (2 yr) testing of some commercial two-part, room temperature–curing RTV adhesive-sealer products on vapor-degreased–only and unprimed 6061-T6 bare aluminum adherends. The durability obtained was considered compar;able to that found using many two-part, room temperature–curing commercial epoxy systems, although the joint strength levels with the RTV products were generally less than 1000 psi (6.89 MPa) as compared with 2000–3000 psi (13.78–20.67 MPa) for the epoxy candidates. It can only be determined by specific durability testing, however, which of the currently produced RTVs have these responses. (More information on the technology of two-part RTV silicone rubber formulations can be found in Ref. 1358, and specific durability test results will be offered in Chapter 10.)

If after having evaluated a certain silicone adhesive-sealant/aluminum combination for adhesion, it may still be shown that the adhesion levels are not as satisfactory as desired, it may be possible to further enhance this adhesion by use of a special primer recommended by the formulator for use with a specific sealant. These primers are generally low-solids, solvent dispersions of silicone resins, silanes, or other organic resins or some combination of these materials. Primers recommended for specific use on aluminum surfaces may differ from those recommended for ceramic or plastic surfaces. All must be separately evaluated using the specific primer, RTV silicone adhesive-sealant, and the specific aluminum surface alloy and pretreatment using elements of the weathering to be found in the service environment.

Cured silicone rubbers offer a wide range of resistance to both chemicals and environmental weathering. Ultraviolet light, ozone, salt spray, or a combination of these generally have little or no effect. They will, however, undergo some swelling when immersed in various kinds of oils. The extent of this swelling can be modified significantly by formulation variations. For the ultimate in chemical, gasoline, and solvent resistance, RTV adhesive-sealants have been formulated around silicone polymers with fluorine substitutions. These fluorosilicones have extremely good resistance to hydrocarbon fuels even with high levels of aromatic organic compounds present. Such fuels will normally completely destroy standard dimethylsilicone adhesive sealants.

IV. THERMOPLASTIC RESIN ADHESIVES

Thermoplastic resins are employed in compouding many industrial adhesives. They can have excellent adhesion properties to aluminum; especially where the chemistry of the resin structure indicates the presence of highly polar chemical linkages, which have the potential for interaction with the polar aluminum oxide. Since these resins are heat fusible, they are not usually employed in an unmodified state for structural bonding purposes. Combinations of thermoplastic and thermosetting resins are, however, well accepted as effective raw materials for commercial structural adhesive products.

A. Hot-Melt Adhesives

Hot-melt adhesive usage is one of the fastest growing application areas in the adhesives industry, because of the possibility for fast application and curing by only cooling to ambient temperature. The lack of any solvent to disperse is certainly an additional plus for any manufacturing plants focusing on lowering atmospheric pollution. In addition, hot-melt formulations will generally require lower energy expenditures than the typical waterborne adhesive systems, which would be the main competition where minimizing atmospheric pollution is an important factor. Even the capital investment to install a hot-melt coater line will generally be less than the large ovens required to remove water or other solvents from solvent-dispersed formulations.

A partial listing of some of the main application areas for hot-melts is offered in Table 41. In Table 42, the actual volume usage broken down by application industry is offered over the period from 1975 to 1981, permitting comparison trends with time.

Most hot-melts are based on polyolefins, the best-known being ethylene vinyl acetate (EVA) copolymer. A variety of hot-melt formulations based on EVA can be reviewed in *Adhesives in Manufacturing* (1359). More recently, additional changes have been made by Dupont in their Elvax II family of EVAs in which the ethylene copolymers contain a substituted alkyl group plus an acid functionality. Both the original Elvax and the Elvax II products are thermally stable and compatible with virtually all the materials currently used in hot-melt formulations. The acid functionality serves to improve adhesion to aluminum, as well as enhance mechanical properties such as toughness. Finally, the Type II materials are cross-linkable through the acid groups. Eastman (1360) published a 1983 article in the journal *Adhesives Age* which summarizes these improvements as compared to the older EVA materials.

Table 41 Where Hot-Melt Adhesives Are Used

Industry	Application areas
Packaging	Case sealing, bags, trays, labels, tapes cartons, composite cans, corrugated boards, fiber drums, pouches, 3-piece beverage cans
Bookbinding	Paperbacks, catalogs, hardcover books, magazine, directories, calendars
Textiles	Interliners, hemming, waistbands, labels, cuffing, pillows, aprons, curtains, towels, patches, basting
Disposables	Diapers, hospital pads, sanitary items, hospital garments, surgical pads
Furniture	Edge banding, cabinet assembly, overlay lamination, door panels, stereo cabinets, window frames
Footwear	Lasting, sole attaching, box toes, counters, folding
Automotive and related industries	Door panels, hand windshield washer, tail-light assemblies, electronic controls, vinyl roofs, trunk and window seals, carpet installation, outboard motor housing, fire-wall insulation, headliners, seating
Product Assembly	Vinyl-clad window frames, envelope batteries, thermal- and noise-insulation pads, mobile-home components, mirror assembly, utensils, electronic assemblies, flexible ducting
Appliances	Air conditioner components, coffee pots, tape cassette holder, refrigerator components, television components, calculator display windows, slowcook-pot heaters
Carpets	Backing, seaming tapes
Tapes (hot-melt pressure-sensitive adhesives [HMPSA]	Electrical, binding, sealing, nylon- or glass-reinforced
Filters	Oil (automotive), air (automotive and residential), exhaust (gas scrubber)

Source: From Ref. 1359 (p. 332, section VI).

Table 42 Hot-Melt Adhesives Usage by Industry

Industry	Year (usage in millions of lb) 1975	1977	1979	1981[a]
Packaging	145	175	200	220
Bookbinding	20	22	24	25
Textiles	15	20	25	30
Disposables	15	10	24	28
Carpet seaming tape	4	5	6	7
Pressure-sensitive tape (hot-melt)	6	8	12	20
Furniture	10	12	15	17
Footwear	5	6	6	6
Automotive	4	7	8	10
Appliances	5	6	7	8
Filters	2	2	2.5	2.5
Product assembly	10	14	16	19
Consumer	0.5	1	1.5	2
Electronic	1	1.5	1.5	2
Other	2.5	1.5	1.5	2
Total	245	300	350	400

[a]Assuming no serious disruption in the volume growth curve during 1980 and subsequent years.
Source: From Ref. 1359 (p. 331, Table 2).

Most recently, Filbey et al. (1361) investigated the cross-linking of modified EVA adhesives with resultant improved high-temperature properties on aluminum. They compared the peel strengths on aluminum using both the unaltered and cross-linked EVA.

A typical hot-melt formulation using EVA copolymers is shown in Table 43 and the overall list of possible components of a hot-melt are reviewed in Table 44.

1. THERMOPLASTIC-BASE POLYMERS

These adhesives can be defined as 100% nonvolatile thermoplastic materials that are heated to the melt condition and applied as a mobile fluid to the aluminum adherend. The bond joint and maximum joint strength will only be developed when the fluid again solidifies. Historically, the oldest

Table 43 Typical Hot-Melt Formulation Using EVA Copolymers

	Parts by weight
EVA copolymer	40
Phenolic resin tackyfier	15
Polyterpene tackyfier	12.5
Synthetic polyterpene tackyfier	12.5
Wax for viscosity control	20
Antioxidant	0.1

Applied at 170–200°F (77–93°C).
Source: From Pocius Lecture 4 on Chemistry of Adhesives, ACS Audio-Teleconference Course on Adhesion Science.

Table 44 Components of a Hot-Melt Adhesive

1. Relatively high molecular weight thermoplastic polymer (provides basis for the adhesive's strength)
2. Tackifying or wetting agent (provides tack during the bonding process; e.g., terpene or polyterpene resin, chlorinated hydrocarbons)
3. Flexibilizer (plasticizer) (provides thermal and mechanical shock resistance; e.g., phthalates, polybutenes, tricresylphosphate)
4. Antioxidant (to prevent excessive oxidation when adhesive is melted; e.g., hindered phenols)
5. Fillers (to reduce cost, opacity, modify flow of the melt; e.g., clay, talc, TiO)
6. Waxes (reduce viscosity, alter surface characteristics; e.g., microcrystalline wax, paraffin)

Source: From Pocius Lecture 4 on Chemistry of Adhesives, ACS Audio-Teleconference Course on Adhesion Science.

example of a hot-melt might be the bitumen (asphalt) used to set the stones in the Tower of Babel in the bible. A more widely used early hot-melt, however, would be the sealing wax used by man for many ages to seal writings and documents. Types of thermoplastics used in compounding hot-melts are listed in Table 45.

Hot-melt adhesives today are usually elaborate mixtures of thermoplastic polymers, tackifying agents, flexibilizers (plasticizers), antioxidants, fillers, and waxes, each of which has a specific contribution to the overall adhesive properties. The types of thermoplastics used also have considerable variety and the relative concentrations of polar groupings present in these resins undoubtedly have a distinct correlation with their abilities to bind to the polar aluminum oxide. In addition, the tackifier may also contribute adhesive polarity as exists in a wide variety of hot-melt formulations where phenolic resin was used as the tackifier. Possibly the best overall summary and practical application reading for the general manufacturer on hot-melt technology can be found in *Adhesives in Manufacturing* (1340) (see also Refs. 1362–1369).

It should be clear that flow and wetting characteristics of hot-melts are directly correlatable with temperature change. Even in the elevated temperature state, however, the structural hot-melt compositions are viscous fluids, restricting ready flow and wetting of aluminum adherends. Also, the high heat conductivity of aluminum is an ever-present negative factor in achieving maximum flow and wetting. This relates to the fact that the hot-melt that is in immediate contact with the aluminum surface can readily lose heat in the interfacial area to the aluminum heat sink. The melt flow at the interface will accordingly diminish when it is most critical for wetting that the spread and flow be maximum. In effect, the hot-melt in the contacting layers will slide up the curve to higher modulus levels before intimate wetting has occurred. Thus, it is absolutely essential in bonding aluminum with hot-melts to monitor the surface temperature of the aluminum adherend throughout the manufacturing

Table 45 Types of Thermoplastics Used in Compounding Hot-Melts

Used in compounded state (formulation design)	Used in uncompounded state (molecular design)
1. Low-density polypropylene	1. Polyesters
2. Ethylene vinyl acetate	2. Polyamides
3. Ethylene ethyl acrylate	3. Polyurethanes
4. Paraffin waxes	
5. Polypropylene	
6. Styrene-butadiene block copolymers	
7. Phenoxy resins	

Source: From Pocius Lecture on Chemistry of Adhesives, ACS Audio-Teleconference Course on Adhesion Science.

process to be sure optimum flow and wetting occurs. It should also be clear that any surface contamination that can adversely affect wettability needs to be removed before application of a hot-melt.

If the aluminum surface passes a simple wettability test, i.e., continuous flow of water demonstrated over the surface, and the surface is initially heated to minimize heat loss to the aluminum heat sink by the heated adhesive, it is probable that satisfactory adhesion can be created. Because the majority of commercial hot-melts are not structural, as contrasted with the thermoset resin-type adhesives, there is usually no need for the general manufacturer to produce an etched, anodized, or chemically converted aluminum surface for bonding purposes. The use of a so-called structural hot-melt adhesive, however, might benefit significantly from the use of a good deoxidizing-type aluminum surface treatment, especially where the service environment has hot-water or high-humidity soaking conditions to resist in the service environment.

A new technological development for hot-melts has been their application by robotics (1370). This application method has been extremely useful in making gaskets and seals.

2. HOT-MELT FOAM ADHESIVES

An additional area of interest for the future has been the development of foamable hot-melts (1371, 1372,1373). Possible advantages could be (1) better control of glueline deposition, (2) extended open times, (3) better void filling for poorly mated parts, and (4) reduction in adhesive consumption. Studies have included the hot-melt adhesives: (1) polyethylene, (2) polypropylene, (3) polyamide, (4) EVA, and (5) polyester. Foamability potential varies according to the adhesive choice. For example, polyethylene falls into the excellent-choice category; polypropylene, polyester, and 12-polyamide (Nylon 12) all foam in the very-good category. Styrene-butadiene and EVA adhesives foam only in the good category, whereas dimer acid polyamide is only a fair foamer. Hot-melt foams could have aluminum applications in appliances, automobiles, cans and various other enclosures, and metal building fabrication elements.

3. HIGHER TEMPERATURE–RESISTANT HOT-MELTS

Kettleborough (1374), Eichhorn and Reiner (1375), and Annighofer (1376) pointed out the higher temperature–resistance properties of melts produced from polyesters and polyamides as compared to more widely used rubbery hot melts. These so-called high-performance hot-melts are semicrystalline polymers on the basis of polyester or polyamide (1376). Schaaf (1377) obtained a copolyamide hot-melt by polymerizing caprolactam, laurinlactam, and several aliphatic diacids and diamines as early as 1974. Blocked isocyanate technology (1378) has been combined (1379) with copolyamide technology to produce cross-linkable hot-melt resistant to a temperature of over 428°F (220°C) even though the initial melting point was 212°F (100°C). The idea is to improve the heat performance without losing the favorable hot melt properties. Bugel et al. (1380) bonded aluminum adherends with a thermoplastic structural adhesive based on phenoxy resin. A strength of 3300 psi (22.74 MPa) was developed and this bond was maintained under a variety of environmental conditions. The idea is to improve the heat performance but without losing the favorable hot-melt properties.

An area of future development would be in the direction of combination thermoplastic and thermoset resin formulations. St. Clair has utilized radiation cross-linking of HMPSA tapes (1368). Another possible route is combining two flows of hot-melt materials which can react with each other. Finally, Hinterwaldner (1365) and Eddy (1369) have discussed exotic, high-performance polymers for aerospace applications which might be adaptable for use as adhesives.

Minford (1381) has undertaken evaluations on aluminum adherends of very high-temperature–melting hot-melts classified as structural adhesives. Pointed toward the direction of use for repairing aluminum heat-exchanger tubing, he evaluated the bonds in hot-humidity and various water-soaking conditions with less than hoped for success. Although it must be acknowledged that no surface pretreating other than solvent cleaning and abrading was used in these evaluations, since the facilities for deoxidizing aluminum surfaces would not be considered practical for many field repair operations.

Very recently, Wielinski (1382) reported on investigations to formulate hot-melts with more heat and oil resistance. Copolymerization of ethylene with vinyl acetate (VA) to produce ethylene vinyl acetate (EVA) resins was the basis for a variety of hot-melt systems but more heat and oil resistance was often desired. These important property improvements have been gained by substituting n-butyl acrylate (nBA) for the VA. These resins now constitute a new family of ethylene copolymers for compounding hot-melts. Improvements in heat stability, low-temperature flexibility, and oil resistance are significant.

One of the problems that many hot-melts have when bonding to polar surfaces like metal oxides is that the backbone polymers, such as the polyolefins, are nonpolar. It is fortunate, however, that bonds can still be developed, since the adhesive is being applied from the melt condition. Schreiber (1383) delivered a recent paper which covers discussion of this adhesion and processing of the polyolefin-type of hot-melt. A similar problem of possible poor wetting on a polar oxide surface would be expected for block polymer-based hot-melts, which are mainly nonpolar in character. Dillman and Sanders (1384) presented a 1989 solution to the problem which considered development of functionalized SEBC polymers which offer the presence of polar functionality for improving performance of such hot-melts over polar surfaces.

B. Pressure-Sensitive Adhesives

Undoubtedly, one of the largest markets for adhesive in our modern world is based on the need to secure temporary attachment of one surface to another. Because of the outstanding esthetic character of aluminum in its many finished forms and its vapor-barrier property, there has been a particularly great need to temporarily attach aluminum as a surface laminate to many different manufactured products. Aluminum with printed information and product identification information has been coated with pressure-sensitive adhesive (PSA) on the backside and bonded to a wide variety of product surfaces. Larger areas of aluminum overlays can also be attached to offer vapor-barrier or corrosion-prevention characteristics to surfaces lacking such properties.

Frost and Sullivan (1385), in a 1978 market study, showed the main PSA polymers had the following percentages of use: natural rubber (46%), acrylates (34%), SBR rubber (10%), and block copolymers and all others (10%). Most PSAs originally were applied from organic-solvent solutions; however, Hamar et al. (1386) have discussed the attempts to develop alternatives. One such alternative, emulsion technology, has made considerable advances (1387). Another, the hot-melt PSAs (HMPSAs) have advanced at the expense of solvent-types (1388). Finally, the waterborne PSAs as the major alternative (1389). (Information about many applications can be reviewed in Refs. 1340, 1390, and 1391.)

1. PSAs ON ALUMINUM ADHEREND SURFACES

The formulation of a pressure-sensitive adhesive which has just detachment, when desired, has been a carefully worked out science. Obviously, the secret is to choose the most appropriate combination of an elastomer polymer, tackifying resin, filler, antioxidant, stabilizer (if necessary), and cross-linker. By altering the ratio between rubber, resin, and cross-linker, a balance is obtained between tack and shear or peel strength. Elastomers and resins used constitute a basis for naming types. These include (1) natural rubber, (2) styrene-butadiene-styrene block rubber, (3) styrene-isoprene-styrene block rubber, (4) butyl rubber, (5) polyisobutylene rubber, (6) acrylic resin, (7) vinyl ether polymer, and (8) silicone.

Obviously, use of pressure-sensitive adhesives on aluminum adherends must be confined to low non–steady-stress service conditions. The use of pressure-sensitive adhesive is found in literally thousands of products where aluminum attachment fulfills an esthetic or practical property need as mentioned above. In most of these situations, the aluminum is either foil or in thin sheet form. This permits the peel or shear strength per unit area of adhesive (though relatively small) to be more than adequate to prevent delamination due to the static load imposed by the dead weight of the aluminum attachment. While it may be virtually impossible to fail such thin aluminum adherends in pure shear due to the relatively large bond area, much lower resistance of the bond in peel across a line of stress permits relatively easy detachment of the foil, sheet, or tab when that is desired. More discussion will follow about pressure-sensitive adhesives when discussing adhesives in film form.

It would seem inappropriate to evaluate weather resistance of PSA adhesion to aluminum surfaces, since the adhesion of these products are generally intended to be more or less temporary. This is often not the case, for many uses of PSAs on aluminum must resist exterior weathering for long times. Minford (1392) has conducted long-term durability testing of the situation where a PSA-type adhesive backing has been used to attach aluminum trim to other body surfaces. Separate testing needed to be conducted where the aluminum surface may be a vapor degreased mill finish, abraded surface with regrown oxide, anodized oxide, or a prepainted aluminum surface. Another use of PSA has been on the backside of plastic films used to protect finished aluminum surfaces while being stored or transported for later field erection or after field erection until the building is completed. Minford (1393) found special testing had to be conducted to determine strippability of such protective plastic films after exposure to accelerated weathering conditions in a laboratory weather-o-meter, in hot-condensing humidity or directly to the natural atmospheric weathering for periods of weeks to months. An alarming number of these commercial protective films showed a total altered adhesion history and become virtually unstrippable after weathering exposure. Finally, a new generation of PSAs has been introduced recently which are actually semistructural in adhesion strength and formulated to resist external weathering conditions. Minford (1394) evaluated in long-term weathering the ability of PSA products like 3M's Isotac pressure-sensitive acrylic transfer tape to join both painted and anodized aluminum-facing sheets for trailers or mobile home applications with encouraging results. The data will be discussed in more detail in Chapter 10 on joint durability.

2. *EVALUATION OF PROPERTIES OF PSAs*

Based on use of the Polyken Probe Tack Tester, Hammond (1395) has plotted the effects of increasing temperature, contact time, and contact pressure on the adhesion of PSAs involving both smooth and rough surface conditions. Regardless of which of these parameters is considered, the adhesion values for the rough surfaces were significantly lower. Magunov (1396) published a 1989 report on the wetting of slightly nonuniform solid surfaces which can be relevent. It was also verified (as might be expected) that adhesion strength diminished with increasing temperature and increased with increasing contact time or pressure regardless of whether rough or smooth surfaces were investigated. The rate of joint separation with the tack tester showed a dramatic rise in measured tack for the smooth surface up to a value of 1 cm/sec, reaching a plateau extending through 2 cm/sec. With a rough surface, the rate of separation was very gradual, never reaching more than 20% of the plateau value obtained with a smooth surface. Bates (1397) developed a method for measuring tack by determining the energy dissipated during debonding at constant strain rate.

In straight peeling testing, Satas determined a maximum value was found with a peel rate of about 0.3 cm/sec. Thereafter, increasing the peel rate to about 0.8 cm/sec showed decreasing peel strength. However, strength then rose slowly to about the same peel force again after about 30 cm/sec as achieved after 0.3 cm/sec. Thereafter followed a second steady decline up to the extremely fast rate of about 800 cm/sec. These results indicate a very complex surface adhesion response for values above the 0.3 cm/sec level. When Satas conducted peel testing at a lower range of 0.1–100 cm/sec, he showed three different responses from three different formulations. Product A peel adhesion rose steadily but slowly over total range. Product B peel rose rapidly to a threefold higher peel adhesion at 25 cm/sec rate and then declined to a twofold higher value. Meanwhile, PSA-C rose in peel at about the same rate as B to the 1 cm/sec peel rate condition and then moved along a plateau to highest rate at 100 cm/sec. Thus, generalizations about different formulations remain difficult if not impossible to propose. (Other reading about the nature and response of PSAs can be found in Refs. 1398–1401.)

The shear resistance of PSAs have also been evaluated. Aubrey (1402) proposed that it is necessary to increase the concentration of network or gel so that a PSA can sustain high shearing force. Hagan et al. (1403) pointed out the importance of molecular weight, glass transition temperature, and polarity factors on the tack and adhesive and cohesive characteristics of PSAs.

Recent works elucidating properties of PSAs include publications by Zosel (1404) on specific mechanics or viscoelastic characteristics of adhesion performance, including deformation; Urahama (1405) on the effect of peel load on stringiness phenomena; Tsukanatani et al. (1406) on bonding processes in tack; and Dae et al. (1407) on viscoelastic behavior, thermodynamic compatibility, and phase equilibria in block copolymer-based HMPSAs. Sparks (1408) discussed the important properties of acrylic HMPSAs, which include good chemical stability, color stability, clarity, and excellent adhesion. He also included the promise of acrylic latexes for PSAs because of their ultraviolet (UV) and thermal stability, resistance to plasticizer migration from vinyl adherends, and lower formulating and manufacturing costs. Hickman (1409) considered the design of S/B latex PSAs, whereas Oldack and Bloss (1410) reviewed compounding of natural latex in water-based PTAs.

The study of the dynamic mechanical properties of silicone PSAs was reported by Copley (1411) in 1984, whereas Class and Chu (1412) used the same techniques to study the viscoelastic properties of rubber-resin PSAs (1413).

C. Waterborne Adhesives

The desire to replace solventborne systems with waterborne adhesive systems in general manufacturing in recent years has promoted increasing research in developing such systems. The changeover, however, has not been made without some difficulties (1414). Some of these problems have been related to (1) poor coating quality, (2) excessive foaming in the bondline, (3) insufficient water resistance, and (4) some corrosion and clean-up problems. Despite such difficulties, some waterborne systems have been adopted for manufacture of PSAs, contact-type cements, and laminating adhesives. Because of their relatively lower strengths and lower resistance to aggressive weathering, they have found only a few applications in manufacturing aerospace, automotive, and over-the-road vehicles.

In general, waterborne adhesives can be divided into two broad categories: water-soluble and water-dispersed products. The water-solubles include animal glues, vegetable glues, and other natural systems (1415). The water-dispersed adhesives include those based on natural rubber latices, synthetic-rubber (polymer) latices, and postformed-rubber (polymer) latices.

The water-soluble adhesives had not played a major role in early bonding of aluminum surfaces. However, with the introduction of the various synthetic latices, a variety of properties became available that made such adhesives important for laminating thin gauges of aluminum to wood, paper, and plastic adherends. The synthetic latices are aqueous dispersons of polymers obtained by emulsion polymerization. Synthetic polymers used include chloroprene (neoprene), butadiene-styrene (GRS, SBR), butadiene-acrylonitrile (Buna N), vinyl acetate, vinyl chloride, and acrylates. Earlier in this chapter, discussion was offered for the various means of bonding aluminum with this kind of adhesive. Under thermoplastic resin material latices, it would now seem appropriate to continue with the adhesive subject areas of the polyvinyl acetates, polyvinyl chlorides, and acrylate polymers. Evans and Krajca (1416) reviewed the waterborne adhesive tackifiers and Jones (1417) the elastomer and resin modifiers for water-based adhesives. Growing use of waterborne adhesives, as of 1981, was reviewed by Fries (1418).

1. POLYVINYL CHLORIDE ADHESIVES

These adhesives are classified as synthetic latices obtained by making aqueous dispersions of polymers through emulsion polymerization. Molecular weights range from 10,000 to 1,000,000 with particle sizes from 0.001 to 10 μm. Since the majority of these solution-type adhesives remain water soluble after drying, their use has been restricted to those application areas that do not require water resistance for the final bond. However, they cannot be redispersed in water after drying, so bonding must be part of a continuing lamination process after adhesive application. In the wet state after application, particles of polymer, resin, and other modifiers exist as individual particles. These particles coalesce into a homogeneous film while drying. There will be a uniform loss of water during the operation until particles begin to contact and coalesce. The rate of water loss then begins to diminish until the particle coalescence becomes irreversible. The drying process can be readily accomplished in general manufacturing by the use of forced-air drying. The most consistant theory of film formation of water-dispersed adhesives, as described above, was proposed by Vanderhofft et al. (1419) in 1973.

2. POLYVINYL ACETATE ADHESIVES

In most adhesive applications, vinyl chloride is copolymerized with vinyl acetate for compounding in waterborne adhesive formulations. These combination adhesives have been used to bond vinyl plastics to both metal or glass. Such adhesives, as well as polyvinyl acetate solution adhesives, are often used to laminate thin aluminum foils to papers to constitute an economical moisture-barrier film for packaging many foods and commodity materials, like cigarettes. Volume of adhesive used in these types of composite-laminate products is staggering.

Rosenblum (1420) discussed an important way of increasing the adhesion of vinyl acetate and EVA adhesives to metal and polymer surfaces by carboxylating the conventional emulsion polymer. Dried adhesive films are highly resistant to oil, grease, water, acids, alkalies, and UV. Carboxylation offers the possibility for cross-linking and reaction with aminoplasts, phenolics, and epoxy resins, giving even greater water and creep resistance.

Obviously, these laminating operations are mostly conducted at high speed for economical reasons. It is absolutely necessary to provide sufficient drying of water from the bondline before mating to obviate the possibility for acidic (hydrolytic) reactions from taking place later owing to water trapping. The products of such hydrolytic reactions can actually strip away the natural protective oxide from the aluminum foil surface, leaving it unprotected against disasterous pitting perforation. Even a few such perforations destroy the vapor-barrier nature of the foil, which was the basis for using the aluminum in the first place. The author has observed manufacturing situations in which the desire to accelerate production by decreasing time in the drying tunnel led to such disaster. Simply slowing down the speed through the drying tunnel restored the integrity of the laminate for the application.

These kinds of waterborne latices do not have sufficient strength or permit any use for structural bonding applications of aluminum. Also, they could not be used effectively for laminating two vapor-barrier aluminum surfaces. At least one adherend surface must ofer an escape route for the last traces of water in the bondline.

3. ACRYLIC ADHESIVES

Acrylic latices are also obtained by emulsion polymerization and usually consist of copolymers or terpolymers. Many types of properties are possible depending on the particular monomers used. Adhesives produced can accordingly range from those with PSA-type properties to some moderate high-performance adhesives. With the appropriate attention to monomer to charge ratio, it is possible to incorporate cross-linking mechanisms to regulate tack, shear strength, and cohesive strength of the PSA-type systems.

Acrylics have fundamental properties in the area of UV resistance, thermal stability, and plasticizer resistance. Enhancement of these properties may be obtained by using a variety of cross-linking or curing mechanisms. Monomers such as acrylamide, N-methyolacrylamide, glycidylmethacrylate, and dimethlaminoethyl acrylate can be used to provide internal curing sites, whereas zinc acetate, urea-formaldehyde, sodium carbonate, and sodium sesquicarbonate have all been used as external cross-linkers. For example, Sparks (1408) developed a new acrylic family for bonding to low-energy surfaces by using a sodium sesquicarbonate alkaline catalyst.

As contact-type adhesives, acrylic latices compare favorably with both solvent- and water-based neoprenes. Their lap-shear adhesion can actually be superior to that of a neoprene latex product, with accompanying equal freeze-thaw stability, nonflammability, ease of handling, and ease of clean-up. As a group, these waterborne acrylates are widely used in manufacturing flexible packaging and industrial laminates.

In 1979, Oldack and Bloss (1410) considered the basis for higher-performance acrylic latices by using acrylate polymers with T_c in the range of 32–50°F (0–10°C) and tackifying them to give more acceptable green strength, open time, adhesion to nonporous adherends, and higher temperature resistance.

D. Modified Acrylic Structural Adhesives

Curable acrylic adhesives are reactive cross-linking structural adhesives that cure by free radical initiation. They are based on methacrylate monomers which must be toughened with elastomeric polymers to offer better properties for truly structural bonding. Their chemistry is quite different from anaerobic, cyanoacrylate, and acrylic solution adhesives and emulsions. First developed in West Germany in the late 1960s, the first materials were low molecular weight solutions of polymethylmethacrylate dissolved in the monomer and subsequently cured by peroxide initiation. Aluminum windows and door bonding constituted one of the earliest applications. Research on the development of many additional sophisticated systems of this family proceeded rapidly in the United States since this early time, ending in the present formulation of what has been referred to as "high-performance" acrylic systems.

The modified acrylic adhesives cure by addition polymerization reactions. Formation of free radicals initiates a very rapid chain reaction resulting in final cure. As compared with condensation-type curing of epoxy and urethane adhesives, this cure chemistry is significantly more rapid. At the same time, however, there is relatively little polymerization in the early stages of the reaction, which still permits positioning of assemblies without concern about disturbing ultimate cure. The fact that this rapid setting is possible at room temperature is also unique to the acrylics as compared to the slow change in viscosity at this temperature offered by epoxy and urethane adhesive formulations. In addition, a variety of formulating procedures and different polymerization initiators offer potential for the development of specific adhesives fitting particular manufacturing conditions. Two concerns that can be present, however, are (1) the tendency to autopolymerize without the use of externally added accelerators or hardeners, making shelf stability a problem; and (2) the characteristic odor which can be quite objectionable even though it is not necessarily a health problem.

Damico (1271) provided a 1990 review of the progress made in recent years in developing a variety of modified structural acrylics and described overall chemistry in some detail. Generally, the method of initiating a cure is known as redox (reduction and oxidation) reaction. Redox reactions for acrylics typically involve an aromatic amine, like dimethyl aniline, in the resin portion of the adhesive which can interact with a "mix-in" peroxide paste. A resulting free radical is capable of interacting with double bonds in a methacrylated monomer, initiating a chain reaction that finally leads to the formation of high molecular weight polymers. Details involve initiating and controlling chain reactions in such adhesives to ensure final polymer molecular weight is sufficiently high and also useful as an adhesive. Formulators such as Lees (1421), Skoultchi (1422), Bachmann (1423), and Kishi et al. (1424) have developed various patented combinations for generating free radicals in these adhesives. Other investigators such as Owston (1425), Briggs (1426), Howard and Owston (1427), Wolinski (1428), Dawdy et al. (1429), and Charnook (1430) have secured patents on various elastomeric polymers used as modifiers. It can be assumed that any fully formulated acrylic adhesive in today's marketplace will have some level of elastic modification present.

1. CONVENTIONAL OR FIRST-GENERATION STRUCTURAL ACRYLICS

These formulations generally contained methacrylated monomers, some type of polymer modifier, aromatic amine initiator, and methacrylic acid. Such adhesives found early utility in bonding thermoplastics but they were also recommended for bonding aluminum. Many of these early products appeared as two-part systems under the Chemlok trademark of Hughson Chemicals. First data sheets made some claims for being capable of bonding with structural strength and good joint durability to "poorly prepared aluminum surfaces." Claims were also made for having certain advantages in handling, which included three different methods of initiating the build-up of bondline strength. First, accelerator lacquer could be preapplied to aluminum by roller, brush, spray, or any conventional coating process. Once activated, this surface could be immediately used for final bonding or stored for periods up to 12 months before closing in a final bonding combination with a second part of the formulation. In use, this made the adhesive function as a single-component adhesive capable of curing at room temperature. Second, if the lacquer method was not appropriate, then this same acrylic adhesive could be supplied as a conventional two-part package. The author found the high ratio of resin to catalyst in these early formulations required very close attention for weighing and mixing procedures (especially in making small batches). Method three was referred to as the "two-component, no-mix method." This procedure was less commonly used, since it required that the ingredients in the second package be premixed with a catalyst up to 12 hr before use and applied to the second adherend surface. (More information can be found about these early modified acrylic formulations in Refs. 1431–1433.)

Minford (1434) was among the first investigators to subject such adhesives to long-term aluminum joint durability testing under a variety of accelerated weathering conditions. By using an alloy with an existing heavy heat-treat oxide film on the surface (6061-T6), a more severe test was made of how well these new products could durably bond to a "poorly prepared" aluminum oxide surface. Evidence from these early tests of several years' duration clearly demonstrated that the advertised ability of these adhesives to bond durably to even a mill-finish aluminum surface was lacking. However, when similarly bonded joints were made with chromic/sulfuric acid–deoxidized 6061-T6 aluminum adherends, the performance could be judged approximately equivalent to that achieved with a variety of two-part epoxy adhesive products on the same kind of aluminum surface.

2. *SECOND-GENERATION MODIFIED ACRYLIC FORMULATIONS*

A so-called second-generation line of structural acrylic formulations was introduced in the early 1970s which was based on the carrier polymer chlorosulfonated polyethylene and acrylics. These adhesives have received the designation DH-type products, since the developer, Dupont, utilized their hypalon polymer in the first formulations (1435,1436). Gilch and Piestert (1437) describe a member of such a family of formulations. They are similar in handling characteristics to the first-generation products, but differ in use of peroxide as part of the adhesive itself, with use of an amine as the mix-in hardener or accelerator lacquer. Second-generation acrylics have also been described as "reactive acrylics" or "toughened acrylics." They retain the advantage of rapid curing but with more tolerance for adherend surface oil contamination, considerable flexibility with higher peel strength, and significant impact resistance. Their heat resistance may be termed good accompanied by good enough low-temperature properties for general manufacturing. Their lap-shear strengths on aluminum surfaces (even without cleaning or etching) can range up to 4000 psi, which compares favorably with the 4000- to 6000-psi (28- to 41.3-MPa) range for heat-curing epoxies.

Minford (1438) also conducted several long-term aluminum joint durability investigations under accelerated weathering conditions with these products using the same 6061-T6 surface conditions as employed earlier with the first-generation materials. Their abilities to demonstrate good performance potential without need to specially treat the surface could be readily demonstrated, although there appeared to be considerable performance differences using different commercial formulations. (More information about development of these second-generation products can be found in Refs. 1439–1442.)

a. Aerobic Acrylic Products

Improved versions of the reactive acrylics have been developed more recently, which have been termed "aerobic acrylics" (1443–1447). These aerobic acrylics have diminished sensitivity to air inhibition, enabling cure between two surfaces regardless of the presence of air. One of the main negatives about the acrylics in general, i.e., a distinctly disagreeable odor, is significantly reduced in these products. Patented compositions have been awarded both Loctite and DuPont.

3. *HIGH-PERFORMANCE ACRYLIC PRODUCTS*

It was a couple of years later that a number of additional patents appeared which specified the use of a variety of methacrylated and acrylated acid phosphate esters as key ingredients (1448). Such formulations were designed to achieve the same degree of adhesion to unprepared adherends as had been so advantageous in the DH types but without the need for an oily accelerator. While bonding very effectively to bronze, lead, nickel, magnesium, copper, steel (mild and stainless), and aluminum oxide surfaces, these products have weak bonding tendencies toward zinc. This has limited their use in the automotive industry where there has been increased efforts in recent years to increase the use of the galvanized steels.

In Table 46, Graham's data are presented on the comparative joint strengths obtained with the above three different classes of modified acrylics on steel adherends. The author obtained data that approximately equivalent joint strength values can be obtained with aluminum adherends providing equivalent stiffness aluminum adherends are used in fabricating ASTM D-1002 lap-joints. It is observed in this table that highest strength properties were actually obtained with the oldest "general purpose-type" formulations at lowest temperature test conditions. The second-generation (DH and high temperaturence [HP] products permitted significantly higher strength potentials at temperatures of 250°F (121°C) and above. Finally, 40% higher ambient temperature strength was achieved with the HP-type product joints.

More data on the performance of HP-formulated joints made with unprepared metals were published by Zalucha (1449). He also considered their application for plastic joints (1450). Lees (1451–1453) discussed the advances made in formulation and use of modified acrylics for manufacturing purposes in the late 1970s and early 1980s. Minford (1454) evaluated the comparative durability of 3003-H14 and 6061-T6 aluminum alloy joints using HP products. While superior durability was achieved with deoxidized surfaces, the performance on vapor-degreased–only surfaces was vastly improved over that achieved much earlier with the original "general-purpose"–type products.

4. *HIGH-IMPACT ACRYLIC ADHESIVES*

The development of acrylic adhesives with improved low-temperature properties can be observed in adhesives of the type patented by Dawdy (1455) in 1988. These adhesives contain olefinic elastomers further reacted with monoisocyanates to eliminate any residual hydroxyl functionality. These adhesives still bond at room temperature to oily metal surfaces but with very good low-temperature performance. Good adhesion and impact resistance to temperatures as low as −13°F (−25°C) has been obtained with use of a urethane-modified, olefinically terminated liquid elastomer.

5. *HYBRID EPOXY-ACRYLIC ADHESIVES*

Additional patents of Dawdy (1456,1457) further expanded the high-impact (HI) technology by achieving improved elevated-temperature performance. Of special interest is patent 4,457,071 (1457), describing

Table 46 Heat Resistance of Different Types of Structural Acrylic Joints as a Function of Temperature

		General purpose	Second generation	High performance
Adherend surface prep.		SAE 1010 cold-rolled steel gritblasted	SAE 1010 cold-rolled steel—none	SAE 1010 cold-rolled steel—none
Test temp.				
°F	(°C)	(psi)	(psi)	(psi)
−40	(−40)	2600	800	2000
0	(−18)	3000	3400	3400
75	(24)	3800	3900	5500
150	(66)	1900	2400	4400
200	(93)	1400	1800	3600
250	(121)	800	1400	1500
300	(149)	—	600	800

The joint strength values obtained by Minford using equal stiffness aluminum adherends would be very comparable. However, a chromic acid–deoxidizing would replace the gritblasting when using the general-purpose formulations.
Source: Data from Ref. 1441 (Table 4, p. 215).

the use of large amounts of epoxy resin in conjunction with the urethane-modified elastomer described above for the HI formulations. Prolonged resistance of at least 30 min at 400°F (204°C) was obtained with optimum ratios established by Dawdy (1457). Such adhesives might be considered in automotive metals bonding where the bondline resistance to paint bake-oven temperatures is required.

6. *EQUAL-MIX ACRYLIC ADHESIVES*

Another variation of special acrylic technology has been named "equal-mix" or "no-mix" (honeymoon-type). Polymerization is achieved in these products when component A is applied to one adherend and B applied to the other. A good example of this technology is offered in the patent of Kishi et al. (1424), which employs a cobalt salt–cure accelerator in one component and hydroperoxide and aromatic amine in the second component. Formulations are completed with incorporation of reactive monomers, elastomers, fillers, and stabilizers. Such adhesives have been useful in bonding magnets in electrical motors.

7. *SURFACE-ACTIVATED ACRYLIC ADHESIVES*

Although directed at bonding of galvanized steel surfaces rather than aluminum, surface-activated acrylic adhesives are worthy of mention because of the recentness of their development and the employment of the specific surface of the adherend as the catalyst to promote the curing. Damico (1458, 1459) and Bennett and Damico (1460) secured patents relative to this technology. In one sense, such adhesives might be considered more like anaerobic or cyanoacrylate-type adhesives, since they are only one-part formulations and begin to cure when brought into contact with a specific adherend face; in this case, galvanized steel. They are, however, considerably lower in cost than either anaerobic or cyanoacrylate products.

8. *PHOTOCURABLE ACRYLIC ADHESIVES*

This separate class of products cure by exposure to photons such as UV wavelengths below 400 nm and to a lesser extent the visible range from 400 to 700 nm. These materials are less important for opaque adherends, like aluminum, but are important for bonding glass and clear plastics. Specific photoinitiators such as benzoin ethers, benzophenones, and similar compounds are incorporated to develop the necessary triggers for the curing process. Other than replacement of a methacrylate monomer with an acrylate monomer and the substitution of the photoinitiator for a redox technique for generating free radicals, the composition of these adhesives is quite similar to those described earlier. An early overview of photoinitiators has been compiled by Gatechair and Wostratsky (1461), whereas Pasternack (1462) discussed the fundamentals of radiation and electron beam curing. (For other references of pertinence to the general field, see Refs. 1463–1467.)

V. THERMOSETTING RESIN ADHESIVES

The adhesives which depend for their structural strength on some form of curing mechanism to thermoset the resins in the formulation are available in a variety of physical forms, as enumerated in Table 47. The most widely used structural adhesive groups are epoxies, polyurethanes, modified acrylics (discussed above), cyanoacrylates, and anaerobics. Also considered are the hybrid structural adhesive combinations where the above thermoset families are combined with thermoplastic resins and the combination forms special structural adhesives with generally greater impact resistance properties.

An excellent review of general information and the chemistry of a wide variety of these structural adhesives can be found in 1973 and 1983 publications by Bolger (1468,1469). Curing mechanisms of thermosets has been discussed by Cope (1470) in the book *Advances in Adhesives* edited by Brewis and Comyn (1471). In addition to their range of physical forms, as shown in Table 47, structural adhesives can be discussed as being chemically reactive, evaporation, or diffusion types, hot-melt types, delayed-tack types, pressure-sensitive types, or electrically and thermally conductive types.

Table 47 Physical Forms of Uncured Structural Adhesives

1. 350°F (177°C) curing films[a]
2. 250°F (121°C) curing films
3. 250°F (121°C) curing pastes
4. Room temperature–curing pastes (two-part systems)
5. Primer/liquid combinations (room temperature curing)
6. Room temperature–curing liquids
7. High temperature–adhesive films and pastes

[a]Adhesives deposited over carrier materials are called tapes and can be available for curing at various temperatures.

For further reading on these classifications, see Refs. 1472–1476. Millard (1477,1478) offered two excellent chapters dealing with the specific application of structural epoxy adhesives to aluminum adherends in Thrall and Shannon's *Adhesive Bonding of Aluminum Alloys.*

A. Phenolic Resin–Containing Structural Adhesives

Phenolic resins can be bonding agents as neat resin adhesives or as part of special formulations as phenolic modifiers. It is mostly in this latter role that we find phenolic resins as important components of adhesives for bonding aluminum. They do have good adhesion to polar surfaces, because of their own polar chemistry. As a neat resin, the phenolics are primarily used as bonding agents in fiber-bonding and friction, abrasive, and foundary applications. In the friction area, bonding can also exist to aluminum, since the applications have included bonding fraction elements to metal backings, for example, automotive brake linings, disk pads, and transmission elements. Other related areas could include bonding to aluminum in air filters, oil filters, and as modifiers in a wide range of formulated coatings, sealants, and adhesives.

Phenolic modifiers, while constituting a smaller volume of phenolic than used as a major constituent in phenolic-base adhesives, are used in developing many structural adhesive formulations used to bond aluminum to itself and a wide variety of other materials of construction. The latter has already been discussed in earlier sections of this chapter describing the use of heated neoprene-phenolic contact cements for laminating aluminum-faced building panels. There are two general types of phenolic modifiers depending on whether the phenolic resin is functioning as a tackifier or as a cross-linking agent. As tackifier, its function is to enhance adhesion and improve wettability to relatively nonpolar adherend surfaces. Such adhesives are distinguishable from true structural-type systems, which rely on cross-linking ability to provide both additional chemical and thermal stabilities. More specific information describing the general form and use of phenolic adhesives and modifiers in bonding processes can be found in the work on that subject by Young and Tandrede (1479).

For structural adhesives, the modifying phenolic resins are resols or hexa-modified novolac phenolic resins. Strength, durability, and heat resistance of elastomer and thermoplastic resin formulations are greatly enhanced by combining the cross-linking properties of phenolics with synthetic polymers like polyvinyl butyral and acrylonitrile copolymers or epoxy resins (1468,1469,1477,1478, 1499).

1. VINYL-PHENOLIC ADHESIVE PRODUCTS

This class of structural adhesives is actually the first successful candidate adhesives developed in England during World War II for joining and assembling aluminum aircraft. Tensile-shear strengths in the range of 3000–4000 psi (20.67–27.6 MPa) and T-peel strengths of 5–35 pli can be achieved on aluminum adherends depending on the particular formulation. Earliest formulations were referred to as the "redux-type" adhesive systems and consisted of first spreading a resol phenolic solution on the surface to be bonded followed by sprinkling powdered polyvinyl formal resin over that coated surface. After solvent evaporation, the bondline was closed and cure affected under heat and modest high-pressure conditions. Variations of this development still exist today in the form of modern REDUX structural film adhesives produced by CIBA. These formulations can generate outstanding joint strengths with good bond durability potentials, which has been the basis for their continued use in the bonding of European aircraft (1480). Details about the applicability of vinyl-phenolic adhesives for aluminum bonding can be found in various publications (1468, 1469, and 1481–1488).

Eickner and Scholwater (1489,1490) made some of the earliest evaluations in the United States on different aluminum preparations using vinyl-phenolic adhesives of the 1950 vintage. Olson et al. (1486) followed with more investigations in the early 1960s, also at the Forest Products Lab. Minford (1481) offered extensive long-term (3 yr) environmental testing in various accelerated laboratory weathering conditions with commercial formulations like American Cyanamid's FM-47 and Narmco's 105 over 6061-T6 adherends. Solomon (1487) also provided a number of literature references illustrating this generally accepted excellent performance on aluminum. Comyn (1491) cited the possible presence of covalent bonding through phenolic oxygens to aluminum ions to help explain the outstanding durability of such aluminum bonded joints to water exposure conditions.

2. *NITRILE-PHENOLIC ADHESIVE PRODUCTS*

As expected from their name, the nitrile-phenolic adhesives are a blend of a nitrile rubber with phenolic novolac resin (sometimes called thermoplastic phenolics). For some time, these formulations in tape or film form accounted for the largest poundage of tape and film adhesives used in the structural bonding of aluminum products. Minford (1481,1492–1494) has shown the exceptional strength retention of chromic acid–deoxidized 6061-T6 joints exposed for 2 yr in various hot-water and condensing-humidity conditions. With natural atmospheric weathering conditions, testing was continued up to 8 yr with equally good results. Although these adhesives formulations do contain nitrile rubber, they are not particularly flexible, because the phenolic characteristics predominate. The phenolic component does, however, make the primary contribution to the respectful strength values shown in Table 48. Similarly, it is the significant factor in enabling the aluminum/nitrile-phenolic joints to withstand temperatures up to about 350°F (176°C).

Table 48 Tensile Shear Strength of Nitrile Phenolic Adhesive Film/Aluminum Adherend Joints

Adhesive identification	Mgf.	Test temp. (°F)	Test temp. (°C)	Shear strength (psi)	Shear strength (MPa)
AF-31	3M Company	75	24	4430	30.54
		300	149	2550	17.57
		350	176	2350	16.20
		400	204	—	—
		500	260	1650[a]	11.37
FM-238	American Cyanamid	75	24	4300	29.64
		300	149		
		350	176	800	5.51
Metlbond 402	Narmco	75	24	4500	31.02
		300	149	2200	15.16
		350	176	1500	10.34

[a]Data developed on steel adherends after 2 hr at 500°F (260°C) using a postcure of 20 hr at 500°F (260°C).

Source: Data from Ref. 1478 (Table 3, p. 137).

Bodnar and Wegman (1495) evaluated the proprietary nitrile-phenolic aerospace film AF-30 on aluminum adherends in the jungles of Panama, obtaining excellent joint strength retentions. Bethune (1074) has confirmed the very satisfactory performance of these kinds of aluminum joints in Boeing aircraft for test periods up to 25,000 flying hours. Modified epoxy systems of the 1974 period offered sporadic disbonding in comparable testing. It should be noted, however, that the PAA pretreatment for preparing aluminum surfaces had not yet been used commercially and the data obtained today for current epoxy formulations would be vastly superior to these earlier epoxy formulations. Wolfe et al. (1496) showed superior fatigue resistance with nitrile-phenolic aluminum joints in 1981. Even with the many epoxy developments of recent years, it is probably true that no other structural adhesive formulations have demonstrated any superiority over the past nitrile-phenolic aerospace films in maintaining higher levels of initial bond strength after extended exposures to water, salts, or other corrosive media.

Unfortunately, it has been a manufacturing disadvantage for these adhesives that they require a higher bondline pressure while curing at approximately 50–100 psi (0.34–0.70 MPa) along with 300°F (149°C) or higher curing temperatures. These bonding demands have been counter to the desired trend in general manufacturing of encouraging the use of structural adhesive products which require only fixturing pressure on the bondline and rapid curing at temperatures of 250°F (121°C) or below. These latter conditions have been conveniently offered in the newer epoxy technology products.

While the technology for these adhesives was largely developed by the major U.S. rubber companies at the end of World War II, there have been additional improvements more recently. These mainly came in the early 1950s patent of Reynolds (1342) incorporating small amounts of carboxyl groups. Other investigators like Doolittle and Powell (1343), Brown and Anderson (1344), and Smarnook and Bonotto (1345) have also reported this positive effect of carboxyl group additions on enhancing the adhesion to aluminum adherends.

B. Epoxy Resin Adhesives

1. *GENERAL CONSIDERATIONS*

A variety of good texts can be referred to the reader's attention with regard to the chemistry and composition of epoxy resin–based adhesives such as those of Lee and Neville (1497) and Lewis and Saxon (1498). Important chapters which consider epoxy chemistry and various modifications to form structural adhesives are those of Bolger (1468,1469), Millard (1478), Dannenberg and May (1499), and Nielsen (1500). Most recently, Mulhaupt and Powell (1501) have considered the subject at the Adhesion '90 meetings at Cambridge and Behn and Cannon (1267) have produced a chapter for the 1990 ASM's *Engineered Materials Handbook*.

The basic resin used in most epoxy-formulated adhesives can be readily and rather economically synthesized. There can be little doubt that for the majority of aluminum structural bonding applications that exist today, epoxy-type resin formulations play a major role. Epoxy resin may be used as main adhesive component, or in combination with various elastomer or other resin raw materials. Unlike vinyl- or nitrile-phenolic structural adhesives that must be heat cured to develop true structural character, epoxy-type adhesives may be formulated to cure at room temperature. These are two-part systems where the curing agent must be kept physically separate until mixing prior to application. As a one-part system, the means of affecting cure must be premixed and packaged with the epoxy resin, but some heat-triggering mechanism must be provided to affect final cure with the adhesive already present in the bondline. One-part epoxy adhesives have a distinct advantage over elastomer-phenolic structural adhesives because they can be cured at lower temperature and without special bondline pressure. With epoxies there is no need to exert this pressure, which is necessary in the case of phenolic formulations to control the volatiles released during the normal curing mechanism. It will be confirmed later in Chapter 10, however, that there is a tendency to develop some higher joint durability responses using one-part epoxies when curing in the 350°F (177°C) temperature range rather than at 250°F (121°C), which constitutes the lower end of the practical heat-curing range for one-part structural epoxy adhesives.

Two-part epoxy formulations are only available as pastes, liquids, or combinations of the two. One part-systems are also available in paste form for many general manufacturing applications, whereas the aerospace industry has historically favored the use of preformed films and tapes because of the better control of adhesive thickness in the bondline. Pressure requirement on bonding either one- or two-part epoxies is most related to achieving a uniform contact by the adhesive without any movement in the bonding while curing is ongoing. This versatility of being able to use a wide variety of forms of epoxy plus offering a wide range of curing conditions has obviously provided a means of finding a set of bonding conditions to fit most any manufacturing operation involved in structurally joining aluminum.

2. *EVALUATION OF PROPERTIES OF EPOXY ADHESIVES*

While most of the fundamental properties of epoxy adhesives have been investigated for many years and can be reviewed in older texts referred to above, there also has been a continuing interest in developing more detailed data in recent years using the earlier simpler formulations. At the same time, there has also been continual development of newer formulations whose properties have to be investigated as soon as possible to properly recommend them in the marketplace.

a. General Properties

In 1985, Gupta et al. (1502) reported temperature-dependence of some mechanical properties of a cured epoxy resin system and Lau et al. (1503) studied the cross-link density of cured epoxy resins. In 1989, Sharon et al. (1504) continued studies of the hygrothermal properties of epoxy film adhesives. Also in 1989, these authors (1505) reported effects of loading rate and temperature on the mechanical properties of tape forms of modified epoxies with a carrier present. Kulawik et al. (1506) made determinations of glass transition temperature, thermal expansion, and shrinkage of selected epoxy products. Garton et al. (1507), in 1989, investigated means of modifying the free volume in epoxy formulations, whereas Rudoi et al. (1508) made surface modifications with reactive fatty acid monoepoxy esters.

In 1990, Sastri and Tesoro (1509,1510) explored new approaches to reversible cross-linking of epoxies and Engleberg and Tesoro (1511) offered measurements of the mechanical and thermal properties of epoxy resins with reversible cross-links. In the same year, Bidstrup and Macosko (1512) considered chemorheological relationships for various epoxy-amine cross-links. Lee and McKenna (1513) measured viscoelastic property responses of epoxy glasses that had been subjected to varying thermal treating. Nakamure et al. (1514) made internal stress measurements on epoxy resins which had been modified with acrylic polymers and polymerized with in situ UV radiation. Massingale et al. (1515) conducted flexibility and adhesion studies of epoxy resins for can and coil coatings. Also in 1990, Ryan et al. (1516) considered networks resulting from fast epoxy polymerization followed in 1991 by Glad and Kramer's (1517) study of microdeformation and network structures in epoxies.

Investigation about mechanical and viscoelastic properties of epoxy networks cured by aromatic diamines was issued by Grillet et al. (1518) in 1991. Physical aging effects on epoxies were investigated by Jo and Ko (1519) and Truong and Ennis (1520) in 1991 on the thermal and mechanical properties and fracture behavior, respectively.

b. Evaluating Cure of Epoxy Adhesives

A review of most important types of curing agents used with epoxy resins will be addressed later in separate discussions about both one- and two-part commercial types of epoxies. The following remarks are intended to describe some of the aspects of curing epoxies which have been reported most recently in the literature. Gillham (1521) has shown that when cure temperature of an epoxy adhesive is significantly below the Tg, it may vitrify before full development of cross-linked networks. Further, Rozenberg (1522) demonstrated that the glassed resin may remain in this chemically uncured state for a considerable time, since chemical reaction in the glassy state is very slow (being diffusion controlled). Sinclair (1523), in 1990, investigated this matter further by following the epoxide depletion with FTIR of adhesives cured at low to high temperatures. He was concerned with identifying some of the properties of the glass/undercured state, including the cure kinetics, dynamic mechanical response, and tensile adhesive strength in bonded joints, under liquid uptake conditions and creep conditions.

In 1987, Jones et al. (1524) conducted a radiochemical study of kinetics of epoxide cure on several model epoxies using various imidazole catalysts. Although phthalic anhydride has been used for curing epoxies for many years, usually with the production of higher temperature resistant polymers, Steinman (1525) only in 1990 investigated more details about this rather old subject matter. Also in 1990, Hodd (1526) discussed various novel curing agents for use with epoxies for adhesive jointing. Lee et al. (1527), in 1990, discussed the dielectric behavior data on epoxies while undergoing cure. Adolf et al. (1528) evaluated both structural and viscosity changes that occur in epoxies when near their sol-gel transition point. Vercheve et al. (1529) concentrated studies in 1990 on network build-up in an epoxy-cycloaliphatic amine system which also included the study of kinetics, vitrifications, and gelation relationships. Kaplan et al. (1530) studied the use of lanthanide-imidizole complexes as latent curing agents for structural epoxy adhesives. Again showing necessity to conduct more detailed investigations of older epoxy systems, Plazek and Frund (1531) obtained more data on curing and physical aging process effects on diglycidyl ether of bisphenol A (DGEBA). Even the effect of the aluminum surface on the curing reaction of an epoxy/amine adhesive was investigated by Zuchas et al. (1532) using differential scanning calorimetry (DSC) data. The same year, these authors issued a paper titled "Model Adherend Surface Effects on Epoxy Cure Reactions" (1533). Their work confirmed significant shifts in the cure exotherms for resin mixtures in the presence of aluminum oxide where fillers were present as compared to situations in which no filler was employed. Earlier Comyn et al. (1534) reported from IETS studies that epoxy/amine reactions were not observable at an aluminum oxide surface owing to absorption of amine by the surface.

In papers issued in 1987 and 1990, Dillingham and Boerio (1535,1536) used XPS and RAIR spectroscopy to show an aluminum oxide surface could actually catalyze the cure reaction of an epoxy/amine system. In 1990, Boerio and Ondrus (1537) reported a different situation when nadic methyl anhydride (NMA) curing agent and benzyldimethylamine accelerator were used with Epon 828 epoxy resin. Using the same RAIR spectroscopy procedures with copper or aluminum adherends, they found NMA curing agent seemed adsorbed onto the aluminum oxide to form a surface carboxylate. With copper oxide present, a vigorous reaction ensued between the anhydride and copper oxide, forming a carboxylate salt accompanied by significant surface etching. This was reported in more detail in 1991 by the same authors (1538). Garton et al. (1539) used IR spectroscopy and differential scanning calorimetry (DSC) procedures to show an acceleration of epoxy/amine systems curing in the presence of acidic carbon surfaces.

Hazony et al. (1540) used a dynamic mechanical analysis technique (tortional braid experiment) to study the curing mechanism of epoxy resins. Lee and Oliver (1541) used the same technique 9 yr later in similar investigations. Levy and Ames (1542), in the 1980s, reported the use of a fluorescence probe for monitoring epoxy cure kinetics.

A number of reports appeared in 1991 about epoxy curing studies, including those of Gilbert et al. (1543) on the mechanism of the dicyandiamide/epoxide reaction, Kaplan et al. (1544) proposing a model reaction for an epoxy resin cured with an aromatic amine, Trappe et al. (1545) on anhydride curing of epoxy resins via chain reaction, and Forster et al. (1546) on the effect of different hardeners on the aging of cross-linked epoxy resins.

c. *Flexibilizing and Toughening of Epoxy Adhesives*

The trend in recent years has been toward toughening the class of epoxy resin adhesives with elastomers and choosing catalyst systems that can cure the epoxy resins at relatively lower temperatures in shorter times. Lewis and Saxon (1498) offered comparisons showing the significantly enhanced tensile-shear and T-peel values for a rubber-modified epoxy as compared to an unmodified epoxy product, as shown in Table 49. Note in particular the fivefold increase in T-peel strength obtained under room temperature test conditions.

Table 50 contains data on typical physical properties of aluminum/aluminum joints bonded with (1) rubber-flexibilized film epoxy adhesive, (2) CTBN rubber-toughened epoxy film adhesive, and

Table 49 Comparison of Aluminum Joint Bond Strengths of Epoxy Adhesives With and Without Rubber Addition

Test	Rubber-modified			Without rubber		
temperature	Tensile	Shear	"T"-peel	Tensile	Shear	"T"-peel
−40	4900	33.76	48	2100	14.47	2
0	5100	35.14	47	2500	17.23	3
25	5200	35.83	44	2700	18.60	3
75	3600	24.80	22	2200	15.16	4
100	2800	19.29	21	3000	20.67	5

Source: Data from Ref. 1498.

(3) a nylon-toughened epoxy adhesive as developed by Pocius (1547). The particularly wide range of lap-shear and T-peel values that can be obtained with different modifier/epoxy combinations is quite dramatic. It might be expected that those modifications producing highest overlap-shear and T-peel would be automatically preferred above others producing distinctly lower values. However, the highest strength formulations are seldom those offering the highest level of durability response, so that compromises are usually sought. The data in Table 51 also come from the same source and covers descriptions of how the "flexibilizing" and "toughening" processes alter glass transition, modulus, and peel strength properties. The differences are decreasing glass transition and modulus as produced by flexibilizing, whereas toughening procedures create little change in these particular property values. Both procedures, however, produce significant increases in peel strength which are a direct reflection of increased toughness and fracture resistance. The effect of incorporating rubbers was recognized more than 20 years ago, as described in 1968 by McGarry and Willner (1548) and Lewis and Saxon (1498) in 1969. Many additional investigations followed in the early 1970s, such as those of Rowe et al. (1549), Siebert and Riew (1550), Sultan et al. (1551), Sultan and McGarry (1552,1553), Siebert and Rowe (1554), and Riew et al. (1555). Important advancement in properties is even more dramatically presented by comparing the range of higher lap-shear and peel strengths shown in Table 49 with the 1000–3000 psi (6.81–20.67 MPa) shear strengths measured by Minford for most conventional unmodified, two-part, room temperature–curing epoxies on variously pretreated 6061-T6 aluminum adherends. More details about the character of the separated phase particles and their effect on the strength and fracturing properties of the rubber-modified epoxies will be considered in Chapter 7 on the subject of joint failure.

The proliferation of investigations about the means of toughening and flexibilizing epoxies, determining fracture properties and unique phase separation, and curing conditions continued through the 1980s. Some more recent investigations include those of Vazirani (1556) on various ways of flexibilizing epoxy systems for bonding with metals and plastics, as potting compounds, and binders for composites. Considerable effort was extended in the early 1980s at Goodrich to develop a series

Table 50 Typical Physical Properties of Aluminum/Aluminum Adhesive Bonds Using 250°F (121 °C) Curing Epoxy Films

	RT overlap shear		RT T-peel	
	(psi)	(MPa)	(piw)	(N/25 mm)
Rubber-flexibilized film adhesive	3700–5500	(25.5–37.9)	22–90	(96.3–394)
CTBN rubber-toughened film adhesive	4000–6000	(27.6–41.4)	25–75	(109–328)
Nylon-toughened film adhesive	5500–7200	(37.9–49.6)	80–130	(350–569)

Source: From Pocius Lecture 4 on Chemistry of Adhesives, ACS Audio-Teleconference Course on Adhesion Science, May 1, 1986.

Table 51 Epoxy-Flexibilizing vs Toughening

Flexibilizing	Toughening
1. Decrease in Tg	1. Slight change (if any) in Tg
2. Decrease in modulus (shear)	2. Slight change (if any) in modulus (shear)
3. Increase in peel strength (fracture resistance)	3. Increase in peel strength (fracture resistance)

Source: From Pocius Lecture 4 on Chemistry of Adhesives, ACS Audio-Teleconference Course on Adhesion Science, May 1, 1986.

of model epoxy formulations containing conventional amine-terminated butadiene/acrylonitrile (ATBN) based on N-aminoethyl piperazine (1557,1558). Chung (1559), in 1984, investigated cure mechanisms of a modified nitrile-epoxy, whereas Bishop (1560), in the same year, considered novel surface and interfacial analysis techniques as aids in the development of new high fracture toughness film adhesives. Drake (1561) offered some formulations for epoxy adhesives with amine-reactive polybutadiene/acrylonitrile liquid polymers to a conference in 1984.

In 1985, Takamura et al. (1562) developed data on the dynamic mechanical properties and adhesive strength of epoxies modified with ATBN rubber and Sanjane and Kupchella (1563) made dynamic mechanical analyses of rubber-toughened epoxy. Goodrich investigators Drake and Tolley (1564) offered discussion of some of these systems in 1986. In 1986, Siebert et al. (1565) discussed formulating epoxy adhesives with carboxy-terminated polybutadiene/acrylonitrile liquid polymers. Kandyrin et al. (1566), in the USSR, discussed the structure of epoxy-rubber films prepared from solution in a mixture of solvents.

Inoue et al. (1567), in 1988, described a cocontinuous or interwoven two-phase structure for formulations described as ATBN/fatty polyamide/epoxy adhesives. Microscopy studies showed connected globules epoxy-rich phases dispersed in a rubber-rich matrix. Pulliam et al. (1568), in 1989, claimed much greater formulating versatility was achieved with amine-reactive modifiers in epoxy formulations as a result of different compatibilities with major groups of amine-cure agents and resulting lower viscosities of the formulations. In 1989, Shaw and Tod (1569) updated the data on the effect of cure conditions on rubber-modified epoxy adhesives. In 1990, Sankaran and Chanda (1570) stated their mechanical, thermal, and microscopic studies of epoxies toughened with hydroxy-terminated polybutadiene-acrylonitrile. Lee and Kyu (1571) obtained data on the phase separation dynamics of rubber/epoxy mixtures the same year. In the area of fracture mechanics investigations, Hunston and Mizumachi (1572) cooperated in their 1990 investigations about micromechanics of fracture in toughened epoxies, whereas Ricco et al. (1573) studied problems of the fracture-mechanics characterization of rubber-modified glassy polymers using double tortion procedures.

The year 1991 showed a continuing flood of papers related to this subject area. Chartoff et al. (1574) reported on the morphology and mechanical properties of rubber-modified aromatic-ether bismaleimide matrix resins. Verchere et al. (1575,1576) continued papers covering the influence of the morphology and mechanical properties and the influence of cure schedule and rubber concentration on generated morphology. Iijima et al. (1577) considered toughening of epoxy resins with aromatic polyester modification and Chen and Jan (1578) the effect of rubber/matrix interfacial modifications on properties of rubber-toughened epoxy resin. Gilbert and Buchnall (1579) worked on epoxy toughening with various thermoplastics, whereas Levita et al. (1580) discussed the relation between the cross-link density and the fracture toughness of epoxy systems. Zhang et al. (1581) discussed rubber toughening of glass fiber-epoxy filament wound composites. Sue (1582) issued paper I in a series of studies of rubber-modified brittle epoxies. This first paper considered measuring fracture toughness using the double-notch four-point-bend method. Finally, Levita et al. (1583) discussed toughness of epoxies that had been modified by preformed acrylic rubber particles.

C. Room Temperature–Curing Epoxy Adhesives

These systems generally consist of a liquid or paste epoxy resin (component A) which must be thoroughly mixed with an aliphatic amine or a polyamide (component B) in a ratio designed to reach a desired level of polymerization of the epoxy resin. Various pigments or fillers like aluminum powder, silica, or carbonate in desired particle size are added by the formulator to component A to control mix viscosity and color and reduce shrinkage of adhesive during the curing process.

Polyamide-cured formulations have found greater favor among manufacturers because of the lower skin and respiratory irritant action of these curative agents compared with the aliphatic amines. Polyamides permit more nearly equal resin to catalyst ratios, making it possible to develop easy to measure equal volume mix formulations. Many later failures with aluminum joints bonded with two-part epoxies have been traced directly to weighing errors which either prevented the development of full curing or accelerated the curing rate to the degree that the structural integrity of the final bondline was altered.

Another possible manufacturing advantage for polyamide-type curing agents stems from their slower catalytic effect. This offers a longer pot-life situation which favors better control of manual assembly operations. Finally, polyamide-type curing formulations cure with less overall rigidity, which means less sensitivity to fracture failing in the bondline. The degree of rigidity can be controlled to a significant degree by altering the resin to curing agent ratio. Regardless of which type of hardener is selected, there is a general relationship of hardener to overall epoxy strength which needs to be considered. Mahoney (1584) pointed out that excessive amounts of hardener obviously remain unreacted and may cohesively weaken areas within apparently cured epoxy resin. Heat resistance will be consequently lowered, whereas permeability to water will increase. Plotting the rate of development of tensile strength against hardener will show a rapid development of strength to a maximum value (hardener content for stochiometric full cure) then a rapid decline and an approach to a minimum, which could be as little as 25% of the maximum value.

Minford (1585–1590) evaluated a large number of different proprietary two-part, room temperature–curing epoxies in an attempt to assay variability in joint strength and joint durability that was present in the range of products available to the general manufacturer. These kinds of data should be of particular interest to many aluminum structure manufacturers, since the product data sheet descriptions for this kind of product all seem remarkably similar. This similarity could create the general conclusion that all epoxies are members of a common family of materials and, thus, should be expected to be relatively interchangeable. Nothing could be further from the truth, as shown by the joint durability investigations of Minford over many years. It should be remembered that Minford chose to use 6061-T6 aluminum alloy adherends for developing this initial strength and long-term durability data, which further deviates from the general manufacturer's data sheets where aircraft alloys like Alclad 2024-T3 were more generally the adherends of reference. The heavy heat-treat film on 6061 alloy could discriminate further bond strength and durability differences as compared with the data on manufacturer's data sheets. This would be especially the situation in which general manufacturing conditions did not permit the use of more expensive surface pretreatments as used in aerospace applications. Significant differences in the ability of different two-part epoxies to adequately wet these surfaces would also exist owing to inherent viscosity differences and the possible introduction of some heating to accelerate the time to achieve full cure conditions.

In order to obtain the full potential of aluminum joint strength possible with two-part paste epoxy products, it is necessary to explore their abilities to chemically wet deoxidized aluminum surfaces and/or include accelerated curing with external heating. Many investigators have shown different epoxy formulations can respond with wide variation to elevated temperature curing. Minford (1591), in this regard, investigated the effect on cure rate and lap-joint strength development using a commercial two-part epoxy (Amicon T-518) on vapor-degreased–only 6061-T6 adherends. The data in Table 52 show that as the curing proceeds for periods from 0 to 120 min at room temperature, the lap-shear strength increased up to 580 psi (4 Mpa) with essentially 100% cohesive failure in the adhesive. In an additional 60-min curing to 180-min total curing time, the strength continued to rise to 880 psi (6.06 MPA); however, the joint now failed 100% adhesively. It is generally understood that a joint showing cohesive-type failure in the bondline is usually going to result in maximum joint strength

Table 52 Rate of Development of Shear Strength in 6061-T6 Lap-Joints Bonded With Amicon T518 Two-Part Epoxy Adhesive

Curing conditions						Lap-shear strength		Visual-type failure
Step 1			Step 2					
Time (min)	Temp. (°F)	Temp. (°C)	Time (min)	Temp. (°F)	Temp. (°C)	(psi)	(MPa)	
60	75	25	—	—	—	60	0.41	100% cohesive
120	75	25	—	—	—	580	4.00	80% cohesive
180	75	25	—	—	—	880	6.06	100% adhesive
1[a]	250	121	([b])	75	25	5	0.03	100% cohesive
1[a]	250	121	([b]) + 10	75	25	30	0.21	100% cohesive
1[a]	250	121	([b]) + 25	75	25	180	1.24	100% cohesive
2[a]	250	121	([b])	75	25	140	0.96	100% cohesive
2[a]	250	121	([b]) + 15	75	25	920	6.34	70% cohesive
3[a]	250	121	([b])	75	25	1200	8.27	30% cohesive
3[a]	250	121	([b]) + 15	75	25	2400	16.54	15% cohesive
3[a]	250	121	([b]) + 60	75	25	2260	15.57	80% cohesive
5[a]	250	121	([b])	75	25	3160	21.77	100% cohesive
5[a]	250	121	([b]) + 15	75	25	3000	20.67	100% cohesive

Source: Data from Ref. 1591 (Table I, p. 6).
[a]Total resident time in a recirculating air oven at indicated temperature.
[b]Few minutes time allowed to cool to room temperature before tensile testing.

for that joint. Further, an adhesive-type failure usually represents an inferior wetting performance by the adhesive on the adherend. This clearly is not applicable to the above observation where an adhesive-type failure accompanied a higher joint strength failure. The explanation is apparent if one analyzes the above observation as being due to slow strength build-up in the adhesive through 120 min, with the stress level to fail the adhesive cohesively being less than that necessary to fail the overall adhesion of the adhesive to the adherend. After 180-min curing, however, the cohesive strength within the bulk adhesive now exceeds the stress level to fail the adhesive at the interface and the mode of failure changes from cohesive in the adhesive to adhesive at the interface. For this to take place, the ability of the paste epoxy adhesive to wet the vapor-degreased adherend at room temperature had to be relatively poor. Now when similar joints were placed in 250°F (121°C) air for as short a time as 2 min and cooled to room temperature and held for only 15 min, the joint strength exceeded that obtained after 180 min at room temperature. Further, the joint failure was 70% cohesive, indicating much superior wetting had been achieved during the brief heating period. With only 5-min heating, these joints could be cooled to ambient temperature and immediately tested to produce a 100% cohesive-type failure at a joint strength that was almost four times higher than achieved after 180 min at ambient temperature. Obviously, even though the epoxy paste did not appear to go through a lowered viscosity and better flow state during the 5-min heating period, a much superior chemical wetting of the vapor-degreased 6061-T6 surface must have been achieved. It will be reported in Chapter 10 that the durability of these heat-cured joints under water-soaking exposure conditions was significantly improved.

It needs to be pointed out that this same degree of improvement in joint strength with heat curing of two-part epoxies is not necessarily present in all formulations. In Table 53, more data are reported by the author using two high-technology, two-part, room temperature–curing epoxies formulated for aerospace applications. Notice that the joint strength advanced about 25% with heat curing the EA-9320 adhesive over vapor-degreased surfaces. With abraded surfaces, the advance was only about

Table 53 Effect of Heat-Curing on Two-Part Epoxy/Aluminum Initial Joint Strengths[a]

Curing conditions	Adhesive[b]	Surface treatment	Lap-shear strength (psi)	Lap-shear strength (MPa)
7 da at RT	EA-9320	Vapor degrease	2860	19.71
60 min at 180°F (82°C)	EA-9320	Vapor degrease	3640	25.08
7 da at RT	EA-9309	Aloxite 60 abrasion	3640	25.08
60 min at 180°F (82°C)	EA-9309	Aloxite 60 abrasion	4100	28.25
7 da at RT	EA-9320	Aloxite 60 abrasion	4100	28.25
60 min at 180°F (82°C)	EA-9320	Aloxite 60 abrasion	4360	30.04
7 da at RT	EA-9309	Acid etched	3640	25.08
60 min at 180°F (82°C)	EA-9309	Acid etched	5700	39.27
7 da at RT	EA-9320	Acid etched	5620	38.72
60 min at 180°F (82°C)	EA-9320	Acid etched	5700	39.27

[a]All joints made with 0.125-in (3.18 mm) 6061-T6 aluminum adherends.
[b]Both adhesives were two-part epoxy products from Hysol specially formulated to offer exceptionally high lap-shear and peel strength without heat-curing of the bondline.
Source: Data from Ref. 1592.

10%, whereas there was virtually no strength advance when acid-deoxidized surfaces were employed. With a different EA-9309 epoxy from the same adhesive formulator, the strength advancement with heat curing was 56% even with deoxidized 6061-T6 alloy adherends. The conclusion must be that each different epoxy formulation must be individually tested to determine the special effect of heat curing. It should be pointed out that the data sheets for these two adhesive formulations from the same manufacturer failed to hint that this strength performance differential might be present.

As just mentioned, the highest lap-shear strength values were generally obtained with two-part epoxies by heating even though the manufacturer's data sheet may only refer to room temperature–curing conditions. A further gain, especially in joint durability, will result from the use of specially pretreated surfaces. Regardless of the optimization of results achieved using heat curing and surface pretreating, there can still be a wide divergence of joint strength and joint durability properties among the available commercial epoxies not found in data sheets. For example, Minford (1592) has shown that the two aerospace-type, two-part, room temperature–curing epoxies shown in Table 53 could have significantly different durability performances, though the information about these two products on the data sheets failed to include any differentiating information. When the author was asked to make a recommendation between the two for a structural aluminum application requiring simultaneous stress and aggressive weathering in service, some lengthy test program had to be undertaken. The data in Table 54 were developed from this program (1593) showing the distinction that could be made when joints were soaked in 100% RH at 125°F (52°C) conditions accompanied by stress levels equivalent to 2.5, 5, 10, and 20% of the average joint strength. Distinctly, the average joint survival times ranging from 25 to 19,600 min for the EA-9309 joints were significantly lower than the 20,160–691,200 min average for the corresponding EA-9320 joints. Without such specifically developed information, it would have been impossible to recommend this adhesive from the data available from the adhesive formulator. Subsequently, this adhesive was recommended for fabricating large aluminum satellite dishes which have been in service for many years in many remote areas.

Blommers et al. (1594) reported on a large aerospace study to develop a low-pressure, room temperature–curing epoxy for field repair of aluminum airframes capable of meeting the MIL-A-5090B testing requirements. However, by showing a pot-life of only 40 min at 176°F (80°C), it missed meeting all parts of the specification. The formulation called for alloying epoxy resin with polyamides or polysulfide rubbers to give a tough, impact-resistant quality. In 1962, Dickstein et al. (1595) gave an expanded review of room temperature–curing adhesives for aluminum structures. In 1965 and 1972, Tanner (1596,1597) reviewed the evaluation data on a variety of adhesives studied at Picatinny

Table 54 Different Stress/Humidity Durability Response of Two Epoxy Adhesives With Indistinguishable Manufacturer Data Sheet Properties for Bonding Aluminum Adherends[a]

Adhesive	Curing conditions: Time (min)	Temp. (°F)	Temp. (°C)	Surface pretreat	Stressing conditions (%)	(psi)	(MPa)	Time to failure (min)
EA-9309	60	180	82	VD[b]	20	848	5.84	25
RS-9309	60	180	82	VD	10	424	2.92	195
EA-9309	60	180	82	VD	5	212	1.46	1,250
EA-9309	60	180	82	VD	2.5	106	0.73	19,600
EA-9320	60	180	82	VD	20	728	5.02	20,160
EA-9320	60	180	82	VD	10	364	2.51	43,200
EA-9320	60	180	82	VD	5	182	1.25	691,200

[a]All joints fabricated with 0.125-in (3.18 mm) 6061-T6 aluminum alloy adherends and exposed to 100% RH at 125°F (52°C) under stress.
[b]Vapor-degreased–only pretreatment.
Source: Data from Ref. No. 1593.

Arsenal which included a number of two-part, room temperature–curing epoxy products on aluminum adherends. Bodnar and Wegman (1598) published their outdoor-aging test results on aluminum joints combined with some commercial two-part epoxy adhesives in 1968. Earlier Schlies (1599) reported on the long-time stability of aluminum/epoxy joints under ambient laboratory conditions.

a. *Wettability of Aluminum Adherends*

The ability of two-part epoxy paste adhesives to wet aluminum surfaces needs to be especially discussed, since differently treated surfaces produce different strength joints, as shown in Table 55. Typical commercial formulations can require as long as 7 days at room temperature to develop maximum strength potential and the author typically allowed this curing time in all his studies of two-part, room temperature–curing epoxies before attempting to measure strength values or change in joint property values. The relatively viscous nature of these adhesives (especially those heavily filled) makes it quite difficult to adequately wet a mill-finish or even a vapor-degreased–only aluminum adherend. This is evident from the nature of the typical lap-joint failure observations after failing in a tensile test machine. Even after the 7 days of curing, these adherend joints will usually show very little cohesive-type failure, indicating the adhesive would have been capable of generating a stronger joint if better wetting conditions had existed. When aluminum adherends are specially pretreated by abrading or chemically deoxidizing the surface, then cohesive failures will generally result with the same epoxy. It is also interesting to note that the author has obtained the same type of cohesive failure even with vapor-degreased–only aluminum adherends when heat-curing the same two-part epoxy. Even more surprising has been the author's observation that duplicate joints to those showing purely adhesive-type failure after 7 days of curing could be converted to 100% cohesive-type failures by simply heating the joints for a few minutes at 250°F (177°C). Considering that the epoxy adhesive after 7 days is totally thermoset, it seems remarkable that such a short exposure to heat can change the nature of the failure-mode as though better surface wetting were being achieved without the surfaces or interfacial area being directly exposed.

b. *Effect of Pot-Life on Adhesion*

The serviceable pot-life of most commercial two-part epoxies has been in the range of 30–60 min at room temperature providing that high exotherm is not developed, as would result from mixing large

Table 55 Initial Joint Strengths of 6061-T6 Aluminum Alloy Joints With Various Abraded, Etched, Conversion Coat, Pain, and Anodize Surface Pretreatments Bonded With One- or Two-Part Epoxy Adhesives

		Curing		Avg. joint strength	
Adhesive	Surface pretreatment	Time	Temp. (°C)	(psi)	(MPa)
2-Part epoxy	Vapor degrease	7 da	25	2131	14.68
1-Part epoxy	Vapor degrease	15 min	204	4308	29.68
2-Part epoxy	Belt-sanding	7 da	25	2030	13.99
1-Part epoxy	Belt-sanding	15 min	204	4869	33.55
2-Part epoxy	Silica gritblast	7 da	25	2017	13.90
2-Part epoxy	Chromic acid deoxidize	7 da	25	2830	19.50
1-Part epoxy	Chromic acid deoxidize	15 min	204	5329	36.72
2-Part epoxy	Alodine 1200	7 da	25	1130	7.79
1-Part epoxy	(conversion coating)	15 min	204	2470	17.02
2-Part epoxy	Epoxy-phenolic (baked	7 da	25	2930	20.19
1-Part epoxy	paint primer)	15 min	204	5155	35.52
2-Part epoxy	Alumilite 215[a] (H_2SO_4)	7 da	25	2530	17.43
1-Part epoxy	(anodize at 0.8 mil)	15 min	204	3270	22.53

[a]Alcoa trademark.
Source: Data from Ref. 774 (Table I, p. 488).

batches. This would imply that sufficient viscosity change can occur by this time that effective wetting of the aluminum oxide has become highly improbable. The physical difficulty of spreading such a viscous mass also exists. These pot-life estimates are based on the use of a certain bulk of mixed adhesive. As the mass increases, the amount of heat-exotherm increases rapidly and correspondingly decreases pot-life. Manufacturing operations employing automated mixing has distinct advantages over such manual mixing and application methods. Recently, epoxy formulators have sought to find curing systems for two-part epoxies which can create fixturing strengths in 5–10 min. Obviously, any so-called pot-life for such formulations using manual mix procedures is so short as to circumvent practical manual useage. The rapid development of significant heat exotherm in such adhesives is highly significant even with small batches. While the initial strength that can be achieved with these accelerated cure, two-part exposies is quite comparable with the longer pot-life products, there is a sacrifice of joint durability performance in water-soak conditions. Minford (1600) has shown there is a highly significant sacrifice of overall aluminum joint durability as compared to many of the older longer pot-life formulations. It would seem a more practical compromise for achieving more acceptable joint permanence with faster production potential might be achievment of 85% of ultimate joint strength overnight at room temperature.

c. *Relationship Between Type of Joint Failure and Joint Durability*

While the above observation of finding some correlation between increasing amounts of cohesive failure (better wetting) and joint durability potential can be demonstrated for joints made with the same adhesive in the bondline, the same relationship does not necessarily hold when comparing cohesive failure percentages and durability performances using different epoxies cured at the same temperature. The author in evaluating thousands of aluminum/two-part epoxy joints found no satisfying correlation between the presence of initial 100% cohesive failure and the joint's ability to survive in accelerated weathering conditions. Many times when evaluating a current crop of new commercial two-part epoxy adhesives, the author found that a particular adhesive which happened to record the highest initial joint strength (accompanied by an apparent initial 100% cohesive-type joint failure) might manifest the poorest response in the group for maintaining strength in adverse weather. Another proprietary adhesive in the group with half the initial strength level, meanwhile, might show no strength decline, though showing the same kind of initial appearing failure response. Thus, unless the manufacturer's data sheet has furnished specific durability test results that fit all the requirements of the customer's service conditions, the best adhesive choice can only be proven by separate testing. Certainly in the situation just recounted, a manufacturer would have made a grevious error by making a selection based only on the manufacturer's data sheet and observing the initial 100% cohesive nature in the absence of any weathering or accelerated laboratory testing.

Many durability investigations of two-part epoxy formulations have been conducted, often with some significant variability in test results. Eichner (1489,1490), Minford (1356,1357), and Bodnar and Wegman (1598) all claimed good durability in jungle sites of joints fabricated with two-part, room temperature–curing epoxies. Cotter (851), however, obtained relatively poor performance from this same class of adhesives at a jungle site in Australia. It would seem more likely in the author's view that this different response was more a manifestation of the variability in performance of different commercial epoxies than the presence of some unique deteriorating factor in the Australian jungle as contrasted with the jungles of Panama used by Eichner and Bodnar and Wegman or the Surinam jungle in South America employed by Minford.

Although Minford has found credible durability performance could be achieved in various water-soaking exposure conditions (especially where the aluminum surface is deoxidized), there is one type of accelerated weathering condition that is not well tolerated. Reference is made here to continuous or intermittent salt water exposure conditions. Even with an FPL-type deoxidizing treatment, the joint survival times are limited to a few weeks of direct exposure owing to the rapid progression of pitting corrosion proceeding from the joint edges across the interface. Minford (533,534,774) has shown that the durability of two-part epoxy joints can still be quite acceptable if CAA, SAA, or PAA anodizing pretreatments have been used. To a lesser degree, this has also been demonstrated with some chromate conversion coating surface preconditioning. Unfortunately, many general manufacturing operations that might wish to use two-part epoxy adhesives because of their lower cost than aerospace film and tape products will find it to expensive to use these pretreatments. It may be necessary to shift to a one-part, heat-curing epoxy adhesive giving generally higher initial joint strengths and probably better opportunity for superior surface wetting even in the absence of special aluminum pretreatments.

In 1965, Sharpe (1601) evaluated the sensitivity of two-part epoxy/aluminum joints at Bell Labs under simultaneous stress and hot humidity conditions and demonstrated that failure times were only measured in a matter of minutes to hours. Ordinarily, these joints would not delaminate in the hot humidity alone for months to years providing the surface was deoxidized and no external stress was imposed on the bondline. The maintenance of a high percentage of initial strength by such unstressed joints was demonstrated by Minford, as shown in Figure 17 (1602,1603). Minford (1604) has also reported similar observations on the effect of stress superimposed on wet-soaking conditions using a different method of imposing the stress (Fig. 10). Wangsness (1605) has also reported stress/hot humidity aluminum joint durability results for a variety of 3M's two-part epoxy formulated products using still another method of imposing the stress on the joint, as shown in Figure 11. It should be pointed out that only the Alcoa ring fixture offers the ready possibility of easily transporting and stacking the lap-joints under sustained stress among different weathering conditions to study the added effect of cycling weather situations.

Additional references of possible interest might include the following. Garnish (1606) has presented a survey titled "Epoxy Resins as Adhesives: Past and Present." Graham and O'Connor (1607) reported a low temperature–cure adhesive containing an epoxy resin combination with triglycidyl p-aminophenol (TGPAP) and BTDA curing agent. While no other catalyst system is necessary, the solubilization of the BTDA will be aided by warming the mixture. In 1982, Ku and Bell (1608) described a room temperature–curing epoxide made by converting a part of the epoxy function into an episulfide by the addition of N-methyl-benzothiazole-2-thione (NMBTT). The resultant can be cured into a "bond cement" within 5–20 min at RT. Also, in 1982, Brown and McCrea (1609) considered competition changes for growth of the epoxy adhesive markets, whereas Tompson and Ting (1610) studied some epoxy polymers for underwater bonding applications.

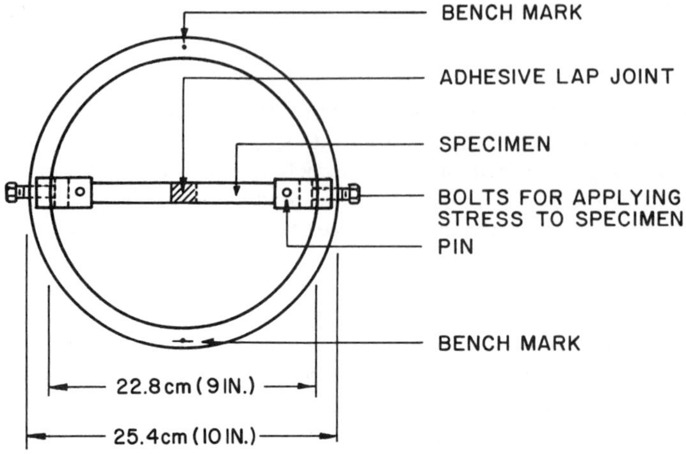

Figure 10 Alcoa Stressing Fixure Details. (From J. D. Minford, Aluminum Adhesive Bond Permanence, Treatise on Adhesion and Adhesives [R. L. Patrick, ed.]. Marcel Dekker, New York, Vol. 5, Figure 3-53, p. 120 [1981].)

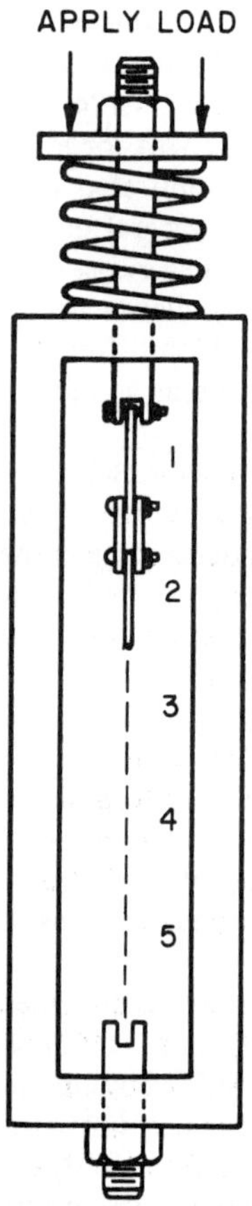

Figure 11 3M's spring-loaded fixure. (From J. D. Minford, Aluminum Adhesive Bond Permance, Treatise on Adhesion and Adhesives [R. L. Patrick, ed.]. Marcel Dekker, New York, Vol. 5, Figure 3-52, p. 118 [1981].)

4. *ELEVATED TEMPERATURE–CURING EPOXY ADHESIVES*

The two-part, room temperature–curing epoxy formulations are undoubtedly produced and consumed in larger quantities by general manufacturing processes than the one-part, heat-curing types. However, when restricting our view to the higher-technology industries like aerospace and automotive manufacturing, the opposite is certainly true. Most of the major manufacturers of epoxy adhesive products are constantly exploring the new curing techniques that will permit the formulation of one-component structural epoxies that have long shelf life under room temperature storage conditions but will cure rapidly in the 180–350°F (82–177°C) range.

The stimulus for the above extra research effort can be cited in a variety of factors which affect the manufacturing reproducibility and economic costs. It should be obvious that when two or more reactive liquids or paste combinations must be mixed shortly before use that increased production costs are necessarily involved in the metering, mixing, and degassing of the separate components. Further, as the pot-life of the multiple mix materials becomes shorter, the possibilities increase for premature thickening or gelling before adequately wetting the surface. All these factors can significantly reduce the reliability of finding consistently strong and equally durable joints in service.

A plant producing a large number per day of any bonded items cannot actually afford to wait for any significant length of time for bonded parts to cure before moving them along to the next assembly steps. This is even more unacceptable if the bonded part must be clamped or fixtured for some time until the bondline cures. Furthermore, heat to cure the bondline, while not inexpensive, can often be less expensive than the time and space requirements required for long room temperature cures. Finally, the two-part epoxy systems simply are not capable of matching the better overall performance properties of heat-cured systems, including higher heat distortion, higher electrical service temperature ratings, superior resistance to chemical attack by acids, bases, swelling solvents, and bond deterioration by water. Minford (774) has published evidence showing that this last superior durability in water condition may not always be true. His data showed that vapor-degreased-only 6061-T6 joints that had been soaked for 24 mo in room temperature distilled water actually retained a higher percentage of their initial joint strength fabricated with a two-part, polyamide-cured epoxy than with a one-part, dicy-cured, nitrile-modified, heat-cured epoxy. The final joint strengths were very similar but only because the one-part epoxy joints had a significantly higher initial joint strength. Even when the 6061-T6 surfaces were deoxidized, the actual loss of strength by the heat-cured epoxy joints was larger in magnitude; however, the overall retained joint strength average was 32% higher. Obviously, there can be a fairly wide range of different durability performances shown by different available commercial one-part products. Thus, the author can only report the above relationships to be true for the specific products evaluated under the specified conditions of exposure. In general, it has been reported that the heat-cured epoxies can provide higher combinations of both peel and tensile-shear strengths as well as better tolerance to oily metal surfaces. This subject will be considered in more detail in Chapter 10 on joint durability evaluations.

The best current one-part epoxy paste adhesives are relatively complex mixtures of epoxy resins, flexibilizers, surfactants, colloidal fillers to provide sag resistance, and "latent" curing agents which are inactive at normal storage temperatures but can be activated with heat. Since heat-curing film epoxy adhesives are almost exclusively used in aerospace bonding, it should be obvious that a better array of properties has been demonstrated overall as compared to those generally found in two-part, room temperature–curing paste compositions. The most frequently used forms of structural epoxies are the films and tapes, which will be discussed in more detail later. It would seem appropriate at this time to further discuss the variety of one-part, heat-curing epoxies that have been developed by modifications with elastomer, nylon, or phenolic resins. Since heat-curing film epoxy adhesives are almost exclusively used in aerospace bonding, it should be obvious that a better array of properties has been demonstrated overall for the one-part systems as compared to the two-part, room temperature–curing epoxy alternatives.

a. *Elastomer-Modified Epoxy Formulations*

The trend in recent years toward toughening the one-part epoxy resins by blending in elastomer has been discussed earlier citing comparative benefits in lap-shear and T-peel values as compared with the two-part, room temperature–curing products. That the formulations for different commercial nitrile-modified epoxies can also vary in property values is demonstrated in Table 56. The tensile-shear strengths of eight different proprietary products when measured at −67, 75, or 180°F (−55, 24, or 82°C) are compared. While the low-temperature and ambient-temperature strength values of the Metlbond 1133 are obviously higher than for all others, the elevated temperature strength is somewhat lower than the FM-73M or EA-9601 adhesives. The range of low-to-high values for the overall group at the low-to-high test temperatures are 4800–6700 psi (33.1–46.2 MPa), 4800–6700 psi (33.1–44.1 MPa), and 2890–4180 psi (19.9–28.2 MPa), respectively. This reflects the fact that the lowest testing temperature produces the widest range of difference in values, whereas the highest testing temperature offered the least spread among the different commercial products.

In Table 57, the flatwise tension strength values are cited for the same list of eight different epoxy film products. The Metlbond 1133 joints which showed the highest tensile-shear strength in Table 56, are now passed by the FM-73M, AF-126, and Plastilock 717 joints in flatwise tension value at both the −67 and 75°F (−55 and 24°C) test temperatures. At the highest test temperature, however, the Metlbond 1133 joints now excel, surplanting the EA-9601 and FM-73M joints, which were both relatively higher in tensile-shear strength values. Still further comparisons can be made in honeycomb climbing drum peel strength values, as shown in Table 58. The peel strength of the Metlbond 1133 joints now fall in the center of the group with the AF-126-2 joints (not previously high in either tensile-shear or flatwise tension) and now showing a significantly higher peel strength value. Finally,

Table 56 Tensile Shear Strength of Nitrile Epoxy Adhesive Films

Adhesive identification	Manufacturer	Test temperatures (°F)	(°C)	Shear strength (psi)	(MPa)
FM-123-2	American Cyanamid	−67	−55	5180	35.7
		75	24	5140	35.4
		180	82	3300	22.7
FM-73M	American Cyanamid	−67	−55	6540	45.1
		75	24	5900	40.7
		180	82	4180	28.2
AF-126-2	3M Co.	−67	−55	5070	34.9
		75	24	5200	35.9
		180	82	2890	19.9
AF-163-2K	3M Co.	−67	−55	4800	33.1
		75	24	4800	33.1
		180	82	3420	23.6
EA-9601	Dexter Corp.	−67	−55	5000	34.5
		75	24	5300	36.6
		180	82	4300	29.7
EA-9628	Dexter Corp.	−67	−55	6100	42.1
		75	24	6000	41.4
		180	82	4000	27.6
Meltbond 1133	Narmco	−67	−55	6700	46.2
		75	24	6400	44.1
		180	82	4000	27.6
Plastilock 777	B.F. Goodrich	−67	−55	5530	38.1
		75	24	5260	36.3
		180	82	3980	27.4

Source: Data from Ref. 1477 (Table 1, p. 114).

Table 57 Flatwise Tension (FWT) Strength of Nitrile Epoxy Adhesive Films[a]

Adhesive identification	Manufacturer	Test temperature (°F)	Test temperature (°C)	FWT strength (psi)	FWT strength (MPa)
FM-123-2	American Cyanamid	−67	−55	1267	8.7
		75	24	837	5.8
		180	82	420	2.9
FM-73	American Cyanamid	−67	−55	1565	10.8
		75	24	1200	8.3
		180	82	690	4.8
AF-126	3M Co.	−67	−55	1427[b]	9.8
		75	24	1150[b]	7.9
		180	82	535[b]	3.7
AF-163-2K	3M Co.	−67	−55	—	—
		75	24	1025	7.1
		180	82	560	3.9
EA-9601	Dexter Corp.	−67	−55	—	—
		75	24	1230	8.5
		180	82	—	—
EA-9628	Dexter Corp.	−67	−55	—	—
		75	24	1400	9.7
		180	82	—	—
Metlbond 1133	Narmco	−67	−55	1400	9.7
		75	24	1050	7.2
		180	82	720	5.0
Plastilock 717	B.F. Goodrich	−67	−55	1522	10.5
		75	24	1126	7.8
		180	82	600	4.1

[a]All flatwise tension results are from tests conducted with 0.25-in (6.2-mm) cell honeycomb core.
[b]Tests were conducted using 0.085 psf (412 g/m^2) adhesive film. All remaining values are for specimens prepared using 0.06 psf (291 g/m^2) film.
Source: Test data from Ref. 1477 (Table 4, p. 116).

Table 58 Honeycomb Climbing Drum Peel Strength of Nitrile Epoxy Adhesive Films

Adhesive identification	Manufacturer	Test temperature (°F)	Test temperature (°C)	Peel strength (in-lb/3-in width)	Peel strength (N/mm)
FM-123-2	American Cyanamid	75	24	80	14.0
FM-73M	American Cyanamid	75	24	82	14.3
AF-126-2	3M Co.	75	24	120*	21.0[a]
AF-163-2K	3M Co.	75	24	70	12.2
EA-9601	Dexter Corp.	75	24	75	13.1
EA-9628	Dexter Corp.	75	24	60	10.5
Meltbond 1133	Narmco	75	24	80	14.0
Plastilock 717	B.F. Goodrich	75	24	80	14.0

[a]Tests were conducted using 0.085 psf (412 g/m^2) adhesive film. All remaining values are for specimens prepared using 0.060 psf (291 g/m^2) film.
Source: Data from Ref. 1477 (Table 3, p. 116).

in Table 59, the metal-to-metal climbing drum peel values are cited for the same products. It would appear that the value here for the AF-126-2 joints is much lower than previously shown in Table 58 for honeycomb climbing drum peel. However, it should be noticed that a heavier (thicker) film weight adhesive has been used to make the tests in Table 58. This should teach the potential selector of epoxy film products to always be sure comparative values have employed equal thickness materials. These overall results also clearly indicate that dramatic differences can occur between different commercial products, all of which qualify as being described as elastomer-modified epoxy formulations.

Minford has included the evaluation of the same one-part, nitrile-modified, dicy-cured, epoxy paste adhesive and two-part, polyamide-catalyzed, room temperature–curing paste adhesive over 6061-T6 aluminum adherends for more than 20 yr of durability testing in order to be able to make direct comparisons where other test variables such as curing conditions, surface pretreating, and variations in weathering were being evaluated. Obviously, newer formulations were being developed over that time that might improve any particular durability numbers for each different set of fabricating and testing conditions; however, having a constant base for making comparisons seemed more important for long-term durability test evaluations. Recent testing in programs, like PABST, have confirmed that much higher levels of aluminum joint durability are possible when the highest state of the art is used regarding best surface pretreatment and optimum and precisely controlled curing conditions are employed with the latest aerospace film products. Some of the investigations involving the same one- and two-part epoxy adhesives involving exposure times as long as 8–12 yr at the Alcoa Labs can be reviewed (39,68,531–535,771,774,1611,1612).

Both Minford (1356,1357) and Bodnar and Wegman (1598) have confirmed a superior durability performance of particular two-part, room temperature–curing epoxy/aluminum joints over a one-part, heat-curing, nitrile-epoxy joint of the same time frame in jungle-exposure conditions. It should be pointed out, however, that the sophisticated surface pretreating of the aluminum and the high degree of modification achieved in present-day aerospace film adhesives were not employed. Minford has shown the excellent performance of one-part, vinyl- and nitrile-modified phenolic adhesives with chromic acid–deoxidized 6061-T6 adherends as compared with the earlier vintage one-part, heat-cured, nitrile-modified epoxy adhesives in the corrosive seacoast atmosphere. On comparatively treated 6061-T6 adherends, the nitrile-epoxy joints failed after about 4 yr, whereas the vinyl-phenolic or nitrile-phenolic joints remained virtually unchanged after 6 yr. This could be an indication of superior ability to wet the aluminum oxide surface plus some possible covalent bonding between the oxide and the phenolic portions of the adhesives.

Because of a readily apparent sensitivity of the EC-2086 nitrile-modified epoxy paste joints to long-term exposures to 100% RH at 125°F (52°C) conditions, the author anticipated that this same adhesive joint would show an even greater sensitivity to a combination of stressing and hot-soaking

Table 59 Metal-to-Metal Climbing Drum Peel Strength of Nitrile Epoxy Adhesive Films

Adhesive identification	Manufacturer	Test temperature (°F)	Test temperature (°C)	Peel strength (in-lb/3-in width)	Peel strength (N/mm)
FM-123-2	American Cyanamid	75	24	70	12.2
FM-73M	American Cyanamid	75	24	85	14.9
AF-126-2	3M Co.	75	24	80	14.0
AF-163-2K	3M Co.	75	24	79	13.8
EA-9601	Dexter Corp.	75	24	69	10.5
EA-9628	Dexter Corp.	75	24	80	14.0
Metlbond 1133	Narmco	75	24	90	15.7
Plastilock 717	B.F. Goodrich	75	24	84	14.7

Source: Data from Ref. 1477 (Table 2, p. 115).

Table 60 Change in Solubility During Room Temperature Aging of Two Nitrile-Epoxy Adhesives

	Solubility in Chlorobenzene (wt %)	
Aging time (wk)	Adhesive A[a]	Adhesive B[b]
0	84	88
5	60	88
10	57	80
15	54	63
20	50	52

[a]Carboxylated nitrile rubber-modified DGEBA epoxy adhesive where nitrile rubber and epoxy resin were not prereacted.
[b]Similar to ([a]) except the rubber and epoxy resin were prereacted.
Source: Data from Ref. No. 1617 (Table 3, p. 241).

conditions. Surprizingly, Minford has provided data showing similarly treated two-part epoxy joints stressed to 500 psi (3.45 MPa) in this hot humidity condition failed in less than 100 min. Corresponding joints with the EC-2085 one-part epoxy in the bondline could be stressed to 1000 psi (6.89 MPa) (which was approximately the same stress percentage based on initial joint strength) and survived for approximately 72,000 min. Wangsness (1605), Hughes and Rutherford (1614), Schwartz (1615), and DeLollis (1616) have all similarly provided results showing an acceptable performance of nitrile-epoxy, heat-cured aluminum joints under stress and hot humidity–soaking conditions.

The superior durability potential of some of the latest nitrile-modified epoxy formulations on CAA and PAA anodized aluminum adherends have been recorded in the various PABST publications. Four adhesives in this category of structural aerospace films would be Bloomingdale's FM-73, 3M's AF-55, Hysol's EA-9628, and Narmco's M-133. Unlike Minford's earlier evaluations with 3M's EC-2086 paste adhesive mentioned above, a primer was always employed to maximize wetting and protect against the possibility of surface contamination prior to bonding.

Finally, it might be of interest to mention the recent work of Morris et al. (1617) on the aging of structural nitrile-modified epoxy films. In Table 60, we can note the significant decrease in the solubility in chlorobenzene with increasing aging time, whereas Table 61 shows an increase in epoxide content of the film with increasing aging time up to a maximum aging time of only 20 wk. A relationship between the epoxide content, solubility in chlorobenzene solvent, and the adhesive flow properties are reviewed for a high temperature–curing epoxy adhesive in Table 62. Once again, the inverse relationship of increasing epoxide content and decreasing solubility in chlorobenzene previously shown in Tables 60 and 61 is manifested, whereas the ability of the film adhesive to flow over the adherend is virtually diminished to zero after only 30 wk of room temperature aging.

Table 61 Change in Epoxide Content During Room Temperature Aging of Two Nitrile-Epoxy Adhesives

	EEW (g/eq)	
Aging time (wk)	Adhesive A	Adhesive B
0	425	442
5	516	446
10	535	526
15	572	676
20	598	817

Source: Data from Ref. 1617 (Table II, p. 241).

Table 62 Change in Epoxide Content, Solubility, and Flow During RT Aging of a High-Temperature–Curing Epoxy Adhesive

Aging time (wk)	EEW (g/eq)	Chlorobenzene solubility (wt %)	Flow (%)
0	229	66	370
10	245	65	410
20	284	60	310
30	474	38	0

Source: Data from Ref. 1617 (Table I, p. 237).

b. Nylon-Modified Epoxy Adhesives

When the nylon-modified epoxy structural adhesives first appeared on the commercial scene in the early 1960s, the 80–130 piw peel strength and the near 7000 psi (42.23 MPa) shear strengths combined with exceptional fatigue and impact properties seemed like a quantum jump forward as compared to the existing one-part, heat-curing epoxy formulations. It had been anticipated that a combination of the toughness and high-tensile properties of the nylon family of polymers with the inherent strength potential of the epoxy resin family should produce an adhesive with remarkable physical properties. Unfortunately, the various crystalline nylon polymers first available proved to be quite incompatible with the epoxy resins. The situation changes, however, with the development of noncrystalline nylons which were soluble in alcohols and other solvents, melted at temperatures below their decomposition temperatures, and could be blended with epoxy resins to yield curable solid films, tapes, or solvent-based adhesives. Smith and Sussman (1618) described how these low-melting, semicrystalline nylons could be made by melt-blending mixtures of crystalline polymers to obtain random copolymers by chain segment interchange in 1963. Commercial examples of adhesive-grade nylons include BCI's 600 and 800 series nylons and Dupont's Zytet 61. One of the most commonly employed nylon-modified epoxy formulations over many years for laminating helicopter rotors and honeycomb core-to-skins in aerospace consisted essentially of Zytel 61, resorcinol diglycidyl ether, and an alkyl-substituted melamine catalyst.

The tensile-strength values for two important nylon-modified epoxy commercial products (including the widely used FM-1000) can be seen in Table 63. Notice, however, that there can be a 47% decline in joint strength when the testing temperature is elevated from ambient to 180°F (82°C). In Table 64, the flatwise tension strength values for one of these adhesives (Meltbond 406) is shown at −67, 75, and 180°F (−55, 24, and 82°C) and compared with that of FM-1000 at ambient temperature. The values appear to be very similar to those obtained with the nitrile-epoxy adhesives shown

Table 63 Tensile Shear Strength of Nylon-Epoxy Adhesive Films/Aluminum Joints

Adhesive identification	Manufacturer	Test temperature (°F)	Test temperature (°C)	Shear strength (psi)	Shear strength (MPa)
FM-1000	American Cyanamid	−67	−55	7400	51.0
		75	24	7000	48.3
		180	82	3670	25.3
Metlbond 406	Narmco	−67	−55	5000	34.5
		75	24	5000	34.5
		180	82	3200	22.1

Source: Data from Ref. 1477 (Table 5, p. 117).

Table 64 Flatwise Tension (FWT) Strength of Nylon-Epoxy Adhesive Film/Aluminum Joints[a]

Adhesive identification	Manufacturer	Test temperature (°F)	(°C)	Flatwise tension strength (psi)	(MPa)
FM-1000	American Cyanamid	−67	−55	—	—
		75	24	1200	8.3
		180	82	—	—
Metlbond 406	Narmco	−67	−55	1200	8.3
		75	24	900	6.2
		180	82	325	2.2

[a]All flatwise tension results are from tests conducted with 0.25-in (4.7 mm) cell aluminum honeycomb core.
Source: Test results from Ref. 1477 (Table 8, p. 118).

in Table 57. However, the metal-to-metal peel value shown in Table 65 is obviously significantly higher than the corresponding nitrile-epoxy adhesive joint values shown in Table 59, as might be expected. It is interesting to note in Table 65 that the peel strength value for the Metlbond 406, which was determined by the T-peel test procedure instead of the climbing drum procedure, gives a number that is only one-third as high. This again emphasizes the necessity to make sure that a testing procedure has been employed to develop the number offered on an adhesive manufacturer's data sheet before drawing final conclusions about its significance.

After some employment in bonding aircraft, however, it was demonstrated in durability testing that these magnificent initial performance numbers were considerably reduced by a significant sensitivity to moisture in the service environment. Even today, except for improvement provided by special primers on the aluminum, no effective solution has been found to circumvent this inherent sensitivity to water. Many of the efforts to offset this moisture sensitivity involved the use of resol-phenolic resin-based primers. While these primers did prolong the service life, they did not raise the performance levels to those of the best nitrile rubber-toughened epoxies.

In 1985, Millard (1619) presented his conclusions that these epoxy-polyamide adhesives (he prefers this nomenclature for the nylon-modified epoxies) are very useful in secondary bonding operations such as applying rub strips, local doubler reinforcements, and the application of small fittings. They also have performed excellently when bonding aluminum honeycomb-producing well-formed fillets to the honeycomb cells which provide high peel strengths at ambient temperature. They are not recommended for bonding primary structure in aircraft as are the nitrile-epoxy films. Millard

Table 65 Metal-to-Metal Peel Strength of Nylon-Epoxy Adhesive Films

Adhesive identification	Manufacturer	Test temperature (°F)	(°C)	Peel strength (in.-lb/3-in width)	(N/mm)
FM-1000	American Cyanamid	75	24	200[a]	35.0
Meltlbond 406	Narmco	75	24	70[b]	12.3

[a]Climbing drum peel test method.
[b]T-peel test method.
Source: Data from Ref. 1477 (Table 6, p. 118).

also points out that this adhesive class can be modified with a variety of fillers and thixotrophy-modifying compounds, which can produce a wide range of adhesive characteristics. For example, the addition of glass or phenolic microballoons can produce light-weight syntactic foams useful for space-filling applications.

Millard (1619) has also stated that these nylon-modified epoxies, while normally used with adhesive primer, do not necessarily need one. These adhesives were employed prior to the introduction of corrosion-inhibiting (CIAP) primers and have continued to be used with the non–corrosion-inhibiting types of primers. Unlike many other epoxies, these heat-curing, epoxy-polyamide systems require 350°F (177°C) to cure, yet do not perform well above 180°F (82°C). This puts them in direct competition with later-developed nitrile-epoxies which can cure as low as 250°F (121°C). In practice, the very high shear strengths of these adhesives cannot always be utilized by the design engineer, which again makes them quite similar in performance characteristics to the lower temperature-curing nitrile-epoxies.

Both DeLollis (198) and Minford (1620) have presented data showing the sensitivity of this class of adhesive to hot humidity exposure, which will be discussed in more detail in Chapter 10 on joint durability. Minford has also presented data showing the effect of varying stress levels on FM-1000/6061-T6 aluminum joints exposed to condensing humidity at 125°F (52°C) or cycling in corrosive salt water. Also included was the effect of widely varying surface pretreating of the aluminum adherend. Kuno (1621) has studied the effect of higher service temperature on Metlbond 406 nylon-epoxy/aluminum joints.

While one-part, epoxy-polyamide adhesive systems have been in aerospace service for a number of years with generally good results, there have been enough delaminations in service to cause concern where moisture has found egress into the bondline. While there has been some evidence that degradation in bondline strength has often been recoverable when the moisture was eliminated or removed by some natural or artificial means, it is likely that a gradual phrase-out of these film systems will take place in the aerospace industry. The liquid systems, however, may still find a useful application in semistructural bonding and syntactic foam formulations.

c. *Epoxy-Phenolic Adhesive Systems*

Epoxy-phenolic–type structural adhesives are actually a fairly specialized group of film adhesives aimed at the high temperature end of the service temperature range for bonded aerospace structures. They were primarily developed in the 1950s to obtain service at temperatures higher than achievable with the vinyl- and nitrile-phenolics used at the time. Their use has normally been confined to aircraft parts exposed to high temperature during service such as engine nacelles, auxillary power units, and areas generally subject to kinetic heating. A few liquid versions are available which are formulated for use as core splice adhesives. They cure at 350°F (177°C) and can perform well at 300–350°F (149–177°C) for several thousand hours and for short periods up to 500°F (260°C).

Black and Blomquist (1622–1625), at the Forest Products Lab, were the first investigators to combine newly available, high molecular weight epoxy resins with the resol phenolics to generate this new class of epoxy-phenolics. As a substitute for the polyvinyl butyral (PVB) or polyvinyl formal (PVF) resins used in the earlier vinyl-phenolic class of structural adhesives, the epoxy backbone polymer seemed to offer a major improvement in thermal stability. At the same time, however, there was an expected sacrifice in flexibility, elongation, and peel properties in these new formulations. While the maximum tensile-shear strength at room temperature for this class only range from about 2500 to 3500 psi (17.23 to 24.12 MPa), they do have the ability to retain about 2000 psi (13.78 MPa) at temperatures as high as 500°F (260°C), as shown in Table 66. As regards flatwise-tension strength (see Table 67), they tend to form slightly weaker joints at low and room temperature conditions than shown by the nitrile- or nylon-modified epoxy joints in Tables 57 and 64. Their flatwise tension strength at 500°F (260°C) can be equal to or even exceeded by 100% that of similar nylon-epoxy joints at only 180°F (82°C). We might expect because of the high rigidity of their polymer structure to find lower honeycomb peel strength and that is confirmed in Table 68.

Bodnar and Wegman (1495) evaluated HT-424 epoxy-phenolic adhesive on aluminum adherends in three different atmospheric conditions. After 3-yr exposures, the joints sustained higher losses

Table 66 Tensile Shear Strength of Epoxy-Phenolic Adhesive Film/Aluminum Joints

Adhesive identification	Manufacturer	Test temperature (°F)	Test temperature (°C)	Shear strength (psi)	Shear strength (MPa)
HT-424	American Cyanamid	−67	−55	3225	22.2
		75	24	3550	24.5
		300	149	2760	19.0
		500	260	2000	13.8
Meltlbond 302	Narmco	−100	−73	2300	15.9
		75	24	2300	15.9
		300	149	2150	14.8
		500	260	1900	13.1

Source: Data from Ref. 1477 (Table 9, p. 119).

Table 67 Flatwise Tension Strength of Epoxy Phenolic Adhesive Film/Aluminum Joints

Adhesive identification	Manufacturer	Test temperature (°F)	Test temperature (°C)	Flatwise tension strength (psi)	Flatwise tension strength (MPa)
HT-424	American Cyanamid	−67	−55	760	5.2
		75	24	615	4.2
		300	149	485	3.3
		500	260	290	2.0
Metlbond 302	Narmco	−67	−55	925[a]	6.4
		75	24	900[a]	6.2
		300	149	—	—
		500	260	600[a]	4.1

[a]Values obtained from specimens using 0.1875-in (4.7-mm) cell aluminum honeycomb.
Source: Data from Ref. 1477 (Table 11, p. 119).

Table 68 Honeycomb Climbing Drum Peel Strength of Epoxy Phenolic Adhesive Film/Aluminum Joints

Adhesive identification	Manufacturer	Test temperature (°F)	Test temperature (°C)	Peel strength (in-lb/3-in width)	Peel strength (N/mm)
HT-424	American Cyanamid	75	23	33	5.8
Metlbond 302	Narmco	75	23	25	4.4

Source: Data from Ref. 1477 (Table 10, p. 119).

than those made with a simple unfilled, polyamid/epoxy system like Epon 828/Versamid 140. Of particular concern was a loss of 22% in strength in Yuma, Arizona (a desert atmosphere), where the similarly tested polyamide-epoxy joints suffered no losses.

Minford (1626), however, found an apparent higher degree of bond durability performance using EPON 422 (epoxy-phenolic formulation) over chromic acid deoxidized Alclad 2024-T3 aluminum adherends. After 3 yr of continuous immersion in 145°F (63°C) water, these joints averaged retention of 81% of initial strength values. Kuno (1621) has reported the good high-temperature resistance of both HT-424 and EPON 422 epoxy-phenolic bonded aluminum joints. Since the epoxy-phenolic adhesives are formulated to operate at 300°F (149°C) plus, the supporting carrier fabric in tape products is normally of woven glass rather than nylon or polyester, as used in many other structural tape products.

E. Unmodified Epoxy Adhesives

There are a large number of epoxy adhesive systems which must be considered to be "unmodified" in nature. These adhesives can be liquids or film in form and may consist of a number of epoxy polymers in combination with various curing agents.

The liquid combinations can be cured at ambient or elevated temperature depending on the manufacturing facilities available and the particular end use in service of the bonded product. So-called "general-purpose" types of formulations are commonly cured at ambient temperature employing diethylenetetramine (DTA) or triethylenetetramine (TETA). Good bonding can be obtained to properly prepared aluminum adherends, although peel strengths will be relatively poor because of the overall rigidity of the cured adhesive.

The unmodified film adhesives utilize curing at elevated temperature, often utilizing chemical agents such as m-phenylenediamine (MPDA) or methylenedianiline (MDA). Both of these curing agents can produce good elevated-temperature performance but without the addition of some special modifying polymers can only provide modest to poor peel strengths. The peel strength may be raised to some modest degree by combination with certain fillers but only at the sacrifice of some of the elevated-temperature strength. The use of dicyanodiamide and diaminodiphenyl sulfone as curing agents also should be mentioned.

Generally, the ummodified epoxy films are formulated for service at 250°F (121°C) or above but special formulations can resist 350°F (177°C) for several thousand hours or even a few hundred hours at 400°F (204°C). The tensile-shear strength values for aluminum joints formed with three commercial unmodified epoxy film adhesives are reviewed in Table 69 for test temperatures from room temperature up to 400°F (204°C). As can be seen, the strengths can approach 2000 psi. The general rule still prevails that the higher the elevated temperature performance, the poorer the peel strength available regardless of resin/catalyst choice. In 1985, Millard (1619) reproted that unmodified epoxy films had been developed to the degree that they could perform very well at quite high service temperatures, yet still provide moderate peel strengths that could be adequate for many aerospace structures. In the aerospace industry, these films would always be used in conjunction with CIAP primers in the bondline; thus providing excellent resistance to hostile environments. Most of these films are cured at 325–350°F (163–177°C) for 60 min and can develop adequate tensile-shear properties for many design purposes. When used to bond aluminum honeycomb structures, the flatwise-tension strengths are quite good but the honeycomb peel values still may require some design changes to minimize peel loading as much as possible.

In the intermediate range of temperature applications, the unmodified epoxy film adhesives may be the best all-around performing epoxy adhesives, since the best are capable of very long-term durability (tested up to 15,000 hr) at 300°F (149°C) without any significant loss in mechanical properties. This level of performance, however, is likely to be the ultimate development possible for the unmodified epoxy formulations.

The testing of some of these unmodified epoxies was conducted by Olson and Lulling (1627) at Forest Products Laboratory at temperatures from −100°F (−38°C) up to 1000°F (538°C) as early as 1962. Minford (1628) conducted evaluations of 6061-T6 aluminum joint durability with several

Table 69 Tensile Shear Strength of Unmodified Epoxy Adhesive Film/Aluminum Adherend Joints

Adhesive identification	Manufacturer	Test temperature (°F)	(°C)	Shear strength (psi)	(MPa)
FM-150	American Cyanamid	75	24	2255	15.54
		300	149	2580	17.78
		350	176	2240	15.44
EA9649	Dexter Corp.	75	24	4190	28.88
		300	149	3080	21.23
		350	176	2880	19.85
		400	204	1860	12.82
AF147	3M Co.	75	24	4500	31.02
		300	149	2200	15.16
		350	176	1750	12.06
		400	204	1500	10.34

Source: Data from Ref. 1477 (Table 1, p. 136).

unmodified epoxies under ambient water-soak, hot-humidity soak, and hot-humidity soak/freeze/thaw cycling conditions in 1973 and encountered considerable performance variation.

F. "Modified-Epoxy"-Type Adhesives

We have discussed the employment of nitrile rubber, nylon, and phenolic resin components as modifiers for epoxy resins earlier in this chapter. However, the adhesive customer is not always permitted to know what particular form of modification has been used in a specific commercial product, since the manufacturer's data sheet only provides the general description "modified-epoxy" type. The number of literature reports dealing with the general subject of modification of epoxy resins to formulate commercial adhesives show a wide variance of properties, so that only a small percentage of these reports can be offered as examples. Most of the early efforts in aerospace were involved in the challenge to develop formulations with even higher temperature resistance. This had to always be balanced by the simultaneous need to offer good retention of physical properties. We will try to offer a chronological cross section of some of these progress reports.

In 1958, Klopfenstein and Lee (1629) offered an early discussion of the thermal aspects of epoxy resins. Overnall (1630) discussed aspects of the electron spin resonance of free radicals in epoxide resins in 1963 along with Lee's (1631) paper on thermal stability and degradation schemes for epoxy resins. In 1964, Lee and Neville (1632) offered data on the use of dianhydrides in high-temperature epoxy adhesives to a SAMPE conference, and Owen (1633) recounted his development work on a specific high-temperature epoxy for aerospace applications at the same meeting. In 1972, Grimes (1634) considered the long-time storage effects on mechanical properties of products available to the aerospace industry. In 1975, Griffith and co-workers (1635) started a series of investigations involving fluorinated variations of epoxy modification starting with a fluoroanhydride-curing agent for already heavily fluorinated epoxy resins. In 1980, Griffith and O'Rear (1936) reported on silicone amine-cured fluoroepoxy resins, and Griffith (1637) reviewed the progress on epoxy resins containing fluorine in 1981. Also in 1980, Kourtides et al. (1640) reviewed their data on the thermophysical and flammability characteristics of phosphorylated epoxy adhesives. In 1981, Jerome and co-workers (1641) further reported progress in the area of modifying epoxy resins, whereas 1982 saw publication of the data of Hawthorne et al. (1642) on the good performance of tris(hydroxylphenyl)methane–based epoxy resins.

Later publications include the 1987 paper of Zhi-Lu et al. (1643) on the study of the properties and structure of polyester-sulfone–modified epoxies. In 1989, Jozavi and Sancaktar (1644) discussed the effects of cure temperature and time on the so-called "stress whitening" behavior of structural epoxies and Garton and Haldanker's (1645) conducted investigation of the modification of the free volume in epoxies.

In the 1960s, Minford (1646) conducted some 3-yr durability testing of 12 different commercially produced aerospace film-type structural epoxy adhesives. Of the 12 products evaluated, more than half were in the category of being only described as modified-epoxy type. After the 3-yr exposure to hot-water soaking and aggressive natural atmospheric weathering conditions, the performances of the 6061-T6 joints (Alcoa A3 deoxidation pretreatment) can only be described as widely varying. This would indicate that the data sheet terminology of modified-epoxy was not discriminating in predicting some level of performance in service. Wegman (1647) also conducted durability evaluations with some modified-epoxy formulations available at the time in 1977 with similar variable performances.

G. Cyanoacrylate Adhesive Formulations

The cyanoacrylates are single-part, solvent-free adhesives that can form strong bonds to a wide variety of adherends, including aluminum. They cure at room temperature by anionic polymerization initiated either by adsorbed surface moisture or sites of alkalinity on the adherend. First developed by Tennessee Eastman Company in the early 1950s, the first commercial offering was in 1958 for special industrial applications. Today, the cyanoacrylates are used in a wide variety of application areas, including assembly of electronic components; attachment of gaskets and strain gauges; repair of appliances; locking of nuts and bolts; and construction of jigs, fixtures, displays, and scale models. Most recently, these products have been widely marketed to the home consumer with adjectives describing their use as "instant," "magic," "miracle," and "crazy-glue" products. They do form strong bonds with most rubbers, metals, hardwoods, plastics, and nonporous materials where smooth, close-fitting surfaces are available.

More information can be obtained from Thomsen (1648) in *Adhesives in Manufacturing*. Thomsen and Von Bramer (1649) reported their bonding characteristics at the 14th SAMPE Symposium in 1968. Coover (1650) has furnished information in the *Handbook of Adhesives*. Coover et al. (1651) also presented an early literature assessment in 1959. Most recently, Schoenberg (1272) has reported his findings in the 1990 *ASM Engineered Materials Handbook*. In 1981, Lee (1691) edited a publication titled "Cyanoacrylate Resins—The Instant Adhesives."

The 2-cyanoacrylate esters are the basis of the present best-known commercially available products with the methyl and ethyl members of the family used most commonly because of their low price, high-bonding strength, and availability. Some typical physical properties of the cured methyl and ethyl cyanoacrylates can be viewed in Table 70. It can be seen that the methyl-polymerized polymer has a significantly higher softening point, flexural modulus, and heat-distortion temperature. However, the decomposition temperature of the ethyl is higher. Rooney (1652) has described the

Table 70 Typical Physical Properties of Cured (Polymerized) Cyanoacrylates

Property	Methyl 2-cyanoacrylate	Ethyl 2-cyanoacrylate
Softening point	329°F (165°C)	259°F (126°C)
Index of refraction	Approx. same as glass	Approx. same as glass
Flexural modulus (psi) (ASTM D-790)	457,000	300,000
Heat distortion temp. at 264 psi	246°F (119°C)	156°F (69°C)
Temp. of decomposition (temp. at which 10% is decomposed)	388°F (198°C)	415°F (213°C)

Source: From Ref. 1648 (Table 3, p. 307).

curing as being based on the instability of the 2-cyanoacrylate esters in the presence of amines resulting in an anionic polymerization. An excess of cyanoacetate ester used in the Knoevenagel reaction acts as a chain transfer agent, reducing the molecular weight of the polymer and lowering the reaction viscosity. Hawkins (1653) secured a patent on the process for manufacture of high-purity alpha-cyanoacrylates in 1969 which may be of interest. The typical joint shear strengths of both methyl- and ethyl-type cyanoacrylates with various types of metals, plastics, rubbers, and wood adherends is reviewed in Table 71. Because of the inherently higher stiffness of the metal adherend joints, they test significantly higher than the weaker adherend plastics, rubbers, and woods. For the metals, the methyl-cyanoacrylate product seems to offer a higher metal-to-metal strength than the ethyl product. With stainless steel adherends, this advantage can be more than threefold. Similarly, as expected, the strength of the aluminum/nonmetallic dissimilar materials joints is more than twofold lower than the stiffer aluminum-to-aluminum joint structures. In most cases with plastic, rubber, and wood adherends, the joints while much lower strength than for metal-to-metal structures, were still sufficiently strong to achieve failure in the adherends when pulled in a tensile tester.

While cyanoacrylates can actually polymerize by either an anionic or free radical mechanism, the former is the mechanism of bond formation in the adhesives. Free radical polymerization, on the other hand, may be a possible cause of instability during long-term storage (particularly at elevated temperature). Almost any electron-donor compound known as a nucleophile (for example, water, alcohols, amines, phosphines, carboxylic anions, inorganic bases, halide ions, epoxides, mercaptans, and sulfoxides) can initiate an anionic polymerization, as mentioned by Eromosele and Pepper (1660), Cronin and Pepper (1661), Wicker and Shearer (1654,1655), and Donnelley et al. (1656,1657). Rasoul and Hall (1658) and Wicker and Shearer (1655) point out that electron-rich olefins like styrene or vinyl ethers can react spontaneously with cyanoacrylates to produce alternating copolymers.

Almog and Gabay (1662), in 1986, described the use of a special modified "super-glue" for fingerprint development work in forensic science. Although cyanoacrylate adhesives are relatively expen-

Table 71 Typical Shear Strength Values of Cyanoacrylate Adhesive Bonded Joints

		Bond strength psi (MPa) after 48 hr	
Type of bond	Adherends	Methyl 2-cyanoacrylate	Ethyl 2-cyanoacrylate
Metal/metal	Steel	3200 (22.05)	2900 (19.98)
	Aluminum	2400 (16.54)	1700 (11.71)
	Stainless steel	2500 (17.23)	800 (5.51)
Metal/plastic	Aluminum/nylon	900 (6.20)	700 (4.82)
	Aluminum/phenolic	1000 (6.89)	900 (6.20)
Plastic/plastic	ABS	([a])	([a])
	Acrylic	([a])	([a])
	Nylon	500 (3.45)	700 (4.82)
	Phenolic	([a])	([a])
	PVC	([a])	([a])
Rubber/rubber	Butyl	([a])	([a])
	EPDM	50 (0.34)	60 (0.41)
	Neoprene	([a])	([a])
	Nitrile	([a])	([a])
Wood/wood	Maple	600 (4.13)	500 (3.45)
	Yellow pine	90 (0.62)	110 (0.76)
	Oak	500 (3.45)	([a])
	Particle board	([a])	([a])

([a]) Indicates that the bond held while the adherend broke.
Source: From Ref. 1648 (Table 3, p. 312).

sive on a weight basis, their economical use must be based on having a number of special properties such as (1) very sparse application, (2) automatic dispensing equipment is readily adaptable, (3) cure times are very short, (4) versatility to bond a wide variety of adherends, (5) no need for curing ovens, (6) no need for premixing, (7) no solvents to be recovered or burned, and (7) no requirement for expensive fixturing. The cost and projected sales figures were reviewed in a 1989 article in the journal *Adhesives Age* (1663).

Only modest heat resistance in the earlier cyanoacrylate adhesives triggered modification developments in this area. Kotzev and co-workers (1664,1665) have progressed in this area through the use of allyl-cyanoacrylate, which undergoes a secondary cure by thermal polymerization of the allyl groups. This produces a cross-link polymer with superior stability and heat resistance. Drain et al. (1666,1667) have concentrated attention on the moisture- and aging-resistance properties of cyanoacrylates, since moisture can migrate into the interface and form a weak boundary layer. Humidity resistance on plastic surfaces, however, is quite good providing the plastic has some solubility in the cyanoacrylate before curing. Coover and McIntire (1668) have shown that cyanoacrylate joints may undergo a gradual loss of strength after several years of long-term aging. Bonds on glass deteriorate very rapidly because of the alkaline nature of the surface. Surfaces that are easily corroded, such as steel, can oxidize and form weak metal-to-metal oxide interfaces. Polymerized cyanoacrylates can be removed with certain polar aprotic solvents, as discussed by Donnelley and Pepper (1669).

Cyanoacrylate adhesives usually also contain one or more additives such as (1) inhibitors, (2) thickeners, (3) adhesion promoters, (4) activators, (5) plasticizers, (6) fillers, or (7) dyes. Thickeners are especially needed to control the rheology to prevent the adhesive from being absorbed into the surface or running off the adherend. Two disadvantages of particular importance are the relatively low impact and peel strength properties. Petrov et al. (1677,1678) and Gleave (1674) discussed addition of elastomeric materials or toughened plastics to improve these properties. Litke (1676) prepared thixotropic cyanoacrylates with hydrophobic silicas.

Considerable research has been conducted over the years to find chemical agents capable of acting as adhesion promoters for the cyanoacrylates over metals. O'Sullivan and Mebody (1680), in 1974, found adhesion promotion attributed to the anhydrides of maleic and itaconic acids and received a patent, as did Konig (1670) in 1976. Schoenberg (1671) obtained a U.S. patent on the promotion abilities of esters of gallic acid in 1978, whereas Krall (1681) attributed adhesion improvement to the same kinds of materials in a 1981 East German patent. Schoenberg and Ray-Chaurhuri (1672) obtained a patent on the promotion properties of monocarboxylic acids also in 1978. The attention was specifically focused on the effectiveness of acetic acid as a promoter in 1987 by Kotzev et al. 1683). They investigated steel, stainless steel, brass, copper, and duraluminum as adherend surfaces and established that adhesion promotion with acetic acid was dependent on the nature and surface treatment of the various adherends, with selectivity being shown toward steel, stainless steel, and aluminum. The results of their preferred surface pretreating for the different adherends using 0.25% by weight acetic acid are shown in Table 72. They concluded that acidic etching was best for all the metal adherends; however, the choice of acid and temperature conditions could differ. The stainless steel surface responded best to strong etching in hydrochloric acid at room temperature, whereas mild steel was hot etched in sulfuric acid and the surface acidity subsequently neutralized. Brass,

Table 72 Surface Treatments for Various Adherends

Substrate	Chemical treatment
Steel (0.2% C)	Immersion in 17% H_2SO_4 at 80°C for 10 min followed by neutralization in Na_2CO_2 and CH_3OH
Stainless steel	Immersion in 30% HCl for 15 min
Brass, copper, and duraluminium	Immersion in aqueous solution of 27.5% H_2SO_4 and 7% $Na_2Cr_2O_7$ at 20°C for 5 min

Source: From Ref. 1683.

copper, and aluminum were all etched in a bath at elevated temperature with a chromate/sulfuric acid mixture similar to an FPL etchant. In Table 73, the same investigator's results are given comparing the bond strengths secured with and without the pretreatments shown in Table 72. The aluminum surface seemed to benefit the most from the hot chromate/sulfuric acid etching of any of the metallic adherends investigated. Suetaka (1684) was able to correlate the actual bond strength with the carbonyl group orientation using infrared spectroscopy techniques.

It was further established that certain chelating agents might be capable of decreasing the adhesive setting time without appreciably reducing stability. Motegi et al. (1675) used crown ethers for this purpose. Plasticizers are occasionally added to increase the flexibility of a cured cyanoacrylate and to improve impact resistance. Allies and Zimmerman (1673) have promoted the use of esters of dicarboxylic acids such as phthalic and sebacic acids; however, high concentrations of plasticizer can reduce both cure rate and bond strength. Such lower bond strength can sometimes be used to advantage when an adhesive is desired which has a temporary function and will need to be broken easily. Finally, electrically conductive fillers such as finely divided silver has been used to compound electroconductive cyanoacrylates, as described by Chorbadjiev and Kotzev (1679).

At a time when cyanoacrylates were being promoted by Loctite for bonding aluminum side moldings on automobile bodies, O'Connor at Zimmerman (1685) at Loctite encountered a problem of obtaining considerable variation in the durability of the preferred commercial product. This prompted a rather extensive investigation of the variability that can be found in the sealed bright anodized surfaces on aluminum trim extrusions. As a background for their investigations, the technical papers of Garwood (1986), Walton (1687), and Spooner and Forsyth (1688–1690) are worthy of inspection. The final conclusion was reached that there was enough variation in the bright anodized surfaces being produced by different suppliers to the automotive industry to account for the wide-ranging performance in the field. The author conducting concurrent evaluations at the Alcoa Labs did not encounter the variation experienced in the field, since all adherend surfaces being produced for experimental bonding purposes were capable of obtaining and sustaining excellent adhesion. This

Table 73 The Effect of Adhesion Promotion, Substrate Nature, and Surface Treatment on Lap-Shear Strength of Joints Bonded with ECA[a]

			La-shear strength (MPa)	
Substrate	Chemical treatment[b]	ECA	ECA with 0.25% acetic acid	ECA, surface activated with acetic acid
Steel	No	9.7	19.1	11.7
	Yes	17.1	6.8	9.0
Stainless steel	No	11.3	18.0	12.5
	Yes	23.1	6.3	10.8
Copper	No	15.1	10.0	10.7
	Yes	6.4	5.8	2.2
Brass	No	12.2	6.7	9.3
	Yes	2.3	2.6	1.8
Duraluminium	No	7.8	8.4	6.5
	Yes	6.4	13.6	6.6

[a]ECA contains 110 ppm hydroquinone and 60 ppm SO_2.
[b]Chemical treatments identified in Table 72.
Source: Data from Ref. 1683.

situation does point to the variation in surface preparation that can creep unknowingly into a manufacturing operation where some of the variables that are important to control are not being controlled within the final customer's manufacturing facility.

As regards the manufacturer's viewpoint for using cyanoacrylate adhesives, the SME Tech. Paper titled "Cyanoacrylates—A Practical Guide to Their Chemical and Physical Properties" by Blomquist (1692) can be consulted. Also, the SME Tech. Paper titled "How to Dispense Adhesives" by Cotter (1693) should be helpful for setting up a manufacturing line. Finally, the paper in the journal *Adhesives Age* by Peace (1694) titled "Evaluating Cyanoacrylates For Product Assembly" should be of benefit to the manufacturing engineer with a charge to use these kinds of adhesives on an assembly line.

The availability of aluminum joint durability data with the cyanoacrylate family of adhesives is rather sparce. Thomsen and Von Bramer (1649) have furnished some Eastman Lab results shown in Table 74 which can be discussed. The residual lap-shear strength found in SBR rubber-to-SBR rubber, SBR rubber-to-aluminum, and aluminum-to-aluminum joints after up to 2-yr soaking in water were compared. A dramatic loss in strength down to a low plateau value occurred with the aluminum-to-aluminum joints after only 1 mo. This did not occur when exposure was at 50% RH and ambient-temperature conditions, indicating the accelerated effect of water soaking. While no change in noting a low-strength failure in the rubber occurred over the 2-yr period, the strength retained even in the aluminum-to-aluminum joints was sufficient that rubber failure should be expected in a dissimilar materials aluminun-to-rubber joint pulled in a tensile tester after 2-yr exposure to water soaking.

H. Anaerobic Adhesive Products

The origins of commercial anaerobic products is most closely allied to work performed by Burnett and Nordlander (1695) in the form of a patent to General Electric in 1953. Additional sources of general and specific information on the subject can be found in the 1985 work of Hauser and Haviland (1698) and in the 1990 work of Melody (1270).

The first anaerobic formulations were prepared by bubbling oxygen or air into dimethacrylate monomers; however, continuous air bubbling was necessary for acceptable stability. The commercial breakthrough came when it was discovered that a stable, yet reactive, one-part formulation could be prepared by simply mixing certain peroxy compounds like cumeme hydroperoxide into the meth-

Table 74 Moisture Resistance of Methyl 2-Cyanoacrylate Bonds

Type joint and exposure conditions	Avg. lap-shear strength (psi)		
	1 mo	4 mo	2 yr
SBR rubber/SBR rubber 50% RH, 73°F (23°C)	120[a]	105[a]	110[a]
SBR rubber/SBR rubber 75°F (24°C), underwater	110[a]	85[a]	80[a]
SBR rubber/aluminum 50% RH, 75°F (24°C)	120[a]	110[a]	110[a]
SBR rubber/aluminum 75°F (24°C), underwater	110[a]	100[a]	110[a]
Aluminum/aluminum 50% RH, 73°F (23°C)	1320	1100	1010
Aluminum/aluminum 75°F (24°C), underwater	390	390	230

[a]Indicates adherend failure.
Source: Data from Ref. 1649 (Table 5).

acrylate monomers. Murray et al. (1696) have described these early operations. A further early description of how to use anaerobic adhesives was provided in 1971 by Drafone (1697). Progress has continued to be made in discovering new and better formulations until in 1990, a typical formulation can comprise some or all of these components: (1) monomer and/or resins, (2) initiators of free radical polymerization, (3) accelerators (for initiation process), (4) stabilizers, (5) thickeners and form modifiers, and (6) performance modifiers (such as plasticizers and adhesion promoters. The final cured adhesive properties are, of course, determined to a major extent by the selection of monomer. Most formulations use methacrylates but dimethacrylates of polyalkoxy glycols have also been useful bases for formulation. Baccei (1699) has pointed out that methacrylate-capped urethanes can vary from being relatively simple in nature to very sophisticated oligomers containing the hard and flexible block moieties typical of cured polyurethanes. Acrylate monomers are seldom used in today's anaerobic products.

The initiators used most broadly are hydroperoxides, of which cumene hydroperoxide is often the most widely accepted example. This component provides stability in the formulations, yet is still suitably reactive (sometimes along with accelerators) when contacting various adherends in the absence of oxygen. As compared with the larger commercial family of "acrylic adhesives," the anaerobics must exist as "one-part" adhesives incorporating a fine balance of stability versus reactivity. The choice of initiator is an important factor in this careful balance. Baeder (1701) has reviewed the more complex "initiators" employed, the function and nature of accelerators desired for rapid hardening, and the function and nature of the stabilizers, of which oxygen is obviously the most important. Lees (1700) has discussed the potential to form a peroxy derivative of methacrylate monomers as a further factor complicating the mechanism of curing.

The so-called "form modifiers" in anaerobic formulations can include dyes, fluorescing agents, thickeners, thixotropes, and pigments. The function of the dyes, pigments, and fluorescing agents is to facilitate on-part inspection and sometimes certain strength characteristics. Thickeners and thixotropes, as anticipated, can promote desired viscosity and flow characteristics desired for the particular application (water-thin compositions for wicking into preassembled structures or thixotropic pastes for flange bonding and sealing functions).

The "performance modifiers" employed can be of the plasticizing or adhesion-promoting varieties. The former are used to generate lower compressive strengths. Other performance modifiers have included the use of lubricants like polytetrafluoroethylene (PTFE) powder and fillers like mica. Malofsky (1702) has described heat-resistance enhancers in products such as the use of imides containing reactive double bonds. Baldwin et al. (1703) have described incorporation of low/medium molecular weight elastomers to offer enhanced toughness properties in the cured adhesive. Finally, the use of photoinitiators has added the convenience of fixturing of anaerobic bonded joints by ultraviolet curing of adhesive fillets.

A wide variety of technical papers are available to the reader covering virtually all the areas of anaerobic technology. For example, Stamper (1704) has discussed the overall curing characteristics of the anaerobic sealants and adhesives. Fulda (1705) has reviewed the procedures for impregnation of porous metal components, which is especially important for processing porous aluminum castings. Pearce (1706) presented an early 1972 lecture on processing anaerobics at a Stevens Institute of Technology Symposium. Hulstein (1707–1709) offered three technical papers on the application, assembly, and automatic application of anaerobic formulations in 1976, 1978, and 1979, respectively. Minford (1710) has investigated the overall durability of anaerobic formulations over aluminum adherends over the 1970s showing a steady progress toward achieving strong and durable joining.

Lee and Bruce (1711), in 1980, discussed the thermal stability and aging of the current crop of formulations, whereas Drake and Siebert (1712) included the use of reactive butadiene/acrylonitrile liquid and solid elastomers for formulating anaerobic adhesives in 1984. Hauser and Loft (1714), at Loctite, described the bonding of anaerobics and the range of potential adherends in 1980, whereas Malofsky and Baccei (1713) discussed improving thermal resistance with maleimides in 1984 and Helmstetter (1715) described the advantages of anaerobic adhesives.

The role of the anaerobics in bonding fasteners has been discussed in a variety of papers (e.g., 1716–1719, 1721). Mention should also be made of a technology report from Loctite (1720). Of great interest to the production line engineer should be the 1984 paper on robotics by Pilarski (1723).

Some data from Hauser and Haviland (1698) reviewing some of the typical property data for five different types of anaerobic structural adhesives is shown in Table 75. For best joint tensile-shear strength, impact resistance, and T-peel values, the data indicate the preferred type would be the formulations described as being the tough, activator-cured general-purpose formulations. For gap-filling, the lowest visocisty and smallest gap line values obviously come from formulations designated as fast activator-cured types. The same investigators also supply data on the typical properties of anaerobic gasket-type sealants in Table 76. The highest-strength joints were made with the so-called "unitizing" types with lowest strength achieved by the general purpose flexible variety. The special type characterized as "high-temperature" shows an intermediate joint strength with temperature resistance up to 204°C as compared to 120 and 149°C for the flexible and "unitized" type formulations. Where elongation is important, only flexible-type general-purpose formulations can be considered.

I. Polyurethane Adhesive Products

Polyurethanes were first developed by Otto Bayer and co-workers at I. G. Farbenindustrie in Germany in the early 1940s (1725,1726). Initially, the work was actually directed toward synthetic fiber development; however, it was soon realized that this chemistry also lent itself to polymers which could be considered elastomers, coatings, and adhesive materials (1727). Hess et al. (1728) cites early use as adhesives in 1954, whereas Dombrow (1729), in 1957, cited the use of polyurethanes in World War II for bonding life rafts and inflatable vests. The excellent water resistance of the polyurethane adhesives remains an important advantage yet today.

Table 75 Typical Properties of Anaerobic Structural Adhesives

Type anaerobic	Fixture (sec)	Tensile strength (psi)	Tensile strength (MPa)	Impact (J/cm²)	T-peel (kN/m)	Viscosity (Pa-sec)	Gap line (mm)	Temp. range (°C)
Tough, activator-cured, general-purpose	180	4354	30	7	8	18	1	−54–135
Environmentally resistant, activator cured	300	2464	17	2	1.7	20	<1	−54–177
Fast activator cured	15	2758	19	2.7	—	12	<0.5	−54–107
Surface tolerant, bonds many materials without surface prep.	120	2758	19	2.3	4.8	70	<1	−54–120
Fast, UV light-cured	10	3048	21	2	[a]	20	<1	−54–100

[a]Not applicable for transparent substrates.
Source: Data from Ref. 1698 (Table 7, p. 297).

Table 76 Typical Properties of Anaerobic Gasket-Type Sealants

Name	Viscosity (Pa-sec)	Tensile strength (psi)	Tensile strength (MPa)	Maximum gap (mm)	Temperature range (°C)	Time to gel (hr)	Elongation (%)
General-purpose (flexible)	800/200 thixotropic	798	5.5	0.3[a]	−54–120 flexible	2[a]	50
High temp.	950	2903	20	0.08[a]	−57–204	40	2
Unitizing	6.5	5515	38	0.3[a]	−54–149	12	1

[a]Activator or heat cures can quadruple the possible gap and decrease time to gel by a factor of 4.
Source: Data from Ref. 1698 (Table 3, p. 289).

Generation of a urethane linkage product by reacting diisocyanates and polyols is the general basis for urethane adhesive technology. The resulting polymers can be viewed as a series of interconnecting, soft and hard segments (crystalline) that are quite unlike polymers used for all other structural adhesives, which are either hard or soft segment variety but never both. The ratio of soft-to-hard segments (domains) can, in turn, be varied to produce an unusually wide range of physical properties. The viscoelastic behavior and toughness of urethane adhesives can be attributed to the coexistance of these domains, as reviewed by Gosnell and Smith (1730) in 1962 and Dollhausen and Warrach (1731) in 1981. More extensive reviews have been offered by MacIver and Thompson (1732) in 1983 and Edwards (1733) in 1986. Zimmer and Morphy (1734) recently considered the employment of polyurethane adhesives as a solventless approach to adhesive compliance. Most recently, a 1990 review of the subject has been offered by Kreider (1269). The industrial synthesis of urethane adhesives has been considered in some detail by Dollhausen (1735) in a product literature bulletin of Bayer AG.

It should be pointed out in regard to the general use of polyurethane adhesives that they should be expected to have excellent resistance to water because of the good stability of the urethane linkage itself to water. This anticipated excellent resistance to water while very often present, does not guarantee the universal good integrity of polyurethane adhesive joints in water-soaking or hot-humidity conditions. This is related to the fact that the wetting of various adherend surfaces can be quite variable; and thus, water debonding may result in water service conditions simply because of such poor or marginal wetting of the adherend surface with the bulk adhesive relatively unaffected.

Polyurethane adhesives can be classified according to physical form (thermoplastic or thermoset) and whether one or two component. The environmental resistance required in the adhesive or coating will usually dictate the type, since thermoset materials generally have higher thermal and chemical resistance than their thermoplastic cousins. Whether a moisture-cure (one-part) or two-component system is chosen will be mostly dictated by the application method and the relative porosity of the adherends. For laminating large areas like plastic films or aluminum foil, the water- or solvent-carried formulations will only function properly if at least one interface is porous, permitting escape of last traces of the water or other solvent.

Frisch et al. (1736) have provided a very recent description of the many variations of polyurethane adhesives available in the marketplace and how they may be generally employed. While the broad selection is between a one- or two-part package, there are many additional subdivisions which can be considered. Use of polyurethane polymers in waterborne form was reviewed by Dietrich and Rieck (1737) in 1978, whereas Dietrich (1738) summarized the various methods of making such dispersions using acetone solvent as well in 1981. Prepolymer mix and melt-dispersion was the main consideration of a 1972 patent by Guise and Jackson (1739). Self-dispersion of solids via a ketimine process was covered in a 1978 patent of Pedain et al. (1740). A variation ketazine process was covered by the patent of Nachtkamp et al. (1741).

Table 77 Properties of Cured Polyurethane Structural Adhesives

Property	Test temp. °F	(°C)	Adhesive formulation 1[a]	2	3	4	5	6
Tensile shear strength	−40	(−40)	2871	6090	7324	1302	2166	
	Ambient		2496	3246	5655	1112	825	3040
	180	(82)	582	622	667	217	825	
	250	(121)	266	607	405	164	340	
Elongation (%)	−40	(−40)	73	17	7	15	254	
	Ambient		52	39	7	105	200	100
	180	(82)	15	10	11	24	90	
	250	(121)	12	30	15	18	141	
Young's modulus	−40	(−40)	15,687	73,209	70,936	23,633	8281	
	Ambient		13,834	19,856	66,591	1535	1491	43,600
	180	(82)	4831	5779	11,514	924	922	
	250	(121)	3737	4381	5275	776	629	
100% modulus	−40	(−40)				1078	591	3040
	Ambient							
	250	(121)					294	
Hardness shore			65 (D)	80 (D)	90 (D)	73 (A)	72 (A)	61 (D)

[a]Description of different formulations given in Ref. 1732.
Source: Data from Ref. 1732 (Table 1, p. 203).

The variation in aluminum lap-joint strength that can be measured at −67°F (−55°C), 75°F (25°C), and 180°F (82°C) for some of the current polyurethane structural adhesive products is reviewed in Table 77. Bondline thickness can also be quite influential on joint strength, as shown in Table 78. With a bondline thickness of 10 mil (0.025 cm), the lap shear strength on primed steel in Table 78 was shown to be similar to that obtained with a good-quality two-part, room temperature–curing epoxy on aluminum by the author. However, when the bondline thickness was increased to 30 mil (0.076 cm), then the joint strength declined by 24%. A similar thickness variation with a two-part epoxy paste adhesive product on aluminum adherends would hardly be noticed. Actually, 10-mil bondlines on aluminum adherend surfaces are not always easy to achieve without elaborate jigging

Table 78 Effect of Bondline Thickness on Bond Joint Strength Using a Polyurethane Adhesive

Bondline thickness (in)	(cm)	Lap shear joint strength (psi)	(kPa)
0.010	0.025	2540	17,000
0.030	0.076	1940	13,367
0.050	0.127	1632	11,244
0.070	0.178	1496	10,307
0.090	0.229	1352	9,315
0.110	0.279	1304	8,985

Tested according to ASTM D-1002 with cold-rolled steel primed with Pliogrip 6025 except with a 1 × 1 in bond area.
Source: Data from Ref. 1732 (Table 3, p. 205).

unless preformed 10-mil film- or tape-type epoxy adhesive is employed. As mentioned earlier, a wide range of different structural urethane properties can be formulated into structural polyurethane adhesives. This can be seen for the six different formulations shown in Table 77. For example, the tensile strength values for these six adhesives could vary by as much as 560% at −40°F (−40°C), elongation by 360%, and the Young's modulus value by 884%.

In general, polyurethane adhesives have been used more frequently to bond elastomers, fabrics, fibers, and thermoplastics to one another than for the joining of metal to metal. A significant technology has been developed, however, for joining polyester fiber glass parts to steel parts in the automotive industry and these procedures are equally applicable to aluminum-to-polyester fiber glass structures. However, it must be realized that a special epoxy primer must be used to ensure reasonable bond durability under aggressive exterior weathering conditions. Obviously, this is a matter of enhancing the degree of wetting through the use of a separate primer which does have the proper ability to wet the aluminum oxide. Minford (1742) has conducted some extensive longer-term, epoxy-primed aluminum/two-part polyurethane adhesive testing in the laboratory using a simulated accelerated tropical-type exposure. While no effective, durable bonding could be achieved in the absence of the primer, the degree of joint durability achieved with the recommended primer present was deemed quite acceptable. Primers have also been developed for the aerospace industry by North American Rockwell for bonding polyurethane adhesives to metal adherends. The results have been discussed in detail by Constanze (1743).

Relatively recently, Goodyear announced the development of a one-part, heat-curable polyurethane adhesive for bonding metals to fiber glass. Having established the good performance of their two-part formulations for aluminum, the author (1744) tried to repeat his earlier investigation substituting the newer one-part polyurethane formulation with its different (vinyl) primer. Durability response proved to be totally unacceptable with failure at the primer to aluminum interface. The one-part urethane could, of course, be acceptable if a better, more compatible, and water-resistant primer could be made available. The author has not pursued this matter further.

1. CRYOGENIC TEMPERATURE SERVICE

Claims have been made by Twiss (1745), Smith and Sussman (1746), and Vaughan and Sheppard (1747) about the superior strength and toughness of polyurethane adhesives at cryogenic temperatures. A comparison of their unusual shear strength properties in aluminum joints at temperatures from −400°F (−240°C) up to 200°F (93°C) with the rubber-phenolic and modified epoxy structural adhesives can be seen in Figure 12. It is clear from their exceptional strengths at cryogenic temperatures why their application for sealing at such temperature deserves attention. Frost (1748) and Hertz (1749) also conducted polyurethane adhesive studies under cryogenic service temperature conditions in the aerospace industry. Delmonte and Sarna (1750) have reported high-strength polyurethanes exist with lap-shear strengths in the area of 5000 psi (34.45 MPa) and peel strengths in excess of 60 pli. However, when these joints were fabricated with deoxidized aluminum adherends and exposed to 95% RH at 75°F (24°C), they lost 48% of initial strength after only 1 mo. Thus, in the author's opinion, there is little evidence that such structural polyurethanes are competitive with a wide variety of epoxies under water-present weathering conditions. For example, Minford (1751) has shown that the simple two-part, polyamide-curing epoxy he employed over 25 yr of testing as a control adhesive on deoxidized 6061-T6 aluminum could be expected to retain almost 90% of initial joint strength after 2-yr soaking in 100% RH conditions at 125°F (52°C). Two years after the published results by Delmonte and Sarna (1750), Delmonte (1752) again published data on bonding metals with polyurethane adhesives, stating that they showed considerable promise. Recently, Mobay (1753) has reported development of a very high peel and tensile-shear strength polyurethane composition on unprimed steel and aluminum with excellent moisture resistance. However, at only 176–212°F (80–100°C), the tensile shear strength had already been reduced to half the room temperature strength value. This fact might have to be considered in view of the fact that a black painted aluminum surface in solar exposure can have a surface temperature above this range even in temperate zone exposures.

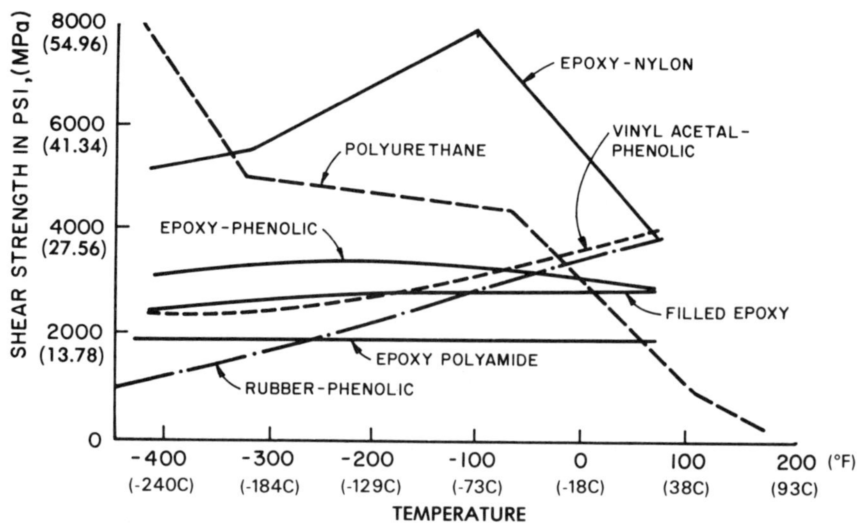

Figure 12 Properties of cryogenic structural adhesive systems. (From J. D. Minford, Aluminum Adhesive Bond Permanence, Treatise on Adhesion and Adhesives [R. L. Patrick, ed.]. Marcel Dekker, New York, Vol. 5, Figure 3-41, p. 104 [1981].)

2. APPLICATION AREAS

Polyurethane adhesives are used mostly in footwear, plastics processing, packaging, apparel, automotive, and building industries. The use of these adhesives with aluminum would most likely be found in the packaging, automotive, and building industries and in some selected applications in the aerospace industry (especially as sealants).

Polyurethane sealant formulations have also found a definite place both in automotive and industrial building product assemblies. Eagle (1754) has reported the substitution of room temperature-curing polyurethanes as replacements for the earlier employed heat-cured vinyl plastisols. It has been further claimed that the paint baking temperatures might be lowerable with such formulations present with still good enough adhesion to satisfy the auto body warrenties. Minford (1348,1349) has evaluated a wide variety of commercial polyurethane sealant products on various alloy aluminum adherends using bare, painted, and anodized surfaces. Acceptable joint durability was obtained after very long-term natural weathering exposures. Both chemically curing, two-part formulations, and one-part, moisture-curing formulations were included as well as the added effect of using special primers specifically recommended by the sealant formulators.

Although conducted in 1967, the investigations of Koch (1755) relative to the processing variables necessary to apply and use urethane adhesives on the Saturn Booster Program can still be of reader interest. Three relatively recent summaries of the technology of urethane adhesives can be reviewed in the publications of Kimball (1756), Dollhausen and Warrach (1757), and Edwards (1758).

The range of polyurethane structural adhesives/sealants use in the automotive industry now includes bonding automotive windshields, truck roofs, doors, lift gates, fenders, head lamps, rear lighting accessories, supports, and grill openings. In the construction market, the bonding of building panel components involving polyurethane foam and core insulations has also been well accepted as described earlier by the author in discussing the fabrication of Alply Building Panels.

J. High-Temperature-Resistant Adhesives Other Than Epoxies

The development of epoxy-phenolic formulations in the 1950s and 1960s by Black and Blomquist (1622–1625,1759,1760), Naps and Hopper (1761), Krieger and Politi (1762), Stewart et al. (1763), and Suffredini and McCann (1764) offered adhesive candidates that were able to operate at the maximum operational temperatures of aircraft of that vintage. However, the need for adhesives for aircraft with even longer survival time at temperatures up to 555°F (288°C) started to take place and the epoxy-phenolics were limited for these conditions. As the space age dawned, designers began to talk about the need for adhesives which would be suitable for both cryogenic temperatures as low as −421°F (−253°C) and high temperatures of 600°F (316°C). DeLollis (1765), St. Clair and St. Clair (1766), Meade (1767), and Nakao (1768) have offered reviews on this subject of high-temperature adhesives in 1977, 1981, 1983, and 1985, respectively.

By the late 1960s, an international effort began to develop a new generation of polymers for sustained use above 600°F (316°C). Levine (1769) provided one of the earliest comprehensive reviews of high-temperature adhesive development in 1962 at the Picatinny Arsenal Symposium on "Adhesives for Structural Applications." Levine pointed out that there were already a variety of materials that were capable of operating at temperatures in excess of 500°F (260°C), but their thermal stability alone was only one of many important criteria for use as adhesives. Some of the most well-known heat-stable materials such as ceramics, glasses, metals, and cermats had great limitations in brittleness, opacity, inelasticity, and high density. Although the organic polymers of that time could be tailored to have toughness, clarity, elasticity, and lower density, they were still failing to meet the increasing heat-stability requirements.

The use of rockets and missiles had begun to involve the possible use of adhesives in situations of short exposure times to very high temperatures. Procedures had to be developed to produce very high temperature rises in very short times even for test evaluations. In this regard, Allison et al. (1770) utilized quartz lamps to control heating to 1500°F (816°C) in 1 min. Meanwhile, Janis et al. (1771) and Levine and Boram (1772) were working on semi-inorganic high-temperature-resistant adhesives at Narmco which were showing some promise on stainless steel for 1000 hr at 500°F (260°C). Levine (1773) developed an adhesive with improved properties at 500°F (260°C) a year later in 1960 but optimum properties were only achievable with a cure schedule of 3 hr at 600°F (316°C) followed by a 24-hr postcure at 550°F (288°C). Development progress reports started to be issued from a variety of sources such as Thomas (1774) at General Dynamics, Ruetman and Levine (1775) and Ruetman et al. (1776) from Narmco, Roseland (1777) from Douglas, and Hergenrother et al. (1778) and Hergenrother et al. (1779) from Narmco in 1964. In the same year, Kausen (1780) presented a review of all progress to date. Additional reviews were offered by Humphreys (1781) in 1965, Mark (1782) and Oldham (1783) in 1979, Cassidy (1784) in 1981, and Hendricks and Hill (1785), Hergenrother (1786), and Lee (1787) in 1984.

1. HISTORY OF DEVELOPMENT

At first, the efforts mainly resolved about attempts at making improvments to fluorocarbons or silicone elastomers. Next, new heterocyclic aromatic polymer families such as polyimide (PI), polybenzimidizole (PBI), and polyquinoxaline (PQ) began to appear in the literature. A comparison of the heat-resistance and relative thermal aging properties of these adhesives with the modified-phenolic and modified-epoxy adhesives up to 800°F (427°C) can be reviewed in Figure 13. A listing of the superb level of joint strength that can be achieved at temperatures up to 700°F (371°C) by two commercially available polyimide adhesives used in aerospace bonding can be seen in Table 79. The temperature range of these adhesives, however, carries them beyond the feasible use range for aluminum as the primary structure in aircraft. Riel and Levine (1788) reviewed development efforts for structural adhesives for use above 700°F (371°C) as early as 1962, whereas Whitney (1789), in 1962, considered the possible use of high-temperature ceramic adhesives.

Levine was active in development work at Narmco in the early 1960s, producing various technical reports in 1960 (1790), 1961 (1791,1792), and 1962 (1793,1794). In 1962, Levine specifically mentions use of polybenzamidizole adhesives (1795) and S-Triazine polymers (1796). Marvel (1797,1798),

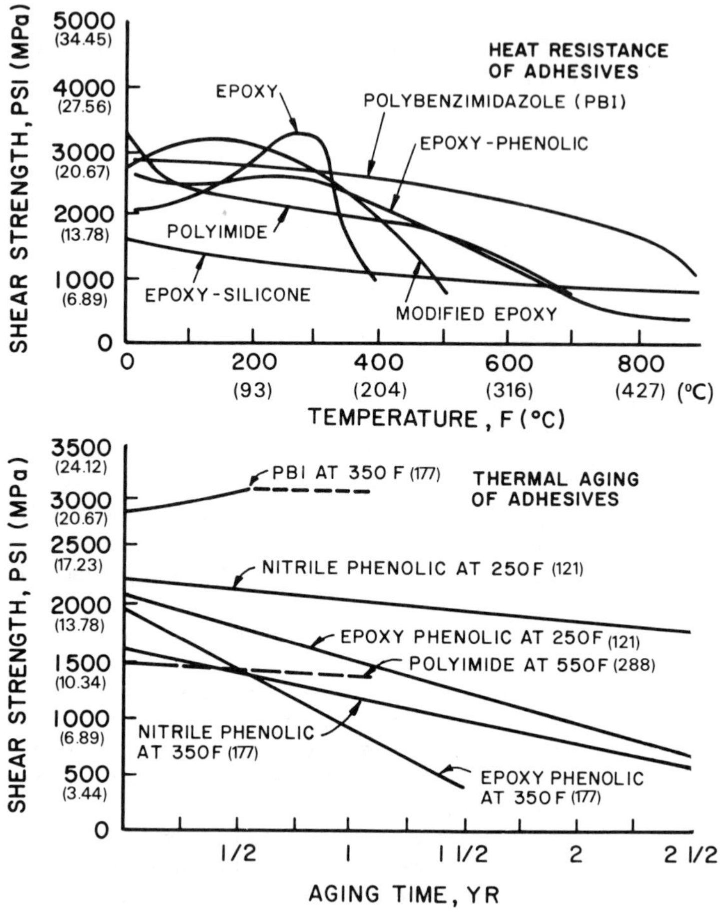

Figure 13 Comparison of heat-resistance and thermal-aging properties of adhesives considered for use in aerospace applications. (From J. D. Minford, Aluminum Adhesive Bond Permanence, Treatise on Adhesion and Adhesives [R. L. Patrick, ed.]. Marcel Dekker, New York, Vol. 5, Figure 3-40, p. 103 [1981].)

Table 79 Tensile Shear Strengths of Polyimide Adhesive Film/Titanium Adherend Joints

Adhesive identification	Manufacturer	Test temperature (°F)	Test temperature (°C)	Shear strength (psi)	Shear strength (MPa)
FM-34	American Cyanamid	75	24	2850	19.64
		300	149	2220	15.16
		400	204	1750	12.06
		500	260	1600	11.03
		600	315	1300	8.96
		700	371	1150	7.92
LARC-13	NASA LaRC	75	24	2800	19.30
	(Langley Research	400	204	2200	15.16
	Center)	600	315	1800	12.40

Source: Data from Ref. 1477 (Table 4, p. 138).

in the same year, reviewed the chemistry of polyaromatic candidates for possible high-temperature adhesive applications followed in 1964 with an SPE J. article on thermally stable polymers with aromatic recurring units. Also in 1964, Shenker et al. (1799) looked at adhesives for bonding thermal reentry space vehicle shields. These polymers definitely showed the ability to retain useful joint strengths on metal adherends like aluminum after extended exposure times in the range from 600°F (316°C) to 1000°F (538°C), where all earlier developed adhesives suffered rapid thermal or oxidative degradation.

Kausen (1780), in 1964, classified adhesives into four classes as regards their temperature resistance: (1) pyrogenic group (retaining shear strength of at least 1000 psi [6.89 MPa] above 1000°F [537°C]), (2) high-temperature group (retaining 1000 psi [6.89 MPa] between 350°F [177°C] and 1000°F [537°C]), (3) transition group (suitable for service between −67°F [−55°C] and 350°F [177°C]), and (4) cryogenic group (retaining at least 1000 psi [6.89 MPa] at temperatures from room temperature down to −423°F [−253°C]). A 1984 report issued by the National Research Council (1800) chose an alternative procedure of arranging according to a use temperature and a time classification, as shown in Table 80. Lee (1801) utilized this classification in 1987 to review the development of different adhesives which would best qualify to fit the range I, II, and III performance levels as described in Table 80. His conclusions are generally summarized as follows. The PBIs and PQ-type polymers have both offered useful bond strengths in the highest use temperature range (Type I). St. Clair and St. Clair (1802) have considered the use of polyimides (PIs) for the Type II service conditions, whereas Hergenrother (1803) has added the polyphenylquinoxalines (PPQs) to this same service group. In the range III group are nadic-terminated or addition-type PIs and acetylene-terminated PPQs. There has been some problem of brittleness in these ATPs and ATPPQs which has been countered by St. Clair et al. (1804) by using rubber reinforcement. Mention should also be made of an excellent 1979 review of the relative stability of high-temperature polymers by Arnold (1805).

2. *POLYIMIDE CLASS OF HIGH-TEMPERATURE-RESISTANT ADHESIVES*

The most widely developed of the present crop of high-temperature-resistant adhesive systems has been the polyimide adhesive products. The first commercial products were introduced in the early 1960s as represented in the patents of Edwards (1806,1807) and Lavin et al. (1808,1809). Since that time there has been a considerable development effort on many systems, as reviewed by English (1810) in 1986 and Rossi (1276) in 1990. In a 1985 discussion of elevated-temperature adhesives, Millard (1811) stated that few or no commercial applications had been utilized up to that time. However, English (1810), in 1986, mentioned applications as a high-temperature adhesive for aerospace structural parts and electronics/microelectronic end uses. Millard cited the current principal development materials as (1) LARC-13, developed by NASA, and (2) FM34, developed by American Cyanamid. The latter system has supposedly had some success in volume-production applications involving bonding of aluminum, steel, titanium, and polyimide glass and graphite composite structures. It is acknowledged by the author that the use of such materials on aluminum as an adherend probably will remain less extensive, since the upper operating range of such polyimide systems can be beyond the structural capability use range for aluminum for extended periods of time.

Table 80 Classification of Adhesives and Sealants According to Use Temperature and Time

Range	Use temperature (°C)	Use time
I	538–760	Seconds to minutes
II	288–371	Hundreds of hours
III	177–232	Thousands of hours

Source: From Reliability of Adhesive Bonds Under Service Environments. National Materials Advisory Board, National Research Council, USA, December, 1984.

A number of articles in the literature offer varying degrees of review on various aspects of aromatic polyimide chemistry and processing of which the articles by Sroog (1812), Serlin (1813), Scola (1814), Landis (1815), Bessonev et al. (1816), and Scola and Vontell (1817) may be representative. From a chemical viewpoint, aromatic polyimides can be either thermoplastic or thermoset in nature. Rodriguez (1818) discussed thermoplastic types as early as 1970, with continuing reporting in 1984, 1987, and 1989 by Cotts and Volksen (1819), Bessonov et al. (1820), and Lee (1821), respectively. Hergenrother (1822) reviewed the so-called "reactive oligomer" approach for formulating the thermoset varieties. The Lubowitz patent (1823,1824) covers production of the bisnadimides (BNIs) and bismaleimides (BMIs) reactive oligomers, which were among the first available. Stenzenberger (1825) and Varma (1826) have also discussed variations of the processing reactions leading to these types of materials. The most popular BNI formulation has been LARC-13, which has been mentioned above and discussed by Hergenrother (1822) in 1984. However, the BNI system PMR-15, as discussed by Wilson (1827), should also be mentioned. The BMIs are closely related chemically, as discussed in the 1986 by Ninan et al. (1828). Shaw and Kinloch (1829) have discussed the formulation of BMIs with chain extenders, reactive diluents, and elastomer resins to overcome the brittleness inherently present in these thermoset materials. Progar (1830) conducted an adhesive evaluation of LARC-TPI and a water-soluble version of LARC-TPI in 1986.

A number of processing problems can be identified. Quite often they require either a step cure or postcuring of some complexity. Volatiles are released during the curing cycles which are very difficult to handle where large surface areas are being bonded. Accordingly, full vacuum processing is recommended with only perforated honeycomb cores as practical internal adherends. The chemical structure of the particular polyimide is related closely to the performance temperatures that can be tolerated. In general, they will perform well at temperatures of 400–500°F (204–260°C) with fall off beyond 500°F (260°C) out to their limit of about 625°F (329°C). Good aging performance for several thousand hours may be expected at 500°F (260°C).

In 1971, Roper (1831) demonstrated a broad temperature range for maintaining high shear strength at elevated temperature with a polyimide system designated BR34/FM34. Critchley and White (1832), in the following year, first sought to incorporate flexible bridging units into either the diamine or dianhydride polyimide monomers to impart mobility into the otherwise rigid backbone. The technology of using acetylenic groups as the reactive moities in a polyimide oligomer was introduced by Landis (1832) in 1974. Other investigations of acetylene-terminated polyimide oligomers were later reported in 1980 by Hergenrother (1833) and 1982 by Landis and Naselow (1834). St. Clair and St. Clair (1835), in 1980, investigated aluminum ion–containing polyimide adhesives, whereas 1981 saw the same investigators reporting of a multipurpose thermoplastic polyimide adhesive (1836,1837). The year 1982 saw Burks and St. Clair (1838) describing the synthesis and characterization of a melt processible polyimide with St. Clair and St. Clair (1739) describing addition polyimides containing various kinds of end groups. In 1984, Scola (1840) delivered a special lecture on the addition-type bismaleimides and related imide polymers at a polymer conference at the State University at New Paltz in New York.

The development of water-soluble thermoplastic PIs was announced by St. Clair (1841) at a 1984 Program/Review Workshop at Virginia Polytechnic Institute. St. Clair and Yamaki (1842) reviewed the development of a polyimidesulfone system in 1984 along with reports of the continuing work of Burks and St. Clair on incorporating flexible bridging units into polyimide monomers. Harris et al. (1843) reported success in incorporating bulky side groups into the monomer which improved processibility. McGrath and co-workers (1844–1846), in the period 1985–1987, published a series of papers relating to achieving enhanced solubility and much improved weatherability of polyimides by incorporation of flexible siloxane segments. In 1987, Johnston and Hergenrother (1847) offered a review of the state of the art of high-temperature thermoplastics.Progar and Pike (1848), in 1988, attempted an evaluation of the water-soluble LARC-TPI as a usable adhesive system. In 1989, Progar et al. (1849) investigated the LARC-TPI 1500 series controlled molecular weight polyimide system, whereas Arnold et al. (1850) showed that the incorporation of siloxane segments into polyimides could significantly reduce the brittleness.

In 1984, Startsev et al. (1851) described the nature of the adhesion interaction and destruction of the polyimide-aluminum foil contact during hardening. Buchwalter (1852), in 1990, overviewed the subject of adhesion of polyimide to metal and ceramic surfaces, and Wilkensen (1853), in 1991, reviewed NASA Lewis developments on high-temperature thermosets.

3. *OTHER HIGH-TEMPERATURE–RESISTANT ADHESIVE FAMILIES*

Some of the earliest investigations on the joint strength of known adhesives for high-temperature screvices were conducted by Eickner and co-workers (1866–1868) in the early 1960s at the Forest Products Lab using aluminum and stainless steel adherends. The evaluation of the durability potential of these adhesives had been mainly predicted from their property values measured at the elevated temperature in the form of lap-shear joints. Later, the measurement of retained joint strengths after varying lengths of exposure time at the elevated temperatures was the better measure of joint durability potential (resistance to high-temperature oxidation). It was assumed that the retained tensile strength of the cured bulk adhesives would directly reflect the thermal aging sensitivity after varying lengths of exposure time. However, Black and Blomquist (1891), at Forest Products Lab, found an alarming acceleration in joint strength decline with the same epoxy-phenolic adhesive on stainless steel as compared to Alclad aluminum 2024-T3 alloy, as shown in Figure 14. The stainless steel joints having degraded to failure in 100 hr as compared to 500 hr to achieve the same end with the aluminum joints. No loss of joint strength occurred with duplicate stainless steel joints when heated in nitrogen. It must be concluded that both the condition and chemical composition of the metal adherend surface and the oxidation result at high temperature can be quite variable. It was assumed that deterioration of such adhesives in the presence of high humidity or condensed water was only of interest for evaluating the storage service times. The latter could be of special interest in evaluating adhesives for missile applications where ordnance must be stored for long periods of time before the short service time involved in its actual usage occurs.

A number of literature references exist for reader study relative to several other important families of high-temperature adhesives in addition to the polyimides studied previously. Tsou et al. (1869), in a contract to Picatinny, studied polyxylenols, polyxylenol ethers, and triazine polymers in the early 1960s. Thomas (1774) studied a large number of high-temperature–resistant adhesives of late

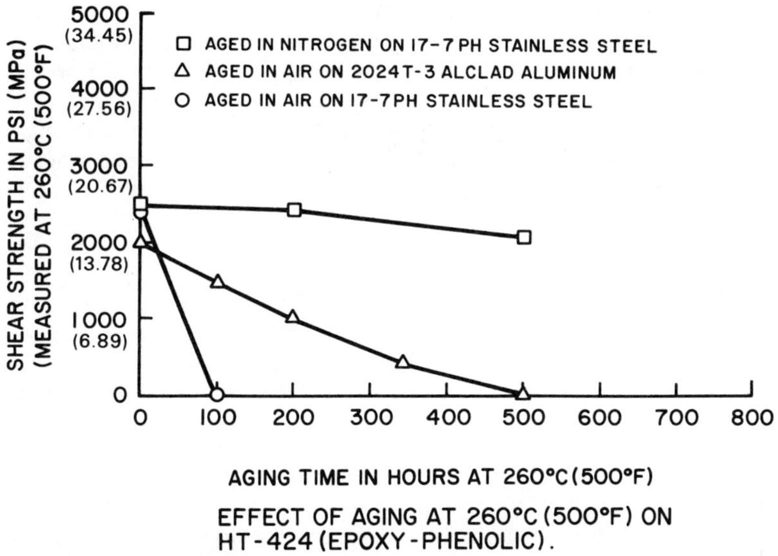

Figure 14 Interacting effect of adhesive and metal surface on thermal aging of bonded metal joints. (From J. D. Minford, Aluminum Adhesive Bond Performance, Treatise on Adhesion and Adhesives [R. L. Patrick, ed.]. Marcel Dekker, New York, Vol. 5, Figure 3–42, p. 104 [1981].)

1950s vintage at Convair. Also, as stated earlier, Ruetman et al. (1776) were developing high-temperature adhesives in 1962 with Hergenrother et al. (1779), citing work on the polyphenyl-as-triazines and polyphenylquinoxalines. Complex curing mechanisms of virtually all very high temperature-resistant adhesives have continued to pose problems in their actual bonded joint configurations. Large metal-to-metal faying surfaces being especially difficult to handle owing to the presence of voids from entrapped volatiles. Elaborate venting has, therefore, had to be employed. Bonding of honeycomb is still generally not recommended with the current state of the art adhesives according to Millard.

a. Polyphenylquinoxalines

Polyphenyquinoxaline adhesives, though not readily available in commercial adhesive formulations, must be considered because of the promise they provide for the most harsh end-use conditions. Hergenrother (1870) provided a most comprehensive 1988 review of polyquinoxalines, including the polyphenylquinoxalines. Wentworth (1277) also provided a 1990 discussion of the subject. They are linear heteroaromatic polymers containing phenylquinoxaline units in the chain. They are linear amorphous thermoplastics, which is a basis for both advantages and weaknesses. Their thermoplastic nature must cause some concern about elevated-temperature creep and substantial loss of properties above the glass transition temperature. Reduction of elevated-temperature plasticity has been a major goal of further research on these materials. An important technique has been to process them as a high molecular weight linear polymer and then achieve some cross-linking in the finished bondline.

Hergenrother (1871–1875) has been the most prolific investigator over a period from 1974 up to the present. An alternative synthetic pathway to produce these materials was reported in 1988 by Hedrick and Labadie (1876) and Connell and Hergenrother (1877). Because use temperature of these materials precludes their use with aluminum, the literature largely reports on bonding titanium and advanced composites, as discussed by Hergenrother (1878), Hergenrother and Progar (1879), and Hendricks and Hill (1880). In 1990, the only commercial supplier of polyphenylquinoxaline polymer products is Cemota, a French firm represented in the United States by IFP Enterprises.

b. Polybenzimidazoles

This is a heterocyclic family of polymers that has received extensive evaluation as adhesives and composite matrices from the 1960s. The first preparation was reported in the patent of Brinker and Robinson (1881) in 1959. Vogel and Marvel (1882) reported additional preparation procedures in 1961. Some publications arranged chronologically include Levine (1883), Jones (1884), Korshak (1885), Frazer (1886), Cassidy (1887), Powers and Serad (1888), Buckley et al. (1889), and Ward (1890). These papers can serve for further amplifications about the structure-property relationships of the neat resin, composites, and fiber mechanical properties of the PBIs. Recent summary of the progress has been reviewed by Connell (1278).

K. Cryogenic Adhesives

Aerospace adhesives and sealants must be equally resistant to extreme low service temperatures because the space vehicle that heats to an extremely high temperature when leaving and reentering the earth's atmosphere, also must function in the low-temperature conditions of space itself or the extremely low temperature ambient conditions of the moon or other planet surfaces. For example, adhesives for cryogenic engines should maintain their mechanical performance between the limits of −299°F (−184°C) and −320°F (−196°C) even though this seems beyond the limits of polymer capability. For such needs, it is likely that a composite or blend of adhesives in a multilayer form needs to be developed. Recently, Burford (1892) performed some interesting fracture toughness measurements on cross-linked polybutadienes which appear to be contrary to the common belief that the fracture energy of such polymers would decrease to a very low level at cryogenic temperatures. He showed, surprisingly, that fracture energy actually increased under the lower temperature conditions. It was speculated that this performance could be due to a craze formation. It might be expected then that common structural adhesives reinforced with rubbers (rubber-toughened systems) could survive extremely low environmental conditions. Keimel (1893) offered an example of a 3M's polysulfide adhesive showing brittle properties at −40°F (−40°C), yet elastomeric properties with attendant good bond strength in liquid helium. Keimal also reported a MOCA-cured polyurethane adhesive provided equal peel test results whether tested under room temperature or liquid helium temperature conditions. These results suggest that a wide range of elastomers might perform well at cryogenic temperatures.

A technical summary of properties of various adhesives under both high and extremely low temperatures by Kausen (1894) serves as a review for the technology to develop adhesives for service down to −423°F (−253°C). Hertz (1895–1899) was a very active investigator of adhesives for possible service under cryogenic temperature conditions with the indicated papers published between 1959 and 1962. He has focused considerable attention on the nylon-epoxies and polyurethanes for aerospace applications. Hiza and Barrick (1900) and McClintock and Hiza (1901) focused their attention on the properties of a variety of epoxy adhesives at cryogenic temperatures. In addressing a symposium on "Plastics and Adhesives in the Space Age," Pascuzzi and Hill (1902) in 1964 reported test results for a variety of structural adhesives under cryogenic conditions. Frost (1903) methodically measured the strength of 10 structural adhesives at temperatures down to −423°F (−253°C), whereas Claus et al. (1904) published an evaluation of the behavior of a variety of materials under conditions of space. Some of the quite dramatic changes in aluminum joint strength that were found with a variety of adhesive polymer families under cryogenic conditions can be seen in Figure 12.

Other investigators whose publications may be of interest include Smith and Susman's (1746) report on a NASA contract for development of very low temperature adhesives in 1963. A NASA contract by Sandler et al. (1905), in 1964, specified development of improved adhesives for use at −423°F (−253°C), with additional reports from the same Borden Chemical authors issued in 1965 (1906). In 1965, Roseland (1907) discussed his evaluation of some cryogenic adhesive usage.

Although a separate section on adhesives as sealants will be offered next in this chapter, it is difficult to divorce that discussion from this subject of cryogenic adhesives, since many investigations have involved development of sealants which to some degree serve both functions well at cryogenic temperatures. Schollenberger (1908) mentioned this connection in 1962 in the *Adhesives Handbook* under isocyanate-based adhesives. In 1964, Saunders and Frish (1909) discussed this idea under the overall subject of chemistry and technology of polyurethanes. The subject again surfaces when a number of reports were written about studies with elastomer polymers as sealants at cryogenic temperatures. Reference is made to a 1961 report by Whetzek et al. (1910), a 1963 Gosnell report (1911), and two 1964 reports by Ludtke (1912) and Wilson and Feldman (1913), and Robinson and Schreihans (1914) report on RTV silicones as both sealants and adhesives for cryogenic applications. Other general papers on development of sealants for cryogenic applications have been issued by Zelman et al. (1915) and Akawie (1916) at Hughes Aircraft in 1961 and 1964 NASA contracts.

VI. SEALANT FORMULATIONS

A. General Background

In sealant applications (outside of the extreme temperature service demanded for some aerospace structural applications), conventional sealants (such as the polysulfides, flexible epoxies, silicones, polyurethanes, elastomerics, and acrylics) do not require very low temperature flexibility according to Klosowski's (1917) 1980 review. Flexibility performances at intermediate low temperatures to −14.8°F (−26°C) tend to suffice in temperate zones where the highest volume of sealants are employed in construction. Among this list of commonly employed sealants, the polysulfides, silicones, polyurethanes, and nitrile elastomers have shown themselves to be quite versatile in sealing between two aluminum surfaces. Panek and Cook (1918) pointed out in their 1984 book, however, that superior water-resistance and aging properties of silicones tend to make them the more acceptable in recent years.

Pertinent references dealing with sealants and sealant technology arranged chronologically are by Rochow (1919), Prost (1920), Damusis (1921), Seymour (1922), Sealers and Waterproofers Institute (1923), Flich (1924–1926), Frisch (1927), Klosowski (1928), ASTM (1929), *DATA Digest*

(1930), Meyer (1931), Keimel (1932), Dunn (1933), Beers and Brassard (1934), Elias et al. (1935), and Skeist (1936).

Minford (1348,1349) conducted long-term atmospheric weathering tests of adhesion retention and physical property changes of most families of conventional sealants used for aluminum architectural applications. This included commercial products identified by their formulators as polysulfides, polyurethanes, elastomerics, acrylics, silicones, and the most common, lowest-cost organic sealers classed as being oleoresinous in nature. These products were extruded on the aluminum alloy surfaces representing a variety of surface conditions, including mill-finish, vapor-degreased/mill-finish, etched (deoxidized), anodized, and painted. The degree of adhesion and retention of various physical attributes was measured as frequently as every 6 mo over a total exposure period of 10 yr. Other accelerated laboratory weathering tests were also conducted, including soaking in hot condensing humidity at 125°F (52°C) or cyclic exposure conditions in commercial weather-o-meter cabinets. The most significant overall conclusion was that it was dangerous to conclude that all products of a given chemical class would be expected to perform equally. While lowest-cost, oleoresinous-type sealants might be expected to generally perform more poorly in adhesion retention and to offer least possibility of maintaining physical properties when exposed over aluminum adherends, even here a considerable range of performance was observed. Overall, there is no substitute for individual direct test evaluations. Generally, Minford found that the silicones as a group performed best; however, polysulfides, nitriles, acrylics, and more recently formulated urethane sealants could offer quite satisfactory substitution at a lower price. Separate proof must also be obtained whether a primer is necessary for optimum durability performance and whether such use is economically acceptable. In some situations where a significant movement of the aluminum in the joint is anticipated, as in aluminum gutter jointures, a good-quality nitrile rubber-based formulation was often preferable at a modest cost. More information on these tests will be summarized in Chapter 10 on joint durability.

Fedor and Brennan (1937) conducted the most recent comparison studies of what correlations can be found between natural and accelerated weathering of sealants. Specimens were fabricated with aluminum adherends by volunteer companies included in ASTM Committee C-24 on Building Seals and Sealants. Sealant types included one- and two-component polyurethanes, polyvinyl acetate latexes, thermoplastic rubber, black and white silicones, solvent-based acrylic, and latex acrylic proucts. Natural exposure to Florida weathering and two accelerated testing devices were employed. Some difference in test results was explained by different damaging forces of UV light, moisture, and temperature afforded by the different exposure methods. Although accelerated testing is often used to predict product durability, there seemed to be definite limitations compared to natural weathering. In this study, the most obvious limitation was the inability of the laboratory test conditions to reproduce the microorganism growth seen in Florida. Laboratory devices also were unable to reproduce degradation by air pollution and windborne contaminates. Nevertheless, accelerated weathering testing can still be considered a useful tool. Best agreement between accelerated testing and Florida atmospheric results was with regard to surface changes (i.e., cracking, pitting, flexibility, and hardness [durometer]). Different generic types of sealants did seem to exhibit varying acceleration factors owing to either form of test condition.

B. Effect of Water and Means of Altering Sensitivity

Conventional construction sealants, i.e., polysulfides, polyurethanes, epoxies, elastomerics, and acrylics, are all sensitive to moisture to varying degrees. Strecker and Gettys (1938) demonstrated that the major failure problem of insulated glass units was due to penetration of water vapor through sealant into sealed units. Spetz (1939), in 1985, demonstrated that fabricators could significantly reduce these failures by simply improving their techniques and eliminating poor workmanship. Berenbaum (1940) established as early as 1969 that after desiccant inside a sealed unit has been consumed, the condensed and entrapped water can chemically hydrolyze and swell the polysulfide sealant being investigated. Thus, chemical action can eventually cause bond separation between two aluminum adherends owing to the creation of stresses not initially present. Other manifestations of this problem have been actual breakage of glass and the development of a fog inside the initially sealed units.

The first approach to solving this moisture-sensitivity problem might be a design change in the sealant to one more hydrophobic or lower in surface energy. Griffith (1941) showed a lower water absorption existed where fluorinated forms of epoxies were employed, whereas Griffith et al. (1646) extended the benefit by using fluorinated curing agents. The well-established nature of fluorinated polymers to show lower surface energies makes them less wettable and, thus, more hydrophobic. A more economical (yet still 50% improvement in hydrophobic properties) solution was achieved by Johncock and Tudgey (1942) using other halogenated epoxies and curing agents. Gibbs and Breder (1943) suggest fluorinated polyimides should also be less sensitive for high-temperature sealing applications.

Another approach to improving moisture resistance involved introduction of siloxane linkages into polymer chains. St. Clair (1944) and Maudgal and St. Clair (1945) made such copolymers of the polyimide class of materials. Ryang (1946) obtained a 1983 G.E. patent involving incorporation of siloxane in a segment of the main chain. Siloxanes offer similar low surface energies and low water wettability as found with the fluorinated polymers mentioned above. While main chain flexibility and water resistance are both enhanced by these alterations, the thermal stability is adversely affected.

Rathbun and Schuessier (1947) focused on the need to keep moisture from penetrating into electronic-type sealants. They found the best treatment was to surface modify the sealant by plasma with hexamethyldisilazane (HMDS) to yield a 100-nm thick coating. Schuessler (1948) listed polymers like siloxyimides, fluorosilicones, fluoroacrylics, phenylated silicone, and silastyrene as potential electronic sealants with more hydrophobic properties than the regular hydrocarbon polymers. The silastyrene seemed to resist any moisture absorption but even its hydrophobic nature is probably not absolute.

C. Sealants in Aerospace Applications

While many of the investigations mentioned previously involved the use of sealants of the aerospace variety, their inclusion under the effect of water heading seemed appropriate when talking about that particular subject. Additional investigations can be included under this broader subject area, because a significant number of formulations have been completed in an attempt to meet all the especially tough specifications of the aerospace market.

Schmitt (1949) reviewed the use of sealants in spacecraft and aircraft in specific application areas of sealing fuel tanks, channel filleting, and for producing aerodynamic smoothing in general. Meyer (1950,1951) meanwhile considered the widespread use of very special polysulfides for use between various metals in aerospace applications. Conventional polysulfides for the building and construction industries have been based on disulfides, but a new family of sulfur-containing sealants has been developed by Singh et al. (1952) at Products Research based on thioethers. This has permitted an elevation of service temperature by 79–88°F (27–31°C) with a lower temperature limit which still conforms to a −76°F (−60°C) specification. Additional benefits of better hydrolytic stability and general chemical resistance over the disulfide compositions should also be mentioned. Hutt and Singh (1953) described the latest of these thioether products in 1964 at an SPE "Plastics and Adhesives in the Space Age" meeting showing its good performance at −94°F (−70°C). Klosowski (1928) also discussed several advanced new aerospace sealants for this −94°F (−70°C) service.

Some other special sealant types that can be mentioned are cyanosilicones, fluoroalkylarylenesiloxanylenes (FASIL), phosphonitrilic fluoroelastomers (PNF), and flexible polyimide fuel tank sealants. Singh (1954) discussed the member of the cyanosilicone family that was used in one of the earliest space shuttles. However, for some unknown reason, these structurally interesting cyanosilicones have been withdrawn from the marketplace. FASIL was a development of the Air Force Materials Lab in the late 1970s and two different classes are reported in publications of Rosenberg and Choe (1955,1956). These formulations demonstrate excellent adhesion to both aluminum and titanium with good hydrolytic stability and resistance to JP-4 aviation fuel. The PNF sealant composition has been reported in two publications by the developer, Firestone (1957,1958). These polymers contain an amazing 55% fluorine that imparts outstanding resistance to jet fuels, oils, and most hydraulic fuels and chemicals. The unique mechanical properties include low compression set, high modulus, good fatigue properties, and good abrasion resistance. The service temperature range appears to range from −85°F (−68°C) to 347°F (175°C). The polymer has been available from the Ethyl Corp. as EYPEL-F.

Jones and Casey (1959) reported the development of a flexible polyimide sealant from a synthetic chemical approach based on the Diels-Alder addition procedure. The special properties reported consisted of (1) high-temperature fuel stability to 500°F (260°C), (2) low-temperature flexibility to −50.8°F (−46°C), (3) excellent JP fuel resistance, (4) high adhesion to aluminum and titanium, (5) facile cure and processability, (6) availability in 100% solids formula, and (7) excellent tensile properties. In comparison with late 1970s developments reported above, the reader may still find the 1958 review by Berg (1960) of seal materials for guided-missile applications and the 1962 and 1963 reports of Lee (1961) on seal and sealant materials development for the supersonic transport still of significant interest. The reader is also reminded of the earlier publications under the cryogenic adhesives section of special development of elastomeric, polyurethane, and silicone seals for those low-temperature conditions (1908–1916).

In 1990, Gooch and Daher (1962) published a comprehensive investigation of how to reduce corrosion between aluminum surfaces in Air Force aircraft and weapons systems with elastomeric conductive sealants. While corrosion obviously creates structural weaknesses, the nonconductive products resulting from corrosion processes also destroy the nuclear-hardening capacity of the structure. Previously, only Type 3, MIL-C-5541 chemical conversion coating of the aluminum was the only protective finish meeting electromagnetic pulse (EMP)/electromagnetic interference (EMI) requirements. Even here, trapped moisture ends in enough corrosion to make the impedence high enough to destroy the nuclear hardness of the corresponding aircraft and weapons. 7075 aluminum alloy test joints with top-candidate sealing materials were weathered in a salt fog environment and the dc resistance and shielding effectiveness were measured before and periodically with increasing time. Control joints had no sealant present. As a result of these extensive tests, these authors concluded that aluminum filler materials should be avoided. Instead, either silver-coated aluminum or aluminum/nickel fillers should be employed with galvanic potentials of about −740 mV with an appropriate matrix such as RTV silicone or polysulfide. Such conductive sealant materials can improve long-term effectiveness (both structural and electrical) of metal-to-metal bonds typically found on exposed surfaces of military aircraft and weapons systems.

D. Sealants in Nonaerospace Applications

A wide range of generic-type sealant formulations are widely used outside the aerospace market. Obviously, the total market volume is many times as great as employed in aerospace, since the number of applications is spread across a great many different types of products. In general, performance requirements are less critical, although the need for good very long term performance is still quite vital for satisfactory performance of many manufactured products. While the primary function of sealing materials is to seal holes and gap areas between various other materials of construction, there is a secondary bonding function that must also be provided if the sealing is to be totally successful. The term *adhesive/sealant* formulations is, therefore, quite apropos for describing many of the present products in the marketplace. The polymeric families currently employed for use in the transportation, construction, electronics, industrial maintenance, and consumer markets are mainly covered in discussing: (1) oil-base caulks, (2) latex acrylic sealants, (3) polyvinyl acetate caulks, (4) solvent acrylics, (5) butyl sealants, (6) polysulfides, (7) polyurethanes, (8) modified silicones, and (9) silicone sealants, including RTV adhesive/sealants.

1. OIL-BASE CAULKS

These are the lowest-cost, lowest-performance, general-purpose sealants formulated mostly from oils and various fillers. In general, they are truly only crack- and gap-fillers and are usually not recommended for exterior applications where some degree of movement of parts is anticipated. As they age in natural weathering, they become increasingly rigid, making the presence of any significant movement disasterous. Minford (1348,1349) always included one of the better-grade oil-base caulks in his long-term weathering studies on variously treated aluminum surfaces with the results described above as the invariable result after a few years of service time.

2. LATEX ACRYLIC SEALANTS

Introduced in the 1960s, these products are based on latex acrylic polymers and cure by evaporation of the water present. Lomax (1963) offered a short paper describing such acrylic polymer caulks at the January 1990 Adhesives and Sealants Council Meetings, whereas Johnson (1285) issued a 1990 review of their chemistry, properties, major suppliers, costs, and application parameters. New products are constantly under development with general acceptance because of their good handling and durability characteristics. Other advantages include medium-durability weathering resistance in exterior atmospheric services, water cleanability, low toxicity, low cost, and low odor with no flammability. Their ability to be painted over after a few hours has been quite influential in residential construction. Sensitivity to rain washout and the need to be applied above 40°F (5°C) is a C-834 specification (1964) to which they must perform as well as Standard Practices specification C-790 (1965) that is pertinent to their application and use in the field. Other standard testing procedures include Federal Specification TT-S-00230C (1966) and Architectural Aluminum Manufacturers Association Specification 802.3 (1967).

3. POLYVINYL ACETATE SEALANTS

A rather low-performance product, these sealants have been in the marketplace for many years and are based on the same technology used to make white glues. An inherent lack of flexibility is not totally overcome in the formulations containing a variety of plastizers. They have a very high modulus change with temperature flucuation, an excessive shrinkage in the bondline with the adherend, and a generally high modulus which makes them generally unacceptable for use in many natural atmospheric services.

4. SOLVENT ACRYLIC SEALANTS

These sealant compositions offer a superior ability to wet even unprimed aluminum surfaces; however, they exhibit a curing mechanism which releases disagreeable odors along with some toxity and some flammability. Their use can circumvent the need for any costly surface treatment. These sealants do have a low recovery in the event of a cohesive failure and this was discussed by Cook (1968) in 1970. Their formulation as copolymers of acrylic and/or methacrylic acid esters and other monomers is shown in BASF patents (1969,1970). Factors which influence their physical properties were reviewed by Brendley (1971), Amstock (1972), and Reid (1973). The most recent review of solvent acrylics is by Baghdachi (1284). Their general properties make them most appropriate for outdoor service conditions. They can be considered a moderate-cost product, yet give service outdoors which may only be second to the silicones. Some disadvantages include low joint movement capability after a year or so with more modulus change with changing temperatures than desirable. Minford (1348,1349) showed relatively good performance with these sealants in aluminum joints in up to 10-yr exposure testing in the natural industrial atmosphere at the Alcoa Laboratories.

5. BUTYL SEALANTS

These products are available in the form of solvent-release, deformable preformed tapes, hot-application types, and polyisobutylene sealants. In 1990, Higgins and Jagisch (1282) produced a fairly detailed summary of the chemistry, properties, testing, forms, applications, and suppliers of butyl-based sealant products.

Both butyl rubber and polyisobutylene are used as the primary binders, tackifiers, and modifiers in a wide range of adhesive and sealant formulations. Major difference between the two materials is that the former is a copolymer of isobutylene with a small amount of isoprene, whereas the latter is a homopolymer. Additional information on the former is also found in a 1987 publication of Fusco and Hous (1974), whereas a 1987 publication on isobutylene by Kresge et al. (1975) furnished good information for the reader on that subject.

Among the advantages of butyl sealants are their good flexibility, good adhesion even without primers, and resistance to UV and water. Their extremely low moisture vapor transmission rate has resulted in their wide use in insulated window units. Long used in a variety of automotive applications for keeping out water, butyl sealants have lost some of this market to newer polyurethane formulations. A 1989 appraisal of their use as preformed tapes in glazing is offered by Bellance (1976). Berejka (1977) and Higgins (1978) supplied 1967 publications about the butyls and the same authors collaborated in 1973 on an updating (1979). In 1968, Berejka (1980) described testing methods with architectural tape sealants. Callan (1981) discussed cross-linked butyl hot-melt use in 1974, whereas Jagisch (1982) wrote about the use of the polyisobutylene polymers in sealants in 1978. In 1983, Quade (1983) included butyls in his discussion of insulating glass.

6. POLYSULFIDE SEALANTS

Entering the market in the early 1950s, these sealants were probably the first high-performance products available. They can be found in three forms: (1) two-part manually mixed, (2) two-part mechanically mixed, and (3) one-part. Their chemistry, functional-types, markets, application suppliers, and properties have been discussed by Giangiordano and Sorensson (1281). These authors also discuss recent advancements in modifying polysulfides which show improved thermal and chemical resistance, lower permeability to inert gases and fuel vapors, and superior physical properties as well. Other variations include newer mercaptan-terminated polymers and new polymers containing sulfur in the form of a thioether.

These sealants have been widely used in aerospace, defense, and construction industries and when in contact with aluminum in these industries have shown good durability performance. Minford (1984,1985) has evaluated both the one- and two-part formulations of polysulfides, both with and without recommended primers for aluminum. Lasting for periods up to 10 yr under external atmospheric weathering, these investigations have generally shown excellent results. However, as has been mentioned many times earlier, no specific commercial formulations can be assumed to be capable of good performance simply because they are described as a polysulfide on the manufacturer's data sheet. Some testing needs to be conducted to confirm that the product selected can perform like the best possible formulations available with this generic description.

7. POLYURETHANE SEALANTS

Polyurethanes had their origin in the work of Bayer (1986) (for other selected book and chapter references, see, e.g., Refs. 1987–1998.)

In the sealant industry, the polyether hydroxy-terminated compounds have been more widely accepted than the corresponding polyesters because of lower price. These materials are prepared by base-catalyzed addition of alkylene oxides to difunctional or polyfunctional alcohols or amines. Active hydrogens present in these compounds can react with a variety of isocyanates, creating a primary urethane chemical linkage. This is the situation only in two-part systems where a polyol-rich component reacts with a second isocyanate-rich component with a nominal 1:1 equivalent ratio producing a urethane rubber. In one-component systems, reaction of the excess isocyanate with moisture generates an amine which reacts with the terminal isocyanate group in the polymer to form a disubstituted urea. In still another type, one unit is an amine which reacts with a second isocyanate unit offering urea formation with a primary polyurea linkage. Polyurethane sealants are of varying molecular weight with corresponding varying hard and soft segments to suit most any application area. Particularly in two-part formulations, end groups can be modified to be hydroxyl, amine, epoxide, mercaptan, or isocyanate to give different specific properties appropriate to varying applications. Urethanes, in general, can accommodate expansion, compression, vibration, UV radiation, and moderate heat, all of which are favorable characteristics for sealants.

Polyurethanes, as with polysulfides just discussed, come in three forms: one-part, two-part manually mixed, and two-part meter-mixed formulations. They are high-performance sealants with excellent adhesion, durability, and long life expectancy. The newer formulations have lower modulus and higher movement potentials than earlier with relatively high-tear strength. Main disadvantages have been loss of adhesion to glass under UV exposure, some higher than desirable modulus changes with temperature fluctuations, and an upper temperature use limit of about 180°F (80°C). It must be remembered that temperature on a black aluminum can range beyond this temperature in jungle or desert atmospheric exposure. Minford (1348,1349) included the testing of both one- and two-part polyurethane sealants with and without special primers on various aluminum surfaces for periods up to 10 yr at a natural industrial atmospheric site. While some early formulations had questionable long-term adhesion, newer formulations in the late 1960s and 1970s performed acceptably and sometimes exceptionally well.

8. *SILICONE-FORMULATED SEALANTS*

This family of sealants seem clearly to be the highest-performance sealants available for most applications in today's marketplace. Their simultaneous ability to maintain stability at high temperature with excellent flexibility at low temperatures is unequaled. At the same time, they have good weathering resistance, modest structural strength like adhesives, can cure quickly if necessary, show excellent adhesion to a wide variety of adherends (including aluminum), and are available in the widest choice of modulus values. Negatives include above-average surface dirt accumulation, relatively low tear strength, nonpaintability, and higher price than most other sealant formulations. They are available in either one-package (ready to use) or two-part manual or mechanical meter-mix forms. The former generally take moisture from the air in order to react to form an elastomeric material with accompanying by-product elimination.

The most common ingredient of silicone sealant is a polymer known as a silanol-terminated polydimethylsiloxane which is produced by the hydrolysis of dimethyldichlorosilane (1999). Resultant hydrolyzate can be polymerized by a number of either basic catalysts (2000,2001,2002) or acidic catalysts (2003–2011).

The molecular weight of the resulting polymer will be dependent on the amount of moisture (chain stopper, establishing the terminal silanol groups) retained in the system. For speciality sealants as used in aerospace applications, other organic groups may be introduced in place of the pendant methyl groups on the polydimethylsiloxane mentioned above.

Some additional readings of value may be the discussion of Klosowski and Gant (2012) on the attractive flow characteristics of silicone sealants even at −40°F (−40°C), discussion of their superior long-term weathering resistance in the construction industry by Klosowski (2013), and a fast-growing application for special building and highway expansion joints by the same author (2014). The most recent summary on the silicone sealant chemistry, forms, properties, and applications is by Beers and Klosowski (1286). Minford (1348,1349) evaluated a wide variety of silicone sealants on aluminum surfaces under long-term atmospheric weathering conditions with variable responses from very poor to excellent adhesion retention. It should be mentioned that specially formulated primers have sometimes been recommended for use under some silicone sealants over the years, so that specific testing needs to be conducted to demonstrate whether the additional use of a primer is necessary. The so-called RTV silicone sealants of recent years have been successfully employed by Minford (2015) to protect aluminum joint edges from corrosive pitting penetration in seacoast exposures for periods up to 8 yr without the benefit of any primer.

E. Preformed and Formed-in-Place Gasketing

The use of gasketing materials in contact with aluminum requires that the chemical compatibility of the preformed gasketing material be assured under all possible service conditions and that, additionally, the long-term adhesion of the formed-in-place gasket be maintained. Because the physical and chemical properties of these gasketing materials need to be very similar to the properties of adhesives, they are manufactured from the same or closely related chemical families of polymers. Perkins and Glarum assembled a booklet for NASA on the subject, *Adhesives, Sealants, and Gaskets—A Survey*, which is available from the Government Printing Office, that may still be of significant benefit to the reader. These authors provide data on the comparative effectiveness of different gasket materials.

The main industrial markets for the preformed type of gaskets have consisted of use in insulating glass, for in-house glazing, and in sealing metal buildings. Today, there are over 2000 companies alone in the United States which have functioned as insulated glass unit manufacturers. Elias et al. (2016) cited forms of design of insulated glass construction in their survey article on sealant markets and application areas. Preformed tapes employed are thermoplastics and butyls that can be used as bedding, end sealers, vision lights, or with panels of aluminum, steel, or porcelain. Extruded gaskets used in the various dry glazing systems are typically neoprene or ethylene propylene diene monomer (EPDM) with lesser use of vinyl or silicone formulations. These gaskets are employed in so-called "wet/dry" sealing systems along with conventional acrylic, butyl, polyurethane, and silicone extrudable bead sealant formulations. Aluminum and other metal building markets often use acrylic and butyl sealants at window and door perimeters and panel joints. Butyl tapes and/or pumpable butyl sealants are used for sealing lap-joints in aluminum roofing panels as well as sealing wall flashings. Minford (2017) evaluated a number of different commercial butyl sealers for durable sealing of aluminum rain-carrying gutters and down spouting with good results under relatively long-term accelerated laboratory weathering conditions (hot humidity and weather-o-meter evaluations).

In recent years, there has been an ever-widening use of liquid sealants that cure in place to constitute formed-in-place gaskets. Many failures have occurred but are not usually related to the materials potential but rather to badly designed joints. It is an error to attempt to make a formed-in-place gasket with these liquid sealants over a flange system previously designed for a solid gasket. Also, there has been an accelerated use of these liquid sealants with application by robotic devices to increase manufacturing rates. If such devices are not constantly monitored, some minor malfunctioning can cause the small variance in the applied bead geometry that can cause either instant or later leaks. Even the incorporation of air bubbles in the sealant can cause quite negative results.

The need to specially design flange areas for liquid sealant application can be explained by the fact that the functioning of a preformed and formed-in-place gasket is very different. While the conventional preformed gasket maintains its sealing capability by direct compression sealing (tightening of fasteners), a formed-in-place gasket does not cure until after assembly. It must depend on metal-to-metal contact with the cured gasket and flanges, which, with the fasteners, carry all the tension. Even when the tension is lost that would fail the preformed gasket seal, the adhesion of the formed-in-place gasket can still prevent failure of the system. Thus, flange systems must be specifically designed for or modified for optimum use with the cured liquid sealant.

Two types of liquid sealant in common use are anerobics and medium-modulus RTV silicones. Once again, flange and gap design are important with the anaerobic cures requiring relatively rigid flanges and narrow gaps that prevent joint movement with the relatively rigid anaerobic product. Where joint movement is anticipated, the medium-modulus RTV silicone will be the desired candidate sealer. The flange design must have sufficient gap, however, to permit entrance of atmospheric moisture to affect the full cure.

VII. THERMALLY OR ELECTRICALLY CONDUCTIVE ADHESIVES

A. Thermally Conductive Adhesive Technology

The need for development of high-thermal conduction of heat by an adhesive has been especially accelerated by an increasing number of miniaturization of electronic circuitry applications. Excessive heat-development problems, which are a natural consequence of this technology, causes premature failure of electronic components when their optimum operating temperature is exceeded. Small metallic-finned heat sinks have been used to carry this heat passively from the power device (heat generator) whose efficiency depends highly on interface of the heat generator and heat sink. Mechanical fixturing of the two still results in about a 20% metal-to-metal contact area, so a more efficient method has been to provide a thermally conductive alternative pathway, as discussed by Frauenglass et al. (2018) in 1984. Batson (2019) has also contributed a discussion of this subject. Metallic (electrically conductive) or nonmetallic (electrically insulative) powders are blended into the adhesive formulation to make a high-viscosity, pasty adhesive. The most popular conductive systems are formulated with epoxies, silicones, anaerobics, acrylates, and methacrylates. Spacer particles can also be added to these adhesives to provide a definitive gap between the two adherends to ensure electrical isolation. There have been some disadvantages, since metering and mixing are often required and it is essential that there be no air bubbles entrapped in the mixture. A special packaging permitting mixing of two parts within the same enclosure has been one answer to this trapping of air problem. Some other single-component thermally conductive formulations must be thermally cured. Anaerobic/aerobic systems of heavily filled resin and a preapplied surface activator has also been introduced, as described by Hastie (2020). Thermally curable-conductive tapes and films are also available which circumvent most of the handling difficulties mentioned above.

B. Electrically Conductive Adhesive Formulations

To gain desired electrically conductive properties in an adhesive it is necessary to add electrically conductive fillers like silver, nickel, or conductive carbon. The latter two with their significantly lower conductivity potential are most used in grounding of circuits, electrical static discharge protection, electromagnetic interference/radio frequency interference (EMI/RFI) protective coatings, gaskets, and adhesives. Silver additions achieve the highest level of conductivity available in the marketplace and yet can be cost effective. The adhesive bases used are mainly in the epoxy, silicone, or acrylate polymer families. One- or two-part epoxies are most commonly produced.

A new technology recently introduced has involved UV curing of the electrically conductive formulation but it is still in the introduction stage by several large companies for future use. Tape adhesive products are also in the marketplace but are mainly used only where the convenience outweighs the need to minimize the electrical resistance. Batson (2019) offered a good summary of adhesive applications as applied to electronics where this kind of adhesive is predominantly used.

Use of electrically conductive adhesive with a good electrical conductor like aluminum might seem like an ideal duo; however, some investigations have shown that some surface changes can proceed while in humid service conditions, which are not thoroughly understood. Thus, it will be necessary to assure long service life potential under such conditions to conduct some thorough testing under controlled laboratory conditions designed to simulate or accelerate the service environment conditions.

Minford (2021) repeatedly demonstrated that a bondline corrosion problem can arise in the presence of a condensing humidity condition at 125°F (52°C) when these electrically conductive adhesives are sandwiched between aluminum adherends. This corrosion can be induced in a relatively short exposure time of days to weeks. While it would be obvious that such conditions must be protected against in service, these tests were employed at the Alcoa Labs in the standard manner to establish the relative compatibility of various adhesives and adherend materials over many years. The author has not further pursued the obvious question of what, if any, level of environmental humidity can be tolerated and for what time period? Just as these more noble metals (used in formulation of conductive adhesives as the oxides) can plate out on aluminum in conductive water solutions and create localized pitting corrosion on the surface. It could be shown that leaches from such adhesives might function similarly. In this regard, it was not necessarily anticipated that the type of electrically conductive epoxy containing conductive carbon particles would create the same incompatibility (corrosive sites) problem as those containing silver and nickel oxides. It proved to do so, however, a longer exposure time to humidity was necessary to create the same level of corrosive action in the bondline.

VIII. PRIMERS FOR ADHESIVE-BONDED ALUMINUM JOINTS

Primers are various materials which have the ability to enhance the probability of obtaining and retaining good adhesion; thus, their function may be considered to both adhesion promotor and protective coating over the bondline. A closely related material is the coupling agent, although its only function is to create some chemical bonding between the adhesive and the adherend surface. This distinction is not always clear, however, since a silane coupling agent might equally fulfill the first definition above, since it is the means of obtaining the excellent adhesion and also the means of retaining it.

Often there is an accompanying increase in joint strength when a primer is used, although that is not always true. For example, there is the possibility that the cohesive failure strength of the primer coating could be lower than the cohesive failure strength of the structural adhesive in the bondline above it. In this case, it would appear to be an example of the old principle of finding bond failure in an unexposed adhesive joint in the weakest linkage part of the overall joint; in this case, a cohesive failure in the cross section of the primer. While this can be the situation when testing the lap-shear strength of the joint with no exposure to weathering, a duplicate of this joint could have significantly enhanced durability when exposed to varying weathering conditions. This response can occur because the primer has achieved a significantly higher degree of chemical wetting of the adherend surface than would have been possible using the structural adhesive alone. This superior wetting is often achieved because the primer employed has been a lower viscosity dilution of the structural adhesive in the bondline, permitting superior flow characteristics as compared with the same undiluted adhesive. In many other situations, the primer polymer and main adhesive polymer can be quite different chemically, yet the same condition of superior wetting by the primer will be expected. For example, the wettability of an aluminum oxide surface by a thixotropic two-part polyurethane adhesive, as employed in the automotive industry, can be quite marginal in aggressive weathering service. A two-part, flowable epoxy primer has been recommended by the polyurethane formulator to serve as a primer in this instance with a consequent significant improvement in the overall durability of the joint. The primer concept thus becomes one of achieving a degree of optimum wetting at the interface in a situation where this optimum wetting cannot be achieved by the main structural adhesive that has been chosen for its other good physical characteristics.

There is a secondary reason for using a priming technique when attempting to optimize the strength and durability of aluminum adherend joints that can be highly significant and might even be considered by some to be the major factor in some manufacturing situations. This is the ability of the primer to protect the adherend from contamination or other changes in the chemical constitution of its surface. It should be recognized that there is almost always a need to clean or clean and pretreat an aluminum surface prior to applying the adhesive and closing the joint. An interruption in this process can obviously provide a time period in which contaminates can deposit from the plant atmosphere or the treated surface can chemically alter from its most favorable state for being "wetted" by the adhesive. A primer applied as soon as possible after any surface cleaning or cleaning and pretreating will obviously find optimum conditions for wetting. At the same time, it also can provide a more responsive surface for later wetting by the main adhesive even with a storage period in between. This immediate application of a primer to aluminum adherends has been a response that the aerospace industry has followed implicitly in all their manufacturing operations. When the author has discussed with aerospace manufacturers the high degree of aluminum joint durability he achieved without use of a primer, their response has been that the immediate use of a primer to protect the surfaces from fallout contamination has been sufficient reason in itself to justify the procedure. The author would like to add his observation that the use of a primer might also serve as an important aid in permitting the highest degree of joint reproducibility as compared with no primer presence.

Bascom (2022) has furnished a 1990 summary discussion of the use of primers and coupling agents in which he points out another function of primers as the superior ability to penetrate surface roughness (mechanical pretreatments) or microporosity deliberately generated by various chemical pretreatments. It should be obvious that the typical paste or even film structural adhesives would lack the ability to perform this function as readily. Bishop et al. (2023) have recently examined aluminum/epoxy joints with and without primer by sectioning across the bondline with an ultramicrotome and viewing the sections by TEM. When no primer was used, there was clear evidence of trapped air in rough aluminum profiles produced by methods ranging from light mechanical abrasion to the microporosity produced by acid-dichromate etching. Much of this was eliminated when a low-viscosity primer was used. The probable ability of primer solutions to dissolve low levels of organic contamination that might remain at the adhesive/adherend interface as a weak boundary layer should also be mentioned. Reinhart (2024) has summarized a number of U.S. Air Force efforts to develop improved primers for both aluminum and titanium, including water-based primers, electrodeposited primers, and more effective corrision-inhibiting primers. Primers are rarely used for bonding adherend surfaces of polymers or polymer matrix composites, though Kinloch and Taig (2025) and Petrie (2026) issued reports relating to these matters in 1987 and 1989.

While it might be expected that the use of primers might be encouraged in most manufacturing for the above reasons, it should be pointed out that it is an additional processing step with some significant attendant costs which can influence most industries outside of aerospace. It should not be unexpected to find that there has been a determined effort in recent years for the adhesive formulators to try and find adhesives which can circumvent the need for use of primers in general manufacturing. There seems to be two levels of performance that have been considered. First, there are new formulated adhesives which carry the information on the manufacturer's data sheet that the adhesive can be used with untreated aluminum adherends. Some adhesives of this type have been formulated by modified acrylic adhesive manufacturers where the claim is made that the accelerator can be brushed on an untreated aluminum adherend followed by application of the acrylic adhesive to the second surface with immediate closing of the bondline. After a modest fixturing time, handling strength is developed and the joint has acceptable resistance to weathering service. The author has evaluated a number of such products over many years with variable results but it must be acknowledged that a rather amazing degree of joint durability has been found with some of this family of products. More data will be found in Chapter 10 on joint durability.

The second level of performance has revolved about the formulation of adhesives that not only can bond to untreated aluminum surfaces but also to surfaces which have layers of lubricant or other oils on the surface. Wegman (2027) in his publication on surface preparation techniques has reviewed the use of such adhesives on oily steel in the automotive industry. Obviously, the surfaces, whether steel or aluminum, should be free of corrosion and any thick layers of oil might be better removed or at least reduced. All the adhesives which have been most effective in bonding over significant oil have been heat curing, as reviewed by Siebert et al. (2028). They used CTBN-modified epoxies which could improve adhesion to oily steel by as much as 65% based on lap-shear and peel tests. The author has conducted literally thousands of aluminum joint durability tests using heat-cured, modified-epoxy, and vinyl plastisol adhesives with varying levels of lubricant contamination on the surface, which will be discussed in Chapter 10.

A. Elastomer or Contact Cement-Type Primers

Because primers are generally relatively low percentage solid solutions and dispersions of polymer resin formulations, they have the ability to penetrate porous structured adherend surfaces. This has obviously been a major benefit in improving the strength and durability of bonded wood jointures and can also play an important role in enhancing aluminum-to-wood bondments. The importance of using a primer on the wood in such jointures is readily demonstrated by the achievement of a higher percentage of wood fiber tearing when such joints are tested both before and after weathering. After exposure, testing will also show clear evidence of a higher retention of initial bond strength as compared with no primer usage. There are a variety of laboratory accelerated and natural weathering environments that can be used to demonstrate the importance of priming such as continuous exposure in water or high humidity at various temperatures (including short-term evaluations in boiling water), repeated cyclic exposures to water soaking or hot humidity, freezing, and thawing (sometimes with accelerated drying conditions), exposure to the cycles in a standard weather-o-meter test machine, or exterior exposure to various natural weathering atmospheric conditions. The author has routinely used all these kinds of test exposure procedures for evaluating aluminum-to-wood bonded structures over many years at the Alcoa Labs.

Quite often, the contact adhesive primer employed is simply a more diluted version of the adhesive to be used for making the final bondline. This can be especially beneficial for bonding non-etched or marginally cleaned aluminum surfaces to wood with priming of both surfaces. The solvent-rich characteristics of such primers can permit dissolving or dispersing modest surface soils, creating a new bonding interface which can be more readily wet by a higher solids version of the same adhesive. This degree of bonding durability would necessitate thorough aluminum surface cleaning where 100% solids, mastic-type adhesives are to be used.

B. Characteristics of Primers for High-Strength Structures

A freshly deoxidized aluminum surface followed by adequate water rinsing and thorough drying will generally be receptive to high-strength bonding. It will also be an ideal surface for immediate priming. Because it has not always been logistically possible to make an immediate bond to this surface in the aerospace industry, it has been a universal procedure to apply a primer as soon as practical after proper pretreating of the aluminum surface. Primer application can be accomplished through a variety of methods, including (1) spraying, (2) dripping, (3) brushing, (4) roller coating, and even (5) curtain coating. Spraying, roller coating, and curtain coating are usually more economical, especially when large surface areas must be covered. After priming, the aluminum part may then be more safely transported and stored until a more appropriate time for making the final assembly bondment.

The primers used in aerospace aluminum structures were first designated for compatible use with the 250°F (121°C) curing structural film adhesives. These included both epoxy- and phenolic-based primers, since American aircraft were primarily being assembled with modified epoxy adhesives and nitrile-phenolic adhesives. At the same time, European aircraft manufacturing employed a preponderance of bonding with variations of the vinyl-phenolic family of products. Up until the development of phosphoric acid anodizing as an aluminum surface pretreatment, American aircraft were being mainly assembled with an FPL (sulfuric acid–dichromate etching) or some modification of that process. At the same time, European aluminum aircraft manufacturers were favoring variations of the chromic acid anodizing procedure (CAA). In recent years, the majority of American assemblies have been featuring variations of the phosphoric acid anodizing (PAA) procedure developed at Boeing.

Peterson and Arnold, in the early 1980s, studied the chemical characterization of aerospace adhesive primers after their analyses of failure modes of 350°F (177°C) primers over both FPL and PAA pretreated surfaces had indicated the failure mode was more dependent on the intrinsic properties of the adhesive or primer than the method of surface preparation. They detected three different mechanisms for toughened epoxy primers. One, the addition of an "incoherent" elastomeric additive to the primer could form phase separated rubber particles which had poor mechanical linkage with the matrix resin. Second, the addition of a "coherent" elastomeric additive to the primer could form phase separated rubber particles which had superior mechanical linkage to the matrix resin. And three, the formulation of the primer with a matrix resin with higher ductility could lead to both higher shear strength and toughness values.

Pocius (2029), in 1981, reported on a detailed 3M's study of the effect of cured primer thickness on the joint peel strength. He concluded that the floating roller peel strength was apparently inversely proportional to the primer thickness but the T-peel values seemed unaffected by thickness variation. With the aid of measurements made by varying the angle of peel, rate of peel, temperature, adhesive thickness, and scrim constitution, Pocius was able to develop a phenomenological description of the peel process even in the presence of a highly cross-linked primer.

It has been quite important for a number of years to impart aluminum corrosion-inhibiting properties into primers for aerospace applications. Considerable progress in this regard had been made by the beginning of the 1970s, as shown by Marshall's (2030) development of an epoxy primer to conform to MIL-P-233377B requiring superior corrosion-inhibitory properties. The range of inhibitors evaluated included alizarin/strontium chromate, cadmium phosphate chromate, or chromium chromate. Croke et al. (2031) have described specific testing procedures to determine the corrosion resistance of adhesive/primer systems. In the development of durable corrosion-resistant aluminum joints for the highest levels of performance, it becomes necessary for the developer of primers and compatible adhesive systems to determine and publish the kinds of data shown in Table 81 for a combination of BR-127 corrosion-inhibiting primer and FM-123-5 aerospace film adhesive (2032). While these kinds of data can go a long way toward enabling the customer to predict what level of initial joint strength might be expected under some precisely stated testing conditions, they do not offer confidence as to the actual long-term durability response of the adhesive/primer system in critical field service conditions. For these kinds of testing and data results, the kinds of investigations reported in the PABST Program need to be studied and reproduced.

While bonds incorporating corrosion-inhibiting adhesive primers (CIAPs) have considerably increased bond durability potential over non-CIAP primed joints, the CIAP primer products have the disadvantage that they must be baked prior to cure of the primary bond. Offsetting this disadvantage, however, is the fact aluminum pieces with the CIAP coating can be stored for long periods of time if properly protected from sunlight and dirt. This offers the option of stocking precleaned and

Table 81 Typical Mechanical Properties of BR-127 Primer with FM-123-5 Adhesive Film Tested in Conformance with Federal Specifications MMM-A-132 Type I, Class 2

Test no.	Property and test condition	Spec. avg. requirement	Avg. results with film wts 0.030 psf	0.60 psf	0.85 psf
1	Tensile shear, psi Normal temp. 75° ± 5°F	2500	5020	5755	6360
2	Tensile shear, psi 10 min at 180° ± 5°F	1250	3600	4025	3920
a	Tensile shear, psi 10 min at 250° ± 5°F	None	1300	1430	1500
7	Tensile shear, psi 10 min at −67° ± 5°F	2500	5815	6050	6380
8	Fatigue strength Normal temp. 75° ± 5°F	750 psi 10 cycles	No failure	No failure	No failure
a	Fatigue strength 1200 psi at 10 cycles	None	No failure	No failure	No failure
9	Creep rupture 1600 psi at normal temp. 75° ± 5°F	192 hr, 0.015 in max deform	0.00056	0.00060	0.00061
			0.00093	0.00070	0.0020
10	Creep rupture 800 psi at 180° ± 5°F	" "			
13	Tensile shear, psi Normal temp. 75° ± 5°F after 30 da salt water spray	2250	5380	5750	6520
14	Tensile shear, psi Normal temp. 75° ± 5°F after 30 da at 120° ± 5°F	2250	5280	5655	6175
15	Tensile shear, psi Normal temp. 75° ± 5°F after 7 day immersion in JP-4 fuel (MIL-J-5624)	2250	5275	6160	6315
	After 7 da immersion anti-icing fluid (MIL-F-5566)	2250	4895	6260	6750
	After 7 da immersion hydraulic oil (MIL-H-5606)	2250	5300	6315	6505
	After 7 da immersion in std. test fluids (TT-S-735)	2250	5142	6220	6540
	After 30 da immersion in tap water	2250	5400	5750	6650
a	After 30 da immersion in Skydrol 500 at 120° ± 5°F	None	4675	5465	—
16	T-peel lbs/in Normal temp. 75° ± 5°F	15	29	40	45
17	Blister detection tensile strength, psi Normal temp. 75° ± 5°F	2250	4175	4650	4455

Table 81 Continued

Test no.	Property and test condition	Spec. avg. requirement	Avg. results with film wts 0.030 psf	0.60 psf	0.85 psf
[a]	Bell peel, lbs/in Normal temp. 75° ± 5°F	None	—	76	—
[a]	Metal-to-metal climbing drum peel, inch lbs/in Normal temp. 75° ± 5°F	None	63	85	90

BR-125 primer 10% solids cured 60 min at 250°F with cured primer thickness 0.0002 to 0.0005 in.
[a]Data in addition to requirements of MMM-A-132.
Source: Data from American Cyanamid Co. publication 4-3-69 titled BR-127 Corrosion Inhibiting Adhesive Primer.

primed aluminum sheet material at a repair facility where it can be cut to size as required and merely solvent-wiped prior to final bonding. Where the additional time required to bake CIAP primer is not permissible in repair of adhesive-bonded aluminum structures, the use of non–corrosion-inhibiting primers may still need to be employed. Shannon (1077) evaluated a number of CIAP primers under 250°F (121°C) curing adhesive films using PAA-treated aluminum adherends with satisfactory performance indicated for all. However, not all primers will perform exactly the same and there is still a need to determine the level of performance potential for each product separately under the particular service conditions deemed appropriate.

While those primers specially developed for use with most 350°F (171°C) curing adhesive films have mostly proved to be compatible with the PAA-treated aluminum surface, a few rubber-containing primers have exhibited some suboptimal behavior, as reported by Marceau (1094). After joint failures were found, it was observed that a thin, rubberlike layer, approximately 100 nm thick, remained on top of the aluminum oxide.

C. Silane (Coupling Agent) Primer Products

The term *coupling agent* originally referred to a surface coating that resulted in distinct improvement in the moisture resistance of glass fiber–reinforced polymer composites. Principal agents in this respect were the silanes, which are distinct adhesion promotors between glass and resins (like epoxy and polyester), retarding the bond failure potentials when these combinations were exposed to water, especially under hot-wet conditions. As early as 1972, Bascom (2033) established the fact that these silanes from strongly adsorbed polysiloxane films on glass and metals. It was also shown that the mechanical and chemical conformation of these films and their thickness, cross-link density, and reactivity are highly dependent on the application conditions.

The silanes have been the subject of much research and speculation about how they operate. The most accepted theory has proposed that trialkoxysilane chemically bonds to the adhesive by reaction of the organofunctional group with the particular resin molecules in the adhesive. The bonding to the aluminum surface (hence the proper use of the terminology of coupling agent) appears to be an interaction of the alkoxy groups of the silane with the inorganic aluminum oxide. It was suggested as early as the 1970s that the adsorbed polysiloxane film has an open network structure that should permit interdiffusion of the adhesive molecules to form an interpenetrating network. Boerio (1117) wrote a 1989 review which summarizes a great number of subsequent investigations, providing both circumstantial and some direct evidence that supports this view. However, while it should not be difficult to envision penetration of relatively small molecules like epoxy monomers into this film, it stretches the concept of interpenetration to include success shown by the use of silane coatings to improve adhesion of various high molecular weight thermoplastics to a variety of other adherends. Diffusion of long polymer molecules into such a network would not be expected. For fur-

Table 82 Effect of Pretreatments Involving Silanes on Lap-Shear Strength of Aluminum/FM-73 Joints[a]

	Lap-shear strength (MPa)			
Surface treatment	Ambient	176°F (80°C)	194°F (90°C)	212°F (100°C)
Chromic acid etch	38.7	26.3	24.7	22.1
Epoxy silane	36.8	23.8	21.8	17.2
Methacrylate silane	36.7	20.9	18.5	11.2
Chloropropyl silane	36.5	19.8	18.3	12.6
Amine silane	35.6	22.5	22.4	18.3
Diamine silane	35.4	23.1	20.6	13.8
Vinyltriacetoxy silane	34.4	20.5	18.1	12.5
Vinyltrimethoxy silane	31.1	19.6	16.3	12.5
Degrease	31.2	19.6	14.6	10.2
Degrease/sandblast	27.9	17.6	15.0	11.4

[a]Lap-shear joints all fabricated with Alclad 2024-T3 aluminum sheet.
Source: Data from Ref. 1155 (Table III, p. 314).

ther study of the development of adhesion through silane coupling agents, the reader is referred to References 1109–1112, 1120–1125, and 2034–2036. Other investigations of specific interest in studying the interaction of polysiloxanes with aluminum oxide are those of Kaul, Sung, Chin, and Sung (2046–2048).

Where the adhesion between a particular RTV silicone adhesive/sealant and an aluminum surface has proven to be inadequate in water-soaking or high-humidity service conditions, special silanes have been recommended to improve the adhesion and overall water resistance of the bondline. Companies who have specialized in formulating and selling silicones and attendant silanes like Union Carbide and General Electric have routinely furnished lists of which silicones may require silane priming on metals along with suggestions of the appropriate silane for particular metal and silicone combinations.

Primers currently available are generally low-solids, solvent dispersions of silicone resins, silanes, organic resins, or some combination of these materials. Primer recommendations for metal and ceramics with oxides present on the adherend surface will usually be quite different from those suggested for primer use on plastic surfaces. Curing is generally completed at room temperature through solvent evaporation followed by hydrolysis of remaining resins with moisture from the air. Sometimes a postcuring operation at elevated temperature may further improve adhesion. Usually, the silane-primed surfaces need to be protected from atmospheric moisture if storing is contemplated before final bonding.

Table 83 Physical Forms of Uncured Structural Adhesives

1. 350°F (177°C) curing films[a]
2. 250°F (121°C) curing films
3. 250°F (121°C) curing pastes
4. Room temperature–curing pastes (two-part systems)
5. Primer/liquid combinations (room temperature curing)
6. Room temperature curing liquids
7. High temperature adhesive films and pastes

[a]Adhesives deposited over carrier materials are called tapes and can be available for curing at various temperatures.

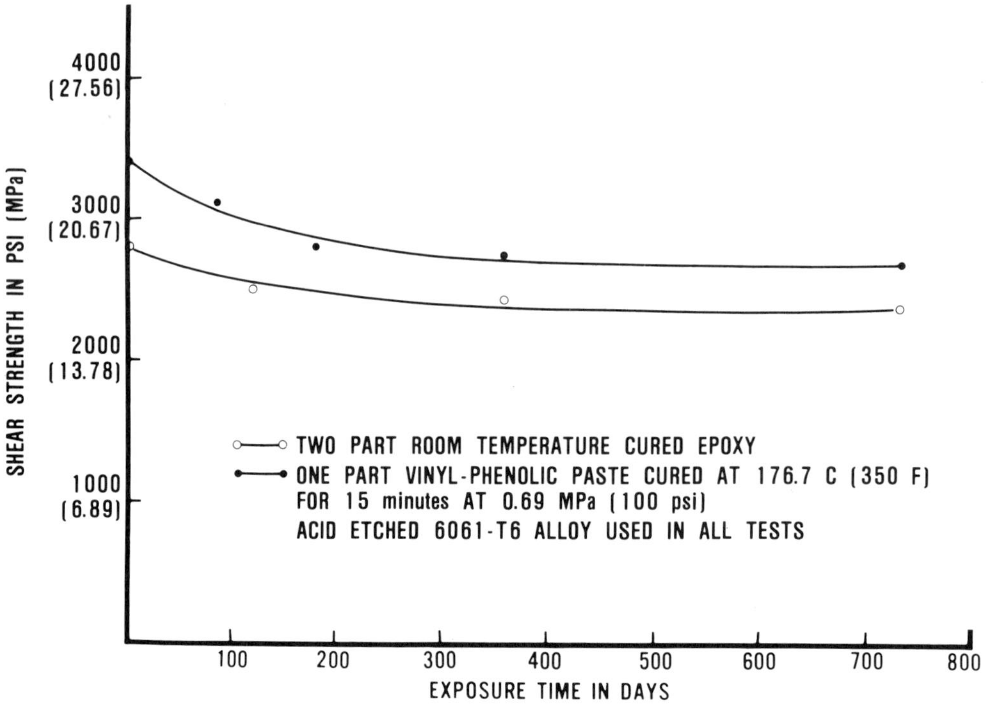

Figure 15 Unstressed specimens in 100% relative humidity at 51.7°C (125°F).

Gettings and Kinloch (212,213) have investigated a range of silane primers on steel adherends and have suggested the same mechanism augmenting adhesion would likely serve for other metals like aluminum as well. Boerio and Gosselin (717) and Kaul et al. (2054) have conducted similar investigations using aluminum adherends with similar results. Minford has employed silane primers on 6061-T6 aluminum adherends bonded with epoxy adhesives to enhance joint durability but with varying degrees of success. It would appear that most silanes available in the marketplace still need to be carefully screened with testing in comparable or even accelerated weathering conditions as compared with the recognized service environment for a particular manufactured structure.

Comyn (2037) has published a 1986 overview of silane coupling agents (2038). Comyn et al. (2039) focused their 1989 study on some aminosilane coupling agents over aluminum oxide surfaces using the IETS spectroscopic technique. Aminosilanes are of particular importance because the amine group can react with epoxide groups, making the coupling agent covalently bonded to the adhesive. Earlier IETS investigations of silanes on aluminum oxide arranged in chronological order are Diaz et al. (2040), Furukawa et al. (2041,2042), Van Velzen (2043), and Brewis et al. (2044). Other spectroscopic analytic techniques have also been employed to study aminosilanes such as the reflection absorption infrared spectroscopy procedures used by Boerio et al. (1117) and Boerio (2045) using 3-aminopropyltriethoxysilane (APES) on aluminum oxide.

Cassidy et al. (2049) issued a comprehensive study of coupling agents with urethane and epoxy adhesives as early as 1970. Kaul et al. (2047) have studied in detail the failure analysis of aluminum/polyethylene jointures under water-soaking conditions where the aluminum was silane treated. Liang and Dreyfuss (2050), also in 1984, published their studies of the effect of silanes on the durability of glass or metal adherend jointures using a polybutadiene-polyurethane adhesive. In 1989, Comyn et al. (2051) reported IETS results on the 3-glycidoxypropyltrimethoxysilane/aluminum oxide interaction. This was followed in 1990 by another paper on the interaction of silanes containing carbon-carbon double bonds with aluminum oxide as investigated with the IETS technique (2052).

The use of silanes as coupling agents to promote more durable aluminum-to-aluminum joints will be discussed in more detail in Chapter 10 on the permanence of aluminum-bonded joints.

D. Epoxy-Based Primers

Epoxy-based primers have been especially successful with structural aluminum adherend joints based on the fact that epoxy-based structural adhesives have constituted the majority of candidate adhesives for American aircraft production. This includes the acceptable joining of primary aircraft production. This includes the acceptable joining of primary aircraft structures as confirmed in the PABST investigation conducted by Douglas for the U.S. Air Force (75,76,77). These primers can be classified as (1) room temperature–curing primers, (2) 250°F (121°C) curing primers, and (3) 339°F (171°C) curing primers. The room temperature–curing epoxy primers would, of course, benefit adhesion when used under 100% solids, paste, two-part, room temperature–curing epoxy products. They are generally liquid primers and, thus, offer a superior opportunity for flowing and wetting the treated oxide surface. This kind of primer has also been employed successfully under two-part polyurethane adhesives for bonding aluminum to plastics like polyester/fiber glass in the automotive industry. Without such a primer, the author demonstrated that there was no chance to achieve a durable aluminum/polyurethane bond using the same polyurethane adhesive. Epoxy adhesive primers are not really mandatory when using the typical two-part commercial liquid epoxy systems for general manufacturing as long as the aluminum surface has been effectively cleaned of obvious soiling. In fact, all of the long-term (up to 12 yr in either the industrial atmosphere or Surinam jungle) aluminum joint durability investigations using a commercial polyamide-cured, two-part epoxy paste adhesive conducted by the authors did not employ any room temperature–curing epoxy primers.

The 250°F (121°C) curing epoxy primers have generally been designed for compatibility with the various 250°F (121°C) curing epoxy film adhesives as used in aerospace structures. Shannon et al. (1077) has shown the satisfactory performance of a number of these epoxy primers with added CIAP properties on aluminum. Scardino and Marceau (853) have illustrated the particular importance of using the CIAP-type epoxy primers when hot sulfuric acid/dichromate deoxidizing (FPL) was used as the aluminum surface pretreatment. Even with some anodizing surface pretreatments, they found greater crack growth in "wedge specimen" tests when the CIAP primer was not used. When PAA pretreatment was employed, however, the crack growth values were quite small regardless of whether the CIAP primer was used or not. The mode of joint failure seemed to remain interfacial in nature regardless of presence or absence of CIAP primers. Both the 250°F (121°C) and 350°F (177°C) curing primers are supplied in liquid form, ensuring a high degree of surface wetting. They are generally a low percentage of resin solids dispersed or dissolved in a single or combination of organic solvents. They may be applied by spray, roller, or flow-coater to a precisely controlled thickness ranging (depending on the film adhesive) from 0.0001 in (0.1 mil) to 0.0025 in (0.25 mil). The primer has to have complete compatibility with the film epoxy, enabling an enhancement of mechanical and/or physical properties in the cured film. It has been the contention of the aerospace manufacturers that the use of a primer is always justified even if they offer little improvement in overall bondline properties, because of their ability to protect freshly deoxidized or anodized surfaces from degradation due to atmospheric exposure.

The 350°F (177°C) curing epoxy primers (described in the earlier classification as the 339°F [171°C] type) have been similarly designed to be used with the same temperature-curing range of aircraft epoxy film adhesives. When a rubber was added to such epoxy primers to reduce the primer rigidity, the ultimate lap-shear strength values were significantly reduced. Marceau (1094) reported finding a thin, rubberlike layer on top of the oxide in failed joints of this kind using PAA-pretreated aluminum adherends. It was presumed that the rubber particles could prevent entry of the primer into the microscopic surface structure of the PAA-treated surface. Rather surprising, however, was the fact that the stress durability of these same joints proved to be excellent at stress levels of 1500 psi (10.34 MPa) or below. Schwartz (2053) when evaluating similar type joints at stress loadings above 1500 psi (10.34 MPa) in hot humidity reported poor durability results. It would appear that some sort of limiting physical stress level had been reached at about this value.

The discussion of Millard (2054) on epoxy-primer compatibility in his chapter on epoxy adhesives should be of further benefit to the reader. Reference might also be made to the SAE Aerospace Recommended Practice for surface preparation and priming of aluminum alloy parts for high-durability structural bonding (2055). In 1986, Dodiuk et al. (2056) issued a paper titled "Chemical Interaction System Aluminum Oxide–Primer–Adhesive" dealing with this general subject. In 1990, Berbezier et al. (2057) investigated the microstructural changes induced by aging in a study of adhesion primer properties, and Xue et al. (2058) reported on the chemical reactions of an epoxy-functional silane in aqueous solution in 1991.

E. Chelation Primers (Complex Formation)

An unusual type of primer has been advocated by Brockman based on selections from chemical agents which can offer a chelating potential with the aluminum oxide surface. Brockmann choose to experiment with 8-hydroxyquinoline or 2,4,5,7-tetrahydroxyflavonene as primers to enhance aluminum joint durability in water-exposure conditions. He offered successful demonstrations using either epoxy-phenolic or nitrile-epoxy adhesives on aluminum adherends.

Yang and Zhou (2059), in 1984, reported no bond separation of aluminum/polyethylene joints after 3-yr immersion in water through the use of a fumarato coordination compound designated Volan 82. For some reason, Dupont no longer markets this coordination compound.

F. Primer Thickness Effects

Sell (1098) has evaluated the effect of different thicknesses of primer on aluminun joint surfaces with a relationship being demonstrated between increasing strength and increasing primer thickness up to the point where the primer thickness can cause the primer to act as a separate layer in the bondline. As mentioned by Kinloch (1099), it should be at some such point that the primer can become the weakest site in the overall joint system. The work of Pocius (2029) was also mentioned earlier on the effect of cured primer thickness on the peel strength of 2024-T3 aluminum joints. T-peel values seemed unaffected, but the floating roller peel values proved to be inversely proportional to the primer thickness.

Spencer (2060) investigated the role of primer in bonding aircraft and space structures as early as 1972, whereas Aker (2061), in the same year, also discussed their function in aircraft bonding. Falcone and Miller (1088) were among the earliest to discuss the special use of primers in making repairs on rotor blades in helicopters.

G. Heat-Cured Paints as Primers on Aluminum for Bonding

Minford (2062) has considered the application of cured paint coatings on aluminum surface should function in a similar manner to similar chemistry adhesive primers. This should be true, of course, only if the paint adhesion durability itself had been guaranteed by being applied over a well-prepared aluminum adherend surface and fully cured before adhesive application. A vast array of commercially prepared prepainted aluminum materials already exist in the marketplace for such bond durability evaluations. This has resulted from the fact that the aluminum manufacturers have extensively studied the best means of painting aluminum surfaces for use in long-term weathering-resistant building applications such as aluminum commercial buildings and siding for residential properties. These investigations have clearly led to methods of mass producing prepainted aluminum sheet with a paint adhesion capable of withstanding 20 or more years of direct atmospheric weathering. Thus, excellent long-term durability of adhesive-bonded joints made by bonding with adhesives that are members of the same or similar polymer families to these already paint-primed surfaces might be considered to have equal long-term service potential.

Minford (774) has attempted to evaluate and further economize the paint primer concept by applying heat-curing paints over a simple vapor-degreased 6061-T6 adherend sheet as compared with the chemical conversion coated aluminum sheet preparation usually employed to manufacture commercial prepainted aluminum sheet products. The results will be considered in detail in Chapter 10 on joint durabilities. However, it can be stated here that this testing proved that the cost of separately deoxidizing and conversion coating the sheet might be preempted because of the low viscosity and high degree of flowability (wetting) obtained when using certain epoxy-urea or epoxy-phenolic paint formulations. Further, the wetting was further positively affected by the heat curing of these paint coatings. Briefly reviewed, the results confirmed that direct exposure of such joints for 8 yr in natural weathering did not significantly degrade the joints made with either a two-part, room temperature–curing epoxy or a one-part, heat-cured, nitrile-modified epoxy formulation when using the epoxy-phenolic paint as a primer under these adhesives. In contrast, thermoset acrylic paints which have been highly successful top coatings for aluminum siding applications did not perform as durably in the same weathering conditions.

Finch (2063) produced a 1986 paper on the use of zinc-rich paint primers for protecting metal adherends. Such surfaces might be bondable providing tests demonstrated good adhesion to the primer.

H. Surface Inhibitors

Ahearn and Davis (38) have demonstrated that they could achieve significant improvement in aluminum joint durability in "wedge" specimen testing in 100% RH at 140°F (60°C) conditions by applying literally monolayers of phosphonic acid inhibitors. This could qualify these materials as being classifiable as some sort of primer. Specifically, the FPL-etched aluminum adherends were simply immersed before bonding in an inhibitor solution containing just 10 ppm of nitrilotris (methylene) phosphonic acid (NTMP). After treatment and bonding, the joint durability reportedly approached that found with PAA-pretreated aluminum adherends. Hardwick et al. (527) have even included reference to further improving the durability of aluminum PAA-pretreated joints through use of these surface inhibitors. Again more discussion will be provided on the subject in Chapter 10.

I. Metal Alkoxide Primers

Pike (2064–2067) and Pike and Lamm (2068) have demonstrated that inorganic primers such as the metal alkoxides can improve the durability of aluminum/epoxy joints. When sec-butyl aluminum alkoxide, for example, was applied to either porous or nonporous aluminum oxide surfaces, the durability of bonds made with a 350°F (177°C) cured epoxy film adhesive improved 4–10 times over that of the organic primers under comparison. Pike (2067) further demonstrated the same enhanced durability using epoxies cured at only 250°F (121°C). The resulting primer layer is smooth, which would certainly restrict the use of mechanical interlocking as an explanation for the exceptional joint stability.

Pike and co-workers (2066–2068) have proposed a number of possible mechanisms to explain this enhanced durability, including (1) formation of a stable amorphous boehmite (AlOOH) after hydrolysis of the primer, (2) a chemical interaction between primer and epoxy, and (3) reduction of chances for a weak boundary layer formation. The last supposition is really based on Brockman's (2069) contention that low molecular weight fragments can segregate and change the cross-link density at the interface where the surface is porous. The smoothness of the alkoxide primer surface would minimize this effect. Pike (2065,2067,2068) also found that the bond performance of the sec-butyl aluminum alkoxide–primed joints varied with the coating thickness, with the thicker coats performing more durably. The locus of bond failure was indicated to be within the alkoxide layer, since primer was usually found on both sides of the failed joints. Pike (2070) reviewed the effect of metal surface treating and formation temperature of inorganic adhesive primers in 1986.

Filbey and Wightman (2071) later performed similar testing with Ti-6Al-4V adherends and tetraisopropyl titanate (TIPT) or sec-butyl aluminum alkoxide primers. The failure in the joints primed with the TIPT appeared to be interfacial between the primer and epoxy, indicating weaker interactions between titanium alkoxide primer and epoxy than for the aluminum alkoxide and epoxy. Pike (2072) offered a review of the present state of the art of water-based inorganic primers at the Adhesion '90 meetings at Cambridge.

IX. FILM VERSUS TAPE ADHESIVE CONSIDERATIONS

A. Structural Bonding Types

Adhesive films are widely used because of their convenience in precisely covering a fixed area of adherend and providing uniform adhesive bondline thickness. They are actually made by preblending one or more high molecular weight polymers with appropriate fillers and other compounding ingredients. This operation is followed by extruding, calendering, or casting of the mixture into a thin film. Typically, these films are 5–10 mil (0.13–0.25 mm) in thickness. If the mixture has been cast or calendered onto a mesh support, such as a woven or nonwoven mesh of glass, polyester, nylon, or other fiber, then the resulting supported film adhesive is referred to as a "tape." More information for the reader can be found in publications by Bolger (2073,2074) and Millard (2075). Either film or tapes can be soft and tacky or stiff and dry depending on the particular blended combination of adhesive ingredients. Some film or tape adhesives were formulated to cure at room temperature for aerospace manufacturing, but the problems attendent to keeping these adhesives in dry ice or low-temperature freezers during shipping and storage periods at the manufacturing facility have made them increasingly less acceptable. Most films and tapes today are cured at elevated temperatures if they are truly structural in nature, although the use of overaged film adhesive must still be guarded against. Because of the extensive use of epoxy-based films in more recent aerospace structures, the need for generating the higher bondline pressures required earlier for properly curing the vinyl- and nitrile-phenolic film products has diminished and fixturing with only modest bondline pressures is now predominant.

The distinguishing compositional feature of tape and film adhesives, as contrasted with the typical 100% solids paste or liquid structural adhesives, is that they contain a high proportion of a high molecular weight polymer instead of only low molecular weight fragments which must be subsequently polymerized to higher molecular weight moieties. Thus, molecular weights of 20,000 or more are quite common. The network polymers that can be generated from these high molecular weight linear polymers can be expected to lead to much tougher and more resilient bondlines with more recoverable elongation than those resulting from the curing in low molecular weight paste and liquid adhesives. These higher property potentials have permitted the tape and film adhesives to replace or supplement welding, riveting, and other assembly methods in aerospace manufacturing. Actual usage per aircraft unit has increased over the years from several hundred pounds of tape and film adhesives on the early Boeing 707 jet to several thousands of pounds of film and tape products on the Boeing 747. On the Boeing 727 fuselage, the combination of bonding with riveting of fuselage splices and reinforcements has provided a variety of advantages such as (1) better overall fatigue characteristics, (2) more uniform load distribution, (3) higher achievable strength to weight ratios, (4) smoother skin profiles, and (5) sealing as well as fastening potential, making it possible to provide better air-tight and liquid-tight pressure cabins, fuel, and hydraulic systems.

Although a large listing of tape and film products can be found in the marketplace, they are all composed of a relatively small number of combinations of polymers and resins, as shown in Table 84. Their three essential components can be described in general terms as listed in Table 85. The adhesives are named according to the identity of the components as described in components (1) and (2)

Table 84 Most Important Tape and Film Adhesives

Designation	Backbone polymer	Cross-linking resin	Catalyst	High-pressure cure
Nylon-epoxy	Soluble nylon	Liquid epoxy	Dicy-type	No
Elastomer-epoxy	Nitrile rubber	Liquid epoxy	Dicy-type	No
Nitrile-phenolic	Nitrile rubber	Phenolic novolac	Hexa, sulfur	Yes
Vinyl-phenolic	PVB or PVF	Resol phenolic	Acid	Yes
Epoxy-phenolic	Solid epoxy	Resol phenolic	Acid	Yes

Source: From Ref. 2074, p. 151.

Table 85 List of Components Comprising Tape and Film Structural Adhesives

1. A high molecular weight "backbone" polymer which can provide the good elongation, toughness, and peel properties desired in such adhesives
2. A low molecular weight crosslinking resin invariably either an epoxy or phenolic type
3. A curing agent (catalyst) for the cross-linking resin

Source: From Ref. 2074, p. 150.

of Table 85. Their overall class designations are as listed in column 1 of Table 84. Because two resins or polymers are usually used in the naming of these structural film and tape adhesives, they have been referred to as "two-polymer" adhesives.

Any of the adhesives in Table 84 which contain phenolic cross-linking resins will also produce some volatiles while heat-curing and, therefore, their use in manufacturing bondlines will require some modest to high pressure on the bondline. Some strictly phenolic film adhesives have been used in bonding wood-to-wood structures. Minford (2076) has evaluated the durability of aluminum/wood laminates using some of these hot press–cured films but found the extreme rigidity of the final bondline a significant detriment for applications requiring some accommodation to the stresses of cycling temperature service conditions. Meanwhile, there is no actual pressure requirement for heat-curing epoxy films beyond that of maintaining alignment conditions while securing significant handling strength in the bondline. This has been of considerable economic significance in manufacturing structural joints, since lower cost heating fixtures can be employed.

Since epoxy-based film adhesives contain both base epoxy prepolymer, toughening agents, and appropriate curing agents in their uncured state, they should be stored at much lower than ambient temperatures in plants where they are not immediately used in production. In recent years, there has been extensive study of what might occur with respect to final cured bondline properties when structural adhesive films are aging in storage and moisture pickup is a possibility. Prominant investigators in this area have been Morris et al. (1617) in Australia and Dodiuk and co-workers in Israel. That moisture pickup in the film under storage conditions is quite important has been investigated by Dodiuk et al. (2077–2081). Several mechanisms have been suggested to explain the property decline, including hydrogen bonding between water molecules and the polar epoxy moieties, (2) hydrolysis of epoxide groups, and (3) homopolymerization in the presence of water. These same authors have also shown that vacuum drying of the uncured epoxy adhesive may reverse some of the adverse effects to the point of regaining properties equivalent to those of moisture-free compositions. However, a moisture uptake level of about 0.3% seems to be a threshold value beyond which vacuum drying no longer works. Most of the work has been conducted with epoxy film products meeting performance requirements at 250°F (121°C) or 350°F (177°C). More recently, epoxy formulations designed for thermal service to 400°F (204°C) has been achieved by incorporation of aluminum powder, which Dodiuk et al. (2081) showed was similarly adversely affected by moisture. With electron microscopy and FTIR techniques, these authors have shown a correlation between macromechanical properties and the microanalysis comprising failure morphology and infrared spectroscopy. The effect of drying time on the strength and weight change of aluminum-filled FM-400 structural epoxy film adhesive can be reviewed in Table 86. An extension of this work by Dodiuk and co-workers involving the effects of rehumdification followed by drying is shown in Table 87.

The effect of freeze-thaw cycling on some structural film adhesives was reported by Grabovac et al. (2082) in 1984.

B. Comparison of Film and Tape Properties

While the film or tape adhesives both have the handling and reliability advantages of no mixing, no degassing, and no possible errors due to incorrect catalyst additions, the use of a support mesh has better served to control the adhesive bondline thickness. This can avoid thin, adhesive-starved bone-

Table 86 The Effect of Drying Time on the Strength and Weight Change of FM-400 Film Adhesive

	FM-400[a] (1 yr)		FM-400[a] (6 yr)		FM-400[b] (1 yr)		FM-400[b] (6 yr)	
Drying time (hr)	LSS (psi)[c]	Weight change (%)	LSS (psi)[c]	Weight change (%)	T-peel (lb/in)[d]	Weight change (%)	T-peel (lb/in)[d]	Weight change (%)
0	2920 ± 52	—	3080 ± 28	—	9.4 ± 0.6	—	9.6 ± 0.1	—
1.0	3150 ± 26	0.26	3164 ± 40	0.26	7.0 ± 0.4	0.08	10.9 ± 0.5	0.50
2.0	3104 ± 34	0.28	3262 ± 43	0.43	12.8 ± 0.8	0.10	11.0 ± 0.7	0.56
2.5	3170 ± 38	0.29	3188 ± 57	0.78	8.5 ± 0.4	0.26	9.8 ± 0.7	0.42
3.0	3100 ± 50	0.29	3110 ± 32	0.78	16.2 ± 0.5	0.25	11.4 ± 0.3	0.69
3.5	3104 ± 26	0.32	3651 ± 21	0.55	16.3 ± 0.1	0.28	16.6 ± 0.8	0.40
4.0	3084 ± 19	0.32	3577 ± 142	0.73	7.8 ± 0.6	0.23	22.9 ± 0.6	0.56
4.5			3436 ± 30	0.56				

[a]Catalogue value 3780 psi.
[b]Catalogue value 8 lb/in.
[c]Lap-shear strength at 25 °C, average of five test specimens ± standard deviation.
[d]T = 25 °C, average of five test specimens ± standard deviation.
Source: Data from Ref. 2081 (Table I, p. 324).

lines where curvature or external pressure on the bondline might create a possible excessive thinning of corresponding composition film products. Consequently, various investigators have attempted to make direct data measurements on the same adhesive formulation.

Certain carriers can actually improve the properties of the adhesive in the bondline, especially in the more brittle adhesive formulations. Peel strength has been shown to be strongly influenced by the substitution of different carrier materials and the form of these materials. Hughes and Rutherford (2083) showed a reduction of effective tensile modulus in cured adhesives with carriers present, whereas Seago (2084) related the presence of a carrier to a lowered resistance of the tape adhesive to slow-cycle fatigue. Moloney (2085) found that both nylon and polyester fiber carriers were actually too low in mechanical strength to act as effective adhesive reinforcements. The cured adhesive strength seemed to be reduced by an amount equal to the volume fraction of carrier present. Brewis et al. (2086) showed that the incorporation of either nylon or polyester carriers into a particular modified epoxy had a negligible effect on the durability of aluminum joints exposed at 125 °F (52 °C) and 100% RH for periods up to 10,000 hr. This would seem to be in agreement with the extensive long-term durability evaluation of a number of film and tape aerospace adhesives on 6061-T6 aluminum by Minford (1416). Table 88, Brewis et al. (2087) show data obtained on the effect of two different weaves (a tight versus an open knit) under soaking conditions. The data show the highest joint strength was measured with no carrier present, whereas an open knit carrier version showed a 12% reduction and a tightly knit version a 37% lower strength. It would appear that the opportunity for the adhesive to penetrate and impregnate the carrier was quite influencing on initial joint strength. The relative resistance of these different adhesives with different carriers was also of interest. Starting at 37% lower strength, the tightly knit carrier joints degraded an additional 40% in approximately 1000 hr. Meanwhile, the open-knit joints starting only 12% lower declined an additional 38%. Pure FM-1000 film joints declined 33% from their highest initial level. In Table 89, the same investigators have presented data on the effect on peel strength resulting from use of the same tightly knitted nylon carrier. The perhaps surprising result was to find no significant difference between the peel strength values of the respective film and tape version joints.

The possibility has been advanced frequently over the years that the carrier backbone of a tape adhesive might wick moisture into the interior of the bondline at a faster pace. The additional question

Table 87 The Effect of Humidifying/Drying Cycles on Lap-Shear and T-Peel Strengths of FM-400 Film Adhesive

	Lap-shear strength (psi)[a]				T-peel strength (lb/in)[a]			
Adhesive (age)	FM-400 (1 yr)		FM-400 (6 yr)		FM-400 (1 yr)		FM-400 (6 yr)	
Moisture added (%)[c]	1st RC[b]	2nd DC[b]	1st RC	2nd DC	1st RC	2nd DC	1st RC	2nd DC
0.0	3170 ± 38[d]		3651 ± 21[d]		16.3[d]		16.6[d]	
0.1	3005 ± 69	3026 ± 36	2933 ± 10	3500 ± 81	11.1 ± 0.2	8.9 ± 0.4	15.6 ± 1.0	18.5 ± 0.4
0.2	2827 ± 13	2892 ± 35	3040 ± 39	3475 ± 22	12.1 ± 0.4	11.8 ± 1.4	13.6 ± 0.2	18.0 ± 0.7
0.3	2885 ± 80	2980 ± 60	3061 ± 34	3357 ± 55	11.9 ± 0.3	6.7 ± 0.0	13.9 ± 1.5	13.4 ± 1.2
0.4	2538 ± 24	3068 ± 24	3069 ± 45	3328 ± 38	10.9 ± 0.4	11.9 ± 1.3	13.3 ± 1.5	13.8 ± 1.5
0.6	2518 ± 11	2951 ± 46	3226 ± 31	3257 ± 35	10.9 ± 0.5	12.0 ± 0.7	15.2 ± 1.5	17.0 ± 1.0

[a]Average of five test specimens ± standard deviation.
[b]Code: RC = rehumidifying cycle, DC = drying cycle.
[c]Weight gained in 1st rehumidifying cycle.
[d]1st drying cycle (3.5 hr, 3–5 mmHg).
Source: Data from Ref. 2081 (Table II, p. 326).

Table 88 Effect of Carriers on Joint Strength

Adhesive	Conditions	Time (hr)	Mean failure load (kN)
FM 1000[a] (unstressed)	Initial controls	0	42.4 + 1.3
	Laboratory stored	1003	42.7 + 0.7
	50°C, 97–100% RH	1003	28.0 + 1.5
FM 1000 (stressed to 20% of initial dry strength)	Initial controls	0	42.2 + 0.9
	Laboratory stored	1009	40.8 + 0.5
	50°C, 97–100% RH	1009	28.5 + 0.9[d]
FM 1000/EP15[b] (unstressed)	Initial controls	0	26.9 + 1.3
	Laboratory stored	1007	26.4 + 0.6
	50°C, 97–100% RH	1007	16.1 + 3.1[e]
FM 1000/OKN[c] (unstressed)	Initial controls	0	37.5 + 0.9
	Laboratory stored	1003	37.3 + 0.8
	50°C, 100% RH	1003	23.2 + 0.7[f]

[a]No carrier present, apparent interfacial failure is observed with initial and laboratory stored joint with crack jumping from one adherend surface to other.
[b]Tightly knit nylon carrier present in FM 1000. Failure occurs along the carrier fabric with torn elements half adhering to each surface.
[c]Open-knitted nylon carrier incorporated in FM 1000. Failure about 50% cohesive and 50% apparent interfacial with fibers visually appearing to each adherend surface.
[d]Apparent interfacial failure for first 500 hr; thereafter cohesive failure around the edges started to be increasingly observed.
[e]Failure occurred straight along the carrier as for the unexposed specimens in [b].
[f]Failure was 50% cohesive and 50% apparent interfacial as observed with unexposed specimens in [c].
Source: Data from Ref. 2087 (Table 1, p. 346).

might be whether this moisture in the carrier could be a hindrance to drying out after prolonged soaking. Brewis et al. (2087) in Table 90 have presented data in this respect where FM-1000 film joints were compared to a tight-knit carrier nylon form designated FM-1000/EP15. As shown from the earlier Table 88 data, the tight-knit carrier contributed to a significantly lower initial strength as well as a larger percentage decline in strength after 1000 hr of soaking. With 1000 additional hours of drying at 122°F (50°C) imposed, the film FM-1000 joints regained to 89% of their already higher initial strength. Meanwhile, the tape version also regained to 85% of its lower initial strength. This would indicate no significant negative factor due to the presence of the carrier beyond its depression of the initial strength. Brewis and co-workers went one step further in Table 90 by storing these dried

Table 89 Effect of Carriers on the Peel Strength of Aluminum Joints Using FM 1000 Adhesive

Adhesive	Peel strength (kN/m)		
	Maximum	Minimum	Mean
FM 1000	22.4 + 0.5	17.3 + 2.0	20.0 + 1.2
FM 1000/EP 15[a]	21.0 + 0.5	19.3 + 0.8	20.8 + 0.4

[a]FM-1000 with a tightly knitted nylon carrier.
Source: Data from Ref. 2087 (Table 4, p. 347).

Table 90 Recovery of Joint Strength on Drying After Exposure of Aluminum Joints With and Without Carriers in the Bondline

Adhesive	Conditions	Time (hr)	Mean failure load (kN)
FM 1000	Initial controls	0	42.3 + 0.7
	50°C, 100% RH	1004	27.2 + 0.7
	50°C, 100% RH	1000	
	then 50°C drying	1006	37.7 + 2.0
	Laboratory stored	1004	
	then 50°C drying	1006	40.3 + 0.5
FM 1000/EP15[a]	Initial controls	0	27.6 + 1.0
	50°C, 100% RH	1000	16.1 + 3.1
	50°C, 100% RH	1000	
	then 50°C drying	1000	25.9 + 0.5
	Laboratory stored	1000	
	then 50°C dried	1000	31.9 + 0.3

[a]FM 1000 adhesive with a tightly-knitted nylon carrier in the bondline.
Source: Data from Ref. 2087 (Table 2, p. 346).

joints for an additional 1000 hr in the laboratory then repeating the 1000 hr of drying at 122°F (50°C). While the film adhesive joints gained additionally to 95% of their initial high-strength values, the tape version joints also continued to gain strength up to a value that was 16% stronger than their initial values. The end results of these kinds of tests leads to the conclusion that the magnitude of how different carrier materials can affect initial strength, water-soak strength, and joint strength recovery with heating needs to be carefully studied by direct testing, although trends based on closeness of the weave even with the same carrier material are somewhat predictable.

Romanko et al. (2088–2091) have been important contributors to analyzing the properties of aluminum joints with scrim cloth in the bondline. While the scrim's main function has been to control the bondline thickness, it may also contribute some negative aspects. Where an aerospace adhesive is present with a scrim cloth, it is likely that the scrim cloth is not really bonded to the adhesive or at best only weakly. This conclusion has been based on SEM photographs of fracture surfaces obtained after slow static testing of thick adherend aluminum joints (comparison of neat [FM-73u] adhesive and regular FM-73 tape with scrim cloth present). The fatigue life of the former was greater by a factor of four. Such nonbonding or weak bonding may additionally provide a pathway for preferential moisture transmission under service weathering conditions. However, as mentioned above, both Brewis (2087) and Minford (1481) found little evidence of overall inferior durability of tapes compared with film forms of aerospace adhesives in long-term water soaking conditions.

C. Nonstructural Types of Film Adhesive

1. Hot-Melt Film Adhesives

Large volumes of hot-melt blends are used in film form where the use of a thermoplastic resin permits simple heat and light pressure on the bondline to create adhesion and subsequent cooling will solidify the resin to produce the ultimate joint strength. If some modest structural strength and durability in weathering is demanded under the known service conditions, then considerable testing of the hot-melt set between the aluminum adherends will be required to demonstrate this capability.

The low polarity of many typical hot-melt raw materials make the thorough chemical wetting of an aluminum surface more critical. This wetting may be especially difficult to achieve on many lower-cost manufactured products where "as received" or only rudimentarily cleaned aluminum adherends might be contemplated for economical reasons. For example, there should be a real question of whether an aluminum adherend surface which lacks potential for being water wettable would be readily wetted by the typical low polarity hot-melt formulation. The fact that the hot-melt must be flowed over the surface while hot (in routine manufacturing) should have some positive effect on accomplishing some wetting. However, the high thermal conductivity of an aluminum adherend can have a distinct negative effect, since it can cause a rapid radiation of heat from the interface unless special measures are taken to prevent or minimize this heat-sink effect. This has after resulted in the need to take some countermeasures such as preheating of the aluminum to assure the critical melt temperature is present at the time of contact with the flowing hot-melt adhesive. The final product needs to be evaluated in any case to assure that the bond strength and durability meets the service standards.

As mentioned earlier in the section on Hot-Melt Adhesives, additional information of general reader interest can be found in References 1259 and 1362–1364. Of possible additional interest might be a number of early technical papers by Brantley and co-workers on the science of adhesion of films on surfaces like aluminum oxide. Brantley and Jones (2092) produced a report under government contract studying the nature of the forces of adhesion between films and adherend surfaces in 1952. Brantley et al. (2093) produced yet another report the next year. Brantley et al. (2094,2095) reported 2 years of additional investigations using the adherometer for reproducibly measuring the stripping force of films from aluminum oxide surfaces.

a. Hot-Melt Pressure-Sensitive Adhesives (HMPSA)

By comparison with other methods of applying PSAs, the hot-melt method is the most energy-efficient method as appraised by de Jager (2096). In addition, it offers the highest known production rates, the lowest capital costs for new lines, and with no solvents present, the absence of air pollution problems. Only some odor due to thermal decomposition of the molten polymer is sometimes a sort of pollution factor to the operator. There are, however, some product deficiencies as compared to the best-performing PS tapes and labels that can be produced from solvent-based and water-based PSA adhesive formulations (2097). A concise review of the history of HMPSA development, with emphasis on the basic polymers and compositions, has been published by Meletsky and Villa (2098). About 90% of commercial HMPSA compositions are based on block and graft polymerization technology, as discussed by Holden (2099) in 1973. Harland and Petershagen (2100) presented an excellent review of the characteristics of the thermoplastic A-B-A block copolymers in adhesives in 1977. Other products used to a lesser degree in HMPSAs include ethylene vinyl acetate (EVA), as discussed by Brinker (2101) and Duncan and Bergerhouse (2102), ethylene ethyl acrylate (EEA), reported by Eifi (2103), and amorphous polypropylene patented by Schmidt and HOlmes (2104). Acrylics, while viewed as superior PSAs, have been difficult to utilize in hot-melt technology.

The second surface of the aluminum foil or sheet to which the PSA adhesive has been applied may be clean mill-finish, anodized, painted, with or without additional decoration by direct printing, etching, or embossing. This would generally end up as the exterior-viewed surface in the final product. As such, it will be called upon to furnish weathering protection and vapor barrier protection. In addition, these specially prepared surfaces may furnish product or manufacturer identification or even some more detailed information about the product and how it should be used.

These PSA formulations generally start out as a solvent or water dispersion of a tacky resin, which can be readily applied by roll bar, roller coater, or curtain coater directly over the aluminum or other adherend surface at a high rate of speed. Some solvent dissipation process (usually hot air) is immediately started after the coating process, resulting eventually in a 100% solids permanently tacky layer. Because of the necessity to handle such a precoated surface for later attachment to a second surface, a release surface attachment must be pressed in place. The adhesion to this surface must be only marginal, at best, and always much lower than that between the PSA adhesive and the aluminum. Low-energy plastic surfaces are mostly employed for these release liners.

One additional type of controlled adhesion product needs to be mentioned in which a PSA formulation must have the properties of marginal adhesion to any specially prepared aluminum surface that needs to be protected from the environment. In this case, the adhesion to the carrier film surface must be superior, since it will eventually need to be peeled away, leaving the aluminum surface unaffected. These materials have the general name of "temporary protective films or papers." Such temporary protective films are designed to protect a specially prepared surface (aluminum or otherwise) during shipping, storage, handling, and finally field erection on a building in the case of architectural aluminum sheet or casting products. Such protection has been shown to be especially beneficial to protect anodized or painted aluminum from the chemical deterioration effects of noncompatible building materials like lime, mortar, or cement.

Minford (2106) has evaluated many commercial strippable coatings designed to protect anodized or painted aluminum building surfaces during building erection. The problems that have arisen apparently relate to the ability of ultraviolet radiation in the atmospheric weathering to change the degree of adhesion from one of easy detachment to a tenacious adhesion, often prohibiting any easy stripping even after only modestly long exposure times. The author has also compared the relative acceptability of a variety of such protectant papers or films by determining the relative number of cycles in a commercial weather-o-meter cabinet necessary to destroy the easy detachment of the protective product from differently prepared aluminum surfaces. The most reliable testing procedure, however, remains exposure to various natural atmospheric weathering conditions to determine the practical service life of the protective product.

Yet another application of pressure-sensitive adhesives has been a two-sided coated carrier tape product where the typical carrier materials have ranged from neoprene foams to various plastic foams such as vinyl or polyurethane foam materials. In this product, there must be a release liner rolled with the two-sided tacky tape to enable it to be used. The liner can be used to attach one tacky side to some adherend surface then peeled away to permit attachment to a second adherend surface. Thus, the bondline between the two adherends is now constituted by the combination of the two tacky bondlines and their foam carrier. Minford (2107) has evaluated commercial forms of such PSA products involving each of the three carrier materials mentioned above when used to join various aluminum commercial surfaces that they might contact. The adhering strength of the PSA formulations must be carefully formulated so that the bond to the neoprene or plastic foam surfaces and to the aluminum or other adherend surfaces is strong enough to force a failure in the foam carrier when the bond is deliberately destroyed by peeling or shearing forces. More details and results obtained in the testing of these products will be given in Chapter 10. It should be sufficient to say here that aggressive hot-humidity conditions altered the weakest linkage in the bondline from a cohesive failure in the foam carrier to an adhesive-type failure between the foam and the PSA adhesive.

D. Radiation-Curable Adhesives

Closely related to the hot-melt and PSA film adhesives just discussed is the technology of radiation curing. The potential energy sources include (1) ultraviolet light (UV), (2) electron beam (EB), (3) visible infrared (IR), and (4) microwave sources. The UV-curing process typically involves the exposure of some reactive liquid that already contains a photoinitiator to UV radiation at wavelengths between 200 and 400 nm. Conversion to a solid can occur in as short a time as 60 sec. Electron beam curing processes differ by requiring special generation of electrons with subsequent velocity accelerations to create energies ranging from under 100 keV to more than 1 billion keV. The SME publication titled "Radiation Curing" (2139) points out, however, that the reactive liquids involved in EB curing do not require addition of a photoinitiator. Obviously, the fast-curing capabilities of such processing permits a significant advantage in bonding to heat-sensitive adherends. Also, the need to fixture parts can often be eliminated with such processing speed, thereby greatly increasing production rates. As a consequence, the use of these systems has enjoyed a rather rapid growth in both industrial and consumer markets in the last 10 years, as reviewed in 1987 and 1991 by Woods (2127, 2128).

It is important to understand that the UV-type processes are based on photoinitiated polymerization of functional oligomers and monomers into a crosslinked polymer network. The process is mediated by a photoinitiator system that absorbs and converts UV-visible light into reactive intermediates such as free radicals and radical ions and/or long-lived intermediates such as acids or bases (which subsequently initiate or catalyze the polymerization process). Recent reviews are available on photoinitiated polymerization by Pappas (2129), photocross-linking by Pappas (2130), photoinitiators by

Pappas (2131), functional oligomers/monomers by O'Hara (2132), and characterization of cross-linked polymer networks by chain growth polymerization by Kloosterboer (2133). The overall interrelationships of UV-cure rates, photoinitiator concentration, and film thickness have been discussed in some detail by Pappas (2134). There has been recent attention on the use of some strategy to use auxiliary dark cure processes in order to reduce some of the inherent limitations of UV curing, as discussed by Grant (2175). These processes can be classified as using one of the four following types of adhesives: (1) UV-thermal cure adhesives, (2) UV-moisture cure adhesives, (3) UV-oxidative cure adhesives, and (4) UV-anaerobic cure adhesives.

1. UV-THERMAL CURE ADHESIVES

There are several reasons for combining UV and thermal curing which include (1) more rapid "immobilization" and "fixturing," (2) diluents can be eliminated and use of solvents reduced for achieving proper application viscosity, and (3) a long-lived catalyst can be UV generated for subsequent thermal cure utilizing a significantly lower temperature than is often required to thermally activate latent catalysts or curing agents. Peeters et al. (2165) and Berner et al. (2164) have utilized the latter approach in coating applications involving photogeneration of strong acids. Pappas and Feng (2166) have discussed the problems attendent to creating a one-part package with moderate curing temperatures using this processing. Crivello (2167) has listed photoinitiators, however, that produce free radicals for ambient cure and acid for thermal curing. Green and Zahir (2168), Gervay (2169), and Morgan (2170) have produced interesting compositions containing epoxies, phenols, acrylated resins, and thiolenes with appropriate photoinitiators present. An interesting modification suggested by Allen et al. (2171) uses a perester capable of absorbing UV-visible light sufficiently to function as both the photo- and thermal initiator.

2. UV-MOISTURE CURE ADHESIVES

Such products consist of a photoinitiator, functional oligomers, and monomers with polymerization and cross-linking through photogenerated initiators (or catalysts) and moisture. No secondary heating step is a main advantage over the previously discussed UV-thermal cure systems. However, there is a distinct challenge to achieve adequate storage stability by minimizing reaction with inadvertant water which often is associated with oligomers, fillers, and pigment materials.

Moisture-curing rates obviously depend on both humidity and diffusibility of water in the polymer system. Gozzelino et al. (2173), in 1989, reported studies on one such system in which the trimethoxysilane groups are incorporated into the photocross-linked network in situ via an acrylate or thiol functionality. Nakos (2172) has secured a Loctite patent on yet another current formulation.

3. UV-OXIDATIVE CURE ADHESIVES

The requirements include a photoinitiator and functional oligomers and monomers that polymerize and cross-link with a photogenerated initiator (or catalyst) and oxygen as described by Peeters et al. (2165). Cobalt carboxylate salts (together with auxillary salts or [driers]) are typically required to affect an oxidative cure. Nativi and Kadziela (2174) have secured a patent incorporating a dual UV-oxidative curing mechanism which would be representative of this technology from as early as 1984.

4. UV-ANAEROBIC CURE ADHESIVES

The combination of UV with an anaerobic-type curing process is relatively easy to accomplish. Ultraviolet curing compositions based on the free radical polymerization of meth(acrylate) resins can have continued dark cure capability with addition of an anaerobic cure system. Alternatively, a photoinitiator for free radical polymerization can be added to a typical anaerobic cure composition to provide added UV cure capability. Grant (2175) has made applications to electronic assemblies, whereas Siga et al. (2176) has extended use to optical equipment assemblies.

5. JOINING OPAQUE ADHERENDS

This must be accomplished by exposing UV-curable adhesives prior to assembly with subsequent dark cure following assembly. The best potential might be achieved through photogeneration of long-lived catalysts such as acids or bases with ultimate cure controlled by subsequent dark reactions, or photogeneration of free radicals that are convertable into more latent or longer-lived initiators. Rooney et al. (2177), Conway et al. (2178), and Woods and Coakley (2179) have all developed relevent patented compositions of this type adhesive. A visible light-activated adhesive for opaque adherends based on photogeneration of free radicals has been patented by Gonzales et al. (2180) at Henkel.

6. GENERAL DISCUSSION OF UV AND EB PROCESSES

Marino (2135) reported how the more intense the UV, the faster the cure. Unlike the UV curing, the EB curing can be selective and the depth of penetration can be controlled as described by Pinkus (2136,2137). Lewis (2138) has also pointed out the very high cost of EB equipment ($400,000–800,000) as a limitation as compared to UV equipment; however, UV-curing materials can be more expensive because of the need for an initiator. A detailed description of radiation curing can be found in a 1984 SME publication (2139). The materials used for most UV/EB adhesives are acrylic acid esters of various forms or combinations of acrylates with aliphatic or aromatic epoxies, urethanes, polyesters, or polyethers, as summarized by Woods (2140), Smoluk (2141), Morris (2142), and Steuben (2143). Barker (2144,2145) has forecasted significant gains in market sales and spread in applications through 1993.

In making a PSA, a properly formulated fluid composition is applied to an appropriate adherend (for example, mylar) and polymerized by a few seconds' exposure to a radiation source. Thus, there is a conversion from a low-strength composition to a tacky, higher-strength adhesive tape with little loss of volatiles. Obvious advantages are (1) no disposal of solvents into the atmosphere, (2) extremely low energy requirements, (3) high line speeds, (4) compact operation, (5) heat-sensitive adherend carriers may be employed, and (6) improved adhesion to adherend obtained due to grafting. Schwartz (2146) reviewed the advantages of radiation curing in 1976.

Nakano (2147) produced an early review paper in 1973 followed by a literature review by Steuben (2148) in 1977. In 1979, Taso and Ketley (2149) obtained a W.R. Grace patent on the addition of acrylate to epoxy using radiation procedures. Crivello (2150) and Crivello and Lam (2151) discussed the nature of photoinitiators for the cationic polymerization of radiation-curable adhesives in two 1980 publications. Perkins (2152) summarized new developments in photoinitiated cationic polymerizations in 1981. In the same year, Ledwith and co-workers (2153,2154) also discussed new findings related to the polymerization mechanism involving cation radicals. Finally, in 1984, separate discussions were written by Steuben (2155), McGuinness (2156), and Crivello (2157) on ultraviolet-cured PSAs, formulation design characteristics for radiation-curing systems, and latest progress in photoinitiated cationic polymerization, respectively.

The process of forming a bond with a hot-melt adhesive is admittedly one of the most rapid of all techniques, since neither solvent loss nor chemical reaction is involved. As a direct result of their characteristics, however, most hot-melt adhesives have relatively poor resistance to heat as compared to the cross-linking polymer-based adhesives. Solvent resistance also may be comparatively poorer. However, improvements in such properties can be achieved in some cases by post-cross-linking with irradiation, as reviewed by Torrey (2158), St. Clair (2159), and Ganslaw (2160). Most recently, Smith (2161) has reported on the chemistry and applications of UV-cured adhesives at the Adhesion '90 meetings at Cambridge and Fernando and Kinloch (2162) have discussed the adhesion of photopolymers at the same meetings.

Bluestein (2108) has provided a review article which can be of benefit to the manufacturing engineer. Other publications of general interest include ones on photochemistry (2109), radiation processing (2110), radiation of elastomers (2111), UV curing (2112), radiation curing (2113,2114), safe handling of radiation processes (2115), and a handbook of radiation curing (2116) and reactive curing (2117). Other authors of pertinent technical papers include Harries (2120), Mahoney (2121), Kardashian (2122), and Friedman (2118).

X. FIRE- AND FLAME-RESISTANT ADHESIVES AND SEALANTS

While fire resistance is generally more important in selecting materials of structures, both fire and flame resistance in adhesives can be important for certain applications. An example would be found in aircraft interior panel assemblies for flame-resistant adhesives. Flame-resistant sealants would be important where they are applied to smooth or cover the exterior surfaces of aircraft parts. The necessity to develop such materials has been called to mind by the results of a variety of airplane accidents in recent years.

Of particular importance in modern electronic technology has been the need to provide as much fire and flame resistance as possible to printed circuit boards. The materials used in the manufacture of these boards (with the exception of metal core boards and injection-molded boards) has been cooper-clad laminates or organic dielectrics, which include phenolic-paper, epoxy-glass, polyimide glass, and some others of lesser note. Although polyimide is being used increasingly because of its high-temperature stability and excellent handling properties, it is epoxy-glass and it is its variants that are most widely used. Nakahara (2181) and Tummala and Rymaszewski (2182) have discussed the range in complexity of printed boards and details of their fabrication, respectively. Markstein (2183) has particularly focused on the fire-retardant nature of the epoxy-glass laminates. Schlack (2184) has discussed the elevation of the T through the addition of epoxy novolac resin. The use of an organosilane coupler has been shown by Antoon et al. (2185) to be essential to the maintenance of the insulation resistance of the dielectric when making epoxy-glass products.

In the past, additives such as phosphorus, halogen compounds, or antimony oxides have often been employed to enhance the flame resistance of structural materials in general. The difficulty in uniformly dispersing these kinds of ingredients has long been recognized as a major concern for obtaining best performance. This is especially difficult in the formulation and application to adhesive and sealant products. One alternative has been to synthesize the adhesive or sealant with phosphorus or halogens as actual parts of the chain polymer structure. Kourtides et al. (2186) have developed a phosphorylated epoxy with the designation BGDPMPO with a higher oxygen index than the regular epoxy, thus earning it the description of being "flame resistant." The fact that some of these chemical additives or polymer modifications might produce toxic fumes when subjecting them to heating must also be given proper consideration.

Lee (2187) has reported receiving information by private communication from Fohlen regarding investigations with phosphorylated polyimides. For example, maleimide derivatives have been prepared from bis(m-aminophenyl)methyl phosphine oxide and tri(m-aminophenyl)phosphine oxide. These maleimides can be cured by an addition reaction where no volatiles are evolved. These materials do not burn in the presence of oxygen even when heated to 572°F (300°C).

Another promising family of flame-resistant materials are the phosphazenes and cyclophosphazines, as described by Kumar et al. (2188). The cyclotriphosphazene has both outstanding fire and heat resistance as demonstrated by its "char yield" at elevated temperatures. Values of 82% at 1472°F (800°C) in nitrogen and 81% at 1292°F (700°C) in air have been measured. Synthesis by a new synthetic route was mentioned to Lee (2187) by Fohlen which would enable manufacturing of this product at an approximate cost of $20.00/lb.

XI. MICROBIAL CONTAMINATION OF ADHESIVES

The possibility of microbial contamination of adhesives has been a problem associated with the use and application of adhesives for as many years as they have been utilized by humans. Starches, gums, and proteins, in particular, serve as excellent hosts for microorganisms. The development of more complex adhesive products in recent years has not really lessened the microbial problem.

Microorganisms exist everywhere and cannot, because of their universal presence, be kept out of manufactured products without resorting to quite expensive and exotic sterilization procedures. In order to reproduce, microorganisms require only a food/energy source, water, and a reasonably hospitable environment. If these criteria are met, then bacteria and molds under ideal conditions can double every 30 min. However, service conditions are usually too dry, too hot, too cold, or too toxic for this to occur.

Sharpell (2189) has produced a review article on this matter which cites a number of important considerations. We are reminded that even distilled water can have growing bacteria present that apparently get their nutrients from absorbed gases. The microbial contamination of jet fuels has been a problem in aircraft from water condensation inside the tanks unless a biocide is present. Phenol, which is present in many structural adhesive formulations, can actually be used in low concentrations as a carbon source by bacteria even though this chemical itself is generally considered quite a toxic material when used in various cleaning solutions. Kostenbader and Flecksteiner (2190) have even described the deliberate use of bacteria to degrade the phenolic by-products of coke plant wastes. In the recent war in the Persian Gulf, it has been mentioned that specific microorganisms might be employed to remove and clean-up the largest oil spill experienced in the world to date. The close relationship between chemicals from oil and materials present in various adhesives is well known to all.

Waxes are used in significant amount in many hot-melts for viscosity control (see Table 39). Sharpell (2191) has produced a paper on how to control the microbial contamination of such materials in manufacturing plants. Very large quantities of emulsion-type adhesives are used for laminating in a number of industries, and Wedderburn (2192) has pointed out that water used to batch chemicals is often the worst offender for providing a microbial source of contamination. He has further discussed the relationship between the degree of hygiene in a manufacturing plant and its effect on the preservation of emulsion-type products. The author remembers that it was necessary to throw out many test samples of various emulsion-type adhesives on a regular basis owing to contamination obviously introduced as soon as the can was opened for use. Upsher (2193) has determined the fungal resistance of modified epoxy structural film adhesives. He points out that while certain ingredients in epoxy formulations have a natural resistance to microbial attack, the diversity of ingredients that can be present makes it impossible to be specific about these adhesives as a group.

Sharpell (2189,2194) has provided an extensive list of biocide preservatives by trade name, active ingredients, and suppliers. Lederer et al. (2195) have offered a complete evaluation of a particularly effective preservative (1,2-dibromo-2,4-dicyanobutane), including its specific activity against particular organisms that are commonly associated with known adhesive failures in the field. Although not initially planning to examine aluminum/epoxy joints for the effect of microbial growths on aluminum/epoxy joints in a secluded jungle exposure, Minford was presented with that opportunity after up to 12-yr exposures in the Surinam jungle. One group of joints was deliberately exposed only under foliage that screened the test joints from ever drying out under sun-drying conditions. Under these continuously hot/wet relatively dark conditions, there was a vigorous growth of mosses and molds on the underneath side of these exposed joints. Examination of the joints before and after clean-up and before and after deliberate failure in a tensile tester failed to disclose any variation in attack of the fillet or edge areas as compared with joints exposed in open jungle where such organic growths did not accumulate. This would serve to indicate that structural aluminum/epoxy joints might not be as sensitive to microbial attack atmospheric exposure conditions as might have been anticipated in the absence of such direct observations.

6

Design of Aluminum-Bonded Joints

I. INTRODUCTION

A. General Considerations

In the preceding chapters, we have considered the physics and chemistry of the aluminum oxide surface, how it may be modified by special surface pretreating, and what adhesive candidates we might consider for adhering to these surfaces. Having specified the surface finish and the adhesive candidate, it still remains to consider the geometry of the joint in a structure and how it will react under a load. The bond width, length, part dimensions, adhesive thickness, and the types of stresses that might be encountered in service all need to be considered carefully. An excellent review of the mechanics of a wide variety of joint designs used in bonded aluminum structures has been assembled by Kinloch (2196). Various reference works on designing adhesively bonded structures have been published (901,2197–2215).

In creating joints it must be remembered that the challenge is to properly design joints for bonding rather than simply substitute adhesive bonding for joints previously employed for other means of joining. Properly designed bonded joints will generally depart significantly from conventional mechanically united joints, making it necessary to conduct both a valid stress analysis of the adhesive in the joint and reliable tests to determine the allowable stresses present in the adhesive, as discussed by Krieger (2204). Prediction of the bond performance is achieved by comparing the calculated stresses with the allowable stresses to create a safe design. Krieger presents a series of design geometries based on the linear stress analysis parameters (2204). These geometries show by example and discussion how stress analysis is used to provide aluminum joint designs with maximum efficiency at minimum weight. This stress analysis is shown to be valid when the aluminum adherends are in tension, compression, or shear. Thus, it appears that the indispensible property of the adhesive for stress analysis is stiffness in shear, which was first obtained in 1973 using a Krieger-designed extensometer and a skin-doubler aluminum specimen (2216–2218).

Hart-Smith (2205) proposed an additional method of rating and comparing structural adhesives in aluminum joints by the use of a new specimen, which he described as an inverse skin-doubler specimen. It is actually a geometric inversion of the conventional double-lap–shear joint that permits freedom from all the parasitic peel stress effects ordinarily present in testing the double-lap joint configuration. Two modes of failure are possible with this new joint: either the continuous outer adherends will break if the adhesive is sufficiently strong or the discontinuous inner member will disbond if it is stronger than the adhesive. The effective shear strength of the adhesive layer is established by performing such tests with various thicknesses of adherends. For final acceptance, the tests need to be repeated throughout the range of environments anticipated in service.

Regardless of the form of test vehicle, there is additional need to properly analyze the adhesive bond. Penado and Dropek (2206) have summarized the state of the art in this regard, discussing both closed-form and finite-element methods. Numerical analysis techniques such as the finite-element method (FE) can be coupled with the outstanding capabilities of today's computers to provide very useful information to the bonded joint analyst, permitting stress or strain to be calculated either at the local detail or even with large bonded assembles. (For chronological reviews of the closed-form methods, see Refs. 2219–2232.) The closed form methods are limited, however, by the fact that only the simplest geometries and boundary conditions can be accommodated.

For the more complex situations, the use of approximate numerical solutions such as finite-difference (FD) or finite-element (FE) procedures become a necessity. The FD procedure does indeed provide for accommodating more complex boundary conditions than achievable with the closed-form solution methods. However, problems are encountered with curved or other irregular boundary conditions. The FE method has the advantage of permitting division of the region of interest into an assemblege of smaller subregions or elements, making it attractive for solution of jointures with very complex geometries. Also, this method offers the ability to interface elements with different material properties, making it well suited for analysis of bonded joints. (For more complete descriptions of the theory behind the PE method, see Refs. 2233–2237.) Some commercially available computer codes utilizing the method are NASTRAN, ADINA, and ANSYS for special applications like truss and beam elements, two-dimensional solid plane stress or plane strain elements, plate and shell elements, axisymmetrical shell or solid elements, three-dimensional solid, and crack-tip elements. FE analysis results for single- and double-lap joints using the MSC NASTRAN code have been demonstrated by Penado and Dropek (2238).

Most recently, Reddy and Roy (2239–2241) reported their work on finite element joint analysis. The technical discussion comes largely from the same authors' ONR reports of 1985 and 1986 (2241, 2242). They also have furnished a review of the background investigations on finite element analyses of many joint geometries (2227,2228,2240,2243–2270).

In addition to calculating the state of stress in bonded joints, the design engineer also needs to predict load failures and margins of safety for the assembly. Methods for predicting adhesive joint strengths must be followed by some failure-prediction approach to accomplish this end. The most important methods for predicting joint strengths can be included in the list of (1) the average stress method, (2) the maximum stress method, and (3) the fracture mechanics method. Each method has some distinct limitations which must be recognized. The average stress method, unfortunately, will only produce adequate predictions if test coupons have the same loading, geometry, and material properties under same temperature and moisture service conditions. In the absence of precisely comparable conditions, the method may still offer useful qualitative comparative strength conditions among several adhesive candidates. Anderson et al. (343) have discussed the limitations of the average stress method in some detail. More recently, Anderson and DeVries (2271) have shown the inability of the method to account for variations in strength due to bondline thickness.

The maximum stress method improves on the average stress method by considering the variation of stresses in the adhesive along the bondline and across the thickness. Also, the test specimens do not need to be of identical geometry to the final assembly. Limitation stems from an inability of the method to account for points of stress singularity present when joining dissimilar materials. The presence of such stress singularities in the region of the bond might indicate that the various fracture mechanics methods might be more suitable for these situations. One method is applicable to the extension of existing cracks of known size (2272–2276). Supporting experimental data will be required, however, of the location and size of cracks or flaws present in the adhesive. The fracture mechanics approach, which considers predictions based on propagation of inherent flaws or imperfections on a more microscopic scale, will be considered in more detail in Chapter 7.

As noted above, after the use of a method to predict adhesive joint strength, there is a further need to employ some failure-prediction approach. Penado and Dropek (2277) have provided one practical joint failure criterion. This FE methodology accounts for the presence of singularities and minimizes the inconsistencies between test specimens and the final structural part and consists of the following steps: (1) determine elastic moduli of the adhesive, (2) determine failure loads from simple analogue joint specimens (e.g., lap-shear and butt-tension), (3) determine stress or strain allowables from the failure loads in (2) using FE analysis, and (4) conduct the appropriate FE analysis (linear or nonlinear) on the structure being considered to obtain the maximum, normal, or von Mises stress under the applied loads. The same authors (2278) also have provided three detailed examples using such a procedure with (1) single-lap-shear specimen with and without an adhesive fillet, (2) small-diameter graphite-epoxy tube with a bonded aluminum end fitting, and (3) joint analysis with metallic fitting and graphite-epoxy composite tube.

Another important consideration in the overall estimation of how strong and durable a bonded aluminum joint might be is based on the ability of the manufacturer to ascertain the presence, nature, and extent of flaws (effect of defects). In this regard, Chang (2208) has reported work on flaw detection (effects of defects). This subject will be considered in greater detail in Chapter 9. (Refs. 2279–2284 on the subject of detecting flaws in bonded structure may be reviewed as a general background.)

After experimentally determining data on the various elements enroute to a final structural design, it remains for a critical experimental verification of the analytical prediction to be conducted. This will likely be used to validate the production method and the process procedure at the same time. A wide range of procedures have been used to validate designs ranging from simple test coupons to a full-scale evaluation of the product. Regardless of which procedure is employed, the critical load conditions, specimen size, and service environment all remain critical issues in deciding the final testing plan. Johnson (2209,2209a) has presented an overall review of the subject beginning with reference to DeHavilland's bonded Mosquito bomber of World War II (2285). Examples of early aluminum-bonded primary aircraft structure in the 1964 Fokker F-27 and F-28 and 1976 Boeing YC-14 (2286) are provided. Anniversario et al. (2287) discussed primary bonded structure in the Boeing 737. Other examples of composite bonded primary structure are found in the Bell/Boeing V-22 (2288), the Grumman/Boeing A-6 Intruder (2289), and the current Boeing 757 commercial passenger plane. Also mentioned, is the U.S. Air Force massive testing program PABST reported in many publications in the late 1970s, of which the details on design was reported by Thrall et al. (2290) in 1977. Johnson (2209) has chosen to discuss design proof testing by considering the following categories: (1) testing rationale, (2) environmental considerations, (3) coupon testing, (4) subcomponent testing, (5) component testing, and (6) full-scale testing. More detail on the testing rationale can be found in the NASA report of Johnson et al. (2291). More about the considerations relating to the environment can be found in the Boeing report of Aniversario et al. (2287) and the AFWAL 1988 report by Horton et al. (2292). More information related to the matter of coupon testing has been offered in the 1981 and 1982 reports of Hart-Smith (2293,2294), whereas details of full-scale testing are described by Aniversario et al. (2287), Chovil et al. (2295), and McCarty (2296).

Designing aluminum joints for use with sealants involves some quite different considerations as compared with the mostly structural aluminum joint designs considered above. Thompson (2210) has offered a recent 1990 review of the subject. Generally speaking, sealant joints between two aluminum adherends experience considerably more movement in service than allowed for in structural joint designs, and consequently the formulations must allow accommodation to multitudinous cycles of such movement without detaching from the adherend or failing in cohesion as described by O'Connor (2297). The variety of application areas for sealants has been reviewed (1921,2298,2299). The strain caused by shear forces on sealant joints must be considered in designing joints and values have been calculated by Shishler and Klosowski (2300).

In practice, the uses of elastomeric sealants can be roughly grouped into those joints requiring thin bond lines and thick bond lines. Included in the thin-line variety are liquid containment applications using a mechanical fastener such as automotive valve and differential cover seals. The earliest gaskets in these areas were cut gaskets from cork, paper, metal, or mineral compositions. This eventually gave way to the form-in-place gasket (FIPG) procedures which early were made from natural oils and plastiziers under the name of gasket dressings. Beginning about 1960, high-performance materials like one-part moisture-curing silicone sealants were used for such gasketing. This market is reviewed in General Electric's bulletin CDS-4128 (2301). Similar applications exist for the anaerobic sealants (2302).

Another type of gasketing has been termed CIPG, which stands for cured-in-place gasketing. These joints call for some special design, because the cured sealant beads need not develop actual adhesion to the two sides of the joint. Instead, the seal is accomplished by compression of the cured bead. This joint design (see Ref. 2210) is particularly suitable for joints requiring frequent disassembly. Finally, there are the sealant joints which require many cycles of movement accompanied by complete sealing against weathering properties. Most often these formulations are silicones, polyurethanes, or polysulfides designed to be highly elastic, yet with retention of excellent adhesion to aluminum for many years in direct weathering. (For some pertinent references, see Refs. 2303–2305.)

In discussing how single aluminum joints will function in structures, we need to consider the factors of load, stress, strain, and the yield strength of the chosen alloy. By load, we refer to the total actual load that is being supported, whereas the joint stress will be some variation that depends on the extent of bond area. For example, if a lap-joint is supporting 1000 kg and the bond area is 1 cm wide and 4 cm long, then the joint stress is only 25% of the load, or 250 kg/cm. Meanwhile, the strain is a measure of the reaction of the bonded assembly to the stressing level. If the above-mentioned stress elongated the joint by 0.05 cm, the strain would be described as 0.0125 cm/cm (i.e., 0.050 cm/4 cm). The most frequently used terminology is to cite the strain as 1.25% elongation. Of course, if the bond undergoes continuous elongation to failure, then the correct terminology to the reaction has become creep, or creep resistance factor. Where deformation can take place in a bonded structure, we need to distinguish that which is temporary, i.e., movement undergoing elastic deformation and recovery, from that which is permanent, where the yield strength of the adherend is actually being exceeded.

The kinds of materials that have been formulated into an adhesive are distinctly related to the nature and extent of creep that can take place. Some adhesives will continuously creep until failure under any load which even approaches the yield strength of the cured adhesive. This is especially true of adhesives with significant percentages of elastomer resin present. Adhesives that are thermoplastic in nature will generally be unable to resist even modest stressing in any high-temperature service conditions. In contrast, highly thermoset adhesive alloys, or high molecular weight polymers like polyimides, can offer service at temperatures up to 1000°F (538°C) for significant times.

The main stresses to which aluminum joints can be subjected can be designated mechanical, thermal, or chemical (environmental) in nature.

II. STRESS FACTORS

A. Mechanical Stresses

Five forms of mechanical stress can be present in joints depending on the geometry of the design. These common stresses are designated tensile, compressive, shear, cleavage, and peel, all of which can be reviewed in Figure 16.

1. TENSILE AND COMPRESSIVE FORCES

The first two of these forces, i.e., tensile or compressive, are found in the reaction of butt-type joints to a load that is acting perpendicularly to the bondline, as shown in Figure 16. The joint is said to be reacting to the stress of tensile-load forces when the stress is tending to extend both the adhesive and the adherends. In contrast, when the loading is such that the adhesive and the adherend are both being directed inward, then the term *compressive stress factor* is obviously appropriate. Certainly, one of the most difficult conditions for a bonded joint to resist will be when tensile and compressive loads are alternating or when the respective loads are cycled on and off. It should be obvious that tensile and compressive joint strength values will be directly proportional to the bond area. In practice, however, manufactured joint perfect alignments are rarely achieved. Thus, other mechanical stresses (particular in the cleavage mode) may complicate accurate prediction of the load-carrying capacity

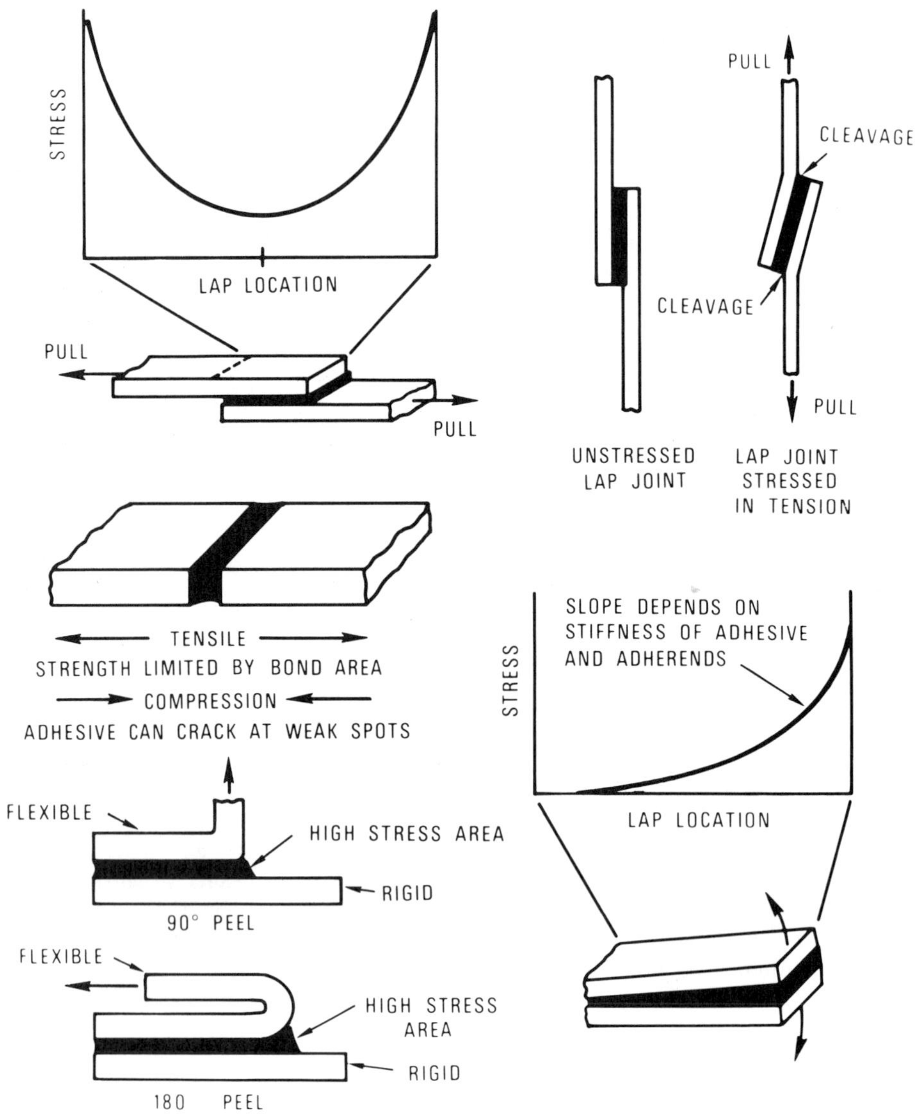

Figure 16 Common stresses met in adhesive joints. (From J.D. Minford, Aluminum Adhesive Bond Permanence, Treatise on Adhesion and Adhesives (R.L. Patrick, ed.). Marcel Dekker, New York, Vol. 5, Fig. 3-22, p. 84 [1981].)

of a given joint (2306). It is equally true that the specific construction loading in service will most likely not be applied exactly perpendicular to the bondline. This will introduce appreciable cleavage stress, so that the highest potential joint strength cannot be achieved for this joint that was planned to be stressed primarily in pure tension. A number of investigators have analyzed the stress distribution that occurs in butt adhesive joints (2307–2314).

2. *TENSILE-SHEAR FORCES*

Lap-shear stresses are those that are most commonly found in single-lap-joints, as shown at the top left in Figure 16. The idealized relative stress pattern across this joint configuration is shown above the drawn joint. These configurational joints are also the most commonly employed joint design for fabricating general manufactured structures. Because of this fact and their simplicity of fabrication in the laboratory, they have been universally employed to establish the general database in the literature that distinguishes the performance levels for the different adhesive families. The loading on such lap-joints has generally been in a tensile mode; i.e., a loading of the adherends along the bondline axis. The net effect is to achieve a sliding or shearing action; hence, the name lap-shear-type test specimen. Because the two adherends of the joint are not actually in direct alignment in the lap-shear joint, the stress across the bondline is not uniform and the joint strength does not vary directly with the relative size of the overall bonded area. Rather there is a higher actual localized stress at the edges of the bondline than would be calculated using the whole bond area. This means that the stress will be diminishing over the adjacent bond areas progressing toward the center joint area. Quite often the idealized representation of the joint stresses are served by drawing a smooth U-shaped curve. However, in real life, this smooth line would be replaced by a jagged-appearing, saw-tooth contoured line caused by the fact that there would likely be some variable distribution of microdefects in the bondline interface. Sharpe has more accurately referred to such a situation using the term *interphase*. This term allows for the various defects that can arise as a result of the involved interaction between the adhesive and adherend.

The stress analysis of the single–lap-joint has received attention from a large number of investigators over a period beginning with Volkersens's work in 1938 until the present. Volkersen described the stresses arising from the differential straining of the bonded adherends (2219) but failed to take account of the tearing stresses generated in the adhesive. The classic work of Goland and Reissner (2220) first recognized and attempted to account for this effect. Other investigators (2227,2315–2327) further extended and confirmed the agreement of Goland and Reissner's analytical solutions with experimental results. Adams and co-workers (24,2227,2320,2321,2328) have examined the transverse stresses as well as the longitudinal and concluded for metal-to-metal joints that the transverse shear stresses can be as much as one-third of the minimum longitudinal shear stresses. All of the above analyses considered that the adhesives and adherends behaved as purely elastic materials.

Most adhesives, however, may exhibit plastic as well as elastic behavior. Adams and co-workers (24,2328) and Penning (2329) have accordingly made allowances for the same. The most recent extensive studies, however, have been conducted by Hart-Smith (2330–2332). His studies show that inclusion of adhesive plasticity under shear loadings may decrease the stress concentration significantly as compared to using predictable stress concentrations based solely on elastic analysis. This means that the predicted joint strength would be higher than previously calculated from theory.

Kinloch and Shaw (2333), in 1981, reported the fracture energy was not independent of the joint geometry. In 1984, Adams and Harris (2334), using a finite element method (FEM), predicted with reasonable accuracy the strength and locus of failure with a wide range of adhesives. Adams and Harris (2335) published an interesting paper in 1987 on the influence of edge geometry on the strength of aluminum lap-joints. They focused particularly on the effect of filleting and radiusing the corners of square-edged joints developing a finite element analysis for predicting joint strengths. We note in Table 91 the significant increases in joint strength experimentally measured for filleted and filleted and radiused joints. In Table 92, we can observe the magnitude of lower stress realized by filleting and radiusing, whereas in Table 93 the close comparison obtained between their predictions and the actual experimentally measured joint strengths can be compared.

Table 91 Experimental Joint Strengths (kN)

	Square-edged	Filleted	Filleted and radiused
Median	15.9	20.4	24.5
Lowest	15.4	19.2	23.8
Highest	16.4	21.0	24.7

From Ref. 2335.

Table 92 Peak Values of Adhesive Maximum Principal Stress From Elastic Analysis with 16kN Applied Load

Joint type	Peak maximum principal stress (MPa)
Square-edged	258[a]
Filleted	147[a]
Filleted and radiused	83[a]

[a]From modified joint models.
From Ref. 2335.

3. CLEAVAGE FORCES

The most easily recognizable form of a true cleavage specimen would be that shown in the bottom right corner of Figure 16. Failing this specimen in the manner indicated would produce a stress curve pattern, as shown above the joint as drawn. This kind of stress is also a prominant contributor to the failure of single–lap-joints, as seen in the upper right hand corner of Figure 16 and is dependent on a number of factors, as will be discussed. Depending on the relative yield strength and thickness of the aluminum adherends, and modulus of the cured adhesive, the cleavage forces arise when the adherends deform to become coplanar. This deformation places an extremely high stress on the edge area of adhesive-adherend jointure. It should be obvious that these cleavage forces can lead to joint separation at significantly lower stress levels than would be calculated from the total stress divided by total bond area value. It should also be clear that the use of stiffer adherends (a higher tensile strength alloy) would lessen the amount of deformation at any particular stress level and thus raise the breaking stress value.

The joint designer can reduce these cleavage stresses by beveling (tapering) the edges of the single-lap adherends. Another variation is the scarf-joint, which can actually be interpreted as a tapered adherend variation of the butt-joint. This design has the effect of considerably reducing the end stress concentrations in the butt-joint as well as lowering the tearing stresses in the adhesive. Reduction of these cleavage forces by tapering edges can be reviewed (see Refs. 2317, 2326, and 2337). While shear-stress concentration is lower, there is no significant decrease in eccentricity for the beveled single–lap-joint. Thamm (2326) has shown that tapering to about half original adherend thickness is practically valueless. The substantial increase in joint strength is only achieved with virtually complete tapering.

4. PEEL FORCES

Peel stresses are similar to cleavage stresses with the stress being concentrated at the failing edge of the bondline. They are more extreme forces than cleavage stresses, because in many joint configurations one or both adherends is likely to be relatively more flexible. Tests to evaluate the magnitude of such peel forces are often conducted by using a specimen consisting of one rigid and one flexible adherend. The flexible adherend is pulled uniformly at an angle of either 90° or 180° to the rigid

Table 93 Comparison of Joint Strength Predictions (kN) with Experimental Values

Joint type	Predicted	Median of experimental
Square-edged	14.5	15.9
Filleted	21.2	20.4
Filleted and radiused	25.3	24.5

From Ref. 2335.

adherend from an adhesive layer which is being supported by the more rigid adherend in a manner as shown in the lower left portion of Figure 16. The former will be referred to in the literature as the 90° peel strength of the adhesive and the latter as the 180° peel strength value. It is, of course, imperative to relate this value only to the particular set of test materials and fabricating procedures employed to conduct the test. Changing any factor of joint test geometry, adherend composition, surface pretreatment, adhesive choice, adhesive curing, or adhesive thickness could alter the peel force value that will be found. Other types of peel test procedures also exist and will be discussed Chapter 9. It should be obvious that the opportunity for peel forces to exist in bonded structures should be minimized as compared to designing joints with primarily tensile- and shear-stressing components.

Some of the peel strength investigations of the 1950s included works by Spies (2338), Kaelble (2339), and Inoue and Kabatake (2340). Considering the adhesive and adherend as elastic materials, these investigators attempted to calculate the tensile and shear stresses in the adhesive layer. Kaelble (2341, 2342) has confirmed experimentally that the distribution of cleavage stresses is a highly damped harmonic function involving alternating regions of tension and compression. A summary of the work through 1968 was developed by Bikerman (2343), and through 1982 by Wake (27). Most recently, Crocombe and Adams (2344,2345) have reexamined the problem employing an elastic, but large displacement, finite element analysis approach. They report that the initial peel stress failure is likely caused by the principal tensile stresses in the adhesive driving a crack toward the interface between the adhesive and the flexible adherend. Propagation of the crack will then occur at some critical bending moment, which is directly related to the choice of adhesive and adherend. This reaction seems to be relatively independent of the angle of peeling.

5. *MEASUREMENT AND APPLICATION OF STRESS/STRAIN DATA IN BONDED JOINTS*

As has been mentioned earlier in this chapter, the stress analysis of a single–lap-joint has been studied and attempts made to predict the magnitude since as early as 1938. By 1944, the classic efforts of Goland and Reissner (2220) had already been published and latter investigators have tried to further fine-tune their equations. The focus on making actual measurements of stress under tension, shear force, and bending moments was reported by Pahdja (2346) in 1972. Ishai and Gali (2347) measured the interlaminar stress distribution within an adhesive layer in the nonlinear range in 1978.

A number of additional investigations conducted in the 1980s bear on this subject of which the following can be cited for the reader. In 1981, Gali et al. (2348) used the stress/strain concept for the mechanical characterization of structural adhesive bonding. Krieger (2349) described a model joint in 1986 which enabled valid stress analysis of adhesively bonded 2024 aluminum alloy joints. He also presented a linear formula for calculating the maximum stress on the joint using two extensometers of his invention designated KGR-1 and KGR-2. Gillibert et al. (2350), in the same year, determined the mechanical behavior of a double-lap A-U4G aluminum alloy joint by attachment of strain gauges located at the outer surfaces of the adherends, taking into account the nonlinear behavior of the bond. Finally, in 1986, Alvazzadeh et al. (2351) published on special mixed finite elements for interfacial stress analysis. They presented results with special triangular and rectangular mixed finite elements for the static analysis of the joints. Liechti and Hayashi (2352) analyzed the uniformity of stresses in certain adhesive deformation specimens in 1989. Carpenter and Barsoum (2353) offered an investigation of two finite elements for modeling adhesive in bonded configurations, whereas Bigwood and Crocombe (2354) suggested an elastic analysis and engineering design formulae for bonded joints in that same year. Gent and Lin (2355) offered comparisons between peel and lap-shear bond strengths for elastic joints, both with and without residual stresses present. Amijima and Fujii (2356) investigated the extension of a one-dimensional finite element modeling program for analyzing elastic-plastic stresses during the progressive failure of bonded joints, whereas Sawa et al. (2357) conducted a three-dimensional stress analysis of adhesive butt-joints of solid cyclinders while subjected to external tensile stresses. Groth (2358) has offered a calculation of stresses in bonded joints using the substructuring technique.

Most recently, Ritter (2359) at the Adhesion '90 meetings in Cambridge described evaluating the adhesion of thin polymer coatings under contact stresses. At the same meetings, Hu et al. (2360)

utilized a boundary element method for analyzing stress in adhesive-bonded structures, and Zhao and Adams (2361) described a new approach to determining the bending moment factor in single lap-joints. In 1990, Bigwood and Crocombe (2362) wrote two articles on a nonlinear adhesive-bonded joint design analysis, and Davies and Khalil (2363) described design and analysis for bonded-double containment corner joints. Chen and Cheng (2364) have evaluated stress distributions in plane scarf- and butt-joints, and Wycherley et al. (2365) proposed a method for analyzing the shear stress-strain of adhesives. Suzuki (2366) has proposed strain analysis methods for cemented orthotropic lap-joints. Finally, Dike and Johnson (2367) have used acoustoelasticity for determining the residual stress left in fabricated joints.

B. Thermal Stress

The second major stress factor in joints consists of thermal stresses, which are primarily present because of expansion and contraction forces generated by changes in the service temperature. When there is a significant difference between the coefficients of linear expansion of the adhesive and the adherend, these stresses can become severe. Hughes and Rutherford (2083), Dannenberg (2368), and Chen and Nelson (2369) have investigated residual joint stresses due to differences in thermal expansion coefficients between adhesive and adherends. The range of different thermal coefficients that can be present in joints can be gauged from the values found for different choices of adherend and adhesive materials, as shown in Table 94. It is possible, for example, for aluminum and an anaerobic adhesive to have a fivefold difference in response to thermal heating. When the aluminum is thermally stimulated to move 0.01 mm, the adjacent anaerobic adhesive will try to move 0.05 mm. This

Table 94 Approximate Coefficients of Thermal Expansion for Various Adherends and Adhesives

Adherends	Expansion coefficient ([cm/cm/°C] × 10)
Steel	1.2
Aluminum	2.5
Copper	1.7
Brass	1.9
Stainless steel	1.7
Glass	1.2
Polyethylene	10–25
Acrylonitrile butadiene styrene (ABS)	5–10
Nylon (unfilled)	8–90
Polyvinyl chloride (rigid)	5–20
Polymethylmethacrylate	5–9
Phenolics	2–4
Epoxies	1–4
Urethanes	10–20
Epoxy (unfilled)	5–6
Epoxy (filled)	2.5–4
Urethanes	10–20
Phenolics	2–4
Silicones	20–25
Anaerobics	12–14
Acrylics	13

Data from Ref. 2201 (Table 2, p. 74).

could create a high enough stress to rupture the adhesive or cause dramatic interfacial separation. By comparison, a phenolic adhesive or a filled-epoxy would expand almost identically as the aluminum adherend. Of course, if the two adherends are dissimilar (with corresponding varying coefficients of thermal expansion), it may be of importance to choose an adhesive whose thermal expansion coefficient lies between that of the two adherends. Alternatively, an adhesive can be chosen with a very low modulus enabling it to actually function as a stress-relief interlayer.

C. Chemical Stress

By chemical stresses, we mean those combined effects of all the chemicals in the joint environment which influence its durability. The service environments of many bonded structures may be primarily interior exposure conditions, which means that the more aggressive chemical actions of high concentrations of water, heat, ultraviolet light, salts, and noxious gases are mainly absent. The chemical action of most of the aggressive weathering conditions that exist in exterior weathering and can result in joint disbonding will be considered in more detail in Chapter 10 on durability performance of bonded aluminum structures. The specific details about the special influence of water on joint survival will be considered in Chapter 8.

III. GEOMETRY FACTORS

We generally refer to width, length of bond overlap, and thickness of adhesive and adherend as the critical geometry factors. While the strength of a lap-joint increases linearly as the width increases, the increase is nonlinear with increasing length of overlap. It should be remembered that these effects cannot be studied in isolation from the adherend yield strength and thickness, since these adherend factors are highly influential in establishing the level of overall stiffness of the joint. Stiffness, in turn, strongly influences the load level necessary to initiate the high cleavage stresses which can initiate deformation failure. It is a good design principle, therefore, to make sure the yield strength level and thickness of the aluminum alloy adherends constitute a combination which will ensure that initial failure of the unexposed joint will always occur cohesively in the bondline.

A. Single-Lap-Joint

DeBruyne and Houwink (2370) made an early analysis of the relationship between bondline thickness, overlap, and stress and originated a relationship called the joint factor. They defined it as the ratio of the square root of the adherend thickness to the length of the overlap. As the adherend thickness increases and the overlap decreases, the ratio approaches a limiting value which will be characteristic of the adhesive used. The relationship between the mean failing load of a joint and the joint factor values is essentially linear for joint factors up to about 0.3.

Rolf et al. (2371), at the Alcoa Labs, conducted an extensive series of tests with varying aluminum geometries to measure the shear strengths that might be achieved in single-lap-joints. Their data relating the joint strength to lap length, lap width, and the aluminum adherend thickness is shown in Table 95 and should be of particular interest to design engineers. It needs to be remembered that these are comparative joint strength values using a particular epoxy structural adhesive and surface pretreatment condition, which would not necessarily represent the maximum strength or long-term durability response that is possible with the highest state of the art bonding in aerospace manufacturing. These joints were fabricated by the author for these investigators using chromic acid deoxidized 6061-T6 aluminum adherends without the use of a primer or the present best surface pretreatment of phosphoric acid anodizing (PAA). Also, these joints were bonded using a 1950 vintage relatively rigid two-part, room temperature-curing epoxy adhesive which would lower the potential strength levels significantly as compared to the present-day aerospace, heat-cured film adhesive products. The values still offer good comparisons, however, for adhesive bonding in general manufacturing. Comparisons relating to adherend thickness alone can be found in the data from Anderson et al. (2372) in Table 96.

Table 95 Shear Tests of Aluminum Adhesive Bonded Joints With Varying Geometries

Lap length		Adherend thickness		Lap width		Joint strength[a] Breaking load		Avg. shear strength	
(in)	(cm)	(in)	(cm)	(in)	(cm)	(lb)	(kg)	(psi)	(MPa)
0.75	1.91	0.125	0.32	1.0	2.54	2180	989	2900	19.98
1.0	2.54	0.25	0.64	3.0	7.62	13300	6039	4500	31.01
1.5	3.81	0.25	0.64	3.0	7.62	15800	7173	3500	24.12
3.0	7.62	0.25	0.64	3.0	7.62	23100	10487	2500	17.23
2.0	5.08	0.5	1.27	3.09	7.62	19500	8853	3200	22.05
3.0	7.62	0.5	1.27	3.0	7.62	25400	11532	2800	19.29
6.0	15.24	0.5	1.27	3.0	7.62	25100	11395	1400	9.65
6.0[b]	15.24	0.5	1.27	3.0	7.62	59800	27149	3300	22.73
				Double lap joints					
0.077	1.67	0.25	0.64	5.1	13.02	53400	22244	6400	44.10
1.625	4.13	0.25	0.64	5.1	13.02	58600[c]	26604	3500	24.12
1.5	3.81	0.25	0.64	3.0	7.62	34900[c]	15845	3900	26.87

[a]Each value is the average of triplicate tests.
[b]Pair of 0.25-in (0.64-cm) aluminum antipeel bolts at each end of lap.
[c]Tensile failure of main plate; no bond failure.
From Ref. 2371.

While the single–lap-joint, as described in ASTM D-1002, has been the most widely employed joint to describe the properties of adhesives in bonded joints because of its simplicity, the average shear strength that is determined is not a unique material property characterizing the adhesive. Instead this value is a rather vague property that depends strongly on the particular geometry of the joint being tested. Even if the joint geometry and adhesive material is left unaltered, the switch from an aluminum to a steel adherend or a greater thickness of either adherend can dramatically change the apparent joint strength. Hart-Smith has analytically interpreted data showing the stronger and stiffer the adherend material, the more the measured "strength" of the adhesive. Where the adherend thickness was doubled, the adhesive strength of long-overlap joints was doubled because failure of the adhesive is preceded by yielding of the adherend. The pioneering analysis of the single–lap-joint was completed by Goland and Reissner (2220) with further refinement by Hart-Smith (2228).

Table 96 Lap Shear Test Data Relating Effect of Adherend Thickness

Adherend thickness		Average stress at debond		
(in)	(cm)	(psi)	(N)	Reduction (%)
0.0625	0.16	1530	10500	63
0.125	0.32	2440	16800	40
0.25	0.64	3600	24800	12
0.375	0.95	4200	29000	0
0.50	1.27	4010	27600	0

Data from Ref. 2372 (Table II, p. 282).

It can be shown that for the particular overlap-to-adherend ratio of 8:1 designated for the ASTM D-1002 lap-joint (0.063 in [1.6 mm] thick) aluminum adherends, the main cause of failure of the adhesive is usually the induced peel stresses with some contribution from the applied shear stresses. No shear-dominated failure occurs anywhere near the geometry used for this test coupon. However, quite high structural efficiencies can be achieved with this coupon provided that the adherend thickness is limited and the overlap is made long. Thus, in contrast, if the above ratio is made 80:1 (10 times the ASTM recommended ratio), Hart-Smith (2373) has indicated this type of joint was successfully used in the skin splices of the adhesively bonded fuselage in the successful PABST program. When using a double-shear-joint, typical overlaps for bonding aluminum aircraft structures in subsonic transports employ 30 times the skin thickness ratio.

IV. MORE COMPLICATED JOINT DESIGNS

Hart-Smith (2205) has discussed the many disadvantages of using the ASTM D-1002 simple lap-joint for assessing the strength of adhesives. In 1990, he discussed the rating and comparison of structural adhesives using the simple ASTM lap-joint and proposed a new method. This specimen has been given the name *inverse skin-doubler* specimen and consists of three uniformly bonded plates which are loaded in tensile. Each outer plate member needs to be at least half as strong as the inner plate member so the strength is controlled by the central member. The load transferred through the adhesive layers is then proportional to the strength of the inner member. It is appropriate to make each adherend the same thickness and compensate analytically for any associated minor effects from stiffness imbalance. The strength of the adhesive layers in such a specimen is independent of overlap when the overlap is long. The adhesive shear stresses and strains are precisely the same as would occur in a structural double joint or double–strap-joint. Hart-Smith also discussed the use of more complex stepped–lap-joints to develop higher bond strengths and the effects of various tapering of the ends of double–strap-joints and a skin doubler test specimen.

Also, of interest is the 1990 work of Penado and Dropek (2206) on numerical design and analysis of joints by finite-element analysis. These authors offered graphs of the FE results for the stresses in the adhesive along the overlap of single-, double-, stepped- and scarfed–shear-joints. Distinctions are made between normal stresses, shear stresses, maximum principal stresses, and von Mises or equivalent stresses. Penado and Dropek (2206) also have provided examples and test correlation data for single–lap-shear specimens with and without fillets, small-diameter graphite-epoxy tube bonded to aluminum end fittings, and a complex jointure involving bonding of an aluminum fitting to a graphite-epoxy composite tube.

A. Double–Lap-Joint

A double–lap-joint, as the name might imply, is like an adherend with a single–lap-joint on each side. The symmetry of this configuration results in a much reduced normal stress across the adhesive layer. Some comparative strength values and variation due to geometry changes can be seen in the data from Rolf et al. (2371) in Table 95. Some stresses obviously still exist as calculated by Volkersen (2374). Adams and Peppiott (2227) have employed an elastic finite-element analysis to estimate the stress across an equivalent single–lap-joint (half physical dimensions of the double lap) to be 1.85 times greater. Actual experimental observations measured a value of 1.92 showing an excellent agreement between the calculated and experimentally verified values. Hart-Smith (2375) published a 1973 NASA report describing many aspects of the adhesive-bonded double–lap-joints. A number of investigators in recent years, such as Adams (24), Hart-Smith (2330–2332) and Penning (2329), have shown that allowance for adhesive plasticity affected a further reduction in the calculated stress concentrations. In this group, Hart-Smith has established that there is a plateau strength, i.e., a set of conditions where no further strength can be gained by increasing the overlap of the double–lap-joint configuration. Hart-Smith has also extended consideration to the tearing stresses developed across the adhesive layer. He suggested that these increase as the thickness of the outer adherend increases. Beyond a certain degree of outer adherend thickness, however, these stresses may even exceed the peak shear stresses in magnitude. Adams (24) has plotted the shear stress concentration values he found using elastic analysis for typical epoxy-bonded aluminum joints in the single-lap, double-lap, and double-butt-strap configurations (see Fig. 7, p. 623, in Ref. 2196).

In the double-lap-joint, the integrated strength of the adhesive bond is independent of the total overlap for most all possible designs except impractically short overlaps. Thus, if the overlap is made long, the adhesive in the middle of the overlap cannot creep even under the worst circumstances. Any shorter overlap, however, might fail by creep-rupture, whereas any longer overlap would not add to either the strength or durability of the bonded assembly. In a good joint design process, it should be ensured that the integrated shear strength of the adhesive is at least 50% greater than the strength of the adherends outside the joint. Also, it must be noted that the strength of the bond must exceed the strength of the adherends even where the design ultimate load is much less. In the absence of the latter condition, there could be no tolerance to damage in either the bond or the adjacent adherend. In this case, Hart-Smith points out the adhesive is actually performing as a weak-link fuse. The pioneering elastic analyses of the double-lap-joint are those of Volkersen (2219,2376) and DeBruyne (2377). Hart-Smith (2378) has discussed the double-shear-joint in aluminum as regards the design and analysis of joints for aluminum aircraft repair. Hart-Smith has analyzed double-lap-joints and has concluded that there is always an upper limit on the thickness of the adherends that can be safely used and that the ends of the adherends must be tapered to realize the full potential shear strength and avoid premature failure due to induced peel stresses.

B. Scarf- and Stepped-Scarf-Joints

The scarf-joint is an example of a joint design with low stress concentration and also zero weight penalty (24,2330,2332,2379). With this joint modification to other joints, the stress concentrations at the joint ends can be considerably reduced accompanied by a lowering of the tearing stress in the adhesive. The effects are particularly marked in the case of the scarfing of single-lap-joints where the eccentricity of the loading path is greatly reduced. (For discussion of the stepped-scarf-joint, see Refs. 2330,2329, and 2380.) Penado and Dropek (2381) have made finite element calculations and plotted the results for the stresses in the adhesive for single-lap-, double-lap-, stepped-lap-, and scarf-joint designs as regards normal stresses, shear stresses, maximum principal stresses, and von Mises or equivalent stresses. Rolf et al. (2371) have published shear test strength results of bonded scarf joints with varying geometries, as shown in Table 97.

Table 97 Shear Tests of Aluminum-Bonded Scarf-Joints With Varying Geometries

Lap length		Adherend thickness		Joint width		Joint strength[a]					
						Breaking load		Average stress			
								Shear		Tension	
(in)	(cm)	(in)	(cm)	(in)	(cm)	(lb)	(kg)	(psi)	(MPa)	(psi)	(MPa)
0.75	1.91	0.125	0.32	1.0	2.54	4250	1930	5900	40.65	1000	6.89
1.5	3.81	0.125	0.32	3.0	7.62	25500	11577	5900	40.65	1000	6.89
0.75	1.91	0.5	1.27	3.0	7.62	13600	6171	4500	31.02	3000	20.67
1.5	3.81	0.5	1.27	3.0	7.62	19400	8808	4300	29.63	1400	9.65
3.0	7.62	0.5	1.27	3.0	7.62	40100	18205	4500	31.01	800	5.51
6.0	15.24	0.5	1.27	3.0	7.62	51800	28057	3500	24.12	300	2.07
9.0	22.86	0.5	1.27	3.0	7.62	63100	28647	2400	16.54	100	0.69

[a]Each value is the average of triplicate tests. Stresses based on actual dimensions.
Data from Ref. 2371.

C. Single–Strap-Joint

The single–strap-joint can be essentially described as a single reinforcement to a butt-joint. Obviously, the backup plate will have the overall effect of rigidifying the butt-joint, and thereby, a better normal stressing alignment of the joint structure should be possible. However, because the reinforcement is not symmetrical, at some stress level, a deformation will be initiated and cleavage forces will begin to build. Nevertheless, the concept of bonding single-strap splices over butted jointures is widely used in commercial bonded structures, including aircraft (2382). The problem with the single–strap-joint is that reinforcing the splice plate(s) to increase the bending strength also increases the eccentricity in load path, and hence the bending moment. In aircraft, the real concern about this kind of flat joint is that it is inherently weaker than the skins being bonded together. It therefore lacks the excellent damage tolerance of double-strap spices and has the potential for complete failure either in the splice plate as observed under test or in the adhesive. In tests conducted by Smith, a narrow strip of adhesive peeled apart immediately adjacent to where the skins butted together on the test panel.

Study of single-strap bonded joints indicated that increasing the overlap was a powerful technique for alleviating the effects of the eccentricity in load path; however, such joints would be much heavier. The simple design rules of thumb are that the total splice width should be about 150–200 times the thickness of the sheet being spliced and that the thickness of the tip of the splice should not exceed about 0.04 in. Hart-Smith and Thrall (2382) also provide experimental evidence on tapered single-strap splices and fatigue testing of the same.

D. Double-Strap Bonded Joint

The double-strap bonded joint is probably the most desirable joint for transferring loads between two aluminum members. The loads in such joints in aluminum aircraft structures may be tension, compression, or in shear. Ideally, each splice strap should be precisely one-half as thick as the adherend parts joined. Only then will the adhesive be equally effective at each end of the overlap. In fatigue testing of aircraft joints, the failures consistently occurred in the splice plates, making it practical to increase their thickness to obtain an added measure of service durability. Detailed studies of this joint have been undertaken by Hart-Smith (2330), Hart-Smith and Thrall (2383), Adams and co-workers (2328), Wright (2384), and Sage (2385). By adding the factors of tapering and radiusing the outside ends of the straps, Sage ensured the failure site in the adherend completely outside the joint. The efficiency for strength for such a joint is virtually 100%; however, a weight penalty must obviously be paid for such outstanding performance.

E. Joints Between Angles and Gusset Plates

Data have been developed involving these matters by Rolf et al. (2371) using aluminum adherends, as shown in Table 98.

Table 98 Shear Tests of Aluminum Adhesive Bonded Joints Between Angles and Gusset Plates

			Joint strength[a]			
	Lap length		Breaking load		Average shear stress	
Type of joint	(in)	(cm)	(lb)	(kg)	(psi)	(MPa)
Single-angle	6.0	15.24	25400	11535	1700	11.71
Single-angle	12.0	30.48	34000	15441	1100	7.58
Double-angle	6.0	15.24	34300	15577	1200	8.27
Double-angle	12.0	30.48	106000	48138	1800	12.40

[a]Each value is the average of triplicate tests.
Data from Ref. 2371.

Table 99 Tests of Aluminum Adhesive Bonded Tube and Sleeve Joints[a]

			Joint strength			
	Lap length		Breaking load		Shear stress[b]	
Type of test	(in)	(cm)	(lb)	(kg)	(psi)	(MPa)
Tension	2.25	5.72	30800	13983	970	6.68 (avg)
Bending[c]	2.25	5.72	13500	6129	2100	14.47 (max)
Bending	4.0	11.43	26200	11895	2000	13.78 (max)
Bending	6.75	17.15	24900	11305	1300	8.96 (max)
Bending[c]	0 (plain tube)		33900	15391		

[a]Loads causing failure of adhesive; plain tube failed by plastic buckling.
[b]Shear stress in bending tests assumed to be proportional to distance from neutral axis.
[c]Single test; all others duplicates.
Data from Ref. 2371.

F. Aluminum-Bonded Tube-and-Sleeve Joints

Appropriate to this subject are the data from Rolf et al. (2371), as shown in Table 99. The author has conducted many tests over the years using these types of joints to evaluate various adhesives for possible use in gas and liquid transmission lines.

G. Aluminum-Bonded Tube-and-Socket Joints

Miller et al. (2386), at Alcoa Labs, have studied the effects of adhesive ductility and bondline geometry on tube-and-socket joint performance, which could be very pertinent to bonding lighter-weight automobiles. Miller (2387) and Bresko (2388), at the same laboratory, have developed unpublished data reports on the evaluation of adhesive-bonded extrusion to simulated casting joints and the behavior of aluminum tube-and-socket adhesively bonded joints, including mechanical fasteners, respectively.

V. FINAL DESIGN GUIDELINES

The following recommendations made by the American Iron and Steel Institute (2389) should be appropriate for all metals: (1) always use the largest possible area for bonding in keeping with reasonable cost, (2) align the bondline (in the structure) so that it is stressed in the strongest direction, (3) maximize shear and minimize peel and cleavage loads, (4) design subassemblies so that only one bonding operation is required for the total assembly, (5) remember to anticipate stresses (particularly peel or cleavage) which may be encountered during assembly or handling, and (6) avoid parts with complex curvatures.

A. Practical Joint Geometries

A variety of corner and tee design joints have been used in general manufacturing. Based on factors of the degree of symmetry and relative likelihood of peel and cleavage forces arising in service, these joints can be classified as being poor, fair, or good candidates for long-term service potential. In general, the cost of making these joints will also vary from low to high in the same order. Consequently, economics and the desire to fabricate the best product possible must be weighed by the design engineer. Unlike aerospace adhesive-bonded joints where machining operations and costs are numerous and expensive, these various practical joints for general manufacturing generally require only cutting operations. A number of these commonly used corner and tee joint designs have been illustrated with an assessment of their successful adhesive-bonding potential on page 80 of Schneberger's book titled *Adhesives in Manufacturing* (Marcel Dekker, 1983).

B. Thin-Walled Tube Assemblies

Adhesive bonding provides a convenient method of assembling structures containing thin-walled aluminum tubing. Various jointures may be envisioned such as a larger aluminum sleeve bonded over a butt-joint between two tubes (tube-in-sleeve joint). It is a variation of the concept to relieve cleavage stresses in a butt-joint by mechanical restraint. Still another variation of tube jointure is called the flared-tube type. One tube end is flared to receive a second tube which has not been so expanded. Adhesive is applied in the gap between the two sections and cured in place. A variation is to insert and bond in-place aluminum tubing selected to fit into cast or extruded aluminum receptacles. If desired, the cast or extruded sections may accept multiple sections of tubing. Stress distribution in such joints has been considered by Adams and Peppiatt (2227) and Lubkin and Reissner (2390). Adams and Peppiatt (2391) and Nagaraja and Alwar (2252) have both employed finite-element analysis techniques on such designs. In an attempt to improve performance with both axial and torque loading conditions, Adams and Peppiatt added partial tapering to form a tubular scarf–lap-joint, but concluded the stress reduction obtained did not justify the significant additional preparation costs.

In the category of thin-walled aluminum tubing may be considered that used for carrying refrigerant gases for refrigerator and air conditioning applications, as well as water coolant systems for vehicles, cooling towers, and even smaller instrument tubing for a wide variety of commercial applications. An additional stress present in tubing carrying refrigerant gases is the high internal pressures generated by low-boiling refrigerant gases when operating on the hot side of the systems. This creates an outward bursting pressure on the joint which must be considered in joint design. Also, the requirement for special adhesive properties must be satisfied both as to their structural requirements, resistance to chemical degradation due to the strong solvent properties of the refrigerant gases, and resistance to long-time operation conditions at elevated temperatures. At Alcoa, many investigations were conducted by the author (2392) over many years to ascertain the best commercially available adhesives for making such tubular connections.

C. Box-Section Beam Assemblies

It is envisaged that box sections and "top hot" stiffener configurations that are adhesive bonded will play an increasing role in the future manufacture of automobiles, transport containers, bus coaches, commercial vehicle cabs, and railroad carriages. Beevers and Kho (2393) have described a convenient box-section beam specimen for studying this problem, as shown in Figure 1, p. 629, of their publication. They made a direct comparison between 0.9-mm steel and 1.2-mm aluminum sheet when this box section was assembled by riveting, spot-welding, or adhesive bonding with either a two-part, room temperature–curing epoxy or a one-part, heat-cured epoxy. The summary data on relative values of flexural stiffness, tortional stiffness, and ultimate flexural strength is cited in Table 100. It appears from these data that the flexural stiffness values for the steel versus the aluminum construction were quite similar regardless of assembly method. Tortional stiffness of the riveted aluminum and steel box sections were very similar; however, the steel structural stiffness was distinctly superior when spot-welding or assembling with the rigid two-part epoxy. This situation reversed with the one-part, heat-cured epoxy where the aluminum box was 35% stiffer. This would indicate that there is considerable interaction of the structural materials with different adhesives because of the wide variability in the properties of different cure adhesives. The aluminum box beams assembled by riveting and spot-welding were distinctly higher in ultimate flexural strength but essentially equivalent to the steel structures when assembled with either form of epoxy.

These investigators compared ultimate compression load strengths of 0.9-mm and 1.2-mm aluminum and 0.9-mm steel boxes assembled by the same three methods and with varying flange bend radii, as shown in Table 101. The adhesive-bonded specimens always had the highest ultimate comparative strength regardless of the gauge of material, type of material, or flange radius situation.

Table 100 Flexural Stiffness, Tortional Stiffness, and Ultimate Flexural Strength of Steel and Aluminum Box Beams Assembled by Different Methods of Joining

Beam material	Joining method	Flexural stiffness (kN/mm)	Tortional stiffness (kNm/rad)	Ultimate flexural strength (kN)
0.9 mild steel	Rivet	1.8	2.5	3.4
	Spot-weld	1.9	9.5	3.6
	Two-part, RT-cure epoxy adhesive	2.9	11.6	7.7
	One-part, heat-cure epoxy adhesive	4.2	8.2	6.9
	Toughened acrylic adhesive	2.85	11.0	5.0
1.2 mm aluminum alloy (2.5% Mg)	Rivet	1.6	2.6	7.0
	Spot-weld	1.9	5.3	6.4
	Two-part, RT-cure epoxy adhesive	2.8	6.7	8.2
	One-part, heat-cure epoxy adhesive	3.1	11.1	8.5

Data from Ref. 2393 (Table I, p. 631).

Table 101 Comparative Compressive Load Strengths of Bonded, Riveted, and Spot-Welded Aluminum Beam Sections

Sheet materials	Thickness (mm)	Flange bend radius (mm)	Ultimate comparative strength: Adhesive bond (kN)	Rivet (kN)	Spot-weld (kN)
Aluminum alloy (2.5% Mg)	0.9	2.5	1.03	0.51	0.51
		1.5	1.18	0.53	0.73
		1.0	1.47	0.72	0.80
	1.2	2.5	3.12	1.09	1.12
		1.5	2.87	1.19	1.49
		1.0	2.80	1.63	1.63
Mild steel	0.9	3.0	1.89	0.54	0.59
		2.5	1.94	0.58	0.76
		2.0	2.94	1.11	1.25

Data from Ref. 2393.

Table 102 Calculated Stiffness of Box Section Beams

Theoretical beam model	Sheet thickn (mm)	I mm × 10	Theoretical stiffness (kN/mm): Aluminum	Steel
Homogeneous continuous connection	0.9	0.227	2.15	6.39
	1.2	0.303	2.84	8.53
No connection	0.9	0.078	0.72	2.14
	1.2	0.112	0.94	2.83

From Ref. 2393 (Table 2, p. 633).

Table 103 Comparison of Experimental Test Results with Theoretical Models of Aluminum Beam Stiffness Using Various Methods of Joining

Method	Comparative stiffness ratio
Theoretical beam continuous, homogeneous joint	1.00
Hot-cured, one-part epoxy beam joining	1.02
Toughened acrylic joined beam	0.98
Spot-welded	0.71
Riveted	0.56
Theoretical beam, no joint	0.33

Aluminum alloy in 1.2 mm thickness employed.
Data from Ref. NO. 2393 (Fig. 3, p. 633).

When comparing 0.9-mm steel with equal gauge (physically weaker) 0.9-mm aluminum sheet, the ultimate strengths were about equivalent when the aluminum was employed with a 2.5-mm flange as compared to a 3.0-mm flange with the steel with riveting or spot-welding. While the comparative strength increased overall for both materials when the flange dimension was decreased, the rate of increase in strength was obviously higher for the steel box. With equal gauge assemblies, all steel box values for adhesive-bonded constructions were higher; however, with 1.2-mm aluminum sheet adherends the highest strength box was obtained with a 2.5-mm flange.

Beevers and Kho (2393) made theoretical calculations of the projected stiffness of their box sections made with both 0.9-mm and 1.2-mm sheet using a homogeneous continuous connection model as compared with a no connection model, as shown in Table 102. The theoretical stiffness of both metal sheets with a homogeneous continuous connection model ranged from 2.95 to 3.02 times stiffer than where the no connection model was employed. In Table 103, the authors made comparisons of these values with actual comparative stiffness values using 1.2-mm aluminum sheet. With the theoretical continuous beam value used as 1.00, a continuously joined box with the heat-cured epoxy measured 1.02. With a toughened acrylic adhesive the value was 0.98. The relative stiffness values for spot-welding and riveting dropped to 0.71 and 0.33, respectively.

7

Mechanisms of Bond Joint Failure

I. INTRODUCTION

Obviously, there is no single mechanism of aluminum-bonded joint failure, since there is a possibility for disrupting the bond in each of the separate parts of a joint. If the cohesive strength of the adherend is less than the stress imposed, then a failure will undoubtedly occur in that cross-sectional area. On the other hand, a joint failure will be similarly forecast if the imposed stress exceeds the bulk cohesive strength of the adhesive. Then there is the exceedingly complex chemistry of the interphase to deal with, where the primary forces of ionic, metallic, or covalent natures can be contributing to the jointure. Also, the secondary forces of van der Waals are always present to make their particular contribution to establishing the bond. Then there are the factors of overall joint strength that relate specifically to the morphology of the adherend surfaces and the rheological properties of the adhesive materials as cured. Finally, there is the distinct possibility that the various joint parts have been attacked and diminished in properties by the weathering environment, which may be further complicated by stress simultaneously imposed on the joint.

II. POLYMER SURFACE DEGRADATION AND POLYMER STABILITY

It should be obvious from the above comments that the physical and chemical properties of both adherends and polymer adhesives can be important factors in precipitating joint failures and, thus, need to be thoroughly studied in order to formulate various mechanisms of joint failure. As regards the degradation and relative stabilities of polymers, Jellinek (2394) has produced an overview of surface degradation and stability of polymers in which he discussed the surface degradation produced by thermal radiation, photodegradation, and oxidative degradation. What might at first thought be considered extensive polymer surface modification by oxidation procedures can be distinctly advantageous for improving the adhesive properties. Publications by Willis and Zichy (2395) and Brewis and Briggs (2396) are quite descriptive in regard to describing enhanced wettability produced on nonpolar surfaces by procedures like surface grafting, alkali or acidic treatments, halogenation, sulfonation, and cross-linking in the presence of fluorine. Oxidation of such surfaces may either remove potentially weak boundary layers or actually strengthen them through cross-linking procedures. The latter has also been achieved by the use of an activated species of inert gases (the CASING process). Schonhorn (2397), a major contributor to this process, also reported, in 1978, on the environmental effects and mechanisms for producing improved polymer adhesion. Briggs et al. (2398), in 1979, described the surface properties of a nonpolar polymer like polyethylene after various surface pretreatments.

The employment of a number of different instrumental surface analysis techniques are required to understand the conversion of polymer surfaces. Jellinek has described some 14 of these procedures (2394). Some review articles dealing with these various methods have been produced by Dilk (2399) and Dwight and Thomas (2401) and can be found in the book titled *Polymer Surfaces* edited by Clark and Feast (2400).

III. ELASTIC AND VISCOELASTIC ADHESION CONSIDERATIONS

The mechanisms that make things adhere to one another are obviously intimately related to their failure in that the favorable conditions for adhering are somehow in reversal. The mechanics of adhesion between surfaces has usually included an attempt to account for both surface energy and surface topography contributions. Briggs (2402) offered a review of such mechanisms in 1983. First studies involved the simple case of an elastic sphere in contact with a rigid second surface and a fracture mechanics analysis seemed to offer the same result as advanced by energy balance–type arguments. Contact between realistic surfaces was generally described as a statistical average of simple spherical elastic asperity contacts. For a rubber, considerations were also extended to the adhering and peeling mechanism from a rough surface and a viscoelastic second surface.

In 1881, Hertz (2403) first made an analysis of the mechanics of contact between two smooth spheres and his work stood essentially unmodified until Johnson et al. (2404), in 1971, showed larger contact area measurements than predicted by the Hertz theory. This was presumed to be due to lack of inclusion of certain surface forces. Greenwood and Johnson (2405), in 1981, used fracture mechanics to consider the unaccounted for forces. In effect, they proposed that the edge of the contact zone between the surfaces might be regarded as a crack tip with an accompaning stress-intensity factor.

Archard (2406), in 1974, considered the surface topography effect as one that needed to include a function indicating the vertical distribution of height along with some horizontal scale features. Greenwood and Williamson (170), in 1966, had showed that if this vertical distribution was Gaussian, then the standard deviation factor could be employed. Reference can be made to a Whitehous and Archard (2407) publication in 1970 for a description of the horizontal features.

Experimental investigations of elastic sphere contact was undertaken using molded silicone rubbers against concave lenses by Fuller and Tabor (2408) in 1975. The effect of surface topography in terms of a ratio (alpha) was derived by these authors. Briggs and Briscoe (2409,2410), in 1976 and 1977, extended the Fuller and Tabor work by studying other values of the ratio in an attempt to explain some observations not previously satisfactorily explained by those authors. It seemed that the asperites have to be a certain right size for the effect of roughness to exert an important effect. The observed effect of very low roughness surfaces on adhesion continues to be especially difficult to explain.

Maugis and Barquins (2411), in 1978, studied viscoelastic peeling of optically smooth polyurethane from glass. Using fracture mechanics, they deduced a fracture energy value for each different peeling angle studied. More recently, Greenwood and Johnson (2405) have analyzed the mechanics of adhesion to open a viscoelastic crack. The value at low peeling speeds (as compared to high speeds) involved multiplying the former value by a factor k. This k factor proved to be the ratio of the instantaneous modulus to the relaxed modulus. Such values compared quite favorably with empirical values found by Maugis and Barquins (2411) 3 yr earlier.

In 1991, Maugis (2412) discussed the work to cause the adhesion in a joint and the work to break the bonds. The latter can be considered equal to the Dupre energy of adhesion (or thermodynamic work of adhesion); i.e., the surface energies of the two solids minus the interfacial energy. While adherence depends on the Dupre energy of adhesion, it is not always directly proportional to it.

One needs to distinguish between adhesion and adherence. The latter being the force necessary to separate two solids which are in contact, and which is dependent on a number of parameters such as geometry of the solids, the stiffness of the measuring apparatus, and the rheological properties of the solids.

The basis for understanding the service failures of structural bonded joints has been mainly rooted in the science of continuum fracture mechanics. Fracture mechanics provides a fundamental and quantitative means of describing material failure, including the effects of environment. Failure begins with the formation of one or more minute flaws or cracks usually at the surface. Stress applied to the material is magnified at the crack tip until the crack begins to propagate. The propagation usually begins slowly but may be followed by an abrupt increase in the rate of extension and can reach a terminal velocity of about 40% of the speed of sound. With a properly designed test specimen and method, this crack growth can be controlled and the energy expended measured. From this information, a fracture toughness can be computed which is a characteristic property of the material. Excellent investigations and reviews of the procedures for evaluating the cohesive fracture of materials can be cited in chronological manner through various publications (2413–2430).

IV. INTERACTION OF ALUMINUM SURFACE WITH POLYMERS

The interaction of the aluminum surface in joints has been discussed by Smith and Kaelble (2431). All metals, except gold, are thermodynamically unstable at room temperature when in contact with oxygen at atmospheric partial pressure causing the formation of oxide surface layers. Thickness of this oxide will vary widely, as already discussed in Chapter 3 on aluminum surface oxides, depending on environmental conditions. Aluminum is sufficiently surface active to displace hydrogen from water to form hydroxyl groups; however, this reaction fails to occur at any appreciable rate. The oxide formed is insoluble, however, and tends to be very protective of the underlying metal even in thin film form.

The theory of oxidation at the metal/oxide interface can be reviewed in an article by Kubaschewski and Hopkins (2432). Bolger and Michaels (147) have suggested that the arrangement of water and oxide layers on aluminum results in three basic assumptions: (1) the oxide layer is sufficiently thick that the underlying metal surface exerts no influence on reactions or electrostatic interactions at the oxide surface; (2) while it is possible for the oxide surface to undergo dehydration at elevated temperatures, conditions for normal bonding operations involve hydrating to form a high density of hydroxyl groups; and (3) these hydroxyl groups can retain additional molecular layering of water. This situation was shown in Figure 4 together with the additional possible overlayering of various polar organic, nonpolar organic, and adsorbed gas layers which further complicate the bonding situation. It was further stated that the spacing and activity of the hydroxy groups will usually depend on the spacing and valency of the metal atoms in the oxide. It might be expected that the use of a hot chromic acid–deoxidizing pretreatment of the aluminum surface should remove these hydrated oxides, oxidize and remove most surface impurities, and thus increase the valency of the surface metal ions.

Considerable investigation of the surface layers on aluminum were reviewed in Chapter 3, including the conclusion of Keller et al. (2433) that the oxide formed by anodizing in sulfuric, chromic, or phosphoric acid electrolytes all consist of similar hexagonal cells whose dimensions and porosity will vary, creating different sealing conditions to alter corrosion resistance. As might be anticipated, the PAA with greatest porosity and least ability to be sealed against corrosion of its surface can form the highest strength and most failure-resistant structural bonds that can be made with aluminum surfaces. Other complications for best bonding may result from findings by investigators like Liechti and Treadwell (2434) that sulfuric anodize layers can contain as much as 13% sulfur trioxide as well as a few percent of bound water. Regardless of these observations, it is still necessary to seal SAA films in boiling water for the development of optimum corrosion resistance. The question of whether this sealing of CAA and SAA oxide layers to increase their overall resistance to corrosion is beneficial or negative in its effect on joint durability has been argued at length in the literature for many years. Bijlmer (815) has offered some evidence that sealing of CAA is a negative factor for joint durability based on European experience, yet Rogers (820) discusses good bonding experience with sealed CAA surfaces in American experience. Minford (2435) has plotted results of stress/humidity testing of both sealed and nonsealed SAA oxide surface joints which show a lesser survival time for the sealed joints. Harrington and Nelson (2436) and Phillips (2437) have showed that the film produced by the sealing process is boehmite in nature, since abrasion of such a film removed the electron diffraction pattern of boehmite. Minford (39) has deliberately heated freshly deoxidized 6061-T6 aluminum in both boiling tap and distilled waters to establish a boehmite surface oxide layer for bonding both two-part, room temperature–curing and one-part, heat-curing epoxy adhesives. The joint durability was generally very acceptable in a wide range of aggressive weathering conditions, including stress combined with hot condensing humidity. Mechanisms of joint failure thus must deal with the failures that occur at the weakest link of the aluminum/adhesive boundary. An additional technical problem is related to identifying the locus or loci of failure, i.e., in the primer, at the primer/oxide interface, in the oxide, or at the oxide/metal interface, or some combination.

Some examples of special studies of oxide properties may be of interest to the reader for better understanding of the bondability. Grosskreutz (2438,2439) has both determined mechanical properties of aluminum oxide films and their effect on dislocation-surface interactions in aluminum. With Benson (2440), he offered a 1968 special Air Force Material Lab Report on the means of suppressing fatigue cracking through control of the aluminum surface conditions. Other investigators such as Bascom (2441) have focused their attention on the surface of the adhesive. In 1972, Bascom showed that the interfacial region of the adhesive could be vastly different from the bulk adhesive, making the effectiveness of different surface treatments for promoting good permanence dependent on their in-depth interaction with this altered form of adhesive. Wegman (2442) meanwhile had studied adhesive joint failures using evaporative rate analysis and carbon-14–tagged radioactive chemicals to provide evidence that both adhesive and adherend surface failures were often in evidence when studying ruptured adhesive-bonded aluminum joints. Brinson (2443), in 1984, lectured on the viscoelasticity of adhesives and composites, including both his understanding of failure and fracture concepts at the Symposium on Fundamentals of Adhesion Theory at New York University at New Platz. In this lecture, reference was made to earlier studies on adhesion chemistry which might also be of reader interest, including 1975 and 1979 publications by Lee (2444,2445) on current advances in adhesion and adhesive studies and the relationship of adhesion and adsorption properties of polymers. Kinloch (2446) completed his very comprehensive review of investigations of adherend surfaces and the interfacial aspects of joints in 1980, which was followed by his equally comprehensive review of the mechanics of adhesive failure (2196) in 1982. Of special interest for study is the early 1973 mechanics of failure publication by Andrews and Kinloch (315) mentioned earlier.

More recent publications about analysis of thin films are 1987 investigations of Sampath and Wilber (2447) describing a new rapid technique for characterizing microstructures of films produced by ion beams with TEM, and Ishida's (2448) description of quantitative surface analysis of polymer surfaces with the FT-IR spectroscopic technique. Publications in 1990 include Gilbert's definition of different performance capabilities when films of intermediate and high film build-up are employed over metallic adherends using electrocoating procedures (2449). Meanwhile, Zimba and co-workers (2450) used FT-IR spectroscopy in studying thin polymer films, and Ager et al. (2451) analyzed the Raman intensities and interference effects found with thin films absorbed on metallic surfaces.

The remarks of Russell (2452) would seem pertinent in regard to identifying the origin of fracture in a bonded metal joint. He suggests study of the fracture surface, especially in the area of the adhesive fillet utilizing surface markings on the fracture which indicate the direction of crack propagation. Further, the mechanism is elucidated by studying the surface of the fracture to detect the presence of aluminum oxide on the adhesives and the bare metal surface. If the oxide is present, then the failure likely resulted from the oxidation of the metal in the interface between the metal and the adhesive. This would suggest failure as a result of a low load surface under warm, humid conditions. Where the fracture surface is completely covered by the adhesive, the indication is that service was under relatively dry conditions. However, it can be shown that a high-load situation under warm, humid conditions may also result in a fractured surface completely covered by cohesively failed polymer. Where the surface is partially covered with adhesive, failure was probably due to service under warm humid conditions accompanied by moderate loading. The author has noted in his many studies under stress and humidity conditions that relatively short-time survival joints, as fabricated from deoxidized 6061-T6 surfaces and nonaerospace-type, room temperature–curing epoxy adhesives, usually show no evidence of adhesive on the aluminum surface. This is in spite of the fact that a duplicate joint pulled in a tensile tester might show 100% cohesive failure in the absence of such exposure and stressing. How such dramatic loss of total adhesion can be induced by the addition of several minutes of stressing still seems amazing to observe, yet it seems consistently reproducible.

V. MECHANICS ANALYSIS

A. Mechanics Analysis of Adhesive

An overall mechanics analysis of a joint should be expected to yield information on the influence of the mechanical properties of the adhesive, the geometric configuration of the joint design, and the type of joint loading. The mechanical properties of the adhesive are reflected mainly in the stress-strain law which includes the effects of time and temperature (separately or together). Thermal strain may be assessed through a combination of the coefficient of thermal expansion and the range of temperature change. Or the magnitude of the adhesive creep deformation can be measured at essentially constant temperature. The time (rate) and temperature effect predictions about many adhesives have been considered mathematically in an equation formulated by Williams et al. (2453), better known as the WFL shift factor. Essentially, it is a semiempirical correlation between adhesive viscosity and the free volume of the viscoelastic adhesive employed. This equation also incorporates another material property known as the glass transition temperature. Other fundamental adhesive properties entering into mechanical analyses are the stiffness (modulus of elasticity) and the incompressibility value (Poisson's ratio).

A variety of papers were introduced in the middle 1980s with respect to the study of material properties involving the use of fracture mechanics and crack propagation of both bulk polymers and adhesives in joints. In 1984, Gent and Pulford (2454) reported on the micromechanics of fracture in elastomers; Enderle et al. (2455) discussed the irreversible deformation of macromolecular networks; Carpenteri (2456) the size effects in material strength due to crack growth and material nonlinearity; and Ravi-Chander and Knauss (2457) an experimental investigation of the microstructural aspects induced by dynamic fracture. In 1975, Relland and Broutman (2458) offered a fracture mechanics analysis of surface embrittlement present in ductile polymers; Sinclair (2459) discussed some inherently unreliable practices in fracture mechanics as practiced at the time; and Cho and Xu (2460) the mixed mode ductile fracture condition using the strain-energy density criterion. Karami and Fenner (2461) made analyses of mixed-mode fracture and crack closure using the boundary integral equation method. Melin (2462) discussed the conditions for crack growth under Mode II conditions. In 1991, Gent and Wang (2463) looked at the fracture mechanics and cavitation in rubberlike solids and Termonia (2464) multiaxial deformation of polymer networks.

Jangblad et al. (2465) presented a method in 1986 for determining the elastic constants E (Young's modulus) and V (Poisson's ratio) for a general adhesive in situ using aluminum adherends. In the same year, Crews et al. (2466) conducted an elastic stress analysis using an aluminum DCB specimen and finite-element methods. An attempt was made to identify the important parameters that can influence stresses advancing ahead of the delamination front. Stringer (2467), in 1985, compared the shear stress-strain behavior of some structural adhesives, and in 1986, Atkins and Mai (2468) measured residual strain energy under elastoplastic adhesive and cohesive fracture conditions. Also, in 1986, Post et al. (2469), using thick adherend aluminum joints, measured deformation and strain values and concluded that moire interferometry seemed especially suited for such studies.

In 1990, Temma et al. (2470) reported a two-dimensional stress analysis of adhesive butt-joints which were under cleavage loads and Temma et al. (2471) a similar analysis where a circular hole defect was present in the adhesive and the joint was subjected to external bending moments. Also in 1990, Temma et al. (2472) conducted a three-dimensional stress analysis of butt-joints with disbonded areas and spew fillets. Corson et al. (2473) conducted peel stress distribution testing between adherends with varying curvature mismatches.

Other 1990 investigations of possible reader interest are Chao's (2474) study of the viscoelastic properties of glass-forming polymers; Korb et al. (2475) reporting on the nuclear relaxation and fractal structure of cross-linked polymers; Parvin and Knauss's (2476) description of the damage-induced constitutive response of a thermoplastic adhesive related to both composites and adhesive bonding; and Cantwell et al.'s (2477) examination of the processes of deformation and fracture using a silica-filled epoxy resin. In 1991, van Hartingsveldt and van Aartsen (2478) investigated the strain rate dependence of interfacial adhesion in particle-reinforced polymers.

B. Analysis of Joint Geometry

Various geometrical configurations of adhesively bonded aluminum joints are used to develop engineering properties. The single- and double-lap-shear joints, T-peel, climbing drum peel, and flatwise tension specimens have all been mentioned in Chapter 6 and will be considered in more detail in Chapter 9 on testing. The thick adherend and double-cantilever beam-type joints are those most commonly employed when considering mechanisms of failure due to cracking induced by various stressing factors like static and fatigue loading. The bulk fracture testing of the adhesive can be conducted with single-edge notched, compact tension, or tapered double-cantilever beam specimens as described by Bitner et al. (1333) and Kinloch and Young (2479). The results of such tests can be converted to fracture energies using standard procedures, as discussed by Bitner et al. (1333), Kinloch and Young (2479), and Hunston et al. (2480). Use of the TDCB specimen permits the bulk fracture determination over a wide range of rates and temperatures, as described by Kinloch et al. (2481) in 1982. In 1984, El-Senussi and Webber (2482) reviewed the use of the double-cantilever beam technique for studying crack propagation in adhesives with Zhurkov et al. (2483) elucidating the principles of the kinetic approach of fracture prediction, and Fischer (2484) reviewed brittle fracture criteria using either statis or cyclic mixed mode loading in the same year. In 1985, Jozavi and Sancaktar (2485) investigated the effects of cure temperature and time on the bulk fracture properties of a structural adhesive. Zeng et al. (2486) observed microstructure, mechanical properties, and fracture behavior using liquid rubber-toughened thermoset adhesives in 1991. In addition to fracture tests, the basic mechanical properties of adhesives can be determined by measuring the dynamic mechanical and stress relaxation properties, as described by Hunston et al. (2487).

Another important geometrical factor that has importance in affecting the failure of joints is thickness (h) of the adhesive interlayer, as discussed by Anderson et al. (2488) and Williams (2489), which will be discussed in greater detail later in Section IX.B.

The loci of fracture has been mentioned as often being difficult to determine. Smith and Kaelble (2431) have considered the situation where the aluminum oxide itself may be fractured. In Table 104,

Table 104 Oxide Thickness Before and After Fracture SET[a] and BET[b]

	Bond history	Thickness (Å)
FPL etch		
SET (24 hr)	Before bonding	623
BET (264 hr)	After fracture	195
FPL etch		
SET (24 hr)	Before bonding	576
BET (0 hr)	After fracture	433
FPL etch		
SET (24 hr)	Before bonding	530
BET (0 hr)	After fracture	273

[a]SET = surface exposure time.
[b]BET = bond exposure time.
Data from Ref. 2431 (Table 4–10).

the data of these authors are presented for aluminum surfaces which were FPL etched then exposed either as the surface itself or after bonding to 129°F (54°C) and 95–100% RH conditions. It was concluded that the aged FPL-etched surface oxide in either situation obviously failed within the aluminum oxide layer.

Not mentioned earlier, but of demonstrable importance in altering bonded joint failure in practical situations, is the presence or absence of fillets at sharp corners. The dramatic effect of sharp corners in homogeneous materials as the points from which fractures could propagate had been recognized and studied for many years. Two individual examples were offered by Sneddon (2490) as early as 1946 with additional generalization given by Williams (2491) in 1957. Linear elastic analyses of plane problems yield solutions for stresses near a sharp notch. Hein and Erdogen (2225), in 1973, further treated the situation for a plane problem in which one material was bonded at an angle to a semi-infinite sheet of a second more rigid material.

C. Mechanics of Loading Analysis

The third consideration of mechanics analysis deals with the manner of load orientation and how it is applied with respect to the total structure or on some individual component of the structure. The load may be either mechanical or thermal in nature and is generally imposed in some time-dependent way from one or more directions. Examples of time-dependent stress variations could be steady, repeated or oscillatory fatigue loadings, or simple sudden shock impact. (For some pertinent articles about appropriate methods of testing see Refs. 2492–2498.)

The mechanics analyst characterizes a crack normal to the primary direction of applied load as Mode I loading (tensile-opening mode). Mode II loading is when the applied load is applied parallel to the crack and the adherends are slipping in the direction of the crack growth (inplane shear). The Mode III situation is similar to the Mode II except that the direction of the adherends is perpendicular to the crack growth (antiplane shear). Two investigations by Jemian and Ventrice (2499) and Jemian and Wilcox (2500) showed the presence of Mode I loads using photoelastic modes of double-lap-shear joints. Only normal stresses (Mode I) were acting at the loading edge or at a crack tip, whereas the maximum shear stress (Mode II) occurred in advance of the edge or crack tip. They concluded that only normal stresses, acting in Mode I, were responsible for the joint failure.

Ripling et al. (2501–2503) continued their studies on the application of fracture mechanics to elucidate the mechanisms of aluminum joint failures in 1963 and 1964. Investigations by Mostovoy et al. (2504) and Mostovoy and Ripling (2505) followed in 1965 and 1966, respectively. In 1967, Mostovoy et al. (2506) issued their report describing their progress in evaluating the joint failure problem using crackline-loaded specimens. Mostovoy and Ripling closed out their 1960s' evaluations with yet another report in 1969 (2507).

In the decade of the 1970s, Ripling and Mostovoy and their co-workers continued to produce publications at a fast pace on all aspects of the problem of fracture failure of joints. A publication with Bersch (2508) followed in 1970. Another publication with Bersch (2509) followed in 1971 with a publication with Corten (2510). Two more publications followed in the same year authored by Mostovoy and Ripling (2511) and the same authors with Smith and Lingwall (2512). The year 1972 produced yet another Mostovoy and Ripling (2513) publication along with two summary reports (2514, 2515) relating to specific government contracts on fracturing characteristics. Another contract report on fracturing characteristics of aluminum joints covered work in 1974 (2516). A chapter was produced by these authors in *Adhesion Science and Technology* in 1975 (2517) as well as a summary of yet another fracturing contract (2518). Yearly summaries of contract work on fracturing characteristics for the years 1976 (2519) and 1977 (2520) were followed by yet another article in 1979 (2521).

In 1972, Bascom and Cottingham (2522) described the joint failure in an aluminum/modified epoxy joint as resulting from a propagating crack being deflected at the interface. Liechti et al. (2523) made comparisons between actual measurements and finite element predictions of crack opening displacements near the front of an interface crack in 1987. Barsoum (2524) investigated asymtotic fields in adhesive fracture in 1989. Some 1990 investigations in matters of investigating the crack propagation failure of joints include Needleman's (2525) analysis of decohesion along imperfect interfaces; Reynolds et al.'s (2526) studies on the mechanics of the interface crack using finite element procedures; Schom-Rode and Berger-Bocker's (2527) detections of microcrack accumulations in tensile-loaded specimens; Charalambides et al.'s (2528) investigation resulting in the development of a test method for measuring mixed-mode fracture resistance of bimaterial interfaces; and Ortiz and Blume's (2529) results of a study of the effect of a test method for measuring mixed-mode fracture resistance of bimaterial decohesion and sliding on bimaterial crack-tip fields. In 1991, Crosley and Ripling (2530) added yet another refinement to the crack study field.

Williams (2531) projected the point of view in 1984 that there needed to be more collaboration between chemists and physicists in working on adhesive fracture from the point of view of continuum mechanics. As discussed by Williams (2531), Anderson (2488,2532), and Kinloch (2196,2446), the amount of new surface created in fracture is sensitive to the loading mode. The analytical treatment assumes that the total surface energy created is split into three separate specific fracture energies by mode, each multiplied by the projected new fracture energy. The existence of different height, depth of hills and valley are not recognized in this treatment.

The second area of collaboration is recognized in the development of an "interaction matrix" as set forth by Williams and Kelly (2533). The matrix consisted of a vertical listing of properties as studied and measured by chemists such as (1) cross-link density, (2) chain stiffness, (3) monomeric friction coefficient, (4) solubility parameter, (5) molecular weight, (6) heterogeneity index, (7) molecular weight between entanglements, (8) degree of crystallinity, (9) volume fraction of filler, and (10) volume fraction of plasticizer. A corresponding horizontal listing included those categories measured and used by mechanical property investigators; e.g., glassy modulus, rubbery modulus, and glassy temperature. The final goal being to be able to obtain mechanical-chemical associations for all the intersections in the matrix. Extensions of the theories to cover adhesive joint failures have been reviewed (2196,2488,2534).

VI. FRACTURE MECHANICS OF ADHESIVE JOINTS

The listing in chronological order of references on the general subject of the fracture of materials, including polymeric adhesives, might furnish a proper background for this subject. Such a list might include publications by the following authors: Tetelman and McEvily in 1967 (2603), Yokobori in 1968 (2604), Liebowitz (2605), Andrews in 1968 (2606), Jenkins in 1972 (2607), Kausch et al. in 1972 (2608), Kinloch and Young in 1983 (2609), Williams in 1984 (2610), Kanninen and Popelar in 1985 (2611).

Kinloch (2196) reviewed the application of fracture mechanics for assessing the strength of adhesive joints by both the energy-balance and stress-intensity factor approaches in 1982. Though the values obtained, G for the former and K for the latter, are related through the Irwin treatment (2580); the use of G has generally been preferred for expressing the adhesive strength. Difficulties still may exist according to Wang et al. (2547), however, because of significant differences in the stress field from those estimated for a homogeneous material outside a short distance in front of the crack tip.

Fracture energetics of brittle and glassy polymers have been discussed in terms of fracture energy, G, and the stress energy or fracture toughness, K, by Lee in 1983 (2566) and 1984 (2567). Also in 1984, Lee discussed the fracture energetics of both the ductile- and viscoelastic-types of polymers. For such ductile polymers, Lee points out that it is necessary to employ the critical J-integral (ductile fracture energy) derived by Rice (2417) in 1968 as a substitute for G. Two 1984 papers with amplifications about the J-integral were by Schapery (2612) and Schindler (2613). In 1985, Dodde and Read (2614) published on experimental and analytical estimates of the J-integral for tensile panels containing short center cracks.

For viscoelastic-type polymers, the tear energy, T, is the value more frequently determined in terms of G. In 1985, Gent and Jeong (2615) published on the contribution of bending energy losses to the apparent tear energy values. In the generalized theory of fracture mechanics as developed by Andrews (2423), the tear failure energy value, T, is claimed to also include the terms G_c, T_c, and J_c.

Actually, the physical properties of a modern structural adhesive can be determined both in the form of bulk specimens as well as adhesive joints by a variety of testing techniques, which will be described in more detail in Chapter 9 on testing procedures. Kinloch (2196) has offered a summation of the physical property values for a specific rubber-toughened epoxy adhesive as measured in bulk and via testing of various adhesive joints taken from unpublished data of Atkins, Kinloch, and Shaw. These results, shown in Table 105, illustrate the wide variety of property factors that need to be considered and demonstrate how much the strength depends upon the joint design and can vary with different adherends such as steel, aluminum, and carbon fiber–reinforced polyester (CFRP). The higher 90-degree peel strength shown for the aluminum alloy adherend certainly reflects the extensive yielding that occurs when the aluminum is peeled away. In contrast, the steel adherend will remain elastic. Also, note the higher single–lap-joint strength with the higher modulus steel adherend. The effective fracture surface energy and fracture energies of an elastomer-modified epoxy has been compared with the unmodified epoxy and a nylon-modified epoxy by Bascom et al. (2680) in Table 106. Notice that while the elastomeric modification creates a significant boost in fracture energy over the unmodified epoxy, its fracture energy is still significantly lower than for the nylon-modified adhesive.

A. Fracture Mechanics Approach to Analysis of Adhesive Joints

The fracture mechanics approach to studying and predicting the failure load of joints is well accepted, since adhesive failure parameters are characteristic of the bond itself. The fracture parameters obtained, however, are empirical quantities which are not explicitly related to either the surface or mech-

Table 105 Physical Properties of Rubber-Toughened Epoxy Adhesive Measured in Bulk and in Various Adhesive Joints

Test	Adherends	Comments	Property
Bulk properties[a]			
Flexure test			E_a = 2.8 GPa
			σ = 74.5 GPa
			= 2.7%
			248°F (120°C)
Glass transition temperature (°T)			
Adhesive joints[a,b]			
Tortional shear	Aluminum	t = 0.1 mm	= 61 MPa
Axially loaded butt-joints	Steel	t = 0.5 mm	= 58 MPa
Single–lap-joint, loaded in tension	Aluminum	t = 0.5 mm	= 9 kN
		d = 1.6 mm	= 28 MPa
		1 = 12.7 mm	
	Steel	As above	= 12.3 kN
			= 38 MPa
Double–lap-joint, loaded in tension	CFRP/steel	No taper; 1 = 80 mm	= 24 kN
			= 6.9 MPa
	CFRP/steel	Reverse taper: 1 = 80 mm	= 80 kN
			= 19.7 MPa
Peel joints (1) 90-degree peel	Steel		$F_{a/H}$ = 0.6 kN/m
(floating roller)	Aluminum		$F_{a/H}$ = 5 kN/m
135-degree peel	Aluminum		$F_{a/H}$ = 4 kN/m
Precracked, tapered double–cantilever beam	Steel		G_{IC} = 0.9 kN/m
			G_{IIC} = 2.3 kJ/m

[a]Tests conducted at 73°F (23°C) and a moderate rate.
[b]All joints failed by cohesive fracture through the adhesive except for the CFRP/steel double–lap-joints which failed in the CFRP.
From Ref. 2196 (Table V, p. 638).

Table 106 Effective Fracture Surface Energies and Fracture Energies of Elastomer-Modified Epoxy Resins

	Γ (Calc)				G_c			
	Cleavage (I)		Cleavage + Shear (I,II)		Cleavage (I)		Cleavage + Shear (I,II)	
Adhesive resin	J/m^2	10^5 erg/cm^2	J/m^2	10^5 erg/cm^2	J/m^2	10^5 erg/cm^2	J/m^2	10^5 erg/cm^2
Unmodified epoxy (HHPA-DGEBA)	58.0	0.58	68.0	0.68	116.0	1.16	136	1.36
CTBN-epoxy (Piperidine-DGEBA)								
10% elastomer	1750	17.5	55.0	0.55	3500	35.0	110.0	1.1
30% elastomer	1100	11.0	55.0	0.55	2200	22.0	110.0	1.1
Commercial elastomer epoxy	1150	11.50	435	4.35	2300	23.0	870	8.7
Nylon-epoxy	3050	30.5	375	3.75	6100	61.0	750	7.5

Data from Ref. 2680.

anical properties of the adhering phases. Actually, the use of generalized fracture mechanics (GFM) procedures permits a detailed analysis of the fracture process, since separation of the surface and bulk contributions to the adhesive failure energy can be distinguished in terms of the interfacial energy magnitude and the mechanical hysteresis properties of the bulk adhesive phases. It can be further shown that the adhesive fracture energy is dramatically controlled by various rate, temperature, and environmental factors acting on the adhesive joint structure.

The fracture mechanics approach to evaluating joint failure utilizes adhesive-bonded sections which can fail by the propagation of a brittle crack through the bondline, as extensively investigated by Jones and Kaelble (2536). The premise would seem practical, since it is readily observable that real adhesive joints are invariably blemished by some sort of flaw of a variety of types and sizes; e.g., bubbles, dust particles, voids, or unbonded areas. Progressive separation arising from the growth of these flaws will basically control the strength of the structure. From the durability evaluation of as near-perfect joints as can be assembled in the laboratory, it also is clear that overestimates may be expected of the total service life potential of routinely manufactured bonded structures. Finally, we are becoming increasingly aware that, as the sophistication of our analytical observations of the joint interphase area advances, many subtle changes can and do occur in the bulk adhesive and adherend surface interface which escaped earlier detection.

Anderson and DeVries (2537), in 1987, pointed out the need for a fracture mechanics approach by making comparisons between the previously employed method of obtaining bond strength data from laboratory-size specimens using a specific adhesive and surface preparation and a fracture mechanics approach. The former procedure using average stress values for design purposes is not valid because it does not account for variations in stress within the bonded joint. Other factors must be accounted for, including the loading mode (direction of load application with respect to a crack or debond termination line), loading rate, joint temperature, adhesive thickness, joint geometry, residual stresses, moisture content and distribution within the adhesive, adherend stiffness, and adhesive stiffness and compressibility (Poisson's ratio). Each one of these factors, according to these investigators, could alter the load carrying capability by as much as a factor of two. Overall the load carrying capability could be in error in excess of an order of magnitude.

In theory, it might be expected that the failure of any bonded joint occurs when the stress reaches some critical value which exceeds the strength of the weakest point in the joint bondline. Such a maximum stress resistance value might be termed the bond stress capability. The joint requirement would then consist of a bondline which could resist at all its points the maximum stress that might be exerted in actual service. The amount by which the joint capability exceeded the joint requirement value could be considered the safety margin of the manufactured joint. The procedure of evaluating the stresses at each point in the bondline to determine the safety margin under maximum service stressing can be very useful when dealing with homogeneous materials except when a notch or crack (360-degree notch) is present.

When linear elastic analyses are employed with notch or crack presence, the stresses will not be defined at the tip location but rather this tip will constitute a point of stress singularity. A fracture mechanics approach is then relied upon to predict load-carrying capability for this geometry situation. Williams (2538), Hein and Erdogan (2225), Dill et al. (2539), and Anderson et al. (343) have offered discussions of the many bond termination geometries that are points of singular stress in bonded joints. Even in the absence of any adhesive voids or debonds, a debonding can be initiated; for example, at bond edges. However, bond failures do not always initiate at a point of apparent stress singularity. A failure criterion is desired then which can be applicable to both singular and nonsingular points. Anderson and DeVries (2540) have chosen to determine an "inherent" flaw size from which energy release rates can be calculated regardless of whether the failure initiates within a bonded joint or at the edges. The four adhesive testing configurations of blister test, cone (cylinder) test, peel test, and butt-joint test were employed with specified loading modes (343,2274,2541,2542).

B. General Considerations

There are two different formulations of fracture mechanics, which can be labeled (1) linear elastic fracture mechanics (LEFM) or (2) nonlinear, finite-strain fracture mechanics (NLEFM). Williams (2421) has considered the matter in detail dating back to 1973.

1. *LINEAR ELASTIC FRACTURE MECHANICS (LEFM)*

The LEFM defines a stress field parameter, K, which reflects the overall intensity of the stress field around the crack. Joint failure will be indicated when K achieves some critical value, which has been designated K_c. The formula for K will be dependent on the type of test specimen configuration. A second fracture parameter also provided for by LEFM has been designated G and is the energy required to create unit area of fracture surface. Once again, there is a critical value, G_c, which corresponds to that energy required to begin crack propagation. Both K and G values are only defined for linear elastic solids, which is a complication, since many materials in joints are nonlinear and inelastic. Metals like aluminum, while presenting linear properties at low strain rates, can have pronounced plasticity around a crack tip where high stresses can develop. With plastic and elastomerlike materials, there will be significant viscoelasticity property development rendering linear elastic assumptions invalid.

2. *NONLINEAR, FINITE-STRAIN FRACTURE MECHANICS (NLEFM)*

When the strains are large or the stress-strain curves are nonlinear in form, it is obvious that the LEFM assumptions become invalid and an alternative formulation of fracture mechanics has to be developed. The earliest alternative can be found in the 1953 publication of Rivlin and Thomas (2413) dealing with rubber materials. They investigated the energy release from such test specimens as a result of crack growth. They proposed that as the crack begins to propagate, the energy consumed per unit area should equal the energy released from the stress field. There is, however, an implication that no energy is dissipated in the recovery of strain for the material with true elastic behavior. This is not in general a valid assumption, so that the fracture parameter developed here contains contributions from energy dissipated remote from the crack itself. This could lead to the further supposition that the parameter value can cease to be independent of the specimen size and shape (configuration).

A second approach can be attributed to Rice (2417), in which he defined a path-independent integral around the crack utilizing the parameter designation, J. For nonlinear elastic solids, the integral ends up being equivalent to the energy released by the stress field per unit propagation of the crack;

thus making it essentially the same as the empirical parameter derived by Rivlin and Thomas (2413). Such nonlinear fracture parameters can handle the large nonlinear deformations found in elastomeric and plastic materials, but like the LEFM parameters, they ignore energy losses. Such losses do obviously occur and must be lumped into the NLEFM parameter values even though they may occur at points quite removed from the crack.

3. NEGATIVE ASPECTS OF CONVENTIONAL FRACTURE MECHANICS

While LEFM defines a true fracture parameter when the material is perfectly elastic and linear, it does not, as mentioned above, have applicability to nonlinear systems. Also, the parameter is not specimen independent if significant inelasticity is present making the equations for calculating K and G values invalid. The NLEFM theory equations can handle the calculation for nonlinear materials but fail to satisfy for inelasticity characteristics. Thus, all energy losses must be included in the fracture parameter regardless of where they may occur in the specimen and regardless of whether they are or are not directly associated with the crack-propagation process.

An additional limitation of both LEFM and NLEFM equations is their lack of application to the propagation of a crack over some time period. These equations only seek to define conditions under which the stationary crack is caused to move. Once the crack velocity moves off zero, the equations no longer apply, since the equations are based on the assumption that the stress field increases monotonically until the critical point of propagation is attained. This is the critical point described by the K, G, and J values. When the crack begins to move, then there will be unloading in some regions of the specimen and conventional fracture mechanics does not consider unloading stress fields. The exception being the special case of perfectly elastic solids accompanied by very slow crack-propagation rates. In spite of the above considerations, investigators routinely publish data showing critical fracture parameters as functions of crack velocity. Andrews (2543) has pointed out, however, that such data may be perfectly valid in the empirical sense that the fracture parameter defined for the loading situation may still be measured under dynamic conditions which can show its dependence upon crack velocity. However, such a dynamic value is nothing more than an empirical quantity with no guarantee that it possesses the characteristics of a specimen-independent material property.

4. GENERALIZED FRACTURE MECHANICS

The purpose of generalized fracture mechanics (GFM) has been to overcome some of the problems mentioned above by addressing (1) nonlinear and inelastic materials, (2) steady-state crack propagation, and (3) expression of critical fracture parameters in terms of the physical properties of the materials involved. Its first introduction by Andrews and Walker (2418) was to explain the fatigue properties of plastics. Andrews and Kinloch (315) followed 2 yr later in 1973 with further development in the context of adhesive failure. The full theory was proposed the next year by Andrews (2423) and the reader is referred to these papers for more details. The outline of the theory considers a preexisting crack propagating steadily through an infinite solid or along an interface. A volume element of the material at a point will undergo certain deformations during the crack propagation; first loading as the crack approaches then unloading as the crack recedes. With perfectly elastic material, no energy should be dissipated and an energy release rate can be obtained by simple integration over the whole specimen. When energy is dissipated in the volume elements (unloading of an element returns only a fraction of the input energy), then the specimen must be divided into loading and unloading regions and integrated separately over these two domains. This new parameter is thus the energy truly available for the creation of a crack surface. If the crack is propagating (at a certain speed), this available energy must also equal the energy required to create a crack surface in the absence of radiated stress waves. This has been designated the failure energy, T, and is said to embrace G, T, And J components. In practice, it is difficult to classify an adhesive joint as being exactly brittle, ductile, or viscoelastic. For most technologies, it then becomes easier to differentiate only rigid from flexible joint geometries.

5. MODES OF CRACKING

The fracture mechanics parameters of G, K, T, and J have been indicated to have critical values which are further designated G_c, K_c, T_c, and J_c. In addition, there are three recognized modes of cracking, which are designated as Modes I, II, and III depending on the direction of the stressing. These three modes have been designated tensile-opening, inplane shear, and antiplane shear, respectively. The combination of these subscripts with the parameter letter designations offers a specific designation of what parameter is being tested and how it is being stressed. For example, the designation G_{Ic} indicates that the fracture energy is being evaluated by testing the specimen in the tensile-opening mode. Correspondingly, G_{IIc} would indicate that the fracture energy was being evaluated with the specimen subjected to stressing in the inplane shear mode. The remaining K, T, and J letter identifications would indicate some mode of stressing employed to evaluate the "stress intensity" or fracture toughness parameter, elastomeric fracture energy, or ductile fracture energy, respectively. It is also possible to consider evaluation of fracture mechanics parameters of the bulk adhesive or as present in a bonded joint. For example, the value of K_c (bulk) would be the magnitude of the fracture toughness value for the solid "as cured" adhesive used as the test material. Where making measurements of the same adhesive as employed in a joint, the value would be indicated by K_c (joint).

The tensile-opening or cleavage mode is technically the most important, since it is the most commonly encountered and most likely the one which has resulted in the observed bond failure. Thus, the adhesive joints with Mode I crack growth have been well characterized. Recent work has led to a qualitative understanding about the failure behavior of adhesive joints as regards the magnitude of the crack-tip deformation (1333,2480,2535,2545).

Bascom et al. (2546) have offered a crack-opening displacement model for explaining fracture failure behavior. Further Bascom et al. (222) have conjectured that the mechanism of joint reinforcement extended by the addition of a rubber phase to epoxy adhesives manifests itself as a means of increasing the plastic zone size at the crack tip due to the presence of the rubber particles. It has become evident over the years that this plastic zone developed by the rubber-modified epoxy adhesive formulations can largely determine the crack resistance of such aluminum structural joints. Thus, the fracture energy value, G, can be attributed to the size of the plastic zone developed at the crack tip of a particular joint under evaluation.

The work of Wang et al. (2547), in 1978, pertained to the use of finite-element analysis calculations for revealing the unique crack-tip stress pattern in the adhesive bond using a double-cantilever beam specimen. In 1981, Kinloch and Shaw (2535) used this background to offer arguments about an observed dependence of G_c (joint) on the bondline thickness (2196). This will be discussed in detail later in Section IX.B. For details about the general subject of the use of finite-element analysis on adhesive joints, the reader is referred to the 1991 work of Reddy and Roy (2239). Also important are their 1985 and 1986 reports to the Office of Naval Research (2241,2242) on this subject. The reader is also reminded of the extensive list of references provided by Reddy and Roy that reviews the background of developing the science of using finite-element analysis for bonded joints (2243–2270).

6. OTHER FACTORS THAT CAN AFFECT THE FRACTURE CHARACTERISTICS

A number of investigators have showed an effect of temperature on the value of G_c (joint) (1333, 2334,2480,2544,2545,2548,2549). Ting and Cottington (2549) have employed two different thermosetting resins in structural joints and demonstrated that the G_c (joint) decreased as the temperature was raised. This is contrary to the fracture behavior of bulk resin where the G_c (bulk) for the same adhesives increased with increasing temperature. The proposed explanation has been that the G_c (joint) value is a function of both G_c (bulk) and the ratio between adhesive yield stress and Young's modulus. At higher temperatures, both modulus and adhesive yield stress is decreasing; however, the rate of change of yield stress is much higher than for the modulus. This results in a significant decrease in the overall ratio and, consequently, the G_c (joint) value trends opposite to the G_c (bulk) value movement. Muscati and Lee (2550), in 1984, made an elastic-plastic finite-element analysis of thermally loaded cracked joints.

Bitner et al. (1333), Hunston et al. (2480), Kinloch and Shaw (2535), and O'Connor and Brinson (2551) have all evaluated the changes in fracture energy values that can be affected by changing the testing rate. Once again, the G_c (joint) may not follow the same trend as the G_c (bulk) value as regards the effect of loading rate changes. Whereas the G_c (bulk) usually decreases with increasing loading rate, the G_c (joint) can be either an increasing or decreasing function depending on the temperature (1333,2480,2544). Khalil and Bayoumi (2552), in 1991, also evaluated the effect of loading rate on fracture toughness of joints. Similar variational differences can be observed with changes in bondline thickness (1333,2535,2551). Related to the above is the difference noted between static and dynamic testing or service life conditions of a joint. In 1984, Woo and Chow (2553) made fatigue crack-propagation measurements using aluminum adherends and PMMA adhesive, and in 1986, Kobayashi et al. (2554) reported their dynamic fracture toughness studies.

Obviously the fracture qualities of different types of adhesives in joints can vary significantly. For example, ductile failures have been evaluated by several investigators. Thomson and Hancock (2555), in 1984, published on ductile failure by void nucleation, growth, and coalescence. In 1986, Lorenzo and Hehn (2556) evaluated the effect of ductility on the fatigue behavior of epoxy resins and Kinloch et al. (2557) proposed a mechanism for ductile crack growth in epoxy polymers. It also would seem logical to expect that a rigid inclusion in the bulk adhesive would affect cracking rate, and Belgacen and Brethear (2558), in 1985, provided a paper titled "A Local Study of the Mechanical Interaction Between a Rigid Inclusion and the Surrounding Plasticized Matrix."

Kinloch (2196) offers the data given in Table 105 showing the tests that might be necessary to thoroughly characterize the physical properties of a single particular rubber-toughened epoxy in bulk and joint situations using aluminum, steel, and carbon fiber-reinforced polyester (CFRP) adherends. With such data it should be possible to show how much the strength of an adhesive joint depends upon the design employed, the degree of yielding, and the modulus of the adherend materials.

Ripling et al. (2559) investigated the effect on pure Mode I and mixed Modes I–III loading of adhesives FM-73 and FM-300K, which are aerospace structural tape adhesives with open knit polyester carriers. Their results using a modified 7075 aluminum alloy tapered double-cantilever beam (TDCB) test specimen indicated fracture laws for adhesive joints are undoubtedly more complex than previously believed.

B. The Energy-Balance Approach

The fracture mechanics approach to study the degree of potential to resist failure in bonded joints is based on early work of Griffiths (2560,2561) on glasses. That the energy to cause crack growth will be greater than the surface free energy for metals and cross-linked elastomers was examined by Orowan (2562) in 1948 and Rivlin and Thomas (2413) in 1953. This was assumed since for real materials, some inelastic deformation mechanisms operate, including plastic, viscoelastic, and other energy-dissipative processes. Such losses are magnified by relatively high strains occurring around the crack tip. Rice (2563) and Gent et al. (2564) both contributed discussions about analyzing materials which exhibit non-linear-elastic behavior in 1968 and 1975; however, Andrews (2423,2430, 2565) developed a generalized theory for applying fracture mechanics to such materials as early as 1974. Other readings from Kinloch (2196) and Lee (2566,2567) are also recommended about this subject.

Kaelble (122,2568,2569) developed a modified form of the fracture energy criteria in which the fracture surface energy is redefined in terms of surface energetics arguments. This enabled Kaelble to analyze the effects of varied vapor and liquid phase immersion upon the energy requirements for interfacial crack propagation. He further postulated that there was a potential zone of spontaneous environmental debonding that existed between two solid phases even with zero external stress. The model developed was specifically addressed to failure criteria at adsorption interfaces.

Lee (2570) has suggested a theoretical model for failure based on adhesive joints that should be mentioned. The main concept is that the adhesive fracture is controlled by the plastic zone developed at the crack tip. He found a quantitative representation to relate adhesive fracture energy to the bulk resin properties of fracture toughness, yield stress, and Young's modulus. The temperature and loading rate dependence of adhesive joint fracture can be well interpreted by this in situ failure model.

It should be remembered that Bascom et al. (222) 10 years earlier had theorized that the benefit of rubber reinforcements in epoxy formulations could be explained by the extra deformation they provided around the crack tip.

Krieger (2216–2218,2571) conducted a number of investigations regarding stress analysis of metal-to-metal bonded joints including their durability in hostile environments. Of practical importance was his development of an extensometer (KGR-1) capable of economically developing stress-strain curves and establishing a value of G for actual adhesive bondlines in shear. The entire procedure has been discussed by Hughes et al. (2572) and Kreiger's (2204) in their publications. In addition to the use of this extensometer for establishing G values as noted above, Krieger described the development of a second extensometer, KGR-2, which is capable of measuring values of K. In brief, the stress analysis approach settles on the fundamental problem of adhesive shear at the tip of a doubler bonded to a skin. This produces a shear distribution beginning with a maximum peak at the doubler tip and curving down and away to zero at some distance. This data can be developed for any adhesive in any environmental condition that can be reasonably reproduced in the laboratory. Krieger (2218) has furnished test data and analyses demonstrating the accuracy of data developed with the KGR-2 device.

The energy-balance approach has been developed with some success utilizing both rigid and elastomeric (high-strain–type) adhesives. (More details and information for the reader about the use of the energy-balance approach for studying mechanisms of failure in flexible joints can be found in Refs. 258,315,422,423,426,2196,2340,2342,2343,2573–2578.)

C. Stress-Intensity Factor Approach

Westergaard (2579) first developed a mathematical solution relating the local concentration of stresses at the crack tip to the applied stress. Irwin (2493,2580–2582) extended the concept to include the stress field around the crack, which was given the designation stress-intensity factor K. G and K values can be related for linear elastic systems through the Irwin treatment but they are not numerically identical. Many investigators have preferred to use the term *K* for fracture toughness value of a bulk or in joint material; i.e., K_c (bulk) or K_c (joint) depending on which is being measured. As with G values, allowance is made in nomenclature to indicate the direction of stress application; i.e., Mode I, II, or III. Thus, the fracture toughness of a particular aluminum/epoxy joint would be designated K_{Ic} (joint) when being stressed in tension.

Wang et al. (2547) reviewed the major difficulties in obtaining K values directly and accurately. Many investigators have calculated results for stress-intensity factors for an interfacial crack between two bonded materials dating from the early 1960s to the present (2583–2592).

Ikemami and Kamiya (2593) investigated the strength of metal-to-metal bonded joints with a flaw in the interface applying linear elastic fracture mechanics to their experimental results. Formulas of Erdogan and Gupta (2586) were employed to calculate stress-intensity factors in terms of flaw size and loading. Fracture stresses with the flaw were predicted at critical values of the stress-intensity factors. Strength of joints without a flaw were correlated with the stress-intensity factors by use of a concept of "effective flaw size."

Anderson and DeVries (2594,2595) measured both Modes I and II stress-intensity factors in lap-shear joints. They show in Table 107 that both these factors increase with decreasing aluminum adherend thickness in the test specimen. Thus, it is expected that thicker adherend aluminum joints should resist much larger loads in service. For all adherend thicknesses, however, the Mode I stress intensity factor can be demonstrated to be greater than that for Mode II.

As mentioned earlier, the energy-balance approach has achieved success utilizing both rigid and elastomeric (high-strain) adhesives. The stress-intensity approach, in contrast, has been more or less limited to studies with more rigid adhesives such as phenolic and epoxy resin variations. For the more general interest reader, the 1975 article by Rowe and Riew (2596) titled "What Failure Mechanisms Tell About Toughened Epoxy Adhesives" may be interesting for explaining the qualitative relationships between the adhesive and its type of failure. A 1989 paper by Liechti and Freda (2597) describes the testing procedure for determining the various types of fracture properties using laminated beam specimens.

Table 107 Stress Intensity Factors for Lap-Shear Tests

Adherend thickness		Shear			Tensile			
		K/τ			K/τ			
cm	(in)	$cm^{0.34}$	$(in^{0.34})$	$K_{II}/K_{II0.5}^{a}$	$cm^{0.34}$	$(in^{0.34})$	$K_I/K_{V0.5}$	K_I/K_{II}
0.16	(0.063)	0.448	(0.326)	2.59	0.914	(0.666)	1.77	2.04
0.32	(0.125)	0.272	(0.198)	1.57	0.660	(0.481)	1.28	2.43
0.64	(0.250)	0.189	(0.138)	1.10	0.560	(0.408)	1.09	2.96
0.95	(0.375)	0.178	(0.130)	1.03	0.526	(0.383)	1.02	2.95
1.27	(0.500)	0.173	(0.126)	1.00	0.516	(0.376)	1.00	2.98

[a]$K_{II0.5}$ is the value of K_σ for an adherend thickness of 0.5 in.
Data from Ref. 2595 (Table 3.2, p. 78).

A number of investigations relating to the stress-intensity factor can be cited in the period from 1984 to 1986. In 1984, Grilitskii et al. (2598) published their experimental determination of the stress-intensity factors in structures with cracks. In the same year, Lem and Williams (2599) issued a paper titled "The Effect of Contact Stress Intensity Factors on Fatigue Crack Propagation," whereas Wah (2600) reported stress-intensity factors determined by use of Westergaard's stress functions. In 1985, Foote and Buchwald (2601) published an exact solution for the stress-intensity factor for the double-cantilever beam. Ruiz and Epstein (2602) in 1986 discussed the variation of the stress-intensity factor along the front of a surface.

D. Ductile Polymer or Viscoelastic Polymer Considerations

The J values for some ductile polymers have been determined by Ferguson et al. (2616). Sumpter and Turner (2617) have shown that, in general, the J values will be greater than the G values for ductile polymers, because J_c accounts more inclusively for the plasticity component. Very recently, Saleemi and Narin (2618) have investigated the plain strain essential work of fracture as a measure of the fracture toughness of ductile polymers.

For viscoelastic polymers, the tear energy also needs to be considered for which the term T_c has been introduced by Rivlin and Thomas (2413). In practice, however, it is difficult to classify a particular joint as being brittle, ductile, or viscoelastic. This follows from the fact that most successful adhesives are mixtures of a variety of polymers, resins, and special organic and inorganic entities. Even when the properties of some of these ingredients are known in detail, it is unlikely that the specific properties of the final formulated adhesive can be exactly predicted in the cured state. The most ready differentiation that is used by technologists is usually rigid or flexible. Lee and McKenna (2619) very recently investigated the viscoelastic response of epoxy glass polymer subjected to different thermal treatments. Also in 1990, Ricco et al. (1573) have described the problems in fracture mechanics characterization of rubber-modified glass polymers using double-torsion study techniques.

E. Mixed-Mode Considerations

It needs to be mentioned that crack propagation is constrained within the adhesive layer in bonded joints regardless of the method of loading. Thus, a mixed-mode fracture needs to be considered. In most practical adhesive polymer applications only G_{Ic} and G_{IIc} fracture values are present because of the peel and in-plane shear stresses that predominate in lap-joint geometries. Thus, creating tensile and shear forces on the adhesive crack; i.e., a mixed-mode cracking. The sum of G_{Ic} and G_{IIc} is generally considered as a new single fracture energy value. There are three basic specimen geometries for the mixed-mode testing of adhesives in joints: (1) cracked lap-shear (CLS) (2620–2622), (2) scarf-joint (2623), and (3) independently loaded mixed-mode specimen (ILMMS) (2503). Of these, only the last type permits the calculation of G_{Ic} and G_{IIc} without the need for finite-element analysis. The originally proposed geometry has been substantially altered more recently by Sancaktar and Tang (2624) and Sancaktar (2625,2626) to minimize any interference between the two modes (cleavage loads and so on) and the effects of stress discontinuities.

Johnson and Mangalgiri (2627) predicted that the fracture of brittle polymer resins with low G_{Ic} values are controlled by the G_{Ic} component; however, the fracture values are quite elevated for the tough polymer resins. They reasoned that these higher G_{Ic} values were achieved because there was a larger free volume in rubber-modified epoxies, plasticized epoxies, or epoxies in general at high temperatures. At the same time, there is only a limited improvement in G_{IIc} values because shear deformation does not require volume dilatation. They also reported that the G_I/G_{II} ratio will vary with the matrix modulus which, in turn, will vary with moisture and temperature.

F. Cure Temperature Considerations

Jozavi and Sancaktar (2628) attempted in 1985 to distinguish the effects of cure temperature and time on the bulk fracture properties of structural adhesives and were able to show some distinct differences based on the length of time and temperature of the cure. The optimum fracture toughness values were obtained with a modified nitrile-epoxy structural adhesive under low-temperature and long-time curing conditions. Also, the optimum toughness could not be achieved in tape forms of adhesive with carrier cloths until the carrier was removed. For a given set of cool-down conditions conducted slowly, the fracture toughness and tensile strength values could be used to determine the average crack-tip plastic zone radius. Such values, in turn, could be used with available tensile rigidity values to evaluate the optimum fracture energies for a number of different curing conditions. In contrast to those conditions above indicated for obtaining optimum fracture toughness values, the optimum fracture energy values were obtained at high temperature with short-time curing parameters. However, the same slow overall cooling conditions need to follow. Once again, no carrier was permissable if optimum fracture energy values were to be achieved. Since a cohesive-type joint failure is usually considered to be most desirable for long service life, it should be expected that the conditions for curing the adhesive would strongly influence the fracture properties and ultimately the joint durability.

G. Experimental Versus Theoretical Fracture Energy Values

Considerable evaluation of the fracture energy values for thermoplastic polymers have been conducted since the early 1960s (2629–2633). All investigators have commented on finding measured fracture energy values well above their theoretical estimates. For example, Berry found the fracture energy of polymethymethacrylate (PMMA) to be 2×10^5 erg/cm^2 at 25°C compared to a calculated energy of 500 erg/cm^2. The assumption as made in these calculations that carbon-carbon bonds were being broken and polymer chains were oriented perpendicular to the direction of crack propagation with close packing achieved. It is generally now agreed that these high measured energies are due to inelastic deformation processes at the crack tip. Sometimes this deformation has been visibly evident on the fracture surface as evidenced by interference colors (see Refs. 2629,2630,2634). Berry (2630) and Wolock and Newman (2634) have also reported both "dimple" markings and craze formation on such fracture surfaces. Other indications of inelastic deformation are temperature dependence (see Ref. 2631) and rate sensitivity (2635). Similar discrepancies (though less) were reported for diglycidyl ether bisphenol A (DGEBA) epoxy (2505,2511).

Recent publications concerned with fracture toughness matters include Springarn's (2636) use of Chevron-notched 2024 aluminum alloy specimens for measuring adhesive fracture toughness. Making Mode I fracture toughness measurements on a nylon-modified epoxy, he found toughness value increased slightly with mouth opening rate C or crack velocity. It decreased with adhesive thickness and remained fairly constant with increasing temperatures up to T_g. In 1987, Herrera et al. (2637) advanced a simple method for fatigue precracking specimens for fracture mechanics testing. Most recently, in 1990, Hinckley and Mung (2638) made fracture toughness measurements on new polyimide film polymers, whereas Adams et al. (2639) reported impact fracture toughness testing of various polymers.

VII. FLEXIBLE JOINT STUDIES

The most common test method for studying the energy-balance approach to flexible joint failure has involved the peeling of an elastomeric adhesive away from a rigid adherend support. Such a support for the purposes of this chapter could well be an aluminum sheet or plate of selected thickness depending on the rigidity required. Where the adhesion failure of the adhesive to the aluminum adherend surface is to be studied, the aluminum needs to be provided with the appropriate surface pretreatment. The surface preparation on the aluminum and the adhesive application and curing conditions (if any) must be identical to those to be used in the manufacturing process if the maximum adhesion potential is to be accurately evaluated. In making a laminate for optimum peel performance, it would also be necessary to use a bonding procedure that can produce maximum surface wetting by the elastomeric adhesive.

The earlier investigations of Gent and Petrich (2577) and Gent and Kinloch (2640) evaluated the resistance to separation by peeling and the breaking stress (in tension, pure shear, or simple extension) for a simple viscoelastic adhesive spread on a rigid adherend. Both the peel resistance and the breaking stresses proved to be variable with the rate of adhesive deformation and the temperature in agreement with the WLF rate-temperature equation of Williams et al. (2453). This led to the conclusion that the strengths seemed to be reflecting a viscoelastic property rather than a thermodynamic property like the heat of wetting of the adherend. However, there was a striking difference between the two different measures of adhesion; i.e., peel resistance and breaking stress. Peel resistance proved to be greater as the adhesive layer thickened, whereas tensile strength moved to lower values. In addition, the peel strength increased with a high extensibility (or ductile flow) in the elastomeric adhesive, while the tensile strength value was less influenced by such a change. This would indicate that a local critical strain energy density is required to cause bond failure (2577). This criterion can be expressed as release of energy due to fracture plane growth being sufficient to supply the energy requirements for new fracture surface. Relatively, this quantity of energy turns out to be some orders of magnitude higher than the thermodynamic surface energy due to the additional energy consumed in irreversible processes around the fracture front (2562,2581,2641).

Greensmith et al. (2642) and Gent and Henry (2643) have demonstrated that the fracture energy for viscoelastic solids depends highly on the speed of fracture propagation and the temperature. Gent and Petrich (2577) have further shown that the failure energy is much smaller if the adhesive is prevented from deforming. The failure energy criterion mentioned above applies only to test methods which involve a similar degree of local deformation of the adhesive in the vicinity of the separation zone. Thus, it seems that the deformation energy is the dominant term in the observed failure energy for an elastomeric peeling joint. Further, the energy value measured is valid for any particular adhesive irrespective of mode of deformation and energy dissipation within the adhesive. Peeling experiments by Gent and Kinloch (2640) did show that as the rate of separation failure is decreased to extremely low values, the measured failure energy is not much larger than would be expected on equilibrium thermodynamic grounds. As the rate of failure is gradually increased from such low values; the measured failure energies do increase by some measures of magnitude. Presumably, this is a reflection of the fact that energy dissipation is accompanying the deformation process as it proceeds under nonequilibrium conditions. In summary, Gent and Kinloch (2644) and Andrews and Kinloch (315,2645,2646) have demonstrated that a major feature of the fracture mechanics approach is that the fracture energy for a given joint, when tested at a stated rate and temperature, is independent of the test geometry. While these investigators used a cross-linked styrene-butadiene rubber for adhering to the rigid adherend, Andrews and King (2647) were successful in extending the study to the adhesion of epoxy resin adhesives above their glass transition temperatures by using metallic adherends.

Another frequently used test configuration involving crack growth study can be used for both flexible and rigid joints. It has been referred to as the pressurized blister-test. Originally a suggestion of Dannenberg (2574), there have been further refinements by Malyshev and Solganik (2575), Williams and co-workers (340,2648–2658), and Anderson et al. (2488).

Recent investigations pertaining to failure of elastomeric adhesives have included the 1987 study of Maugis (2659). Tsukatani et al. (2660) reproted on the bonding and debonding processes in tack of pressure sensitive adhesives, and in 1989 Urahama (2661) followed with a study of the effect of peel load on stringiness phenomena and peel speed of pressure-sensitive adhesive tapes. Weissburg and Arcan (2662) have contributed a paper in 1990 on comparison of mechanical evaluation of structural adhesives using stiff versus flexible adherends.

VIII. RIGID JOINT STUDIES

As noted earlier, both the energy-balance approach (determination of the adhesive fracture energy) and the stress-intensity factor approach (determination of the fracture toughness of the adhesive) can be employed to determine failure values in rigid joints. Kinloch and Shaw (2535) and more recently Kinloch (2196) have presented reviews on this subject.

A. Mode of Cracking

Pioneering work in this area was conducted by Ripling et al. in the early 1960s (2501–2503). Continued expansion of these early studies has been ongoing by Mostovoy and Ripling together with a number of workers at the Materials Research Laboratory up until the present time (2504–2521). They have considered each of the three different modes of crack growth as previously described (Modes I, II, and III). These investigators chose to employ aluminum double-cantilever beam test specimens with special focus on Mode I and II forces. Later studies were mainly conducted with a tapered modification, as in Figure 17. The testing of this tapered specimen has, however, been strictly oriented to Mode I-type failure. For structure design pruposes, it would also be of interest to evaluate joint fracture energy under other fracture modes or combinations of modes. (Specimens for measuring Mode II fracture values can be reviewed in Refs. 2501,2502,2663–2665.) Mode III measurements were made by Brussat et al. (AFML Report TR-77-162 [1977]), whereas specimens representative of combinations of Mode I and Mode II have been studied by Bascom and Orashnik (2666).

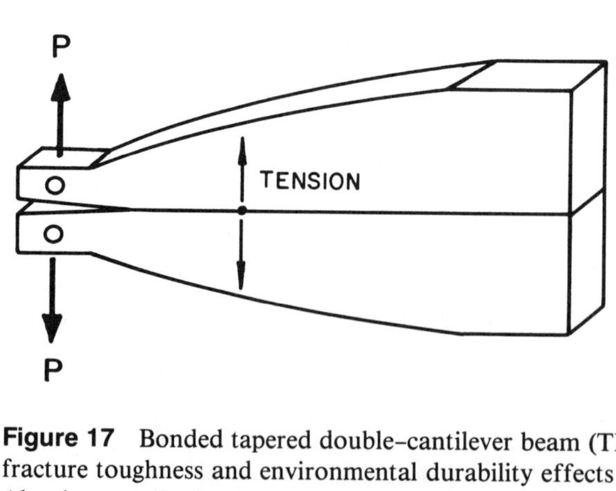

Figure 17 Bonded tapered double-cantilever beam (TDCB). Used for quantitative evaluations of fracture toughness and environmental durability effects on bonded systems. (From J. D. Minford, Aluminum Adhesive Bond Permanence, Treatise on Adhesion and Adhesives (R. L. Patrick, ed.). Marcel Dekker, New York, Vol. 5, Figure 3-46, p. 111 [1981].)

B. Other Factors than Mode of Cracking

The fracture energy values obtained from the failure testing of structural adhesive joints are dependent on a variety of other factors that are separate from the mode of the crack propagation. Since the principal adherend material used in the investigation of these other factors has been aluminum, these discussions have particular relevance to a book on the technology of aluminum joining.

1. ADHESIVE COMPOSITIONAL CHANGES

These changes have been studied by Mostovoy and Ripling (2505) and Gledhill and Kinloch (2548) and Gledhill et al. (2667). One of the most widely studied compositional changes has been the addition of a rubber which phase separates to toughen cross-linked epoxy systems. Reasons for such an addition have been investigated by Bitner et al. (1333), Kinloch and Young (2609), Hunston et al. (2480), and Hunston and Bascom (2668). Papers on the toughening mechanism have been produced by Bitner et al. (1333), Kinloch et al. (2669), Hunston et al. (2670,2671), and Bascom and Cottington (2545). Linear viscoelastic properties were measured by Hunston et al. (2671) and Rowe et al. (1549). Rokhlin (2672) and Rokhlin et al. (2673) have reported the application of interface waves for investigating the curing and viscoelastic properties of structural adhesives which polymerize over several hours after the addition of a solidification agent. The velocity and attenuation of the interface wave depend on the elastic and dissipative properties of the adhesive film and half spaces. In particular, the maximum on the transmission loss curve and the point of inflection on the velocity curve of the interface wave characterize the entire system (film and adherend). Sultan et al. (1551) showed fracture energy could be increased by more than an order of magnitude by elastomer modification of epoxy resins. Wang et al. (2547) conducted a stress analysis using an elastic finite element analysis. (For other publications about toughening effects see Refs. 1552,2544,2674–2677.)

Reinforcement of a brittle adhesive with an elastomer increases the fracture energy value as compared with the unmodified adhesive. The concept of toughening thermoset resin adhesives through dispersed rubbery particles, as a discreet second phase, has been described by McGarry and Willner (2676) and Sultan and McGarry (1552) at the time of their 1969 and 1973 investigations as well as by Bolger (2678,2679) in discussing the range of structural adhesives available to the manufacturer. Bascom et al. (2680) provided examples of effective fracture surface energy values, as shown in Table 106 on p. 245, using an epoxy that was reinforced with a nitrile rubber. More recently, Ting and Cottington (2549,2681) published a series of G_{Ic} values for the very high-performance adhesive types. These data demonstrate the definite positive elastomer reinforcement effect on modified epoxies as well as a similar effect when elastomer is added to high molecular weight polyimide and polysulfone thermoplastic adhesives. Bascom has postulated that the reinforcement effect may be due to the increase obtained in the plastic zone site at the crack tip due to the presence of the separated rubber particles. Generating a quantitative analysis of the plastic deformation zone involving time-temperature–dependent viscoelastic properties, however, would seem to be too complex for illustrative purposes. Accordingly, an elastic-plastic model has been assumed. This crack opening displacement model has been characterized and discussed by Bascom and co-workers (2544).

The kinds of fracture energy evaluations described above have provided a practical approach to developing modified epoxy adhesives which can be expected to perform better in aerospace applications. One such group has been referred to as ITCF (internally toughened controlled flow) adhesives. As new candidate adhesives were formulated, mechanical property studies were undertaken to optimize the products. Then fracture toughness, fatigue resistance, and stress corrosion test results can enable valid comparisons to be made regarding the service life potential of the product. Some mechanical properties data from B. F. Goodrich on their ITCF adhesive PL 717B is shown in Tables 108 and 109. Mostovoy, Ripling, and co-workers used such a battery of test procedures to demonstrate that a particular 250°F (121°C) modified epoxy adhesive was second in projected joint durability only to a 350°F (171°C) curing nitrile-phenolic aerospace adhesive.

Since G depends on so many variables, it cannot be used as a single parameter for predicting long-term stability of a joint. The insertion of the plastic zone concept, however, might be usable for some such predictions.

Table 108 Mechanical Properties of ITCF Adhesive (PL 717B)

Tensile strength (psi [MPa])	7800	(53.74)
Elongation (%)	4–5	
Notched Izod impact (ft-lb/in)	1.6	
Gardner impact (ft-lb)	1	
Flexural modulus (psi [MPa])	380,000	(2618)
Fracture toughness (lb/in)	12–15	
T-peel at RT (lb/3 in)	125	
Lap-shear (psi [MPa]) at RT	6150	(42.37)
(psi [MPa]) at 180°F (82°C)	4100	(28.25)

Data from B.F. Goodrich published information on PL 717B.

2. *TEMPERATURE CHANGE AND/OR STRESSING RATE CHANGE*

The particular involvement of the variables of testing temperature and stressing rate have been studied by many investigators (2513,2548,2682–2687). Pertinent reviews have been published by Roberts and Thomas (2686) and Andrews and Stevenson (2687).

3. *ENVIRONMENTAL EXPOSURE CONDITIONS*

It would seem obvious that environmental exposure conditions could have significant degrading effects on adhesive and adherend in the interfacial area, which can lead to alteration of the fracture energy response. There have been a number of investigations in this subject area (278,2535,2688–2707). The general procedure employed after pretreating the aluminum and fabricating the joints was to determine an initial fracture energy value then retest for a final value after environmental exposure.

Mostovoy, Ripling, and co-workers (2505,2509,2708–2710) spent more than a decade conducting adhesive bondline evaluations with tapered double-cantilever beam aluminum test specimens measuring the Mode I opening fracture values. When such test specimens were exposed to water or air at various temperatures and humidities and a static stress imposed, the data derived could be classified as being stress corrosion data in nature. Since the effect of water is so universally present in most service environments that employ bonded joints, and its effect so traumatic in nature, the detailed discussion of such environmental effects has been considered as a separate subject in Chapter 8.

4. *REPRESENTATIVE FRACTURE ENERGY VALUES*

A listing of some typical values for different kinds of structural aluminum joints is shown in Table 110. It is clear from these values that the fracture energy values for even structural adhesives can be quite low. However, there are methods for evaluating these values so that the effect of adhesive modifications can be directly determined. For example, the fracture energy value for an epoxy can

Table 109 Typical Thick Adherend Data for ITCF Adhesive (PL 717B)

	−67°F (−55°C)		RT		180°F (82°C)	
Yield strength (psi (MPa])	9382	(64.64)	5700	(39.27)	2871	(19.78)
Yield elongation	0.006		0.003		0.001	
Failure strength (psi [MPa])	9286	(63.98)	6122	(42.18)	4733	(32.61)
Failure elongation (mil)	0.008		0.014		0.016	
Work to failure (ft-lb/in)	4.86		6.35		4.42	

Data from B.F. Goodrich published information on PL 717B.

Table 110 Typical Values of Adhesive Fracture Energy, G, for Some Structural Adhesive Joints (All Tests Conducted at Room Temperature and at Moderate Test Rates)

Joint[a]	G_{Ic} (kJ/m^{-2})	G_{IIc} (kJ/m^{-2})
DGEBA-TEPA-aluminum alloy	0.07	1.45
DGEBA-TEPA-aluminum alloy (in water)	0.015	
DGEBA-tert. amine-aluminum alloy	0.10	
DGEBA-HHPA-aluminum alloy	0.10	
DGEBA-PIP-CTBN-aluminum alloy	3.40	3.55

Note (for comparison): G_{Ic} (polystyrene) = 0.4 kJ/m^{-2}; G_{Ic} (high-impact polystyrene) = 16 kJ/m^{-2}; G_{Ic} (7075 aluminum alloy) = 20 kJ/m^{-2}.

[a]Resin: DGEBA, diglycidyl ether of bisphenol A. Curing agents: TEPA, tetraethylene pentamine; tert. amine salt; HHPA, hexahydrophthalic anhydride; PIP, piperidine. Rubber toughening agent: CTBN, carboxyl-terminated-butadiene-acrylonitrile rubber.

Data from Ref. 2196 (Table II, p. 632).

be significantly elevated by modification with a toughening agent like a CTBN rubber. Even after such modification, however, the value must still be classified as being comparatively low as compared to other adhesives without thermoset polymer components. It is presumed that the high cross-link density and high interchain attractive forces that can exist in such (modified or unmodified) epoxies create an overall high modulus and low creep system which more readily cracks and fractures. However, even the low fracture energies of the unmodified epoxies can still be about 100 times greater than the energy actually required to break covalent bonding. It has been postulated that this extra energy which is required to propagate crack extensions in actual structural adhesives may be that required to overcome viscoelastic and plastic deformation processes taking place at the crack tip. Patrick and co-workers (2700) and Wilcox and Jemain (2698) both identified the presence of such processes at the crack tip by examining the fracture surfaces of failed aluminum joints using scanning electron microscopy (SEM).

IX. EFFECT OF JOINT GEOMETRY

Joint geometry might be expected to affect the fatigue and fracture failure values of bonded joints and many investigators have obtained test results confirming this. Because of the better control that can be exerted in forcing a failure in axially loaded butt-joints, this joint geometry has been widely employed by many investigators for their studies of joint geometry effects; however, more recent investigations have attempted to similarly evaluate lap-joints and scarf-joints.

A. Stress Analysis in Single-Lap-Joints

Beginning with the stress analysis of Goland and Reissner (2220), reference can next be made to other modifications and simplifications of this analysis by Hart-Smith (2228), Ojalvo and Edinoff (2230), and Renton's investigation where the loads are not eccentric (2711). Sancaktor (2712), in 1979, showed that due to the nature of the joint that bending still occurs. Although a number of investigators (2089, 2247,2712,2713) used numberical finite element analysis on single-lap-joints, the thinness of the bondline and singular stress fields occurring at the bond edges have created conputational difficulties. Accordingly, Delale and Erdogan (2714,2715) have stated that the numerical analysis involve such severe convergence problems that the accuracy of the results must be highly questionable.

It should be pointed out that the various analyses of the single-lap-joint requires that values be furnished for Young's modulus and the shear modulus of the adhesive layer for calculation of stress distribution. Young's modulus values are mostly obtained from tensile testing the bulk adhesive and the shear modulus value from a lap-shear–type specimen. Several investigators (2716–2718) have shown only small success relating the bulk tensile properties to the lap-shear properties, which makes it doubtful that bulk tensile properties really have any meaningful relation to the elastic properties needed to evaluate accurately the stress distributions. Also, it has been demonstrated that the use of shear properties measured from a lap geometry using joint elongation to calculate strain properties are usually in error. Renton (2713) and Sancaktar (2712) chose to measure the deformation over only a small portion of the joint using special small-gauge–length extensometers. Renton (2713), Nirajan (2719), Everett (2720), and Padgilwar and Sancaktar (2721) have all stated that the use of photoelasticity may be an excellent method to determine the distributions of stresses in the single–lap-joint. However, the method requires relatively thick adherends and adherents which may not always be representative of practical bonding applications. Yet another problem in making stress analyses of lap-joints stems from interface effects, since the overall joint is actually a combination of two adherends with an intervening adhesive. This has introduced the possibility of modeling of the interface as a separate material. It must be acknowledged that the quality of this interfacial bond has to be of prime consideration in considering other methods of studying the surface chemistry and interfacial effects (2722,2723). Mention should also be made of the good potential shown by Sancaktar and Brinson (2718) in studying interfaces using neutron radiography.

Possibly the most critical area for initiating the failure of most well-bonded joints lies in the changes taking place in the interfacial zone. In 1984, Sugama et al. (2724) investigated the nature of interfacial interaction mechanisms between polyacrylic acid macromolecules and oxide metal surfaces. In 1985, Knollman (2725) discussed the variation occurring in shear modulus through the interfacial zone of an adhesive and Knollman and Hartog (2726) published in the same year a paper titled "Experimental Determination of the Variation in Shear Modulus Through the Interfacial Zone of the Adhesive." In 1990, Clough et al. (2727) studied fiber and interface fracture in single-crystal aluminum silicon carbide fiber composites, whereas Napolitane and Moet (2728) considered the dissapative processes in interfacial failures.

It should not be unexpected that many flaws can exist in bondlines, since so many variables are involved in applying and curing adhesives. Consequently, the use of fracture mechanics to analyze joint failures has been a major focus of many investigators (340,2273,2480,2487,2488,2544,2680, 2729). Different approaches have been taken such as: (1) The fracture toughness is determined from the testing of tensile specimens with scarf-joints, tapered double–cantilever beams or other joints with known crack geometries. The inherent flaw size is estimated and critical loads are determined using the measured fracture toughness. (2) The assumption can be made that the single–lap-joint is an infinite body with a double-edge crack from which joint loads can be assessed directly. (3) The strain energy of a single-lap system using the Goland and Riessner analysis can be equated to the strain energy necessary for crack propagation. It should be noted in procedures (2) and (3) that all information is derived from a single-lap specimen without the need for actual fracture testing.

While it has been recognized that adhesives are viscoelastic in nature and can be considered fracture-dominated materials, most of the continuing experimental and analytical investigations have continued to be based on elasticity concepts. Attempts to relate the properties of lap-shear geometries to stress-rate and creep properties of bulk tensile adhesives clearly shows the shear deformation in lap-joints is significantly larger than would be predicted from these bulk tensile results. Delale and Erdogan (2714,2715) have performed a linear viscoelastic stress analysis of a single–lap-joint using assumptions similar to those of Goland and Reissner which show substantial variations of both peel and shearing stresses with time.

Actually, time-dependent failure analyses have not been extensively investigated, although the work of Brinson and co-workers (2717,2718) and Padgilwar and Sancaktar (2721) can be reviewed for the technique used to predict creep ruptures of both tensile and lap-shear specimens. Cartner and Brinson (2730) also have conducted analyses using the nonlinear viscoelastic models of Schapery in a similar manner. Time-dependent fracture mechanics analyses have also been performed by O'Connor and Brinson (2551) and Bitner et al. (1333). Liechti and Knauss (2731) have conducted one of the few studies involving slow crack growth resulting in long-term fracture. They observed the inherent nonlinear nature of time dependent crack propagation. The proposal has been that the viscoelasticity of the surface layer is considerably different from that of the same adhesive in either lap or bulk forms. Also, the viscoelastic nature of the adhesive can alter failure modes from adhesive to cohesive or vice versa as compared with the type failure obtained in short-term static testing.

The prediction of ultimate joint life under known loading conditions for very long times would be most productive but is seldom practical, so that an alternative procedure to predict joint life from short-term testing is always being investigated. Brinson et al. (2732) and Anderson et al. (2488) have both used higher temperature as a means of accelerating the deformation process with the assumption that the mechanisms might remain essentially the same at both low- and high-temperature conditions. This process has been called the time-temperature superposition principle (TTSP). The references cited above can be used for more reader information. This principle is definitely inadequate for handling nonlinear systems; however, Griffith et al. (2733) have described a companion time-stress superposition principle which might be applicable in this situation. The concept proposes that events which would take a long time at a low stress level can be accelerated by testing at a higher stress. Again the assumption must be made that the deformation mechanisms will remain the same.

B. Adhesive Bondline Thickness

Byrant and Dukes (2694) have shown the stress necessary to create fracture in axially loaded butt-joints with relatively thin adhesive bondlines is apparently independent of the adherend diameter used to fabricate the joint. At least, this was true over a wide range of diameter butt-joints that they evaluated. This same conclusion seemed to be obtainable whether the adhesive was rigid like an epoxy or rubbery like a silicone elastomer. Increasing the thickness of the adhesive, however, decreased the stress-to-fracture energy (2690,2692,2695,2705). (For more details about an explanation for elastomeric adhesives based on the fracture mechanics approach, the reader is referred to Refs. 2534,2701.) Related studies using structural epoxy/aluminum joints were carried out by Hilton and Gupta (2699) using a stress-intensity factor approach, who concluded that the fracture toughness may be considered to be independent of adhesive thickness, but they acknowledged that the adhesive joint strength should increase as the adhesive thickness decreases. Bascom and co-workers (2544,2545) also showed a strong dependence of fracture behavior on bondline thickness, whereas Kinloch and Shaw (2535) calculated optimum bond thickness at particular temperature and testing rates.

Tanaka and Taniyama (2734), in 1975, evaluated the effect of the adhesive layer thickness on the failure pattern of joints, whereas Nakayama (2735), in 1978, studied the effect of the adhesive layer on the static tensile and tensile-fatigue resistance of bonded joints. Minford (2736) has obtained a decrease in adhesive joint strength as the adhesive thickness increased with aluminum joints bonded with a one-part, heat-curing epoxy adhesive modified to especially bond in the presence of oil or lube on the surface. The effect mentioned seemed to be equally present whether the aluminum adherend surface was mill-finish or heavily-lubed mill-finish condition. Actually, low-strength joints that would be unacceptable for most automotive structural applications were produced unless relatively thin bondline thicknesses were employed. It was clear from inspection of the failed joints, however, that the decreasing failure load with increasing bondline thickness in this case was due to the increasing porosity in the bondline produced by the high temperature curing which had been designed to simulate the high heat input from passage through paint-bake ovens used to cure the paint coatings on automotive surfaces.

Later work by Adams and Coppendale (2737) using a considerable range of aluminum alloy adherends and epoxy adhesives suggested that the situation is probably more complex than summarized by the Hilton and Gupta investigation (2699). Adams concluded that the ductility of the adhesive played a primary role in determining the strength of an aluminum butt-joint. Adams and Peppiatt (2227) have graphed the theoretical values for failing stress versus adhesive thickness using the equations of Volkersen, Goland, and Reissner, and their own finite element analysis model. According to the plot utilizing the Goland and Reissner formulas, the failing load was predicted to rise significantly over the thickness layer range up to 0.6 mm. With the Volkersen equations, the increase in failure load up to about 0.25 mm coincided with the Goland and Reissner curve but thereafter started to level toward a plateau value. The finite-element produced curve proceeded in a very gradual rise through the maximum bondline thickness value calculated of 0.45 mm. Meanwhile, the actual experimental curve rose slightly to 0.15-mm thickness, then declined slightly to the 0.3-mm thickness level. The decline in failure load beyond this point was almost insignificant. Other workers (2738–2741) also showed the actual breaking load does not increase with increasing bondline thickness and may even slightly fall.

Kline (2742) has shown with his model published in 1984 that there can be a significant difference in edge peel stress operating through a thick adhesive layer. However, the edge shear strength between two such separated bondlines can be barely recognizable. Most recently, in 1990, Liechti and Liang (2743) have further studied the effect of the adhesive layer thickness on interfacial crack initiation. While the effect of adhesive layer thickness on the cohesive toughness of the adhesive had been well studied under Mode I fracture conditions by Bascom and Cottington (2544) and Kinloch and Shaw (2535), the extent to which the adhesive layer thickness affects the interfacial crack growth has not been adequately investigated according to Liechti and Liang (2743). These authors have recently reviewed their results to determine this relationship using bimaterial and trimaterial strip blister specimens. The former provides a varying failure mode mix, whereas the latter gives essentially pure Mode I conditions for all crack lengths. The data furnished by these authors need to be reviewed directly from their publication for best comprehension.

1. STRESS ANALYSIS AND ADHESIVE BONDLINE THICKNESS

For many years the classic analysis of Goland and Reissner (2220) has been used to analyze the stresses developed in overlap-joints. The basic assumption was that the adherends were assummed to deform as thin plates coupled by an elastic adhesive layer. Particular attention was paid to a resulting shear stress from joint loading and an appreciable transverse tensile stress or peel stress which also developed. Also, for single-lap-joints, the eccentricity of load application produced a significant bending moment at the joint edges. Later work by Chang (2744) extended the research to include the effect of plasticity in the adherends. Adams and Coppendale (2737) and Hart-Smith (2228) investigations included the effect of plasticity in the adhesive layer. Weitsman (2745) and Delale and Erdogan (2714) considered the effect of viscoelastic adhesives, and Renton and Vinson (2324) the effect of anisotropic adherends. A major drawback to the above analyses was the assumption that the adhesive layer was very thin in comparison with the adherend layer. There could be, therefore, some situations where adhesive thickness effect could be important. Ojalvo and Eidinoff (2230) dealt with this situation by including a linearly varying stress distribution through the adhesive layer. Even here a deficiency existed because the peel stress was modeled as being invariant in the adhesive layer. This would leave the equilibrium relations for the adhesive layer unsatisfied. A further modification by Carpenter (2746) still failed to satisfactorily resolve the problem. Kline (2742) has also shown with his model that increasing the adhesive thickness tends to reduce peak stresses providing that the adhesive thickness is much lower than the adherend thickness (see Figs. 11 and 12, pp. 607 and 608, respectively, in Ref. 2742). For adhesive thicknesses comparable to or greater than the adherend thickness, increasing the adhesive thickness increased the peak stresses as a result of the increase in the bending moment.

While addressing the question of stress effects due to variation in bondline thickness, it may be interesting to note experiments of Bryant and Dukes (2747,2748). These investigators inserted copper wires in the center of the bondline of joints to precisely control the bondline thickness and found no negative effect on the joint strength as compared to equal-thickness bondlines without the wires. It might have been expected that such an inclusion would have introduced some altered stress effects changing the measured joint strengths.

Also of possible interest to the reader could be some 1990 investigations involving residual stresses and fracture aspects of the adhesive in the bondline. Bandyopadhyay (2749) offered a review of the microscopic and macroscopic aspects of fracture in unmodified and modified epoxy resins. Low (2750) studied the effects of residual stresses on the failure micromechanics in toughened epoxy systems, whereas Srivastava and Shembekar (2751) investigated the tensile and fracture properties of epoxy resin filled with fly ash particles. Wu et al. (2752) discussed the structural heterogeneity that can exist in epoxies, whereas Wu (2753) reviewed the chain structure, phase morphology, and toughness relationships in polymers and blends. Finally, Lee (2754) has discussed the thermomechanical properties of polymeric materials and the related stresses that can exist.

C. Overlap Length Considerations

The general relationship for aluminum lap-joints between the magnitude of the failure load and failure stress with increasing overlap has been investigated by DeBruyne (2370,2740). The adhesive used was one being employed in aircraft bonding during World War II. It can be shown that the breaking load increases up to an approximately constant value, whereas the average applied stress at failure decreases as the overlap length of the lap-joint increases. The overlap dimension chosen for the standard ASTM lap-joint was accordingly chosen at a point where the relationship between overlap length and breaking load was still linear in nature. Kline (2742) has plotted the variations in edge peel and shear stress as the overlap dimension changes using this stress analysis model (see Fig. 10, p. 606, in Ref. 2742). The edge stresses were significantly changed by altering the overlap distance by a ratio of 1:2:4 times.

D. Adhesive Modulus Considerations

The joint may employ relatively lower or higher modulus adhesives and this factor in itself may produce different levels of edge peel and shear strength results. The differences are especially significant in these stress levels as the distances change from the edge. Kline (see Fig. 8, p. 604, in Ref. 2742) has plotted situations where a low- and high-modulus adhesive has been employed in the bondline. A higher edge shear stress, but lower interior joint area stress level, was found to be present with a higher-modulus adhesive. The difference between the peel stress patterns for high- versus low-modulus adhesives in the bondline was much less apparent. A clearer picture of the high significance that adhesive modulus can produce was shown by Kline's plot of results using his model for adhesives differing in modulus by an order of magnitude. In this case, there was a particularly rapid rise in both peel and shear edge stress in moving from the middle when very high modulus adhesive was in the bondline.

E. Adherend Thickness Effects

Changing the adherend thickness in a lap-joint has a relatively minor effect on the peak peel or shear edge stresses, as shown by the plot of Kline (see Fig. 9, p. 605, in Ref. 2742). It should be expected that the thickening of the adherend should induce an overall added stiffness to the joint and, thus, reduce stress concentrations. This appears to happen only up to some limiting value. Among the factors considered to this point, we could summarize that in descending order of effectiveness for influencing edge peel and shear stresses are adhesive modulus, joint overlap, and thickness of adherend. We might also mention that the factor of adherend thickness should be considered in conjunction with adhesive thickness. Kline (2742) has offered his calculated edge peel and shear stress changes where adhesive bondline thicknesses of 0.24, 1.27, and 2.54 mm were incorporated with either 2.54- or 6.35-mm adherends. He concluded that increasing adhesive thickness tended to reduce peak stresses only where the adhesive thickness was much less than the adherend thickness values.

X. PEEL JOINT CONSIDERATIONS

The static stress distribution that results when a flexible member is peeled away from a rigid adherend with an attached adhesive layer has been investigated by a number of investigators (435,2339, 2341–2343,2755–2760). Such analyses have considered the adhesive and adherends as elastic materials and deduced the tensile and shear stresses in the adhesive layer. Assumption is also made that the tensile or cleavage stress is constant throughout the adhesive thickness and across the width of the joint. Kaelble (2342) and Kaelble and Ho (2755) confirmed that the distribution of cleavage stress is a highly damped harmonic function involving alternating regions of tension and compression. More recently, Crocombe and Adams (2344,2345) re-examined the problem by employing an elastic (large displacement) finite-element analysis approach. They stated that the initial joint failure was caused by the principal tensile stresses in the adhesive driving a crack toward the interface between the adhesive and the flexible adherend. Continuing crack propagation will then occur at a critical applied bending moment for each different combination of adhesive and adherend which will be independent of the peel angle. Finally, it was shown that the load measured by the peel test was not proportional to the actual strength of the adhesive, since a small increase in adhesive strength could produce a much larger proportionate increase in the applied peel load.

Aluminum adherends have not been prominately featured in these classic studies of peel joint failures; however, the use of aluminum foils as the flexible adherend and sheet, plate, extrusion, or forgings as the rigid adherends in structures experiencing peeling failure stresses can certainly be found in the real world. The effect of adhesive layer thickness, flexible adherend thickness, and the angle of peel have been reviewed by Gent and Hamed (2761). Separate theoretical equations have been evolved using the stress analysis or energy-balance approaches for predicting the failing peel force for particular joints. Both approaches predict that the failing in peel force should be independent of the thickness of the adhesive layer. It has been the common observation that the peeling failure force increases up to a certain point as the adhesive layer thickness increases (2317,2757,2762–2765). The data of Gent and Hamid (2761) showing the plateau point at about 2 mm of bondline thickness can be seen in their paper or in Table 15 of Ref. 2196. The result is mostly attributed to additional energy dissipation within the bulk of the adhesive acting up to the point where the increasing thickness prohibits the dissipation process from involving the entire layer of adhesive.

The effect of the thickness of the flexible adherend seems most pronounced when the level of adhesion is high and the peeling strip is relatively thin. Interacting strongly with these conditions will be the relative rigidity of the adherend and whether any significant plastic deformation of adherend surface can occur. While the former factor can be greatly varying depending on the aluminum alloy of choice, the latter condition would not be of great significance with flexible aluminum adherend peeling situations. With a strong, rigid aluminum adherend backing, the bending stresses with a thick peel strip will not be great enough to cause yielding, so the peel forces will decrease back to the value that would relate to zero thickness. Thus, the peel forces at both zero and large thickness of peeling adherend should be equal to the value of the peel failure load in the absence of plastic yielding of the flexible adherend.

Using energy considerations and stress analysis, a conclusion has been drawn that the peel force should be inversely proportional to (1-cos α) where α is the angle of peel. Discrepancies have appeared between investigators regarding this matter (258,2339,2341,2342,2578,2755,2759,2766–2769). For example, at angles below 30, the measured values have tended in general to be less than those predicted by Kaelble (2341) but greater than those predicted by Lake and Stevenson (2769). At high angles of peel (above about 150), the peel force values have usually been greater than predicted. Gent and Hamid (2761) and Lake and Stevenson (2769) have attributed these discrepancies to energy-dissipating processes which take place to different extents.

XI. COMPARISON OF DIFFERENT JOINT DESIGNS

A. Axially Loaded Butt-Joints

Bryant and Dukes (2694) have demonstrated that the fracture stress of axially loaded butt-joints is independent of the diameter of the adherend, as mentioned earlier. This applies for thin bondline joints employing either a rigid epoxy or rubbery adhesive. A large number of investigators (2688–2692,2694,2695) have all agreed that the fracture stress will increase as the adhesive layer becomes thinner. Wake (2693) originally attributed this to higher internal stresses, whereas Meissner and Baldauf (2688), Bikerman (2692), and Dukes and Bryant (2695) reasoned it was the statistical probability of larger flaws in relatively thick adhesive layers. Gent (2534,2701) and Hilton and Gupta (2699) have presented their more recent evidence that a fracture mechanics approach can provide essentially a quantitative mechanism for this effect. Gent (2701) investigated butt-joints with elastomeric adhesives bonded to rigid adherends, whereas Hilton and Gupta (2699) concentrated on structural butt-joints with epoxy adhesive and aluminum adherends. The work of Adams and Coppendale (2737) have suggested that the solution may be more complex and that the relationship between the strength of a butt-joint (aluminum adherends) and that of a bulk specimen of the adhesive is highly dependent upon the ductility of the adhesive used.

B. Lap-Shear Joints

A significant discussion of the stress analysis of single–lap-joints has already been offered in Section IX.A and the reader is referred to that section for details. It might be appropriate, however, to cite the work of DeBruyne (2739) and Mylonas and Debruyne (2771) in establishing and developing data for the general concept that the breaking load increases up to an approximately constant value, whereas the average applied stress at failure decreases as the overlap length is increasing (see Fig. 13, p. 635, in Ref. 2196).

C. Peel-Type Joints

A significant discussion of the stress analysis of peel-type joints has been included earlier in Section X and will not be repeated here.

D. Other Types of Joints

Gent (2701) has employed a fracture mechanics analysis approach to predict joint strength using different joint designs, relatively rigid adherends, and a typical rubbery adhesive. For more details about the various assumptions and value parameters, the reader needs to consult Gent's paper. Joint strengths calculated by Gent using various joint geometries are presented in his paper (see also Table IV, p. 638, in Ref. 2196). It seemed clear that the strength values prove to be highly dependent on the Young modulus of the adhesive, an intrinsic flaw size selected, and local stress concentration effects. It can also be observed in Gent's results that the main tensile breaking stress for butt-joints is dependent on a radius/adhesive thickness factor which has been called the shape factor.

While a relatively large amount of research has been done on adhesive joints connecting adherends having similar widths and thicknesses, relatively little attention has been directed to joints connecting elements of different geometries such as thin membranes to relatively thicker and narrower strips of laps. Such joints actually have a wide range of practical application, because they provide one of the most suitable means of loading thin-walled structures. The load can be transferred from the lap onto the membrane with the aid of a flexible adhesive layer capable of producing lower concentrations of stresses in the membrane than achieved with rivets or welds. Czarnocki and Piekarski (2772, 2773) used a finite-element investigation to study this matter and have demonstrated that the stress state in the membrane, as well as in the adhesive layer, depends on the mode of loading and the shape of the lap. Cold-rolled 6061-T6 alloy was used for the laps and the membrane bonded with Hysol's EA-9320 aerospace-type epoxy. The three-dimensional finite-element analysis employed has been described in a 1986 publication of the same authors (2268). They found that these joints with different geometry situations showed different failure patterns than those investigated by Gali and Ishai (2251), where lap-joints connecting adherends of the same width had been employed. A more favorable stress state in the membrane was obtained by appropriately shaping the laps. Wider and shorter laps produced a less severe stress state in the membrane than narrower and longer laps.

Wang (2774,2775) produced comprehensive reports on the effect of stress distribution on the fatigue behavior of a variety of aluminum joints, including lap-joints, scarf-joints, and butt-joints. Both a vinyl-phenolic and a nitrile-phenolic aerospace adhesive were compared. The influence of the stress distribution variation caused by the interaction of different adhesive mechanical properties, joint geometry, and type of loading on fatigue strength has been most recently reviewed by Sancaktar (2776). Chen et al. (2777), in 1987, reviewed early information on fatigue data of bonded joints ranging back to 1957. Adams and Wake (2778) reviewed knowledge available in the aircraft industry as early as the 1960s regarding the fatigue resistance superiority of aluminum-bonded joints as compared to riveted aluminum joints. For example, these authors mention a 63% higher fatigue strength-to-ultimate strength ratio by Argyris in 1962. Sancaktar and Lawry (2779), in 1980, demonstrated as much as a 71% increase in load-carrying capacity for bent adherend lap-joints, as did Johnson (2780) in 1981. In 1983, Johnson and Mall (2781) modified the end taper of a lap-joint from 90 to 5 degrees to increase the no-growth threshold nondamaging stress range by about 50%. In 1986, Adams et al. (2782) tripled the monotonic strength of double-lap bonded joints by tapering adherends and using an adhesive fillet at the overlap edge. Jablonski (2783), in 1980, described crack growth rates in aluminum

(TDCB) specimens using AF-163 and EA-9649 rubber-filled solid film adhesives and varying film thicknesses showing cracks propagated at lower onset threshold values for thicker bondline specimens.

Also of interest as another type of joint is the 1986 study of Miller et al. (2386) which attempts to understand the effects of the adhesive ductility and bondline geometry using aluminum tube-and-socket–type jointures.

In 1990, Matsui (2784,2785) contributed two papers evaluating the effect of size on shear and tensile stresses on adhesive-bonded rectangular or tubular joints. One paper measured ultimate shear stresses under tension loading and the other paper the norminal ultimate tensile stresses under bending or peeling load. Groth (2786) evaluated the viscoelastic and viscoplastic stress analysis of various adhesive joints. Hashim et al. (2787) described design and assessment methodologies for adhesively bonded structural connections, whereas Crocombe et al. (2788) reported a general article on the analysis of structural joints for failure. Fay and Suthurst (2789) described their efforts on the redesign of adhesively bonded box beam sections for improved impact resistance.

XII. TESTING TEMPERATURES AND RATE CONSIDERATIONS

A. Testing with Respect to the Glass Transition Temperature

Two general areas can be considered where the testing temperature is either above or below the glass transition temperature of the adhesive.A number of investigators have been engaged in measuring the joint strengths using elastomeric adhesives over a wide range of temperature and rate conditions (205,315,2577,2644,2646,2701,2790–2792). The data which have been developed can be plotted as a single master curve when normalized to a reference temperature by means of the Williams-Landel-Ferry rate/temperature equivalence for viscoelastic materials (2453). Various master curves for the adhesive fracture energies and tensile breaking stresses of rubbery adhesives can be reviewed in Kinloch's excellent review publication (2196).

The curves for relating the peel strength of un–cross-linked elastomer adhesives are more complex (2577,2793). Gent and Petrich have offered an approximate relationship between peel strength and tensile-stress strain behavior of the bulk adhesives using an empirically determined parameter they have chosen to call the interfacial bond strength.

Typical strength relationship to testing temperature values for some tests conducted mostly below the glass transition temperature have been cited by Cotter (2794). The epoxy-phenolic, polyimide, and polybenzimidazole adhesives exhibited the best high-temperature strengths of those reported. Long-term use of epoxy-phenolics is apparently limited to temperatures of 347°F (175°C) and below, but polyimide and polybenzimidazole polymers can have significant service lives as high as 797°F (424°C). Polybenzimidazole formualtions, however, seem more susceptible to some oxidative degradation above 500°F (260°C).

Yeager (2795) and Sandler and Berg (2796) have considered and reported on the relationship between performance at low temperatures and the structure and formulation of adhesives that can offer realistic service at these temperatures.

B. Testing Rate Considerations

Adhesive joints can creep and suffer stress rupture under either static or dynamic loadings. For the former, the reader can refer to investigations over the decade from 1969 to 1979 (2797–2805).

Investigations involving applied dynamic loadings for the period 1972 to 1981 can be reviewed in various publications (2806–2812). Gledhill et al. (2684) have shown epoxy adhesives (when tested in TDCB configuration specimens) do not appear to suffer at all from relatively high static loads. Kinloch (2813) and Kinloch and Williams (2814) have proposed the explanation that severe crack-tip blunting can occur resulting in a reduction of the stress concentration at the crack tip below that necessary for further crack extension. In some instances where the capacity for crack-tip blunting in the adhesive is limited during a static fatigue testing procedure, delayed failure will eventually be observed. Gledhill and Kinloch (2804) studied one epoxy adhesive composition which showed a linear relationship between the applied fracture energy and the logarithmic failure time with the failure time decreasing as the applied fracture energy value was increasing. The time to failure represented an incubation period in which the crack was not observed to propagate until the instant of fracture when the propagation became catastrophic. These investigators reasoned that the plastic-zone size at time of fracture and crack-opening displacement at fracture could be related to both time-dependent fracture energy and time-dependent modulus value. Further, they conducted static fatigue experiments in which the plot of one of these functions against the other was linear over the whole scale. Until critical values of these two factors are attained, no crack extension will occur.

Fracture mechanics studies have been carried out on rubber-toughened epoxy adhesives by Bascom and Mostovoy (2811) and Mostovoy and Ripling (2815). They found the Paris equation (2816,2817) could be used to describe the fatigue crack growth that they observed. In brief, the crack-growth per cycle equaled a constant (A) times the applied adhesive fracture energy raised to the power of another constant (n).

XIII. OTHER BONDLINE CHARACTERISTICS AFFECTING JOINT FAILURE

In addition to the many characteristics of joint failure considered above, there are also some additional situations that can exist in the area of the bondline which can significantly contribute to the mechanics of joint failure. Minford (2818) considered some of these possibilities in his discussion of factors which affect aluminum bond permanence. A brief review of some of these factors is offered here.

A. Modulus of Elasticity of the Bondline Adhesive

We have already discussed the situations where the modulus of the adhesive or the modulus of the adherends separately were varied to show their effect on both the peel or shearing stresses in a joint based on the modeling of Kline (2742). We have not mentioned their interaction, however, especially where the coefficients of thermal expansion may be quite different. The worst situation can be expected to exist where two adherends of widely varying coefficients of thermal expansion have been assembled into the same structural joint. If a highly cross-linked, high-modulus adhesive is used in the bondline to obtain a strong structural joint, then the stress forces acting on the interfacial area when the structure is being exposed to fast thermal cycling will be quite difficult to resist. Some manner of introducing some stress relief must be sought if very long-time service potential is to be realized. Use of a lower modulus adhesive in the bondline is one commonly employed answer. For example, an aluminum joint fabricated with a rigid epoxy structural adhesive and subjected to cyclic fatigue stressing will demonstrate a limited service life until failure. When the rigid epoxy is replaced with a rubber-modified epoxy, a significant boost in service life cycles would be expected. A further enhancement in service life may be possible if an even lower modulus heat-cured nitrile-phenolic adhesive is substituted in the bondline. Obviously, there may be other tradeoffs such as achieving adequate bondline strength or heat resistance which could limit the level of stress relief potential of a substitute adhesive.

B. Interfacial Imperfections

DeBruyne (2819) and Bascom (2820) both furnished examples in the literature of situations where trapped air bubbles in the adhesive bondline furnished sites for high localized stress situations resulting in premature crack initiation and propagation to joint failure. Such stresses can be additive with the desorption forces inherently present in an exterior weathering environment leading to a significant reduction in overall joint permanence.

It has been mentioned earlier in the work of Jozavi and Sancaktar (2628) that the curing temperature and cooling rate after cure of a polymeric adhesive could be influential in changing the bulk properties of a heat-cured adhesive. It should not be unexpected then that some degree of interfacial imperfection can arise as a result of employing different bondline curing procedures and temperatures.

Unlike the situation studied by Jozavi and Sancaktar, where only the bulk adhesive was investigated, the opportunity now exists for the adhesive to interact with the adjacent adherend surface, making the possibility for altering joint performance and permanence much more likely. Furthermore, the additional opportunity for controlled and uniform heat curing of the adhesive throughout the bondline area, as compared to curing the bulk adhesive outside a joint, becomes much more unlikely.

Another matter of concern is development of localized stresses in the adhesive due to shrinkage when cured in the presence of the adherend. Again the stress development potential is further magnified by adherends of different coefficients of thermal expansion in dissimilar material–type jointures. Plueddemann (1109) listed this factor of development of localized stresses in the bondline as among the principal factors responsible for general failure of joints. It should also be pointed out that most adhesives cure by one of the several mechanisms of solvent evaporation, cooling (hot melts), or chemical polymerization with or without heating. The degree of wetting achieved will thus be variable for reasons discussed earlier under that subject in Chapter 2. To the degree that this overall wetting of the adherend is uniformly achieved during the curing process, the joint failure potential will be correspondingly affected.

A still different situation can develop if the bond interface is altered owing to the swelling of the adhesive. This is particularly a critical factor for thermoplastic-class adhesives in a service environment where organic solvents can be present. A volume change of sufficient magnitude can be generated to create a highly significant stress in the interfacial area. Of course, such a chemical invasion of the bondline adhesive could also create actual chemical degradation of the bulk adhesive as well. Metal adherends like aluminum create the worst possible scenario, since they neither absorb solvent or undergo any similar compensatory change in dimensions. The most common solvent in most weathering situations remains water, and many thermoplastic-based adhesives will actually swell as water is absorbed from the service environment. The loss of adhesion due to physical swelling of the adhesive can also be accompanied by a water-produced desorption of the adsorbed adhesive at the adhesive/metal interface. The failure mechanisms attendant to the presence of water in the service environment will be the subject of Chapter 8.

C. Adherend Surface Corrosion Processes

Mahoon (2821) has pointed out that while bond failures of aluminum joints under moist corrosive service conditions are often clearly due to corrosion of the aluminum surface in the interfacial area, few investigations have been conducted on the mechanisms and modes of failure. Rather most of the corrosion studies have been empirical in nature, assessing either a particular surface treatment or relative performance of different adhesive or adhesive/primer systems (2822–2825).

Stevenson (2826,2827) pointed out in 1985 that the role of electrochemical potentials in causing polymer-metal joint failures had only recently been subjected to detailed scientific study even though the role of electrochemical effects on metal corrosion itself had been well established much earlier. The debate about whether corrosion or bond failure comes first under corrosive weathering conditions has repeatedly appeared in the literature. However, there should be little question about what is happening when a metal adherend joint is immersed in an electrolyte and the adherend is electrically connected in some deliberate fashion to become a cathode. The electrochemical corrosion of the metal will be suppressed, yet the rate of bond failure may be enhanced. Clearly corrosion must be considered to be an after effect rather than the primary cause of bond failure in this situation. However, an intimate relationship still exists between the mechanisms of corrosion and cathodic bond failure. Leidheiser et al. (2828,2829) have pointed out the analogous problems of the bond integrity of organic coatings on metals to adhesive joint bond maintenance. Although the details of the mechanisms may differ, the role of electrochemical potentials are certainly similar. It is necessary, in order to conduct a meaningful investigation of the role of cathodic disbondment, to use adhesive joints that have excellent inherent durabilities in water exposure. Where the durability is sufficiently poor that the bondline would fail by some other mechanism, the role of electrochemical effects would be unclear when failure occurs. Sedillot and Stevenson (2830) have worked with structural joints on oil platforms and subsea pipe coatings on steel that need to function for 20–30 yr and thus have had to consider this problem in detail.

Stevenson (2827) has offered some detailed descriptions of experimentation over recent years to understand and limit the sensitivity of steel-bonded structures in salt water situations where a high cathodic potential was present. The physical effects of hydrogen evolution have been considered the primary cause of bond failure at such high potential levels. Bulloch et al. (2831), in 1988, discussed attempts to enhance the chemistry of adhesive systems to offer increased resistance to cathodic disbondment. None have been successful as yet in completely resisting cathodic disbond. Adhesive joint design can also be important. McIntosh and Stevenson (2832) reported in 1990 on good performance in data of composite flexjoints used as a critical structural bearing to provide flexibility in deep-water compliant oil platforms. (Other publications include Refs. 2833–2836.)

Pourbaix (2837) listed the anodic and cathodic reactions that can occur with the ordinarily stable aluminum oxide in or near neutral pH solution causing it to dissolve when the pH is lowered below 4 or elevated above 8 (the famous U-shaped curve depicting pH dissolution of aluminum oxide in various environments). Mattison (2838) has discussed the pitting of aluminum in outdoor atmospheric weathering conditions and especially the severe pitting and crevice corrosion that can attend in the marine atmospheric conditions. Richardson and Wood (2839) added discussion of how the presence of free oxygen or oxidizing ions (like the chromates) not only increase rate of oxide formation but also reduce the rate of aluminum dissolution. Bijlmer (770), Cotter and Hockney (2840), and Bethune (1000) have all contributed studies on the failure of aluminum joints due to corrosion; however, a clear understanding of the mechanism was not produced. Failure mechanisms have included discussions about microcracks or flaws causing failure close to the interface in weak boundary layers, stress hydrolysis of primary covalent bonds due to humidity, or the possibility of water desorbing the adhesive owing to its greater affinity for the aluminum surface through the formation of hydrogen bonds (2670).

On the aluminum side of the bondline, Smith (2841) proposed a structural transformation in oxide as the main contributor to failure of aluminum joints. Bascom et al. (2842) stated the energy necessary for failure of bonds between aluminum and unmodified epoxy can vary considerably from 2 J/m to 20 J/m. This fracture energy is typical of metal oxides and silicate glasses according to Coble and Parikh (2843), which implies that the fracture can be due to the variable break up of aluminum oxide or corrosion products produced on its surface. Since normal fracture energies encountered are within 500–1000 J/m^2, however, there would appear to be considerable inelastic energy dissipation occurring in the adhesive, and the failure could actually be located near the interface.

The claims of Riel (2844,2845), Greer (2846), and Rogers (2847,2848) that Alclad aluminum alloy joints experience faster corrosive failures under aggressive weathering conditions than nonclad alloy joints has been attributed to the sacrificial electrochemical properties of the clad alloy as compared to the core alloy properties. The fact that a very extensive and comprehensive investigation of this matter conducted by McFarlen (2849) failed to confirm these claims indicates there can be other interacting factors that can be important enough to obviate the earlier observations. The author has speculated for some time about how different sets of competent aerospace investigators could arrive at such different conclusions. The best explanation may be that there are significant differences between the wettability potentials on aluminum oxide afforded by newer adhesive formulations. Thus, when they are cured against either Alclad or bare aluminum surfaces (for example, Alclad 2024-T3 compared with bare 2024-T3), the degree of surface wetting is such that approximately the same degree of resistance to undercutting corrosion of the bondline in a seacoast atmosphere or laboratory salt-spray cabinet exposure is achieved. This explanation fits in with the author's observation over many years that there was an apparent good correlation between the degree of wetting achieved by heat-cured epoxy adhesives on aluminum adherends and the survival time of such joints in salt-water exposure conditions.

When aluminum alloy surfaces are exposed to water soaking or hot humidity soaking conditions, initially bright reflective surfaces will begin to develop increasing amounts (though quite variable) of darker hued appearance. Depending on the initial inherent resistance of the alloy to general corrosion, this water-staining condition may be rapid or relatively slow. Usually, the higher the overall purity of the alloy, the slower the visual change takes place. As other alloying elements are added, their presence on the surface produces a rapid development of dark water stain over the surface. For the higher-strength aerospace alloys (which are most highly alloyed), this kind of observation is often missed, since the surface can be quite dark initially with a heavy heat-treat oxide in place. It should be remembered that the Alclad aerospace alloys are somewhat of an exception, since the cladding alloy is usually a relatively high-purity alloy as compared to the core alloys; i.e., 2024-T3 and 7075-T6.

In a general way, the author has observed that this could be a factor in explaining why the lower-strength (lesser alloyed) aluminum surfaces may tend to generate more environmentally stable joints, especially with regard to joint durability in water-soak weathering using only mill-finish, solvent cleaned, or mechanically abraded surface conditions. Where more extensive surface pretreating can be afforded, then other factors are affecting the wetting which can be of greater importance than the inherent properties of the native oxide surface. While the visual appearance of water stain on aluminum indicates some dramatic surface degradation has occurred, it may be surprising to note how little actual loss of aluminum has occurred in terms of experimentally measured weight loss. Also, the surface while microscopically roughened, shows only modest evidence of gross surface roughness as found after deliberate mechanical abrading. The surface produced by corrosive pitting in salt water will generally appear much rougher. Under these latter weathering conditions there also will usually be much more evidence of corrosive metal weight loss before the joint actually delaminates. Nevertheless, it has been proposed by Walton and Cowling (2850), Gledhill and Kinloch (2851), and Kinloch (2852) that the first mechanistic step leading to joint failure under these salt-water corrosive conditions is the displacement of the adhesive from the aluminum oxide by moisture. These authors contend that the extensive corrosion that is observed after bond failure is a following step in the bond failure process.

The author has conducted many hundreds of tests involving thousands of single–lap-joints in both water soaking at room temperature, soaking in 100% condensing humidity at 125°F (52°C), cycling in water or condensing humidity followed by freezing and thawing temperatures, intermittent corrosive salt-water or salt-spray cycling, and very long time exposures to the natural industrial or seacoast atmospheres. Added in the years since the middle 1960s has been the combination of simultaneous exposures to hot condensing humidity and varying levels of stress on the bondline. A number of observations made during this time has made it extremely difficult to totally accept the simplicity of the mechanism calling for total water debonding before gross corrosion occurs on the aluminum adherend. For example, consider the observations made when vapor-degreased Alclad 2024-T3 joints were bonded with a room temperature–curing, two-part epoxy and the joints were soaked continuously in either distilled water or 3.5% salt water made up with distilled water. After the first 6 mo of continuous soaking at room temperature, no joints of either test set had failed, and the retained joint strengths of triplicate specimens of each group continued to be very comparable. During the next 6 mo, however, several joint failures began to occur in the salt-water soak group, indicating some additional degradation of the bondline in the presence of the salt. While these observations might fit the theory that a certain level of adhesive displacement had been necessary to enable the salt water to additionally accelerate joint failure, the overall rate of acceleration had been modest with the absence of any gross corrosion of the adherends. The substitution of a cycling exposure procedure, which consisted of taking duplicate joints out of the 3.5% salt-water bath daily for drying in ambient air for 8 hr produced a most dramatic change in joint survival. Within a few days to weeks' time, all the cycled in salt-water test specimens had totally fallen apart with total gross pitting corrosion of the surface visually evident. When the failed adherends dried out, they were totally covered with a heavy white corrosion powder. If one presumed that the water had even less opportunity (33% less soaking time) to replace and debond the same adhesive on the same adherend, then what was responsible for the vastly different aggressiveness of the exposure conditions. The author believes that the aggressive pitting corrosion which began to be evident along the bondline edges under these wet/dry cycling conditions was able to literally chunk out the aluminum oxide surface, thus removing one of the key bridges in the bondline. By moving the specimen in and out of the water daily, there could be a literal washing of corrosion product from the bondline permitting new exposure surfaces for the process to be repeated until there was insufficient bridge left to physically maintain the joint.

Minford (2853) more recently published an investigation which was proposed to prove that a standard exposure test ASTM B1117 involving continuous exposure to 5% salt-water fog at 95°F (35°C) was not as readily predictive of aluminum joint durability under aggressive corrosive service conditions as either of two Alcoa-developed alternate immersion-type salt-water exposure procedures. A test specimen was used which deliberately employed a specially compounded, heat-curing, one-part epoxy adhesive which could bond in the presence of surface lube. Thus, it was presumed that a relatively poor condition for good surface wetting was present, since the aluminum surface was untreated and an automotive-type metal-forming lubricant was present. After 50 da, no evidence of any corrosion was observable on the aluminum adherend interfacial surfaces when joints were forcibly delaminated after continuous exposure to the salt fog. Meanwhile, specimens exposed to a single daily wet/dry cycle in 3.5% salt water for only 30 da were already showing initiation of gross corrosion on failed adherends. When 24 daily wet/dry cycles were employed in a paddle wheel–type exposure tank (specimens moved once each hour under salt water for 20 min), significant gross corrosion started to be picked up within only 20 da. It seems evident again that increasing the opportunity for corrosion product to be chunked out of a bondline correlated with instituting faster joint failures, and that the situation where water soaking was most continuous was actually less able to initiate and accelerate joint failures.

Dickie and co-workers at Ford (2854–2858) extensively used XPS techniques to study interfacial phenomena in paint coatings on metal adherends. Most systems studied involved painted galvanized steel specimens, where it was shown that a corrosion-induced deadhesion occurred which actually involved a cohesive failure of the paint resin close to but not at the paint/metal oxide interface. It would appear that a degradation occurred in these near interfacial areas of the paint polymer that could also be described as generation of weak boundary layer conditions. Van Ooij (2859) and Van Ooig and Kleinhesselink (2860,2861) also obtained evidence for the formation of weak boundary layer (WBL) conditions when studying rubber-to-metal adhesion. In the case of rubber-to-copper bonding, failure of the system was clearly in a copper sulfide film that formed during the curing process. Smith (2862) has employed a variety of experimental techniques to determine the failure mechanism of epoxy bonds to both aluminum and titanium. It could be shown that mapping of the adherend surfaces prior to bonding could reveal distinct areas with a predisposition to fail interfacially.

8

Water in the Service Environment

I. INTRODUCTION

Water must be considered one of the most important factors in determining the permanence of aluminum adhesive-bonded joints in weathering exposure service conditions. In Chapter 7 on mechanisms of joint failure, we have primarily considered the response of the bulk adhesive or the adhesive in the interfacial area to physical stressing situations. No allowance was generally made in such cases for the fact that the adhesive joint, as originally fabricated, has also to exist over long time periods in the presence of the degrading chemical factors of the environment. Tetelman and McEvily (2863) have reported that basic continuum fracture mechanics only requires a small crack before a substantially decreased failure time or diminished joint strength should be observed. However, this failure can indeed become much more catastrophic and can be actuated by a lower stress loading when water is also present in the environment. Althof (2864), Wake (2865), and Comyn (2866) have contributed recent discussions concerning the influence of water on the stability and durability of bonded joints, which can be of special interest to the reader. These investigators have focused on the fact that water can attack the adhesive (both in the bulk phase and in the interphase region) as well as the adherend directly. Since most adhesives have polar sites in their chemical structure, it is natural that highly polar water can literally be attracted into their bulk and permeate toward the interfacial area. Such water entry can cause physical swelling and deformation, as will be discussed in more detail later. At the interface or interphase area, water can actually act to displace some of the bonding already in place between the adhesive and the adherend. The excellent resistance to water debonding of phenolic-modified structural adhesives in aluminum joints was early reported by DeLollis and Montoya (2867), Bolger (2868), Buck and Hockney (2869), and has been explained by the strong covalent bonding that can arise between the oxide and phenolic configurations resisting rupture by water. Finally, water can directly chemically attack the aluminum via a corrosion mechanism and undermine the bondline structure. In this regard, the method of aluminum surface pretreatment can be most important as established in the author's investigations (532–534,774).

II. MECHANISMS OF FAILURE

A. Interface Stability

A number of theories have been advanced over the years relating to the mechanisms of failure of the water-soaked joint, including consideration of the thermodynamic properties of the interface. Kinloch (2870) has used the thermodynamic approach to predict stability, but it must be recognized that any chemisorption or interdiffusion of water across the interface is not provided for. Also, this theory does not account for interfacial adhesion forces arising from primary or secondary bonding forces or mechanical interlocking or stressing of the joint in some hostile environment. Nevertheless, values of the work of adhesion for different interfaces in either inert medium or water has been calculated by Gledhill and Kinloch (2851) and Kinloch et al. (1160). Levi et al. (2871) combined thermodynamic analysis with a stress-biased activated rate theory to help predict joint durability in hot water soaking with stress imposed.

Different investigators have utilized various procedures for studying the decay of the wet strength of aluminum/epoxy joints. Both Falconer et al. (2872) and DeLollis and Montoya (2867) made their early observations emphasizing the change in the type of joint failure observed; i.e., progressing from an initial 100% cohesive-type failure to 100% adhesive-type at the time of total loss of joint strength. Gettings et al. (2873) studied the locus of bond failure using AES and XPS analytical procedures. Kerr et al. (2874) continuing the earlier investigation added the concept of the probable breaking of hydrogen bonds at the interface due to water infusion. Gledhill and Kinloch (2851) still later added the concept of complete displacement of the adhesive at the interface with water molecules. Butt and Cotter (2875) still later considered the swelling and plasticization of the adhesive caused by direct water uptake, whereas Kerr et al. (2876) investigated the concept of direct hydrolysis of the adhesives.

Kinloch (2877) in discussing the kinetics of failure of epoxy/mild steel joints determined an activation energy of displacement of epoxy by water of 32 kJ/mol. This value was remarkedly similar to the 16–38 kJ/mol reported by Bardeleben (2878) and Manson and Chiu (2879) for the diffusion of water through an epoxy resin. This suggests that the rate of interface debonding may be controlled by the availability of water at the interface. Assuming Fickian, two-dimensional diffusion, the mathematics due to Cranks (2880) permits calculating the rate of water penetration into the joint. More recent investigations involving water diffusion into polymers include a 1984 report by Wilson et al. (2881) and Majerus et al. (2882) discussing experimental measurements and Monte-Carlo simulation of water diffusion into epoxy matrices. In 1986, Johncock and Tudgey (2883) investigated some effects of structure, composition, and cure on the water absorption and glass transition temperature of amine-cured epoxies. Dodiuk et al. (2884), in 1990, reported on the hygrothermal properties of adhesively bonded joints and their correlation with bulk adhesive properties, and Whitaker et al. (2885), in 1991, considered the influence of temperature and hydrostatic pressure on moisture absorption in polymer resins.

B. Nature of the Oxide on the Adherend

Venables et al. (36), in 1980, considered the hydration, oxidation, and/or corrosion of the aluminum adherend surface in the overall process of joint strength decay. Brewis et al. (1106), in the same year, observed that a linear relationship exists between joint strength and the water content of joints; however, Nakamura and Maruno (2886) observed that adhesion strength diminished very little with time under low humidity conditions. In 1987, Nakamura et al. (2887) formalized the quantitative relationship between joint wet shear strength and adhesive moisture content as a continuation of earlier investigations by Nakamura and Kondo (2888) in 1973 and Nakamura (2889) in a separate 1974 investigation. Noland (838) noted the reconstituted oxide produced after chromic acid deoxidizing an aluminum adherend hydrated to what he termed a "gelatinous boehmite." Although failure appeared to be adhesive-type, XPS analysis revealed the locus of failure was in the weak oxide layer. Riel (2844) noted aluminum joint failures with Alclad aluminum adherends which he attributed to the sacrificial nature under water-exposure conditions of the more anodic cladding compared with the underlying core alloy. Bascom and Patrick (2890) also expressed concern for the durability of any adhesive/metal combination where the surface of attachment was anodic with respect to the rest of the system. Cotter (1012), Scardino and Marceau (853), Shannon and Thrall (77), Riel (2844, 2845), Greer (2846), and Rogers (2847,2848) have all recommended preference of bare over clad aluminum alloy surfaces for aerospace bonding for this reason. The author believes the choice of a better adhesive (with a better surface wetting potential) and the use of proper protection against corrosive water entrance can still provide very acceptable jointures even using clad materials.

Although Bascom (2891) was specifically investigating the effect of water on the glass/resin composite in this particular investigation, there is a great deal of information which is translatable to the aluminum/resin situation as well. This follows from the fact that water is adsorbed similarly on the surface of nonhydroscopic oxides which would include SiO_2, Fe_2O_3, and Al_2O_3. The adsorption occurs in the form of hydroxyl groups (– M – OH) and as molecular water held by hydrogen bonding to the surface hydroxyls. There could also be some adsorption of molecular water on nonhydroxyl sites due to weak dispersion force interactions. Despite these similar features, however, the hydrated oxides can still differ greatly among themselves owing to the M – O bonds having different degrees of ionic-covalent character. Infrared spectroscopy can give direct information about the molecular state of this adsorbed water (2892,2893). The M – O bond involving Al_2O_3 is highly ionic, whereas that with SiO_2 is essentially covalent. From infrared spectra it also appears that clusters of water molecules seem to develop over the surface and there seems to be a continual diffusion of water molecules from one such cluster to another according to Dacey (2894) and Resing (2895).

Bowden and Throssell (2896) conducted early studies of water adsorption on aluminum foil in an attempt to confirm Stromberg's earlier observation that polar solids adsorb water films that are tens of molecular layers thick. This fact, if confirmed, would imply that the attractive force of the surface could be relayed through successive layers of water. Henniker (2897) has cited other authors' discussions as evidence for long-range surface forces. Bowden and Throssell (2896) actually found that this apparent heavy adsorption of water was apparently due to contamination of the oxide with some hydroscopic ions which could, in turn, be thermally desorbed at higher temperature. After such treatment, the water absorption was hereafter confined to no more than two monolayers even at partial pressures close to saturation. If the foil was subsequently rinsed in ion-containing tap water, the adsorption of many more molecular layers again became possible. From ellipsometer studies, these authors further demonstrated that only a small amount of hydroscopic contamination was required to promote this multilayered adsorption.

Water is universally present in all natural weathering conditions but obviously to widely varying degrees. These conditions can range from the extremely low humidity conditions of the desert areas to the near 100% humidity conditions of the tropical rain forest. It may also range from the rare exposure to any rain in the desert to the drenching almost daily rain conditions in the tropics. In between we could have the daily early morning condensed dew conditions of many temperate zone regions. At the atmospheric exposure station maintained by the Alcoa Labs near Pittsburgh, Pennsylvania, the presence of daily wet/dry cycling early morning conditions are present for most of the year. Finally, water can be present in the aggressively corrosive form of salt water in the coastal and marine atmospheric areas of the world. Thus, water in some form is the predominate form of weathering that can be experienced by bonded joints over the whole planet. The questions then become those of how fast and how much water will invade the joint structure. It can strike by moving directly through permeable adherends to the interface. Alternatively, with nonpermeable adherends like aluminum, the pathway must be by some combination of diffusion through the adhesive or via the interfacial zone through an edgeline.

C. Fracture Mechanics Mechanisms of Failure

This approach is based on the application of continuum mechanics which considers that the strength of a material containing a flaw (mostly considered as an ellipical crack) will be governed by the rate and magnitude of stress which can develop around the crack. Two different approaches have been made, as described in the terms G for fracture energy and K for stress-intensity factor or fracture toughness. The former has been further designated the energy-balance approach, since the criterion arising from Griffith's (2560) and Orowan's (1296) works supposes fracture occurs when sufficient energy is released from the stress field through crack growth processes to equal the requirements for creating the new fracture surfaces. In like manner, the latter is predicated on the discovery of Irwin that the stress field around the crack could be defined by a stress-intensity factor, K, which must be exceeded to initiate crack propagation. The critical value in each case has been written in Chapter 7 as G_c and K_c. Also, as mentioned earlier, there can be separate critical values depending on the mode of stressing. The corresponding tensile opening-mode values now become G_{Ic} and K_{Ic}.

The pioneering work of Mostovoy, Ripling, and co-workers in making fundamental measurements of these values through the use of a TDCB-type specimen has continued over many years (2200,2492, 2501–2503,2504–2521,2708–2710,2815). The work of Corten (2898) may also be of review interest. The TDCB is a constant-compliance geometry specimen (fracture energy is independent of crack length), making it suited for environmental studies where crack velocity will be a function of both applied load and the environment. They found that in water immersion, the interfacial debonding of the loaded joint occurred near the original cohesive starter crack tip. This new interfacial crack eventually propagated along the adhesive/metal oxide interface, permitting the development of a plot of adhesive fracture energy versus crack velocity. From such a plot it can be readily seen that the G_c value required in water is much lower than that required in a drier environment. These values are, however, much higher than the corresponding values for the thermodynamic works of adhesion. Kinloch (2899) has stated this is because mechanical strain-energy is available under an applied load to assist environmental crack propagation. This is further reflected in inelastic energy-dissipative processes like plastic flow which can occur in regions of the adhesive around the crack tip. Obviously such processes are missing from any thermodynamic consideration.

Mostovoy and Ripling (2507) further denoted a G term below which slow crack growth would not occur in an aqueous environment. This implies some stress level below which no environmental attack occurs. Conflicting evidence has been obtained, however, probably for several reasons. First, the time-scale employed in such investigations (2507) has not been sufficiently long to really be sure when such a G value has been attained. Second, it has been established by thousands of unstressed water-immersion tests by the author (such as reviewed in Refs. 532–534,774,976) that aluminum joints still suffer environmental attack even when no externally applied stress is present. On the other hand, however, it is also obvious from the work of Sharpe (1601), Minford (2900), Weitsman (2901), and many others that the presence of both applied and swelling stresses can significantly accelerate the environmental decay mechanisms whatever they may be.

Kinloch et al. (1160) and Gledhill and Kinloch (2804) have conducted some interesting aluminum TDCB tests with tertiary amine-cured epoxy bondlines exposed to either a relatively dry (50% RH, 73°F [23°C]) or wet (water immersion, 73°F [23°C]) environment. In the former case, the locus of joint failure was cohesive in the center of the adhesive layer with a linear relationship between the G_{Ic} and logarithmic time to fracture. There was also an incubation time until crack propagation began with a modest crack growth rate of about 20 m/sec. Modeling the crack as an elastic-plastic material with a plastic zone at the crack tip, there appears to be a critical value of the plastic zone size at which the fracture occurs. Gledhill and Kinloch (2548,2804) calculated a value of about 16 μm for the critical plastic zone size for this particular adhesive. In the presence of water, the second set of identical joints now demonstrated a departure in failure mode by changing from the center-moving cohesive-type failure to an apparent interfacial-type failure. Some time was required for water to diffuse in order for this change to take place. Accordingly, when the applied adhesive fracture energy was relatively high with time to fracture times of about 300 sec or less, the failure behavior more resembled that of the drier exposure joints.

Kinloch (2902) calculated the thermodynamic work of adhesion as being a positive value of about 232 mJ/m^2, whereas the value for a similar joint with adsorbed water at the interface became a negative -137 mJ/m^2. This predicted instability was manifested in the test joints by the observed interfacial-type failure observed. Apart from the exceptions noted above, the mechanism of failure seemed to be a penetration of water followed by displacement of the epoxy/aluminum oxide interface.

It should be pointed out, however, that the geometry and testing procedures using the constant load-type TDCB specimen has proved too complex and costly for most industrial test purposes. Accordingly, Shannon and Thrall (77) have discussed two constant displacement-type specimens (double-cantilever beam [DCB] and wedge-type joints). Unfortunately, since stress relaxation can occur in the adhesive and diminish the effective stress at the tip, the observed crack-growth rate can diminish even with constant loading (2667). Bascom et al. (2842) have discussed the nonlinear bending displacements of the Boeing wedge specimen, making it difficult to stress analyze. However, its economy, ease of use, and speed of testing have resulted in its widespread use for discerning between different surface treatments on aluminum adherends. There will be more discussion of wedge testing in Chapter 9 on test methods.

III. THE ADHESIVE PATHWAY

As mentioned above for aluminum adherends, the water pathway to the interface must be via the bulk adhesive, through an edge area, with subsequent migration along the interface or more likely a combination of the two. Very often there is a competition between two mechanisms depending on the degree of adhesion that has been achieved in the "as manufactured" joint. Where no aluminum surface treatment has been used, or is minimal at best, i.e., poor conditions for wetting by the adhesive, the debonding of the adhesive at the interface will likely occur through edge moisture migration along the interface. This should occur long before significant moisture penetration of the cured bulk adhesive can reach any critical level. Recently, Takahashi (2903) has reported a study of interfacial moisture ingression as measured by AC impedence measurements. At the same 1987 meeting, Lefebvre et al., (2904) reported investigation of the influence on strain and boundaries of the diffusion of gases and vapors in polymers, especially as it applies to the durability of adhesive joints. Lefebvre et al. (2905) have introduced the concept that the available free volume within a cured adhesive structure should be highly important in determining the rate of water diffusion through the adhesive and applies to the durability of adhesive joints. In 1990, Brewis et al. (2906) discussed the concept of whether there is a critical relative humidity for the weakening of joints in wet air, whereas Lefebvre et al. (2907) reported their investigation of the abrupt loss of epoxy adhesion to inorganic oxide adherends when the epoxy is equilibrated in air whose humidity exceeds some critical value. These investigators concluded that the critical relative humidity is associated with (1) a sudden increase of water solubility in the epoxy, (2) a corresponding increase in the volume of the epoxy, and (3) a strong decrease in the mobility of absorbed water. They also proposed reconsideration of the earlier concept of "water in clusters" in the polymer at sites which become available above the critical humidity. Finally, they postulate a basis for the pronounced irreversibility of solubility and swelling as due to a reaction of the absorbed water with residual oxirane groups resulting in the formation of diols.

In contrast, when the best types of aluminum surface pretreating have been employed, i.e., FPL-etching, conversion coating, or anodizing, the chemical stability of the interfacial area will be tremendously enhanced. This means that sufficient time is now available for slow moisture migration to continue into the bulk adhesive until plasticization or swelling of the adhesive can become another significant factor in contributing to overall joint failure.

Bascom (2891) has published his investigation of how water can degrade the strength of both the adhesive and the aluminum adherend. He points out that fracture mechanics can provide a fundamental and quantitative means of describing material failures by various environmental factors including water. While failure may begin with the formation of one or more minute flaws or cracks in the interfacial area, it is the application of stress at some higher level at the crack tip which inaugurates the propagation. The action of an adsorbate like water could act to further significantly alter the bond strength at the crack tip effecting a reduction in the cohesive energy of the adhesive. Another possibility is that the presence of water may promote an embrittlement of the material around the crack tip, thus restricting the amount of deformation that would ordinarily occur ahead of the crack. Under conditions where the crack is propagating rapidly, this would be an unlikely mechanism, since it would require a rapid infusion of water to the crack tip and into the deformed area which would seem impossible. Finally, water condensing in a crack could exert considerable capillary pressure, acting to open the flaw (2908).

Minford (2909) has plotted his data in Figure 15 showing the very slow 2-yr decline in joint strength of chromic acid–deoxidized 6061-T6 aluminum adherends bonded with a polyamide-cured, two-part, room temperature–curing epoxy while soaking in 100% RH at 125°F (52°C). This would confirm that whatever the water incursion rate for the water vapor might be, it was not enough of a detriment to significantly lower the overall joint strength over a 2-yr period. In contrast, in Figure 18, a rapid catastrophic bond failure did occur with duplicate joints within a matter of minutes when external stress was place simultaneously on the bondline. The significance of the water was still quite evident, however, since other similar joints could be stressed to the same levels at ambient laboratory humidity levels without any apparent detrimental effect on the adhesion. Only the combination of the high humidity and stress was capable of precipitating the catastrophic failures. After joint exposures of 2 yr in the hot condensing humidity alone, the forced joint failures in a tensile test machine could be termed 100% cohesive in nature. Within minutes of exposure to the simultaneous conditions of stress and hot humidity, the joints catastrophically separated in the stressing fixture and the failure condition had to be termed 100% adhesive in nature. While it was never possible to predict which adherend in the joint would end up retaining all the adhesive initially present in the bondline, all the adhesive in these particular tests always remained as an intact attachment to only one of the adherends. It would seem that the short survival time of the joints would not permit an explanation based on moisture transfer via the bulk adhesive. It is even difficult to promote a totally satisfying explanation of total water replacement of the adhesive at the bondline in the instant of catastrophic failure. Also of interest should be the fact that a chromic acid–deoxidizing pretreatment of the 6061-T6 surface was able to provide a wetting condition of the surface which provided a total prevention of significant damage from water incursion either through the bulk adhesive or along the interfacial area for a full 2 yr.

Comyn (2910) has reviewed the subject of the kinetics of water entry into the bulk adhesive and the interfacial area. He makes reference to additional pertinent publications by Crank (2880) and Carslaw and Jaeger (2911). (For the mathematics involved in the calculation of the distribution of water in the adhesive using Fick's first law, see Ref. 2910.) The general procedure concerns treating the adhesive in a lap-joint as a section of a rectangular prism formed by the volume of the intersection of two slabs at right angles. Fijita (2912) first showed the uptake of water by cured adhesive

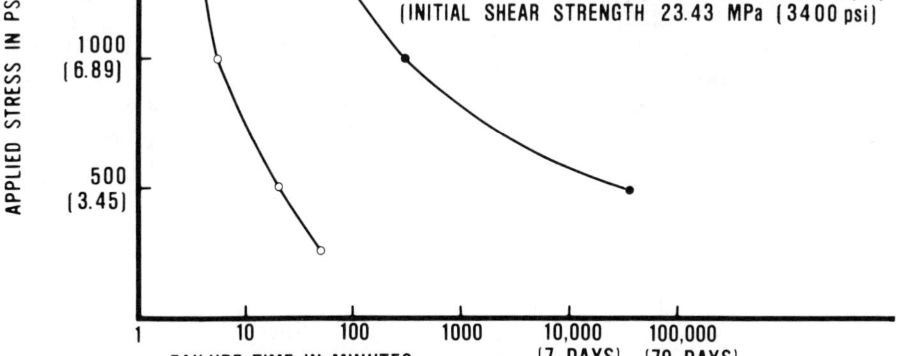

Figure 18 Exposure in 100% relative humidity at 51.7°C (125°F). (From J. D. Minford, Aluminum Adhesive Bond Permanence, Treatise on Adhesion and Adhesives (R. L. Patrick, ed.). Marcel Dekker, New York, Vol. 5, p. 122, Fig. 3–58 [1981].)

films could agree with Fick's first law and, hence, could be termed Fickian. Brewis and co-workers (2913) at Leichester have examined the liquid water uptake of a series of cured epoxy resin adhesives by exposing thin film samples to water at 77°F (25°C), 113°F (45°C), and 158°F (70°C). The mass uptake of liquid water at 113°F (45°C) was as observed in Table 111. The data show an initial rapid uptake of water followed by a long plateau. This type of diffusion has been termed Fickian and is typical of diffusion in a polymer in the leathery state; i.e., above it glass transition temperature. In general, glassy polymers have shown non-Fickian S-shaped uptake plots which makes the data plot from Table 111 rather unexpected. However, that this is more the rule than the exception for glassy epoxy adhesives has been repeated in the literature quite often. Diffusion coefficients for the uptake process for any adhesive showing this pattern can be calculated from the slope of the linear region of such graphs. For example, Narmco 5208 (an epoxy-novolac adhesive) has had its glass transition temperature reported at 482°F (250°C) by Kaelble and Dynes (2914) and 421°F (216°C) by McKague et al. (2915). However, Fickian uptake plots were developed both for the neat resin and as a matrix resin at temperatures considerably below these quoted T_c values by Shirrell and Halpin (2916), McKague et al. (2917), Cook et al. (2918), and Long (2919). Another example has been furnished by Brewis et al. (2086) using once popular aerospace polyamide-epoxy film FM-1000. This adhesive has a T_c when dry of only 93°F (34°C) but is further depressed in room temperature water to −25°F (−32°C). Nevertheless, these investigators measured a Fickian uptake curve at only 77°F even when the adhesive material changes from glassy to a leathery state under such soaking conditions. An alternative method for calculating diffusion coefficients was employed by both Daniely and Long (2920) and Marom and Broutman (2921) using the actual changes measured in a length of water swelled adhesive.

Thermodynamic law would predict that the equilibrium uptakes of the same adhesive film in liquid water as compared to its saturated vapor should be the same. Saturation of the infinitesimal surface layer should be instantaneous, which should lead to identical uptake rates. Antoon and Koenig (2922) have shown with composites, the moisture uptake from liquid water will generally be greater than from exposure to saturated humidity conditions, which may be somewhat surprising, since most investigators would consider the latter weathering environment to be more accelerating for laboratory testing. A possible explanation offered has been that there is an augmented capillary attraction for liquid water or that there is great difficulty in obtaining truly saturated vapor. At the Alcoa Labs, for example, the author employed specially constructed aluminum humidity cabinets which constantly pressured water vapor into the enclosure to maintain truly continuously condensing conditions; i.e., 100% RH exposures.

Table 111 Uptake Plot for the Epoxide DGEBA-DDM from Water at 212°F (100°C)

Time ($t^{1/2}l^{-1}\cdot 10^6 s^{1/2} m^{-1}$)	Fractional uptake (M_t/M_∞)
25	0.13
50	0.26
75	0.36
100	0.38
200	0.42
300	0.42
500	0.47
1000	0.60
2000	0.83
3000	0.95
4000	0.95

Data interpreted from Ref. 2910 (Fig. 6, p. 97).

There seems to be a definite problem in establishing any predictive general rules based on the choice of adhesive about whether Fickian or non-Fickian water uptake will be found. Brewis et al. (2923) have also investigated the dependence of the diffusion coefficient (D) on water concentration by performing uptake experiments using water vapor at different pressures. The results clearly indicate that D was independent of concentration for the particular adhesive system studied (DGEBA/DAPEE). Augl and Berger (2924) had earlier showed a similar response for some other epoxy systems. While such a constancy of D over a considerable concentration range greatly simplifies the calculation of water distribution in joints, it unfortunately is not always true. Instances of reporting that D can actually increase with increasing water concentration also must be dealt with in the publications of Perera and Heertzes (2925) and Mozisek (2926). Daniely and Long (2920) also showed an increase in the overall water uptake with longer adhesive curing times, which they attributed to an accumulation of hydroxyl groups as a result of curing. Although the authors did not conduct water diffusion tests with the thicker bondlines of highly modified epoxy joints cured at automotive paint-baking temperatures, as mentioned earlier, they did show significantly reduced joint strengths. There would be good reason to expect that the extreme porosity of the adhesive would also contribute to finding a higher diffusion rate for moisture under such conditions.

As indicated earlier, many adhesive seem to show an equilibrium uptake of water (see Table 111) ending with a plateau at the end of the uptake curve. Unfortunately, for ease of making generalizations, different adhesives can show several additional types of water uptake behavior. Brewis et al. (2913), for example, have developed the same-appearing data for DGEBA-DDM epoxy as shown earlier in Table 109 for short exposure times; a second slower rise in moisture uptake rate can proceed with this same adhesive after longer exposure at 212°F (100°C) (see Fig. 6, p. 97, in Ref. 2910). Possible explanation has been offered for the second uptake phenomenon stating that further cross-linking of the already water-plasticized adhesive can occur at either a higher temperature or longer curing time. A second uptake phase is then possible because of the new presence of cracking and crazing in the adhesive.

Althof (2927) has studied water uptake by commercial aerospace structural adhesive FM-73 which shows a single Fickian uptake to a plateau, as shown in Table 111, when testing is conducted at 68°F (20°C), 104°F (40°C), or 122°F (50°C) in 95% RH conditions. However, when the testing temperature was elevated to 158°F (70°C), then a distinct second sorption state occurred. To further demonstrate the variation possible in uptake rates, we can cite the situation found with FM-1000 film adhesive as tested by Brewis et al. (2913). Previously it had been mentioned that a typical Fickian uptake curve was found at 77°F (25°C), though the material actually changed from a glassy to a leathery state under this condition. When these investigators elevated the testing temperature to 122°F (50°C), instead of the plateau response that was expected, there was a persistent weight loss after the plateau was attained. If this adhesive was then dried out and retested at the same temperature in saturated water vapor, the same original Fickian response was repeated at room temperature. The authors offered the explanation that the weight loss under the liquid water-soaking conditions at 122°F (50°C) had been due to leaching out of some adhesive component (perhaps some unreacted component or an earlier product of hydrolysis). When this material was removed in the liquid sorption process, a re-exposure after drying could then repeat the earlier response with the temperature again at 77°F (25°C). From such examples it seems quite clear that it may be necessary to separately evaluate each different adhesive and its processing conditions in order to establish the overall nature of its sorption process.

Wong and Broughman (2928) found that the diffusion through an incompletely cured glassy epoxy can be distinctly non-Fickian. They theorized that it was due to the continuing cure of the epoxy with direct water sorption in the resin. Also in 1985, the same authors (2929) proposed a new diffusion model in which the diffusion of water is through two regions in a glassy epoxy resin with varying densities present. The subsequent implication was that the older diffusion theory equations really apply only at relatively low concentrations of water. Regardless, the sorbed water can cause direct chemical changes in adhesives and hydrolysis reactions have been established in the literature by Pearce et al. (2930) for epoxies, Abbott and Brumpton for polyurethanes (2931), and Droin et al. (2932) for cyanoacrylates.

Schneberger (2933) in discussing bulk polymer structure has stated that the cross-linked polymer is in a constant vibration movement with the greatest vibrations occurring across the hardener linkage points. Because of this motion and imperfect packing, there is a certain amount of free volume which constitutes as much as 2–3% of the total volume. This volume can increase with increasing temperature, thus increasing the affinity that a liquid penetrant can show for that particular polymer. General Dynamics (2934) issued a 1975 AFML report in which a finite difference solution to Fick's law was employed for predicting the concentration and distribution of moisture in a resin system with good correlation with laboratory test results. A 1977 report from the same company offered a procedure for determining the diffusion coefficients for neat adhesive FM-73M in a study of the fatigue behavior using thick-adherend aluminum joints (2935). In 1979, Kinloch (2852) published an excellent review of some investigations designed to study the deleterious effect of water. As discussed earlier, Kinloch et al. (1160) had pointed out that thermodynamics takes no account of interfacial adhesion forces from primary or secondary bonding or mechanical interlocking. Thus, no information should be expected from such calculations regarding the expected service life of joints that have been stressed while undergoing simultaneous water infusion. Rather it is necessary to combine the thermodynamic data with either a stress-biased activated rate theory as developed by Zhurkov and Tomashevsky (2936) and Zhurkov and Korsukov (2937). Such theories were utilized by Levi et al. (2871) in joint fracture studies. Alternatively, a continuum fracture mechanics approach can be employed. Reference should also be made to the 1977 publication of Brockmann (373) on the effect of interface reactions on the long-term properties of metal bonds. Brockmann bonded variously treated 6-mil aluminum foil with a phenolic adhesive as a peel specimen which could be peeled after aging in 275°F (135°C) water. The peel test proved to be the only method where adhesional failures were always obtained regardless of the surface pretreatment used.

More recent 1990 studies have related to joint hydrolytic stability effects such as those of Bowditch (2938) on the relationship between hydrolytic stability of adhesive joints with equilibrium water content; Small and Fay's (2939) investigation of the creep of adhesive lap-joints in dry and high-humidity environments; Bianchi's et al. (2940) study of water sorption effect on toughened acrylic adhesives, and ElSa'ad et al.'s (2941) investigation titled "Moisture Absorption by Eposy Resins: the Reverse Thermal Effect."

A. Comparisons of Predicted and Measured Water Uptakes

A number of cured aerospace adhesives have been tested for water sorption. Althof (2942,2943) has made measurements using BSL 313A and FM-73 (modified epoxy types), AF 126-2 and FM 123-5 (nitrile-epoxy types), and Redux 775 (vinyl-phenolic) adhesives cured between aluminum adherends. The FM-73, AF 126-2, and FM 123-5 adhesives would also be considered different from the other mentioned adhesives because of the presence of a fibrous carrier to assist in the handling and control of bondline thickness. Althof (2943) has measured the moisture distribution across sandwich specimens using FM-73 adhesive and made comparisons with moisture content values calculated from a diffusion coefficient obtained from film uptake experiments. It seemed clear that different moisture concentrations arose from the two different procedures and that the maximum deviation seemed to occur at the most interior areas from the edge. Althof has also compared the actual diffusion coefficients measured from mass uptake of the film experiments with those from sandwich-type specimen experiments using four different adhesives, as shown in Table 112. Not only were the values of D obtained with the two different specimens not identical but they could be vastly different. Althof observed the opening of some of his sandwich specimens along edges, and sometimes even observed some corrosion of edge areas. He presumed this could account for some of the larger differences observed.

Brewis et al. (2944) have also employed aluminum/adhesive sandwiches and immersion in isotopically labeled water for up to 6 wk at 122°F (50°C). Both adhesives with and without carriers were included in the investigation. They measured and calculated what water might be expected to be absorbed in 25 equal area parts of the bondline after the immersion, as reported in Table 113. It seems clear that moisture was still gradually moving from the edge areas toward the center based on finding as little as 20% in the center areas as compared to edge areas. However, they found no difference

Table 112 Comparison of Diffusion Coefficients Measured From Mass Uptake by Adhesive Films (f) and Sandwiches (s) of Adhesive Between Aluminum Adherends

Temperature		RH	$D/10^{-13}\ m^2s^{-1}$ Adhesives FM-73		BSL313A		AF126-2		FM123-5	
(°F)	(°C)	(%)	f	s	f	s	f	s	f	s
68	20	70	2.8	3.6	1.4	1.7	1.7	2.2	—	—
68	20	95	1.7	3.9	0.7	1.8	—	—	—	—
104	40	70	8.0	10.3	1.9	3.9	5.0	6.1	—	—
104	40	95	4.2	4.1	0.9	28	—	—	—	—
122	50	95	1.2	1.5	1.1	5.0	—	—	1.4	7.5
158	70	95	—	—	0.8	3.6	—	—	—	—

Data from Refs. 2942 and 2866 (Table 1, p. 99).

in the magnitude of the moisture pickup in the outer perimeter areas whether the same adhesive compositions had carriers present or not. In the inner areas of the sandwiches, however, they found significantly less water than they would expect by calculation. The measured water levels toward the center of the joint were lowest compared to calculated values where no carrier was present. Apparently, the combination of carriers and adhesive in different adhesive formulations can produce varying effects with respect to hindering the entrance of water. These authors also conducted some experiments with FM-1000 formulation with and without a carrier which definitely showed hindrance to water passage with a carrier present. Contrastingly, the movement of water in the tape version (carrier present) of BSL 312/5 adhesive did not seem to be hindered as compared to the corresponding film version of the adhesive. All organic polymers are, of course, permeable to water and values of permeability coefficients (P) and diffusion constants (D) have been compiled (2945). In such lists it can be seen that epoxy and phenolic materials rank among those adhesive materials with the lowest diffusion and permeability constants; however, it still must be remembered that other properties such as wetting/adhesion characteristics, processability, toughness, and cost have still to be considered as well.

Some concern has been expressed that carriers might act as water-entry points causing a strength decline in joints exposed to saturated air or liquid water in service. Brewis et al. (2944) have supplied data showing no alteration of aluminum joint durability in saturated humidity exposures for either BSL 312 or FM-1000 joints for up to 10,000-hr exposures with carriers present in the bondline. This was the result even though these same two adhesives in tape forms had shown the opposite responses, as just mentioned above as to whether the carrier seemed to hinder water entry.

B. Effects of Water on the Adhesive

Most cured adhesives studied in water-soaking conditions do seem to reach an equilibrium condition. Kerr et al. (2874,2876) and Kerr and Orman (2946) reported a number of investigations of the

Table 113 Fractional Uptakes of Isotopically Labeled Water in Segments of an Aluminum Alloy FM 1000 Sandwich[a] After Exposure to Vapor for 100 hr at 122°F (50°C)

A	(1.05)	B	(1.06)	C	(1.07)	B	(1.06)	A	(1.28)
D	(0.80)	E	(0.70)	F	(0.61)	E	(0.75)	D	(1.08)
G	(0.62)	H	(0.20)	I	(0.20)	H	(0.42)	G	(0.88)
D	(0.56)	E	(0.34)	F	(0.38)	E	(0.50)	D	(0.80)
A	(0.41)	B	(0.28)	C	(0.52)	B	(0.71)	A	(0.91)

[a]Overall size of sandwich was 12.5 mm × 25 mm.
Data from Ref. 2943.

effect of water on aluminum/epoxide bondments in the late 1960s showing leveling off of joint strength decline which was presumed to result from such an equilibrium state. These results would seem to indicate the rate of adhesive deterioration might not necessarily continue to accelerate with increasing exposure time. Overall, the epoxy adhesives studied in water exposure with specially treated aluminum adherend surfaces have generally shown good resistance results. For example, reference can be made to the slow decline to what appears to be a stable joint strength over a 2-yr exposure in 100% RH conditions at 125°F (52°C) using a chromic acid–deoxidized 6061-T6 adherend, as in Figure 15. Although these data from the author would tend to support the joint stability present in an equilibrium situation, the author has also obtained every conceivable manner of aluminum joint failure under steady soaking conditions depending on different choices of adhesive, alloy, and surface conditions. Data of these sorts will be considered in more detail later in Chapter 10 on evaluation of aluminum joint durability.

Antoon et al. (2947) and Antoon and Koenig (2948) have used FTIR spectroscopy to follow possible chemical changes in the DGEBA/MNA epoxy adhesive system, including exposure in 176°F (80°C) water for periods up to 90 da. In unstressed joint specimens, there was no hydrolysis of ester groups noted; however, hydrolysis did occur in increasing amounts as the stress level was advanced from 0 to 19 to 26 to 54 MPa. This was actually a contradiction to earlier work by Levy et al. (2949), who subjected the material system TGDDM/DDS to various stress levels, elevated temperatures, and moisture in an autoclave at 275°F (135°C). These investigators also employed the FTIR analysis technique but concluded that the extent of bond rupturing appeared to be independent of stress level. However, they did claim finding a relationship between mechanical stress and the swelling stress in water.

The formation of cracking or crazing in adhesives due to exposure to varying climatic conditions may be the more important factor affecting joint debonding as compared to hydrolytic splitting of the chemical structure. This has been shown experimentally in aluminum aircraft bonded joints by simulating the thermal spike condition present under flight conditions; i.e., a return to below zero temperature within a few minutes after the aircraft skin has reached above 302°F (150°C) during a sudden supersonic maneuver. Epoxy/fiber composites have furnished especially good materials for such studies, and Browning (2950,2951) has demonstrated equilibrium water uptake in such composites can significantly increase when experiencing such spiking conditions. The obvious explanation is that water can now enter the microcracks which have been initiated by the spike cyclic experience. The long-term exposure (11 yr) of epoxy composites to either salt water or distilled water by Mazor et al. (2952) indicated the presence of possible microdamage effects. They based this on the fact that water-diffusion gains could be shown after drying out the specimens and re-exposing them to the same water conditions. Farrar and Ashbee (2953) and Nicholas and Ashbee (2954) showed cracks or crazing occurring in cured epoxy resins when small amounts of electrolytes were present in the water in which specimens were immersed. Still other investigations that have revealed changes in water uptake or diffusion coefficient values attributed to microcracks or cavity production have been conducted (2923,2955,2956–2958).

Ripling et al. (2509) have evaluated the short- to long-term effects of water soaking on epoxy adhesive toughness. They observed that toughness could actually increase for the short-term exposure conditions; however, toughness can begin to decrease as the water-soaking time is lengthened.

1. PLASTICIZATION AND SWELLING

The general effects of water on adhesives is to depress their glass transition temperatures and lower their modulus and cohesive strength. Actions which accompany these changes can be labeled plasticization and swelling. Some of the early theories which attempted to explain why water ingress might depress the T_c were referred to as "free volume theories" as enunciated by Fox (2959) and Kelley and Bueche (2960) in the late 1950s and early 1960s. In order to use the pertinent equations proposed by Kelley and Bueche (2960) for predicting this effect, it was necessary to establish a value for the glass transition temperature of water itself. There seemed to be good agreement among McMillan and Los (2961), Suisake et al. (2962), Rasmussen and MacKenzie (2963), and Frank (2964) that this value was between −209.2°F (−134°C) and −216.4°F (−138°C). However, when these values were employed by Morgan and Mones (2965) in 1979 on the TGDDM-DDS-water system, the actual measured glass transition temperatures of the wet resins were well below those calculated using these values. Browning and co-workers (2950,2951,2966) found very good agreement between the measured and predicted values of these same resins when using a value of 39.2°F (4°C) for the T_c of water, which corresponds to the temperature at which water is known to have its maximum density. The further justification for using this value could be that it is a better measure of water's free volume contribution to a polymer/water mixture.

As expected, all T_c values of adhesives were depressed when first absorbing water as was shown by Brewis et al. (2913) (Table 114). However, if the exposure in water was prolonged, some values could actually increase above the T_c of the dry adhesive, making plasticization unlikely as a satisfactory explanation for the long-time durability results for many joints. Undoubtedly, plasticization can still be considered a contributory factor in certain situations. Shalash (2967) has shown one case where plasticization does seem to be highly contributory in his evaluation of aluminum/FM-1000 joint specimens. When plotting the strength-temperature curves of wet and dry joints, Shalash found the wet curve was shifted to lower temperatures by about 86°F (30°C) to 122°F (50°C), corresponding to an amount similar to the water-induced depression of T_c.

Water may also cause swelling in a joint which may have the effect of inducing weakening stresses in a joint. Studies have been made by Gazit (2968) and Shirrell and Halpin (2916) which included examples of showing the swelling strain in neat resins as being proportional to the moisture content. Weitsman (2969) analyzed water swelling in rigid composite adherend joints fabricated with 3501 epoxy adhesive. He measured higher compressive edge stresses with increasing moisture diffusion times and demonstrated that the maximum tensile stress present in the adhesive moved into the interior of the joint with increasing diffusion time. After an initial rise in these values, however, the stress concentrations began to degrease, making it seem unlikely that any long-term weakening of the joint would occur.

Sargent and Ashbee (2970) measured normal swelling strains induced in joints consisting of thick adherends bonded to thin sheets of glass. They graphically differentiated the normal displacement profiles and applied equations by Love (2971). Jesson and Sargent (2972) outlined an even better estimate of the distribution of normal swelling stress in 1982 by using the same Sargent and Ashbee (2970) experimental technique but employing a "theory assisted" fit for the normal displacement profiles.

Gledhill et al. (2705) proposed in 1980 that the outer water-containing zone of the adhesive might constitute a crack. However, this might infer that even lower humidity exposures, such as ambient laboratory conditions (usually around 50% RH) should have some significant negative effect over

Table 114 Glass Transition Temperatures of Wet and Dry Adhesives Formed from DGEBA with Various Hardeners

	Measured T_g (°C)			T_g (°C) from
Hardener	Dry	After equilibration	After 10 mo in water	Fox equation after equilibration
DAPEE	67	37	49	44
TETA	99	86	111	76
DAB	161	143	157	139
DDM	119	110	130	92
DMP	68	51	54	47
BF_3 $C_2H_5NH_2$	173	155		151

Abbreviations: Code for hardeners: DAPEE = di(1-aminopropyl-3-ethoxy)ether; TETA = triethylenetetramine; DAB = 1,3-diaminobenzene; DDM = 4,4′diaminodiphenylmethane; DMP = tris(dimethylaminomethyl)phenol.
Data from Ref. 2866 (Table 5, p. 109).

Table 115 Dependence of Joint Strength upon Water Uptake for Aluminum Alloy Double-Lap-Joints Bonded With FM 1000

Water uptake (M_t/M_∞)	Joint strength (kN)	Water uptake (M_t/M_∞)	Joint strength (kN)
0.00	42	0.87	23
0.25	40	0.94	20
0.50	31	0.97	13
0.75	26	1.00	6

Data from Ref. 2866 (Fig. 14, p. 107).

long periods of time. In this regard, DeLollis (2973) has showed that there was no apparent loss of aluminum/epoxy joint strength after an ambient laboratory exposure for up to 11 yr. Gledhill et al. (2705) exposed steel/epoxy butt-joints for 2500 hr at 55% RH at 68°F (20°C) with no joint strength weakening, whereas Brewis et al. (1106,2085) found no significant weakening of aluminum/FM-1000 joints (with or without carrier present) after 10,000 hr in 45% RH at 68°F (20°C). It should be pointed out that based on a knowledge of water sorption isotherms for other adhesives, it would be expected that such joints would absorb significant quantities of water even under such low humidity conditions.

Boutilier et al. (2974), Weitsman (2975), Gazit (2968), Sargent and Ashbee (2970), and Althof (2927) have studied the generation of stresses in bonded adhesives due to moisture absorption. One measure of the dimensional stability under such conditions is the change in creep behavior. Hughes and Rutherford (2976) particularly studied the effect of thermally induced stresses on creep behavior, whereas Hughes et al. (2977) studied the combined effect of thermally and moisture-induced stresses.

Table 116 Linear Relationship Between Joint Strength and Fractional Water Uptake for Aluminum Alloy Joints with Five Different Surface Pretreatments and Bonded with DGEBA-DAPEE

Solvent degreasing		Sandblasting		Chromic acid etching	
Water uptake (M_t/M_∞)	Joint strength (kN)	Water uptake (M_t/M_∞)	Joint strength (kN)	Water uptake (M_t/M_∞)	Joint strength (kN)
0.00	2.0	0.00	2.6	0.00	6.4
0.25	1.6	0.25	2.3	0.25	5.8
0.50	1.3	0.50	2.0	0.50	5.1
0.75	1.1	0.62	1.85	0.75	4.5
		0.75	1.80	0.87	4.2
		0.80	1.75	0.95	3.9

Anodizing in chromic acid

Method I		Method II	
Water uptake (M_t/M_∞)	Joint strength (kN)	Water uptake (M_t/M_∞)	Joint strength (kN)
0.00	2.8	0.00	4.8
0.25	3.2	0.25	4.3
0.50	3.5	0.50	4.15
0.75	3.9	0.75	4.0
0.87	4.0	0.85	3.9

Data from Ref. 2866 (Figs. 12 and 13, pp. 106 and 107, respectively).

They concluded that the tortion-shear creep of both filled and unfilled epoxy adhesives was increased by moisture, with the greater susceptibility being present in the unfilled systems. Also, it can be shown that the creep rate at any given applied stress increases with increasing humidity.

Brewis et al. (1106) have observed a linear dependence of joint strength of aluminum double-lap/FM-1000 joints upon total water uptake, as shown in Table 115. Measurements made on aluminum joints with five different surface pretreatments similarly showed linear dependence, as shown in Table 116. Good's (2978) work, "Combined Temperature-Moisture-Mechanical Stress Effects on Adhesive Joints," might be of reader interest. He further cites the early 1965 discussion of Fox et al. (2979) on the concept of molecular free volume in the adhesive as a basis for understanding water diffusion and subsequent degradation of adhesives.

IV. EFFECTS OF WATER ON THE INTERFACE

Metal and glass adherend structural joints have high-energy surfaces on the adherends which demand best-candidate adhesives be polar in nature (27). Even so, the use of such joints in water-containing service environments has some definite limitations which need to be discussed. Since water is a highly polar molecule that can permeate polymers, it is practically impossible to prevent it from migrating to the interface, where it can interfere with the bonding. Establishing the long-term joint durability potential under such service conditions is a key issue in extending the application potentials both within and beyond the aerospace industry.

Water can arrive at the interface by some combination of diffusion through the adhesive and wicking along the interface from a joint edge. Bethune (1000), Brewis et al. (1106), and Gledhill and Kinloch (2851) have all demonstrated a preference in epoxy/metal joints of water diffusing through the epoxy rather than passage along the interface. It must be remembered that special surface pretreatments promoted good wetting of the adherends in these investigations which limited the potential for water wicking along the interface. Where such procedures might not be employed in manufacturing outside aerospace industries, the faster pathway to the interface might very likely be by wicking.

Water as a relatively small molecule with strong association through hydrogen bonding can develop strong localized interactions with the polar groups in organic polymer adhesives (2980). Though many examples of Fickian water sorption have been offered earlier in this chapter, water diffusion in a polymer is typically non-Fickian (2981). Thus, the diffusion coefficient is not constant but may depend highly on the concentration and the sample thickness. Further, the overall rate of water penetration can increase with increasing temperature and the presence of internal and external stresses (including the swelling stress). The publications of Lefebvre and associates (2904,2905,2907) mentioned earlier constitute a rigorous analysis of the kinetics of water diffusion in bonded joints.

Kinloch (1099) has indicated that a relatively lower diffusion rate shown in both epoxy and phenolic structural adhesives can help explain their lower susceptibility to moisture attack as compared with some other adhesive families. Kinloch (1099) has proposed that the water front advances toward the joint center with the entire joint eventually attaining an equilibrium moisture content.

Water can be present in the environment as liquid or vapor at some varying relative humidity. Little difference in threat should exist between liquid or 100% RH conditions; however, the equilibrium moisture uptake should be diminished substantially as the humidity is successively lowered. Enns and Gillham (2982) have developed typical sorption isotherms showing that the equilibrium moisture uptake is reduced by increasing the cross-linking in a cured epoxy. Kinloch (1099) has reported an estimated 1.35% water uptake in an epoxy to be a critical water concentration level for that system. Exposure to a humidity below 50% apparently failed to cause any permanent damage to the joint strength. Bethune (1000), Brewis et al. (1106), Brockmann (2983), and DeNicola and Bell (1114) all stated that prolonged water exposure will shift the failure mode from cohesive to adhesive in nature. This indicated to Kerr et al. (2984), Noland (838), and Butt and Cotter (2985) that interface weakening rather than bulk adhesive degradation was responsible for the joint strength decline. Butt and Cotter (2985), for example, showed the water-affected epoxy suffered a decline in tensile modulus and glass transition temperature. The changes proved to be reversible, however, when the joint was subjected to dehydration conditions. The lowered permeability of adhesive formulations with inert mineral or metallic fillers can be reviewed in the *Handbook of Fillers and Reinforcements for Plastics* (2986). Griffith and Bultman (2987) have shown the equilibrium moisture uptake could be reduced by 85% by highly fluorinating an epoxy.

Obviously the water can also exert some possible changes in the inorganic oxide side of the interface which can be disasterous. Venables and co-workers (33,34,754) have proposed that moisture can convert the initially present oxide in aluminum joints to aluminum hydroxide with a chemical composition between that of boehmite (the monohydrate) and a pseudoboehmite (a dihydrate). The cohesive strength of this converted oxide layer is quite weak, so that the adhesive side of the failed joint is covered with a portion of this oxide. The aim of the joint manufacturer, therefore, is to provide a surface pretreatment with enhanced oxide stability in water such as that provided by the PAA process (30). Ahearn et al. (2988) have additionally described the use of a monolayer of hydration inhibitor to further improve aluminum joint durability.

Finally, we can apply a primer (coupling agent) solution to the adherend surface prior to bonding to improve wet strength and durability. The organosilanes (1110,2989,2990) have been the most common type used. Gettings and Kinloch (212,213) have conducted durability testing with epoxy/steel joints and Boerio and Gosselin (717) evaluated aluminum adherend joints with interface inspection. Gettings et al. (2991) determined by AES and XPS that failure in such cases occurs through the polysiloxane layer, making it the weakest link in the joint system. Sung and co-workers (2046–2048) have showed both the structure and thickness of the polysiloxane are important variables. Calvert et al. (2992) have employed organotitanates on metal adherends as well. Bolger and Michaels (147) have discussed the good metal joint durability responses in water exposure with phenolic-based primers.

A. Interface Stability in Terms of the Thermodynamic Work of Adhesion

Each adherend has a surface free energy in the absence of water which can be calculated as the summation of the polar and dispersion components of the overall surface free energy. The former is due to dipole-dipole interactions and hydrogen bonding, whereas the latter is due to London dispersion forces. A compilation of the surface free energy values for a variety of adherends, adhesives, and water compiled by Kinloch can be viewed in Table 117. Kinloch and co-workers (1099,1160,2851) have considered the thermodynamic work of adhesion, W_A, required to separate a unit area of two

Table 117 Surface Free Energies (mJm^{-2}) of Adherends, Adhesives, and Water

	γ^{o}		γ^{p}		γ
SiO_2	78		209		287
Al_2O_3	100		538		638
Fe_2O_3	107		1250		1357
CRFP (heavily sanded)	27.4	5.3	30.6	5.3	58
CRFP (HNO_3-treated)	12.9	3.5	60.6	9.3	73.5
Carbon fiber	25.9	1.5	25.7	3.3	51.6
Epoxide[a]	41.2	2.4	5.0	0.8	46.2
Modified epoxide[b]	37.2	3.1	8.3	2.0	45.5
Cycloaliphatic epoxide[b]	28.9	3.2	25.0	4.7	53.9
Polymethyl methacrylate[c]	35.9		4.3		40.2
Water	22.0		50.2		72.2

[a]DGEBA cured with 9.4 wt% of 2,4,6-tris(dimethylamino)phenol and 2-ethylhexanoate.

[b]Cured under dried nitrogen.

[c]PMMA representing a reaction-setting acrylic adhesive

Data from Ref. 2866 (Table 6, p. 117).

Table 118 Work of Adhesion in Air and in Water

Interface	W_A (mJm^{-2})	(W_{AL} (mJm^{-2})
Epoxide/SiO	178	−57
Epoxide/Al O	232	−137
Epoxide/Fe O	291	−255
Epoxide/CFRP	88–90	22–44
Epoxide/CFRP (heavily sanded)	92	18
Epoxide/CFRP (HNO treated)	95	−20
PMMA/Al O	216	−102
PMMA/Fe O	270	−167

Data from Refs. 1099 and 2866 (Table 7 on p. 118).

phases forming an interface which is related to the surface free energies by the Dupre equation. For adhesive/adherend surfaces in an inert atmosphere, the term W_A is employed but if the separation is to take place in a liquid then the term W_{AL} is employed. As can be seen in Table 118, W_A usually has a positive value; however, in the presence of a liquid such as water, the value may be negative (1099). When the work of adhesion value is negative, it indicates an unstable interface is present which could dissociate under water-soaking conditions. Kinloch (2852) has also provided information comparing the values of work of adhesion for various interfaces with the physical evidence of what happened to such joints after their immersion in water in the unstressed condition, as shown in Table 119. In addition, Kinloch makes comparisons of the effect of shifting the immersion medium from water to various other liquid environments. For example, in ethanol instead of water, the value for work of adhesion for epoxy/ferric oxide joints rises from −255 mJm in water to 22 mJm in ethanol. This would predict no joint failure in ethanol immersion and definite joint failure in water soaking and that is what is observed in actual experimentation.

Table 119 Values of W_A and W_{AL} for Various Interfaces and Environments

	Work of adhesion (mJm^{-2})			Evidence of interfacial bond after immersion of unstressed joints
Interface	Inert medium (W_A)	Liquid medium (W_{AL})		
Epoxy/ferric oxide	291	Ethanol	22	No
		Formamide	−166	Yes
		Water	−255	Yes
Epoxy/aluminum oxide	232	Water	−137	Yes
Epoxy/silica	178	Water	−57	Yes
Epoxy/carbon fiber–reinforced plastic	88–90	Water	22–44	No
Vinylidene chloride copolymer/polypropylene	88	Water	37	No
		Sodium n-octyl sulfate (3.5%)	1.4	No
		Sodium n-dodecyl sulfate (0.5%)	−0.9	Yes
		Sodium n-hexadecyl sulfate (0.05%)	−0.8	Yes

Data from Ref. 2196 (Table VI, p. 641).

Table 120 Effect of Surface Treatments on Different Alloy Aluminum Joints Using a One-Part Heat-Curing Epoxy Adhesive

		Alloy 2036-T4			Alloy 6151-T4			Alloy X5085-H111		
Exposure condition		Mill-finish	Vapor-degrease	Alodine 1200[a]	Mill-finish	Vapor degrease	Alodine 1200[a]	Mill-finish	Vapor degrease	Alodine 1200[a]
Int. strength	(psi)	1850	1930	2130	2550	2690	2530	2250	2270	2110
	(MPa)	(12.75)	(13.30)	(14.68)	(17.57)	(18.53)	(17.43)	(15.50)	(15.64)	(14.54)
85% RH[b]		380	313	2200	1350	1550	2410	1860	1890	1910
75°F (24°C)		(2.62)	(2.16)	(15.10)	(9.30)	(10.68)	(16.60)	(12.82)	(13.02)	(13.16)
100% RH[b]		20	0	690	920	1050	1100	1260	1520	1230
125°F (52°C)		(0.14)	0	(4.79)	(6.34)	(7.23)	(7.58)	(8.68)	(10.47)	(8.48)
5% salt spray[c]		0	0	1110	80	150	2090	690	530	1210
95°F (35°C)		0	0	(7.65)	(0.55)	(1.03)	(14.40)	(4.75)	(3.65)	(8.33)

[a]Commercial chromate-phosphate conversion coating solution.
[b]Exposure time was 6 mo.
[c]Exposure time was 3 wk.
Data from Ref. 859 (Table 3–2, p. 82).

Where positive W_A values in inert medium become negative in water, a driving force has had to be provided for progressive displacement of the adhesive at the interface. It is likely as a result the joint strength will first be reduced then the locus of joint failure will gradually change from cohesive to adhesive in nature. Evidence of this is clearly offered in Gledhill and Kinloch (2851) and Gettings et al. (2873) technical reports. Notice in Tables 118 and 119, however, that unlike the aluminum/epoxy interface, the carbon fiber–reinforced plastic interface with epoxy retains a positive work of adhesion value even under water-soaking conditions. We further note in Table 119 that this agrees with the visual observation that no interfacial bond failure occurs under water-soak conditions. In spite of all these observations, it still needs to be recognized that the adhesive failure energies required for the range of crack growth rates normally encountered in the real world are much higher than any of these thermodynamic work of adhesion calculated values. Kinloch has offered the explanation that such disparity between the values is probably because under an applied load, mechanical strain-energy is available to assist environmental cracking or debonding. This is then reflected in inelastic energy dissipative processes.

Gent and Schultz (2993), Mai (2703), and Kinloch and Shaw (2333) have all shown the expected reduction in measured adhesive failure energy for interfacial crack growth where a liquid is present in the environment as compared to air. It needs also to be emphasized that the mathematical equations dealing with thermodynamics do not take into account any interfacial adhesion forces that can arise in some jointures from primary bond interactions or mechanical interlocking situations. Another disadvantage is that the thermodynamics approach implies that the joint strength will fall to zero. In practice, however, we know that the declining aluminum/epoxy joint strength in water-soaking conditions, in many cases, seems to have reached some limiting value as shown by Orman and Kerr (2994) in Figure 19. Also, the thermodynamics approach does not permit the concept of recovery of lost joint strength after water soaking followed by drying out, as shown in Figure 20. If the duration of exposure and drying out times are controlled by the rate of water diffusion, then it might be expected that a similar time period to attain a certain level of water intake will be that which is required to remove it. Most sorption/desorption cycles conducted on aluminum joints in

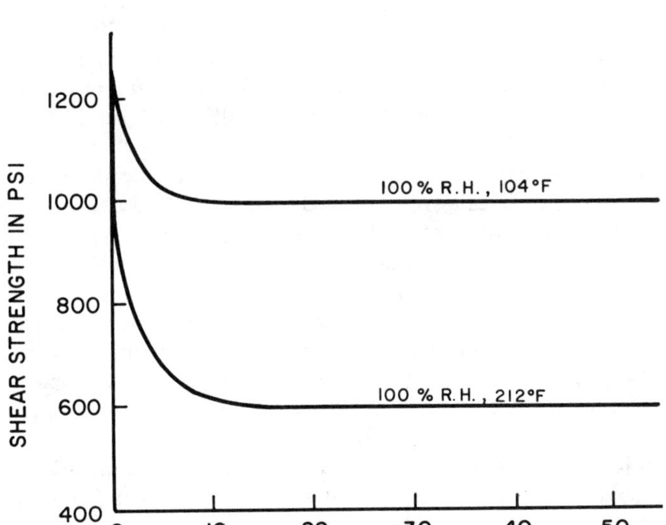

Figure 19 Typical bond strength decay curves of aluminum/epoxy bonds. (From J. D. Minford, Aluminum Adhesive Bond Permanence, Treatise on Adhesion and Adhesives (R. L. Patrick, ed.). Marcel Dekker, New York, Vol. 5, p. 64, Fig. 3–9 [1981].)

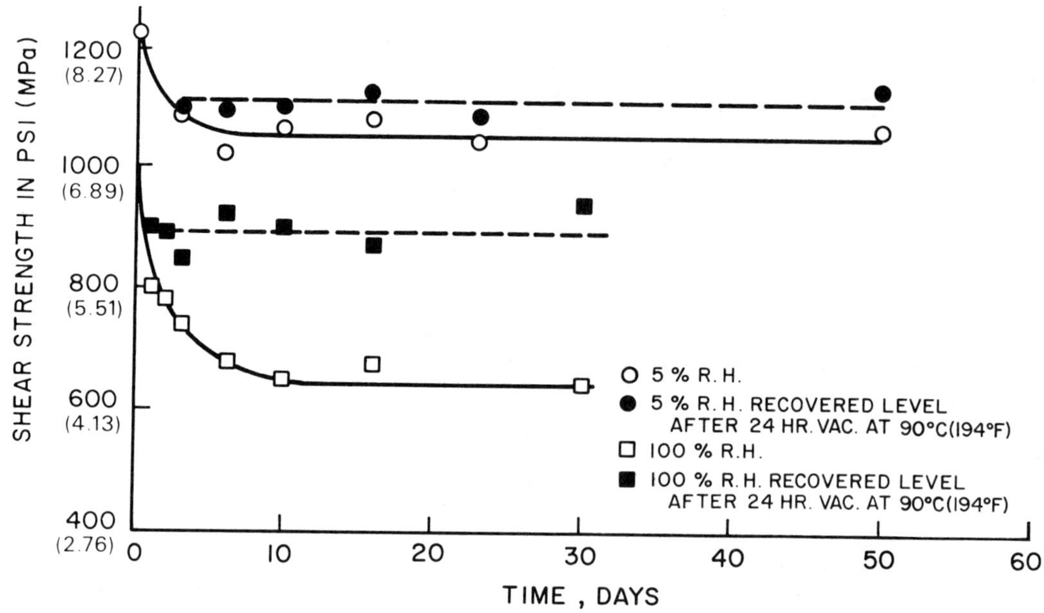

Figure 20 Exposure of aluminum/epoxy bonds at 90°C (194°F). (From J. D. Minford, Aluminum Adhesive Bond Permanence, Treatise on Adhesion and Adhesives (R. L. Patrick, ed.). Marcel Dekker, New York, Vol. 5, p. 65, Fig. 3-11 [1981].)

the literature have consisted of much longer sorption times, as indicated in Figure 19, where the soaking time was 50 da compared to 24 hr of drying out. Comyn and co-workers (2085,2995) have exposed FM-1000 and BSL 312 bonded aluminum joints to water for 1000 hr followed by 1000 hr drying out but could only conclude that some strength recovery was possible but never full recovery. Ishii and Yamaguchi (2996) exposed aluminum/epoxy-polyamide joints to 140°F (60°C) and 212°F (100°C) and nitrile-phenolic joints to 212°F (100°C) wet conditions with the expected declining joint strength results in each case. Upon subsequent exposure to drying at 104°F (40°C), however, the strength recoveries were only slightly noticeable. Once again, as noted earlier for the water-uptake plots, a great variation in performance will apparently be found with different adhesive types. The author has even shown great variation between different formulations of the same chemical type of adhesive in water-soaking durability studies with aluminum joints.

Kerr et al. (2874) compared the effect of dry ethanol or water on cured epoxy adhesive alone and in an aluminum adherend joint with the results shown in Figure 21. The alcohol obviously had virtually no effect on the interfacial joint strength, yet degraded the tensile strength of the cured epoxy adhesive significantly. In contrast, the tensile strength of the cured epoxy was little affected by water-soaking conditions, yet the joint strength was seriously degraded.

Other potential differences between joints that are not allowed for in thermodynamic considerations are: (1) values for W_A and W based on air-cast adhesive samples may be different from the actual situation of curing the same adhesive against the adherend surface, (2) there may be special orientation or preferential adsorption of one adhesive component as compared to other component ingredients, or (4) mechanical interlocking of the adhesive and adherend may occur (2997). Finally, thermodynamics provides no information about the actual service life potential of stressed joints in water environments which can be so dramatic, as shown by comparing Figures 15 and 17.

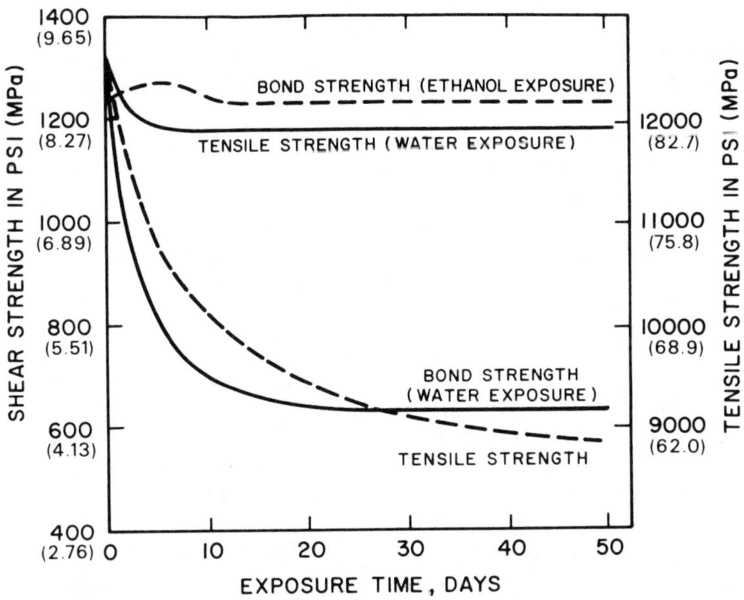

Figure 21 Bond strength of aluminum/epoxy joints and tensile strength of epoxy resin after exposure to water and ethanol at 90°C (194°F). (From J. D. Minford, Aluminum Adhesive Bond Permanence, Treatise on Adhesion and Adhesives (R. L. Patrick, ed.). Marcel Dekker, New York, Vol. 5, p. 64, Fig. 3-8 [1981].)

B. Presence of Primary Interfacial Bonding

We have already considered the fact that if secondary force interactions across the interface are presumed present, then it is possible from thermodynamic calculations to derive the required values of dispersion and polar force components of the surface free energy to produce adhesive-adherend interface stability in the presence of the environment. Kinloch (2998) has graphically presented such data for a polyethyleneterephthalate adherend and an iron oxide surface in a water environment. Further, it could be shown that certain adhesives such as an amine-cured epoxy resin should form an environmentally water-stable interface with this polyethyleneterephthalate adherend but an unstable interface with the iron oxide. This demonstration from thermodynamic considerations that water will virtually always desorb an organic adhesive from a metal oxide surface, i.e., if only secondary interfacial bonding is present, must lead to the conclusion that stronger interfacial forces need to be present if the interface is to resist rupture in water.

One method of establishing a primary interfacial bond between metal oxide surfaces and epoxy adhesives has been to link these surfaces via an organometallic primer such as gamma-glycidoxypropyltrimethoxy silane (213,2999). The increased durability is graphically shown in both Kinloch publications (2196 and 2999). It would, of course, be of great interest to be able to quantify the contribution of such primary interfacial bonding to the stability of such an interface in water; however, at present, it is impossible to calculate the contributions exactly. Nevertheless, Kinloch (2998) has offered an approximate indication by taking the interfacial, chemical bond energy and assuming a certain coverage per adsorbed site. The intrinsic work of adhesion value obtained would make it unlikely that water would displace such a chemisorbed primer layer.

Bolger and Michaels (147) have suggested that the known excellent durability of aluminum/phenolic-modified epoxy and vinyl adhesives in water environments might be explained by the possibility for interfacial, covalent bonds to exist. While the fact that the reaction conditions required to cure such adhesives can be considered to be those favoring the formation of ether linkages between the metal oxide and the phenolic resin, the existance of such interfacial bonds have best been evidenced indirectly through the use of inelastic tunneling spectroscopy evidence by investigators such as Lewis et al. (214).

Comyn (3000) has also proposed the good durability of phenolic-modified aluminum joints to presence of covalent bonding across the interface. He further suggests such bonds might be formed by the coordination of phenolic oxygens to aluminum ions as demonstrated in the form of complexes between aluminum and 8-hydroxyquinoline by Brockmann (373) or alizarin dye attachments (see Refs. 374 or 375). Comyn (3000) also discussed the possibility for ion-pair bonding at aluminum interfaces with epoxy adhesives based on the combination of aluminum preparation procedure and mechanism of the epoxide curing itself. He has further included his speculations about the possible interaction at the aluminum interface with primary, secondary, and tertiary amines which often function as curing agents for the epoxies. Finally, he suggests that ion exchange site production on aluminum surfaces may be particularly pertinent to the acid pretreatment procedures used to promote more effective wetting and bonding to aluminum surfaces.

Knoch and Locke (3001) have also stated that organic polymers in adhesives can react with metal oxide surfaces (including the hydroxides) through chemical linkage. Brockmann (373) has demonstrated this by his studies of adsorption and desorption on aluminum adherends or by chromatographic methods he employed using alumina as the adsorbant. Brockmann (3012,3013) and Kinloch (879) have considered the initial adhesional strength between polymers and polar groups and metal oxides in terms of surface energy theories, adsorption models, acid-base interactions, and straight chemical reactivity.

V. ALUMINUM ADHEREND STABILITY CONSIDERATIONS

Good aluminum surface oxide stability certainly seems to be a definite requirement for ensuring long-term joint durability in the weathering service of bonded aluminum structures. Good oxide stability can also be obtained with accompanying microporosity through various deoxidizing and deoxidizing plus anodizing pretreatments. It was Hartman who first showed the important of micromorphology on an aluminum oxide surface by etching away the aluminum after curing phenolic reson against it. He was able to show that the phenolic resin now had the exact replica surface of the oxide morphology to which it had first been adhering. Twenty years later, Hennemann and Brockmann (3013) showed that the European Chemoxal pretreatment for aluminum and the American developed FPL treatment had similar surface morphologies and, consequently, similar aluminum joint durability potentials. Pretreatment of the aluminum adherend seems most often responsible for establishing the good resistance to debonding in water. The reader is referred back to the section on different surface pretreatments for more details about these procedures. The durability of aluminum joints so pretreated will be considered in detail in Chapter 10 on the durability of aluminum joints in various weathering conditions.

In spite of the extensive investigations that have been conducted over many years on the special pretreating and choice of adhesive in aluminum bondlines, it still may not be possible to precisely define all the exact chemical and physical parameters that may be required to fabricate the joint having the maximum resistance to all kinds of environmental conditions. For example, the particular oxide which offers the maximum opportunity for physical interaction (mechanical interlocking) may not be the most wettable. While the oxide with highest wettability may not offer the maximum opportunity for interlocking with the adhesive, in like manner, the oxide established by one pretreatment offering excellent wettability may not be the most stable in some particular type of service weathering. Perhaps the classic example of this latter situation was that manifested in the exposure of new-design aluminum-bonded joints in naval aircraft in the Vietnam War. In this case, the currently employed aluminum surface pretreatment was the FPL deoxidation procedure with ongoing efforts to maximize the durability experience by creating a so-called "modified FPL" procedure. In prior aircraft, the joint design had offered relatively good opportunities to protect the bondline from salt-water invasion, so that there was no question raised about the proficiency of the existing pretreatment for bonding naval aircraft that would be operating from carriers at sea or close by naval seacoast stations. With the newer joint designs built into a new wave of aircraft, the exposure conditions changed so that the jointures could be more easily contacted directly by the corrosive environmental conditions. The end result was a sudden rash of aircraft joint failures that was both unexpected and alarming to the military. Without the benefit of joint design playing a role in the overall durability evaluation, the author had been proceeding to conduct voluminous testing of FPL-type joints with a wide variety of aerospace-type adhesives present but no primers in the bondlines or sealants overlaying the joints. One of the five types of general weathering conditions employed at the Alcoa Labs involved the use of wet/dry cycling in salt water in the laboratory or to the related wet/dry salt-water cycling conditions encountered at a natural seacoast exposure station. The extreme sensitivity of FPL-treated joints to premature debonding under either set of weathering conditions could be readily demonstrated as compared with excellent resistance to such debonding shown by CAA or SAA pretreated joints. Unfortunately, this Alcoa information had been developed and was being communicated only to industry outside of aerospace. It was presumed that the extensive government funding of the aerospace manufacturers had ensured that their technical expertise in manners of pretreating and durably bonding aluminum was the most comprehensive available. Nevertheless, it must be concluded that the information should have been already available before manufacturing was instituted that the FPL procedure created vastly inferior joints for seacoast exposure situations to those using some form of anodize pretreating. The preference for an anodizing pretreatment instead of a straight chemical deoxidation in present-day aerospace craft is confirmation of this.

Now the PAA is used almost exclusively in American aircraft and the CAA pretreatment in their European counterparts. Some unusual situations have even been found using PAA oxide at the interface. For example, Knoch and Locke (3001) conducted a case study of water sensitivity of two PAA oxides formed at anodize temperatures of 62°F (16.7°C) and 77°F (25°C). They established that load transfers in the joint under wet conditions seemed facilitated by relatively large pore/oxide volume conditions at the oxide/polymer interface. If the oxide generated by one set of anodizing conditions has more tapering at the upper oxide surface, then the load can be transferred better into the oxide depth, resulting in a cohesive failure. In the alternative situation without a tapered oxide condition present, a "slick-type" joint failure could be found. Minford (533,534) has shown a modest difference in performance between three sets of PAA anodizing conditions as compared with that employed by Boeing. The best joint durability performance in aggressive weathering or a combination of stress and hot humidity was shown by the Boeing conditions, which also seemed to have the largest pore/oxide volume statistics.

We need to remember that with metallic adherends, the water in the service environment may either attack the interfacial oxide directly or become involved with corrosion processes acting on the bulk metallic adherend itself. In the former situation, the presence of water in the interfacial area can be the single most degrading factor in converting the oxide from a desirable to a less desirable form. It is for this reason that the differences in the chemistry and morphology of the oxides produced by the different surface pretreatment procedures do directly affect the joint permanence. The speed and degree with which water can convert the existing oxide to a less stable form translates directly to the level of bond durability achievable. A lower oxide stability can also be related to its lower inherent resistance to stressful rupture. Thus, the joint can fail cohesively at a much reduced joint strength level in a rupturable plane of the interfacial oxide. This could occur, in the case of particularly unstable oxides, at a time when the loss of cohesive strength in the adhesive layer due to water diffusion was still relatively insignificant.

Venables (3002,3003) stated that moisture intrusion at the interface in an aluminum joint can cause the oxide (whether naturally formed or created by some special pretreatment) to eventually convert to an hydroxide. This will obviously be accompanied by a drastic change in morphology. The resulting hydrated layer can show poor adhesion to the underlying aluminum, so that bond strength will be severely diminished. (See Ref. 33, p. 461, for Venables' proposed failure model.) Venables has also observed a correlation between the time to failure of aluminum "wedge-joint" specimens and the incubation times determined for the oxide-to-hydroxide conversion process to take place (this test procedure will be considered in greater detail in Chap. 9). For those aluminum surface pretreatment oxides that require longer incubation times, the wedge test results will be correspondingly better.

Working in the same laboratory, Davis et al. (525) have developed a graphical device (similar to a phase diagram) which represents the reaction between a surface, an adsorbate, and the environment. It was shown by this technique that the oxide produced by PAA had a much greater stability in water than that produced by the FPL-type etching procedure. These authors claim this good stability may be due to the presence of aluminum phosphate on the anodized surface. The phosphate supposedly continues to act as a chemical inhibitor until it is itself hydrated and lost by dissolution.

The second method of joint debonding involving the direct corrosion of the aluminum adherend is also quite important. Certainly, it is the primary mechanism of failure when aluminum joints are exposed to the seacoast or marine salt-water weathering conditions. In our present highly industrialized society, we also might need to consider the possible effect of so-called acid rain or other corrosive industrial pollutants. The significant acceleration of joint failure that can occur with an electrolyte like sodium chloride is clearly evident by comparing the survival times of joints under water-soaking conditions with those under wet/dry cycling salt-water exposure. Minford (39,68,72,533,534,772–774) has routinely made such comparisons between thousands of bonded aluminum joints that attest to the shorter survival times of similarly fabricated joints in salt fog, wet/dry salt-water immersion or natural seacoast exposures as compared to ordinary water immersion or condensing humidity conditions. One exception to this overall generalization has been the exposure of aluminum joints to ambient salt-water conditions with simultaneous imposed stressing as compared to hot-humidity exposures with added stressing. The latter situation has always produced the shorter-life joint response using duplicate joints. For example, the short joint survival times obtained by Minford (see Fig. 17) using FPL-type pretreated joints and a two-part, room temperature–curing adhesive under stress and hot condensing humidity conditions would always be significantly longer if the same stressed joint had been placed in ambient temperature salt water. If similar joints were exposed under the same stress in distilled water, then about the same joint survival times would be recorded as found for those using salt water. It would appear that the stress/hot humidity environment is so destructive that any salt-water effect is a longer time response stimulus to joint failure.

While ordinary water in contact with aluminum can cause a form of surface corrosion and gradual oxide conversion resulting in eventual debonding, the wet/dry salt-water contact is more clearly evidence of a severe, deep-pitting electrochemical type of attack. When an aluminum adherend surface has not been specially treated to form a resistant form of oxide to this type of attack, i.e., special conversion or anodized oxide layers, the aluminum edge surface can rapidly experience a pitting attack in which the surface in contact with adhesive can be undermined. Cycling the joint in such conditions leads to an ungoing process of (1) pitting with corrosion product development, (2) washout of corrosion product under cycling exposure conditions, and (3) exposure of a fresh metal surface for more subsequent attack and repeat of the process.

It can be shown that even pure aluminum (99.99% Al) can develop pitting to some degree in salt-water exposures. However, as more alloying elements are added to develop stronger aluminum alloys, the relative sensitivity to pitting corrosion increases. This means that the need to employ the strongest alloys in load bearing structures carries with it the penalty of using the most inherently sensitive aluminum adherends to saltwater corrosion. The greatest need thus exists to better protect such joints by special surface pretreating and sealing the bondline edges as much as possible. This effect can be clearly seen in the data presented by Minford (3004) in Table 115, where the comparative retention of joint strength in 5% salt spray exposure is directly related to the order of respective inherent corrosion resistance of the adherend alloys to salt water; i.e., X5085-H111 > 6151-T4 > 2036-T4.

The corrosion sensitivity is based on the relative ease with which the chloride ion in water can disrupt the aluminum oxide layer present and create an anodic pit area. This creates a condition whereby the more anodic metal present will be locally sacrificed in order to protect the more passive cathodic area around the pit. Although not commonly recognized, Donnelly and Cohen (3005) have shown that even so-called inert polymers like polyethylene, polypropylene, polystyrene, and polymethyl methacrylate can be degraded in undersea exposure conditions. It has been shown using stressed double-cantilever beam aluminum joints that a 3-mo exposure in 5% salt-water can be more severe than a 3-yr exposure in semitropical conditions. McMillan (2707) has made comparisons of aluminum joint strength losses found in various weathering conditions and has shown the most degrading weathering conditions to be a 5% salt spray. Of course, Minford (2853) has more recently directly compared the 5% salt-spray condition to two different wet/dry salt-water cycling procedures from Alcoa, and has clearly shown the significantly more severe degrading effects of the cycling type of salt-water exposure as compared to continuous salt-spray conditions. The effect of added stressing in the presence of salt water has also been studied in detail by Minford (773) using a wide variety of aluminum surface conditions. This will be considered in more detail in Chapter 10. It might be theorized that the diffusion constant of water should increase exponentially with added stress as pointed out by Fahny and Hart (3006). However, Gilat and Broutman (3007) have shown that water uptake sorption does not follow such a pattern. It appears that whether the existing stress affects the joint strength during exposure to humid environments depends greatly on the type of adhesive present. For example, Minford has shown in Figure 17 that while the effect of stress and humidity on the epoxy-polyamide/aluminum joints was highly adverse, the joint survival times of a vinyl-phenolic adhesive on a similar aluminum surface were significantly longer. Minford has studied the stress/water-soak response of nylon-modified epoxy FM-1000 film adhesive on aluminum-etched 6061-T6 adherends, as shown in Figure 22. He found the stress effect was considerably less drastic than for either the two-part epoxy or vinyl-phenolic bonded joints using similarly prepared 6061-T6 adherends. Actually, this response was rather unexpected considering the usual high sensitivity of nylon-modified epoxies to water-soaking conditions alone.

Venables (34) has proposed that a primer is often of value in improving the moisture resistance or preventing metal corrosion of the aluminum adherend. It has always been the general practice of the aerospace joint manufacturing procedures to employ priming of the specially pretreated aluminum surface. The extensive testing of baked paint primers on aluminum by Minford (774) which were applied over only vapor-degreased surfaces can also be cited and will be discussed in more detail in Chapter 10. Hardwick et al. (527), in 1984, found a new aluminum oxide hydration inhibitor which

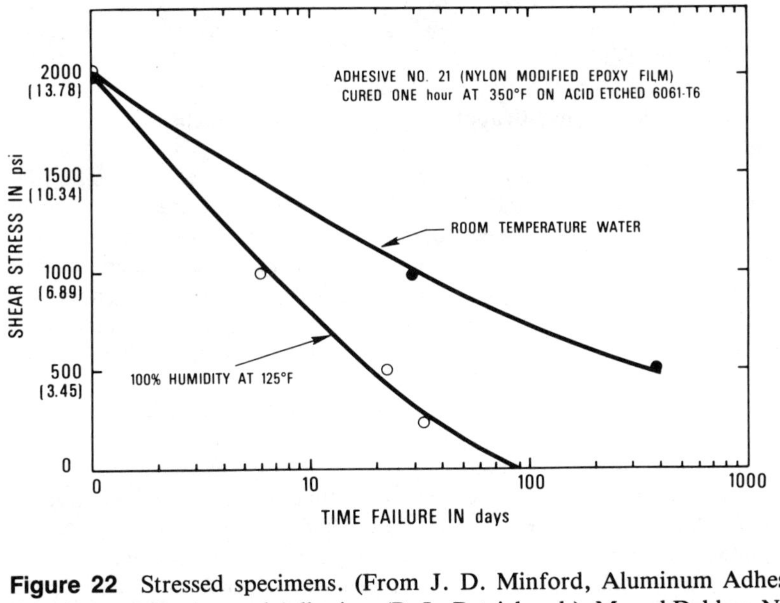

Figure 22 Stressed specimens. (From J. D. Minford, Aluminum Adhesive Bond Permanence, Treatise on Adhesion and Adhesives (R. L. Patrick, ed.). Marcel Dekker, New York, Vol. 5, p. 124, Fig. 3-60 [1981].)

can be considered a form of priming to promote greater aluminum bond durability. They proposed a three-step hydration process consisting of (1) reversible physisorption of water, (2) slow dissociation of the inhibitor followed by a rapid hydration of freshly exposed Al_2O_3 to boehmite (AlOOH), and (3) further hydration of the monohydrate to bayerite ($Al[OH]_3$). Their inhibitor NTMP was found to adsorb onto the alumina surface via POH bonds of the phosphoric acid groups, resulting in a displacement of water that would normally adsorb on the surface. Thus, the eventual formation of the weakly adhering bayerite form of oxide is circumvented.

The investigation of Bailey (3008) might be mentioned as one in which various conclusions were drawn from the failure analysis study of bonded aluminum honeycomb structure where significant corrosion of the aluminum surface had occurred. The characterization of oxidized aluminum alloy 1100 surfaces when exposed to hygrothermic environments was a 1987 study of Dodiuk et al. (3009), whereas Just (3010), in the previous year, had investigated the time-temperature-moisture effects of environmental exposure on the fracture failure of thick adherend aluminum 6061-T6 alloy joints bonded with FM-73M and FM-300M structural aerospace adhesives. Brewis et al. (3011), in 1990, added to their many previous investigations of the effect of moisture on aluminum/epoxy joints in a paper simply titled "The Effect of Humidity on the Durability of Aluminum-Epoxide Joints."

In the above discussions relating the presence of water and the thermodynamic relationships at the interface, it had to be presumed first that water could reach the interface via the adhesive and/or travel successfully along the interface from the edge areas. Brockmann (373) has pointed out that the polymer adsorbed on anodized aluminum oxide was still separable in water, indicating that chemical bonding alone between polymers and metal oxides cannot account for the stable adhesion in water demonstrated by many investigators using anodized aluminum adherends [Minford (533,534, 773,774)]. Brockmann et al. (523) have demonstrated that very small differences in the parameters of the aluminum surface pretreatment leading, in turn, to relatively small differences in surface morphology can still significantly change the water stability of the boundary layer in the joint. The transition zone between the oxides and various primers also was investigated by using a wet-peel testing procedure. Ultrathin cross sections from this boundary zone before and after delamination of the joints were studied by TEM, whereas the actual failed adhesive surfaces were analyzed by ESCA. As a result of the water exposure, a zone was found very near to the oxide layer whose properties were quite different from those of the bulk adhesive. In many cases, relatively large amounts of the adhesive remained on the oxide after failure, but no metal oxide was detectable on the failed adhesive layers. It was concluded that, at least partly, the adhesional failure is actually cohesional failure in very weak boundaries of the adhesive. The properties in this area have undoubtedly been influenced by the state of the anodized aluminum oxide surface. Other Brockmann papers relating to the effects of surface morphology on aluminum joint strength and durability have been co-authored by Kollek (3012) and Hennemann (523,1046), whereas investigations by Bijlmer (3014) and Ahearn et al. (3015) would have related value.

Many investigators have stated a possible relationship between the rate of water diffusion into a joint and the environmental joint failure (1106,2535,2085,2086,2929,2930). The predictions made by these investigators seem reasonably accurate, especially around the periphery of the joint where the initial debonding occurs. Brewis and co-workers (1106,2085,2086) have made water sorption measurements in adhesives in joints to demonstrate that a linear relationship can exist between the loss of joint strength and the total water content of the ahdesive layer. Obviously, when comparing Figures 15 and 17 involving the stress and humidity environmental testing of aluminum/epoxy joints by Minford that imposing additional stress simultaneously with entering water can precipitate total and sudden catastrophic failure long before any significant total water uptake could occur.

Gledhill et al. (2705) have developed a model for predicting joint durability in water exposure which is based on a combination of thermodynamics, water diffusion data, and continuum mechanics. The agreement they obtained between predictions and actual failure data using their model was termed good by the authors. The fact that they employed mild steel adherends and tensile butt-joints (instead of tensile shear lap-joints) should not make their data any less pertinent for offering explanations with aluminum/epoxy lap-joints exposed to the same water contact. While the two different metallic oxides may give different absolute numbers for debonding purposes with different epoxy adhesives in the bondline, the fact that both metallic adherends are equally impervious to any water entry should make the water pickup similarly difficult and quite comparable. Kinloch and co-workers have stated that the concentration profiles for water ingressing into the joint as a function of time and temperature should be quite pertinent to predicting overall joint durability. The kinetics for failure of the butt-joint configuration they believed should be governed solely by the rate of water diffusion (obviously using unstressed specimens). They were able to arrive at a constant critical water concentration for debonding, which could be used to deduce the interfacial environmental crack length, a, as a function of time in water at a given temperature. Finally, the crack length could be combined with the independently measured values of G_c and E_a of the adhesive to predict failure stress when the test specimen was fractured in a testing machine after exposure.

Ripling et al. (2492,2503) determined with their tapered double-cantilever specimen that the crack extension force, G, was about the same for a particular adhesive polymer whether the adherend was glass or aluminum. With water-induced joint failures, the induced fracture appeared to have propagated at the polymer/adherend boundary. Failure seemed to occur in a zone that was much greater than one or two molecular layers. This might indicate that there is a rather wide water-sensitive zone. This zone could be a hydroscopic oxide on the adherend or some component of the polymer that might have accumulated near the interface.

9

Test Evaluation Standards

I. INDUSTRY SPECIFICATIONS AND STANDARDS

All the testing procedures that have been developed have mainly been centered around the evaluation of adhesive properties, adherend properties, or how these materials function together in a bonded joint structure. Most influential in developing specific testing procedures for general comparative use has been the American Society of Testing Materials (ASTM) (3016) through the activities of Subcommittees D-14 and C-24 dealing with adhesive and sealant technologies, respectively. ASTM Committee D-14 on adhesives is further subdivided into 11 subcommittees covering the development of material specifications and methods for testing for adhesives and adhesive/sealant materials. The C-24 subcommittee on building seals and sealants is further split into 20 additional committees whose interest covers both the development of materials specifications and test methods but also the promotion of knowledge and stimulation of research for sealing of buildings and traffic decks. The approximately 235 specifications and standards for adhesives and sealants are contained in Volumes 15.06 (3017) and 04.02 (3018), respectively. ASTM standard specifications are typically listed, as shown in Table 121. Also functioning as a national source for testing specifications is the British Standards Institution (BS series of publications).

Other sources for industrial standards have been developed by the Society of Automotive Engineers (SAE) (3019) and the Technical Association of the Pulp and Paper Industry (3020), typically designated as shown in Tables 122 and 123. More than 200 SAE specifications can be found on plastics, adhesives, elastomers, and related materials in the AMS Index of Specifications.

Various federal and military Specifications also exist for the evaluation of materials used in various forms of industrial bonding and assembly of military hardware.

There is also the International Organization for Standardization (ISO) which is headquartered in Geneva, Switzerland. Copies of the ISO standards and other industry standards can be obtained from the American National Standards Institute (ANSI) at 1430 Broadway, New York, NY 10018.

Table 121 Typical ASTM Standard Specifications and Test Methods Relavant to Adhesives and Sealants

Designation	Title
ASTM C-557	Specification for Adhesives for Fastening Gypsum Wallboard to Wood Framing
ASTM C-920	Specification for Elastomeric Joint Sealants
ASTM D-905	Test Method for Strength Properties of Adhesive Bonds in Shear by Compressive Loading
ASTM D-950	Test Method for Impact Strength of Adhesive Bonds
ASTM D-1151	Test Method for Effect of Moisture and Temperature on Adhesive Bonds
ASTM D-2235	Specification for Solvent Cement for Acrylonitrile-Butadiene-Styrene (ABS) Plastic Pipe and Fittings
ASTM D-2559	Specification for Adhesives for Structural Laminated Wood Products for Use Under Exterior (Wet Use) Exposure Conditions
ASTM D-2851	Specification for Liquid Optical Adhesive
ASTM D-3933	Practice for Preparation of Aluminum Surfaces for Structural Adhesive Bonding (Phosphoric Acid Anodizing)
ASTM D-4317	Specification for Polyvinyl Acetate-Based Emulsion Adhesives
ASTM F-607	Test Method for Adhesion of Gasket Materials to Metal Surfaces

From Ref. 3021 (Table 1, p. 62).

Table 122 Typical SAE Specifications and Test Methods Relevant to Adhesives and Sealants

Designation	Title
SAE AMS-1320	Decal Adhesive Remover
SAE AMS-3107	Primer, Adhesive, Corrosion Inhibiting for High Durability Structural Adhesive Bonding
SAE AMS-3374	Sealing Compound, One-Part Silicone, Aircraft Firewall
SAE AMS-3375	Adhesive/Sealant, Fluorosilicone Aromatic Fuel Resistant, One-Part Room Temperature Vulcanizing
SAE AMS-3376	Sealing Compound, Noncuring, Groove Injection, Temperature and Fuel Resistant
SAE AMS-3681	Adhesive, Electrically Conductive, Silver-Organic Base
SAE AMS-3686	Adhesive, Polyimide Resin, Film and Paste, High-Temperature Resistant, 315 Degrees C or 600 Degrees F
SAE AMS-3695	Adhesive Film, Epoxy-Base for High Durability Structural Adhesive Bonding
SAE AMS-3704	Adhesive, Contact Chloroprene, Resin-Modified
SAE ARP-1843	Surface Preparation for Structural Adhesive Bonding Titanium Alloy Parts
SAE ARP-4069	Aerospace Recommended Practice of Sealing Integral Fuel Tanks
SAE J-1523	Recommended Practice for Metal-to-Metal Overlap Shear Strength for Automotive-Type Adhesives
SAE J-1525	Recommended Practice for Lap-Shear for Automotive-Type Adhesives for Fiber Reinforced Plastics (FRP) Bonding

From Ref. 3021 (Table 2, p. 62).

Table 123 Typical TAPPI Technical Information Sheets and Test Methods Relevant to Adhesives and Sealants

Designation	Title
TAPPI T-540	Determination of Polyethylene Adhesion to Non-Porous Substrates
TAPPI T-814	Peel and Shear of Hot Bonds at Elevated Temperature
TAPPI TI—0305-25-82	Buying, Storing, and Handling Adhesives
TAPPI UM-562	Adhesiveness of Gummed Paper Tapes (McClaurin Test)
TAPPI UM-816-82	Gluebility of Linerboard

From Ref. 3021 (Table 3, p. 62).

II. GOVERNMENT STANDARDS

The federal specifications and standards are listed in the Index of Federal Specification and Standards (FPMR 101-29.1) issued annually by the General Services Administration, Federal Supply Service (GSA-FSS) and obtainable from the Superintendent of Documents, Government Printing Office, Washington, DC 20402. Each acquisition item in the index has a federal supply classification number (FSC), which is 8040 for adhesives and 8030 for sealants. Some representative typical federal specifications are listed in Table 124. The GSA-FSS also issue some simplified forms of specification designated as the A-A item in Table 124 which can apply to both adhesives and sealants.

The most useful sources of information used by the military, however, come from the Department of Defense Index of Specification and Standards (DoDISS) which is available from the Naval Publications and Forms Center (NPFC 106), 5801 Tabor Avenue, Philadelphia, PA 19120. Part 1 of the index is alphabetical, whereas Part 2 permits identification in document-number sequence. In addition to the military specifications, DoDISS contains listings of military handbooks, federal specifications, commercial item descriptions, and the DoD-adopted industry standards (principally ASTM and SAE-AMS standards). Typical listings are shown in Table 125 for military specifications and handbooks. Military standards establish engineering and technical requirements for processes, practices, methods, and materials. Some additional information can be found in Nardone's article (3021) in the 1990 *ASM Handbook*.

Table 124 Typical Federal Specifications

Designation	Title
MMM-A-121	Adhesive Bonding Synthetic Rubber to Steel
MMM-A-134	Adhesive, Epoxy Resin, Metal-to-Metal Structural Bonding
MMM-A-180C	Adhesive, Polyvinyl Acetate Emulsion
MMM-A-1931	Adhesive, Epoxy, Silver-Filled, Conductive
SS-S-210A	Sealing Compound, Preformed Plastic for Expansion Joints and Pipe Joints
TT-S-227B	Sealing Compound, Rubber-Base, Two-Component (For Caulking, Sealing, and Glazing in Building Construction)
TT-S-1732	Sealing Compound, Pipe Joint and Thread, Lead-Free General Purpose
A-A-1556	Sealing Compound (Elastomeric Joint Sealants)
V-V-190	Sealing Compound, Dipcoating

From Ref. 3021 (Table 4, p. 63).

Table 125 Typical Military Specifications and Military Handbooks for Adhesives

Designation	Title
MIL-A-83376	Adhesive Bonded Metal-Faced Sandwich Structures Acceptance Criteria
MIL-A-47040	Adhesive Sealant, Silicone, RTV, High Temperaturc
MIL-A-46864	Adhesive, Epoxy, Modified, Flexible, Two-Component
MIL-A-48611	Adhesive System, Epoxy, Elastomeric, for Glass-to-Metal
MIL-A-24179	Adhesive, Flexible, Unicellular-Plastic, Thermal Insulation
MIL-A-87135	Adhesive, Non-Conductive, for Electronic Applications
MIL-C-2399	Cement Liquid, Tent Patching
MIL-G-46030	Glue, Animal (Protective Colloid)
MIL-P-47279	Primer, Silicone Adhesive
MIL-S-12158	Sealing Compound, Noncuring, Polybutene
MIL-S-46897	Sealing Compound, Polyurethane Foam
MIL-S-11388	Sealing Material for Metal Container Seams
MIL-S-8802	Sealing Compound, Temperature Resistant, Integral Fuel Tanks and Fuel Cell Cavities, High Adhesion
MMM-A-132A	Adhesive, Heat-Resistant, Airframe Structural, Metal-to-Metal
MIL-S-81733C	Sealing and Coating Compound, Corrosion Inhibitive
MIL-S-83430	Sealing Compound, Integral Fuel Tanks and Fuel Cell Cavities, Intermittent Use to 360°F
MIL-S-29574(AS)	Sealing Compound, Polythioether, for Aircraft Structures, Fuel and High-Temperature Resistant, Fast Curing at Ambient and Low Temperatures
MIL-HDBK-337	Adhesive Bonded Aerospace Structure Repair
MIL-HDBK-691	Adhesive Bonding
MIL-HDBK-725	Adhesives-A Guide to Their Properties and Uses

From Nardone Ref. 3021 (Tables 5 and 6, p. 63).

III. ADDITIONAL GENERAL INFORMATION SOURCES

Landrock (3022) has furnished a very complete review of such sources in regard to the available information about adhesives, sealants, and the technology of their use and application in the 1990 *ASM Handbook*. Some of the categories covered include (1) literature from manufacturers; (2) journals, trade magazines, and periodicals; (3) books, handbooks, and monographs; (4) short courses, seminars, and conferences; (5) ASTM committees; and (6) databases.

IV. GENERAL DEVELOPMENT OF TESTING PROCEDURES

Representative scientists and engineers from most industrial firms whose major business or related activities can be connected to the development, evaluation, or manufacturing of adhesives and structures bonded with adhesives or protected with sealants must be major contributors. These individuals plus corresponding scientists and engineers from the federal and military agencies interested in such matters are invited by ASTM and other test-developing organizations to regularly meet in organized subcommittees whose special interests cover the entire range of adhesive and sealant technology and applications. In addition to the development of new testing procedures, they also need to provide continual updating of all existing procedures. Having such standardized procedures available for use all over the world has enabled comparative judgments to be made across the whole field of adhesion science. The testing conditions must, of course, be rigorously adhered to if such comparisons are to be valid. After the experimental data have been accumulated, the proper mathematical procedures for drawing the right statistical inferences must be followed (3023).

The aerospace industry has directly adopted the majority of the ASTM and BS procedures for developing adhesive bonding pertinent to their industry. Because they work so closely with the government in many of their activities, however, the typical aerospace laboratory must also be equipped to conduct any of the many pertinent procedures described in a number of federal and military specifications such as those listed earlier in Tables 124 and 125. While there are many more specifications than listed in these example-type tables, the author will attempt primarily to emphasize those that have more direct relationship to usage involving use of aluminum adherends. Included are a number of procedures involving nonstructural-type adhesives which might be employed in general manufacturing of aluminum products utilizing foil and thin sheet. Some of the most important testing procedures used in aerospace have been considered in a review by McMillan (833) for a 1979 AGARD lecture. Koski and Schneberger (3024,3025) have produced two published articles on testing adhesives and adhesive joints for general manufacturing. The most detailed technical analyses of the existing testing procedures have been published by Anderson and DeVries and co-workers (2488, 2594,2595,3026–3028).

Most recently, the 1990 *ASM Engineered Materials Handbook* has included a number of review articles under Section 5, which is titled "Testing and Analysis." In this series, Dillard and LeFebvre (3029) begin by outlining the scope of the articles to follow. In consecutive order are reviews by the following authors on their specialty subjects. Osswald and Rietveld (3030) on "Measuring Constitutive Properties," Adams (3031) on "Failure Strength Tests and Their Limitations," Liechti (3032) on "Fracture Testing and Failure Analysis," Sancaktar (3033) on "Static and Dynamic Fatigue Testing," Allen (3034) on "Special Tests for Membranes and Miniature Components," Lefebvre (3035) on "Special Tests for Sealants and Elastomeric/Foam Materials," Drzal and Herrera-Franco (3036) on "Composite Fiber-Matrix Bond Tests," Packham (3037) on "Microstructural Analysis," Roylance (3038) on "Thermal Properties and Temperature Effects," Takahashi (3039) on "Electrical Properties," and Schmueser (3040) on "Evaluating Test Geometries."

It would seem that the most organized procedure for presenting the multiplicity of testing methods available would be to consider specific groups of testing procedures that relate to (1) the adhesive itself, (2) the adhesive combination with the adherends in the various joint designs, (3) the nondestructive testing of joint strength, and (4) the testing procedures for evaluating the permanence of joints. Another aspect of testing has been the development of methods to determine the sensitivity of adhesive materials to fracture, a sense of the alteration of this property in the presence of the adherend, and the elucidation of specific relationships found in the interphase area of the joint. These latter types of testing procedures are not really identified by specific ASTM procedure numbers, which is the major focus of this chapter on testing methods. While they have been discussed in some detail as part of the means of determining the mechanisms of joint failure and water deterioration of joints in Chapters 7 and 8, there have been even more recent literature contributions to extending these general investigative procedures. We will seek to mention these before proceeding with the outline for reviewing the more standard types of testing. These will include the identification of pertinent and new publications on spectroscopic, radioactive tracer, Iosipescu shear, and miscellaneous thin adhesive film procedures.

A. Recent Publications Concerned with Spectroscopy Procedures

Romand et al. (3041) discussed recent developments in the spectroscopic characterization of modified surfaces for adhesive bonding or painting in 1987, whereas Tokutake et al. (3042) reported on scanning Auger electron microscope resolution as determined by a quantitative AES method. Petrakian and Renucci (3043) were concerned with theoretical analysis of AES depth profiling in multilayers in the same year. In 1989, Bishop et al. (3044) suggested applications with a high spatial resolution combined AES/SIMS instrument.

Five XPS spectroscopy-related papers appeared in 1990 from Pertsin and Pashunin (3045) on differential charging in XPS studies of polymer/metal interfaces, Atanasoska et al. (3046) on XPS studies of chemical bonding at polyimide/metal interfaces and semiconductor overlays, Turner (3047) on XPS and AES surface analysis, Watts and Gibson (3048) on the determination of acid-base properties of inorganic surfaces using XPS, and Akhter et al. (3049) reporting on the distribution of compositional defects in cast polyvinyl alcohol films using angular dependent XPS.

SIMS-related papers in 1987 were issued by Wittmaack (3050) on charge compensation in SIMS while analyzing polymer foils using negative secondary ions, and Save et al. (3051) on SIMS studies on spin-deposited submicrometer films of organic molecules. In 1990, Hook et al. (3052) studied secondary ion formation from functional polymer systems using static SIMS.

Two 1987 FT-IR spectroscopy studies were conducted by Kodema et al. (3053) on ESCA and FT-IR studies on boundary-phase structure between blend polymers and polyimide adherend and Xue et al. (3054) on in situ studies of coatings on metal wires. Three 1990 papers included studies by Zimba et al. (2450) on FT-IR studies on thin polymer films, probing organic-inorganic interactions and curing processes using photocoustic FT-IR by Urban and Salazar-Rojaz (3055), and chemisorption of linear and cyclic polymethylsiloxanes on alumina by Cosgrove et al. (3056).

IETS study of silane coupling agents on aluminum oxide was reported by Comyn et al. (3057) in 1988, followed by additional publications on the same subject in 1989 (2039,2051) and in 1990 (2052). Wertheim (3058), in 1989, described a new method for bulk analysis by ESCA procedures.

Six papers on nuclear magnetic resonance (NMR) spectroscopy were issued in the period 1988–1990. In 1988, Koenig (3059) studied interpenetrating polymer networks in the interface of glass-reinforced epoxy composites using NMR, whereas Cheng (3060) characterized two adhesive systems (an acrylic resin and a rosin ester and a hot-melt formulation) using a variety of procedures including NMR, FT-IR, ultraviolet (UV) microscopy, and x-ray diffraction. In 1989, it was Nieminen and Koenig (3061) studying NMR imaging of epoxy adhesive joint interfaces. Finally, in 1990, Blum (3062) reported on magnetic resonance of polymers at surfaces, Kinsey's (3063) report on the solid state NMR of elastomers, and Kuhn and Eichoff's review of the limitations and applications in material science of NMR spectroscopy (3064). Reference is also made to the 1990 overall review article of Siegbaum (3065) titled "From X-ray to Electron Spectroscopy and New Trends."

No mention has been made previously about electrochemical impedance spectroscopy but two 1990 references can be cited. McDonald (3066) has cited advantages and pitfalls of this investigative method, whereas Titz et al. (3067) have reported characterization of organic coatings on metal substrates using such a procedure.

B. New Investigations Utilizing Radioactive Tracer Techniques

Marceau and Firminhac (3068,3069) first studied surface treatments and environmental effects on aluminum adherends using radioactive tracers in 1969 and 1970, respectively, whereas Westerdahl and Hall (3070) attempted to find the location of adhesive bond failures using radioactive tracers. In 1987, Jones et al. (3071) conducted a radiochemical study of the kinetics of epoxide curing, whereas Kelson (3072), in the same year, discussed delayed desorption as a new utilization of radioactive atomic nuclei for study of adherend surfaces.

C. Iosipescu Shear Test

The Iosipescu shear test procedure has not been mentioned previously but some recent publications can be cited. Adams and Walrath (3073) published a 1987 current status report on the method, whereas Barnes et al. (3074), in the same year, discussed both theoretical and experimental aspects. In 1990, Pindera et al. (3075) made Iosipescu shear characterizations of both polymeric and metal metrix composites.

D. Miscellaneous Recent Reports Involving Testing Procedures

Two 1987 reports involved testing procedures with thin films. Allen, Howe and Senturia (3076) used microfabricated structures for the in situ measurement of residual stress, Young's modulus, and ultimate strain of thin films, whereas Sampath and Wilbur (3077) suggested a new rapid technique for characterizing microstructures of films produced by ion beams in the TEM. Liechti and Freda

(3078) first elaborated on the use of laminated beams for determining pure and mixed-mode fracture toughness of adhesives in 1987 and followed with an elaborating paper in 1989 (2597). Also in 1987, Herrera et al. (3079) suggested a simple method for fatigue precracking specimens for use in fracture mechanics evaluations. Crawley and van Schoor (3080), in the same year, discussed material damping procedures for aluminum and metal matrix composite structures.

Budziak and Neumann (3081) produced a 1989 report describing automation of the capillary rise technique for measuring contact angles for bonding. Also, Spelt (3082) utilized contact angle data to determine solid surface tension values through the use of thermodynamic models.

Another thin film analysis procedure was described in 1990 by Ager et al. (3083) involving Raman intensities and interference effects for thin films adsorbed on metals. Sura and Rhinehart (3084) utilized an indentation test procedure for adhesion measurements of polyimide films. Ritter (3085) has discussed the overall question of evaluating the adhesion of thin polymer coatings under contact stressing conditions at the 1990 meetings of the Plastics and Rubber Institute. Klotz et al. (3086) has discussed imaging polymer interfaces by element specific electron microscopy and electron-energy loss spectroscopy. Finally, Manoharan et al. (3087) suggested an additional procedure for obtaining combined Mode I/Mode III fracture toughness values.

V. TESTING OF ADHESIVE FORMULATIONS

The most commonly used tests for determining the properties of adhesive formulations measure (1) viscosity, (2) percent of solids, (3) shelf life, (4) pot-life, (5) tack, and (6) curing rate. It is virtually mandatory for the adhesive manufacturer to conduct such testing in order to characterize both the raw materials for formulating adhesives and the final formulation itself. It is also recommended that the adhesive customer learn such techniques in order to be able to check on each separate batch of adhesive as it is delivered to the manufacturing plant.

A. Viscosity Testing

Viscosity measurements for free-flowing Newtonian or near-Newtonian adhesives are described in ASTM Std D-1084. The viscosity number obtained will be a direct indication of how easily the formulation will be to pump and spread over the adherend surfaces. The same measurement can often be used by the adhesive manufacturer to detect manufacturing errors in the production of already established formulations. The customer and the adhesive manufacturer can both use this test as a means of determining if the flow or pumping properties have been affected by excessively long storage times. Viscosity levels can vary tremendously between different products and the magnitude of the viscosity may even dictate the particular testing procedure which is preferable. Fast-flowing (low viscosity) formulations are generally measured with so-called volume flow cups, where the time in seconds for the adhesive to flow from a specially numbered cup is the measured number. For higher viscosity products in the range from 50 to 200,000 cP, the rotating spindle method should be used, as described in ASTM Std D-2556. Many structural adhesive formulations actually are made available in the form of two-part 100% solids or nonflowable one-part pastes. These require an extrusion-type of measurement, where a specified nozzle attachment is used on a standard extrusion cartridge. A specified air pressure (90–95 psi [630–665 kPa]) is applied to extrude a bead of the test product. The desired value is given in terms of the rate in grams per minute that extrusion is accomplished based on the average of three 10-sec extrusion trials (also see Ref. 1263).

B. Measurement of Percent Solids

Solids content needs to be checked by the formulator before shipping to ensure that the formulation meets the desired specification. Upon receipt, the customer needs to repeat and confirm the manufacturer's measurement for final acceptance. Improperly sealed adhesive containers can obviously cause unrecognized solvent losses both in shipment and storage conditions which may not be otherwise detected before using the adhesive in manufacturing. For maximum application efficiency and even best joint durability potential, the user will often find a particular bondline thickness, or at least a minimum thickness, must be maintained. With these bondline relationships having been previously established through specific durability testing, close attention should be paid to the relationship between the control of the viscosity, the speed of adhesive deposition on the adherend by the applicator, and the percent of solids content of the adhesive that relates to proper thickness of adhesive deposition. ASTM Std D-898 is one standard method of determining the applied dried weight of adhesive per unit of bondable area.

Where the adhesive is water dispersed, the nonvolatile content of the adhesive can be determined by ASTM Std D-1489.

C. Shelf-Life Considerations

The shelf life of an adhesive is not as readily evaluated as the two measurements indicated above. If the adhesive contains sufficient solvent to make its viscosity easy to measure, then any increase in viscosity may be one of the best indications that overaging has already proceeded to the degree that a lowered shelf life is present. Usually, the adhesive formulator has established certain time and storage temperature requirements which should not be exceeded by the customer. This necessitates that reliable records be kept of dates of receipt of adhesive shipments, and routine procedures developed to further check on time of storage at temperature conditions. Another manifestation of the adhesive being past usable shelf life may be its inability to achieve the originally established bond strength level. Additionally, this condition may be picked up by conducting some accelerated aging and/or weathering tests to find some indication of the existing joint durability potential. There is, however, no universally accepted degree of decline in initial joint strength which fits all occasions and automatically affects an adhesive rejection. Any decline in potential service life, as indicated by some form of accelerated weathering test, should certainly constitute a good basis for rejection. This is an especially critical type of screening and pertinent where a certain guaranteed level of performance must be met.

Obviously, the measurement of pourable viscosity has no real meaning with respect to the bondability of heavy mastic or 100% solids adhesives, which lack this quality from the beginning. For these kinds of adhesives, the rate of extrudability is much more meaningful. A decrease in this extrudability rate below that deemed acceptable from the adhesive formulator can often be correlated with a change in the final curing properties. With aluminum adherends, we should expect to find poorer adhesion (lower wettability) and lower initial bond strength as the spreadability decreases. Lowered performance in any accelerated weathering tests should also be expected. This is often reflected by shifting from observation of a cohesive-type failure to an adhesive-type failure when a lap-joint is pulled apart in a tensile testing machine. A higher viscosity will usually correlate well with lower flow and chemical wetting potential on any adherend surface as compared with freshly processed and flowable mixes of the same adhesive formulation. Adhesives which have a relatively short storage life can be evaluated for their storage life by measurement of consistency tests (Procedure A) and bond strength tests (Procedure B) as described in ASTM Std D-1337 and ASTM Std D-1338.

D. Pot-Life Considerations

The definition of pot-life is that time period during which the adhesive remains usuable in production. Some manufacturing operations require that major holding strength be achieved in a matter of seconds, whereas many other structural-type adhesives must be held in the bondline overnight under ambient conditions, like casein-neoprene cements used for laminated aluminum-to-wood structures. Most of the highest-demand structural applications have curing properties that can only be accelerated to full cure through the use of combined heating with either pressure or at least some fixturing pressure to hold the parts until holding strength levels are attained. With such a wide variation in the consideration of what may constitute the acceptable pot-life, it is not unexpected that there is no single testing procedure that can be applied by the adhesive formulator or structure assembler that covers all possible manufacturing situations.

Methods of determining pot-life based on either viscosity or bond strength are described in ASTM Std D-1338, which has already been mentioned under shelf life. It should be clear that a practical basis for using change in adhesive viscosity or bond strength is equally valid for determining when the end of pot-life is approaching. In particular, it may be necessary for any adhesive formulator to devise a battery of tests to determine the most acceptable pot-life for all manufacturing alternatives. It is even possible that different adherend surfaces will respond so variably to wetting by the same adhesive that a pot-life number may need to be specifically indicated for each different property adherend surface. An example might be the use of an initially solvent-rich contact cement over a porous wood surface to penetrate the surface as compared to an aluminum surface where such a low-viscosity adhesive would only create the need for longer setting or heating periods to remove the extra solvent. Even this example is not totally true, however, because the situation could exist where bonding to a poorly cleaned or even slightly contaminated aluminum surface might be enhanced by having extra solvent present when the contact adhesive is being applied. These kinds of factors have to be especially considered when bonding dissimilar adherends whose surface characteristics may vary widely from one to the other. The major basis, for example, for achieving a high degree of fiber tearing in a wood surface may depend heavily upon the degree of mechanical interlocking that has been achieved in the upper layers of the bulk adherend. For this reason, the author often recommended to Alcoa customers engaged in laminating aluminum-to-wood that the wood be separately primed with a diluted version of the same contact cement to be used in the final lamination processing at some higher solids level. Thus, two different viscosity variations of the same adhesive would be used to make the strongest and most durable aluminum/wood structure.

Whether either a one- or two-part epoxy mastic adhesive is overage can be quite important in the bonding of aluminum adherends with these adhesives. In the latter situation, the epoxy formulator has designated some preferred mixing ratio on the data sheet, which also requires an accurate measurement and thorough mixing at ambient temperature. As soon as the mixing is completed, the counting of the pot-life time must commence. Obviously, the formulator has established these data for the customer based on original testing using the freshest ingredients. At some point, the mixed adhesive becomes too viscous to apply, and the degree of wetting it can achieve on the adherend surface is questionable for achieving optimum strength and durability joints. Another related factor that needs to be recognized is a relationship that will exist between the batch size and the length of acceptable pot-life time. The formulator will also designate a maximum batch size that relates directly to the pot-life time specified on the data sheet. The larger the bulk mass of mixed adhesive, the shorter will be the pot-life, since the normal curing accompanies a heat exotherm development that is much faster developing the larger the batch of adhesive being cured. The most practical (though also expensive) answer to this possible manufacturing danger is to employ special two-part mixing and dispensing machines. Even here, however, there is both the initial cost to consider and the extra expense of cleaning and maintaining the intricate mechanical parts. A final factor that can significantly alter the pot-life times using the same adhesive is the ambient temperature of the surroundings. The basic chemical fact is that the mix reaction rate accelerates with increasing temperature. Since heat is already developing from the exotherm from the curing reaction, it is often beneficial to keep the mixed batch of mastic externally cooled to at least guarantee that the practical pot-life will be at least as long as mentioned on the formulator's data sheet. This factor is often overlooked as being a seasonal variation in manufacturing, where a favorable pot-life of 30–60 min in the winter plant temperature conditions becomes an unacceptably short pot-life under summer ambient plant conditions.

With respect to the situation with one-part, structural mastic adhesives that are overaged, the adhesive manufacturer should be consulted about the range of bonding results that might be expected after varying periods of storage at various temperatures. These data should be pertinent for the customer to develop guidelines for handling and storage conditions that are practical for the particular manufacturing situation.

E. Tack Measurements

Tack is that characteristic where contact causes easy sticking of an adhesive-coated surface to some other adherend surface. Pressure-sensitive adhesives in particular must have a continuous property with this characteristic to justify the use of this name. Many so-called contact-type adhesives will also have this property at some time during their use in manufacturing. During this time of high tackiness, the manufacturer must affect the bond between his adherend materials in order to achieve the highest degree of surface wetting, highest final bond strength, and finally best accompanying bondline durability.

A number of devices have been contrived to measure the force that is required to detach or separate a surface that comes in contact under controlled conditions with a coating of the adhesive to be measured for tackiness (3095). It is absolutely necessary that the conditions for making the measurement be precisely described and exactly duplicated each time a measurement is made if valid comparisons are to be made between two different operators.

ASTM Std D-2979 is one method that has been employed to measure the pressure-sensitive tack of adhesives. It is applicable to those adhesives which form a bond of measurable strength rapidly upon contact with an adherend surface. It involves bringing the tip of a cleaned probe of defined surface roughness into contact with the adhesive at a controlled rate under a fixed pressure and for a short dwell time at a given temperature. Subsequently, the bond must be broken at some controlled rate with the tack measured as the maximum force required in the total breaking of the bond.

ASTM Std D-3121 is another procedure that was originally developed by Douglas Aircraft for testing low-tack adhesives. It involves the release of a steel ball from the top of an incline to produce a certain known degree of momentum as it moves onto a level plane of the adhesive. The distance the ball can travel across the horizontal surface of the tacky adhesive is the comparable number used to describe the magnitude of tackiness present. It would seem logical that certain variables like (1) adhesive thickness, (2) bonding of the adhesive to some backing, and (3) the rigidity of the backing need to be carefully controlled for satisfactory comparisons to be made.

Johnston (3088) has published a 1983 article that discusses rolling-ball methods for measuring the tack of pressure-sensitive adhesives. Urushizaki et al. (3089) have also evaluated f (rolling friction coefficient) by analyzing the motion of a rolling ball. Mizumachi and Saito (3090), in 1986, further suggested a more unified theory involving both the rolling process and the rollout distance. Mizumachi (3091,3092) pointed out that f of a pressure-sensitive adhesive can more easily be determined by a pulling-cylinder method. Finally, Mizumachi (3093) proposed yet another model theory whose validity was confirmed by Mizumachi and Hatano (3094) in 1988.

In evaluating the tack level on aluminum sheet coated with a neoprene-phenolic contact adhesive, Alcoa has utilized an internally developed device for panel manufacturing (3096,3097). A simple spring-loaded stainless steel disk of prescribed weight and diameter could be dropped from some fixed height onto the adhesive-coated surface under the line conditions used to laminate adhesive-coated aluminum for building panels. The force in pounds necessary to just detach the weight was registered on a calibrated scale. While this simple device may not have the analytical quality of some laboratory measurement methods, it was found to be very adequate for comparing adhesive tackiness in varying thickness coatings and at different line operating conditions. Most practically, it could be used as a direct measuring instrument for comparing different lots of adhesive and project possible quality of different lots of panels before shipping. Further, it was established by Krubsack et al. (87) and Minford and Vader (89) that a certain minimum acceptable level of tackiness based on these readings under production conditions would produce acceptable panels which could be expected to survive for periods commensurate with their use for sheating commercial buildings.

F. Cure Rate Determinations

How fast an adhesive can develop maximum strength potential is certainly of considerable economical interest to every manufacturer who is trying to mass produce an adhesive-bonded product. ASTM Std D-1144 offers one procedure for conducting a planned experiment to establish bond strength development at two-fifth, three-fifth, and four-fifth time periods based on the recommended full cure time listed by the adhesive formulator. The test specimen may be one suggested by the adhesive formulator. The test specimen may be the one suggested in ASTM Tensile Test Method D-987. Results are commonly reported as a graph of average breaking load versus cure time.

The rapid curing rate even at room temperature of the cyanoacrylate and anaerobic adhesive family formulations has obviously been a boon to their widespread use in many mass-produced products.

In contrast, the ultimate bond strength of contact adhesive-bonded joints made at ambient temperature between aluminum adherends would likely require days to weeks even when the joining was affected just as the last vestiges of solvent needed for tackiness were almost dissipated. Of course, as mentioned earlier in Chapter 5 on adhesive selection, the best chance for fast bond development here would be to heat reactivate a dried adhesive bondline under static or momentary contact pressure or both.

Full-strength attainment with a two-part, room temperature–curing epoxy with aluminum adherends can require as long as 7 da for many proprietary commercial products. There is a direct relationship here between curing rate and rate of production, since extremely short pot-life products that may be available also have excessively shorter pot-lives, making their use risky under many conditions of general manufacturing. Thus, many manufacturers still would prefer the 30- 60-min pot-lives of the longer-curing time products for accommodating a wider range of manufacturing variables. It should also be mentioned that the author has obtained very poor joint durability responses using these fast-curing, two-part epoxy adhesive products demonstrating that there is a distinct trade-off of durability potential for being able to so dramatically shorten the cure time to attain full strength. Whereas the two-part epoxy/aluminum joints curing to full strength in 10 min failed to survive for more than 30 da in water soaking, the two-part epoxy/aluminum joints with formulations that were cured for 7 da before exposure could survive up to 720 da with more than 90% joint strength retention.

The structural acrylic adhesive products also present a wide range of times to secure full cure. The earliest acrylics from Hughson (so-called first-generation acrylics) were distinctly two-part systems that had to be mixed and applied to the adherend like the two-part epoxies just discussed. The newer second-generation products are similarly two parts (resin and accelerator) but self-mix on the adherend can be achieved when the two parts are separately applied. Mating the adherends is able to produce mixing, enabling joint strength to cure rapidly to full strength in a few hours.

The highest strength and most durable aluminum joints that can be fabricated are those made with the general class of heat-curing, one-part paste, tape, and film products developed for the aerospace industry. Full curing is only achieved at temperatures in the range of 250°F (121°C) to 350°F (177°C), generally with curing time ranging from 40 to 60 min. The author has demonstrated that further accelerating the curing rate by higher temperature curing conditions may or may not alter the durability response. For example, he found no difference between aluminum/nitrile-modified epoxy joints cured at 350°F (177°C) or those cured for only 15 min but at the higher temperature of 400°F (204°C). This was based on long-term durability evaluation in accelerated weathering for periods up to 2 yr. However, joints made with an epoxy modified to wet the aluminum surface in the presence of a lubricant were distinctly more durable in water-soaking exposure than similarly prepared joints cured at 425°F (218°C).

Finally, we might mention that several families of adhesives like the hot-melts and high-temperature thermoplastics have no additional curing time to consider. It is obviously their nature to develop essentially full-strength potential while cooling in the bondline back to room temperature conditions.

G. Bulk Adhesive Testing

Bulk adhesive test methods cannot determine all the adhesive performance parameters; however, they can contribute useful data on the adhesive proper. The fracture behavior of epoxy-based polymers in bulk, as adhesives, and as matrix resins in fiber-reinforced composites, has assumed considerable importance for structural components in aircraft, automobiles, ships, and even housing. Where the component is load bearing, the strength, stiffness, and toughness of such resins are critical to the overall reliability of the structure. Some other important aspects of failure criteria for adhesive joints beyond peel and lap-shear testing data seemed necessary for designing joints, since these characterizations of adhesive strength are much too dependent on specimen geometry factors. Fracture toughness values seemed an appropriate criterion, since it could be shown structural joints with these adhesives, as a whole, fail in a linear elastic "brittle" fashion. Publications of Mostovoy, Ripling, Patrick, Corten, and Bersch (2501–2521,3098) resulted from a U.S. Navy–sponsored program to develop the methodology of adhesive fracture testing from 1963 to 1976. From 1975 to 1981, Bascom and co-workers (222,1332,2522,2544,2545,2666,2674,2680,3099–3103), at the Naval Research Lab

(NRL), using the earlier techniques as well as some newer modifications. They also investigated the micromechanics of adhesive fracture in unmodified epoxies, rubber-modified epoxies, and various commercial structural adhesives. Also, included were the matrix resins in fiber-reinforced composites. Most recent work has been a continuation of the earlier efforts by Hunston and co-workers (1333, 1334,1572,2480,2481,2668–2671,3104–3108), starting out at the National Bureau of Standards. Also significant is a 1983 paper by Kinloch et al. (3109) and a 1984 publication by Hunston et al. (3110).

Jemian (3111) has looked at the joint as a composite structure composed of the adherend and adhesive layers with variation of chemical and structural properties. As a result one can expect precursors to joint failure to exist which need to be studied. Mechanical characteristics of the bulk adhesive which could offer valuable information about loading responses and adhesive behavior would include flow (stress-strain), creep, and stress relaxation curves, as discussed by Cottrell (3112) in 1964. In 1972 and 1973, Wilcox and Jemian (3113,3114) attempted to depict the nature and sequence (progressive deformation changes) in an aluminum/epoxy joint under loading. Masubuchi and Keith (3115) demonstrated that the mechanical actions are transferred through the bulk structural components to the region of the interface producing both normal and shear strain. Phases of markedly different stiffness will produce stress fields that are locally complex that include concentrated stress components. The deformation mechanisms they described were later confirmed by the finite element structural analysis of Grimes (3116), photoelastically by Jemian and Ventrice (2499), and a closed-form mathematical solution by Renton and Vinson (3117). Gent (3118) found an important effect of applied stress was to mechanically enhance flow properties. In 1967, Alfrey (3119) reviewed the role of bulk properties of the adhesive. It was his contention that the interfacial forces play their crucial role during joint formation, whereas the bulk properties are of great significance during the subsequent mechanical loading to joint failure. Buchnall (3120), Kambour (3121), Rabinowitz and Beardmore (3122), and Argon et al. (3123) have concluded that the principal deformation processes in glassy polymers (like epoxies) are shear bond formation and crazing.

Decreasing adhesive thickness has been reported by Bikerman (2343), Gardon (3124), Bryant and Dukes (3125), and Wake (3126) to increase the fracture strength. Bascom and Cottington (2545), however, report deviations to this and more overall detailed explanation of the effect. Diem et al. (3127) have reported on the influence of the adhesive layer thickness and internal stresses on metal-to-ceramic joints. Cottrell (3112), Buchnall (3120), and Friedel (3128) have discussed void formation as an included process in the deformation and fracture of both adherend alloys and polymer adhesives. Donatelli et al. (3129) made a 1984 assessment of the fracture toughness of elastomer-modified epoxy adhesives, whereas Ripling et al. (3130), in the same year, reported on the fracture of composite-adhesive-composite systems.

Some specific test procedures are enumerated in the Index of ASTM Standards, Part 22, under the general heading of "cracking," which certainly relate to fracture testing. ASTM Std D-3166 describes fatigue testing of metal-to-metal adhesive joints in shear by tension loading. Fracture testing is also involved in testing aluminum "wedge" test specimens for joint durability potential, as described in ASTM Std D-3762. Fracture testing of adhesive bonds in cleavage is also described in ASTM Std D-3433. Finally, the rupture/rupture strength of nonrigid adhesives is described in the practice set forth in ASTM D-3983 using thick adherend tensile-lap specimens.

1. DOG-BONE SPECIMEN TESTING

The dog-bone specimen, as used for tensile testing metals, can be used to generate bulk adhesive data like Young's modulus, Poisson's ratio, yield stress, and ultimate strain values. Neat specimens in "dog-bone" shape are generally tested in uniaxial tension at several strain rates and temperatures. The neat material may be "as-cured" or conditioned for various times in, e.g., humidity. Secant moduli and stress-strain data to failure are determined from the constant strain rate testing. (Further details can be found in Refs 3131 and 3132.)

2. TORSION SHEAR TESTING

A napkin-ring type of specimen, consisting of two relatively thin-walled tubes bonded together end-to-end, was selected by Hughes et al. (3132) for evaluation. In test, the adhesive bond is stressed by rotating one adherend relative to the other about an axis that is coincident with the longitudinal axis of the assembly. The resultant loading is pure shear with no tensile or cleavage loading on the bond. Because of the relatively small wall thickness to radius ratio, the stress across the adherend face can be considered constant. Special precautions must be taken, however, with respect to adherends preparation, tortion apparatus design, and the strain-measuring system employed (3132). The data generated can be used to determine the torsion shear modulus and microyield data; e.g., precision elastic limit, microyield stress, irreversible work done in a load-unload cycle, the fracture stress and strain of the adhesive, and time-dependent properties (creep). Typical torsion shear properties of Metlbond 329 aerospace adhesive can be seen in Table 126. This adhesive, which is a 100% solids, modified epoxy adhesive film supported on a synthetic fiber carrier, is typical of the high strength–low ductility structural adhesives. Tests with a typical ductile structural adhesive like FM-73 show a significantly lower average fracture stress value of 5400 psi compared with the 9,900 psi value for Metlbond 329 in Table 126. As might be expected, the average fracture strain value is much higher for the ductile adhesive, showing a value of 615 compared with 110 for the Metlbond 329 adhesive.

Complete stress-strain curves to failure can be recorded with the torsion shear apparatus. When ductile adhesives are being studied, the sensitivity of the test may have to be reduced to obtain the full stress-strain curve. This can be accomplished by (1) increasing the capacitor plate separation, (2) decreasing the proximity meter sensitivity, or (3) decreasing the voltage sensitivity of the X-Y recorder.

3. *TORSION PENDULUM TEST ON CAST (NEAT) ADHESIVES*

This method of testing is the one standardized in ASTM Std D-2236 (3134). Its aim is to determine the shear modulus and the logarithmic decrement of the adhesive at different temperatures. A plot of these two at the various temperatures will indicate the thermomechanical characteristics of the adhesive. A specimen of cast adhesive (neat) is stimulated to oscillate around its longitudinal axis and the free decay oscillations are recorded. The test can be performed at various temperatures which cover a range including (1) the freezing range (high and constant shear modulus with low decrements; (2) the softening range (shear modulus decreases steeply while decrement increases and passes through a maximum; (3) the rubber elastic range (occurs with cross-linked resins with shear modulus and decrement low and remaining approximately constant with increasing temperature); and (4) the plastic range (non–cross-linked resins showing shear modulus decreasing and decrement further increasing with temperature).

4. *THICK ADHEREND–LAP-SHEAR TESTING*

Where adhesive is used as a fastener for joining two pieces of metal, the load transfer is shear in the plane of the bondline. Unfortunately, as has been mentioned earlier in Chapter 6 on design of joints, the shear stress distribution in the bondline of such structures is not uniform. The relative stiffness of the adherends and the typical structural adhesive produce a peak or maximum stress at each end of the splice jointure. Theoretically, the stress at the center of a joint could be zero if the joint were long enough.

Table 126 Typical Torsion Shear Mechanical Properties of Metlbond 329

Bondline thickness (in)	Test temp. (°C)	Elastic shear modulus (psi)	Precision elastic limit (psi)	Microyield stress (psi)	Avg. fracture Stress (psi)	Avg. fracture Strain ($\times 10^{-3}$)
0.005	24	370,000	930	1800	9900	110
	191	130,000	120	450	4600	170
	−55	472,000	—	—	10,200	72

Data from Ref. 3132 (Table 1, p. 155).

The data needed to determine this stress distribution is the shear modulus (ratio of shear stress to shear strain). It is theoretically possible to obtain this value if measurements could be made of the shear displacement, bondline thickness, and the shear stress with a specimen having uniform shear stress distribution. However, we need to answer the questions of what kind of specimen would have this quality and how can we measure the displacement factor? A best specimen should provide stresses that are comparable to those that can be calculated if prediction of service performance is to be possible. Hughes et al. (3133) have considered these matters in their discussion of the "skin-doubler" concept resulting in the use of a thick adherend short-lap specimen which minimizes the presence of tension in the bondline. Aluminum alloy bare 2024-T3 has become the standard material for constructing such specimens, because it is a representative aluminum aircraft alloy which does not have the complications offered by the use of a dissimilar metal Alclad-type alloy. It may be feasible, however, to use steel for adhesives with extremely high modulus values. Krieger has designed his KGR-1 extensometer to obtain the required data using this specimen. It measures shear displacement of the bondline in this thick adherend–lap-shear specimen. The KGR-1 device is attached and a deflection curve made for a particular test load cycle. It should be pointed out, however, that test values of stress-strain data using the above three methods have produced measurably different values using the same adhesive.

VI. TESTING THE PROPERTIES OF JOINTS

Methods of testing the adhesive joint are mainly relied upon by both aerospace and general manufacturing to provide the bulk of adhesive performance data. The performance of the adhesive, constrained as it is in the bondline between two adherends, appears to be more readily analzyed with data from testing performed on bonded joints, at least for the present. There is, however, the element of constant interdependence between adherend surface and adhesive-bonded joint methodology that frequently obscures the actual performance of the adhesive layer itself. While this interdependence is always present in any actual bonded assembly, it is often difficult to isolate the causative factor for service failures when neither of these values can be effectively evaluated separately.

Anderson and DeVries (3146) have pointed out that more than 50 bond strength test methods have been accepted by ASTM in regard to the development of data about the adhesive joint properties. They conducted a poll in 1983 among 38 different companies who routinely conduct some form of adhesive joint testing with the results shown in Table 127. Tensile, lap-shear, and peel tests were apparently the primary test techniques used by 92% of the companies, so the primary emphasis in this chapter will be on those particular kinds of procedures.

A. Tensile Testing Procedures

The type of joint geometry used for evaluating the tensile strength of adhesive joints is the axially loaded butt-joint. The general procedure for measuring the tensile properties of adhesive bonds is

Table 127 Adhesion Tests in Use (1983)

	Percent of respondents[a]	
Test	Primary test method	Some degree of use
Peel	46	93
Lap shear	22	74
Tensile adhesion	24	63
Other	8	33

[a]Forty-three total respondents representing 38 companies.
Data from Ref. 3146 (Table 3.1, p. 70).

set forth in ASTM Std D-897 (British BS 5350: Part C3: 1979). Some of the materials used as adherends are aluminum bar, rod, or wire, as described in ASTM B-211. Many parameters must be carefully controlled in this procedure and need to be reported along with the results. These items might include (1) the alignment of the grips, (2) thickness of the adhesive layer, (3) the speed of testing, and (4) the conditioning procedure used for the parts, including the testing room conditions. Other pertinent testing procedures such as ASTM D-2094 describe methods of preparing the bar and rod specimens (including unlike adherends). Also, ASTM D-2095 should be consulted for means of conducting the actual tensile strength tests using the prescribed bar and rod specimens. For samples that cannot readily be fashioned into one of the above forms, the configurations described in ASTM C-297 or D-1344 may be useful.

While the geometry for the standard tensile adhesion test is visually simple, it has been known for some time (2402,2539,3135) that the stresses can become singular at the bond edges when there are differences in the mechanical properties of adhesive and adherend. (See Ref. 3136 for additional specific details.)

A specific procedure for tensile testing rubber-to-aluminum joints can be found in ASTM Std D-429. ASTM Std D-1344 is a test procedure for determining the comparative strength of adhesives through the use of a cross-lap assembly with specific reference to the conditions of adherend pretreatment, test temperature, and the testing machine speed. The use of this procedure would be especially recommended where glass is one adherend and the other aluminum.

Some additional discussion on testing axially loaded butt-joints or poker-chip joints has been offered by Kinloch (2196) as part of his review of the mechanics amd mechanisms of failure of joints. Messner (3137) made an early analysis of the stress distributions using the poker-chip specimen in 1963. Lindsay (3138) came up with an additional stress analysis in 1967. His solution was only valid, however, if the adherend can be considered rigid as compared to the adhesive and the adhesive itself is virtually incompressible; i.e., the adhesive Young's modulus/bulk adhesive modulus is nearly zero. The effect of adhesive thickness can also be quite significant, as shown in Table 128. As the adhesive thickness increases compared to the bond area diameter, the stress concentration factor at the bond center decreases, and for sufficiently thick bonds becomes less than unity; i.e., the local axial stress at bond center becomes less than the average stress.

1. TENSILE TESTING LAMINATED ALUMINUM-SANDWICH BUILDING PANELS

In response to the need to evaluate the bond strength between aluminum faces and back-up sheets and insulation cores in the Alcoa Alply laminated building panel, a standard bench-model tensile

Table 128 Effect of Adhesive Thickness on Average Failure Load for Button-Type Specimens Pulled in Tensile

Average thickness		Average failure load	
mm	in	N	lb
6.38	(0.251)	286	64
6.17	(0.243)	138[a]	31
0.71	(0.028)	1430	322

[a]The lowered resistance to failure for a joint of approximately equal bondline thickness was shown to be the result of a flaw with approximately twice the area. Assuming this flaw size, the investigators predicted a failure load of 156 N (35 lb) which was in good agreement with the 138-N (31-lb) value actually measured in test.

Data from Ref. 3136 (p. 103).

Figure 23 Alcoa lab tester for determining tensile-cleavage bond strength of laminated aluminum faced architectural building panels. (From unpublished work of J.D. Minford at Aluminum Co. of America.)

tester of the type shown in Figure 23 was specially modified. The standard tensile test grips were altered to pull against the facing in a tensile mode. Special thick steel holding devices were constructed to grip and add significant stiffening to the faces to minimize the recognized peeling forces that could operate around the periphery of the panel as it was being pulled at the center point of the back-up plate nearest the hydraulic ram. The test values obtained were best described as being a combination of tensile and peel stresses and, hence, were labeled tensile-peel tests. While not corresponding to any value that might be secured from use of some standard ASTM test procedure, these values could be conveniently compared internally at both the Alcoa Research Laboratory and the Alcoa Alply manufacturing facility. They could also be used for developing long-term durability data by recording an initial tensile-peel laminate strength value and repeating this determination on other specimens removed from the same panel after exposure to varying periods of accelerated weathering conditions in the laboratory or under natural exterior atmospheric exposures.

Mention should also be made of an Alcoa-developed quality control tensile test procedure for evaluating the durability potential of actual day-to-day manufactured lots of panels. The different procedures and the passing test performance levels established by laboratory testing of carefully prepared full-size panels are as shown in Table 129. Small 2 × 2 in-test specimens were statistically selected with respect to location over the panel facing and sawed-out for tensile-rupture testing under the testing conditions shown in the left-hand column of Table 129. It could be shown that a panel selected from a particular day's manufacturing run could produce a series of test specimens that were representative of the overall quality of the panels being produced during that day's production. The final test specimen was produced by slightly undercutting the backside facing of the small test specimen so it could be dropped through a commensurate sized hole in a horizontal holding plate. The specimen was then being totally supported by the core material bond to an area similar to the drop-through dimensions. The prediction of bond quality was then based on the dead-load tension loading necessary to totally separate the core from the front facing. Progressive loading to failure times under room temperature or 180°F (82°C) facing temperature conditions (separate thermocouples were

Table 129 Stress-Rupture Testing[a] to Evaluate the Service Durability Potential of Alcoa Laminated Alply Building Panels with the Maximum Facing Temperature Induced Under Solar Exposure

Test description	Established requirements (1)
180°F (82°C) progressively loaded tensile testing of outer facing adhesion	Progressively dead-load panel facing to failure. Acceptance only with resistance to stress-rupture level of 15 psi or higher
180°F (82°C) with a 5-psi dead-load applied in tensile to the outer facing	Stress-rupture must be resisted for 60 min or longer
180°F (82°C) with a 10-psi dead-load applied in tensile to the outer facing	Stress-rupture must be resisted for 2 min or longer

[a]Most important test to determine acceptability of a neoprene-phenolic contact adhesive candidate for laminating exterior weathering aluminum-faced building panels. In manufacturing, this testing is conducted after room temperature–aging of the bondline for 7 da, or after an artificial aging in a hot room for 2 hr at 140°F (60°C) following 1 da of room temperature–aging.
From Ref. 89.

attached to each facing under infrared lamp heating conditions) were recorded for at least triplicate test specimens. Tensile testing with static loads of 5 and 10 psi with the facing temperature at 180°F (82°C) was considered especially critical in terms of predicting the ability of the freshly manufactured panel to be erected and immediately resist solar heating conditions in the field. A panel whose test specimens showed the ability to resist delamination at 180°F (82°C) for 10 min under a 10-psi static tensile loading was shown to be able to resist solar heating atmospheric conditions for an indefinitely long time. This form of testing proved to be especially amenable to evaluation of aluminum-faced/foamed polystyrene or foamed polyurethane insulation cored panels bonded with the best neoprene-phenolic contact cement. It was also shown that this type of test procedure could be employed as a screening test for acceptance of best neoprene-phenolic adhesives in the marketplace. Minford showed that less than a dozen commercial neoprene-phenolic contact adhesive products consistently could pass such a test out of more than 200 products that were screened. In regard to the manufacturing procedures for such building panels, it was shown that the aluminum/foamed plastic bondline had to be heated to at least 150°F (66°C) during the lamination processing to pass this quality control testing.

B. Lap-Shear Test Evaluations

1. *SINGLE–LAP-SHEAR SPECIMEN*

The single–lap-shear test specimen is the most commonly employed overlap joint for determining the initial and after exposure bondline strength. The procedure is covered in ASTM Std D-1002 (British 5350: Part C5: 1976). It is also used in aerospace as the preferred specimen in Specification MMM-A-132. The main incentive for its use has been its relative low cost to produce and test, and yet there is some validity also from the fact that it is a type of jointure that is present in many manufactured bonded structures. The elastic bending of the adherend, and the resultant peel at the end of the joint (previously discussed in Chap. 6) reduces the reliability of the shear data generated. ASTM D-3163 describes an almost identical test specimen except for thickness to better accommodate the evaluation of rigid plastic joints. Although the specification calls for the specific use of 2024-T3 aluminum sheet or plate (ASTM B-209), it should be mentioned that any other aluminum alloy can be used providing there is an attempt only to make data comparisons between adherends of similar specimen stiffness.

At the Alcoa Labs, for example, Minford preferred to make most of his evaluations over many years using 6061-T6 alloy of approximately double-thickness adherends as compared to the 2024-T3 joints specified in ASTM D-1002. This choice was based on the desire to obtain practical data on the alloy most widely employed outside of aerospace; i.e., 6061-T6 sheet and extrusions. Obviously, the appropriate joint strength and durability data for clad or bare 2024 or 7075 aluminum alloy adherends was being adequately furnished through government-subsidized aerospace research contracts; however, similar data for the lower-strength but still structural 6061-T6 alloy continued to be unavailable to the general manufacturer. Clearly, the aluminum alloy selected for strength and durability studies should always be dictated by what alloy will be projected for the end-use application. When making comparisons between different sets of data, the same alloy, adherend thickness, surface pretreatment, adhesive choice, curing conditions, and bondline thickness need to have been employed.

Many analyses have been made of this lap-joint geometry, including the early treatments of Volkersen (2219) and Goland and Reissner (2220). It was Plantema (3139) who also included the shear effects in this system. A finite-element method was first used in 1971 by Wooley and Carver (3140) and Adams and Peppiatt (2227) also used a finite-element procedure, deriving data agreeable to that of Goland and Reissner but they added consideration of a triangular fillet.

Some investigators have suggested modifications of the standard ASTM procedure. Adams and Peppiatt (2227) tested joints with and without triangular adhesive fillets showing decreased maximum stress values and less overall testing variability with the fillets present. Renton (2711) suggested thickening the adherend to reduce cleavage stress in the lap-joint. His work determining adhesive strain led to suggested optimum specimen designs for various adhesive/adherend combinations. Guess et al. (3141) showed evidence that the failure in the standard lap-joint is often controlled more by the tensile than the shear strength of the adhesive. River (3142) tried methods to reduce the stress concentration at the ends of the overlap to obtain more accurate shear property data.

The principles of fracture mechanics have been applied by many investigators using lap-shear test specimens (2488,2511,2545,3143). Of special interest for the lap-shear joint has been the application of a Griffith-type energy-balance procedure by DeVries et al. (3144). They noted an apparent dependence of the adhesive fracture energy on the mode of stress at the crack tip. Wang and Yau (3145) studied the effects of relative stiffness of both adherends and adhesives and also included debonds of various lengths. It seemed that the component of Mode I stress-intensity factor is always greater than the Mode II component except when the adherend and adhesive moduli are nearly equal. When the adherend thickness was large, the tensile component in the adhesive could be several times larger than the shear component. (See Ref. 3146 for more discussion of the properties of the lap-joint.)

ASTM D-3165 describes how a specimen can be prepared to determine the strength properties of adhesion in shear by tension loading of laminated assemblies. Compression shear tests are also commonly used. ASTM D-2182 describes a geometry similar to a lap-shear specimen and the necessary compression-shear test apparatus. ASTM D-1759 describes a similar test and fixture for conducting shear block tests for quality control of adhesive in scarf-joints. For wood testing in shear D-2718 and D-2719 measure adhesion in plywood in rolling shear and the strength of plywood in shear through the thickness. D-902 offers measurement of shear properties of adhesives in plywood-type construction by tensile loading (similar to D-1002).

2. *BLISTER DETECTION SPECIMEN AND THICK ADHEREND SPECIMEN*

This form of specimen is described in ASTM Std D-3165, as shown in Figure 24. The joint configuration closely simulates the actual joint configuration of many bonded assemblies and can be used for developing design parameters for such assemblies. In practice, the laminated assembly is either made oversize and test specimens removed from the real product or a certain percentage of the actual manufactured assemblies are destructively tested. It has also been called a thick adherend test, since two strips are laminated to make the specimen whose thickness can be varied to give whatever specimen stiffness is desired.

Yurek (3147), at 3M's, used this kind of aluminum joint test specimen with multiple specimens strung together (like links of sausage) in a fixture culminating in a calibrated spring on one end and a fixed attachment at the other, as shown in Figure 25. This unit could then be immersed in a thermostated recirculating hot-water bath chamber to conduct wet/stress testing. Alternatively, such a unit could be enclosed in a condensing humidity cabinet for stress/humidity evaluations. Brewis (3148) has also reported the results of strung-together lap-joints under sustained loading. Minford (3149) made a direct comparison of the load-to-failure values between blister specimens and single-lap-joints with identical 6061-T6 adherends and a two-part relatively rigid epoxy adhesive and showed significantly

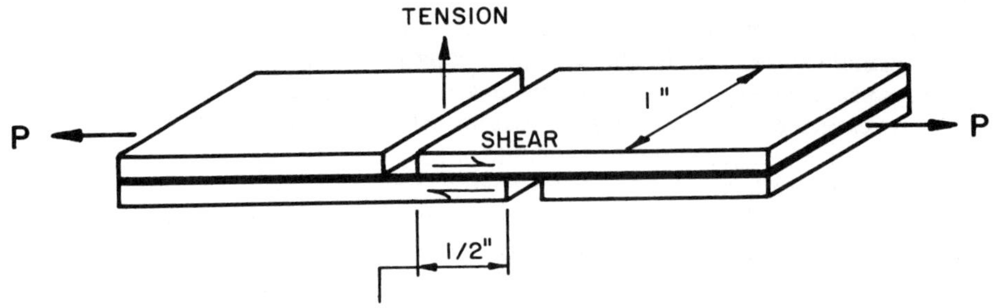

Figure 24 Blister detection lap shear (MMM-A-132) time to failure recorded. (From J.D. Minford, Aluminum Adhesive Bond Permanence, Treatise on Adhesion and Adhesives (R.L. Patrick, ed.) Marcel Dekker, New York, Vol. 5, p. 118, Fig. 3-51 [1981].)

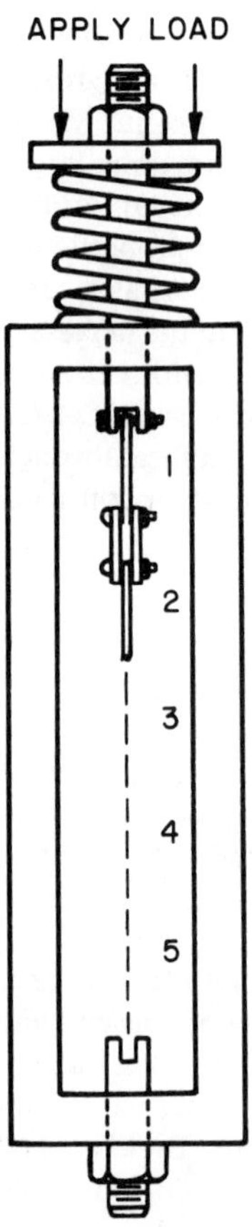

Figure 25 Spring-loaded fixture for evaluating a series of bonded lap-joints. (From J.D. Minford, Aluminum Adhesive Bond Permanence, Treatise on Adhesion and Adhesives (R.L. Patrick, ed.). Marcel Dekker, New York, Vol. 5, p. 118, Fig. 3-52 [1981].)

different joint strengths as determined in a tensile tester. He found a statistically different lower value population for the blister joints, which was rather unexpected considering the greater asymetrical geometry of the single lap when pulled in aligned grips. Undoubtedly, the nature of the adhesive must play some significant role in altering the difference in the measured initial strengths when comparing the two different joint configurations.

The use of a thick adherend shear specimen is described in ASTM Std D-3983 to permit making accurate determination values of the shear modulus and shear strength. This is important for determining the property values for bonded building structures using nonrigid adhesives with modulus ranging up to 100,000 psi (700 MPa). Frazier (3150) described the use of a thick adherend shear test specimen in 1970 which had the advantage of reducing the peeling action always present with thinner adherend joints. He obtained mechanical properties with a computer assist. Seago (3151) extended this work at Bell to include analysis of test data obtained from different production runs of adhesive, eliminating the separate shear testing of each different roll of adhesive as received.

Anderson and DeVries (3146) have tested the effect of varying adherend thicknesses using steel/epoxy structural lap-shear joints with thicknesses ranging from 0.0625 in (0.16 cm) to 0.5 in (1.3 cm). The adhesive was a fairly rigid two-part epoxy (Hysol EA-934). All testing was conducted at 72°F with a displacement rate of 0.05 in/min (0.13 cm/min). For an adherend thickness greater than 0.25 in (0.64 cm), a fairly constant plateau value was reached. As shown in Table 130, these authors' data showed the breaking load had diminished by some 63% when the steel adherend was thinned to the ASTM D-1002 recommended adherend thickness of 0.625 in (0.16 cm). It should be remembered that the ASTM D-1002 specification was originally designed to measure the lap-shear strength of aluminum alloy (relatively lower stiffness per unit of thickness than steel) joints and 2024-T3 aluminum alloy in particular.

Special ASTM standards exist for testing single-lap, metal-to-metal joints at elevated or low temperatures. ASTM Std D-2295 should be consulted for testing at elevated temperatures, whereas ASTM Std D-2557 pertains to the testing procedures at low temperatures.

3. DOUBLE-LAP-SHEAR SPECIMEN TESTS

The manner of assembly and the testing procedure for double-lap-shear aluminum joints is described in ASTM Std D-3528 or BS 5350: Part C5: 1976. The method covers the determination of the tensile-shear strengths of adhesives for bonding metals when tested in an essentially peel-free standard specimen. It represents the shear distribution which should be representative of typical low-peel production-type structural joints. Volkersen (2374) showed some normal stresses still exist across the adhesive layer. Adams and Peppiatt (2227) calculated the maximum stress ratio of a single- to double-lap-joint was about 1.85, which agreed well with experimental observations of about 1.92. While the double-lap-joint will give the shear result desired, it is not an economical joint to fabricate for general testing.

Table 130 Lap-Shear Test Data Relating Effect of Adherend Thickness

Adherend thickness		Avg. stress at debond		
in	cm	psi	N	Reduction (%)
0.0625	0.16	1530	10500	63
0.125	0.32	2440	16800	40
0.25	0.64	3600	34800	12
0.375	0.95	4200	29000	0
0.50	1.27	4010	27600	0

Data from Ref. 3146 (Table 3.3, p. 82).

Also, any differential in bondline thickness between the two separate bondlines in the overall joint can cause a moment in the specimen to develop during loading.

The double-lap-shear joint has the advantage of greatly reducing the bending stresses as compared to those in the single-lap version. It has, therefore, received much attention for those structural applications that demand a high level of structural integrity as required in the aerospace industry. The finite-element technique for calculating stresses in these kinds of joints has been utilized by Amijima et al. (3152) and Sen and Jones (3153,3154). The former were able to claim agreement between their predictive joint strengths and experimentally observed values. The latter considered a viscoelastic adhesive and reported that the magnitude of the maximum adhesive stress decreased as the ratio of the adhesive modulus to adherend modulus decreased from unity. Their evaluation of the effect of adhesive thickness and length of overlap on stress distribution along the joint showed that decreasing the adhesive thickness resulted in a distinctly more uniform stress distribution. Measurement of the exact stress in any adhesive joint configuration has always been difficult and reference should be made to the work of Sharpe and Muha (3155) in this regard. They employed a laser technique and reported stresses in good agreement with the results of Goland and Reissner. Their method is only pertinent, however, to the use of transparent linearly elastic adhesives in the bondline.

A number of investigators (24,2319,2320,2330–2333,2375) have considered the effect of adhesive plasticity in this joint type with reported reductions in the calculated stress concentrations as plasticity advanced.

4. SCARF TEST SPECIMEN

In most structures, it is a rare occurrence to find a structural joint that is designed to experience pure Mode I loading. Generally, a bondline and flaws within it will likely cause some combination of Modes I, II, and III loadings. Irwin (3156) and Erdogan and Sin (3157), in 1957 and 1963, respectively, determined that Mode I was the lowest-energy fracture mode making the crack want to propagate along a path normal to the direction of maximum resolved tension. This means that a crack in any isotropic plate should propagate in Mode I fracture regardless of how the initial flaw is oriented with respect to the applied load. However, this is not necessarily the case in adhesive fracture situations because the flow propagation can be constrained to the resin layer if tough adherends are employed. This situation can force the crack to propagate in the bond direction at or near the interface regardless of the bondline orientation. Thus, some effort must be made to adhesive fracture in combined stress situations, since bulk sample measurements may be of little use in predicting this adhesive behavior in real-world structural joint assemblies.

Accordingly, in 1963, Ripling et al. (2501) first determined the Mode II adhesive fracture values for a simple unmodified epoxy (DGEBA-TEBA) using the ILMM specimen. This was followed in 1971 by the use of a scarf-joint design by Trantina (2274,2275,3158) capable of measuring a mixed-mode adhesive fracture in epoxy/aluminum joints. In 1975, Bascom et al. (222) used this same scarf specimen to study the mixed-mode adhesive fracture behavior of two other unmodified epoxy resins (DGEBA-HHPA and DGEBA-piperidine).

The scarf-joint (literally a tapered butt-joint) thus has both shear and tension stresses acting on the adhesive in the bondline. This specimen while useful for conducting combined load testing, requires a three-dimensional analysis in order to obtain shear or tension data. Also, it is quite difficult and expensive to fabricate, and extremely complicated local measuring devices are required for making strain measurements. More detailed statements and references for further study have been mentioned under the earlier discussion of the design of joints in Chapter 6 and the mechanisms of joint failure in Chapter 7.

One form of scarf-joint has been especially valuable in studying the fracture of epoxy-based adhesives contained between aluminum adherends (3159). Both shear and tensile forces are applied by this specimen, and the bond angle is usually at 45 degrees for loading. The ratio between the Modes I and II components will change with change in this bond angle. Trantina (2274,2275,3158) developed the finite-element stress analysis for a scarf-joint from which the adhesive fracture energy can be calculated. How the crack growth proceeds can be readily observed and is important in explaining the fracture energy data obtained. For example, the initial precrack will tend to proceed across the bondline from the centerline during initial loading, which is largely mode I fracture. As it approaches the adherend/adhesive interface, however, the crack will arrest. A higher specimen loading will then be required to force the crack to continue to grow along the interface. This latter pure shear fracture value could be more than a factor of 10 greater than the Mode I or mixed-mode values present in the initial loading phase. Observations by Ripling et al. (2503) and Bascom and Timmons (3160) suggest that failure in the shear mode may not be simple Mode II fracture. Instead, it seems to proceed via the formation throughout the length of the bond of microcracks across the bond in the Mode I direction, and these microcracks then can join by failure of the material in between.

Bascom et al. (222) also have shown that roughness of the adherend surface in these scarf-joints can also strongly influence the mixed-mode fracture values. The explanation advanced lies in the proposal that the crack tip can be trapped in the surface roughness, causing extra energy to be needed to divert it out. The visual failure of these scarf-joints is apparently interfacial; however, Bascom and Oroshnik (2666) have used radioactive tracer procedures to show that the failure is actually deceptively close to the interface rather than constituting a true interfacial failure.

While the mixed-mode fracture behavior of the unmodified epoxies using this scarf specimen are at least understood qualitatively, the behavior of the CTBN elastomer-modified epoxies constitute a very different picture. Investigators have obtained quite diverse results with Mostovoy and Ripling (2517) and Brussat, Chiu, and Mostovoy (3161) obtaining mixed-mode energies higher than Mode I values. Bascom et al. (222,2666,3101,3105,3162) obtained lower values for mixed-mode fracture, whereas Johnson and Mall (2622,3163) showed little difference, at least until the Mode II component was very high. The possible explanation for such diverse results will likely require more detailed research; however, some available results can be mentioned. The investigations of Hunston and Bascom (3159) and Bascom et al. (3162) inserting precracks at different angles to the applied load to obtain different types of mixed-mode loading should be studied. Also, it should be pointed out that the tests giving higher mixed-mode fracture energies by Mostovoy and co-workers (2517,3161) involved a scrim cloth in the tested adhesive, producing a center of bond failure along the scrim cloth. All the investigations showing lower mixed-mode values meanwhile were conducted with adhesive bondlines without scrim material present.

5. OTHER SHEAR-RELATED TESTING PROCEDURES

This may be the most appropriate time to introduce the subject of three-point bend test specimen for evaluating adhesive joints, or alternatively, as it has been called the short beam shear test specimen. McDevitt and Baun (3164) stated that attempts to correlate data from accelerated laboratory experiments with actual service life by Rogers (3165,3166), Srinath (3167), and Bascom (3168) using peel, "wedge opening," or lap-shear mechanical test results have been only marginally successful. Actual in-service failures of adhesive-bonded metal structures have often been reported to occur in the region of the adhesive/oxide interface, although Baun (3169,3170) has more recently shown failure sites occurring at the aluminum oxide/metal interface.

A basis for using a test like the three-point bend was explained by McDevitt and Baun:

> Unless a large amount of degradation has occurred in the interfacial region of a metal-to-metal adhesive bonded joint, the T-peel and "wedge opening" tests provide only Mode I (tensile) information on the bulk adhesive. However, thickness of the adherend must be carefully considered in order not to have a large amount of Mode I failure. When these tests failed consistently to find defects that were incorporated into the interfacial region of test specimens, the three-point bend test geometry was evaluated.

The three-point specimen visually indicates the basis for anticipating that information obtained might be relatable to both failure sites involving the adhesive-to-oxide and oxide-to-metal interfaces. These investigators have also considered the importance of corrosion processes when considering aluminum surface joints. They further suggest that this type of test specimen could be of interest in dealing with the thin films that would be involved with the early stages of corrosion in the oxide/metal interfacial area.

For a more complete description of the test specimen and testing procedure, the reader should review the McDevitt and Baun (3164) work mentioned above. The authors show the distinctive load-displacement curves that are obtained for (nonbonded) and (bonded and cured) aluminum joints. Also, distinct patterns of joint test specimens with especially low-strength interfaces as compared with joints showing complete interfacial failure. From the linear portion of these curves, they could evaluate what they described as bondline stiffness (BLS) which could be compared with the directly measured yield strength of the interface (YSI). Data comparing aluminum bonded specimens of varying PAA oxide thickness and exposed to varying environmental conditions from these authors are shown in Tables 131 and 132.

A final test procedure described in ASTM E-229 is titled standard test method for shear strength and shear modulus of structural adhesives. Torsional shear forces are applied to the adhesive through a circular speciment which produces a peripherally uniform stress distribution. A special adhesive tortional shear jig permits application of a torsional shear load without inducing bending, peeling, or transverse shear stresses in the bondline. The basic material properties obtainable can be used in the theoretical equations for designing joints and in the evaluation of new adhesives.

C. Static Versus Dynamic Testing

Test methods described above are used primarily to determine static properties of adhesives; however, the dynamic properties can also be evaluated by use of similar-type specimens. The performance of adhesive joints have been evaluated under repeated loading conditions for many years in the laboratory to attempt to simulate cyclic stress conditions which occur naturally in many service environments. Early testing of this type was often accompanied by actual first failure in the adherends, since many plastics, and to a lesser degree metals, are subject to fatigue-type failures. The majority of the better structural adhesives available today can be combined with proper surface pretreatment and adhesive curing to produce conditions that may see fatigue failure of the adherend before

Table 131 Specimen Preparation and Exposure Conditions Employed to Develop Data Shown in Table 132

Anodization Conditions for PAA Treatment of Aluminum Adherends

Surf. prep. no.	Voltage	Time in bath (min)	Avg. oxide thickness (Å)
1	10	0.5	225
2	10	2.0	1000
3	10	10.0	3200
4	10	16.0	3600
5	40	2.0	1200
6	40	5.0	3000

Types of Curing and Exposure Used Before Testing Specimens

Test ident.

A—Specimens cured then tested within 24 hr.
B—Specimens cured then postcured at 220°F for 8 hr.
C—Specimens cured then stored in desiccator under ambient conditions for 90 da.
D—Specimens from test C subjected to dry heat, 100°C, for 63 hr.
E—Specimens from test C subjected to SO environment at RT for 2160 hr.
F—Specimens from test C subjected to SO environment at 100°C for 72 hr.
G—Specimens from test C stressed to 1000 lb for 2 min then subjected to 100% RH and 100°C for 8 hr.

From Ref. 3164 (Tables 1 and 2, p. 383).

Table 132 Yield Strength of the Interface (YSI) and Bondline Stiffness (BLS) Data Utilizing Three-Point Bend Testing

	Control Specimen Data					
	Curing and Exposure Conditions[b]					
	A		B		C	
Surface[a] prep. no.	BLS	YSI (lb)	BLS	YSI (lb)	BLS	YSI (lb)
1	1.53	2825	2.06	3050	1.84	2758
2	1.59	3012	2.15	3552	1.80	2793
3	1.55	3200	2.12	4120	1.86	3222
4	1.58	3175	2.12	4202	1.92	3202
5	1.53	3025	2.01	3725	1.84	2900
6	1.59	2875	1.98	3650	1.85	2940

	Environmentally Exposed Specimen Data[b]							
	D		E		F		G	
	BSL	YSI (lb)	BSL	YSI (lb)	BSL	YSI (lb)	BSL	YSI (lb)
1	1.53	3100	1.76	2520	—	NIF*	1.62	2455
2	16.2	2960	1.82	2760	1.76	2475	1.70	2590
3	1.60	2780	1.81	2945	1.70	2820	1.73	3045
4	1.65	3210	1.82	3145	1.70	2900	1.53	2500
5	1.59	3000	1.88	2910	1.70	2515	1.56	2640
6	—	NIF[c]	1.82	2400	1.69	2690	1.65	2847

[a]No interfacial failure.
[b]See Table I in Ref. 3164, which describes the conditions for pretreating these test specimens.
[c]See Table II in Ref. 3164, which describes the curing and exposure conditions employed prior to testing these specimens to failure.
Data from Ref. 3164 (Tables III and IV, p. 391).

the joint. Usually such failures are observed to occur adjacent to the actual bonded area. Thus, information is not only obtained about the projected long-term durability of the bondline but also could indicate that a stronger alloy or thicker adherends of the same alloy should be used. For checking the longevity of thinner adherends, the use of reversed load cycles or shear loads which pass through zero may not always be feasible. However, with the introduction of thick adherend aluminum specimens, the evaluation of the fatigue strength even of high-performance adhesives with reversing shear loading does become possible. In addition, with the further incorporation of low mass zero length extensometers, or other displacement measuring devices, it is possible to record the displacement of the adherends at the joint during load cycling. Even the phase relationship between the applied load and the adhesive response can be followed with proper electronic devices. From these kinds of data, the response of the adhesive joint to load rates can be accurately determined and pared to the data calculated from bulk adhesive testing.

For example, Althof and Brockmann (3171), at Deutsche Forschungsanstalt fur Luft und Raumfahrt (DFVLR) Institut fur Structurmechanik, have determined adhesive shear properties by measuring the shear strains of short bondlines between adherends of high stiffness; i.e., thick adherends. A few years later, Althof (3172) proved that the stress-strain relationships of such short bondlines are representative of the adhesive sections in long bondlines as well. These investigations established that adhesives deform linear viscoelastically only at low stresses, and only under this condition can the time-dependent stress-strain relationships be described by simple mathematics as discussed by Althof (3173) in his publication on the effects of low cycle loading on shear stressed adhesive bondlines.

Gilibert and Verchery (3174) also employed low cycle loading of special specimens with two double-lap-joints to determine the influence of surface roughness on the mechanical properties of joints. Although these investigators chose to use low carbon steel adherends, the same procedures could be used with comparable aluminum roughened surfaces. They presented the effects of roughness on the three typical mechanical parameters: elastic limit (crack initiation threshold), crack propagation threshold, and ultimate strength. Their general conclusions stated that fine grinding produced better mechanical properties than coarse grinding. Sandblasting, however, improved properties better than shotblasting or pure grinding. Finally, sandblasting offered the best behavior when the total depth of roughness was equal to the mean diameter of the dispersed particles in the resin.

D. Combined Shear and Tension Effects

Although it may be preferred and planned to have a structural bonded joint be loaded in pure shear, this situation is very rarely achieved in practice. Instead the joint field experience will involve varying combinations of shear and tension. This has always been known from the beginning of adhesive bond structural technology, and a favorite countermove by the adhesive bonding expert has been to find and use the adhesives with the highest possible peel strength combined with high toughness. While the actual value of this extra peel strength was not directly used in any calculation of allowable adhesive stress, the very high peel value was considered to have some positive contribution for compensating for the tension element of the load on the joint. This assumption had merit, but led the adhesive formulators to incorporate large amounts of elastomers or plasticizers in their adhesive compositions to meet the high peel requirements demanded for many types of field service. Unfortunately, these incorporations usually adversely affected the overall durability of the joint.

An outstanding historical example was the period of aerospace bond manufacturing when the incorporation of nylon with epoxy formulations yielded adhesive products with exceptionally high initial strength combined with apparent high toughness and joint peel strengths. These adhesives appeared to be like an ideal structural adhesive for aircraft. That is, until it became known that these initial favorable properties were accompanied by amazingly fast drops in bondline strength when water could egress into the bondline. Thus, the rationale changed to selection of adhesive formulations where the combined load testing data would indicate the adhesive had met certain lower minimum peel requirements but also could survive in the real service environment without any penalty from poor resistance to water or combinations of stress and water. In further exposition of the subject of combined loading effects on adhesive joints is the work of Krieger (3175) using a simulated skin doubler test coupon in combination with a specially designed extensometer.

E. RABB Specimen Testing

More recently, a modified blister-detection joint has been extensively used in aerospace testing which is really like three blister-detection specimens in one continuous specimen. The overall length is specially chosen so that three 0.5-in overlap bonds can be fabricated in the overall length by notching three locations on either side of the thick adherend laminate. In addition, a 0.25-in diameter hole is drilled in the center of each overlap area. The specimen width is only 0.5 in as compared to the 1.0-in width dimension of the standard lap-joint described in ASTM Std. D-1002. This hole is employed so that water entering from an edge only needs to travel a fraction of the distance it traverses in the standard lap-joint. This so-called RAAB specimen was developed in the American Cyanamid Lab and is described in detail as to form and use by Marceau and Thrall in *Adhesive Bonding of Aluminum Alloys* (3176; see Fig. 6, p. 187). This volume covers all aerospace bondments in considerable detail. While the thick adherend-type test joint design may be more favored in aerospace testing, the standard ASTM D-1002 type lap-joint is most commonly employed in general manufacturing. Even thick adherend-type specimens fall short of acting as a duplicate of production-type aerospace joints like the double–butt-strap production joint. Hart-Smith (3177) offers a comparison list of characteristics for the thick adherend single-lap configuration joint as compared to this aerospace production-type joint (see Table 133).

Table 133 Thick Adherend Versus Aircraft Configuration Bonded Joints

Thick adherend joint characteristics (thick adherend single-lap configuration)	Aircraft joint characteristics (double-butt-strap configuration)
1. Uniform shear strain in adhesive	1. Nonuniform adhesive shear strain
2. Adhesive creep occurs everywhere	2. Adhesive creep only occurs at end of overlap—none in elastic trough
3. Only weak adhesive locked-in stresses to restore to original state	3. High locked-in metal stresses to restore to original configuration
4. Joint cannot recover during unload periods, short or long	4. Joint recovers during long unload periods
5. Environmentally degraded	5. No evidence of in-service degradation, except at interface with poor surface treatment
6. Useful test to investigate adhesives	6. Most failures in metal
7. Results not necessarily indicative of configured joints	7. Necessary to verify environmental durability

From Hart-Smith lecture on mechanical testing and design of adhesive bonds. ACS Teleconference Course on Adhesion and Adhesives.

F. Peel Joint Testing

A peel joint test involves measuring the force necessary to peel a flexible adherend away from either a rigid or second flexible component. Most common specimen configurations are the 90-degree tests (T-peel), as described in British Std BS 5350: Part C10: 1976 and BS 5350: Part C14: 1979, and 180-degree test from ASTM Std D-903 (British Std 5350: Part C11: 1979). Where the adherends are both flexible, the specimen and manner of testing with a T-peel specimen is described in ASTM Std D-1876. The definition of flexible adherend that applies is an adherend that permits bending through any angle up to 90 degrees without breaking or cracking.

For aircraft testing, a climbing drum peel test is described in ASTM Std D-1781 and is used for evaluating skin/sandwich assemblies (3178). This measures relative peel between a relatively flexible and a rigid adherend. The method can be used with a variance in specimen to determine the relative peel resistance of adhesives in sandwich structures providing the facings are relatively flexible. Prior calibration is required in this method to determine the actual load beyond the actual peeling strength that is necessary to overcome the resisting torque of the drum and clamps and counterweights usually present. Since the climbing drum peel strength values are often cited for aerospace structural film and tape adhesives by their manufacturers, it should be remembered that comparative testing by users should always involve the same angle of peel. Variation of adhesive thickness can also be quite influential and needs to be carefully controlled when making direct comparison tests.

Another frequently used industrial procedure has been designated floating-roller test procedure, as described in ASTM Std D-3167 (British BS 5350: Part 9: 1978). This is another variation of the 90-degree peel test involving peel separation of a flexible from a rigid adherend. It is very important to specify thickness of the adherends. The name is derived from the fact that the 1-in diameter rollers over which the flexible adherend is peeled must roll freely; hence, the term *floating roller*. Rubber-to-aluminum peel strength measurements are described in ASTM Std D-429.

While any of the above test procedures could be used to obtain initial and after-exposure peel strength results for a particular combination of adherends and adhesives, Carter (3179) was the first to report as early as 1966 that a dramatic change in peel strength would be expected if the peel testing was conducted in a totally wet environment, such as immersion under water or in a hot humidity cabinet. More details on a test procedure designed to assay peel joint durabilities in water can be found in ASTM Std D-2918. It is suggested that this practice could be especially useful as an accelerated screen test for assessing durability in aggressive exterior weathering for the design engineer.

Locke (3180) has reported internally at Boeing that in his testing, the wet-peel test procedure has proved to be even more discriminating of acceptable bond strength and durability situations than the use of the widely used Boeing "wedge test" method, which will be discussed under cleavage testing. Of course, it clearly states in the beginning of the wedge test ASTM D-3762 specification that the procedure is for making discriminations between surface pretreatments rather than showing actual prediction of the degree of joint durability in service.

Of course, it should be obvious that peel evaluations will be especially important for evaluating pressure sensitive–type adhesive products. ASTM Std D-3654 considers the holding power of such adhesives under contact loading, whereas their 180-degree peel adhesion strengths are discussed in ASTM Std D-3330.

G. Cleavage Strength Testing

While it is preferred to design adhesive-bonded structures so that the joints are primarily stressed in tensile or shear, in practice, there are often significant cleavage stressing components under anticipated service conditions. ASTM Stds D-1062 and D-3433 are both designed to evaluate the cleavage strength of metal-to-metal joints. This is accomplished by designing a special form of off-centered butt-joint which can be pulled in tensile with the tendency to cleave open the butt-joint from one side. Through the use of two other types of off-centered joints (described as parallel or tapered double-cantilver beam joints (see Fig. 17), Mostovoy and Ripling have been able to accurately measure the actual fracture strength in cleavage of aluminum bonded joints. The method is described in detail in ASTM Std D-3433, which also contains references to 10 publications on the subject by these investigators. The use of these specimens by Mostovoy and Ripling was discussed earlier in Chapter 7 on mechanisms of joint failure. The use and description of parallel double-cantilever beam (Mode I), special shear specimen in cleavage for measuring Mode II, and tapered double-cantilever beam (Mode I) specimens are also pictured and discussed by Kinloch (3181).

1. WEDGE-TYPE TEST SPECIMEN

Another important test specimen using the measurement of cleavage conditions in a bondline is described in ASTM Std D-3762 under the title "Adhesive Bonded Surface Durability Of Aluminum-Wedge Test." The specimen employed is that shown in Figure 26 as it was originally devloped by Boeing. More background information can be provided from investigations at Boeing Aircraft by Marceau and Moji (1015,1017). Marceau et al. (3182), and Scardino and Marceau (2824). For this type of test to be reliable, it is necessary to use relatively thick aluminum adherends, and the ASTM specification suggests aluminum aircraft alloy 2024-T3 or 7075-T6. If data are to be developed for any lower-strength aluminum alloys, then it would be necessary to increase the thickness of the adherends in order to achieve corresponding adherend stiffness. While there have been reports in the literature in which this testing procedure has been used to select a best adhesive for use with aluminum adherends, the procedure is only recommended by the ASTM D-14 committee for comparison of the joint durability produced by varying surface pretreatments on aluminum. Actual in-service failures

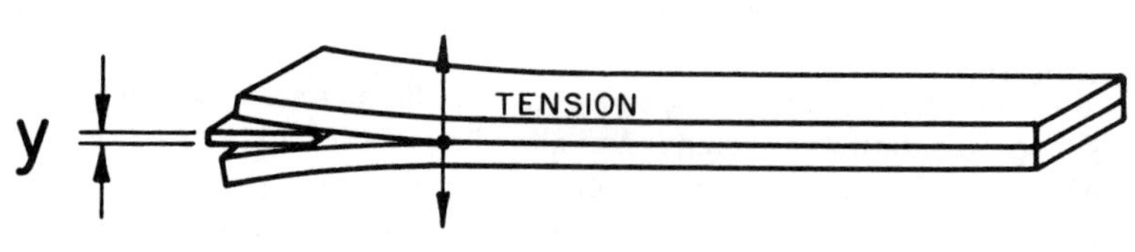

Figure 26 Thin adherend uniform DCB (wedge test). (From J.D. Minford, Aluminum Adhesive Bond Permanence, Treatise on Adhesion and Adhesives (R.L. Patrick, ed.). Marcel Dekker, New York, Vol. 5, p. 112, Fig. 3–48 [1981].)

in aluminum aircraft joints have been reported in the region of the adhesive/oxide interface, and more recently by Baun at the oxide/metal interface (3169). It would seem, however, that very little information is really generated by either the standard peel or "wedge opening" testing procedures about the subtle changes that may occur in the interphase region. More information of this nature can be found in lap-shear testing, but only when most of the peel forces have been eliminated.

Having previously conducted long-term durability testing with standard lap-shear aluminum joints for many years, Minford (3183) has conducted wedge tests to obtain some comparative estimates of aluminum joint durability between those long-term known results and the short-term exposure testing using the wedge test. Both a rigid, two-part, room temperature–curing epoxy and a more flexible, but higher-strength, one-part, heat-cured, nitrile-modified epoxy were used for making these comparisons. A room temperature, hydrochloric acid etching procedure was used with the rigid epoxy, and Alcoa A3 etch (equivalent to FPL) was used with the higher-strength modified epoxy. Some comparisons were attempted with the wedge test durability data in Boeing's AFML-TR-78-104 report titled "Anodize Optimization and Adhesive Evaluations for Repair, Dec. 1976–July 1978." Factors evaluated were (1) fabricating wedge specimens by sawing individual specimens from a larger laminate versus precutting the individual specimens, (2) desmutting or not desmutting the HCl etched surface before bonding, and (3) comparing crack-propagation rates during the first 24 hr with 7–30 da wedge testing times. Some conclusions were (1) the relative brittleness of the adhesive was quite influential in addition to the type of surface pretreatment; (2) the number of physical flaws in the bondline could be a significant factor when introduced by methods of curing and type of adhesive curing mechanisms; (3) the maximum spread in crack growth values among identically prepared specimens usually would occur between 4 and 24 hr test times; (4) the evidence was not convincing that sawing specimens produced different results from individually preparing the specimens; (5) unexpectedly, desmutting did not change the results even though a visually weak boundary layer seemed to be present before curing the adhesive (a dark smut layer could be wiped away with finger pressure yet was not apparently acting to produce lowered adhesion responses); and (6) good durability agreement was found between the "wedge" test results and the long-term lap-shear tests using the two-part epoxy adhesive over the HCl etched 6061-T6 adherends; however, the agreement between the higher-strength, modified, one-part, heat-cured epoxy joints was much less acceptable. However, the long-term durability results had been obtained using 6061-T6 adherend surfaces as compared with 2024-T3 alloy having been employed to fabricate the corresponding wedge-type specimens in order to conform to the ASTM specification. Since no stressing was involved in the long-time weathering exposure testing of the 6061-T6 alloy adherends, the difference in yield strength between 6061-T6 and 2024-T3 should not invalidate the comparisons between overall durability prognostications.

Recently, Cognard (3184) has produced a paper using the wedge test to make quantitative measurements of the energy of fracture. This requires careful design of the joint. Cognard showed the relationship between the length of fracture and the geometrical parameters of the joint and the variation of the resistance of fracture in the presence of humidity. Also, he confirmed the fracture length follows a Weibull statistics allowing for qualtity control. Earlier Cognard estimations (3185,3186) that the crack growth should be stable after 1 da in a desiccator were confirmed.

2. *DOUBLE-CANTILEVER BEAM SPECIMEN*

The idea of a monolithic double–cantilever beam (DCB) was originated by Obreimoff in 1930 with a simple analyses proposed in 1957 and 1959 by Benbow and Roesler and Gilman, respectively. The two bonded adherend beams are separated by initiating a crack and treating as a cantilever beam built-in at one end. Only the bending of the cantilever part was considered in the early energy analyses. Kanninen modeled each half of the DCB as a beam partly free and partly supported by an elastic foundation in 1973. In 1978, Gates further generated the Kanninen model accounting for shear stresses noticing that this effect is significant in the case of a short DCB when the uncleaved ligament is comparable to the beam height. The DCB specimen used so extensively by Ripling, Mostovoy, and co-workers (2501–2521) to evaluate the fracture energy of structural adhesives with aluminum adherends was further refined by these investigators by special tapering or contouring to produce a specimen known as TDCB. Bascom and co-workers (2674) and Kinloch and Shaw (3187) observed the dependency of the fracture energy on the properties of the adhesive (elastic modulus and Poisson's ratio) and adhesive bondline thickness.

3. OTHER PROCEDURES REPORTED IN THE LITERATURE

Another testing procedure was used with aluminum joints in the early 1960s by Eickner and co-workers at the Forest Products Lab (3188,3190) with the adhesive bondline under a cleavage force, as shown in Figure 27. Obviously, a continuous cleavage stress will be present on the bondline as long as the adhesive does not creep sufficiently to relieve the stress or there is debonding. Actually, as employed, there was no quantification of the actual amount of stressing imposed; however, one would anticipate that such stressing in the natural weathering conditions where these joints were exposed would significantly decrease the average joint survival time as compared with similar nonstressed lap-joints. Because Eickner did not report significant decreases in joint survival time due to the use of this specimen would lead to the conclusion that the degree of stressing must have been quite modest. It may be recognized by the reader that this type of test is the same as used to determine the relative resistance of metal bar test specimens to stress corrosion. For example, the comparative resistance of various aluminum alloys to stress corrosion could be conducted by bending aluminum sheet specimens between specifically spaced restraints to provide a known bending stress and exposing the specimens to various kinds of accelerated weathering conditions such as salt-water soaking, intermittent salt spray, or the natural atmosphere. Thus, this test using adhesive lap-shear joint specimens would appear to constitute merely a bonded joint substitution for the single-coupon specimen used for metal stress-corrosion evaluations.

Still another improvised type of cleavage test for bonded joint evaluations, not included to the author's knowledge in any ASTM standard, is that used by Carter (3191) in the early 1960s. Carter was attempting to evaluate the effect of simultaneous stress and outdoor weathering on aluminum joint durability. As shown in Figure 28, a large single-lap aluminum joint was anchored to a stationary support and allowed to overhand with a large static weight as the load attachment to the overhanding end. Unlike the earlier testing of Eichner in Figure 27, the stress load could be precisely determined and data point gathered is the time period to produce a joint failure. Under these very severe cleavage loads, Carter found catastrophic failures of room temperature–curing aluminum/epoxy joints even in a natural atmospheric weathering exposure. It should be pointed out, however, that the load imposed is being resisted entirely in a pure cleavage mode, a condition that would never be knowingly employed by a design engineer. Carter was also among the first to vertically suspend a string of single-lap-joints and attach a static load to the bottom while exposing to outside weathering conditions. In this case, any cleavage stress component would be quite secondary to the more predominant tensile-shear forces acting on the bondlines of the string of joints.

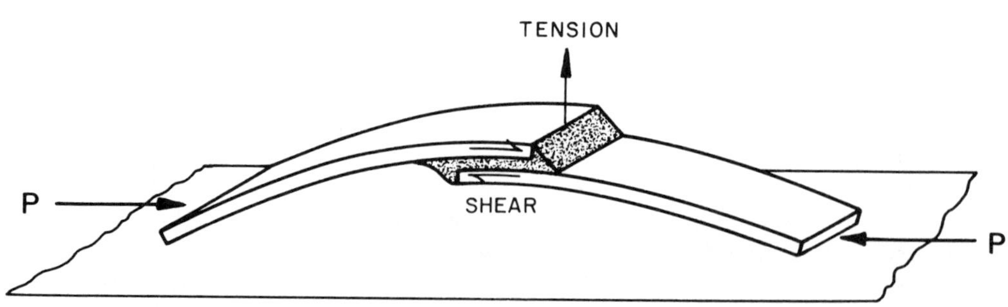

Figure 27 Standard lap-shear in bending. Specimen, 2.54 cm (1-in) wide or entire assembly, is bent and time to failure is recorded. (From J.D. Minford, Aluminum Adhesive Bond Permanence, Treatise on Adhesion and Adhesives (R.L. Patrick, ed.). Marcel Dekker, New York, Vol. 5, p. 112, Fig. 3-47 [1981].)

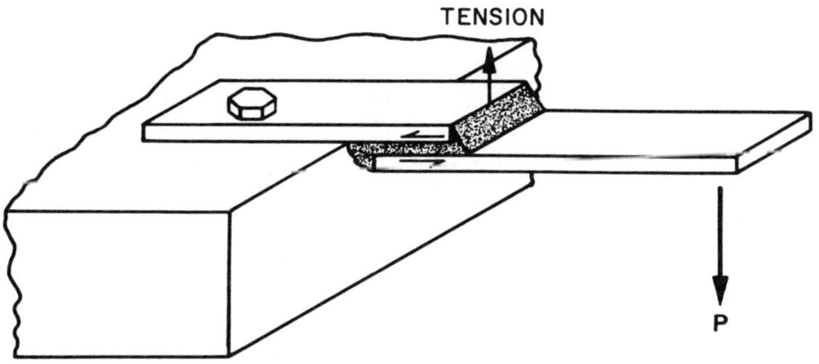

Figure 28 Cleavage lap-shear time to failure recorded. (From J.D. Minford, Aluminum Adhesive Bond Permanence, Treatise on Adhesion and Adhesives (R.L. Patrick, ed.). Marcel Dekker, New York, Vol. 5, p. 117, Fig. 3–50 [1981].)

H. Impact Strength Testing

Impact testing measures the force required to break an adhesive bond when one adherend is rigidly held and the other adherend is impacted. The description of the most commonly employed testing procedure is offered in ASTM Std D-950. Typically, a pendulum-type striker is used and the magnitude of the energy absorbed in producing failure of the specimen is recorded. Reproducible results are generally difficult to achieve, and typically the results from at least 10 test specimens are averaged. Thickness of the adhesive can obviously be an important factor in the amount of energy absorption that can be tolerated, and test data presented must include bondline thickness measurements to within 0.025 mm. Also, the method of making such measurements needs to be carefully described in detail, so that comparative results can be obtained at any second testing site.

Harris and Adams (3192) found the existing ASTM D-950 procedures were not suitable for their studies of the impact strength of adhesive joints and instead chose the single–lap-joint for impacting. They employed a rubber-toughened epoxy and varying strength aluminum adherends. The adhesive was separately tested in bulk form to show its mechanical response when tested in shear over a wide range of strain rates. There were large changes in response of the bulk adhesive under quasistatic loading versus impact conditions as anticipated. However, the strength of lap-joints were only slightly reduced by impact loading as compared to the static loading, as shown in Table 134 for three different strength aluminum adherends in the approximate strength ratios of 4:2:1. In regard to energy

Table 134 Comparison of Single–Lap-Joint[a] Strength Between Static and Impact Loading for Various Adherend Materials

Adherend	0.2% proof stress level (Nmm^{-2})	Failure loads: Static (kN)	Failure loads: Impact (kN)	% reduction due to impact testing
Aluminum alloy 2L73	430	15.75	14.25	10
Aluminum alloy BB2hh	220	10.25	9.70	5
Aluminum alloy BB2	112	7.80	7.49	4

[a]Single-lap specimens manufactured to dimensions specified in ASTM D1002 using a rubber-modified epoxy adhesive with glueline thickness maintained at 0.125 mm.
Data from Ref. 3192 (Figs. 6 and 7, pp. 620 and 621, respectively).

absorption, it was small when high-strength aluminum adherends were used; however, lower-strength aluminum adherends could absorb much more energy because of their significant plastic deformation. Based on the above, these investigators concluded that aluminum adherend joints could be quite suitable for the joining of structures from which high-energy absorption is required. However, for any given adhesive, the energy absorption capability is strongly dependent on the aluminum adherend strength. Since fracture energy of the adhesive itself is relatively small, significant energy absorption can only be achieved through plastic deformation of the adherends. The same investigators (3193) examined the feasibility of using adhesives in structures in which joints may be required to be highly deformed or crumpled under an impact load. (For a description of the results, see Ref. 3192, p. 626.) Kinloch (3194) provided a 1987 presentation to the Adhesion Society titled the "Impact Performance of Structural Adhesive Joints."

I. Fatigue Strength Testing

Fatigue tests are usually carried out using standard fatigue-testing machines and single–lap-joints fashioned according to ASTM Std D-1002. The overall procedure is described in ASTM Std D-3166. The mode of stressing is cyclic axial or bending. Important variables that need to be recorded when reporting data are cycling frequency, amplitude of the motion, testing temperature, and type of stressing. Cycles to failure and corresponding load levels are generally plotted for best interpretation. The intersection of the minimum stress points line with the 10 million cycle line is generally referred to as the fatigue strength value for the specific type of joint tested.

In order to quantify the fatigue behavior of any bonded joint, it would be obvious that an accurate analytical representation of the stress-strain distribution in the adhesive interlayer needs to be established as a function of the applied loading force. Two of the main properties of interest in this regard are the tensile creep compliance and Poisson's ratio (a constant for each different material designated as the ratio of lateral contraction over unit elongation when the material is stretched). It has been considered that these two values are interrelated in the so-called constitutive relations of any material. These relations may be time, temperature, and moisture dependent in general. Of course, there are other known material properties that can also influence stress-strain state such as moisture absorption (expansion), moisture desorption (contraction), thermal coefficient of expansion, and glass transition temperature, among others.

Chen et al. (2777) reported fatigue strength data on adhesively bonded joints from as early as 1957. Adams and Wake (2778) pointed out in 1984 that knowledge was already in existence in the aircraft industry about the fatigue strength superiority of the adhesive joint as compared to any riveted connections. Measurements by Argyris in 1962 suggested a 63% higher value for the fatigue strength–to–ultimate strength ratio obtained under monotonic loading conditions. Wang (2775) produced extensive fatigue testing results at the Air Force Materials Lab in 1963. The early 1970s showed the results of investigations by Mostovoy and Ripling (2511,2518) in 1971 and 1975.

In 1975, Roderick et al. (3195) reported their cyclic debonding data on composite/metal joints, whereas Brussat et al. (3161,3196), in 1976 and 1977, demonstrated from use of fracture mechanics principles that the strain-energy release rate could be useful for correlating cyclic debond growth rate. These strain-energy release rate calculations do require specimen analysis with a finte element program such as GAMNAS (see Ref. 3215).

Brussat et al. (3161,3196) showed cyclic failure rate could be further resolved into I (opening), II (sliding), and III (tearing) debonding modes using a cracked lap-shear joint. It should be pointed out that in practice only the peel and shear stresses exist near the debond front during fatigue loading. In 1980, Romanko and Knauss (3198) further extended the concepts to fatigue failures occurring under various environmental conditions. In 1981, Jacobson (3199) demonstrated that the various operational modes of aircraft (combinations of thermal and acoustic conditions) can subject the bonded aircraft structures to fatigue failure leading to the early conclusion that sonic fatigue considerations should govern the design of any advanced aircraft structures. Everett (3200), in 1982, concentrated on the evaluation of the role of peel stresses in cyclic debonding of joints. Mall et al. (3201) produced a 1984 report on cyclic debonding which described a combined experimental and analytical study of cracked lap-shear specimens subjected to constant-amplitude cycle loading. More recently, Sancaktar (2776) has reviewed the field of fatigue and fracture mechanics in the 1990 *ASM Engineered Materials Handbook.*

McMillan (833), in his AGARD Lecture, reviewed the evidence that cyclic stressing testing of laboratory-fabricated structural joints gave similar type failures to aircraft in service; i.e., interfacial. Further, he notes that the combined effect of environmental exposure and slow-cyclic fatigue stressing are especially damaging to aluminum joint survival as compared with much higher frequency cyclic testing alone. Some of the most comprehensive studies involving simultaneous cyclic stressing and weathering have been those reported by Frazier and Lajoie (3202) at Bell Helicopter under government contract. They also found that a question arises as to the selection of a load/unload periodicity. This occurs because adhesives will vary as to the rates at which they creep under load and relax when the load drops off.

Althof (3173) has studied the effect of varying cyclic loading on shear stressed adhesive bondlines. This is especially pertinent where organic polymers like adhesives have viscoelastic properties. In this case, the adhesive will undergo a displacement in the bondline under mechanical loading which is time-dependent. Obviously, if the load is constant, then the displacement may continue to increase with time, and we find ourselves dealing with the creep properties of the adhesive. The load required for a constant displacement can decrease with time, because the adhesive is showing accommodation by gradual relaxing. Under cyclic loading conditions, however, we should expect that such time-dependent displacements will be the typical bondline response. Of greatest importance to the practical service life of a bonded structure will be its response to the loading and unloading conditions on the bondline. The primary importance of the viscoelasticity of adhesives in these matters has prompted many extensive investigations of the viscoelastic properties of adhesives using cured neat adhesive coupons.

Romanko (3203), in his 1979 AGARD lecture, presented a good review of the investigative work performed in determining the fundamental mechanisms of fatigue degradation in structural aluminum joints and the dominant fatigue mechanisms with various service environmental regimens, including cyclic mechanical load, temperature, and humidity. Additional pertinent data has been developed under a number of Air Force–sponsored programs, including Schjelderup's 1977 compilation of mechanical properties of FM-73 at Douglas (3204), Renton's 1978 AFML study of the structural properties of adhesives (3205), General Dynamics' 1976–1980 study of joint fatigue failures (3206), Vought Corporation's 1976–1978 investigation of the structural properties of adhesives (3207), and Douglas Aircraft's PABST study relating to this subject (3208).

Althof and Brockmann (3171) have added measurements of adhesive shear properties by determining the shear strains of short bondlines between aluminum adherends of very high stiffness. Althof later proved that the stress-strain relationships of these short bondlines are also representative of the adhesive sections in long bondlines. Matting and Ulmer (3209) reported as early as 1963 that there is an accumulation of residual shear displacements in adhesively bonded joints when subjected to sinusoidal cyclic loads. Quantitative values could not be offered, however, since methods to measure shear displacements in thin adherend bondlines were not available. Romanko and Knauss (2812) almost 20 years later reported extensive test results on shear displacements using cycled thick adherend aluminum joints but still without quantitative measurements of bondline displacements. Althof (3173) has made some of these bondline shear strain, time-dependent measurements using cycled thick aluminum specimens. Althof was able to (1) analyze with respect to a mathematical description, (2) estimate the bondline shear stresses at which strains are negligible with respect to a long-life bondline, (3) compare the effects of squarewave and sinusoidal cycled bondlines, and (4) estimate the criteria for bondline failure under cyclic loading. His studies included shear stress load levels ranging from 20 to 40 MPa and five cycling conditions, which are described in Table 135. The test results obtained with thick adherend FM-73 adhesive bonds has been summarized in Table 136.

In these investigations, a number of characteristics could be observed such as (1) when a plastic strain could be induced in the bondline during the first cycle, then the joint would survive only a few cycles; (2) whereas shear stresses below the quasi-elastic stress limit could cause low strains, the strains could continue to increase continuously, if slowly; (3) the creep was always shown to be maximum during the early cycles as compared to later cycles; (4) the residual strain remaining after a cycle appears to be due to the lower recovery during the unload period as compared to the creep experience

Table 135 Times of Load and Unload Periods During Squarewave Cycling

Load period (min)	Unload period (min)	Abbreviation
60	15	60′/15′ "Boeing mission"
15	5	15′/5′ "Douglas mission"
4	2	4′/2′
1	1	1′/1′
0.5	0.5	0.5′/0.5′

From Ref. 3173 (Table I, p. 662).

during the loading period; (5) the bondline will fail when residual strains build up to a critical strain level; (6) the shorter the periods of loading and unloading, the greater the number of cycles required for bondline failure, i.e., slow cycle fatigue conditions are the most damaging; (7) the accumulated failure strain from load cycling will correspond to the failure strain measured for the static load situations; (8) mathematical descriptions only seemed possible under conditions of very low stress where the adhesive could deform linear-elastically; (9) the observed shear strain behavior at medium and high shear stresses indicate the effective mean load causes an increase of the total strain, which is analogous to the strain behavior of static creep tests under sustained loads; and (10) the above observations were believed to be strictly valid only for short bondlines with homogeneous shear-stress distribution in the bondlines. In real-world bonded structures, the adhesive strain behavior would undoubtedly be more complex but there would be a similar dependence on time and adhesive strengthening factors. Althof concluded that this work demonstrated that the failure criterion could not be fatigue failure caused by a progressive crack front as proposed earlier by Romanko and Knauss (2812).

Table 136 Test Results of FM-73 Adhesive Bondlines, Squarewave Cycled on Different Times of Cycle Periods and Different Shear Stresses

Shear stress (MPa)	Time of Load Period/Unload Period (min)									
	60/15		15/5		4/2		1/1		0.5/0.5	
max	N_f	γ_f	N_f	γ_f	N_f	γ_f	N_f	γ_f	N_f	γ_f
40	7	1.48	6	1.48	5	1.46	29	1.48	30	1.50
			5	1.45	25	1.60				
37.5	8	1.40	11	1.45	62	1.41	27	1.43	28	1.38
	5	1.43	13	1.52	26	1.50				
35	15	1.50	35	1.48	47	1.36	164	1.52	72	1.42
			35	1.53	56	1.40				
					171	1.45				
30	33	1.50	61	1.60	49	1.48	1765	No failure		
	52	1.38								
25	120	No failure	1626	No failure	975	No failure	1640	No failure		
20	1270	No failure								

N_f = numbers of cycle at failure, γ_f = shear strain at failure; on static loading, the bondline fail at γ_f = 1.50.
Data from Ref. 3173 (Table II, p. 666).

In addition to the large aerospace reports summarizing fatigue investigations during the period from 1975 to 1980 mentioned above, there were a number of other fatigue-related investigations of bonded joints in the period ranging from about 1975 to 1978. Some examples from 1975 included publications of Kuhbander and Apoyni (3210), Klapprott et al. (3211), and May and Adsit (3212). Three 1984 fatigue-related publications included those of Hashemi and Williams (3213) on the size and loading mode effects in fracture toughness testing, Sewaki et al. (3214) on determination of fatigue fracture toughness by conventional rotary bending specimens, and Chaplya and Gvozdyuk (3215) describing a new method of investigating fatigue crack growth in lamellar composite materials.

Additional 1978 publications on fatigue were issued by Wentz and Wolfe (3216), Johnson et al. (3217), Anderson et al. (3218), Ratwani (3219), Brussat and Chiu (3220), Pfeiffer and Alic (3221), and Mulville et al. (3222). Parida et al. (3223) offered a pertinent discussion of the subject of probability, confidence, and sample size in fatigue testing in 1990.

J. Creep Strength Testing

The ability of different cured adhesive formulations to creep under an applied stress has been extensively investigated and the results are of particular interest to the design engineer. Obviously, a structure bonded with any adhesive that can creep under a service loading condition cannot sustain itself under any steady stressing situation of that magnitude. ASTM Std D-1780 test procedure is titled "Conducting Creep Tests of Metal-to-Metal Adhesives" and is the recommended practice for determining the amount of creep of aluminum-to-aluminum bonds due to the combined effects of temperature, tensile-shear stress, and time. The primary use of such a creep test is to provide basic data for the choice of safe working stresses for applications, in which allowable deformation within the service life of the structure is the criterion of failure. Creep is recognized as an extremely sensitive index of strength; however, it does not usually vary as a linear function of stress but rather depends on the material itself, the stress present, the temperature, and the time of stress impression. Separate measurements are indicated for creep, instantaneous strain, and instantaneous recovery using this standard with bonded test panels of 7.5 × 12 in (190.5 × 305 mm) dimensions. A variation procedure involving the loading in shear by tension loading using a single–lap-joint is offered in ASTM Std D-2294. The British equivalent is BS 5350: Part C7: 1976.

Temperature is obviously an important parameter in such testing as indicated by the common situation where a structure designed to operate safely at a lower temperature will inexorably creep to failure at some higher service temperature or rapidly fail at some very high temperature without extensive chemical decomposition of the bulk adhesive.

Constant-stress creep testing is the simplest procedure and can in its simplest context be carried out by applying a constant dead-weight load to the bottom of a suspended ASTM Std D-1002 single–lap-joint. It has also been mentioned that ASTM D-1780 practice permits measurements of instantaneous strain, instantaneous recovery, and total recovery. The increment of time used to measure these values must be specifically designated, since it is nearly impossible to obtain strain reading at the instant of loading or decrease in strain that occurs immediately. The recovery value will be the total decrease in strain minus the instantaneous recovery value. As also noted elsewhere, the rates at which various cured adhesive formulations will creep can vary widely. Also, the extent of recovery after stress release.

Another form of creep test can be conducted when this same single–lap-joint is fastened into a fixture which is spring loaded, as described in ASTM Std D-2919. The specific aim of this specification, however, is to provide a sustained stress on the bondline while the specimen is simultaneously exposed to some specified hot humidity weathering condition. This test procedure was evolved from important investigations conducted by Sharpe at the Bell Labs in the 1960s showing the significant acceleration of time to failure of aluminum lap-joints bonded with room temperature–curing epoxy adhesives. More will be said about this subject when discussing the overall joint durability problem in Chapter 10. A later developed stressing fixture constructed entirely from aluminum at the Alcoa Labs by Jomboch, as shown in Figure 29, was used by Minford (976) for the first time in a 1972 report to develop aluminum joint survival times with both room temperature– and elevated temperature–curing epoxy adhesives under variable stressing and weathering conditions. This stressing device

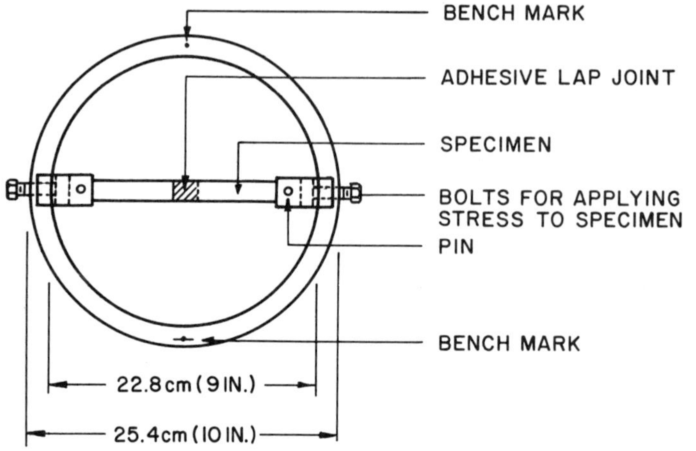

Figure 29 Circular stressing fixture details. (From J.D. Minford, Aluminum Adhesive Bond Permanence (R.L. Patrick, ed.). Marcel Dekker, New York, Vol. 5, p. 120, Fig. 3-53 [1981].)

had the advantage over the ASTM D-2919 stressing fixture of being much lower cost, lighter, and more conserving of exposure space requirements. Thus, it could be readily used even in conducting daily cyclic test exposures with steady stress imposed on the bondline. With all parts made from aluminum, with the same coefficient of thermal expansion, it could accommodate significant temperature changes during varied exposure conditions while maintaining the same alignment features. The use of anodized aluminum surfaces also permitted the use of the fixture in corrosive salt water or salt-fog as well as condensing hot humidity, freshwater or wet/dry salt-water cycling in the seacoast or marine atmospheres.

Hughes et al. (3224) have demonstrated the relationship between the humidity of the joint environment and the level of creep stress on the rate of creep, as shown in Table 137. From these data we can observe that the creep rate at the same creep stress could be 28-fold higher with the environmental humidity at the 70% RH level as compared to the 20% RH level.

Finally, there is a compression loading test method for evaluating creep in aluminum-to-aluminum joints designated ASTM Std D-2293. The use of a compression loading specimen requires a special spring-loaded device which is described in the specification.

1. TORSION SHEAR TESTING OF CREEP

The torsion shear technique has been described in some detail by Hughes et al. (3225). A napkin-ring type of specimen, consisting of two relatively thin-walled tubes bonded together end-to-end are used. The adhesive bond is stressed by rotating one adherend relative to the other about an axis that is coincident with the longitudinal axis of the assembly. The specimen has the advantage that the loading is in pure shear with no tensile or cleavage loading on the adhesive bond. In making the specimen, the ratio of the wall thickness to the diameter of the adherends can be deliberately selected to minimize the variable strain in the adhesive. The technique can ordinarily be used to determine the tortion shear modulus, generate microyield data (precision elastic limit, microyield stress, and irreversible work performed in a load-unload cycle), and the fracture stress and strain of adhesives. Creep (as a time-dependent property) can additionally be studied using this technique.

K. Other Special Testing Procedures

1. PIN-AND-COLLAR SHEAR TEST

In recent years, there has been a considerable increase in the use of curing-in-place liquid adhesives like anaerobic adhesives for retaining and locking aluminum cylindrical assemblies. This has generated

Table 137 Effect of Atmospheric Humidity on Creep Rate of a Cured Adhesive

Humidity level (%) at RH	Creep stress		Creep rate ($\times 10^{-8}$/sec)
	psi	MPa	
20	500	3.45	2
	1000	6.89	13
	1500	10.34	48
55	500	3.45	5
	1000	6.89	48
	1500	10.34	150
70	500	3.45	65
	1000	6.89	250
	1500	10.34	1350

Data interpreted from Ref. 3225 (Fig. 2, p. 131).

a new ASTM Std D-4562 procedure. By using a specially fabricated pin-and-collar assembly, the shear strengths of such assemblies can be accurately determined as the adhesive is being cured in place.

2. EXTENSOMETER TESTING PROCEDURES

Krieger (2216–2218,2349,2571) proposed evaluation of structural bonded aluminum joints through the use of a device called an extensometer. In 1973, Krieger first published a paper titled "Evaluating Structural Adhesives Under Sustained Load in Hostile Environment" which initiated a series of papers extending the use of an extensometer test approach. In 1975, a paper followed discussing the stiffness characteristics of adhesives in hostile environments, whereas 1976 and 1977 generated publications dealing with the shear stress-strain properties under similar conditions. In 1978, Krieger (3226) discussed the techniques he had developed in the journal *Adhesives Age* for more general reading, and again, in 1985, in the same publication, with additional comments. This last article can be recommended for surveying this important work. Finally, in 1985, Krieger (3227) composed a summary article of progress to date in the journal *Adhesives Age* which can be recommended for surveying this important work. In 1986 (2349), he offered another application of his techniques for the testing of bonded structure. The most important advantage of the procedure is that Krieger was able to make predictions about performance in bonded structure without the benefit of the expensive peel, lap, and fatigue testing of actual parts which had been usual procedure before. Most recently, Krieger (2204) has provided a 1990 review of his technique for the part of the *ASM Engineered Materials Handbook* that relates to "Designing with Adhesives and Sealants."

Hughes and Rutherford (3131) have employed high-sensitivity extensometers based on earlier microstrain test methods developed by Bossler et al. (3228) and Hughes et al. (3229). With this instrumentation, they were able to develop values for the modulus, elastic limit, microyield stress, and energy loss. They reported that stress decreased the modulus but raised the microyield stress and energy loss. Similarly, stress increased the energy loss with load/unload cycles. Some property losses were restored with time, however, in the absence of applied load. Hughes and Rutherford (3131) employed these procedures to evaluate the potential for developing satisfactory weld bonding. At the same laboratory, Shen and Rutherford, using similar techniques, have studied and determined various time-dependent properties of the adhesives themselves.

More recently, Miller et al. (2386), at the Alcoa Labs, have utilized the Krieger extensometer (KRG-1) to determine adhesive shear behavior in illuminating the effects of adhesive ductility and bondline geometry on aluminum tube-and-socket joint performance. Three 1991 papers related to creep include Kim et al. (3230) presenting a novel specimen geometry for double shear creep experiments, McDonald and Hartwig (3231) describe testing under cryogenic environmental conditions, and Elster and Honerkamp (3232) discuss a modified maximum entropy method and its application to creep data.

3. TORSION PENDULUM TEST ON CAST (NEAT) ADHESIVES

This method is one standardized in ASTM Std D-2236 (3134). This is a testing procedure that is aimed at determining the values for the shear modulus and the logarithmic decrement of the adhesive at varying temperatures. The material for making specimens from adhesive film products is fabricated by taking a number of layers of film, compressing them in a mold, and curing the material in a heated press under the equivalent time at temperature conditions to be used to manufacture the final structure. If the adhesive is a liquid, then it must be cast into a mold. A bubble-free test specimen is then cut from this cast and cured material for fixing to a flywheel mass of known inertia. The sample must be stimulated to oscillate and the free decay values are recorded for each different testing temperature.

The individual values of shear modulus and logarithmic decrement are calculated for each temperature and each is plotted versus temperature in a semilogarithmic diagram. The different characterizations for an adhesive can be described and the values compared for various adhesive conditions. In the so-called freezing range for the adhesive, the expectation is that a high and constant shear modulus should be present and will be accompanied by low decrements. This would mean the adhesive would have the properties of being hard and brittle with reversible deformations in the bondline. Creep rate will be small but cracks may occur under dynamic loading conditions like fatigue testing.

In the softening temperature range for the adhesive, the shear modulus will decrease sharply with the decrement increasing and passing through a maximum. This so-called dispersion range has other gradations, which are described along with the method details and data by Hughes et al. (3233). The occurrence of a number of these dispersion ranges can be characteristic of copolymers. In such a softened state, the bondline may deform irreversibly when loaded, with its strength decreasing as it creeps.

The rubber-elastic range is one which is observed in the case of cross-linked resins. The shear modulus and the decrement are both relatively low and will remain approximately constant with increasing temperature. Mechanical loading is not recommended with the adhesive in this state, since adhesive in the bondline is very ductile and even with relatively small loads will creep.

The final range is referred to as the plastic range and is found when the resins are not cross-linked. The shear modulus will decrease and the decrement consequently increase with increasing temperature. Again loading of the bondline is not recommended, since any deformations of the adhesive in the bondline will be irreversible.

Hughes et al. (3234) employed the torsion shear creep test as a function of temperature, humidity, and applied stress to determine the effect of moisture on two bisphenol A-type epoxy adhesives in 1984.

4. BLOW-OFF ("BLISTER") PROCEDURE

The blow-off ("blister") test for evaluating the adhesion of adhering layers was originally proposed by Dannenberg (3235) for analyzing the debonding of paint or thin films in 1961. Williams (340), in 1969, and Andrews and Stevenson (341), in 1978, used the pressurized blister test to measure the fundamental specific adhesive fracture energy. Gent (3236), in 1987, offered a reconsideration of the blister for measuring the work required to detach a unit area of a material from an adherend. Also in 1987, Dillard (3237) used the constrained blister test as a novel way for measuring adhesive fracture toughness. Allen and Senturia (3238,3239) reported on the use of an island blister test for peeling thin films in 1987 and thin films adhering to thermosetting adhesives in 1989. Moet (3240), in 1987, considered the thermodynamics of the blister test in 1987. In 1990, Dillard (3241) reported on current developments in blister tests and with Lai (3242) discussed an elementary plate theory prediction for strain energy-release rate using the constrained blister test in 1990. Finally, Fernando and Kinloch (3243) reported on the use of the "inverted blister" test to study the adhesion of photopolymers.

VII. NONDESTRUCTIVE TESTING PROCEDURES (NDT TECHNIQUES)

A. General Background

Interface defects are not always detectable by even state of the art NDT methods, so test coupons are still often processed with production parts for subsequent laboratory evaluation. Schliekelmann (3244,3245) and Bijlmer and Schliekelmann (779) conducted a variety of investigations about the ideal aluminum oxide configuration for best adhesion using inspection with an electron microscope. Another physical parameter that was used early as a basis for NDT of aluminum oxide surfaces was the surface potential difference (SPD), or Volta-potential. This is the difference in work function between a reference surface and the surface of study. The extreme sensitivity of this technique to changes in the oxide and to the presence of surface contamination was reported in 1963 by Matting and Ulmer (3246). The SPD value is obtained with a vibrating capacitor technique originally ascribed to work of W. Thomson (Lord Kelvin) (3247). Based on this principle, Bijlmer (3248) helped develop one of the first NDT instruments; i.e., the Fokker Contamination Tester. Bijlmer and Schliekelmann (779) and Bijlmer (3248) reported finding correlations between the SPD values and bondability of chromic-sulfuric acid deoxidized aluminum joints (FPL-type).

When this work was extended to the study of anodic aluminum oxides, however, the situation was obviously more complicated (3249). Nevertheless, Bijlmer (3248) was able to monitor with SPD values the effects due to adsorption and desorption of water on such surfaces. Michaelson (3250) and Fromhold and Kruger (3251) explained the effect in terms of oriented dipoles or a residual space charge. Smith and Lindberg (3252) also studied the relationship between SPD measurements and adsorped dipole layers, the detection of surface organic contaminants, and anodic oxide aging. Kwakernaak et al. (3249) concluded that the SPD values seemed to depend mainly on the anodizing voltage and to a much lesser degree on the time and temperature associated with the anodizing. They also seemed to provide information about the rinsing and aging of the oxide after anodizing.

Another predictive measurement made on anodic surfaces has been surface impedance. Englehart and George (3253), at Alcoa, pioneered the use of impedance measurements as a simple method to measure the quality of sealing the anodize surface. Birtel and Leute (3254) and Ott (3255) have both reported the use of impedance for this purpose in industry. The method was used by Schliekelmann (3256) to discriminate the relative aggressiveness of different alkaline cleaning solutions on aluminum prior to deoxidizing the surface for bonding. Kwakernaak et al. (3249) concluded that the surface impedance seemed to be largely influenced by the anodizing voltage and to a lesser extent by the processing time in the bath. Rienka (3257) presented a 1976 paper reviewing the use up to that time of the resonance/impedance and the Volta-potential methods for the NDT testing of bonded joints.

It is the obvious goal of manufacturers of bonded structures to determine the strength and durability potential of the structures they manufacture. Equally important is the evaluation of the day-to-day quality of the product coming off the assembly line. For this latter purpose, there has been a continuing search for better testing procedures to make this determination without having any detrimental effects upon bondlines or adherends. The designation NDT has been applied to the total umbrella of procedures that have been developed for this purpose. In 1987, Adams et al. (3258) discussed the known concepts in NDT testing involving defect, including porosity, surface unbonds, zero volume bonds, and poor curing conditions.

Because of the critical nature of the aircraft industry's bonded structures in regard to human safety, there has been the highest possible incentive to develop NDT practices. Further, because of the vested interest of the military in such matters, there has been the strong financial support necessary to both develop, purchase, and maintain such equipment. In the future, it might be expected that other industries such as the automotive, trailer, and railroad car industries might develop their use of adhesive bonding to the level that they will want to similarly inspect their production. Certainly, the successful use of bonding in such vital areas of the car as hem flanges on hood, decks, and doors has been significant. Advancements in bonding dissimilar materials like steel and aluminum to polyester fiber glass with adhesives was discussed in 1980 by Meyer and Chapmen (3259).

The most prominent nondestructive-type tests fall into the major categories of sonic, ultrasonic, radiographic, eddy current, thermal, acoustic emission, and acoustic holographic varieties. Regardless of which technique is employed, the success is dependent on observable response differences to a physical stimulus applied to well-bonded versus poorly bonded areas. Schliekelmann's (52) excellent review of the state of the art just prior to 1980 should be quite helpful to the reader desiring the larger overview of the subject.

Schliekelmann's personal theoretical considerations about this matter were reported as early as 1964 (3260). In 1967, the same author published another general article on NDT (3261). In 1966, a defense conference on NDT was held in Boston in which Lennon and Wang (3262) looked at the implications for inspecting adhesive-bonded metal helicopter blades, whereas Vogel (3263) described the experimental development of an infrared multiple scan system for inspecting bond integrity. Meanwhile, in 1968, Jarvis (3264) reported the development of a process specification at American Aviation in Cleveland to inspect metal-to-metal and aluminum-faced aluminum honeycomb-bonded assemblies using specified adhesives.

In 1991, Nieminen and Koenig (3265) reviewed the use of macroscopic and modern microscopic nondestructive evaluation methods (NDE) for adhesive-bonded structures.

B. Sonic Testing Procedures

The idea behind this type of procedure is quite elementary in its concept and simplistic in its operation. For this reason it usually fails to imbue much confidence in its diagnostic abilities for finding even modest-sized debonds. An example of a sonic procedure would be the use of an aluminum rod or a large metallic coin as tapper over the bondline area desired to be checked. It must be admitted, however, that investigations have shown that disbonds must be of about 38-mm diameter dimensions for even experienced testers to detect the sound difference between the clear sound from solid bonds as compared to the dull sound from voided bonds. Every inspector using such a method has worked constantly to be able to perform this job better. Unfortunately, the method offers no opportunity for predicting bond quality other than the opinion of the operator. Obviously, we would only expect the use of this method primarily to detect possible disbonds within large area bonded structures like laminated building panels or very large parts on aircraft. Kraska and Kann (3266) summarized some of their experience in 1970 with sonic inspection of bonded aircraft honeycomb structures, as had Botsco and Roopenian (3267) in 1968.

Adams et al. (3268), in 1985, considered increasing the sensitivity of the coin or hammer-tap sonic test method by further processing of the force-time histories. Another variation of a sonic vibration procedure has been referred to as the mechanical impedance method. Lange and Teumin (3269) discussed their efforts on sonic testing in the Soviet journal *NDT* in 1971, whereas Schrooer et al. (3270), a year earlier, investigated a procedure which they described as being an acoustic-impact technique. Cawley (3271) issued a recent report on the subject in 1984.

Schliekelmann (52) has considered the Fokker Bond Tester Type I under sonic testing techniques because of the low frequency of operation. A transmitter and receiver probe are used with the transmitter introducing vibrations into the top bonded sheet (known as white noise). The meter reads the ratio between the transitted and received energy. Areas of detection must produce resonance frequencies within its frequency band of 0.5–10 kHz. Geometry and edge effects play a particularly important role because of these low frequencies, which was a stimulation by Fokker and others to move to higher-frequency techniques. Schliekelmann (52) also has listed the Sondicator tester as an improvement over Fokker Tester I because of a significant boost in frequency range. However, reports have indicated it suffers from the same general limitations of any low-frequency resonance systems.

Schliekelmann (52) also considers under sonic testing procedures the more sophisticated methods, still using the same basic principle of eddy-current–producing devices. The Harmonic Bond Tester developed by Boeing and manufactured by Shurtronics Corporation being one commercial example. The probe in this case contains an electromagnetic coil which maintains an oscillating field which induces eddy currents in the conductive upper sheet. The interaction of the original and induced fields causes the sheet to be driven at a frequency double the coil frequency. The response of the laminate is detected by a wide-band microphone (3272). There is still a serious limitation for detecting small voids for all of these low-frequency resonance systems. Schliekelmann (52) has calculated that if 30 kHz is considered to be the highest frequency detectable for an upper sheet that is 0.4 mm thick, then the void size must be in the order of 7 × 7 mm to be detectable.

C. Ultrasonic Testing Procedures

The aircraft industry has probably more widely employed this general procedure than any other. It is based on the general principle that there is an impedance of sound in various materials. This impedance can play a particularly important role at the interface between a metal and the adhesive. It is dependent on recognizing and measuring a change in frequency and amplitude which can occur when an ultrasonic transducer is liquid coupled to a bonded aluminum containing discontinuous areas. Claims have been made that bondline voids and porosity in the bondline can be detected in an early stage of failure while they are still in the precursor steps leading to ultimate disbonding. It is still quite true, however, that no single NDT method or procedure is known which is truly capable of finding all the bondline flaws that can occur in manufacturing. Thus, it is important to know what procedures are available and what their particular limitations may be.

Curtis (3273) has stated that the ultrasonic equipment used for bonded joint inspection is similar to that used for inspection of solid metals. He refers to two methods used on bonded joints; i.e., through-transmission and pulse-echo ultrasonic resonance techniques.

1. THROUGH-TRANSMISSION TECHNIQUES

The through-transmission procedure uses a separate transmitter and receiver on each side of the joint. Piezoelectric transducers are mainly employed for both stations. Such devices act to transform electrical into mechanical vibrations and vice versa. Simplest devices use probes on opposite surfaces in direct contact and alignment so the ultrasound from the transmitter can reach the receiver if the bond in intact. Water or an oil of low viscosity must be used as a couplant to ensure acoustic contact. An obvious difficulty would be maintenance of probe alignment while the probes are being slid over the surface in scanning. If the received signal is very small compared to that transmitted, then most of the energy has been reflected owing to the low impedance of an intact adhesive layer. Further interpretations about the quality of cohesion must come from analysis of the amplitudes of the received signal.

Improvement had to be sought in such devices on account of difficulties in maintaining transducer alignments. A major improvement came with the introduction of the immersion technique, where both transducers are submerged with the water acting as the couplant. A variation has been the so-called water-jet technique which uses a water jet as a pathway for ultrasound. A device for moving the probes via a guide-rail mechanism has permitted making simultaneous use of an x-y recorder resulting in production of what has come to be known as a C-scan record. Yet another development of through-transmission utilizes a search wheel where the two transducers are mounted in the hub of a wheel.

2. PULSE-ECHO TECHNIQUE

This procedure is designed to obviate the technical difficulties associated with the above probe systems. The same transducer is now used as both transmitter and receiver. The single transducer receives a signal (echo) back from the bond interface. Obviously, there would be an impossible to deal with interference if a continuous signal were to be transmitted, so a pulsed signal must be transmitted instead. Energy can thus be received during the dead periods between short pulses.

Several variations of this technique are possible. First, where thickness is large enough compared to the pulse length, it is possible to distinguish the reflected signals from each interface. The amplitudes of the echoes from the metal/adhesive interface (particularly from the lower interface) are parameters of special interest. Where echoes are missing from the lower interface there could not have been a firm bond to that surface. Thin adherend sheets can cause some problems even here, because of interference between the reflected and transmitted pulse. Special probes capable of producing very short pulses must be used in such cases. A reverberation of signals can occur within the top sheet which is known as "ringing." Because the difference in impedance at each dictates the

Table 138 Generic Flaw Types and Flaw-Producing Mechanisms

Flaw producing mechanism	Generic flaw type				
	Metal-to-metal	Metal-to-core	Core	Surface	Adhesive
1. Disbonds, internal	X				
2. Disbonds, part edge	X	X			
3. Disbonds, high core	X				
4. Porosity	X	X	X		X
5. Unremoved protective release film from adhesive	X	X			
6. Foam adhesive in film adhesive bondline	X	X			
7. Cut adhesive	X	X			X
8. Adhesive gaps	X	X			
9. Missing adhesive	X	X	X		
10. Weak bonds	X	X	X		
11. Extra layers of film adhesive	X	X			
12. Foreign objects	X	X	X		X
13. Double drilled or irregular holes	X			X	
14. Disbonds, low core		X			
15. Void or gap, chemical milled land		X			
16. Void or gap, doublers		X			
17. Missing fillets		X			
18. Voids, closure-to-core		X			
19. High-density inclusion, (chips, etc.)	X	X			X
20. Voids, foam joint		X	X		
21. Disbond, shear ties	X	X			
22. Lack of sealant at fasteners		X			
23. Thick foam adhesive		X			
24. Broken fasteners	X	X			
25. Crushed core			X		
26. Wrinkled core			X		
27. Condensed core			X		
28. Distorted core			X		
29. Blown core			X		
30. Node bond separation			X		
31. Missing core (short core)		X	X		
32. Cut core		X	X		
33. Water in core			X		
34. Cracks				X	
35. Scratches				X	
36. Blisters				X	
37. Protrusion				X	
38. Indentations (dents/dings)				X	
39. Wrinkles				X	
40. Pits				X	

Data from Refs. 3274 and 3276 (Table 1, pp. 339–342).

amount of reflected energy, conclusions about cohesive properties of a cured layer may be drawn from observations of the pattern on the CRT. The higher the adhesive quality, the smaller is the number and amplitude of reflected signals. The pulse-echo can be adapted for immersion testing. There is also a reflector plate technique where a sound pulse, after passing through a bonded component, bounces back from a glass or metal plate behind or below and passes again through the component, being received again by the same transducer that transmitted the pulse.

From the preceding discussion, it is easy to see how so many investigators have been involved over many years in trying to establish a particular variation of technique that fits their bondline situation.

Recognition of the wide variety of flaws or discontinuities which can occur in practice can be viewed from Clark's data in Table 138 (3274). Further, Clark itemized the relative frquency of flaws in aircraft in Table 139, showing that metal-to-metal voids and disbonds could account for 74% of the overall total. Defects can be divided first into three broad categories; i.e., metal bond defects, adherend defects and honeycomb sandwich defects. The first includes (1) voids (2) unbonds or disbonds, (3) porosity, (4) porous or frothy fillets, (5) lack of a fillet, (6) fractured or gouged fillets, (7) adhesive flash, and (8) burned adhesive (from drilling or cutting operations). Second category-type defects include (1) fractures or cracks; (2) double-drilled or irregular holes; (3) dents, dings, and wrinkles; and (4) scratches or gouges. Finally, honeycomb panel defects include (1) water in the core cells, (2) crushed core, (3) condensed core, (4) node separation, (5) blown core, (6) voids in foam/adhesive joints, and (7) protective films left on adhesive. The variety of above possible defects seems overwhelming and it is easily apparent why no single technique has been successful for pinpointing all the different situations.

Hagemaier and Fassbender (3275), in 1978, and Hagemaier (3276), in 1985, considered the various means of inspection. Hagemaier (3276a) has characterized the ultrasonic methods as being in several different general categories of (1) contact echo-pulse, (2) contact through-transmission, or (3) immersion method procedures. Babb (3277), in 1966, characterized the main ultrasonic procedures as being either (1) pulse-echo reflection, (2) pulsed through-transmission, or (3) sweep frequency resonance. The latter could be utilized using hand-held probes, semiautomatic scanning with part water immersion, or search unit wheels. Lanier (3278) had the opportunity in 1969 to review the pulse-echo methods that had been used on the production of the C5 military transport plane. From the European perspective, Althof (3279) offered a result summary utilizing ultrasonic pulse-echo instruments. In the same year, Beal (3280,3281) reviewed the use of an ultrasonic emission detector by NASA, including water-coupled impedance in the same year. Clotfelder (3282), in 1966, reviewed the use of the eddy-sonic testing procedure using composite structures.

Arnold (3283) and Arnold and Vincent (3284), in 1957 and 1959, rspectively, reported on developments of nondestructive testing, including ultrasonics. Immersion methods have been widely employed, especially in forms like the Ultrasonic C-Scan immersion system. Schmitz and Frank (3285)

Table 139 Frequency of Rejectable Flaws in Adhesively Bonded Assemblies

Defect	Number of defects	Percent of total
Metal-to-metal voids and disbonds	378	74
Skin-to-core voids and disbonds	19	3
Gap in core-to-closure bond	9	2
Lack of foaming adhesive or voids in foaming adhesives	22	4
Difference in core density	6	2
Lack of fillets	1	1
Crushed or missing core	32	6
Short core	40	8

Data from Ref. 3276 (Table 2, p. 342).

issued a 1965 comprehensive NASA report on this subject which is well illustrated with figures and tables of data. Kammerer (3286), in 1966, described an automated ultrasonic bond inspection procedure used in project Apollo. Even earlier in 1961 and 1962, Reese and Boruff (3287) and Miller and Boruff (3288) had issued summary reports on the Martin Company's efforts to thoroughly evaluate various ultrasonic test devices available at the time. In 1966, LoPilato and Carter (3289) recorded their observations on unbond detection using ultrasonic phase analysis. It is usual to keep on permanent file the C-Scan recorder records of any aircraft bonded parts that are compatible with the technique. They are then available for later comparison when any field problems occur or questions are raised about the soundness of the bonded components.

Schijve (3290), Arnold (3283), Arnold and Vincent (3284), Gonzales and Cagle (3291), and Smith and Cagle (3292) were among the earliest to relate ultrasonic procedures to the evaluation of bond quality. Schliekelmann (52,3244), at Netherlands Fokker, produced numerous progress reports on this matter in 1972, 1975, and 1979. He has certainly been one of the most prolific contributors to the field both in equipment design and interpretation of results. He has even established a correlation between ultrasonic measurements and the actual lap-shear strength. Rose (3293) and Rose and Meyer (3294) in the early 1970s also sought to predict the level of aluminum joint strength utilizing ultrasonic techniques. Tattersall (3295), in 1973, talked about predicting poor bond quality potential from ultrasonic instrumentation. He described what he labeled "slackness" in the bondline, which was something less severe than an actual disbond. About the same time, Lechodziejewski (3296) sought to use ultrasonics for evaluating the quality of bonded track-shoe pads on military vehicles. Adameter (3297), in the same period, extended use to evaluating elastomer-to-metal bondments and identified certain fatigue modes in the jointures.

Meyer and Rose (3298,3299) issued several papers on the general subject of determining aluminum joint strengths using ultrasonic procedures. They considered several modeling concepts for studying the interaction of ultrasonic waves with adhesive bonds. A variation in the interfacial bonding condition of a bond seemed to be most noticeably detected by amplitude change of the interfacial echo. As the quality of the interfacial bonding decreases, the amplitude of the reflections from the adhesive/aluminum interface should increase. Computer programs by Rose and Meyer (3300) have been developed that can be used to calculate the theoretical ultrasonic reflection from the adhesive layer as a function of modeled bond parameters and ultrasonic input wave characteristics. The need to assure bond integrity in the space program obviously intersected with developments in ultrasonic science. Gonzales and Merscheer (3301) described their use of ultrasonic inspections of Saturn II tank insulations, whereas Kammerer et al. (3302) developed some automated ultrasonic bond quality inspections for the Apollo space program. Hagemaier and Fassbender (3303), in 1977, related the use of nondestructive techniques, including ultrasonics for inspections in the primary adhesive bonded structure (PABST) technology program. Earlier, in 1975, Chang et al. (3304) wrote about the means of determining adhesive bond quality using ultrasonic techniques.

Additional research by Chang et al. (3305,3306), at General Dynamics, studied the transmission frequency spectra of ultrasonic waves through multilayer media and correlation of NDE parameters with the bond strength in multilayered structures. In 1976, Thompson (3307), at Rockwell, developed an interdisciplinary program for quantitative bond flaw detection. In 1974, Dukes and Kinloch (3308) published their adhesion science viewpoint on nondestructive testing including ultrasonic use. They mentioned that the mere detection of voids was not a good enough method of quality assurance. Instead, the aim of NDT needed to be to obtain a direct correlation with the failure property which was being predicted. In 1982, Li et al. (3309) reported on the detection of small voids and disbonds using ultrasonic impedance and resonance testing procedures. They further reported that these methods did not seem generally to detect smaller defects, such as porosity, which might be detectable with the domain ultrasonic techniques.

Gonzales and Merscheer (3310) reviewed the use of ultrasonic testing procedures for inspecting the tank insulation bonds on the Saturn S-II tank structures. Specifically, a copy of the testing procedure for evaluating the tanks at liquid hydrogen temperature was described. Newschafer (3311) provided additional information on the use of NDT evaluations in the Saturn production program in 1969. In 1978, Hagemaier and Fassbender (3312) produced a review of the state of the art of testing bonds nondestructively. Rienke (3313) reported comparative testing of the Fokker Bond Tester in 1976, citing its greater reliability in quantitatively measuring the bond strength as related to voids and porosity in the bondline. Some of the NDT commercial test devices have not found general acceptance and have subsequently disappeared from the marketplace. Some reports of note in this regard are Miller and Boruff (3288) on the Coinda Scope and Stub Meter, Bostco (3272) on the Sonic Resonator, and Schroeer et al. (3270) on Arvin Acoustic-Impact Tester. Hagemaier (3314) has offered a review of the present non-destructive testing procedures in the 1990 *ASM Engineered Materials Handbook* and presents a listing and discussion of the operations and use of the commercial ultrasonic testers in current use. The list includes (1) Fokker Bond Tester, (2) NDT-210 Bond Tester, (3) Shurtronics Mark I harmonic bond tester, (4) the Sondicator, (5) Bondascope 2100 (NDT Instruments, Inc.), and (6) The Nova Scope.

Chang et al. (3315) have presented a description of the principles of ultrasonic spectroscopy in the NDT of adhesive bonds, whereas Hagemaier (3316) discussed the ultrasonic maintenance inspection of aircraft structures in the same year. Earlier, Hagemaier (3317) had published a review of ultrasonic applications in the aerospace industry. Also, in 1974, Thompson (3318) discussed the requirements for and advances in nondestructive evaluation. In 1975, Palmer (3319) analyzed the effect of water-path variations on ultrasonic through-transmission sensitivity for adhesive-bonded aircraft honeycomb structures. Alers et al. (3320), in 1977, discussed ultrasonic techniques for measuring the strength of bonded joints. Flynn (3321) offered a means of predicting cohesive bond strength using ultrasonics. Martin (3322) discussed the interference effects present in the ultrasonic pulse-echo technique. Gericke and Monagle (3323) investigated the detection of delaminations using ultrasonic techniques in 1976. Smith (3324) came up with the unusual title "NDT Techniques for the Prediction of Adhesive Failure Loci Prior to Bonding" for his 1975 paper.

Rose et al. (3325) published their results in 1983 on a feasibility study on the NDT evaluation of metal-to-metal bonded joints using the ultrasonic pulse-echo approach. Matzkanin et al. (3326) reported further improvements in the NDT testing methods in manufacturing the C5A cargo plane in 1974. Clark (3327) has reviewed the definition and detection of critical adhesive flaws which would need to be NDT evaluated, whereas Lowe et al. (3328) attempted detection of bondline flaws in multiple-layered bonded structures.

A wide variety of ultrasonic bond testers have been developed over the past 25 years which have greatly contributed to the widespread interest of general manufacturing in this type of nondestructive testing. Many independent studies have been conducted with the result that all the instruments were shown to have varying degrees of capability in regard to establishing the quality of a bonded joint. None, however, have been totally adequate for establishing the adhesive quality of a bond that is defective due to (1) poor adhesive surface preparation, (2) bondline porosity, (3) extra-thick adhesive, and (4) corrosion at the interface. The ultrasonic bondtesters in present-day use in the aerospace industry include (1) The Fokker Bond Tester, (2) NDT-210 Bond Tester, (3) Sondicator, and (4) Bondscope 2100. The Bondscope 2100, an NDT Instruments Inc. product is of particular interest, being an advanced microprocessor-based device operating on the principle of ultrasonic impedance analysis. It offers a response capable of discovering unbonds at different depths in a multilayered adhesive laminate. The relative sensitivity of instruments inspecting bonded laminates of increasing total thickness has been established by testing at Douglas Aircraft, as shown in Table 140. Fassbender (3329) has investigated the NDT-2100 bond tester at Douglas.

A great many investigations of the ability of the Fokker Bond Tester have been undertaken which also may be of interest. Early work coming out of Fokker were investigations by de Jonge and Hartman (3330), de Jonge (3331), and Schjederup (3332). Also in 1962, Gonzales and Cagle (3333) conducted a Hughes investigation, and meanwhile Regalado (3334) had been conducting his own separate investigation at Norair, which he reported a year earlier in 1961. Hertz (3335) investigated bond deterioration with the Fokker Tester in 1964, and Phelan (3336) a 1967 evaluation of the Fokker Bond Tester System covering recent developments and the permanent records available.

Schliekelmann discussed the properties and observations made on the tester in 1975, pointing out it was the only commercially available instrument at the time which attempted to measure cohesive properties of the adhesive in the joint. Also, it was claimed that it would reliably detect small voids and disbonds at different depths in a multilayer joint. The quality of adhesion from any Fokker Bond Tester result may be accurately determined by comparing the instrument reading with estab-

Table 140 Relative Sensitivity of Instruments Inspecting Bonded Aluminum Laminates of Increasing Total Thickness

Laminate thickness (two sheets) (in)	Minimum defect size (in)			
	0.5	0.10	0.15	0.20
Instruments				
Sondicator (contact method)	0.7	1.1	1.5	1.9
harmonic bond tester	0.7	1.1	1.5	1.9
Fokker bond tester	0.4	0.8	1.2	1.7
210 bond tester	0.4	0.8	1.2	1.7
Sondicator (through transmission method)	0.5	0.7	0.8	1.0
Ultrasonic pulse echo	0.5	0.7	0.8	1.0
Ultrasonic (through transmission)	0.3	0.35	0.4	0.45

Data interpreted from Ref. 3276 (Fig. 28, p. 381).

lished correlation curves. In practice, the acceptance limits are based on the load or stress requirements of the adhesive for each joint. The accuracy of the prediction of quality depends mostly on knowing the manufacturing variables and accuracy of the nondestructive and destructive tests conducted in accordance with MIL-STD-860 (3337). Clemens (3338), in 1970, investigated application of NDT on Northrup planes. However, Curtis (3339) demonstrated some difficulties with the instrument in predicting cohesive properties and bond strengths because frequency shifts are so very small on account of any change in cohesive properties or bondline thickness. Uniform bond thickness is necessary, therefore, to obtain true measures of cohesive properties. Fassbender (3340) has also investigated the Fokker Bond-tester at Douglas. In this regard, reference should also be made to a technical handbook from Fokker (3341) and a Fokker testing manual (3342).

A newer technique of ultrasonic spectroscopy has been described by Lloyd and Brown (3343) for measuring cohesive properties but has other difficulties associated with correlating features of the spectrum with adhesive properties and thickness. Mention was made earlier of the detection of delaminations with ultrasonic spectroscopy by Gericke and Monagle (3323) and discussion of the principles and application by Chang et al. (3315) at General Dynamics in 1976. Relatively recent ultrasonic investigations are those of Pilarski (3344) in 1986 on evaluating adhesion degree in layered joints and Rehbein et al. (3345) on fatigue crack characterization by ultrasonic inspection.

D. X-ray Radiography Testing

Radiographic tests require that an x-ray photograph of the bondline be prepared and evaluated. Voids and porosity in the bondline can especially be detected providing the adhesive used is x-ray opaque. Patton (3346) has studied the incorporation of additives into adhesives for radiographic inspection of bonded honeycomb. While x-ray radiography is an effective method of inspection, and offers a permanent film record for interior components of bonded honeycomb structure, it is quite expensive and offers some potential radiation hazards to those in the vicinity. Well-trained personnel should always be sought to conduct such testing.

Hagemaier (3347) has stated that some adhesives like AF-55 and FM-400 are x-ray opaque, which allows voids and porosity to be detected in metal-to-metal joints. This is extremely advantageous where complex geometry joints must be tested for integrity.

E. Neutron Radiography Testing

Similar to x-ray or gamma-ray radiography in depending on attenuation to achieve object contrast, neutron radiography is less influenced by the density or thickness of a material or component. Neutrons, unlike x-rays, will readily penetrate lead, yet will be readily absorbed by the hydrogen atoms in an adhesive. Thus, hydrogeneous (adhesive) materials can be delineated from other elements in many cases where x-ray radiography has been inadequate. Neutron radiography should be used for evaluations where the adhesive is not x-ray opaque. The necessary neutrons for the method must come from accelerator, radioisotope, or reactor sources, so operational costs are high. Thus, the method overall does not appear to be cost effective for any kind of routine inspection of adhesive-bonded structures. More details may be found in a technical paper by Dance and Peterson (3348).

F. Acoustic Emission Testing

There are some situations where acoustic emission techniques have been more effective than x-ray or conventional ultrasonic methods in detecting internal aluminum surface corrosion and areas of moisture-degraded adhesive. The method is based on the observation that physically or mechanically transformed materials will emit sound or stress-wave signals. It is likely that the boiling of moisture may develop the acoustic signal from a corroded aluminum bondline, since drying out the joint before testing will circumvent that response. The same explanation would also explain detection of moisture-degraded bondline adhesive.

One promise of acoustic emission techniques has been to be able to detect adhesion failure prior to a fracturing in the bondline. In order to achieve this, however, the joint generally has to be loaded to approximately 50% of its failure load (3151). While this technique is partially destructive, there are few alternatives if adhesion strength is to be monitored. Schmitz and Frank (3285) investigated acoustic emission in a 1965 NASA report as a function of bond stress. Under stress conditions these investigators found ultrasonic emission to be a reliable indication of bond strength. Their report is quite comprehensive with many illustrations and tables. Brockmann (3350), in 1977, discussed the use of acoustic emission analysis as a method of testing adhesive metal joints, whereas Hill (3351), in the same year, described the use of acoustic emission for characterizing adhesive joint failures.

Rogers and Moore (3353,3354) have made such detections in a number of adhesive-bonded aluminum honeycomb structures. A technical article on the subject also appeared in *Aviation Engineering Maintainence* magazine in April of 1979 (3355) with the appropriate title of "The Sign of a Good Panel is Silence." Clotfelter (3356) used the acoustic emission technique for evaluation of adhesively bonded composite in the development of the *Saturn* vehicle. Matusik (3357) has furnished proof that this technique can provide accurate measurement of bond adhesion. Holt (3358), in 1969, described the use of a Schlieren system using a precision optical and mechanical device for visually studying the effects of an acoustic wave propagating in a transparent media.

Moisture egress to the core of a honeycomb assembly with resultant corrosion of the core has been a continuing problem for many years. This is possible if moisture moves along the bondline to individual cells of the core. If the core is perforated, then such progression is even more rapid. Fortunately, this moisture problem is detectable by NDT inspection methods. Generally, water may be detected by both x-ray radiography or acoustic emission procedures using a hot-air gun or heat lamp to cause boiling (cavitation of the water). The use of acoustic emission technology for this purpose was described in detail in two 1980 reports by Rodgers and Moore (3353,3354).

Three 1984 investigations were those of Leroy and Claeys (3359) on an acousto-optic NDT procedure, Padmanabhan et al. (3360) on analysis of acoustic emissions generated during stress corrosion cracking, and Arora (3361) on acoustic emission characterization of corrosion reactions in aluminum. In 1985, Scruby et al. (3362) made characterizations of fatigue crack extension by quantitative acoustic emission.

G. Holography Test Evaluations

Usually referred to as holographic nondestructive testing (HNDT), this technique can be further separated into two subfields of holographic interferometry and acoustical holography. In holographic interferometry, a hologram of the test specimen is first recorded by means of a laser light reflected from the specimen surface and superimposed on a mutually coherent reference beam in the plane of a high-resolution photographic emulsion. The hologram provides a complete record of the entire illuminated surface of the specimen, including phase as well as amplitude of the reflected wave front.

The specimen is then stressed in one of several ways such as heat, pressure change, evacuation, or acoustic vibration. Obtaining a surface displacement by this action of only a few microinches between a defective and good-quality bond is adequate to produce a holographic image. Differential surface displacements caused by subsurface anomalies can be observed or recorded in one or more of the following HNDT techniques: (1) real time, the stress-illuminated specimen is viewed through its developed hologram (made in a different state of stress); (2) time lapse, in which two holographic exposures are made on the same plate, with the specimen in two states of stress and then reconstructed; or (3) time average, in which the hologram is recorded during many cycles of sinusoidal vibration of the specimen. Wells (3363) has described the procedures, including the use of a fluid-gate photographic plateholder which enhances the speed of operation. In 1985, Hagemaier (3364) also provided a description and discussion of the procedures. Schliekelmann (52) also reports the technique in his 1979 AGARD-NATO Lecture and his 1975 publication (3365).

In acoustical holography, the method is altered to visually provide observation of the composite laminate or adhesive-bonded joint using ultrasound to obtain three-dimensional information on interior structure. The acoustical hologram is the converter, or recorder, which allows acoustic information to be visualized much as film is the recorder for light in holographic interferometry. Holosonics Inc. manufactures acoustical holography systems for a variety of applications. Collins (3366) has described these systems and their application in reactor pressure vessels. One acoustical holography system, for example, uses a liquid surface which acts as a dynamic film for momentary storage of the hologram while it is converted to a visual image through the use of coherent laser light. Another employs a scanning technique to construct the hologram. The hologram is then recorded on either transparency film or a storage oscilloscope, providing a permanent record for later study. A sample of acoustical holography test images have been provided by one of the equipment developers; i.e., Holosonic Inc. (3367).

In general, investigators have reported finding void defects in aluminum skin-to-core bonds, delamination area, crushed aluminum core, and bondline corrosion areas which would be difficult to identify by any other technique. Reviewing the different HNDT techniques for observing or recording differential displacements caused by subsurface defects is too technical to serve much purpose in this general review of NDT procedures. (Additional reading of possible interest can be found in Refs. 3368–3372.)

Other articles of possible interest include those of Vallat et al. (3373) on holographic interferometry, Haflinger et al. (3374) on thermal expansion coefficient measurement of diffusely reflecting samples by holographic interferometry, and two articles by Jones (3375) and Jones and Biji (3376) as early as 1974 on strain distribution and elastic-constant measurements using holographic interferometry and effects associated with the four-point bending technique.

H. Nuclear Magnetic Resonance Imaging (NMRI)

Nieminen and Koenig (3381–3385) have furnished 1988 and 1989 publications on the use of NMRI for investigating adhesives and bonded joints. NMRI can be considered an extension of conventional nuclear magnetic resonance spectroscopy in which the homogeneous magnetic field is modified by linear magnetic gradients to change the local field in a systematic way (see Ref. 3172). In this process, the resonance frequencies of the nuclei are charged. Finally, a reconstruction of a two-dimensional image is achieved and a two-dimensional Fourier transformation calculates the final image. This procedure permits a slice to be selected in any plane without removing the sample. Slice thickness and its location in the sample can be easily varied electronically.

Nieminen and Koenig (3381) have recently demonstrated the ability of NMRI to show the spatial disribution of heterogeneous resin/resin or resin/curing agent mixes in epoxy bondlines. Liu et al. (3386,3387), in 1989, contrasted pairs of polyvinyl acetate emulsions having the same viscosities but different solids, or vice versa, in 200- 300-μg gluelines. (For other earlier relevant papers for review, see Refs. 3377–3380, and for recent 1990 publications, see Refs. 3388–3392.)

I. Thermal or Infrared Testing Procedures

A variety of NDT techniques fall in this category which includes (1) infrared radiometrics, (2) thermochromic or thermoluminescent coatings, and (3) liquid crystal (cholesteric procedures). In infrared radiometer testing, a combination of a hot detector and a moving heat source are used to produce and record variations in heat absorption or emission from the studied structure. No physical contact need be made to the thermal source surface, so the observed surface is not physically disturbed. The thermal pattern of the surface is observed on a cathode ray tube, storage (memory) tube, or x-y recorder, since an electrical signal is generated by the radiometer (scanner). The bonded areas will conduct more heat than unbonded areas, because of the good thermal conduction from the skin to the aluminum honeycomb below when testing.

1. INFRARED RADIOMETER TECHNIQUE

For radiometer testing, the detector employs a moving heat source and records variations in heat absorption or emission while scanning the part surface. The radiometer generates an electrical signal exactly proportional to the incident radiant flux. Since no contact is required, the observed surface is not disturbed. The thermal pattern may be observed on a cathode ray tube, storage (memory) tube, or x-y recorder. Typically, the test panel is painted black before heating to assure uniform surface emissivity. (Further readings can be found in Refs. 3263 and 3393–3401.)

2. THERMOLUMINESCENT OR THERMOCHROMIC PROCEDURES

An ultraviolet-sensitive coating containing a thermoluminescent phosphor that emits light under excitation by ultraviolet radiation (black light) is sprayed and dried on the surface to be tested. Coating thickness is controlled to 3–5 mil (75–130 μm). Defect areas appear as darkened areas when the panel is heated to 140°F (60°C) and viewed under black light. Defect areas can be marked with a felt-tip pen over the dried film left after the coating was dried to the above specified thickness. (More information can be found in Refs. 3393, 3395–3397, 3400 and 3402.)

Thermochromic paints are composed of mixtures of temperature-indicating materials which have the ability to change color when certain temperatures are attained. Originally these paints were developed in Germany and sold under the trade name Detecto Temp. Thirty-six materials are available covering the temperature range from 104 to 2962°F. Obviously, only the very low end of the group can be used for color-change purpose over bonded panels. The lowest-temperature member is initially light green, becoming vivid blue at 104°F. The color change will extend (in the defect area) for about 15 min based on the relative humidity. Original color can be restored with application of steam, and the test can repeated many times without destroying the paint properties. The same reading references apply at listed above for the luminescent-coating technique.

3. LIQUID CRYSTALS (CHOLESTERIC PROCEDURE)

These crystals are a mixture of so-called cholesteric compounds that change color when their temperature changes as little as 1.5°F. They will always attain the same color at a given temperature for a specific crystal composition. After suitable surface cleaning, a thin coating of is applied by brush or spray to the test surface. When the surface is heated using a heat lamp or hot-air gun, the defects are shown by color differences developed. The defects must be marked as they appear. (Additional readings can be found in Refs. 3395–3397 and 3403–3406.) Simcox (3407) has reported General Electric's investigation of the thermal sensitivity of liquid crystals for detecting flaws in the Apollo system structures. Cohen (3408) compiled a bibliographic listing of publications in 1966, giving details on the equipment and procedures developed to that date for applying, testing, and photographically recording the results of liquid crystal thermographic testing.

J. Other Methods

A wide variety of nondestructive evaluation methods have been used for adhesive-bonded structures, as discussed above. Good reviews of the general subject area have been forthcoming from

Treca (3409), Schliekelmann (3410), Hagemaier (3411,3412), Adams and Cawley (3413,3414), and for metallic structures, the article "Nondestructive Evaluation and Quality Control" found in Volume 17 of the *Metals Handbook*, 9th ed. (ASM International, 1989). For the evaluation of composite joints by nondestructive means the most recent 1990 summary has been offered by Bar-Cohen and Mal (3415).

1. LEAKY LAMB WAVE TECHNIQUE

Prominant among the other methods not discussed above has been the introduction by Mal et al. (3416) of the development of the leaky lamb wave technique, which is particularly applicable to evaluating composite materials, since the wave reflection is from layered solids. In 1977 and 1978, respectively, Hagemaier (3417) and Bar-Cohen et al. (3418) had reported on the use of the conventional ultrasonic pulse-echo and through-transmission methods of testing bonded structures. These methods, however, are all based on a normal incidence of the ultrasonic waves to the specimen. The leaky lamb wave (LLW) technique is associated with an angular insonification of the tested joint. The wave is refracted as well as mode converted in this instance to induce useful guided waves. The procedure and examples are to be found in the above paper by Mal et al. (3416) and the *ASM Handbook* (3415). The LLW technique is particularly useful for the NDE of defects in laminates involving graphite-epoxy materials. Theoretical analysis of wave propagation in anisotropic multilayered media by Mal (3419) and Chimenti and Nayfeh (3420) in 1988 has allowed LLW modes to be predicted for bonded and unbonded laminates hitherto too complex to interpret for multilayered laminates.

2. ULTRASONIC MICROSCOPY AND COMPUTED X-RAY TOMOGRAPHY

The majority of methods that have been discussed above can be considered macroscopic in nature; the widespread use of adhesives in most engineering fields has created interests in various microscopic methods as well. Arnold and Reiter (3421) and Thaker and Burton (3422) in 1984 and 1987, respectively, offered pertinent comments for one of these methods designated ultrasonic microscopy. Persson and Oestman (3423) and Persson (3424) in 1985 and 1988, respectively, have investigated yet another method designated computed x-ray tomography. The NMRI method discussed earlier would also fall into this category of microscopic methods. These methods can produce abundant information about the qualtity and state of an adhesive joint with good resolution. For example, they can be used to study curing processes and water- or solvent-penetration mechanisms in complex joints containing various materials.

3. CORRELATIONS OF DESTRUCTIVE AND NDI RESULTS

While not constituting a new method as such, it may be informative to show how the use of more than one procedure can offer additional support for conclusions and may even be required where a variety of defects can be present as in honeycomb assemblies. Hagemaier (3412) cites results in this regard when examining a metal-to-metal joint involving a bonded doubler and longerons and frames that had been rejected for numerous voids and porosity. The Fokker Bond Tester had made the initial discovery. This was confirmed using the NDT-210 tester. Excellent correlation was further obtained with these results and a thermal neutron radiography examination. Ultrasonic scanning using the through-transmission technique again offered excellent correlation. Finally, the aluminum adherends were chemically milled away, revealing thicker adhesive in and around the defect areas as the cause for NDT rejection.

K. Additional Pertinent References on NDT Procedures

A number of Military Standards describe in detail related testing methods and requirements for acceptance and inspection, of which the following is a partial listing.

1. MIL-A-83377, Adhesive Bonding (Structural) for Aerospace Systems, Requirements for.
2. MIL-A-83376, Adhesive Bonding Aluminum Honeycomb Sandwich Structure, Acceptance Criteria.
3. MIL-I-6870 (ASG), Inspection Requirements, Nondestructive: For Aircraft Materials and Parts.
4. MIL-STD-410, Nondestructive Testing Personnel Qualification and Certification
5. MIL-STD-401, Sandwich Construction and Core Materials; General Test Methods.

The Air Force Materials Lab has issued a report titled "Adhesive Bonded Aerospace Structures Standardized Repair Handbook" which contains a special section on nondestructive inspection (3425).

Various aircraft companies have written specific reports on different types of testing, of which the McDonnell-Douglas specification on "Acceptance Criteria for Nondestructive Testing of F-15 Bonded Assemblies" is typical (3426). Similarly, Douglas' (3427) report on the "Eddy-Sonic Inspection of Adhesive Bonded Metal Assemblies" should be relevent. Hagemaier (3428) has discussed correlations of destructive and NDI results in his 1990 summary article in the *ASM Engineered Materials Handbook*.

VIII. JOINT STRENGTH RETENTION PROCEDURES

This subject does not really introduce any different testing procedures beyond those already mentioned. It does, however, provide amplification where a manufacturer might desire to know the rate at which a specified bonded structure or a given set of adhesive and surface treatment conditions might be projected to lose joint strength under a given set of exposure or special service conditions. For many years, these kinds of investigations have been actively pursued by those who sought to establish what the best choice of adhesive or surface pretreatment conditions might be for a wide variety of aluminum-bonding applications.

In order to establish some basis for making practical commercial judgments, it is often necessary to resort to use of test conditions which would be considered accelerated as compared to the service conditions known to exist for many general manufacturing products. For example, in Figure 30, we see picture of one of the specially built aluminum cabinets employed at Alcoa Labs to establish continuously condensing 100% RH exposure conditions at 125°F (52°C). designated as an accelerated tropical-type laboratory exposure, this type of environmental exposure was developed from extensive earlier testing at Alcoa to determine a set of conditions which could predict chemical compatibility relationships between aluminum surfaces and other nonmetallic manufacturing materials. Thus, this type of weathering should also be amenable to determining the longevity of organic adhesive/aluminum oxide surface jointures.

At the same time, it might also be pertinent to initiate very long-time exposure-type evaluations of aluminum-bonded joints in natural exterior weathering conditions. Such data could not only directly relate to predicting how long such bondments might survive in natural exterior weathering conditions, but also offer comparisons for determining the rate of acceleration possible with accelerated weathering conditions produced in the laboratory. For these long-time exposures, Alcoa Labs constructed specimen-holding racks in the natural industrial atmospheric conditions at their New Kensington Lab site, as shown in Figure 31. Similar racking for seacoast exposure conditions were installed on a finger of land protruding into the Atlantic Ocean at Point Judith, Rhode Island, as shown in Figure 32. In this latter regard, it had been shown that such seacoast sites provided a very severe testing condition for both aluminum adherends and the adhesive bondline. A similar weathering condition was provided at the laboratory for wet/dry cycling in salt water both by cyclic immersion in 3.5% salt water or in a specially provided 5% salt-fog condition.

Also, there was consideration of the fact that bonded structures are often subjected to cyclic natural weathering conditions which can include water entry into the structure followed by alternate freezing and thawing. For this purpose, an Alcoa Laboratory testing procedure was evolved in which

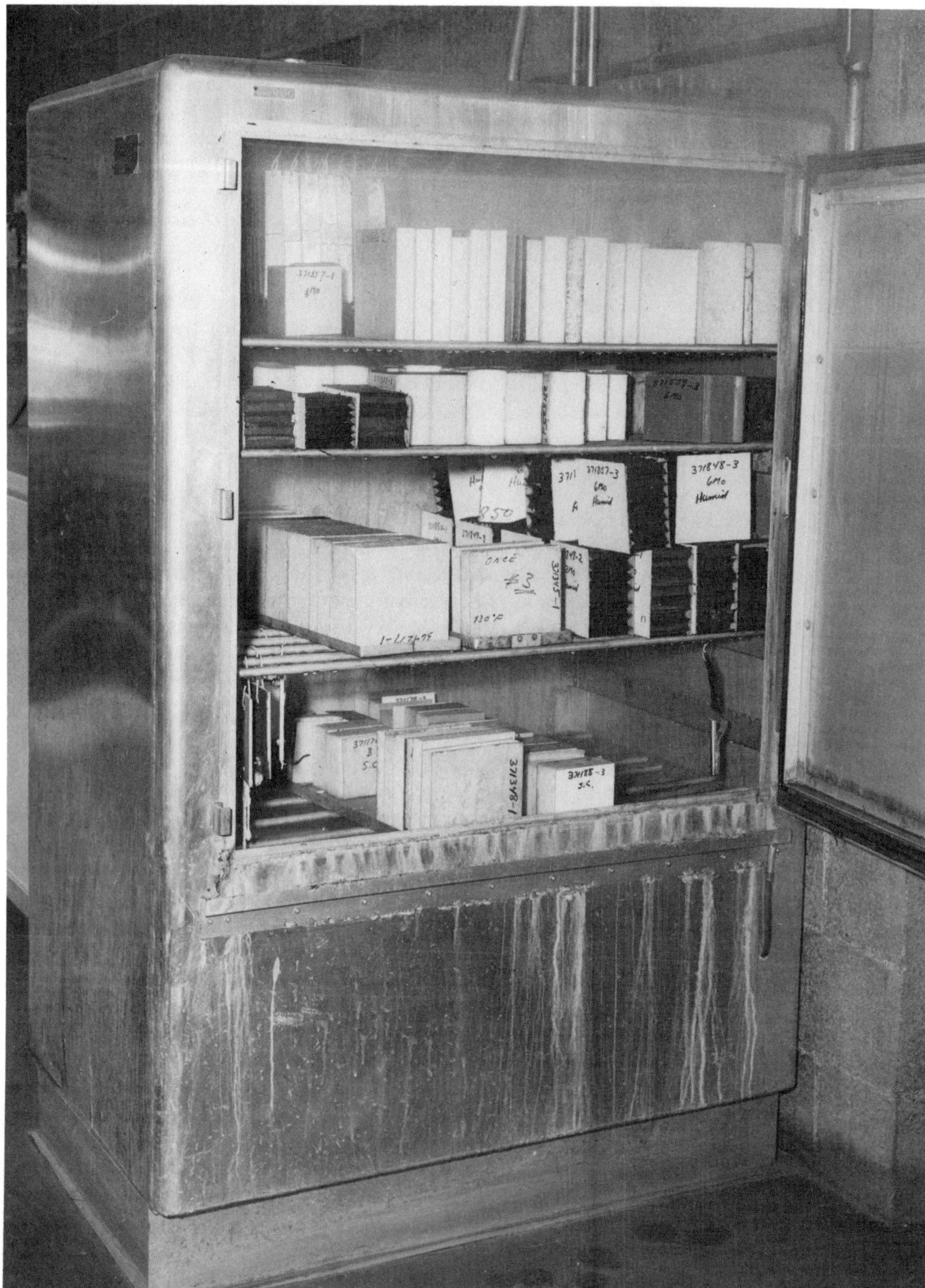

Figure 30 Humidity cabinet operated at 125°F.

Figure 31 Industrial atmospheric exposure station at Alcoa Labs in New Kensington, PA. (From unpublished work of J.D. Minford, Aluminum Co. of America.)

Figure 32 Seacoast atmospheric exposure station at Point Judith, RI. (From unpublished work of J.D. Minford, Aluminum Co. of America.)

the test specimens might be subjected to designated consecutive time periods in hot-soaking condensing humidity, freezing, and hot-air drying conditions. Such accelerated laboratory weathering conditions would be appropriate for evaluation of both aluminum-to-aluminum bonded joints and bondlines of aluminum laminated to other materials of construction. Finally, a way was provided to place various stress levels on an aluminum joint through use of the Alcoa Stressing Fixture (see Fig. 28).

A. Water-Exposure Testing Procedures

Soaking bonded aluminum joints in liquid or vapor water at temperatures from ambient to above the boiling point (using autoclaves) has been widely employed to establish the overall sensitivity of the adhesive, adherend, and the interfacial area to water in the service environment. The test specimens are usually fabricated according to ASTM standard procedures in order to be better able to make cross-comparisons between different laboratories. The test specimens (the simple ASTM D-1002 lap-joint specimen is most commonly employed) can be subjected to tensile, shear, peel, or cleavage mode stressing depending on which is most relatable to the product in its service environment. The average initial joint strength (usually obtained from testing triplicate specimens) can be used as the baseline from which any joint degradation can be measured. The total number of test specimens in these classic-type exposure test procedures is usually determined from the number of exposure periods planned and the number of specimens desired for reliability purposes. Either the absolute average retained joint strength or the percent of initial joint strength retained can then be plotted against the exposure time to obtain a joint strength decay curve, as shown in Figure 15. At the same time, other similarly fabricated joints can be undergoing exposure to longer times in natural atmospheric weathering such as in the desert, rural areas, industrial, seacoast, or actual marine conditions. Many thousands of such tests were conducted at the Alcoa Labs by the author over a 25-yr span of time to attempt to predict the best adhesives for bonding aluminum adherends which might experience a wide variety of service conditions. Some of the results obtained in a continuing effort to help establish confidence levels for manufacturers fabricating aluminum products outside the aerospace industry can be reviewed in publications by the author (39,68,72–74,81,84,85,87,89,531–536,771–775,978, 979, 3429).

1. *EFFECT OF MOISTURE AND TEMPERATURE ON ADHESIVE JOINTS*

The ASTM Std Test Method D-1151 defines certain conditions of moisture and temperature which are recommended for exposure of bonded joint specimens. The test specimens can be prepared by any method recommended by the manufacturer of the adhesive employed, although they should conform in detail with the requirements prescribed in various ASTM methods which cover the desired strength property. For example, tensile properties would be measured by specimens as described in ASTM D-897. Peel or stripping strength as described in ASTM D-903, or shear by tension loading as described in ASTM D-1002 for metal-to-metal joints. Twenty-one standard test exposure conditions of temperature and humidity are designated in ASTM D-1151 ranging from −70°F (−57°C) to 600°F (316°C). Conditions of 50 and 88% RH and immersion in water are specified at some temperatures, whereas the humidity in the testing ovens is generally uncontrolled above 180°F (82°C). The joint strengths after exposure are compared with an initial average strength.

a. Effect of Different Temperature Waters

It might be generally predicted that increasing the temperature of the liquid or vapor water would elevate the chemical aggressiveness of the testing conditions, since there is a general doubling of many chemical reaction rates for each 10°F rise in the environmental temperature. In this regard, Peters (3430), at Alcoa, investigated the effect on joint durability of both vapor-degreased and acid-deoxidized Alclad 2024-T3 aluminum joints bonded with either a one- or two-part epoxy adhesive when exposed to 40°F (4.4°C), 75°F (24°C), 145°F (63°C), or 165°F (74°C) distilled water for up to 12 mo. The data can be reviewed in Tables 141 and 142 for a two-part, room temperature–curing and a one-part, heat-curing epoxy adhesive. In Table 141, we observe that both vapor-degreased

Table 141 Durability Patterns of Two-Part, Room Temperature–Curing Epoxy Joints[a] with Vapor-Degreased or Chromic Acid–Deoxidized Alclad 2024-T3 Adherends in Different Temperature Waters

Water exposure temperature (°F [°C])	Surface pretreatment	Initial strength (psi)	% Retained joint strength after indicated exposure time in months[b]		
			3	6	12
40 (4.4)	Vapor degrease	1400	102	76	86
	Acid deoxidize	2130	119	105	100
74 (24)	Vapor degrease	1400	124	112	91
	Acid deoxidize	2130	102	108	109
145 (63)	Vapor degrease	1400	134	138	134
	Acid deoxidize	2130	107	107	97
165 (74)	Vapor degrease	1400	134	121	34
	Acid deoxidize	2130	94	92	63

[a]All joints fabricated with M688/CH-16 two-part, room temperature–curing epoxy from National Starch. All joints were cured at room temperature for 7 da before exposure.
[b]All values are the average of triplicate test specimens.
Data from Ref. 3430 (Tables 1–4).

Table 142 Pattern of Durability Test Results in Different Temperature Water Exposures with One-Part, Heat-Curing Epoxy Adhesives Bonded to Vapor-Degreased or Chromic Acid–Deoxidized Alclad 2024-T3 Adherends

Water exposure temperature (°F [°C])	Surface pretreatment	Adhesive	Int. joint strength (psi)	% retained joint[a] strength after indicated months of exposure		
				3	6	12
40 (4.4)	Vapor degrease	1[b]	3070	108	76	55
	Acid deoxidize	1	3270	91	102	89
	Vapor degrease	2[c]	3930	88	85	69
	Acid deoxidize	2	3830	97	102	96
75 (24)	Vapor degrease	1	3070	79	56	56
	Acid deoxidize	1	3270	82	67	88
	Vapor degrease	2	3930	70	58	41
	Acid deoxidize	2	3830	90	81	84
145 (63)	Vapor degrease	1	3070	120	86	66
	Acid deoxidize	1	3270	83	97	81
	Vapor degrease	2	3930	23	5	3
	Acid deoxidize	2	3830	54	37	4
		1	3070	90	71	63
165 (74)	Vapor degrease	1	3270	80	64	41
	Acid deoxidize	2	3930	0	0	0
	Acid deoxidize	2	3830	30	8	2

[a]All values are averages for triplicate test specimens.
[b]Adhesive #1 was an acid anhydride-curing, one-part epoxy powder impacted into a rod form for melting on a hot metal surface. The adhesive was cured after melting in the bondline and closing the lap joint for 15 min at 400°C (204°C).
[c]Adhesive #2 was a nitrile-modified, one-part epoxy paste with a Dicy-curing mechanism. The joints were cured for 15 min at 400°F (204°C).
Data from Ref. 3430 (Tables 1–4).

and acid-deoxidized joints maintained nearly 100% of initial strength at water temperatures up to and including 145°F (63°C). The further elevation of the temperature to 165° (63°C), however, affected a significant drop-off in joint strength to only 47% after 12 mo. At this temperature, benefit of acid-deoxidation surface pretreatment could be clearly observed. In Table 142, similar comparative results are shown with a one-part, acid anhydride–catalyzed heat-curing epoxy and a one-part, nitrile-modified, heat-curing epoxy adhesive. The significant lowering of retained joint strength even at 40°F (4.4°C) and 75°F (24°C) is now apparent as compared with the room temperature–curing form of epoxy. At temperatures of 145°F (63°C) and 165°F (74°C), the nitrile-modified epoxy joints were completely degraded, whereas more rigid, anhydride-cured epoxy joints retained significantly higher joint strength under equivalent soaking conditions. It should be pointed out these tests were performed with nonaerospace structural adhesives, and, thus, the apparent sensitivity to water shown by the heat-cured epoxies, in particular, would not be anticipated had modern-day aerospace film epoxy adhesives and appropriate aerospace pretreatments been employed.

2. EFFECT OF ATMOSPHERIC EXPOSURE ON ADHESIVE-BONDED JOINTS

It should be obvious that the range of humidity in the natural atmospheric service environments can range from a relatively low value in the desert atmosphere to near 100% in the tropical atmosphere. ASTM Std Practice D-1828 defines the procedure for directly exposing adhesive joints and structures to natural atmospheric environments. The same general comments about the range of specimens that can be used in atmospheric exposures has been cited above for ASTM D-1151 and moisture exposures. In the case of metal adherend joints, the practice is to include exposure site areas that include (1) rural-pure atmosphere; (2) industrial-sulfurous gases atmospheres; (3) marine-seacoast atmospheres, where chlorides are actually deposited on the joints; and (4) tropical or southern Florida with higher heat and humidities present. It should be expected that atmospheric conditions can vary greatly even from year to year at the same site, making it more difficult to obtain the degree of reproducibility present in laboratory-aging conditions. Also, weather can vary appreciably at two locations only a relatively short distance apart. Thus, it must be established to the satisfaction of the tester that no significant error will be introduced by using climatological data from a site other than the actual service exposure–type site.

The serious nature of testing in the seacoast atmosphere will be considered in detail in Chapter 10 on evaluating the durability of aluminum-bonded joints.

3. STRESS AND WATER EXPOSURE CONDITIONS SUPERIMPOSED

As has been mentioned earlier, the simultaneous exposure of all manner of adhesive-bonded joints to a combination of stressful condition and water soaking can lead to a greatly accelerated rate of degradation in the bondline, reducing the joint survival times significantly, as shown in Figures 15 and 18. ASTM Std Practice D-2919 describes one method for evaluating the effect of stress and humidity using a special ASTM durability test apparatus that had previously been used at the Bell Labs by Sharpe. The test specimen for this and other related devices, like the Alcoa Stress Ring, has been the simple lap-joint described in ASTM D-1002. The specimen, with its selected stress level, can be used to determine joint durability while stressed in (1) air, (2) air in equilibrium with certain solutions, (3) water, (4) various aqueous solutions, or (5) various environments like natural atmospheres or salt-water corrosive cycling conditions. More details of test procedures and results obtained with the Alcoa Stress Ring can be found in publications by Minford (3431–3434) and Brewis (3435).

Marceau and Scardino (3436) conducted a rather complete survey of test methods for determining the durability of adhesive-bonded joints under stress and adverse environments. Their report provided many illustrations along with a discussion and summary table giving evaluation criteria and the advantages and disadvantages of the various procedures. Krieger (2204) has also written a number of papers which are concerned with the environmental durability of structural adhesives bonding metals together. He has stated that stress-strain curve data for structural adhesives cannot solely predict the environmental durability of these adhesives in joints. However, this can be achieved with his skin-doubler concept with accompanying stress analysis and test data.

The prediction of failure time for joints using a reaction rate method developed by McAlbee and Levi (3437–3440), mentioned above, is also related to this discussion of combining the effects of stress and humidity on the bondline. Different reaction rate equations were developed for three decreasing relative humidity situations. Under 90–95% RH conditions, the equations were relatively successful in predicting joint failure times for AF-126 film/aluminum joints. Under 50% RH conditions, the data scatter was considerably greater. With 20% RH conditions, the derived equations were simply not useful. It was reasoned that the lower humidities do not promote significant weakening of the bonds, and, hence, random flaws in the bondlines probably play a more predominant role in reducing joint failure.

4. SPECIAL ALCOA TESTING PROCEDURES WITH WATER-EXPOSURE CONDITIONS

After conducting testing of thousands of aluminum joints using the above-described procedure of generating joint strength decay curves, there seemed a need for a more economical type of testing that would still provide screening of the proliferation of new adhesives, surface pretreatments, methods of adhesive curing, and fabrication procedures pouring into the marketplace. To offer one answer, Minford (1038) reported the development of a type of screening procedure that could be more economical for the average manufacturer seeking to establish the most practical combination of adhesive, surface pretreatment, and set of manufacturing conditions for his bonded product. The statistical proof was offered that the same general conclusion about bond durability potential might be achievable utilizing only a fraction of the number of individual lap-joint specimens used to generate joint strength decay curves. Simply stated, the procedure is based on similar exposure of ASTM Std D-1002 joints to whatever weathering conditions are deemed most appropriate for anticipated service conditions; however, instead of automatically scheduling failure of each joint in a tensile tester, the joint is stressed only to some percentage level of the established initial strength (usually 50%). As soon as no joint failure occurs at this level, the loading on the tensile tester is dropped to zero and the specimen is returned for longer weathering exposure.

In order to establish the validity of a testing procedure where the same test joint is repeatedly retested at 50% of initial strength after increasingly longer exposure times, Minford developed data similar to that shown in Table 143 which demonstrated that the imposing of 50% stress every 15 da during 1 yr of soaking in a 5% antifreeze/water solution at 185°F (85°C) did not produce a significantly different joint strength retention response than for duplicate joints soaked continuously for 1 yr without the intermittent stressing. It should be noted that this particular antifreeze solution was chosen based on earlier investigations involving adhesive-bonded aluminum radiator parts. In these investigations, it had been shown that an antifreeze/water mixture which gave inadequate protection against aluminum corrosion also constituted an unusually aggressive weathering exposure for an aluminum-bonded joint. A situation was uncovered, however, where this alternate test procedure was not applicable, as seen in Table 143. This occurred where a very thick oxide (Alodine 1200) was built-up on the aluminum adherend and a thinner (less stiff) 6061-T6 adherend used to fabricate the joints. The higher degree of bending of the adherend in combination with a relatively rigid two-part room temperature–curing epoxy adhesive apparently could create some fracturing in the thicker oxide layer which led to joint strength reductions within 60 da equivalent to that produced by 1-yr exposure without the intermittent stressing. Even here, however, it was shown that using a thicker (stiffer) 6061–T6 adherend could make the methods workable. Since a 0.125-in (3.18-mm) thick 6061–T6 adherend had always been the standard test specimen for Alcoa joint durability evaluations, this method could then be adopted for future screening of the durability of new bonding combinations with the opportunity for comparison with all previous published results.

It may also be considered that this method of screening may have the positive effect of being considered a possible underestimate of joint durability, since some minor bondline damage rather than interface strengthening would be anticipated by the periodic stressing. It should be mentioned that this periodic stressing to 50% of initial strength would in most cases greatly exceed the level of static stressing for which bondlines are designed for service. At the same time, using fewer test specimens

Table 143 Effect of Periodic Stressing on 6061-T6 Aluminum Alloy Joints Exposed to Antifreeze/Water Solution at 185°F (85°C) for 1 Yr

			Avg. lap-shear strength psi(MPa)		
Adherend thickness[a]	Surface preparation	Adhesive[b]	Initial	No periodic stressing	With periodic stressing[c]
3	Alodine 1200 (10mg/ft)	A	1700 (11.71)	430 (2.96)	435 (3.00)[b]
3	Alodine 1200 (25mg/ft)	A	1070 (7.37)	400 (2.76)	345 (2.38)[d]
3	Aloxite 120 abrasion	A	1596 (11.00)	1605 (1106)	1655 (11.40)
4	Alodine 1200 (10mg/ft)	A	2334 (16.08)	770 (5.31)	860 (5.93)
4	Alodine 1200 (25mg/ft)	A	2380 (16.40)	1405 (9.68)	1445 (9.96)
4	Aloxite 120 abrasion	A	2160 (14.88)	2200 (15.16)	2225 (15.33)
3	Alodine 1200 (10mg/ft)	B	1210 (8.34)	580 (4.00)	410 (2.82)[d]
3	Alodine 1200 (25mg/ft)	B	706 (4.86)	165 (1.14)	245 (1.69)[d]
3	Aloxite 120 abrasion	B	810 (5.58)	2045 (14.09)	2085 (14.37)
4	Alodine 1200 (10mg/ft)	B	1620 (11.16)	1365 (9.40)	1337 (9.21)
4	Alodine 1200 (25mg/ft)	B	2232 (15.38)	1360 (9.37)	1220 (8.41)
4	Aloxite 120 abrade	B	1060 (7.30)	2600 (17.91)	2615 (18.02)

[a]Two thicknesses of 6061-T6 alloy adherends: 3 corresponds to 0.064 in (1.63 mm) and 4 corresponds to 0.125 in (3.18 mm).
[b]A = National Starch and Chemical two-part, room temperature–curing, polyamide-cured epoxy paste adhesive. Cured 7 da at ambient lab temperature with a mix ratio of 100/32. B = Hysol Chemical Epoxi-Patch Kit 6C adhesive cured for 7 da at ambient lab temperature with a mix ratio of equal parts.
[c]Condition of periodic stressing consisted of stressing to 50% of initial average joint strength every 15 da during one yr of immersion unless failure occurred earlier. Other set of comparable joints was in the antifreeze/water immersion for one yr and only stressed to failure after that time.
[d]Joints degraded in strength to indicated level after only 60-da exposure.
Data from Ref. 1038 (Table IV, p. 39).

to evaluate each set of overall bonding conditions might enable many more variables to be evaluated at the same level of effort and expense.

5. *CHEMICAL COMPATABILITY EVALUATIONS INVOLVING WATER EXPOSURES*

A final matter of importance regarding the effect of water concerns its ability to influence compatibility between variously joined materials. Certainly the bond durability can be very adversely affected if the adhesive is incompatible with the adherend with water in the service environment. ASTM Std Recommended Practice for Determining Corrosivity of Adhesive Materials (D-3310) is intended to make this determination whether the adhesive is cured or uncured in contact with a metal. It is intended as a screening method to rule out all those materials that give visible signs of corrosion.

Compatability of aluminum should be investigated with any material with which it might come in direct contact. Thus, compatibility of wet wood with contacting aluminum is a matter of concern when predicting the long-term durability of such composites. The general compatibility is first established by mechanically holding the aluminum against the other material in a wet exposure condition for a prolonged time. At the Alcoa Labs, many test conditions were investigated before settling on the soaking of the contacting materials in continuously condensing humidity at 125°F (52°C). Even under these conditions, it was shown that a minimum exposure time of 6 mo might be required to make a final judgment about long-term compatibility. In Table 144, Alcoa data by Vader (3441) show a commercial 100% RH cabinet failed to demonstrate an incompatability problem with mill-finish

Table 144 Effects of Exposure to 100% RH at 125°F (52°C) on Aluminum/Hardboard Laminates Employing Different Protective Paint Primers on the Aluminum-Bonded Surface

Paint primer	Type exposure[a]	Exposure time (mo)	Bond[b] quality	Type bond failure[c]	Delamination (%)	Corrosion noted
None (mill-finish)	A	6	VS	100% H	0	None
None	B	6	VS	100% H	0	Mild
None	B	18	VS	25H/45A/30C	10	Mild
Acrylic (clear)	A	6	VS	70H/30C	0	None
Acrylic (clear)	B	6	P	100% A	60	None
Acrylic (clear)	B	18	P	100% A	90	Severe
Alkyd (clear)	A	6	VS	75H/25C	0	None
Alkyd (clear)	B	6	VS	15H/85A	15	None
Alkyd (clear)	B	18	P	100% A	90	Severe
Epoxy	A	6	VS	100% H	0	None
Epoxy	B	6	P	100% A	75	None
Epoxy	B	18	P	100% A	80	None

[a]A designates exposure in a commercial 100% RH cabinet and B designates exposure in a specially constructed Alcoa 100% RH cabinet designed to achieve continuous condensation.
[b]The code is: VS, very strong bond; S, strong bond; F, fair bond strength; P, poor bond strength.
[c]The code is: H, cohesive failure in hardboard; A, adhesive failure to aluminum facing; C, cohesive failure.
Data from Ref. No. 3441.

aluminum surfaces bonded with a contact neoprene-phenolic adhesive to a particular hardboard. However, the duplicate laminates exposed in a special Alcoa cabinet designed to assure continuously condensing conditions (see Fig. 29) showed evidence of some mild corrosion. An important point to remember is that the ability of any adhesive to act as an insulator against corrosion under continuously wet conditions is not always guaranteed as often intimated by the adhesive supplier. Because some reconstituted wood products like various hardboards may create occasional corrosion sites on aluminum under soaking conditions, Alcoa conducted testing over many years designed to evaluate the possibility that some organic paints might offer protection. With a clear acrylic paint on the same aluminum with the same adhesive and hardboard exposed to the same hot condensing humidity, it took 18 mo to show any significant corrosion. With an epoxy paint coating, no corrosion was found in 18 mo but some significant loss in laminate strength was sustained. The important conclusion from many years of evaluations by the author was that wide variances in protection are possible with different coatings on the aluminum and specific testing is required to find the best protective situations.

Another problem that has arisen in evaluating aluminum building panels is some polyurethane foams used for filling between aluminum facings can create a corrosion problem under these same hot humidity soaking conditions. More details will be offered in Chapter 10, but it needs to be suggested that testing, as described above, will be necessary to find a protective paint, primer, or surface pretreatment after it has been established that a possible incompatibility problem exists when foaming against bare aluminum and exposing to water soak.

Appleman (3442), in 1990, produced the results of a survey of various accelerated testing methods for anticorrosive coatings' performance, which certainly relates to adhesion problems where corrosion must be guarded against. Also in 1990, Simpson et al. (3443) investigated the electrochemical monitoring of organic coating degradation during atmospheric or vapor-phase exposure conditions.

Exposure of metal-bonded joints to the marine atmosphere has already been mentioned in discussing atmospheric exposure conditions. It might be considered that exposure of metal joints to seawater or laboratory salt-spray testing can be classified as one type of chemical incompatibility problem. This subject will be considered in appropriate detail in the Chapter 10 on durability of bonded joints. Additional readings on the subject can be found in Nakahara's chapter on corrosion control of aluminum joint durability (3444). Stevenson (3445), has reviewed some of the test specimen geometries used to study electrochemical and corrosive effects on adhesive joints in water-contact situations.

6. CHEMICAL EFFECTS ON ADHESIVE JOINTS

Reinhart (3446) has discussed this subject in the 1990 *ASM Handbook on Adhesives and Sealants* considering the most often encountered chemical agents and the testing of various types of adhesives and adherend materials to such chemicals. He lists the applicable ASTM testing methods as including (1) D-1002, (2) D-1780, (3) D-1781, (4) D-1876, (5) D-2182, (6) D-2294, (7) D-2918, (8) D-2919, (9) D-3166, and (10) D-3167.

B. Radiation Exposure Testing

ASTM D-1879 is the standard practice for exposure of adhesive specimens to high-energy radiation. The purpose is to define conditions for the exposure of polymeric adhesives present in bonded specimens to high-energy radiation prior to determination of radiation-induced changes in physical or chemical properties. Forms of radiation discussed include gamma or x-ray radiation, electron or beta radiation, neutrons, and mixtures of these such as reactor radiation. Only the conditions of irradiation are included in the specification and the choice of specimen and bond property to be studied is left to the discretion of the investigator. The exposure conditions are further classified into exposures at ambient conditions, at controlled temperature conditions, in media other than air or under a load situation. Kern et al. (3447) have considered the methods of evaluation in the 1990 ASM Handbook on *Adhesives and Sealants*.

C. Special Testing Procedure Examples

The evaluation of the relative aluminum joint durability in various weathering conditions such as water-soaking or hot humidity conditions is obviously of great importance for estimating service life of bonded joints in exterior weathering service. However, for test service potential in marine or seacoast conditions, it is mandatory to substitute some form of salt water. The standard testing procedure offered by ASTM B-117 procedure consists of continuous soaking in a salt-fog cabinet at 95°F (35°C) for either 500 or 1000 hr.

It became evident from early Alcoa investigations that several types of intermittent cycle exposures in 3.5% salt water that had been developed for evaluating the salt-water corrosion resistance of aluminum alloys appeared to be much more discriminating. Minford (2853) was able to demonstrate this in a specific investigation designed to make direct comparisons between exposures of identical aluminum joints under three different salt-water contact procedures. The data developed in a 1440-hr exposure period has been summarized in Table 145 and Figure 33. It clearly demonstrates the greater severity of the Alcoa cycling wet/dry procedures. It was also demonstrated that the greater the num-

Table 145 Comparative Corrosion Testing of 6009-T4 Aluminum-Bonded[a] Joints in Continuous 5% Salt Fog, Alternate Immersion in 3.5% Salt Water (10 min/hr), or 3.5% Salt-Water Immersion and Ambient Lab Condition Drying Daily Cycle

Exposure condition	Exposure time (da)	Avg. final joint strength[b] psi	MPa	Observations of joint failure
None	0	1120	7.72	100% thin cohesive
5% continuous fog[c]	10	1140	7.85	100% thin cohesive
Alternate immersion[d]	10	1160	7.99	100% thin cohesive
Alternate immersion[e]	10	1280	8.82	100% thin cohesive
5% continuous fog[c]	20	1220	8.41	100% thin cohesive
Alternate immersion[d]	20	470	3.24	75% adhesive, 20% corrosion area
Alternate immersion[e]	20	1240	8.54	100% thin cohesive
5% continuous fog[c]	31	1140	7.85	100% thin cohesive
Alternate immersion[d]	31	455	3.13	100% corrosion area spec. 1, 90% thin cohesive joint
Alternate immersion[e]	31	930	6.41	20% corrosion area
5% continuous fog[c]	39	1400	9.65	95% thin cohesive
Alternate immersion[d]	39	288	1.99	30% corrosion area spec. 1, 95% corrosion duplicate joint
Alternate immersion[e]	39	460	3.17	20% corrosion spec. 1, 100% corrosion on duplicate
5% continuous fog[c]	49	1320	9.09	100% thin cohesive
Alternate immersion[d]	49	325	2.24	80% corrosion area
Alternate immersion[e]	49	0	0	80% corrosion area
5% continuous fog[c]	60	1110	7.65	5% corrosion area
Alternate immersion[d]	60	40	0.28	80% corrosion area
Alternate immersion[e]	60	415	2.86	60% corrosion area

[a]All joints were fabricated with mill-finish 6009-T4 aluminum adherends bonded with Cybond 4533 one-part, heat-curing epoxy modified to bond in the presence of oil contaminated surfaces. All joints were cured for 1 hr at 400°F (204°C) before exposing.
[b]The values and observations were made on duplicate or triplicate joints which demonstrates some of the individual variation in performance that can be found in practice.
[c]This exposure and conditions for testing are described in ASTM B-117 procedure. It provides for continuous exposure in a salt fog cabinet operating at 95°F (35°C) with 5% sodium chloride solution.
[d]This alternate immersion procedure consisted of exposing the joints on racks attached to a revolving paddle wheel whose timed revolutions permitted 10 min of every hour immersion under 3.5% salt water.
[e]This is a simple testing procedure set up by Minford at the Alcoa Labs consisting of immersing test joints in a 3.5% sodium chloride solution for 18 hr overnight and drying the joints for 8 hr under ambient lab conditions each day.
Data from Minford No. 2853.

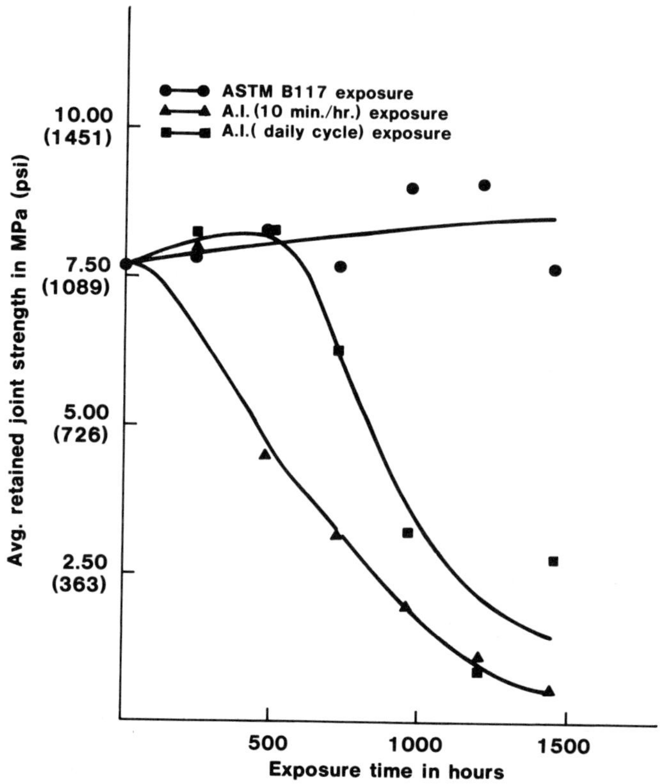

Figure 1 Joint durability results in various salt-water exposure conditions. (From Ref. 2853.)

ber of daily wet/dry cycles in salt water, the faster the bond-destroying corrosion could delaminate identically prepared joints. For example, an alternate immersion cycle procedure wherein the test joints were immersed for 10 min of each hour in a 3.5% salt-water immersion tank initiated visible corrosion within the bondline within 20 da. In a related Alcoa test where these specimens were immersed 8 hr and dried under ambient laboratory conditions for 16 hr each day, the evidence of debilitating corrosion was initiated in about 30 da. With continuous soaking in a 5% salt-spray cabinet (ASTM B-117), the first trace of bondline corrosion became evident only after about 60 da.

Even though a procedure using one daily wet/dry cycle soaking was apparently less aggressive than a 24 daily cycle procedure, the author found it to be much more economical to employ the former, since it could be conducted without any special mechanized equipment or special cabinet. Very extensive screening programs involving varying adhesives, alloys, surface pretreating conditions, and curing procedures could be conducted using this simple test procedure at minimal expense. Another Alcoa procedure extensively used to evaluate bonded joint durability has consisted of overnight exposure in a 3.5% salt-fog cabinet followed by a daily drying. These cabinets were deliberately designed so the salt fog could be shut off during the laboratory day and the cabinet opened to drying in the ambient laboratory conditions. The aggressiveness of this procedure was about equal to that found with the 24 daily drying cycles even though there was only one daily drying cycle. It was presumed that the salt-fog condition substitution for simple salt-water immersion was the basis of some acceleration.

The variety of tests that can be devised to evaluate the wide variety of commercial products available can be virtually limitless. However, a few examples may serve to illustrate with respect to bonded aluminum products. Some tests deal with products that are so widely manufactured that their overall quality is under the control of the main producers. A good example is the control of manufacturing and testing of vinyl/aluminum laminates, which are tested according to the Vinyl Laminator Std CS 245-62. In this standard, the requirements and methods of testing are specified for acceptable bond strength, degree of flammability, abrasion resistance of the weathering surface, light stability of the surface, and finally the stain resistance requirements. More than just initial adhesion strength also must be evaluated in terms of bond strength levels maintained after various exposure times to salt spray or boiling water.

Another example is Alcoa's development of specific test criteria for their laminated aluminum-faced building panels. This included the development of specially prepared and uniquely designed test specimens which could provide data as to (1) how long the bond could be expected to last in continuous soaking humidity; (2) how many cycles of soaking, freezing, and thawing could be endured; (3) bondline survival prediction for continuous or intermittent service at the highest solar heating temperature present on an exterior face, and (4) how can extrapolations be made from the testing of small selected specimens to include the performance of the largest-sized panels that could be practically manufactured.

For the prediction of how long the panels might be expected to resist exterior weathering conditions, a series of soaking tests in 100% RH at 125°F (52°C) were conducted with tensile-peel testing of the bondline strength after soak periods of 1, 3, and 6 mo. In Figure 22, we see a paper-honeycomb-cored panel ready for testing. The test specimens were 6 × 6 in pieces sawed from statistically selected locations representing conditions at edges and interior areas. For each different set of bonding conditions, separate testing was required. Variables could include differences in aluminum alloy (bare, anodized, painted, or surface texturing), adhesive selection, core insulation selection, and type of manufacturing procedure. During the initial laboratory investigation of choice of adhesive, more than 200 different commercial neoprene-phenolic contact cement formulations alone were evaluated to insure best candidates for final production.

The effect of alternate cycles of wet/freeze/dry conditions needs evaluation, since it exists in natural weathering. This was investigated using similar statistically selected specimens exposed for up to 6 mo in a daily cycle of exposure to 100% RH at 125°F (52°C) for 16 hr overnight and 2 hr in a freezer chest followed by 6 hr in a hot-drying oven at 180°F (82°C) during each day. It was shown that the heating condition prevented as extensive development of bondline deterioration from the water soaking as originally anticipated. A more severe bondline deterioration test resulted when the hot-oven-heating exposure was replaced by a simple daily acclimation to ambient drying conditions in the laboratory. Later, the potential for greater degrading in the bondline was provided by simply extending the daily cycle periods to three equal 24-hr periods. This permitted more gradual encroachment of water into the overall panel with greater damage from the freezing cycles as well as the opportunity for more chemical imcompatibility in the form of initiated aluminum corrosion processes.

The potential for panel survival when aluminum/polystyrene/aluminum– or aluminum/polyurethane foam/aluminum-laminated building panels might be heated by solar means to 180°F (82°C) was evaluated by using smaller 2 × 2 in specimens similarly sawed from statistically selected locations. Both progressive tensile loading to failure and time to failure values under sustained loads of both 5 and 10 psi were utilized for predicting performance. For establishing control norms, a panel had to be produced for testing which could not be failed by any extended period of simulated solar heating when fixtured in the same manner as would be used in erecting the panels in building sidewalls. Constructing panels by bonding aluminum facings to core insulations with two-part epoxy adhesives offered this opportunity. The stress-rupture requirements desired for either 5- or 10-psi loading at 180°F (82°C) were those achieved with aluminum/polystyrene foam bondlines sawed from panels that could perform like these epoxy-bonded control panels. While the 5-psi dead-load tensile test value continued to be included in the overall quality acceptance program for day-to-day production, it became recognized that the 10-psi dead-load stress-rupture time was the more critical for accepting production panels as being ready for shipping to the customer and immediate field erection.

A final example is offered to demonstrate how it is often necessary for the adhesive user to improvise a test procedure which may better satisfy a special purpose than employ a less applicable ASTM standard. This occurred with the author at Alcoa (3448) when the question of repairing aluminum automotive body sheet with organic body solder materials arose. The final test procedure developed consisted of impacting the aluminum sheet with an Olsen Cup Tester with a 2.22-cm diameter ball to a depth of 0.64 cm to produce a reproducible ding configuration. The depth of penetration could vary, however, depending on the yield strength of the particular aluminum alloy being evaluated. Different preparations also could be evaluated; however, the most practical choices usually had to conform to those that could be realistically employed in a typical auto repair facility. The other main variable was the choice of organic body solder to fill this standized indentation. In order to obtain a measure of the degree of initial and final adhesion after varying exposure conditions, a flat-head 0.78-mm aluminum machine screw was embedded head down in the uncured body solder then cured in place. Obviously, careful attention had to be directed to maintaining the screw precisely perpendicular to the plane of the sheet during cure. Also, it was necessary to remove any excess solder squeeze-out before curing, and carefully grind away any residuals after curing to assure minimum variability between duplicate test specimens. With special fixturing to accommodate the specimen to a tensile test machine, the plug repair area could be pulled in tensile from the supported sheet. In addition, other duplicate specimens could be placed in any kind of accelerated weathering conditions deemed appropriate for specified times or numbers of cycles and tested for direct comparison in terms of retained tensile strength of the repair patch. It would seem that this specially designed testing procedure, thought not directly comparable with any specific ASTM standard, would likely develop more pertinent data for selecting best repair procedures.

IX. SOME CRITICISMS OF STANDARD ASTM TESTING METHODS

In recent years, some investigators have tried to present more detailed critiques of some of the standard ASTM procedures described elsewhere in this chapter. They point out some factors which can easily be overlooked by the average investigator who might use these testing procedures in "cookbook" fashion.

Anderson and DeVries (2488,3026–3028) have been particularly specific in this regard, and it is suggested that their comments be reviewed. They point out that most of the tests proposed can be classified as being in one of two general groups. The first group can be labeled the "average stress criteria" type. The inherent assumption here is that failure of the particular joint configuration is controlled by the magnitude of the actual stress at which failure occurs. Most standard ASTM tests that have been considered earlier, such as ASTM Std D-897 (tensile), ASTM Std D-1002 (single–lap-shear specimen), and ASTM Std D-3165 (shear) fall into this group reporting average joint strength. Consider that for all the tensile and shear test procedures, the results are given as force per unit area needed to debond. This is an average failure stress, however, and this value may or may not be closely related to the actual failure stress that exists in commercial structures employing the same adherends, adhesives, and surface preparation procedures. We can say this because the stress on the joint is not uniformly distributed over the total bond area. What we are actually recording is more likely the lowest stress capable of initiating a cleavage failure at some point in the edge area. Hein and Erdogan (2225), Anderson and DeVries (2488), and Nakayama et al. (2735) have all discussed the final fast cleavage separation that occurs after the initial cleavage failure in testing such joints. Thus, the rest of the joint failure (after initial cleavage) is not a measure of the force necessary to disrupt the remaining bonded area in either the tensile or shear mode testing. The failure is not a function of the average shear stress over the whole joint but rather the induced cleavage stress. Nevertheless, the data developed from such standard tests continues to be reported as tensile and shear strength values, as though the values represented all possible aluminum manufactured structures.

More attention should be directed toward control of the adhesive bondline thickness and how thickness is involved in establishing the so-called joint strength in these ASTM standardized tests (see Refs. 2315 and 2373). However, the effect can be reversed depending on the test selected. If aluminum butt-joints are tensile tested, then joint strength will undoubtedly decrease as the adhesive bondline thickness increases. In contrast, joint strength in aluminum single–lap-joints using the same adhesive could likely increase as the adhesive thickness increases in the situation where the bondline stiffness can be increased by such a thicker layer. In the design of joints section, we previously discussed the significant effect of the adherend moduli. If the same aluminum adherend thickness is always used in these ASTM tests, the significantly higher tensile strength of the aircraft-type alloys (as compared with those more commonly employed in general manufacturing) will make a significant contribution to overall stiffening of the test specimen. This will immediately be reflected in a relatively higher tensile-shear joint strength using the same adhesive. When the same-thickness aluminum specimens are used to bond specimens with the same adhesive and equal bond thickness using lower strength alloys like 3003, 3004, and 5052, the tensile-shear strength values will be distinctly lowered. Even the heat-treated 6061-T6 aluminum alloy used in many structural applications in general manufacturing requires some change in adherend specimen thickness as compared with Alclad and bare 2024-T3 and 7075-T6 aircraft alloys. The author found that by substituting a thickness of 0.125 in (0.316 cm) for the ASTM D-1002 recommended 0.064 in (0.163 cm) approximately equivalent joint strengths could be achieved for the 6061-T6 alloy joints as compared to the Alclad and bare 2024-T3 and 7075-T6 alloys suggested for specimen preparation in that specification. In this way, the joint strength data which the author published for aluminum joints for more than 20 yr using aerospace-type film adhesives seemed very comparable to other data being developed in the aerospace industry. The fact that this relationship is not always recognized in the literature is demonstrated by the citing of strength and durability test results using different-strength aluminum alloys and the same adhesive without explanation of the obviously different test values obtained. One major adhesive formulator reported and made direct comparison of initial and retained joint strength values following years of exposure to exterior weathering conditions without regard for the fact that some joints had been fabricated using 2000 (aerospace-type) alloys, whereas joints for direct comparison had been fabricated from the 5000 series (general manufacturing-type) alloys. In such a case, the initial joint strength values were decidedly lower where the lower stiffness 5000 alloy joints were involved. In the same publication, the fact was also ignored that the surface characteristics of heat-treatable and non-heat-treatable aluminum alloys are vastly different, making direct comparison of durability data relatively questionable.

Undoubtedly, the recommended bonding of larger sheet adherends in ASTM Std D-1002 followed by subsequent sawing into individual specimens could also offer a distinct possibility for reducing the chance for misalignment errors as would be possible when separately fabricated individual test specimens are involved. However, it would seem that most manufactured joints for field service are individually formed and later combined into larger unit structures. In addition, the likelihood of duplicating individual errors in fabrication must be diminished by bonding a larger structure and then subdividing it. Of course, any positive or negative effect of heating at the edge of the bondline due to the cutting operation is not present when individual test coupons are fabricated. Because it might be considered closer to practical manufacturing methods, the author has always chosen to fabricate individual test coupons using the ASTM lap-joint dimensions and making compensation for the different strength aluminum alloys that might be employed in specific test evaluations.

Finally, we must recognize the striking effect that imperfections such as cracks or voids in the bondline can have, even though the bondline thickness has been carefully controlled. These voids, which can be crack initiation sites, can go unrecognized during joint fabrication. For example, nonwettable sites on the adherend, mixing air pockets into the adhesive, or having such pockets created by volatiles trying to escape from the bondline under curing conditions all may show themselves as a basis for causing lower stress cleavage values than predicted. Usually, close observation of failed joint interfaces may offer distinct explanations of why particular joints fail at significantly lower than anticipated average values. Such bondline imperfections may also be the highest contributor to the wide range of bond joint strengths shown by particular combinations of adhesive, adherend, and curing conditions.

In determining the tensile strength of joints according to ASTM Std D-897, the misalignment problem is particularly significant. Anderson and co-workers (2488,3026–3028) tested joints as recommended in this standard with duplicate joints tested with special care to achieve the best symmetry loading with a special fixture. Such a test modification was able to produce an increase of 80% in apparent load-carrying capability. In addition, the coefficient of variation using the refined fixture was lowered to only 6% as compared to 36% when following the standard ASTM procedure. This apparently startling fact, however, was not as influencing on manufacturing engineers as might be expected, since they also had to consider that they can never hope to achieve such refined application of service load even in their best-designed structures. As a result, the lower confidence level created by the use of the ASTM test, as written, may actually be more practical and certainly more representative of what the marketplace designer can expect to manufacture.

Of course, one of the most accepted designer rules is that the design load should preferably load the joints in as much shear as possible. For this purpose, the designer will most likely use data produced through the use of ASTM Stds D-1002 and D-3165. Since the critical value which initiates cleavage is what is being measured rather than the average load to break each unit of bonded area, any means of raising this critical value will correspondingly raise the reported lap-shear value. This means that the adhesive can be highly influential, because of its relative ability to absorb energy without cracking. Either the substitution of a more flexible adhesive in the bondline or increasing the stiffness of the aluminum adherend can act in the direction of raising the critical value to initiate cleavage and, hence, the average stress to failure.

10

Durability or Permanence of Aluminum-Bonded Joints

I. INTRODUCTION

The definition of the word *durability* in Webster's dictionary is having the quality of lasting or enduring, not easily worn out, whereas the word *permanence* adds the concept of having continued existence. Accordingly, the main emphasis in this section will be on those bonded structures manufactured for long-term service. It will be important from a practical viewpoint, however, to also consider those multitudinous joined materials, which are more economically adhesive bonded, and have lesser requirements to resist weathering and stress factors.

In an earlier chapter, it was mentioned that the degree of adhesion achieved could often be explained in terms of a combination of valence forces and interlocking mechanical action. However, we also pointed out that the processes of making and breaking joints do not seem to bear any direct fundamental relationship to one another, although the breaking strength of an adhesive joint is, in general, dependent on how it is made. Further, it was pointed out that while the strength of adhesive joints might be related to these specific energy of attractions arising along the border between adhesive and adherend (molecular theory of adhesion), there is a rheological theory that states this strength is usually limited by having a weak boundary layer condition existing somewhere that constitutes the weakest element of the overall joint construction. Thus, the specific adhesion forces can actually account for a greater potential strength than found in the real world in manufactured joints. Nevertheless, we do observe distinct relationships between the potential for long-term survival of joints and the potential they have for good primary and secondary chemical adhesion at the interface, plus the added opportunity for any mechanical interlocking of the adhesive to take place with the adherend surface in the interfacial area.

The special chemical pretreatments of adherends to promote better adhesive wetting and flow, and rough surface geometries to physically entrap the flowing adhesive, constitute important steps toward promoting desirable long-term joint durability characteristics. The most easily understood concept for generating the highest strength and durability in bonded joints is that we need to establish the maximum area of molecular contact between the adhesive and adherend. In all improperly manufactured joints, this goal is not achieved, with the final result that a variety of bondline inperfections can be present such as voids, cracks, and layers of contaminant that prevent optimal chemical and physical interactions from taking place, as well as offer sites for initiation of structural failures at such defects under stressful service conditions.

Brockmann (373,905,1313,3350,3449–3454) has published a number of technical papers on the subject of durability of adhesion between metals and polymers, including a 1989 paper (504) which reviews the present best picture. In his earlier works, focus was on the fact that chemical bonds of high energy, i.e., high strength, exist in the boundary layer between polymeric adhesives and metals, playing a most important role in durability. The situation changed, however, in the first half of the 1970s, when it became more obvious that bonded joints exposed to a combination of mechanical and environmental stresses were losing strength due to diffusional processes in the adhesive layer, as well as "adhesive rupture" at the adhesive/metal interface. As a result, the phenomenon of adhesion between metals and adhesives has been intensely tested and studied in recent years under a variety of water-soaking and exterior-weathering service conditions.

Brockmann (504) has summarized steps leading to joint failure in metal/polymer bonds as (1) a largely reversible weakening due to moisture incursion into the polymer layer near the metal/oxide interface (structure of this layer differs from that of the bulk and is influenced by the chemical and physical properties of the adherend surface); (2) slow transformation of the oxide layer by hydration and diffusion of oxide constituents into the polymer (this process is irreversible and is influenced by the state of the surface and chemical properties of the polymer); and (3) a fast deterioration of the oxide by primary corrosion usually initiating at an unprotected edge but occasionally arising within the body of a joint.

References of interest to the study of joint durability were issued in 1964 by Blomquist (3455) on the importance of evaluation of joint permanence, DeLollis and Montoya (3456) in 1964 on mode of joint failure, Bodner's compilation of 32 papers from the 1965 Symposium at Steven Institute Technology (3457), and a series of 8 papers on service life and durability testing in ASTM STP Publ. No. 401 from 1966 (3458). More recently, Zwei et al. (3459) issued a 1984 research report titled "Durability of Structural Adhesive Joints." In 1985, Carre and Schultz (3460) investigated the effect of a liquid environment on polymer-aluminum adhesion, and Omar et al. (3461) presented a report in 1986 on the adhesive-failure wear of metals. Harris (3462) talked about damage tolerance of adhesive joints at the 1990 Plastics and Rubber Institute Meetings at Cambridge. In 1991, Hand et al. (3463) offered a paper with the general title "Effects of Environmental Exposure on Adhesively Bonded Joints," and Minford (3464) composed a rather extensive chapter on the subject titled "Durability of Adhesive Bonded Joints."

II. MODELS BEARING ON ADHESIVE JOINT PERFORMANCE

A. General Considerations

Lewis and Gounder (3465) have discussed many of the important considerations that can affect the permanence, endurance, and reliability of structural bonded joints in their 1981 chapter. They considered both the subjects of failure criteria in materials (see Ref. 3466) and in adhesive joints (see Refs. 3098 and 2510). Their summary of the service life variables for adhesive joints is shown in Table 146. Consideration was also extended to the complex interactions that can occur between the adhesive and adherend in the boundary layer area. The phases in such a boundary layer might be considered as distinct states of matter which have their own physical, chemical, and mechanical behavior apart from the contiguous polymer phase of bulk adhesive and the aluminum oxide surface layer. It should be pointed out that adhesive/adherend couples are almost always characterized by a modulus mismatch, and if two different adherends are joined, then a three-material modulus "mismatch" must be present.

B. Stresses in Adhesive Joints

Certainly one of the most important intensive variables in establishing the endurance performance of a bonded aluminum joint is the stress environment in which it must operate. Whether the inherent

Table 146 Some Typical Service-Life Variables for Adhesive Joints

Intensive—primary	Intensive—specific
A. Mechanical stress (static, dynamic impact)	A. General industrial-ambient temperature, CO, H_2O, oil, vapor, others
B. Temperature	B. Aerospace fuel, radiation, vacuum, atmosphere (moisture, freeze-thaw cycling), and others
C. Physicochemical environment	C. Hydrospace-hydrostatic pressure, hydrodynamic forces, marine corrosion and fouling, others
Extensive	Random
A. Adhesive material	A. Careless workmanship
B. Adherend material	B. Voids in bondline
C. Joint design	C. Improper surface preparation
D. Extent bondline	D. Improper cure of adhesive
E. Bondline thickness	E. Poor joint alignment
F. Interfacial considerations	F. Murphy's law variables
G. Others	G. Others

From Ref. 3465 (Table 7-1, p.317).

mechanical properties of the joint are one of low strain or high strain will contribute to the fracture property characteristics. Joint design and methods of curing the adhesive and assembling the joint will also interact in the scheme of service life prediction.

1. MECHANICAL RESPONSE OF POLYMER MATERIALS

The response aspects of polymer materials can be related to adhesive response in joints, although many additional factors interrelate in the joint, as compared to bulk polymer response alone. Analyses of polymeric responses by Kelley and Williams (3467) might be reviewed in this regard. Their data relate to the deformation response of a polymer over a broad time or temperature scale interpreted in terms of the relaxation spectrum of the polymer. These authors have considered a so-called "interaction matrix" involving 5 polymer response parameters and 10 polymer "microstructural descriptors," as shown in Table 147. The important factors in influencing modulus are molecular

Table 147 Engineering Properties of Polymers: Interaction Matrix

	Modified power law parameters[a]				
Molecular characteristics	E	E		n	T
Cross-link density	N	S	N	M	N
Chain stiffness	N	N	U	M	S
Monomeric friction coefficient	U	N	S	U	S
Solubility parameter	M	N	U	U	S
Molecular weight	N	S	N	N	S
Heterogeneity index	N	N	M	N	M
Molecular weight between entanglements	N	S	N	N	N
Degree of crystallinity	N	S	S	S	N
Volume fraction of filler	N	S	M	S	M
Volume fraction of plasticizer	N	S	S	N	S

[a]N = negligible interaction, S = strong interaction, M = moderate interaction, U = unknown response.
From Ref. 3465 (Table 7-2, p. 320).

weight, cross-link density, molecular weight between entanglements, degree of crystallinity, volume of filler, and the amount of plasticizer. The creep behavior of adhesive joints in service will obviously have a strong relationship to these kinds of parameters.

2. *FRACTURE OF ADHESIVE POLYMERS*

The fracture of polymers has been the subject of many investigators. Review articles by Manson and Hertzberg (3468) and Andrews (2430) may be of interest for studying polymer fracture concepts. The Griffith crack model (adapted from Ingles [3469]) is still the foundation of all present-day theories of fracture of materials. The tensile strength of a brittle material, e.g., a cured structural adhesive, is dependent upon the existence of inherent cracks and flaws in its cross section. Its strength is considered to be proportional to the energy required to propagate a crack, flaw, or other concentrator of stress when under an overall imposed stress. Further modification of the Griffith work by Orowan (2562) extended the tensile failure description to include more ductile materials like adhesives. An additional analysis of the Griffith equation was made by Berry (3470). The approach to handling more ductile materials was to separate the fracture energy terms into two elements: (1) energy to create new surface, and (2) the energy for plastic deformation. It was Irwin's (2493,2580,2581,3471) extensive study of this matter that derived the important concept of a critical value for the stress concentration. This can serve as a definition of the toughness of an adhesive polymer. For viscoelastic adhesive materials, the critical stress concentration is found to depend on temperature, strain rate, and failure mode, since crack propagation can occur under tension or shear stressing. The energy required to initiate a crack is certainly proportional to, but may not be identical to, another term coming out of the Griffith fracture theory called the critical energy for crack extension. The most common method of measuring fracture toughness in joints has been to employ the cleavage beam procedure employed most extensively by Ripling and Mostovoy, as reviewed in Figure 17.

3. *STATIC FATIGUE OF ADHESIVE JOINTS*

An adhesive joint may often have a complex response to stress in its service environment because the stressing may be random in both its periodicity and intensity. For purposes of analysis, these imposed stresses may be considered static or dynamic in nature, depending on the time scale within which the stress is applied. A typical response to a steady load may be deformation. If the deformation is permanent (tensile, shear, or compression set), then the term *static fatigue* is employed to describe the result. In the case of an oscillatory stress, as in vibration or random or periodic impact, the preferred term is *dynamic fatigue*. Both situations obviously are encountered in adhesive joint service conditions and must be properly interpreted and understood to explain the endurance behavior of bonded aluminum structures.

Lewis (2799,3476) has advanced a qualitative model describing the failure mechanism in materials by characterizing the response to static stress in four regimens of behavior. At very low stress levels, he proposed the term *hookean deformation*, which is reversible. As the stress level rises further, the adhesive polymer deformation becomes irreversible with crack opening with this stress applied, and crack stopping when the stress is relaxed by a blunting process. A lowering of the system modulus can then occur by a result similar to the mullins effect observed in rubber; i.e., a form of stress annealing. As the stress is elevated still further, irreversible crack openings result with a continuing steady cumulative damaging of the defect structure. An incipient failure zone is the result. At or near the measured breaking stress of the joint, the cracks grow more rapidly and coalesce, making joint failure an imminent event. A summary of these events is shown in Table 148. Obviously, in moving to any of the damage stressing conditions, there will be an accumulation of the damages that can result in those situations transversed. This means the rheological damage found in a joint in service will be a record of the joint's previous history. The viscoelastic properties of many adhesives further means that the speed of stress build-up can be an important consideration in joint longevity. If the viscoelastic adhesive enters the strength damage region rapidly via a fast-accelerating stress condition, then the terminal or fracture condition may be attained more rapidly than under an equilibrium accommodating loading rate. The latter permits molecular relaxation. Hata (3472) has proposed some viscoelastic models of mechanisms of failure in adhesive joints which consider the dependence of

Table 148 Rheological Response of Adhesive Material Under Static Stress

Designation	Description	Effect of damage on ultimate strength
O—Hookean	Reversal elastic process	None
I—Modulus damage	Modulus lowers, self-strengthening	None or slight increase
II—Strength damage	Irreversible crack opening continues to grow	Lower strength
III—Terminal damage	Catastrophic failure, cracks coalesce	Drastic strength lowering

From Ref. 3465 (Table 7-3, p. 324).

adhesive strength on the rate of adhesive separation, temperature, and type of failure observed. These models still remain gross oversimplifications of the actual viscoelastic behavior of adhesives in real-life joints.

4. STRESS ENDURANCE LIMIT OF ADHESIVE JOINTS

Lewis et al. (2798) have modified the method of Prot (3473) in an experimental study of static strength and the endurance limit of lap-shear aluminum joints. By definition, the joint endurance limit is that stress level that the joint can continue to operate at for some indefinitely long service time. An idealized plot of a stress endurance curve is shown in Figure 34. Any joint operating under stressing conditions, which exceed this endurance limit value, will suffer damage that can be considered proportional to the amount which the stress exceeds the limit value. Notice the idealized curve has an exponential function character. In order to develop some specific data, Lewis et al. (2798) loaded their lap-joints in an universal tensile test machine at a constant programmed rate until failure occurred to find a stress endurance value. They compared this value with the failure loads obtained when increasing the stressing level continuously to obtain short-term strength values. They employed an equation derived by Loveless and co-workers (3474,3475) which offered a useful interpretation of the endurance limit concept (3476). Their endurance limit (EL) and short-term lap-shear strength (LSS) values

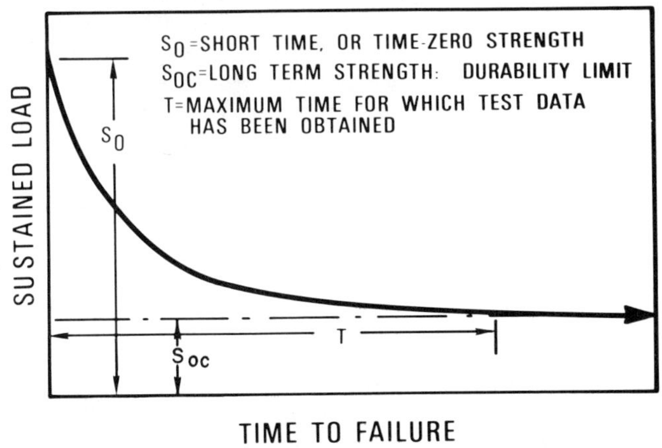

Figure 34 Durability limit curve in shear mode. (From J.D. Minford, Aluminum Adhesive Bond Permanence, Treatise on Adhesion and Adhesives (R.L. Patrick, ed.). Marcel Dekker, New York, Vol. 5, Fig. 3-54, p. 121 [1981].)

for aluminum joints fabricated with nylon-epoxy, nitrile-phenolic, epoxy-novolac, and nitrile-epoxy adhesives are given in Table 149. They found that the ratio of endurance limit strength to short-term lap strength was a near-constant ranging from 0.42 to 0.45 for four different adhesive candidates from four different chemical adhesive families. Based on additional testing, they established the apparent lower range of endurance limit to be about 0.25 of the short-term test strength. This kind of information is of interest to the design engineer who wishes to incorporate a realistic margin of safety into a specific adhesive-bonded structure. Caution that these values may not be directly translatable to all sorts of design structures needs to be given, since the data were developed with the specific joint described in ASTM D-1002, which is not directly used by most design engineers. Also, the data were developed under controlled temperature and humidity conditions (73°F [23°C] and 50% RH) which would not simulate the much higher temperature and humidity conditions that may occur in real-life service exposures. Furthermore, for adhesive joints, the sustained load tests did not wholly support the basic assumption of the Prot equation that plots of stress level versus time to failure should follow a hyperbolic law, as shown in Figure 34. McDowell (3477) has confirmed this in unpublished work at the Lord Corporation, and Minford (3478) has also shown this repeatedly in hundreds of stress tests of aluminum/epoxy joints in 125°F (52°C) 100% RH weathering conditions. More recently, Allen et al. (3479), in 1985, reviewed the concept of an endurance limit for adhesive joints.

5. *DYNAMIC FATIGUE RESPONSE OF ADHESIVE JOINTS*

The manifestations of dynamic fatigue in adhesive joints is damage and failure that is similar to that discussed for static fatigue. The response is time dependent and is caused by various modes of stress application described as sinusoidal, squarewave, or randomwave types. A high-velocity impact mode of stress application would also be considered in this case of response. These forces may be superimposed upon an existing static load as the most complex situation. The time application factor causes the adhesive polymers to respond viscoelastically. Various amounts of elastic energy may be stored or lost during these cycling stress conditions, and the publications Bueche (281) and Lenk (3480) can be used for additional reading on the subject. Part of the supplied energy that is nonrecoverable will manifest itself as heating in the strained adhesive. At relatively low strain levels, the adhesive may absorb a maximum of mechanical energy, which over long-time fatigue conditions can cause chemical changes resulting in rheological damage and ultimate polymer failure. At higher strain levels, Watson (3481) reports that mechanochemical reactions may take place in a polymer. Winspear (3482) has reported on the massive heating that elastomer polymers can undergo under severe straining conditions. Obviously, a very complicated picture is presented when these higher strain levels coexist with a static stress and/or thermal and chemical service life weathering conditions.

Experimentally, the fatigue of materials is normally represented by the so-called S/N-type curves which relate the peak stress failure values to the number of cycles to failure. These curves have a similarity to the stress endurance curve for adhesive joints, since the number of cycles to materials failure begins to extend dramatically as the stress approaches a minimum level for crack initiation to occur; i.e., endurance limit.

Table 149 Comparison of Endurance Limit and Short-Term Strength of Aluminum Lap-Shear Joints

	Shear strength		Endurance limit		Endurance limit/shear strength
Adhesive type	psi	MPa	psi	MPa	
Nylon-epoxy	6258	43.12	2826	19.47	0.45
Nitrile-phenolic	4164	28.69	1728	11.91	0.42
Epoxy-novolac	5052	34.81	2181	15.03	0.43
Nitrile-epoxy	5482	37.77	24.55	16.91	0.45

Data From Ref. 3465 (Table 7–4, p. 331).

Paris and Erdogan (3483) suggested an equation in 1963 which relates the crack-propagation rate to the stress-intensity factor range, ΔK. The ΔK in this equation is raised to a power that is related to the volume of the plastic zone in the area of the crack tip. The original value of this power factor was predicted by Paris to be 4; however, Barsom has suggested a value of 2 (3484). Regardless of its precise value, there seems to be little question that it is a function of frequency, environment, mean stress, temperature, and materials variables (3468). It may also be logical to assume that the rheological damage may be accumulating at a faster rate as the last cycles to failure occur. The dynamic mechanical effects of viscoelastic materials used in adhesive joints have been studied by Mostovoy and Ripling (2516). The suggestion has been made that testing frequency should be restricted to below 3 Hz to avoid nonlinear effects due to nonequilibrium nonisothermal crack growth conditions. In general, an adequate experimentally measurable damage function, or even an explanation of the true nature of damage, has not been developed. Thus, the linear cumulative damage analysis approach is left as a purely statistical identity. One attempt was made by Lepie and Adicoff (3485) to explain damage in terms of rheological parameters using solid rocket propellent; i.e., highly filled elastomeric polymers. Using a molecular primary bond rupture calculation, they concluded the major contribution to damage was that of rupture of the load-bearing molecular chains.

6. EFFECT OF TEMPERATURE ON ADHESIVE JOINTS

Cohesive failures in aluminum joints can also be related to the effect of temperature on the polymeric adhesive. The phenomenon of phase transitions in polymers has been well studied by Boyer (3486) in 1963 and Bever and Shen (3487) in 1974. In theory, an increase in the thermal energy of a polymer should cause an increase in molecular motion resulting, in turn, in producing higher-amplitude vibrations. For example, in passing through its glass transition temperature, a glassy amorphous polymer goes from a cohesively strong brittle or tough solid to a cohesively weak, creep-prone, deformable rubber. It needs to be remembered, however, that most structural adhesives are cross-linked polymer systems with significantly lower mobility under thermal stressing than for purely thermoplastic polymers. The slight softening that they show at high temperatures is generally through a chemical degradation mechanism which is generally irreversible. The rubber-modified epoxy adhesives may exhibit low-temperature transitions due to the fact that they are two-phase systems. This low-temperature transition reflects the glass transition temperature of the secondary rubber phase (3488). The strength of polymeric adhesive materials at various temperatures then depends on a number of factors: (1) the T_g of the base polymer and/or resin modifiers, (2) the degree of crystallinity (if any), and (3) the degree of cross-linking (if any).

The glass transition temperature of an adhesive polymer can be an important factor to consider when predicting the permanence and endurance of adhesive joints. It depends on composition in a number of ways: (1) stiffness or geometry of backbone polymer, (2) the amount of plasticization, (3) the degree of crystallinity, and (4) the degree of cross-linking. From diagrams of thermal temperature/strength states representing various polymer types, it would appear that adhesive joints are structurally durable only when the adhesive exists in its glassy adhesive state below T_g under the expected service conditions. With the highly cross-linked polymers such as phenolics and polyimides used in higher-temperature service, the T_g is usually obscured by the stiff network structure. Some adhesives can exhibit a thermal response that is a combination of the different thermal temperature/strength states; i.e. glass, transition, rubbery, and flow conditions.

As compared to the thermal durability of the bulk polymeric materials separately, the problem of thermal durability of bonded joints is complicated by being a complex material system consisting of two homogeneous material phases and at least one interfacial phase. When the adherends are vapor barriers (e.g., aluminum), then oxygen diffusion to the adhesive can alter the thermal oxidative rate. The geometry of the joint can influence the overall thermal durability as well. The thermal stability of polymers and some structural adhesives was reviewed by Nieman et al. (3489,3490). Krieger and Politi (3491) have developed some quantitative results showing the degradation of structural adhesives under vairous thermal exposure conditions. A distinct acceleration of thermal degradation rate for the same adhesive was observed when using stainless steel adherends as contrasted with aluminum adherends. The chemical composition of the surfaces appeared to be the main influencing factor (see Fig. 14).

Temperature can also influence the durability because adhesive/adherend couples are almost always characterized by a modulus mismatch. The high coefficient of thermal expansion of aluminum can even be a problem in direct sunlight service involving large-area laminated sandwich panels. A compromise in the strength properties of adhesives for such applications may be considered in which some of the aluminum thermal growth can be accommodated by adhesive creep. The stress in the actual bondline must also be kept as low as possible by allowing the overall panel to bow through the attachment design to the rigid framing. The ability of contact cement–type adhesives like the neoprene-phenolics to creep with the aluminum sheet under solar heating of the facing sheet has been a major reason for the use of such adhesives for laminating large aluminum-faced architectural building panels. At the same time, the adhesive tensile strength at the solar-heating temperature must resist the separation of facing from the layer of insulation in the panel cross section. The selection of best adhesive candidates through specific laboratory testing for the Alcoa Alply building panel can be reviewed (87,89). In addition, the adhesive properties and performance at elevated temperature must also be carefully considered with the different materials of construction that might be present in any aluminum-faced sandwich panels. A main characteristic of such building panels is the insulation value that they offer to the building. The most common insulation materials, therefore, are various plastic foams. Durability potential will be significantly lowered, for example, if the plastic foam is chemically degraded by any strong solvents in the adhesive. While this solvent must be dissipated before handling and final load-bearing strength can be developed, there is the matter of keeping any solvent away from sensitive plastic foam surfaces until handling strength can be developed. In the Alcoa manufacturing process, this was accomplished by never allowing the adhesive-coated aluminum sheet to come into contact with the foam surface until all solvent had been thermally volatilized. Alternatively, there are also the type of aluminum structurally bonded panels in service that are rigid and designed to resist the thermal stress situation totally. In this case, the combination of adherend and adhesive must act together to resist any thermal stress anticipated in normal service. It is amazing to the bonding novice to observe how fast the stiffness of a structure can be improved by adhering several layers of bonded weaker individual strength materials. Nowhere is this more evident than with a laminated building panel, where a low-modulus adhesive can combine a relatively thin, high-modulus sheet of aluminum with a low-density polystyrene beadboard to end up with a panel that is amazingly load bearing. The extreme examples of high stiffness are, of course, provided by the use of high-modulus adhesives with high-modulus thick aluminum sheet or extrusion adherends to create a primary structure in aircraft.

Finally, as the anticipated service temperature for aluminum-bonded aerospace structures rose significantly or were lowered into the cryogenic range, adhesive formulations with more extensive cross-linking or more flexibility were necessary. The selection of durable adhesives for such service has been considered earlier when discussing the selection of best adhesive for a particular application (see Figs. 12 and 13).

7. PHYSICOCHEMICAL ENVIRONMENTAL EFFECTS ON JOINTS

Several models have been suggested which relate to separation of an adherend/adhesive interface (3499–3523). Additional investigations relating to service-life prediction of polymers and metal/polymer joint systems might include those found in references 2804, 3525–3543. The effects of liquid water and high humidity on the durability of bonded joints has been widely studied (39,68,72–74,80, 85,89,531–535,772–775,2705,2707,3464,3492–3498).

The first model accounts for the effect of the presence of a liquid phase on the thermodynamic work of adhesion. Fowkes (389,1240) first suggested that the free energy of a liquid may be considered to be made up of dispersion (London forces) and polar contributions. Owens and Wendt (316) developed a method of measuring these dispersion and polar contribution components of surface free energy for a solid. Consequently, Owens (3499) concluded the condition for spontaneous separation of an adhesive/adherend interface must be a negative thermodynamic work of adhesion. If the work of adhesion is positive, it means that work must be done on the joint to separate the solids (e.g., the separation will not be spontaneous). Schonhorn (3500) and Schonhorn and Frisch (3501) have cited an example using the aluminum/polyethylene/aluminum joint with or without a stearic acid monolayer between the aluminum oxide and polyethylene. Work of adhesion for the aluminum/

polyethylene joint (without monolayer), and in the absence of water, was 120 erg/cm^2. With water present, the value was -312 erg/cm^2. Thus, spontaneous separation would be predicted in water. With a monolayer at the interface, the work of adhesion was calculated as 101.8 erg/cm^2, and no spontaneous separation was predicted, since it was prevented by the introduction of the monolayer.

III. ALUMINUM AHDEREND SURFACE CONDITIONS THAT AFFECT DURABILITY

A. Macroscopic Alterations

It is easy to understand and visualize how the high-strength and toughened high molecular weight polymers produced by modern synthetic chemistry can generate strong joints by becoming physically entangled with a macroscopically roughened aluminum surface. The initial strength of such joints will usually be determined by the cohesive strength of the polymer, since it must be ruptured in order to achieve joint separation. It would be expected further that the durability of such joints could be improved if the adhesive has good increasingly better capacity for chemically wetting the aluminum adherend and flowing into and attaching to the roughened surface geometry.

In some adherends, such as wood and ceramics, the presence of macroscopic surface porosity can be a major factor in enhancing joint strength and durability. When bonding aluminum to such surfaces to make multimaterial constructions, it follows that the final joint durability is dependent on the bondability of the adhesive to each adherend. Thus, it behooves the manufacturer of aluminum-bonded dissimilar material structures to be just as careful to select and specially prepare the wood or ceramic surface, especially if high durability is to be achieved. Unfortunately, aluminum sheet and extruded surfaces (as received) lack surface pores on a macroscopic scale. The same low-viscosity adhesive that penetrates readily into porous wood will find no mechanical interlocking with aluminum unless specially surface pretreated. In Table 150, Minford (3545) has shown evidence that the greater opportunity for forcing adhesive into wood pores when laminating with a dwell-time in

Table 150 Comparison of Bond Durability Between Aluminum and Plywood Adherends Joined by Hot-Nipping Versus Hot-Press Fabrication Conditions[a]

Adhesive application	Aluminum foil thickness (mil)	Curing conditions	Test procedure	Cycles completed	Comments
Brush coat on marine plywood and foil	8	Rotary nip at 250°F (121°C)	Cyclic boil	1	40% cohesive failure
Brush coat on marine plywood and foil	4	5 min at 250°F (121°C) in heated platen press under 100 psi	Cyclic boil	25	0% delamination
Brush coat on 0.75-in firboard and foil	4	Same as above	Cyclic boil	25	0% delamination
Brush coat on 0.75-in firboard and foil	8	Same as above	Cyclic boil	25	5% delamination

[a]Laminate joints were made with a water-solvent contact cement recommended for joining aluminum sheet to plywood. After closing the bondline, the bonds were allowed to acclimate for 72 hr under ambient lab conditions before subjection to the boiling water cycle test recommended for bond durability evaluations by the lumber industry.

Data from Ref. No. 3545.

a hot platen press can show up in enhanced bond durability. Mechanical abrading by methods such as sanding, wire brushing, gritblasting, Scotch-Briting, or turning on a lathe can produce scratches and grooves in the aluminum surface that may offer some mechanical component for elevating the overall joint strength and perhaps even extend the durability potential.

While the initial aluminum joint strength can be significantly elevated by abrading a mill-finish surface aluminum, the increase in durability of the joint in aggressive weathering or water-soaking conditions is often modest at best. Minford (3546) has conducted some testing of the durability of two-part epoxy-bonded 3003 aluminum alloy joints, which were pretreated by abrading with either 80, 120, or 240 grit Aloxite papers. As might be expected, the most mechanical interaction was possible with adherends pretreated with 80 grit paper, producing highest joint strengths. The initial joint strength of the 120 and 240 grit-abraded adherend joints were very similar, indicating that the degree of abrasion had to be fairly coarse in order to influence joint strength via macroscopic interlocking. In subsequent water-soaking tests, the more coarsely abraded surface joints also afforded highest durability. Another observation from this investigation was that there appeared to be a greater variability in both initial joint strength and long-term joint durability between similarly abraded surface joints than between similarly etch-pretreated joints whose joint strength and durability were more keyed to microscopic interactions.

B. Microscopic Alterations

The microscopic alterations produced on aluminum surfaces tend to produce higher-strength joints in spite of the fact that there is less direct visual evidence for any opportunity to supplement the chemical surface interactions with physical interlocking. The secret of obtaining such extra strength and joint durability lies in creating a better surface condition for additional molecular contact because of wetting and flow of the adhesive into the microscopic roughness of the chemically etched aluminum surface (see Fig. 1). The microscopic alteration due to chemical etching, or a combination of etching plus conversion coating or anodizing, can so enhance the strength and durability of aluminum joints that bonding aircraft is now routine. In the case of anodizing, the PABST programs have proven the feasibility of producing aluminum-bonded joints for primary structure in aircraft.

The chemical attributes of adhesion cover a wide and subtle range of primary or secondary bridging interactions across the interface. Primary bridging can result from direct chemical reaction between the adhesive and aluminum oxide surface, whereas secondary bridging relates to the residual electrical forces which surround the respective molecules. It is assumed that these fundamental energy relationships and their special association with the "wetting" phenomena have been adequately covered in the chapter on theories of adhesion. A wide variety of technical readings were referenced in that chapter. It should be sufficient here to state that better wetting of the aluminum oxide by the adhesive shows a good correlation with the relative durability of joints in their real-life service environment.

The forces of mutual attraction that can be developed between the specially formulated high molecular weight polymer adhesives available today and a pretreated aluminum alloy surface are certainly spectacular in themselves; however, more is needed than this to assure the most durable joint. The cohesive strength of the adhesive and the aluminum oxide cross section must both be sufficiently high to carry and transmit the highest applied loads to be encountered in service. Some additional safety margin for performance is also necessary to support manufacturers' guarantees. It is a primary goal of the design engineer to build as much extra safety factor into the product as can be economically afforded.

It must also be kept in mind that the service life of any bonded structure is dependent on the accumulated damage effects of the incremental deteriorations that are proceeding with time. The problem then of the evaluation engineer is to develop some means of testing which can simulate this degree of damage in some shorter evaluation time. Having established a testing procedure that can be convincing about predicting the long-term durability potential of the structure, the evaluation engineer must then proceed to establish some quality control procedure which assures that each manufactured unit does meet this standard.

IV. FAILURE SITE VARIABILITIES

When evaluating the durability of a bonded structure, the modes of possible joint failure must be recognized, and the role the adhesive or adherend plays in creating a weakest link in the overall structure must be identified. The variety of possible mechanisms of joint failure has already been considered in some detail in Chapter 7.

A. Cohesive-type Failure Observations

The adhesive can affect the joint failure in a wide variety of ways relating to the relative presence or absence of physical flaws or chemical deterioration of its surface layers or bulk structure. Thus, an apparent cohesive-appearing failure may occur as a result of testing a lap-joint for initial strength, and yet identically fabricated joints often show a disturbing range of initial failing values. This variation in strength using the same adhesive and adherend surface pretreatment must be due to the presence of near microscopic cracks, crazes, or voids which are nonuniformly distributed. It is to minimize the development of such adhesive bondline flaws that the aerospace industry has employed such close control and slow uniform bondline heating and cooling rates.

DeBruyne (3547), Bascom (3548), and Bascom and Cottingham (2522) show examples where imperfections like trapped air bubbles can be the site for high localized stress. These built-in stresses can add significantly to the effect of any desorption agent like water in the service environment, further reducing the joint durability. Even with normal curing, there may not be complete freedom from some degree of interfacial imperfection. This is because many adhesives (such as the neoprene-phenolic contact cements) only develop handling strength after most of the solvent has been driven from the bondline, and the volume of the adhesive in the bondline has accordingly shrunk. Other adhesives may undergo some shrinkage when cooling from the heat-curing temperature, or it could even be a result of chemical polymerization enroute to full cure. Plueddemann (1109) has listed stresses due to shrinkage processes combined with coefficient of expansion differences as among the principal factors responsible for bond failure.

A different situation can develop if the bond interface is altered on account of swelling of the adhesive, which may particularly occur when organic solvents or petroleum derivative fuels or oils are in the service environment. Thermoplastic-type polymer adhesives are generally more sensitive to solvents in this regard, and the resultant swelling can cause significant interfacial stress. Any type of metal adherend offers the worst possible situation in this regard, since metal adherends neither absorb solvent nor change in volume to compensate for volume change in the adhesive. Water can also serve to swell some adhesives as it penetrates into the bondline in service. For example, Minford (3549) has observed the significant swelling of a neoprene-phenolic contact cement exposed to 100% RH at 125°F (52°C). While wet this adhesive could be easily scrapped from the aluminum surface. It had very little adhesion at this time and resembled the crumbling of rubber particles when using a rubber eraser on paper. If the wet swollen adhesive was dried out, however, the adhesion was restored to the degree that it could not be readily scrapped from the same surface even with a spatula. Obviously, the initial adhesion had been destroyed and then restored by simply changing the environmental exposure conditions.

It may be generally agreed that a bonded aluminum joint that shows no initial cohesive failure will likely have little chance of achieving long-term survival in aggressive weathering conditions. However, after evaluating literally thousands of aluminum joints under water-exposure conditions, Minford has concluded that observing cohesive failure initially is not a reliable prediction of longevity. For example, aluminum joints of high initial joint strength may loose that joint strength more rapidly and fail faster than similar alloy and surface pretreated joints with low to modest initial strength adhesives present. DeLollis (197,198) offers examples of both these situations with aluminum joints when comparing the performance of nylon-modified epoxies and nitrile-phenolics. While the initial strength of the former adhesive joints can be three times higher than the latter, they also could fail within 3 mo when exposed to high humidity conditions. However, the author has evaluated more modern nylon-modified epoxy aerospace film adhesives and obtained much better durability responses, though not near as durable as aluminum joints fabricated with the best nitrile-phenolic aerospace film adhesives. DeLollis reported both types of joints initially showed 100% cohesive failure; however, the nitrile-phenolic joints continue to show 100% cohesive failure even after several years in hot condensing humidity exposure.

B. Adhesive-type Failure Observations

When addressing the subject of joint durability, the failure site most frequently observed may be visually between the adhesive and the adherend. As incremental deterioration of the joint interfaces continues to accumulate with increasing simulated or real service exposure conditions, the percentage of initially observed cohesive-appearing failure begins to decrease with a corresponding increase in adhesive-appearing failure. We often observe no visual evidence that any adhesive had ever been attached in these areas, and thus conclude that the adhesive has been totally detached. If water was the deteriorating agent in the environment, the surface of the adherend would appear to be covered with water, and it would be easy to conclude that water had desorbed the adhesive at the interface. Highly magnified viewing, however, has shown that such a situation rarely, if ever, occurs.

Even before such sophisticated viewing of the adhesive-type failed surfaces was possible, some adhesion scientists such as Bikerman (361,3550) and Sharpe and Schonhorn (295) had stated that there was only the degree of cohesive failure to consider in failed joints, since the attractive forces involved at the interface are invariably greater than the cohesive strength of the adhesive. Bikerman (3551,3552) pointed out bond rupture so rarely occurs in the area between adhesive and adherends that such events, i.e., failure in adhesion, need not be considered in any theory of adhesive joints. He has furnished more examples of noninterfacial failure more recently (3553). Whether the failure path in such adhesive-apparent failures is truly at the interface, or in the surface oxide layers of the aluminum, or in some weakened boundary layer of the adhesive, or in a primer layer (if present) is still a matter of debate and controversey. Regardless of where the exact path may be, we need to concentrate our attention on the interphase region if we are to understand how to minimize the effect of the environment, and hopefully extend the service life potential. Good (3554) has also concluded that true interfacial separation seems highly improbable, particularly in those situations where a true chemical wetting of the adherend surface has occurred. Good (3555,3556) has offered a critique of Bikerman's views with regard to the "weak boundary layer" concept, and concluded that what Bikerman calls a weak boundary layer is actually the interphase between the adhering phases. The use of AES, SIMS, and ESCA to define the locus of failure by workers like Baun (3557) and Wyatt et al. (3558) has been very helpful in understanding these concepts better.

Sharpe introduced the concept of an interphase being present consisting of layers which intervene between adhesive and adherend which can be distinctly different propertywise from the original adhesive or adherend. These altered property layers could lower the breaking stress resistance owing to the presence of weak boundary layers (WBL). Thus, if a joint fails at or near the interface at relatively low applied stress, then a weak boundary layer must certainly be involved. There would seem to be no doubt that a weak boundary layer could be the cause of many joint failures that appear to be adhesive related. Whether they exist universally or there are other situations leading to joint failure remains difficult to answer specifically. The reference by Sharpe (295) is recommended for further considerations about evidence for WBL existence, its nature where isolable and identifiable, its development, control of WBL properties, and its influence on joint behavior. For metals like aluminum, Sharpe has contended that the intrinsic boundary layers are the oxides whose properties depend on the history of the particular batch of aluminum and the environmental conditions. That is, boundary layer properties will be dependent on the surface preparation conditions. Minford (534) has investigated the principle of anticipating decreasing joint strength with increasing oxide thickness, as shown in Table 8. The opportunity for flaws to exist should increase as the planes of oxide build-up, and consequently bond fracture can occur at lower and lower stressing levels. There must be an acknowledgment of some significant contributions to joint strength through mechanical interlocking with the adhesive; however, the joints listed in Table 8 do show a relationship between joint strength and oxide thickness which is undeniable.

The matter of the development of weak boundary layer conditions in organic polymeric adhesives is undoubtedly more complex than in the aluminum oxide owing to the wider variety of functional, structural, and morphological factors that can be present. Sharpe's main point is that a complex of interacting factors enter into the mechanical response of any composite structure under an applied load to which the author would add some sort of weathering condition. It is certainly not clear how these ideas about WBL and the changed material properties in the interphase exactly interact with the various other bond-failure theories.

Acceptance of these ideas, however, could possibly offer some complication to the acceptance of the bond desorption by the invading water theory that is espoused by a number of investigators observing the bond failure between epoxy adhesives and silicon dioxide surfaces; i.e., glass adherends (3559–3563). Confusion also exists about water debonding since Laird (3564) measured the diffusion rate of water along an epoxy/glass interface as potentially 450 times faster than through the adhesive. Yet, more recent work by Gledhill et al. shows a distinct relationship between the failure of steel/epoxy joints and the water diffusion rate through the adhesive. At the same time, Minford (3565,3566) has shown certain aluminum/epoxy joints in water-soaking conditions could fail in a matter of minutes when a modest stress was applied simultaneously. In this latter instance, the water has obviously not had time to diffuse through even a short distance of the edge adhesive. Many investigators, such as Mazor et al. (3567), have referred to the observable damage to the adhesive imposed by the presence of water in the service environment.

The concept that true interfacial polymer unbonding is not possible has not gone unchallenged. Huntsberger (3568), in 1963, demonstrated the regeneration of intact free surfaces using selected bonded adherends to prove that interfacial bonding does exist. However, Kaelble (1308) has pointed out the special importance of rheology in regard to Huntsberger's example by suggesting that the opposite result, i.e., cohesive failure, could have resulted if a regularly increasing temperature or a diminished peel rate had been present to change the bulk response. Earlier experiments of Bright (3569) had shown such a rate- and temperature-dependent transition between adhesive and cohesive failure when the temperature of peel was controlled. Between the low-temperature branches of the peel curves (showing interface failure) and the high-temperature branches (showing cohesive failure) was a transition region of combined adhesion-cohesion type failure.

Finally, some failures classed as purely adhesive even after examination by an ISS/SIMS technique were shown to be situations where the adherend surface was obviously not wet by the adhesive. Closer examination of such cases have shown that a thin layer of contaminating elements caused the adherend to be most accurately described as "dirty." No question of whether the failure was truly adhesive can be raised, since real bonding never really occurred.

Admittedly, it is difficult to determine after bond failure whether an apparent adhesive failure occurred at the original interface or at some new interface leaving behind a thin layer of adhesive. In comparing surface features after bond failure with the original adherend surface, the maximum resolution of about 100 Å for scanning electron microscopes may not always be sufficient to detect a thin film of adhesive closely reproducing the original surface profile. Brett (3570) has attempted to use optical and staining methods to determine the presence of such films, but since the optical technique uses an interference phenomenon, it is really only applicable to fairly thick films. Films a few angstroms thick are, therefore, still largely undetected. The use of AES, ISS, and SIMS in the early 1970s greatly improved the opportunity for deducing surface compositional changes and, hence, the locus of failure in failed joints as described by Baun et al. (563). Even when surface films are only on the order of atomic dimensions, or when the failure occurs near the original interface and includes parts of both the adhesive and adherend, useful information on the locus of failure can be provided. Benninghoven (567), Schubert (569), Schubert and Tracy (568), and Werner (570) have all made contributions to increasing the potential of using SIMS for such evaluations.

Venables (33) and Brewis (3571) have pointed out the importance of the aluminum surface morphology in developing initial bond strength and long-term durability. Even the PAA oxide will eventually convert to a poor adhesion oxide with prolonged exposure to humidity (529). Kinloch et al. (781) have suggested local chemical variations within the pretreated oxide layer can greatly affect durability. A precise determination of the final locus of bond failure has been difficult, however, even with sophisticated techniques as discussed by Baun (3572), who used the best analytical procedures known in 1978. Good (3556) has emphasized that it is critical to understanding the adhesion between two materials to determine precisely the locus of bond failure between them.

Brockmann et al. (524) have identified transitional layers in the interfacial regions that are variations in both structure and chemistry from the bulk adhesive or the prepared aluminum oxide surface. Crompton (3573) has developed special techniques for studying these layers in the interfacial region of the bond and published his observations in 1989. His work showed the failure locus of joints (visually labeled as occurring at the bond interface) actually occurred within the adhesive adjacent to the oxide surface. A thin coating of polymer less than 200 Å remained on that surface. This was interpreted as support for the proposition that it is the mechanical properties and stability of this transitional region of a bond which is important in determining bond strength and durability. Hennemann (1085) has recently suggested that differential straining may take place between these different boundary components.

Investigators like Baun et al. (3574) and Baun (3169) have utilized AES, ESCA, and SIMS to improve the chances for discovering loci of joint failure. McDevitt and Baun (3575) have employed these analytical techniques to investigate the surfaces of 6061 and 7050 aluminum alloys. McDevitt and Baun have used these kinds of techniques to show the relationships between deoxidizing treatments and growth of anodic barrier films. Their studies include where the failure may have occurred near the original interface, and could include parts of both the adhesive and adherend. In addition, these tools could provide useful information on the actual locus of failure. When examined with these tools, the majority of apparent adhesive-type failures do show cohesive failures in both adhesive and adherend. It is still necessary, however, to use an elemental analysis method to confirm whether the failed surface contains both adhesive and adherend. The AES has proved to be less useful, because of heating effects causing decomposition of adhesive surfaces. Some of the other important analytical tools have been x-ray emission spectroscopy (XES), x-ray photoelectron spectroscopy (XPS), high-resolution scanning electron microscopy (XSEM), and inelastic electron tunneling spectroscopy (ITES). For more details, the reader is referred to various technical articles on the subject of surface analysis (465–467,588). Also, a variety of individual papers was discussed in Chapters 3 and 4 on characterizations of the aluminum adherend surface and methods of surface pretreatment.

V. ADHESIVE FACTORS AFFECTING DURABILITY

A. Modulus of Elasticity

Adhesives can cure and develop a wide range of modulus values which, in turn, will interact with adherends of varying modulus. The property differences for some adherends and adhesives that would seem appropriate to the discussion are shown in Table 151. High modulus in the adhesive is particularly destructive if coupled with dissimilar adherends of significantly different coefficients of thermal expansion. The worst situation for long-term joint stability under these conditions would be under service conditions which included wide-ranging and cycling temperatures. The thermal-induced stresses at the interface could be quite high and impossible to resist in the bondline. Where adherends of significant difference in their coefficients of thermal expansion have been used, it is usually beneficial to employ an adhesive of low modulus in the bondline, such as a heat-cured nitrile-rubber-modified phenolic. This family of adhesives can combine adequate strength and outstanding joint durability with significant stress relief. Obviously, the best adhesive choice for these situations must be a compromise between adequate strength to support the bonded structure and some significant ability to withstand the generally alternating high bondline stresses that can develop as the temperature environment fluctuates. Aluminum must often be combined, in this regard, with other materials of construction which have different thermal coefficients. Thus, the rubber-modified cements, or even some of the adhesive/sealant products derived from modern synthetic chemistry, can be considered for these special service conditions where the thermal stresses are cyclic and continuing indefinitely. Although the joint strength of such aluminum joints would be considered low to modest at best, this low strength may be sufficient where the stress loads in service are known to be rather

Table 151 Property Differences For Some Adherends and Adhesives

Adherend	Modulus (10^3 MPa)	Linear coefficient of expansion (10^{-6} m/m°C)
Stainless steel	199.96	17
Aluminum	72.40	22–28
Alumina	344.75	8
Adhesive		
Epoxy (unfilled)	3.45	50–60
Epoxy (filled)	13.79	25–40
Nitrile-phenolic	0.03	50+

From Performances, Durability and Testing of Structural Metal Joints. Technical paper given by J.D. Minford before Epoxy Formulators Association Meeting.

low. An outstanding example is the use of flexible sealants to caulk aluminum window products, where the differential expansion between frame and window must be tolerated for many years.

B. Interfacial Imperfections

This subject has been mentioned and considered in some detail previously in this section under the subjects of cohesive and adhesive-appearing failure sites. The important consideration is that the uniform structural integrity of all elements in the interfacial area is a goal of all adhesive joint fabricators but is rarely, if ever, achieved in practice. If multiple-part adhesive formulations are employed, then the possibility for error in weighing and blending of the components starts with the adhesive manufacturer. It is for this reason that the listing of various test procedures was outlined under the testing section, which could be conducted to give assurance to the adhesive manufacturer of the quality of the product. Further testing would then be recommended on the part of the adhesive customer to establish that the adhesive product, as received, meets all the data sheet specifications.

By interfacial imperfections, we also mean to include more than physical voids, pockets of volatile reaction products, and cracks in the cured adhesive cross section. We must include all those variations in final adhesive structure that can arise from improper or incomplete mixing. The greatest volume of adhesive used in general manufacturing falls in the classification of polymer and solvent blend products. These range from very low solids content products to thick, viscous mastics in the 70–85% solids category. As viewed in cans of freshly opened paints, the blend of solids and solvents will begin to segregate as soon as motion ceases after packing. Thorough stirring is absolutely necessary before each use period in production. It is also characteristic of organic solvent adhesives that the solvent will have a fairly high degree of volatility. Thus, the concentration of solvent after mixing needs frequent monitoring if the adhesive dispersal on the adherend surface is to be applied uniformly at some controlled thickness. All manufacturers need to be ever mindful of the fact that it will be virtually impossible to operate their manufacturing line without some periodic down-time periods. It is necessary then to make close observations and sometimes conduct some testing to be sure that no significant operating factor has begun to vary from predetermined levels. Most pieces of adhesive application equipment will be subject to some down-time situations. The equipment will then need to be adjusted to allow for the property of environmental differences that exist as compared with the standards previously set for production.

All of the above situations for creating nonuniformity in the bondline are further complicated by variations in operator performance and variation in "as received" adherends. This variation can be quite pronounced when comparing different mill-finish aluminum surfaces, which will subsequently be bonded without additional mechanical or chemical pretreating. For example, the author has observed from years of bonding heat-treat aluminum alloys that they can have quite variable wettability even though their chemical compositions were maintained within the alloy composition limits established by the aluminum industry for these alloys. It was also demonstrable that some variations in pretreatment schedules might be required in order to achieve equal joint strength bonding. This could be due to significant differences in the thickness of the heat-treat oxide. It was also observed that different lots of the same alloy material could show similar initial joint strengths in the mill-finish state, yet show distinct joint durability differences in later accelerated stress and weathering testing. These remarks apply mainly to manufacturing situations where essentially no cleaning, or only solvent cleaning, precedes the bonding operations.

Those adhesives which initiate a joint through contact bonding between two adherends are particularly vulnerable to interfacial imperfection problems because of the variety of different bonding situations that can occur in production. Contact cement bonding adhesives usually fall into one of two general types. First, there are the low solids adhesives, where the solvent concentrations can range from 70 to 80% of the adhesive weight. The most general method of use is to brush, roll, spray, or curtain coat a layer of solvent-rich adhesive on each adherend and then wait until the last solvent is escaping from at least one surface. At this time, the maximum immediate bond can be achieved by contacting the two surfaces and applying some pressure to affect intimate contact. If too much solvent is present on either surface, then insufficient immediate bond strength will be developed to permit handling on the production line. Handling with insufficient green strength in the bondline can also cause a slippage or displacement of joint mating and be the basis of lower service durability. If too little solvent remains in both coated surfaces to create a tacky surface condition, then both marginal handling and final strength and lower durability can result. Both situations arise in a contact bond made between aluminum and another material that is sensitive to solvent. This occurs because no adhesive can be applied to the sensitive surface. This was the situation when Alcoa manufactured aluminum/polystyrene/aluminum laminated architectural building panels. Polystyrene foam cells would be immediately collapsed by the strong solvent used in a neoprene-phenolic contact cement. Thus, when the final bondment to polystyrene was made, no solvent could be left in the adhesive. All solvent had to be removed by volatilizing the solvent from the coated aluminum before mating the two adherends. Further, any heat used to remove the solvent needed to be sustained to create sufficient tackiness to make the final lamination in combination with external pressure from heated press and/or a rotary press.

The second general class of contact cements are those with a high solids content (50–85%). Imperfect bondlines usually stem from difficulties associated with the procedures used to remove enough solvent to retain sufficient tackiness but stop short of the creation of a solvent-starved adhesive coating with marginal tackiness. Because of the thickness of adhesive deposited by a heavy mastic adhesive, only one of the two surfaces to be joined is usually coated with adhesive. This places a heavy responsibility on the ability of the adhesive to achieve good wetting on the noncoated adherend. If the open time before joining is beyond a relatively short time, the surface tackiness will rapidly diminish and joint strength and durability may be inadequate or marginal at best. Poor service potential could likely result. This kind of adhesive is often employed to laminate corrugated aluminum to wood structurals for prefabricated buildings. Adhesive is extruded as a series of beads on the flats of the corrugated aluminum and pressed against the wood. A number of these laminations are then stacked up for further development of handling strength as the additional solvent escapes from the adhesive bondline into the porous wooden surface. Attempts to handle such structures too soon will usually cause localized areas of nonbonding and interfacial imperfections to develop, which result in poor resistance to weathering conditions.

There are some generalizations about the negative effects which contact cements and mastics can generate on joint durability: (1) a variety of solvents can be used to disperse such adhesives, which can have different time rates in escaping from the bondline; (2) water dispersed contacts and mastics tend to develop handling strength and surface tackiness slower than organic solvent dispersed products; (3) the thicker the coating, regardless of what type of solvent is used, the longer must be the open-time to achieve handling strength and uniformity in the bondline; (4) because of the lower strength present in these adhesives, thin bondlines may be inadequate for long-time service with any stress present; and (5) attempts to accelerate solvent removal with heating to improve production rates must be controlled through specific testing of the final products to demonstrate bondline imperfections have not been unknowingly introduced.

Another interfacial factor which can significantly affect overall joint durability is the presence and relative concentrations of nonadsorbable or nondesorbable contaminating films at the interface. The most common contaminates found on aluminum are grease films, forming lubricants, and anticorrosion-type coatings. Minford has observed a manufacturer's attempt to bond aluminum sheet to itself or other materials where the contaminant layer of oil was so thick that a name could be traced on the surface with a finger. Obviously, bonding with such gross contamination conditions makes bonds that may not even survive in mild weathering. Unfortunately, aluminum manufacturing and fabrication procedures often require the use of special forming oils and lubricants. In such cases, the bonder must be particularly careful to remove this contaminant to assure highest durability performance. Minford (68,72–74) conducted a number of joint durability investigations in the early 1970s to attempt to establish how much forming lubricant could be tolerated on aluminum auto body sheet surfaces in manufacturing bonded aluminum joints for that industry. It was confirmed that virtually no contaminating oils or lubricants could be tolerated with the typical two-part room temperature–curing epoxy cured at room temperature (3576). It was demonstrated, however, that one-part, heat-cured epoxies or one-part, heat-curing vinyl plastisol formulations could bond to varying degrees in the presence of modestly high concentrations of surface lubricant. This will be discussed later in the chapter under deliberate bonding in the presence of metal lubricants on the formed aluminum surface. In the 1980s, adhesive formulators have attempted to develop heat-curing epoxies that can effectively bond to oiled-metal surfaces with remarkable resistance to exterior weathering conditions. The author (3577) evaluated the comparative durability of mill-finish and variously lubed autobody sheet alloy adherends as lap-joints in room temperature water immersion, 100% RH at 125°F (52°C) and in an autoclave at 250°F (121°C). Similar testing was performed in corrosive salt water (3578) and for stress-endurance performance in 100% RH at 125°F (52°C) or corrosive salt water (3579).

C. Adhesive Heat-curing Variabilities

The heat-curing of adhesives can affect the quality of aluminum joints in both positive and negative ways. The shrinkage of the adhesive in a bondline during curing has already been mentioned as a negative factor, because it can lead to the production of bondline stresses. When 100% solids paste or film adhesives are cured at elevated temperature, however, it can be a definite positive factor on durability. This occurs because the adhesive enroute to cure proceeds through a state of lowered viscosity. This can promote an improved flow and wetting stage for the adhesive over the aluminum surface as compared with the wetting which would occur under room temperature conditions. Even with soiling present on the aluminum, the lower viscosity of heat-curing adhesive might absorb, dissolve, disperse or desorb the soiling. This certainly is one explanation for Minford's observation (3576) that it is virtually impossible to achieve strong or durable aluminum joints with any nonmodified two-part ambient curing epoxy formulation over an oiled or specially lubed aluminum surface. As noted above, special heat-curing epoxy pastes were developed in the 1980s which could both develop strong initial bonding and acceptable bond durability even in aggressive water-soaking exposures (3577–3579). This could be an explanation for Minford's observations (3580) that curing a typical two-part, epoxy paste adhesive for 7 da at room temperature over degreased 6061-T6 adherends often resulted in development of only modestly strong joints. Such joints usually showed 100% adhesive-type failure when tested at that time in a tensile tester. In contrast, however, when duplicate joints were heated to only 200°F (93°C) for 5–10 min in an oven after the 7 da RT cure, the joint strengths more than doubled. Furthermore, the mode of joint failure had now become 100% cohesive and the joint survival times in hot humidity could increase severalfold. Going back to the earlier mentioned "attachment site theory" of adhesive joint strength, we might conclude that the effect of heating the room temperature–cured bondline was somehow able to create and consummate a number of additional new attachment sites.

Sell (1098) showed that the higher the cure temperature for the structural adhesives he evaluated on aluminum adherends, the greater the tendency to find a higher order of joint durability response. He has also shown that within a given family of adhesives, the joint durability would increase as the creep resistance increased, and the creep resistance increased as the heat curing was advanced. The heating could obviously generate significantly greater cross-linking in the same adhesive polymer, which could then increase the resistance to creep.

We need to constantly keep in mind that higher-temperature curing of an adhesive does not automatically ensure higher joint durability. Minford (3581) has shown that, for example, in the data presented in Table 152, two (one-part, heat-curing epoxy) products demonstrated quite different durability responses when longer than recommended heat-curing was employed to fabricate 2036-T4 aluminum joints. In product A when the heat-curing was essentially doubled, the joint durability in hot humidity diminished by 75%. In another epoxy, designated B from the same manufacturer, a similar elevation in heat-cure time and temperature showed no deleterious effects on the durability in hot humidity or salt-spray exposure. It also could be quite significant in Table 152 that epoxy A had both the temperature and time elevated over the recommended values, whereas only the time was doubled for epoxy B.

D. Bondline Pressure

Adhesives vary greatly in their requirements for pressure on the bondline, but in most all situations, the net effect is positive for increasing joint durability. (*Note*: An exception to this statement is the situation where excessive pressure on the bondline can create an adhesive starved-bondline.) The simplest case would be simple fixturing where two adherend surfaces are being held in contact and proper registration while the adhesive is curing. In others, the necessity exists for a higher order of bondline pressure to reduce the bondline imperfections that can be caused by volatiles generated by the curing adhesive. Finally, there is the purely physical factor of more adhesive being forced into the structure of a porous adherend surface by high pressure on the bondline during manufacturing. (See Table 150 for the positive effect on durability of using dwell time in a heated platen press as compared to only passage through a rotary press for laminating aluminum-to-plywood as determined by the author [3545].) Information on the effect of laminating pressure on bonding to porous wood adherends can be found in a recent publication of River et al. (3583).

E. Contact Cement-type Adhesive Responses

Pressure exerted on the surfaces to be joined by a contact adhesive would seem to be particularly vital, since it is mandatory that each adhesive-coated adherend be intimately joined to its counterpart.

Table 152 Variation in Bond Durability of 2036-T4 Aluminum/Epoxide Joints with Variation in Heat Curing

Epoxy	Curing conditions	Initial shear strength: psi	Initial shear strength: MPa	Retained joint strength (%): 100% RH at 125°F (52°C) (4 wk)	Retained joint strength (%): 5% salt spray at 95°F (35°C) (3 wk)
A	1 hr at 300°F (149°C)[a]	1670	11.50	100	0[b]
A	2 hr at 350°F (177°C)[c]	1900	13.09	25	0[b]
B	1 hr at 300°F (149°C)[a]	1930	13.30	64	55
B	2 hr at 350°F (177°C)[c]	1950	13.44	61	58

[a]Manufacturer's recommended cure condition.
[b]All joints failed within 7 da.
[c]Simulation of cure cycles used for commercially priming and painting aluminum.
Data from Ref. 3581.

Where both surfaces are vapor barriers (as with two aluminum adherends), it can be possible to end up with an adhesive-starved glueline if the applied pressure is too high. On the other hand, if aluminum is being laminated to a porous adherend like wood, there is a certain level of pressure, which may force adhesive into the porous surface of the wood resulting in a higher bond strength and joint durability due to generation of more mechanical interlocking components. For this reason, hot laminated aluminum-to-wood products will always show enhanced durability when held under pressure in a static press or passed through a rotary press followed by dwell time in a static heated press. Adherend surfaces of marked roughness could also show some such benefit from pressure on the bondline, especially where a high-solids adhesive is being used to fabricate the joints.

Although the two-part, room temperature–curing, thermosetting resin adhesives like the epoxies are not usually included under the heading of contact-type adhesives, they may be considered such, since they only require simple surface contact between two adherends to achieve a joint. The effect of extra pressure on such aluminum bondlines may be positive on durability only if it somehow can create more contact sites. It can be definitely negative if the pressure is much higher than required for simple fixturing, leading to a starved bondline. However, unlike the contact rubber–type cements, the epoxy mastic adhesives generally show higher and more durable joint strength in thinner bondlines. This is probably due to more uniform adhesive distribution and greater freedom from voids and air pockets.

F. Heat-curing Adhesive Responses

Some adhesive formulations (such as phenolics) must be cured under heat and modest to high pressures to achieve good weathering resistance in service. Products produced by the heat-triggered polymerization without accompaning pressure on the bondline may allow creation of significant porosity in the bondline and, hence, reduce durability potential in the joint. One of the reasons that epoxy adhesive formulations have been preferred in bonding American aircraft has been their low production of volatile reaction products in the bondline, which is a very positive factor when using vapor-barrier adherends like aluminum. In contrast, the heat-curing vinyl-phenolics that are more widely employed in bonding aluminum aircraft in Europe do require higher bondline pressures to restrict concentration of voids. In spite of this fact, the vinyl-modified phenolics have not been displaced by epoxies overseas. This stems from their undeniably long-term proven performance in aluminum aircraft operated under all types of weathering service.

A word of caution needs to be given, however, about the possibility of overheating (overcuring) an adhesive in a bondline and promoting poorer durability than may be indicated by the data sheet of the manufacturer. This is definitely one situation where more of the same may not be beter. The bulk structure of adhesives may be significantly altered by both the degree and speed of heating in-place. However, Minford (3584) has been able to demonstrate a one-part nitrile-modified epoxy paste adhesive (recommended for curing 1 hr at 350°F [177°C]) could be cured at 400°F (204°C) for 15 min with no recognizable change in 2-yr durability test performance. This could not be determined by simply observing the appearance of the adhesive after the two curing conditions. It had to be established by direct side-by-side durability testing using identical exposure conditions. The intent of finding these two widely different, but equally acceptable, curing times was to demonstrate the possibility that a much faster rate of production could be achieved without sacrificing product quality in service. Quite often, however, the appearance of an adhesive after an overcure can be indicative of poor durability potential. Minford (3585) showed the appearance, strength, and durability potential of aluminum joints could be drastically affected when another adhesive recommended for curing at 350°F (177°C) was cured for the same time at 250°F (121°C), 300°F (149°C), 400°F (204°C) and 425°F (218°C). At the lowest temperature, the adhesive barely developed any measurable joint strength and durability was totally unacceptable. At 300°F, the joint strength was less than stated by the adhesive manufacturer, but water durability results were relatively good. At 350 or 400°F, the curing produced maximum strength joints and essentially the same good durability response. At 425°F, the cured adhesive was noticeably darker and more friable (probably due to an excessive high heat exotherm) and the durability results shown in water-exposure conditions were significantly lowered along with the initial joint strength level. In spite of this lower strength, the joint failed cohesively showing a significantly lowered strength in the bulk adhesive.

G. Effect of Fillers in Adhesives

The effect of fillers in adhesives on bond durability has been the subject of several papers. Bodnar and Wegman (1495) have shown the superior resistance to various natural weathering conditions of filled polyamide-epoxy (Epon 828/Versamid 140)/aluminum joints compared to joints made with the unfilled version. Minford (3586) obtained similar results with 6061-T6 joints and filled versions of Epon 815/Versamid 125 adhesive subjected to long-exposure times in accelerated laboratory water-soaking testing conditions. Minford (3586) has also detected significant durability differences between different filler choices. For example, aluminum joints with either an aluminum-filled or a china clay–filled version of the same polyamide-epoxy successfully resisted 8-yr exposure to direct industrial atmospheric weathering, whereas a carbonate filler modification-type joint delaminated several years earlier. The fact that aluminum powder as a filler in Epon 815 (60%)/Versamid 125 (40%) adhesive mixtures produced joints with twice the survival time in the corrosive seacoast atmosphere of corresponding joints made with the unfilled epoxy was shown by Minford (3587). Spathis et al. (3588) have suggested that a positive effect from the use of filler in the adhesive might result from the filler acting as a spherical exclusion resulting in a stiffening of the matrix and increase in average strength of the polymer. They found the adhesion of the filler to matrix seemed strongly dependent on the diameter of the particles. The volume content of the filler seemed to be of much less significance.

The effect on durability of aluminum powder filler in nitrile-modified heat-curing epoxies was studied both by Wegman (1495) and Minford (3589). Wegmen's exposures were conducted at military exposure sites under jungle, desert, and industrial atmospheric conditions and Minford used accelerated water soaking procedures. Neither could find a special durability benefit, however, from the incorporation of the aluminum powder filler. It should be pointed out that the nitrile modification, while quite successful in flexibilizing the heat-cured epoxy, achieves it at great expense to the overall resistance to water-exposure conditions. This aids in explaining that the rate of deterioration of joint strength for some one-part, nitrile-modified, heat-cured epoxy/aluminum joints has often proved to be faster than found when using a good-quality two-part, room temperature–curing epoxy bonded to the same aluminum adherends. It is not unusual to find that an unmodified (more rigid) one-part, heat-curing epoxy can produce more durable joints in water-soaking exposure conditions. The flexibility-producing rubber, however, may still be a positive factor for achieving better stress-humidity–resistant joints and, indeed, produces significantly higher survival-time joints than can be achieved with most any commercial two-part rigid epoxy.

Minford (3586) has investigated the effect of several different filler additives to two-part epoxies based on durability testing in hot-water–soak/freeze/thaw cycling for 2 yr. It was shown that both the ratio of polyamide curing agent and the choice of aluminum oxide filler could be highly influential on strength and durability. An equal weight mix of Epon 815 epoxy and Versamid 125 polyamide curing agent plus an addition of 140 parts of Albron 120 powder gave highest strength and durability. Lowering the Albron addition to 100 parts and changing the resin/curing agent ratio to 40/60 gave lowest strength and only 40% of the strength retention value. Data obtained with Albron additions to other commercial two-part epoxies showed a very wide variability response depending on choice of epoxy resin, curing agent, and relative concentration of filler. Addition of a second aluminum flaking powder to provide a more metal-like appearance to the adhesive significantly lowered the water resistance, indicating a complex interaction effect.

H. Adhesive Thickness in Bondline

Earlier references have been made to the effect of bondline thickness on initial joint strength as developed by Bryant and Dukes (2694,2695,2748). There has not been a large volume of data available in regard to the effect on joint durability in weathering conditions. It might be suspected, however, that increasing the thickness of the bond would generally lead to more opportunities for finding non-homogenuity and, consequently, more chances for weak boundary conditions somewhere. Wolfe et al. (1496) have evaluated joint durability in terms of fatigue resistance and found noticeably shorter joint fatigue life as bondlines thickened. Minford (3590) has looked at the effect of increasing bondline thickness in aluminum joints made with controlled lube concentrations on the adherend surface and with heat-curing highly modified epoxy adhesives designed to bond through the lube on the surface. Several factors seemed to be operating to significantly decrease the joint durability. As the mass of uncured adhesive was increased in the bondline, the heat-curing conditins may need to be altered to prevent the development of significant void areas and the development of more friable fracture properties. This seemed to be due to a significant rise in heat exotherm in the bondline, producing volatiles which could not escape. Presumably either a slower rate of heating to the final curing temperature or a longer curing time at a somewhat lower than recommended temperature might change the durability potential significantly.

Wake (3591) has proposed that the weakness of the thicker bondlines is largely due to stresses developed by the volume contraction of the adhesive or by differential contraction between adhesive and adherend after curing at elevated temperature. This implies a linear relationship between strength and bondline thickness, which is what has been usually found. It occurs to the author that we might also consider how the weathering environment might have a differential effect on the adhesive in different parts of the bondline. If we consider that the main pathway for water to enter the bondline is through the adhesive by diffusion, then the tendency to find more porosity in the center of a thick bondline could mean a faster diffusion rate in that area and a faster degradation of the bulk strength in that area and early fracture under any sustained or oscillating stress. Finally, we might also state that the cured adhesive in proximity to the aluminum adherend appears to be a higher density and definitely has less porosity as compared to that in the center of a thick bondline. The lowered resistance of this center area to fracture would explain the observations by Minford that the joint failures in stress exposure of these joints was largely in the center of the bondline, leaving a thick layer of adhesive on each adherend surface.

I. Environmental Temperature Effects on Durability

The nature of the joint-failing process associated with its temperature change has been studied by Foulkes et al. (3592). While thermosetting-resins have softening points at high temperatures, they do show a change in strength which is independent of any thermal degradation process. These authors gained evidence that while the adhesive fails by a brittle mode at low temperatures, many begin to fail by a ductile mode at room temperature. Thus, the recorded bond strength decrease with increasing temperature may be due to a plastic flow phenomenon.

Since temperature is a common factor in all service environments, the effect of temperature on aluminum joint durability needs to be considered. This is especially true since service temperatures of interest in recent aluminum aerospace bonded structure applications can range from as low as −421°F (−253°C) up to 1000°F (538°C). Providing the adhesives have chemical stability at these temperature extremes, bond permanence is based on the result of stress concentration and gradients developed within the joints. Such stress concentrations may be triggered by thermal expansion coefficient differences between adhesive and adherend, adhesive shrinkage during curing, modulus of elasticity and shear strength differences between adhesive and adherend, thermal conductivity differences, or residual stresses in the adherends due to a release in bonding pressure.

Functioning aluminum joints at the lower cyrogenic temperatures will experience intensified stress concentrations, because of the magnified differences in the physical and mechanical properties of adhesives and the inorganic aluminum oxide. For example, a low-modulus–type adhesive which can readily stress relieve itself at room temperature by deformation, can increase in modulus of elasticity at cryogenic temperatures to the degree that no adequate stress relief is now possible. If we further increase the area of adherend and joint under these conditions, the stress situation can actually multiply. This was apparently the situation that occurred during the development of bonding for the space shuttle. When testing was made at the cryogenic temperatures of space on scale-model specimens, the results could not be properly extrapolated to the performance obtained with full-size constructions. The supposedly adequate test results using scaled-down joints simply failed to simulate the greater magnitude responses in full-sized structures.

Obviously, bonded joints can be dstroyed at high temperatures as well, but now the most significant factor becomes resistance of the adhesive in the bondline to thermal oxidation. It became imperative to find adhesives for bonding aluminum aircraft that were increasingly resistant to thermal oxidation in the early 1950s. This matter was reviewed earlier in the section dealing with the selection of an adhesive for particular environmental circumstances. Black and Blomquist (1622–1625,1759,1760, 1891) provided acceptable candidates with the epoxy-phenolics, but service response was still limited to about 500°F (260°C). They also provided the evidence that metal surface chemistry was highly influential in accelerating the rate at which thermal decomposition might take place. Olson and Lulling (1627) pushed the testing temperature to 1000°F (538°C) by 1962 and demonstrated that the rate of heating to test temperature could also be important. For example, to go to peak temperature within 10 min proved to be more destructive than slower heating rates. Krieger and Politi (1762) pushed performance still further by 1965. The technical papers of St. Clair and St. Clair (1766,1835–1837) in the 1980s, however, represent more recent developments in high-temperature–resistant adhesive technology. Rossi (1276), Wentworth (1277), and Connell (1278) have each presented summaries of such developments in the 1990 ASM Handbook.

J. Effect of Natural Weathering on Aluminum Joint Durability

Beginning in 1955 and continuing through 1960, Eickner (964–966) began to report on aluminum joint permanence in various exposures to exterior natural weathering conditions as compared to laboratory-type accelerated weathering conditions. At this time, the majority of laboratory-type conditions were those specified under military specifications MIL-A-5090B or MIL-A-8431. Eickner and co-workers specifically choose two sites known for their aggressive environmental conditions; i.e., a site 50 yd from the seashore in Miami, Florida, and another a 0.25 mile from the ocean in the Panama Canal Zone. In 1959, the Royal Airforce Establishment issued a report on the tropical durability of metal adhesives as known at that time (3739).

General good performance of unstressed aluminum joints was found after several years exposure where the aluminum surfaces had been prepared by hot chromic acid deoxidation. The next logical step was taken to introduce some stress in the joint using the test specimen shown in Figure 27 and data from 11 different adhesives as shown in Table 153. Some correlation between laboratory-accelerated weathering testing and atmospheric exposures was sought by testing stressed and nonstressed

Table 153 Percentages of Control Strength Retained by Stressed and Unstressed Aluminum Joints After 36 Months of Exposure at Florida Site

	Percentage of control strength	
Adhesive	Unstressed	Stressed
Bondmaster M24B	88	21
Redux 775 film	96	60
Swedlow 371W	78	19
Griffobond P262-A	23	0
Bloomingdale FM-47	77	0
Scotchweld AF-6	72	56
EC 1471	63	0
AF 5930	64	28
EC-1469	56	0
Epon VIII	54	0
Epon VIII (painted)	96	92
Epon VIII (primed)	86	83
Epon 422 (on steel)	64	0

Data from Ref. 1490.

joints in 30-da in salt-water spray, 30, 60, or 90 da in 120°F (48.9°C) in 97% RH, or 15 da at 165°F (73.9°C) in 97% RH. Stressed joints (admittedly these had relatively low stressing levels employed based on the modest decline in joint survival time) did tend to show some lower levels of retained joint strength. The more corrosive accelerated salt-water spray and the Miami seacoast exposures were generally noted as most deteriorating. This relatively short exposure time to 165°F (73.9°C) at 97% RH did not generally cause failure of any of the Eickner fabricated aluminum joints even under stress conditions employed. This also would indicate that the level of stress in these Eickner investigations must have been low. It should be compared that Minford's testing conditions in 100% RH at 125°F (52°C) were arbitrarily set up to run for 2 yr when he initiated his first series of durability investigations in the early 1960s.

Important early natural atmospheric aluminum joint durability investigations were conducted at Picatinny Arsenal by Wegman et al. (3740) using exposure sites in tropical Panama, desert in Yuma, Arizona, and an industrial site at Picatinny Arsenal. Separate reports were issued after 1, 2, and 3 yr for making comparisons. No stressing was involved in any of these exposures. The Panama site was described as being a "moderate" jungle site by the investigators. Fourteen commercial structural adhesives were used with curing temperatures from ambient to 350°F (177°C). One RT-curing epoxy series of aluminum joints and one 250°F (121°C) curing modified epoxy series were used, which could be comparable to the two control epoxy adhesives used by the author during the same period. Wegman used mechanical and acid-etching pretreatments on 6061-T6 alloy, whereas Minford has employed as many as 25 different aluminum surface treatments with subsequent exposure of 6061-T6 joints for 8 or more years in the industrial and seacoast atmospheres. Minford always included one series of joints using only vapor-degreased 6061-T6 adherends and one series of joints using a hot Alcoa A3 (FPL-type) deoxidizing pretreatment as control joints in all investigations.

After 1 yr, Wegman concluded that the hot, dry Yuma exposure failed to cause deterioration and could elevate initial strength as compared with control joints exposed for 1 yr in the laboratory. Some slight decline in joint strength was noted with similar exposure at Picatinny Arsenal. All joints in the Panama exposure showed deterioration to some degree with some affected drastically. Duplicate joints were exposed to MIL-STD-304 conditions (cycling from low air temperature to elevated temperature at 95% RH). While the results from this MIL-STD tended to predict the trends noticed for joints exposed to jungle high humidity, the individual values showed wide variance. Similar joints in accelerated laboratory environments showed as much as 450% higher to 35% lower joint strength retentions than found in Panama. Treating the aluminum adherends with hot chromic acid deoxidation resulted in highest initial and highest joint strength retention values when exposed in the Picatinny atmospheric conditions.

A survey study of the effect of 1-, 2-, or 3-yr natural weathering in south Florida was conducted by Hause et al. (3741) in 1965. Ambient curing two-part and heat-curing one-part epoxy paste adhesives were evaluated as well as heat-cure nitrile-phenolics, nylon-epoxies, modified epoxies, and vinyl-phenolics in film form. All joints were fabricated with acid-deoxidized aluminum adherends. Overlap-shear, honeycomb peel, and honeycomb beam flexture testing was conducted with the general conclusions drawn that the bond strength for most of the adhesively bonded joints were generally unaffected after the maximum 3-yr exposure time.

Possibly the largest number of long-time aluminum joint durability test investigations in natural industrial and seacoast atmospheres were performed by Minford beginning in the early 1960s and continuing to the 1980s. First publication was in 1972 (976) followed by a large summary of tests, including 27 different adhesives from a wide variety of different chemical families in 1973 (531). The typical testing schedule specified that sufficient joints would be exposed that triplicate joints could be deliberately failed in a tensile tester to determine average joint strength values initially and after 1, 2, 4, and 8 yr of atmospheric exposure. In Figure 35, it can be seen that some exposures were continued in the industrial atmosphere for 12 yr to correspond to 12-yr exposures that were conducted in the tropical jungle in Surinam, South America, using the same adherends and adhesives.

Additional reporting of the effect of outdoor exposure on both stressed and nonstressed aluminum joints was issued by the Royal Airforce Establishment in 1970 (3742). In 1977, Cotter (851) reported on some weathering trials conducted at three sites. The weathering at the Royal Airforce Establishment at Farnborough was representative of most outdoor conditions for the temperate zone, whereas

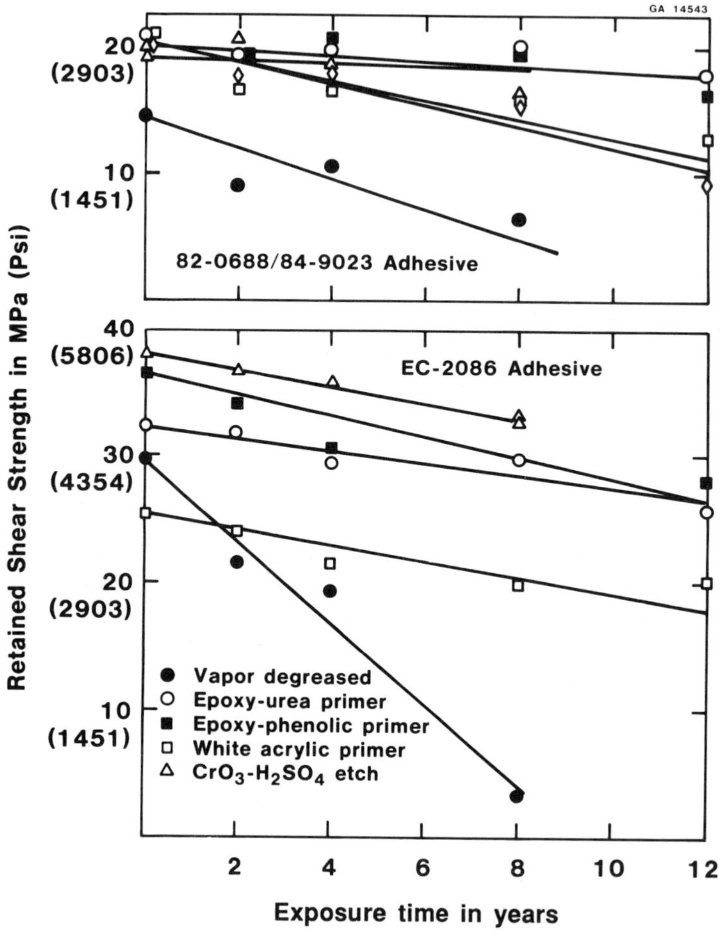

Figure 35 Long-term durability testing in the natural industrial atmosphere. (From J.D. Minford, unpublished work, Aluminum Co. of America.)

the Innisfall site in Australia represented (wet/hot) jungle conditions and Cloncurry a dry/hot Australian site. Tests at the jungle site involved chromic-deoxidized clad aluminum joints. As unstressed joints, epoxy-Novolac, nitrile-phenolic, and vinyl-phenolic adhesives all showed excellent durability, but the polyamide-cured epoxy showed drastic loss of strength. This is in contrast to the results of Minford, where a polyamide-epoxy adhesive showed good resistance to 12-yr jungle exposure using chromic acid–deoxidized Alclad 2024-T3 adherends. With stress present, the epoxy-novolac and nitrile-phenolic adhesives continued with excellent response, but the vinyl-phenolic joints showed rapid loss after only 2 yr (851).

Of special interest are the very long-time (12 yr) durability tests by Minford (1356) using one- and two-part epoxy-bonded Alclad 2024-T3 and 6061-T6 aluminum adherends in the jungle in Surinam, South America. Just a few degrees off the equator, this environment offered both secluded and open jungle sites with corresponding high-temperature and near 100% condensing humidity weathering conditions. The experimental variables offered the opportunity to study the performance differences between (1) Alclad 2024-T3 (aircraft alloy with high-purity clad surface) and 6061-T6 (general manufacturing sheet alloy with a heavy heat-treat film on the surface), (2) vapor-degreasing–only or hot chromic acid–deoxidized surface pretreatments, (3) two-part, room temperature–curing or one-part, heat-curing epoxy adhesive-type joints, (4) silicone sealant edge protection or no edge protection, and (5) exposure to site characterized as open jungle weathering conditions and a different secluded-type jungle site. These data have been plotted in a number of graphs, which can be reviewed in more detail in the author's publication (1356).

Although the clad surface on Alclad 2024-T3 is relatively high-purity aluminum, the joint durability performance during the 12-yr exposure was slightly poorer than for corresponding joints fabricated with a more highly alloyed and heat-treated oxide surface on 6061-T6 adherend joints. This was observed with either a vapor-degreased or acid-deoxidized treatment with the one-part, heat-cured, nitrile-modified epoxy adhesive. Also observed was the fact that no significant improvement in durability could be noted from use of a silicone sealant to screen out the destructive jungle weather. This was in great contrast to the effective protection that the same RTV silicone had provided using similar joints exposed to the seacoast for 8 yr. The author has explained this lack of protection in the humid jungle as evidence of the poor ability of such a sealant to act as an effective vapor barrier. However, from the data obtained at the seacoast exposure station in Rhode Island, it does seem to be able to function as an effective barrier to passage of liquid water into the joint. The author cannot offer any reasonable explanation as to why the more alloyed surface on the 6061-T6 alloy joints offered more durable bondability than the higher-purity Alclad 2024-T3 surface. Whether these data can be considered as offering some evidence of support for a preference for using bare alloys as compared to clad alloys is still difficult to answer. It must be pointed out that these joints (unlike those fabricated for airplane bonding) did not employ the anodize surface pretreating or a primer to affect better surface wetting as would be mandated in aerospace manufacturing.

The much higher resistance of the relatively lower-strength and more rigid 2-part epoxy in the bondline has been mentioned elsewhere by Minford. For example, while the correspondingly treated one-part epoxy joints in this study were 75% higher in initial strength, they degraded to only 6% of this value after 12-yr jungle exposure. At the same time, the lower strength two-part epoxy joints, even on vapor-degreased–only surfaces, exhibited virtually no joint degradation after 12-yr exposure. The corresponding 6061-T6 joints with their heavy heat-treat oxide degraded to about 25% of initial strength, indicating the distinct need to move beyond a simple vapor-degrease treatment. This was accomplished by vapor degreasing followed by a hot chromic acid–deoxidizing pretreatment to achieve more durable performance.

This investigation by Minford (1355,1356) also demonstrated the significant difference in joint durability that can occur with exposure to different jungle locations. There was sufficient high humidity in either of these open or secluded sites to cause comparable joint strength decline with vapor-degreased–only Alclad 2024-T3 or 6061-T6 joints bonded with the more hot-vapor–sensitive one-part nitrile-modified epoxy adhesive. There was a clear distinction, however, with the corresponding two-part epoxy joints. Those vapor-degreased–only 6061-T6 joints which showed high-to-moderate resistance to high humidity at the secluded site showed a precipitous decline in joint strength in an open jungle site. Minford has offered the explanation that we are experiencing a combination variable temperature and wet/dry cycling condition (sun in and out situation) at the open jungle site which makes it a more severe exposure than continuously humid conditions. By acid deoxidizing the 6061-T6 adherends, however, the relative joint durability is sufficiently improved that no difference can now be distinguished for these joints whether exposed to the open or secluded-type jungle exposure.

We might expect that the joint durability test which should come most close to predicting the service life potential for most general manufactured products would be the industrial atmosphere, since that simulates general exterior weathering. Wegman showed earlier that very little joint degradation was encountered in desert or dry and semiarid weathering found at Yuma, Arizona. We might expect that another recognizable form of atmospheric weathering might be labeled as agricultural, or rural. We would expect higher humidity than the semiarid, but relatively devoid of much of the industrial pollutant fall-out. The predicted overall aggressiveness of various atmospheres would be seacoast > jungle > industrial > farm or rural > desert or semiarid. All of Minford's atmosperhic durability data from 20 yr has clearly established these relative relationships using the whole range of commercial adhesives, including water and solvent contact cements, viscous mastics, 100% solids pastes, and the aerospace films and tapes.

The relationship of surface pretreatment to joint durability in the industrial atmosphere is clearly seen in Table 154. For a two-part room temperature–curing epoxy over 6061-T6 adherends, the 43% joint retention of vapor-degreased–only joints is significantly poorer than for any of the other 11 surface pretreatments, which show an overall average of 93% retention. As mentioned earlier, strength retention of similarly treated adherends with EC-2086, one-part, nitrile-modified, heat-curing epoxy paste is significantly poorer in any direct weathering situation involving water-soaking conditions. For this adhesive, the average joint strength retention for vapor-degreased–only adherend joints has dropped to 10% after 8 yr in the industrial atmospher;e. Among the other 11 pretreating situations, however, the lowest retention was still 59% and the overall average of all was 78%. It would seem that the overall benefit of additional surface treating beyond degreasing is more striking for the one-part

Table 154 Comparative Durability Responses in 8-yr Exposure in the Industrial Atmosphere of 6061-T6 Aluminun Alloy Joints with Varying Surface Pretreatments

Pretreatment	Adhesive	Average joint strength retention		% of Initial strength
		psi	MPa	
Vapor degrease	2-part epoxy	907	6.25	43
Vapor degrease + grit-blast	2-part epoxy	2030	13.99	100
Ridoline 53 (hot alkaline cleaner)	2-part epoxy	2230	15.32	85
Caustic + desmut (hot)	2-part epoxy	2665	18.36	100
Acid paste cleaner	2-part epoxy	2030	13.99	85
Deoxidine 526 (RT) (H_3PO_4 + alcohol)	2-part epoxy	2130	14.68	79
Alum-Surf-Prep (RT Acid Cleaner)	2-part epoxy	2840	19.57	100
HNO_3 + H_2SO_4	2-part epoxy	2360	16.26	94
H_3PO_4 + H_2SO_4	2-part epoxy	2305	15.88	97
CRO_3 + H_2SO_4 (RT)	2-part epoxy	2645	18.22	100
CRO_3 + H_2SO_4 (hot)	2-part epoxy	2370	16.33	84
HCl + $K_2CR_2O_7$ (RT)	2-part epoxy	2520	17.36	96
Vapor degrease	1-part epoxy	450	3.10	10
Belt-Sand + vapor degrease	1-part epoxy	3400	23.43	70
Ridoline 53 (hot alkaline cleaner)	1-part epoxy	4330	29.83	82
Caustic + desmut	1-part epoxy	2715	18.71	59
Acid paste cleaner	1-part epoxy	3670	25.29	79
Belt-sand + acid paste cleaner	1-part epoxy	3730	25.70	76
Deoxidine 526 (RT) (H_3PO_4 − alcohol)	1-part epoxy	4300	29.63	87
Alum-Surf-Prep (RT acid cleaner)	1-part epoxy	4160	28.66	82
HNO_3 + H_2SO_4	1-part epoxy	3470	23.56	68
H_3PO_4 + H_2SO_4	1-part epoxy	2950	20.33	62
CRO_3 + H_2SO_4 (RT)	1-part epoxy	4585	31.59	83
CRO_3 + H_2SO_4	1-part epoxy	4730	32.59	89
HCl + $K_2CR_2O_7$ (RT)	1-part epoxy	4595	31.66	93

Data from J.D. MInford, unpublished work, Aluminum Co. of America.

epoxy adhesive. This would seem to indicate that the expected better wetting potential using the heat-curing paste adhesive epoxy is not as important in establishing the level of joint durability as providing a superior surface to enhance wetting potential.

The most widespread differences in durability performance occur in the seacoast atmosphere, where degradation is due to the combination of washing in salt water accompanied by alternating wet/dry cycling. This can create a corrosion undercutting of the bondline, whose effect is shown in Table 155, where exposures of variously pretreated 6061-T6 joints were exposed for periods up to 8 yr in this type of weathering. The extreme severity of the exposure, even in the presence of many

Table 155 Comparative Durability Responses in 8-Yr Exposure to the Seacoast Atmosphere of 6061-T6 Aluminum Alloy Joints with Varying Surface Pretreatments

Pretreatments	Adhesive	Average joint strength retention: psi	Average joint strength retention: MPa	% of initial strength
Vapor degrease	2-part epoxy	0	0 (80 da)*	0
Vapor degrease + grit-blast	2-part epoxy	1350	9.30	67
Ridoline 53 (hot alkaline clean)	2-part epoxy	0	0 (290 da)	0
Caustic etch + desmut (hot)	2-part epoxy	0	0 (394 da)	0
Acid paste cleaner	2-part epoxy	0	0 (275 da)	0
Deoxidine 526 (RT H_3PO_4 + alcohol)	2-part epoxy	0	0 (270 da)	0
Alum-Surf-Prep (RT acid cleaner)	2-part epoxy	0	0 (580 da)	0
HNO_3 + H_2SO_4 (hot)	2-part epoxy	0	0 (674 da)	0
H_3PO_4 + H_2SO_4 (hot)	2-part epoxy	0	0 (644 da)	0
CrO_3 + H_2SO_4 (hot)	2-part epoxy	0	0 (760 da)	0
CrO_3 + H_2SO_4 (RT)	2-part epoxy	2290	15.78	89
HCl + $K_2Cr_2O_7$ (RT)	2-part epoxy	2560	17.64	97
Vapor degrease	1-part epoxy	0	0 (270 da)	0
Belt-sand + vapor degrease	1-part epoxy	0	0 (760 da)	0
Ridoline 53 (hot alkaline clean)	1-part epoxy	0	0 (390 da)	0
Caustic etch + desmut (hot)	1-part epoxy	0	0 (749 da)	0
Acid paste cleaner	1-part epoxy	0	0 (1440 da)	0
Belt-sand + acid paste cleaner	1-part epoxy	0	0 (1450 da)	0
Deoxidine 526 (RT H_3PO_4 + alcohol)	1-part epoxy	0	0 (1440 da)	0
Alum-Surf-Prep (RT acid cleaner)	1-part epoxy	0	0 (1449 da)	0
HNO_3 + H_2SO_4 (hot)	1-part epoxy	0	0 (768 da)	0
H_3PO_4 + H_2SO_4 (hot)	1-part epoxy	0	0 (1524 da)	0
CrO_3 + H_2SO_4 (hot)	1-part epoxy	0	0 (1460 da)	0
CrO_3 + H_2SO_4 (RT)	1-part epoxy	4300	29.63 (4 yr)	78
HCl + $K_2Cr_2O_7$ (RT)	1-part epoxy	4300	29.63 (4 yr)	86

Data from J.D. Minford, unpublished work, Aluminum Co. of America.

surface pretreatment situations, is dramatically indicated by the predominance of 0% joint strength retention values shown in the last column of Table 155. Only the anodizing and cold chromic acid–deoxidizing pretreatments were able to prevent the undercutting pitting corrosion action on the aluminum interface for the one- and two-part epoxy joints employed. Another exception using the two-part epoxy was the silica gritblast-pretreated surfaces, which has been explained by Minford as being due to a corrosion-inhibiting silica surface film. Minford tested at least 12 tape and film adhesives of the aerospace variety with hot chromic acid–deoxidized pretreatment (see Ref. 531, p. 108) and showed 10 survived more than 4 yr, whereas the two-part paste epoxy joints generally failed in 2–4 yr. This indicated the better flow and wetting achieved with the aerospace film adhesives.

Very recently, Fedor and Brennan (3743) have reported a paper on correlations they obtained between accelerated and natural weathering of sealants.

VI. ADHEREND EFFECTS ON JOINT DURABILITY

A. Choice of Aluminum Alloy Effect

Aluminum alloys have a wide variety of alloying compositions, tensile strengths, and processing conditions that can affect bond joint permanence. In particular, Minford (3593–3595) has shown a relationship between tested joint permanence and the general corrosion resistance of the aluminum alloy. Brewis (3596) has also reviewed the subject, and Kinloch and Smart (663) and Kinloch et al. (781) have demonstrated that durability decreased with increasing concentration of surface magnesium. Although this relationship can be readily confirmed for various alloy joints in water-soaking conditions, it is most dramatically demonstrated in marine or seacoast weathering conditions or laboratory testing conditions like exposures to continuous or intermittent salt-fog or spray salt-water conditions. The higher strength aluminum alloys like 2024, 2036, or 7075, which are used in the manufacture of very high-stress structures, are unfortunately also the most sensitive to the destructive effects of freshwater and salt water. This correlates to their greater sensitivity to corrosive conditions in general. Somewhat less sensitive and lower-strength alloys like 6061 can actually form structural joints with the same adhesive system, which can survive up to four times as long in corrosive weathering conditions. As the number and concentration of other elements in an aluminum alloy are further diminished, the alloys strength is correspondingly decreased but with a consequent increase in the inherent corrosion resistance shown by the alloy. Thus, high-purity aluminum itself has the highest degree of corrosion resistance, which was the original basis for its use in a variety of manufactured products.

The most widely used compromise alloy products for strength and relatively good corrosion resistance are the 3000 and 5000 series of alloys. Because aluminum foil alloys are generally in the category of higher-purity aluminum compositions, they have good resistance to corrosion and, thus, have found wide application for sealing containers and creating vapor barriers over large surface areas. Higher-purity aluminum alloys are also used as the cladding material for higher-strength core alloys in an attempt to extend the service life potential for such alloys. Originally, such clad alloys were routinely recommended for aerospace bonding applications with the idea in mind of helping extend the long-term serviceability. More discussion on this subject will be given later in the section dealing with the effects of cladding on joint durability.

Minford (3595) has investigated the relative joint strength declines and survival times of 2024, 6061, and 5052 alloy joints all fabricated with the same adhesive and surface pretreatment. This testing was deliberately conducted in salt-spray exposure conditions to obtain the most rapid definition of difference. The relative inherent general corrosion resistance of these alloy is in the order 5052, 6061, 2024, and this fact was reflected in the relative survival times of the joints in that same order. While the order of general bond durability in corrosive salt-water exposures has proven to be 5000 series alloy joints, 6000 series alloy joints, 2000 series alloy joints in a number of different Minford investigaitons, one exception has been shown in a 1982 investigation of Minford (3600) comparing the relative durability of 2036-T4 and 6009-T4 joints made with surface contamination in the form of a particular automotive metal-forming lubricant. In this investigation, the durability order apparently reversed with 2036-T4 oiled surface joints, performing distinctly more durably than the corresponding fabricated 6009-T4 joints. While it is not understood why the contaminated surface joints have had a different relative durability response from earlier results, it clearly means that existing durability data based on performances with mill-finish surfaces may require re-evaluation when dealing specifically with lubed surfaces and especially with specifically modified adhesives designed to adhere through such oil or lube contamination layers.

Poole and Watts (3597), in 1985, published on the effect of alloy composition and surface pretreatment on the durability of aluminum joints, whereas Price et al. (3598), in 1991, specifically investigated durability of Al-Li-Cu-Mg alloy joints.

Stifel (3601) has showed a remarkable difference in response of 2024 aluminum in two different heat-treatment forms to FPL deoxidation and subsequent joint durability. The 2024-T3 alloy retained a bright metallic luster after the acidic deoxidation, whereas 2024-T81 acquired a darkened, smutted appearance. Joint durabilities were compared using FM-400 adhesive (without a corrosion-inhibiting primer present) under 1500 psi (10.34 MPa) in 140°F (63.3°C) and 100% RH. Whereas the bright-etched–appearing 2024-T3 joints survived an average of only 14 da, the darkened-appearing 2024-T81 joints averaged 83-da survival. Quite unexpectedly, when the testing temperature was raised to 160°F (71°C), the survival time for the 2024-T3 joints declined only to 12 da; however, the 2024-T81 joints also declined to about the same 12-da survival time from their significantly elevated average survival time of 83 da. An elevation of only 20°F would not be expected to create such a difference of response, with the two alloys essentially reversing their relative durability responses. With a relatively thick application of BR-400 corrison-resistance primer on the etched 2024-T81 surface, the joint survival times were increased dramatically by 275% with a corresponding 15% decline in initial strength. Thinner layers of the inhibiting primer led to lower levels of stressed durability, so the inhibiting primer thickness seems to be playing a critical role in establishing the level of durability.

Minford (3594) obtained more expected differences when evaluating three different possible aluminum candidates for fabrication of aluminum joints for aluminum automotive assemblies, as shown in Table 156. Comparison can be made between the durability performance of the three alloy joints with three different surface pretreatments while being exposed to either of three different weathering conditions; i.e., 85% RH at room temperature, 100% RH at 125°F (52°C), or 5% salt spray at 95°F (35°C). The alloys in this investigation were members of the same aluminum alloy families already investigated as 2024, 6061, and 5052 joints. They should be expected to have the same relative order of general corrosion resistance and relative retention of joint strength after exposure. This conclusion was verified by the results which show relative durability in order of X5085-H111, 6151-T3, 2036-T4.

Table 156 Effect of Surface Treatment[a] on Bonded Joints Using a One-Part, Heat-Cured, Nitrile-Modified Epoxy Adhesive and Three Different Aluminum Alloy Adherends

	Alloy 2036-T4			Alloy 6151-T4			Alloy X5085-H111		
Exposure	A	B	C	A	B	C	A	B	C
Initial strength (psi)	1850	1930	2130	2550	2690	2530	2250	2270	2110
(MPa)	(12.75)	(13.30)	(14.68)	(17.57)	(18.53)	(17.43)	(15.50)	(15.64)	(14.54)
85% RH at	380	313	2200	1350	1550	2410	1860	1890	1910
75°F (23.9°C)	(2.62)	(2.16)	(15.16)	(9.30)	(10.68)	(16.60)	(12.82)	(13.02)	(13.16)
100% RH at	20	0	690	920	1050	1100	1260	1520	1230
125°F (52°C)[b]	(0.14)		(4.75)	(6.34)	(7.23)	(7.58)	(8.68)	(10.47)	(8.48)
5% salt spray at	0	0	1110	80	150	2090	690	530	1210
95°F (35°C)[c]			(7.65)	(0.55)	(1.03)	(14.38)	(4.75)	(3.65)	(8.34)

[a]Surface treatments: A = mill-finish, B = vapor degreased, C = chromate conversion coating.
[b]3-mo exposure period.
[c]3-wk exposure period.
From Ref. 535 (Table 3-2, p. 82).

Kinloch (3602) examined the uppermost atomic layers of three British commercial degreased aluminum alloys for magnesium content using XPS. The highest purity alloy (E1C), nominally containing no magnesium, showed no magnesium on its bonding surface. Alloys NE4 and NE8 (similar to Alcoa alloys 5052 and 5054) showed similar amounts of surface magnesium though the latter had twice the bulk magnesium concentration. When bonded with epoxy adhesive and immersed in water at 140°F (60°C), the E1C (magnesium-free surface) joints retained more than twice the joint strengths shown by the NE4 and NE8 joints after 1500 hr. With different surface pretreatments, the order of magnesium left on the surface was solvent degreased, grit blasted, chromic acid etched, PAA. This appears to say that durability may be expected to be higher as the amount of magnesium on the surface is decreasing. However, since magnesium-free degreased E1C joints were distinctly inferior to that of PAA-treated E1C joints (where some small amount of magnesium showed up), a low level of magnesium cannot be the sole criterion for producing good-durability aluminum adherends. (Additional discussion of the effect of surface magnesium on aluminum joint durability can also be found in Ref. 3603.)

The use of x-ray photoelectron spectroscopy (XPS) to investigate the chemistry of aluminum alloy surfaces has been most recently discussed by Davis (3604,3605). Wagner (3606), as early as 1972, had suggested a method for making the surface analysis of metals more quantitative by applications of appropriate sensitivity factors. However, Seah (665) contended that there were possible systematic errors which would still be operating and present in the XPS analytical procedure itself. Sun et al. (985) used AES to analyze the surface of PAA-anodized aluminum, finding phosphate present on the surface as did Kinloch et al. (781) using XPS analysis. A negative effect of surface copper on the durability of aluminum joints was established by Sun et al. (985), whereas Bijlmer (1008) had earlier shown the poor bonding results with excess quantities of copper in the FPL-type etching bath. McMillan (3607) and Chen et al. (227) both have discussed the negative effect of fluoride ions on aluminum joint durability.

Another way that bond permanence can be affected by alloy selection is based on their variable tensile strength as adherends and special processing conditions that may vary according to their production requirements. The importance of alloy strength is observed by using different strength alloys to fabricate equal thickness lap-shear joints using the same adhesive. With the higher tensile-strength alloys, the adherends are correspondingly more stiff, and the degree of cleavage force necessary to fail the joints in a tensile test machine is increased with a subsequent higher recorded shear strength. Thus, if the aluminum alloy families are in the following order of tensile strength, i.e., 2000 alloys, 6000 alloys, 5000 alloys, 3000 alloys, then the lap-shear joints made from those alloys will be in the same order in regard to their initial strengths. In order to secure a joint of equal lap-shear strength with the same adhesive, it will be necessary to select a thicker sheet of any lower tensile-strength alloy. The presumption is made in this discussion that the two different strength alloys have been pretreated to remove any different surface conditions causing differential wetting of the two different alloy surfaces. For example, Minford (3608) has shown that it was necessary to use 0.625-cm sheet thickness of a high-purity aluminum alloy like 1100 (lower tensile-strength alloy) as compared to 0.317-cm sheet thickness of a medium-strength alloy like 6061 to achieve lap-joint strength equal to that measured for a 0.163-cm thick 2024 (aircraft alloy) adherend. The adhesive can further interact, however, depending on whether it is brittle or quite yielding. The stiffer the adherend, the better the match will be with a brittle adhesive for producing higher lap-shear–strength aluminum joints. Because such a joint would be able to better resist a particular stress level imposed on a structure in service, the apparent survival time would be greater with a higher-strength alloy (under stressful exposure conditions) than with a lower-strength alloy joint under equivalent exposure.

The effect of specimen stiffness on joint durability when exposed to water soaking and occasional stressing was shown by Minford (1038) using the same 3003-H14 aluminum alloy in two different adherend thicknesses. Lap-joints were fabricated with 0.163- and 0.317-cm 3003-H14 sheet which had been chromate conversion coated as a pretreatment. A relatively rigid, two-part, room temperature–curing epoxy adhesive was used to fabricate the lap-shear joints. Periodically, the joints were removed from a hot humidity cabinet and stressed up to 50% of the strength of the unexposed control joints. After some modest exposure time, the thinner adherend joints failed in stressing at only 50% of original strength, whereas the thicker adherend (same alloy joints) did not fail throughout the 1-yr planned exposure time. The author's explanation was that the higher degree of flexing experienced in the periodic stressing of the thinner adherend joints created a fatigue response in the thick conversion coating oxide layer. Meanwhile, the thicker adherend joints did not permit enough crack damage to develop to institute sudden failure under stressing at up to 50% of initial joint strength.

The effect of manufacturing and processing on the aluminum alloys can also be significant. Of particular effect can be the variation in oxide build-up, which will result from the special heat treatments which are necessary for highest-strength development in certain alloys. The more commonly employed 3000 and 5000 series aluminum alloys used in general manufacturing are usually hot rolled to final sheet or plate thickness. In contrast, the 2000, 6000, and 7000 series higher-strength alloys must be soaked for some considerable times in high-temperature furnaces. As a result, the former alloys have significantly lesser films of oxide present on their surfaces, and durable bonding can be achieved with less surface pretreating. Thus, in commercial bonding applications where mill-finish, solvent, or vapor-degreased pretreatments are all that can be economically afforded, it will be more practical to employ aluminum alloys which do not have the heat-treat films. These alloys often may lack sufficient strength, however, to produce the level of structural requirement of a particular designed product. The aluminum alloys that require extensive heat treating also happen to be alloys which have more extensive alloying compositions. The 2000 series of aircraft-strength alloys, for example, have significant additions of copper to pure aluminum, and even higher than bulk percentages of copper may end up resident on the surfaces after heat treating. The dramatic effect of acid etching on these heat-treat film alloys can easily be followed visually. The dark and nonuniform appearance of the heat-treat oxide is replaced with the bright and uniform appearance of the higher-purity aluminum alloys. Only when these thick and complex oxides of the heat-treated alloys are thoroughly removed can the adherend surface be effectively wetted and durable joints produced. This dark oxide surface can be practically removed by uniform mechanical abrasion processing to provide a higher order of joint durability in moderate weathering. Minford (532–536,771–775) has provided results from many different investigations over a 20-yr period to demonstrate the significant improvement in bond durability provided by abrading away the nonuniform "as recieved" oxide. The improvement in service life extension provided by acidic deoxidizing and deoxidation followed by development of new conversion coatings and anodized coatings is even more highly significant.

B. Comparison of Clad Versus Bare Aluminum Alloy Adherends

A final aluminum choice that has been shown to affect joint durability has been whether a clad or bare aluminum alloy has been chosen as the adherend. Clad aluminum alloys were developed at Alcoa with the intent to improve the corrosion resistance of bare aluminum alloys without adhesive bonding being a considered factor. After their development, the data were easily obtained to establish their benefit in extending the resistance of bare aluminum to all manner of aggressive weathering. Their employment as bondable adherends for aircraft followed. The principle of using a metallurgically bonded sandwich of two different aluminum alloys in a manner that one thinner layer will be electrochemically sacrificing to the thicker layer has been universally accepted and proven amazingly effective for extending service life potential for freely exposed aluminum surfaces to corrosive weathering. The beneficial nature of this cladding is usually due to the spreading of a relatively shallow pitting corrosion attack over the total surface of the cladding layer, whereas attack of the less corrosion-resistant core alloy is retarded until the cladding layer is electrochemically sacrificed.

If the cladding is operating satisfactorily, as described above, and the cladding actually constitutes the adherend surface in contact with adhesive in the bondline, it would seem plausible to expect this built-in sacrificial anode might be a negative factor in retaining long-term adhesion in the bondline. This would especially be true where corrosion-inducing factors like water and/or chloride ions of salt-water infiltrate the bondline. In practice, however, aluminum-bonded joints incorporating FPL-deoxidized, clad-aluminum surfaces were employed in commercial and military aircraft for many years following World War II without expressed concern about the specific use of Alclad as compared with nonclad alloys. While corrosion was being encountered in aluminum aircraft joints under severe weathering exposure conditions after long service, the degree and catastrophic nature of such debonding only became alarming with some joint design changes during the Vietnam War. Apparently the demand of the military for newer aircraft designs coincided with moving the joint design closer to peripheral surfaces, thus exposing the joint edges (and ultimately the interfaces) to direct electrochemical attack by the seawater conditions always present on aircraft carriers and at coastal aircraft stations. Riel (2844,2845), Greer (2846), and Rogers (2847,2848) all issued investigative reports in the period from 1968 to 1973 which concluded that the significant corrosion failures being experienced in naval carrier and seacoast air station aircraft in Vietnam might be attributed to the use of clad alloys. They recommended that corresponding bare alloys be used. Data as shown in Fig. 36 seemed to confirm their concern. Minford, from past Alcoa experiences, offered the view that the apparently altered and unacceptable performance of these joints might not be totally due to the use of the Alclad aluminum, but the change in joint design permitting freer access of corrosive elements into the area of the bondline. Data available, but not published by the Alcoa Labs, had earlier indicated an extreme sensitivity of FPL or Alcoa A3 pretreated joints to the seacoast-exposure conditions. Unfortunately, it was more recognizable with newer joint designs. Now that both American and European aircraft are being manufactured with anodized surface preparations and better joint sealing, the negative effect of cladding is rarely reported.

Actually, the European aircraft manufacturers never adopted any real condemnation of the use of Alclad alloys. Cotter (851), in 1977, reported that durable joints oculd be obtained in the European experience with clad alloys providing (1) correct pretreatment was provided (CAA anodizing in this case), (2) a corrosion-inhibiting primer was used, (3) a suitable adhesive chosen, (4) the bondlines are sealed, and (5) mechanical fasteners are avoided through the bondlines. Albericci, similarly speaking from European experience, suggests that some difference may be anticipated based on the different response of clad versus bare alloys to the acidic deoxidation. The clad layer of higher-purity

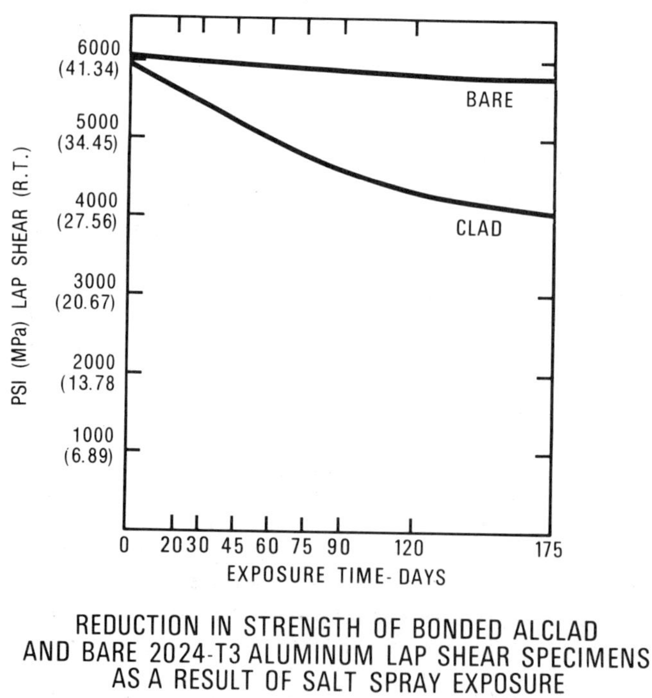

Figure 36 Durability testing of the effect of cladding in aluminum alloys. (From J.D. Minford, Aluminum Adhesive Bond Permanence, Treatise on Adhesion and Adhesives (R.L. Patrick, ed.). Marcel Dekker, New York, Vol. 5, Fig. 3-32, p. 92 [1981].)

aluminum will etch much more slowly and probably with a different surface morphology presented to the adhesive. Of course, when proceeding to anodizing as a next step in the surface pretreatment, this argument has less significance. However, the argument has complete validity based on direct bonding to the deoxidized surface and bonding directly to that surface.

McFarlen (2849) was among a number of investigators who tried to obtain more data to resolve the controversey. He reported his extensive comparative durability study of clad and bare aluminum adherends and found that no significant differences in durability could be found. The "optimized FPL" etch pretreatment was used, so surface pretreatment cannot be the sole explanation for these different results from those shown earlier by Riel et al. After carefully studying the McFarlen data, the author can only conclude that the combination of primer and adhesive must have offered superior wetting at the cladding interface, which reduced the amount of interfacial corrosion of the cladding on the adherend surfaces. Thus, it might be possible to find no significant difference between the durability performance of clad versus bare joints. It is possible that there are a number of bonding variables that can be highly influential in this matter; i.e., surface treatment, adhesive type, curing conditions, joint design, and protective measures taken to screen the joint interface from the environment.

It should also be pointed out that there are two different commercial approaches to furnishing protection by a cladding. Alclad 2024 aircraft alloy is one example where the cladding chosen is actually a high-purity form of aluminum (1230 alloy), which can offer electrochemical protection to the 2024 core, since 2024 alloy is strongly cathodic with its high concentration of copper. In the case of Alclad 7075 aircraft alloy with a high concentration of zinc and a lesser amount of copper, a different cladding alloy is employed which contains 1% zinc (7072 alloy). Many investigators using the anodized surface pretreatments now agree that whether the joints are fabricated with Alclad or bare 2024-T3 alloys is of minor or questionable difference. However, the durability evaluation of Alclad and bare 7075 joints (with the sacrificial zinc-type cladding) could still be showing some negative effect on durability with the clad-type alloy joint as compared to the joints fabricated with the bare 7075 alloy surface.

VII. EFFECT OF ADHESIVE CHOICE ON JOINT DURABILITY

We have already described the wide range of adhesive products that can be considered for bonding aluminum joints in an earlier section. It remains, however, for the different adhesives to be discriminated among in regard to the durability potential of their joints with different alloys.

A. Nonstructural-type Aluminum Joints

A nonstructural aluminum joint is one that is not intended to resist any significant stress under intended service conditions. Included in this category could be the situations where aluminum foil has been bonded as a decorative surface, a vapor barrier, or between two adherent surfaces. Stresses can exist, even here, when the coefficient of thermal expansion of the foil is a mismatch with the attached adherend and the overall product is subjected to thermal cycling conditions. It might also be possible for some stress to develop owing to the vapor barrier acting to restrict the loss of liquid or vapor. Usually, relatively low-strength and nonrigid adhesives can be used for bonding aluminum foil, since large bonded areas are mostly employed. Such adhesives can offer some stress relief when necessary. A preponderance of nonstructural adhesives are used for bonding aluminum overall in general manufacturing both as to square footage and quantity of adhesive.

1. PRESSURE-SENSITIVE ADHESIVE ALUMINUM JOINTS

Pressure-sensitive film-type adhesives are often used to attach aluminum foil, since such adhesives make it relatively easy to bond large expanses of thin aluminum foil in fabricating laminated products. The permanently elastic properties of a pressure-sensitive adhesive can act as a stress-relief interlayer in products where cycling service temperatures are present. The same properties may not be acceptable, however, where the thermal stress is steady and continuous creeping to failure might result. Under conditions where intermittent weathering includes the presence of water, as in exterior weathering service, the gradual encroachment of water into the bondline may be the main driving force for eventual failure. In more recent years, pressure-sensitive film adhesives have been developed, which border on being load-bearing. They may even be used to bond broad expanses of overlap trailer or mobile home sheet in vehicle construction. Minford (1392) has conducted extensive testing of lap-joints, which might be representative of the various surfaces and adhesive products used in such applications with good bond-durability results. Such fabricated joints could resist continuous soaking in 100% RH at 125°F (52°C) for periods up to 2 yr with little evidence of water degradation in the bondline. This is a very creditable performance, especially considering that no special surface preparations beyond using clean aluminum needed to be considered. Also, immediate handling strength in the bondline can be anticipated in contrast to the curing times required for contact cements or epoxies. It has also been shown by Minford (1394) that semistructural pressure-sensitive adhesives (some of the newer acrylic formulations like 3M's Isotac products) can actually resist some low levels of sustained stress (such as supporting the weight of an aluminum adherend on a vertical surface) while being continuously exposed to hot humidity weathering.

The use of pressure-sensitive adhesives for attaching aluminum trim, decorative patterns, or identification logos consumes a very large volume of adhesive, and this application is part of the manufacturing of the widest possible range of industrial products. Minford (1392,1394) has evaluated bond durability of both pressure-sensitive films and transfer tapes for bonding foil or thin sheet in the "as received," solvent cleaned, painted, conversion coated, or anodized forms. While the relative durability is generally in the same order of finish as mentioned above, the biggest difference is almost always between the "as received" surface joints and the rest as a group. Foil products are usually received by the customer in a condition that can be attached by pressure-sensitive adhesives with satisfactory resistance to many service conditions. Thin-sheet aluminum, however, may require special cleaning if it has been pretreated by the manufacturer to resist or prevent corrosion under plant storage conditions.

Although not involved in structural joints as just discussed, a considerable volume of pressure-sensitive adhesives are used to coat plastic films which are then applied as protective films over painted or anodized aluminum surfaces. The purpose is to offer some protection against weathering in storage and after field erection. To offer such protection, the adhesive needs to maintain perfect adhesion through any time period demanded by the user. When required, there is an equal need to be able to peel away the protection without blemishing the surface underneath. Minford (1393) found that a wide range of performances was to be found in the existing commercial products which needed to be discriminated by conducting exposure tests in laboratory weather-o-meters or condensing hot humidity cabinets for up to 6 mo or direct exposure in a natural atmosphere for periods up to 1 yr. Special peel test specimens had to be devised for making direct comparisons. In the atmospheric exposures (probably involving the direct effect of sun radiation), some pressure-sensitive coated films could actually increase the degree of adhesion, prventing any easy peeling away of the protective film from the treated aluminum surface. The author had to deal with many customers complaints that under a combination of factors operating in the natural atmosphere these products would move from easy to peel protective films to virtually impossible to remove protective films. Several years of testing of many proprietary protective films over bare, painted, and anodized aluminum led the author to conclude that none could withstand extended weathering and ultraviolet light. The author summarized protective film performance as falling in separate categories such as those offering (1) good protection and with good potential for removal without ultraviolet light for periods up to 1 yr (warehouse storage type), (2) protection only in the absence of high humidity and ultraviolet light for up to 2 yr exposure, and (3) protection with both high humidity and ultraviolet light for at least 6 mo (outdoor exposure type).

2. HOT-MELT ADHESIVE ALUMINUM JOINTS

Originally, the use of hot-melt adhesives for nonstructural bonds was based on the selection of very low-cost raw materials for manufacturing the adhesives. The primary adherends bonded with early hot-melts tended to be lower-cost and -strength materials like cardboard and paper products, where all the strength required was that necessary to exceed the tensile strength of the wood surface. Today, the state of the art hot-melt technology is such that adhesive strengths up to that required for the structural bonding of aluminum can be formulated. Strictly speaking, we might even consider that the highest temperature-resistant adhesives used to bond aluminum aerospace structures are more like highly sophisticated hot-melts in the sense that the adhesive must be flowed in the bondline by very high-temperature melting.

Minford (1381) has investigated the long-term durability under aggressive weathering conditions of structural hot-melt adhesives used for affecting repairs on aluminum heat-exchanger structures. Since hot-melt adhesives are totally thermoplastic polymers, their highest strengths are only developed after the bondline cools to room temperature. At the same time, the maximum wetting and flow over the adherend surface can only be achieved when the hot-melt polymer is at the highest melt temperature. To satisfy both these conditions requires careful planning and control of the total bonding operation. The question of achieving maximum wetting (for best durability performance) must take the high thermal conductivity of aluminum into account. If the melt polymer is cast against an unheated aluminum surface, there is the distinct possiblity that the heat will be so rapidly abstracted from the layer in immediate contact with the aluminum that the degree of wetting will be suspect. For this reason, preheating of any aluminum adherends is recommended when using hot-melt adhesive joining. Proper fixturing of the joint to prevent movement in the bondline while the adhesive is cooling has to be another factor that must be carefully controlled if durable bondlines are to be achieved.

3. RUBBER MASTIC CEMENT ALUMINUM JOINTS

Certain rubber mastic cements might be considered as bonderline cases for fabricating nonstructural aluminum joints. In particular, we refer to adhesive formulations based on reclaimed or GRS-type rubber elastomers, as have been commonly used to attach aluminum-facing sheets to wood or wood derivative–type building materials for prefabricated housing sidewalls. The requirement for adhesive strength is mainly that required to hold the weight of the aluminum in vertical alignment against the stiffer and thicker wood back-up members. In most cases, nails may be employed as randomly spaced mechanical fasteners, with the adhesive needed to hold the facing flat in the other areas. For additional facing stiffness, the aluminum sheet will often be corrugated. The corrugation has the additional benefit of permitting thermal movement of the facing under solar heating conditions via an accordionlike movement. The adhesive is applied to the flat areas only, since these will be the only areas of direct contact with the structural members. Because there are significant open gap areas on either side of the flats, water in vapor or liquid form from the surroundings can readily enter the bondlines and cause premature delamination if good wetting by the adhesive has not been achieved. It is suggested that the manufacturer conduct specific durability testing using specimens representative of any prospective adhesive candidate and use the precise manufacturing procedures contemplated. At Alcoa, it was assumed that such testing would not only include soaking tests in hot humidity conditions but also some sort of soaking, freezing, and drying out cycling conditions. In an attempt to find most acceptable candidates for bonding aluminum-to-wood structures for prefabricated homes, a wide variety of commercially available rubber mastic cements were screened at Alcoa using exposure to 100% RH at 125°F (52°C) conditions for periods up to 6 mo as a standard test procedure (3609).

B. Structural Aluminum Joints

The adhesive bonding of aluminum for aircraft, automotive components, consumer appliances, process industry structures, and sporting equipment would all likely fall into the structural joint category. Aluminum building products may or may not fall into this category depending on their designed use in the building.

1. RUBBER-PHENOLIC MASTIC ALUMINUM JOINTS

This category of adhesive (especially neoprene-phenolics) has been widely employed to bond aluminum curtainwall structures. The typical panel may consist of exterior facing/back-up stiffening board/plastic foam insulation/interior facing. Quite often the panels are made symmetrical in design to better balance their response to thermal cycling conditions. The best description of details about the durability testing of such panels can be found in the publications of Minford and co-workers (87–89) dealing with the development of the Alcoa Alply building panel. It should be of interest in regard to durability evaluation that Alcoa Labs evaluated over 200 commercial adhesives in this chemical classification over the development years of the Alply panel. Although all were deemed worthy of consideration by their respective formulators, less than a dozen were entirely satisfactory based on the total battery of Alcoa testing procedures. Another important aspect of evaluating such adhesives is consideration of their effect on the differing materials of construction that can be used in architectural panels. The overall panel durability is not only dependent on the ability of the adhesive to wet and bond to the aluminum but also to all other components it contacts. The range of other materials that could be part of an Alply panel included plywood, fiberboards, cement asbestos board, gypsum board, polystyrene beadboard, styrofoam, polyurethane foam, foamglass, foamed vinyl, particle board, paper-honeycomb core, and inorganic aggregate boards. Each change in combination of materials for a panel should be separately evaluated to show long-term durability potential. The most discriminating test for such contact adhesives is that they are sufficiently strong and heat resistant at the maximum exterior skin temperature in service. The procedures for determining this have been discussed in some detail in Chapter 9.

There are three factors that need to be especially considered when conducting tests on large laminated aluminum-faced panels in regard to their sensitivity to delamination under solar heating conditions: (1) test panels of maximum overall size to evaluate highest service stress levels, (2) test panels with color facing that will absorb highest thermal energy (Alcoa used flat black painted aluminum facings), and (3) use the same physical attachment system for mounting panel to frame that will be employed in field erection. It should be remembered that the use of smaller panel sizes, lighter paint colors, or a method of attachment that permits greater panel bowing with solar heating could underestimate what conditions the adhesive must pass under the worst set of service conditions. The difficulties associated with passing such rigorous tests was the basis for only being able to find a few contact adhesives of 200 tested that could be used in production. The same adhesives were also tested for their long-term weathering resistance by exposing portions to continuous soaking in 100% RH at 125°F (52°C) for a minimum of 6 mo. Production panels were also exposed in an experimental building to natural industrial weathering conditions for indefinitely long periods of time. Periodically panels were removed and evaluated in simulated thermal stress and tensile testing of the bondline. Other duplicate panels were exposed to a wet/freeze/thaw cycling exposure as the common wall between two chambers whose temperature and humidity could be exactly controlled. Finally, quality control test procedures were developed which predicted for each lot of production panels whether the same quality of bonding had been achieved that was equivalent to those in laboratory panels which had passed all the contrived longer-time exposure testing. The details of how these quality control tests were conducted has been described earlier in Chapter 9.

2. THERMOSET STRUCTURAL ADHESIVES

The range of adhesive formulations that might qualify under this general classification would be too numerous to describe completely. They can exist as solids, liquids, or spreadable pastes. The solid forms can be further broken down into film types (unsupported), tape adhesives (supported), and solid powders or preforms. The film and tape types have been particularly important in aerospace manufacturing, since they permit excellent control of the thickness of adhesive in the bondline. Also, they guarantee uniformity of coverage in the interfacial area. In general manufacturing, economics dictate widespread use of so-called "100% solids" adhesive paste formulations. This group may subsequently be subdivided into two-component adhesives that cure at room temperature and one-part paste adhesives that cure only at elevated temperatures.

The evaluation of the long-term durability of a wide variety of these thermoset adhesives on aluminum adherends was undertaken by Minford in the 1960s and was published in 1973 (3610) and in 1983 (3611). The attempt was made to conduct 2- 8-yr laboratory and natural weathering evaluations of commercial adhesives representative of two-part polyamide or amine-cured epoxies with a variety of different fillers, dicy and nondicy curing one-part epoxies in paste, rod, or powder form, and nitrile-modified epoxy pastes. Also undertaken was the evaluation of thermoset film and tape form adhesives, including modified nitrile-phenolic unsupported film, vinyl-butyral elastomer-modified phenolic tape, nitrile-phenolic and modified epoxy two-sided adhesive-coated tape, phenolic-nitrile unsupported tape, high-temperature–resistant elastomer-modified epoxy film, thermoplastic-modified nitrile-phenolic unsupported film, vinyl-phenolic film, nylon-modified unsupported epoxy film, and other forms of modified-epoxy unsupported films (3612,3613).

The testing procedures included continuous immersion in room temperature water, continuous soaking in 100% RH at 125°F (52°C), or exposure to a hot-water–soak/freeze/hot-air–dry cycle up to 2 yr. Also, exposure to the natural industrial or seacoast weathering for 4 yr, which was later extended to 8 yr. Examples taken from a later Minford publication (3614) shows the dramatic effect of adhesive choice on the long-term durability of aluminum structural joints exposed to either the aggressive exposure conditions of a hot-water–soak/freeze/hot-air–dry cycle (Fig. 37) or the corrosive seacoast atmosphere in Figure 38. In Figure 37, some interesting variations in performance can be seen which need to be discussed. No relationship can be observed between the initial level of joint strength and the relative joint durability, since the one-part, nitrile-modified epoxy joints with the highest strength lost up to 73% of that strength within the first 90 da of the exposure. In contrast,

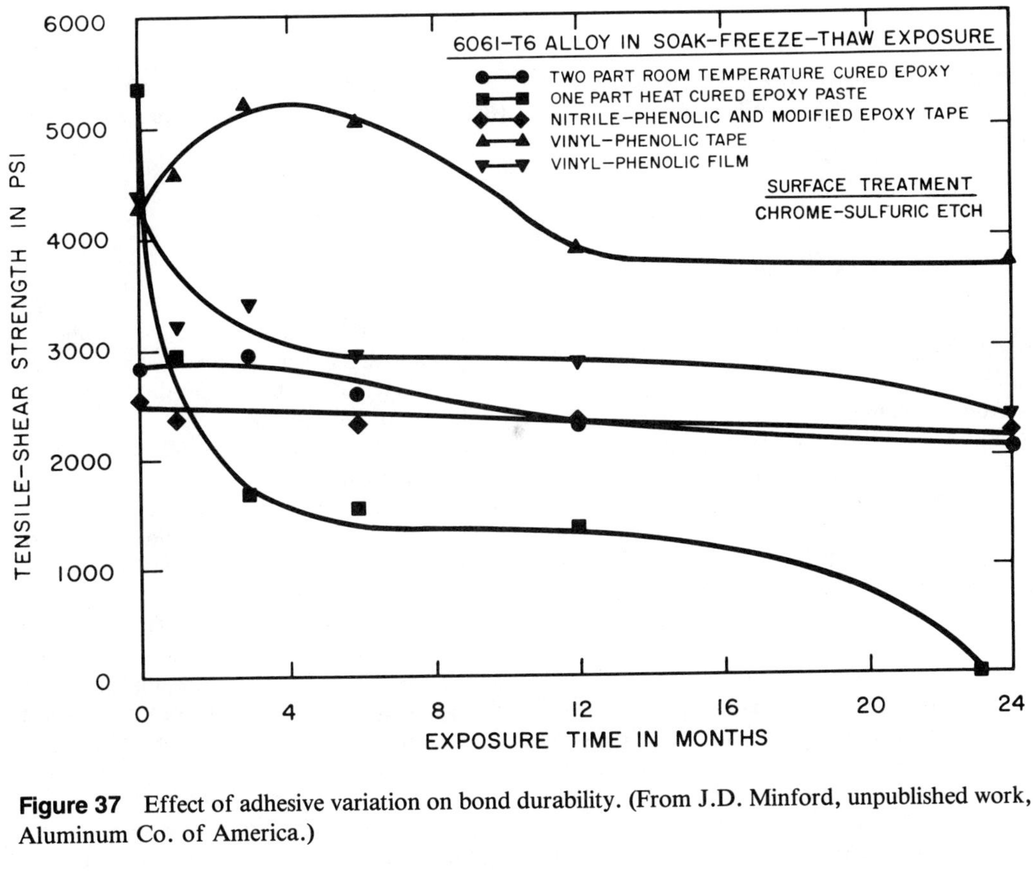

Figure 37 Effect of adhesive variation on bond durability. (From J.D. Minford, unpublished work, Aluminum Co. of America.)

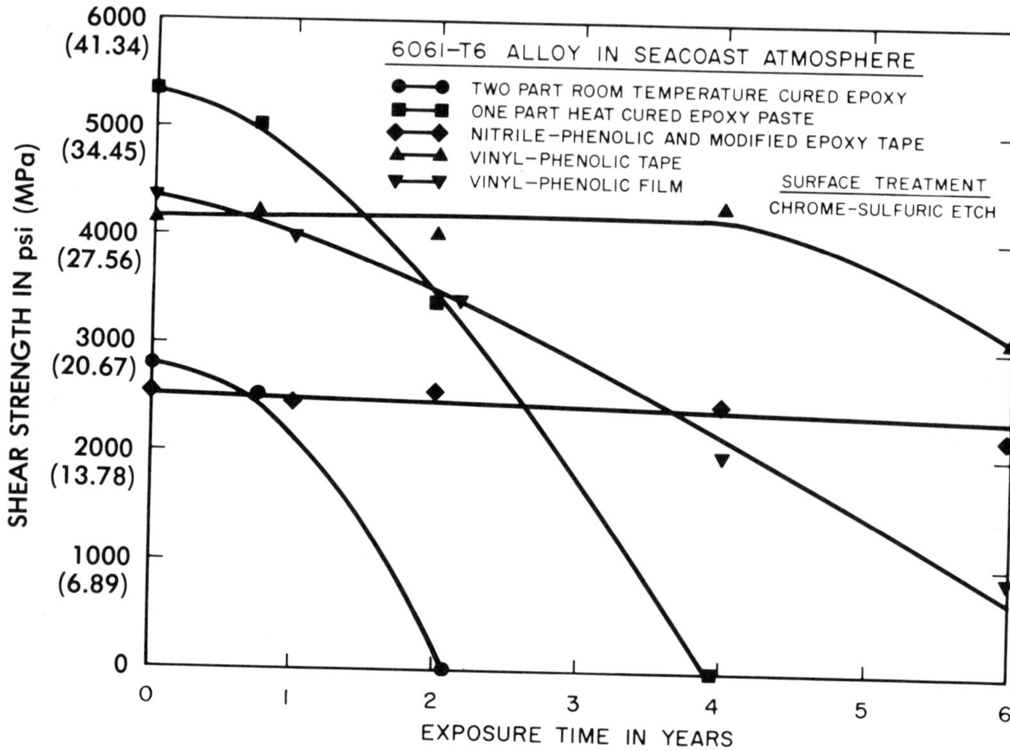

Figure 38 Effect of adhesive on bond durability. (From J.D. Minford, Aluminum Adhesive Bond Permanence, Treatise on Adhesion and Adhesives (R.L. Patrick, ed.). Marcel Dekker, New York, Vol. 5.)

the two-part, polyamide-cured epoxy joints start with only 50% of that strength, yet show little or no decline during a subsequent 2-yr exposure time. Earlier in Chapter 8 (joint decay in water exposure conditions), aluminum/epoxy joint strength decay curves in water soaking were shown where a rapid initial loss in joint strength was followed by a lower-strength plateau within 60 da (see Fig. 19). In Figure 37 (wet/freeze/thaw cycling), we note a variety of different responses with different modified epoxy/aluminum joints using the same type of aluminum-pretreated surface and exposure time extended to 2 yr. The two-part epoxy joints show no initial significant strength decline but a plateau throughout the total exposure time. As indicated above, the one-part, nitrile-modified epoxy paste adhesive joints show the dramatic initial decline in joint strength for about 90 da followed by a plateau effect for the next 270 da. Finally, there is a slow decline in strength till total delamination failure after about 23 mo. It is the vinyl-phenolic film/aluminum joints which show the typical decay to a plateau performance shown earlier in Figure 19. The only other reaction possible (an actual initial rise in joint strength) is shown by the heat-cured vinyl-phenolic tape commercial adhesive. After 4 mo of rising joint strength, these joints then repeated the earlier pattern (see Fig. 19) of joint strength decay for 8 mo followed by an apparent plateau response. It would seem an appropriate conclusion to state that literally any kind of joint strength response that can be imagined was found in these tests depending on the combination of adhesive, adherend, surface pretreatment, and type of environmental test conditions.

An even wider joint-durability response using the same adhesives is shown in Figure 38, where the added effect of corrosive salt water was a factor. It is the two-part epoxy joints that are now most vulnerable to this kind of weathering and all these had debonded by about 2 yr. While starting at twice the initial strength, the one-part epoxy paste joints decline at about the same rate but require twice the time (4 yr) to completely debond. The vinyl-phenolic film adhesive joints barely survive 6-yr exposure, whereas the vinyl-phenolic tape adhesive joints show only a modest 25% decline over the period of 4–6 yr of exposure. It is interesting to note that the vinyl-phenolic tape formulation performs the most durably regardless of whether water soaking with thermal stress cycling or corrosive salt-water deterioration is involved. Meanwhile, the two different kinds of epoxy formulations flip-flopped in their performances. The preferred adhesive, however, might be the combination nitrile-phenolic/modified epoxy duplex tape aluminum joints whose joint strength was unaffected throughout the total 6-yr exposure period.

a. Phenolic Resins in Structural Adhesives

Although two types of phenolic resins exist (resol and Novolac), structural adhesives usually contain the alkylated (methylated or butylated methyol group) resol variety. Such resins are compatible with elastomers like nitrile rubber, vinyl thermoplastic resins, and epoxies which form separate families of structural adhesives used for bonding aluminum adherends (see Refs. 3615–3618).

(1) NITRILE-PHENOLICS. These adhesives were among the earliest used to durably bond aluminum aerospace structures but had the limitations of lower bonding strength ranges and a requirement for controlled higher pressure on the bondline than necessary for bonding the epoxy structural formulations. For many years, this class of adhesives accounted for the largest poundage volume of tape, film, or solution structural adhesives. As can be seen in the long-term durability testing of Minford (see Ref. 3613, Figs. 20 and 21), nitrile-phenolic bonded 6061-T6 joints retained essentially all their initial strength after 2 yr of continuous immersion or soaking at room temperature or hot condensing humidity conditions. In Figures 39 and 40, we can see the relative durability of nitrile-

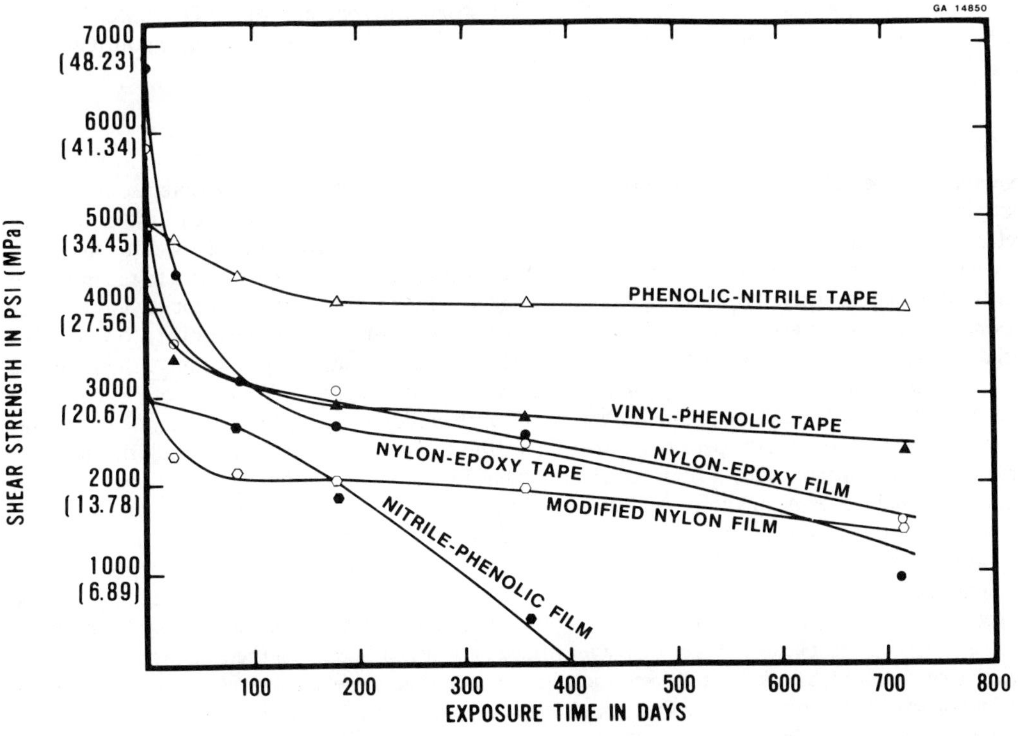

Figure 39 Tape and film structural adhesives—water immersion cycle. (From J.D. Minford, Adhesives, Durability of Structural Adhesives (A.J. Kinloch, ed.). Offl. Sci. Publ., London, Chap. 4, Fig. 22, p. 180 [1983].)

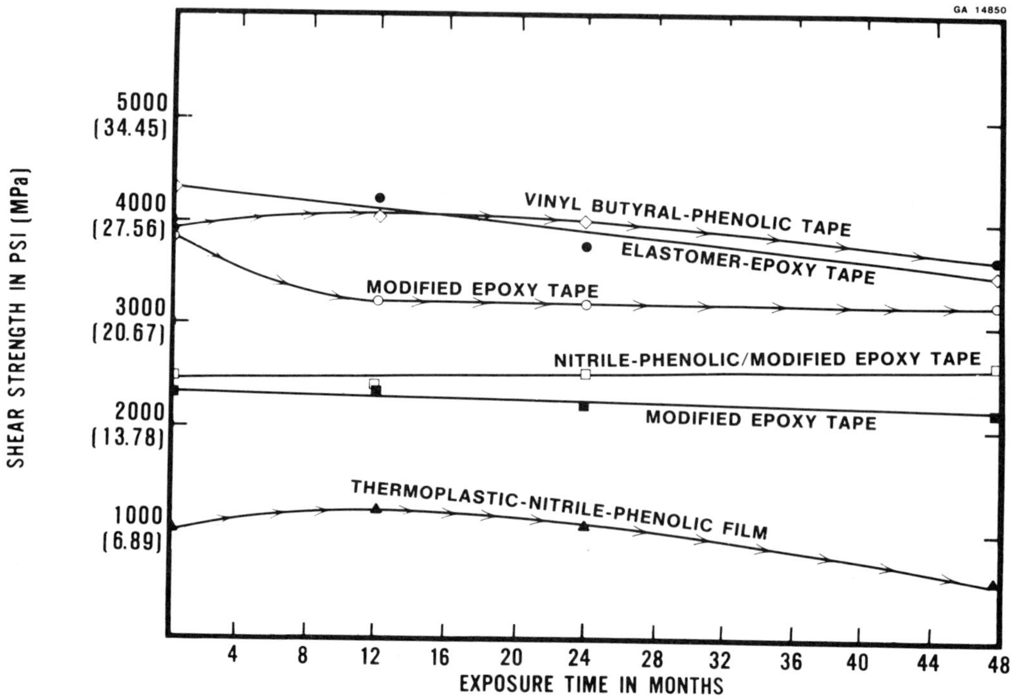

Figure 40 Tape and film structural adhesives—industrial atmosphere. (From J.D. Minford, Adhesives, Durability of Structural Adhesives (A.J. Kinloch, ed.). Offl. Sci. Publ., London, Chap. 4, Fig. 22, p. 180 [1983].)

phenolic-bonded aluminum joints during a 2-yr exposure to an aggressive hot-water-soak/freeze/hot-air-dry cycle or 4 yr in the industrial atmosphere. We might expect to find excellent joint strength retention of what we have chosen to designate nitrile-phenolic adhesive joints. However, one adhesive with that terminology attached to it has also performed the poorest in the group tested in Figure 45. It must be observed that this was a thermoplastic film version. Included are various nylon-type adhesives previously characterized by De Lollis (197) as adhesives with notorious sensitivity to water. We need to conclude that nylon-modified adhesives of more recent vintage seem to have the potential to resist water soaking to a much greater degree than hitherto reported. We also note in Figure 39 that an adhesive classified by the manufacturer as a phenolic-nitrile tape has by far the best durability response. This should focus our attention on the fact again that without testing the adhesive thoroughly, we cannot make a prior assumption that by calling an adhesive a member of a certain chemical family does not assure performance similar to all other commercial adhesives with that same family name. In this instance, however, the fact that the adhesive formulator chooses to use the word *phenolic* as the first part of his generic description indicates that phenolic is the more major constitutent. Hence, the water resistance of this adhesive in its final form might be expected to be superior. In contrast, the other formulator using the terminology on the data sheet of "nitrile-phenolic" may also be correct, because this adhesive may have been formulated with the emphasis on developing the best possible flexibility properties in the adhesive with the elastomer constituent as the primary portion of the composition. This could mean a greater sensitivity of the bondline to failure in water. In contrast, in Figure 40, in a milder exposure environment, even the thermoplastic-nitrile-phenolic film adhesive joints, declined only about 40% after a 4-yr exposure. The performance in a corrosive seacoast atmosphere, previously shown in Figure 38, demonstrated that the nitrile-phenolic/modified epoxy duplex tape-bonded aluminum joints were unaffected in exposure for 6 yr.

Minford has exposed the same phenolic-nitrile tape and thermoplastic nitrile-phenolic film adhesive joints shown in Figures 39 and 40 to a seacoast atmosphere for 4 yr with excellent retention of joint strength by both. This supports the contention that nitrile-phenolic adhesives have excellent wetting potential for a deoxidized aluminum surface, creating a high degree of resistance to undercutting corrosion of the bondline in the presence of chloride ion in salt water. It needs to be noted that the common aluminum surface pretreatment for all these 6061-T6 alloy joints was the same hot chromic-sulfuric acid etch solution that leads to a high susceptibility to undercutting corrosion of the bondline when two-part, room temperature–curing epoxies are used for bonding. In support of this statement, notice the data of Minford plotted in Figure 38, where all the comparable joints bonded with a two-part epoxy adhesive all failed within 2 yr. The importance of surface pretreatment is shown in the Minford data of Figure 41, where this same two-part, room temperature–curing epoxy was bonded to 6061-T6 adherends pretreated by blasting with silica grits, developing a chromate conversion coating, or undergoing a sulfuric acid anodizing (SAA). While vapor-degreased–only 6061-T6 joints failed within 3 mo, the other three pretreatment procedures all permitted significant joint strength retention for 4 yr.

The ability of solvent-diluted nitrile-phenolic contact cement adhesives to prevent undercutting corrosion of aluminum adherends in corrosive salt-water exposures has prompted their use as primers for increasing the long-term durability potential of aluminum aerospace structural joints. A number of investigators have reported a distinct adhesion benefit was achieved when organic additives containing carboxyl groups were added to nitrile-phenolic adhesive combinations when bonding aluminum surfaces (1342–1345).

Bethune (1074) has published Boeing data showing satisfactory service of nitrile-phenolic adhesives in aircraft during more than 25,000 hours of flying. By contrast, room temperature–curing epoxy joints of the same vintage had failed in a few thousand hours, and even 121°C (250°C) curing modified epoxies had shown sporadic disbonds under similar service conditions. All these adhesives had met the U.S. Military Spec. MMM-A-132 which was required for aerospace use at that time. Wolfe and co-workers (1496) demonstrated superior fatigue resistance of nitrile-phenolic adhesives as compared to a nitrile-modified epoxy for structural joints in aerospace. It was their conclusion that the lower modulus of elasticity and higher strain-to-fracture qualities of the nitrile-phenolic should be expected to withstand better oscillating peel stresses.

Bodnar and Wegman (1495) evaluated AF-30 aerospace nitrile-phenolic adhesive in aluminum joints in Yuma, Arizona, Picatinny Arsenal at Dover, New Jersey, and in the jungles of Panama with excellent durability response.

(2) EPOXY-PHENOLICS. As mentioned earlier under the section of adhesive selection, the epoxy-phenolic family of structural adhesives were among the first developed when higher operating temperatures became the operating requirements for 1950s vintage aircraft. At that time, the polyvinylbutyral- or polyvinylformal-modified phenolics used for highest-temperature service would begin decomposing with long-time thermal exposures above 250°F (112°C). The substitution of the epoxy backbone polymer offered a major improvement in thermal stability but with an expected sacrifice in flexibility, elongation, and peel properties. Bodnar and Wegman were among the earliest to publish durability tests with HT-424 epoxy-phenolic aluminum joints after 3-yr exposure under the three atmospheric conditions mentioned in the previous paragraph (1495). Their report that the joints sustained higher losses than found when a simple unfilled polyamide-epoxy–like Epon 828/ Versamid 140 was used under identical conditions was not encouraging. Of particular concern was a 22% joint strength loss in the desert exposure, where the polyamide/epoxy two-part joints suffered no loss. This poor performance was in contrast to the durability test data secured by Minford (1626) at the Alcoa Labs where Epon 422 (epoxy-phenolic) Alclad 2024-T3 aluminum joints (chromic/sulfuric acid deoxidized) retained as much as 80% of their initial strength values after 3-yr continuous soaking in 145°F (63°C) water. The good resistance of either HT-424 or Epon 422 epoxy-phenolic aluminum joints to high temperature was reported by Kuno (1621).

(3) VINYL PHENOLICS. The earliest member of this family of adhesives (Redux) was actually used for bonding the De Havilland aircraft in World War II, and variations of the Redux technology continues in use until the present day in joining European aircraft. In its original concept, a resol-phenolic solution was spread on one adherend with a powdered poly(vinyl formal) resin sprinkled

over the top. After evaporating any solvent present, the joint was closed and heat and pressure were applied to effect cure. This toughening of a phenolic resin through co-curing with a high molecular weight linear polymer gave joint strength previously unobtainable with a highly acceptable level of durability performance in service. There have been many demonstrations of the excellent durability of aluminum joints bonded with this family of adhesives starting with the 1952 comparative tests by the Royal Airforce Establishment. In 1965, Salomon (3620) cited a number of references with excellent results with aluminum. Minford has investigated the long-term durability of the proprietary aerospace vinyl-phenolics (FM-47 and Narmco 105) using 6061-T6 aluminum adherends. With chromic acid–etched adherends, the Narmco 105 adhesive–produced joints showed no significant loss in joint strength after 2 yr in room temperature water immersion, soaking in condensing humidity at 125°F (52°C), or a quite challenging soak/freeze/thaw cycle environment. The bond strength declines with the other commercial adhesive product (FM-47) were moderate to highly significant in the same exposures. This clearly reflects the need to separately evaluate each new adhesive even though it is a member of a chemical family which has earlier performed acceptably. In a 4-yr (milder exposure) in an industrial atmosphere, however, the adhesives could not be distinguished. Finally, in a 4-yr exposure to the corrosive seacoast weathering, the superiority of the Narmco 105 joints again was manifested. Reference should also be made to Figures 37–40 for data showing evidences of the good durability of aluminum/vinyl-phenolic adhesive joints in soak/freeze/thaw cycle and the industrial and seacoast atmospheres.

The vinyl-phenolic structural adhesives have been especially important in the development of bonded aluminum aircraft structures in Europe using the chromic acid–anodizing surface pretreatment. Catchpole (1480) reviewed some of these application areas in 1965, whereas Koetsier (1484) reviewed their excellent performance in Fokker aircraft in 1975. DeLollis (1485) reviewed the stressed and nonstressed durability data of Olson et al. (1486), confirming the good durability responses of Redux 775 and FM-47 vinyl-phenolic aluminum joints. Hockney (1483) has also furnished evidence of the good durability of vinyl-phenolic aluminum joints under natural weathering conditions.

As can be seen in Figures 37–39, the vinyl-phenolic film and tape-type commercial adhesives evaluated by Minford performed very durably in the very corrosive wet/dry salt-water cycling of the seacoast and the challenging cycling between hot-water soaking, freezing, and hot-air thawing on a 3-da cycle basis.

The question of why the phenolic-modified structural adhesives bond so durably to aluminum oxide has been widely speculated about. DeLollis and Montoya (1201), Bolger (3615), and Buck and Hockney (3619) all have suggested the older phenolic-based European structural adhesives have superior metal joint durability in water to later epoxy-based adhesives; however, better peel strength and lower cure temperatures and pressure requirements generate more acceptance for the epoxies. In 1983, Albericci (3622) offered two explanations. First, the Redux system as originally used could be considered to function in the manner of a primer with separate low-viscosity and surface-wetting advantages. Second, the phenols as a class can form chelate-linked compounds with metal ions, giving a more stable aluminum-to-resin bond. In 1987, Comyn et al. (3621) made comparisons between aluminum-bonded to some phenolic adhesives and to a modified epoxy in moist-air exposure at 122°F (50°C).

b. Two-part Room Temperature–curing Epoxies

Most commonly, these adhesives consist of a liquid or paste consistency epoxy resin (component A) plus an aliphatic amine or polyamide (component B). Fillers like aluminum powder, silica, or carbonate usually need to be added to reduce viscosity and reduce shrinkage in the cured bondline. Pigments are often used to identify the separate parts and to help recognize thorough blending of the parts when the adhesive is mixed. If the viscosity is too high, because of using a very high molecular weight epoxy resin, then certain low molecular weight diluents (such as butyl or phenyl glycidyl ether) may be added when mixing the two other components. While the simple aliphatic polyamines can effectively cure the liquid epoxies, they are less frequently employed in manufacturing because of skin- and respiratory-irritant activity on workers. In addition, the mix ratios in such formulations is high resin-to-low catalyst, making it necessary to very accurately weigh out small amounts of catalyst (except when mixing very large batches). In contrast, polyamide-cured epoxy adhesives often can be close to 50/50 by weight or volume, making it easy to proportion accurately with even small batches for the laboratory or production. An obvious advantage of using a polyamide-curing agent is that a less rigid and more toughened adhesive can be secured; however, this advantage is often only obtained at the expense of some water resistance.

From the data of Minford (3623), it is seen that a variety of unfilled and filled room temperature–curing, two-part epoxy joints had excellent durability in 2-yr RT water soak or hot humidity. This performance is predicated, however, on the use of a good surface pretreatment; in this case, hot chromic-sulfuric acid etching. In Figure 42, we see six of seven commercial epoxy adhesive–produced joints showing no change in joint strength after 4 yr in an industrial atmosphere. That joints with one commercial adhesive had debonded after 4 yr again reinforces the need for concern to evaluate each adhesive separately. Based on the manufacturer's data sheet, there was no reason to suspect that this adhesive would not perform as well as the others or that it might only start to decline rapidly after 2-yr exposure.

In publications as early as 1955, Eickner (964–966) had claimed good performance for clad aluminum/epoxy joints (polyamide-catalyzed epoxy) in the Panama jungle. Later, he reported some stressing of similar joints with a fixturing, as shown in Figure 27. We have mentioned earlier that his levels of stress must have been extremely low, since no difficulty was noted in durability response. More recently, Minford (1355,1356) reported a very satisfactory durability response with chromic-sulfuric–etched 6061-T6 aluminum joints exposed for up to 12 yr at open or more secluded jungle exposure sites in Surinam. Duplicate joints were exposed which had been fabricated with a one-part heat-curing nitrile-modified epoxy adhesive, and a definitely inferior durability response was shown. This finding was in agreement with the results of Bodnar and Wegman (1495), who had also showed greater durability in the jungle for a polyamide-cured epoxy adhesive in aluminum joints over duplicates made with a nitrile-epoxy heat-cured adhesive. Cotter (851), however, obtained his best aluminum joint performance in an Australian jungle site with an epoxy-Novolac or a nitrile-phenolic adhesive with a polyamide-epoxy adhesive showing the poorest resistance to the weathering. Minford (1481) has evaluated a number of different two-part epoxy adhesives in 6061-T6 aluminum joints with an exposure for 2 yr in a repeated 3-da cycle consisting of 24 hr in 165°F (74°C) water immersion, 24 hr in −30°F (−34°C) air chest, and 24 hr in 180°F (82°C) recirculating air in an oven (see Ref. 1416, Fig. 25, p. 86). Four of the adhesives produced joints which showed significant increases in joint strength instead of strength declines for periods in excess of 200 da. Two of these four adhesives showed their joints holding an elevated plateau strength level for the rest of a 2-yr exposure. A third adhesive showed a steadily declining joint strength after 400 da with complete debonding after 600 da. Another group member showed joints declining in strength after 150 da but still retaining about 30% of initial strength after test completion at 730 da. A fifth epoxy adhesive produced joints which lost strength for 400 da then leveled off. A final epoxy adhesive group of joints lost only about 5% of initial strength spread over the 2-yr test period. These kinds of data certainly demonstrate that literally any type of joint durability response that is possible can occur. No response can then be said to be representative for all two-part epoxy adhesives available. Though manufacturers' data sheets may read nearly identically, each different commercial two-part, room temperature–curing epoxy must be separately evaluated in a testing environment that is like that in which the joints must perform during their anticipated service life.

Minford (3614) has shown that the poorest bond durability response for the room temperature–curing variety of epoxies with aluminum adherends seems to be in the corrosive seacoast or marine environments. Minford has developed data showing a wide variety of responses of a whole group of commercial two-part epoxy adhesive/aluminum joints in the seacoast atmosphere. Some adhesive candidate joints were undercut with corrosion within 16 mo, whereas some other candidate joints lasted until 30 mo (see Ref. 3614, Fig. 28, p. 88). This variation and relatively short overall average joint survival performance was shown in spite of the fact that all the 6061-T6 adherends had been thoroughly deoxidized in hot chromic acid. In a later durability testing evaluation, one of these epoxies with adherends that were pretreated in room temperature chromic-sulfuric acid was evaluated. Very unexpectedly, as shown in the data of Table 155, there was no debonding after 8 yr at the seacoast. Rather amazingly, this room temperature deoxidizing (from a visual appearance) had not seemed to remove or have much visual effect on the heavy heat-treat film. Yet, somehow, the adhesion

Table 157 6061-T6 Aluminum-Bonded Joints Exposed to Up to 8 Yr at the Seacoast with Various Abraded, Etched, Conversion Coated, Paint Primed, and Anodized Surface Pretreatments

Surface pretreatment	Adhesive type[a]	Avg. joint strength after indicated exposure time (MPa)				% strength retention after longest exposure
		1 yr	2 yr	4 yr	8 yr	
Vapor degrease	A	0	All failed by 71 da			0
Vapor degrease	B	0	All failed by 270 da			0
Belt-sanded	B	11.99	All failed by 760 da			0
Silica-gritblast	A	16.74	12.40	10.82	9.30	67
Chromic acid etch	A	17.43	All failed by 760 da			0
Chromic acid etch	B	34.66	23.22	Failed by 1440 da		0
Alodine 1200 (conversion coat)	A	7.43	4.58	2.76	All failed by 1475 da	0
Alodine 1200	B	16.74	14.68	16.26	17.78	100
Epoxy-phenolic (paint primer)	A	—	11.09	5.17	All failed by 1497 da	0
Epoxy-phenolic	B	—	31.90	30.04		71
Alumilite 215 (H_2SO_4 anodize to 0.8 mil thick)	A	15.64	14.81	14.29	13.94	80
Alumilite 215	B	19.29	14.67	16.26	17.77	79

[a]Adhesive type A is a two-part, polyamide-cured epoxy; adhesive type B is a one-part, nitrile-modified epoxy.

Test results from Ref. 774 (Table VI, p. 498).

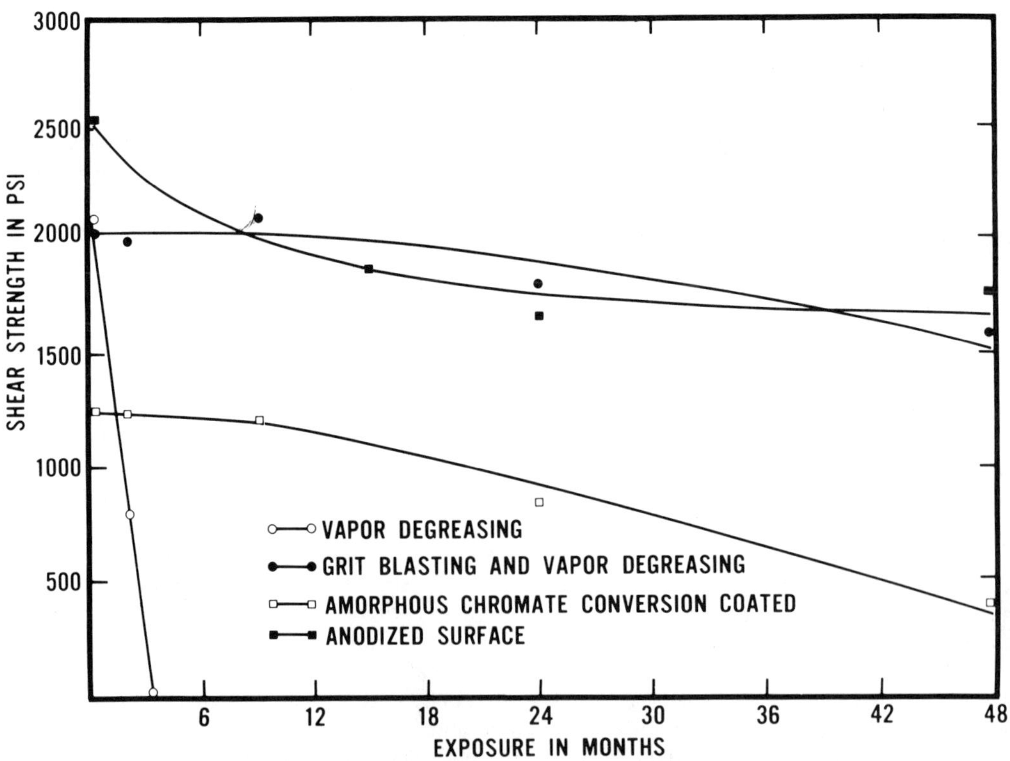

Figure 41 Two-part epoxy, seacoast atmosphere, abrasion, conversion coated, and anodized 6061-T6 aluminum adherends. (From J.D. Minford, unpublished work, Aluminum Company of America.)

to this surface highly resisted any undercutting corrosion of the bond interface. In yet another separate investigation summarized in Table 157, Minford (774) showed no debonding in 8 yr at the same seacoast exposure where the 6061-T6 adherends were pretreated by silica gritblast (Fig. 41), priming with epoxy-phenolic heat-cured paint, or anodizing. Another set of weathering exposure conditions of considerable severity for evaluating bonded joints is a hot-water soak/freeze/hot-air cycling sequence. Minford (see Ref. 774, Table IV, p. 493) showed chromic-deoxidized 6061-T6 joints with

Table 158 Comparative Durability Responses in 2-yr Soaking in RT Water of 6061-T6 Joints[a] Involving 26 Types of Surface Pretreating

Pretreatment	Adhesive	Average joint strength retention[b] psi	Average joint strength retention[b] MPa	% of initial joint strength
Vapor degrease	2-part epoxy	1130	7.79	53.2
Vapor degrease + grit-blast	2-part epoxy	2130	14.68	105.0
Ridoline 53 (hot alkaline cleaner)	2-part epoxy	2200	15.13	83.7
Caustic (hot) + desmut	2-part epoxy	2130	14.68	90.1
Acid paste cleaner	2-part epoxy	1970	13.57	82.3
Deoxidine 526 (RT) (H_3PO_4 + alcohol)	2-part epoxy	2570	17.71	95.3
Alum-Surf-Prep (RT acid cleaner)	2-part epoxy	1870	12.88	66.8
HNO_3 + H_2SO_4 (hot)	2-part epoxy	2330	16.06	93.2
H_3PO_4 + H_2SO_4 (hot)	2-part epoxy	1970	13.57	83.2
CrO_3 + H_2SO_4 (hot)	2-part epoxy	2670	18.40	94.3
CrO_3 + H_2SO_4 (RT)	2-part epoxy	2270	15.64	88.3
HCl + $K_2Cr_2O_7$ (RT)	2-part epoxy	2530	17.43	96.2
Vapor degrease	1-part epoxy	1170	8.05	27.1
Belt-Sand + vapor degrease	1-part epoxy	2170	14.93	44.6
Ridoline 53 (hot alkaline cleaner)	1-part epoxy	2370	16.33	45.0
Caustic (hot) + desmut	1-part epoxy	2200	15.16	47.5
Acid paste cleaner	1-part epoxy	2470	17.02	53.4
Deoxidine 526 (RT) (H_3PO_4 – alcohol)	1-part epoxy	2830	19.50	57.0
Alum-Surf-Prep (RT acid cleaner)	1-part epoxy	3870	26.66	76.4
HNO_3 + H_2SO_4	1-part epoxy	3000	20.67	58.8
H_3PO_4 + H_2SO_4	1-part epoxy	1500	10.34	31.8
CrO_3 + H_2SO_4 (RT)	1-part epoxy	3760	25.91	66.8
CrO_3 + H_2SO_4 (hot)	1-part epoxy	3530	24.32	66.4
HCl + $K_2Cr_2O_7$ (RT)	1-part epoxy	3800	26.18	76.5

[a]Two-part, room temperature–curing epoxy was National Starch M688 (100 parts) mixed with 32 parts of CH-16 polyamide-curing agent and cured in the bondline for 7 da before exposing. One-part, heat-cured epoxy was 3M's EC-2086 nitrile-modified epoxy cured in the bondline for 15 min at 400°F (204°C) before exposure. Lap-shear joints were tested according to ASTM D-1002.

[b]Initial joint strength values can be found in Table 157.

Data from J.D. Minford, unpublished work, Aluminum Co. of America.

the same two-part epoxy adhesive that failed in 2 yr at the seacoast now survived with 75% joint strength retention. However, the silica gritblast–treated joints which totally survived 8 yr at the seacoast (see Table 157) now are totally degraded within 2 yr.

The durability results for a specific two-part epoxy (M688/CH16, National Starch and Chemical Product) in three weathering environments (room temperature water, soak/freeze/thaw cycling, and the natural industrial atmosphere) can be seen in Tables 154, 158, and 159. The specific results will be discussed in more detail later in this chapter under the specific aluminum surface pretreatments employed.

Table 159 Comparative Durability Responses in 2-Yr Exposure to a Hot-Water/Freeze/Hot-Air Dry Cyclic Environment of 6061-T6 Aluminum Alloy Joints with Varying Surface Pretreatments

Pretreatments	Adhesive	Avg. joint strength retention after 2 yr or indicated time		% of initial strength
		psi	MPa	
Vapor degrease	2-part epoxy	120	0.83 (90 da)	6
Vapor degrease + grit-blast	2-part epoxy	40	0.28	2
Ridoline 53 (hot alkaline clean)	2-part epoxy	500	3.45 (180 da)	19
Caustic etch + desmut (hot)	2-part epoxy	1065	7.34	45
Acid paste cleaner	2-part epoxy	600	4.13	25
Deoxidine 526 (RT phosphoric acid)	2-part epoxy	1166	8.03	43
Alum-Surf-Prep etch acid cleaner)	2-part epoxy	0	0	0
HNO_3 + H_2SO_4 (hot)	2-part epoxy	1850	12.75	74
H_3PO_4 + H_2SO_4 (hot)	2-part epoxy	1200	8.27	50
CrO_3 + H_2SO_4 (hot)	2-part epoxy	2100	14.47	74
CrO_3 + H_2SO_4 (RT)	2-part epoxy	2270	15.18	88
HCl + $K_2Cr_2O_7$ (RT)	2-part epoxy	2170	14.95	82
Vapor degrease	1-part epoxy	500	3.45	12
Belt-sand + vapor degrease	1-part epoxy	0	0	0
Ridoline 53 (hot alkaline clean)	1-part epoxy	600	4.13 (180 da)	11
		0	0 (360 da)	0
Caustic etch + desmut (hot)	1-part epoxy	0	0 (180 da)	0
Acid paste cleaner	1-part epoxy	900	6.20 (360 da)	19
		0	0 (2 yr)	0
Deoxidine 526 (RT) H_3PO_4 + alcohol)	1-part epoxy	0	0 (360 da)	0
Alum-Surf-Prep (RT acid cleaner)	1-part epoxy	1870	12.88 (360 da)	37
		0	0 (2 yr)	0
HNO_3 + H_2SO_4 (hot)	1-part epoxy	675	4.65 (180 da)	13
H_3PO_4 + H_2SO_4 (hot)	1-part epoxy	805	5.55 (180 da)	17
CrO_3 + H_2SO_4 (hot)	1-part epoxy	1330	9.16 (360 da)	25
CrO_3 + H_2SO_4 (RT)	1-part epoxy	770	5.31	14
HCl + $K_2Cr_2O_7$ (RT)	1-part epoxy	2330	16.05 (360 da)	47

Data from J.D. Minford, unpublished work, Aluminum Co. of America.

Some kind of special surface pretreating of heat-treated aluminum alloys is usually required with room temperature–curing epoxy adhesives for acceptable survival in exterior weathering service conditions. If the aluminum adherend has a high-purity cladding (like Alclad 2024-T3), however, the requirement for special etching, conversion coating, or anodizing could be somewhat diminished. Minford (1356) has showed that only vapor-degreased Alclad 2024-T3 adherends could be bonded with a simple two-part, polyamide-cured epoxy paste adhesive and show acceptable joint strength retention after a 12-yr exposure in the equatorial jungles of Surinam in South America. Corresponding joints made with 6061-T6 that had only been vapor degreased showed significantly poorer durability performance under identical tropical exposure conditions. The durability response of 6061-T6 adherends that were pretreated by a wide variety of methods and exposed to various water-soaking and natural atmospheric weathering conditions using this same two-part epoxy or a one-part, heat-cured nitrile-modified epoxy paste adhesive can be reviewed in a 1984 publication of Minford (774).

Schlies has published data on the long-term storage of two-part epoxy/aluminum joints at ambient temperature, whereas Olson (1486) and Cotter (851) exposed stressed and nonstressed two-part epoxy joints to natural weathering conditions. In 1963, Meckelburg and Althof developed a test procedure for determination of the fatigue strength for long lifetime by measuring the creep deformation under elevated time-phase loads for a corresponding shorter lifetime. The estimated rupture time point under weaker loads is determined graphically by extrapolating the log-log plotted straight-line processes, McAlbee and Levi (3624,3625), at Picatinny Arsenal, have used a hot-water soak reaction rate method for evaluating the durability of two-part epoxy/aluminum joints when exposed to a combination of stress and hot and high humidity. The evaluation of two-part epoxy/aluminum joints were included in a wide-ranging program at Picatinny Arsenal on the relative durability to be expected from different adhesive families. Tanner reported some of these results in 1965 and again in 1972. Wangsness, at 3M's, reported results from the monitoring of the long-term durability response of a variety of their structural adhesives for bonding aluminum, including a number of two-part, room temperature–curing epoxies.

c. *Elastomer-Epoxy Adhesives*

The most important development in recent years in formulating structural adhesives with extra toughening has been the addition of elastomeric materials to epoxy resins. This has been discussed earlier in some detail under the selection of adhesive section and the works of a number of investigators need to be reviewed for more detailed explanation of the effects on the properties of such modified epoxy formulations (1498,1548–1555,1561,1564,1565,2679,3626,3627).

For example, the fracture energy of an unmodified epoxy can be increased 30- to 40-fold by the addition of a rubber such as a carboxyl-terminated polybutadiene/acrylonitrile (CTBN) elastomer to the epoxy resin. Below about 15% addition, the elastomer precipitates and cures as spherical particles, which appear in the fracture surfaces of the modified epoxy adhesive. Above this level of additive, the CTBN forms a more of less homogeneous blend, with the fracture energy declining sharply. Of special interest, is the fact that the gain in toughness occurs with little loss in tensile strength, modulus, or thermal mechanical resistance of the epoxy matrix resin. Also of benefit from an operating-cost point of view is effective curing at relatively low bondline pressures and short elevated temperature dwell times. Such curing properties have been helpful in assembling ever larger wing, tail, and fuselage sections in aircraft with lower steam pressures, and thinner-walled autoclaves that can be employed in manufacturing. Lewis and Saxon (1498) have published data indicative of the dramatic improvements in tensile-shear and T-peel strengths that are obtained with adding nitrile rubber to an epoxy (see Table 48). Lees (3628) has evaluated elastomer-modified products developed in the early 1980s. Bascom et al. (222) have explained the improvement in properties through the role of the elastomer particle in permitting a much larger volume for plastic deformation at the crack tip.

The durability data for several nitrile-modified epoxy, one-part, heat curing adhesives of early development have been cited by Minford (3629). Data obtained with chromic-sulfuric acid–etched 6061-T6 adherends in room temperature water or 100% RH at 125°F (52°C) for 2 yr has been plotted in (see Ref. 3629, Figs. 6 and 7, p. 97, and Ref. 3611, Fig. 30, p. 188). Of particular note is the wide range of initial joint strength from 2000 to 5500 psi which was found present in a variety of one-part, heat-cured epoxies. While all the tested commercial adhesives with etched 6061-T6 gradually declined in strength by about 25% in 2-yr RT water soaking, the similar adhesive joints in hot 100% RH conditions tended to decline much more severely, losing as much as 80% in the same time period. One dicy-cured, epoxy powder-form adhesive only lost 17% joint strength; again indicating a distinct need to evaluate each different adhesive of the same generic family. The joint durability results with duplicate joints exposed to aggressive hot-water-soak/freeze/hot-air-dry cycle exposure for 2 yr is shown in Figure 42. While two of five adhesive candidates could not survive the total exposure time, the same dicy-cured powdered epoxy mentioned above again showed little change in joint strength throughout the exposure. The 4-yr exposure of similar joints to an industrial or seacoast atmosphere have been reported (see Ref. 3629, Figs. 9 and 10, p. 100, and Ref. 3611, Figs. 32 and 33, pp. 189 and 190, respectively). All adhesive candidates performed without joint decline over the 4-yr exposure to the industrial atmosphere; however, none of the dicy-cured nitrile-modified epoxy adhesive joints survived in the seacoast beyond 4 yr. The best performer was an anhydride-cured, nondicy, nonnitrile epoxy powder candidate. The aluminum joints fabricated with the dicy-cure, epoxy powder adhesive (which had shown the most durable performance in the water-exposure conditions) showed the shortest survival time in the corrosive seacoast conditions. It is recognized that while an elastomer addition imparts some undesirable sensitivity to water soaking (as compared to the two-part epoxy adhesives), the comparative performance in the seacoast is improved. The author attributes this to the better wetting from heat-curing the adhesive rather than any mechanical effect of the elastomer addition. Minford (531–536,771–775) has continued to use the same nitrile-modified commercial epoxy (EC-2086) over many years in order to be able to make direct comparisons of the effect of other variables, such as different alloy adherends, weathering conditions, and surface pretreatments.

Bodnar and Wegman (1495,3630) have evaluated EC-2086 bonded aluminum joints in natural weathering conditions in the desert, industrial plant atmosphere, and jungle. All joints failed in the

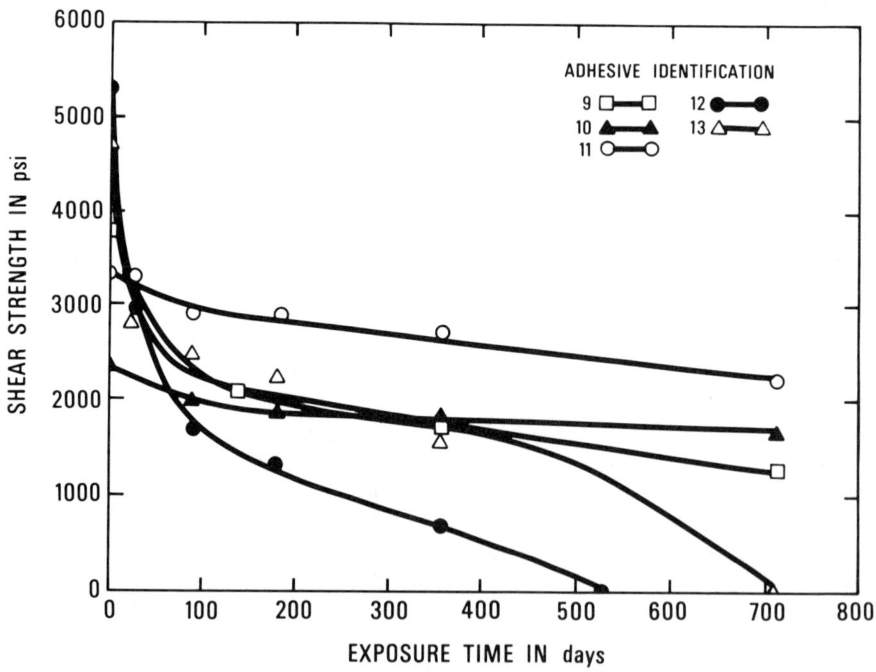

Figure 42 One-part epoxies—water immersion cycle. (From J.D. Minford, Adhesive Bonded Aluminum Joints, Treatise on Adhesion and Adhesives (R.L. Patrick, ed.). Marcel Dekker, New York, Basel, Vol. 3, Chap. 2, Fig. 8, p. 98 [1973].)

time frame between 2 and 3 yr in the jungle, whereas joint strength losses in even the dry desert averaged 15%. These results indicated a significantly lower order of durability than found for the corresponding vinyl- or nitrile-phenolic aluminum joints evaluated under the same conditions. Wangsness (1605) reported the durability of various 3M's commercial nitrile-epoxy paste and film adhesives under sustained load and hot high humidity, whereas Hughes and Rutherford (1614) tested Metlbond 327 (nitrile-epoxy) compared to 3M's aluminum powder–filled EC-2214 nitrile-epoxy under similar testing conditions. Grimes (3631) has investigated the tension-tension fatigue performance of another aerospace nitrile-epoxy adhesive in film form (AF-126-2). Schwartz (1615) has compared the stressed durability performance of a nitrile-epoxy requiring 350°F (177°C) cure with two modified-epoxy aerospace film adhesives cured at 250°F (121°C) and included data involving varying aluminum surface pretreatment. Finally, DeLollis (1616) has evaluated FM-123-2 and EC-2214 nitrile-epoxy aluminum joints in continuous immersion in water or high humidity–soaking conditions.

Luckyram and Vardy (3632) have published some interesting data comparing the fatigue performance of two toughened epoxy adhesives considered for possible use as heavy structural engineered joints in bridge construction. Rather unexpectedly, it was shown that the single-part, heat-cured member of the duo performed better in fatigue than a corresponding two-part, heat-cured toughened epoxy which actually had a much greater measured fracture toughness. The one-part epoxy joints also showed no signs of crazing due to fatigue as shown by the two-part adhesive joints and were more highly resistant to crack initiation as well as crack propagation. While bonding of steel plates was involved, the data could also be of benefit for bonded heavy aluminum structures.

Reference can be made to earlier work by Jones et al. (3633) and MacDonald and Calder (3634) with data for the bonding of steel plates to concrete beams. Albrecht (3635) and Martin (3636) have demonstrated superior fatigue resistance of adhesively bonded cover plates on the tension flange of a steel girder and adhesive bonded web stifferers, respectively, using epoxy adhesives. Desirable characteristics and properties of useful adhesives in this regard have been compiled by Lark and Mays (3637) and Lark (3638), respectively. The toughened epoxies stand out as the best candidates for such structural uses.

d. Nylon-Epoxy Adhesives

The first nylon-modified epoxy adhesives introduced into the aerospace industry appeared to be an ideal solution for the airplane designer. These adhesives showed significant boosts in peel, shear, fatigue, and impact resistance over the adhesives currently in the marketplace. When the durability of these adhesives was extensively investigated in a variety of service conditions where water was present, however, serious durability limitations became evident. In addition to the sensitivity to moisture problem, the nylon-epoxies also demonstrated a significant reduction in peel strength at low temperature as well as a poor resistance to creep. We mentioned earlier that DeLollis (198) documented this sensitivity to moisture, showing aluminum joints with the nylon-epoxy adhesive lost 80% of initial strength in high humidity within 2 mo compared with 18% loss by nitrile-phenolic joints in the same exposure after 18 mo. Minford (772,773) compared the relative durability results of Alclad 2024-T3 and 6061-T6 aluminum adherends pretreated in a variety of ways and bonded with an aerospace-type, nylon-modified epoxy film adhesive. Joint strengths ranging from 6750 psi (46.51 MPa) down to 3700 psi (25.49 MPa) could be obtained depending on choice of alloy and surface pretreatments, as shown in Table 160. Using hot chromic acid–deoxidized 6061-T6 adherends and the same nylon-modified epoxy, Minford (3639) compiled joint strength decay curves in both RT water and 100% RH at 125°F (52°C) (for results, see Ref. 3639, Fig. 32). It was found that the curves are quite different with the higher-temperature humidity environment degrading the joint strength much more rapidly over the first 100 da of exposure as might be expected. However, a flat plateau occurred in the hot humidity exposure curve after a decline of about 65% from initial values after 100 da, and no subsequent significant strength decline occurred through a 2-yr exposure time. In contrast, the decline in joint strength in ambient temperature water was gradual until about the same magnitude of joint strength decline (intersection with the hot humidity curve) was reached at 450-da soaking. Unlike the hot humidity–exposed specimens, however, a somewhat slower decline in joint strength continued to the 15% of initial strength level after 2 yr. An important point to be mentioned here is

Table 160 Initial Bond Strength Values for Different Alloy Aluminum Joints[a] Fabricated with Different Surface Pretreatments

Alloy	Surface pretreatment	Average initial shear strength	
		psi	MPa
Alcoa 2024-T3	Deoxidine 670 etching	6000	41.34
Alcoa 2024-T3	Mill-finish	5750	39.63
Alcoa 2024-T3	Chromic acid deoxidizing	5520	38.03
Alcoa 2024-T3	Vapor degreased	5500	37.89
Alcoa 2024-T3	Alodine 1200 (conversion coat)	5190	35.76
Alcoa 2024-T3	Alumilite 215 (H_2SO_4 anodize)	4400	30.32
Alcoa 2024-T3	Silica-gritblasting	3700	25.49
6061-T6	Deoxidine 670 etching	6750	46.51
6061-T6	Chromic acid deoxidizing	6120	42.17
6061-T6	Alodine 1200 (conversion coat)	5650	38.93
6061-T6	Vapor degreased	5100	35.14
6061-T6	Mill-finish	5050	34.79
6061-T6	Alumilite 215 (H_2SO_4 anodize)	4350	29.97
6061-T6	Silica-gritblasting	4300	29.63

[a]All joints fabricated with FM-1000 aircraft-type, nylon-modified epoxy film adhesive cured between the adherends in a platen press for 60 min at 350°F (177°C) under a pressure of 10 psi (0.069 MPa).
Data from Ref. 772 (Table 1, p. 85).

that many water-soaking durability performance evaluations in specifications are not designed to go beyond 30–90 da. Thus, the lower plateau performance of the nylon-modified epoxy in the hot humidity soaking would have gone undiscovered. However, it also must be admitted that a loss of 30% of initial joint strength within 30 da or the 70% loss in 90 da would be a realistic basis for rejecting this adhesive regardless of its eventual achievement of a lower plateau strength value.

Minford (3639) (see Figure 33, p. 118) has also developed the stress endurance curves for similarly prepared nylon-modified epoxy aluminum joints in the same ambient and 125°F (52°C) water-soaking conditions (see Ref. 3639, Fig. 33, p. 118). In hot humidity, as might be expected, the joint survival times were significantly lower. Joint survival times of about 20 da were found with the stress at 500 psi (3.45 MPa). In contrast, the survival time at this stress level could be extrapolated to 1300 da when the water temperature was lowered to ambient. Another interpretation could be that the joint when stressed in ambient temperature water could survive as long at 1000-psi stress as one stressed to only 500 psi exposed to hot humidity conditions. An attempt was made to offset this water sensitivity through the use of water-resistant resol-phenolic resin-based primers. While such primers did prolong the service life to some degree, they were not able to elevate the overall joint durability to the performance levels established for unprimed joints using nitrile- or acetal-modified epoxy adhesives.

Minford (772,773) has studied the effect on durability of varying surface pretreatment on Alclad 2024-T3 and 6061-T6 aluminum adherends using the aerospace nylon-epoxy film adhesive FM 1000 in the bondline (see Table 160 for initial joint strengths). The evaluated surface pretreatments included mill-finish, vapor degreasing, silica gritblasting, Deoxidine 670 acid etching, hot chromic-sulfuric–acid etching, Alodine 1200 conversion coating, and Alumilite 215 (architectural thickness of sulfuric acid anodizing). A common factor in each of the separate investigations was the study of the effect of stress at 2000 psi (13.78 MPa), 1000 psi (6.89 MPa), and 500 psi (3.45 MPa). The exposures were either continuous soaking in 100% RH at 125°F (52°C) (772) or alternate immersion cycling in 3.5 salt water (773). To compensate for the different strengths of the two alloys, the 6061-T6 alloy joints were twice as thick. The significant effect of varying the surface pretreatment conditions on joint strength is clearly evident. High joint strength, even with the mill-finish adherends, must be attributable to the effective surface wetting achieved with the high-temperature curing. It was not expected that the mill-finish (as received) Alclad 2024-T3 joints would be as strong as the chromic acid–deoxidized joints. However, it must be remembered that the cladding on 2024-T3 alloy is relatively high purity. The lower comparative strength of the mill-finish 6061-T6 joints would be expected with the heavy heat-treat oxide present in contrast to the high-purity nature of the 1260 alloy clading on the 2024-T3 surface. It should be noted that gritblasting to provide the possibility for much more physical mechanical locking potential actually had a negative effect on initial strength as compared to all the other surface pretreatment conditions. This is one illustration of the greater benefit of microscopic rather than macroscopic surface conditions on joint strength and durability. The excellent bonding results achieved on steel by gritblasting does not seem to be paralleled in the bonding of aluminum adherends. Possible explanations for the very low joint strengths found with the sulfuric acid–anodized surfaces and only modest strengths of the Alodine 1200–pretreated joints needs to be suggested. Alcoa Alumilite 215 anodize layer thickness is recommended for maximum corrosion resistance to natural weather. This is only achieved by building up an oxide layer that is many times thicker than the chromic acid or phosphoric acid–anodized layers most used to bond aluminum aircraft structures. In these investigations, the Alcoa Labs was intending to demonstrate that architectural thickness anodized aluminum could be effectively and durably bonded with the same kinds of structural adhesives as used in aerospace. While this proved to be true, it was recognized that the joint strengths would be significantly lower with the much thicker oxide present, since some weaker boundary layer in the thicker oxide could fracture at some load below that producing pure cohesive failure in the cured adhesive. The stress-durability time to failure values in these tests will be considered in more detail when the subject of durability under simultaneous stress and exposure conditions is discussed in a later section.

Kuno (1621) has studied the effects of higher temperature on the durability of aluminum joints fabricated with the nylon-epoxy Metlbond adhesive.

e. Modified-Epoxy Adhesives

Often the adhesive formulator of structural epoxy adhesives is not willing to divulge class or type of modifying resin used. In these cases, the general description of "modified-epoxy" type is indicated on the data sheet. In an extensive program of screening the durability of a number of proprietary tape and film aerospace-type adhesives, Minford (1626) found that 6 of 12 commercial products were only identified in this manner. These included Metlbonds 324 and 329, FM-61 (one side of this duplex tape was modified epoxy, whereas the other side was nitrile-phenolic), FM-96, Plastilock 677, and Aerobond 3030. The resultant chromic-sulfuric–etched 6061-T6 adherend joints were evaluated in room temperature water immersion, soak in 100% RH at 125°F (52°C), hot-soaking/freeze/thaw cycling, industrial atmosphere, and the seacoast atmosphere. The range of the durability response in these exposures was so wide ranging that no conclusion could be reached regarding what level of performance might be anticipated from only a formulator's description of "modified epoxy"-type structural adhesive. For example, four categories of relative durability performance were used to describe the results: (1) joints would actually increase in strength, (2) joints show no significant change in strength, (3) joints show a modest decrease in strength, and (4) there is a highly significant decline in strength. After 2-yr room temperature water immersion, four of the six "modified epoxies" were in categories (1) and (2), whereas the other two were in category (3).

When exposed to the more degrading hot humidity soaking, two adhesives remained in categories (1) and (2) with three adhesives now showing up with category (3) performance. The final adhesive now fell into the significant strength decline category (4). Under the still more aggressive cycling weathering conditions for 2 yr, only two adhesives qualified in the no change category. Joints of only one adhesive declined moderately, whereas half of the adhesives evaluated showed significantly deteriorated joints. Even in the natural atmospheric exposure after 4 yr, there was a greater performance variation than anticipated from extensive prior testing of many two-part epoxy adhesives. Half of the adhesives constituted joints showing no decline; however, two showed some significant decline. Finally, in the normally most aggressive weathering, i.e., corrosive seacoast weathering, the unexpected occurred with all the modified epoxies showing unaffected joints after 4 yr. Minford has speculated that perhaps the most important factor in resisting undercutting corrosion in aluminum joints is the degree of chemical wetting achieved before and during any bondline heating. Since all of these adhesives were cured at 350°F (177°C), the amount of flow and wetting was adequate to resist undercutting corrosion, yet water could still gain entry if continuously present in the service testing environment. This explanation does not seem adequate to explain the situations where some joint declines were noted in the natural industrial weathering where continuous water contact would not be anticipated. There was the opportunity at the test site in Pittsburgh, Pennsylvania, however, for many wet/freeze/thaw cycles, which also seem to be notoriously able to penetrate and damage the bonding of particles in various types of road construction.

Peterka (3640) obtained good performance with paint-overcoated and heat-cured epoxy joints after 10-yr exposure to the central European natural atmospheric weathering processes. Wangsness (1605) included five so-called modified-epoxy 250°F (121°C) curing adhesives with aluminum adherends in his 1977 sustained load testing under hot humid soaking conditions study. Hughes and Rutherford (1614) evaluated modified-epoxy ADX 41.2 under stress/weathering conditions. Schwartz (1615) employed both stressed blister-type aluminum lap-joints and aluminum wedge-type specimens to evaluate two 250°F (121°C) cure modified-epoxy adhesive formulations. Cotter (851) employed a modified-epoxy heat-curing adhesive with aluminum adherends in his high-humidity jungle exposure in Australia and offered comparisons with room temperature–curing two-part and nitrile-epoxy adhesive joints. Grimes (3631) used Metlbond 1113 modified-epoxy joints to compare in constant cycle fatigue with a nylon-epoxy and a nitrile-epoxy structural adhesive with aluminum adherends.

Askins and Schwartz (3641) have evaluated FM-73 and FM-400 modified-epoxy adhesives for bonding aluminum sandwich panels. The same investigators (3642) also evaluated the durability under adverse weathering of FM 400 for graphite epoxy to aluminum joints. There were four commercial adhesives chosen for extensive testing in the PABST program by Shannon and Thrall (76–80, 3643,3644). All (M1133, AF-65, FM-73, and EA-9628) were designated by their formulaters as being modified epoxies requiring curing at 250°F (121°C). The aluminum adherends were all primed with BR-127 corrosion-inhibiting primer as part of the overall preparation before bonding. A variety of weathering conditions under various sustained loads were employed to evaluate the total durability picture. McMillan (3607) has attempted to find the best relationships between surface preparations on 2024 and 7075 aluminum adherends and various 250°F (121°C) curing modified-epoxy adhesives. Wegman (1647) has also conducted aluminum joint durability evaluations with four modified epoxies (ADX 6562, EA 9628, M1113, and PL 717B). Comparisons were made in stress/hot humidity and fatigue testing conditions. Finally, Scardino and Marceau (853) have conducted extensive wedge-type test evaluations of a number of commercially available 250°F (121°C) curing, modified-epoxy adhesives.

f. Structural Acrylics

These adhesives are actually two-part systems and in that regard tend to resemble the room temperature–curing epoxies. However, they may or may not have pot-life problems like the epoxies depending on how the curing is affected. The older generation of acrylics had a resin and curing agent which had to be premixed. Pot-life would be a consideration with these adhesives. The more recent commercial formulations (second-generation types) can cure rapidly without premixing. It is only necessary for the activator solution to be preapplied to the adherend, followed by an application of the acrylic resin.

The earliest published durability test results involving aluminum adherends was by Zalucha in 1972 (1431), whereas Minford had also been accumulating extensive weathering data at the Alcoa Labs on the same Hughson Chemical adhesive formulations (1438). Zalucha tested steel and aluminum adherends for 1000 hr in room temperature water, hot condensing humidity, and 5% salt spray, finding at least 60% strength retention for all the adherends except SAE-101 CRS in the hot humidity. From the 1960s, Minford had been testing the available commercial products obtaining a wide-range durability response in 100% RH at 125°F (52°C). With minimal preparation such as vapor degreasing, the response was generally poor. Where the aluminum adherends were chromic-sulfuric acid etched, the durability response was equal to that of the average commercial two-part epoxy. Shorter curing times of the acrylics could still be considered a possible advantage for production, but the ratio of curing agent was usually quite small and often fractionally based on the amount of acrylic resin specified. This made the need to accurately weigh the activator critical and its thorough mixing imperative. When testing under combined conditions of hot humidity and moderate stressing, Minford found the acrylic joined aluminum joints to be definitely superior to typical two-part epoxy-bonded joints using etched adherends.

The present generation of modified structural acrylic adhesives are recommended for bonding to unprepared aluminum surfaces, and Minford (3645–3650) has tested a variety of these so-called second-generation products for bonding aluminum-to-aluminum and aluminum-to-plastics such as rigid vinyl, ABS, polycarbonate, acrylics (such as Plexiglas), and butyrate. It was found that these products are capable of developing durable bonding to mill-finish aluminum surfaces as well as to some deliberately oiled adherends. Bonding to the latter surfaces was not possible with any unmodified two-part, epoxy products. Acceptable durability levels were shown after 2-yr liquid water immersion or soaking in hot condensing humidity using vapor-degreased 6061-T6 adherends. There was also clear evidence that these acrylic-bonded joints might offer superior resistance to undercutting corrosion of the bondline in corrosive salt-water exposures (as compared to two-part epoxies). Even more durable performance in all these exposures was shown by Minford using plastic-to-plastic adherends, as mentioned above. Minford also evaluated acrylics for bonding aluminum/dissimilar materials joints (polyester-fiber glass) (3651) and engineering-grade styrene and cross-linked styrene (3652). In these investigations, Minford compared the strength and joint durability potentials of the reactive acrylic adhesives with various solvent-type dope cements, synthetic resin contact-type cements, and two-part, room temperature–curing epoxy commercial adhesives, as in Table 161 using rigid vinyl adherends. It should be noticed that the initial joint failures (without involving any weathering) occurs almost exclusively in the plastic adherend. This is generally true, as will be shown in more detail later, regardless of whether the plastic adherend is rigid vinyl, ABS, polycarbonate, acrylate, or butyrate.

Table 161 Initial Joint Strength of Vinyl-to-Vinyl and Aluminum-to-Vinyl with Varying Adhesive and Surface Pretreatments

Type[a] joint	Adhesive[b]	Surface preparation	Initial lap-shear strength		Type failure
			psi	MPa	
V-V	Weld-On 1001	Alcoa wipe	750	5.17	In vinyl
A-V	Weld-On 1001	Vapor degrease—A Alcoa Wipe—V	620	4.27	In vinyl
A-V	RA-0018	Vapor degrease-A Alcoa wipe—V	720	4.96	In vinyl
A-V	EC-4475	Vapor degrease—A Alcoa wipe—V	500	3.45	100% cohesive
A-V	EC-2216	Abrade—A Alcoa wipe—V	820	5.65	In vinyl

[a]A = aluminum adherend was 0.32 in (0.81 mm) 3003-H14 alloy; V = vinyl adherend was 0.062 in (1.57 mm) rigid PVC.

[b]Weld-On 1001 is a water-clear, solvent-type cement from Ind. Polychemical recommended for bonding plastic adherends. RA-0018 is a reactive acrylic from H.B. Fuller which cures with a chemical accelerator preapplied to adherend surface. EC-4475 is a clear synthetic resin contact-type adhesive from 3M's. EC-2216 is a two-part nitrile-flexibilized epoxy from 3M's.

Data from J.D. Minford, unpublished work, Aluminum Co. of America.

Table 162 Durability Performance of High-Performance Acrylic[a] on 3003-H14 and 6061-T6 Aluminum Adherends with Varying Surface Pretreatments and Exposure Conditions

Adherend[b]	Surface pretreatment	Exposure condition	Exposure time (days)	Avg. retained joint strength	
				psi	MPa
3003-H14	Vapor degrease	None	None	1320	9.09
3003-H14	HCl (RT) etch	None	None	4130	28.46
6061-T6	Vapor degrease	None	None	4300	29.63
6061-T6	HCl (RT) etch	None	None	4420	30.45
3003-H14	Vapor degrease	RT water	500	1050	7.23
3003-H14	HCl (RT) etch	RT water	500	2550	17.57
6061-T6	Vapor degrease	RT water	180	0	0
6061-T6	HCl (RT) etch	RT water	500	2470	16.77
3003-H14	Vapor degrease	100% RH—125°F	360	330	2.20
3003-H14	HCl (RT) etch	100% RH—125°F	450	2990	16.81
6061-T6	Vapor degrease	100% RH—125°F	60	0	0
6061-T6	HCl (RT) etch	100% RH—125°F	450	1050	7.23
3003-H14	Vapor degrease	Int. salt-fog	90	10	0.07
3003-H14	HCl (RT) etch	cycling	450	3190	21.98
6061-T6	Vapor degrease	Int. salt-fog	60	0	0
6061-T6	HCl (RT) etch	cycling	450	3100	21.43
3003-H14	Vapor degrease	Industrial	540	1270	8.75
3003-H14	HCl (RT) etch	atmospheric	540	3400	23.43
6061-T6	Vapor degrease	exposure	540	2070	14.26
6061-T6	HCl (RT) etch	conditions	540	4350	29.97

[a]All joints were bonded by brushing Accelerator No. 4 on both adherend surfaces and allowing 5–10 min for solvent to volatilize from each interface. Hughson Chem. Co. B-3327-6 high-performance acrylic structural adhesive was then spread on one adherend and the lap-joint was fixutred and allowed to cure under ambient lab conditions for 24 hr before exposing.

[b]Both adherends were 0.125 in (3.18 mm) thick. This would mean the 6061-T6 adherend would be stiffer and would be capable of bonding with somewhat higher lap-shear strengths with the same adhesive and curing conditions if no other factor was involved.

Data from Ref. 3654 (Tables I, II, III, V, and VI).

Table 163 Effect of Adherend Surface on Curing in Air and Underwater Using a Two-Part Toughened Acrylic Adhesive

Curing environment	Initial Lap-Shear Strength (MPa)			Aluminum/fiber glass epoxy
	Steel/steel	Steel/aluminum	Aluminum/aluminum	
Air (5 min)	8	6	7.5	8.2
Air (30 min)	11.3	9.1	11.0	13.0
Underwater (5 min)	7.0	6.0	6.8	7.1
Underwater (30 min)	10.5	8.0	9.0	10.0

Adherends were untreated 1020 steel, Alclad 2024-T3 aluminum, and fiber glass/epoxy.
Data from Ref. 3655.

Zalucha (1449,1450) reported using so-called high-performance acrylic adhesives to unprepared aluminum adherends and the special bonding of plastics to metal for solar energy applications. Minford conducted extensive testing of mock-up full-size aluminum/plastic solar-heating panels using circulating heated water (3653). The specific acrylic adhesive selection was not as critical as other structural problems which could occur. Martin (1442) reviewed the developments in the entire field of acrylic adhesives in the late 1970s. Lees (1453) found good durability results with toughened acrylic Flexon 241 using etched aluminum in 1982. This was based on finding only a 3% loss in joint strength after 1000 hr in 100% RH at 113°F (45°C).

Minford (3654) used Hughson B-3327-6 high-performance acrylic for bonding 3003-H14 and 6061-T6 adherends with vapor degreasing or HCl RT-etched surface treatment and exposed comparative joints to RT water soaking, 100% RH at 125°F (52°C), intermittent salt-fog cycling, or an industrial atmosphere, as shown in Table 162. It seems quite clear from such durability test results that the performance on vapor-degreased–only adherends is much inferior to that found on acidic deoxidized adherends even with the present state of the art high-performance designated acrylics. Even so, the usual adhesive formulator of acrylics continues to recommend that these structural acrylics can bond effectively to noncleaned aluminum joint surfaces. Allen and co-workers (3663,3664) evaluated acrylic adhesives on aluminum adherends in the 1980s.

Recently, Angelovici et al. (3655) have referred to the earlier works of Drisko et al. (3656), Clark et al. (3657), Allen et al. (3658), Sharp (3659), Drisko (3660), Hegedus (3661), and Oliver (3662) on various investigations of underwater repair by adhesive bonding. Obviously, the adhesive wetting of adherends and displacing the water from the interface for bonding immersed structures has been a major problem. While the most commonly employed adhesive has been an epoxy type, the long-curing time with special support while curing has been a major disadvantage. Angelovici et al. (3655) have investigated the bond curing rate and ultimate joint durability of lap-joints made with mild steel, Alclad 2024-T3 aluminum, and fiber glass/epoxy using rapid-curing toughened acrylics. Joint-strength development values after 5 or 30 min in air or the same times underwater can be seen in Table 163. The speed of curing appeared to be only slightly slower under water. These authors found poor durability results with untreated aluminum joints in both moisture and water-immersion conditions. Further, galvanic corrosion occurred on the aluminum adherend side of the joint when cathodic metals like steel or copper were present in the joints. Marine paint application or chromic acid anodizing of the aluminum before bonding and application of the paint over the steel surface in bimetallic joints afforded reasonable durability, as shown in Table 164. Anodizing both aluminum adherends' surfaces afforded similar joint durability after water immersion for 8 mo.

g. *Cyanoacrylates*

The extremely rapid development of high bond strength that can be achieved with these adhesive polymers would lead one to expect that extensive durability testing had been conducted. Their brittleness and water sensitivity under stressful conditions with aluminum adherends has tended to make information rather sparse. One should not conclude, however, that this kind of adhesive has no place

Table 164 Durability of Toughened Acrylic Joints Cured Underwater and Aged in Sea Water for 8 Mo

Joint adherend combinations	Surface pretreatments	Joint strength (MPa) after 8-mo exposure
Steel/aluminum	None/CAA	0.4
Steel/aluminum	Marine paint/CAA	7.0
Aluminum/aluminum	Marine paint/CAA	5.3
Copper/aluminum	None/CAA	0.8
Fiber glass epoxy/aluminum	None/CAA	6.2

Data estimated from Ref. 3655 (Fig. 10, p. 48).

in bonding to aluminum surfaces. However, it does mean that they must be individually evaluated in each separate potential application area using some accelerated condition that would be present in the service environment.

The most detailed investigations of the overall durability question was conducted by Eastman-Kodak in the early days when their Eastman 910 product almost stood alone in the marketplace. As far as the author knows, they usually used natural outdoor weathering conditions instead of accelerated laboratory weathering test condition. No actual aluminum-to-aluminum test joints appeared to be employed, but they did evaluate shear-strength changes produced by 4- to 7-yr outdoor weathering in dissimilar material joints like rubber-to-aluminum or steel. The worst situation seemed to be aluminum-to-phenolic plastic joints, which lost 62% of initial strength in 2 yr. A number of elastomeric-to-aluminum joints seemed to fare much better than this.

(1) HEAT RESISTANCE. The durability of structural joints made with cyanoacrylate adhesives must be of some concern under elevated temperature conditions, since most polymerized cyanoacrylates are thermoplastics with relatively poor heat resistance compared to thermoset adhesives like the epoxies. In addition, they may be susceptible to depolymerization at elevated temperatures. Also, those with the highest softening points like poly(methyl 2-cyanoacrylate) will be especially brittle in the bondline; consequently, they can fail owing to temperature cycling because of the differences in the coefficients of thermal expansion between these adhesives and the adherend. Obviously, the joint situation involving two dissimilar adherends with different coefficients of thermal expansion could be a particularly difficult problem for such brittle adhesives.

An upper temperature limit for sustained use of cyanoacrylates in structural joints might be considered to be about 160°F (70°C), although short exposure can be tolerated to about 212°F (100°C). Kotzev et al. (3665) considered cross-linked polymer produced by allyl cyanoacrylate with a higher heat resistance. However, Kotzev et al. (3666) point out that further polymerization may continue in elevated temperature service, resulting in eventual joint strength decline due to further embrittlement. Any joint stressing during this period will only hasten the time to failure.

(2) MOISTURE AND AGING RESISTANCE. Drain et al. (3667) have identified moisture migration into the interface of cyanoacrylate-bonded metal joints as formation of a weak boundary layer. The same investigators have demonstrated the much superior resistance of cyanoacrylate bonds to plastics under the same conditions, especially where the plastic has some solubility in contact with the adhesive. Coover and McIntire in Skeist's Handbook have reported observing a gradual loss in strength of cyanoacrylate-bonded joints over several years. In the presence of high-energy adherend surfaces, however, the water can cause chemical changes and more rapid joint failure. Bonds on glass deteriorate especially rapidly in water exposure as do metals that can be easily corroded.

(3) SOLVENT RESISTANCE. The polymerized cyanoacrylates can be removed with certain polar aprotic solvents. Different types have different solubility parameters, making them more or less soluble as solvents of increasingly higher polarity are employed. Donnelly and Pepper (1669) have offered a discussion of this subject.

h. Polyurethane Adhesives

One of the first known uses of polyurethane adhesives was for the bonding of rubber-to-metal tank treads by Germany during World War II. While this application has distinct structural aspects to it, the preponderance of polyurethane adhesives used commercially have been for bonding elastomers, fabrics, fibers, and thermoplastics in considerable less structural capacity. Two-part polyurethane adhesives are currently employed to bond polyester fiber glass to steel parts structurally in the automotive industry and could similarly be used for bonding aluminum adherends.

Minford (1742) has shown that the two-part epoxy primer used for durably bonding steel with the polyurethane adhesive must also be employed over aluminum based on long-term water-soaking testing. MacIver and Thompson of Goodyear (1732) have provided some data in Table 165 on the effect of various environmental conditions on the tensile strength of polyurethane bonds between various types of sheet moulding (polyester fiber glass composition) compounds, cold rolled steel, and aluminum adherends. Although the exposure conditions were for relatively short times, they seem to indicate that durable joints could be made to all three adherend surfaces.

Table 165 The Effect of Various Environmental Conditions on the Tensile Strength (psi) of Polyurethane Bonds Between Various Substrates

Testing conditions	Sheet moulding compound (low profile)	High glass sheet moulding compound	Directional glass sheet moulding compound	Cold rolled steel	Aluminum
-40°F	646	956	2500 (C)	3205 (C)	1638 (C)
	740	1232	2744 (C)	2550 (C)	1050 (C)
180°F	480 (C)	471 (C)	690 (C)	794 (C)	700 (C)
250°F	280 (C)	293 (C)	304 (C)	525 (C)	580 (C)
14 da at 190°F	852	952	2610 (C)	2520 (C)	1650 (C)
14 da water immersion at 72°F	834	1158	1958 (C)	2395 (C)	1624 (C)
14 da at 100°F at 100% RH	620	1076	1970 (C)	2170 (C)	1604 (C)
500 hr weatherometer	798	1250	2736 (C)	2557 (C)	1043 (C)
Flex fatigue (cycles)	4×10^6	2×10^6	2×10^5	1×10^5	—
Mode of failure	None	Plastic	Plastic	Stress cracks steel	
Slam test	Passes	Passes	Passes	Passes	Passes

All bonds resulted in substrate failure except where (C) denotes cohesive break.
Data from Ref. 1732 (Table 2, p. 204).

More recently, Goodyear has added a one-part, heat-curing product to their list of available polyurethanes which is recommended for the same market applications. Minford (1744) has evaluated this product on 6061-T6 aluminum in the same manner as used for evaluating the two-part product above (1742). The recommended vinyl primer was not able to sustain good adhesion under the accelerated weathering conditions, so that the durability of the polyurethane could not be compared with the two-part material. Twiss (1745) and Smith and Susman (1746) demonstrated the superior shear and toughness of the polyurethane adhesives at cryogenic temperatures. However, they can rapidly lose this strength above 176–212°F (80–100°C), and this has been a serious limitation to their use in structural bonding. For example, Minford (3668) measured black aluminum surface temperatures on laminated building panels in desert areas above this temperature. In 1990, Krieder (1269) reported that specially formulated products could have maximum service temperature range as high as 300–325°F (150–163°C).

Mobay Comporation has been very active in the development of polyurethane chemistry and reported a very high peel and tensile-shear strength polyurethane-type adhesive with excellent moisture resistance even on unprimed steel and aluminum. However, the tensile strength was shown to diminish by 50% from the room temperature value when the temperature reached 212°F (100°C).

Delmonte and Sarna (1750) discussed some high-strength polyurethane adhesive reporting lap-shear values exceeding 5000 psi (34.45 MPa) and peel strengths exceeding 60 lb/in. Their overall assessment was that these adhesive were more than competitive with the epoxies of the same period. However, durability testing actually showed that the adhesives on etched aluminum adherends produced joints which lost about 48% of initial strength in 95% RH at only 75°F (24°C) in 1 mo. This would be unacceptable for most metal-bonded joints even in interior service application conditions. Delmonte (1752) produced an additional report 2 yr later that still failed to be convincing about accepting polyurethane as a replacement structural adhesive for most epoxies. DeLollis (1765) also conducted some durability testing of polyurethane-joined adherends using both water-immersion and soaking in hot high humidity. The environmental concern about using MOCA-curing agents with polyurethane compositions has resulted in the search for other means of curing. Lauer and

Boyaner (3669) have reported the formulation of non-MOCA (4,4′-methylenebis[2-chloroaniline])-cured polyurethane with good properties development for −300°F (−184°C) service.

i. Anaerobic-type Adhesives

The uniqueness of this type of adhesive stems from its ability to be kept in the unreacted state for indefinitely long periods; i.e., as long as oxygen is excluded from the environment. This is a very appropriate property to have where two vapor-barrier adherends are to be joined such as aluminum to itself or another metal. The general reaction, even in the absence of oxygen, is inherently relatively slow. However, this problem has been overcome by rapidly accelerating the polymerization process through the use of special accelerator chemicals and elevated temperature. The Redux catalysts have proved to be sensitive to the kind of metal ion present, so different curing rates can be found when different adherend surface are employed for adhering.

The main markets for such adhesives seem to be where the joint once made will be impervious to oxygen migration into the interface. This situation can be present whether thin or thick metal adherends are involved. Some forms of anaerobics have proved to be uniquely appropriate for sealing porosity in aluminum castings. The anaerobic formulations are available in very low-viscosity versions which can penetrate very fine pore structure. The anaerobics do sell for unit prices that can be an order of magnitude higher than for other adhesives like the structural epoxies. However, the anaerobics are often the only material available; for example, where adhesive penetration is required into very close tolerance fits. Also, the thin bondlines that result combined with their excellent spreading capacities can make even the overall costs much more competitive than might first be anticipated based on cost of anaerobics on a purely weight basis.

Pearce (1706) gave a 1972 review of the successful use of anaerobics, including the necessity to exercise special care to achieve the required level of joint durability. There needs to be some concern about adhesive fracturing under stress and cycling service conditions, since these materials are relatively brittle in the joint. Also, there can be a special durability problem when trying to join materials of differing coefficients of thermal expansion. Thus, sudden thermal shocks that can cause debonding by fracturing must certainly restrict the application range under some service conditions.

However, probably no class of adhesives in the marketplace has shown such increasing improvement in recent years. The chemistry of these materials is so unique that countless modifications have been possible that have resulted in substantial joint durability potential even under the most aggressive weathering conditions. Minford (3670) has evaluated two of these newer products under the trade names Superbonder 324 and 325. The former is considered a more general-purpose product, whereas the latter has been formulated for better resistance to high temperatures above 250–275°F (121–135°C) and water-soaking service. Under testing conditions involving simultaneous exposure to stress and humidity, both adhesives proved to be superior to general-purpose two-part epoxies but not to the best second-generation modified structural acrylics. The water resistance of these products seemed much improved over earlier developed anaerobic products and compared favorably with the performance of the typical room temperature–curing commercial epoxies. On higher-purity aluminum such as 3003 alloy (as compared to more structural 6061-T6, 2024-T3, and 7075-T6 alloys) both Superbonder products formed excellent bonds to vapor-degreased–only surfaces. These joints survived more than 1 yr in corrosive intermittent 3.5% salt spray exposure conditions. Minford also investigated 6061-T6 adherends which required some acid deoxidation to effectively remove the heavy heat-treat oxide film in order to assure similar durability.

A comprehensive review of the technology, as it existed in 1977, was offered by Murray (3671). Lee and Bruce (1711) offered data on the thermal stability and aging characteristics of anaerobics in 1980. The same year, Hauser and Loft (1714) discussed what kinds of adherends can be bonded effectively, whereas in 1984, Malofsky and Baccei (1713) offered the possibility for anaerobic products with improved thermal resistance by formulating with maleimides. Melody (1270) has offered the most recent review of the technology in the 1990 ASM Handbook but does not offer any information on any investigations involving the overall joint durability potentials under various weathering conditions.

j. Adhesive Use at Cryogenic and High Temperatures

As the space age dawned, designers began to talk more about the need for adhesives that would function to bond structures that had to operate at both the low cryogenic temperatures of space (−424°F [20°K]) and temperatures that are developed on the metal skins of aircraft flying at supersonic speeds. Most demanding was the very high temperature that could develop on spacecraft returning to the atmosphere from space. From the early 1960s, an international assault was launched to develop new families of polymers for use above 600°F (316°C). Initial efforts concentrated on improvements in fluorocarbons and silicone elastomers, but new heterocyclic aromatic polymers like polyimides, polybenzimidizoles, and polyphenyquinoxalines soon appeared. Such efforts produced products which retained useful joint strength after extended exposure times in the temperature range of 600–1000°F (316–538°C). These kinds of exposures were capable of destroying even an epoxy-phenolic aluminum joint within 1–10 hr.

Durability evaluations of these high-temperature and cryogenic service adhesives proceeded differently than outlined previously for the various adhesive polymer products discussed to this point. Measurements of their properties are often quite relatable to their service potential. Important data on their overall durability might include (1) how long they can survive at some extremely low or high temperature?, and (2) at how high a temperature can they maintain their structural strength properties? How strong they are at any given temperature is also a direct reflection of how well they have wet the adherend surface. Getting them to effectively wet adherend surfaces has certainly been one of the problems with using these adhesives, since they must flow over the surface at very elevated temperatures in the bond formation procedure. It would seem obvious that the effect of water on such adhesives is less important, since it cannot even exist in the environment at these extremely high service temperatures. However, the effect of liquid or vapor water might be a factor in situations where these adhesives have been employed on structures which are stored in buildings or silos until they are used as missiles or spacecraft.

Data on properties under extreme temperature conditions has been provided by a number of investigators over the years. For example, Wrasidlo (3672) reported as early as 1974 on the measured thermal properties of a variety of such adhesives, including polyimides, polyphenylene ethers, and polyphenylquinoxalines. He also included information on their transition and relaxation properties. At the same SAMPE symposium, Hergenrother (1870) offered information on polyphenyl-as-triazines and polyquinoxalines as either adhesives or in composite matrices. Aponyi and Delano (3673) added studies of the polyamidazoquinazolines in high-temperature–reinforced composites. The ability of BR34/FM34 polyimide joints to maintain high shear strength over a broad temperature range was measured by Roper (1831). Vaughan and Jones (3674) added that P4/A5F polyimide showed significant property and processing advantages over the BR34/FM34 system. Further testing of both systems was undertaken by Vaughan and Sheppard and reported in 3 yr along with corresponding results from a Boeing polyphenyquinoxaline adhesive and DuPont R1508 polyimide. The typical testing included (1) measurements of static strength and thermal shock from −423°F (20°K) to 600°F (589°K), (2) aging effects and 50% stressing at 400°F (477°K) to 600°F (589°K). At higher temperatures, P4/A5F joints were the strongest of all those evaluated, indicating superior resistance to adhesive oxidation. With stress and increasing aging, the polyphenylquinoxaline proved superior. The strength test results at the cryogenic level showed BR34/FM34 as being superior. Obviously, the best compromise adhesive may vary depending on the particular service condition envisaged.

Another problem involved with these very high molecular weight polymers has been the development of a significant number of voids in the bondline due to volatiles produced enroute to full curing. St. Clair and St. Clair (1836) subsequently developed a linear thermoplastic polyimide which can be imidized and freed of volatives at the very low temperature of 446°F (230°C) and then handled in the bonding operation as though it were truly thermoplastic. At the same time, these linear polyimides have proved to be very practical, because they exhibit unusual toughness with attractive flexibility over a wide temperature range. The same investigators (1837) studied in detail the effect of adding elastomers like the amine-terminated acrylonitrile-butadiene copolymers, or silicones. As might be expected, some high-temperature stability had to be sacrificed following the rubber elastomer additions, whereas silicone elastomer additions improved the aging resistance at 450°F

(505°K). The ambient- and elevated-temperature fracture toughness of such modified polyimides was improved over three- to fivefold with rubber additions. Silicone additions had little effect, if any, on the fracture toughness. Thus, we might interpret that the joint durability could be advanced at higher temperature service by the silicone additions, but under intermittent impact stresses a compromise with some rubber addition might be the most acceptable.

Rossi (1276) has provided the most current review of polyimide technology. He points out that there is a distinct competition involving the development of various polymeric systems as high-temperature adhesives. In addition to the polyimides whose development he considers in some detail, reference is made to disclosures about benzocyclobutenes by Kirchhoff (3675,3676) in 1985 and 1986 and polybenzimidazoles and polyquinoxalines by Chih-Ching (3677) in 1988 and Lee (3678) in 1987. He further speculates that it is not likely that future trends will involve new monomers and polymers because of the difficulty and cost of developing and bringing new polymers into the marketplace. It is more likely that combinations of existing systems in partially interpenetrating and full interpenetrating polymer networks (IPNs) and other compatible blend forms will be used to compensate for individual system deficiencies (3679–3683).

As expected, the main application area for the use of such adhesives has been in the bonding of titanium-to-titanium joints. Hergenrother (1878) has evaluated polyphenyquinoxaline adhesives for joining titanium to itself, titanium-to-graphite/polyimide composite, and the composite to itself as early as 1971. There was a surprising superiority of joint strength retention at 600°F (316°C) for the composite joints, which as explained as a partial softening of the composite material offering a less brittle joint behavior with better stress distribution. Studies of a variety of polyphenyquinoxaline adhesives listed chronologically from the 1990 review article of Wentworth (1277) include Hergenrother (1803,3684,3685), Hendricks and Hill (1880), Hendricks et al. (3686), and Havens et al. (3687). The most current review of polybenzimidazoles is that of Connell from NASA Langley (1278).

The development of durable structural foams for the use at cyrogenic to ultrahigh temperatures has been a continuing effort of a number of investigators, including Allen and Yates (3688), Segal (3689), and Kimmel (3690), all as early as 1974.

C. Sealant Durability Evaluations

1. *GENERAL BACKGROUND*

The service-life potential of many aluminum-bonded structures is related to the ability of the design engineer to keep the degrading effects of weathering from entering the structure and ultimately the interfacial area. The most widely employed procedure for keeping weathering out of structures has been to use a variety of the same general types of organic polymers as used in adhesives but in the different formulated compositions known as sealants. These materials can be extruded or troweled over the jointures between the adherends to fill the gap areas that exist in most assemblies or between assemblies. While the sealant compositions do not generally have sufficient tensile-shear strength to be considered for making the structural joints themselves, they do have good adhesion potential to form a seal which must maintain that adhesion if sealing properties are to be maintained.

Their primary function is to fill gaps with the subsequent development of sufficient adhesion and cohesive strength through various curing mechanisms to form a long-lasting barrier to the weather. In making evaluations of their potential to accomplish these purposes, it is necessary to measure their initial adhesion strength to any material surface that is to be protected. Any changes that might occur between initial physical and chemical properties and the existent properties after simulated or actual periods of exposure to pertinent service conditions must be measured to predict long-term durability potential. Each separate combination of particular adherend surface and particular sealant formulation must be separately evaluated because data involving one particular combination is not applicable to another. For aluminum, it is necessary to consider that a different sealant response might be found depending on whether the adhesion was being generated against a bare, abraded, etched, conversion coated, anodized, or painted (pigmented or clear) aluminum surface. For bare (mill-finish or mill-finish plus degreased), abraded, or deoxidized aluminum surfaces, there could be a different response from different alloys.

Wide variation can also exist within general chemical families of sealants because of the variation in kind and concentration of special additives such as plasticizers, fillers, thixotropic agents, or other additives used to impart specific properties. In the latter category, we find additives to influence adhesion and/or package stability. In waterborne sealants, there might be extra additives to protect against bacteria, freezing (antifreeze solvents), or surfactants to disperse the filler, aid in extruding, and generally improve the surface wettability potential. In still other formulations, there may be special adhesion-promoting agents or chemical-coupling agents (like silanes or titanates). Protection against UV light requires special antioxidants and/or UV absorbers. Also, there can be special catalyst additives which can improve curing characteristics. With this range of additives superimposed over the basic characteristics of the wide range of chemical polymers, it is clear that specific testing must proceed for each combination of sealant and adherend.

Four classes of sealant performance are recognized as being (1) low, (2) medium, (3) high, and (4) advanced performance (3691). This refers to the main requirements for performance in terms of the movement capability of the sealant. The "advanced-performance" sealants in aerospace would add the requirement of high resistance to extremely high and low temperatures as present in aerospace service. The low-performance types actually may not even have a distinct curing mechanism, being commonly based on air oxidation of unsaturated carbon-to-carbon linkages present in fatty oils. The only curing results when solvent or water has evaporated from the sealant bead. Their low cost is balanced by the presence of relatively low flexibility and tendency to crack even under modest joint movement.

In the medium-performance category, we find the solvent-base acrylics and butyl rubber sealant formulations. The loss of solvent creates some undesirable joint shrinkage which must be compensated for by the application. Other variations in this category have been the latex sealants based on vinyl acetate homo- and co-polymers, vinyl-acrylics, and acrylics. Being able to paint such sealants has helped to improve the durability performance. More recently, the joint-movement capability of the latter products has been improved to the point where they can virtually meet the high-performance-type specifications.

In the high-performance category are the polyurethanes, polysulfides, and silicones (3692). Their curing mechanisms provide for little shrinkage after cure with excellent joint movement potential. For hot-melt sealant application procedures, the styrene block copolymer types fall into this overall performance category. Foster (3692) describes the transparent acrylic latices as high-performance products with good weatherability. There are still relatively few commercially available aerospace sealants that have service temperatures in the desired ranges. FASIL has been mentioned earlier to perform in the service range of −66–500°F (−54–260°C). Also mentioned were the phosphazenes from Ethyl Corp., which perform in the −85–345°F (−65–175°C) range. Usmani has provided an excellent table of comparable sealant performance (see Ref. 3691, Table 5, p. 675 for additional study).

2. *SEALANT-DURABILITY EVALUATION*

It is resistance to the weathering elements that usually determines the lifetime of a sealant in a particular service application. Useful life can also be influenced by degradative processes occurring during polymer synthesis and the sealant processing to service-life limits. Durability requirements can actually vary quite widely, since stresses may be imposed on the sealant and its attendant adhesion by many processes. These include (1) mechanical processes, (2) heat and cold, (4) radiation (UV, x-ray, nuclear), (5) humidity, (6) gases (oxygen, ozone, sulfur dioxide), (7) solvents (jet fuels, oils, hydraulic fluids, and water), (8) corrosive chemicals, and 9) bacteria. There can actually be distinct physical and mechanical property change in the sealant polymer due to chain scission, depolymerization, substitution reactions, and elimination and cross-linking changes. Usmani has offered an excellent visual schematic linking the various degradations sources to polymer bond scission and ultimately degraded polymer (see Ref. 678, Fig. 9, p. 678).

There are two major sources of potential sealant failure that can be broadly characterized as (1) the ability to withstand the overall expansion and contractions imposed by low to high-temperature change on the overall sealed structure, and (2) the ability to withstand the combined effects of moisture (water), temperature, radiation, and chemicals in the service environment. The former is further affected by the factors of stress, temperature, rate of deformation, humidity, light, air, adherend surface, liquid water, and other chemicals. Thus, tensile testing under controlled conditions along with some mathematical manipulations may be able to predict a sealant's performance and overall durability.

Water effects need to be considered separately, since polymers, in general, are permeable to moisture or liquid water which subsequently provides water to the interface and the surface of the adherend as well. A significantly lower water-permeability property is shown by the fluorine and silicon-containing polymers. Further, most sealants are polar in nature, which while aiding in promoting adhesion to polar metal oxides, can promote moisture diffusion in the bulk. Among the construction sealants, polysulfides, polyurethanes, and acrylics are all moisture sensitive to some degree. This has been offered as the major cause for failure of insulated glass units along with poor workmanship in production. Condensed trapped water will swell and hydrolyze a polysulfide sealant, for example, and sealant cohesive failure may result. Of course, any physical swelling of any sealant will automatically induce stresses at the interface, causing a bond failure.

In the late 1950s and early 1960s, Minford initiated a continuing long-term durability test program ending with the evaluation of some 67 sealants, including candidates from the oleoresinous, butyl, nitrile, polysulfide, acrylic, polyurethane, Hypalon, and silicone families. Periodic observations and measurements were made for periods out to 10 yr and relative ratings assigned to describe different observations on (1) the degree of adhesion retention, (2) consistency of the sealant as joint filler, (3) degree of hardness, (4) degree of shrinkability, and (5) degree of soiling accumlation. The first three factors were considered of greatest importance in extending service life and were appropriately weighted in calculating the final evaluation averages. The overall summary observations after 7–10 yr of exposure in the natural industrial atmosphere outside Pittsburgh can be seen in Table 166. Because many

Table 166 Summary of Performance[a] of 67 Sealants and Caulking Products on Bare Aluminum After 4–10 Yr of Weathering in the Industrial Atmosphere

Chemical family type	Adhesion	Resilience		Shrinkage	Soil accumulation
		Consistency	Hardness		
Oleoresinous	Fair	Poor	Poor	Moderate	Light to moderate
Butyls	Fair	Good	Good	Slight	Heavy
Polysulfides	Good	Excellent	Excellent	None to slight	Light
Nitriles	Fair to good	Excellent	Fair to good	Slight	Light
Acrylics	Good	Excellent	Fair	Excessive	Moderate
Urethanes	Inadequate sampling for meaningful averaging				
Hypalons	Range of performance to broad for meaningful averaging				
Silicones	Poor[b] to excellent	Excellent	Excellent	None to slight	Light

[a]A system of judging relative performance of sealants on aluminum was based on periodic observation over periods from 4–7 yr in real-life industrial atmospheric weathering with separate qualitative visual judgments made about each of the five properties listed above. For more quantitative judgments, relative numerical values were assigned to the above descriptive terms for each individual sealant. Averages for each separate sealant or each separate family of sealants could then be made.

[b]The silicone sealants tested here were among the earliest commercial products and the improvements made in recent years have been outstanding. While the silicone sealant manufacturers originally advised the use of special primers for their products on aluminum, formulations are available today for use without priming which offer excellent adhesion. However, any sealant (regardless of type) should be tested in some fashion before acceptance for long-term service.

Data from J.D. MInford, unpublished work, Aluminum Co. of America.

aluminum products were not anodized or painted for early aluminum building applications, the sealant adhesion was first tested only against a bare 3003 alloy surface. This was, of course, the most commonly employed alloy for roofing and siding applications and, thus, offered more directly relatable data for field application of sealants.

Polysulfide sealants appeared to be the most durable as a group, with selected butyl, nitrile, and acrylic sealants also showing similar service potential responses after 9–10 yr of exposure. Silicone sealants, of course, were equally durable providing good initial adhesion was found. More recently, the RTV silicone sealants have proved to have excellent initial adhesion without the need for surface priming; however, 40 yr ago when these initial tests were being conducted, it was typical to have to employ a specific primer for a specific silicone on a particular aluminum surface. It will be evident that polyurethane sealants have been omitted from Table 166, but again, as with silicone sealants, the early polyurethane sealer's performance was yet to be optimized for aluminum. In a later investigation, Minford (3693) evaluated a two-part polyurethane sealant on aluminum for 7 yr in the same industrial atmosphere with fair adhesion retention, retention of a soft and pliable resilience, only moderate volume shrinkage, and no objectionable soiling accumulation. Similar performance was obtained with a one-part package polyurethane except there was a light accumulation of soiling.

Other Alcoa-accelerated weathering-type tests also have been conducted with sealants on various aluminum product surfaces by exposing sheet specimens with center-located sealant beads to 500 hr in a Twin-Arc Weather-O-Meter. Duplicates could be hung vertically in a continuously condensing humidity cabinet held at 125°F (52°C). Further duplicates could be inclined on exposure racks to the

Table 167 Performance of Commercial Sealants and Caulks on duPont 1234 Lacquered Alumilite 400 and Duranodic 312S Anodized Aluminum Surfaces Exposed 9 Mo to Industrial Atmospheric Weathering

	Alumilite 400		Duranodic 312S			
Sealant	Adhesion	Type failure	Adhesion	Type failure	Sealant shore hardness	Soiling condition
2-Part polysulfide	Excellent	Cohesive	Excellent	Cohesive	25	Moderate
(+ primer)	Excellent	Cohesive	Excellent	Cohesive	25	Moderate
Different 2-part polysulfide	Poor	Adhesive	Poor	Adhesive (lacquer soft)	50	Moderate
(+ primer)	Excellent	Cohesive	Excellent	Cohesive	50	Moderate
1-Part polysulfide	Excellent	Cohesive	Excellent	Cohesive	50	Moderate
(+ primer)	Excellent	Cohesive	Excellent	Adhesive	50	Moderate
Different 1-part polysulfide	Poor	Adhesive	Poor	Adhesive	60	Slight
G.E. silicone + primer	Excellent	Cohesive	Excellent	Cohesive	50	Moderate
Dow silicone + surface conditioner	Excellent	Cohesive	Excellent	Cohesive	30	Moderate
Butylseal	Fair	Adhesive	Poor	Adhesive	30	Heavy
Acrylic	Excellent	Cohesive	Excellent	Cohesive	50	Heavy
Tremco Dymeric + Primer No. 1	Excellent	Cohesive	Excellent	Cohesive	40	Moderate
Tremco Dymeric + Primer No. 2	Poor	Adhesive (lacquer soft)	Poor	Adhesive (lacquer soft)	40	Moderate
1-Part polyurethane	Excellent	Cohesive	Excellent	Cohesive	45	Moderate
Vinyl acrylic	Excellent	Cohesive	Excellent	Cohesive	40	Heavy

[a]Items 1 and 3 were classified as the same chemical type of sealant but from two different manufacturers.
[b]Lacquer was aged for 4.5 mo under ambient laboratory conditions before sealants were applied and specimens placed in atmospheric weathering.
Data from J.D. Minford, unpublished work, Aluminum Co. of America.

natural industrial atmospheric conditions. In the latter two test environments, sufficient specimens would be employed to permit peel testing after periods of 3, 6, and 12 mo. Testing in the natural environment would often be extended out to periods as long as 10 yr, as mentioned earlier. Test results from one of these Alcoa studies is shown in Table 167 (3694), where some 11 commercial sealants and some recommended primers representing 7 different sealant chemical families were extruded over one commercial surface of interest (a clear, air-dried lacquer over an anodized finish recommended by Alcoa for aluminum monumental building sheathing). Adhesion failure could occur in a number of locations such as (1) adhesive between lacquer and anodized oxide, (2) cohesive in lacquer, (3) adhesive between sealant and lacquer, or (4) cohesive in sealant. The data can be readily interpreted as to which sealants showed the most long-time durable adhesion but several other points are worth mentioning. For example, one of a pair of one- or two-part polysulfide sealants might show excellent adhesion retention, whereas another of the same chemical family might only be able to establish the same performance with a special primer. A butyl rubber sealant which had previously proven valuable as a sealant in the field for durably sealing bare aluminum gutters was totally ineffective owing to solvent attack on the particular air-dried lacquer coating. A solvent acrylic provided excellent adhesion under identical weathering conditions. Some sealants also developed a significant rigidity (increased shore hardness), which would undoubtedly contribute to earlier cracking and probable adhesion failure under increased thermal cycling in the service environment.

Minford (3695) further employed special exposures to hot condensing humidity in special cabinets or 1-yr exposures to an industrial atmosphere to evaluate adhesion durability of 12 different sealants, including silicone, acrylic, polysulfide, butyl, polyurethane, and oleoresinous chemical families. The data shown in Tables 168 and 169 were obtained with aluminum adherends prepainted with four different commercial heat-cured paint coatings used for long-term service on architectural buildings,

Table 168 Sealants Adhesion to Prepainted Aluminum Adherends Based on Exposure to 100% RH at 125°F (52°C) Conditions

	PVF paint 1[a]				PVF paint 2[a]				Silicone polyester[a]				Water-based acrylic[a]			
Sealant	A	B	C	D	A	B	C	D	A	B	C	D	A	B	C	D
2-Part (No. 1) polysulfide	X				X				X				X			
(+ primer)		X			X				X				X			
2-Part (no. 2) polysulfide			X		X				X				X			
(+ primer)	X				X				X				X			
1-Part (No. 1) polysulfide	X						X				X				X	
1-Part (No. 2) polysulfide			X				X					X				X
Acrylic (No. 1)	X				X						X		X			
Acrylic (No. 2)	X				X						X		X			
Silicone (No. 1) + Suf. conditioner				X	X				X						X	
Silicone (No. 2)	X						X			X			X			
Silicone (No. 2) (+ primer)				X			X				X				X	
Butylseal	X				X				X					X		
Tremco Dymeric			X				X				X		X			
1-Part polyurethane	X					X			X						X	
Oleoresinous	X				X				X				X			

[a]A, Peel strength of sealer either increased, was unaffected, or decreased by 20% or less. B, Peel strength declined more than 20% but was still 10 lb/in or higher. C, Peel strength decline exceeded 20% but was less than 10 lb/in. D, Peel strength declined to 0 during exposure.

Data from J.D. Minford, unpublished work, Aluminum Co. of America.

including aluminum siding. The paints represented two varieties of polyvinyl fluoride (PVF) commercial paint formulations, one silicone-polyester formulation, and one water-base acrylic paint formulation. The test specimens were designed to permit peel testing after periodic exposure times (multiple bead specimen test panels). An arbitrary comparative durability evaluation procedure placed each sealer performance in a relative classification depending on which category of retained peel strength applied. These categories were (1) decline to less than 89% of initial peel strength but with >10 lb/in retained peel strength, (2) decline to less than 50% of initial peel strength but with >10 lb/in retained strength, and (3) failure in adhesion during exposure. An examination of the data indicated the overall performance of a particular sealant could be quite different depending on the weathering conditions employed to evaluate. Second, the same sealant can offer significantly different adhesion retentions on different paint formulations though from the same chemical family; i.e., the first two-part polysulfide sealant in Table 168 shows excellent adhesion retention to all four painted surface variations after hot humidity–soaking conditions. The same sealant formulation in Table 169 meanwhile showed excellent adhesion performance after 1 yr in the atmospheric weathering to two paint surfaces (one PVC and the water-based acrylic formulation), whereas the adhesion was totally lost to the other PVC paint which had fared well in condensing hot humidity. The other result noted was the total loss of bond by the oleoresinous sealant to all four paint finishes before 1 yr in the natural atmosphere, yet demonstrating the highest peel strength retention to the same four surfaces throughout the hot condensing humidity exposure. The author might offer the speculation that the increase in durometer of the oleoresinous sealant under the simultaneous ultraviolet exposure conditions of the atmosphere plus the temperature cycling conditions of the atmosphere exposure had to contribute greatly to the end result. Other similar specimens were also placed in the weather-o-meter cabinet for 500 hr with the result that the majority showed peel-strength increases instead

Table 169 Comparative Ratings of Adhesion of Various Sealants to Prepainted Aluminum Surfaces Based on Exposure for 1 Yr to an Industrial Atmosphere

	PVF paint 1[a]				PVF paint 2[a]				Silicone polyester[a]				Water-based acrylic[a]			
Sealant	A	B	C	D	A	B	C	D	A	B	C	D	A	B	C	D
2-Part (No. 1) polysulfide	X							X			X		X			
(+ primer)	X					X			X				X			
1-Part (No. 2) polysulfide			X				X				X				X	
(+ primer)			X				X				X				X	
1-Part (No. 1) polysulfide	X							X	X				X			
1-Part (No. 2) polysulfide	X							X	X				X			
Acrylic (No. 1)	X				X				X				X			
Acrylic (No. 2)	X				X				X				X			
Silicone (No. 1) + Surf. conditioner				X				X			X				X	
Silicone (No. 2)				X				X	X							X
Silicone (No. 2) + primer	X						X		X							X
Butylseal	X				X				X				X			
Tremco Dymeric			X		X					X			X			
1-Part polyurethane	X					X			X				X			
Oleoresinous				X				X				X				X

[a]A, Peel strength either increased, was unaffected, or decreased by 20% or less. B, Peel strength declined more than 20% but was still 10 lb/in or higher. C, Peel strength decline exceeded 20% and was less than 10 lb/in. D, Peel strength declined to 0 during exposure.

Data from J.D. Minford, unpublished work, Aluminum Co. of America.

of decreases. Thus, depending on the particular weathering conditions chosen for the durability comparison, the investigator might arrive at some different conclusions concerning acceptance of these products for the industrial application of sealing painted aluminum joints.

3. COMPATIBILITY EVALUATIONS OF SEALANTS

Another factor that can affect the bond durability of either an adhesive or a sealant may be its general chemical compatibility with the adherend surface. For example, Minford (3696), at the Alcoa Labs, investigated the compatibility of 17 different sealants when sandwiched between bare 3003-H25 aluminum sheet faces or between one sheet of aluminum and one of mild steel. An electrical couple was made between the bimetallic facings further to study whether an additional negative factor could be the resultant electrochemical couple. After 6-mo hot humidity soaking, most sealants proved to offer complete chemical compatibility with the aluminum. However, one silicone building sealant produced severe pitting and perforation of the aluminum sheet even in the uncoupled specimen. Another silicone produced a mild pitting corrosion situation, whereas a third silicone from still a different formulator produced no visual effect. Meanwhile, a neoprene-Hypalon mix showed only a small area of mild "worm-track-type" corrosion on the aluminum. In the duplicate specimen coupled to steel, this mild corrosion was somewhat accelerated. The corrosion produced on the coupled specimens with the two incompatible silicones appeared to be actually milder in nature, demonstrating that aggravation of the corrosion by the presence of a deliberate electrical couple did not necessarily occur. It should be pointed out that a direct relation did seem to exist between the degree of corrosion that did occur on an aluminum surface, the decrease in bond strength, and the subsequent service potential of a structural joint or sealant in contact with that surface.

In addition to the possible incompatibility of the sealant with the adherend materials, the service life of a laminated aluminum structure could be affected if the sealant can deteriorate other components of the laminated sandwich. For example, Minford (3697) evaluated the effect of solvents present in a variety of sealant formulations with the foamed polystyrene or polyurethane core of Alcoa laminated building panels. While none of the tested sealants (acrylics, nitrile rubbers, vinyl-acrylics, or neoprene rubbers) had any adverse chemical effect on the thermoset foamed polyurethane cores, they did significantly attack and seriously degrade typical foamed polystyrene insulating cores. However, it could be easily demonstrated that an extruded-form of polystyrene insulation (such as Styrofoam) was much less sensitive to such solvent-containing sealants than the expanded bead form of polystyrene.

There is a distinct need to know that any organic sealant which is to be recommended for use to seal joints in aluminum gutters is chemically compatible with the aluminum under continuously wet conditions. Minford (3698) evaluated seven commercial sealant candidates which had been recommended for this application by their formulators. The author found that all seemed to be acceptable during an evaluation in 6-mo continuous hot humidity soaking at 125°F (52°C). This was true, however, only if there was no steel electrically coupled to the aluminum in the overall rain drainage system. There was sufficient variation in retention of overall adhesion among the commercial sealer products evaluated by the author over many years to conclude firmly that some measure of comparative testing must be conducted before acceptance by any manufacturer for a particular product. It is certainly critical to reject any sealant which can generate aluminum surface corrosion, since that condition would automatically lead to a poor adhesion retention and service-life potential for any manufactured product requiring sealing of jointures.

4. SEALANT ADHESION TO GASKETING

A number of manufactured aluminum structures have been designed over the years which incorporate other materials as gasketing; for example, neoprene window gaskets or four-way patches over intersections between multiple building panels. While it is usually implied that a gasket should be designed separately to keep out weathering, it is often necessary to supplement the sealing with additional extrudable mastic-type sealants. Minford (3698) has shown the necessity to evaluate the durability of the adhesion between such sealants and a neoprene overlay surface, in addition to the evaluation of the separate adhesion between this sealant and bare, painted, or anodized aluminum panel surfaces. Sealant beads were deposited on the neoprene sheeting and exposed vertically in a continuously condensing humidity cabinet at 125°F (52°C) for periods up to 6 mo. Duplicates were also set out to weather in the natural atmosphere for periods up to 1 yr or longer. Only a one-part polyurethane commercial sealant candidate retained good adhesion to the neoprene under these service conditions, whereas best selection silicone, acrylic, polysulfide, and Hypalon sealant compositions based on other earlier investigations only showed poor-to-fair adhesion retention.

VIII. JOINT PERMANENCE AS A FUNCTION OF SURFACE PRETREATMENT

A. General Relationships

The effect of surface pretreatment on the aluminum oxide surface is critical and intimately related to obtaining a high degree of aluminum joint permanence. It remains difficult to distinguish adequately the benefit of different surface treatments unless the differently treated joints are exposed to some environmental condition representative of what the joints will experience in service. Aluminum joints often show complete cohensive failures when initially tested in a tensile machine, yet show a relatively short survival time even under modest active weathering. This is even more likely to occur where some moderate to high level of stress is simultaneously imposed. Thus, extended exposure of test joints in natural or accelerated laboratory weathering (with or without additional stressing) has remained the only realistic procedure for gaining confidence that the right adhesive, adherend, and surface treatment have been selected for the manufacturing of a product.

The relative order of effectiveness of different pretreatments on aluminum can vary depending on the weathering condition and stressing level chosen for evaluation. The usual acid-etching treatment can often rate relatively high when the test environment is water oriented alone. In corrosive salt-water exposure, most acid-etching treatments do not perform as well as might be desired. The author has found two exceptions to this generalization, which will be discussed later in this chapter, i.e., longer-time etching in RT chromic acid solution or short-time etching in RT HCl solution. At the same time, aluminum adherends treated by acidic deoxidation followed with conversion coatings or anodizing conditions have demonstrated superior adhesion retention by epoxy adhesives even in corrosive salt-water conditions. It is for this latter reason that most of the corrosive failures in military aircraft, which were prevalent using the FPL and optimized FPL treatment in manufacturing aircraft, were significantly reduced when the aircraft industry changed to CAA and PAA pretreating procedures.

Stressing plus environmental weathering can produce still different ratings among surface treatments as well as distinguish between different adhesives over the same pretreatment. For example, while aluminum/two-part, RT-curing epoxy joints can often exhibit very acceptable durability in hot condensing humidity exposure with chromic acid–deoxidized aluminum adherend surfaces, duplicate joints exposed to very modest stress levels can show relatively short survival times. At the same time, the same epoxy bonded to an ambient temperature, HCl-etched aluminum surface may be capable of relatively long-time service at the same stressing level. At high-stress levels, the magnitude of the stress seems to be a most critical test variable even with the best surface pretreatments. For example, the testing of the best joints possible with modern technology will rarely be tested at stresses above 50% of the initial strength when soaking in hot humidity conditions up to about 165°F (74°C). Even at relatively low stressing levels, the form of weathering can still be quite important and eventually produce joint failure after sufficient time has elapsed for the deleterious elements of the weather to encroach into the bondline.

In recent years, there has been increasing interest in a variety of nonaerospace structural aluminum products in fields like automotive and over-the-road trailer manufacturing. In such industries, the economics often preclude the use of the special surface preparations developed and used in aerospace. Because Alcoa recognized that many manufacturers would have more interest in bonding mill-finish, vapor-degreased, emulsion-soaked, abraded, or room temperature–deoxidized surfaces, the author undertook to evaluate the long-term service potential of a wide variety of lower-cost pretreatments using both room temperature- and elevated temperature–curing epoxy adhesives. The author decided to use 6061-T6 alloy as the preferred adherend, since the aerospace industry had mainly concentrated its efforts on clad and bare 2024-T3 and 7075-T6 alloys as employed in aerospace structures. 6061-T6 alloy, on the other hand, is structural enough to be considered for many manufactured structural aluminum products and is available in both sheet and extrusion forms.

Minford (1022–1024) has accumulated data on the long-term durability potential of as many as 30 different surface pretreating conditions using 6061-T6 aluminum adherends bonded with both a two-part, room temperature–curing polyamide-cured epoxy and a one-part, heat-cured, nitrile-modified epoxy adhesive. The data shown in Table 170 demonstrate the full range of initial joint strengths

Table 170 Initial Variability in Aluminum Joint Strength with Different Surface Pretreating of the Aluminum Adherend with a RT-Curing or Heat-Curing Epoxy Adhesive

	Lap-shear strength			
	2-part epoxy		1-part epoxy	
Procedure	psi	MPa	psi	MPa
Vapor degreasing	2130	14.67	4330	29.83
Vapor degreasing + silica-gritblast	2030	13.99	—	
Vapor degreasing + gritblast + applied EC-1597 primer	2030	13.99	—	
Belt-sand + degrease	—		4870	33.55
Belt-sand + degrease + applied EC-1597 primer	—		1970	13.57
Vapor degreasing + RT etch in CrO_3/H_2SO_4 solution	2570	17.71	5500	37.90
Vapor degreasing + hot etch in CrO_3/H_2SO_4	2830	19.50	5330	36.72
Vapor degrease + etch in hot HNO_3/H_2SO_4	2500	17.23	5100	35.14
Vapor degrease + etch in hot H_3PO_4/H_2SO_4	2370	16.33	4730	32.59
Etch in R.T. 25% Arcal Alum-Surf-Prep	2800	19.29	5070	34.93
Vapor degrease + carbide Y2102 silicone primer + bake	553	3.81	—	
Vapor degrease + abrasive cleanser + RT etch in CrO_3/H_2SO_4	2600	17.91	5200	35.83
Immerse in hot alkaline cleaner (Rido-line 53) + hot etch in CrO_3/H_2SO_4	2630	18.12	5430	37.41
Vapor degrease + Alcoa A1 hot caustic etch + 40% HNO_3 dip	2370	16.33	4630	31.90
Etch in hot Alcoa A1 + hot etch in CrO_3/H_2SO_4	2530	17.43	5370	37.00
Vapor degrease + RT immersion in $HC1/H_2O/K_2Cr_2O_7$ solution	2630	18.12	4970	34.24
Anodize in H_2SO_4 (Alcoa Alumilite 215)	2530	17.43	3270	22.53
Solvent wipe + cleaning with Hughson B-727-6 paste	2400	16.54	4630	31.90
Chromate conversion coating (Alodine 1200)	1130	7.79	2470	17.02
RT etch in 25% solution of Deoxidine 525 (H_3PO_4/alc.)	2700	18.60	4970	34.24
Vapor degrease + application of Dow Surface Conditioner A	645	4.44	2030	13.99

Table 170 Continued

Procedure	Lap-shear strength 2-part epoxy psi	MPa	1-part epoxy psi	MPa
Hot cleaning in alkaline cleaner (Ridoline 53) + hot tap water rinse	2630	18.12	5270	36.31
Organic acid anodizing by Alcoa Duranodic 335S Proc.	2530	17.43	3170	21.84
Vapor degrease + Hinac II org. coat	2470	17.02	4270	29.42
Vapor degrease + immersion for 1 min in boiling water	1830	12.61	4800	33.07
Vapor degrease + application of baked epoxy-urea paint primer	2970	20.46	4667	32.16
Vapor degrease + application of baked epoxy-phenolic paint primer	2930	20.19	5300	36.52
Vapor degrease + application of baked white acrylic paint primer	2970	20.46	3730	25.70
Vapor degrease + application of RT clear acrylic paint primer	2970	20.46	Destroyed by heat curing	
Belt-sand + vapor degrease + clean with Hughson B727-6 paste	—		4900	33.71
Vapor degrease + application of Corricone 1700 corrosion protective coating + air dry	—		3000	20.67
Vapor degrease + application of baked white acrylic paint primer and oven dry	—		3100	21.36

Note: All joints fabricated with 6061-T6 aluminum with heat-treat film initially present. Two-part epoxy was National Starch M688 epoxy resin (100 parts) cured with 32 parts of CH-16 polyamide. Curing time was for 7 days at RT before determining lap-shear strength. One-part epoxy was 3M's EC-2086 nitrile-modified epoxy cured for 15 min at 400 °F (204 °C). From J. D. Minford, unpublished work, Aluminum Co. of America.

that can be obtained using these surface pretreatment variables and the two epoxy adhesives. Note that the range of joint strengths using the two-part epoxy was from 553 psi (3.81 MPa) up to 2970 psi (20.46 MPa). Meanwhile, the joints fabricated with the one-part, heat-cured epoxy ranged from 2030 psi (13.99 MPa) up to 5500 psi (37.90 MPa). These various joints were further exposed to establish their relative durability performances in room temperature water soaking, hot-water–soak/freeze/hot-air–thaw cycling, the natural industrial atmosphere, or the seacoast atmosphere, as shown in Tables 158, 159, 154, and 155, respectively. More recently, considerable interest has been growing in evaluating aluminum joint permanence when varying amounts of lube are present. In the earlier Minford publications (68, 72–74), the ability of commercial epoxy and vinyl plastisol heat-curing adhesives was evaluated over various aluminum automotive alloy surfaces. Most recently, Minford (3577–3579, 3699, 3700) has been investigating the durability potential for newer commercial heat-curing epoxies that have been deliberately formulated to bond in the presence of surface lubricants.

B. Effect of Specific Aluminum Surface Pretreatments

The nature and character of the aluminum oxide surface is directly influential on the survice potential for aluminum-bonded structures. In the most fundamental sense, two atomically clean aluminum atom surfaces should be capable of developing direct adhesion if brought into solid-state contact. If this situation could be realized, then the interfacial bond strength that results should be stronger than the cohesively weaker of the two adherends. This follows from the practical fact that no two adherend surfaces will ever be absolutely identical. The whole concept is purely idealistic, however, since we are unable to produce two such surfaces in real life by any presently known surface pretreating procedure.

1. MILL-FINISH ALUMINUM ADHERENDS

One interpretation might be that a so-called mill-finish aluminum surface is not a surface pretreatment for bonding, since the joint manufacturer does not perform any deliberate prebonding operation. However, the lack of specific pretreatment by the manufacturer does not preclude the fact that the aluminum fabricator earlier performed some specific operations in producing an acceptable product for sale. For reading about the properties and physical metallurgy of aluminum the author can suggest Volume 1 of the *Aluminum* series edited by Van Horne and published in 1967 by ASM. More specifically within this volume, the reader's attention can be directed to individual chapters (3701–3706). A volume issued in 1984 by ASM has been given the title *Aluminum: Properties and Physical Metallurgy* as edited by J. E. Hatch. This is the successor to the Volume I of the 1967 ASM series already described above. This updating was undertaken under the auspices of the Aluminum Association, Inc.

The many details that can affect the aluminum surface as it could be supplied by the aluminum manufacturer in not in the intended scope of this book. However, the range of the chemical constitution of the main alloy families can be cited as one means of demonstrating the enormous potential for surface variability that can be present in the mill-finish surfaces furnished to the bonding manufacturer. For example, the major alloy systems discussed by Brown and Willey (3701) include the following: (1) aluminum-copper systems, (2) aluminum-magnesium systems, (3) aluminum-manganese systems, (4) aluminum-silicon systems, (5) aluminum-zinc systems, (6) aluminum-copper-magnesium systems, (7) aluminum-magnesium-silicon systems, and (8) aluminum-magnesium-zinc systems. Each different alloying ingredient and combination of ingredients can affect both the surface and internal bulk structure as described by Dean (3702).

a. Aluminum Supplier Contributions

The mill-finish condition is generally that produced by the aluminum supplier as a result of those processing conditions which best fit the manufacturing characteristics of the particular alloy chosen by the customer. For example, the surfaces found on aluminum sheet, extrusion, castings or forged products will be different as mentioned above. Further, the surface oxides themselves will vary both because of different alloying ingredients and variances in the processing of the form of the alloy. Not only the chemical properties of the surface but also the physical nature of these different alloy products can be quite different. The surface topography will definitely have a distinct relationship with the nature and design of the surface of the rolls used in rolling down an ingot to final sheet thickness. Further, the customer may dictate differences being present in a particular shipment of a particular alloy, as compared to some other customer, by requesting some special surface mechanical, chemical, or optical property which better fits into the final manufacturing procedures or product quality.

The surface of aluminum that has been extruded through a die will, in turn, be different from a rolled surface, as will a cast surface produced by a mold or a combination of forging hammers and mechanical and hydraulic presses required to create a forging. Another common interfering agent on mill-finish aluminum surfaces for bonding may be the coating of special lube or an anticorrosive coating that the aluminum supplier has applied to the surface to obviate the unexpected development of surface corrosion during handling, shipping, receiving, or storage prior to use by the customer. Thus, if a manufacturer of bonded aluminum structures chooses to bond directly to a mill-finish surface for economic reasons, the manufacturer must be prepared to face a wide range of possible responses depending on type and form of adhesive that has been chosen for use. It now becomes clear why special surface pretreating procedures had to be developed to cover all the situations and needs that can exist for good service-life potential. Finally, it should be evident why data need to be developed to offer guidelines for choice of bonding conditions which will result in acceptable long-term service-life potential.

b. Surface Chemistry Effects

Other oxides such as copper and magnesium oxides have been shown on the mill-finish surface of alloys containing these alloying elements in the bulk microstructure. Sun et al. (850) have indicated that a high concentration of copper at the metal/oxide interface would have an important effect on joint durability. This supports several earlier joint-durability evaluations of the author showing that aluminum adherends with increasingly higher concentrations of alloying copper would debond at shorter and shorter times in cycling corrosive salt-water conditions. This would apply to both situations involving mill-finish or acid-deoxidized surfaces. This latter situation should be anticipated, since alloys with higher copper contents still have higher copper residues after deoxidizing.

Kinloch and Smart (663) have noted a correlation between magnesium concentration in the oxide layer and poorer aluminum joint durability. Kinloch et al. (781) confirmed with specific alloys of increasing magnesium content that durability definitely decreased as the surface magnesium oxide increased. In this same regard, we should appreciate that the purest and most uniform mill-finish surfaces would be those that exist on the highest-percentage aluminum commercial alloys. Unfortunately, these relatively pure aluminum alloys (1100 alloy as commercial purity aluminum is 99% purity metal) are relatively low strength for most structural applications. They are, however, often used as a foil overlay or in some electrical applications where the excellent conductivity is desired. When bonding is employed, the use of elaborate surface pretreating may often not be required because of the superior bondability that can be achieved as compared to the more heavily alloyed (and much stronger) commercial alloys. They are also used as cladding on some higher-strength alloys such as 2024 and 7075 aerospace alloys. For example, the cladding on 2024 alloy is 1285 alloy which is 99.85% aluminum. The clad and nonclad products can even be visually distinguished because of the different chemical action of the processing environments. Thus, the mill-finish surface of Alclad and bare 2024-T3 alloys are distinctly different because of the significant difference in the concentration of alloying ingredient present in each. Consequently, any chemical pretreatments of these two surfaces whether it be FPL etching or CAA or PAA anodizing will have some variation in response. Generally, for a more simple type of pretreatment like vapor degreasing, the greater durability bonds should be generated on the Alclad surface. The possible exception to this statement offered by weathering in salt-water instead of distilled, deionized, or tap-water compositions will be discussed later. This relates to the cladding acting as a sacrificial anode to protect the cathodic core alloy in corrosive salt water.

It is also worthy of mentioning that aluminum alloys can have widely varying amounts of oxide on their surface depending on the inherent chemical properties of the particular alloy and the need to specially heat treat some of the more highly alloyed aluminum types. Thus, while some aluminum alloys will have encountered less total thermal exposure during their manufacturing, others will be exposed to what might be termed intermediate thermal soak, and still others will see the quite high oven temperature soaks used for the so-called heat-treatable aluminum alloys. It is their relatively thick oxide that must be removed by either mechanical (abrading) or chemical means to achieve the highest strength and more durably performing joints. The higher the stress on the final joint and the more aggressive the service weathering conditions, the greater will be the necessity to completely remove this oxide film and regrow one of more uniform composition and environmentally stable configurations.

Minford (3707) has conducted a comprehensive investigation of the effect on bond strength and joint durability of additional customer preheating before bonding of 2000, 3000, 5000, and 6000 series alloys in hot convection ovens. The vapor-degreased surfaces of 2036-T4, 3003-H14, 5182-0, 5657-H25, 6010-T6, Duranodic (anodized) 335, and mild steel were all pretreated by heating at 250 °F (121 °C), 350 °F (177 °C), or 450 °F (232 °C) for 0.25, 2.0, or 4.0 hr before bonding. Durability responses were noted following soaking in RT water or 100% RH at 125 °F (52 °C), soak/freeze/thaw cycling, intermittent spray exposure to 3.5% salt water, or simultaneous stressing with hot humidity soaking for up to 44 wk. Relatively little evidence could be shown indicating any significant change in the normal durability patterns obtained by using unheated vapor-degreased alloy adherends as controls. Actually, there proved to be more evidence of performance differences between different alloys than shown by any of the indicated heat exposures on the same alloy. The most significant difference shown was observation of some elevation of the joint-survival time in stress/humidity conditions by prebond heating, as shown in Table 171. Only preheating of steel surfaces seemed to promote a distinctly negative effect on overall joint durability.

Further, Minford (3708) has attempted to demonstrate whether the bond strength and durability of different lots of 6009-T4 and 6010-T4 automotive body sheet alloy could be influenced by altering the manufacturing line speed of these two products. Production line speeds were altered from 80 to 120 ft/min and epoxy-bonded joints were made with these mill-finish surfaces. Additionally, comparisons were made with 2036-T4 and 5182-0 aluminum auto body sheet alloys as presently manufactured. Changing the line speed did not significantly change either initial joint strength or durability response. Different production lots of the 6009-T4 and 6010-T4 alloys supplied for further line processing altered joint strength values and durability responses to a greater extent than changing line processing speed with a single lot of material. Thus, metallurgical factors seemed to be much more important than line processing variabilities. The relative joint durabilities (especially in corrosive salt spray) for various alloys again fell into a reverse order to their general sensitivities to overall corrosion; i.e., 2036-T4, 6010-T4 = 6009-T4, 5182-0 joints.

If joints are fabricated from mill-finish surfaces, there is an additional criterion that can relate to their durability, as noted above. Reference is to the relative inherent alloy chemical resistance. Minford (3709) has demonstrated this with less reliance on the presence of high concentrations of copper by choosing to compare the performance of joints made with Alclad 2024-T3, 1100-H14, and 6061-T6 adherends. This order corresponds to a decreasing order of aluminum purity in the surface, since 1285 cladding is 99.85% aluminum, 1100 alloy is 99.00% aluminum, and 6061-T6 is 97.95% aluminum. The relative order of inherent general resistance to exterior weathering should fall into this same order as well. Long-term water-soaking tests of these structural joints indeed showed that their decreasing levels of joint strength retention fell into the same order predicted. In yet another investi-

Table 171 Effect of Hot-Air Pretreatment on Aluminum Joint Durability Under Stress and Hot Humidity Conditions

	Surface pretreatment after vapor degreasing			Stress level		Joint failure value times (min)	
	Time	Temperature					
Adherend alloy	(hr)	(°F)	(°C)	psi	MPa	Range	Average
2036-T4	None	—	—	250	1.72	(20–66)	43
2036-T4	2	350	177	250	1.72	(25–83)	54
2036-T4	0.25	450	232	250	1.72	(98–110)	104
3003-H14	None	—	—	250	1.72	(49–65)	57
3003-H14	0.25	450	232	250	1.72	(85–150)	118
6010-T4	None	—	—	250	1.72	(20–25)	23
6010-T4	2	350	177	250	1.72	(27–30)	28
6061-T6	None	—	—	250	1.72	(75–95)	85
6061-T6	4	250	121	250	1.72	(130–160)	145
Mild steel	None	—	—	250	1.72	(134–160)	147
Mild steel	4	250	121	250	1.72	(86–106)	96

Data from Ref. 3707 (Table VI).

gation, Minford (3710) evaluated bare 2024-T3, 6061-T6, and 5052-H32 joints using the same epoxy adhesive and the same kinds of water-soaking conditions. Again it was noted that the order of joint strength retention was 5052-H32, 6061-T6, 2024-T3; the same order as that of decreasing inherent resistance to corrosion in salt water. This same observational relationship between general resistance to corrosion and joint durability will, of course, be noted whether the joints are fabricated with mill-finish, solvent-cleaned, abraded, or chemically deoxidized surfaces.

c. *Preapplied Finishes*

Since any surface on aluminum prepared by the aluminum producer can be considered a mill-finish surface to the bonding customer, some attention needs to be paid to the fact that a large poundage of aluminum is specially prepared by application of a heat-cured paint finish. In performing this operation, the aluminum supplier is confronted with the same situation as the structural joint manufacturer; namely, how must the aluminum be pretreated to gain long-term adhesion when a heat-curing organic polymer formulation in the form of a commercial paint system is to be used. The largest operations of this kind in the aluminum industry are concerned mainly with the painting of aluminum sheet and extrusion. From a practical view, the various chemical families of the most prominant paint formulations have a close relationship with the same polymers as structural adhesives. Usually, the paint compositions are offered in a much lower-viscosity form for best flowing and wetting properties. In general, they need not be more structural than to support their own weight, since there is no second adherend to make an actual joint. They must, however, develop sufficiently high-adhesion and cohesive strength within the paint film to be further bondable and supportive of much greater stresses if they are involved in making a joint by bonding to a second adherend.

Typical processing of the aluminum sheet for commercial painting requires prior cleaning of the mill-finish surface, usually with some sort of commercial hot-emulsion–type immersion procedure. The Alcoa architectural painting recommendations also required a further deoxidizing bath step with a final treatment in a conversion coating bath such as Alodine 1200. Thus, either water- or organic-solvent–dispersed paints can be used. Finally, the paint is cured at some elevated temperature, usually in the vicinity of 450°F (232°C) to 550°F (288°C) to achieve true thermoset properties. When used as a further bondable surface, prepainted aluminum is acting like an adherend which has received a pretreatment very much like it would receive from a customer who has to pretreat a mill-finish aluminum. Further, the cured paint can serve to act as a special primer which has been applied in low viscosity over deoxidized and conversion coated aluminum.

Minford (3711) has demonstrated that there can be some chemical incompatibility problems involved with bonding over one of these pre-applied paint finishes which needs to be evaluated by the manufacturer. For example, Minford showed this by attempting to bond acrylic-painted or epoxy-painted aluminum surfaces with both room temperature–curing modified-acrylic and two-part, room temperature–curing epoxy adhesives. For further delineation of the problem, both a heat-cured thermoplastic acrylic paint and a thermoset-type acrylic paint were employed. In general, the joint strengths attained were quite respectable, with cohesive failures occurring in the paint cross sections. These tensile-shear lap values using the thermoset paints approached the values that would be obtained with the same adhesives over freshly deoxidized aluminum adherends. The thermoplastic acrylic painted joint failures while still cohesive in the paint films, were significantly lower overall than with thermoset coatings. In long-term water-soaking tests, the thermoset painted joints maintained good adhesion strength for periods of months to years using either type of structural adhesive. The thermoplastic acrylic paint film, however, was gradually degraded by the modified acrylic adhesives, eventually failing adhesively to the aluminum oxide surface. No degradation could be noted with 100% solids epoxy adhesives.

In a separate investigation under natural weathering conditions up to 8 yr, Minford (774) demonstrated the amazing durability that occurred when an epoxy-phenolic thermoset paint was applied over a vapor-degreased mill-finish 6061-T6 adherend surface. It must be remembered that the commercially painted aluminum adherend discussed above had received emulsion cleaning, acid deoxidizing, and conversion coating as surface preparation before painting. This experiment was considered to be a more rigorous situation for obtaining good durability performance because the heavy

Table 172 6061-T6 Aluminum Joint Durability Testing in Room Temperature Water for 2 Yr with Varying Abrasion, Etching, Conversion Coating, Paint Primer, and Anodizing Surface Pretreatments

Surface pretreatment	Type[a] adhesive	Average joint strength after indicated exposure time (MPa)				% Strength retention after longest exposure
		3 mo	6 mo	12 mo	24 mo	
Vapor degrease	A	8.96	7.34	8.47	7.79	53
Vapor degrease	B	21.15	13.78	10.82	8.06	27
Belt-sanding	A	12.73	12.88	12.35	—	88
Belt-sanding	B	25.98	23.63	21.36	14.95	45
Silica-grit blasting	A	14.88	—	12.88	14.68	100
Chromic acid etch	A	20.46	—	17.43	18.40	94
Chromic acid etch	B	30.80	30.32	28.73	24.32	66
Alodine 1200 (1)	A	7.85	6.75	5.99	4.69	60
Alodine 1200	B	17.02	14.95	14.95	11.51	67
Epoxy-phenolic paint primer	A	20.88	20.19	20.88	19.29	96
Epoxy-phenolic paint primer	B	34.93	32.38	31.49	30.32	83
Alumilite 215 (2)	A	17.57	17.02	15.16	14.95	86
Alumilite 215	B	17.02	14.59	16.54	16.74	74

[a]Adhesive type A is a two-part polyamide-cured epoxy; adhesive type B is a one-part, nitrile-modified epoxy.

Test results from Ref. 774 (Table II, p. 490).

Table 173 6061-T6 Aluminum Bonded Joints Exposed to 100% RH at 125°F (52°C) with Various Abraded, Etched, Conversion Coated, Paint Primed, and Anodized Surface Pretreatments

Surface pretreatment	Adhesive type[a]	Average joint strength after indicated exposure time (MPa)				% Strength retention after longest exposure
		3 mo	6 mo	12 mo	24 mo	
Belt-sanded	A	10.32	10.80	11.07	—	79
Chromic acid deoxidize	B	14.68	8.75	4.55	6.10	17
Epoxy-phenolic paint primer	A	19.77	18.60	17.91	17.43	86
Epoxy-phenolic paint primer	B	28.73	25.70	22.25	19.50	53
Alumilite 215 (H_2SO_4 anodize at 0.8 mil)	A	14.33	12.40	13.09	15.16	87
Alumilite 215 (H_2SO_4 anodize at 0.8 mil)	B	18.33	16.54	18.12	17.91	80

[a]Adhesive type A is a two-part polyamide-cured epoxy; adhesive type B is a one-part, nitrile-modified epoxy.

Test results from Ref. 774 (Table III, p. 492).

heat-treat oxide on the 6061-T6 mill surface was not removed or even disturbed by the simple vapor-degreasing cleaning procedure. Both a room temperature– and a heat-curing epoxy were employed, as shown in Tables 23, 157, 172, and 173. Retention of 96% of initial joint strength after a 2-yr soak in room temperature water with the room temperature–curing paste epoxy certainly justifies the use of this type of heat-cured epoxy-phenolic paint as an effective primer over a minimally prepared surface (Table 172). While the percentage of retained strength for the corresponding heat-curing, nitrile-modified epoxy was lowered to 83%, the absolute magnitude of joint strength was still 57% higher than that obtained with the room temperature–cured epoxy because of its much higher initial strength level. In Table 173, corresponding joints were evaluated in the normally more severe exposure condition of 100% RH at 125°F (52°C). The strength retention after 2 yr for two-part epoxy joints was only lowered to 86%, whereas the heat-cured nitrile-modified joints decreased from 83% down to 53%, which was still 19% higher than the two-part epoxy joint because of its much higher initial strength value.

We might anticipate that the best demonstration of good joint durability would occur with exposure to the relatively mild natural atmospheric conditions reviewed in Table 23. This is demonstrated by the 91 and 98% lap-shear strength retensions shown by the one- and two-part epoxy adhesive joints after 8 yr. Finally, we might anticipate that the most dramatic joint degradation would occur in the corrosive seacoast exposure, and this is clearly true from the values shown in Table 157. All the joints made with two-part epoxy adhesive degraded to failure within 4.1 yr. With the better surface wetting potential afforded by heat curing the one-part epoxy at 350°F (177°C), the comparable one-part, heat-cured epoxy joints actually retained 71% of initial strength values after an 8-yr exposure. As mentioned several times earlier, the author firmly believes that the ability of the adhesive to wet the adherend surface more effectively is a prime factor in promoting joint longevity in the seacoast exposure.

The performance of the epoxy-phenolic paint primer just discussed was also compared in similar preparation and testing investigations to a commercial epoxy-urea, heat-cured, thermoset paint. This paint primer had, in turn, proven to be distinctly superior in earlier durability tests to a nonpigmented or white-pigmented, heat-cured acrylic paint primer. All these paints had been deliberately applied over a vapor-degreased-only 6061-T6 adherend and, thus, did not have the additional adhesional benefits that would exist in a commercially prepainted aluminum sheet adherend. The overall durability performance ranked as follows: epoxy-phenolic, epoxy-urea, white pigmented acrylic, clear, nonpigmented acrylic. In Figure 43, the comparative durability test results are shown for the first

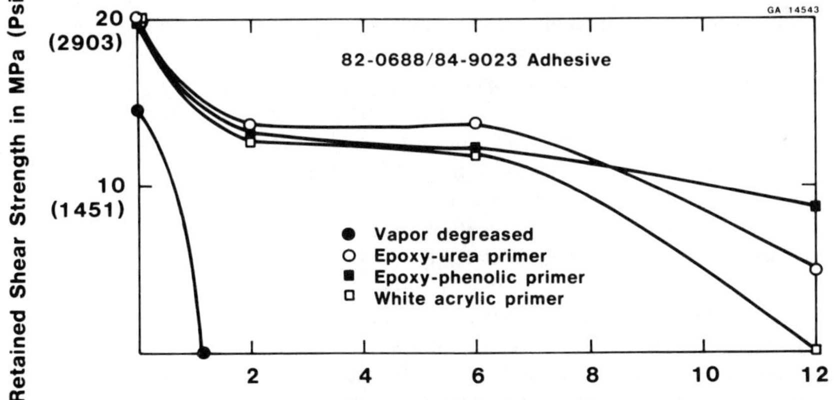

Figure 43 Comparative aluminum joint performance in daily intermittent salt spray of three paint primers and vapor-degreased–only 6061-T6 aluminum. (From J.D. Minford, unpublished works, Aluminum Co. of America.)

three paint primers in an intermittent salt-spray exposure for 1 yr. In Figure 35, comparison is made in the natural industrial atmosphere for 12 yr along with conditions of vapor degreasing only and Alcoa A3 chromic acid deoxidizing. As can be seen, the primers applied over a vapor-degreased surface lift the performance level to that of a chromic acid–deoxidized adherend surface preparation joint where the initially thick oxide film was entirely removed.

Technically, the manufacturer of bonded joints could request aluminum that had a conversion coating or an anodized finish from the aluminum supplier, in which case, such adherends might still be mill-finish in regard to their use for bonding by a manufacturer. The relative durability enhancements afforded by adhesive bonding to either of these kinds of surfaces will be discussed later in this chapter.

2. *SOLVENT-WIPED OR VAPOR-DEGREASED ALUMINUM*

It is usually agreed that an aluminum adherend for bonding should at least be free of nonadsorbable or nondesorbable contaminating films. A possible exception in recent years has occurred in some selected industries where using specially formulated adhesives which must be heat cured has permitted some satisfactory bonding over surfaces which would have been considered too oil contaminated in earlier times. Even in these instances, however, bonding in the presence of contamination can hardly be considered a desired procedure.

The simplest way to produce an acceptably clean surface has been the use of a solvent-wipe or vapor-degreasing procedure. Between these two alternatives, the use of the vapor-degreasing procedure might be considered distinctly more advantageous, though possibly more costly than simple solvent wiping. The main reason for rejecting the use of solvent wiping is that when the operation is repeated using the same rag, the efficiency must suffer. It is usually observed in most manufacturing lines employing solvent-wiping procedures that rags are repeatedly used for economical purposes. Each time more soiling is accumulated until the operator is redepositing as much soiling as has been removed. A vapor degreaser can also experience a build-up of contamination even when properly protected by screens. The solvent also undergoes some slow chemical breakdown due to the continual recycling at elevated temperatures or acceleration of the breakdown by picking up some contamination chemicals which catalyze breakdown. The use of a simple water-wetting test on every adherend surface that is about to be bonded should make the solvent rag-wiper operator or vapor-degreaser operator aware of how the service-life potential of the bonded product may be seriously jeopardized when either system is not employed with maximum efficiency.

For more than 25 yr, the Alcoa Labs investigated the durability potential of thousands of vapor-degreased 6061-T6 lap-joints in the screening of different adhesive products. It was the purpose of these investigations to offer helpful durability data on the use of the widest possible variety of adhesive and sealant materials on solvent-cleaned aluminum adherends. It was believed that there was a distinct opportunity to use adhesives for many manufactured products outside the aerospace industry that could not afford any surface cleaning operation beyond simple degreasing.

This was most easily demonstrated for the use of contact-type cement adhesives manufactured with various elastomers. Especially amendable for long-term service were the neoprene-phenolic contact cements such as used to laminate the Alcoa Alply panel. Some panels produced in the early 1960s have already surpassed the 20–30 yr of service exposed to outdoor weathering that was projected for them from the use of various accelerated weathering tests (87–89). It is acknowledged that the stress per unit area imposed on such panels is relatively low because of their very large surface areas. It was recognized that any adhesive formulation that was originally dispersed in 20–30% strong organic solvents like toluene and methyl ethyl ketone (MEK) could effectively disperse and usually displace traces of oil, lubes, or fall-out soiling on an aluminum sheet surface. Thus, a simple solvent cleaning of the surface should be sufficient to enable a stable, durable bond to be formed.

With the use of 100% solids adhesives like the two-part epoxy commercial paste adhesives, the joint durability of marginal solvent-cleaned surfaces could be suspect because of questionable uniform wetting of the adherend by the adhesive. Also, since the use of a solvent for cleaning does not alter or improve the stability and durability of the aluminum oxide, there is no added contribution to stability in the presence of water at the interface. Minford (3711) has shown that the durability results obtained with vapor degreasing as the only surface preparation on different lots of the same alloy bonded with a room temperature–curing paste epoxy can be significantly different. This became apparent from repeating test evaluations each time a new lot of a particular alloy was obtained for additional bonding investigations. This difference between the surfaces on different lots of the same alloy was sufficiently great that even the more favorable wetting conditions available from heat-bonded epoxies could not compensate in comparative durability tests. Minford obtained durability data provided by three different lots of vapor-degreased Alclad 2024-T3 when immersed in 145°F (63°C) water using the same heat-cured EC-2086 nitrile-modified epoxy. While no joints degraded to absolute failure in less than a year, the average joint strength retention values for triplicate specimens after a year could be widely varying. Data were also obtained using a RT-curing epoxy paste where wettability is even more critical. While the difference between the performance of different lots was distinct in a hot-water soak, it became even more pronounced when the severity of the weathering exposure was increased through the use of a hot-water soak/freeze/thaw in hot-air cyclic exposure. In a much milder soaking in room temperature water, the joint performances became quite similar, indicating that the variability in performance between different lots of aluminum was more easy to detect as the severity of the weathering exposure increased.

Continuous soaking of aluminum joints in room temperature water can provide a simple and economical means of distinguishing durability differences between different aluminum pretreatments. The relative sensitivity of a vapor-degreased surface joint as compared to a deoxidized surface pretreatment using a dicy-cured, nitrile-modified epoxy paste, nylon-modified epoxy paste, and anhydride-cured epoxy in melt-rod form can be reviewed in the Minford data shown in Figure 44. Improvement in joint-durability performance resulting from deoxidizing the surface is clearly established regardless of which adhesive is employed.

In Figure 45 (535), durability data is presented with the added comparison of a belt-sanded aluminum surface preparation by exposure to a mild RT water-soaking situation or more aggressive hot-water

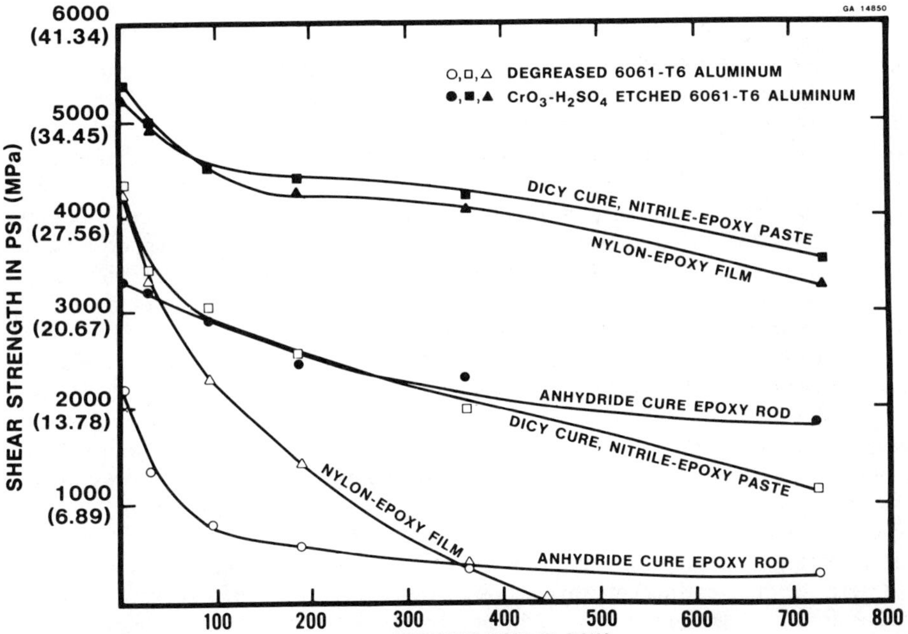

Figure 44 Comparison of durability performance between vapor-degreased–only and chromic acid–deoxidized 6061-T6 aluminum adherends. (From J.D. Minford, unpublished work, Aluminum Co. of America.)

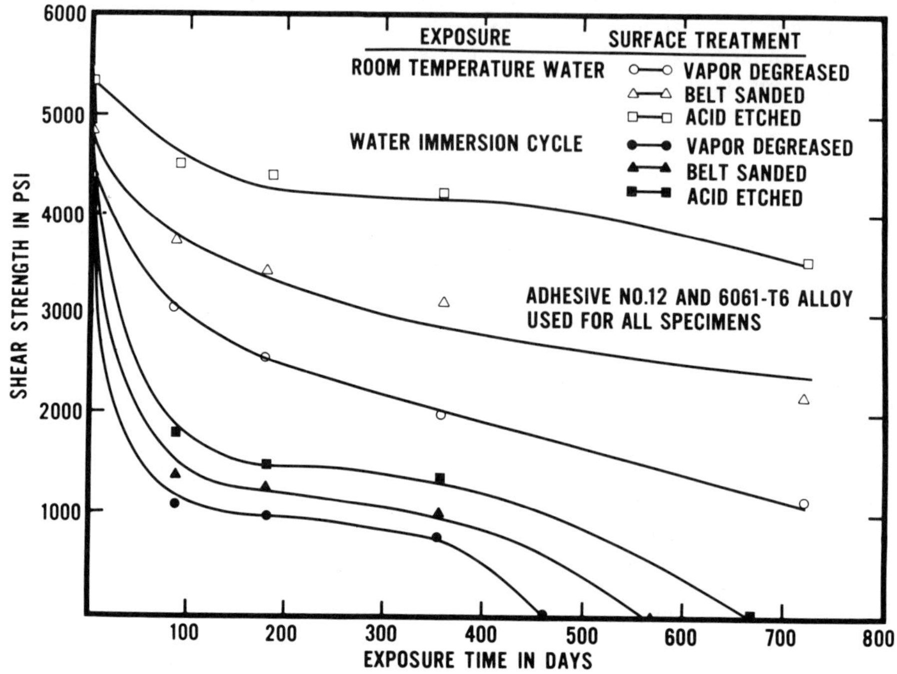

Figure 45 Comparison of durability performance between vapor-degreased–only, belt-sanded, and chromic acid–deoxidized 6061-T6 aluminum adherends. (From Ref. 535 [Fig. 22, p. 109].)

soak/freeze/hot-air dry cycling test environment. It is immediately observed that the more aggressive weathering significantly lowers all the joint strength retension curves regardless of which surface pretreating is employed as compared with immersion in RT water. The pattern is clearly established, however, for discriminating the acid deoxidation as the most effective surface pretreatment followed by the belt-sanding and vapor degreasing. It also can be seen that the more severe the weathering exposure, the closer is the range of performance between different types. The relative rating of the three different types of surface pretreatment (acid deoxidation, belt-sanding, vapor degreasing), however, remains the same regardless of the type of weathering involving contact with water.

In Figures 23 and 24 (see Ref. 3712, pg. 110), the data show comparative performances of 6061-T6 joints treated with the above surface pretreatments in 4-yr exposures to an industrial or seacoast atmosphere. The same pattern of joint-durability acid deoxidation > belt-sanding ₆> vapor degreasing is clearly shown in the milder industrial atmospheric exposure. This was true whether a one- or two-part epoxy adhesive was used. Since the former joints were significantly stronger initially, their joint decay curves constituted a distinctly higher range of curves. Both types of joints, however, showed only modest declines in joint strength, including vapor-degreased joints only. Acid-deoxidized joints were insignificantly degraded by exposure after 4 yr. The results shown in Figure 24 involving exposure to corrosive seacoast conditions were dramatically different. The author has stated that the degree of surface wetting by the adhesive is one of the most important factors in the bondline's ability to resist the undercutting corrosive action of wet/dry sea-water cycling which exists in any seacoast environment. Therefore, it may not be unexpected to find that the combination of an acid-deoxidized surface with increased adhesive flow potential offered by the one-part, heat-cured epoxy adhesive might perform most durably of the combinations tested in this investigation. The retention of more than 50% of initial joint strength after 2-yr exposure places the performance of these joints above those of any other combination of adhesive and surface pretreatment. Since we are focusing in this section on the relative ability of vapor-degreasing–only 6061-T6 joints to form durable joints, the complete loss of bond in less than 4 mo clearly demonstrates the ineffectiveness of simple solvent cleaning as a treatment for aluminum joints which must see service in seacoast- or marine-type weathering conditions.

Although the main thrust of the Alcoa durability testing was to establish the performance of minimum surface treatment conditions like vapor degreasing, it also seemed necessary to have comparative data using a surface treatment like chromic acid deoxidizing, which was widely used in aerospace manufacturing at that time. The Alcoa A3 process was selected for use because it had been historically employed at Alcoa for surface pretreatment before conversion coatings or anodized coating procedures. As compared with the FPL deoxidizing used in aerospace at the time, the Alcoa A3 differs only in substituting chromic acid for the dichromate salt and carrying the deoxidation out at a slightly higher temperature for a shorter time. Thus, as 30 different surface treatments for aluminum were eventually evaluated over many years, the data for comparing each with vapor degreasing or a chromic acid deoxidizing were always possible. First results were published by Minford in 1972 (976) followed by a larger compendium in Volume 3 of the *Treatise on Adhesion and Adhesives* series in 1973 (531). The emphasis in these earlier publications, however, was more focused on the comparative results using 27 different commercial adhesives in one- and two-part paste, tape, and film forms representing 9 different chemical families with vapor-degreased or chromic acid–deoxidized 6061-T6 aluminum adherends.

3. *MECHANICALLY ABRADED ALUMINUM ADHERENDS*

We need to remember that the emphasis to this point has been on vapor degreasing only, which leaves the mill-finish oxide intact and unchanged. In this section, we will focus on the use of various surface-abrading techniques which do disturb and largely remove the existing mill-finish oxide. As it is removed, however, a new oxide growth is initiated which will be a new base for adhesive wetting and bonding. Unfortunately, this oxide has mostly been identified as being amorphous in nature, as was the original mill-finish oxide.

Accordingly, we should not expect, and do not find, a level of joint durability produced by this treatment which is equivalent to most of the acid-deoxidizing pretreatments. In the latter, regrowth is more uniform and often can be identified as being other than amorphous. We have already noticed in Figure 45 and (see Ref. 3712, Figs. 23 and 24) that the relative order of joint durability did indeed fall into the anticipated order of joint durability that would be predicted from the above statements; i.e., acid deoxidation > abrading > vapor degreasing. It might be noted in comparing these data that the relative durability spread between the abraded and deoxidized joints seems less pronounced for the two-part epoxy paste than for the one-part, heat-curing epoxy paste. The author would relate this to the fact that the greater mobility of heated epoxy adhesive permits a superior degree of wetting to be achieved on a regrowth oxide as compared to the mill-finish oxide. The greater uniformity and superior oxide stability present after acid deoxidation permits superior wetting by the heat-cured epoxy over that adherend surface as compared to a rougher, more diverse oxide form on the mechanically abraded surface. This factor assumes maximum importance where weathering exposure is wet/dry salt-water cycling (clearly shown in Ref. 3712, Fig. 24). Also, we note the mechanical abrasion pretreatment can be highly variable owing to the different efforts of the individual operators in a production line.

Peters and Minford (3713) have investigated the combination of a mechanical abrasion belt-sanding plus acid deoxidation on 6061-T6 joint durability in room temperature water immersion or more aggressive hot-water soaking/freeze/hot-air dry cycle. These results were directly compared to similarly bonded joints which had received only vapor degreasing, belt-sanding plus vapor degreasing, and vapor degreasing plus acid deoxidizing. The opportunity was presented to differentiate between a two-part, room temperature–curing epoxy, a relatively rigid one-part, heat-curing epoxy, and a flexibilized nitrile-modified, one-part, heat-curing epoxy. In general, the relative durability rating for the various treatments was discerned as being belt-sanding plus chromic acid deoxidation = chromic acid deoxidation alone > degreasing and deoxidizing in 25% Deoxidine (RT phosphoric acid/alcohol mixture) > degreasing + a commercial acid paste cleaner for aluminum > vapor degreasing alone. The significantly higher initial strengths afforded by using heat-curing epoxies was not always translated into a superior long-term weathering response. This was especially true for the flexibilized version of the heat-curing epoxy joints, which were more rapidly degraded in the aggressive cycling exposure conditions than those made with the more rigid version.

a. Grit-particle Abrading

The most common technique is to use sanding surfaces which consist of various material grits of a uniform and specified size adhered to some carrier surface. The abrading may be performed manually or by some power-tool receptacle to which the grit paper (or grit particles on a cloth base) can be attached. It should be self-evident that a solvent cleaning still needs to follow any surface abrasion processing to remove the grits and surface debris left behind by the operation. Regardless of the method, there is opportunity for considerable nonuniformity owing to operator variability. It is virtually impossible to find individual operators who will exactly duplicate one another's work when manually sanding. Variation is only further exaggerated through the use of power equipment. In the latter instance, such factors as the angle the power tool is held against the workpiece and how much pressure is exerted by the operator against the workpiece can be highly influential. This can be better understood when it is remembered that the joint strength can be highly influenced on rough profile surfaces by the relative opportunity to achieve some significant degree of mechanical interlocking with the adhesive. Even durability can be affected because the abrading will probably not be uniform between two different operators, and this could affect the amount, distribution, and uniformity of adhesive flow and wettability.

Finally, one might expect that deeper abrasive pathways for the weathering to travel might increase the joint survival time to some extent that can only be measured by actual joint durability testing. Minford (3546) has evaluated the relative abrasion effects on aluminum joint durability by manually abrading bare 3003-H14 aluminum alloy sheet with Aloxite 80, 120, and 240 particle size papers and bonding with a two-part, room temperature–curing epoxy paste adhesive. There was relatively little difference in the durability response in room temperature water immersion or soaking in 100% RH at 125°F (52°C). However, in wet/dry cycling exposure in 3.5% salt water, the greater durability was distinctly achieved using the most coarse particle (Aloxite 80) abrading. Presumably, the corrosion pitting process was being slowed down by a circuitous pathway as compared with the more shallow abraded adherend joints.

An inherent problem exists when using bonded grit-particle paper that is similar to that mentioned for solvent wiping. In like fashion, contamination from the workpiece begins to build up on the abrading material surface and cannot be readily removed to regain the degree of original cleanliness. Accordingly, as soon as a grit paper is not being replaced frequently by the operator, the degree of contamination on the surface becomes intolerable and is further trapped by a rough surface profile. While it is hoped that the use of a solvent-wiping or vapor-degreasing process after abrading will remove all such soiling, this is unlikely, since the pressure required to affect the abrading also affords great opportunity to embed the debris in the surface. The entrapment of this debris in the interface and even in the bulk adhesive can increase the opportunity to cause joint failure prematurely as compared to a similar surface with all debris removed. It should be clear that a highly mobile, heat-cured epoxy adhesive would have a much greater opportunity to wet a rough adherend surface effectively and fill the valleys produced by the pretreatment procedure.

A question of particle compatibility with the aluminum surface also needs to be addressed. The incompatibility that can arise is largely electrochemical in nature and can arise when any abrasive material is transferred and embedded in the aluminum adherend surface. Of course, the most aggressive incompatibility will be attained when the abrasive grit particle itself can function as a strong cathode in the presence of a conducting medium such as salt water. If the basic electrochemical incompatibility of the grit particle is not recognized, then the debonding hazard may not be recognized until the joint is placed in a service environment containing liquid or vapor water. In either case, the water will eventually encroach into the interface and generate a conducting medium between the anodic aluminum and cathodic grit material areas. As the anode, the aluminum will sacrifice itself electrochemically and corrosive debonding will take place. The main concern for aluminum abrading is to omit the use of any metallic grits like steel whose particle residues embedded in the aluminum surface would be deadly in a water weathering situation. Fortunately, one of the most successful and widely used abrasive grit materials is aluminum oxide itself (such as the Aloxite abrasive products). Silicon carbon grits also appear to be quite acceptable, as well as any sand or silica grit materials.

b. Manual or Power Wire Brushing

Similar concern must be taken about the continued accumulation and build-up of surface contamination using power abrading tools. Unlike the more or less continuous nature of the grit paper or cloth surfaces, the separation of the wires on wire-brushing devices can create more centrifugal loss of the debris while the operation is proceeding. However, this same vigorous action on debris also goes along with a tendency to produce abrasion by literally chunking and gouging pieces out of the treated surface. Opportunity is presented, however, to be able to clean such brushes with whatever frequency is demanded to maintain relatively debris-free treated surfaces. This could be performed by swishing the brushes in buckets of solvent or even suspending them in a vapor degreaser.

The same obstacles remain, as noted above, in regard to controling the uniformity of the operation between different operators. The electrochemical corrosion dangers exist, and under no circumstances should ordinary steel brushes be employed. The use of stainless steel bristles, however, can be considered with aluminum surfaces, since the electrochemical relationships between stainless steel and aluminum are not nearly so dangerous.

c. Grit- or Shotblasting Procedures

The final method used for treating large surfaces is to shot- or gritblast the aluminum surface. One main advantage is obviously the rapidity of completing the abrading. The very fact of this speed, however, also carries with it a penalty, because so much oxide and underlying metal can be removed with virtually any dwell time in one location. Significantly, this can, in turn, create compressive stresses in the actual surface. All these factors add up to demanding an experienced operator as compared to the previous abrading methods. Again, the warning must be sounded against using any steel grit particle when abrading aluminum.

Washed silica sand may be the most commonly employed abrasive medium for gritblasting aluminum. The method was used by Minford (532,774) to develop 6061-T6 aluminum joint studies using a two-part, room temperature–curing epoxy adhesive. The most unexpected result occurred when comparing similar joints with a chronic acid–deoxidizing (Alcoa A3) surface pretreatment in a seacoast exposure for up to 8 yr. It was shown in (see Ref. 3712, Fig. 24) that the belt-sanded and chromic acid–deoxidized joints made with this same adhesive had failed after about 2 yr. Amazingly, the silica-gritblasted 6061-T6 joints with this adhesive in a different investigation retained 74% of initial strength after 2 yr and 67% after 8 yr at the seacoast. This performance proved to be very comparable to corresponding joints made with anodized surfaces (SAA). In Table 157, comparison between these silica-grit joints and anodized surface joints can be seen. Only the anodized surface joints showed a higher durability using a two-part epoxy. Unfortunately, Minford did not conduct testing using a one-part, heat-curing epoxy, but it is presumed the performance should be at least as good and possibly better based on achieving better surface wetting with the heat-curing epoxy paste. An explanation for this amazing performance in the seacoast atmosphere needs to be advanced, since duplicate joints failed before 2 yr in an exposure to hot-water/freeze/hot-air dry cyclic conditions. Minford has proposed that the resistance of the bondline to undercutting corrosion in wet/dry salt-water cycling is due to the deposition of a layer of silica particles in the surface which perform as an effective corrosion inhibitor.

Comparative test results with 2-yr exposure to room temperature water in Figure 46 can be compared to the results for duplicate joints exposed to hot-water soak/freeze/hot-air dry cycle in Figure 47. Essentially no degradation on either silica-gritblasted or SAA-anodized surface joints was noted in RT water soaking. However, as mentioned above, the silica-gritblasted joints proved to have less than acceptable resistance to the aggressive soak/freeze/thaw cycling, failing before 2 yr. In contrast, the SAA-pretreated joints declined only about 40% in the same period. In an industrial atmospheric exposure, joint durability performances of similarly prepared epoxy joints with vapor

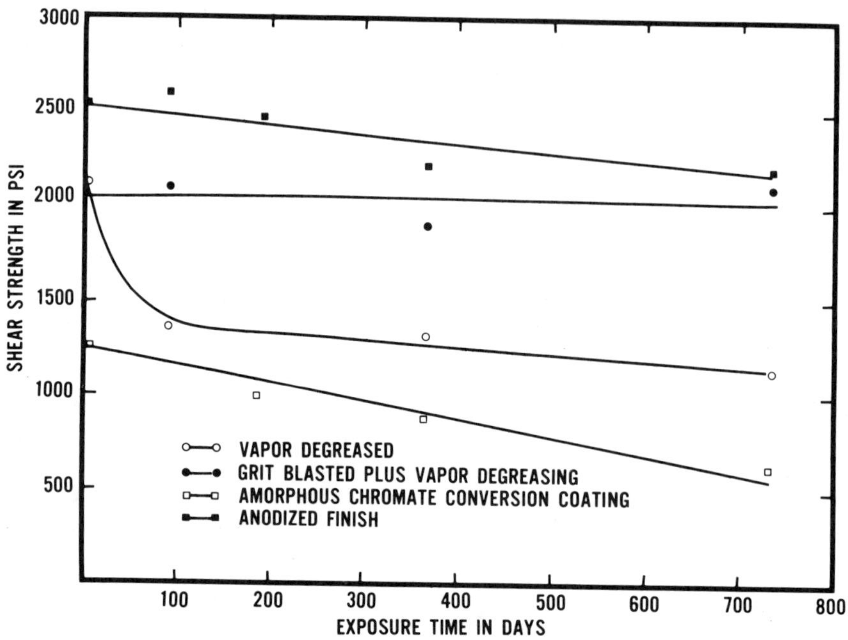

Figure 46 Comparison of durability performance between vapor-degrease-only and grit-blast, conversion coating, and SAA anodize 6061-T6 pretreatments in room temperature water soaking. (From J.D. Minford, unpublished work, Aluminum Co. of America.)

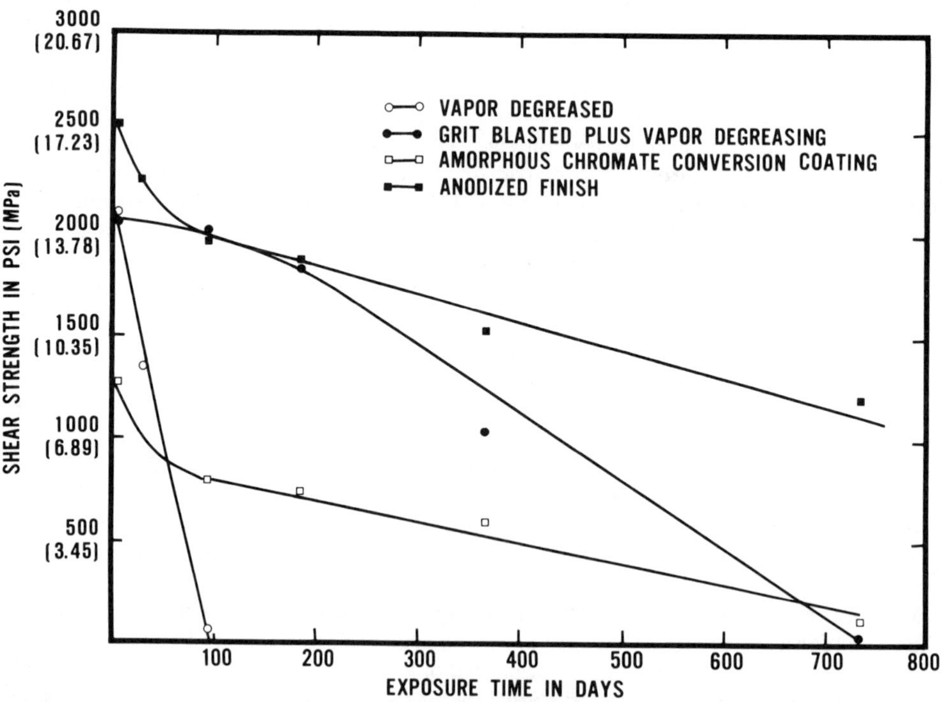

Figure 47 Comparison of durability performance between vapor-degrease-only and gritblasted, conversion coated, and SAA anodize 6061-T6 pretreatments in soak/freeze/thaw cycling weathering. (From J.D. Minford, unpublished work, Aluminum Co. of America.)

degrease only, silica-gritblasting, or anodizing treatments were similar over a 4-yr exposure time. It should be noted in both Figures 46 and 47 that similar joints made by room temperature curing a 2-part epoxy over an amorphous chromate conversion coating had significantly lower overall initial strength and gradually lost additional strength in each exposure over a 2-yr exposure. In the exposure of similar joints for 4 yr in the industrial atmosphere, this did not occur. This will be considered in more detail when discussing this kind of surface pretreatment later.

4. CHEMICALLY DEOXIDIZED SURFACES

Considerable activity has been undertaken in the aluminum industry about the means of deoxidizing aluminum surfaces due to the need to provide suitable surfaces for durable painted aluminum. The same polymer families and combinations of different polymers that function as adhesives usually can be made into paints. Like adhesives, paints may be room temperature or elevated temperature cured. Because of their general lower viscosity, however, they may be expected to have superior flow and wet-out qualities at room temperature. Regardless of their constitutional differences, the requirement of both adhesives and paints for establishing a high degree of initial and long-term durable adhesion are quite similar. The necessity for a cured structural adhesive to maintain its adhesion, however, is further complicated by the necessity to support a much greater stress than expected for paint coatings. Consequently, there are a number of additional factors that can lead to its failure in service.

a. Alkaline Cleaning Procedures

The lack of any serious discussion in the structural bonding literature about the acceptability of alkaline cleaning alone as a base for durable structural bonding seems to preclude much discussion. However, even an alcohol-phosphoric acid deoxidizing at room temperature has been considered as being acceptable for structural adhesive bonding under exterior weathering conditions. We might, therefore, consider that the use of an alkaline cleaning procedure has no real place in overall bonding, but that is not necessarily true.

It must be pointed out that more manufacturers are actually applying some sort of adhesive to a clean adherend surface than are demanding the highest performance of structural bonding in aerospace. The reason this is possible is because more manufacturers are producing relatively nonstructural products which only require rubber contact cements, rubber latex emulsions, casein-rubber latexes, acrylic emusions, polysulfides, and high solids rubber mastics than to high-strength epoxies and aerospace film and tape adhesives. The ability of these lower-strength adhesives to achieve their optimum strength really only requires that the aluminum surface be cleaned of major amounts of contamination. For example, it was found by the author when evaluating the long-term durability of Alply panels (adhesive is a neoprene-phenolic contact cement), that it was almost impossible to distinguish any performance different between aluminum sheets cleaned by an expensive furnace-cleaning procedure and or just vapor degreasing. Eventually, the extra cost of furnace cleaning bare aluminum facings was discarded for simple vapor degreasing.

Minford (3715) has tested the relative durability potential for pretreating 6061-T6 aluminum sheet in a hot commercial alkaline cleaner (Ridoline 53). By using the same one- and two-part commercial epoxy adhesives used over many years, a direct comparison could be made between different surface pretreatments, since the change in treatment is the only variable in the continuing exposure testing in different weathering environments. From Table 170 we see that the initial joint strengths of 2630 psi (18.12 MPa) and 5270 psi (36.31 MPa) compare favorably with the 2830 psi (19.50 MPa) and 5330 psi (36.74 MPa) strength values obtained with hot chromic acid–deoxidized 6061-T6 aluminum alloy adherends. Durability can be reviewed in Tables 154, 155, 158, and 159. Immediately we observe an excellent example of the deception of concluding from a high initial joint strength that good durability follows, since the overall durability of the alkaline-etched surface joints is much inferior to those made with chromic acid–deoxidized surfaces.

From Table 158 we observe that hot alkaline-cleaned aluminum joints with the two-part, room temperature–curing epoxy adhesive had lost 56% of initial strength as compared to 47% for the vapor-degreased joints in RT water soaking. Meanwhile, corresponding joints treated with hot chromic acid deoxidizing lost only 6%. With a heat-curing, one-part epoxy, 55% loss in joint strength was comparable percentagewise using alkaline-cleaned adherends. However, this adhesive lost 74% of initial strength on vapor-degreased-only joints with the loss on chromic acid–deoxidized joints listed as 34%. In the more aggressive cyclic weathering exposure shown in Table 159, the lower-durability potential for the alkaline-only cleaned adherend joints became much more evident. Here the sustained loss in 2 yr was 89% compared with 94% for the vapor-degreased-only joints. In contrast, the chromic acid deoxidize–treated joints suffered only a 26% loss.

In the relatively mild industrial atmospheric exposure (see Table 154) which is a realistic test for exterior weathering, the data support the possibility of using the alkaline treatment, since overall loss in 8 yr was only 15%. Meanwhile, the comparable vapor-degreased-only joints lost 57% in this period and the chromic acid–oxidized surface joints lost an equal 16%, indicating it was not really superior to the alkaline pretreatment under this kind of mild continuous weathering. In Table 155, the data cover responses in a corrosive seacoast atmospheric exposure. The jump in joint survival time from 80 da for vapor-degreased–only treated joints to 290 da for the hot alkaline-cleaned joints would indicate some significant benefit obtained by alkaline cleaning after vapor degreasing. However, the 760-da survival time for chromic acid–deoxidized surface joints is a significant additional boost in joint durability potential.

The general subject of alkaline cleaning pretreatments would also include any hot deoxidizing of an aluminum surface using strong caustic materials like sodium hydroxide. Alcoa has historically used a rapid deoxidation procedure involving the use of a hot sodium hydroxide solution, referred to as the Alcoa A1 process. Sometimes it has been used as the first step of a two-step process followed by Alcoa A3 hot chromic acid deoxidation. Unlike most acidic deoxidation processes, the caustic etching leaves a considerable smut on the surface which needs to be removed with some desmutting step before acid deoxidizing. Minford (3714–3716) has used a dip in cold 40% nitric acid to remove this smut after hot caustic deoxidizing in order to evaluate the caustic treatment for subsequent bonding of 6061-T6 adherends. It should be remembered that this alloy has a considerable heat-treat film on its mill-finish surface. Tables 154, 155, 158, and 159 list the long-term weathering results. From Table 170 we note that the initial joint strengths achieved with the one- and two-part epoxy adhesives were actually about 10% lower than shown in the previous discussion for the hot Ridoline soaking pretreatment. In Table 158, we note that the hot caustic deoxidize–treated joints are about as durable in the 2-yr water-soaking testing as obtained with the hot acidic deoxidized joints. In Table 159 we note that the retention of 45% of initial joint strength after 2 yr in the soak/freeze/thaw cycling conditions makes them much superior to any vapor-degreased–only, silica-gritblasting, hot-inhibited alkaline cleaner and even some of the acid deoxidation methods. The hot chromic acid–deoxidized joints at 74% retention are still clearly superior. We might expect that these hot caustic-etched joints would be very durable in the industrial atmosphere, and from Table 154 we see that they retain 100% after 8 yr. In Table 155, citing results in the corrosive seacoast exposure, we observe a maximum survival time of 394 da, which is much better than vapor-degreased joints at 80 da and hot Ridoline-treated joints at 290 da. However, it is only 58% as long as the hot chromic-deoxidized 6061-T6 aluminum joints.

Another variation of alkaline pretreatment, the European Chemoxal procedure, was reported in the literature by Brockmann in 1974. It involves acetone wiping, etch for 1 min in RT 10% caustic, desmut in 1:1 nitric acid and water, and 1-min exposure to dilute alkaline buffer solution at 180°F (82.2°C). Minford (3717) has evaluated and compared the relative durability of 6061-T6/epoxy joints treated with three variations of buffered phosphate solution as the final processing step. Comparative durability testing was conducted using Alcoa A3 (FPL equivalent), Alcoa A1 (hot caustic etch plus 40% nitric acid dip), Deoxidine 670 (proprietary mix of alcohol/phosphoric acid), and a hydrochloric acid/dichromate mixture suggested by some earlier experiments of Wegman at Picatinny Arsenal. The comparative data in four different weathering environments can be reviewed in Tables 174–177. The claim by Brockmann of equal aluminum joint durability using the Chemoxal and FPL-type procedures using aircraft aluminum alloys was not confirmed in Table 175 for exposure to 100% RH at 125°F (52°C) conditions by Minford. However, it seems to be relatively true for the other weathering conditions of room temperature water-soaking, hot-water–soak/freeze/hot-air-dry cycling conditions and the seacoast, as shown in Table 174, 176, and 177. Also, the joint durabilities of Deoxi-

Table 174 Comparative Durabilities of Various Etch Pretreated Aluminum Joints in Room Temperature Water Soaking

Surface pretreatment	Exposure time (da)	Average initial joint strength		Retained strength		% retention	Durability ratio[a]
		psi	MPa	psi	MPa		
Chromic-sulfuric (FPL-type)	730	4930	33.96	2680	18.47	54.4	1.00
HCl-dichromate	730	4970	34.24	3800	26.18	76.5	1.42
Deoxidine 670	730	4970	34.24	2630	19.50	56.9	1.06
Alcoa A1 + 40% nitric acid dip	730	4630	31.90	2200	15.16	47.5	0.82
Chemoxal variation No. 1	730	5167	35.60	3127	21.55	60.5	1.17
Chemoxal variation No. 2	730	4960	34.17	2833	19.52	57.1	1.06
Chemoxal variation No. 3	730	4720	32.52	2794	19.25	59.2	1.04

[a]Durability ratio $= \dfrac{\text{Avg. retained strength} \times \text{days of exposure or survival}}{\text{Avg. bond strength of FPL-type joints} \times \text{days exposure or survival}}$

From Ref. 3717 (Table VIII).

Table 175 Comparative Durabilities of Various Etch Pretreated Aluminum Joints in 100% RH at 125°F (52°C)

Surface pretreatment	Exposure time (da)	Average initial joint strength		Avg. strength retention		% retention	Durability ratio[a]
		psi	MPa	psi	MPa		
Chromic-sulfuric (FPL-type)	224	4930	33.96	2060	14.19	41.8	1.00
HCl-dichromate	730	4970	34.24	2800	19.29	56.3	4.43
Deoxidine 670	NA	4970	34.24	NA	NA	NA	NA
Alcoa A1 + 40% nitric acid dip	730	4630	31.90	390	2.69	8.4	0.62
Chemoxal variation No. 1	224	5167	35.60	985	6.79	19.1	0.48
Chemoxal variation No. 2	280	4960	34.17	960	6.61	19.4	0.58
Chemoxal variation No. 3	224	4720	32.52	920	6.34	19.5	0.45

[a]Durability ratio $= \dfrac{\text{Avg. final bond strength} \times \text{no. of days exposure or survival time}}{\text{Avg. final strength FPL-type joint} \times \text{no. days exposure or survival}}$

From Ref. 3717 (Table IX).

Table 176 Comparative Durabilities of Various Etch Pretreated Aluminum Joints in Soak/Freeze/Thaw Cycling

Surface pretreatment	Exposure time (da)	Average survival (da)	Average joint strength		Retained strength		% retained	Durability ratio[a]
			psi	MPa	psi	MPa		
Chromic-sulfuric (FPL-type)		309	4930	33.96	0	0	0	1.00
HCl-dichromate	730	730	4970	34.24	2770	19.08	55.7	>2.36
Deoxidine 670		178	4970	34.24	0	0	0	0.58
Alcoa A1 + 40% nitric acid dip		95	4630	31.90	0	0	0	0.31
Chemoxal variation No. 1		234	5167	35.60	0	0	0	0.76
Chemoxal variation No. 2		347	4960	34.17	0	0	0	1.12
Chemoxal variation No. 3		378	4720	32.52	0	0	0	1.22

[a]Durability ratio $= \dfrac{\text{Avg. final strength of joints} \times \text{no. days of exposure or survival}}{\text{Avg. final strength FPL-type joints} \times \text{days exposure or survival}}$

From Ref. 3717 (Table X).

dine 670 joints (which do not require any heat pretreating) and those treated with the Alcoa A1 (hot caustic plus desmut procedure) seem to be inferior in performance to both Alcoa A3- and Chemoxal-pretreated joints. However, the best overall performance afforded by any of the pretreatments reported in this investigation was shown by room temperature etching in HCl/dichromate solution joints. The ratio of improvement range from a low of 1.45 times better in retained joint strength up to as high a ratio as 5.76 times better.

Table 177 Comparative Durabilities of Various Etch Pretreated Aluminum Joints in the Seacoast Atmosphere

Surface pretreatment	Average survival time (da)	Average initial joint strength		Retained strength		Durability ratio[a]
		psi	MPa	psi	MPa	
Chromic-sulfuric (FPL-type)	784	4930	33.96	4370	30.11	1.00
HCl-dichromate	>2190	4970	34.24	4195	28.90	>5.76
Deoxidine 670	987	4970	34.24	3600	24.8	1.65
Alcoa A1 + 40% nitric acid dip	466	4630	31.90	2350	16.19	0.66
Chemoxal variation No. 1	>720	5167	35.60	4940	34.04	>1.13
Chemoxal variation No. 2	>720	4960	34.17	4580	31.56	>1.05
Chemoxal variation No. 3	>720	4720	32.52	4800	33.07	>1.10

[a]Durability ratio $= \dfrac{\text{Avg. final bond strength} \times \text{no. days exposure or survival time}}{\text{Avg. final bond strength FPL-type} \times \text{no. days of exposure or survival}}$

From Ref. 3717 (Table XI).

(1) COMBINATION ALKALINE CLEANING OR DEOXIDIZING PLUS ACID DEOXIDIZING. Minford (3714–3716) made direct durability comparison between combination alkaline pretreatments plus acidic deoxidizing and the same acidic deoxidizing without alkaline precleaning. For example, 6061-T6 adherends were treated by a two-step process of (1) preclean in hot alkaline Ridoline 53 and (2) deoxidizing in hot chromic acid (Alcoa A3). Another set of test joints were prepared with the two-step process of (1) deoxidizing in hot caustic (Alcoa A1) and (2) deoxidizing in hot chromic acid (Alcoa A3). Both sets of joints were compared to joints which were hot chromic acid deoxidized only. In the order described above, the initial joint strengths with two-part epoxy were 2630 psi (18.12 MPa), 2530 psi (17.43 MPa), and 2830 psi (19.50 MPa), respectively. With a heat-curing, one-part epoxy paste, the corresponding initial joint strengths were 5430 psi (37.41 MPa), 5370 psi (37 MPa) and 5330 psi (36.72 MPa), respectively. Thus, there was no reason to choose any specific treatment based only on average initial strength. In the long-term weathering exposures, the relative durability performances of the two-step process joints did not indicate any advantage from performing the extra step of an alkaline pretreatment before the regular chromic acid deoxidizing.

b. Acid-Deoxidizing Cleaner Types

Since alkaline cleaners are most commonly used on aluminum for simple soil removal, it is usually important to precede an acid deoxidation with an alkaline cleaning, especially where heavy soiling is present. However, in most of the joint fabrication at the Alcoa Labs, it was not found that an alkaline cleaning was necessary where the mill-finish 6061-T6 surfaces were heavily oxidized but not heavily overlaid with soil. It might be necessary, for example, if aluminum material was to be treated which had been inappropriately stored and had serious soiling and even some corrosion on the surface. For the 6061-T6 material received directly from the Alcoa fabricating plants, the author routinely used a vapor degreasing directly before the hot acidic deoxidizing. Even where some mild smutting was evident, it still seemed to be perfectly satisfactory to rely on only the acid deoxidizing without a prior hot caustic etching. With appropriate material, the use of the acidic deoxidizing would invariably produce a uniform, bright-looking wettable surface.

(1) PHOSPHORIC ACID–CONTAINING TREATMENTS. Certain acid solutions, especially those containing phosphoric acid, are able to dissolve not only oxide films but also oxidation products (as from atmospheric weathering). These solutions are often economical where large areas are to be cleaned for subsequent bonding. Although they can be used for shorter times at elevated temperatures, they are more frequently employed at room temperature. The opportunity to use them without heating further permits them to be used in relatively low-cost tankage such as an easily constructed wood framing with an interior fitted vinyl liner. Typical commercial formulations like Deoxidine 526 or 670 contain alcohols such as butyl or isopropyl and can be further diluted to offer whatever level of aggressive deoxidizing is needed. Most Alcoa tests conducted by the author utilized a 25% solution of the Deoxidine concentrate diluted with deionized water with the 6061-T6 adherend immersed for about 5 min at ambient temperature. In Table 170, it can be observed that the initial joint strength levels with the one- and two-part epoxy adhesives are significantly above that achievable by only vapor degreasing and only slightly lower than that attained using the hot chromic acid deoxidizing. From the data in Table 158, we note that the 2-yr soaking in room temperature water produced only a 5% decline in joint strength, which was equivalent to that obtained with the hot chromic acid-pretreated joints and quite superior to the 47 and 16% declines shown by vapor degreasing or hot alkaline cleaning. The hot caustic (Alcoa A1) pretreatment, however, was only slightly poorer with a 10% decline in strength. The soak/freeze/thaw cycling data in Table 159 permit first delineation of a superiority of the hot chromic acid deoxidation over the room temperature Deoxidine deoxidation, with the former joints declining an average of 26% as compared with 57% for the latter. The advantage over vapor degreasing only or hot alkaline soaking is clear, however, since these two pretreatments only offer 6 and 9% joint strength retentions after 2 yr.

In the milder industrial atmosphere after 8 yr in Table 154, the advantage of the ambient phosphoric acid deoxidizing over vapor degreasing is still quite evident; however, the hot alkaline-soaking–pretreated joints now perform as well as either the room temperature Deoxidine-treated or hot chromic acid–deoxidized joints. In the seacoast exposure results in Table 155, the relative maximum joint survival times of 80 da for the vapor-degreased-only joints now compares with 290 da for hot alkaline-soaking conditioning, 270 da for the RT phosphoric acid Deoxidine conditioning and 760 da for the hot chromic acid–deoxidation conditioning. Thus, it appears that a surface pretreatment for optimum economy and acceptable service potential may be selected based on the worst kind of weathering that might be encountered in service.

While the time to complete the deoxidation pretreating in the above Deoxidine-type phosphoric acid/alcohol proprietary etchants can be reduced by heating the solution to temperatures in the range of 120–125°F (49–52°C), the durability test performance would not be significantly changed. Minford (3714–3716) has made evaluations of hot phosphoric acid–type deoxidizer by substituting phosphoric acid for the chromic acid in Alcoa A3 or dichromate in FPL (see procedure 9 in Table 170). The initial joint strength achieved by this substitution was modestly lower than found with the original Alcoa A3 or an FPL pretreatment. However, in the 2-yr water-soaking evaluation in Table 158, there was essentially no difference in the average strength retention of joints treated with hot phosphoric or chromic acid when mixed with sulfuric. Likewise, no performance difference could be ascertained after 2 yr in the more aggressive soak/freeze/thaw cyclic exposure in Table 159. It would be presumed from prior results that the durability performances would be near identical in the relatively mild 8-yr exposure in the industrial atmosphere in Table 154; however, the percentage of actual strength retention was some 13% higher for the phosphoric acid substitute mix. The most discriminating 8-yr planned exposure to the seacoast atmosphere showed very similar maximum joint survival times of 674 da for the phosphoric acid–containing mix and 760 da for the hot chromic acid–containing mix, as shown in Table 155.

(2) ALUM-SURF-PREP PROPRIETARY SURFACE DEOXIDIZER. A variety of fast acting at room temperature, aluminum cleaning and brightening products are commercially available which might be able to substitute as a pretreatment bath for aluminum bonding. One of these products from Arcal Chemicals has been investigated by Minford for this possibility and the joints subjected to the battery of weathering conditions discussed above. The advantage of such a pretreatment would be the fact that it could be easily diluted in water and pretreatment affected by immersion periods as short as 1 min at room temperature. These conditions were used with 6061-T6, which with its heavy heat-treat film should offer a real test of its deoxidizing effectiveness at such a short-treatment time. Again the data from Tables 154, 155, 158, and 159 are most pertinent for making direct comparisons.

Although the initial joint strength in Table 170 was similar to the other acid deoxidizers already discussed, the 2-yr water-soaking results show 67% strength retention compared with 95 and 94% for the room temperature Deoxidine and hot hot chromic acid deoxidizers. In a more severe soak/freeze/thaw environment, the fast etch of Alum-Surf-Prep fails significantly, with no joints surviving the 2-yr testing. Perhaps unexpectedly, the 8-yr exposure in the rather mild industrial atmosphere caused no apparent decline in joint strength, which might indicate that this kind of fast, economical pretreatment should suffice for all sorts of indoor exposure service. From the poor response in the 2-yr water soak/freeze/thaw cycle shown in Table 159, we might expect a very poor durability response in the seacoast exposure in Table 155. However, a maximum joint survival time of 580 da would make this type of pretreated joint rather close in performance to the hot chromic/phosphoric and chromic/sulfuric pretreatments.

(3) NITRIC ACID COMBINATIONS. Nitric acid is a mineral acid that has been used as a desmutting agent on aluminum but with a similar but lower, overall corrosive action on aluminum than the phosphoric acid discussed above. Minford (3714–3716) has made a substitution for the chromic acid in Alcoa A3 and dichromate in FPL with nitric acid. It might also be considered as a substitution of nitric acid for the phosphoric acid, as listed in procedure 9 in Table 170.

From Table 170, we observe a similar initial joint strength as compared with the various deoxidizing procedures such as Alcoa A3 and the FPL-type deoxidizing. From the data in Table 158 relating to 2-yr immersion in water we observe a 93% joint strength retention, which is equal to that of the Alcoa A3 and FPL but slightly higher than when phosphoric acid was mixed with sulfuric acid as deoxidizer. The same relationships were found when the exposure was for 2 yr in the more aggressive soak/freeze/thaw cycling in Table 159; i.e., joint strength retention about equal to the Alcoa A3 but higher than for the phosphoric acid substitution for chromate. In the milder natural industrial atmosphere, a joint retention of 94% was actually 10% better than the joints treated with the hot

Alcoa A3. Finally, maximum joint survival time of 674 da at the seacoast falls in between the 760 da for the chromic/sulfuric deoxidizer and the 644 da for the joints treated with a phosphoric/sulfuric deoxidizer, as shown in Table 155.

(4) HYDROCHLORIC ACID VARIATIONS. This work involves an elaboration by Minford (1026–1028,3718) on a pretreatment mentioned in the literature earlier by Wegman (991) but with no aluminum joint relative durability data provided for comparison with other pretreatments. Wegman suggested that a strong aluminum joint was formed by the simple treatment of brushing on an HCl-dichromate solution. This would relate to a substitution of HCl for the sulfuric acid in the conventional FPL etchant. Starting with this ideal that you needed both the hydrochloric acid and the passivity control afforded from addition of dichromate, Minford ran long-term joint durability evaluations, as described in Tables 154, 155, 158, and 159. Later Minford ran tests with butyl alcohol substituting for the dichromate, which afforded retardation and better control of the etching process. In effect, this mixture represented a substitution of hydrochloric acid for the phosphoric acid used in the proprietary Deoxidine-type deoxidizers. Regardless, the big economical advantage should come from (1) the low cost of the hydrochloric acid compared with any of the other mineral acids, (2) the very short-deoxidizing time, and (3) no need to heat the solution. It was realized early that some extra provision had to be made to remove the corrosive chlorine vapors from the atmosphere near the workpiece. Also on the negative side, was rapid fouling of any treating solution by violent chemical action on the aluminum surface, which virtually mandated fresh etchant solution be used to treat each new batch of material. In contrast, the FPL-type deoxidizing bath can be used to treat multiple batches.

It is immediately observed that even at room temperature a considerable dark smut is formed on the aluminum surface which does not appear to be strongly adhering. In fact, this smut could be easily disturbed and usually removed by simply brushing the surface with a finger or wiping lightly with a piece of cheesecloth. Thus, it was early suspicioned that this loosely adhering layer might constitute a weak boundary layer in a structural joint and should be removed before final rinsing and drying. For some time, Minford proceeded to remove this layer by immersion in a room temperature, 40% nitric acid solution followed by strong rinse in deionized water and air drying. Later, in a durability test program designed more from curiosity than scientific deduction, the author surprisingly established that some surface change took place when bonding an epoxy to this smut layer. The end result was that the same high-bond joint strengths that were achieved when the smut was removed before bonding were still achievable when the smut was not removed. To this date, the author has no rational explanation of why this happens, but it is a repeatable observation from hundreds of additional tests. In the present recommended procedure, the smut is washed away with a forceful stream of water and immediately dried in a blast of filtered air. Before and during the rinsing period, the surface appearance is nonuniformly dark; however, the compressed air stream dries it rapidly to a lighter gray. At the present time, the recommended procedure is to prepare a brushable paste by adding concentrated hydrochloric acid to Cab-O-Sil filler. The paste is brushed over and just beyond the area to be bonded and the etching allowed to proceed until gas evolution ceases. Any residues are flushed away in water and the surface is blown dry. It can immediately be bonded, or the author has allowed the treated surface to stand in the laboratory from hours to days before final bonding without adversely affecting the overall joint durability.

Minford (3718) has compared the relative durability of joints made by the last method noted in the previous paragraph in 3.5% intermittent salt spray with 6061-T6 adherends. Comparison were also made with joints made by immersing the joints in either the HCl/dichromate or HCl/butyl alcohol at room temperature for periods from 15 sec up to 5 min. Best durability performance was achieved with the paste treatment consisting of bottle-strength HCl mixed with the Cab-O-Sil to form a viscous paste. A 1:1 dilution of bottle-strength HCl with distilled or deionized water also could form an effective solution for 1-min immersion treating at room temperature.

Some of the earliest but still important long-term durability results obtained by Minford are shown in Tables 154, 155, 159, 159 with immersion of 6061-T6 adherends into a diluted HCl plus dichromate solution followed by a nitric acid dip to remove the smut and a final forceful rinse in deionized water. This is a room temperature FPL-type etch with HCl substituting for sulfuric acid. It appears in Table 170 that initial strength levels for both a one- and two-part epoxy adhesive were approximately

13 and 7% lower than produced by the hot Alcoa A3 chromic acid deoxidizing. In Table 158, after a 2-yr soak in water, the HCl deoxidized joints and Alcoa A3 two-part epoxy joints both averaged near 95% joint-strength retention. However, the HCl-treated joints bonded with a heat-cured epoxy actually averaged 10.1% higher joint-strength retention. In the aggressive cyclic soak/freeze thaw 2-yr testing data, the HCl joints averaged 8% higher durability with a two-part epoxy and 22% higher with the one-part, heat-curing epoxy, as reviewed in Table 159. As anticipated, the 8-yr exposure to the mild industrial atmosphere was not a challenge to joints deoxidized in either fashion, as shown in Table 154. Finally, in Table 155, the greatest challenge for all deoxidized surface joints in the seacoast atmosphere for 8 yr showed by far the greatest advantage for using the HCl deoxidizing. Whereas the chromic acid–deoxidized joints using one- and two-part epoxies had maximum joint survival times of 1460 da and 760 da, respectively, the comparable HCl-deoxidized joints showed 86% strength retention after 4 yr for the one-part epoxy joints and 97% strength retention after 8 yr using a two-part epoxy cured at room temperature.

(5) CHROMIC ACID PROCEDURES. This subject has been kept to the end of the deoxidation discussion because of its close relation to the earliest main method of aerospace pretreatment; i.e., FPL and optimized FPL etch. As mentioned earlier, the many investigations by the author continued to use the hot Alcoa A3 procedure because it differed from the FPL only in the selection of chemicals, bath temperature, and time to achieve the same etching endpoint.

Some of the earliest aluminum-bond permanence studies were by Eickner (964–966) and Eickner and Schowalter (29,974,975) at FPL Lab involving exposures in Miami seacoast and Panama jungle sites. The investigation was later extended to include a form of stress testing. Wegman et al. (3719) used a moderate jungle site in Panama, an industrial atmosphere at Picatinny Arsenal, and a desert site at Yuma, Arizona, to test the FPL joint durability. As might be imagined, the jungle sites and seacoast sites were more discriminating than the industrial atmosphere and desert conditions. Cotter (1012), in 1977, described some relative durability results obtained at a Royal Aircraft Establishment site as compared to a wet/hot site and a dry/hot site in Australia. While durability normally deteriorated more rapidly with increasing temperature and humidity, anomalies did occur as shown by some stressed joints, with a modified epoxy deteriorating more rapidly on a temperate site than at a wet/hot site. McMillan (2707) has also compared the severity of a semitropical natural environment with various laboratory and in-service conditions.

Use of FPL on Alclad 2024-T3 adherends afforded joints which gave more consistent high-quality and salt-water spray resistance than by the use of any of the abrasive-, solvent-, or alkaline-cleaning methods previously used with aluminum. The etching conditions must be carefully controlled, and, as McMillan (833) reported, the durability was poorest with a fresh solution, which led to finding conditions to better optimize the conditions. Bijlmer (770) established through his special surface studies that the special microetch pitting condition produced by the particular formulation suggested in the FPL etchant was responsible for the good-durability performance. Bethune (1000) found the addition of small amounts of aluminum could improve the etching uniformity and with Smith (1004) that this applied to small additions of copper. The "optimized" FPL resulted from the addition of both ingredients to a fresh bath. Bijlmer (1007,1008) also demonstrated that excess quantities of both aluminum and copper could produce negative effects that should be avoided. Thus, careful analysis and control of FPL etchant baths are demanded for maintaining good durability results. McMillan (2707) confirmed poor-durability results with iron above 1 g/dm. Chen et al. (227) confirmed poor results with both excess chloride and fluoride ions. Scardino and Marceau (853) have evaluated the durability of FPL-treated joints when varying several of the important processing variables.

The strength-retention values obtained from Tables 158 and 159 show a creditable 94 and 74%, respectively, after 2-yr water soak or a hot water soak/freeze/hot-air dry cycling for 2 yr. Also quite acceptable was the 84% strength-retention average after 8 yr in the industrial atmosphere (Table 154). Disappointing is the maximum joint survival time of 760 da shown at the seacoast (Table 155), where an 8-yr exposure had been planned and was successfully resisted by joints with silica-gritblasted surfaces, HCl-deoxidized surfaces, and a room temperature deoxidizing in the Alcoa A3 ingredients. This latter unexpected response will be separately discussed. The early dramatic failure in the seacoast atmosphere of the hot Alcoa A3–treated joints was the basis for the prediction by the author that dramatic joint failures of naval aircraft in the Vietnam War should have been anticipated based on joint designs used and the coastal air station and carrier weathering service in which they were employed.

Minford (3720) conducted an extensive investigation involving 12 different variations of hot chromic-sulfuric and chromic-phosphoric acid and room temperature deoxidizing in HCl variations using 2036, 6061, and 5052 alloy adherends. An intermittent salt-spray weathering condition was used to discriminate the relative joint durabilities. As expected from the earlier discussion about the overall relationship of aluminum alloy corrosion resistance to joint durability, the observed durability of these joints with the same surface pretreatment was 5052-H34 > 6061-T6 > 2036-T4. The effectiveness of the different pretreatments in promoting best joint durability was HCl deoxidizing > chromic-sulfuric deoxidizing plus desmutting > chromic-sulfuric–deoxidizing variations without desmutting > vapor degreasing only. Different treatments did offer maximum durability potentials with different alloy adherends. A two-step hot etching chromic-sulfuric deoxidizing followed by a chromic-phosphoric deoxidizing proved most effective with the 2024-T3 adherends. 6061-T6 adherends were treated with a short room temperature immersion in HCl to produce maximum joint durability. A short-soak treatment in 150°F (66°C) chromic-sulfuric acid produced highest durability with 5052-H34 adherends. It is not really understood how the joint performance in the very demanding weathering of alternate wet-dry salt-water cycling can be so different with 2000 and 5000 alloy surfaces as compared with 6061-T6 surfaces when chromic-sulfuric deoxidized. It does show that many subtle differences exist between different alloy surfaces which can affect the weathering resistance of bonded joints.

In the early 1960s, Minford tried to study the durability of 6061-T6 joints pretreated by immersion in room temperature chromic-sulfuric acid solution. It was immediately evident that the visual appearance belied any evidence of the removal of any significant amount of the heat-treat oxide as was always evident by immersion in the same solution at elevated temperatures. The appearance of the surface, even after 20-min immersion at room temperature conditions, stayed relatively dark and quite variegated in appearance. Nevertheless, the joints were fabricated with the same two epoxy adhesives used previously to compare other pretreatments and exposed to the same accelerated and long-term natural atmospheric exposure conditions. Not much difference between the hot and ambient temperature chromic acid–immersion procedures can be seen based on initial strengths in Table 170. Hot-treated joints showed 10% higher strength with the two-part epoxy adhesive but 3% lower strength joints with the one-part epoxy. After 2-yr immersion in water, the two-part epoxy joints retained 94 and 88% of initial strength with identical 66% retentions using the one-part epoxy, as shown in Table 158.

It was unexpected from the initial absence of any visual evidence of etching to find that after 2-yr water soaking the retained joint strengths were similar to that achieved by the brightly etched hot chromic acid–treated joints. However, it was even more surprising to find that the ambient deoxidized joints after 2-yr soak/freeze/thaw cycling in Table 159 show the highest joint strength retention (88%) of any of 30 different pretreatments evaluated in Table 170.

After an 8-yr exposure to the industrial atmosphere (Table 154), both pretreated types retained joint strengths of 84 and 100%, respectively. The most striking performance was achieved by the ambient deoxidized joints after 8 yr in the very aggressive and corrosive seacoast conditions. In Table 155, we see that whereas the hot deoxidized joints had maximum survival times of 1460 and 760 da using one- and two-part epoxy adhesives, the comparable ambient temperature–treated joints actually retained 78 and 89% of initial strength values after a full 8 yr of exposure. Allen and Smith (3721) also showed good-durability results with aluminum joints pretreated with various paste forms of a cold chromate etch composition in 1985.

Clearfield et al. (3722), in their 1990 discussion of adherend surface preparations report an interesting series of FPL-treated joint-durability results based on the proposition that bond durability should be improved by (1) decreasing the rate of water diffusion through the adhesive or (2) along the bondline (33,37,524,838,3723–3725). Concepts were demonstrated with: (1) a water-wicking adhesive (worst-case scenario), (2) a water-resistant adhesive, (3) an organic hydration inhibitor with a water-wicking adhesive, and (4) a corrosion-inhibiting epoxy primer with a water-resistant adhesive. Figure 10 in Clearfield et al.'s publication (3722) shows crack propagation curves with data taken from Davis et al. (37,525), Hardwick et al. (529), and Shaffer et al. (3726) which prove the relationship between joint durability and slowing down the egress of water. The plotted curves clearly show a steadily decreasing crack growth rate as conditions for making the joint progressed from (1) through (4) above.

(6) VARIATIONS OF NONCHROMATE-TYPE TREATING. In recent years, the U.S. Environmental Protection Agency (EPA) has had definite concerns about the presence of chromates in industrial wastes, which caused a greater emphasis on finding deoxidizing formulations which effectively perform that task but do not contain chromium salts. Russell (767) developed an early nonchromate variation of the FPL etchant at Picatinny Arsenal as early as 1976. Clearfield et al. (3722) mentioned one variation of the FPL for pretreating where the stresses are low: $NaHSO_2$ would be substituted for H_2SO_4. They also refer to the P2 etchant, which is toxic chromate-free, yet can provide the complex aluminum oxide surface morphology that is required for mechanical locking, high strength, and good joint durability. Rogers (1009) described substitution of ferric sulfate as an oxidizer in place of the sodium dichromate in FPL. Desai et al. (1010) have demonstrated the similar oxide morphology to that produced by a broad range of FPL conditions. Long-term durability test data, if conducted, have not come to the author's attention.

Minford (3727) has evaluated a commercial aluminum cleaner without chromate that Wyandotte has formulated and named Pondax. This cleaner is employed at room temperature, which should make it of interest for manufacturers who cannot afford heating and controlling treating solutions. Using the same one-part heat-curing epoxy paste adhesive as for previous investigations, an initial joint strength of about 5000 psi (34.45 MPa) was quite acceptable. The strength retention of joints weathered by a 2-yr water soak was 82%, which would be comparable to other room temperature deoxidizers. The performance in the seacoast atmospheric conditions, however, was distinctly poorer than found with RT HCl or Alcoa A3 pretreating. All joints had failed by 695 da, which was comparable to the 760 da for hot Alcoa A3 pretreating but significantly less than the 8-yr survival with significant joint strength retention of both room temperature–deoxidizing systems (HCl and Alcoa A3). Comparative durability was also checked in 100% RH at 125°F (52°C), where the Pondax joints compared favorably with both ambient and elevated temperature chromic acid–deoxidizing procedures. Since the Pondax joints performed as well at the seacoast as the majority of the hot-acid deoxidizers, its relatively easy use at room temperature might offer a lot to recommend it for general bonding use.

(7) ACID PASTE DEOXIDIZERS. It was mentioned under the HCl treatment discussion that a paste was used on 6061-T6 adherends containing bottle-strength HCl and Cab-O-Sil filler. This form of paste deoxidation has been very effective in the factory production of joints involved in the structural assembly of very large parabolic satellite dishes. There were several earlier attempts by various chemical suppliers to develop some other acid paste–type products. One from Hughson Chemical, named B-727-6 acid paste cleaner, was evaluated by Minford. In Table 170, it can be seen that the overall initial joint strength using the two epoxies previously used for comparative testing at Alcoa was only slightly below that afforded by the hot chromic acid deoxidizing. The joints in Table 158 after 2-yr water soaking were slightly lower in joint-strength retention as compared to the hot chromic acid deoxidized. In the more critical soak/freeze/thaw cycling in Table 159, the hot Alcoa A3 joints with the two-part epoxy survived 2 yr with 49% higher strength retention. Again, a big difference was noted between the two epoxies under this weathering situation, since with the one-part epoxy, the joint strength involving both hot chromic acid– or the acid paste cleaner–treated surfaces had degraded to 19–25%. This paste cleaning (which could be applied in the field where the oxidation on the surface was not too advanced) might be considered for industrial atmospheric exposure, since the average joint strength retention after 8 yr was 76 and 85% for the one- and two-part epoxy adhesives, respectively, as shown in Table 154. However, the lesser durability afforded by the acid paste cleaning pretreatment was most clearly identifiable in the seacoast exposure, where the two-part epoxy joints averaged only 275 da to failure, as shown in Table 155. The choice of adhesive was again quite important, since the corresponding one-part, heat-cured epoxy joints survived an average of 1440 da compared with 1460 da using the hot chromic acid–deoxidizing treatment.

5. CONVERSION COATING PROCEDURES

Conversion coating is used to describe a surface film on aluminum which has been formed by a reaction, in which a portion of the base metal has been converted into one of the film components. Thus, the film is actually an integral part of the metal surface, which must exhibit excellent adhesion properties of its own to an underlying structure. Such coatings can be built-up in a wide variety of ways with a wide range of thicknesses depending on the bath reaction time. As the thickness increases, it has been shown that joint strength can decrease because of the higher probability of developing internal film flaws. Increasing peel forces exerted at the ends of lap-joints, for example, would play an increasing role in affecting joint failure as this layer thickened. This can be shown by making joints with increasing thickness of aluminum adherends using the same conversion coating thickness, and noting the increase in joint strength achieved by creating greater overall adherend stiffness. Increasing the adherend thickness can even produce a higher joint strength with a thicker conversion coating.

There has been interest in the aluminum industry to provide the most stable and inert surface beneath organic paints. Adhesives being similar in chemical composition to paints, should also develop similar adhesion and possible bond durability to aluminum surfaces specially pretreated by conversion coating. These conversion coatings can offer excellent bondline protection against a corrosive environment like the seacoast and marine atmospheres. It should be remembered that they would not constitute a replacement for the excellent adhesion and joint durability that can be achieved with the CAA and PAA anodizing oxide layers. The conversion coatings generally fall into five types: (1) alkaline oxide, (2) crystalline phosphate, (3) amorphous phosphates, (4) amorphous chromates, and (5) boehmite. Most widely employed today under organic paints on aluminum have been the amorphous chromate type, and they accordingly have been extensively studied by the author. They must be applied in acid solution.

First consideration for reacting an aluminum surface is that the surface be thoroughly deoxidized and clean. The variety of methods for achieving this has been discussed in Chapter 4. Since the aluminum joints that have been made from acidic deoxidizing only have shown some very acceptable joint-durability responses under a variety of different weathering conditions, the question might be asked why perform an additional pretreatment operation. The most acceptable answer that can be given might be that the very high resistance of the conversion coating layer to very corrosive conditions may be very desirable in many corrosive weathering service environments.

Alternatively, the deoxidizing step could be followed by one of the several anodizing procedures such as SAA, CAA, or PAA, and the latter two are widely used in aerospace applications for the extra joint durability that they can provide. The higher cost of anodizing as compared with building up a modest conversion coat must also come into the economic picture.

To better understand the precise benefits of these conversion coatings for structurally bonded aluminum joints, joint durability results obtained by Minford after 8 yr at the seacoast can be cited in Table 157. We observe that 6061-T6 joints with an Alodine 1200 conversion coating were bonded with a comparable one-part, heat-curing, nitrile-modified epoxy adhesive. The ability of such a joint to almost stand alone as being able to survive beyond 8 yr with little or no loss in bond strength constitutes a most excellent performance. We have noted on several occasions that a two-part, room temperature–curing epoxy can sometimes show a greater joint durability potential in various aggressive water-soaking conditions than a much stronger one-part, heat-cured product. Also in Table 157, we notice that a reverse situation has occurred with two-part epoxy over Alodine 1200 joints, all failing by 1475 da. The author would suggest that there is a different mechanism for enhancing durability that is generated due to the heating of the conversion oxide as a result of heat curing of a one-part epoxy. This consists of a further conversion change in the surface layers of the coating that is strictly thermally induced. As a result, the water resistance of the conversion oxide layer increases as water is lost from heating. For example, a fresh conversion coating becomes completely water nonwettable when heated. Fortunately, this heating to cure the adhesive takes place after the adhesive has been applied. At the same time as heating is commencing, the epoxy mobility is decreasing, so that wetting is essentially accomplished before the conversion coating can become nonwater wettable. The two-part epoxy joints without such heating apparently have sufficient wetting to show reasonable initial strength. However, the conversion coat layer without heating remains much more sensitive to slow migration of water, eventually producing a weak interface and final joint failure. The amount of water contact provided by the wet/dry cycling conditions of the seacoast is apparently enough to cause this gradual but steady decline in overall joint strength throughout 1475 da of exposure time, as shown in Table 157.

The final category of conversion coating mentioned at the beginning of this section was boehmite oxide. The solution mentioned by George et al. (856) is to use boiling distilled or deionized water. Accordingly, Minford (39) has investigated the joint durability of such films with both one- and two-part commercial epoxies. The heat-treat film on 6061-T6 adherends was removed either by deoxidizing in hot caustic plus dilute nitric acid desmutting or hot chromic acid deoxidizing alone followed by water rinsing. The 6061-T6 joint surfaces were set into a boiling tap or deionized water bath as soon as possible after the deoxidizing and incubated for varying periods of time up to 1 hr to establish boehmite films of varying thicknesses.

The aluminum adherend surfaces generated with the two types of water appeared distinctly different in appearance. The joint strength could vary greatly depending on the boehmite layer thickness. Short immersion times produced thinner coatings with higher final joint strength (see Table 17). A relative joint durability factor was calculated after each different exposure based on the average joint-strength retention value and relative overall exposure time. Results in room temperature water soaking were as shown in Table 178. Very acceptable joint durability could be achieved with exposures to the boiling water as short as 1 min. The thicker films evaluated (up to 1 hr in boiling water) tended to offer even higher durability potentials. The thickest boehmite film (after 1-hr exposure) offered the best overall durability responses; however, the initial joint strength was significantly lower because a point of lower stress fracture existed in the significantly thicker oxide. This is similar to the same observations the author had made earlier about the decreasing joint strengths obtained with increasing thicknesses of SAA anodic oxide.

Table 178 Durability Test Results in RT–Water-Soaking Conditions with 6061-T6 Alloy Joints with Varying Thicknesses of Boehmite Oxide on the Surface

			Average type failure		
Surface condition[a]	% Average retained joint strength	Average exposure (da)	Adhesive %	Cohesive %	Durability factor[c]
Mill-finish	35	182	97	3	6.37
Vapor degrease	40	126	95	5	5.04
Caustic etched	42	532	50	50	22.34
Deionized H_2O (1 min)	40	630	85	15	25.20
Tap H_2O (1 min)	35	518	95	5	18.13
Deionized H_2O (5 min)	63	714	50	50	44.98
Tap H_2O (5 min)	40	560	85	15	22.40
Deionized H_2O (10 min)	69	714	40	60	49.27
Deionized H_2O (60 min)	123	714	NA[d]	NA[d]	87.82

[a]The conditions for producing the different thickness boehmite oxide surfaces are described in the Ref. 39. In essence, the procedure consisted of exposing the caustic-deoxidized and acid-desmutted surfaces to the indicated type waters at the boiling point for the indicated times.

[b]The average initial joint strength of these joints can be found in Table 17.

[c]The durability factor is a calculation designed to weight the percent retained joint strength obtained considering the length of exposure time as follows:

$$\text{Durability Factor} = \frac{\text{\% Retained Joint Strength} \times \text{Days of Exposure}}{1000}$$

[d]Not applicable.

Data from Ref. 39 (Table II, p. 509).

Joints pretreated in boiling deionized water seemed more durable even though Wegman (561) had observed that rinsing an FPL-pretreated surface in 160°F (71°C) deionized water subsequent to bonding with a two-part, room temperature–curing epoxy produced a relatively low joint strength. In the boiling deionized water, a quite different situation obviously develops. Durability results in 100% RH at 125°F (52°C) weathering conditions are shown in Table 179 with lower overall durability factors created because of the higher-temperature soaking condition. The same pattern for a higher-durability response of the deionized water–pretreated joints seemed also to exist as compared to boiling tap water. Again, the highest level of joint durability in terms of percentage of joint strength retention occurred with thickest boehmite coating. Even more spectacular performances were shown in tests of 1 yr in cycling salt fog (see Table 180). Remarkably high-durability factors were shown by all joints which were fabricated with boiling deionized water. For example, an immersion in boiling DI water for only 1 min produced boehmite oxide joints with averaged joint-strength retentions of 78% after 1 yr exposure to this most severe of all accelerated laboratory weathering test environments. In contrast, the corresponding joints produced in 1 min in the local tap water at the boiling point showed joint strength decline to the 28% level. Again, the thickest oxide (1-hr exposure) joints showed highest durability values averaging slightly higher than the initial strength average value. The higher-durability potential of the mill-finish 6061-T6 joints in this severe weathering than those which had been initially vapor degreased only may seem to be unexpected. However, the presence of some thin layer of contamination on the mill-finish surface could slightly retard the initiation of severe pitting as compared to the vapor-degreased adherends with some of this contaminating layer removed.

Table 179 Durability Test Results in 100% RH at 125°F (52°C) Soaking Conditions with 6061-T6 Alloy Joints with Varying Thicknesses of Boehmite Oxide on the Surface

Surface conditions[a]	% Average[b] retained joint strength	Average exposure (da)	Durability factor[b]
Mill-finish	22	140	3.08
Vapor degrease	37	84	3.11
Caustic etch	38	84	3.19
Deionized H_2O (1 min)	45	294	13.23
Tap H_2O (1 min)	31	196	6.08
Deionized H_2O (5 min)	54	266	14.36
Tap H_2O (5 min)	43	224	9.63
Deionized H_2O (10 min)	58	336	19.49
Tap H_2O (10 min)	73	364	26.57
Deionized H_2O (60 min)	91	364	33.12
Tap H_2O (60 min)	54	308	16.63

[a]The conditions of producing the different thickness boehmite oxide surfaces are described in Ref. 39. In essence, the procedure consisted of exposing the caustic-deoxidized and acid-desmutted surfaces to the indicated type waters at the boiling point for the indicated times.

[b]The average initial joint strength of these joints can be found in Table 17.

[c]The durability factor is a calculation designed to weight the percent joint strengths for comparative purposes by incorporating the length of the exposure time as follows:

$$\text{Durability Factor} = \frac{\%\ \text{Retained Joint Strength} \times \text{Days of Exposure}}{1000}$$

Data from Ref. 39 (Table III, p. 511).

Table 180 Durability Test Results in Salt-Fog Cycle Exposure with 6061-T6 Alloy Joints with Varying Thicknesses of Boehmite Oxide on the Surface

Surface conditions[a]	% Average retained joint strength	Average exposure time (days)	Durability factor[c]
Mill-finish	17	28	0.48
Vapor degrease	0.5	70	0.04
Caustic etch	27	266	7.18
Deionized H_2O (1 min)	78	364	28.39
Tap H_2O (1 min)	28	168	4.70
Deionized H_2O (5 min)	77	364	28.03
Tap H_2O (5 min)	45	266	11.97
Deionized H_2O (10 min)	89	364	32.40
Tap H_2O (10 min)	85	364	30.94
Deionized H_2O (60 min)	103	364	37.49
Tap H_2O (60 min)	66	364	24.02

[a]The conditions of producing the different thickness boehmite oxide surfaces are described in Ref. 39. In essence, the procedure consisted of exposing the caustic-deoxidized and acid-desmutted surfaces to the indicated type waters at the boiling point for the indicated times.
[b]The average initial joint strength of these joints can be found in Table 17.
[c]The durability factor is a calculation designed to weight the percent joint strengths for comparative purposes by incorporating the length of the exposure time as follows:

$$\text{Durability Factor} = \frac{\text{\% Retained Joint Strength} \times \text{Days of Exposure}}{1000}$$

Data from Ref. 39 (Table IV, p. 513).

The durability testing under a combination of stressing and hot humidity conditions will be considered in a later section.

Minford (3728) has also evaluated whether any benefit in joint durability might be obtained over vapor degreasing only by boiling vapor-degreased Alclad 2024-T3 and 6061-T6 adherends in tap water for 1 min. As noted above, this was quite beneficial after deoxidizing 6061-T6 surfaces. The oxide produced by a 1-min boil on Alclad 2024-T3 (higher-purity 1285 alloy surface) actually produced joints which were 29% lower in joint strength than similarly processed 6061-T6 joints with their initially heavy heat-treat oxide film. 6061-T6 joints, in turn, were 14% lower initial strength than the vapor-degreased–only 6061-T6 joints, as shown in Table 170. With a one-part, heat-curing epoxy, the 6061-T6 boiled in tap water surface bonded with 11% higher joint strength than with a vapor-degreased–only pretreatment. While there was some modest extension of bond durability of the tap water boiled surface joints over the vapor-degreased–only joints in 2-yr water immersion, 100% RH at 125°F (52°C) or soak/freeze/thaw cycling, there was no detectable improvement in durability after 4-yr exposure to the industrial atmosphere. The greatest benefit was actually observed in the corrosive seacoast exposure, where 1-min treatment in boiling tap water joints survived 3.25 times as long as vapor-degreased–only joints.

In all weathering conditions except for the seacoast, the performance of any joints produced by one of the deoxidizing aluminum procedures afforded superior durability performance.

6. ANODIZING-TYPE PRETREATMENTS

a. General Considerations

When aluminum is made the anode in a suitable electrolyte with a metal or carbon cathode, and current is passed through the cell, the surface is converted to an oxide coating. This oxide is intergral with the surface and has excellent adherence to the underlying surface, thus making an excellent base to which to adhere. How thick these coatings are can be precisely controlled through regulation of anodizing time and choice of electrolyte. Where the electrolytes have little capacity for dissolving the oxide, the coatings will tend to be relatively thin, nonporous, and have a limited thickness barrier layer at the base. Practical electrolytes are those with appreciable solvent action which generate a porous outer portion which has the most favorable properties for bonding the adhesive with both chemical forces and significant mechanical interlocking.

As a preview of the relative effectiveness of the three main anodizing pretreatments (as compared to vapor degreasing only and degreasing plus deoxidizing), the joint durability data from Minford's investigations (533,534) have been cited in Tables 181–183. A 2-yr soaking in room temperature water results are shown in Table 181, with little difference in joint durability shown between the chromic acid deoxidation–treated 6061-T6 joints and various anodized surface joints with about 63–68% average joint strength retentions. It should be noted that all these joints were bonded with a one-part, nitrile-modified epoxy paste, which has been shown earlier in Table 158 to have less actual water resistance than a comparable two-part, room temperature–cured commercial product. With the one-part epoxy, the vapor-degreased-only–treated joints retained a 27% overall average joint strength retention after 2-yr soaking in room temperature water.

In the 100% RH at 125°F (52°C) weathering (considered an accelerated tropical laboratory exposure) shown in Table 182, anodized joints as a group were distinctly superior to acid-deoxidized–pretreated joints with a similar 54–68% average strength retention. In aggressive intermittent salt-spray weathering shown in Table 183, discrimination is only found between the variously anodized joints, since earlier evaluations by the author had showed the vapor-degreased–only joints failed in about 30 da, whereas the chromic acid–deoxidized surface joints with the same adhesive survived about 360 da. It appears that the relative durability of the different anodized joints is PAA > CAA > SAA. It should be noted that this is in reverse order to that which would be found for the overall

Table 181 Relative Durability of Unstressed Aluminum-Bonded Joints[a] with Different Anodizing Pretreatments in Room Temperature–Water Immersion

Surface pretreatment	Initial joint strength		Average % retained lap-shear strength after indicated exposure times			
	psi	MPa	3 mo	6 mo	12 mo	24 mo
Vapor degrease	4330	29.83	70.9	59.4	46.2	27.0
Deoxidine 526 (5 min, 25% conc. at RT)	4970	34.24	83.1	84.5	73.8	56.9
Chromic acid deoxidize	5330	36.72	83.9	82.6	78.2	66.2
Chromic acid anodize	5513	37.98	85.6	83.6	70.9	65.0
Phosphoric acid anodize (Boeing procedure)	6480	44.65	89.8	80.2	74.4	63.0
Sulfuric acid anodize (12 asf, 60 min, boiling water seal)	3550	24.46	69.6	66.8	67.6	68.5

[a]6061-T6 aluminum alloy adherends were employed for all joints bonded with EC-2086 one-part, nitrile-modified epoxy adhesive cured at 400°F (204°C) for 15 min.

Data from Ref. 533 (Fig. 1, p. 20).

Table 182 Relative Durability of Unstressed Aluminum-Bonded Joints[a] with Varying Anodize Surface Pretreatments in 100% RH at 125°F (52°C) Exposure Conditions

Surface pretreatment	Initial lap-shear strength		Average % retained lap-shear strength after the indicated exposure times			
	psi	MPa	3 mo	6 mo	12 mo	24 mo
Vapor degrease	4330	29.83	40	25	20	17
Chromic acid deoxidize	5330	36.72	48	30	20	18
Chromic acid anodize	5513	37.98	70	69	58	58
Phosphoric acid anodize (Boeing procedure)	6480	44.65	70	63	58	54
Sulfuric acid anodize (12 asf, 60 min, boiling water seal)	3550	24.46	77	76	73	68

[a]6061-T6 aluminum alloy adherends were used to fabricate all joints with EC-2086 one-part, nitrile-modified epoxy adhesive cured at 400°F (204°C) for 15 min.
Data from Ref. 533 (Fig. 2, p. 21).

relative resistance of the surfaces to general corrosion. Thus, the differences in the surface morphologies and variations in the sensitivity to hydration of the different oxides must play a vital and predominate role in establishing overall bond durability.

b. Chromic Acid Anodizing

Chromic acid anodizing (CAA) has been in use for many years for treating aluminum surfaces to make them more resistant to general and concentrated pitting. While it is a much thinner oxide than

Table 183 Comparative Durability of Unstressed Aluminum Joints[a] with Various Forms of Anodizing Surface Pretreatment in Intermittent Salt-Spray Weathering Conditions

Type anodizing pretreatment	Exposure time (mo)	% Retention of initial lap-shear strength
Phosphoric acid (110 V for 6 min)	12	96
	24	85
Phosphoric acid (60 V for 18.5 min)	12	93
	24	87
Phosphoric acid (30 V for 30 min)	12	88
	24	84
Phosphoric acid (Boeing procedure)	12	79
	24	77
Chromic acid	12	82
	24	78
Sulfuric acid (12 amp/ft² for 60 min) (unsealed)	12	74
	24	71
Sulfuric acid (12 amp/ft² for 60 min) (sealed in pH 6 deionized water for 30 min)	12	50
	24	45

[a]6061-T6 aluminum alloy adherends were anodizied as described for all joints and bonded with EC-2086 one-part, nitrile-modified epoxy adhesive cured at 400°F (204°C) for 15 min.
Data from Ref. 533 (Fig. 4, p. 22).

the sulfuric-acid–anodized surfaces used most extensively on aluminum curtain wall buildings, it is considered a thick dense coating as comapred to the phosphoric acid anodize oxide (PAA). Venables et al. (837) have shown its solid column geometry which should offer good corrosion resistance. McDevitt and Solomon (841) showed that the thickness of the anodic oxide can depend strongly on the nature of the aluminum alloy chosen. Both Bethune (1074) and Venables et al. (837) have predicted the manner in which the relatively thick CAA oxide layer should be able to protect the metal/oxide layer from environmental attack as compared to the thinner amorphous oxide layers on mill-finish and vapor-degreased aluminum adherends.

Bijlmer (3729) in 1985 discussed CAA as used in Europe. The 1925 Bengough-Stuart process is still one of several used in France, Germany, and Holland. Bijlmer prefers not to seal the CAA coating and believes its excellent joint durability record in Europe justifies that position. He also favors higher voltages in the 40- to 50-V range without sealing as the best overall set of conditions for durable bonding. In the United States, CAA-pretreated joints were first reported on by Eickner (964) in 1955. The good initial joint strengths reported for CAA surfaces prompted subsequent durability testing by Fokker in Europe and by Bell Helicopter in the United States.

Additional aluminum joint durability evaluations, as shown in Tables 181–183, were conducted and published by Minford (533,534) in 1977 and 1978 which confirmed the good potential of CAA coatings. The results for CAA-pretreated 6061-T6 joints bonded with a one-part heat-curing epoxy adhesive show the CAA joints were superior overall to vapor-degreased–only or chromic acid–deoxidized joints in both room temperature water or hot humidity soaking for 2 yr. The greatest improvement of CAA over the other treated joints, however, was found in corrosive intermittent salt spray, where 78% joint-strength retention after 2 yr compared with failing in 30 da for vapor-degreased joints and total failure within 1 yr for chromic acid–deoxidized joints. The Boeing-type PAA-pretreated joints, however, outperformed the CAA joints and the CAA joints outperformed the SAA joints in the author's investigations.

Rogers (817,820–822) has reported on the durability of aluminum joints pretreated with a sealed CAA procedure at Bell Helicopter. A sealing process had been employed by aluminum manufacturers to obtain maximum weathering resistance for aluminum surfaces exposed in aggressive natural weathering conditions. However, Rogers has stated that fully sealed coatings fail to develop suitable bond strengths. Rogers (820) also makes the point that use of CAA should be restricted to those alloys that contain less than 5% copper and/or 7.5% total extra alloying elements if suitable coatings for durable bonding are to be achieved. Special problems apparently arise with alloys containing very high percentages of magnesium or silicon. These statements certainly should have relevence to the subject of overall joint durability. In addition, Rogers discusses the compatibility of sealed coatings with primers, stating both corrosion-inhibiting (CIAP) and non–corrosion inhibiting primers appear to be compatible. Finally, he offers an order of increasing durability in environmental exposures of anodize only < anodize plus noncuring CIAP < anodize plus curing CIAP. It should be pointed out that all the author's investigations were performed without the presence of any primers in order to establish the effectiveness of the different pretreatments without any interaction with other interfacial factors. Also, the author's work had always been intended to furnish data for the general manufacturer who cannot afford all the steps taken to ensure best performance in aerospace.

Rogers (820) concluded the combination of sealed chromic acid anodize and the CIAP products can greatly improve resistance of Alclad 2024 and Alclad 7075 alloy joints as well as the nonclad types of joints. He cautions that the use of high-peel strength aerospace film adhesives may require some sort of modification of the sealing process. He states that the paste and low-peel strength aerospace adhesives can be bonded to fully sealed surfaces, as produced by dichromate or boiling-water treatments after anodizing.

Minford (3730), in 1979, extended his studies of attempting to discriminate between the relative durabilities of CAA- versus PAA-pretreated aluminum adherends by conducting 4-yr testing in 3.5% intermittent salt-spray cabinets. At the end of that time, both kinds of joints still retained joint strength averages of more than 2000 psi (13.78 MPa), which ranged from 30 to 40% of initial joint-strength values, with no evidence of corrosion encroachment into the bondline based on the observations made on the joints after deliberately failing them to determine joint-strength retentions. Actually, this investigation failed to provide a clear superiority for either the CAA or PAA procedure. Direct comparisons have also been made by other investigators such as Brockmann and Hennemann (1046), Brockmann et al. (524), and Arrowsmith and Clifford (3731), who concluded that CAA-treated joints could perform as durably as PAA-treated aluminum joints.

Durability testing with a combination of stress and water-soak conditions will be discussed later.

c. Phosphoric Acid Anodizing (PAA)

The use of PAA for aerospace structural bonding is rather interesting in view of the fact that it illustrates how an aluminum surface preparation process rejected earlier for one reason may later be reviewed and be a near-perfect solution for a different market area. Alcoa made the earliest investigations of phosphoric acid as an electrolyte for anodizing and rejected its use as being unable to generate a protective enough oxide layer against external weathering conditions. The thinness of the layer and the large open pores which would not effectively seal against weathering prompted the emphasis to be placed on CAA and SAA procedures for commercial application uses. The same properties that made it unacceptable for protecting walls on building proved to be the best properties for generating high-strength and most durable aluminum joints under the combination of stress and weathering conditions.

McMillan (833) stated that the so-called optimized FPL etching treatment was proving to be unsatisfactory for bonding primary aircraft structures, and PAA was the best answer to this problem as introduced about 1974. Some immediately recognizable advantages were a lower criticality in its processintg variables compared to careful bath controls needed in the optimized FPL and the fact that its presence could be monitored by polarized light. Again, it was Venables et al. (837,844) who offered a visualization of oxide morphology with its longer "whiskers." Bijlmer (811) showed that the actual thickness of the oxide was somewhat alloy dependent. The hydration resistance of this oxide (established by Hunter et al. [834], Keller et al. [786], and Amore and Murphy [3732]) proved to be important in predicting excellent durability that has been established in these type joints. Noland (838) confirmed this using XPS. Both Hunter et al. (834) and Amore and Murphy (3732) had suggested this resistance to hydration might be due to the actual presence of phosphate ions in the PAA layer. Kinloch (3733), Solomon et al. (3734), and Sun et al. (635) showed location of the phosphorus near the oxide surface. McDevitt and Solomon (841) stated that additional bond stability could come from the presence and concentration of metallic elements at the metal/oxide, thus affecting the stability of the interphase. Kinloch et al. (781) and Sun et al. (850) have indicated detrimental effects of copper or magnesium on joint durability when present on an etched surface, so these elements would need to be excluded for best stability of the joint. Davis et al. (525) have offered a possible three-step mechanism for the hydration of such PAA aluminum–pretreated surfaces.

McMillan reported the results from a comparative durability test program at Boeing involving 1250 wedge-type specimens, each with the PAA treatment or optimized FPL etch. The anodized joints invariably showed the better response and consistency performance. Etching could still offer very good durability results, but the scatter of the results was much greater. Brewis et al. (2086) conducted some durability testing with both etched and PAA surface preparations using the BSL 312 aerospace adhesive. While the initial joint failures for all the joints were clearly cohesive, after exposure to hot humid conditions some apparent interfacial failure was observed with some joints. It could not be determined whether this was failure in the adhesive near the interface or true interfacial failure. Brewis (3735) contends that the oxide stability of the PAA layer is probably the paramount factor in promoting the excellent durability of these types of joints rather than the highly porous structure of the oxide and the attendant mechanical interlocking. Venables (33) and Ahearn and Davis (631) have discussed the effect on bond durability of the degree of mechanical interlocking and the resistance of the oxide to hydration both of which play important roles in PAA joints' overall durability performance. Hardwick et al. (529), Ahearn et al. (540), and Davis et al. (37,525) have reported investigations relating the greatest inherent durability of PAA aluminum joints to hydration resistance of oxide and mechanical interlocking with the microroughness of the surface. They relate this to slower hydration rate involving the same sequence involved in hydrating the FPL-generated oxide: (1) reversible adsorption of water, (2) slow dissolution of present oxide followed by rapid hydration of the surface to boehmite, and (3) final hydration of boehmite to the bayerite form. The crack propagation is said to be mainly connected to the second step of the existing oxide's conversion to boehmite step regardless of whether the oxide is FPL or PAA derived.

Minford has shown in Tables 181 and 182 that the durability of unstressed PAA joints and CAA joints in 2-yr soaking in liquid water or 100% condensing humidity at 125°F (52°C) conditions were quite similar. In fact, in Table 182, it would appear that the performance based on percentage of strength retention was actually best for the sulfuric acid-anodized and 60-min boiling water-sealed joints. However, it should be recognized that these joints suffer a highly significant penalty in initial joint strength because of the significantly thicker oxide produced; i.e., a 45% lower joint strength than developed by the same adhesive with the PAA-pretreated adherends. In Table 183, however, where the variously anodized joints were exposed for up to 2 yr in intermittent salt spray, the Boeing PAA procedure joints retained about the same joint strength as the CAA but performed better than sealed or unsealed SAA joints. It can be seen that when three other sets of PAA conditions were tested against the Boeing PAA-treated joints, their average strength retention in each case was higher by about 10%.

d. Sulfuric Acid Anodizing (SAA)

The most widely used process for several decades for producing thick porous anodic coatings on aluminum for architectural applications employed sulfuric acid as the electrolyte. For rendering maximum protection to exterior weathering conditions, it had proved to be superior to CAA coatings, because much thicker coatings could be developed. While the typical CAA coatings can be about 0.3 mil, the typical SAA coating thickness for buildings has been about 0.8 mil. At the same time, this processing was relatively easy to control in manufacturing and less expensive. The PAA-anodized coating was never even a viable candidate for these exterior weathering applications, because of thinness and resistance to accommodating any practical sealing.

(1) CONVENTIONAL TYPE. Two types of SAA coatings are produced by aluminum manufacturers; i.e., the conventional and the hardcoat varieties. The conventional type was produced in the United States by Alcoa under the general family name of Alumilites. Alumilite 215, which is a relatively colorless and transparent oxide, was first used by Minford (532–534,772–774) to evaluate the bond durability of structural adhesives on anodized surfaces. The considerable work carried forward in the airplane industry on the use of CAA anodizing for structural bonding was already appearing in the literature, whereas literature references on PAA evaluation were not present. Little or no technical information on the durability of adhering structural adhesives to the SAA surface was available for general manufacturing, and that obtained within the aerospace industry was not favorable, because of problems encountered at high peel loads at low temperatures. This could be attributed largely to the much thicker oxides than present with the CAA procedures and the recognized presence of fracturing within these thicker anodic layers. This would not really be as critical a factor for general structural bonding applications operating close to room temperature conditions. It also had to be accepted (because of thicker anodic oxides) that significantly lower initial joint-strength potentials would be present even with the same structural adhesive when comparing with CAA- and PAA-pretreated aluminum adherends (see Tables 7, 181, and 182).

Joint-durability studies were conducted by Minford (533,534,772–774) using 6061-T6 adherends with both one- and two-part epoxies. The results comparing the bonded SAA joints with a variety of other surface pretreatments are shown in Tables 23, 24, 172, and 173. The 86% joint-strength retention for two-part epoxy joints after 2-yr immersion in room temperature water in Table 172 should be acceptable, although somewhat lower than those pretreated with a silica-gritblast, chromic acid–deoxidation, or over an epoxy-phenolic–baked paint primer. However, it is not the resistance to water-soaking conditions but rather the superb durability of the anodized joints in corrosive salt-spray conditions (Table 183) and under simultaneous stress and water-soak conditions that makes the anodizing pretreatments most desirable. A distinct superiority over most other surface pretreatments can also be seen for the SAA-treated joints with a more aggressive water exposure like soak/freeze/thaw cycling conditions, as in Table 24. The most unexpected results are shown in Table 23 after 8-yr exposure to the relatively mild industrial environment, though 63% joint-strength retention is only considered low because a number of other kinds of treated joints produced equal or better numbers.

The corresponding results in Tables 23, 24, 172, 173, and 183 with a one-part, heat-cured, nitrile-modified epoxy paste adhesive show a poorer percentage retention of joint strength; however, the fact that initial joint strength of the SAA joints with this adhesive were significantly higher than with two-part epoxy adhesive must be considered. An exception to this general description of comparison performances between the two adhesives can be noted in Table 23 involving 8-yr industrial atmospheric exposure. Here, the superior wetting affected by heat curing apparently could lift the performance level significantly in the industrial atmosphere exposure with 80% average joint-strength retention.

Minford and Peters (3736) attempted to evaluate the feasibility of producing an SAA-anodized coating over some previously adhesive bonded aluminum structure. This could have commercial interest, where the advantages ordinarily offered by an SAA-anodized oxide layer might be desired after assembly by bonding. Alclad 2024-T3, 6061-T6, and 6063-T5 adherends were bonded with one- or two-part epoxies as lap-joints. These joints were then subjected to the commercial SAA anodize processing capable of building up a 0.8-mil weather-resistant oxide layer. It was shown that such joints showed a wide range of sensitivity to weathering as compared to control joints that were bonded and evaluated without anodizing. A small percentage of joints experienced failure during anodizing, yet duplicate joints might offer the same good-durability performance as control joints. The proportion of failed joints could be altered by switching adhesives indicating a significant performance variation among different commercial adhesives of the same class; i.e., one- or two-part epoxies. With the same adhesive, changing adherend from 6061-T6 to 6063-T5 could greatly reduce the number of joint failures that were experienced when the bonded joints were subjected to the anodizing process. The author attributed this to better wettability on the 6063-T5. Changing length of joint overlap from 0.5 to 3.0 in 0.5-in increments had no significant effect. It was deduced that some significant flaw had to exist in the bondline which permitted the strongly acidic mixture in the bath to enter and rapidly degrade the bond. Obviously, no flaw must exist in those joints which performed like the control joints. The anodizing current also was no factor, since equal numbers of failures occurred without the anodizing current present.

(2) HARDCOAT TYPE. The hardcoat processing results in thick, hard, wear-resistant anodic coatings which range in thickness from 1 to 5 mil. In addition to the change in electrolyte incorporating organic-type acid with sulfuric acid, lower electrolyte temperatures and higher current densities and voltages are employed. This reduces the electrolyte solvent action permitting generation of a harder, denser coating. These coatings are the darker-colored (grayish, gold through bronze and gray to black) architectural-type finishes which have been most widely employed on the majority of aluminum monumental buildings in recent years.

In an earlier publication, Minford (534) had shown an inverse relationship exists between the initial joint strength of variously anodized 6061-T6 adherend/epoxy joints and the thickness of the anodized layer. For example, the oxide thickness relationship was SAA > CAA > PAA, whereas the joints made with the same adhesive showed the joint-strength relationship was PAA > CAA > SAA. Presumably, this might be related to the increasing statistical probability of finding greater numbers of structural faults in increasing thickness of the aluminum oxide layer. Of course, when speaking only of initial joint strength, we must acknowledge that the increasing opportunity for the adhesive to find better potential for mechanical interlocking also is present in the series PAA > CAA > SAA anodic coatings. Since Alcoa Duranodic (hardcoat-type) anodized thicknesses can be approximately 20, 6, or 2.8 times thicker than the typical PAA, CAA, and SAA oxides that have been adhesive bonded, the prediction might be that much lower bond-strength potential might exist with the same adhesive bonded to a Duranodic product surface.

Minford (1086) conducted both initial joint-strength and long-term durability testing using Alumilite 215 (SAA-type) 6061-T6 adherends with typical thickness of 0.7 mil (17.8 μM) and Alcoa Duranodic 335S hard anodize procedure yielding a coating thickness of 2.0 mil (50.9 μM). Unexpectedly, the initial joint strengths of Duranodic adherends using the same one- or two-part epoxies gave a quite similar result to the Alumilite 215 adherend joints. This might be explained by the fact that while the hard coating oxide is significantly thicker, it is also much more dense. This latter factor must be quite significant in operating to prevent the development of the greater number of weak boundary layer conditions that would be expected based on thickness considerations alone.

When the above Duranodic surface–treated joints were compared in the same weathering conditions, however, a difference in performance was noticed that requires some clarification. A greater reduction in retained joint strength was observed for the Duranodic-anodized joints in hot condensing humidity a 125°F (52°C) as compared to the thinner cross-section Alumilite 215 (SAA-type) joints. This appeared for the one-part, heat-curing epoxy joints but not for those fabricated with the two-part, room temperature–curing adhesive. The author offers the explanation that this difference in durability performance is related to the presence of a greater degree of surface crazing of thicker coating while curing at elevated temperature. This lowering of durability potential was not observed with an ambient temperature–cured epoxy adhesive, because no condition for promoting increased surface crazing of the oxide was present.

Either Alumilite- or Duranodic-finish joints presented very similar durability results in both an industrial or seacoast atmospheric exposure condition. The explanation offered was that the amount and degree of direct water contact in either of these atmospheric conditions was significantly less than encountered in the continuously condensing hot humidity situation. Thus, there was no distinction between responses of room temperature- or elevated temperature–cured epoxies. The excellent resistance of the Duranodic surface to corrosive salt-water cycling has been well established in their direct exposure on monumental buildings and in laboratory-accelerated tests. Application of an adhesive over such an inherently corrosion-resistant coating should not show anything but excellent joint durability providing good wetting was achieved with the adhesive and this was observed.

e. *Effect of Anodizing on Previously Bonded Joints*

The question of whether an adhesive-bonded aluminum structure can subsequently be put through an anodizing process should be of some interest to the general manufacturing engineer. Minford and Peters (3736) have conducted a comprehensive investigation at Alcoa Labs to attempt to answer this question. ASTM D-1002 lap-joints were fabricated with Alclad 2024-T3, 6061-T6, or 6063-T5 aluminum adherend using both one-part, heat-cured epoxies and two-part, room temperature–cured epoxy adhesives. In order to develop information that would be of greater interest to general manufacturing applications, the use of higher-priced structural epoxy film adhesives was excluded, and all adherends were pretreated only by vapor degreasing. The anodizing procedure was standardized by using the same three-step process: (1) deoxidation in hot chromic-sulfuric acid followed with a cold nitric acid desmut, (2) anodizing in 16 weight percent sulfuric acid under a current density of 12 asf (amp/ft^2), and (3) sealing in boiling deionized water for 15 min.

It was rather surprising that any loss in joint strength after anodizing failed to serve as a reliable criteria for comparing different commercial epoxy performances. Either complete joint failure by total delamination occurred or the initial joint strength was generally maintained. This observation was always true for the heat-curing epoxy joints; however, room temperature–cured joints could actually gain additional joint strength. Minford has attributed this to some additional cross-linking of the adhesive during the sealing processing in boiling water. The criticality of alloy could also be shown by the fact that the results were significantly different using the same adhesives for the 6061-T6 and 6063-T5 adherends. For the former, 37% of all joints produced with vapor-degreased 6061-T6 joints totally delaminated during anodizing, whereas only 4% of 6063-T5 joints failed under identical fabricating and testing conditions.

IX. PREDICTABILITY OF LONG-TERM JOINT PERMANENCE RESULTS IN NATURAL ATMOSPHERIC WEATHERING BASED ON SHORT-TERM ACCELERATED-WEATHERING TESTING

Minford (3737) has attempted to find some predictive relationship between shorter-time exposure results in laboratory-accelerated conditions and very long-time exposures in natural atmospheric weathering conditions. Nine different methods of aluminum surface pretreatment were utilized on 6061-T6 alloy adherends to observe if any predictive relationships could be influenced by choice of pretreatment with both one- and two-part structural epoxy paste adhesives used to fabricate all the joint specimens.

As might be anticipated, an extremely wide range of durability test results were found depending on the combination of pretreatment and adhesive choice. The strength decay curves could be plotted, since triplicate joints of each different combination of bonding parameters were deliberately failed in tensile-shear after 3, 6, 12, and 24 mo in water-soak exposures and 3, 6, 12, 24, 48, and 96 mo in the industrial atmosphere. By comparing the residual joint-strength averages after 8-yr exposure in the industrial atmosphere with the various points on the joint decay curves from accelerated water exposures, it should be possible to calculate an "acceleration" ratio for each different combination of adherend, adhesive, and surface pretreatment condition and weathering condition. An overall (except for vapor-degrease–only joints) average retention in the 2-yr water immersion environment was 76% as compared with an overall average strength retention of 87.6% for the same collective joints after 8 yr in the industrial atmosphere. In spite of what appeared to be an overall similar deterioration in 2-yr in water immersion as compared to 8 yr in the industrial atmosphere, it was impossible to find a ready basis for predicting a specific length of time in water immersion that would directly predict an anticipated strength loss of differently surface treated joints in 8 yr in the industrial atmopshere. The rather close agreement averagewise between the overall deterioration of all the differently treated joints after 2 yr in water-soaking as compared with 8 yr in the industrial atmospheric exposure did not reflect the wide variance in different degrees of predictability that simply averaged out fortuitously.

The data shown in Table 184 can be studied to observe several difficulties encountered in making a consistent prediction that would apply to all situations. Obviously, one difficulty arises from the wide range of variations in joint degradation that arise with different surface pretreating conditions. This is clearly shown by observing the range of acceleration ratios that exist as the surface pretreatment is varied. For example, the highest acceleration ratio was shown by the SAA-treated joints bonded with the one-part epoxy, where the same rate of joint strength deterioration that took 8 yr in the industrial atmosphere required only 1.25 mo in water immersion. Thus, the acceleration ratio was 96:1.25, or approximately 77 times more degrading in water soaking than in natural weathering. On the other end of the range was one-part epoxy adhesive bonded over hydrochloric acid–etched joints, where it took 6 mo in water soaking to achieve the decline found after 96 mo in the industrial atmosphere. Thus, the calculated acceleration ratio was 96:6, or 16 times. Any attempt to make valid comparisons of the type just described is also complicated by those adherend treatment/adhesive combinations where the joint strength retention after 8 yr in the industrial atmosphere averaged above initial joint strengths. Since all combination joints lost some strength under water-soaking conditions, no acceleration ratio can exist and any prediction attempt becomes meaningless.

The fact that the acceleration ratios for eight different surface pretreatments with the one-part, heat-curing epoxy averaged 31.3 (3 mo in water soak is like 96 mo atmospheric exposure) shows a general higher order of water sensitivity as compared with the two-part epoxy performances. In contrast, an acceleration factor can only be calculated for five pretreatments with two-part, room temperature–curing epoxy with an average of 6.42, indicating a much more durable performance for the two-part epoxy joints in water-soaking conditions. Thus, simply changing to another formulation of the same chemical polymer family can create an entirely different set and range of values.

It might be questioned why the prominant surface pretreatments of CAA and PAA anodizing have not been included, but at the time this investigative program was initiated in the early 1960s only the chromic acid deoxidizing (FPL) was widely employed in American planes.

It should also be pointed out that the use of simple solvent degreasing (no alteration of the surface oxide) created joints which had essentially the same calculated acceleration factors using either the one- or two-part epoxy adhesives; i.e., 1.8 and 1.6. This might be interpreted to mean that no alteration of the mill-finish oxide actually provided the closest approach to projecting about equal joint deterioration in 2-yr water immersion and 8 yr in the industrial atmosphere. Consequently, any change in the oxide provided by various different deoxidizing, conversion coating, or anodizing procedures served to create a larger diversification of interfacial resistance to weathering as found

Table 184 Relative Predictions of Joint Durability in the Natural Industrial Atmosphere Based on RT-Water–Immersion Data

			Retained joint strength					
			2 yr		8 yr			
		Initial joint strength	RT Water		Ind. Atm.		Estimated equivalent exposure time[a]	Accelerated ratio[b]
Surface treatment	Adhesive	(psi)	psi	%	psi	%	(mo)	
Vapor degrease	1-part epoxy	4330	1170	27.1	450	10.4	54	1.78
Vapor degrease	2-part epoxy	2130	1130	53.2	907	42.6	60	1.6
Alcoa A3 (RT)	1-part epoxy	5500	3530	64.2	4585	83.4	2.5	38.5
Alcoa A3 (RT)	2-part epoxy	2570	2270	88.3	2645	102.9	<0	—
B-727-6 acid paste	1-part epoxy	4630	2470	53.4	3670	79.3	2.5	38.5
B-727-6 acid paste	2-part epoxy	2400	1970	82.3	2030	84.6	17	5.6
Deoxidine 526	1-part epoxy	4970	2830	57.0	4300	86.5	2.9	33.1
Deoxidine 526	2-part epoxy	2700	3570	95.3	2130	78.9	37.7	2.5
Alcoa A3 (180°F)	1-part epoxy	5330	3530	66.4	4730	89.0	2.5	38.4
Alcoa A3 (180°F)	2-part epoxy	2830	2670	94.3	2370	83.8	13.7	7.0
Alcoa A1	1-part epoxy	4630	2200	47.5	2715	58.6	11	8.7
Alcoa A1	2-part epoxy	2370	2130	90.1	2665	112.5	<0	—
HCl-dichromate	1-part epoxy	4970	3800	76.5	4595	92.5	6	16.0
HCl-dichromate	2-part epoxy	2670	2530	96.2	2520	95.8	6	16.0
Alodine 1200	1-part epoxy	2470	1670	66.7	2770	112.2	<0	—
Alodine 1200	2-part epoxy	1130	680	60.2	1130	100.0	<0	—
Alumilite 215 (SAA anodize)	1-part epoxy	3270	2436	74.4	2625	80.3	1.25	76.8
Alumilite 215 (SAA anodize)	2-part epoxy	2530	2170	85.8	1585	62.6	>96	>1

[a]Value is the extrapolated number of months on the joint-strength decay curve in RT–water soaking corresponding to the final joint-strength value after 96 mo in the natural industrial atmospheric exposure.
[b]The value obtained by dividing the extrapolated number of months in RT water required to achieve the same degree of joint degradation shown in 96-mo exposure to the industrial atmosphere by that period of time; i.e., 96. This would represent the relative degree of acceleration of joint strength decline in RT water compared to equivalent exposure to the industrial atmosphere.
Data from Ref. 3737 (Table II).

in direct water soak or the modest alternating wet/dry and temperature cycling conditions of the industrial atmosphere.

X. PROTECTIVE COATINGS FOR EXTENDING THE DURABILITY OF ALUMINUM JOINTS

Minford (3738) evaluated a variety of organic coatings or sealants for protecting aluminum joints in (1) room temperature–water immersion, (2) immersion in 145°F (63°C) water, (3) hot-water soaking/freezing/hot-air thawing cycling, (4) intermittent daily cycling in 3.5% salt water, (5) the industrial atmosphere, and (6) the seacoast atmosphere. The coatings and sealants included (1) bitumastic asphaltic-type coating, (2) vinyl air-dry paint, (3) nitrile-rubber sealant, (4) two-part, ambient-curing epoxy caulk, and (5) RTV silicone sealant. In test environments (1–3,5,6), the joints were all initially formed by bonding vapor-degreased Alclad 2024-T3 aluminum with a two-part, room temperature–curing epoxy. In test environment (4), the degree of severity was accelerated by using the less corrosion-resistant 6061-T6 alloy.

After 2-yr soaking in room temperature water, no joint failures were noted and all joints retained above 50% of initial bond values. The fact that unprotected joints showed the highest joint-strength retention of 79% was unexpected. After 2-yr soaking in elevated temperature water, the unexpected continued to occur with unprotected joints actually gaining 5% over initial average strength. The three quality sealants (Gutterseal, RTV 731 silicone, and Pro-Seal Epoxy Caulk) in sharp contrast showed an average decline in joint strength of 37%, whereas the lower-cost Bitumastic 50 and Ucilon 450-10 vinyl painted joints averaged only 1.5% decline. This most unexpected result is indeed difficult to explain. If the evaluation had been conducted for only 1, 6, or 12 mo, the average joint strength declines would have led to the conclusion that the sealants, caulk, and vinyl paint were all having a positive effect by eliminating any decline in joint strength. However, only vinyl paint maintained this same degree of protection after 24 mo exposure. Meanwhile, the nitrile-rubber and silicone sealant protected joints, and the epoxy-caulked joints declined by 37, 58, and 23%, respectively. During the same final 12 mo, the vapor-degreased-only joints actually gained for the first time by 19% and the asphaltic-coated joints remained unchanged. Whatever the mechanism of this degradation in the bondline may be in the presence of the sealants and caulk, it only became evident and accelerating in nature after 12 mo of soaking. This is yet another example of the errors in judgment that can be associated with conducting the short-term type of weathering exposure that is often employed for expediency in many acceptance test procedures.

In the aggressive hot-water–soak/freeze/hot-air–dry cycling only the lowest-cost asphaltic-coated joints retained virtually all initial joint strength after 2 yr. In 1 mo, all joints showed some elevation above initial strength, but by 6 mo all showed some modest level of decline. After 12 mo, there was no basis for seeing any positive effect from any overcoating, sealing, or caulking as compared with the unprotected control joints. However, after 2 yr, all joints but the asphaltic-coated dramatically declined, showing only 21–48% retentions, whereas the asphaltic-protected joints retained 92% of original strength levels. In corresponding 2-yr exposure to the milder industrial atmosphere, protected joints outperformed the unprotected controls throughout the exposure period. Nitrile-rubber-sealed and epoxy-caulked joints showing essentially full protection. Whereas no joints had failed in water-soaking exposure, three joints (one unprotected, two protected) had failed in water-soak cycling, and one unprotected joint in the industrial atmosphere.

There was a highly significant percentage of joint failure in both 2-yr seacoast exposure and 2-yr intermittent salt-spray exposure. After 30 da in the seacoast atmosphere, 75% of unprotected joints failed. This allowed the best opportunity to observe the benefit of protecting the duplicate joints. The most outstanding performer for the first time was the silicone sealant, which permitted 47% of initial strength retention and only one debonding due to corrosion in 2 yr. Each other type of protection permitted four or more corrosive bond failures with no joints remaining after 2 yr to obtain any data point. It was on the basis of this evidence that the author has recommended the use of a silicone sealant for protecting general manufacturing joints which must give service in a marine or seacoast condition. The same conclusions were evident from a 2-yr exposure to intermittent 3.5% sodium chloride–spray conditions. Once again, one joint with silicone sealant failed and 40% overall joint-strength protection was secured after 2 yr. Actually, fewer joint failures and significant joint-strength retentions were still evident for a period out to 12 mo even for the unprotected joints. This would confirm that the natural seacoast exposure can be even more severe than the accelerated laboratory test using intermittent salt-water spray.

In the same investigation by Minford (3738), the attempt was made to determine the relative effectiveness under corrosive intermittant salt-water spray conditions of several primer coatings with additional sealant candidates evaluated after 1-, 3-, 6-, and 12-mo exposure periods. The control, unprotected 6061-T6 alloy joints showed debonding within 24–35 da. A methacrylate sealer 1234 from duPont used as primer under the adhesive only extended the joint survival time range to 79 da. A silicone-alkyd primer (Plaskon ST873) further extended survival time to 112 da, whereas an alkyd coating (duPont 93-508) increased the maximum survival time range to 177 da. Butyl rubber sealant (Durribon 4040) afforded similar extension to 180 da, whereas urethane (Eccocoat OP2) joints only ranged to 110 da. Finally, two sealants with well-established good performance in architectural applications (Tremco Lastomeric polysulfide and Tremco Mono Lastomeric) afforded the best performance by extending joint survival times to 294 and 365 da, respectively.

XI. EFFECT OF PRIMERS ON ALUMINUM JOINT PERMANENCE

Primers must flow over the adherend surface and, therefore, they are generally diluted adhesive formulations. The property requirements for materials for primers under structural adhesives has been particularly demanding. Early primer systems exhibited some serious deficiencies such as (1) poor protection against corrosion, (2) poor toughness, and (3) overall poor durability. This necessitated making a considerable effort to better understand corrosion inhibition, polymer impact resistance, and durability phenomena in general. Lehmann (3744), in 1983, authored a technical discussion in the book titled *High Performance Adhesive Bonding* (3745). This information had been originally developed as a 1978 SAMPE paper by Lehmann et al. (3746). Particular interest was focused on the development of corrosion inhibitors for formulating effective primers. Specific reference was made to the development of soluble corrosion inhibitors such as the macrocyclic polyethers (3747–3752).

Reinhart (3753) also offered an excellent review of new primer technology. Noland (838) conducted extensive primer testing with various heat-cured epoxy adhesives in high-humidity exposures. Ehlers (3754) published an early paper on the correlation between structure and thermal stability for epoxy primers, whereas the patent of Klapprott and Paradis (3755) is typical of many protections sought to use certain formulation types. Rice et al. (3756) have attempted to correlate adhesive primer performance with the chemical structure and bulk properties of the primer. The reduction of the cross-link density has been shown to decrease the thickness sensitivity of primers, but the technique has a negative effect of reducing the heat and solvent resistance properties. Two methods of circumventing these negative effects have been to include a rubbery heterophase particle into the resin matrix. Van Hoorn (3757), Ehlers (3754), and Klapprott and Paradis (3755) discussed results attempting to promote varying chain stiffness with corresponding variable heat resistance. Falcone and Miller (1088) evaluated the benefit of primers on repair of main rotor blades in helicopters. Spencer (2060) investigated primers in both aircraft and space vehicles, as has Aker (2061). Data on durability of primed aluminum aircraft joints has been developed by Shannon and Thrall (3643), Schwartz (1615), and Scardino and Marceau (853).

Nakahara has summarized a great deal of the current aerospace technology about chemical primers and coatings on aluminum for enhancing the durability of structural joints in various environmental weathering. In Table 185, the evaluation results with five protective-finish systems used over aluminum and some dissimilar metal jointures in accelerated weathering conditions are summarized. While no discriminations were possible in adhesion retention in humidity or 5% salt-spray exposure, differences could be seen after acidified salt spray and seacoast beach exposures. It would appear that the epoxy-polyamide primer plus urethane top-coating system was most protective. Nakahara has also offered information on the relative corrosion resistance of aluminum interfaces pretreated by FPL etching, CAA, or PAA procedures and with and without adhesive primers evaluated in 5% salt spray in Table 186. While even the CAA- and PAA-pretreated interfaces could show some pitting corrosion without a protective primer, no corrosion was observed with a primer present. By contrast, scattered surface blisters with attendant corrosion were found with FPL-etched surfaces even with a primer.

Minford (3758) evaluated various baked paints on aluminum as primers for structural bonding to increase the joint resistance. This effort was directed, however, toward finding if different types of heat-cured paints applied over vapor-degreased–only mill-finish 6061-T6 alloy could afford sufficiently good adhesion to serve as primers that would be resistant to both water-soaking and salt-water exposure weathering. It was shown that there was no difference in initial strength of joints made to acid-deoxidized surfaces or heat-cured paint over vapor-degrease–only 6061-T6 surfaces using two-part epoxies. With one-part, heat-cured epoxy, only an epoxy-phenolic paint–primed adherend showed equivalent joint strength to acid-deoxidized surfaces. In the latter case, joints fabricated with epoxy-urea paint primer showed 12% lower strength, whereas white-pigmented acrylic paint–

Table 185 Accelerated Environmental Testing of Protective-Finish Systems for Aluminum Aerospace Bonded Structures

	Protective-finish systems				
Test[a]	Epoxy amine primer	Epoxy amine primer + epoxy topcoat	Epoxy polyamide primer	Epoxy polyamide + primer + urethane topcoat	Wash primer + urethane primer + urethane topcoat
Adhesion	Pass	Pass	Pass	Pass	Pass
Humidity (1 mo)	Pass	Pass	Pass	Pass	Pass
5% salt spray (2000 hr)	No corrosion	No corrosion	No corrosion	No corrosion	No corrosion
Acidified salt spray (1 mo)	—	Fair	—	Fair	Poor
Beach site (12 mo)					
Chalk and gloss retention		Poor	—	Excellent	Excellent
Filiform corrosion resistance		Poor	—	Good	Poor to good
Exfoliation corrosion resistance		Fair	—	Good	Poor

[a]Dissimilar metal panels used in acidified salt-spray and beach exfoliation-corrosion tests.
Data from S. Nakahara, Coatings, Adhesive Bonding of Aluminum Alloys (E.W. Thrall and R.W. Shannon, eds.). Marcel Dekker, New York, Table 3, p. 233 (1985).

primed joints averaged 30% lower initial strength. These differences reflected the varying tensile strengths of the heat-cured paint films, since all joint failures were cohesive in the paint primer.

Weathering tests were conducted in RT water immersion or soaking in 100% RH at 125°F (52°C) for 2 yr, with much more paint degrading evident in the acrylics. A pigmented acrylic was superior, however, to a clear acrylic paint. The hot humidity exposure was more aggressive to the urea-modified–epoxy primer, with 57% average strength decline compared to only 13% decline in ambient water. Epoxy-phenolic paint-primer stability was best with joints made to this surface, only losing 11% strength even in the hot condensing humidity. The comparative performances of the epoxy-phenolic joints and white-pigmented acrylic joints in a soak/freeze/thaw testing was surprising with about equal strength declines. The author speculated that this might be related to the much higher rigidity of the phenolic-modified paint and the consequence of that in a lap-shear specimen test.

Table 186 Corrosion Resistance of Aerospace Aluminum Surface Pretreatments

Surface treatment	Adhesive primer (mil)	5% salt spray (hr)	Observations
FPL etch	None	21	Heavy corrosion with pitting
	0.13	3000	Scattered blisters with corrosion
Phosphoric acid anodize (PAA)	None	141	Heavy corrosion with pitting
	0.15	3000	No corrosion
Chromic acid anodize (CAA)	None	336	Scattered minute corrosion pits
	0.18	3000	No corrosion

Data from S. Nakahara, Coatings, Adhesive Bonding of Aluminum Alloys (E.W. Thrall, R.W. Shannon, eds.). Marcel Dekker, New York, Table 4, p. 235 (1985).

In Figures 43 and 35, the relative performance of pigmented acrylic, urea-modified, and phenolic-modified epoxy paint primers are reviewed in intermittent salt spray and the natural industrial atmosphere. In all cases, the paints had been spray applied over vapor-degreased–only 6061-T6 mill-finish alloy. Clearly the relative joint durabilities in the more corrosive salt spray was epoxy-phenolic > epoxy-urea > pigmented acrylic. In the more mild industrial atmospheric exposure out to 12 yr, the same general order of relative joint durability was found with the same two-part epoxy adhesive. However, with a better wetting potential using a one-part, heat-cured epoxy in the industrial atmosphere, the performance of the epoxy-urea–primed surface was about the same as with the epoxy-phenolic–primed surface. In all these evaluations, the improvement in joint durability affected by the primer paints over the vapor-degreased–only joints was highly effective for all three types of paint primers. Even the poorest primer enabled an increase of 12 times in joint-survival time. The data comparing epoxy-phenolic–primed joints with chromic acid–deoxidized, conversion coated, and anodized 6061-T6 joints in RT water immersion, 100% RH at 125°F (52°C), soak/freeze/heat cycling and industrial atmosphere, can be reviewed in Minford (774) and Tables 23, 24, 172, and 173.

Sung has provided a pertinent discussion on the diffusion of water through various polymers (see Ref. 3759, Table 1, p. 623), including the effect of liquid versus vapor, which should be related to the improvement of joint durability through the use of special primers. Bascom (1087) has provided an article titled "Primers and Coupling Agents."

Bascom (1087) has published data on the lap-shear and flatwise-tension strengths of commercial adhesive applied to aluminum with and without primer. While these data appear quite promising, Bascom (1087) points out that published data on these types of joints under severe field service conditions are quite difficult to obtain.

Although virtually all Alcoa durability testing of anodized aluminum joints have been conducted without a primer in order to establish the level of durability in their absence, Minford (3760) has conducted an investigation of CAA- and SAA-pretreated 6061-T6 joints using AF-6 nitrile-phenolic film adhesive with and without the recommended EC-1290 primer in the bondline. Joints were exposed for 1 yr to a 3.5% sodium chloride intermittent spray to discriminate. No significant difference in corrosion resistance could be detected with a primer as compared with no primer present; however, the primer was able to support a significantly higher initial joint strength potential for both CAA- and SAA-pretreated adherends. While the cost of the primer for boosting joint strength, protecting the treated surface, and increasing the level of stress-endurance could make it economically acceptable in the aerospace industry, there might be various manufacturers outside aerospace who could achieve satisfactory performance levels without the need for these special primers.

Minford (3761) has investigated a vinyl wash primer as used under paints as another possible primer surface for bonding. It was shown that a vinyl wash primer that was heated to the temperature necessary to cure a typical paint used for architectural coatings developed a non–water-wettable condition. A joint made with a water-dispersed casein-neoprene adhesive failed after only 1-mo soak in RT water. Good adhesion was maintained with the same adhesive when the wash primer was applied after the paint bake cycle.

Earlier it was mentioned that the evaluation by Minford of a wide variety of silicone sealants had disclosed a wide range of adhesion performances often requiring use of a separate primer. More recently, additional durability tests of more modern modifications have shown that silicone sealants can maintain good adhesion to a deoxidized aluminum surface over an 8-yr exposure time to direct seacoast atmospheric weathering. Minford has also evaluated the durability of primerless RTV adhesive/sealant products comparing vapor-degreased–only and chromic acid–deoxidized 6061-T6 aluminum adherends in lap-joints. The data from exposures in hot condensing humidity, soak/freeze/thaw cycling, continuous salt spray, and a natural industrial atmosphere for periods up to 2 yr are summarized in Table 187. It is clear that a satisfactory performance was only possible with a deoxidized aluminum surface in the first three aggressive weathering conditions. The performance of the vapor-degreased–only surface joints was reasonably acceptable in the relatively milder industrial atmospheric exposure. In all cases, the initial joint strength attainable was only modestly structural with lap-shear strengths under 800 psi.

Table 187 Durability Evaluation of Primerless RTV Adhesive/Sealant Products on 6061-T6 Aluminum Adherends

Surface pretreatment	Adhesive	Average initial strength (psi)	Exposure conditions[a]	Exposure time (mo)	Average final strength (psi)
Vapor degrease	RTV-154	763	A	1	103
Alcoa A3 (FPL)	RTV-154	586	A	24	120
Vapor degrease	RTV-156	573	A	1	127
Alcoa A3 (FPL)	RTV-156	466	A	24	245
Vapor degrease	RTV-154	763	B	1	107
Alcoa A3 (FPL)	RTV-154	586	B	24	255
Vapor degrease	RTV-156	573	B	3	107
Alcoa A3 (FPL)	RTV-156	466	B	24	300
Vapor degrease	RTV-154	763	C	3	167
Alcoa A3 (FPL)	RTV-154	586	C	24	327
Vapor degrease	RTV-156	573	C	24	187
Alcoa A3 (FPL)	RTV-156	466	C	24	413
Vapor degrease	RTV-154	763	D	24	420
Alcoa A3 (FPL)	RTV-154	586	D	24	560
Vapor degrease	RTV-156	573	D	24	460
Alcoa A3 (FPL)	RTV-156	466	D	24	620

[a]Exposure conditions: A—100% RH at 125°F (52°C; B—24 hr in 100% RH at 125°F (52°C), 24 hr in −0°F (−17°C), 24 hr in 170°F air (77°C), 3 da cycle repeated; C—5% continuous salt spray; D—natural industrial atmosphere.
Data from J.D. Minford, unpublished work, Aluminum Co. of America.

XII. EFFECT OF SILANES ON ALUMINUM JOINT PERMANENCE

An alternative approach to application and curing of paints or organic polymer primers for enhancing durability has been to pretreat a metal surface with a solute solution of a special primer like an organosilane-coupling agent. Originally developed as a pretreatment for silicon dioxide surface of glass fibers for improving the durability of adhering epoxy resins in water-soaking conditions, they can perform similarly on the various metallic oxides, including aluminum oxide. Sterman and Marsden (3762), in 1966, showed the effect of silane-coupling agents on the wet strength of polyester-glass cloth laminate tested in three-point flexure.

Several workers (1140,1160,3763) have investigated the effect of silane primers on durability of bonds during exposure to a humid atmosphere or water. In these investigations, aluminum/epoxy TDCB specimens were prepared, and sophisticated viewing equipment or special analytical spectroscopy was employed to establish where the failure in the bondline was occurring. A static load was applied to force a joint failure. Patrick (3763) found failure occurred near the interface with exposure to water without silane, and it was inhibited in the presence of silane as long as the concentration of the depositing solution was close to 0.01%. At higher or lower concentration, the failure near the interface was still obtained. Kinloch et al. (1160) compared both stress under water and in 56% RH. As viewed by Patrick, failures were observed in the former exposure, whereas failure was cohesive in the latter situation. Thin films produced cohesive-type failure with about two orders of magnitude increase in failure time. Boerio and Ondrus (1140) always found that beams prepared from silane-primed aluminum failed at much longer times. The pH of the primer solutions had little effect on the times to failure, but drying the primer films at 212°F (100°C) before bonding definitely had an effect. Longest times to failure were found with 10-min heating, whereas the time decreased with shorter or longer times. Boerio and Ondrus (1140) related increase in bond longevity with a silane present in water exposure as increasing the induction time to form a boehmite film, as proposed by Venables et al. (36). (Additional information can be found in Ref. 3764.)

Kinloch (3765) evaluated the effect on the durability of an epoxy/steel bond using an epoxy-terminated silane. In 1988, Ripling et al. (3766,3767) made a study of the effect on bond durability in bonding steel bridge components using primer and coupling agent, separately and in combination. A wedge specimen was used. Exposure time at which the crack driving force decreases precipitously was a measure of the adhesive bond resistance to hot-wet conditions. The bonds made without the manufacturer-recommended primer actually showed the least resistance to water. However, a silane coating on the adherends with the primer gave definite improvement. When silane was actually mixed with the adhesive, it gave the overall most effective and consistent results.

Minford (3768) evaluated the effect of primer AP-828-P (diluted form of Union Carbide A-187 silane) on 3003-H14 sheet alloy and Alcoa Brazing alloys CS99 and CP89 bonded with a commercial two-part ambient-curing general purpose epoxy adhesive. In Table 188, we observe results with joints fabricated with this silane applied over these surfaces after vapor degrease or abrading with Aloxite 120 grit paper plus vapor degrease. Bond-strength averages were compared with or without silane. The effect of the silane on initial strength was not clear because of the wide variation between joints with varying adherend, surface treatment, and selection of epoxy adhesive. The presence of silane apparently doubled the strength of one combination of CS99/CP89 joints with Aloxite abrade and Hysol Epoxy Patch Kit 6C adhesive. The same combination of adherends and adhesive with vapor degrease offered half-strength joints as compared to similar joints without silane. In Table 189, similarly bonded joints were exposed to a 50/50 by volume solution of Prestone II in tap water at 275°F (135°C) for 420 da to compare the effect of the silane on joint durability. Results with aluminum brazing sheet alloys should be relatable to making adhesive repairs to aluminum brazed-surface radiators, whereas the results with the 3003-H14 alloy should have correlation with the adhesive repair of aluminum air conditioner coils exposed to exterior weathering conditions behind the radiator in

Table 188 Effect of Silane Adhesion Promoter Primer AP-828-P[a] on Aluminum Lap-Joint Strengths with Different Aluminum Adherends and Surface Pretreatment

Adherend[b]	Surface treatment	Adhesive[c]	Average bond strength ratio for joints with primer/no primer
3003-H14	Vapor degrease	Hysol Patch Kit 6C	0.87
3003-H14	Vapor degrease	Armstrong A31	1.11
3003-H14	Vapor degrease	Hysol Patch Kit 0538	0.50
CS99/CP89	Vapor degrease	Hysol Patch Kit 6C	0.46
CS99/CP89	Vapor degrease	Armstrong A31	1.35
CS99/CP89	Vapor degrease	Hysol Patch Kit 0538	0.51
3003-H14	Aloxite 120 abrade	Hysol Patch Kit 6C	1.42
3003-H14	Aloxite 120 abrade	Armstrong A31	1.60
3003-H14	Aloxite 120 abrade	Hysol Patch Kit 0538	0.94
CS99/CP89	Aloxite 120 abrade	Hysol Patch Kit 6C	2.12
CS99/CP89	Aloxite 120 abrade	Armstrong A31	0.94
CS99/CP89	Aloxite 120 abrade	Hysol Patch Kit 0538	1.13

[a]Identified as Union Carbide's A-187 silane in chlorothane solvent at 2% concentration.

[b]3003-H14 alloy sheet in 0.050-in (1.27-mm) thickness to simulate aluminum air conditioner tube alloy surfaces and same thickness of CP-87 and CS-99 brazing sheet to represent adherend surfaces for bonding in brazed aluminum radiators.

[c]Adhesives used in this study were two-part, room temperature–curing epoxy adhesives recommended for general-purpose bonding.

Data from Ref. 3768 (p. 7).

Table 189 Effect of Silane Primer on Aluminum Joint Durability[a] Involving Aluminum Adherends Representative of Brazed Aluminum Heat Exchanger or Air Conditioner Coil Surfaces

Adherend[b]	Surface preparation	Adhesive[c]	Effect of silane primer on bond joint durability[a]
Brazing sheet	Degreased	Armstrong A31	Highly improved
Brazing sheet	Degreased	Hysol Patch Kit 0538	Not improved
Brazing sheet	Degreased	TPX 1015	Moderately improved
Brazing sheet	Aloxite 120 abrade	Hysol Patch Kit 6C	Highly improved
Brazing sheet	Aloxite 120 abrade	Armstrong A31	Highly improved
Brazing sheet	Aloxite 120 abrade	Hysol Patch Kit 0538	Moderately improved
Brazing sheet	Aloxite 120 abrade	TPX 1015	Moderately improved
3003-H14	Degreased	Hysol Patch Kit 6C	Not improved (good durability with or without primer)
3003-H14	Degreased	Armstrong A31	Highly improved
3003-H14	Degreased	Hysol Patch Kit 0538	Highly improved

[a]Comparative joint-durability performance based on 420-da exposure to a 50/50% by volume solution of Prestone II and New Kensington, PA, tap water. Joints were tested in an autoclave with the antifreeze/water solution operating at 275°F (135°C).
[b]Adherends used to simulate surfaces of aluminum radiator were CP-87 and CS-99 brazing sheets. The 3003-H14 surfaces were representative of bonding aluminum air conditioner coils.
[c]Adhesives Armstrong A31, Patch-Kits 6C and 0538 are typical two-part, room temperature–curing epoxies that might be considered for field repair of automotive radiators or air conditioner coils. The TPX 1015 is a General Mills high-temperature–softening polyamide-type hot-melt adhesive suggested for radiators or air conditioner coil assemblies.
Data from Ref. 3768 (p. 8).

an automobile. Again, the response of using the silane for improving bond durability was variable, ranging from showing no improvement through moderate improvement to highly improved.

It has been typical of silane producers to publish lists of their products with recommendations that specific products are to employed for specific adherends. The author would suggest that, based on the diversity of durability of results he has obtained using different adherends, surface pretreatments, and adhesives, separate durability testing is required to confirm the performance that might be expected with a particular set of bonding parameters and weathering conditions that might be encountered in service.

XIII. BOND PERMANENCE WITH SIMULTANEOUS STRESS AND WEATHERING

A. Sustained Static Load Stressing

Up until 1960 there were few references in the adhesive-bonding literature involving the simultaneous application of a stress load and exposure to water-contact weathering. Most specifications for quantification of adhesives, even in the military, were based on joint-strength retention values for aluminum joints after certain specified periods of exposure to specific environments such as (1) immersion in room temperature water, (2) elevated-temperature liquid water, (3) high or condensing humidity conditions at various temperatures, (4) corrosive salt fog, or (5) various natural atmospheric weathering conditions.

Eickner and co-workers (1489,1490), at the Forest Products Lab, generated early data under various conditions of humidity with some unquantified stressing level, as shown in Figure 27. This type of simple stress fixturing had been used for many years by the aluminum producers to determine the stress resistance of different aluminum alloys under various weathering conditions, including salt-fog cycles. The time to joint failure could be a measure of the degradation in the joint due to the environmental exposure conditions, or the residual joint strength might be measured if no failure had occurred as a measure of the magnitude of joint deterioration caused by the exposure to both the stress and environment. The data of Eickner et al. (3208) shown in Table 153 is indicative of this form of relating the joint durability. Aluminum joints made with 11 different adhesives were exposed to Florida weathering with the results that all joints survived as unstressed specimens for the total 3-yr test period. However, when stressed, only 6 of the 11 adhesives affected joints that could survive the same period of exposure. Of special interest in this investigation is the fact that painting over the joints or using primers on the surface to be bonded could affect strength retention, so that about the same level of joint-strength retention was found whether stress was present or not. The duplicate control without such additional procedures all failed during the same test period.

Certainly among the first investigators to attempt to combine an element of weathering and quantified stress, as far as the literature indicates, was Carter (3179,3191,3769). In a paper in ASTM STP 401 in 1946, Carter demonstrated the much higher criticality for evaluating joint durability when testing the peel strength of joints while the specimens were immersed under water. In 1967, he was able to publish some early durability results on some large aluminum lap-joints held under sustained static stress under outside natural weathering conditions. The manner of stressing is that shown in Figure 28 and clearly places the bondline under a severe cleavage stress situation. While the presence of water is intermittent and variable in an atmospheric exposure, it must still be present in far higher doses than for any interior service conditions. The author noticed at the Alcoa industrial atmospheric site that the surfaces were wet with dew and condensation on an almost daily basis.

The big rush to incorporate added stress to weathering aluminum joints really started after Sharpe's data on aluminum stressed joints held in a special Bell Lab's stressing fixture in hot humidity as reported in 1965 at the Stevens Institute of Technology Symposium on Joint Durability (1601) and in a 1966 paper (3770). The dramatically shortened survival times announced were obviously due to the simultaneous use of sustained static stressing in the presence of an aggressive hot humidity. Obviously, this matter of finding such rapid joint failures needed to be addressed not only in hot humidity conditions but also in the wide variety of other weathering conditions that can exist in various service environments. Structural joints are fabricated for the sole reason that certain levels of stress must be safely resisted in manufactured structures with possible contact with water in the service environment. Thus, the margin of safety in a joint that might be extrapolated from its joint durability in water alone cannot be relied upon when stress is also present. Other articles of related interest can be cited by Steffens and Brockmann (3771) on aging resistance and Althof and Brockmann (3772). Wegman (3773) conducted some aluminum joint durability testing involving environment and stressing and reported in a 1971 Picatinny Arsenal Report. Some of these tests were performed with a modified Carter T-peel specimen in water, whereas others utilized an outdoor weathering frame in which the lap-joints were attached in tandem, multiple-joint strings with an overall calibrated spring supplying the tension at one end.

Minford (976), in 1972, published the results of evaluations of aluminum joints which were stressed in the Jombock-designed Alcoa Circular Aluminum Ring Fixture and stacked in continuously condensing humidity at 125°F (52°C), as shown in Figure 10. The data were plotted as a stress-endurance curve whose idealized form is shown in Figure 34. This was followed in 1973 with more complete test results (3774). In Figures 15 and 18, the change in joint performance for two different adhesives bonded between chromic acid–deoxidized 6061-T6 aluminum adherends can be seen when stressing was imposed in hot condensing humidity. Additional disclosures shown in the above Minford reference (3774) document what change in performance might be expected with similar joints with stressing present with milder weathering conditions like ambient water immersion or the natural industrial weathering atmosphere. The distinctly different joint-strength decay curves for room temperature water-immersion conditions and soaking in a continuously condensing humidity cabinet are compared to corresponding stress-endurance curves for duplicate joints with stressing combined with these two accelerated laboratory weathering conditions (see Ref. 3774, Figs. 32 and 33). Whereas both the room temperature water- and hot humidity-exposed joints approach approximately the same final low plateau of retained joint strength after 1-yr exposure, the hot humidity exposure with combined stress precipitates a much steeper decline in joint strength and significantly shorter joint-survival time values.

In 1975, Minford extended his durability testing to include the effect of stress and hot humidity weathering on 6061-T6 aluminum joints fabricated with vapor-degrease, SAA, CAA, or PAA pretreatments using the same nylon-modified adhesive (FM-1000). The data were published in 1977 (533), with further testing time reported in 1978 (534). The data in Figure 48 indicate the PAA-treated joints not only had higher initial joint strength potential but also could survive for longer times under higher stressing levels than the CAA-treated joints which, in turn, could survive longer than all FPL-pretreated joints. The corresponding SAA-pretreated joints all performed the poorest among the anodized and unsealed joints outperformed those whose surfaces were sealed before adhesive bonding. Meanwhile, the 6061-T6 joints fabricated with only vapor-degreased adherends virtually were incapable of surviving for any significant time at the same stress levels, whereas the CAA, PAA, and FPL-deoxidized joints could survive for periods up to 1 yr. The overall performance among the anodized-surface joints under combined stress/weathering conditions was distinctly PAA > CAA > SAA.

The distinctly poor performance of SAA joints was presumed to be influenced by sensitivity of the thick anodized layers to fracture under stress. The above relationships seem to indicate some dependence on the opportunity for mechanical interlocking. The known morphologies of PAA, CAA, and FPL surfaces would also predict increasing mechanical interlocking in the order PAA > CAA > FPL. Although there might also be greater opportunity for significant interlocking with the SAA (especially unsealed) surface than for FPL, the thickness of the oxide and sensitivity to fracture may be a more negative factor creating lower stress-rupture performance.

Minford has also compared the stress-durability resistance of 6061-T6 adherends PAA anodized using four different sets of anodizing conditions, as shown in Figure 49. While the Boeing set of anodizing conditions produced the highest initial joint strength, the voltages and time conditions could obviously be varied higher and lower without significantly altering the stress-endurance performance. As noted earlier, modification of a heat-curing epoxy with a nitrile rubber tends to lower the joint sensitivity without stress being imposed, as shown in Figure 50. When stress was added, as

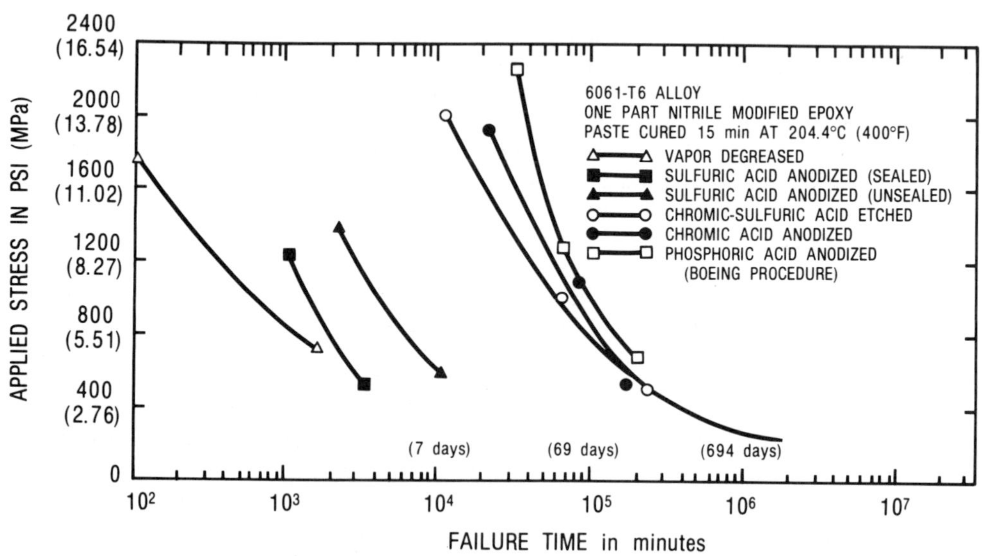

Figure 48 Comparison of stress-durability curves with various surface pretreatment conditions. (From J.D. Minford, Adhesives, Durability of Structural Adhesives (A.J. Kinloch, ed.). Applied Sci. Publs, London, Fig. 5, p. 168 [1983].)

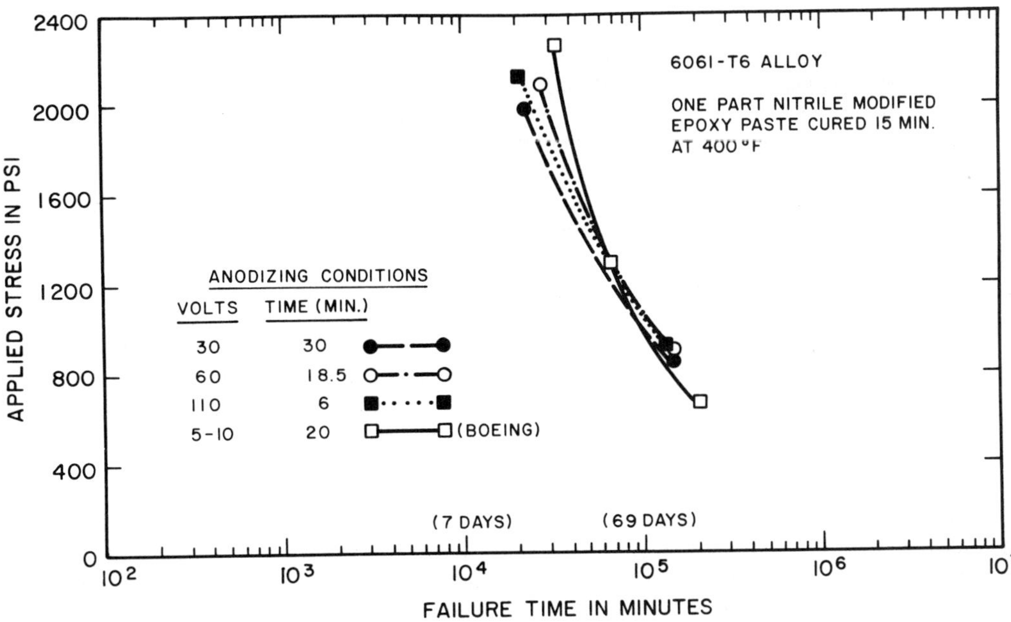

Figure 49 Comparison of stress-durability performance of 6061-T6 adherends with four different sets of PAA-anodizing conditions. (From Ref. 533.)

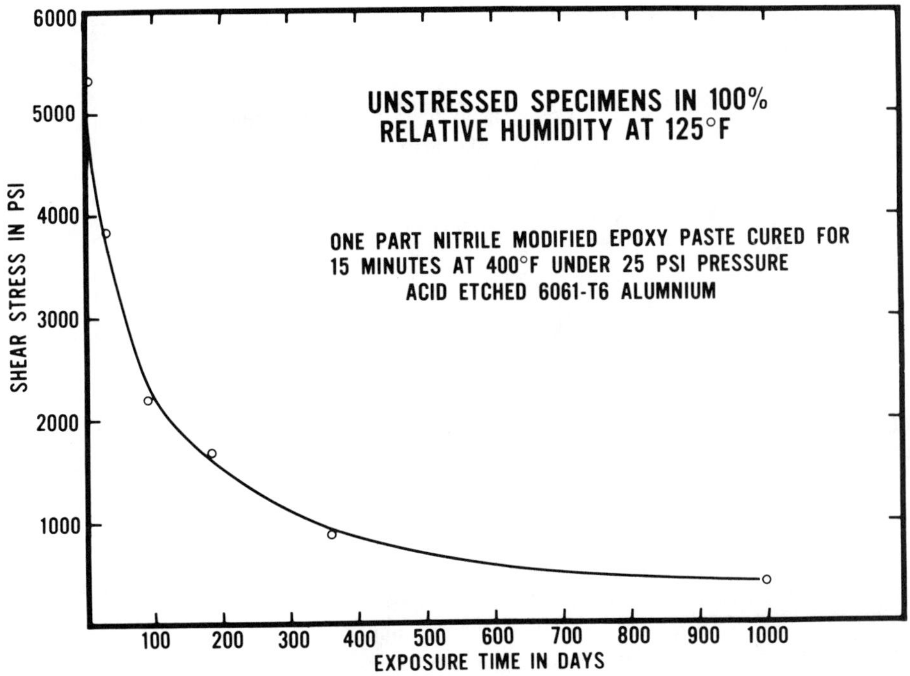

Figure 50 Sensitivity of a nitrile-modified epoxy (EC-2086) to soaking in hot humidity without any applied stress. (From Minford Ref. 3774 [Fig. 27, p. 114].)

shown in Figure 51 with duplicate joints, the stress endurance was much better than anticipated from the unstressed performance, especially compared to the performance of a two-part, room temperature–cured epoxy, as previously shown in Figure 18. Whereas the two-part epoxy with the same pretreated adherends in Figure 18 retained virtually all its initial joint strength after 2 yr in hot humidity, the nitrile-modified, heat-curing epoxy under similar hot humidity exposure conditions in Figure 50 had lost 92% of initial joint strength in the same 2-yr period. Obviously, there are very important factors contributing to overall durability under stressing conditions beyond the simple factor of what is the relative joint resistance to exposure in the same water conditions without the stress.

In 1980, Minford (772,773) reported results of more detailed combined stress and hot humidity evaluations using a variety of different surface pretreatments and the aerospace film adhesive FM-1000. In the first publication (772), the stress was imposed while soaking in a hot continuously condensing humidity cabinet at 125°F (52°C). The second report (773) in the same year reported duplicate joints where the stressing was imposed while the joints were being exposed to intermittent salt-water spraying in specially constructed cabinets. A daily cycle consisted of overnight continuous spraying for 16 hr followed by an 8-hr drying period. The effect of aluminum alloy was included by duplicating all joint variables using the high-purity aluminum (alloy 1285) surface on Alclad 2024-T3 and the heavy heat-treat oxide surface on 6061-T6. A wide-ranging set of initial joint strengths have been recorded in Table 160. Rather surprisingly, the highest-strength joints were produced by the phosphoric acid and chromic acid–deoxidation–treated 6061-T6 surfaces rather than the higher purity. The differences in tensile strength of the two alloys required that the stronger Alclad 2024-T3 alloy joints be fabricated from 0.064-in (1.62-mm) sheet to make the overall stiffness approximately equivalent to 0.125-in (3.24-mm) 6061-T6 joints. The dramatic effect of surface pretreatment on initial lap-shear strength can be observed with the range of strengths found with Alclad 2024-T3 adherends (3700–6000 psi [25.49–41.34 MPa]) and with 6061-T6 alloy adherends (4300–6750 psi [29.63–46.51 MPa]). In Tables 190 and 191, we can compare the relative stress-endurance resistance of both types

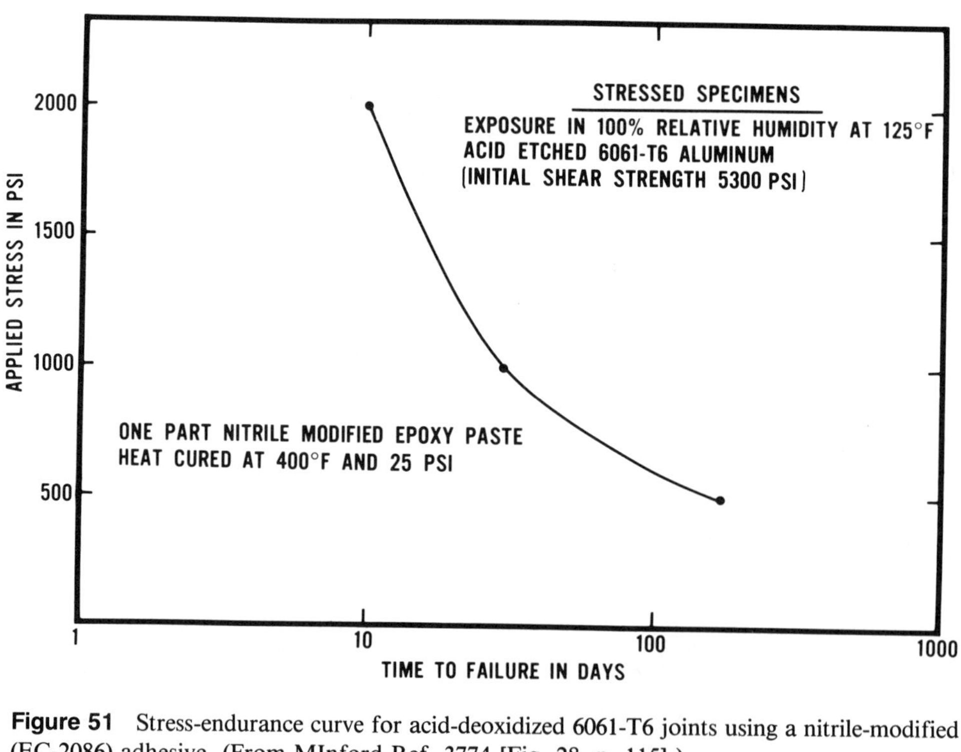

Figure 51 Stress-endurance curve for acid-deoxidized 6061-T6 joints using a nitrile-modified epoxy (EC-2086) adhesive. (From MInford Ref. 3774 [Fig. 28, p. 115].)

Table 190 Stress Endurance in 100% RH at 125°F (52°C) Exposure Conditions for Alclad 2024-T3 Aluminum Alloy Joints[a] with Varying Surface Pretreatment Conditions (Arranged in general order of decreasing joint endurance)

	Average time to failure values (min) for 3 stressing levels		
Surface pretreatment	500 psi (3.45 MPa)	1000 psi (6.89 MPa)	2000 psi (13.78 MPa)
Chromic acid deoxidize	86,400	10,080	1320
Alodine 1200 (chromate/phosphate conversion coating)	19,860	7500	2320
Deoxidine 670 (RT etch, H_3PO_4 + alcohol solution)	3240	780	135
Vapor degrease	2880	390	150
Alumilite 215 (H_2SO_4 acid anodize, 0.8-mil thick)	2880	150	42
Mill-finish	1800	600	180
Silica-gritblast + vapor degrease	460	30	4

[a]All joints were fabricated using FM-1000 aircraft-type, nylon-modified epoxy film cured for 60 min at 350°F (177°C) in a platen press at 10 psi (6.89 mPa).
Data from Ref. 772 (Table 3, p. 86).

of alloy joints with seven different surface pretreatment conditions at three different stress levels of 500 psi (3.45 MPa), 1000 psi (6.89 MPa), and 2000 psi (13.78 MPa). The order of effectiveness of the surface pretreatments based on the average time to joint failure was quite varied, and we must conclude that the different physical and chemical natures of the surfaces has a high degree of significance in establishing the level of stress endurance for joints made with those conditions.

Table 191 Stress Endurance in 100% RH at 125°F (52°C) Exposure Condition for 6061-T6 Aluminum Alloy Joints[a] with Varying Surface Pretreatments (Arranged in general order of decreasing joint endurance)

	Average time to failure values (min) for 3 stressing levels		
Surface pretreatment	500 psi (3.45 MPa)	1000 psi (6.89 MPa)	2000 psi (13.78 MPa)
Alodine 1200 (chromate-phosphate conversion coat)	30,240	9,600	1,250
Alumilite 215 (H_2SO_4 anodize at 0.8-mil thickness)	21,720	45	25
Deoxidine 670 (RT etch in H_3PO_4 + alcohol solution)	20,200	7,000	1,200
Chromic acid deoxidizing	16,800	1,020	28
Vapor degrease	4,320	1,080	240
Silica-gritblast + vapor degrease	4,220	600	25
Mill-finish	120	45	10

[a]All joints bonded with FM-1000 aircraft-type, nylon-modified epoxy film cured for 60 min at 350°F (177°C) in platen press under 10 psi (6.89 mPa).
Data from Ref. 772 (Table 4, p. 87).

We might expect and anticipate that the higher average times to joint failure might be found with the joints made with the higher-purity aluminum surface; i.e., Alclad 2024-T3. Especially since it was reported earlier in this chapter that the joints made with higher-purity aluminum surfaces required less aggressive surface pretreatment to produce more durable joints in water-soaking environments. It may not be surprising to note that the highest individual joint-survival time average under 500 psi (3.45 MPa) stressing was achieved by the chromic acid–deoxidized Alclad 2024-T3 joints with a value of 86,400 min. This Alclad 2024-T3 joint-survival time was only slightly less time than the summarization of survival times for all seven kinds of 6061-T6 joints. Obviously this does not tell the full story, however, since four of the other surface treatments afforded longer joint survival time averages when used on 6061-T6 adherends (SAA [Alumilite 215], Deoxidine 670 [phosphoric-alcohol etch], vapor-degreased [mill-finish oxide], and silica-grit mechanical blast). At the 1000-psi (6.89 MPa) stressing level, all survival times for both sets of joints were significantly lower, as might be anticipated. While the same order of surface pretreatments found at the lowest stressing level were, in general, those existing at the next higher level of stress, there were two exceptions which should be noted. The first was the very dramatic drop-off in the joint-survival time for the SAA-anodized 6061-T6 joints whose average was the same as the lowest performing mill-finish surface joints. Among the Alclad 2024-T3 joints, it was this same mill-finish surface condition that generated a joint performance elevation from next to last at the 500-psi stress level to a level above both the vapor-degrease and Alumilited surface joints at the 1000-psi stressing level. This would indicate that the sensitivity to cracking failure of the anodized 6061-T6 surface oxide had not been as evident at the lower stressing level; however, at the more elevated stressing condition of 1000 psi (6.89 MPa), the cracking sensitivity of the two different anodized surfaces became essentially equivalent. It probably should be pointed out that the overall average joint-survival time for all the different pretreated 6061-T6 joints has now become greater than that achieved by the collective Alclad 2024-T3 joints.

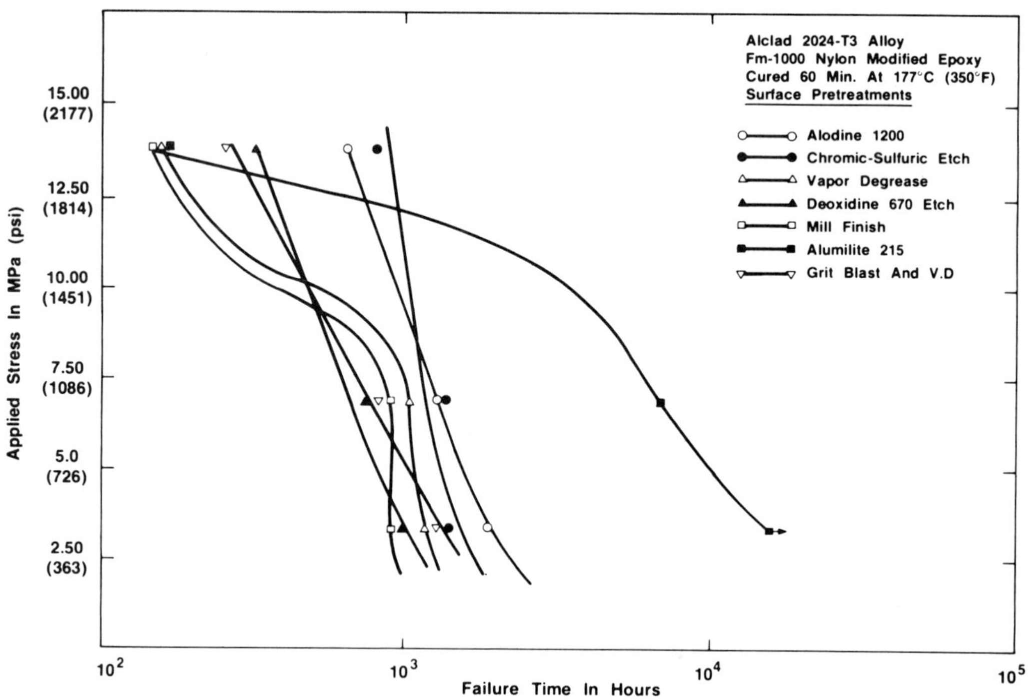

Figure 52 Stress-endurance curves in intermittent salt spray exposure with various pretreatments on Alclad 2024-T3 aluminum adherends. (From Ref. 773.)

At the highest stressing level of 2000 psi (13.78 MPa), no surface condition with either alloy could form a bond with the FM-1000 adhesive which could survive more than a few percent of the time that was shown at the lowest stress level of 500 psi (3.45 MPa). It is interesting to note that the silica-grit–blasted surface prepared joints for both alloys were especially sensitive to early joint failure under combined stress-humidity exposure conditions. It will be remembered that a unique ability was shown by this surface treatment with a two-part epoxy paste adhesive to survive 8 yr in direct exposure to the very corrosive seacoast atmosphere. The explanation for that performance, however, was not based on what surface morphology was produced but rather what benefit could be performed by a unique corrosion-preventive layer physically deposited over the surface.

The data for the stressed joints in intermittent salt-water spray testing (773) is summarized in Figures 52 and 53 for the Alclad 2024-T3 and 6061-T6 joints, respectively. In general, the same three surface pretreatments (chromic acid deoxidizing, Alodine 1200 [conversion coating], and Alumilite 215 [SAA anodizing]) appeared to be the best-performing pretreatments on both alloys. This is what might be predicted based on the earlier investigation employing stress and hot humidity reviewed in Tables 190 and 191. As noted earlier, the mill-finish and grit-blasted adherend joints performed the poorest, whereas the Alodine 1200 conversion coated–type joints overall performed the best under stress and hot humidity exposure conditions. The chromic acid–deoxidized and Alumilite SAA-anodized joints tended to flip-flop in performance, with Alclad 2024-T3 adherends favoring chromic acid etching treatment, whereas the anodized 6061-T6 adherends performed better. In corrosive intermittent salt spray combined with stressing, the mill-finish and grit-blasted–treated joints in both alloys again showed the poorest overall durability performances. The reader might be reminded that the silica-grit–blasted 6061-T6 aluminum joints had earlier showed outstanding durability in the corrosive seacoast exposure, presumably due to deposition of a corrosion-inhibiting layer of silica on the treated surface. Some evidence of this behavior can be seen in Figure 53, where the grit-blasted

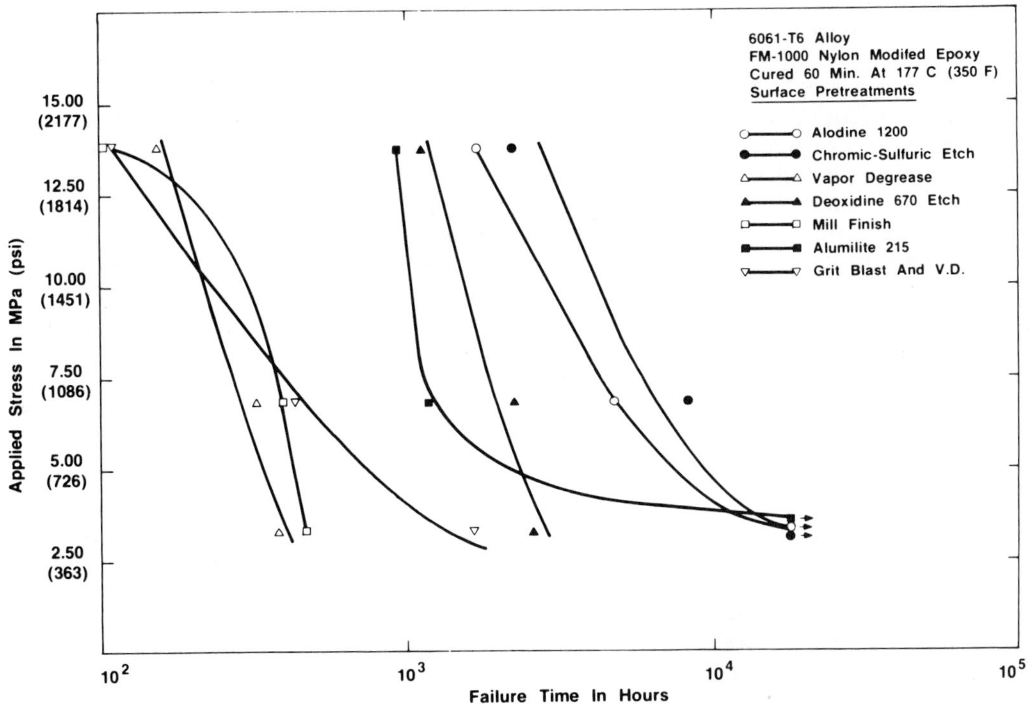

Figure 53 Stress-endurance curves in intermittent salt-spray exposure with various pretreatments on 6061-T6 aluminum adherends. (From Ref. 773.)

joints using the 6061-T6 adherends showed a distinct elevation of joint durability compared to the mill-finish and vapor-degreased–only joints at the lowest stress level. Presumably, at this lowest level of stressing, the ability of the interfacial condition to better resist corrosive undercutting of the bondline could exert a more positive effect on joint survivability.

At the other end of the range of joint survivability under stress and corrosive salt-spray conditions, we can note that the same three surface pretreatments show best performances, as shown in Tables 190 and 191 for the stress and hot humidity tests; i.e., SAA anodizing, Alodine 1200 conversion coating, and chromic acid deoxidizing. While these three surface condition 6061-T6 joints all show almost identical survival times at the lowest stressing level, the SAA-anodized surface on Alclad 2024-T3 alloy creates joints which survive the Anodine 1200 and chromic acid–deoxidized Alclad 2024-T3 joints by an order of magnitude. The benefit of anodizing is almost as great even at the intermediate 1000-psi stress level. However, at the 2000-psi stress level, the greater sensitivity of this quite thick oxide to fracture makes the joint-survival time no better than for the mill-finish and vapor-degrease–only joints.

The reader may question why comparative joint-survival time performances of CAA and PAA pretreatments are not offered in these investigations; however, these studies were undertaken at a somewhat earlier date before PAA anodizing had been established as a better substitute for the chromic acid deoxidizing. Also, it was the author's intention to make comparisons between surface treating conditions which were generally outside those suggested for aerospace and were more amenable for general manufacturing considerations.

Yurek (3154), at 3M's, used their Test Method NO. C-294 for conducting stress/endurance testing of their products in hot humidity. In this method, a calibrated spring device is used to support the stress on a series of strung-together joints, as shown in Figure 25. While all aspects of the mechanism involved in adhesive bond stress corrosion-type failures is not clearly understood, the work of Bascom (3775) nicely summarizes a relatively recent state of the understanding. Bascom has also supplied an appropriate list of references the reader may find of further help.

Marceau and Scardino (3776) have reviewed the literature and surveyed adhesive manufacturers, prime aerospace contractors, and other sources to determine currently used stress/exposure durability testing procedures as of the middle 1970s. Having selected the best candidate procedures, it was their purpose to assess these methods and choose those most appropriate for their Air Force–sponsored programs at Boeing. The four major types of specimens evaluated were (1) thick adherend machined lap-shear, (2) thick adherend double–cantilever beam (DCB), (3) thin adherend DCB specimen (Boeing wedge) test specimen, and (4) thick adherend single–cantilever beam (SCB) specimen.

Patrick (2700) conducted a series of studies using DCB-type stress test specimens and amine-cured epoxy resin joints in water-exposure conditions to develop an explanation of the mechanism of stress corrosion failure. He proposed when the specimen is immersed in water under stress, the water interacts at the highest stress concentration point, causing cracking to initiate by a solvolyis of the bond between the resin and the aluminum adherend. As the crack propagates, water follows into the crack void, interacting with the aluminum oxide. By some complex mechanism of solvating the aluminum oxide, the aluminate ion is formed, and this ion diffuses into the aqueous media. The aluminate ion continues to concentrate until the solution saturation point is reached. Then aluminum hydroxide will precipitate out and more aluminate ions will diffuse. Thus, a steady-state condition is eventually reached. Erickson et al. (3777) earlier showed the rapid diffusion of polyamines like triethylene tetraamine (TETA) and tetraethylene pentamine (TEPA) through epoxy resins. Patrick suggests that such amines, as contaminants on aluminum surfaces, could accelerate formation of bayerite oxide on the aluminum surface under room temperature water-soaking conditions. The fact that polyamine can diffuse in such joints simply increases the probable rate at which the oxide may be hydrated, solubilized, and precipitated as bayerite. Patrick emphasized that the very large concentrations of bayerite on the surface is a postcracking function rather than a causative phenomenon for cracking.

It seems that as the service environment diminishes in aggressiveness of the weathering condition, the bond permanence can rise under identical conditions of stress, as shown by Minford's comparisons of the stress-endurance performance of duplicate joints in the natural industrial atmosphere, room temperature water, and soaking in condensing humidity at 125°F (52°C). For example, note that the relative rate of joint strength decline in these three exposures for chromic acid–deoxidized

6061-T6 adherends bonded with a nitrile-modified, heat-cured epoxy paste (EC-2086) is shown in Figures 50 and 54. It is clear that the relative aggressiveness of these exposure conditions on joint durability is hot condensing humidity > room temperature water immersion > natural industrial atmosphere. It is not unexpected then that the relative order of joint survival time under stressful conditions might be in the reverse order; i.e., natural industrial atmosphere > room temperature water > hot condensing humidity. This would certainly fit the theoretical concept that stress corrosion cracking in joints will be directly related under water-exposure conditions to the rapidity that water can penetrate through the adhesive or along the interface. The data shown in Figure 55 based on stress testing of chromic acid–deoxidized 6061-T6 joints bonded with a nitrile-modified epoxy adhesive by Minford show to what degree the failure times are delayed when the opportunity for water to enter the bondline is that present in the natural industrial atmosphere. Note in Figure 51 that the joint survival time in 100% condensing humidity at 125°F (52°C) for similarly treated joints was slightly under 10 da. However, from Figure 55, the corresponding failure time at 2000-psi stress was approximately 1200 da. Minford (see Ref. 3774, Fig. 30, pg. 116) has established a similar stress-endurance curve for similar joints in room temperature water conditions showing an average joint survival time of about 35 da. In comparing the modest difference in joint strength decay of these type joints in Figure 54 over 2 yr in the natural atmosphere and room temperature water, the author would point out that the degree of acceleration to joint failure under stress in room temperature water was far greater than might have been projected from the difference in the joint strength decay curves between similar joints in these exposures shown in Figure 54. Thus, there must be some other additive factor than relative joint strength decay in the exposure to water that is also operating.

In some of the last investigations performed at the Alcoa Labs by Minford (3778), there was an attempt to identify some additional factors which could affect the joint-survival time under simultaneous exposure to various stressing levels and hot humidity weathering. This coincided with the

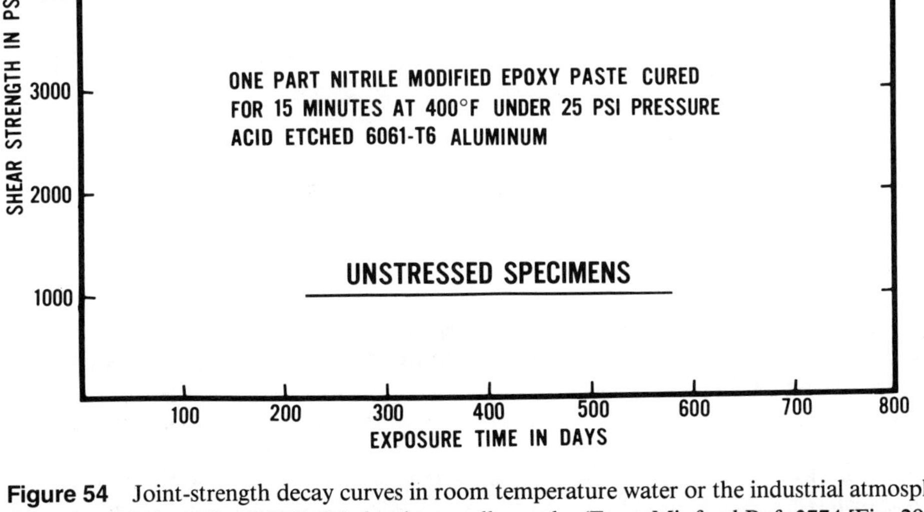

Figure 54 Joint-strength decay curves in room temperature water or the industrial atmosphere using chromic acid–deoxidized 6061-T6 aluminum adherends. (From Minford Ref. 3774 [Fig. 29, p. 115].)

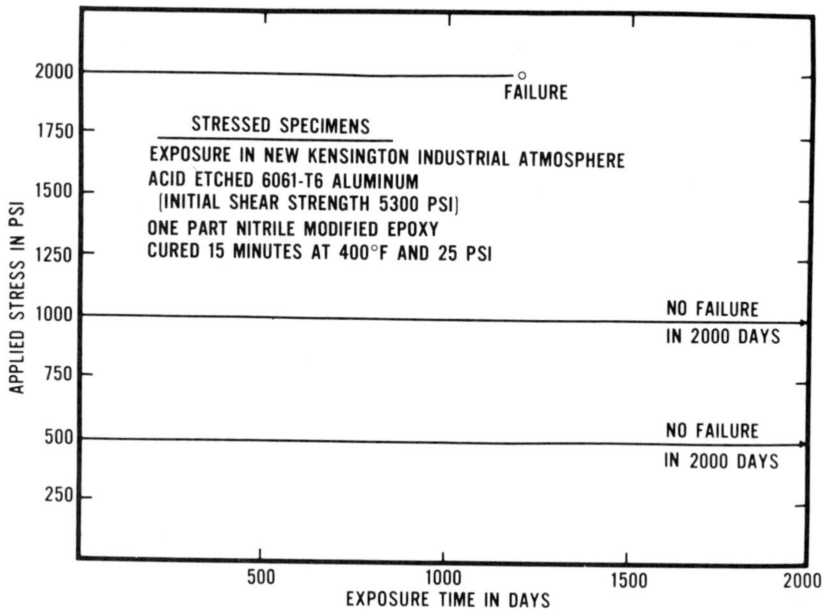

Figure 55 Stressed durability performance of chromic acid–deoxidized 6061-T6 aluminum joints in exposure to the natural industrial atmosphere. (From Ref. 3774 [Fig. 31, p. 116].)

Table 192 Effect of Preheating the Aluminum Adherend Surface Prior to Application of Paste Epoxy Adhesive on Stress/Humidity Durability of Joints[a]

	Adherend surface temperature when adhesive applied			Stress conditions			
Adhesive[b]	°F	°C	Surface pretreat	%	psi	MPa	Time to failure (min)
EA-9320	77	25	Vapor degrease	20	728	5.02	20,160
EA-9320	77	25	Vapor degrease	10	364	2.51	43,200
EA-9320	77	25	Vapor degrease	5	182	1.25	691,200
EA-9320	150	66	Vapor degrease	20	618	4.26	4,620
EA-9320	150	66	Vapor degrease	10	309	2.13	>602,800[c]
EA-9320	150	66	Vapor degrease	5	155	1.07	>604,800[d]
EA-9320	150	66	Vapor degrease	20	618	4.26	12,920[e]
EA-9320	150	66	Vapor degrease	5	155	1.07	>730,080

[a]All joints fabricated with 0.125-in (3.18-mm) 6061-T6 aluminum alloy and exposed to 100 RH at 125°F (52°C) under stress.

[b]Adhesive was a Hysol two-part epoxy formulation specially formulated to provide exceptionally high lap-shear and peel strengths with only a room temperature cure. In this investigation, however, all joints were cured for 60 min at 180°F (82°C).

[c]Joint was removed from stressing fixture and tested to determine retained bond strength after 602,800 min under stress and high humidity. The strength value was 4000 psi (27.56 MPa), which was actually +29% than projected initial strength.

[d]The stress level was raised to 20% after 604,800 min stressed at the 5% level when no joint failure had occurred.

[e]The above joint ([b]) failed after 12,960 min of additional exposure after raising the stress level from the 5 to 20% level. This was actually three times longer time than showed by other joints initially stressed only to 20%.

Data from Ref. 3778.

introduction into the aerospace market of new specially formulated two-part, room temperature-curing epoxy adhesives which could develop exceptionally high lap-shear and peel strength joints. It was desired to employ such an adhesive for bonding large satellite dishes which would need to survive in weathering conditions around the world at remote sites. The data shown in Tables 192–194 are representative of some of the results developed to assure highest durability responses under the conditions of manufacturing available. The question needed to be raised as to whether preheating of the 6061-T6 aluminum adherend might be justified in terms of an enhanced survival time under stress/humidity–exposure conditions.

As pointed out earlier in this chapter, the heat-curing of a typical two-part epoxy paste adhesive in the bondline generally can affect an improvement of joint durability properties because of the opportunity to obtain better wetting of the aluminum adherend. Having settled on heat curing for the projected manufacturing process, a series of tests was conducted (Table 192) to determine whether preheating of the aluminum prior to application of the adhesive might be justified. The data are somewhat conflicting at the 20% stress level, where preheating actually produced shorter time to failure values. At 10% stress, the situation was clearly reversed. It appeared that preheating the adherend could significant extend survival time at 10% stress more than a factor of 14 times, with the retained joint strength actually measuring 29% higher than projected as an initial strength value from earlier data. When similar joints were stressed to only 5% for 507 da (730,080 min) and then the stress raised to 20 psi, the joints again failed after only 9 da. Thus, it might be questionable whether the additional cost of preheating adherends could be justified.

In Table 193, an investigation was conducted to determine the relative survival time of joints with varying aluminum surface treating. Abrading the aluminum with Aloxite 60 grit paper was definitely preferable over vapor degreasing at 10% and lower stress levels. However, further enhancement of

Table 193 Comparison of Vapor Degreasing, Mechanical Abrasion, and Room Temperature Acid-Etching Surface Pretreatments on Aluminum Joints[a] Survival Time Under Stress and High-Humidity Exposure Conditions

		Curing conditions			Stress conditions			
			Temp.					Time to failure
Adhesive	Surface pretreatment	Time (min)	°F	°C	%	psi	MPa	(min)
EA-9320	Vapor degrease	60	180	82	20	728	5.02	20,160
EA-9320	Vapor degrease	60	180	82	10	364	2.51	43,200
EA-9320	Vapor degrease	60	180	82	5	182	1.25	691,200
EA-9320	Aloxite 60 abrade	60	180	82	31	1380	9.51	128,160
EA-9320	Aloxite 60 abrade	60	180	82	20	875	6.03	159,840
EA-9320	Aloxite 60 abrade	60	180	82	10	438	3.02	648,000
EA-9320	Aloxite 60 abrade	60	180	82	5	219	1.51	>777,600
EA-9320	Aloxite 60 abrade	60	180	82	5	219	1.51	>777,600
EA-9320	HCl etch at RT	60	180	82	25	1380	9.51	30,240
EA-9320	HCl etch at RT	60	180	82	18	1000	6.89	648,000
EA-9320	HCl etch at RT	60	180	82	9	500	3.45	>777,600[a]
EA-9320	HCl etch at RT	60	180	82	4.5	250	1.72	>777,600[b]

[a]All joints were fabricated from 0.125-in (3.18-mm) 6061-T6 aluminum alloy adherends and exposed in 100% RH at 125°F (52°C) under the indicated stressing conditions.
[a]Joint deliberately failed after indicated stress/exposure time to determine the retained joint strength. This joint strength was 3300 psi (22.74 MPa), which was 60% of original projected joint strength.
[b]Joint deliberately failed after indicated stress/exposure time with a joint retention of 4800 psi (33.07 MPa); i.e., 87% of initial projected joint strength. This joint differed from ([a]) only in a 50% lower stressing level throughout exposure time.
Data from Ref. 3778.

survival time at a stressing level of about 1000 psi (6.89 MPa) was achieved by etching the 6061-T6 adherends with HCl at room temperature. At the 500-psi (3.45-MPa) level of stress, the joint strength retention after more than 500 da of exposure in hot humidity was still 60% of originally projected strength. When the stress was only 250 psi (1.72 MPa), the retained strength after similar exposure time was 87%, indicating some interfacial degradation had been incurred by doubling the stressing level.

In Table 194, Minford (3779) studied an effect not previously reported in the literature to his knowledge. The question is what is the possible damaging effect on joint survival of subjecting a joint to a relatively low static stress and then moving to a higher static stress later in the joint service. The survival time of individual joints stressed to failure by continual stressing at 10, 18, or 20% levels were used for comparison control values. Alternatively, similar joints were stressed to low levels of 2.3–2.5% for periods of about 417 da and then were restressed to higher levels as used to develop the control data times. Regardless of which surface treatment was used, the survival time values were significantly lowered by prestressing service at a lower level. This indicates that some form of specific damage had been induced in the bondline even at very low stress. This damage then manifested itself by reducing the joint survival time values below those already established as the control values.

Minford (39) has developed stress in hot humidity data for aluminum joints fabricated with varying thickness boehmite oxide films on 6061-T6 joints, as shown in Table 195. Two stress levels of either 15 or 35% of initial joint strength averages were used and stress/humidity–durability factors

Table 194 Effect of Changing Stressing Level on Survival Time of Aluminum Joints[a] While Soaking in Hot Humidity Conditions

Joint number	Adhesive	Surface treatment	Stressing conditions %	psi	MPa	Time to failure (min)
1	EA-9320	Aloxite 60 abrade	20	875	6.03	159,840
2	EA-9302	Aloxite 60 abrade	2.5	110	0.76	>601,920
2	(Same joint [2] with stress increased to 20% after initial 601,920-min exposure)		20	875	6.03	54,720
3	EA-9320	Aloxite 60 abrade	10	438	3.02	648,000
4	EA-9320	Aloxite 60 abrade	2.5	110	0.76	>601,920
4	(Same joint [3] with stress increased to 10% after initial 601,920-min exposure)		10	438	3.02	170,500
5	EA-9320	HCl etch at RT	18	1000	6.89	648,000
6	EA-9320	HCl etch at RT	2.3	185	0.86	>601,900
6	(Same joint [6] with stress increased to 18% after initial 601,900-min exposure)		18	1000	6.89	125,280
7	EA-9320	HCl etch at RT	2.3	185	0.86	>601,900
7	(Same joint [7] with stress increased to 15% after initial 601,900-min exposure)		15	820	5.65	172,800

[a]All joints fabricated from 0.125-in (3.18-mm) 6061-T6 aluminum alloy adherends and exposed in 100% RH at 125°F (52°C) at the indicated stressing levels.
Data from Ref. 3779.

Table 195 Joint Durability Results Under Simultaneous Stress and Hot Humidity Exposure Conditions of Aluminum Joints[a] with Varying Thickness of Boehmite Oxide Surface Conditions

Surface condition	% Stress	Average joint failure time (da)	Calculated stress/humidity durability factor[b]
Mill-finish	4.9	123	30.75
Mill-finish	7.8	118	47.20
Mill-finish	15	1.12	0.86
Mill-finish	35	0.005	0.009
Vapor degrease	5	120	30.00
Vapor degrease	8	104	41.6
Vapor degrease	15	53	0.39
Vapor degrease	35	0.009	0.015
Caustic etch	15	53	49.13
Caustic etch	35	8.5	18.39
Deionized H O (1 min)	15	45.3	38.94
Deionized H O (1 min)	35	9.5	19.05
Tap H O (1 min)	15	115	67.85
Tap H O (1 min)	35	20	34.50
Deionized H O (5 min)	15	62.5	44.5
Deionized H O (5 min)	35	18.5	30.75
Tap H O (5 min)	15	124	110.24
Tap H O (5 min)	35	17	35.28
Deionized H O (10 min)	15	122	74.79
Deionized H O (10 min)	35	18.5	26.47
Tap H O (10 min)	15	135	101.25
Tap H O (10 min)	35	23	40.25
Deionized H O (60 min)	15	135	61.43
Deionized H O (60 min)	35	33.5	34.81
Tap H O (60 min)	15	147	78.79
Tap H O (60 min)	35	42	52.54

[a]6061-T6 aluminum alloy joints pretreated as indicated under the "Surface condition" column. Items after the caustic etch joints were additionally treated in boiling water of the indicated type for the time shown in parentheses. All joints were made with 3M's EC-2086 nitrile-modified epoxy adhesive cured for 15 min at 400°F (204°C).

[b]The durability factor is a calculated number designed to consider the combination of the level of stress and the time to fail the joint under that stress as follows:

$$\text{Durability Factor} = \frac{\text{Imposed Stress in psi} \times \text{Joint Failure Time in Days}}{1000}$$

Data from Ref. 39 (Table V, p. 516).

were calculated which consider the stress level and the joint failure time in days. For comparison of the mill-finished and vapor-degreased–only joints with the same adhesive, the stress levels chosen had to be lowered to the 5–8% range to obtain similar durability factor results. In contrast, with the earlier cited data involving the soaking of unstressed boehmite surface joints in the same hot humidity conditions (where boehmite films created in boiling deionized water had the highest durability responses), we now see that the boehmite films generated in boiling tap water show the higher durability factor responses. It seems that the presence of the various cations and anions in the tap water have somehow become involved in creating a boehmite oxide that is more resistant to failure mechanisms operating under stress and humidity conditions.

Minford (39) has also made comparisons using new and earlier data on 6061-T6 joints fabricated with the same one-part epoxy adhesive using six different pretreatments, including (1) vapor degrease plus chromic acid deoxidize, (2) hot alkaline soak/clean plus hot chromic deoxidize, (3) CAA anodize, (4) PAA anodize (Boeing process), (5) chromic acid deoxidize plus 5-min boil in deionized water, and (6) chromic acid deoxidize plus 60-min boil in deionized water. All successfully survived an arbitrarily selected 70-da (100,000-min) stress in hot humidity at 16% of initial average joint strength, as shown in Table 196.

B. Cycling Stress Loading

All of the above comments on stress testing with weathering present have related to an imposed static load. However, there are many structural aluminum applications, where steady load application does not represent the service condition. Certainly this is the case with aircraft aluminum joints and most vehicle structures, where loads are applied and relaxed at intervals. This has caused a trend in

Table 196 Comparative Durability Performances of Aluminum Joints Pretreated by Acid Deoxidation, Anodizing, or Boehmite Oxide Created in Boiling-Water–Immersion Conditions[a]

Surface pretreatment	Initial strength (psi)	Exposure time (da)			Joint strength retention (psi)			Stress level (lb)	Survival time[c] (min)
		A	B	C	A	B	C		
Chromic acid deoxidation[d]	5330	730[a]	365	N.A.	3624	885	N.A.	800	>100,000
Chromic acid deoxidation[e]	5430	N.A.	364	N.A.	N.A.	1629	N.A.	N.A.	N.A.
Chromic acid anodize	5513	730	365	365	3198	3363	4521	900	>100,000
Phosphoric acid anodize (Boeing procedure)	6480	730	365	365	4212	3551	5119	1050	>100,000
5-min treatment boiling deionized H_2O	4750	714	266	266	2993	2565	2138	710	>100,000
60 min treatment boiling deionized H_2O	2970	714	364	364	3653	2703	3059	700	>100,000

[a]Pretreatment for producing boehmite oxide film includes (1) vapor degreasing, (2) hot caustic etching, (3) desmutting in cold nitric acid, (4) water rinsing and air drying, and (5) immersion in boiling deionized water for indicated period.

[b]Exposure conditions: A—water; B—100% RH at 125°F (52°C); C—salt-fog cycle (daily cycle of 16 hr in salt-fog cabinet and 8-hr drying with cabinet open to ambient lab atmospheric conditions); D—Stressing in Alcoa stressing fixture in 100% RH at 125°F (52°C) for arbitrary time extending up to 100,000 min if no joint failure observed.

[c]Stressing of joints was in 100% RH at 125°F (52°C) conditions.

[d]Treatment conditions were (1) vapor degreasing, and (2) immersion in chromic acid solution at 180°F (82°C) for 5 min.

[e]Treatment conditions: (1) vapor degreasing; (2) immersion in Ridoline 53 alkaline cleaner at 180°F (82°C) for 3 min; (3) rinse in deionized water, (4) immersion in chromic acid solution for 5 min at 180°F (82°C) followed by a final water rinse and dry.

Data from various publications of Minford.

durability testing of aerospace bonds toward procedures in which the load is repetitiously applied and relaxed. This cyclic stressing may be employed under controlled environmental conditions consistent with the weathering factors in the service environment. Some of the most comprehensive studies, in this regard, were undertaken by Frazier and Lajoie at Bell Helicopter. Their work has been summarized in an Air Force Materials Lab Report (3202). In planning any work of this kind, the question always arises as to the selection of a load-unload period, because adhesives vary considerably as to the rates at which they creep under load and relax when the load drops off.

The principal used at Bell can be demonstrated by comparing a relatively low temperature–curing adhesive like FM-123-7 with 350°F (177°C) curing FM-1000. The FM-123-7 adhesive polymer shows a yield point in its stress-strain curve but no work-hardening properties. Creep under stress measurements showed the almost instantaneous deflection was 0.0027 in (0.0069 cm), creep in the next 15 min was 0.0003 in (0.0008 cm), and for the next 45 min was 0.00015 (0.00038 cm). When the load was relaxed, all the displacement relaxation was completed within 2 min. For this adhesive, a load time of 30 min and an unload time of 5 min would seem to be a possible desirable cycle. However, for an extremely tough, high-elongation adhesive like the FM-1000, the creep and relaxation properties are quite different, and another cycle selection must be made. At the end of 1 hr, the FM-1000 polymer was still creeping, although at a slower rate. Initial deflection was similar to the FM-123-7, i.e., 0.0024 in (0.0061 cm), but within 30 min additional creep was 0.0028 in (0.0071), with an additional 0.0008 in (0.0020 cm) of creep after 60 min. The same sequence occurred in reverse when the load was relaxed, with the relaxation still measurable after 15 min. Based on this information, the investigators decided that a maximum bond damage cycle for the FM-1000 joints would likely be 2 hr of loading followed by 1 hr of relaxation. Although these kind of differences can be shown between adhesives, it became necessary for economy of evaluation of a large number of adhesives to set some sort of compromise load cycle such as 4000 psi (27.56 MPa) for 1 hr and unload for 15 min. The selection of a specific load level does, however, create an additional problem in that a curve of durability versus stress level cannot be determined, as has been shown earlier to be the procedure when static load testing is conducted.

Table 197 Comparison of Cyclic Load–Environmental Exposure Data for Three Aerospace Adhesive Systems Using Vacuum-Bag or Autoclave Curing

Adhesive	Load (lb/in)	Type curing	Life cycles
FM-73[a]	900	Vacuum bag	4200
	900	Autoclave	5750
	1200	Vacuum bag	190
	1200	Autoclave	2200
	1500	Vacuum bag	300
	1500	Autoclave	600
EA-9628[a]	900	Vacuum bag	900
	900	Autoclave	1900
	1200	Vacuum bag	700
	1200	Autoclave	1750
	1500	Vacuum bag	200
	1500	Autoclave	550
EA-9309[b]	Less than 1 cycle at 900 lb/in		

[a]One-part, heat-cured adhesives.
[b]Two-part, room temperature–curing adhesive.
Data interpreted from Kuperman and Horton, Adhesive-Bonded Aluminum Structure Repair, Adhesive Bonding of Aluminum Alloys (E.W. Thrall and R.W. Shannon, ed.). Marcel Dekker, New York, Fig. 11, p. 452 (1985).

Althof (3173) has published pertinent data using low-cycle loading of thick adherend specimens bonded with FM 73 adhesive. The range of combinations of load/unload periodicities for testing was shown in Table 135. The number of cycles to joint failure and shear strain at failure are summarized in Table 136. Depending on the load/unload frequency selected, the number of cycles to joint failure could vary widely at the same shear stress level. For example, at 37.5 MPa in Table 136, the cycles to failure could be 8, 11, 62, 27, and 28 for load/unload periods in minutes of 60/15, 15/5, 4/2, 1/1, and 0.5/0.5.

These kind of cyclic loading investigations are used for evaluating bonded joints in the so-called spectrum stress testing of aluminum alloys for aircraft. In testing metallic components, three different conditions of periodic stressing and vibration can be recognized depending on whether the aircraft is landing, in steady flight, or under severe maneuvering conditions. Data from Kuperman and Horton are reviewed in Table 197.

XIV. BOND PERMANENCE OF COMBINATION ADHESIVE AND RESISTANCE SPOT-WELDED JOINTS (WELDBOND)

A. Aerospace Applications

Weldbonding is the combination of incorporating an adhesive in the bondline together with the benefit of a heat-fusion spot-weld. Two approaches have been used in practice: (1) apply an adhesive to the area to be joined, and make one or more spot-welds through the adhesive, (2) make a spot-weld first and attempt to capillary an adhesive into the joint area. It is possible to use either a one- or two-part epoxy adhesive using method 1, since either can be distributed evenly before the spot-welding operation is affected. The adhesive is then cured under whatever temperature conditions are recommended by the adhesive formulator. It is usually necessary in method 2 for the adhesive to be specially formulated to have exceptional flow and wetting and achieve a very low viscosity at some time enroute to full curing.

The Russians initiated the first use of this technology, and they continued to use it more extensively than elsewhere in the world. It was introduced in the United States via a Russian document translation discussing utilization in building machinery in 1967. This led to financing a project at Lockheed-Georgia by the Air Force. The journal *Iron Age* focused on some of Lockheed-Georgia's initial conclusions, including comparative sonic fatigue endurance data (62). Weldbond-durability performance attributed to Lockheed-Georgia stating that weldbond aluminum joints retained much of their initial strength in high humidity was published in the journal *Material Engineering* in 1971 (63). Seventy-day data were cited cycling 16 hr at 100°F (38°C) at 95% R.H. and 8 hr ambient air that reduced strengths by 10% to 2700 psi (18.60 MPa). Weldbonds were also successfully tested under low-cycle flexure in liquid hydrogen without brittle fracture.

In 1971, Lockheed-Georgia issued an interim report (64) describing evaluation of a wide variety of surface pretreatments, primers, adhesives, and weldbond processing variables. Relatively short-term weathering cycle tests and wedge-bond tests after high-humidity soaking were used to discriminate among the surface treatments and adhesives. In 1973, in another official report to the Air Force, Lockheed-Georgia (65) showed significant differences in weldbond joint durability in salt spray depending on the type of surface pretreatment used. It was shown that while the aluminum industry developed spot-weld etching that optimized spot-welds, their joint durability was significantly lower for bonded joints than for FPL-etched weldbond joints.

Mahon et al. (66), at Grumman Aerospace, began work separately on methods to achieve durable weldbonds that needed to operate in 350°F (177°C) service. They concluded that immersion chemical treatment of aluminum with HNO_3/Na_2SO_4 solution at 175°F (80°C) to 190°F (88°C) was preferable to immersion in the FPL-etchant solution at 180°F (83°C) for development of class A spot-welds and generally tough adhesive joints. Stress/durability testing was conducted in spring-loaded fixtures conditioned in 100% RH at 140°F (60°C) in accordance with MIL-A-83377. Joint survival times were compared with corresponding adhesive-bonded joints with the result that most weldbonds lasted longer at 33% higher stressing. The same investigators compared the relative performance of weldbonded versus adhesive-bonded joints fatigued per the general requirements of MMM-A-132.

The comparative fatigue performance of the adhesive-bonded joints were actually higher, but the weldbonds with the same adhesive were dramatically above the performance of conventional riveted joints employing various sealants. Choice of adhesive could be influential, since corresponding joints made with a different adhesive subjected to the same test conditions showed superior weldbond fatigue strength over the corresponding adhesive joint.

More recent technical information about weldbonding has come from another aerospace company in a series of reports by Bowen et al. (760–764) at Northrup. Wu and Bowen (71) sought to establish a production method for aluminum weldbonding which included a particular surface prep, adhesive application, and spot-weld variables to assure environmental durability and consistent weld quality. The previous year, Bowen et al. (70) had focused on the development of improved surface treatments and selection of best adhesives, including elucidation of the mechanism of weld expulsion with spot-weld etch as compared to FPL etching. At first, Bowen and Wu advocated the use of an FPL etch plus a 60-min sodium dichromate seal as the best surface pretreatment for aerospace weldbonding. Later, an improved treatment was developed consisting of degreasing, hot-alkaline cleaning, deoxidizing in nitric acid, and anodizing in phosphoric acid/sodium dichromate solution at room temperature. Based on durability in salt-water exposure and minimum crack growth in wedge test specimens in 100% RH at 120°F (50°C), they suggested Goodrich 0500PE130 adhesive with phosphoric acid/sodium dichromate anodizing at 1.5 V for 20 min. Goodrich continued to develop many new adhesives after this time, so this particular adhesive may not be available now, and testing would be required to show superior properties of new products.

B. Nonaerospace Applications

Both truck and trailer manufacturers have developed interest in weldbonding aluminum applications in recent years as an alternative to the extensive use of riveting and sealing for joining. Weldbond and adhesive joint durability comparisons were undertaken by Hall (69), in 1974, referring to current interest by truck manufacturers but still using the aerospace FPL etch as surface treatment for the weldbonding and adhesive bonding. No difference in bond durability was noted between the adhesive and the weldbonded joints in 100% RH at 160°F (71°C). Both, however, declined steadily in strength during a 6-mo exposure period. Some significantly higher bond-strength retention by the weldbond joints was shown after 10-mo exposure to a daily cycle between 60% RH at 80°F (27°C) and 100% RH at 160°F (71°C).

It seemed to Minford, at Alcoa, that comparative durability test information seemed to be needed involving no special surface treatment of aluminum. This should be especially pertinent for operations in the truck and automotive industries whose operations simply could not economically support the use of aerospace adhesive materials and surface preparation techniques. A technical paper in 1975 by Minford et al. (72) compared relative permanence of weldbond and adhesive-bonded joints. This work was actually an extension of an earlier 1974 paper by Minford and Vader (68) designed to offer information to the automotive industry on the strength and durability of bonds made to unprepared aluminum body alloy. Data are shown in Table 198 which demonstrate that the weldbond joint made with a vinyl plastisol (as already used in auto jointures) not only could provide a substantial boost of about 60% in initial strength over a similarly bonded joint but also retained a higher percentage of initial strength in hot humidity. As often occurs, in Table 199, we observe that properties of the adhesive can have a distinct effect. For example, a 2036-T4 alloy weldbond joint fabricated with a high-strength, heat-curing epoxy paste actually was 21% weaker than the corresponding adhesive joint. Evidence was presented that the strength of the alloy can also be important. In this case, the weaker X5085-H111 alloy adherends used in a weldbond joint with the same epoxy adhesive created joints with the same strength as the adhesive-only joints. Another difference with epoxy adhesive substitution was finding equally degrading percentages of adhesive-only and weldbond joints in hot humidity–exposure conditions. In this combination, the weld nugget was not functioning as though it was having an overall negative effect. Duplicate joints were also exposed to a corrosive salt spray (Table 200), which shows a more rapid joint deterioration by pitting corrosion penetration into the interfacial areas than present from hot humidity testing. After 3-wk exposure, some durability benefit of weldbonding could be seen with vinyl plastisol/2036-T4 combination joints.

Table 198 Effect of Weldbond on Durability of Aluminum Joints in High-Humidity Exposures Using Heat-Curing Vinyl-Plastisol or One-Part Epoxy Adhesives

Type joint	Average initial bond strength, lb (kg) 2036-T4	X5085-H111	% overall strength retention of aluminum joints[a] 2036-T4 alloy 85% RH at 75°F (24°C)	2036-T4 alloy 100% RH at 125°F (52°C)	X5085-H111 alloy 85% RH at 75°F (24°C)	X5085-H111 alloy 100% RH at 125°F (52°C)
Adhesive[b]	910 (409)	860 (387)	76	70	78	70
Weldbond	1450 (652)	1510 (680)	99	97	89	93
Adhesive[c]	1610 (725)	1220 (549)	100	59	94	76
Weldbond	1270 (572)	1210 (545)	100	64	95	74

[a]Exposure time in all tests was 90 days.
[b]Adhesive used was a one-part, heat-curing vinyl-plastisol paste.
[c]Adhesive used was 3M's EC-2086 one-part, nitrile-modified epoxy paste adhesive.
Data from Minford No. 535 (Table 3–4, p. 97).

Table 199 Effect of Weldbond on Durability of One-Part, Heat-Cured Aluminum Joint in High-Humidity Exposure[a]

Type joint	Average initial joint strength, lb (kg) 2036-T4	X5085-H111	% strength retention 2036-T4 alloy 85% RH at 75°F (24°C)	2036-T4 alloy 100% RH at 125°F (52°C)	X5085-H111 alloy 85% RH at 75°F (24°C)	X5085-H111 alloy 100% RH at 125°F (52°C)
Adhesive	1610 (724)	1220 (549)	100	59	94	76
Weldbond	1270 (571)	1210 (544)	100	64	95	74

[a]Exposure period was 90 da.
Data from Ref. 535 (Table 3–5, p. 97).

Table 200 Effect of Weldbond on Joint Durability in Salt-Spray Exposure Conditions

Type joint	Average initial bond strength, lb (kg) 2036-T4	X5085-H111	Overall strength retention, % 2036-T4 alloy 1 wk	2036-T4 alloy 3 wk	X5085-H111 alloy 1 wk	X5085-H111 alloy 3 wk
Vinyl plastisol	910 (409)	860 (387)	85	35	99	94
Vinyl plastisol (weldbond)	1450 (652)	1510 (679)	96	75	95	93
One-part epoxy	1610 (724)	1220 (549)	96	78	84	86
One-part epoxy (weldbond)	1270 (571)	1210 (544)	100	77	84	86
Polysulfide-modified epoxy	700 (315)			69		
Polysulfide-modified epoxy (weldbond)	1175 (529)			47		

Data from Ref. 535 (Table 3–6, p. 98).

The superior corrosion resistance of X5085-H111 alloy proved to be so significant that it created equally durable adhesive-only or weldbond-type joints. There was little difference in final joint strength joints with 2036-T4 adherends and the polysulfide-modified epoxy adhesive. However, it must be remembered that this adhesive was cured at room temperature, whereas the plastisol and one-part epoxy may have had a better wetting opportunity. As we will seen later, the main advantage of the lower-strength polysulfide-modified epoxy is its greater ability to withstand cyclic fatigue.

It has been demonstrated by Nordmark (3780), Minford et al. (72), and Vader et al. (3781) that weldbonding can generally improve long-life fatigue strengths by 100–500% even in the absence of the more sophisticated surface pretreatments recommended by the aerospace industry. At the same time, it has been demonstrated by Minford et al. (72) that depending on the particular adhesive selection and the severity of the simulated weathering used for exposure, the joint strength and survival time can be increased. Nordmark (3780) has shown a similar effect on the fatigue strength of similarly fabricated weldbond and rivbond joints. Manufacturers of trailers, vans, and trucks have primarily used riveted construction, because they could not afford the more sophisticated and costly adhesive-bonding techniques as developed in the aerospace industry. The automotive industry has mostly favored spot-welding. While it was known from aerospace experience how more durable bonded aluminum joints could be fabricated, the use of these methods in general manufacturing could raise the price of the products beyond what the typical consumer could afford.

Joint fatigue testing had been undertaken at Alcoa with joints assembled by all these methods over many years, and Figure 56 is representative of data relating spot-welds, adhesive-bond joints, and weldbonded joints using a two-part relatively rigid, polylamide-cured epoxy. With this relatively rigid adhesive, it can be seen that either adhesive-bonded or weldbonded joints can be an order of magnitude greater than spot-weld joints alone. However, the adhesive-bonded joints with a rigid adhesive tend to be better at higher-load cycling conditions than where the nugget is present with

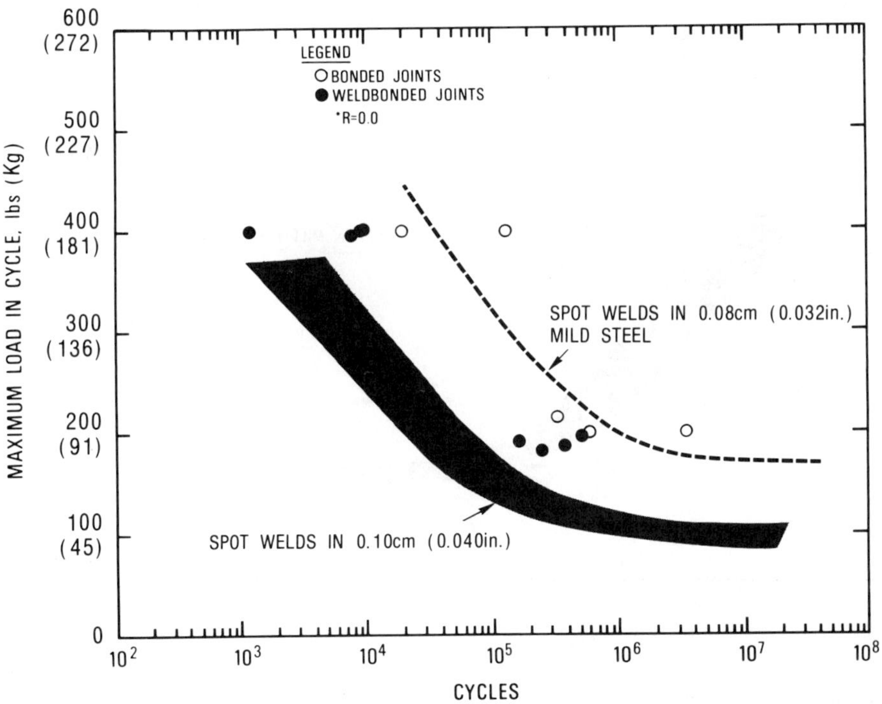

Figure 56 Fatigue strength of lap-joints in 0.10-cm (0.040-in) 2036-T4 sheet polyamide-modified epoxy. (From Ref. 535 [Fig. 3–37, p. 99].)

adhesive. Performance of the weldbond and adhesive-only bonded joints tend to come closer together as the load level decreases. Characteristics of the adhesive can be important in contributing to performance, as is demonstrated in Figures 56–58. In Figure 57, a high-shear, high-peel, one-part, heat-cured epoxy was substituted for the more rigid two-part epoxy in Figure 56. By comparison, we immediately note the improvement in fatigue strength of the adhesive-only and weldbond joints has improved over plain spot-weld joints to two magnitudes of difference. The fatigue strength of both these kinds of joints are quite similar. Finally, in Figure 58, the heat-cured epoxy is replaced by a polysulfide-modified epoxy cured at room temperature. Performance of the weldbonded joints is further enhanced, whereas corresponding performance of the adhesive-only bonded joints is somewhat diminished. Such weldbond joints can be as much as 2.5-fold greater than spot-wels in fatigue strength with load cycling at the 200- to 300-lb level. This is in spite of the fact that the polysulfide-modified epoxy adhesive itself is significantly weaker in tensile strength than either of the other adhesives in Figures 56 and 57. This demonstrates that several factors can interact to alter the overall fatigue strength of weldbond joints.

In determining the potential of weathering for decreasing the fatigue resistance of aluminum weldbonded joints (see Ref. 3780), the joints were subjected to either 100% RH at 125°F (52°C) for 3 mo or 5% continuous salt spray for 3 wk. Nordmark summarized his investigation with the following five conclusions: (1) effect of simulated service ranged from complete adhesive failure of the bond to apparent improvement in fatigue life, (2) flexible adhesives generally appear to be better able to resist the combination of environmental and repeated loadings, (3) sheet alloy selection was shown to affect performance significantly, (4) oil presence on the sheet during bonding will likely lower the

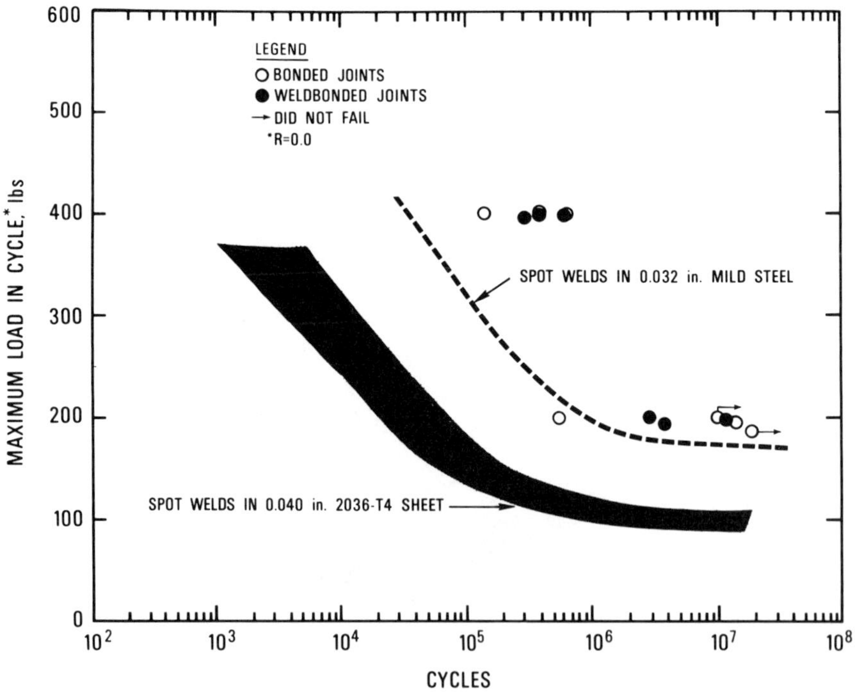

Figure 57 Fatigue strength of lap-joints in 0.040-in 2036-T4 sheet high shear–high peel epoxy. (From Minford Ref. 535 [Fig. 3-38, p. 100].)

resistance to exposure, and (5) in selecting an adhesive, it is important to evaluate it in typical service conditions, including both manufacturing procedures and exposure conditions. Some typical data are shown in Table 201.

Minford (3782) evaluated the comparative durability of three adhesives in weldbonded joints that were being considered for use in building an aluminum Naval Surface Effect Ship. In addition to hot condensing humidity, soak/freeze/thaw cycling, cycling in 3.5% salt spray, and in continuous 5% salt spray (ASTM B117), the factor of simultaneous stressing in the first three accelerated weathering conditions was also considered. A modest-strength polysulfide-modified epoxy with some significant flexibility actually outperformed the other two more stronger and more rigid epoxy candidates under all these aggressive testing conditions.

In 1987, two technical papers were presented at the SAE International Congress and Exposition Meeting which considered the evaluation of aluminum weldbond joints. Marwick and Sheasby (3783) reviewed their durability test data using a manufacturing scheme for weldbonding aluminum vehicles developed at Alcan. They employed the previously well-accepted testing of lap-joints by exposure to salt spray or using combined stress and hot humidity conditions. Good durability results were obtained with certain heat-cured, oil-tolerating epoxy paste structural adhesives, but extensive testing of a number of formulations was necessary to obtain most acceptable performance. In the same year, Selwood et al. (3784) produced a paper titled "The Evaluation of an Adhesively Bonded Aluminum Structure in an Austin-Rover Metro Vehicle" which related directly to the same type of Alcan assembly and joining technology. The conclusion that the proposed manufacturing system for this aluminum vehicle was quite commercially feasible seemed well established.

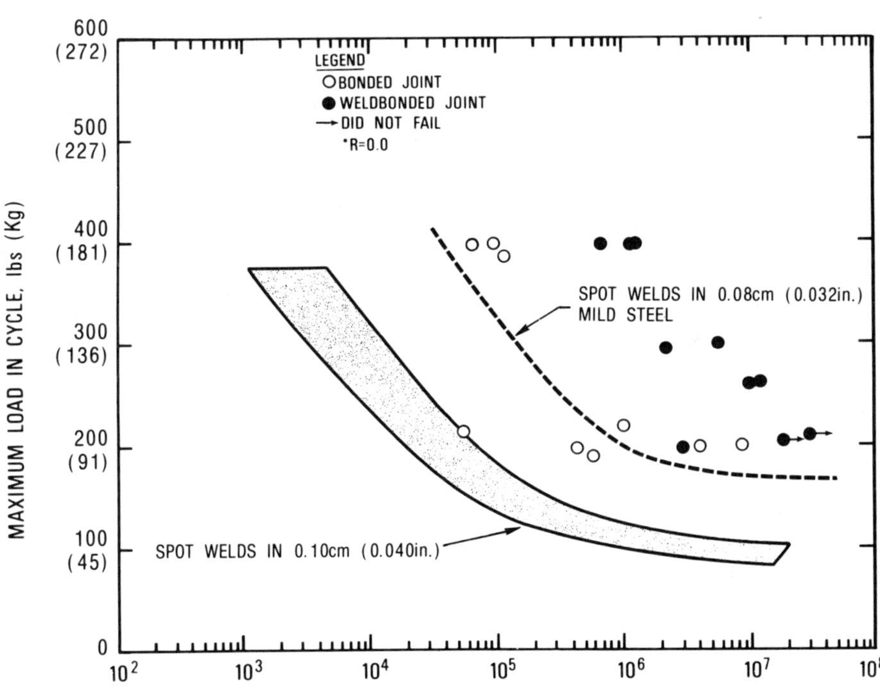

Figure 58 Fatigue strength of lap-joints in 0.10-cm (0.040-in) 2036-T4 sheet room temperature-curing polysulfide-modified epoxy. (From Minford Ref. 535 [Fig. 3-39, p. 100].)

Table 201 Fatigue Test Results Using Weathered Weldbonded Aluminum Joints

		Exposed to condensing humidity				Exposed to salt fog			
			Maximum load				Maximum load		
Alloy	Adhesive	Spec. no.	N	lb	Life cycles	Spec. no.	N	lb	Life cycles
2036-T4	Epoxy A	8	1780	400	265,000	20	1760	400	67,000
		11	1310	294	1,003,100	15	1310	294	305,500
		2	1330	299	1,884,000	13	1330	299	344,500
		5	1330	299	31,998,500	19	1330	299	599,200
						17	1100	247	6,167,666
2036-T4	Epoxy B	8	1310	294	17,100	15	1330	299	30,600
		14	894	201	2,307,500	16	885	199	41,000
		13	894	201	9,928,700	19	894	201	656,800
2036-T4	Vinyl-plastisol D	6	1330	299	883,800	10	1330	299	1,300
		7	1330	299	1,550,800	12	1310	294	114,800
		8	1120	252	1,317,000	17	894	201	114,100
						13	894	201	964,800
5182-0	Vinyl-plastisol D	B7	1330	300	1,076,300	B18	1760	395	1,484,400
		B5	1330	300	>29,813,700	B14	1340	302	1,804,200
		B2	894	201	>26,246,100	B22	1340	302	3,535,700
						B9	894	201	>26,511,100
6009-T4	Vinyl-plastisol D	B38	1770	397	457,000	B43	1790	397	319,200
		B34	1780	400	763,700	B41	1780	400[a]	332,500
		B32	1330	300	1,592,900	B42	1340	302	740,900
		B39	1330	300	1,655,200	B46	1330	300	741,300
		B36	1330	300	11,392,200	B45	1340	302	1,650,500
		B31	903	203	>25,080,100	B48	1105	249	>35,363,900
						B41	894	201[b]	>25,703,500
6010-T4	Vinyl-plastisol D	B64	1780	400	586,300	B77	1790	492	286,700
		B66	1760	395	688,200	B72	1760	395	321,700
		B65	1530	345	1,089,200	B74	1540	346	779,100
		B69	1540	346	1,375,200	B76	1560	350	1,217,100
		B62	1330	300	>25,127,000	B78	1550	345	1,476,200
						B71	1340	302	>18,026,700

[a]Specimen previously loaded to 894 N (201 lb) for 25,703,500 cycles.
[b]Load raised to 1780 N (400 lb).
Data from Ref. 3780 (Table I, p. 5).

XV. BOND PERMANENCE OF COMBINATION ADHESIVE AND RIVET OR METAL CLINCHING JOINTS (RIVBONDS AND CLINCHBONDS)

Combination of adhesive bonding with a mechanically inserted rivet has been of interest for manufacturing activities where riveting alone has been a standard method of assembly. The possible advantages would include both an increase in overall joint strength and a significant increase in fatigue life. Low fatigue life potential have always been recognized as one of the important shortcomings of riveted metal structures. One successful method of increasing the fatigue life of riveted structures has been to provide separation and insulation in joints to minimize sensitivity to metal fatigue failure.

Table 202 Fatigue Test Results of Weathered Rivet and Rivbond Aluminum Joints

		Exposed to 100% RH at 125°F (52°C)							
		Spec.	Maximum load		Life	Spec.	Maximum load		Life
Alloy	Adhesive	no.	N	lb	cycles	no.	N	lb	cycles
2036-T4	None	15B	1750	393	451,800	18B	1760	397	94,900
		16B	1330	300	3,203,600	19B	1330	300	1,113,700
2036-T4	Epoxy M	15C	1790	402	644,200	19C	1780	400	517,100
	(polysulfide	17C	1760	397	1,100,000	20C	1780	400	636,800
	modified)	16C	1330	300	11,465,000	18C	1310	294	>29,612,500
2036-T4	Epoxy A	15C	1740	391	490,300	18C	1760	397	354,900
	(heat-cured)	16C	1300	292	3,365,300	19C	1330	300	944,200
		17C	1100	247	>36,031,100	20C	894	201	>32,581,500

Data from Ref. 3780 (Table 2, p. 6).

Thus, replacement of a nonbonded insertion with a bondable adhesive should represent a further enhancement of that approach.

Minford and Vader (3785) investigated comparative strength and durability of aluminum joints fabricated by either riveting or rivbonding which showed enhancement of both joint strength and joint durability when combining adhesive with a rivet.

Nordmark (3780) provided data showing the relative fatigue resistance values for rivbond and riveted-only 2036-T4 aluminum joints bonded with the same adhesives as tested above by Minford. Representative data are provided in Table 202 showing the significant improvement in fatigue resistance obtained with a rivbonding procedure.

XVI. BOND PERMANENCE WITH CONTAMINATED ALUMINUM SURFACE JOINTS

The main basis of adhesive bonding aluminum to obtain the highest degree of joint durability has historically been to make sure the adherend surface is clean and free of contaminating soils and to follow that processing with some special surface preparation known to offer excellent resistance to service weathering. It would then seem inappropriate to consider the subject of deliberate bonding to a contaminated surface, but it is a subject that must be addressed in the real world because of economic necessity. As has been mentioned, aluminum manufacturers use materials (such as rolling lubricants) which must be considered contaminants for any adhesive bonding. Additionally, varying amounts of commercial forming lubricants may be used in the forming and shaping of parts prior to their assembly into a final structure. This is especially true in the automobile, truck, trailer, and transit car industries. Finally, there are unplanned contaminations such as the situation where handling has caused contamination or industrial atmospheric fallout has occurred during handling or storage even including occasional corroding of the adherend surface.

If these contaminated surfaces are subsequently exposed to solvent or emulsion cleaning followed by deoxidizing, then there need not be any negative effects to consider. However, if these surfaces are the ones to be directly bonded, then a wide range of possible joint durabilities may be expected. What the final answer to the question is, however, will only be ascertained by specific testing designed to evaluate the effect of all the factors that can be operating throughout the manufacturing and service life of the product.

The need to create wetting of the surface by the adhesive is always necessary to achieve any reasonable level of joint strength and acceptable joint durability. This can be achieved to some degree if the adhesive solids have been dispersed in a large volume of solvent, as in formulating the contact-type adhesives. If the solvent is water, i.e., water latex or emulsion adhesives, then there will be little tendency to achieve real wetting in the presence of any surface contamination. However, if the solvent is an organic type (and preferably a strongly polar solvent), then surface contaminants might be dissolved, dispersed, or desorbed depending on the nature of the contaminants. The relative acceptable level of bond-joint durability that can be achieved by bonding with a 25–30% solids neoprene-phenolic contact-type cement over even a mill-finish aluminum surface has been discussed in an earlier section. It has also been mentioned (under the earlier discussion of high-solids mastic adhesives) that as percentage of organic solvent present is reduced, the degree of wetting achieved to a contaminated surface, or a mill-finish surface, might diminish with consequent lowering of joint durability in service weathering. When the adhesive is 100% solids (as a two-part epoxy paste adhesive), then it has (in the past) been impossible to expect that any acceptable level of joint durability under exterior weathering conditions will occur using uncleaned adherends. One exception (using a 100% solids type of adhesive) has been found with the second-generation acrylics, which do have a rather unique ability to bond even to unclean aluminum with some expectation of decent joint durability. However, as with any situation that we might face, there undoubtedly is some level of gross contamination that simply cannot be overcome.

Finally, there is the situation posed by the 100% solids, heat-curing adhesives which can overcome some surface contamination owing to the low viscosity melt state they go through enroute to total curing. While such adhesives do not have the ability to actually remove the contaminants directly, they can have the ability to assimilate some levels of contaminants into their bulk, and consequently, achieve some degree of wetting, respectable joint strength, and worthwhile durability response. Adhesive formulators have worked hard in recent years to develop some very special adhesives which can even show a combination of assimilating contamination and providing for some displacement of the contaminants to provide better direct contact between adhesive and adherend surfaces.

Before discussing the durability results of aluminum joints whose surfaces have been deliberately contaminated with an oil or forming lube before bonding, it may be beneficial for the reader to review some results obtained from uncontaminated surface joints with varying surface pretreatments. Minford (535) obtained the joint-durability test results for 6061-T6 aluminum alloy adherends that had been vapor degreased or chromic acid deoxidized involving both a one- and two-part epoxy commercial adhesive paste as a basis for later comparisons. The joint-strength retention values have been compiled for these two surface-treating conditions when exposed to room temperature water or a hot-water soak/freeze/hot-air thaw cycle for 2 yr, as well as 8-yr exposures to both the industrial and seacoast atmospheres. The data employing a nitrile-modified, one-part, heat-cured epoxy paste adhesive are shown in Table 203. Data from similar joints with a two-part epoxy paste adhesive substituting for the one-part, heat-cured epoxy in Table 203 are shown in Table 204.

Some of the important points for comparison involving each of the four different weathering conditions can be briefly reviewed. Overall, it is clear that the relative order of ability to degrade and destroy adhesion across the whole spectrum of tests is seacoast atmosphere > cyclic soak/freeze/thaw > room temperature water soak > natural industrial atmosphere. In the natural atmospheric weathering (least severe-type weathering), only the vapor-degrease pretreated joints are seriously degraded in 8 yr. This is, of course, the only surface pretreatment considered which has only removed obvious surface contaminants with no oxide-removal or oxide-regeneration process in effect as would be the case with any of the other seven different pretreatments considered. Of course, these other surface pretreatments perform oxide removal and oxide regeneration with such different results that each needs to be separately compared with all the others in each different weathering situation.

The varying performance in water-soaking conditions is most evident by comparing the vapor-degreased, belt-sanded, and chromic acid–deoxidized type of joints in Table 203 using the one-part epoxy adhesive. The vapor-degreased surface joints again (as with the industrial atmosphere weathering) degrade most significantly. In fact, the retained joint-strength averages are reduced by about 50% each year of exposure. The abraded (belt-sanded) joints meanwhile show the anticipated better

Table 203 Relative Effectiveness of Different Surface Treatments on 6061-T6 Aluminum Bonded with a One-Part, Heat-Cured, Nitrile-Modified Epoxy Adhesive

	Exposure Conditions						
	RT water Retained strength (psi [MPa])			Cyclic wet/freeze/thaw Retained strength (psi [MPa])			
Surface treatment	Initial	1 yr	2 yr	Initial	1 yr	2 yr	
Vapor degrease	4330 (29.83)	2000 (13.78)	1170 (8.06)	4330 (29.83)	820 (5.65)	0	
Belt sand and vapor degrease	4870 (33.55)	3100 (21.36)	2170 (14.95)	4870 (33.55)	1050 (7.23)	0	
Chromic acid etching	5330 (36.72)	4270 (29.42)	3530 (24.32)	5330 (36.72)	1330 (9.16)	0	
	Industrial atmosphere				Seacoast atmosphere		
	Initial	2 yr	4 yr	8 yr	2 yr	4 yr	8 yr
Vapor degrease	4330 (29.83)	3100 (21.36)	2090 (14.40)	450 (3.10)	0	0	0
Belt sand and vapor degrease	4870 (33.55)	4130 (28.46)	3670 (25.29)	3400 (23.43)	0	0	0
Chromic acid etching	5330 (36.72)	5300 (36.52)	5010 (34.52)	4730 (32.59)	3370 (23.22)	0	0
Chromate conversion coating	2670 (18.40)	2600 (17.91)	2730 (18.81)	2700 (19.09)	2665 (18.36)	2730 (18.81)	2580 (17.78)

From Ref. 535 (Table 3–8, p. 110).

Table 204 Relative Effectiveness of Different Surface Treatments on 6061-T6 Aluminum Bonded with a Two-Part RT-Curing Epoxy Adhesive

	Exposure conditions							
	Joint strength in industrial atmospheric exposure [psi (MPa)]				Joint strength in seacoast atmospheric exposure [psi (MPa)]			
Surface treatment	Initial	2 yr	4 yr	8 yr	Initial	2 yr	4 yr	8 yr
Vapor degreased	1970 (13.57)	1530 (10.54)	1264 (8.72)	907 (6.25)	1970 (13.57)	0	0	0
Gritblast and vapor degrease	2030 (13.98)	2060 (14.19)	2030 (13.99)	1930 (13.30)	2030 (13.98)	1800 (12.40)	1580 (10.89)	1350 (9.30)
Chromate conversion coating	1260 (8.68)	1230 (8.47)	1198 (8.25)	1130 (7.79)	1260 (8.68)	835 (5.75)	430 (2.96)	0
Anodized (SAA)	2530 (17.43)	2130 (14.68)	1868 (12.87)	—	2530 (17.43)	1660 (11.44)	1690 (11.64)	—
Chromic acid	2830 (19.50)	2900 (19.98)	2731 (18.82)	2370 (16.33)	2830 (19.50)	0	0	0
Phosphoric acid/ alcohol	2660 (18.33)	2470 (17.02)	2450 (16.88)	2130 (14.68)	2660 (18.33)	0	0	0
Acidic paste	2500	2300	2099	2030	2500	0	0	0

From Ref. 535 (Table 3–7, p. 109).

durability produced from grossly removing original mill-finish oxide and having some oxide regrowth. In this instance, the initial joint strength average decrease to about 50% of initial required an additional year of water soaking. Even the preferred treatment by deoxidizing in hot chromic acid showed some sensitivity to degradation in room temperature water soaking, suffering a 33% decline in joint strength after 2-yr exposure.

The second most aggressive weathering condition was noted as the cyclic soak/freeze/thaw exposure in Table 203 involving the one-part, heat-cured, nitrile-modified epoxy. Of special interest is the fact that none of the surface pretreatments (vapor degrease, abrading, chromic acid deoxidation) can produce joints which are capable of surviving 2 yr of such exposure. The author calls attention to the earlier discussed Figure 47, where only the SAA anodize-pretreated joints survived a 2-yr exposure in this weathering with about 75% overall joint strength retention. Admittedly, these data were obtained with a two-part room temperature–curing epoxy; however, by comparing the relative performance of the vapor-degreased and abraded surface joints in Figure 47, we can observe that these types of joints were rapidly degraded when bonded with the same adhesive as shown in Table 204 using a one-part, heat-cured epoxy.

Finally, in both Tables 203 and 204, we can make many relative comparisons between a wide variety of surface pretreatments where bonding was affected in the absence of contamination and exposure was to the very corrosive seacoast condition. It should be pointed out that McMillan (833), in evaluating the relative aluminum joint strength loss caused by different environments, came up with the order shown in Table 205, which places 5% salt-spray exposure as the most severe environment. This would be in complete agreement with anticipating that the seacoast exposure should be the most severe among those shown in Tables 203 and 204. Five of the eight different surface-pretrated types of aluminum joints in these tables (regardless of whether one- or two-part epoxies were employed) failed to survive the first year of exposure. Included in this group was vapor degreasing, belt-sanding, phosphoric acid/butyl alcohol deoxidizing, acid paste–deoxidizing, and chromic acid–deoxidized joints. From Table 155, we can observe the actual survival time in days for all these pretreatments with both one- and two-part epoxy adhesives. The two-part epoxy joints, in general, will have the poorer durability response, since the ability of the modestly viscous paste two-part epoxy cannot as readily wet the aluminum surface as the more flowable (lower-viscosity) one-part epoxy paste while being heat cured. For the former adhesive, the survival times for vapor-degreased, phosphoric acid/butyl alcohol, acid paste cleaner, and chromic acid–deoxidized joints are 80, 270, 275, and 760 da, respectively. For the latter (one-part epoxy), the survival time for correspondingly treated adherend joints are 270, 1440, 1440, and 1460 da. The significant overall elevation of survival time for the one-part, heat-cured epoxy joints as compared to room temperature–cured two-part epoxy joints has been an important bit of evidence for the author to conclude that the relative degree of wetting by the adhesive can relate directly to long-term durability in a corrosive salt-water exposure.

Minford (68) issued a first paper on bonding aluminum automotive body sheet which included the relative ability of certain adhesives such as elastomeric and vinyl-plastisol antiflutter-type adhesives to bond to different types of aluminum and steel adherend surfaces as they might be used to form hood and deck cover components. In Figure 59, the author's data have been plotted where four different aluminum alloy adherends and automotive body steel sheet were used to make lap-joints with a vinyl-plastisol and an automotive elastomeric antiflutter product. It can be seen that the vinyl-plastisol could develop joints that were three times stronger; however, the range of initial strength values were significantly more varied depending on the particular metal adherend employed.

Table 205 Relative Aluminum Joint Strength Loss Caused by Different Environments

Mild flight service environment	<	Semitropical natural weather environment	<	Ground-air-ground cycle	<	Condensing humidity at elevated temperature	<	Severe flight-service weather	<	5% salt-spray exposure

From Ref. 833.

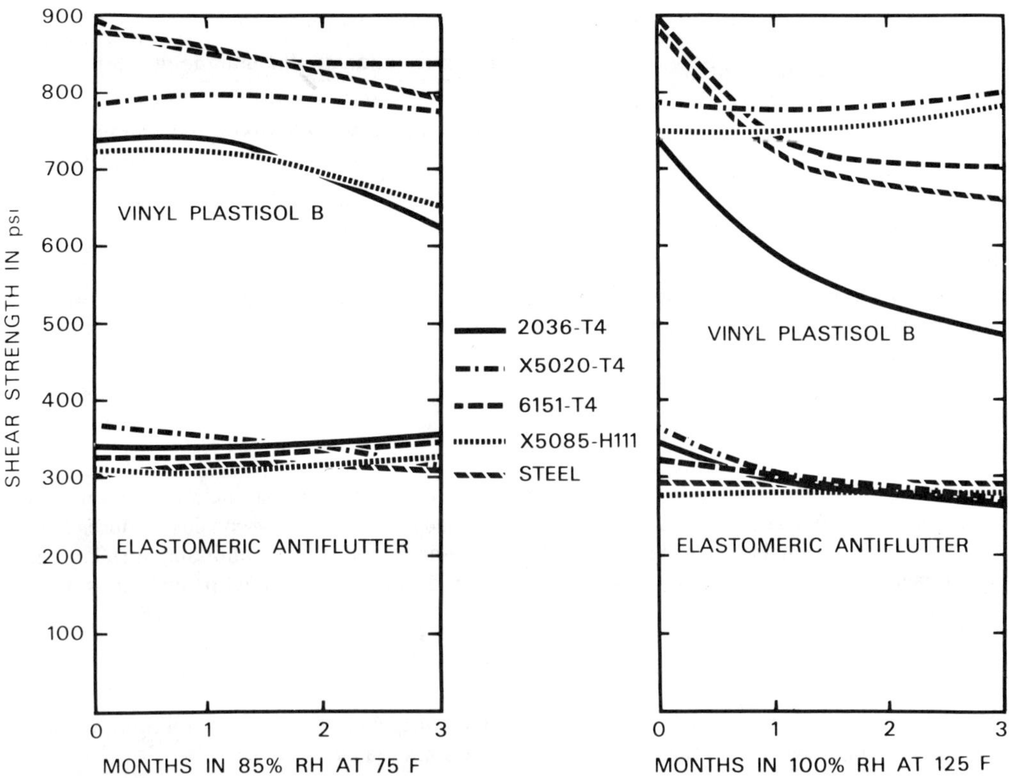

Figure 59 Performance of low-strength adhesives on various alloys in high-humidity exposures. (From Ref. 68.)

In contrast, the joint-strength values using the elastomeric antiflutter were almost identical regardless of which adherend was chosen. Also, it is clear from Figure 59 that the sensitivity of the bond made between different adherends and the vinyl-plastisol adhesive was much more varied when exposed to a deteriorating weather environment like 85% RH at room temperature or 100% RH at 125°F (52°C).

Additional investigations were made using a different vinyl-plastisol, designated A, and durability evaluations were extended to the very aggressive weathering of intermittent exposure to 5% salt spray for 500 hr. With the same mill-finish surfaces used in Figure 59, the data from all surfaces were averaged for each of the three different antiflutter adhesives, as shown in Table 206. It can be seen that the joints fabricated with the different vinyl-plastisol A adhesive averaged 26% lower than those made with the previously tested vinyl-plastisol B and 71% higher strength than joints made with the elastomeric C adhesive. All conditions of joint showed a retention of 90% + when exposed to 85% RH conditions, but only the joints made with vinyl-plastisol A remained in that range when weathered in 100% RH at 125°F (52°C). However, this may be somewhat misleading when considering the significantly higher initial average strength of the vinyl-plastisol joints. Even with a 13% overall lower joint-strength retention, the vinyl-plastisol B joints remained 18% higher in actual joint strength retention after the 3-mo exposure to hot condensing humidity. In the corrosive salt-spray exposure, the overall average percentage of joint-strength retention ranged only from 52 to 69%, indicating much less sensitivity than found with the more structural epoxy A and slightly more structural vinyl-plastisol used and reported in Table 207 using only the 2036-T4 aluminum adherend.

Table 206 Comparative Durability Performance on Aluminum Adherends of Antiflutter-Type Adhesives Used in Automotive Hood, Deck, and General Sealing Applications

Adhesive	Average initial joint strength		Overall strength retention of aluminum joints (%)[a]			
	psi	MPa	85% RH 75°F (24°C) (3 mo)	100% RH 125°F (52°C) (3 mo)	5% salt int. spray (3 wk)	Overall average
Elastomeric C	352	2.43	97	85	69	84
Vinyl-plastisol A	602	4.15	98	96	61	85
Vinyl-plastisol B	810	5.58	90	84	52	76

[a]Data includes averages obtained with joints prepared with mill-finish 2036, X5020, 6151, and X5085 aluminum auto body–type alloys, and carbon steel.
Data from Ref. 68.

However, it can be seen from Table 207 that substitution of an epoxy E adhesive offered equivalent high percentages of joint-strength retention to those found with the lower strength antiflutter adhesives used in Table 206. The overall conclusion that must be reached by these considerations is that a very wide range of joint performance exists when bonding to mill-finish aluminum surfaces with different chemical classes of adhesives and even between different commercial formulations within the same chemical class.

Table 207 Effect of Oil Contaminant on the Strength and Durability of 2036-T4 Aluminum Automotive Alloy Joints Involving Several Automotive-Type Adhesive Candidates

Adhesive	Lube	Surface oil conc.	Initial strength		Retained joint strength (%)		
			psi	MPa	85% RH 75°F (24°C) (3 mo)	100% RH 125°F (52°C) (3 mo)	5% salt spray 95°F (35°C) (3 wk)
Epoxy A	None	Vapor degreased[a]	2500	17.20	61	44	5
	#1	Neat[b]	950	6.55	33	20	0
	#1	10%[c]	2470	17.02	60	45	9
	#2	Neat[d]	2070	14.26	43	0	0
	#2	10%[e]	1600	11.02	63	17	27
Epoxy E	None	Vapor degreased	3250	22.39	71	76	41
	#1	10%	1940	13.37	58	54	53
	#2	10%	1870	12.88	63	57	56
Vinyl plastisol	None	Vapor degreased	930	6.41	99	96	23
	#1	10%	750	5.17	100	84	39
	#2	10%	790	5.44	92	91	22

[a]Mill-finish surface was vapor degreased before bonding.
[b]Lube #1 neat condition = 3.0 mg/in².
[c]Lube #1 10% condition = 0.34 mg/in².
[d]Lube #2 neat condition = 2.24 mg/in².
[e]Lube #2 10% condition = 0.20 mg/in².
Data from Ref. 68.

Because use of heat-curing vinyl-plastisols had existed for some time in certain automotive applications where the surfaces may have lubricant contamination, Minford added this class of adhesives to the heat-curing epoxy types included in his earlier 1974 (68) and 1975 (72) durability evaluations using aluminum autobody sheet. In addition to looking at relative durability performance with mill-finish, vapor-degreased-only, and chromate-phosphate conversion coated conditions, Minford also investigated the effect of preapplying lube oils in the neat (full-strength) and diluted to 10% conditions using two different commercial metal lubricant products and three different adhesives, as shown in Table 207. The specific variations in initial joint-strength and durability responses caused by these variable surface preparations and adhesive-bonding candidates can be readily seen. With an adhesive labeled epoxy A (an unmodified one-part, heat-curing epoxy), the presence of lube at the neat concentration decreased initial joint strength by 62%, whereas applying a 10% dilution of the same lube enabled 99% of the strength obtained with a vapor-degreased adherend to still be developed. With this particular lube designated as #1, the relative joint durabilities in subsequent 85% RH at 75°F (24°C) or 100% RH at 125°F (52°C) exposure conditions for 3 mo were diminished according to the degree of surface contamination present; i.e., neat > 10% > no lube.

When the same tests were repeated with a different commercial lube product (designated #2), different results were achieved, as shown in Table 207. The neat lube concentration now only produced a 17% decrease in initial strength but 100% strength loss in 3-mo exposure to 100% RH at 125°F (52°C). When diluted to only 10%, the initial strength average was unaccountably lower than present with the neat concentration of lube; however, the joint-strength retension average in hot humidity was still 39% of that achieved when no lube was present on control joints. The durability responses in a 5% salt spray at 95°F (35°C) (ASTM Std B-117) were difficult to interpret, since the control joint themselves only retained 5% of initial strength. It did seem logical to expect then that the joints made with a neat lube concentration should completely delaminate. Unexpectedly, the joints made with 10% of lube #2 on the surface retained 27%, yet initially showed lowest initial joint strength of any of the variously lubed surface joints. It might be interpreted that a modest amount of lube could offer some protection against pitting corrosion, i.e., offer mild anticorrosion film protection, at least as compared with the mill-finish condition. Later investigations by Minford evaluating the bonding of second-generation-type acrylic adhesives on deliberately prelubed aluminum surfaces also seemed to offer some confirmation of this idea.

The same 2036-T4 adherend and lube products were employed in Table 207 with a different heat-curing epoxy product that had been suggested by another formulator for possible automotive structural bonding. Both lubes applied at 10% dilution showed about a 40% decline in initial joint strength. The better durability of joints made with this different epoxy candidate with water soaking as compared to the epoxy A joints just discussed is very evident. The average joint strength loss with either lube in 100% RH at 125°F (52°C) proved to be only 28% lower than the retained joint strength of the nonlubed control joints in the same exposure. In fact, the joints with 10% lube actually performed better than their controls, allowing for the fact that controls were 40% higher in initial strength level. Even more impressive was the fact that both types of lubed joints with epoxy E outperformed the controls in exposure to the aggressive salt-spray exposure. Once again, indicating that the presence of some lube in the bondline could actually serve to promote some inhibition of the pitting corrosion which always leads to final bond failure in such an exposure.

Finally, we might anticipate the good joint resistances shown by some heat-cured vinyl-plastisols in auto manufacturing might predict even better comparative performance from the use of such an adhesive as compared with epoxies A or E. Such a vinyl-plastisol did show only a 19% lower initial strength, and there was little difference from the vapor-degreased control joints in the hot 100% RH exposure. In the corrosive salt-spray testing, the vinyl-plastisol joints over lube #1 proved to be superior to the control surface joints and with lube #2 surfaces about the same performance. These data were published in a report issued by the Aluminum Association in 1975 (3787). It should be pointed out that these earlier-formulation epoxy adhesives were the best available at the time, but they have now been replaced by a whole generation of epoxies that have much superior ability to wet through such lubed layers on the aluminum surface.

Minford (3788) conducted a series of durability tests designed to determine the relative effect on contaminated surface joints of water exposure at different temperatures. The exposure temperatures were 250°F (121°C), 125°F (52°C), and 75°F (24°C). For exposure at 250°F (121°C), it was necessary to place the test joints in a heavy-walled bomb with some deionized water and keep in a 250°F (121°C) air oven. This joint was not only being exposed to the effect of water at 250°F (121°C) but also to the pressure generated by water at a temperature above boiling. This is the same sort of test environment that has been used for many years by wood bonders to determine the relative durability of structural wood joints. The 125°F (52°C) water exposure was conducted by suspending aluminum adherend joints in a specially designed aluminum cabinet designed to create continuously condensing conditions by constantly forcing new heated water into the cabinet to assure such condensation. It must be remembered that this is the same kind of cabinet used earlier to determine chemical incompatibility between aluminum and wood when a commercially purchased 100% humidity cabinet failed to generate sufficiently soaking conditions to generate the desired corrosion situation. Finally, the 75°F (24°C) test environment consisted of simple immersion in a water bath at ambient laboratory conditions.

The investigation was set up so a number of different variables were being simultaneously tested. There were three different surface conditions for bonding: (1) mill-finish condition, (2) 0.51 mg/cm² preapplied DC-4251 lube over mill-finish, and (3) 0.82 mg/cm² DC-4251 lube over mill-finish. Two different aluminum adherends were employed (2036-T4 and 6009-T4) representative of 2000 and 6000 series automotive sheet characteristics. Two different Cybond heat-curing epoxies were used which were specially formulated to wet and adhere in the presence of lubricant. Three different sets of adhesive-curing conditions were also employed covering the range from possible undercuring to possible overcuring. The latter condition could actually be present in manufacturing as a summation of all the possible paint-curing temperatures that might be used on an automobile involving primer, finish pigmented-paint coats, and a possible clear final topcoat.

Because of the greater severity of the 250°F (121°C) water exposure, the exposure time was limited to 500 hr, whereas the other two water-exposure times were both 6 mo. In Table 208, the data have been selected to consider the effect of the three levels of surface lube contamination. The average initial strength reflects the overall performance involving the two different aluminum adherends and the use of Cybond 4533 and 4543 adhesives and three different sets of adhesive-curing conditions. The average joint strength with a mill-finish surface for bonding of 1223 psi (8.43 MPa) thus reflects an average for all such joints with the other variables present. As expected, the joint-strength average

Table 208 Effect of Surface Contamination on the Durability of Aluminum Joints Immersed in Water at Varying Temperatures Using Heat-Curing One-Part Epoxy Adhesives Modified to Bond in the Presence of Surface Lubricants

Surface contamination	Average initial joint strength		Average retained joint strength in water at the designated temperatures[a]					
			250°F (121°C)		125°F (52°C)		75°F (24°C)	
	psi	MPa	psi	MPa	psi	MPa	psi	MPa
Mill-finish	1223	8.43	1150	7.92 (94%)[b]	1464	10.29 (139%)	1221	8.41 (107%)
0.51 mg/cm² (DC-4241 lube)	1039	7.16	977	6.73 (94%)	1304	8.40 (124%)	1227	8.45 (109%)
0.82 mg/cm² (DC-4251 lube)	769	5.30	869	5.99 (113%)	973	6.70 (135%)	1013	6.98 (129%)

[a]Exposure times were 500 hr in 250°F (121°C) water, and 180 days in 125°F (52°C) or 75°F (24°C) water.
[b]Percent of average initial joint strength. Pooled data used to calculate average value included separate investigations using two different adhesives (Cybond 4533 and 4543), two different aluminum alloys (2036-T4 and 6009-T4), and three different sets of bond curing conditions.
Data from Ref. 3788.

was highest with no contaminant present. It would not necessarily be expected to find the average joints strengths with significant oil contamination present to be showing such a close average strength to the mill-finish joints after soaking in an autoclave at 250°F (121°C). In fact, the joints with most contaminate on the bonding surface actually gained strength on average after 500 hr in pressurized-water exposure at 250°F. Still it was clear from the data in the other water temperatures that they were less severe after 180 da than the 500 hr at 250°F. It was expected from extensive earlier Alcoa data involving various water-exposure conditions that 125°F (52°C) humidity soaking might be more degrading than soaking at room temperature, but this is not supported by the data in Table 208.

In Table 209, the data are presented to make a judgment about possible effect of different adhesive-curing schedules. With possible undercuring at 60 min at 250°F (121°C), the average joint strengths were only about 50% as high as for the two elevated temperature–curing conditions of 70 min at 400°F (204°C) and 120 min at 350°F (177°C). It proved to be unexpected to find that the effect of the 500 hr of exposure of the apparently undercured joints to 250°F (121°C) water only served to further extend the degree of curing and elevate the average joint strength by 56%. The effect of the 125°F (52°C) humidity with the undercured joints surprisingly produced the highest-strength joints found under any exposure conditions, with the RT-water–soaked joints showing 35% lower strength. Any satisfactory explanation for this joint strength elevation in RT-water–soak conditions is not readily apparent. All these events only serve to illustrate the extreme complexity of the possible interactions that can take place at the interface when specially formulated adhesives are interacting with lube and various weathering environments.

In Table 210, Minford has attempted to choose data designed to judge the relative effect of using different aluminum alloy surfaces; i.e., 2036-T4 and 6009-T4 alloys. In the 250°F (121°C) pressurized-water exposure, there does not seem to be any significant difference between the joint strengths retained regardless of which adhesive was employed. The overall average of joint-strength retention seems to be higher with the lower temperature water exposures, as might be anticipated. However, there was one set of 6009-T4 joints in room temperature water with Cybond 4543 adhesive which had the overall lowest average retained joint strength. Probably the most important overall conclusion that can be drawn from all these tests is that the resistance to water (at temperatures from 250°F down to 75°F) is quite amazing for joints fabricated under what would be generally considered as totally unacceptable condition just a few years earlier.

Table 209 Effect of Heat-Curing Conditions on Aluminum Joint Durability in Water Immersion at Various Temperatures with One-Part, Heat-Curing Epoxy Adhesives Modified to Bond with Lubrication Contamination on the Adherend

Curing condition	Average initial joint strength		Average retained joint strength in water[a] at the indicated temperatures: 250°F (121°C)		125°F (52°C)		75°F (24°C)	
	psi	MPa	psi	MPa	psi	MPa	psi	MPa
60 min at 250°F (121°C)	671	4.62	1046	7.21	1532	10.56	997	6.87
70 min at 400°F (204°C)	1148	7.91	907	6.25	986	6.79	1124	7.74
120 min at 350°F (177°C)	1233	8.50	1023	7.05	1232	8.49	1285	8.85

[a]The data are based on average values including use of two different adhesives (Cybond 4533, Cybond 4543); two aluminum alloys (2036-T4 and 6009-T4); and three levels of surface contamination (mill-finish, mill-finish + 0.51 mg/cm² lube, mill-finish + 0.82 mg/cm² lube). The water-exposure times were 500 hr in 250°F (121°C) water or 180 da in either 125°F (52°C) or 75°F (24°C) water.

Data from Ref. 3788.

Table 210 Effect of Alloy and Adhesive on Aluminum Joint Durability in Water Immersion at Various Temperatures Using One-Part, Heat-Curing Epoxies Modified to Bond in the Presence of Lubricant Contamination on the Adherend Surface

		Average retained joint strength in water[a] at the indicated temperatures					
		250°F (121°C)		125°F (52°C)		75°F (24°C)	
Alloy	Adhesive	psi	MPa	psi	MPa	psi	MPa
6009-T4	Cybond 4533	1055	7.27	1216	8.38	1093	7.53
2036-T4	Cybond 4533	926	6.38	1484	10.22	1395	9.61
6009-T4	Cybond 4543	1015	6.99	1154	7.95	825	5.68
2036-T4	Cybond 4543	998	6.88	1134	7.81	1264	8.71
	Average	998	6.88	1247	8.59	1144	7.88

[a]The averages include data using three different curing conditions: 60 min at 250°F (121°C), 70 min at 400°F (204°C), 120 min at 350°F (177°C); three levels of surface contamination using DC-4251 lubricant: mill-finish sheet, mill-finish + 0.51 mg/cm² lube, mill-finish + 0.82 mg/cm² lube. The water exposure times were 500 hr at 250°F (121°C) or 180 da in 125°F (52°C) or 75°F (24°C) water-soaking conditions.
Data from Ref. 3788.

In 1982, Minford (3789) published data on the ability of Cybond 4533 (heat-curing epoxy formulated to bond in the presence of surface lube) to bond in the presence of increasing concentrations of an automotive metal lubricant on aluminum body sheet alloy 2036-T4. In Table 211, it can be seen that effective high-strength bonding (equal to that obtained on the mill-finish surface) could be found in the presence of lube contamination levels up to 0.82 mg/cm². A dramatic dropoff of 46% was noted when that contamination level was elevated only 13% to 0.93 mg/cm². At that same time, a sharp rise in percentage of adhesive-type bond failure also occurred. In Table 212, the joint durability averages using duplicate joints are shown after soaking in 100% RH at 125°F (52°C), which is an accelerated weathering. The author's previously published procedure of returning to exposure all

Table 211 Initial Strength of 2036-T4 Adhesive Bonded Lap-Joints[a] with Variable Forming Lubricant on the Adherend Surface

Lube conc. mg/cm²	Range of joint shear strength (psi)	Average initial shear strength		Observations on joint failure	
		psi	MPa	Adhesive (%)	Cohesive (%)
None	1600–1700	1675	11.54		100
0.11	1600–1850	1690	11.64		100
0.28	1700–1900	1800	12.40		100
0.35	1500–1650	1590	10.96	10	90
0.51	1400–1750	1575	10.85	15	85
0.62	1480–1500	1495	10.30	10	90[b]
0.71	1300–1900	1475	10.16	15	85
0.82	1550–1700	1620	11.16	5	95
0.93	800–900	875	6.03	40	60
1.05	800–1050	925	6.37	90	10

[a]All joints fabricated with Cybond 4533 one-part structural epoxy adhesive cured 2 hr at 350°F (177°C).
[b]Thin residual adhesive visually remaining on one surface.
Data from Ref. 3789 (Table 3, p. 215).

Table 212 2036-T4 Aluminum Joint Durability Performance with Varying Surface Lubricant Concentrations in 100% RH at 125°F (52°C)

Lube conc. mg/cm²	Average initial shear strength psi	MPa	% Retained joint[a] strength after indicated days of exposure[a] 90	120	180	210	240	280
None	1675	11.54	<50	<50	<50	<50	<50	105[b]
0.11	1690	11.64	<50	<50	<50	<50	<50	111[b]
0.28	1800	12.40	<50	<50	<50	<50	<50	105[b]
0.35	1590	10.96	<50	<50	<50	<50	<50	108[b]
0.51	1575	10.85	<50	<50	<50	<50	<50	102[b]
0.62	1495	10.30	<50	<50	<50	<50	<50	109[b]
0.71	1475	10.16	<50	<50	<50	<50	<50	46[b]
0.82	1620	11.16	31[c]					
0.82	1620	11.16	<50	<50	<50	<50	<50	86[b]
0.93	875	6.03	46[c]					
0.93	875	6.03	<50	<50	<50	<50	<50	96[b]
1.05	1020	7.03	22[c]					
1.05	1020	7.03	<50	<50	47[c]			

[a]All joints fabricated with Cybond 4533 one-part structural epoxy adhesive (specially formulated to bond in presence of surface lubricants) cured 2 hr at 350°F (177°C).
[b]Average of duplicate joints which survived 280 da.
[c]Retained strength of single joint of duplicate test pair.
Data from Ref. 3789 (Table 10, p. 224).

Table 213 2036-T4 Aluminum Joint Durability Performance with Varying Surface Lubricant in Salt-Fog Cycle

Lube conc. mg/cm²	Average initial shear strength psi	MPa	% Retained joint strength after indicated days of exposure[a] 30	60	90	120	150	180
None	1675	11.54	>50	>50	2.4			
None			>50	>50	1.8			
0.11	1690	11.65	>50	>50	>50	>25	>25	36[b]
0.11			>50	>50	31			
0.28	1800	12.40	>50	>50	>25	>25	>25	19
0.28			>50	29				
0.35	1590	10.96	>50	>50	>50	>50	>50	0
0.35			>50	>50	>50	>50	>50	0
0.51	1575	10.85	>50	>50	>50	46		
0.51			>50	>50	>50	>50	>50	0
0.62	1495	10.30	>50	>50	>50	1.3		
0.62			>50	>50	>50	>25	>25	11
0.71	1475	10.16	>50	>50	>50	>25	>25	20
0.71			>50	>50	>50	>50	5.4	
0.82	1620	11.16	>50	>50	>50	>50	>50	23
0.82			>50	>50	>50	>50	>50	17
0.93	875	6.03	>50	>50	>50	>50	26	
0.93			>50	>50	>50	>50	>50	15
1.05	925	7.03	>50	49[c]				
1.05			>50	47[c]				

[a]Weathering consisted of daily exposure of the joints in a 3.5% sodium chloride salt-fog cabinet for 16 hr followed by drying-out under ambient lab conditions for 8 hr.
[b]Joint was deliberately failed to establish level of strength for comparison purposes.
[c]Only 10% of bond area shows corrosion undercutting as compared to 100% for control nonlubed exposed joints after only 90 da.
Data from Ref. 3789 (Table 11, p. 225).

joints which could be successfully stressed to 50% of initial average strength was employed. Those joints which passed such screening after 90, 120, 180, 210, and 240 da were deliberately failed after 280 da to obtain real average joint-strength retention values. Quite surprisingly, the joints fabricated with surface lube levels up to 0.62 mg/cm² were still sustaining equal bond strength to those made with mill-finish surfaces. The first significant dropoff of joint-strength retention seemed to occur just above this level at a contamination level of 0.71 mg/cm². At any higher lube level on the surface, the joint durability response seemed unpredictable.

The most aggressive weathering condition used was daily cycling in a salt-fog cabinet. As noted in Table 213, the response was not always predictable, since variable inhibiting of pitting due to the presence of varying surface lube seemed to occur. While no precise level of lube contamination could be associated with a predictable good performance, some unusual conclusions can be speculated about. First, the shortest bond survival time responses occurred either with the mill-finish or max lube contamination. Second, the variation in durability performance of all the lube contamination levels from 0.11 to 0.93 mg/cm² were essentially unpredictable. For example, the lowest-level contamination joints might produce the highest level of joint-strength retention of 36% after 180 da in one joint and failure within 90 da at 31% in a duplicate. Or one of the highly contaminated joints could last 180 da with 15% joint-strength retention and a duplicate retain 26% after 150 da. Although the subtlety of the interactions going on to produce such variable durability cannot be enunciated, it is clear that several different mechanisms of behavior involving lube as a corrosion inhibitor, wettability by the adhesive in the presence of lube, and the relative capacity of the adhesive to assimilate or displace lube at the adherend surface must all be operating simultaneously.

XVII. PERMANENCE OF BONDMENTS INVOLVING ALUMINUM CAST ALLOY ADHERENDS

The amount of published data on durability testing of aluminum casting joints has been quite sparse as compared with bonding sheet and plate. There is no question that the "as cast" surface is quite different in some of the chemical and physical properties it possesses, as are the base properties of the bulk casting. The differences in the chemical composition of cast alloys as compared with sheet and plate alloys can obviously account for some chemical differences, whereas the intimate contact in a cooling mode against a die surface can obviously produce a different surface profile for contacting the adhesive. (More information on the various properties of castings can be found in Ref. 3704.) Basically it is some form of aluminum oxide that should not offer enough difference, as compared to rolled sheet, that the same chemical families of adhesives that bond effectively to sheet will probably perform similarly with cast alloys.

Minford (3790) conducted only some limited investigations of aluminum castings at Alcoa Labs when the subject was of interest for a special market. A full-scale investigation of the permanence of cast alloy joints fabricated as lap-shear specimens was never really undertaken.

Four investigations from outside Alcoa have come to the authors's attention; all conducted in the time period from 1965 to 1968. Kron (3791) conducted a 1965 investigation, and his data are reviewed in Table 214. It is noted that EPON 915 adhesive used by this investigator is a one-part, heat-curing epoxy paste, which has been reported for use on aluminum sheet and plate alloys as well. Because this kind of adhesive has the opportunity to go through a low viscosity, highly flowable state enroute to curing, it should be able to flow readily into the very irregular profile surface of the typical casting. The good macroscopic mechanical interlocking that can result should benefit both higher joint strength and bond durability. In spite of this imagined advantage, water-immersion conditions on untreated cast surfaces so bonded became disbonded after 150 da. In contrast, the pretreating of a casting with an Alodine 1200 conversion coating offered 89% strength retention after soaking 400 da. The anodize pretreatment of a cast surface was less effective in extending durability than the commercial Alodine 1200 pretreatment. Another interesting comparison is the relative durability performance of untreated casting joints with Alclad aluminum sheet alloy joints bonded with the same adhesive and immersed in motor oil. While there was no strength deterioration when bonding to Alclad sheet, the casting alloy joints lost 47% of initial strength within 150 da. It should be pointed

Table 214 Environmental Testing of Aluminum Castings Bonded with Shell Epon 915 Adhesive

Adherend surface	Exposure condition	Days of exposure	% Retained joint strength after indicated days of exposure
Original die casting	Water	40, 100, 150, 300, 400	20, 10, 0
Alodine 1200 treatment	Water	40, 100, 150, 300, 400	94, 93, 92, 91, 89
Anodize treatment	Water	40, 100, 150, 300, 400	93, 88, 82, 60, 48
Hinac II treatment	Water	40, 100, 150, 250	101, 100, 98, 95
Original die casting	Motor oil	40, 100, 150	92, 80, 63
Alclad aluminum	Motor oil	40, 100, 150	98, 100, 101
Original die casting	Ethylene glycol	40, 100, 150	60, 55, 50
Original die casting	10% NaCl	40, 100	20, 0 at 65 da

Data interpreted from graph in Ref. 3791.

out that in the case of the Alclad surface, the alloy is much less complicated with various alloying constituents. Finally, a high sensitivity of the casting joint interface to corrosion undercutting in 10% salt-water exposure led to complete debonding in 10% salt-water exposure within 65 da. Although Kron did not provide data on the permanence of conversion coated or anodized cast surface joints in salt water, the author would speculate that a significant improvement in joint durability should be expected based on his extensive bonding studies with sheet and plate surfaces.

Work was performed in 1965 by Anderson (3792) at Picatinny Arsenal which demonstrated a modified epoxy adhesive (heat-curing, one-part, 3M's EC-2214) was capable of bonding die-cast aluminum to steel with sufficient strength to resist 4000 psi in tension, which happened to be demanded by a particular application. A comparison was made in 1965 at McDonnell Aircraft by Wieland (3793) of chemical versus mechanical cleaning of aluminum sand castings which found no significant difference for adhering an epoxy protective finish. While not a direct test of an adhesive in a joint, this information could still have application for bonding organic epoxy structural adhesives to sand cast aluminum surfaces.

Manor (3794) developed special joint-strength data resulting from four methods of joining thin-walled aluminum tube to aluminum or zinc die castings. The joining systems compared included welding, threading, shrink fitting, and adhesive bonding. The final application product was a high-reliability sonobuoy. Final results showed the adhesive-bonded assembly joints exceeded the strength of welded or threaded joint accompanied by an overall cost savings. However, the criticalness of proper metal cleaning was clearly established, and an appropriate chemical or mechanical cleaning procedure had to be developed.

XVIII. BOND PERMANENCE OF ALUMINUM-TO-PLASTIC ADHEREND JOINTS

The joining of a metal to a nonmetal adherend must require equal consideration of the particular properties of both types of surfaces and their interactions with adhesive organic polymers. It may be considered an advantage that the plastic surfaces are organic in nature, like the adhesives, and may be closely related chemically; and often are members of the same chemical families. Also, the plastic adherend materials, like the organic polymers in various types of adhesive products, will usually fall into a classification of being referred to as thermoset or thermoplastic. On the other hand, plastics have quite different surface properties than the typical metal oxides. They can be very different in coefficient of thermal expansion, their heat conductivity properties, and tend to have lower tensile strengths and inherent stiffness in equal thicknesses. All of these latter properties of plastics can be highly influential in the response of dissimilar material joints consisting of a plastic and a metal joint adherend. Also of considerable importance when dealing with the long-term durability potential of dissimilar joints, is the lack of any sensitivity of plastic adherends to corrosion pitting in the sense that it takes place with most metals. However, we need to remember that the use of metals in bimetallic joints offers outstanding resistivity to chemical degradation in many chemical and strong solvent exposures and good continued service potential and resistance to general deterioration under many service temperature conditions that are realistically present in many manufactured product areas.

A. Low-Density Plastic Materials of Construction

Considerable attention has been directed earlier in this book to bonding or laminating of aluminum foil or sheet faces to low-density plastic materials like polystyrene beadboard or extruded boardstocks, polyurethane foamed boardstocks or directly foamed-in-place polyurethane, or various board forms of polyethylene foam, polyvinyl foam, epoxy foam, phenolic foam, and isocyanurate foams. In most of these situations (except for polyurethane and isocyanurate foaming-in-place operations), there is the requirement of using a separate adhesive to join the aluminum sheet product surface to the plastic core material. Regardless of which combination of aluminum facing-to-plastic adherend is considered for bonding, there is a need to consider the specific separate chemical and physical properties of each adherend.

There is also a need to evaluate the durability potential of the combination in a series of exposure tests which include all possible factors operating in the service of the manufactured product. Hundreds of different combinations of aluminum adherend, foam plastic, commercial building slab and sheet products, and different solvent and mastic, contact-type adhesives were evaluated at the Alcoa Labs by the author over a period of more than 20 years. These data were used in the development of the Alcoa Alply Building Panel, which had the potential to have a very large number of different compositions depending on the individual desire of different architects and all those who sit on boards which make final decisions about what buildings should look like or contain within their wall construction.

B. Bonding Aluminum Surfaces to Engineering-type Plastic Adherends

The increasing use of combinations of metals and nonmetals in composite-type structures has been evident in recent years in aerospace, automotive, boat, and general manufacturing industries. The nonmetallic portion may be a relatively homogeneous material such as an organic polymer or an inorganic material in an organic polymer matrix. A study of the literature disclosed that there are relatively few detailed reports on long-term durability of aluminum-to-engineering–grade plastic joints. While more advanced concepts involving the bonding of these materials to aluminum adherends are undoubtedly being developed in the aerospace and automotive industries, the specific end results are not always shared as was the PABST program performed for the Air Force. After consulting available handbooks on bonding like those of Skeist (3795), Cagle (1263), and Cordura Publications (1259,1260), a manufacturer could still be very uncertain of just what adhesive to select or how to manufacture a structure with specific combinations of plastics and metals. The actual joint durability performance of a particular recommended adhesive candidate by a formulator for a specific dissimilar combination would likely be unavailable. Even where actual initial joint strengths may be known, the relative longevity under the particular type of weathering that may be most likely prevalent in actual field service conditions will likely be unavailable.

For example, the 1979 Cordura Publications' books (1259,1260) only attempt to organize systematically literally thousands of individual adherend/adhesive combination data points for ready reference. The lack of generally available information about dissimilar joints of the aluminum/engineering plastic variety is shown by the fact that only two references to bonding fiber glass to itself with a cyanoacrylate, five for bonding fiber glass with acrylics, and eight for bonding polyester fiber glass with epoxies was found included. Only one reference appears on the initial strength of an aluminum-to-fiber glass sheet with a structural adhesive. Although it may not be available, more general information about bonding metals to plastics appeared in a 3M's publication titled *Adhesive Answer Book for Product Assembly* issued in 1976. The 3M's Company has been particularly active in developing special adhesives for bonding various plastics over the years, which is why the author, as the editor of Volume 7 of the *Treatise on Adhesion and Adhesives* series, secured the efforts of 3M's laboratory to compose Chapter 3 on the 1991 state of the art on bonding all the commercial varieties of plastic materials of construction.

1. BONDING ALUMINUM TO POLYESTER FIBER GLASS

Polyester fiber glass is itself a composite-type material consisting of a blend of polyester resin and various fillers with inorganic glass fibers. Many commercial forms and compositions exist in the marketplace. We can be reasonably sure that adhesive bonding and bond durability tests conducted with any single product form will not necessarily be exactly representative of all possible forms of the material listed under the generic term. For this reason, Minford (3651) employed several forms of the commercial materials to determine general bondibility. Three types of electrically grade sheet were selected which initially bonded to 3003-H14 aluminum sheet with similar lap-shear strengths of about 1000 psi (6.89 MPa). Differences were definitely noted later between the water resistance and atmospheric weathering resistance which would be reflected in the overall long-term durability and potential service life. While it is possible to combine aluminum and plastic sheeting by drilling holes and mechanically bolting or screwing, the tendency of such structures to propagate cracking failures at the pierced locations makes adhesive bonding quite desirable. At the same time, all of the typical high-temperature joining methods that can be used to join aluminum and other metals are ruled out because of the thermal sensitivity of the organic polyester matrix material. The low tensile strength

Table 215 Initial Joint Strengths of Aluminum-to-Polyester Fiber Glass–Bonded Adherends

Joint tyupe	Adherends[a]		Preparation[b]		Adhesive ident[c]	Shear strength		Type[d] failure
	No. 1	No. 2	No. 1	No. 2		psi	MPa	
1	3003	GPO-2	MW	AB 320	Fuller	1300	8.96	P
2	3003	GPO-1	AB 120	AB 320	Fuller	1260	8.68	P
3	3003	GPO-3	AB 120	AB 320	Goodrich	1200	8.27	AA
4	3003	GPO-3	MW	AB 320	Fuller	1060	7.30	AA, AP
5	GPO-2	GPO-2	AW	AB 320	Goodrich	1050	7.23	P
6	3003	GPO-2	AB 80	AB 240	3M's	1050	7.23	AA
7	2036	GPO-3	VD	AW	Hughson	1050	7.23	P
8	3003	GPO-1	AB 120	AB 320	Goodrich	1040	7.17	AA
9	GPO-1	GPO-1	AW	AB 320	Goodrich	1020	7.03	P
10	3003	GPO-2	AB 120	AB 320	Goodrich	1010	6.97	P
11	GPO-3	GPO-3	AW	AB 320	Goodrich	1010	6.96	AP
12	3003	GPO-3	N	AW	Fuller	1000	6.89	P
13	3003	GPO-1	MW	AW	Morton	840	5.79	AA
14	3003	GPO-1	N	N	Fuller	820	5.65	AP
15	3003	GPO-1	N	AB 320	Fuller	800	5.51	AP
16	5182	GPO-3	VD	AB 320	Hughson	740	5.10	P
17	3003	GPO-1	MW	AW	Armstrong	700	4.82	AA
18	3003	GPO-1	VD	AW	G.E.	180	1.24	AA
19	3003	GPO-1	AB 120	AB 320	G.E.	160	1.10	AA

[a]Alloys (2036-T4, 3003-H14, 5182-O); fiber glass (GPO-1, GPO-2, GPO-3).
[b]AB (abrade) (Aloxites 80, 120, 240, 320); AW—alcohol wipe; MW—ketone wipe; N—no surface preparation; VD—vapor degreasing.
[c]Two-part epoxies (Goodrich, A1236B; Morton, S8003; Armstrong, A-31, 3M's EC-2216); modified acrylic (Fuller, RA-0018; Hughson, B3327-6); silicone (G.E., 1200 construction sealant).
[d]Failure code: AA—adhesive to aluminum; AP—adhesive to plastic; P—failure in plastic.
Data from Ref. 3651 (Table I, p. 1146).

of polyester, even with the fiber glass reinforcement (as compared to aluminum sheet) makes the initial failure of unexposed dissimilar joints in a tensile test always fail in the plastic. Surface treatments for plastics like fiber glass are usually quite simple compared to the complicated and costly pretreatments for aluminum bonded in aerospace applications. However, it is recognized that some good pretreatments for preparing low-energy plastic surfaces for bonding also can be expensive.

With polyester fiber glass, it is not usually necessary to use any preparation beyond (1) solvent wipe to remove chemical release agents, (2) mechanically abrade by sanding, and (3) solvent wipe again to remove abrading debris. It should be mentioned that the automotive industry has engaged in bonding steel-to-polyester fiber glass for the Corvette sports car for a number of years using a two-part, room temperature–curing polyurethane adhesive. For this purpose, both adherends after cleaning are primed with a two-part epoxy primer which is air dried and then bonded with the mixed polyurethane adhesive. Aluminum body sheet would be similarly processed for similar automotive applications.

Minford (3651) conducted durability testing of aluminum-to-polyester fiber glass of several varieties, as mentioned above, using simple surface preparations that might be affordable in general manufacturing. In Table 215, we note a fairly wide range of initial bond strengths and sites of joint failure noted when dissimilar lap-joints were pulled in a tensile tester. However, abrading either or both adherend surfaces before bonding did not necessariuly guarentee highest initial joint strength. In Table 216, duplicate joints were exposed to either RT-water immersion or soaking in 100% RH at

Table 216 Aluminum-to-Polyester Fiber Glass Joint Durability Tests in RT Water or 100% RH at 125°F (52°C)

Joint	RT Water Immersion[b]				100% RH at 125°F (52°C)[b]			
type[a]	A	B	C[c]	D[d]	A	B	C	D
1	66	720	48	AA,AP	46	720	33	P
2	60	570	34	P	40	720	29	P
3	50	30	1.5	AA	49	720	35	P
4	82	720	59	P	24	540	13	P
5	55	720	40	P	51	375	19	P
6	78	720	56	P	46	720	33	P
7	40	600	24	AA	38	270	10	P
8	29	420	12	AA	36	720	26	P
9	44	720	32	AP	49	180	9	P
10	8	600	5	AA	54	720	39	P
11	70	720	50	P	46	720	33	P
12	100	720	72	P	50	720	36	P
13	—	—	—	—	48	270	13	P
14	46	720	33	P	44	90	4	AP
15	79	420	33	P	72	720	53	P
16	—	—	—	—	86	360	31	P
17	—	—	—	—	44	90	4	AP
18	—	—	—	—	22	30	0.7	AP
19	—	—	—	—	44	90	4	AP

[a]See description of joints and their fabrication parameters in Table 215.
[b]Code for column heads: A—% strength retention; B—days of exposure; C—durability factor; D—failure site.

[c]Durability Factor $= \dfrac{\text{\% Joint Strength Retention} \times \text{Days in Exposure}}{1000}$

[d]Code for failure sites: AA—adhesive to aluminum; AP—adhesive to plastic; P—failure in plastic.
Data from Ref. 3651 (Table II, p. 1148).

125°F (52°C) for periods up to 2 yr. The length of exposure time was actually governed by the time required to fail the adhesive at less than 50% of initial strength value. A durability factor was calculated which consisted of multiplying the percent of joint-strength retention times days in exposure divided by 1000. A higher durability number average for the water soak at RT verified the expectation that it should be a less severe weathering than hot humidity soaking. Two exceptions to that conclusion can be noted and no ready explanation can be offered. An important overall observation can be made that the hot humidity soak is more degrading to the plastic-to-plastic joints as shown by these joints greater tendency to fail at lower than initial strength levels in the plastic cross section, indicating fiber glass thermal/water degradation.

In Table 217, similar joints were exposed to two conditions of weathering generally deemed to be more aggressive to the interface. Certainly many durability results with aluminum-to-aluminum joints previously discussed have demonstrated the hot-water soak/freeze/hot-thaw cycling and intermittent exposure to 3.5% salt water are more aggressive than either water soaking alone. From Table 217, it is somewhat unexpected then to observe that the joints exposed to wet/freeze/thaw conditions are not degraded any more than in the hot humidity soak in Table 216. However, the wet/dry saltwater spray exposure is quite degrading, because the modest surface preparations used in this investigation are unable to produce a corrosion-resistant aluminum interface. It can be seen that where plastic-to-plastic bonds were made much longer joint survival times were possible, since pitting corrosion was not a factor in establishing the joint-survival time.

Table 217 Aluminum-to-Polyester Fiber Glass Joint Durability Tests in Hot-Soak/Freeze/Hot-Air Dry Cycle or 3.5% Intermittent Salt-Fog Cycle

Joint	Soak/Freeze/Thaw Cycle[b]				3.5% Salt-Fog Cycle[b]			
type[a]	A	B	C[c]	D[d]	A	B	C	D
1	43	720	31	AA,P	14	260	5	AA (60%) (1)
2	50	720	40	P	17	180	3	AA (100%)
3	42	720	30	P	33	30	1	AA (80%)
4	46	720	33	AA,P	17	360	6	AA (50%)
5	46	720	33	P	68	540	37	P
6	57	720	41	P	1	180	0.2	AA (100%)
7	47	720	34	P	33	60	2	AA (20%)
8	39	720	28	P	0	30	0	AA (100%)
9	36	720	26	P	48	540	26	P
10	40	720	29	P	—	—	—	—
12	65	720	47	AA,P	48	540	26	AA,P
13	60	720	43	P	24	90	2	AA (10%)
14	33	180	6	AP	51	90	5	AA
15	57	720	41	P	39	180	7	AA (40%)
16	63	420	38	P	30	270	8	AA (20%)
17	60	720	43	P	32	30	1	AA (40%)
18	33	60	2	AP	76	540	41	AP
19	56	180	10	AP	85	540	46	AP

[a]Refer to Table 215 for description of all the joint parameters for each of the joint types under this column.
[b]Code for column headings: A—% strength retention; B—days of exposure; C—durability factor; D—failure site.

[c]Durability Factor $= \dfrac{\text{\% Joint Strength Retention} \times \text{Days in Exposure}}{1000}$

[d]Failure site code: AA—adhesive to aluminum; AP—adhesive to plastic; P—failure in plastic.
Data from Ref. 3651 (Table III, p. 1152).

2. BONDING ALUMINUM TO ENGINEERING-GRADE AND CROSS-LINKED STYRENE

Relatively little information appeared to be available in the literature when this question needed to be answered at the Alcoa Labs in the early 1980s. Skeist (3795) provided a list of possible adhesives for bonding styrene which included unsaturated polyester-styrene cyanoacrylates, polyurethanes, and epoxies. Cagles (1263) added the possibility of bonding styrene with solvent cements, acrylics, epoxies, and elastomeric-base compounds. The dearth of information is emphasized by the fact that the two-volume data bank set of Cordura Publications (1259,1260) in 1978–1979 listed lap-shear data on over 4000 combinations of adhesive/adherend combinations without any data on bonding styrene to any metallic adherends. 3M's Research Labs offered a publication titled *Adhesive Answer Book* which was a source of suggestings for bonding styrene adherends. From these various sources, Minford choose a variety of adhesive candidates for bonding aluminum-to-styrene, including representatives of the solvent cement, acrylic, epoxy, silicone, and polyurethane families of commercial adhesive products which were used to evaluate the long-term durability of aluminum-to-styrene dissimilar material joints. The results were published in 1983 (3652) and are reviewed below.

Styrene adherends were pretreated by washing with detergent solution, wiping with methyl alcohol, or abrading with Aloxite 320 fine grit paper followed by a methyl alcohol wipe. The aluminum alloy employed was 3003-H14, since it was highly likely that this kind of common commercial aluminum

Table 218 Initial Joint Strengths of Aluminum-to-Styrene Plastic Adherends Bonded with a Variety of Adhesives

Joint no.	Adherends[a] No. 1	Adherends[a] No. 2	Surface[b] preparation No. 1	Surface[b] preparation No. 2	Adhesive ident[c]	Average shear strength psi	Average shear strength MPa	Type[d] failure
1	CLS	3003	AB 320	AB 120	Goodrich	480	3.31	P
2	CLS	CLS	AB 320	AW	Goodrich	400	2.76	P
3	S	3003	AB 320	VD	Goodrich	340	2.34	P
4	S	3003	AW	E	Goodrich	340	2.34	P
5	S	3003	AW	AB 120	Goodrich	340	2.34	P
6	S	3003	AW	VD	Fuller	340	2.34	P
7	S	3003	AW	VD	3M (epoxy)	330	2.27	AA, AP
8	S	3003	AB 320	VD	Fuller	330	2.27	P
9	S	S	DW	DW	Cadillac[e]	320	2.20	P
10	S	S	DW	DW	Cadillac[e]	316	2.18	P
11	S	S	DW	DW	Cadillac[e]	310	2.14	P
12	S	3003	AW	AB 120	Weld-On	310	2.14	C
13	S	S	AB 320	AB 320	Weld-On	300	2.07	P
14	S	3003	AB 320	AB 80	3M (urethane)	240	1.65	C
15	S	3003	AW	AB 120	G.E.	20	0.14	AP

[a]Adherends are 3003-H14 aluminum; S—high impact polystyrene; CLS—cross-linked styrene sheet.

[7b]Code for surface preparation: AB—abrasion with Aloxite nos. 80, 120, 320; AW—alcohol wipe; DW—detergent wash; E—HCl etch at R.T.; VD—vapor degrease.

[c]Adhesives included: Goodrich A1236B 2-part epoxy; Fuller RA-0018 reactive acrylic; 3M's XD-3591 fast-cure epoxy; 3M's EC-3549 two-part polyurethane; Ind. Polychemical Weld-On 1359 heavy-bodied solvent cement; Cadillac Plastics SC-2508 light-bodied solvent cement; G.E. 1200 silicone adhesive/sealant.

[d]Failure code: AA—adhesive-type failure to aluminum; AP-adhesive-type failure to plastic; P—failure in plastic adherend.

[e]Solvent dope cement adhesive applied to both adherends and assembled like a contact cement would be employed.

Data from Ref. 3652 (Table I, p. 1166).

alloy might be used for bonding to styrene adherends rather than 6061-T6 or aircraft-type alloys. No deoxidizing pretreatment needed to be employed, since this alloy does not have the heavy heat-treat oxide film present on the higher-strength alloys. Therefore, similar surface pretreating was employed on the aluminum consisting of either vapor degreasing, alcohol wiping, or abrading with Aloxite 120 grit papers followed by alcohol wiping.

In Table 218, the initial joint strengths were significantly lower than recorded earlier for the aluminum-polyester fiber glass joints recorded in Table 215. This was essentially due to the lower tensile strength of the styrene sheet material as compared to the polyester fiber glass adherends, since the lap-shear failures were mostly cohesive in the plastic adherend in either case. The strengths noted for the aluminum/cross-linked styrene joints were slightly higher because of the higher tensile strength of that adherend as compared to the engineering-grade regular styrene. The exceptions to finding plastic adherend cross-sectional failures can be noted in joint types 7, 12, 14, and 15, where the adhesion to the aluminum or styrene was marginal or the cohesive strength of the adhesive was the weakest link in the overall jointure.

In Table 219, the joint durabilities were compared when immersed in room temperature water or soaked in 100 RH at 125°F (52°C) for times approaching 2 yr. A calculated durability factor was employed to make the comparisons which included consideration of how much initial joint strength had been retained and the length of exposure time. On the basis of these comparative factor values, it would be concluded that the overall performances were quite similar. The several poor-durability–type joints shown appeared to result from either poor adhesion or low resistance to hot humidity. The relative joint durabilities in corrosive intermittent cycling in 3.5% sodium salt fog can be reviewed in Table 220. Since the 3003-H14 aluminum adherends did not receive any of the special surface pretreatments which can promote effective resistance to corrosive salt mist, the main basis for

Table 219 Aluminum-to-Styrene Joint Durability Tests in RT Water or 100% RH At 125°F (52°C)

Joint	RT water immersion[b]				100% RH at 125°F (52°C)[b]			
type[a]	A	B	C[c]	D	A	B	C	D
1	100	720	72	P	100	630	63	P
2	100	720	72	P	—	—	—	—
3	29	210	6	AA	89	630	56	P
4	97	720	70	AP,P[d]	100	630	63	P
5	90	720	65	AA	100	630	63	P
6	—	—	—	—	100	630	63	P
7	16	30	5	AA	2	30	0.1	AA
8	—	—	—	—	100	630	63	P
9	100	720	72	P	71	630	45	P
11	98	720	71	P	82	630	52	P
12	73	720	53	AA	39	270	10	C
13	100	720	72	P	80	630	50	P
14	100	720	72	C	23	90	2	C
15	[e]	30	3	AP	25	30	0.8	AP

[a]See description of different types of joints in Table 132.
[b]Code for column headings: A—% joint strength retention; B—days in exposure; C—calculated durability factor using results in A and B; D—failure site of joints.

$$^{c}\text{Durability Factor} = \frac{\%\ \text{Joint Strength} \times \text{Days in Exposure}}{1000}$$

[d]Duplicate joints had different failure sites.
[e]Structural joint strength did not develop to the styrene adherend surface with the silicone construction adhesive.
Data from Ref. 3652 (Table II, p. 1168).

Table 220 Aluminum-to-Styrene Adherend Joint Durability Tests in 3.5% Intermittent Salt-Fog Cycle

Joint type[a]	% Strength retention	Days of exposure	Calc. durability factor[b]	Failure site[c]	% Surface corrosion
1	2	30	0.06	AA	95
2	80	680	54	P	NA[d]
3	18	30	0.5	AA	55
4	90	680	61	P	0
5	18	30	0.5	AA	90
6	57	680	39	AA, P[e]	90,0
7	1	30	0.3	AA	90
8	50	450	22	AA	30
9	90	680	61	P	NA[d]
12	5	270	1.4	AA	95
13	100	680	68	P	NA[d]
14	14	90	1.3	AA	95
15	[f]	60	6.0	AP	0

[a]See description of joints in Table 132.

[b]Durability Factor $= \dfrac{\text{\% Strength Retention} \times \text{Days in Exposure}}{1000}$

[c]Code for failure site: AA—adhesive to aluminum; AP—adhesive to styrene; P—failure in styrene adherend.

[d]Not applicable, since both adherends are styrene.

[e]Duplicate joints had different failure sites.

[f]Structural strength did not develop between silicone construction adhesive and styrene adherend surface.

Data from Ref. 3652 (Table III, p. 1171).

failure of dissimilar styrene/aluminum joints was destruction of the aluminum interface bonding. It should not be unexpected then that the four types of joints showing the highest durability factors (nos. 2, 4, 9, and 13) all were styrene-to-styrene adherend joints.

3. BONDING ALUMINUM TO RIGID POLYVINYL CHLORIDE

A number of industrial applications have arisen over the years in which the composite structure has contained aluminum-to-rigid polyvinyl chloride (PVC) joining. Mechanical jointures have often sufficed where rigid PVC corrugated sheets have been employed as roofing material over an aluminum framework porch enclosure. Obviously, it is necessary to seal any holes punched through the rigid PVC to keep out water. For this purpose, a variety of commercial sealants have been employed. There has also been wide experience in bonding PVC to itself in the form of extruded pipe for conveying everything from rain water to quite corrosive wastes from industrial plants.

There was, however, relatively little published information on the direct bonding of aluminum and PVC, especially in regard to the strength of overlap joints and their durability in various weathering conditions. While the practice of bonding the PVC with so-called vinyl dope–type cements was well established, the use of such cements for joining metals to PVC was relatively undocumented. Minford (3796) has investigated the bondability and relative joint durability of a variety of adhesives for bonding aluminum-to-PVC sheet lap-joints using a vinyl dope cement, a reactive acrylic structural adhesive, a synthetic resin contact cement, and a two-part room temperature–curing epoxy paste adhesive. Data have been summarized for exposure to room temperature water, 100% RH at 125°F (52°C), intermittent salt-fog spray, the natural industrial atmosphere, and the natural seacoast atmosphere in Table 221. The procedure employed was to subject the exposed joints to periodic stressing

Table 221 Comparative Durability of Vinyl-to-Vinyl and Aluminum-to-Vinyl Joints in a Variety of Weathering Conditions

			Average retained lap-shear strength (Avg. no. days in weathering exposure[d])				
Type[a]	Adhesive[b]	Surface[c] treatment	A	B	C	D	E
V/V	Weld-On 1001	1	>34% (510)	35% (120)	>50% (450)	>50% (360)	>50% (360)
A/V	Weld-On 1001	2	7% (420)	38% (225)	18% (60)	1.6% (360)	0% (180)
A/V	RA-0018	2	>50% (510)	>50% (450)	12% (135)[e]	>50% (480)	>50% (360)
A/V	EC-4475	2	>50% (510)	44% (180)	22% (270)[e]	>50% (360)	>50% (360)
A/V	EC-2216	3	22% (420)	>50% (450)	12% (180)	>50% (360)	0% (360)

[a]A—0.32-in (0.81-mm) 3003-H14 aluminum adherends; B—0.062-in (1.57-mm) rigid PVC adherends.
[b]Weld-On 1001 is a water-white, clear solvent cement for bonding plastics from Ind. Polychemicals; RA-0018 is a reactive acrylic adhesive from H.B. Fuller utilizing a chemical accelerator separately applied to each adherend; EC-4475 is a clear synthetic resin contact cement from 3M's recommended for bonding plastics; and EC-2216 is a flexibilized two-part epoxy paste from 3M's.
[c]Code for surface pretreatment: 1—vinyl adherends both wiped with alcohol; 2—aluminum adherends are vapor degreased only and vinyl adherends are alcohol wiped only; 3—aluminum adherends are abraded with Aloxite 120 grit paper and vinyl adherends are alcohol-wiped only.
[d]Code for weathering: A—immersion in room temperature water; B—soak in 100% RH at 125°F (52°C); C—intermittent salt spray; D—industrial atmosphere; and E—seacoast atmosphere.
[e]Joint failures were 100% adhesive to the aluminum adherend due to pitting corrosion of the aluminum surface.
Data summarized from Ref. 3796.

to 50% of the measured initial joint strength potentials until the joints failed below this value or the length of exposure time was considered long enough to be considered in the range of good-durability performance. The data indicate that the vinyl solvent dope cement does not form as generally durable dissimilar material joints between PVC and aluminum adherends as can be affected with a second-generation acrylic, synthetic resin–contact cement or an epoxy. In the corrosive intermittent salt-fog or seacoast atmosphere there was little difference, since the corrosion of the aluminum was the predominant factor establishing joint-strength retention and joint-survival time.

For the aluminum-to-PVC bonding situations, the reactive acrylic RA-0018 and EC-4475 clear synthetic resin contact adhesive both created relatively durable joints in three of the five weathering conditions (corrosive salt-fog and seacoast atmospheric weathering being the exceptions). Meanwhile, the two-part epoxy joints failed to perform as well as might be desired in three of the five environments even with abraded surface preparation. Again, as with the styrene adherends in the previous section, the aluminum interface was the site of preferential attack in dissimilar material joints rather than the PVC interface. If a prepainted, adhesive-primed, conversion coated, anodized, or room temperature deoxidized by HCl or chromic acid solution aluminum adherend had been used to form the dissimilar material joints, then the performance of the reactive acrylic, syntheric resin-contact, or two-part epoxy adhesives would likely have been acceptable in all the weathering conditions.

4. *BONDING ALUMINUM TO POLYCARBONATE ADHERENDS*

Information from the literature about joining polycarbonate plastic in the late 1970s only indicated plastic solvent cements, cyanoacrylates, polyurethane, and epoxy adhesive as potential candidates. No durability data seemed to be available about bonding to metals.

Minford (3797) undertook a durability testing program in 1978 designed to give some insight into effective aluminum/polycarbonate bonding. Since cyanoacrylate adhesives generate rather brittle aluminum joints with questionable durability in water, this class of adhesives was not included in the survey. Those tested included (1) two different plastic dope cements recommended for bonding polycarbonate; (2) four different epoxies, including both polyamide-curing and mercaptan/extra-fast-curing epoxy; (3) two free radical-curing structural acrylics; (4) two two-part polyurethanes; and (5) RTV silicone construction adhesive. The durability evaluations included exposure in (1) room temperature-water immersion, (2) 100% RH at 125°F (52°C), (3) hot-water soak/freeze/hot-air dry cycling (3-da cycle), (4) industrial atmosphere, and (5) seacoast atmosphere.

It became clear that a special technique would be needed to adequately bond with the dope-type solvent cements. The method eventually used consisted of preapplication of the cement with air drying to serve as a primed surface. After application of a second coat of cement to the aluminum, the joint was closed against the polycarbonate surface and set aside jigged until strength developed unber ambient temperature condition. Even after 7-da curing time, the lap-shear strength of such dissimilar material joints had only risen to 40 psi (0.28 MPa). Amazingly, such joints could then be placed in any water-soaking exposure conditions during which exposure the joint strengths continued to rise. After 2-yr immersion in RT water, the average joint strength rose to 520 psi (3.58 MPa). In the 100% RH at 125°F (52°C) soak, the joint strength rose to a similar level then began to decline, reaching 270 psi (1.86 MPa) after 2 yr. The limiting factor governing the joint strength in such dissimilar material joints was always the degree of adhesion to the aluminum surface, since the polycarbonate-to-polycarbonate joints similarly bonded developed joints in excess of 1000 psi (6.89 MPa), with most of that strength maintained after 2-yr water expose.

The silicone construction mastic meanwhile bonded modestly to the aluminum but failed to develop any real adhesion to the polycarbonate. The two room temperature-curing polyurethane adhesives bonded the dissimilar lap-joints with about the same strength as the epoxy adhesive candidates but lost a significant portion of that strength within the first 30–90 da in hot humidity exposure. Mercaptan rapid-cure epoxy joints were actually 44% stronger than the corresponding joints made with the two-part, polyamide-curing epoxy candidates. They subsequently degraded faster to failure than any other composite-type joints regardless of adhesive used. The reactive acrylic adhesive candidates produced higher initial strength joints than any of the other adhesives, and they subsequently also showed the best overall durability response.

There appeared to be a general relationship between level of durability response and severeness of exposure conditions that can be seen in the following data involving the acrylic-bonded joints. After 720-da exposure in RT water, the average retained strength was 880 psi (6.06 MPa), and after 180 da in hot condensing humidity, the strength average was reduced to 440 psi (3.03 MPa), whereas after 60 da in the aggressive soak/freeze/thaw cycling, the decline was to 475 psi (3.73 MPa). When using the plastic dope cement adhesives, the durability of all the polycarbonate-to-polycarbonate joints were superior in all exposures to any of the dissimilar material joints. All joints performed relatively durably in the natural industrial atmospheric weathering; however, the corrosive seacoast exposure was highly discriminating among the composite joints owing to the aggressive penetration of corrosion across the aluminum interfaces. It must be mentioned, however, that no attempt was made to provide the aluminum adherends in these tests with any special surface pretreatment able to offer good protection against the corrosive seacoast conditions. In addition to room temperature deoxidizing in HCl or chromic acid conversion coating, anodizing, or prepainting, the use of a corrosion inhibiting primer, as used in aerospace bonding, would greatly enhance the durability potential in the seacoast exposure.

5. *BONDING ALUMINUM TO BUTYRATE PLASTIC ADHERENDS*

Because of the use of transparent butyrate sheet in architectural and transport markets for viewing and light transmission, there should be some need for developing a joining technology between aluminum and butyrate surfaces. Data on this subject are very scarce. For this reason, Minford (3798) attempted to develop some structural joint data for making aluminum-to-butyrate lap-joints. The 1977 edition of Skeist's publication mentions only cellulose nitrate solvent cement, polyurethane, or cyanoacrylate adhesives as being recommended. The *Modern Plastics Encyclopedia* mentions the largest end use for solvent-cast butyrate as being a laminate to the two sides of 0.35-mil (0.0089-mm) foil. The adhesive employed was not mentioned. It is likely that the butyrate might be cast against a very clean aluminum foil. Also mentioned as an aluminum foil/butyrate laminate was a tough moisture-proof packaging material for packing coffee, penicillin, or effervesant salts. In regard to structural joining to aluminum sheet, there seems to be no information about initial strengths or durability potentials.

The variety of adhesives chosen for testing included (1) two water-clear plastic dope solvent cements suggested for bonding Plexiglas to other acrylate surfaces, (2) three commercial reactive acrylics with nonmix accelerators, (3) mercaptan-cured, 2-part epoxy, and (4) silicone construction mastic. As in the testing of other plastics with aluminum just discussed, the weathering exposure conditions were room temperature water, hot condensing humidity, soak/freeze/thaw cycling, and natural and seacoast atmospheres.

While forming strong and durable joints with butyrate surfaces, the general-purpose monomer-base solvent cements were not effective bonders to an aluminum surface. Both room temperature-curing acrylics and epoxy adhesives formed quite strong initial joints with aluminum. The strength in composite joints with butyrates was always determined by the cohesive failure strength of the butyrate sheet adherend. As with other plastics, the silicone construction mastic failed to develop any significant strength to the butyrate. There was a variable positive effect of abrading the butyrate and solvent wiping before bonding; i.e., often, but not always, higher initial strength results over no abrade. Again, mercaptan-cured two-part epoxy developed strong initial joints which failed in as short a time as 30 da in water soaking. Although it took almost 2 yr, the mercaptan-cured epoxy composite joints even failed in the relatively mild industrial atmospheric exposure. Each of three structural reactive acrylic adhesives formed equally strong initial bonds but wide variability in their individual resistance to the various weathering exposures. It is quite likely that some of this variability was due to variable degradation of the butyrate by the strong solvent action of the adhesive. Those joints so affected always failed via a cracking failure in the butyrate adherend when tensile tested. The crack was sometimes located adjacent to the actual bondline but could also initiate and proceed directly across the actual joint area. Overall, the durability of one of the three acrylics was distinctly superior to all the adhesives evaluated with joints surviving 2 yr in all three accelerated water–contact wathering exposures. It was necessary, however, to abrade and solvent wipe the butyrate surfaces before bonding. When tested after exposure, the joint failure was always to the abraded butyrate rather than the vapor-degreased–only aluminum adherend.

A final observation relating to long-term durability was concerned with the probable ultraviolet degradation of the bond to butyrate even in the usually mild industrial atmosphere. This observation was not made in a seacoast atmosphere, probably because of the relatively short time to joint failure. The limiting factor governing the survival time for these dissimilar materials joints was always the level of resistance of the aluminum interface to undercutting corrosion. Since no special corrosion-resistant pretreatments were used in fabricating these joints, the joints failed relatively early owing to undercutting corrosion.

XIX. BOND PERMANENCE OF ALUMINUM-TO-WOOD JOINTS

Because of voluminous use of aluminum/wood laminates in a wide variety of manufactured products (especially architectural), thousands of durability evaluations of these kinds of joints were conducted for more than 20 yr at Alcoa Labs by Minford and co-workers. During this time, it was the purpose of most evaluations to help provide both initial strength, long-term durability, and chemical compatibility data for Alcoa customers and adhesive formulators. The data were also of considerable value in developing methods of laminating aluminum to the wide variety of wood-related backups used in Alcoa Alply Building Panels, and establishing a basis for projecting the overall service-life potential of such structures.

The situation is complicated by the fact that wood adherends, unlike metallic adherends, are subject to having widely varying natural chemical properties depending on the soil in which they grow. Thus, a pine wood indigenous to the Pacific Northwest can have a different compatibility and bondability response to aluminum surfaces as compared to a pine that is grown in the southern United

States. (For more information about the characteristics of wood which might influence the bonding process and bond quality, see Ref. 3799.) Some factors can lead to basic problems of relative chemical incompatibility with the aluminum surface. The possibility of finding incompatibility with the aluminum surface is often related to the chemical effects of leaches from the wood on the aluminum oxide. The same leaches or extractives can interfere with the adhesion obtained with different adhesives. For example, Mizumachi (3800) studied the effects of 18 species of wood with varying amounts and types of extractive contents on the activation energy of urea-formaldehyde–curing reaction. Extractives from five hardwoods decreased the gel time even added in small amounts. Narayanamurti et al. (3801) showed a case where insoluble nonextractives could actually cause more adhesion problems than the soluble extractives with the same adhesive-solvent system.

Alcoa Labs spent many years developing a test method for determining the relative chemical compatibility of various woods with aluminum. The general mechanism for incompatibility is to provide the right environment to extract any materials (usually acidic in nature) which will cause fracturing or thinning of the protective oxide layer, making it possible for anodic areas to generate. In the presence of electrolyte provided by extractives and water, pitting corrosion can thrive at the anodic sites. Or overall etching corrosion may occur in areas where oxide is dissipated and a ready source of oxygen for oxide repair is screened out by the contacting wood adherend.

This test consists of forming unbonded aluminum/wood/aluminum sandwiches using the particular alloy and type of wood for the particular application. The interfaces should be held together as tight as possible so that extractives are not readily flushed out of the interface. Eight sandwiches are vertically suspended in a special cabinet designed to produce continuously condensing water conditions thermostated at 125°F (52°C). After years of testing to determine the minimum time necessary to assure good compatibility, it was decided at least 6-mo soaking in the hot-wet condition was necessary to establish the safety of using the specific aluminum/wood combination. For best efficiency, duplicate sandwiches would be removed after 1 and 3 mo and disassembled and checked for any evidence of corrosion on the aluminum surface. If no corrosion was evident after 6 mo, then the wood could be declared completely compatible and acceptable for long-term bonding applications. As a further means of condensing the overall evaluation time, the author usually conducted the screening test for compatibility with aluminum and the actual bond strength and bond durability evaluations simultaneously. Thus, if the wood was termed compatible after the 6-mo test period, the first 6 mo of the 12 mo or more desired to evaluate the bonded joint durability would also have been completed. If corrosion was observed during the 6-mo soaking period, then an evaluation was already ongoing which could answer the question of how successful the adhesive might be in acting as an inhibitor against the corrosion that would occur between the direct-contacting aluminum and wood surfaces under similar weathering conditions.

The most discriminating set of environmental circumstances for establishing relative durability of either wood/wood or aluminum/wood joints involves some soaking in water. For the wood-to-wood bond, the durability investigator may often resort to subjecting the bonded joints to boiling water or a boil/dry cycling test for some prescribed time period followed by joint shear testing to determine the decline in joint strength. This procedure is often criticized as being too accelerated for wood structures that only need to survive in natural atmosphering weathering. This testing procedure does have value when trying to discriminate between the abilities of different adhesives and bonding conditions. The use of a boiling-water/dry cycling procedure for aluminum/wood joints is especially devastating for several reasons. First, the significant differences in coefficient of thermal expansion of the two adherends is going to place high demands of accommodation on the adhesive in the bondline. Second, while the aluminum structure will be basically unaffected (except for thermal expansion), the wood fibers will likely be conditioned to swell significantly, adding to the stress in the bondline. In the drying phase of any testing cycle (depending on the drying time), the fibers could reverse their swelling effect by partially shrinking back. Third, the effect of boiling water on the wood may be to generate a much higher concentration of extractibles at the bondline with a possible negative effect on the adhesive and/or the adjacent aluminum interface. While it is widely believed that adhesive layers can screen out most undesirable weathering factors around a bondline, experience teaches it is not always true. At least, it cannot automatically be counted upon to offer the desired amount of protection. It is fortunate in this latter regard that the exposure times used in most boiling/dry cycling tests are still sufficiently long that the negative chemical factors can exert their negative influence on the adhesive bondling. The relative abilities of different adhesives to offer such barrier properties is one positive factor for selecting a particular adhesive candidate.

In order to evaluate the relative durability of aluminum/wood joints, the author choose to use the much longer exposure conditions of soaking in room temperature water, in hot condensing humidity, or cycling between various soaking, freezing, and drying conditions. The ultimately selected procedure (as used for determining chemical compatibility described above) was to expose at least duplicate test specimens (either lap-shear joints or 6 × 6 in laminated panels) to varying periods of exposure such as 1, 3, 6, and 12 mo. By testing frequently with increasing exposure time, it was possible to further discriminate the relative durability performances and better predict service-life potential in real-life applications. Some interesting observations were made about relative aggressiveness of a wet/freeze/dry cycling procedure which are unique only to the aluminum/wood type of jointure. For example, the author started out with the premise that the most aggressive such test would involve using the widest possible cycling temperatures as had been demonstrated for aluminum/aluminum joints. This included soak in 165°F (74°C) water followed by freeze and dry out in a 170°F (77°C) recirculating air oven for consecutive 8-hr periods and then repeating the same cycle for periods up to 6 mo. While this procedure was generally physically quite destructive to the wood structure, it never produced any corrosive-type failure on the aluminum as found in the continuous soaking in water procedures. It was later established by the author that acceleration of joint failure from both thermal stressing and aluminum surface corrosion was produced by lowering the drying cycle temperature to room temperature conditions. This prevented total drying out of the water in the interface, making the exposure one of continuous contact of the aluminum with wet wood and its extractives. The interruption of this condition by drying out the aluminum/wood jointure with hot recirculating air, as noted above, apparently prevented the inauguration of the corrosion and pitting condition on the aluminum surface.

Abstrated from many Minford (3802) reports in Tables 222 and 223 are results on evaluating some responses of water-soaking conditions on aluminum extrusion-to-fir plywood laminates. In Table 222, the data involve the use of seven different commercial high-solids rubber mastic adhesives used to bond the 6063-T6 aluminum extrusions. The range of initial joint strength values from 63 to 95 psi (0.43 to 0.65 MPa) reflects the very wide range of initial tensile-peel strengths found with a variety of commercial mastic rubber products that had been recommended by their formulators. It should also be remembered that tensile-peel values must generally be multiplied by two to five times to obtain the corresponding lap-shear values for the same aluminum/wood-bonded interface. Thus, the numberical range in terms of lap-shear would have a much wider divergence. All the mastics, however,

Table 222 Durability Evaluation of Aluminum Extrusion to Fir Plywood Bonds in Long-Term Hot Humidity Soaking

Exposure time in 100% RH at 125°F (52°C) (mo)	Tensile-peel strength retention (psi)
0	63–95 (initial strength range for seven test groups)[a]
1	24–38
3	13–28
6	9–25

[a]Seven different commercial high-solids mastic adhesives were used to bond 6063-T6 aluminum extrusions to plywood and allowed to cure under ambient laboratory conditions for approximately 28 da before exposed to soaking. The range of initial strength values reflects the wide range of bond strengths that can be found with typical commercial products.

Data from Ref. 3802.

Table 223 Response of Aluminum Extrusion-to-Fir Plywood Laminate Bonds to Periods of High-Humidity Exposure Followed by Acclimation Periods in Ambient Laboratory Conditions

Exposure time (da)	Tensil-peel strength (psi) after indicated cycling time in high-humidity or ambient lab conditions	
	100% RH at 125°F (52°C)	50% RH at 75°F (24°C)
0	75	—
3	58	—
6	32	—
0	This period of testing starts with the above 6-da weathered specimens to determine bondline ability to recover bond strength at the lower temperature and humidity conditions	32
8		40
10		50
14		58
16		55
32		57

6063-T6 aluminum extrusions bonded to fir plywood with B. F. Goodrich high-solids rubber mastic adhesive.
Data from Ref. 3802.

did have the common property of steadily declining in tensile-peel strength over the test time of 6 mo under hot condensing humidity conditions.

The experiment was conducted in Table 223 to determine the extent of possible joint strength recovery that might be induced by simply switching from a 100% RH at 125°F (52°C) exposure to a 50% RH at ambient temperature condition. After only 6-da exposure, the tensile peel strength had declined 67% for these relatively poor-durability joints. By switching these deteriorated joints to an exposure condition of 50% RH, there was a distinct joint strength recovery of 81% after 14 da. This appeared to represent some sort of equilibrium condition, since there was no additional change when acclimated for an additional 18 da.

XX. FORMULATION OF OIL-ACCOMMODATING ADHESIVES

As mentioned earlier in Section XIII, it has become necessary in recent years to try and develop adhesives which can bond directly to metal adherends in the presence of recognized contaminates to bonding such as oils and forming lubricants. Structural adhesives have been increasingly used under such conditions in a variety of industrial manufacturing situations, especially in the manufacturing of such products as automobiles, trucks, and trailers. Two mechanisms have been proposed for formulating such oil-accommodating adhesives: (1) the adhesive must be capable of actually absorbing the oil and diffusing to the metal surface, or (2) the adhesive must be capable of displacing the oil from the surface, allowing the adhesive to adhere in its place. Of course, the most ideal oil-tolerating adhesive would be one that had been successfully formulated to perform both these functions.

The ability of certain solvent-containing adhesives, such as the elastomer and elastomer/phenolic contact cement–type adhesives, to bond effectively over unprepared or deliberately contaminated aluminum adherends is based on the oil- and grease-dissolving properties of these adhesives; i.e., by mechanism 1 above. A discussion of this subject was offered by Baumann (3803) in 1967. This same mechanism is also the one operating with the heat-cured vinyl-plastisol formulations (3804). In principle, the plasticizer that is present in such formulations can be considered to be acting like a solvent.

Of course, the grease-dissolving ability of such adhesives is further improved by the need to cure these formulations at high temperature. For the same reason, it can be shown that any heat-cured epoxy adhesive can offer a distinct advantage over the same adhesive cured at room temperature. The author has attempted many times, but never with any realistic success, to form really durable joints on lubed aluminum adherends using the typical proprietary two-part, room temperature-curing epoxy adhesive. Hot-setting thermoplastic hot-melt adhesives can often show some promise based on this better grease-dissolving ability at elevated temperature. (Some discussion on this matter can be found in Refs. 3805–3807.) However, Ruhsland (3808), in 1984, was quick to point out that such high-temperature-curing conditions alone are not sufficient alone to make the desired high-strength joints along with acceptable water weathering resistance. (For other related publications, see Refs. 3809–3815.)

Ruhsland (3808) wrote a chapter in 1984 in "Adhesive Joints—Formation, Characteristics, and Testing" which made reference to his 1977 paper with Winkler (3816) on "Adhesive Bonding of Metals Without Surface Treatment of Bonding Parts." Further reference was made to an adhesive Epasol FV/ZIS 939, which is a modified cold-setting, two-component epoxy which, because of its composition, is able to satisfactorily wet oil surfaces and to dissolve and to absorb oil films already present. More details involving the development of this adhesive are offered in Ruhsland and Winkler (3816) and Ruhsland (3817–3820). Some Ruhsland data showing the significant advantages of the Epasol FV/ZIS 939 adhesive in developing high joint strength as compared to an unmodified two-part epoxy are reviewed in Table 224. The added ability of this modified epoxy to attain equal joint durability over an oil-contaminated surface as found using degreased or even deoxidized aluminum adherends under water-soak weathering is shown in Table 225.

Ruhsland (3808) also has described (3821) the possibility of joining contaminated surface adherends using a technique referred to as "vibrational adhesive" bonding. This term signifies an adhesive bond method according to which a "hard" filler is added to the adhesive composition. Subsequent bonding is affected in combination with a mechanical relative movement of the bonding parts by subjecting them to ultrasonic energy. The energy causes cleaning and simultaneous roughening of the bonding surface in the presence of the filled adhesive. To accomplish this, it is necessary for

Table 224 Tensile Shear Strength of Epasol FV/ZIS 939[a] Adhesive Joints Fabricated with Greased Aluminum Adherends

Adhesive[b]	Degree of degreasing[c]	Tensile shear strength (N/mm^2)
Ummodified epoxide resin adhesive	06	12
	6	8
	12	7
	18	6
	24	5
Modified epoxide Epasol FV/ZIS 939	0	27
	6	25
	12	24
	18	23.5
	24	23

[a]Epasol FV/ZIS 939 is a modified cold-setting two-component adhesive based on an epoxide resin which, because of its composition (GDR Patent WP C 09j/123672), is able to satisfactorily wet oily surfaces, to absorb oil films present, and activate the joint component surfaces.

[b]Both adhesives were cured for 24 hr at 68°F (20°C) + 2 hr at 194°F (90°C).

[c]The degree of greasing is in terms of arbitrary units where 10 units = approx. grease layer of 1-μm thickness.

Data interpreted from Ref. 3808 (Fig. 1, p. 260).

Table 225 Water-Soaking Durability Testing of Aluminum Joints Fabricated with Epasol FV/ZIS 939–Modified Epoxy[a] Adhesive Over Greased Adherends

Adhesive[b]	Surface pretreatment	Degree of greasing[c]	Exposure time (mo)	Retained tensile shear strength (N/mm²)
Epasol FV/ZIS 939	Degreased and pickled	0	2	20
			4	18
			6	16
			12	16
Epasol FV/ZIS 939	Gear oil GL 125 applied	24	2	17
			4	16
			6	16
			12	16

[a]Modified cold-setting two-part modified epoxide to permit wetting of greased adherend surface.
[b]Adhesive was cured 24 hr at 68°F (20°C) + 2 hr at 194°F (90°C).
[c]The degree of greasing is in terms of units where 10 = approx. 1 μm grease layer thickness.
Data interpreted from Ref. 3808 (Fig. 4, p. 263).

Table 226 Effect of Vibrational Adhesive Bonding on Tensile Shear Strength of Greased Surface Aluminum Joints

Adhesive	Degree of greasing	Tensile shear strength (N/mm²)
1. Unmodified epoxide resin[a]	None (adherend degreased)	9.5
	None (adherend degreased and pickled)	12
	3[b]	9.5
	6	5.5
	12	4.5
	18	3.5
	24	3.0
2. Adhesive 1 + 40% filler[c]	3	16
	6	16
	12	16
	18	16
	24	16
3. Adhesive 2 + 8 silane adhesion promoter	3	19
	6	19
	12	19
	18	19
	24	19

[a]Adhesive cured 24 hr at 68°F (20°C) + 2 hr at 194°F (90°C).
[b]10 degrees of greasing = 1 μm grease layer thickness.
[c]40% of a filler added which has a hardness greater than the aluminum adherend surface. Ultrasonic treatment time for joints made with adhesives 2 and 3 was 2 sec.
Data interpreted from Ref. 3808 (Fig. 6, p. 265).

the hardness of the selected filler to be greater than that of the adherend material to be bonded and any contaminants present. Results comparing the strength of joints made with an unmodified epoxy, epoxy plus 40% of preferred filler, and filled adhesive plus a silane adhesion promoter are reviewed in Table 226.

XXI. MISCELLANEOUS TECHNICAL REPORTS

Although the most prominent theme of the technical reports appearing in *Adhesives Age Magazine* relates to the applications of adhesive bonding for manufacturing specific products, occasional articles relating to the testing and durability determinations about adhesive-bonded joints also appear. Some of these have been separated out for the reader's review by the author and are cited according to their chronological appearance in *Adhesives Age Magazine.*

In 1960, an unauthored article appeared about "Coated Valves Fight Corrosion" (3821) followed in 1961 by an article by Niconchuk (3822) on "Testing Adhesives for the Vulcanized-in-Sole Process." Four articles in 1962 reported on testing and durability by Yurenka (3823) on "How to Test Structural Adhesives," Burrows and Crowe (3824) on "Bonding Contaminated Surfaces," Smith and Susman (3825) on "Structural Variations of Polyurethanes: Their Effect on Metal Adhesion at Various Testing Temperatures," and an unauthored article on "P-S Label Withstands Acid Washings" (3826). In 1965, an article from Picatinny Arsenal by Wegman and Tanner (3827) described "Effects Upon the Strength Epoxy Adhesives When Stressed to Failure in Miliseconds," whereas Bijorksten (3828) considered "Preventing New Product Failure." In 1966, Donovon (3829) described "How to Test the Strength of Bonded Joints," and an unauthored article appeared on "Resin-Based Adhesive Helps Coat Rivets with Anti-Corrosion Liquid" (3830). Three articles in 1967 included the paper of Carlson and Sapetta (3831) on "Stresses in Assemblies Bonded with Thermosetting Adhesives," the paper by Bodnar et al. (3832) on "How Weathering and Aging Affect Bonded Aluminum," and an unauthored article titled "Epoxy Resurfacing Eliminates Corrosion" (3833). Finally, in 1969, Carter's investigation appeared entitled "Low Cost Durability Testing of Stressed Lap-Shear Adhesive Joints" (3834).

From the decade of the 1970s, a nonauthored report was issued by *Adhesives Age Magazine* titled "Developing Testing Standards for Building Joint Sealants: The Job of ASTM Committee C-24" in 1973 (3835). In 1977, Althof and Brockmann (3836) suggested a "New Test Method for the Prediction of Environmental Resistance of Adhesive Bonded Joints," and Marceau et al. (3637) described "A Wedge Test for Evaluating Adhesive-Bonded Surface Durability." Four articles followed in 1978 by Minford (3838) on "Durability of Structural Adhesive Bonded Aluminum Joints," Marceau et al. (3839) on "Cyclic Stress Testing of Adhesive Bonds," Wegman et al. (3840) on "A New Technique for Assessing Durability of Structural Adhesives," and Jackson (3841) on "Improving Adhesion by Gas-Plasma Contaminant Removal." Finally, in 1979, Bascom (3842) issued his distinctive conceptions about "Stress Corrosion of Structural Adhesive Bonds," whereas Rutherford and Hughes (3843) discussed "Creep in Adhesive-Bonded Metal Joints."

The 1980s began with an article by Baker et al. (3844) titled "Characterization of a Hot-Melt Adhesive System for Severe Environments." A 1982 article by Clark (3845) followed with the title "Modifying Pressure-Sensitive Testing for Weather Stripping." Joneja and Newaz (3846), in 1985, investigated "Evaluating SMC Bonds Using a Wedge Test." Four technical papers followed in 1987 including Rantz's (3847) disclosure titled "Proper Surface Preparation: Bonding's Critical First Step," Haviland's (3848) general-interest article on "Challanges and Opportunities in Adhesive Engineering," Lovald's paper titled "Influencing Adhesive Evaluation: Developments in Substrates," and Wightman's discussion titled "Surface Analysis Examines Fundamental Adhesion Questions" (3850).

Finally, in 1990, an article by Fedor and Brennan (3851) considered "Correlation of Accelerated and Natural Weathering of Sealants," whereas 1991 produced an *Adhesive Age* report by Silva and Spindel (3852) describing "Testing Elastomeric Sealants to Predict Performance."

11

Applications of Adhesives in Bonding Aluminum Structures

I. INTRODUCTION

In Chapter 5, Selection of an Adhesive, the full range of adhesive-type materials has been considered which might be used in some manufacturing circumstance to bond or seal to an aluminum surface. For any applications not mentioned in this section, reference should be made to Chapter 5 to decide if an appropriate adhesive is presently being marketed.

A number of general-information articles exist which can provide some pertinent information to the manufacturing engineer about to embark on an adhesive assembly project. Probably the most profound and all-encompassing recent publication of this nature has been Volume 3 of the *Engineered Materials Handbook* series from ASM (3853). The "Glossary of Terms" section by Sharpe (3854) can itself provide an excellent source for generally reviewing the terminology and materials that will be encountered in applying adhesion science to manufacturing. Hagquist et al. (3855) offer brief reviews in the handbook covering the major market areas, with mention of some important applications in each. Another important part of the handbook for the manufacturing engineer is the "Guide to General Information Sources" provided by Landrock (3853), as well as Landrock's 1985 *Adhesives Technology Handbook* (1472).

Kardashov (3856) edited a volume of basic information on bonding agents, including phenol-formaldehyde, epoxy, polyurethane, polyamide resins, silicon-organic compounds and their combinations with other polymers in 1963. The book was designed for a broad circle of engineering and technical workers in planes, design offices, and scientific research organizations. General broad subject areas included the theory of bonding and joint strength, modern synthetic adhesives, bonding procedures used in aviation engineering, and methods of testing and inspecting joints. In 1964, Merriman (3857) produced a general review article for selecting adhesives based on an extensive listing of their characteristics. Minford (3858), in 1967, produced the chapter in the four-volume ASM Series on *Aluminum* on the general joining of aluminum with adhesives. McIntyre et al. (3859) produced a series of reports on the effect of varying process parameters on adhesive-bonded aluminum structures at the Picatinny Arsenal. Report No. 4162 was specifically titled "Production Methods."

Other general-information sources not mentioned earlier, arranged chronologically, include the "Design Guide—Assembling with Adhesives" (3860) published in 1966 by *Machine Design*; "Bonding Aluminum Parts" by Hall (3861) in *Production Engineering*; "Metal-to-Metal Adhesives for Structural Applications at Elevated Temperatures" by Conway (3862) in 1968 in *Adhesives Age*; *Adhesives for Metals-Theory and Technology* by DeLollis (3863) in 1970; and Rider's "Principles and Applications of Adhesives—A Prototype Module" (3864) in *Materials Science and Engineering* (August 1978). A number of review articles dealing with application areas can be cited such as Layman's 1982 review in *Chemical & Engineering News* (3865) of the types of adhesive polymers described as the speciality adhesives. He points out that the present use of synthetic adhesives accounts for more than 85% of today's sales, with the market share of the older natural product adhesives continuing to steadily decline. Ellerhorst (3866), in 1982, described how the traditionally used solvent-borne systems will be gradually replaced by water-borne, hot-melts, nonvolatile solid or liquid formulations, two-part systems, radiation-curable adhesives, reactive-liquid adhesives, and powder adhesives. Bittence (3867), in 1983, projected a steady rise in the opportunities to employ the new "high-technology" formulations in aerospace, vehicle manufacturing, and general manufacturing. Brief and Skeist (3868), in 1983, reviewed 25 years of growth in the adhesives industry, especially the growth of the use for the classes of polyvinyl acetate, polyolefins, styrenic block copolymers, acrylics, cyanoacrylates, anaerobics, polyurethanes, and epoxies. Giudice (3869), in 1990, projected the rise in adhesive and sealant sales would reach $28.6 billion annually by 1998. This translates into a 10.6% rise per year in the decade of the 1990s.

The use of an adhesive for a specific application can be closely related to the conditions for curing the adhesive. The reader is referred to a Society of Manufacturing Engineers' technical paper by Haviland (3870), which discusses the manufacturing aspects of the six main adhesive cure systems, including the whole gamut of potential advantages and disadvantages. The extensive application of adhesives for the bonding of aluminum structures was reviewed by Minford (see Ref. 3858, Table 3, p. 572). The major industries cited in 1967 at the time of this publication include (1) aircraft and military, (2) appliances, (3) automotive, (4) building products, (5) process industries, and (6) sports equipment. It would appear that the major new addition to this listing would be in the electrical and electronic manufacturing industries.

The manufacturing engineer also needs to know what the thick and thin, thixotropic, and dilatant properties of the uncured adhesive might be, and also, the stiff, strong, brittle, flexible, and weak variations of adhesives after curing. Finally, he or she must make the decisions on storing, handling, dispersing, curing, and testing of the differently cured systems available in the marketplace.

II. ADHESIVE CURING AND MANUFACTURING SITUATIONS

A. Reactive Chemical Curing

Haviland (3870) has summarized the manufacturing aspects which arise from having six different types of adhesive-curing procedures. The cure system for structural bonding used most prominantly in manufacturing is probably that described as an interaction of reaction chemicals. These are mostly the two-part room temperature–curing epoxy, acrylic, and polyurethane-type products. Elaborate heating and controlling devices are obviously not required, so capital investment costs can be minimal. Since the adhesive can be cured in large bulk, the presence of unusually large gaps between adherend surfaces can be accommodated. The variation in heat exotherm developed during cure must, however, be recognized; and its relationship to the concept that the larger the bulk of curing adhesive, the greater the likelihood of generating self-destructive temperatures has to be considered. Tensile strengths in the 13.8- to 21.0-MPa range can be developed which are quite adequate for structural designs outside the aerospace industry. Material costs are relatively modest compared to truly aerospace-type tapes and films, and solvent resistance and aging properties will be acceptable. Minford (3871) has evaluated two-part room temperature–curing epoxy/aluminum joints after 12 years' open exposure to the industrial atmosphere and found very minor changes in the cured adhesive strength properties in the presence of a good surface-deoxidizing surface pretreatment. Similarly, he has exposed similar joints to 12 years' exposure in the tropical forest and in an open weathering area in Surinam (1355,1356) near the equator. There were variable performances depending on factors like alloy, surface treatment, and openness or seclusion of the exposure site which need to be reviewed here. Also, the inability of a good weather-resistant sealant around some joints to keep out some degrading effects of the high humidity need to be remembered. Overall, however, the two-part epoxy showed an ability to even outperform a one-part, heat-cured, nitrile-modified epoxy adhesive, which could be important to a design engineer. The author must advise the reader, however, that his long-term durability data had to be obtained using the same adhesive candidates over many years in order to obtain directly comparable data. The data, therefore, may represent only the minumum performance possible, since more durable adhesive products may have been formulated in the meantime. The data do offer a base, however, for making general comparisons with any present-day products using the same test evaluation methods.

On the limitation side for these kinds of adhesives is the need for precise metering and mixing of the resin and reactant. Since most of these adhesives are in paste form, the control of viscosity will greatly influence extruding rate and flow over aluminum surfaces. The overall manufacturing operation must be completed comfortably within the pot-life limitations, never on the ragged edge of that time. Material beyond pot-life must be discarded. It should never be added to fresh material with the anticipation that this mix will somehow still have the same chemical response as a fresh batch.

B. Heat-curing Types

Included in this group are the one-component material adhesives in the anaerobic, epoxy, acrylic, and polyurethane commercial formulations. Heating not only cures these adhesives but can significantly improve the surface-wetting potential, especially as the viscosity is lowest just prior to initiation of chemical cross-linking at peak temperature. This enhanced wetting seems particularly important where the service environment is corrosive, such as in seacoast or marine exposures. Minford (see Table 16) has shown the superior durability of a heat-cured, one-part, nitrile-modified epoxy aluminum joint at the seacoast as compared to the viscous two-part epoxy paste joints. This occurred in spite of the fact that the inherent water-soaking resistance of similar room temperature–curing epoxy joints could be shown to be superior in RT–water-soaking conditions.

Other advantages of heat-cured over RT-cured epoxies are their acceptable shelf life even with catalysts present, and higher shear joint strengths in the 21- to 34-MPa range. The excellent durability of some of these heat-cured epoxies and vinyl-phenolics for joining aircraft structure testifies to their high-performance potential in manufacturing in general. Actually, no other heat-curing adhesives have shown as outstanding long-term durability as some of the earlier nitrile-phenolic aerospace formulations. Also of importance is the precise application and bondline thickness control available with such adhesives in their tape and film forms.

It should be mentioned that attempts have been made from time-to-time to incorporate a means of electrical resistance curing of adhesives formulated in sheet form. Marshall et al. (3872), for example, worked on the development of high-temperature adhesive systems which contained intergral electrical resistance heating elements for use in fabricating metal honeycomb sandwich structures in 1967. Minford (3873) evaluated the durability under hot condensing humidity conditions of aluminum sheet sandwiches combined with a Hughson Chemical–developed structural adhesive film with incorporated electrical resistance screening. The durability of the adhesive itself left much to be desired in this particular product. Althof (38974,3875) produced two technical reports in 1973 and 1974 in an attempt to increase the overall heat resistance of bonded aluminum joints by actual combinations of two different adhesives.

The special features of the heat-curable anaerobics need to be mentioned for structures where their unique properties are applicable. They offer unlimited shelf life, 21- to 28-MPa shear strength, lower curing temperature range of 199°F (93°C) to 300°F (149°C) (as compared to one-part epoxies), and can cure at 300°F (149°C) in as short a time as 30 sec. They also are easy to disperse liquids in the uncured state with a low overall "system" cost. Their environmental resistance can still be termed good.

C. Anaerobic Curing

The adhesives originally called "anaerobics" were stable formulations only as long as oxygen in minute quantities was present. If oxygen was excluded, such as confinement between two aluminum adherend surfaces, then a free radical in the mixture could start a polymerization process. The polymerized end product is truly thermoset and insoluble in all common solvents. An obvious advantage over the heat-curing anaerobic formulations is that no energy is required to affect the curing. Since it is a single-component product, there is no mixing or pot-life problems (as in the two-part systems), and no volatile or nonreactive solvents are present (as in the solvent-dispersed cements, latex, or emulsion adhesives). Clean-up is easy, since no curing can occur outside the bondline. They function as effective sealants where porosity is present in a structural part. They can be easily applied by screening and rolling procedures or can be vacuum impregnated into a porous adherend. Their main limitations are still their high rigidity in the cured state, inability to cure in gaps over 5 mil (0.13 mm) unless specially activated, and slow curing on certain adherend surfaces without some heating.

D. Surface-Activated Curing

This class of curing is related to the anaerobic type of adhesive and it is a means of overcoming a main disadvantage of "pure" anaerobics. These materials are specially formulated to respond to steel, copper, brass, aluminum, and to special chemical activation on other adherend surfaces. Advantages are those commonly enjoyed by the anaerobics in general, such as 15-sec to 3-min fast fixture curing without heat. Cure can be obtained through thicker gaps of 30 mil (0.8 mm) as compared to the "pure" anaerobics. Some cleaning of an adherend can be achieved while surface priming to create chemical activation. A limitation is present because the activator solutions do contain solvents which need to be flashed from the surface. Finally, the humidity, heat, and solvent resistance of the final bonds are definitely inferior to those achieved with heat-curing anaerobics.

E. Ionic Curing

This method of curing primarily relates to the cyanoacrylate adhesives which can be cured by the presence of water vapor and/or an adherend surface that is chemically basic. The employment of cyanoacrylates has really been triggered by those needs in manufacturing to achieve a fast fixture in the shortest possible time (5–60 sec) with modest joint strength of about 17.2 MPa. It must be admitted that materials most commonly bonded are the plastic and elastomer-type adherends rather than metals like aluminum. This level of joint strength is considered sufficient for such adherends which are weaker materials. These adhesives have some limitations for aluminum bonding including (1) maximum temperature resistance of 180°F (82°C); (2) low resistance to solvents, moisture, and general weathering; and (3) low-impact strength. Other general unfavorable characteristics in manufacturing are an irritating odor and possible sticking together of skin parts of workers. For this reason, polyethylene gloves on handling personnel are helpful, but beware the use of cotton gloves, which can cause a fast, high-exotherm reaction.

F. Ultraviolet Light and Electron Curing

Ultraviolet curing is a relatively newer technology of the adhesives industry compared to most of the situations discussed above. It is based on the incorporation of stabilizers in fast-curing, clear liquid adhesive formulations. The triggering mechanism is a destabilization affected by 365-nm wavelength light at some appropriate milliwatts per square centimeter of intensity. The main adherends used in manufacturing have been glass and thermoset plastics. Advantages have included (1) fast cure, (2) one-component adhesive, (3) low safe triggering reaction, (4) easy cleanup of excess adhesive, (5) cures well with any gap dimension, and (6) generally safe to use adhesive with no fumes. For aluminum bonding, obviously, the other adherend must be transparent to UV light. The adhesives are generally anaerobic in nature, so the fillets will not cure in the presence of any oxygen. However, if this is desired, it can be accomplished through the use of high-intensity mercury vapor lights.

The use of energetic electrons for the initiation of free-radical curing of adhesives and coatings was pioneered as early as the late 1950s by Chapiro (3876). The commercial success has been overshadowed by the more widespread use of UV for curing inks in graphics (3877) and for metal decoration (3878). McGinnis (3879) pointed out in 1976 that the free radical–initiated chemical systems were quite similar chemically whether used with UV or electron-type curing. The increasing concern of environmentalists for more pollution controls on thermal curing of adhesive systems could further increase the use of electron curing in the future. The features of modern electron processors have been described by Nablo (3880) in 1973 and Hoffman (3881) in 1977. An Occupational Safety and Health Standard (3882) requires that radiation shielding be an integral part of any electron processor. Nablo and Tripp (3883) summarized the state of the art as of 1979. The publications of Tripp and Nablo (3884), Dowbenko et al. (3885), and Chu et al. (3886) have provided reviews for data on typical formulations that can be electron cured along with their expected properties. It must be acknowledged that the most widely used and advanced adhesives and coatings for this purpose are generally guarded as proprietary product secrets.

Bluestein and Cohen (3887) have discussed the use of visible light–curing adhesives. While these adhesive products are quite similar to the UV- and electron beam–curing products, they are of recent 1986 origin and are more expensive. Dental restorative procedures have been an immediate market along with the attachment of false fingernails in the cosmetics industry. However, interest should grow in any industry where repeated exposure of a manufacturing operator to UV is of concern. Expansion of application might apply to manufacturing of electrical and electronic component assembly, fiber optics joining, jewelry mounting, arts and crafts items, field installation of security systems, and repair work on a variety of devices which involve some bonding. While these kinds of applications may not consume large volumes of adhesives, they do require some very exacting adhesive property requirements.

G. Curing Initiation After Solvent Removal

The preponderance of all general manufacturing operations which involve adhesive bonding of aluminum adherends probably involve the use of some sort of contact-type cement. In such adhesives, the initiating factor for curing involves the loss of the last traces of solvent (water or various organic solvents) from the bondline. The advantages of such adhesives include (1) ease and speed of application; (2) ability to bond very large areas economically; (3) development of adequate strength for most nonstructural applications; (4) choice of a large variety of different chemical formulations with varying strength properties; (5) choice of a wide price range of products to fit different economic situations; (6) lack of specific pot-life problems, as long as solvent is contained in storage, or in the bondline; and (7) available formulations which can cure at room temperature but whose cure can be significantly accelerated and durability response increased by the addition of heat if available.

These adhesives can generally be applied by spraying, roll coating, curtain coating, flow coating, brushing, or knife coating. Bonding techniques include procedures of wet bonding, open-time bonding, contact bonding, and solvent reactivation.

1. WET BONDING

When wet bonding is used with aluminum, the other surface must be porous in order to permit solvent loss from the bondline for curing initiation. Aluminum bonding to wood is an example of extensive commercial use. It is really only necessary to apply the contact cement to one of the adherends; however, it can be considered more advantageous to apply the adhesive to the porous adherend. In this instance, the adhesive solids have maximum penetrative power into the porous structure when the viscosity of the adhesive is the lowest. This can also enable more solvent to be lost from the bondline before contact is made with the vapor-barrier aluminum, thus, minimizing the time to achieve handling strength in the bondline. To achieve maximum aluminum-to-plywood bond strength and durability as well, the author has often applied and dried a prime coat of contact adhesive on the wood surface followed by a separate application of adhesive to the aluminum. This will serve to eliminate the possibility of a more adhesive-starved bondline, since the porous surface of the plywood is already filled with adhesive solids. While a longer time for curing in the bondline should be expected, this can be controlled by some form of heating the bondline before mating the two adherends.

2. OPEN-TIME BONDING

In this technique, the adhesive is applied to both surfaces, and they are allowed to stand (open) until suitable tack has been achieved. The technique works best when at least one surface is porous or semiporous. The simplest testing procedure for ascertaining right tack is to touch a finger to the surface, and if the adhesive feels sticky but does not transfer to the skin, it is ready to bond. Some latex adhesives go directly from the wet to a dry state without developing a tacky range. Their use with this technique might even be negatively influenced, since more solvent could remain trapped in the bondline. The author has found that there is a need only to control and test the tacky condition, as described above, on one of the surfaces, and preferably on that surface which is the greater vapor barrier. It seems very adequate to precoat and dry the adhesive on one adherend and then concentrate on obtaining the right degree of tack in the second surface by the finger test in manual assembly operations or by some sort of automated tack tester for higher production rate manufacturing. Laying one adherend over the other and passing the laminate through a static or rotary presss operation, or both, will ultimately produce a very strong and durable bondment. Obviously, this procedure is not applicable where one of the two adherend surfaces is chemically incompatible with the solvent in the adhesive.

3. CONTACT BONDING

In this procedure, both surfaces must be coated, and the solvent is permitted to dissipate under natural air-drying or forced air-drying conditions until only a slight tackiness remains in at least one surface. Pressing together can create an almost immediate ultimate bond strength. The principle being employed is that even with only slight tackiness, the adhesive can be very responsive to wetting another similar surface. Such bonding is usually preferred for joining two nonporous adherends like aluminum to itself. The adhesive type of preference is usually a neoprene-latex like the casein-neoprene family of contact cements or a neoprene-phenolic cement, as used to laminate large aluminum-faced building panels (87,89). The drying time necessary to remove most solvent will depend on the nature of the solvent and whether auxilary heat is employed. In general, water solvent is more difficult to flash out of an adhesive coating, and drying at ambient temperature could require as long as 30 min. This time may be drastically reduced with additional heating, but discretion must still be used as to the intensity of this heating, since the adhesive can develop a surface-erupting situation where the vaporization of the water is too rapid. Organic solvents, on the other hand, can usually be expected to volatilize more rapidly under any atmospheric or heating condition. The attendant manufacturing danger when using contact bonding is that the level of residual tackiness may have inadvertently fallen below that sufficient for adequate surface-to-surface wetting. This may be evidenced by a lowered tensile or shear tensile bond strength, or it may be evidenced only by observation of a faster joint-strength decline in a weathering environment. If extra solvent is still retained within the bondline following its closing, the bond strength will be lowered from it potential maximum; however, with the later loss of the last residual solvent, the bond strength can still increase to maximum. Also, in this latter situation, the maximal durability potential may still be realized following loss of this last residual solvent with additional aging under ambient conditions.

4. SOLVENT REACTIVATION

In this operation, the adhesive is applied and permitted to dry under either ambient or heated conditions. When a joint is to be made, the adhesive is reactivated to a tacky condition by wiping or misting with solvent or by placing the coated surface on a solvent-impregnated pad. It is theoretically possible to bond such a tacky surface to another uncoated adherend surface providing a sufficient degree of wetting of the initially dry adhesive is achieved. The more surely successful joint is usually achieved by mating two correspondingly reactivated surfaces. This method has a potential for bonding aluminum-to-aluminum since most of the solvent has been removed from the adhesive, which can only lose solvent through the edge areas of the bond. This method can rarely be used for bonding large surface areas, because the tack range is very short when solvent reactivation is properly carried out. The other side of the coin is when excessive solvent may have been used; in which case, the green strength of the adhesive will be too low to adequately hold the adherends in proper registration while curing. The operator now must strive to remove some solvent to get the surface condition back to a proper tack state to develop high-bond strength so precise registration and retention of that orientation of the adherends can be maintained while handling and shipping.

The author would like to offer a word of caution about such bonding in the presence of chemical incompatibility with a second adherend to be joined to aluminum. For instance, in laminating aluminum to a polystyrene beadboard with a neoprene-phenolic contact adhesive, it only requires trapping a very minute amount of the solvent in the bondline to cause collapse or other chemical degradation of the beadboard. In the production of Alcoa Alply panels with polystyrene insulation board cores, it was always necessary to remove solvent totally before joining the aluminum facing to the solvent-sensitive insulating polystyrene beadboard. This would not be the case when a water-solvent cement was employed which had no chemically incompatible problem.

5. *HEAT REACTIVATION*

This involves the application of a thermoplastic adhesive to one or both surfaces and drying the surfaces for later bonding. The advantage of storing such coated parts for several weeks or longer is often of great manufacturing advantage. To bond, the surface is heated by heat lamps, recirculating hot air, or even in a static heated oven to produce a soft and tacky surface. The final bond is then made under pressure (any variety of laminating rolls or static press) and cooled to ambient conditions. For neoprene-phenolic contact bonding, the temperature at which the final pressing is made will strongly affect the ultimate elevated temperature stress-rupture resistance. For this reason, Alply panel laminates had to be hot pressed at a temperature of approximately 185°F to assure resistance to solar heating in service. Such reactivation has an obvious manufacturing advantage, since it can be employed in a continuous in-line manufacturing operation.

H. Adhesive Application Variables

Delmonte and Fullerton (3888) provided an early 1966 study of the influence of application variables on the properties of an epoxy adhesive. They included such factors as concentration of curing agent, application time after mixing, open assembly time, cure time at varying temperatures, and effect of adhesive thickness. Cocanour and Anderson (3889), in 1980, discussed the application of extruded thermoplastic materials in nonwoven web form as heat-bondable adhesives for a broad range of industries. Four basic polymer compositions were involved, including polyamides, polyurethanes, polyolefins, and polyesters. Such adhesives were especially well suited for bonding flexible adherends such as aluminum foil, plastic foams and films, and fabrics. Some appealing features are that no solvents are involved and no waste or a minimal amount of burnable waste is generated. These authors also discussed specific application methods such as the laminating transfer printer, two-roll calender, and two-roll laminator (especially useful when using hot-melts).

Haviland (3890) has produced a publication available from the Loctite Corp. which summarizes the state of the art of automating the use of adhesives in manufacturing. Examples of automation procedures are presented for use with epoxies, polysulfides, anaerobics, and cyanoacrylate products. The particular items of equipment are described and their procurement sources from 23 different equipment suppliers are furnished. Also, the procedures and heating methods used to cure these kinds of adhesives in place are compared as to relative costs and production efficiencies.

A recent 1990 summary of the state of the art of metering and mixing equipment for adhesive and sealants has been offered by Devlin (3891). He considers the details about (1) manual weighting and mixing, and (2) preweighed and packaged kits, and bulk meter/mix equipment. The latter subject he further subdivides into discussions about (1) material transfer pumps, (2) metering systems and mixing systems. Petterborg (3892), in 1988, discussed a so-called barrier/injection kit packaging concept and how its use can reduce costs and production time on the manufacturing line. In 1989, Drake

(3893) wrote about the impact of encapsulation compounds on the selection of dispersing equipment for electronics manufacturing processes.

De Vries (3894) has contributed a 1990 summary of the state of the art for dispensing and application equipment for adhesives and sealants. He points out that for the purpose of specifying dispensing equipment, adhesives and sealants can first be separated into the general categories of (1) heat-processed products, and (2) room temperature–applied products. Types considered under category 1 would include the hot-melts and so-called warm-melts, weld-through sealers, weldable/expandable sealers, and single-part epoxies. Category 2 would include plastisols, silicones (RTV-type), one-part epoxies, and the urethanes. This article intentionally does not include the broad category of two-part epoxies, which would obviously be room temperature applied. It may be confusing to note similar types of adhesives and sealants under each of the two above main categories. De Vries attempted to distinguish whether the equipment itself needs to be heated, since some same-type materials require heating to reduce the viscosity to a pumpable value. Other materials require heating to cut through the die lubricants on the adherend surface. Any equipment that is capable of melting or reducing material viscosity by heating will obviously be both more sophisticated and costly than the room temperature applications–type dispenser. The four main components of all dispensing systems are pumps, header systems, dispensing valves or guns, and system controls, which are discussed by De Vries in that order in his review.

Borstell and Wheeler (3895), in 1990, reviewed the factors of (1) preparation for bonding, (2) application of adhesive, and (3) tooling for manufacturing based on their experience at Grumman Aircraft Systems. They attempted to relate this subject to the fact that the manufacturing operation is most closely related to making insert bonds, doubler bonds, or panel bonds. In the first, one adherend is being bonded to another in either a cavity or a very small local area. Doubler bonds are then made when a second layer of materials is bonded locally to a larger sheet of the material in an area of high stress. Last, a panel bond is generated when two facing sheets are bonded to opposite sides of a core material. Overall, the bonding processing is considered in the individual manufacturing steps of (1) the bonding preparation (prekitting of adherend, bondline thickness control, prefit evaluation, and surface pretreating; (2) the adhesive application (techniques for liquids, pastes, or films); and (3) the tooling problem (design and employment of most appropriate fixture, pressure applicator selection, design selection, and design of appropriate adhesive curing equipment). Taricco and Moulding (3896) discussed the curing and curing control of adhesives in terms of appropriate kinds of ovens, autoclave processing systems, and computer-controlled curing systems. Turner (3897) has written a separate review article titled "Automation and Robotics for Adhesives and Sealants Use," Davies et al. (3898) have described the use of robots and adaptive control in automated dispensing of adhesives.

I. Elastomer-to-Metal Bonding in Manufacturing

Rather closely related to the earlier discussion on curing bonds made with various elastomer contact cements is the subject of bonding elastomers to metals. DeBruyne and Houwink (3899) presented one of the earliest reviews on the bonding of elastomers to metals in 1951. They particularly discussed the use of isocyanates alone or in conjunction with chlorinated elastomers for bonding vulcanized elastomers. Chlorination had been discussed earlier in a DuPont report (3900). Gerstenmaier (3901) pointed out as early as 1953 that "no universal method or all-purpose adhesive had been developed so each combination of materials, such as metal-to-elastomer, presents its own unique problem and resultant bonding procedure." In 1955, Alstadt (3902) discussed rubber-to-metal adhesion theories. Brams (3903), in 1958, estimated that organic adhesive bonding of elastomer-to-metal had already replaced 85% of the earlier bonding by a brass-plating technique. Buchan (3904), in 1959, reviewed the principal reasons for this decline in brass-plating popularity.

Painter (3905), in 1960, discussed the method of determining the adhesion of elastomer-to-metal bonds as a 90-degree peel test in ASTM D-429, Method B. The button or tensile test procedure (Method A) had also been used. A dead-load wrap-around peel test (ASTM D-413) can be used to determine adhesion or friction strength. DeCrease (3906) furnished a summary on the known effects of various rubber compounding ingredients on adhesion and suggested the use of the term *bondability index* in

1960. The most easily bondable nitrile rubber was given an index value of 10, whereas the low-polarity butyl rubber placed it at the other end of the scale at 1. He also discussed the negative effects on adhesion maintenance of certain plasticizers, paraffinic or aromatic oils, and copper-bearing adhesive-curing accelerators. Gallagher (3907), in 1961, offered a considerably detailed discussion of elastomer-to-metal bonding with a multitude of typical adhesion data as had been developed at that time.

Bonding of vulcanized elastomers to produce rubber-tearing bonds has always been more difficult than for unvulcanized elastomer/metal bonding. Some of the reasons include (1) a greater difficulty to obtain any degree of chemical cross-linking with a reacted polymer surface, (2) the vulcanized surface is relatively smooth and nonmobile, and, thus, is more difficult to chemically wet, and (3) the surface may be contaminated with mold-release agents, plasticizers, or compounding ingredients which can migrate, bleed, or bloom to the adhesion interface. Cyclizing the vulcanized surface with acid was an early method of modifying the surface to improve bondability.

Hutchison (3908) produced a comprehensive 1978 article covering elastomer and adhesive selection, metal preparation, and molding and bonding. Causes of some elastomer-to-metal bond failures was also considered. The recommended surface preparation for aluminum at the time was indicated as (1) blasting with clean, sharp 40-mesh aluminum oxide grits or sand, or (2) chromate conversion coating treatment.

The most durable elastomer/metal bonds have been formed under heat (150–300°F [66–149°C]) and pressure (200 psi, 1.38 MPa) with accompanying vulcanization of the elastomer. Such bonds have been shown to be durable even in seawater for several years by Stevenson (2826) and Leidheiser et al. (2828). Emphasis was placed on bonding rubber-to-metals like aluminum as early as 1968 by Gallagher (3909) and Spearman (3910). The former provided a review of the technology up to that date, whereas Spearman specifically evaluated the ability of commercially available rubber-to-metal bonding agents for bonding both steel and aluminum for use in military specifications, including the use of both EPDM and butyl-type rubbers. In 1990, Bond (3911) compared the earlier use of organic solvent–dispersed rubber-to-metal bonding adhesives such as Chemlok 205 primer and Chemlok 252 overcoat with newer water-borne adhesives which can eliminate emission and odors. The alternative procedure used water-dispersed Chemlok 802 primer with Chemlok 828 overcoat. Durability results were essentially equivalent, leaving the advantages of freedom from odor, harmful vapors, atmospheric pollution, and flammability hazards as the basis for selecting such water-borne systems.

An important application of elastomer/metal bonding has involved the development of underwater sonar systems. Ting (3912) has reported on the screening of a number of commercially available adhesive systems for this purpose. ASTM D-429 test methods using both conical button and peel-strips were employed. Cutts (3913) reviewed the various testing methods in 1981. A relative classification method for estimating the bulk rubber tear, rubber/adhesive, and adhesive/metal failure that had been developed by Peterson (3914) in 1964 has been employed in many of these investigations. Greatest bond strength reductions seemed to be produced by chemical attack from salt water under stressful conditions. In 1983, Peterson (3915) wrote a chapter titled "Adhesives for Bonding Rubber to Various Metals" in the SME book *High Performance Adhesive Bonding*.

J. Summary of *Adhesives Age* Articles Describing Adhesive Bonding Procedures, Equipment, and Machinery Available for Manufacturing Applications

Reed (3916), in 1961, reported on "High Frequency Curing of Adhesives." In 1962, Jones (3917) offered "Some Tips on Production Line Handling of Contact Adhesives," and an unauthored article was published on "Improving Bonding Accuracy With Thermistor Controllers" (3918). Klinetob (3919), in 1966, discussed "Proper Care of Spreader Reduces Gluing Problems" following an unauthored 1965 article titled "Methods of Bonding Curved-Form Parts" (3920). Another unauthored article of 1967 discussed "How to Prevent Solvent Hazards in Adhesive Plants" (3921) followed by Sebbard's (3922) discussion in 1968 of "Mishandling of Adhesives Reduces Machineability." Another unauthored article in the same year reported "Infrared Oven Solves Laminating Problems" (3923).

Nine pertinent *Adhesives Age* articles were found in the 1970s decade such as the 1972 unauthored article on "Application Equipment For Anaerobic Adhesives" (3924). Another unauthored article occurred in 1973 titled "Application Equipment for Adhesives and Sealants—A Survey of the Manufacturers" (3925). Spearman and Hutchison (3926), in 1974, considered "Post Vulcanzing Bonding Concepts," whereas Reighard (3927), in the same year, discussed "Hot-Melt Labeling Equipment—Present and Future." In 1977, Jennings (3928) explained "Two Techniques for Characterizing Surfaces for Adhesive Bonding." Roobol et al. (3929), in 1978, reviewed "Determining Airless Sprayability of Adhesives." Three articles in 1979 included "Electron Curing of Adhesives and Coatings" by Nablo and Tripp (3883), Hulstein's (3930) paper on "Automatic Application of Anaerobic Adhesives," and Mushel's (3931) review of "Laboratory and Pilot Coater/Laminators."

The number of papers about these matters definitely accelerated in the decade of the 1980s, with the author being able to cite 21 *Adhesives Age* articles. In 1980, an unauthored paper discussed the general review subject of "What's New in Machinery and Equipment" (3932). Herot (3933), in the same year, reviewed "Engineered Dispensing Stations for Adhesives." A further updating of "Whats New in Machinery and Equipment" was published in 1981 (1934) and again in 1982 (3935). Mosher (3936) discussed "The Rectilinear System of Robot Dispensing" in 1982, and Jacobs (3937) reported on "Adhesive Application Equipment for Two-Part Reactive Materials." Four papers in 1983 included two unauthored reports titled "An Approach to Process Control of Adhesive Bonding" (3938) and a newer "What's New in Machinery and Equipment" (3939). Dueweke (3940) rendered an overview titled "Robotics and Adhesives—An Overview" and Stoops and Ferrier (1370) a discussion of "Merging Two Technologies: Robotics and Hot-Melt Adhesives." In 1984, Morat (3941) reviewed "Industrial Bonding and Automation in France." More unauthored reviews titled "What's New in Machinery and Equipment" (3942,3943) followed in 1985 and 1986. In 1986, another *Adhesives Age* unauthored report "What's New in Machinery and Equipment" (3944) was compiled along with an article titled "Company Halves Materials Costs Using Foamed HM Technology" (3945). Still another unauthored report in 1987 considered "Dispenser Uses Less Adhesive to Produce Improved Product" (3946) (involved lipstick cases). In 1988, an unauthored report considered "Visible Light Curing Provides Low Shrinkage and Good Depth" (3947), and Cheng (3948) discussed "Using Analytical Techniques to Characterize Adhesives." A final review of "What's New in Machinery and Equipment" followed in 1989 (3949) as well as Kingsley's (3950) article on "Dispenser Eliminates Solvents and Improves Profitability."

Three 1990 articles include Elliott et al.'s "Hybrid Technology Fulfills Adhesive Application Needs" (3951) and two unauthored articles titled "Versatile Coater/Laminator Aid in Two-Ply Production" (3952) and "Automated Dispenser Increases Production and Reduces Rework" (3953).

Finally, in 1991, the most recent "What's New in Machinery and Equipment" (3954) and McGinnis and Benham's (3955) "Hand-Held Dispensing Equipment Reduces Waste and Saves Cost."

K. Summary of *Adhesive Age* Articles Describing Different Adhesive and Sealant Formulations Available for Use in Various Applications

1. *GENERAL ARTICLES ON THE ADHESIVE INDUSTRY AND ADHESIVE SELECTION*

In 1966, Carroll (3956) described a process for selecting a high-production adhesive system, whereas Wooten (3957), in 1969, discussed "Adhesives for Coupling Large Strains." In 1973, three papers considered selection of best adhesive. Gould's (3958) article was titled "Guide to Adhesive Selection with a Pictoral Checklist," and Weggeman (3959) described "Adhesive Selection Charts." An unauthored technical article discussed "Checklist for Adhesives Selection" (3960). In 1974, Downey (3961) discussed the subject of "Sprayable Adhesive Systems." Also in 1974, an unauthored article titled "Adhesives Replace Fasteners in Shipping Containers" (3962) appeared and was followed by a paper by Green (3963) with the title "Effectiveness of Cell Edge Adhesives." It was in 1979 that Mahoney (3964,3965) issued two papers titled "Structural Adhesives for Fast-Cure Applications—No's I and II." Lees (3966), in 1983, suggested a method of "Selecting Adhesives by Computer," whereas Brief and Skeist (3967) reviewed growth of the industry in an article titled "The Adhesives Industry—25 Years of Growth." In 1983, an unauthored article appeared titled "The Adhesives Industry—25 Years Into the Future" (3967). In 1984, Lees (3968) described his investigations about "Designing and Producing Toughened Structural Adhesives." In 1989, Hermansen and Tunick (3969) described "Formulating Custom-Tailored Thermal Transfer Adhesives," and in 1990, Ludbrook (3970) wrote about "High Performance Adhesives Meet Many Industry Needs."

2. SEALANTS AND THEIR APPLICATIONS

A number of articles in *Adhesives Age* over the years have described various sealants along with some of the main areas of their application. LeFave et al. (3971), in 1962, composed a paper simply titled "One-Part Polysulfide Joint Sealants," whereas Zakim and Shihadeh (3972), in 1965, offered "A Comparative Guide to Sealants and Caulking Compounds." Peterson (3973), in 1967, issued a paper titled "The Sealant Jungle." In 1974, an unauthored article was titled "Silicone Building Sealant Handles Cold Weather" (3974) and Devine (3975) described practical application in an article titled "Sealants: A Comparative Evaluation of Performance in Two Typical Joint Configurations." In 1977, Brady (3976) reported on "Silicone Sealants for Solar Energy Systems," whereas the same author reported the next year on "Silicone Sealants, Types Available and Typical Uses" (3977). Also in 1978, Brower (3978) reported on "Silicone Sealants, General Properties and Application Details."

In 1979, an unauthored review article appeared titled "Sealants: New Product and Application Profiles" (3979). The following year, Oxley (3980) discussed "Elastomers in Sealants" and Kishita (3981) reviewed "Liquid Sealants as Gasket Materials." Finally, in 1989, an *Adhesives Age* review article was issued discussing "Extruded PVC Foam Sealant Yields Six-Figure Savings" (3982).

3. PRESSURE-SENSITIVE ADHESIVES AND THEIR APPLICATIONS

In 1962, Modic (3983) produced a paper titled "Silicone Pressure-Sensitive Adhesives: Their Properties and Applications." Wangman (3984), in 1977, issued his summary of "Pressure-Sensitive Tapes: Types and Applications." The next year, Miron and Skeist (3985) offered their review of "Trends in Pressure-Sensitive and Heat-Seal Materials," whereas in 1979, Hayes (3986) reviewed "Water-Based Acrylic PSA's" and Merrill (3987) his review of "Silicone PSA's: Types, Properties, and Uses." In 1980, Kegley (3988) predicted "Future Trends for PSA's." An unauthored adhesives article in 1987 discussed "Solvent Acrylic PSA's Resist High Temperature and Humidity" (3989), whereas Sobrieski and Tangney (3990), in 1988, wrote about "Silicone PSA's Perform Well at High and Low Temperatures." Finally, Zosel (3991), in 1989, discussed "Physical Properties and Adhesion Performance of PSA's."

4. HOT-MELTS AND THEIR APPLICATIONS

Carlson (3992), in 1962, produced an early assessment of "Bonding with Hot-Melts." Bucksot (3993), in 1977, reviewed opportunities for "Hot-Melt Adhesive Lowers Costs of Pallet Stabilization," and 2 years later, Cobbs (1367) discussed the "Foamability of Hot-Melt Adhesives." Duncan and Bergerhouse (3994), in 1980, reviewed "EVA and VAE Copolymers for Hot-Melt PSA's." Borg and Boutillier (3995) offered their commentary on "Hot-Melt Copolymers Exhibit High Cohesion and Compatibility" in 1986, and in 1991, McBride (3996) discussed how "Ethylene Terpolymer Expands Hot-Melt Technology to PVC."

5. POLYURETHANE ADHESIVES AND APPLICATIONS

In 1966, McClellan and Rausch (3997) discussed "Polyurethane Sealants: Some Improved Systems," and in 1978, Dietrich and Rieck (1737) reviewed "Aqueous Polyurethane Systems: Their Possible Uses." In 1982, Dollhausen and Warrach (1757) offered an overall review of "Polyurethane Adhesive Technology." In 1990, Trippe and Burke (3998) reported "PU Adherends Many Substrates Offers Production Flexibility."

6. OTHER ADHESIVE TYPES

Kelly (3999) offered a 1962 overview of "Neoprene Contact Adhesives: Some Special Advantages and Production Applications," and Martin (4000) issued a 1966 review of "Neoprene-Phenolic Adhesives." In 1969, Kiriyama (4001) discussed "Metal-Containing Epoxy Resins as Adhesives" and Plonchak (4002) offered "Applications Grow for Anaerobic Adhesives and Sealants" in 1972. An unauthored 1973 *Adhesives Age* article discussed "Epoxy-Based Materials—An Aid to OSHA Compliance" (4003). Vazirani (4004) discussed some ways of flexibilizing epoxy systems for use in bonding metals and plastics in "Flexible Epoxy Resins." Finally, in 1991, Briggs et al. (4005) introduced a discussion of structural acrylics in the paper titled "Structural Methacrylates Yield Process, Performance Benefits."

L. Summary of *Adhesives Age* Articles Dealing with the General Subject of Bonding Metals

In 1961, two unauthored papers appeared titled "Leather to Aluminum Bonding" (4006) and "Plastic Letters Bonded to Metal Station Front" (4007). Also, Epstein and Litvak (4008) discussed the "Influence of Some Selected Fabrication Variables on Structural Metal-to-Metal Bonding," and Bush and Scott (4009) described "Epoxy Adhesive Bonds Extruded Sections of Aluminum Light Pole." In 1962, three unauthored papers appeared titled "Bonded Metal Mural" (4010), "Epoxy Adhesive Speeds Production of Bonded Glass-Aluminum Assemblies" (4011), and "Aluminum and Fiberglass Faced Plywood Panels" (4012). Also in 1962, Jackson (4013) reported "Bonding Plastic Laminates Onto Metals" and Valentine (4014) described "How to Assure Successful Metal and Glass Joints." In 1966, an unauthored paper was published on adhering aluminum parts titled "Diffusion Process Offers New Ways to Bond Metals" (4015), Preiss (4016) described "Bonding Polyurethane to Metal," and Dastin (4017) reported on "Bonded Beryllium Structures." In 1967, Lewis (4018) described fabrication of "Post-Formable Adhesively Bonded Metal Laminates" and Yoshino et al. (4019) described "Fabrication Methods for PBI Adhesive Beryllium Sandwich Structures." In 1968, an unauthored paper described "Laminating Method for Metal Panels Allows Post Forming" (4020). In the same year, Litvak (4021) offered a paper on "Polybenzimidazole Adhesives for Bonding Stainless Steel, Beryllium, and Titanium Alloys," Gallagher (3909) reviewed "Rubber-to-Metal Bonding." Conway (4022) discussed "Metal-to-Metal Adhesives for Structural Applications at Elevated Temperatures," and Howe and Radtke (4023) reported their investigations on "How to Join Lead with Adhesives." Finally, Duncan (4024), in 1969, investigated "Adhesives and Methods for Bonding Metal and Plastics to Porous Substrates."

In the decade of the 1970s, there was an unauthored article in 1973 titled "Vacuum Impregnation for Sealing Porosity of Powdered Metal Parts" (4025) followed by Field's (4026) article "Summary of the Weldbonding Process." In 1974, Darmory (4027) reported an "Extreme High Temperature Polyimide Adhesive for Bonding Titanium and Stainless Steel," Dexheimer and Vertnik (4028) described "Metal Bonding with Polyamide Hot-Melt Adhesives," Bascom (4029) reviewed "The Surface Chemistry of Bonding Metals with Polymer Adhesives," Bethke and Ketcharn (4030) reported on "Polysulfide Sealants for Corrosion Protection of Spot-Welded Aluminum Alloys," and Paul and McGivern (4031) investigated "Electrochemical Characterization and Control of Titanium Surfaces for Adhesive Bonding." In 1977, Brockmann (4032) described "Interface Reactions and Their Influence on the Long-Term Properties of Metal Bonds," whereas Vaughan et al. (4033) reported on "Polyimide Adhesives for Weld-Bonding Titanium." In 1978, Wolff (4034) investigated "Effects of Moisture on Mean Strength of Bonded Composite-to-Metal Joints" and Ishai and Girshegorn (4035) reported the "Strength of Bonded Aluminum-CFRP Single Lap Joints."

In 1983, an unauthored article was titled "Relaxation Time, Bond Strength of Epoxy Resin to Aluminum" (4036) followed by Bunk et al.'s (4037) article in 1984 titled "Epoxies for Residential Copper Plumbing." Another unauthored article in 1987 considered "Tape Reduces Production Costs and Aluminum Skin Marring" (4038).

In 1990, Muller (4039) reported that "Strong Flexible Polyester Hot-Melts Bond Metal-to-Metal Systems," and in 1991, Osterndorf and Bonk (4040) considered "Evaluating Adhesives for Aluminum-to-Copper Bonding."

M. Summary of *Adhesives Age* Articles Dealing with the General Subject of Bonding Plastics

In 1961, unauthored articles described "Synthetic Rubber Compounds Seal Plastic Skylights" (4041) and "Bonded Vinyl Patches" (4042). In 1962, Vohralik (4043) reported "Vacuum Covering with Adhesive Coated PVC," Russell (4044) investigated "Bonding Teflon FEP Film with Conventional Adhesives," St. Cyr (4045) issued a paper on "Methods of Bonding Fluorocarbon Plastic to Structural Materials," and an unauthored article considered "Bonding Tedlar to Steel with Acrylic Adhesive" (4046). Two unauthored reports appeared in 1966 on "Better Adhesion to Plastic Bottles with Resin Adhesive Films" (4047) and "How to Safely Bond Facing Panels of ABS Plastic to Sensitive Styrene Foam" (4048). Debnar (4049), in 1967, described "How to Bond Decorative Laminates to Wood." Delmonte (4050), in 1968, reported on "Bonding Thermoplastics with Improved Urethanes." In 1969, Devine and Bodnar (4051) reported on the "Effect of Various Surface Treatments on Adhesive Bonding of Polyethylene," Rauhut (4052) described "Pretreating Polyethylene for Optimum Structural Adhesive Joints," and Lerner (4053) investigated "Plasma Treatment of Delrin for Improved Adhesive Bonding."

In 1972, Bersin (4054) reported on "How to Obtain Strong Bonds Via Plasma Treatment" and Karle (4055) described "Adhesive Bonding of PVC in Water Demineralizing Systems." In 1973, Panek (4056) considered "Polysulfide Sealants for Plastics" and an unauthored article discussed "High Speed Laminating of Vinyl to Wood." In 1974, Bragole (4057) reviewed "Adhesive Bonding of Polyolefins" and Smith (4058) considered "Silicone Adhesives for Joining Plastics."

In 1980, Petrie (4059) reviewed "Joining the "Engineering Plastics—How Adhesives Compare with Other Bonding Methods," and in 1987, Lombardi (4060) described how "Reactive Acrylic Polymers Give Better Plastic Film Adhesion."

III. AEROSPACE APPLICATION BONDING OF ALUMINUM

A. Early Progress

For many years aluminum has been the primary material of construction for airplanes and also a significant part of space vehicle structures. These uses have been based on good high strength, excellent stiffness-to-weight ratio, and resistance to edgewise compression and flatwise bending. There has been an accumulation of evidence that the present state of the art for adhesive bonding of aluminum has reached the stage where it should be satisfactory for primary structural joining in aircraft.

Early important contributions to the structural use of adhesive bonding in aircraft was early work at the Forest Products Laboratory involving measurements on adhesive-bonded aluminum panels. Kuenzi and Erickson (4061), in 1955, reported that the shear stability found in bonded panels was superior to riveted panels. Another contribution was Revilles's (4062) 1955 investigation of the deflection and stresses in uniformly loaded sandwich construction. Norris (4063), in 1956, reported on the development of compressive buckling design curves for sandwich panels with isotropic facings and orthotropic cores. In the same year, a U.S. Printing Office report, designated ANC-23, was issued with the title "Sandwich Construction for Aircraft." Epstein (4064) wrote an overall review in his 1954 book titled *Adhesive Bonding of Metals.* In 1956, Hardroth (4065) made some early comparison of how cracks due to aluminum fatigue were much more effectively arrested by bonded stiffeners on aluminum box beams, especially as compared to integrally machined or riveted stiffeners. Blair (4066), in 1959, demonstrated the increase in life expectancy of helicopter rotor blades from 200 hr with rivets to 2000 hr with adhesive bonding. Belcher et al. (4067) laid the foundations for many investigations by their measurements on stiffened aluminum aircraft panels excited by acoustic noise as early as 1959.

B. Progress Through the 1960s

As mentioned previously, for many years aluminum has been the primary material of construction for airplanes and also has played a vital materials role in the development of space vehicles using significant amounts of adhesive bonding. The overall use of aluminum has been based on aluminum being able to furnish good high-strength, excellent stiffness-to-weight ratios, and resistance to edgewise compression and flatwise bending. There has been an ever-increasing accumulation of evidence that the present state of the art can provide the means of assembly of the primary structure for joining aircraft, which will be discussed next when reviewing the PABST investigations of the middle to late 1970s.

Myers (4068) produced a 1960 paper on adhesives for honeycomb cores. In 1961, Epstein (4069, 4070) issued investigations of the stresses encountered in sandwich construction and six types of bonds that could conceivably be developed in such a unit of construction and also offered discussion of how surface conditions of aluminum adherends, air entrapment in liquid epoxies, primer coatings, and curing pressures can affect the final bondline. In the same year, Evans (4071) reported the development of autoclave bonding of aluminum subassemblies in British aircraft production. He considered the massive tooling requirement for curing complex curved assemblies in flat platen presses leading to the research of methods that made use of fluid pressure. Lunsford (4072), in 1962, produced the first paper on the design of bonded aluminum joints in the B-58 supersonic "Hustler" bomber aircraft. In 1964, Gunter and Sanger (4073) presented a study on elastomeric adhesives as candidates for low-temperature curing to meet 1964 critical aerospace requirements. Emphasis was on adhesives that cure at 250°F (121°C) but provide stable bonds at higher temperatures that might be required for other components. Hilton (4074), in 1966, first described the use of two-part, room temperature–curing, epoxy-polyamide tape adhesives on the Boeing 727 aircraft. In 1967, McMullon and Garnish (4075) added further to the technology for structurally bonding metals in aircraft.

In 1968, Jensen (4076) reviewed the application of metal bonding in glider structures, whereas Keemar and Strong (4077) furnished some bonding airplane information from British Overseas Aviation in *Shell Aviation News*. Finally, Krieger (4078) discussed the advances that had been made in combatting corrosion in adhesive-bonded aluminum structures with aerospace emphasis in 1969.

C. Primary Adhesive Bonding Structure Technology (PABST) and Other Developments of the 1970s and 1980s

The year 1970 brought a paper from Hong (4079) on the subject of advanced bonding for large aircraft, whereas Snogren (4080), in the same year, provided an excellent article for the engineer with minimal knowledge of adhesion theory by furnishing important information about the fundamentals of joint mechanics and adhesive properties as they relate to bonding the widest possible variety of joint configurations. In 1971, Melcon and Moss (4081) discussed the opportunity to control structural fatigue in structures through the use of adhesive bonding.

It was in the mid-1970s that the PABST program was finally initiated. Under the direction of Thrall at Douglas Aircraft, the most comprehensive technical investigative research program on the structural bonding of aluminum was launched. Continuing technical bulletins were issued steadily from 1975 through 1980 (a partial listing can be found in Refs. 4082–4091). These reports covered the entire range of problems that had to be answered in order to develop the level of confidence that could ultimately lead to the bonding of primary structure in aircraft. Some of the kinds of investigations whose results can be found in the listed references will be mentioned. In Bulletin No. 3 (4082), the broad spectrum of possible structural concept types was considered and reduced to six definitive arrangements. The most cost effective of these designs were further reduced to three designs involving selection of a damage-tolerance test specimen, a shear panel specimen, and the tension testing of a bond joint between the skin and the frame/shear tee in Bulletin No. 4 (4083). The third report (4084) was an alert technical bulletin expanding the cause of the adhesive failure of "wedge-crack" test specimens prepared with a production load of PAA adherend material. Bulletin No. 8 (4085) considered the remainder of the PABST program to be investigated, whereas Bulletin No. 9 (4086) was a continuation of the test program to determine the damage-tolerance level of large flat and curved panels using special test fixtures for the application of combined pressure and axial loading. Bulletin No. 10 (4087) came out at the end of 1976 and contained a correlation of NDT results with built-in defects in laminated panels and honeycomb. Results were offered in Bulletin 13 (4088), in 1977, of tests on coatings over PAA-treated aluminum, beach-exposure effect on adhesives and cyclic exposure tests using the special RAAB test specimen developed for the PABST program to evaluate joint durability data. Aerospace tape adhesives were used with both woven and nonwoven carriers present. The design detail of the full-scale demonstration component (FSDC), which was a simulation of the forward section of the C-15 airplane fuselage, was offered in Bulletin 13. The forward end was to be closed off by a steel pressure dome and the aft by a barrel section test fixture. Bulletin 14 (4089) covered the continuing environmental performance of organic coatings on PAA or modified–FPL-etched aluminum alloys, new adhesive modification testing, and development of the NDT test specifications. The analysis of the results obtained during "tear down" of the full-sized bonded fuselage structure after completion of fatigue, residual strength, and ultimate tests on the 42-ft full-scale demonstration component was reviewed in Bulletin 20 (4090). Additional PABST Technical Bulletins were issued (Nos. 1, 2, 5–7, 11, 12, 15, 16, and 17–19) but have not been reviewed here.

By the time the third PABST Technical Bulletin had been issued in 1975, the overall test program was reviewed by Thrall (4091) before a meeting of the American Institute of Aeronautics and Astronautics. The completion of the first 18-mo phase of the 4.5-yr program was reported in *Business Week* magazine in early 1977 (4092). This portion of the work focused on the evaluation of the most promising candidate new adhesives and their bonding procedures on aluminum adherends. The added comment was made that "these new advances in bonding may make possible the nearly rivetless airplane." Two disclosures have been made in *Adhesives Age* during and at the end of the program: first by Thrall and Shannon (4093) in 1977 titled "PABST Surface Treatment and Adhesive Selection" with an additional summary paper in 1979. Shannon (4094) discussed his particular interest in the PABST results in 1979 at a Business Aircraft Meeting. Thrall (4095) also reviewed the results at an ASTM-sponsored meeting discussing the overall future of metal bonding. Thrall and Shannon (4096) have furnished much of this evidence in a 1985 book titled *Adhesive Bonding of Aluminum Alloys.*

Earlier in 1972, a series of three articles appeared in *Aircraft Engineering* on the adhesive bonding accomplishments presented in the Tri-Star Aircraft. The aluminum alloys chosen for the structure were Alclad 2024-T3 and Alclad 7075-T6. The third article in this series was particularly concerned with the adhesive bonding of large fuselage panels (4097). The next year, Dauksys (4098) produced a paper that touched on a variety of research and development problems that still needed to be addressed and were eventually investigated in the major research programs involved in the overall PABST program noted above.

Reynolds (4099) has discussed the impact that adhesive bonding has had on aircraft design through the 1970s. The development was generally gradual but sometimes suddenly accelerated with breakthroughs like swept wings, pressurized fuselages, and jet engines. Even so, by the end of the 1960s, the overall utilization of bonding had increased by 65% (as shown in Fig. 2 of Ref. 4099). Reynolds further illustrated how improved fatigue resistance, fracture touchness, and lowered crack growth rates in adhesives due to bonding had significantly affected weight savings in the newer plane designs (see Fig. 3 in Ref. 4099). In addition, bonded tear stoppers had greatly reduced the weight of fail-safe designs and the weight of secondary structures had also been considerably paired down through the use of bonded honeycomb sandwich construction.

The supressing of noise around the inlets of jet engines has been another job made easier by bonding. In this connection, Arnold (4100) has reported work at Boeing to develop a 350°F (177°C) service-temperature system for bonding jet engine acoustic liners which consisted of perforated aluminum facing sheets bonded to an aluminum honeycomb core. It was necessary to demonstrate aging durability to the 350°F (177°C) environment, resistance to hydraulic fluids, glycols, aviation fuels, and bond toughness to function as a structural component of the engine nacelle while performing as an effective sound attenuator. A direct relationship appears to exist between the efficiency of noise suppression and the low percentage of hole blockage in the honeycomb in the final bonded assembly. On the negative side has been the disbonding and corrosion under severe weathering conditions. The employment of the CAA- and PAA-anodizing pretreatments for bonding of aluminum has provided virtual immunity to this undercutting corrosion of the bondline with attendent help from the development and use of special corrosion-inhibiting (CIAP) primers under the adhesives.

Wegman et al. (4101) published a 1971 report at Picatinny Arsenal on evaluation of adhesive bonding processes in helicopter manufcture in regard to the durability. Wolff and Lemon (4102), in 1972, summarized the state of making reliability predictions about aircraft bonds at the Air Force Materials

Lab. In the same year, Reinhart (4103) attempted to summarize the new adhesive needs and concepts which had been developed up to that time. The problems attendant to satisfactory repair of aluminum structures began to receive more attention in the late 1960s and early 1970s. A number of separate publications from the aerospace industry were being released with the overall goal of publishing a standardized repair book. Representative of such investigations are four General Dynamics reports dealing with the overall testing of the F-111A aircraft in 1969 (4104,4105) and 1972 (4106,4107). Boeing issued a 1976 report of the static testing of the outboard spoiler on the same aircraft (4108).

In 1975, Rudder and Plumbee (4109) issued an Air Force Flight Dynamics Lab (AFFDL) report of their investigation at Wright-Patterson titled "Sonic Fatigue Design Guide for Military Aircraft." This was followed in 1977 by Zimmerman's general application article in *American Machinist* titled "Gluing Airplanes Together" (4110). In 1978, Wentz and Wolfe (4111) published their development work on random fatigue data for adhesive-bonded and weldbonded structures subjected to dynamic excitation. Potter et al. (4112), in 1979, produced their final AFFDL Tech. Report titled "Primary Adhesively Bonded Structure Technology (PABST) Design Handbook for Adhesive Bonding." In 1980, two important USAF AFML/FIBC contract reports were issued at General Dynamics with the titles of "Laminated Wing Structures" (4113) and "Advanced Technology Wing Structure" (4114). Also in 1980, Romanko and Knauss (4115) published an excellent review of their technical work on developing a methodology for predicting aerospace bonded joint life, which was followed in 1981 with a review of current aerospace bonded structures by Liechti and Knauss (4116) titled "Advances in Aerospace Structures." Finally, in 1982, Romanko et al. (4117) combined their efforts in a final report to AFWAL at Wright-Patterson titled "Integrated Methodology for Adhesive Bonded Joint Life Predictions." In 1985, Sanders (4118) reviewed the concept of "Bonding Aircraft Wings Improves Weight and Durability."

Fehrle and McDougal (4119), in a 1979 paper, quantitized the increase in bonding in aircraft from 1952 through 1966 by comparing the degree of bonding in the 1952 C-130 transport plane to the C-141 1962 plane and the 1966 C5A transport plane. From a ratio of only 0.1 bonded to total joined area in the C-130, the ratio in the C-141 aircraft had increased by a factor of 8–10 times. The first big changes came in the 1962 aircraft with the pressure door, nose wheel well panels, and various control surfaces. By 1966, the C5A transport ratio had increased by another threefold with primary structure applications extended to under floor bulkheads, cargo floor, troop floor, and the troop compartment pressure bulkhead. The entire cargo floor in the C5A is a metal sheet stiffened with adhesive-bonded beams. A total of about 25,000 ft^2. of honeycomb bonded structure is incorporated along with about 10,000 ft^2. of metal-to-metal bonded components. Obviously, it was necessary to study and weigh the advantage and disadvantage of a variety of different potentially usable types of construction and fabrication procedures before setting on the final design in each aircraft.

The question of whether to use adhesives or mechanical-type fasteners has often been influenced by the following list of advantages gained through the use of adhesive bonding:

1. Fabrication of smoother parts is permitted, since adhesives do not break through or deform the surface of the component. The end result being an exterior surface of exceptionally low drag.
2. The use of lighter-gauge materials is permitted, since the full strength of bonded thin sheets can be utilized. This is achievable because bonding minimizes the stress concentrations that commonly occur with screws, bolts, rivets, and spot-welds. Uniform stress distribution will provide both greater strength and rigidity in the bonded assembly.
3. Bonded areas serve to dampen vibrations. Better stress distribution also means better fatigue resistance. Structural adhesives are capable of transferring, distributing, and absorbing stresses so that metals often fatigue before adhesives, especially when used in conjunction with rivets and bolts.
4. Dissimilar adherends are more effectively joined with adhesives. The adhesive layer isolates the two materials, thus eliminating corrosion problems. (The author would direct the readers attention to his observations when corrosion testing aluminum-to-steel joints. It was found that the natural rust corrosion product generated from the oxidation of the steel adherends could eventually bridge across the freely exposed ASTM lap-shear joint and permit corrosion to initiate as a galvanic couple was present via the bridging corrosion product. When this product

was removed, the adhesive would continue to serve as an electrically isolating layer between the two adherends.) Thus, in a well-designed final bonded structure, some means of preventing such bridging would need to be made if long-term exposure to a corrosive salt water was a factor in the service environment.

5. Bonding can provide easier fabrication of complex contours. Many components in both aerospace and nonaerospace manufactured products do have complex contoured surfaces that would be very difficult to achieve through conventional fastening methods but can be bonded with relative ease.

D. Repair of Bonded Aluminum Aircraft

While the PABST investigative program was ongoing at Douglas Aircraft, other aerospace companies were involved in other important Air Force–funded programs. An overall review of the many problems which needed to be investigated can be interpreted from the Report of the Ad Hoc Committee on Structural Adhesives for Aerospace Use of the National Materials Advisory Board of the National Research Council. A review of this report with attendant discussion has been reprinted as Volume 4 of the *Treatise on Adhesion and Adhesives* series under the title of "Structural Adhesives, with Emphasis on Aerospace Applications" (4120).

Boeing had been focusing on a five-phase program to develop a standardized handbook for the repair of bonded aluminum aircraft. In August 1975, McCarty et al. (4121) released AFML Report TR-75-158, which had the title *Adhesive Bonded Aerospace Structures Standardized Repair Handbook.* The following year, the same authors released another similarly titled report with the AFML Tech. Report No. TR-75-201 (4122). In 1979, Horton (4123) offered a review in *SAMPE Quarterly* titled "Demonstration of An Improved Method for Repair of Bonded Aircraft Structure." Doetsch et al. (4124) issued a Rockwell AFFDL Tech. Report TR-76-63 in 1977 dealing with *Low Cost Aircraft Structural Repair and Maintenance Study.*

In 1985, Kuperman and Horten (4125) wrote on repair of aluminum aircraft and summarized the overall PABST program and followed this in 1990 with an update in *Engineered Materials Handbook*, Vol. 3 (4126).

A vast array of different specifications have had to be developed in order to support the overall problem of making acceptable repairs to aluminum aircraft. Some of the most pertinent are included in the following listing:

1. *Metallic Materials and Elements for Aerospace Vehicles Structures, MIL-HDBK-5, Military Standardization Handbook*, U.S. Dept. of Defense (May 1989) (4127)
2. *Aluminum and Aluminum Alloy Plate and Sheet, General Specification for QQ-A-250* (4128)
3. *Cloth, Glass Finished for Resin Laminates, MIL-C-9084, Military Specification* (4129)
4. *Polymer Matrix Composites,* Vol. 1, *MIL-HDBK-17, Military Standardization Handbook,* U.S. Dept. of Defense (June 1989) (4130)
5. *Core Material, Aluminum, for Sandwich Construction, MIL-C-7438, Military Specification* (4131)
6. *Core Material, Plastic Honeycomb, Laminated Glass Fabric Base, for Aircraft Structural and Electronic Applications, MIL-C-8073, Military Specification* (4132)
7. *Core, Honeycomb, Fibrous Aramid Base, Phenolic Coated, AMS 3711, Aerospace Materials Specification* (4133)
8. *Adhesive, Film Form, Metallic Structural Sandwich Construction, MIL-A-25463, Military Specification* (4134)
9. *Adhesives, Heat Resistant, Airframe Structural, Metal-to-Metal, MMM-A-132, Military Specification* (4135)
10. *Sealing Compound, Temperature Resistant, Integral Fuel Tanks and Fuel Cell Cavities, High Adhesion, MIL-S-8802, Military Specification* (4136)
11. *Sealing and Coating Compound, Corrosion Inhibitive, MIL-S-81733, Military Specification* (4137)
12. *Sealing Compound, Topcoat, Fuel Tank, Buna-N, MIL-S-4383, Military Specification* (4138)

As might be imagined, most of the repair adhesives and their assembly operations that are considered in the above aerospace reviews of repairing aluminum aircraft involve hot-setting adhesives. In 1981, Chan and Armstrong (4139) offered the results of their investigations involving the use of cold (ambient)–curing type of adhesives such as the two-part epoxy pastes and the structural acrylics.

E. Effect of Compositional Variations in Structural Adhesives

Any manufacturer of adhesively bonded products needs to be aware of the problems that can arise when the adhesive being used has undergone some compositional change from the product that was originally accepted for a particular application. For many of the less critical bonded products, it may only be necessary to use some standard testing method like ASTM D-1337 titled "Storage Life of Adhesives by Consistency and Bond Strength" to determine that the product is or is not acceptable. The consistency testing may only require some means of measuring the viscosity and comparing the value with that published on the data sheet. This measurement would be conducted with new as received material before acceptance. More critical is the situation of the manufacturer who may typically purchase much larger shipments for economy reasons, then have to be assured that the consistency has been maintained after variable periods of storage. What may be even more subtle is the possibility that consistency may be maintained, whereas bond strength properties have declined.

Characterization of compositional variation can be of immense importance to the manufacturing engineer in aerospace. Kibler (4140) has investigated composition and processing variables of a particular structural epoxy from Ciba-Geigy using typical mechanical-type receiving inspection tests, current physicochemical methods, and various performance tests. These results were then used to assess the sensitivity of processing and performance behavior to material variations and guide implementation of physicochemical quality control procedures at General Dynamics. Previously, Hinrichs and Thuen (4141) had shown that conventional mechanical test results involving stress-strain, peel, and flexure, e.g., can be incomplete indicators of the materials' quality because mechanical results are strongly influenced by specimen fabrication and testing technique. Kibler (4136) cites an example by showing a distinct effect of hardener content on joint strength when using double–lap-shear specimens, whereas single–lap-joint specimens using the same adhesive showed no real trends, apparently because of the greater sensitivity of the single–lap-joints to peel components. A number of other programs to investigate aspects of physicochemical detectability of constituents in nominal and varied matrix resins and adhesives can be reviewed in the literature (4142–4145).

The McDonnell Aircraft Division of McDonnell-Douglas undertook to relate the chemical composition of structural adhesive systems to a wide variety of performance parameters. This work by Britton (4146) was performed in order to define chemical acceptance criteria for adhesives in aircraft. With such information, airframe manufacturers could accept or reject material batches with more confidence based on short, simple chemical analyses. A problem that could still arise even after such acceptance testing was exposure of the materials to moisture in storage. When such moisture pickup in some existing aerospace film adhesives was shown to be detrimental to their low-temperature peel-strength properties, Alner and Pocius (4147) at 3M's were able to demonstrate that new candidate adhesives AF-163 and AF-163-2 were able to maintain their good low-temperature peel properties under equivalent storage conditions.

Grimes (4148) reviewed the adhesive-aluminum honeycomb relationship in 1966, whereas Mahoney (4149), in 1972, investigated the effect of adhesive components on the corrosion of aluminum honeycomb. He found some products produced during adhesive curing could produce corrosion. He suggested adding anticorrosion additives as well as the use of anodized or corrosion-inhibitor–primed honeycomb. Locke (4150), in 1977, published an extensive investigation of the problems encountered in the so-called "older" technology for fabricating aluminum honeycomb sandwich panels in the period 1960–1965. The newer technology of 1975 included replacing the clad 7075-T6 facings with bare facings and the optimized FPL pretreatment (BAC 5514) with the PAA pretreatment (BAC 5555). Adding a sealant to the honeycomb edges or an entire core pour coat did not show a measurable further improvement over the described "new" procedure technology.

Having accepted the adhesive and found it to be acceptable in storage, the next procedure to control was monitoring of the cure of the structural adhesive in the bondline. Lockheed Missile and

Space Company undertook this type of investigation, culminating in a report in November 1980 by May et al. (4151).

Brown and McKeegan (4152) have written a section for the 1990 *ASM Engineered Materials Handbook* titled "Raw Materials Quality Control" which is certainly pertinent to the overall problem of assuring batch-to-batch consistency of adhesive products. They point out that both the quantity and/or quality of the raw materials going into liquid, paste, or film manufactured adhesive products can have a significant effect on the user's final bonded assembly. Typically, the adhesive manufacturer has compiled internal specifications for controlling the quality of raw materials as well as whatever time, temperature, mixing or other processing variables might be employed. As the customer, the user still needs to check how well the manufacturer's specifications were followed in formulating the batch of adhesive received. In the beginning, there has to be a series of adhesive product test phases carried out by the manufacturing engineer which culminate in the development of one or more qualification tests that can be agreed upon by the supplier and user. By the determination of a favorable result from such tests, the supplier certifies the adhesive for the job, and generally a repeat test by the user enables him or her to accept the product for manufacturing. These authors supply additional practical information on such subjects as test techniques, alternative techniques, testing considerations, and acceptance alternatives.

Roberts (4153) has also supplied a section in the same ASM Handbook titled "Processing Quality Control." Roberts points out that industry has historically responded to this problem by testing all lots of incoming material prior to use. Unfortunately, after a significant number of tests are conducted on the basis of statistical process control (SPC) and a proven track record has been established, the inclination is to curtail or reduce this degree of testing. Roberts also refers to matters of (1) storage (material age-life history), (2) moisture exposure, (3) control of adhesive thickness, (4) surface preparation control, (5) handling and exposure to contaminants, (6) control of tooling, (7) control of lay-up technique, and (8) factors to be controlled during curing.

F. Fatigue Life of Adhesively Bonded Aluminum Joints in Aircraft

Sonic fatigue failures in aluminum riveted aircraft structures such as skin panels, ribs, spars, stringers, and longerons have often been the basis of unacceptable maintenance and inspection burdens in aircraft operations. Such failures result, of course, from the vibratory response of the structure to high-intensity noise due to jet propulsion, turbulent boundary layers, separated flow noise, flow impingment, and scubbing flow-effects. The fluctuating pressure field produced results in a resonant response of the parts which ultimately can result in a fatigue failure. Benefits of bonding aerospace structures, in terms of a significant increase in fatigue life, were early recognized, although it must be admitted that many of the specific factors that affect sonic fatigue life have not always been analytically predictable. A broad base of general design information using both analytical and empirical approaches has been developed and summarized in an AFFDL Tech. Report No. TR-74-122 (May 1975) by Rudder and Plumbee.

Hartman (4154) produced an early 1966 study of the effect on fatigue strength of Alclad 2024-T3 joints manufactured by a combination of riveting and adhesive bonding. Particular effects studied included testing temperature, type of room temperature–curing epoxy adhesive, pretreatment on the aluminum, frequency of loading, and artificial aging for 1 mo in 98–100% RH at 158°F (70°C).

Wolfe et al. (1496) have investigated the sonic fatigue life of adhesively bonded aluminum structures to obtain data that could be a basis for the design criteria of aircraft structures. Their evaluation program at Wright-Patterson AFB sought to determine the effects of adhesive thickness, primer thickness, and type of oxide on the random bending fatigue life. It was of great significance that the locus-of-failure in fatigued aluminum joints of the PABST vintage showed the failures were essentially all within the adhesive, with only a few cases with failures very close to the primer surface.

In 1986, Fasold (4155) reported the use of a polysulfide adhesive for attaching anti-icing vanes (aluminum-faced/nonmetallic honeycomb panel) to the B1B aircraft. The attachment had to be structurally adequate to resist a severe load spectrum created by the vibroacoustic environment within the nacelle. Also, the anti-icing blankets had to be removable from the vane without incurring damage to either. The polysulfide, based on 2 yr of aircraft testing, was selected over structural film, epoxy paste, and elastomeric adhesive candidates.

An excellent contribution of Sancaktar (2776) to the 1990 *ASM Engineered Materials Handbook* deals with a number of aspects of the fatigue failure of aluminum-bonded joints. Some of the subsections of this article deal separately with subjects such as (1) fatigue fundamentals, (2) joint and loading geometry effects, (3) adhesive microstructure considerations, (4) viscoelasticity considerations, and (5) fatigue fracture criteria threshold levels.

G. Criteria for Damage Tolerance and Fail Safety in Aluminum Aircraft

Conducting testing to determine the above qualities for any particular aircraft need to be meticulously carried forward throughout the time from design to final manufacture. Four kinds of tests can be recognized as having importance during different stages of the plane's development. They are (1) coupon tests, (2) subcomponent tests, (3) component tests, and (4) full-scale tests. The coupon testing is obviously first, since a great deal of material screening and process control data is needed as soon as possible. Rather early in the developmental process, however, subcomponent-type tests must be initiated to provide better guidance for the critical design details. The actual design concepts must be validated by actual component tests which produce the data which is part of the final substantiation process. Finally, the need arises to evaluate full-scale structures in both static and cyclic stressing conditions to demonstrate strength, stiffness, durability (fatigue), and damage tolerance (safety). A major consideration is to determine the test environments that need to be imposed with all sorts of specimens. The most direct and conservative approach has been to assume that all areas of the aircraft should be thermally soaked at the environmental extremes. Sometimes this procedure may result in excessive weight penalties; however, in which case, thermal analyses must be performed to produce a more realistic distribution of thermal environments to the individual areas of the structure. (Some additional references of possible interest are 4156–4159.)

Fehrle and McDougal (4119) presented a 1979 paper to a Business Aircraft Meeting which related to a number of critical matters about bonding in aircraft. They mentioned a new criterion for damage tolerance and fail safety in aircraft had been specified in MIL-A-83444 (*Airplane Damage Tolerance Design Requirements*) and MIL-STD-1530 (*Aircraft Structural Integrity Program, Aircraft Requirements*). In the nonbonded structures, the MIL-A-83444 specification assumed initial flaws of specific size must exist as a result of manufacturing and processing operations in each hole of each element in the structure. By application of fracture mechanics technology and slow crack-growth analysis, the life of the structure with assumed flaws could be assessed. It can be assumed that positive influences on durability over such structures could be established by use of bonded laminates, doublers, or straps. This laminate structure would resist both crack initiation and extension, whereas the straps and/or doublers would break up large areas and stop or contain rapid crack growth.

Anderson et al. (3218) have conducted investigations related to the benefits on crack retardation by adhesively bonded doublers. It is clear from their data that no growth in the back sheet crack was initiated until after 20,000 stress cycles when the back sheet stress intensity factor first exceeded that of the front sheet crack. The advantage from a durability and crack growth consideration of a laminated sheet versus a monolayer sheet is also shown in Fehrle and McDougal's and Anderson's publications. The time to failure of a laminated sheet is approximately three times longer than for the single sheet to the same stress distribution.

Hart-Smith and Thrall (2202) have considered the various possibilities for structurally improving the design concepts of aerospace bondments in their chapter on structural analysis of adhesive-bonded joints in Thrall and Shannon's *Adhesive Bonding of Aluminum Joints* (4096); e.g., they considered two design concepts for structurally improving flush bonding splices (see Fig. 7, p. 248). In Design (1), a doubler area with maximum bending moment is made even more damage tolerant with the addition of another bonded doubler. The inner doubler overlap on the skin is sufficient to transfer load, and the ends of the inner doubler can be additionally tapered to further reduce peel stresses. Design (b) shows how the skin, being spliced, can be chemically milled at the ends to decrease the eccentricity of the load path. The position of the chemically milled step occurs at an area of reduced axial stress. The thin ends of the outside members in this design also decrease the adhesive peel stresses. Increased thickness at the center will also increase the bending strength. In the same chapter (Fig. 10), these authors consider the design considerations for the proportions of stiffeners to be used in bonded stiffeners, which can be important in increasing the tolerance of bonded structure to damage. The typical Z section joint can be subject to problems such as (1) having too narrow a bond width, (2) excessive peel stresses present at the heel of the Z due to skin shear wrinkles, or (3) bonding pressures on the overhang part can make the stiffener roll over. The authors offer three methods of enhancing damage tolerance: (1) utilizing symmetrical legs with thin flexible tapered tips replacing the bottom of the Z section: This promotes low adhesive peel stresses because the stiffener and the skin can deflect together under low loads; (2) the tapered tips can be replaced by flexible tips of skin legs which can promote the same low peel stresses but the overall altered Z stiffener design can be easier to extrude as compared with the tapering operation. The more constant overall thickness of the extruded skin legs also makes for easier bonding of matching details and more effective stopping of skin cracks; and (3) the outer flanges of the stiffener feet can be made wide enough to permit riveted repairs when necessary. In any design it needs to be remembered that damage-tolerance (crack-stopping) ability requires the bond must be large enough when considering the stiffener area and spacing to prevent the unzipping of the bond.

For additional current information, reference can be made to the presentation titled "Design Proof Testing" by Johnson (4160) in the 1990 *ASM Engineered Materials Handbook*, although there is more emphasis on testing involving bonded composite materials structures.

H. Aerospace Aluminum Weldbonding

The aerospace industry and more recently the automotive and truck and trailer industries have considered the application of the weldbonding technique for aluminum structures. As mentioned in Chapter 10, the investigations about weldbonding have been ongoing since its introduction from Russia in about 1967. An *Iron Age* magazine article in 1969 with the title "Spot Weld Through An Adhesive?" first posed the question as to its general use in manufacturing (62). In 1971, *Materials Engineering Magazine* tried to promote interest with an article titled "Glue-Spot Welding Joins Aluminum, Stronger Than Spot, Cheaper Than Fusion" (63). It was shown that a 19% reduction in overlap aluminum weldbond strength occurred after 70-da exposure to a cycle of 16 hr in 95% RH at 100°F (38°C) and 8 hr in ambient air. Unfortunately, no data with a corresponding adhesive-only bonded joint were presented for direct comparison.

Since the major funding in aerospace was by the U.S. Air Force, the major emphasis in these early days of investigation were confined to only aerospace-type aluminum surface pretreatments. The overall improvement in resistance to sonic fatigue as compared to adhesive-only bonded joints was clearly demonstrable. Also, aluminum weldbond joints successfully passed testing under low cycle flexure conditions in liquid hydrogen without brittle failure. (Work conducted at Lockheed-Georgia can be found in Refs. 64 and 65, whereas results of extensive work at Northrup can be found in Refs. 70, 71, and 760–764.) In the 1971 Lockheed-Georgia investigations (64), the evaluation of a variety of surface pretreatments, primers, and adhesives were combined with various weldbond processing. Relatively short-term weathering cycle tests and "wedge-type" specimen testing after high-humidity soaking were used to discriminate between the various surface treatments and adhesives. In the 1973 reporting (65), Lockheed-Georgia showed significant differences in weldbond joint durability in salt spray depending on the type of aluminum surface pretreatment employed. While the spot-weld etch developed by the aluminum suppliers for pretreating areas for subsequent welding was optimal for making spot-welds, the joint permanence of weldbond joints was lower with this pretreatment than for FPL-treated aluminum adherends.

In 1974, Bowen et al. (760–764) concentrated at Northrup on the identification of the best type of aluminum surface oxide for making most serviceable aluminum weldbonds. They concluded that a "boehmite" film of specific thickness was most desirable. When the pretreating was by vapor degreasing, hot alkaline cleaning, or the conventional aluminum industry spot-weld etchant, only "bayerite" oxide was shown to be formed; however, FPL-type deoxidizing did seem to produce the desired boehmite form. In 1975, later publications by Bowen et al. (70) and Wu and Bowen (71) sought to establish a production method for weldbonding aluminum which included optimizing surface preparation, adhesive application, and spot-welding parameters. In the former publication, the authors also elucidated the mechanism of weld expulsion using the conventional etchant and FPL-treated aluminum adherends. At first, Bowen and Wu had advocated the use of an FPL etch plus a 60-min seal in boiling sodium dichromate solution. Later, their improved treatment consisted of degreasing, hot alkaline cleaning, deoxidizing in nitric acid, and anodizing in phosphoric acid–sodium dichromate solution at room temperature (70). Based on durability results in salt-water exposure and minimum crack growth in "wedge-type" specimen joints, Bowen and Wu suggested B. F. Goodrich 0500PE130 adhesive with phosphoric acid–sodium dichromate anodizing at 1.5 V for 20 min as the best surface pretreatment.

Mahon et al. (66), at Grumann, tested different surface pretreatments and found best durability with a treatment in a hot nitric acid–sodium sulfate solution. When Bowen investigated the oxide produced by this treatment, it seemed to be about 80% bayerite oxide, which conflicted with Bowen's own investigations stating best results were with boehmite oxide on the surface. Mahon's testing should have been discriminating, since he used stress testing with a spring-loaded fixture at 140°F (60°C) in accordance with the MIL-A-83377 procedure. Weldbond joints had slightly longer survival times. Adhesive choice was also shown as being quite important. For example, ADX-41 aerospace adhesive promoted higher fatigue strength in an adhesive-only–bonded joint, whereas the use of Reliabond 398 created better performance in the weldbond form of joint. Further investigations disclosed that the thickness of the boehmite layer could be critical, needing to be about or below 700 Å for good weldability. Conventional PAA oxide (boehmite-type oxide) pretreatments created from about 2400 Å with bare 2024-T3 and 7075-T6 alloys to 5200 Å for the clad versions of these alloys. Thus, a modified procedure is needed if PAA anodizing is used for pretreatment for weldbonding.

I. Bonding Aluminum Space Vehicles

The beginning of consideration of adhesive bonding for aluminum space vehicles probably goes back to the early demonstration of the excellent strength of some adhesive chemical families in aluminum joints under cryogenic temperature conditions in the late 1950s and early 1960s by Hertz (1895–1899,4161), Frost (1903), McClintock and Hiza (1901), and Eppinger and Love (4162). Stress concentrations in aluminum parts at very low temperatures can be highly influenced by such factors as (1) difference in thermal coefficient of linear expansion between different combinations of adhesive and adherends, (2) shrinkage of adhesive in curing, (3) trapped gases or volatiles evolved during bonding, (4) differences in modulus of elasticity and shear strength between adhesive and adherends, (5) differences in thermal conductivity of adhesives and adherends, and (6) residual stresses in adherends as a result of the release of bonding pressure after assembly. A low-modulus adhesive at room temperature might readily relieve stress concentrations by deformation, but at sub-zero temperatures, the modulus of elasticity could actually increase, making relief of concentrated stress impossible. It should be readily apparent that advance missiles and space vehicles could both present many possibilities for aluminum-to-aluminum and aluminum-to-plastic bonds which must withstand extreme low- and high-temperature conditions.

Adhesive bonding for the Saturn S-II stage with room temperature–curing adhesives was reported by Whelton (4163) at North American in 1962. He investigated the bondability of thermal insulation on the aluminum surfaces using 18 different adhesives. Testing temperatures ranged from 160°F (71°C) down to −423°F (−253°C) and included exposure to 96% RH at 120°F (49°C). Inadequate results were obtained with room temperature curing which resulted in substitution of a 16-hr cure at 120°F (49°C) under vacuum pressure. In 1964, Hess (4164) supplied a compilation of the various types of commercial structural adhesives available for the space program as well as available types of curing equipment. In 1966, Hill (4165), at Boeing, compared the concept of using a mechanical fastener with bolting for applying pressure while curing aluminum lap-joints with jigging as prescribed in ASTM D-1002. The joint strengths seemed equivalent for test temperatures from 75°F (24°C) to 300°F (194°C). The studies included aluminum-to-aluminum, aluminum-to-laminate, and laminate-to-laminate structures. Kammerer et al. (4166), in the same year, described the method at North American of applying NDT to the honeycomb structures in Project Apollo quality bonds. These authors provided a copy of an ultrasonic test of a Saturn S-II liquid hydrogen tank insulation bond. Bonding with respect to that found on the Apollo command, service, and lunar modules has been discussed by Spiker and Barnes (4167). Almost a decade later, Mulrey and Mazenko (4168) continued to report on structural adhesives for space systems at a SAMPE Technological Conference.

For space, adhesives are required that are stable in the 500–600°F (280–316°C) range for the bonding of aluminum honeycomb sandwich primary load-carrying structure. Other adhesives used in secondary structure had to serve electrical potting and joint-sealing functions. Epstein (4169) also discussed adhesives that might be employed in space environments in a 1970 U.S. Air Force Report. He further reviewed the requirements and applications for adhesives in various space systems in two *Adhesives Age* articles in 1972 titled "Adhesives for Space Systems, Part I—Requirements and Applications" (4170), and "Part II—Launch and Re-entry Systems" (4171). He pointed out that "for every one pound of structure saved in a spacecraft, the savings in launching can be as much as $10–20,000." This definitely favors adhesive bonding, because weight savings dictates the use of thin-gauge aluminum which is generally not amenable to riveting or bolting. Also, the attachment of the ablative heat shield can only be accomplished by bonding, since any metallic fasteners would melt during reentry. Wherever, stiffness or rigidity and low weight is an important design factor, as in space vehicles, bonded sandwich construction is ideal as previously demonstrated in aluminum aircraft construction.

For structural applications in these space vehicles, epoxy-base (including epoxy-phenolic and epoxy-nylon) and occasionally modified phenolic adhesives have been employed. Similar adhesives could be used in nonstructural applications as well. However, more likely adhesive types could include room-temperature vulcanizing (RTV) silicone elastomerics, cyanoacrylates, polyurethanes, rubber-base cements, and even pressure-sensitive adhesives. Stiles (4172) reviews the extensive use of adhesive bonding in the secondary structure category by pointing out over 700 miscellaneous details like simple angle supports to complex machined fittings are bonded with epoxy-base adhesives. It is quite likely that even in the primary bond category, the common aluminum bulkhead in the Saturn second stage could be the largest primary structure ever bonded—33 ft in diameter. Because of its huge size, special techniques were necessary to pretreat the aluminum surfaces and perfectly fit the honeycomb core. The Apollo service module itself consists of an aluminum cylindrical sandwich construction with 2024-T8 aluminum outer skins and 7178-T6 inner skins bonded to 5052-H39 aluminum honeycomb core using an epoxy-phenolic tape adhesive. The spacecraft lunar module adapter that supports and houses the lunar module has similar aluminum construction as does the lunar module itself.

In 1973, Pascuzzi (4173) considered the role of adhesive bonding in the development of the Mariner Venus/Mercury spacecraft, and in 1975, Srinivas (4174) issued a NASA report on the analysis of bonded joining. Further publications surveying the use of structural adhesives for space systems and methods for selecting the best candidates were issued by Murray (4175) and Lees (4176). Most recently, Novak and Comer (4177) have described an extensive test program involving a toughened cyanoacrylate system for bonding the neoprene foam rubber seals to forward aluminum skirt solid rocket booster structure. The advantages included faster cure time, more rapid application, simplified positioning of the seal to the complex geometry of the skirt, elimination of costly and time-consuming repairs, substitution with a safer and nonflammable adhesive and accelerator, and reduction in inventory and cost of adhesive with enhanced shelf life. Distinct improvement in adhesive bond reliability coupled with reduced processing time and material costs also enhanced flow of critical flight hardware for the space shuttle program.

J. Evaluation of Space Vehicle Structures

The need to carefully control the process of bonding aluminum facings to aluminum honeycomb has been a necessity in all airplane and space vehicle construction. Undetected voids in the bondline or improper curing of adhesive in the bondline can be disasterous in service. Pike and Williams (4178) have attempted an ultrasonic assessment of the cure-rate effects in bonded aluminum honeycomb structures. They illustrate how variation in the curing temperature can influence the type and length of the adhesive fillets, the degree of cure in the bondline, and the T_g of the resultant cured adhesive. Their procedure utilized a mathematical method known as recursive structural identification (RESID), which had been developed earlier in 1982 by Williams and Zwieke (4179). This technique had to be coupled, in turn, with ultrasonic scan data from the honeycomb test panel. Other variables affected by heat-up or cure rates of the adhesive had included stress level by Mahoney (4180), rheological behavior by Thuen and Hinrichs (4181), and tensile lap-shear strength by Sancaktar et al. (4182). The combination of NDE testing and RESID has led to successful prediction of a number of aluminum-to-aluminum adhesive bond properties such as (1) bondline thickness, (2) bondline porosity, (3) glass transition tempertures, (4) degree of cure, and (5) level of bond strength.

Mahoney (4180) has studied the influence of adhesive components on the corrosion of aluminum honeycomb by determining if the presence of various curing agent types would affect the amount of aluminum corrosion found as compared with the effect of water alone. He also evaluated how corrosion could be reduced through coating the aluminum with selected anticorrosion additives or anodizing.

The one-part nature of structural epoxy-based film adhesives, as employed in most aerospace applications, can sometimes be a serious disadvantage as a result of aging. The time in transit between manufacturer and user can sometimes be lengthy, making it desirable to develop a detailed knowledge of the effect on the chemical and physical properties with increased aging. Such studies have shown that slow cure, hydrolysis of the resin, and specific interactions between components can all take place. Ultimately, these changes can alter both the strength and serviceability of aluminum joints in structures.

Considerable effort has been expended by various investigators to establish chemical methods of characterizing the epoxy-based adhesives used in aerospace structures (4183–4185). Other investigators have worked on the adverse effects of high humidity on the properties and cure characteristics of these types of adhesives (4186–4188).

In the period from 1980 to 1984, six technical papers were produced on the changes in chemical and physical properties of the one-part epoxy-type adhesives under storage conditions (4189–4193). Particular focus was on changes in 250°F (121°C) curing nitrile-epoxies with service to 212°F (100°C) and curing epoxy films with aerospace service to 392°F (200°C). The results have been summarized in the 1984 chapter by Morris, Pearce, and Davidson in *Adhesive Joints Formation, Characteristics, and Testing*. The Alclad 2024-T3 joint geometries and testing methods were according to the standard aviation specifications described by Pearce et al. (4193). Some of the various factors evaluated were (1) epoxide content, (2) adhesive flow, (3) lap-joint strength and climbing drum peel using aluminum honeycomb panel specimens. The adhesives (stored for varying time periods at 73.4°F [23°C]) were also subjected to chromatographic separation methods with the isolated materials identified by infrared analysis. Over a period of weeks at only ambient temperature, some polymerization (chain extension and cross-linking) did occur. As anticipated, this led to a measurable reduction in the epoxide content and some reduction of the solubility of the adhesive in certain organic solvents. Where a substantial molecular weight increase had occurred, there was a reduced flow of the adhesive under heat-curing conditions. This can be especially important in aluminum honeycomb panel bonding, where the best possible adhesive flow is required for best filleting action and subsequent strongest bond formation.

Pearce et al. (4191) showed in 1981 the considerable susceptibility to degrading by hydrolysis of certain epoxy resins used in high-temperature adhesive formulations. They extended this to the situation with the diglycidyl ether of bisphenol A (DGEBA) resin as well in 1983 (4193). This hydrolysis could lead to epoxide content reduction plus an overall increase in adhesive polarity, which in turn could further increase the moisture-uptake properties as the aging process continues. In the nitrile-modified epoxy systems, the carboxyl terminal group of butadiene acrylonitrile (if not prereacted) can react with epoxide groups during storage. In addition to the epoxide content reduction which results, there is a decrease in adhesive solubility and adhesive flow during cure. The effect on joint strength seems to be more related to the means by which the epoxide content reduction occurs. If reduction under aging conditions is due to a slow normal cure process, the joint strength will be little affected. However, if it is due to non–cure-related factors like hydrolysis or interaction of components (excluding curing), reduced joint strength will likely result, especially at elevated service temperatures. While the reduction of adhesive flow rate could adversely affect the aluminum wetting situation quite significantly, it might also prevent adequate filleting in the bondline between facings and honeycomb. Increasing the bonding pressure could possibly provide some compensation for decreased wettability, but generally reduced flow situations could not be balanced. It should be remembered that many manufacturing situations simply do not permit application of more than some relatively light pressure.

K. Titanium Applications in Aerospace

Although this volume has carefully omitted the use of any adherends not ordinarily bonded to aluminum, it would seem appropriate when speaking of aerospace applications to remind the reader that applications exist today where higher strength properties at elevated temperature service are needed than available with aluminum alloys. For these types of applications, the unique high strength/weight ratios, excellent fracture toughness, corrosion resistance, and retention of strength at very high-temperature properties of titanium need to be considered. Sanders et al. (4194) have reviewed these properties. Steinberg (4195), in 1986, reported titanium components were acceptable for operating temperatures as high as 797°F (425°C). Donachie (4196) meanwhile has reviewed some of the application areas, including (a) airframe structural components, (2) landing gear beam assemblies, (3) gas turbine blades, (4) variety of medical devices like high-strength prosthetic implants, (5) nuclear power generators, and (6) cooling tower structures.

Mahoon (4197) has listed the advantages of bonded titanium structures in the aerospace industry over those produced by mechanical attachment (riveting and bolting) or by fusion processes such as welding, brazing, diffusion bonding, or super plastic forming. They include (1) lighter weight, (2) improved stiffness and compression, (3) higher fatigue resistance, (4) absence of galvanic corrosion, (5) improved aerodynamics, (6) even distribution and dissipation of localized stresses, and (7) better tolerance to flaws.

As discussed at great length in this volume about adhesion to aluminum surfaces, a number of investigators have also demonstrated similar contributions of mechanical interlocking of the polymer and chemical bonding contributions when bonding the titanium oxide surface. (For further detailed discussion, see Refs. 33,463,523,539, and 755.) The most detailed description of the state of the art of bonding titanium has been assembled by Shaffer et al. (1219).

As with aluminum, the most important aspect of promoting good titanium joint adhesion is related to the type of oxide generated on that adherend surface as has been described (33,621,723, 4198–4213). Stable and microscopically rough titanium oxide seems to be the best overall performer in both tension and shear test modes based on durability studies in moderate aqueous environments (4204,4207,4208,4212–4214). As with aluminum-bonded joints, it may be possible to extend titanium joint durability performance by application of coupling agents (see Refs. 723,4201, and 4215) or alkoxide primers (4216–4222).

As mentioned earlier, titanium adherends are substituted for aluminum where service at very high temperatures is demanded. A number of high-temperature adhesives have been especially developed for high-temperature titanium applications. Some representative candidate for such service was developed in LARC-TPI adhesive by Bell et al. (4223). St. Clair et al. (4224) have been actively engaged in such developments. Special studies on the adhesion of titanium and polyimide-type high-temperature adhesives of possible interest are those of Wightman and Diaz (4225).

L. Listing of Articles from *Adhesives Age* Relating to General Subject of Adhesive Bonding of Aerospace Structures

In 1960, an unauthored article appeared with the title "Sealing the X-15" (4226), followed in 1962 by an article by Hahn (4227) summarizing his work on "Lap-Shear and Creep Performance in Metal-to-Metal Bonds" which was directed toward both aircraft and missile production. The three pertinent articles published in 1962 were all unauthored, with the titles "Critical Area of X-15 Sealed with RTV Rubbers" (4228), "Film Bonding Adhesives and the Convair 990 Jet" (4229), and "Bonded Weatherstripping Solves Boeing Problem" (4230). Two unauthored articles followed in 1965 with the titles "Bonding De-Icer Boats to Helicopter Blades" (4231) and "Aircraft Struts and Pontoons-Adhesive Assembly Permits Weight and Cost Reductions" (4232). Three articles in 1966 were similarly unauthored, citing the subject areas "Pre-Fitted Bonded Parts for Aircraft Construction" (4233); "Pressure-Sensitive Aluminium Tape Shields Aircraft" (4234); and "Using Adhesives in a Floating Airfield" (4235). In 1967, the subjects of "Bonded Helicopter Blades Meets Tough Competition in Vietnam" (4236) and "Sealant Kits Speed Airplane Overhauls" (4237) were discussed. Callahan (4238) submitted data on "Building Supersonic Wings and Stabilizers" in 1968, whereas two unauthored articles had the titles "Adhesive Bonding Jumbo Jet Airframes" (4239) and "Sealants Play Major Role in Construction of Military Helicopters" (4240). In 1969, a single article not hitherto mentioned had the title "Bonded Landing Pads Provide Safer Stops for British Crafts" (4241).

Adhesives Age in the 1970s started with an article by Howeth and Cornelius (4242) in 1971 on the subject of "Advanced Composite Production of Wing Trailing Edge Panels." In 1972, Petrino (4243) considered the use of aircraft adhesives and sealants as employed on the L-1011 TriStar aircraft. An unauthored article followed in the same year titled "Helicopter Research Group Verifies Integrity of Industrial Adhesives" (4244). Powis (4245) discussed "Structural Adhesive Bonding of Hovercraft" in 1973. Hart-Smith (4246) offered a general summary of "Designing Adhesive Bonds" in 1978, and Jones and Cassey (4247) in 1979 discussed "Flexible Polyimide Fuel Tank Sealants."

In 1981, Reinhart (4248) summarized the "Use of Structural Adhesives by the U.S. Air Force." An unauthorized article in 1982 was titled "Fuel Tank Bonding Project Cuts Air Force Repair Costs" (4249). Silverman and Norrbom (4250), in 1983, discussed "Polysulfide Sealants in Aircraft Composite Structures," and Novak (4251) wrote about "Wet Epoxy Primer Improves Bondline Shear Strength" in regard to aerospace bonding.

M. Listing of Articles from *Adhesives Age* Relating to the Adhesive Bonding of Space Vehicles

Two unauthored articles appeared in 1961 titled "Pressure Sensitive Tape Helps Catch USAF Space Capsules" (4252) and "RTV Silicone Rubbers Seal Space Capsule" (4253). Gatzek and Feller (4254), in 1962, summarized the range of "Space-Age Adhesive Materials." In another unauthored article in 1965 considerations were discussed about "Adhesives in the Space Race" (4255). In 1966, Roseland (4256) reported his efforts in "Developing Structural Adhesives Used for Space Applications." "Spaceman to Bond Self to Vehicle with Adhesive Capsules" (4257) was the rather unusual title for an unauthored article in 1967. Robertson and Brown (4258) discussed their investigations on "A Bonded Bulkhead for Saturn" in 1968, whereas 1969 introduced an unauthored article titled "Adhesive Plays a Vital Role in Apollo Heat Shield Structure" (4259).

In 1971, an unauthored article appeared which described "Using Weldbonding to Assembly Spacecraft" (4260). A 1974 article discussed "Epoxy Adhesive Used in Spacecraft" (4261), and Singh (4262), in 1978, described his investigation "Cyanosiloxane Sealants Used in Construction of Space Shuttle Orbiter."

IV. SANDWICH PANELS AND LAMINATES

A. Early History

The technology for fabricating aluminum-faced sandwich panels and laminates had its beginning in the time of World War II, when the all-wooden wings of the early days of the war were improved by the addition of adhesive-bonded metal reinforcements. It was the pioneering research and development work of DeBruyne (4263) that led to the acceptance of hot-curing adhesives for metal surfaces. The British De Havilland Aircraft Company saw the benefit of improving the structural efficiency of their all-wooden wings in the historically famous Hornet aircraft by bonding aluminum alloy spar caps to the plywood shear webs and skin of the wings.

It remained for the Fokker Company shortly after the war to begin using the same polyvinyl acetal/phenol-formaldehyde adhesives that had worked so well with wood to bond entirely aluminum structures in both military and commercial aircraft (41). The advantage gained through the use of bonding was to provide resistance to buckling compressive loadings for aluminum sheet structures. When two sheets of aluminum are laminated to opposite sides of some light-weight core material which can transmit the shear load, a sandwich panel with a high degree of bending stiffness and stability under compressive loading is achieved (4264–4266). Minford has compiled a table which offers the comparison between the actual general properties of 16 different possible core materials for making laminated aluminum-faced panels as compared with the properties that might constitute an "ideal" core material. Properties of maximum importance might include density, tensile strength, compressive strength, surface friability, fire resistance, water absorption rate, vapor transmission rate, insulation value, dimensional change soaked in hot humidity, and general chemical compatibility with aluminum under water-soaking conditions (89).

At first, the aluminum-faced structural panels for aircraft had light-weight, balsa wood cores, but aluminum honeycomb core soon proved to be a much better choice, offering the most effective combination of light weight, high stiffness, and high strength. Providing a strong, durable, continuous adhesive attachment between facings and core was obviously vital to the successful production of such a sandwich panel. There are a number of additional reasons in designing aircraft for choosing bonded sandwich structure beyond those mentioned above. For example, the growing demands for close dimensional tolerances on aerodynamic and related parts (e.g., ailerons, rudders, elevators, spoilers, slaps, air brakes, slats, tabs, and vanes) can be effectively fulfilled by these stable light sandwiches. Sandwich construction also has proven of great value for the acoustical damping characteristics they can provide in and around engines and their nacelles. The same properties also lead to good resistance against high sound pressure, which can cause serious fatigue failures in nonsandwich constructions in relatively short time. Information about aluminum honeycomb as used in aerospace applications is given in MIL-C-7483.

Depending upon the strength and stiffness requirements of the final panel, aluminum honeycomb is available in a wide range of standard cell sizes and densities. Core fabricated with the earlier used 5052 aluminum alloy has been replaced more recently with 5056 alloy and its substantial boost in tensile strength. Current repair procedures call for replacement of the 5052 with 5056 core in existing aircraft (see Ref. 4125). The bonding of aerospace panels can be accomplished by either vacuum bag- or autoclave-curing techniques; however, the durability of the final structure will likely not be equivalent. Obviously, the choice of a most appropriate adhesive can be of primary importance regardless of which curing procedure might be employed. Data from Kuperman and Horton (4125) demonstrate cyclic load-environment exposure variations from both curing and adhesive choice variables. Regardless of adhesive selection, the higher-durability performance seemed to be present with autoclave-cured test specimens. It was certainly significant that the room temperature–curing aerospace-type EA-9309 epoxy adhesive joints failed in less than one cycle under the lowest loading of 900 lb/in. Their use, therefore, cannot be recommended for permanent aircraft repairs or even temporary repairs in critical structure areas.

Over the years, a number of articles pertaining to aspects of adhesive bonding sandwich construction have appeared in *Adhesives Age Magazine* which should be cited. In 1960, Myers (4267) reviewed "Adhesives for Honeycomb Cores," whereas Leeser (4268), in 1961, described fabrication of "Bonded Honeycomb-Core Steel Doors," and an unauthored article appeared on "How Adhesives Are Used in Honeycomb" (4269). An article in 1962 provided information on "Seven Layer Bonded Sandwich Provides Skiis with Permanent Camber" (4270). Tipton (4271), in 1965, discussed his work on "Haystack Antenna System Features Honeycomb Sandwich Construction," and the next year, Hochberg (4272) wrote about "Evaluating Adhesive Bonded Temperature-Resistant Sandwich Construction." In 1969, Kantner and Litvak (4273) described "Adhesives for Bonding Large High-Temperature Sandwich Structures" and Lewis and Elder (4274) discussed "Selecting Adhesives for Vibration Damping Metal Laminate Applications." Phelan (4275), in 1974, described methods of panel attachments in an article titled "The Forgotten Corner of Bonding, How Sandwich Panels are Attached." Also in 1974, Berg and Ramsey (4276) discussed "Viscoelastic Adhesive Effects on Sandwich Structure." Finally, in 1987, an unauthored article described "Waterborne Cuts Cleanup and Eliminates Compliance Costs" (4277) (about stress skin and structural panel fabrication).

B. Nonaerospace Aluminum Structural Panels

A wide variety of physical and chemical characteristics can be of importance in providing properties that need to be considered in selecting a core material for bonded aluminum-faced sandwich panels in different manufacturing industries. Mention has been made above about the comparisons made by Minford between 16 different possible core candidates which were considered at the Alcoa Laboratories.

1. ALUMINUM BONDING TO FOAM INSULATION CORES

The low-density, low-cost, highly insulating foam plastic core materials for use in aluminum-faced sandwich panels were discussed as early as 1954 by Rose (4278). Humke (4279) in 1958 and Hemming (4280) in 1969 reviewed molded laminates and sandwich materials. Minford and co-workers (87–89) described a wide variety of such core insulations and their fabrication into aluminum-faced sandwich panels in 1960, 1973, and 1975.

a. Polystyrene Core Considerations

Actually a much larger consumption of both adhesives and aluminum-facing materials has gone into the production of various types of aluminum building panels than used in all aerospace applications. Only rarely are such building panels fabricated with regular aluminum honeycomb, since good thermal insulation is equally important to structural considerations. For this purpose, the various foamed plastic core materials can act as highly efficient insulation cores while permitting laminating to the aluminum facings to produce a multilayered curtain wall or even a load-bearing commercial building panel with appropriate design and attention to proper erection details. In Minford's early Alcoa work, aluminum facings were placed in the bottom of a mold and the cavity was filled with preexpanded polystyrene beads. These contain a propellent capable of completing the expansion when steam is forced into the mixed beads. Before forcing the steam into the cavity, the top facing had to be put in place as an effective seal to expansion of the beads beyond a designated volume. The weight of beads expanding into a given volume will, of course, determine the final density of the beadboard produced. The author early on found that the degree of adhesion exacted against a bare aluminum facing by the expanding thermoplastic form of polystyrene beads was insufficient to form a laminate with adequate initial strength or one able to adequately resist even mild weathering. The degree of adhesion to a prepainted, anodized, or conversion coated aluminum sheet was overall unacceptable for commercial production. However, a panel with adequate strength and bond durability in hot humidity soaking tests could be produced by precoating the side of the aluminum facing to be foamed against with a good-quality contact cement (preferably of the neoprene-phenolic class). The superior wettability potential of the adhesive surface by the hot, expanding, thermoplastic polystyrene beads enabled an effective bonding that resulted in a cohesive rupture of the beadboard both initially and after 6-mo soaking in 100% RH at 125°F (52°C).

The alternative manufacturing procedure adopted by Alcoa was to separately adhesive bond preformed polystyrene beadboard or Styrofoam board to aluminum facings that had been precoated with a neoprene-phenolic contact cement formulated to resist service temperatures exceeding 180°F (82°C). Alcoa developed the largest industrial machine of its time to further expand partially expanded beads to approximately a 2.5 lb density core board. Actually, a 4-ft wide slab of 8-in thickness was expanded in a continuous web and sawed into designated thickness and length boards. Usually, it was found most efficient to mass produce 4 × 8 ft panels with specific combination of facings and saw into any panels of lesser dimensions. The procedure was described elsewhere when discussing the use of contact cements. Briefly, it consisted of applying adhesive by curtain coater to the facing to be bonded and passing through a heating tunnel to remove solvent that would be destructive to the polystyrene upon contact. Upon exiting the heating tunnel, the hot sheet (inner facing) was placed on the heated bottom platen of a press and the insulation core placed on top. Finally, a hot top facing was placed over the core and the press closed for a period of approximately 90 sec with sufficient pressure to achieve adequate adhesion. Upon opening the press, the hot sandwich was passed through a neoprene-roll, rotary press with a line pressure sufficient to slightly compress the sandwich while moving through the rolls but with insufficient pressure to alter the desired panel thickness permanently. Durability potential of panels made of these materials in this fashion was established through exhaustive testing of first laboratory-fabricated and later the commercially produced panels in a wide variety of water-soak, hot humidity soak, hot water, or hot humidity soak/freeze/air-dry cycling and tensile and shear testing at elevated temperature conditions. Some testing specifications that were set are listed in Table 129. Finally, it should be pointed out that the practical performance in service of large bonded structures cannot be completely determined by evaluating small-segment specimens cut from the larger structure. Alcoa found it necessary to build a special testing apparatus that could enclose and support the full-sized panels (up to 4 × 8 ft) under the field erection design conditions. Further, means of uniformly heating the total panel, as it would be heated in natural atmospheric weathering conditions, had to be undertaken with attachment of devices able to measure all movement of the materials and determine any change in dimensions or geometry of the panel under loading and heating. Where extra stiffening or dent resistance was desired, a second back-up sheet material like pressed hardboard, plywood, or even wood particle board could be separately laminated to the backside of the facing sheet, which could then be laminated to the core materials in the fashion described above.

Some articles that appeared in *Adhesives Age* in the middle 1960s and early 1970s were pertinent to such laminated sandwich panel manufacturing. In 1962, an unauthored article described "High Speed Infrared Panel Curing" (4282), whereas another article in 1966 discussed "Applying Facings with Roll-Coating Techniques" (4283). Another article in 1967 reviewed how "Hot Spray Adhesives Speed Production of Sandwich Panels" (4284), and a 1968 article titled "Latex Adhesive Effectively Bonds Insulation to Metal Panels" (4285). Finally, in 1972, Debnar (4286) reviewed his process for "Selecting Suitable Adhesives for Expanded Polystyrene."

b. Polyurethane Core Considerations

The bonding of an aluminum sheet or prebonded laminate to polyurethane insulation to form sandwich panels needs some special discussion because of the unique physical and chemical characteristics of this thermoset-type plastic core material as comapred with the thermoplastic polystyrene just discussed. Polyurethane insulation can be incorporated behind aluminum facings by adhesive bonding prefoamed slabstock as noted above; however, it provides an additional possibility as a direct foaming-in-place liquid mix material.

(1) Foamed-in-Place Aluminum/Polyurethane Panels. The adhesion that can be developed when foaming against solvent-cleaned bare aluminum by a polyurethane mix (unlike the expanded polystyrene beadboard) is capable of resisting considerable water-soaking exposure or natural atmospheric weathering conditions. The adhesion is further enhanced using a clean abraded or prepainted aluminum sheet. The relationship between the exothermic reaction temperature and the aluminum facing temperature (while foaming-in-place) can be a very important factor, because of the natural high heat conductivity of aluminum. Because the rate and uniformity of foam development depends on this exothermic reaction, the foam uniformity and its wetting of the aluminum surface can vary significantly depending on how much the aluminum is acting like a heat sink. For this reason, it is never good manufacturing practice when foaming polyurethane against any aluminum surfaces to start with the metal at room temperature. Rather, the aluminum needs to be heated to at least 130–140°F (54–60°C) and maintained there during the processing. Evidence of the deleterious effect of aluminum altering the heat exotherm at the aluminum interface is usually quite clear by observing an altered foam density and significant void production in the bondline area. This will result in a much lower initial tensile rupturing strength and a lowered resistance to later weathering conditions. The more practical manufacturing condition has been to use a preheated mold in which the inserted aluminum facings can come to thermal equilibrium with the surroundings before the cavity is charged with the mix. Alternatively, a faster panel manufacturing process has been developed by devising a continuous foaming line where foaming is against a heated continuously moving web of aluminum sheet. Theoretically, the length of panel desired can simply be sawed to whatever length is required. This is not easy to achieve, however, since the sawing operation must somehow be isolated from the continuous panel production line.

In addition to the creation of density alteration and void production near the interface between the expanding line of foam and a cooler aluminum facing, the author has also observed the possibility of actual foam collapse back from the interfacial contact area. This can completely destroy the ability of the final panel to uniformly support and resist loading. This subject introduces the differences that can exist between systems of manufacturing panels that can best be described as horizontal versus vertical foam pouring. In the former, the direction of foam rise is in the short direction of the panel. This most closely corresponds to the method of laying up the polystyrene slabstock panels where the facings are laid horizontally under and over the core material. With a foaming-in-place operation, however, the mold can be operated vertically, in which case, the foam is rising vertically through the total height of the panel. It should be clear that this vertical pour foaming is statistically more subject to a dragging effect of the foam on the sidewalls which can lead to more extensive variations in adhesion and foam density variations between different parts of the overall panel. This will lead to more variability in achieving uniform loading potential. Thus, it would seem to be the distinctly preferable manufacturing operation to have the foam rise through the shortest dimension of the panel.

Unfortunately, after all the positive factors for making a foamed-in-place aluminum/polyurethane panel have been employed as described, Minford (4287) found that a basic chemical incompatibility problem could exist between many commercial foam formulations and aluminum facings when subjected to water soaking exposures for varying periods of time. For many years, Alcoa had employed a standard exposure test of 6 mo in 100% RH at 125°F (52°C) as a means of establishing both compatibility and adequate adhesion strength for laminated aluminum panels involving various materials of construction. When this testing was performed on solvent-cleaned or abraded and cleaned 3003 alloy facings and foamed into panels with the earliest polyurethane foam formulations from duPont, a scattered pitting corrosion developed which could lead to actual perforation of the aluminum sheet. This had not been recognized by any of the polyurethane formulators because their testing evaluations with aluminum had not included such water-soaking exposures. It was further observed that these pits could originate in as short an induction time as 30 da or less. The corrosion rarely took the form of an overall uniform etching which would fail the overall bonding but rather developed scattered but highly localized areas of pitting which could actually perforate the sheet from the inside through to the outside surface before actual panel debonding occurred.

Methods had to be investigated to either obtain compatible formulations or find means of inhibiting the action by pretreating the facing side that would ultimately contact the polyurethane. Of course, it should be noted that literally thousands of polyurethane-foamed aluminum products may have been manufactured over the years which may not have had service conditions which ever developed this incompatibility problem. However, the problem has definitely occurred in the field where water could invade and remain at the foam/aluminum interface. Minford (4288) began to evaluate a wide variety of different polyurethane foam formulations to distinguish more compatible foams. Minford (4289) evaluated a wide variety of differently pretreated aluminum facings, including a number of different protective coatings and primers, always using 6 mo in hot humidity as a screening test. All paint coatings commonly employed for coloring architectural aluminum proved, in themselves, to offer a wide range of protection values against the same foam formulation. For this testing, the author deliberately picked a commercial foam formulation which readily created pitting, and even perforation, of bare 3003 aluminum facings with 30-da soaking. The range of paints tested included vinyl wash primers, various clear and pigmented heat-curing vinyls, alkyds, vinyl-alkyds, acrylics, and epoxy paint coatings. These coatings were commercially applied over a chromate conversion coated surface so paint adhesion was never suspect. In general, the various heat-curing epoxy paints with their outstanding good resistance to a wide variety of chemicals proved to be the favored solution to this problem. However, the author would favor the use of a completely compatible polyurethane foam formulation regardless of whether a heat-cured epoxy painted aluminum facing had been employed or not.

Another of these investigations by Minford (4290) was related to explaining the mechanism by which foamed-in-place polyurethane could produce corrosion when contiguous with an aluminum facing under water-soaking conditions. Minford was eventually able to develop this understanding and was also able to reproduce the corrosion on 3003 aluminum alloy in a test tube by combining dioxane (a simple substitute for the polyol in polyurethanes) and Freon 11 (the blowing agent present in all the commercial polyurethane formulations that had promulgated corrosion in hot humidity soak). Apparently, the isocyanate plays no role in the reaction. When no water was present in a mixture of dioxane and Freon 11 with 3003 surfaces present, there was no evidence of any corrosive interaction. However, when a few drops of liquid water was added and the mixture set aside to incubate at the same temperature as used in the Alcoa hot humidity test cabinets, then the same kind of surface pitting corrosion could be observed. In yet another investigation, Minford (4291) was able to show the specificity of the Freon 11 propellent by substituting Freon 12 for the Freon 11 and not finding corrosion. Further, Minford was able to obtain some specially foamed panels from the Freon Products Laboratory using a specially formulated polyurethane with Freon 12 blowing present. When exposed to periods up to 12 mo in 100% RH at 125°F (52°C), there was no evidence of any incompatibility with that formulation. Minford related this difference in compatibility to the much higher resistance of the carbon-to-halogen bond to hydrolytic rupture as compared to the resistance of that bonding in the asymmetrical Freon 11 molecule. Actually, Freon 12 has been employed as the refrigerant gas in aluminum room and car air conditioning units operating at temperatures up to 250°F (121°C) with no problems. On the other hand, the author has encountered aluminum-tubed large building heat exchanger units over the years when even a modest number of parts per million of water has caused some pitting on the surfaces even at much lower operating temperature where Freon 11 was the refrigerant in the system.

The above corrosion problem encountered under wet conditions with Freon 11 blowing agent has not been encountered by the author with the so-called "polyester-type" foam compositions. The difference in performance has been attributed by Minford to the fact that the gas remaining trapped in the cells to furnish good insulation capacity is carbon dioxide instead of Freon 11. It might be questioned then why such polyester formulations should not be universally substituted for aluminum-faced foamed-in-place panels that might be exposed to high humidity. The answer has two important aspects. First, the thermal efficiency of the foam with carbon dioxide in the cells is significantly lower (about twice the k factor). If designing to a designated panel thermal efficiency, this would require both twice the thickness of panel and a corresponding increase in materials cost.

(2) Adhesive-Bonded Aluminum/Polyurethane Slabstock Panels. Fortunately, no chemical incompatibility problem exists between bonded prefoamed polyurethane slabstock and aluminum facings even in the presence of 100% RH at 125°F (52°C) exposure conditions for 6 mo. This is apparently true whether the foam slabstock is mechanically held in tight contact or separately bonded. Minford (4292) has conducted extensive compatibility testing with a wide variety of polyurethane slab stocks submitted by all the major polyurethane foam manufacturers. Unlike the thermoplastic polystyrene core materials which are all sensitive to some degree to most organic solvents used to disperse neoprene-phenolic cements, the thermoset polyurethane slab cores are quite resistant. For example, a neoprene-phenolic contact adhesive mixed with a 50/50 mixture of methyl ethyl ketone and toluene can be sprayed or even curtain coated directly over the polyurethane cellular surface without any significant cell collapsing. This same practice would utterly collapse and destroy any polystyrene-celled surface. Fortunately, this means that both aluminum facing and polyurethane core surface can be adhesive coated and laminated together in a sandwich.

This latter situation became especially important in manufacturing when Minford showed that the same procedure as used to laminate aluminum/polystyrene panels (adhesive applied only to the aluminum) would produce a relatively low strength and durability in weathering panels. This was due to the presence of a very friable layer of ruptured cells on the polyurethane surface which provided a weaker boundary layer condition when bonded with only the single layer of adhesive on the aluminum surface. Both a significant boost in initial tensile-rupture value and retained tensile-rupture value after prolonged soaking in hot humidity was obtained when both the aluminum and slabstock core were coated and subsequently hot pressed and laminated in a rotary press. It was found to be possible for manufacturing expediency to precoat cores of polyurethane and store for latter laminating to aluminum when various orders called for differing forms of aluminum and aluminum-to-a back-up material laminate.

Some *Adhesives Age* articles also have appeared related to this subject. An unauthored article in 1968 was titled "Sprayable Adhesive Improves Urethane Foam Bonding" (4293). Strobech (4294), in 1977, described "Bonding Lightweight Insulated Sandwich Elements with Polyurethane Adhesives" and DeGisi and Smith (4295), in 1978, wrote about "Improved Bondability of Molded Rigid Urethane Foam."

c. Isocyanurate Core Considerations

Isocyanurate foam formulations are really variations of polyurethane chemistry in which the starting materials have been altered to provide a significant increase in fire resistance. Less than desired fire resistance in the conventional polystyrene and polyurethane core materials has prevented their use in areas with restrictive fire codes on the use of burnable material in building panels or where specific values of acceptance for rates of burn exist. The high resistance of isocyanurate foam insulations, even to direct fire exposure coupled with their nonburnability, has made them desirable for use in laminated aluminum building panels. However, this fire resistance is obtained at some expense of flexibility, and the extreme friability of the isocyanurate surface can require some attention to how much pressing pressure can be exerted in lamination without introducing a significant weak boundary layer condition in the bondline. Also, isocyanurates, because of their complex chemistry, have a high resistance to dimensional change in hot humid conditions. Minford (4296) evaluated the bondability and bond durability in various weathering conditions of a variety of the earlier and later commercial formulations offered by the polyurethane foam manufacturers for applications demanding greater heat resistance and nonburnability. He found some significant variation between the products of different formulators, especially as to the degree of surface friability and dimensional stability in long-time hot humidity soaking. As anticipated, the quality of the later formulations steadily improved, so that good bondability and dimensional stability ultimately was achieved.

Even higher fire resistance aluminum-faced panels have been produced by Minford using core materials like mineral fiberboards, ceramic foams (epoxy-bonded fly ash), or foamed glass. None of these materials, however, can begin to match the higher degrees of thermal insulation available from using the previously discussed polystyrene beadboard or polyurethane slabstock or foam-in-place polyurethane structures.

d. Other Organic Polymer Foams

A number of other polymeric foam insulation materials have been evaluated by Minford (4297) such as polyethylene foam, vinyl foam, epoxy foam, phenolic foam, and urea-formaldehyde foam. All of these materials have a higher degree of solvent resistance than the polystyrene cores, but usually require precoating of their surfaces with the adhesive to secure maximum stress rupture bonds and have definite limitations in regard to their density availabilities. As compared to the approximately 2.5-lb density of the blown polystyrene or polyurethane-type cores used for the Alcoa Alply Panel, only the commercial urea-formaldehyde core can match this low value. However, the durability in hot weathering of this density urea-formaldehyde leaves much to be desired. Polyethylene foam has been available in the density range of about 3 lb/ft^3 but generally is most available in a higher-density range for truly structural panel performance. Polyvinyl, epoxy, and phenolic foam slabs are generally available at higher density levels with corresponding higher expense per square foot of panel. Any of these other organic foams can be effectively bonded with a good-quality neoprene-phenolic contact cement to aluminum facings which have acceptable initial and after exposure tensile-rupture values. Some corrosion problems have been noted with the phenolic foams, which produce hydrochloric acid in the blowing reaction. There have been reports in Europe of their successful use, so that some sort of durability evaluation that simulates the worst condition to be anticipated in service is called for.

e. Inorganic Insulation Aluminum Panels

Inorganic insulations can also be an important form of core material to which aluminum facings can be adhesively bonded. The heat-insulative properties that they can provide are usually significantly lower than present with the various forms of foam plastics discussed above. However, their good fire resistance and nonburning characteristics may be an overriding reason for their selection where stringent fire codes are in effect. Minford (4298) has evaluated a wide variety of commercial products of this category, including mineral fiberboard, foamed glass, ceramic foams, gypsum board, and cement/mineral fiberboards. All of these can be bonded with the same neoprene-phenolic adhesive as used for the various organic foams. It is also possible in some less demanding service environments to use relatively high-solids rubber mastic cements, since the presence of temporarily entrapped organic solvents has no degrading effect on such materials.

Mineral fiberboards can have very rough surfaces and are usually of quite high density but are relatively inexpensive. Compatibility studies need to be considered when involving these materials with aluminum facings even though an adhesive is presumed present as a separating layer. Such studies by Minford always involved sandwiching the fiberboard with clamping between two bare aluminum sheet facings and soaking for up to 6 mo in the hot condensing humidity.

Foam glass similarly presents a relatively coarse-textured surface but with a greater degree of surface friability than found with the mineral fiberboards. Its high tensile and compressive strength, excellent fire resistance, low water absorbtion rate, and general compatibility with aluminum under hot-soaking conditions are all positive characteristics making this core worthy of consideration for making structural aluminum panels. Minford has shown that a manufacturing problem can exist owing to the high friability of the glass cellular structure, especially as to its sensitivity to crushing. This was particularly noted where the high line pressure exerted by a rotary press operation was employed. Minford found that the strongest and most weather-resistant aluminum-faced/foam glass panels were produced by introducing a thin high-density neoprene sheet between the metal and glass interface. Under these conditions, it was possible to even use a rotary press operation for the lamination. The initial strength could be governed by the cohesive strength of the neoprene sheeting. The fire resistance of such a bondline would, of course, be suspect.

Ceramic insulation cores for many years had the reputation of being extremely sensitive to high-humidity weathering. Dow Ceramic insulation slabs underwent significant improvements in dimensional stability over many years as based on compatibility testing and panel durability tests conducted by Minford. The final manufacturing essentially duplicated the procedures used for bonding polyurethane slabstock to aluminum using adhesive application to both surfaces, partial removal of solvent and hot flat bed pressing followed by passage through a rotary press. Once again, the ceramic insulation material (usually an epoxy-bonded fly ash) could be friable enough to fracture if exposed to a high line pressure in a rotary press. (More information on such cores can be found in Ref. 4299.) Handling and assembly problems (especially with long panels) due to their quite high densities and only modest resistance to cleavage breakage when handling did require some special attention on the production line. Finally, the difference in the coefficient of thermal expansion between aluminum and the ceramic could cause localized debonding under high solar heat conditions.

Gypsum boards had the problem of bond failure under adverse weathering conditions in the paper facing of the board as manufactured. No matter how strong or weather-resistant an adhesive is used in this panel, the bond can not be stronger than the wet fiber strength of this paper.

Cement/mineral fiberboards would be those replacing the older cement/asbestos board products. Minford (4300) has shown that typical bonding procedures as outlined for the aluminum/polyurethane foam slabstock-type panel can be used to produce durable panels for most water-soaking conditions. It is suggested that chemical compatibility testing be conducted to accept such cement/mineral fiberboards with the appropriate aluminum alloy when mechanically sandwiched between the aluminum and soaked in 100% RH at 125°F (52°C) for at least 6 mo.

It should be recognized that a room temperature–curing epoxy paste oculd be applied to one adherend surface and held together until the epoxy adhesive cured in final panel form using all the above core materials.

f. Aluminum-to-Wood Building Panels

A substantial market has existed for a number of years in manufacturing aluminum-faced prefab homes where the aluminum has been adhesive bonded to a prefab wooden framework. The aluminum facings are generally overcoated with a baked paint finish on the front side and some designated back coating on the side to be adhesive bonded. The aluminum is usually supplied in a corrugated form which permits direct bonding of the flat areas to the wood support. In other designs, the flat aluminum sheet has been directly laminated to a wood backing, which could be a weather-resistant plywood. An inside facing can be aluminum, an aluminum/wood or plastic laminate, or simply prefinished wood. Because this finished structure is to be the total sidewall construction of some building, it is also necessary to insert some thermal insulation in a cavity between the outer and inner wall constructions. This could be mechanically inserted batt insulation, or some foam such as polyurethane could be poured and formed in the cavity. In this latter case, it may be necessary to conduct the overall expansion while the wall structure is being restrained in a jig.

In many of the high-production prefabricated home operations, high-solids rubber mastic cements are extruded in a series of beads on the flat areas of the corrugated sheet and modest pressure is applied to squeeze the adhesive uniformly over the surface with the hope of achieving some squeeze-out along the edges. Since it cannot be afforded to test these panels in the elaborate methods of the aerospace industry, the best assurance of quality may reside in observing such squeeze-out along all edges of the bondline. Minford and Vader (4301) have evaluated a large number of such mastic cements over the years for this kind of application and have recommended the use of neoprene-based mastic adhesives for most durable bonding.

Another factor not previously mentioned is the effect that the color of the aluminum exterior surface could have on the temperature and speed of heating the bondline by infrared lamps in manufacturing. Data were developed by Minford (4302) showing the metal temperatures attained when heating facing colors ranging from white to black with various heating times under a 250-W infrared lamp positioned 12 in above the surface and the back side insulated from rapid heat dissipation. It seemed clear that a bondline temperature above 135°F was not possible with a white painted panel regardless of heating time, whereas this temperature could be exceeded within 2 min with a black surface with a maximum temperature approaching 200°F. With appropriate insulation of the backside (as in a well-insulated aluminum-faced building panel), temperatures as high as 225°F have been registered in severe desert exposure in Death Valley, California. This exterior color can be very influential for another reason related to the coefficient of thermal expansion for aluminum. The critical matter is the degree of thermal bowing induced in the panel by the extra temperature due employment of darker and darker facing colors. One of the most serious delaminations occurring with the Alply Panel in service took place on a school building in Wisconsin where the architect had requested charcoal-colored aluminum facings. At the same time, the size panel employed was the largest that was manufactured at the time (4 × 12 ft) and the manner of attachment was actually predicated on earlier service conditions using smaller 4 × 8 ft panels with dark bronze finish. Unfortunately, the end results of using darker-colored facings and larger panels were both additive predisposing factors toward creating maximum panel bowing. The debonding actually occurred in the polystyrene foam, indicating that a superior adhesive would not have helped the situation. Several solutions were possible such as using a lighter-colored facing and redesigning the panel attachment geometry. Since the customer insisted on retaining the dark color, new panels were manufactured and attached with a redesigned fixture, with the result that no further problems were encountered while the author was at Alcoa Labs.

The varying chemical nature of different wood species can also play an important role in the bonding of wood to aluminum. This chemical difference can have a distinct effect on the adhesive itself, or secondarily, chemical incompatibility may occur with the aluminum. For details about the differences that can be present in different woods, the author would recommend the chapter titled "Wood as an Adherend" in the 1991 Volume 7 of the *Treatise on Adhesion and Adhesives* series by River et al. (Marcel Dekker, New York). There are considerable extraneous organic and mineral materials in the cell wall and cell lumen of the various woods which can be extracted from the wood by fairly gentle procedures. More can be learned of specific extractives in various textbooks on the subject (4303–4305). These extractives can be soluble to varying degrees in various adhesive-solvent systems. For example, Narayanamurti et al. (4306) found that the extractives of teak wood, although largely only soluble in alcohol/benzene solvent mixtures, can still adversely affect the setting of water-based animal or urea-formaldehyde adhesives. Meanwhile, the extractives from acacia wood, which are quite soluble in hot water, did not interfere with the setting of these same adhesives.

Minford (4307) has conducted evaluations of relative bond strength and durability of adhesion between aluminum sheet adherends and different species of woods over many years. In general, it was found that the woods whose extractives are very acidic, such as the tropical woods, oaks, and Douglas fir, are the most difficult to bond to aluminum and show minimal long-term durabilities.

The author can recount one situation where relatively thin aluminum sheet had been used to laminate to local tropical woods in Surinam for porch enclosures which must survive in the tropical atmosphere. The author happened to be visiting in the country for another purpose and was asked to inspect the condition of these aluminum/wood laminates. It was later proved by Alcoa Laboratory testing that the acidic extractives from the tropical hardwood adherend had been able to randomly dissolve the oxide on the aluminum sufficiently to permit initiation of intense localized pitting ending in perforation of the aluminum facing sheet. The manufacturer had used bare aluminum for the sake of economy and as a result suffered total loss of structure. Later work of the author showed a number of heat-cured epoxy back paint coatings would undoubtedly have permitted this product to survive for the potential service time planned by the manufacturer.

Finally, since large quantities of water-solvent contact cements are being increasingly used for laminating aluminum (foil through sheet thicknesses) to all manner of wood and reconstituted wood-type products in architectural applications, Minford (4308) undertook the study of the water wettability of different commercial aluminum surfaces as affected by the processing of the aluminum. This included mill-finish, painted, alodized, and anodized surfaces as they might be supplied by the aluminum producer to the manufacturer of laminated panels. As might be expected, the range of initial water-wettability conditions was so great that any manufacturer using these aluminum surfaces needs always to make specific wettability potential observations on each new lot of material as received and just before use on the production line. The latter operation will catch the occasional situation where in-plant handling or storage has contaminated the surface since receipt from the supplier.

V. AUTOMOTIVE, TRUCK, AND TRAILER APPLICATIONS

A. General Considerations

The use of a wide variety of organic polymer materials which develop some level of adhesion to steel and aluminum adherends in the automotive and related industries has grown in recent years. Many of these materials, however, must simply maintain good adhesion to achieve sealing between surfaces (sealants) or offer durable attachment of various covering materials (contact cements). Others must offer actual structural joining between the load-carrying members of the vehicle. Because of the lower price of steel body sheet as compared to aluminum, most automobiles have been assembled with steel components for many years. However, the historians of the automotive industry are aware of the fact that many of the earliest automobiles and vehicles were fashioned from aluminum. Today, the most expensive automobiles in the world, like the Rolls-Royce and the Ferrari, continue to be fashioned with aluminum body components, because of the better corrosion resistance and the lighter weight of aluminum. It is anticipated, however, that just as aluminum has replaced steel in many specialized car parts in recent years, it may even offer some replacement of body steel in some models of the future.

Also of importance for future automotive applications is the possibility through bonding of combining aluminum with composite and plastics to produce light-weight but unusually strong components. Fabrication of such combinations in high-volume automotive applications is simply not feasible, however, using the traditional joining procedures. A chapter by Beevers and Kho in *Adhesive Joints: Formation, Characteristics, and Testing* edited by Mittel in 1984 demonstrates the kinds of data that are important for evaluating the use of adhesive-bonded, thin-gauge aluminum sheet structures, especially with reference to box-section beams.

In 1987, Kreibich and Mercantonio (4309) discussed the most recently developed epoxy pastes, reactive hot-melts, adhesive tape films, and polyurethane adhesives for structural bonding in the automotive industry. They point out that structural paste adhesives will possibly require a precure stage to obtain joint handling strength and to resist bondline washout in the painting operations. In this respect, the reactive hot-melts and adhesive film tapes have the advantage of being solids both before and after application with lesser toxicity for the workers. Data are offered on the mechanical properties of these newer automobile bonding candidates with respect to lap-shear and peel strength and variation in lap-shear strength with varying bondline thickness. They concluded that continual improvements in handling strengths after short-induction curing, improved toughness, and improved environmental resistance should be possible with these newer solid adhesives. The design of adhesive joints for optimum strength has become the crucial factor for the usage of adhesives in new automotive designs.

An increased use of aluminum has been under intense consideration because of its great potential for reducing weight in vehicle structures, attractive appearance of its many finishes (including a distinct improvement in inherent weathering resistance over steel), and the availability of detailed knowledge about its bondability to itself and other materials, which has developed out of its long and continued employment in the aerospace industries. In addition, there has been a general increase in the use of other aluminum automotive components such as trim, rocker arms, transmission cases, brake drums, suspension parts, oil pans, intake manifolds, and a variety of engine parts, leading to the overall description of many automotive power plants as being primarily aluminum engines. Still new but definitely planned for many future vehicles will be the use of thin aluminum sheet to make rigid box structures. It is also likely that aluminum structures in future vehicles will be designed with more extrusions and castings with new kinds of structural designs involving tube-and-socket-type jointures. It would seem to be important to recognize the quite different shapes of the stress distributions along the overlap of such axisymmetrical joints as compared with the standard lap-shear joints from which most existing bonding data have been accumulated.

Paris (4310) has reviewed the use of anaerobic adhesives in the automotive industries from the late 1950s to 1977. Of particular interest has been the use of the retaining compounds for the mounting of bearings, shaft-mounted parts like gears, pulleys, collars, fans, impellers, fly wheels, and cams. The use of anaerobics for reduction of bearing cost assembling has been discussed in 3M's Adhesioneering publication titled "Bearing Assembly Costs Reduced with Anaerobic Adhesive." Haviland has summarized the facts in a Loctite Corp. report titled "Automating Adhesive Joining," which is about the use of adhesives in manufacturing as the best way to achieve control of the process variables. Haviland presents examples of automation using members of the epoxy, polysulfide, anaerobic, and cyanoacrylate families of adhesives. Included also are particular items of equipment and where they can be procured from some 23 different equipment suppliers. The procedures and heating methods to cure the adhesive in place are compared as to relative costs, production efficiencies, overall advantages and disadvantages, and their reliabilities.

Some *Adhesives Age* articles of general interest regarding the use of adhesives in the automotive field might include two technical papers in 1972 by Wright (4311) titled "Incentives for Using Automotive Bonding Methods" and Gialanella (4312) on "Application Systems for Structural Adhesives in the Automotive and Construction Industries." In 1981, Lees (4313) offered his discussion on the "Use of Adhesives in Constructing Vehicles." Schilkey (4314), in 1983, discussed "An Automotive Case History," and in 1987, Wall (4315) reported on an "Automaker Halves Man Hours with Robotic HM Dispenser."

B. Adhesive Bonding in the Automotive Industry

Schneberger (4316–4319) published a variety of papers during his tenure at General Motors Institute relating the use of adhesives in this industry such as "Using Adhesives Effectively" in 1973 (4316), "Low Energy Curing Adhesives-Their Uses, Advantages, and Limitations" in 1976 (4317), "Fundamentals of Adhesive Selection" in 1975 (4318), "Automotive Adhesives—Why, What, Where, and How" in 1978 (4319), and "Automotive Adhesives—Past, Present, and Future" in 1980 (4320).

Schneberger has reviewed the history of the use of adhesives in auto manufacturing from the early 1950s to 1980. He has summarized that 64 adhesive articles had appeared in the *Transactions of the S.A.E.* (4321) through 1978, whereas 68 automotive adhesive entries were listed in the *Engineering Index* (4322) in the decade of the 1970s alone. While about half of these listing dealt with the design and use of adhesives, only nine actually contained specific data and results usable for practical design.

The early applications of adhesives in the automotive industry included (1) adhesive brake shoe-to-lining bonding in 1949 (4323) (see also Refs. 4324 and 4325); (2) vinyl plastisol use in the 1950s for antiflutter bonding of hood inner-to-outer panel bonding (see Refs. 4326–4328); (3) roof panel to roof bow bonding to incorporate increased structural integrity in 1950 (note that roof bows are not in present automotive designs); (4) introduction of fabric-to-roof bonding with elastomeric resins in 1957; (5) anaerobic adhesive applications, including numerous engine and power train locking applications also in 1957 (see Ref. 4329); (6) accelerated use of hot-melts in the 1960s for a variety of uses, including bonding coil wire ends (for mroe recent information, see Ref. 4330); (7) also in the 1960s, the use of thiokol glass-to-metal bonding (updating of this subject can be found in Refs. 4331–4333); (8) an expanded use of pressure-sensitive adhesive formulations with the appropriate strengths and adhesion durabilities to attach insignias and a variety of different decorative trim (for a related article in *Adhesives Age*, see Ref. 4334); (9) use of truly structural adhesives for joining roof-to-side structure panels and station wagon load floor panel bonding (see Ref. 4335); and (10) silicone glass-to-metal bonding to offer added resistance to vibration and greater stability at high- and low-temperature extremes.

In the 1970s, the main expansion was adhesive involvement for holding applications plus some structural innovations. For example, (1) nylon-phenolic films for window hinges to glass panels, (2) urethanes for structural bonding of back lights and windshields to car bodies and structural bonding of polyester fiber glass exterior body panels to metal structural body members in special cars like the Corvette (for more recent publications about FRP and SMC bonding, see, e.g., Refs. 4336 and 4337); (3) more extensive use of anaerobics for the tacking of wire and interior decoration holding; (4) anaerobic sealants to reduce porosity in metal castings; (5) pressure-sensitive acrylates for exterior trim and drip molding attachment (often with mechanical fastener reinforcement) and bonding of body color plastic strip attachments (for recent developments in tapes comprising an elastomeric foam core with double PSA backing, see a 1988 *Automotive Engineering* article (4338) can be cited; also Gheyara (4339), in 1989, reviewed the mechanism whereby such PSA-coated tapes can increase in peel strength with time by a diffusion process); (6) hot-melt attachment of interior carpet eliminating holes through the body floor and seal plates; (7) use of liquid gasket compounds (typically polyesters or silicones) applied by extrusion and offering excellent flexibility, resistance to drying out, and outstanding resistance to oil and gasoline as solvents (see Ref. 4340); (8) use of retaining compounds for engine plugs; (9) silicone sealing of exterior rear view mirrors to glass; (10) anaerobic adhesive attachment of interior rear view mirrors to glass windshield; and (11) bonding of stiffeners to body, roof, and exterior automotive panels.

A number of articles have appeared in *Adhesives Age* over the last 20 years which relate to the general subject of structural bonding of various car components. Listed chronologically, we can cite an unauthored article of 1972 titled "Automobile Tug-of-War Tests Cyanoacrylate Adhesive Bond" (4341). In 1978, Wilkerson (4342) discussed "Acrylic Adhesives: New Way to Bond Aluminum Auto Parts," and in 1980, Bolger (4343) reported his work on "Epoxies for Manufacturing Cars, Buses, and Trucks." Wilkinson and Tyler (4344) extended their data in 1981 on "Acrylics Improve for Bonding Automotive Aluminum Alloys," and Damico (4345), in 1987, discussed "Bonding Galvanized Steel with R.T. Curing Acrylics" (GM cars). In 1988, Shah (4346) reported on "Epoxy's Rapid Cure Reduces Cycle Time for Automaker," and the next year, an unauthored article related how "Two-Part Acrylic Used to Bond Sensors to Fuel Tank Bottoms" (4347). Also in 1989, Phillips and Broxterman (4348) projected "Adhesives, Sealants in Autos to Grow at 250% of GNP." Finally, in 1991, Chasser and Maklouf (4349) discussed how "Epoxies Reinforce Metals for Automotive Industry."

Some *Adhesives Age* articles not clearly falling in one of the above mentioned categories might include an unauthored article of 1967 titled "Adhesive Weights Balance Wheels" (4350) and Bartrug and Koleg's (4351) 1968 article on "Improving the Adhesion of Fiber Glass Cord to Rubber." A 1976 article described how an "Epoxy Formulation Solves Battery-Production Problems" (4352). Beaver (4353), in 1989, reported "Replacing Clamp with Adhesive Saves Money for Auto Maker," and the following year, an unauthored article announced "Battery Drain Tube Adhesive Reduces Capital Expenditures" (4354).

Adhesives have long been used for noncritical decorative applications in automobiles (4348). Interior trim items include trim panel fabric, door panel fabric, ceiling fabric, carpet adhesive, and weather stripping. Exterior trim items include the body side moldings, wood grain decals, and stripping decals such as weather stripping, vinyl roof, and mirror to metal frame. Schroeder and Drain (4355) offered a 1989 SME review paper of automotive adhesive applications explaining that the increased acceptance of adhesives in recent years has been greatly accelerated by the positive response of adhesive suppliers to correct the limitations of earlier adhesives through formulation of a variety of new rapid-curing and higher-performance adhesive products.

Most recently, Drain and Chandrasekharan have reviewed this subject in an article titled "Automotive Applications for Adhesives" in Volume 3 of the 1990 edition of the *Engineered Materials Handbook*. They focused their discussion on the major subject areas of (1) brake shoe bonding, (2) decorative trim and fabric adhesives, (3) gasketing and threadlocking, (4) automotive body assembly, (5) structural bonding, (6) sound absorption, and (7) autobody repair.

1. ADHESIVES IN AUTOMOTIVE BODY ASSEMBLY

Main adhesive application areas have been the bonding of hemmed flanges in doors, engine hoods, truck lids, and tailgates. These bonds are partly load bearing and the adhesive is used in conjunction with widely spaced spot-welds whether in steel or aluminum structures. These welds are primarily employed to stabilize the adhesive joint while it is undergoing cure. Other body assembly operations have included stiffener panels to roofs, hoods, and trunk lids, and also the bonding of roof rails, drainage ditches, and wheel house flanges. Margolis (4356) has offered a 1980 paper in the *SAMPE Journal* describing the automotive developments in the 1980s relating to the competition between the use of composites versus metals, and Norton (4357) the next year reviewed the range of structural adhesives employed in the automotive industry. Scheidle (4358), in 1989, described how subassemblies comprising large, highly countoured outer panels and internal stiffeners can be assembled in a single bonding operation, making the use of adhesives a much more cost-effective manufacturing method.

Vinyl plastisol (one-part, heat-curing adhesive) products were extensively used in the various hemmed flange operations early on, because the oven cycles for painting operations could be used to cure them. Increasingly, in recent years, one-part epoxy structural adhesives have been in vogue, because of very demanding corrosion specifications; however, Minford (4359) has demonstrated that this same concern about steel corrosion in the presence of a weathered vinyl plastisol is of significantly less concern when aluminum is substituted. Changes in curing methods in recent years have also seen the conventional air convection ovens being displaced by induction ovens with special adhesives formulated to precure or pregell with such electromagnetic induction heating. The heating generated in the metal adherends is secondarily conducted into the adhesive, thus initiating or accelerating curing (see Ref. 4355). Chang et al. (4360,4361) have discussed how induction curing allows for significant welding reduction with consequent improvements in the quality of exterior automotive panels. By designing adhesives that can pregell under induction heating, considerable progress has been made to improve resistance to wash-out of the adhesive while preparing the surface for painting. While not a major issue in hemmed flange joints, it can be critical in more exposured jointures. Information about the induction heating method and materials can be found in a 1935 article by Clamer (4362) titled "The Development of the Coreless Induction Furnace," a 1944 Curtis (4363) book *High-Frequency Induction Heating*, and the 1979 *Induction Heating Handbook* by Davies and Simpson (4364). Schroeder and Drain (4355) also addressed the use of epoxy adhesives in a hot-melt or film tape form to minimize wash-out. An attempt to combine the excellent impact resistance of vinyls with epoxies in a single reactive vinyl plastisol adhesive was discussed by Namiki (4365) as early as 1976.

For bonding metals-to-plastics, either epoxy or polyurethane adhesives have been used for structural joining. Where a subsequent paint processing would subject the panel assembly to a temperature that might degrade a polyurethane, the better heat resistance of the epoxy dictates its use. Recent feasibility studies by a number of automotive manufacturers have been concerned with the use of induction curing to bond plastics with epoxy adhesives and high-frequency curing or microwave curing to cure urethane adhesives (4366). Assembling operations involving plastics has been significantly advanced by the development of primerless bonding systems. This advance combined with an increase in curing speed has been the subject of both Wongkamolsesh (4367) and Shah (4368) in 1988 conferences. Spinu et al. (4369) reported their investigations in 1990 on the durability of aluminum bonding to sheet moulding compound (SMC) using a two-part polyurethane (Pliogrip 6600/6610) with a diisocyante-containing primer on the SMC and with and without primer on the 6061-T6 aluminum adherend. Short exposure times at 140°F (60°C) dry air or in 100% RH conditions showed joint survivals whether the aluminum was primed or not. Minford had investigated this type of bonding many years earlier between aluminum-and-aluminum and showed an epoxy primer was necessary on each adherend for good durability in 100% RH exposure at 125°F (52°C). Some failures were observed by Spinu et al. with 50% loading superimposed under the hot-air conditions, but all joints failed prematurely when stressed at 50% load with the 100% RH conditions. XPS investigations of the failed surfaces were also provided. It was concluded that priming the aluminum did not significantly improve joint durability. An unusual aluminum surface pretreatment for bonding was employed by deoxidizing in hot caustic followed by desmutting in concentrated nitric acid. Minford had employed this treatment for bonding 6061-T6 aluminum to itself many years before and found good but not exceptional durability performance was obtained with both one- and two-part epoxy adhesives.

Various techniques have been used to join the so-called "engineering plastics" in reviews by Petrie (4370,4371) in 1980 in *Adhesives Age Magazine* and 1983 in *High Performance Adhesive Bonding* edited by DeFrayne. How adhesives compare with the other bonding methods as to advantages, disadvantages, and the equipment required are covered in some detail in these publications. The methods compared include (1) adhesive bonding, (2) heat-tool welding, (3) hot-gas welding, (4) resistance wire welding, (5) spin welding, (6) induction heating, (7) ultrasonics, (8) vibration welding, and (9) solvent cementing. A table was provided by Petrie showing which methods can be used with which plastics, including the engineering plastics: (1) ABS, (2) acetal, (3) epoxy, (4) fluoropolymers, (5) nylons, (6) phenolic, (7) polyphenylene oxide, (8) polycarbonate, (9) polyimide, (10) polyphenylene sulfide, (11) polysulfone, and (12) thermoplastic polyester. Finally, a table was provided to review the details about the special surface pretreating for each of the above-mentioned engineering plastics. (For other pertinent references, see Refs. 4372–4375).

Minford et al. (4376) conducted an extensive investigative program at Alcoa designed to evaluate the bondability and weathering resistance of simulated production specimens of aluminum-to-aluminum, aluminum-to-steel, and steel-to-steel automotive-type assemblies. Relative durability was evaluated after exposure to 1000 hr in 5% continuous salt spray (ASTM B117), 30 "cold crack test cycles" (Fisher Body Test Method 45-61-Cycle "B"), and the natural industrial atmosphere for 1 yr. The all-aluminum assemblies had the highest resistance to corrosion and performed without significant loss in joint strength. All steel assemblies, though subject to much more extensive overall corrosion, still remained well bonded. However, the aluminum-to-steel assemblies performed distinctly poorer owing to the opportunity for generating some galvanic corrosion. This corrosion could be effectively mitigated by the use of a neoprene gasket insertion or a two-part polysulfide sealant.

Also, at the Alcoa Labs, Vader (4377) evaluated the relative strength and durability of aluminum-to-aluminum and aluminum-to-steel adhesive bonded (one-part, heat-cured epoxy or vinyl plastisol) scaled-down automotive hood assemblies. Aluminum alloys X5020-T4 and 2036-T4 and 1010 body steel sheets were used for outer hood sections, whereas 5182-0 sheet was used to form all the inner panels. The effect of 6-mo weathering in condensing humidity at 125°F (52°C) or 6 wk in 5% continuous salt spray (ASTM B117) was compared. As anticipated, the bimetallic hood constructions performed more poorly, showing some loss of adhesion and corrosion at the aluminum interface. Priming the aluminum with a zinc chromate primer distinctly improved performance. No primer was required in the all-aluminum mini-hoods, which performed excellently in both exposure conditions.

In 1988, Sobieski and Tangney (4378) reviewed the ways in which silicone PSAs are replacing mechanical fasteners in the aircraft, automotive, and aerospace industries. They described the methods for evaluating performance, including measurements of glass transition temperature, static shear load, peel adhesion, dynamic lap-shear, and the low-temperature lamination potential.

In recent years, there have been advantages to the use of thin-gauge metal in automotive manufacturing; however, such use also carries certain disadvantages which an adhesive can better deal with than the older conventional joining procedures. One of the most important has been the development of epoxy reinforcing adhesives to increase the flexural strength and minimize the damageability of the thin-gauge component parts. Chasser and Makhlouf (4349,4379) reviewed the technology in a 1990 SME Clinic Meeting and a 1991 *Adhesives Age* article.

2. STRUCTURAL BONDING IN AUTOMOTIVE MANUFACTURING

While the above use of various adhesives in body assembly operations has been generally successful and important in vehicle styling and overall manufacturing costs, a fully bonded car body has not, as yet, been achieved. Some laboratory-assembled prototype vehicles do, however, represent significant advances toward this goal. True structural bonding is still limited to windshield bonding, bumper bonding, and metal reinforcement bonding. The 1989 paper of Wigotsky (4380) discusses one such structural bonding of plastic facia to a plastic reinforcement beam mounted directly to the car frame.

Windshield bonding adhesives have evolved over the years from simple water-leakage sealants to products that can contribute 10–50% of the tortional rigidity of the automobile. Cocozzoli (4381) wrote a General Motors' description of the change from polysulfide sealants to moisture-curable polyurethanes based on the application of the Federal Motor Vehicle Safety Standards (FMVSS). According to Sermolins (4382), this has resulted in being able to remove 8–10 lb (3.5–4.5 kg) of metal from the A pillar and headers in some models, whereas preserving an extra margin of safety against roof crush requirements found in FMVSS 216. Sermolin also discusses the application of new high-viscosity "pumpable tape" adhesives for this windshield bonding. As might be expected, there is a great challenge to find an adhesive that would have universal adhesion to the vast array of different paint systems that can be present in today's automobiles. Kirby (4383) has discussed the use of ceramic frits on the windshield to protect the windshield adhesive from UV degradation, diminishing the need for the typical so-called "blackout" primer which had hitherto been used to affect such protection. In 1990, Zabel and Pfeiffer (4384) described a new one-component polyurethane with high initial strength that had been used in Europe to bond automotive glass without the need for a primer.

Chang et al. (4323) have discussed in some detail the use of die-cut pieces of structural epoxy prepeg patches for bonding to obtain metal reinforcement in vehicles. The adhesive is sufficiently pressure sensitive that the patch can actually be applied over oily or otherwise contaminated surfaces while curing is proceeding in the body assembly paint bake ovens. They have the considerable disadvantage, however, of having limited shelf life unless stored at low temperatures. They do, however, provide high structural strength localized reinforcement in quarter panels, roofs, truck lids, and other closure panels.

The absence of accepted accelerated testing methods for structural automotive body shell construction could mean many years before a major production line will move to overall adhesive bonding. The most accepted test specimen to date has been the adhesive performance as characterized by the adhesive bonded box beam. Moody et al. (4385), in 1987, reviewed some of these test results, which indicated the need for still tougher adhesives. In 1988, Marwick and Powell (4386) reported some good correlation of box beam testing and simple impact testing results. Some adhesives with very exciting impact resistance/temperature profiles were included in this discussion. The need for newer types of impact resistance tests are certainly indicated, and Kinloch and Kodokian (4387) discussed some of these possibilities in 1987. Better characterization of the adhesive by correlating the adhesive properties with the bulk properties is now enabling finite-element analysis in support of joint design for structural bonding as discussed by Fischer and Pasquier (4388) in 1989.

While the above investigations may represent the present state of the art in regard to structurally bonding automobiles, there are a great many earlier reports that have furnished the necessary background of information required to bring us to this level of progress. The progress increase in the general use of adhesives in overall automobile manufacturing can be followed chronologically by reviewing the papers on the subject from 1962–1966 by Twiss (4389–4392), three *Automotive Industry* reports by Twiss in 1971 (4393) and DeFrayne and Twiss (4394,4395) and Beck and Yurek (4396) in 1975. Examples of some of the considerable comparative autobody aluminum joint durability data developed by Minford and co-workers at Alcoa Labs has been reviewed earlier in Chapter 10 when addressing that subject. Joint strength decay curves have been plotted by the author involving as many as five different aluminum alloys and body steel using many different elastomeric antiflutter and epoxy-type heat-curing adhesives. The effect of the alloy surface can be of particular importance, especially as the initial strength potential of the adhesive is raised. Thus, the lowest-strength elastomeric antiflutter adhesives showed very similar overall strength and joint durability performances, indicating a lack of the need for special surface properties. The variation in both strength and performance of these different alloy joints could be quite wide, however, as different formulations of one-part, heat-curing epoxies were employed, indicating much more discrimination between the surfaces of the different aluminum adherends.

In 1978 and 1980, Bolger (4397,4398) contributed technical papers involving several cooperative investigative programs carried on between Amicon and some automotive and bus manufacturers. The programs were primarily pointed toward finding the best means of using epoxy adhesives in assembling both school buses and passenger cars. He described several new epoxy formulations and their bonding processes for use in high-volume, low-cost automotive assembly operations. Included were the assembly of a variety of under-the-hood electrical and electronic components, halogen-quartz headlamps, body repair solders, and low-temperature–curing, flexible body joints for school buses using rivetbond construction (4398).

Lees (1452) in 1981 considered toughened structural adhesives based on acrylic and epoxy resins might be capable of replacing welding as a means of constructing motor vehicles. He described their use in a variety of metal structures, especially in their relation to performance as bonded box beams. He further cited comparative values for riveted or spot-welded aluminum and steel structures. Finally, he discussed the lower insensitiveness of such modified adhesives to bondline thickness, which can be especially important where structures are assembled from pressed, drawn, or folded components with their significant dimensional inaccuracies. Another 1981 paper of possible interest for the vehicle designer is that of Murray (4399) describing a new approach to selection of best structural adhesive for machine manufacturing. In 1982, Lees (4400) produced yet another manufacturing related paper titled "The Use of Adhesives in the Manufacturing Industries."

3. SEALING, GASKETING, AND THREADLOCKING APPLICATIONS

These terminologies actually show considerable overlapping when discussing the application of adhesives and sealants in automotive manufacturing.

a. Porosity Impregnation

The use of anaerobic sealing compounds for impregnating porous materials has drastically replaced other earlier sealing compounds such as linseed oil, varnish, sodium silicate, and polyester (epoxy or styrene variations). Their unique characteristic of staying liquid in air and curing without heat when confined in a condition of porosity makes them ideal for a vacuum impregnation process. The porous part is submerged and a vacuum is pulled to remove air from the pores. When the vacuum is removed, the liquid anaerobic fills the voids and cure commences under the anaerobic conditions in the void cavities. The cured formulation can withstand pressures up to 55 MPa. When used to impregnate powdered-metal parts, machinability can be improved by up to 500% because of the elimination of tool chatter. Metal plating of impregnated pore structures is also considerably enhanced, since the plating chemical will not be trapped under the plate surface. Excellent applicability has been shown on iron and brass powdered-metal parts; zinc, aluminum, and magnesium die castings; steel and aluminum welds; and cast iron sections. Parts may be finish machined before impregnating, because of the cleanliness of the processing. In aluminum engine blocks the resin seals porosity that would be present in the cooling system and oil passages. Loctite has suggested in a press release that more than 20 million automotive castings made of aluminum and other light metals will be sealed for microporosity in 1990. Fulda (1705) produced an SME Tech. Paper on impregnation of porous metal components with anaerobic sealants as early as 1977. Hulstein (1707–1709) has written three technical papers on "Application of Anaerobic Adhesives," "Assembling with Anaerobics," and "Automatic Application of Anaerobics."

b. Sealing Cylindrical and Threaded Fittings

Virtually all types of plugs along with cylindrical, tapered, and straight threaded automotive parts are being sealed with anaerobic adhesives. Unlike the porosity filling type of anaerobic formulations which are required to be very low in viscosity to penetrate and fill pores, these formulations for sealing fittings are usually quite thick or thixotropic to make them hold their position while curing. One of the first applications of an anaerobic sealant in the automotive industry was in 1957 when a thin resin was used to lock the screws which held the butterfly flapper to its shaft in a Holley carburetor. Previous staking methods had often caused shaft bending, leading to sticking of the valve. The reader may be interested in reading more about the topics of (1) how threaded fasteners fail, (2) the removal of torque, and (3) the lubrication capacity of anaerobic threadlockers (4401). Stamper (1704) reported in 1983 on the "Curing Characteristics of Anaerobic Sealants and Adhesives," which should be pertinent to this subject.

c. Engine Sealant Applications

Engine sealants in recent years have had to meet increasingly tighter tolerances, precise torque values, and sealing against the surface of new materials not available for inclusion in earlier designs. The range of these sealants includes chemical gaskets as well as the microporosity and thread sealants just discussed. The fact that fluids that leak from engines are also classified as emissions by government standards makes their effectiveness an even greater issue than earlier when such leaking was strictly only a quality issue. The shift from iron to aluminum engines has also been a factor in creating more emphasis on sealing microporosity. Some of the main application areas include transmission cases, intake manifolds, transaxles, power steering housings, pumpcovers, and even the engine blocks.

Chemical gasketing has significantly enhanced the manufacture of today's more reliable and low-service-cost engines. RTV silicones and anaerobics seem to be the dominant products available. In recent years, there have been significant advances in designing and building smaller, lighter, and higher speed (of rotation) engines, which have required continuing research and improvement in the sealant's ability to resist higher temperatures and the action of the internal engine fluids operating at those temperatures. In addition, it has been necessary to build in higher potential levels of adhesion and accompany that change with easier application properties with the use of newer automatic application equipment. Actually, sealant thixotrophy has proven to be a much more important factor in obtaining a proper seal bead than was first imagined. Since about 1980, the changes in the 4-, 6-, and 8-cylinder engines have demanded a much wider range of sealant capacity properties than present in all earlier engines. Turbocharged engines have made special demands in particular for higher-temperature performance along with demands for higher levels of flexibility character. Such sealants had to perform at temperatures up to 700°F (370°C). Often it can be the sensitivity of some other automotive parts that must be considered in formulating these new sealants. For example, the oxygen sensor which adjusts the timing was being fouled by the curing volatiles developed by the silicone sealant as used in the first cars which were able to meet the California emission standards. This necessitated the development of an entirely new generation of low-volatility RTV silicone gasket formulations.

d. Auto Body Sealant Applications

The use of adhesive/sealant materials for various sealing functions throughout the body of an automobile was one of the earliest applications and probably may still involve the largest poundage of materials. Slingerland (4402) attempted a review of the application of elastomeric body sealers in his S.A.E. 1980 paper titled "Past, Present, and Future of Elastomeric Body Sealers in Automotive Applications." He noted the use of progressively better elastomeric sealers for isolating the car passengers from the heat, cold, wind, water, and noise of the care body's service environment. For example, the elastomers used in windshield seals have evolved from the use of SBR elastomer to neoprene, then butyl, and most recently to EP and EPDM compositions in order to meet the increasingly stringent requirements. He also included discussion of the development of more durable methods of weather stripping doors and deck lids and the array of elastomer and molded plastics that have been increasingly involved in recent years.

Adams (4403) in discussing the applications for sealants in the 1990 *ASM Engineered Materials Handbook* has also offered a review of body assembly sealant applications. He makes the observation that the most dynamic changes in such applications has occurred in the 1980s–1990s period with further evolution expected throughout the 1990s. A number of events promoted this acceleration of development, including (1) introduction of more light-weight corrosion-resistant metals, (2) new low-temperature–bake, clear, base-coat paint systems, (3) increasing use of plastics, and (4) introduction of new low-drag coefficient body designs. While major identifiable areas of application continue to exist like seam sealants, windshield sealants, and gas tank sealants, a broader and more diverse range of different sealant types have had to be employed in the newer design applications.

The manufacture of lighter, more corrosion-resistant body and frame assemblies has actually been facilitated by the qualities of some of these newer sealants. For example, many new sealants are now partially cured in the assembly process with full curing during the paint processing. Urethanes, hot-melts, and new vinyl-plastisol products are widely used to achieve both bonding and sealing functions. The use of sealants in conjunction with adhesives is bound to create stronger and more reliable assemblies capable of more evenly distributing the loads and assisting in corrosion resistance. Wider use of galvanized metal surfaces also promotes greater corrosion-resistant auto bodies, and this is further enhanced by using adhesives on galvanized surfaces as compared with spot-welding.

Solid hot-melt sealants assumed early popularity in European automobiles, and has continued to grow there and in the United States. They are now available in a wide variety of preformed shapes that can be easily applied on fasteners and in seams without the need for special application equipment. During the painting operation, these hot-melt forms will melt and flow out and cure. Such preformed sealants are especially appropriate for the low-volume applications.

Windshield sealants, meanwhile, have changed dramatically from those various rubber channels mentioned earlier to polysulfide and polyurethane direct-glaze systems. Such systems have been demanded by the designs, which include flush-mount glass installations on aerodynamic styling. Also, dictating the use of new materials has been the employment of newer rollover safety standards. The sealant technology has actually contributed significantly to passenger safety by enabling manufacturing of windows which flow out and away from the passengers. The added strength provided by a truly bonding sealant has enabled the fixed windows to simultaneously have complete environmental sealing coupled with providing additional body structural strength. The popular "T" tops and sun roofs have been successful through the use of the same sealant technology as used above for the windshields and fixed window sealing. Because of relatively lower-volume production, urethane, RTV silicone, and polysulfide sealants have been primarily manually applied in production of these special roof designs.

The introduction of epoxy and urethane adhesive/sealant compositons has enabled integration of the rear light system into aerodynamical body design. The expensive plastic lens covers that have been popular since 1986 are today assembled entirely without any mechanical fasteners. Two 1989 papers in *Adhesives Age* have described some of the later developments. Grunder and Koehler (4404) relate the development of automated potting of halogen light sources, resulting in reduced costs and lead time over the existing manufacturing procedure. The requirements for the adhesive and machinery, machine operations, and adhesive selection procedure have also been included. DeBarros and Beschle (4405) described the advantages attendant to using a high-performance thermoset adhesive in place of a flame-sealing operation for a variety of parabolic refector lamps. Adhesive development and assembly evolution are also discussed.

More recently, Norwicki (4406) examined the methods and advantages of utilizing liquid polyurethane foams for fabricating prepositioned gaskets. At the same SME adhesives technology meeting, Bachmann (4407) discussed the use of aerobic acrylic adhesives to form structural bonds in assembly processes while functioning as coatings that are more evenly distributed. And finally, the bonded structure provides an effective seal for the passenger compartment against all weathering conditions. The combination of new polymer technology with very precise application methods has thus produced a truly strong, reliable, sealant/adhesive body assembly. From the use of asphaltic and rubber fillers for physically filling body leaks and accommodating poor door and window fits, the sealant of the 1990s actually assists in the body assembly process and lends a new-found structural integrity to the automotive vehicle not previously envisioned. In 1990, Accettura (4408) described a very efficient solution using modern sealant for the old oil pan gasket problem.

4. BONDING RIGID PLASTIC TO RIGID ADHEREND SURFACES

Because of the presence of so many rigid plastics in the modern automobile, there has been a continuing need to examine how to bond various plastics to one another and to metal surfaces. The bonding of the SMC plastic body on the Corvette to the steel frame components was, of course, one of the most important early successes in manufacturing that can be cited. Many earlier references have been made to this success in the sections under "Polyurethane" adhesives in Chapter 5 and evaluating the durability of bonding polyester fiber glass to aluminum by the author in Chapter 10 (3651). With either steel or aluminum adherends, it was considered that a compatible epoxy primer was needed on the metal surface to obtain an acceptable level of bond durability. Minford also evaluated the use of both epoxies and modified-acrylic adhesives for bonding the polyester fiber glass to aluminum with varying degrees of success as reported in (3651). Minford (3652) also reported successes with bonding engineering-grade styrene and cross-linked styrene to aluminum with these same adhesives. In other investigations of unpublished work, Minford conducted similar investigations of the bonding of PVC, polycarbonate, and butyrate plastics to aluminum (3797,3798), with best durability obtained again with epoxies and acrylics.

Leung and Chmiel (4409) discussed the use of a water-borne polyurethane thermoforming adhesive for laminating expandable vinyl plastic to contoured rigid surface parts. Evaluation of both one- and two-part systems was undertaken using performance in automotive aging cycling conditions. Also, in 1990, Powell (4410) reported on the selection and testing (compatibility, impact, and long-term weathering) of silicone elastomeric adhesives for bonding plastic automotive bumper systems. Hauser (4411) recently discussed the application of hot-melt adhesives for bonding polypropylene and polycarbonate surfaces in automotive headlights. Bond integrity was maintained up to 248°F (120°C) with good prospects indicated for performances at even higher temperatures.

Martin (4412) has reported on the substitution of one-part water-based adhesives for solvent-based systems for bonding foam-backed vinyl and polyethylene-backed vinyl to solid surfaces of ABS, Delac, or urethane adherends. These combinations are important in forming automotive interior trim components. The cost savings of such water-based bonding over the current solvent-based systems has been shown to be very substantial.

5. ELECTRICAL AND ELECTRONICS AUTOMOTIVE APPLICATIONS

A relatively recent area of adhesive growth has been that of automotive electronics. In Publication No. 600 from the Delco Electronics Division of General Motors (4413), a partial listing is given of the range of electronic applications involving adhesives or sealants. In the power train area, the list includes engine control modules, transmission controls, and electronic spark timing controls. Inside the body are the instrument panel display electronics. Chassis electronics include suspension and ride controls, electronic brake and traction controls, security systems, and electronic power-assist steering systems. Entertainment and comfort electronic devices include heater and air controls and music system speakers. Finally, under a classification of "hybrid electronics," Delco lists the ignition modules, pressure sensors, and power-switching controls.

The first conventional adhesive applications in the electrical and electronics area were the fixturing of wire harnesses and the bonding of motor magnets. These applications now pale by comparison with those where the adhesive must additionally act as a barrier to the environment; i.e., the potting- or encapsulation-type applications. These abilities are best served with such adhesive types as the epoxy resins, polyurethanes, silicones, and acrylates. The epoxy types need to be highly filled to reduce the coefficient of thermal expansion. In addition, they might also be modified to dissipate heat when encapsulating heat-generating devices. The low elastic modulus of polyurethanes and silicones can compensate for significant degrees of thermal expansion and contraction that can be experienced in complex assemblies. Acrylates have been used much less frequently, but occasionally the benefit of being able to cure them with UV energy has been the basis for some use. In every case, it is necessary for ionic purity to be maintained, which means a very low moisture uptake even under adverse environmental exposure conditions.

The statement is made by Adams (4414) that sealants have probably enjoyed their biggest market growth in the electrical and electronics fields. While only part of this dramatic growth has been involved in direct products for inclusion in automobiles, the overall range of applications in all fields involving electronic devices have many common denominators with the automotive applications discussed above. More discussion will be offered in a separate section to follow on these other types of general electrical and electronics markets applications. The use of aerobic acrylics by Bachmann (4407) for coating and sealant applications also included the use of these adhesives for certain electrical potting applications.

6. *SOUND ABSORPTION AUTOMOTIVE APPLICATIONS*

Obviously, the use of more and more adhesive bonding in the body shell construction will continue to offer significant potential for significant noise reduction. At the present time, a variety of elastomeric noise-control materials are used to reduce vehicle interior noise. Such materials are quite different in properties as compared to the structural adhesives discussed earlier for joining body component parts. They are specifically required to have relatively low-modulus values in order to provide best damping characteristics. The frequency range of concern is from 20 Hz to 20 kHz in the form of the noise contributed by the drive train, tires, suspension system, and the wind. Noise can travel through pillars, floor rails, windows, and various closure panels directly and indirectly from a source to some other surface by vibrational transmission where the energy is radiated as noise. A 1989 paper delivered at a special S.A.E. Noise Vibration Conference (4415) has reviewed this latter situation as noise transmission from engine to the floor panels. Since each different design and assembly of parts that constitute a new model can have its own unique noise signals, the kind and manner of application of elastomeric materials must be specifically devised for each different model.

Distinctly different from the sound-barrier elastomerics discussed above are the adhesive damping materials which reduce the kinetic energy of vibration by transforming it into heat. This is in contrast to the sound-barrier materials which function largely by reflecting and absorbing the energy. The important property attendant to damping sound is the stiffness of a polymeric material, which is a temperature-dependent property. The viscoelastic dampeners are effective reducers of wind, road, and engine noises when applied in roof, door, and floor pan application areas. The performance of viscoelastic dampers has also been discussed in some detail in the above-mentioned paper (4415).

Some means of comparing and evaluating damping performance can be found in the use of the Geiger plate test (SAE J-671) and the more complex modulus test (Oberst Bar, ASTM E-756). Such methods can offer comparisons in both decay rate and loss factor. Klempner (4416), in 1988, suggested use of dynamic mechanical spectroscopy for estimating efficiency as a vibration damping polymer. He suggested efficiency was apparently related to the loss factor peak width measured.

7. *ALUMINUM AUTOMOTIVE VEHICLE DEVELOPMENTS*

An early summary of developments concerning aluminum and automobile manufacturing can be found in the chapter by Keith et al. "Automotive Applications Outside the Engine" (11). These authors pointed out the prediction in 1896 by the *New York Sun* that aluminum would have a large use in horseless carriages because of its lightweight and readily formable nature. An aluminum crankcase was made as early as 1897, and by 1900, aluminum sheet was already in use in passenger car panels and in cast forms for engine, transmission, and body parts. The body of the 1923 Pierce-Arrow (a luxury sedan of the period) was fabricated entirely from cast and wrought aluminum. In general, the use of aluminum for car bodies has been more widely employed in Europe where the high price of gasoline emphasizes the value of weight savings in any type of vehicle.

In more recent times, again aluminum is beginning to be viewed as a replacement for steel for many inherent qualities that made it an ideal material for transportation vehicles. Among these qualities is its one-third the density of steel, high strength/weight ratio, inherently good corrosion resistance, and excellent formability when not pushed to excessively high-strength levels. Among the most successful modern stories of the successful use of aluminum has been in the Land Rover, which has always had an aluminum body structure showing exceptional durability over many years. Similarly, the United States Postal Service has considered the excellent corrosion resistance as a primary basis for acceptance along with recognized weight savings and long-term durability in their parcel service vehicles. A large number of the luxury European cars shipped to the United States over many years, like the Volvo and Diamler-Benz, have utilized aluminum hoods and deck lids to reduce weight, whereas makes like the Porsche and Rolls-Royce have used aluminum in hoods, doors, and even front fenders. The most successful use of extensive aluminum was in its use for the whole body structure in the Dyna-Panhard of the mid-1950s (4417).

Although the early favored material for construction of automobiles, it has been pointed out that the sudden appearance of a much lower cost body steel sheet in the marketplace displaced aluminum as the best candidate for mass producing automobiles. For many years following this major shift in preferred material for auto body manufacturing to steel, aluminum had to slowly grow in overall usage in that industry by replacing steel in various isolated components of the overall car. The so-called "energy crunch" beginning in the early 1970s seemed, however, to regenerate interest in significantly reducing the overall weight of automobiles where possible. In earlier portions of this section on the use of adhesives in automobiles, a significant number of these substitutions of aluminum in various auto components has been considered. However, because of the significantly higher price of aluminum which remained because of its higher-power production costs per pound, the major goal still remained to have aluminum actually replace steel as a major body structure material.

Earlier references in this book have been made to some of the earlier investigations and reports of Minford and co-workers at Alcoa Labs in the period of the middle to late 1970s (68,72–74). This work continued steadily throughout the 1980s up until the author's retirement in 1982, with intense work concentrating on the bonding characteristics of different aluminum auto body sheet alloys with a variety of different lubricants and levels of lubricants on the adherend surface. This work, in turn, was coordinated with the rapid development of improved formulations of heat-cured structural epoxy adhesives which were evaluated in the form of purely adhesive-bonded, rivetbonded, clinchbonded, or weldbonded aluminum joint structures.

In 1983, King (4418) showed that by using finite-element analysis to optimize the gauge and cross section of individual components coupled with the joint stiffness achieved by adhesive bonding that aluminum vehicle structures with a tortional stiffness equivalent to those of spot-welded steel structure could be produced at half the weight. Keeley (4419), in 1985, indicated finite-element modeling of automotive structural frames had shown that, overall, the gauge of the aluminum is generally governed by the stiffness targets rather than by the load-bearing requirements. In 1986, Sheasby et al. (4420) reported that it seemed to be necessary to pretreat the aluminum surface in order to obtain satisfactory adhesive bonding both in terms of initial joint strength and bond durability. Economic considerations would indicate some sort of aluminum coil pretreating would be preferable to separate pretreating of the final components.

In 1985, Kewley (4421) presented an S.A.E. Technology Paper No. 850103 titled "The BL Technology ECV3 Energy Conservation Vehicle" which dealt with the construction of an adhesively bonded aluminum concept vehicle. The paper described an aluminum structural vehicle technology (ASVT) process which was appropriate for volume car production. In 1987, Kewley et al. (4422) presented an S.A.E. paper titled "Manufacturing Feasibility of Adhesively Bonded Aluminum For Volume Car Production" which covered such subjects as (1) pretreatment and lubrication of aluminum, (2) press shop operations, (3) die engineering requirements, (4) body design, (5) body-in-white assembly, (6) spot-welding requirements, and (7) finishing and final assembly. In the same year, Selwood et al. (4423) produced yet another S.A.E. paper titled "Evaluation of an Adhesively Bonded Aluminum Structure in an Austin-Rover Metro Vehicle." Wheeler et al. (4424) considered a number of important factors in their S.A.E. paper of 1987 titled "Aluminum Structured Vehicle Technology—A Comprehensive Approach to Vehicle Design and Manufacturing in Aluminum." They discussed and listed references to a number of relevent subjects such as the use of fluid forming of aluminum for more economy in parts forming (see Refs. 4425,4426); selection of proper lubricants by Marwick and Sheasby (4427), spot-weld contamination by aluminum (see Refs. 4428 and 4429), and favorable manufacturing costs (see Ref. 4422).

Another important subject when discussing potential aluminum structure for vehicles is evaluation of adhesives that might be considered for various designs. In a 1987 S.A.E. paper, Marwick and Sheasby (4431) conducted evaluation of adhesives for aluminum structured vehicles, which concluded that the choice of the individual components in the bonding scheme (pretreatment, press lube, and the adhesive) are all interrelated and have to be chosen to be fully compatible for best overall performance.

Other S.A.E. technical papers relating to the development of aluminum structured vehicles are those of Seed, Nardini, and Cassese (4430) on a Ferrari research vehicle, and Nardini and Seeds (4432) in 1989 titled "Structural Design Considerations for Bonded Aluminum Structured Vehicles." Three 1990 S.A.E. contributions are those of Bull et al. (4433) about "The Application of a Knowledge Based CAD System to Press Tool Design," McGregor et al. (4434) with a discussion of "The Design of Impact Absorbing Members for Aluminum Structured Vehicles," and Nardini et al. (4435) describing the "Analysis and Testing of Adhesively Bonded Aluminum Structural Components."

C. Adhesive Bonding in the Truck, Bus, Tractor Trailer, Transit Car, and Recreational Vehicle Industries

The extensive use of aluminum as a lighter-weight construction material for building truck cabs and the tractor trailers they haul is certainly well known. The savings in being able to haul larger cargo loads for the same fuel costs in these lighter-weight over-the-road haulers is multipliable by a fantastic number of hauls over the service life of the vehicle. This fact has continued to support the cost of the more expensive aluminum metal as the material of construction for these vehicles. As with the airplane industry, the first and most practical method of assembling an aluminum box for use as an over-the-road trailer was to use rivets. However, adhesive materials in the form of sealants have always had to be used, since riveted sheet jointures are not tight fitting enough to be weather resistant. The possibility that a continuous adhesive bondline could furnish both structural integrity and effective sealing in the same operation has always been recognized as desirable, but the question of economic practicality was always a factor for manufacturing any vehicle outside the aerospace industry. Consequently, a number of the major manufacturers, such as Fruhauf, began to investigate modified designs of tractor trailer construction involving the gradual incorporation of adhesive bonding practices as they became more economical. Unfortunately, it had to be conceded quite early that the already-proven methods of adhesive bonding aircraft or spacecraft could never be considered for these same economic reasons. What was left appeared to be limited research programs designed primarily to decide what level of adhesive cost, adhesive strength, and surface preparation of the aluminum adherends could be afforded and then conduct tests on experimental mock-ups to estimate service life potential.

It should be clear that the use of a combination of riveting and adhesive bonding (rivbonding) would enable significant strengthening of the jointure area and also permit the function of sealing against weathering to be simultaneously achieved. Vader (4436), at Alcoa, evaluated the durability of rivbonded aluminum joints as an alternative technique for assembling auto, truck, and special military light-weight vehicles. Comparisons were made between 6009-T4, 6010-T4, and 5182-0 aluminum sheet jointures using a heat-curing, one-part epoxy or a vinyl-plastisol. For manufacturing operations where economics might prevent consideration of a heat-curing procedure, several room temperature–curing two-part epoxies and a recommended room temperature–curing modified acrylic were also employed. When the more rigid heat-cured epoxy and two-part polyamide-cured epoxy were used, there was little difference between the strength of a rivbond and the adhesive-only-bonded joints. In contrast, when lower-modulus adhesives like the vinyl-plastisols or a polysulfide-modified, two-part epoxy were used, the former produced a 90% stronger joint as a rivetbonded joint, whereas the modified epoxy developed a 50% stronger rivetbond jointure. The second-generation–type modified acrylic rivbond joints, however, were the strongest of all the rivbond joints fabricated. They did show some sensitivity to long-term durability testing in 100% RH at 125°F (52°C) conditions, which was a negative factor. The author would like to comment, however, that the continuing improvement in the modified acrylics' durability over many years would make it necessary to repeat such an investigation with later acrylic adhesives before discounting the use of such an adhesive for this type of application. Some confirmation was also obtained in this investigation showing that as stronger tensile strength aluminum alloys were used, the rivbond joints were correspondingly stronger using equal sheet thicknesses.

Bolger (4397,4398) summarized the existing developmental data obtained by Amicon for applying rivbonding techniques for the manufacture of school buses. These methods are being applied to about 98% of all school buses being currently manufactured. A recent Federal Standard requires each rivbond joint to have at least 60% of the tensile strength of the sheet metal to pass. It would require as much as a 300% increase in the number of rivets to equal this strength gain. Linebarrier (4437) has cited the advantages of rivbonding in the assembly of light-weight fighter aircraft and construction of advanced transit cars.

The possible use of weldbonding for assembling aluminum trailers has also been widely evaluated by manufacturers such as Fruhauf. Also, Minford (4438) has investigated several alternative surface pretreating procedures at Alcoa for fabricating trailers utilizing a weldbond for jigging the extrusion over the sheet-splice joint while a two-part, room temperature–curing epoxy was curing. Since prepainted aluminum sheet was desired for this particular trailer construction, pretreated procedural variations included abrading away the paint in the area to be weldbonded with or without subsequent paint priming, and with or without subsequent application of a separate silicone sealant to keep out the weathering elements. To achieve maximum joint strength using a spot-welding procedure, the low-modulus, two-part, polysulfide-modified epoxy adhesive mentioned in the previous paragraph was used. Based on a 1-yr evaluation in intermittent salt spray (the most aggressive weathering situation), a very satisfactory performance was achieved when the abraded sheet was directly bonded to an unprimed extrusion. Elimination of any signs of corrosion in the area of the joint was only affected by the additional use of a silicone sealant bead along the edge of the bondline. The use of a special primer on the extrusion that had been recommended by one trailer manufacturer proved to actually have a negative effect on long-term adhesion which would reinforce the necessity to carefully screen and test any primer rather than project that it should be a positive step toward prolonging the service life.

In addition to the modified designs of existing trucks as mentioned above, there have been a considerable number of entirely new truck design concepts utilizing bonding such as the TX-450, a Leyland Motors truck design with an adhesive bonded aluminum chassis. The bonded chassis uses hollow side- and large-box assembly cross members providing benefits over a conventional steel ladder frame which also incorporates a 30% weight savings. It needs to be recognized that structural aluminum alloys while having comparable yield stresses to steel have only approximately one-third the modulus of elasticity. Therefore, to achieve comparable stiffness, a significant change in section portions was needed consisting of hollow side-members linked by large, box-assembly cross members closed by top and bottom skins. A substantial aluminum casting was used for the critical connection between the cross and side members. A further integration of the floor structure with the chassis produced a combined load-bearing structure. In some joints, some mechanical fasteners were used which helped provide both location and clamping of the joint while curing. A special primer was used in conjunction with degreasing and shotblasting, which seemed to be the easiest and most acceptable form for commercial vehicle volume production.

The adoption of new materials of construction in the bus and coach industry has also triggered more extensive use of adhesive bonding. For example, it was reported that six different adhesives have been used in the manufacture of the Hestair Duple 425 coach vehicle, including a two-part polyurethane for bonding the single-sheet sides of the bus to its spaceframe (used in place of a conventional chassis). Another polyurethane was used to bond the wooden floor to the metal framework, whereas other adhesives were for sealing the windshield and windows, bonding air conditioning components, trunking and luggage racks to the spaceframe, and attaching floor coverings.

The transit car industry has been especially interested in using aluminum in transit car construction, because of its light weight and esthetic appearance in many of its finished forms. In 1972, Minford (4439) evaluated the comparative joint strength and durability potential of 5052 aluminum sheet to 6061-T6 aluminum structurals using either a rigid epoxy, a lower-modulus polysulfide-modified epoxy, or a polysulfide mastic (earlier used as an antiflutter adhesive by Fruhauf in manufacturing trailer roofs). As expected, the initial joint strength values decreased in that same order. Vapor degreasing both adherends afforded acceptable joint durability performance in water immersion and hot humidity-soaking exposures in spite of the heavy oxide still present on the 6061-T6 extrusion surfaces. Abrading the surfaces removed this oxide and correspondingly increased the initial joint strength values but failed to show significant advancing of the joint durability regardless of the adhesive choice.

Adhesive bonding has played a very important role in the mobile home and trailer home industry where the sidewall, floor, and roof areas are mainly assembled from adhesive-bonded aluminum-faced sandwich components. The author distinctly remembers the earliest days of this industry when the Chairman of Winnebago Industries waited on the telephone at the mobile home show to receive confirmation from the author's laboratory that a totally adhesive-bonded sidewall construction had passed long-term adhesion durability tests. At this time, the development of the Alcoa Alply Building Panel was being finalized, so that the methods of manufacturing and evaluating such laminated aluminum-faced panels were well established at the Alcoa Labs. The main question for concern with a manufactured panel for mobile home construction was whether the lesser sophisticated method of manufacture needed for more economical production of such a panel (without the hot press dwell time plus final rotary press used for Alcoa Alply manufacture) would survive similar weathering tests. One positive aspect of the mobile home lamination procedure was the use of an extruded polystyrene insulation core (instead of the expanded beadboard used in the Alply Panel) which permitted application of an organic solvent containing neoprene-phenolic contact cement adhesive by spray application without collapsing the insulating core structure. This solvent-induced tackiness was necessary as an alternative to the higher-temperature–induced tackiness of similar adhesives used to fabricate the Alcoa manufactured building panels. Thus, where well-controlled adhesive application and proper pressure application by rotary pressing has been used in mobile and trailer home construction, the resulting adhesive-bonded sandwich has been successful.

Over the years there have been a number of articles published by *Adhesives Age* magazine which are related to some aspects of manufacturing in the truck, trailer, and over-the-road vehicle industries. An unauthored article in 1962 dealt with the subject of "Epoxy Repairs Casting" (dealing with refrigerator car maintenance) (4440). Three unauthored articles in 1966 were titled "Floor Covering Bonded to Tracked Vehicles" (4441), "Versatile Adhesive Automates Production of Mobile Homes" (4442), and "Truck Bodies Quickly Repaired with Metal-Backed Pressure-Sensitive Tape" (4443). In 1967, a related article titled "Apache Buffalo Camping Trailer Utilizes High-Resistant-Adhesive" (4444) appeared, whereas in 1969, the title was "Metal Foil Tape Fixes Trucks" (4445). "PVC Foam Tape New Sealant for Recreational Vehicles" (4446) was the subject of a 1973 article with "Adhesives for Recretional Vehicles Kept Uniform with Portable Mixers" (4447) following in 1974. Finally, a more recent 1987 article discussed "Adhesive/Sealant Cuts Riveting and Prevents Leaks for Hauler" (4448).

D. Repair Bonding in the Automotive Industry

The use of adhesives in many forms probably can be considered for a wide variety of repair applications in the automotive industry. There is, of course, first the use of an adhesive to affect a repair where the same or a similar adhesive has been used in the original jointure or combination jointure such as rivbond, clinchbond, or weldbond. In most structural situations, it is most likely that the adhesive would be involved in the manufacturing of a new part, which is then employed as a replacement unit in the repair shop. This type of procedure (originally developed by the Opel European Division of General Motors) offers certain advantages over welded repair kits (4449). Examples of this kind of situation would be new bonded brake shoe replacement kits; new body item replacement units like hoods, rear decks, doors or roof panels where adhesive bonded flanges and hems may be present; or replacement of tail lights and back-up lights as factory-manufactured units.

More directly applied adhesive in repair shops may be involved with the replacement of trim or insignia items, direct installation and sealing of replacement windshields and back window lights, and replacement of weathering roof coverings. Others requiring the spot use of adhesives would involve those automotive applications employing retaining, liquid sealant, or gasketing-type sealants as in gasoline and electric motors and power train components. Further research will be necessary before adhesive bonding can be considered for the repair of actual structural members. Where such repair might be considered, the two-part epoxies are most commonly employed, since there generally is a lack of high-temperature paint bake facilities that are used for curing the one-part epoxies in original manufacturing. While some reduction in joint durability might be anticipated with substitution of the ambient temperature–cured products, there is a greater opportunity in the repair shop to specially prepare the aluminum surface in some fashion as compared to the original manufacturing situation where some lube may have even been present. Also, some convenient dual-chamber cartridges have become available in the marketplace which could permit easy mixing and application of such materials as reviewed in a Ciby-Geigy auto repair manual (4450).

Peters (4451), at Alcoa Labs, conducted some evaluations as early as 1967 of the possible application of adhesives for repair operations on aluminum radiators, since such items were being produced for special cars like the Corvette. The scope of the program involved testing nine adhesives on three different aluminum alloy adherends along with four surface applications that might be practical for shop use. Five accelerated testing conditions were employed, including thermal cycling, intermittent salt spray, and hot 5–50% antifreeze/water solutions. The salt-spray exposure was particularly important to simulate the effect of winter road splash on the exterior radiator surfaces and proved to be the most aggressive of the weathering situations. There was a preference for one-part, heat-curing epoxy adhesives over the two-part, room temperature–curing alternatives. A chromate conversion coating surface pretreatment was much superior to solvent clean, abrasion plus solvent clean, or room temperature acidic deoxidizing. However, room temperature etching in HCl had yet to be considered, and more recent investigations by Minford would indicate this kind of pretreatment might offer the best long-term durability response, especially in the corrosive salt-water condition. Harrison Radiator (4452) had previously issued a repair instruction manual for the aluminum radiator in the Corvette recommending abrasion and solvent cleaning plus heat curing an epoxy at a temperature no higher than 350°F (177°C). Three recommended patching materials suggested were glass tape, aluminum screen, or a thin aluminum sheet.

Minford (4453), in 1978, evaluated the potential of adhesives for repairing aluminum air conditioner units used as standard equipment in the automotive industry. Field service conditions were simulated with actual refrigeration tubes, Freon 22 charges, and pressurized operation at 250°F (121°C). The challenge was to develop a repair method for perforated tubes where the repairman would need to make on the spot repairs and pressure test during the repair sequence. Repair patches had to hold internal tube pressures up to 600 lb. Fourteen commercial adhesives were evaluated using both simple adhesive overlays or folded adhesive/foil laminate patches. High-temperature hot-melts or room temperature–curing epoxies failed to hold against the high pressures consistently. Five epoxy formulations cured with a focused heat lamp source passed 28-da screening tests at these elevated-temperature and -pressure conditions.

VI. ELECTRIC AND ELECTRONIC APPLICATIONS FOR ADHESIVES AND SEALANTS

A. Electrical Applications

Buchoff (4454), in 1973, reviewed adhesives in the electronic and electrical industries and added the question of "What We Need?" The most recent review on the application of adhesives in the electrical industry has been offered by Batson (4455). He reminds us that the adhesives for use in electrical applications can differ from those in electronic applications, because the adhesive also needs to involve significant load-carrying requirements. Such electrical applications as rotating electrical devices, motors, alternators and generators, transformers, solenoids, and the processing controls of instruments, relays, and switches fall in this category. Such devices tend to be electromechanical in nature involving movement due to electrical and magnetic force fields.

Devices such as transformers, solenoid coils, rotors, and stators all are dependent on holding laminated components in place either mechanically or adhesively. In transformers and solenoids there is a low-frequency vibration which must be resisted, and very low-viscosity, wicking-type adhesives are required to penetrate and bond the individual laminates. Such adhesives may be dip applied in vacuum or under pressure in tanks or sometimes simply applied to the laminate edges. Some production problems have arisen in producing transformer or solenoid coils securing the cut wire after the winding operation. Various fast-tacking materials like tapes, rubber bands, and waxes have been employed with varying success to combat this problem. Cyanoacrylates plus an overspray accelerator have recently offered fast wire tacking operations in a second or less. Most recently, UV-curing adhesives have been employed and shown to be ideal, since they are of one-component and can be cured to a strong bond in less than 3 sec.

For electrical devices with coils, there is a need to provide security to wires passing out of the body of the device to external electrical connectors (usually pins). Both UV-curing or epoxy adhesives have offered means of bonding these fine wires to the plastic body of the device. Thus, overstressing of the wire wrap or solder joint is restricted. A very high-volume application has been adhesive involvement in the double-insulated motors for the power tools industry. Newest designs have involved manufacture of 100% adhesive-bonded power tool motors replacing the previous mechanical joining techniques. Batson (4455) has described the sequence of assembly operations in this regard. The adhesives employed fall in the class of anaerobic grades of retaining compounds. They are most convenient in bonding any metal mechanical components together as a single component. In addition, they can bond nonmetallic adherends using the proper heat or surface activator technique.

In the automotive industry, as mentioned earlier, adhesives play a prominent role in the manufacturing of small fractional horsepower motors, which are involved in many uses such as the blower for air circulation inside the car and to various window glass areas, windshield wiper operation, power windows, antenna elevation, and even trunk closing. Such motors operate in innumerable timers in appliances and general industrial operations as their power source. For motor magnet bonding, the methacrylates and urethane-acrylate adhesives have been widely used because of their ability to be automated and rapidly cured either by heat or chemical activation. Batson (4455) has also described such magnet bonding in some detail. We have also previously described the use of various adhesives to bond lens materials for both industrial and automotive lighting. Both plastic and glass lenses must be bondable to other plastics, steel, or aluminum parabolic reflectors to achieve a watertight seal with long-time permanent assembly.

In the home entertainment industry, the use of adhesives to bond the many different kinds of speaker cones and various gaskets can be cited. Usually, the cyanoacrylate or methacrylate adhesives are most employed. While these components have little inherent strength in themselves, they absolutely must be securely bonded to any framing involved in their attachment.

B. Application of Adhesives in the Electronics Industry

For an early assessment of the use of adhesives and sealants in electronics, the author suggests the review of DeLollis from his extensive work at the Sandia Labs in the middle 1960s (4456). For a modern update of the overall technology, we cite the 1990 chapter by Charles (4457) titled "Electronic Packaging Applications for Adhesives and Sealants" in ASM's *Engineered Materials Handbook* (Volume 3). The most important and fast-growing segment for discussion would be integrated circuit manufacturing with all other applications thrown into a category of general component bonding.

1. *INTEGRATED CIRCUIT MANUFACTURING*

The growth in the number of applications involving adhesives and integrated circuit technology has been astounding in recent years. The needs of these complex, high-density, high-speed devices has especially generated difficult demands for device protection, encapsulation, and packaging procedures and separate discussions of each will be undertaken. Moore (4458) has discussed the exponential rise in chip device density in 1980, and Sze (4459) has reviewed the corresponding decline in the size of these devices. This has created rapidly changing packaging parameters requiring advanced materials for device interlayer dielectrics, top surface passivation and overcoating layers, die adhesives, encapsulants, package sealants, and circuit interconnections (4460). Wong (4461) has written a 1986 book on the technology of the integrated circuit encapsulants which demonstrates the complexity of solving such problems.

a. Wafer Production Technology

Integrated circuits (ICs) must start with the growth of either single-crystal silicon or gallium arsenide. Czochralski (4462) has offered one technology for accomplishing this process. A refinement of the Czochralski process to prevent arsenic escape has been offered by Metz et al. (4463). Charles (4457) has presented a typical integrated circuit flow sheet for a simple set of circuit elements which uses the wafers. Various additional processing operations have been described by Charles and Clatterbaugh (4464).

After wafer preparation, one of the first IC processing steps is to grow an oxide layer on the silicon. This oxide layer can act as a barrier or as a source for the introduction of dopants (controlled impurities) into the silicon. Glasser and Subak-Sharpe (4465) have further described special procedures to distribute the impurities deeper into the silicon. These authors showed the diffusion followed according to the conventional diffusion laws (Fick's laws).

Silicon devices are typically metallized or interconnected using aluminum metal because of its high electrical conductivity and ability to form a protective oxide layer. The metal makes contact with the silicon only in the openings in the protecting oxide layer. It has one significant problem, however, because of the solubility of the silicon in the aluminum at very high temperatures. Holland (4466) has discussed this under vapor deposition conditions, whereas Chapman (4467) relates it to sputtering conditions. When the silicon diffuses into the aluminum, holes or voids are left in the silicon which, in turn, are filled by aluminum metallization of "spikes" as they are called. Pai et al. (4468) has studied the minimizing of these spikes by reducing the processing times with accompanying rapidly increasing temperatures. Poate et al. (4469), in their 1987 book, have described a process called electromigration which can shrink the interconnect trace dimensions causing an increase in the current densities carried by the circuit traces. It is desirable to improve the electromigration resistance of the aluminum-copper-silicon interconnect systems, and Steppan et al. (4470) have demonstrated the introduction of some titanium can be a positive factor in this respect. It should be remembered that the typical aluminum alloy used for metallization will contain 4% copper and 1–2% silicon. This is the basis for the copper appearing above as part of the interconnect system. Because the silicon-aluminum interface is basically unstable, the development of more stable and uniform metallized contacts are highly desirable. This surface has been improved by using silicide contacts (4471).

The metallization of gallium arsenide is an even more difficult problem. Pugh and Williams (4472) discussed this matter in 1986. Beyers et al. (4473) have shown the effectiveness of using gold-germanium in this respect. Once fabricated, either silicon or gallium arsenide can be electrically tested at the wafer level using either conventional contact probes or noncontact beam (electron and laser technologies) (4474).

b. Device Packaging

The basic operations in device packaging involves placing the IC into a package (die attach), interconnecting the circuit to the packaging (interconnection bonding) either directly or through some intervening adherend (e.g., hybrid, multichip module), and finally sealing or encapsulating the package. Inspections, testing operations, and environmental screening must be used to verify the product performance.

The electrical connection to most ICs is made to a bonding pad on the top surface or face of the chip. Mechanical support must be provided by bonding to the adherend or package using solder alloy attachment and/or organic adhesive attachment. Charles has described this process in considerable detail (see Ref. 4457, pp. 583 and 584). Interconnecting bonding (step 2 above) is accomplished in any of three ways: (1) wire bonding, (2) tape automated bonding, or (3) inverted-chip reflow (flip-chip technology). Wire bonding, in turn, is divided into three categories: (1) thermocompression bonding (ball-wedge), (2) thermosonic bonding (ball-wedge), or (3) ultrasonic bonding (wedge-wedge). In thermocompression bonding, a gold wire is fused to the chip bonding pad, which is typically aluminum (see Ref. 4457, p. 584). Thermosonic bonding is a similar procedure except the heat associated with thermocompression bonding is reduced by the introduction of ultrasonic energy. Finally, in ultrasonic, or wedge-wedge, bonding the bonds are formed by capturing the wire against the chip bonding pad using a wedge-shaped tool. The bond is formed by applying pressure and an appropriate burst of ultrasonic energy (typically 20–60 kHz frequency).

The tape automated bonding process involves bonding ICs to patterned metal or multilayer tape (copper patterned film on polyimide carrier film) using thermocompression bonding techniques. The process is described in detail by Charles (4475). The inverted-chip reflow, or flip-chip, process involves the formation of ball or bumps on the chip bonding pads (either solderable or nonsolderable bumps). The chip is then inverted over appropriate metallization pads and an interconnection is formed by solder reflow or thermocompression and/or ultrasonic techniques. The interconnect can be extremely high density with excellent electrical performance with good adaptability to automation.

c. *Package Sealing*

Packages can be either hermatic or nonhermatic as broad classifications. The latter are either polymer-sealed or encapsulated where moisture will penetrate in a relatively short time, whereas the hermatic packages must be made of metals, ceramics, or glasses. Leedecke et al. (4476) have described metal and glass packaging using a glass/metal sealing. Over 75% of all metal packages are welded using parallel seam welding. Others are solder sealed using gold-tin or conventional tin-lead solder alloys (4477).

The major chip encapsulation (packaging) techniques are cavity filling, saturation (impregnation), and coating (dip, surface, and conformal). Cavity-filling processes are further subdivided into (1) potting, (2) casting, and (3) molding. In the first process, the potting agent is poured around the electronic component in a cavity and cured. All become a integral part of the final assembly. Casting is similar to potting except the curing container is removed after curing is complete. Typically, no heat or pressure will be used in the process, although some vacuum can be used to help outgas the resin. In molding, the component is placed in a mold and the encapsulant is injected with curing in the mold, which is removed after curing. Gee et al. (4478) have discussed various problems found when using the molding process, whereas Charles and Clatterbaugh (4479) have applied finite-element techniques to defining the stresses which can arise from different conditions of molding.

d. *Board Level Packaging*

The processes of joining packages to other assemblies like cards, boards, or other adherends can have major influence on the design, manufacturability, cost, and reliability of electronic systems. In turn, the choice of what board interconnection methodology to use will play a significant role in determining the board configuration and selection of materials. It also follows that the interconnection method will be influenced by the need for removability to replace failed parts or to change a part to further enhance the overall performance. The above will usually lead to selection of a solder joint or sometimes a socket. Charles (4457) has shown the typical basic forms of packages and soldered interconnects (see Ref. 4457, Fig. 16, p. 589). Pin-socket interconnections have been shown typically to provide inferior-performing interconnections as compared to the more permanent solder joints. Other alternatives such as conductive adhesives or direct wire-bonded interconnections have been described by Bodendorf et al. (4480).

A through-hole–type mounting is common where the lead from the package is formed and then pushed through a hole in the printed wiring board. This hole may be plated and the lead is soldered into the hole. The procedure, known as DIP, has been the mainstay of through-the-hole packaging with use for IC packaging going back to the early 1960s. Plastic DIPs have been used also because of their low cost but suffer poorer hermiticity (4481).

A more desirable method as compared to pinned and through-hole packages has been the surface mounting technology (SMT) procedure. Some of the recognized improvements have been increased density, reduced package size and cost, improved board area utilization, better electrical performance, and improved repairability. Two subtypes of SMT are leadless and lead packages. In leadless packages, solder forms the bridge between metallized areas of the package and the board (4482). In the leaded package, a compliant lead extends from the package and forms the interconnection of the package by means of a solder joint to the board. The major distinction between the two is the elastic compliance of the interconnecting lead in the leaded device between the package and the board. Clatterbaugh and Charles (4483) have discussed their analyses of the importance of the elastic compliance under conditions of thermal and power cycling.

e. Circuit Boards

Circuit boards are a complex array of organic and inorganic materials with both internal and external wiring that allows components to be mounted (mechanically unsupported) and electronically interconnected. A variety of thin-film and thick-film inorganic circuit boards or substrate constructs have been described by Charles and Clatterbaugh (4464) and Borland (4484). With this exclusion, there are four basic printed board technologies that can be considered: (1) rigid boards, (2) flexible boards, (3) metal core boards, and (4) injection-molded boards. Mostly, the rigid and flexible board materials are copper-clad laminates of organic dielectrics, including phenolic-paper, epoxy-glass, and polyimide-glass. The polyimide is being used increasingly because of its high-temperature stability and excellent handling properties; however, epoxy-glass and variants are still the most widely used printed board material.

Printed boards range from single-sided to complex, multilayered boards. The interconnection structures can be extended in principle to any number of layers. Nakahara (4485), for example, has demonstrated boards of 40–50 layers. For details of printed board fabrication see Tummala and Rymaszerski's handbook (4486), and Markstein (4487) can be viewed for comments on the most widely used board using FR-4 fire-retardant, epoxy-glass cloth. The method of making the epoxy fire retardant has been discussed by Schlack (4488). Fiber glass is the major reinforcing material in the epoxy-glass cloth and the polyimide-glass boards. A good coupling agent has had to be employed to achieve good adhesion between the fiber glass fibers and the epoxy resin (4489).

The flexible circuit boards are usually one or two signal layers made on thin, flexible boards of polyimide or polyester film rather than the rigid epoxy-glass fabric. The surface wiring may be photolithographed patterned copper foil that has been laminated to the film dielectric with either epoxy or acrylic adhesives. By mounting rigid board elements to a flexible board or stiffening flexible boards in localized regions, a manufacturer can build a fold-up board which becomes a three-dimensional packaging scheme.

Board overcoatings or encapsulations can provide mechanical protection, moisture and hostile gas protection, and dielectric shock protection. If applied prior to soldering, the coating must withstand the soldering process. Coatings can be applied by screen printing, roller or dip coating, evaporation, or the fluidized-bed process. Sometimes a silane treatment may be necessary to obtain the desired degree of long-term durability.

f. Inorganic Passivation

Typical inorganic passivations and/or encapsulants have included silicone dioxide, silicon nitride, and silicon-oxynitride. They need to be deposited in very thin layers of the order of 1–2 μm (0.04–0.08 mil) requiring sophisticated thermal, plasma, or radiation-assisted deposition techniques. Wong (4490) has reported that these inorganic passivations are excellent moisture and mobile-ion barriers, with silicon nitride being superior to silicon dioxide. These materials typically fall in the range of the glasses as to their moisture diffusion.

Other inorganic oxides employed for passivation of semiconductors and metal layers include aluminum oxide produced by anodization of aluminum along with various deposition techniques used on the nonaluminum surfaces. Yamanaka et al. (4491) have described sputtering and radio frequency ion plating techniques used for this purpose. Berry et al. (4492) have considered anodizing or reactive sputtering of tantalum for applications of this oxide, whereas Maissel and Glang (4493) have described the use of titanium oxide for passivation purposes.

Adams (4494) has described the thermal chemical vapor deposition process (CVD) which is used most widely for preparation of silicon dioxide, silicon nitride, silicon-oxynitride, and polysilicon inorganic passivators. Further, Kern and Ban (4495) and Kern and Schnable (4496) have described the atmospheric and low-pressure variations of the CVD process. This thermal process does, however, require relatively high-deposition temperatures (1110–1650°F [600–900°C]) for production of high-quality coatings. Meyer and Lau (4497) have shown the aluminum metallization will tend to form "hillocks" due to the intermetallic diffusion problems existing at such temperatures. Adams and Capio (4498) have described a parallel-plate plasma-assisted CVD which is typically a cylindrical glass chamber with parallel aluminum plates acting as the top and bottom electrodes. Pliskin (4499) describes another variation; i.e., the hot-wall plasma-assisted CVD process.

g. *Organic Adhesives, Passivations, Sealants, and Encapsulants*

The organic sealants and encapsulants are further divided into the broad categories of thermosets, thermoplastics, and elastomeric types. Examples of the thermoset variety include polyimides, epoxies, polyesters, silicone-modified polyimides, acrylates, and silicon-epoxies, all of which cannot be reversed to their original states after curing. The remeltable and flowable thermoplastics can include polyvinyl chloride, polystyrene, polyethylene, various fluorocarbon polymers, acrylics, poly(p-xylylene), and some preimidized silicon-modified polyimides. The long, linear, flexible chained elastomers include silicone rubber, silicone gel, natural rubbers, and polyurethanes. The most useful sealants and encapsulants have been the epoxies, polyimides, fluorocarbon polymers, poly(p-xylylene), polyurethanes, and silicone (including the gel). Charles (4457) has described in some detail the variety of individual materials and their use as sealants and encapsulants. Specifically, he has reviewed epoxies (4500), polyimides (4501), silicones (4502), poly(p-xylylene) (Parylene) (4503), fluoropolymers (4504), and thermoplastics (4505).

C. Miscellaneous Technical Reports Pertaining to Adhesives and Sealants Applications in the Electrical and Electronic Industries

1. *TECHNICAL REPORTING FROM* ADHESIVES AGE MAGAZINE

In 1960, three unauthored articles included "Epoxy-Sealed Marine Radios" (4506), "Bonding Computer Components" (4507), and "Epoxy for Electrical Application" (4508). Rider (4509), in 1961, described "Adhesive Clips Facilitate Telephone Wire Installations" and an unauthored article considered "Adhesive-Coated Copper Simplifies Production of Die-Stamped Circuits" (4510). Snogren (4511) described "Adhesive Bond Heat Sinks to Printed Circuit Boards" in 1968, whereas Hess and Jackson (4512) discussed their investigation about "Designing Adhesives and Interfaces for Optimum Electrical Conductivity."

Armstrong (4513), in 1972, described "New Concepts in Adhesive Bonding (laminated structural materials, paper, foil, fabric lamination, and motor wire bonding), and Selya (4514) reported on "Encapsulation of Electrical Units with Thermosets" in 1973, followed, in the same year, by Holland's (4515) report on "Adhesives for Flexible Printed Circuits." An unauthored 1974 article recounted a "New Silver Epoxy Speeds Electronic Production" (4516). Leech (4517) described "Adhesive Coated Laminates for Additive Circuits" in 1975, along with O'Malley's (4518) "Silicone Pressure-Sensitive Adhesives for Flexible Printed Circuits." In 1979, Heuring (4519) reported on "Bonding RF Absorbent Material to Microwave Enclosure Covers."

Unauthored articles appeared in 1982 titled "UV Adhesive Lowers Magnet Assembly Cost" (4520), and in 1983, describing "Simplified Wiring in Auto Printed Circuit Boards" (4521). Also in 1983, Schultz (4522) described "Potting and Encapsulating Electronic Components." Pandiri (4523), in 1987, wrote about "The Behavior of Silver Flakes in Conductive Epoxy Adhesives," with Moreau (4524) describing in 1988 how "UV Curing Cuts Time and Costs for Solenoid Manufacturer." In 1990, Leone and Noyes (4525) described how "Automated Potting Reduces Urethane Waste and Rework" (in transducers). Finally, in 1991, Clark (4526) related how "Tapes and Structural Bonding Reduce Radar Unit Costs" and an unauthored article was titled "Turn-Key Potting Benefits OEM Electronic Producer" (4527).

D. Additional Technical Publications Mostly Related to Thermally and Electrically Conductive Adhesive Formulation

These papers will be cited chronologically for best historical development purposes. Kraus (4528) developed an information bulletin for Abselstik Labs titled "Adhesives for Microelectronics." In

1984, Kulesza (4529) described how conductive epoxy adhesives can be used to solve surface mounting problems in electronics manufacturing. In 1985, an unauthored article in *Electronics Packaging Production* was titled "Conductive Epoxy is Tested for SMT Solder" (4530), whereas in 1986, an unauthored article in *Adhesives Age* was titled "Thermally Conductive Acrylic Bonds Solid State Relays" (4531). Coleman (4532) wrote on the subject "Sticky Issues in Electronic Assembly" in 1989, whereas Hermansen and Tunick (4533), in the same year, reviewed "Formulating Custom-Tailored Thermal Transfer Adhesives." Finally, Jagt (4534), in 1990, reviewed the overall subject of "Adhesive Concepts for the Fixation of Electronic Components" in a presentation at Adhesion '90 meetings.

VII. CONSTRUCTION INDUSTRIES

It will become clear from a review of the kinds of adhesives that have found application in the construction industries that many have the function of serving the dual purpose of both adhesive and sealant. However, there are also many instances of sealants serving purely that function alone, whereas more structural adhesives may be used internally in a construction, and thus are serving no sealing function of keeping water out of the structure itself. Many adhesives also fall in the area of being low strength as compared with truly structural adhesives but are still capable of maintaining sufficient strength and adhesion to durably attach surface films, like foils or various metallic or plastic films, to stiffer structural surfaces. Reference is made here to adhesives like the solvent (organic or water)-dispersed contact adhesives. Of course, it should be remembered that the author has discussed in detail the production of laminated aluminum-faced building panels elsewhere where the overall assembly with a neoprene-phenolic contact adhesive could be deemed structural even though the measured lap-shear strength of the adhesive would not be structural for the purposes of joining airplanes, automobiles, or other vehicles. In this case, it is the overall combination of multiple laminated layers of structure which create the final stiffness and high-loading potential because of the large total area of the building panel.

A. Buildings and Associated Bonded Structures

For a more practical appraisal of progress in bonding architectural buildings and related components, the author will present a chronilogical review of the pertinent articles appearing in *Adhesives Age Magazine* over a 3-decade period.

In 1960, McKenzie (4535) discussed "Metal Curtain Wall Sealants." Sheppard (4536), in the next year, reported on "Use of Adhesives in Preparing Prefinished Surfaces for the Housing Industry." Three unauthored articles in the same year were titled "Computer Room's Floating Floor Bonded in Place" (4537), "Bonded Panels for a City Under Ice" (4538), and "Ceramic-to-Glass Bonds in 60 Seconds" (4539). During the year 1962, six unauthored subject areas were discussed, including "Epoxy Helps Put Up a Good Front" (redecoration of a store front) (4540), "Unusual Roof Design Requires New Coating Technique" (4541), "Bonding Prefab Panels of Mosaic Tiles" (4542), "Caulking and Sealing Chicago's New Airport" (4543), "Reinforced Bonded Handles for Multiwall" (4544), and "Sealing Stained Glass" (4545). Also in 1962, Ellis (4546) discussed "Adhesives for Thermal Insulation" and Fitzgerald (4547) considered "Adhesives for Ceramic Tile."

Four unauthored articles appeared in 1965 with the titles "Building Panels Strengthened with Wide-Range Adhesive" (4548), "Adhesives for Wallboard" (4549), "Pedestal Floors Constructed with Contact Adhesive" (4550), and "Chemical Bonding of Stained Glass Windows" (4551). The following year two additional unauthored articles discussed "Pavillion Constructed with Bonded Sandwich Panels Proves Successful" (4552) and "Window-Wall Systems Utilize a Variety of Sealers" (4553). In 1967, *Adhesives Age* published four related articles with the titles "Sealants Play a Vital Part in Building the Largest Structure" (4554), "Commercial and Private Buildings Utilize a One-Part Epoxy System" (4555), "Aggregate Epoxy Protects Facade" (4556), and "Experimental Arch House Features Adhesive Bonded Stressed Skin Roof" (4557). Finally, in 1969, the published information appeared in a paper by Valentine (4558) on "Glass and Glazing Joints" and an unauthored report on "New Butyl Rubber Offers Advantages as Window and Construction Sealant" (4559).

Santaniello (4560), in 1972, discussed "Polysulfide Sealant Meets the Needs of the World Trade Center." Four unauthored reports were issued on "Roll-Out Roofing Meets Need of Modular Home Construction" (4561), "Construction Adhesive Saves Time in Two Building Applications" (4562), "Adhesive Dispensing System Cuts Prefab Housing Costs" (4563), and "Silicone Sealant Waterproofs Spires of Musuem Towers" (4564). The next year Meinertzhagen (4565) reviewed "Using Adhesives in Thermal Insulation," Chase (4566) revealed how "Multipurpose Construction Adhesive Increases Structural Soundness," and Lowe (4567) considered "Structural Adhesives and Compounds for Use in the Construction Industry." Two unauthored articles also appeared about "Cross-linked Butyl Solves Roofing Problem" (4568) and "Primer/Conditional Seals Concrete Floor in Space Conversion" (4569). A 1978 report discussed "Manufacturing Dual-Seal Insulating-Glass Units" (4570), and the following year an article by Booth titled "A Guide to Building Construction Adhesives" appeared (4571).

In the decade of the 1980s, the number of articles relevent to adhesives use in building construction began to diminish. Bax (4572), in 1981, discussed "Formulating Solvent-Free, Water-Resistant Adhesives for Ceramic Tiles." In 1982, an unauthored article titled "Sealant Enhances Building Design" appeared (4573). Slautterback (4574) issued a 1983 paper reviewing special bonding on a $32 million dollar aquarium in Monteray, California, specifically bonding fiber glass with a hot melt. In 1984, Bouchey (4575) reviewed the report of "Task Group Examines Glazing for Insulating Glass Units," and in 1986, Sandberg and Klosowski (4576) discussed "Structural Glazing: Behavior Details of Double-Bead Installations." Finally, in 1990, Halbmaier (4577) considered "Fast Cure Hot-Melt Increases Parquet Flooring Assembly," whereas an unauthored article was issued in 1991 about "Waterborne Solves Builder's Production and EPA Problems" (4578).

B. Other Miscellaneous Construction Applications

Among the types of construction applications outside building and architectural just discussed is the use of adhesives in various bridge and highway projects. In 1961, an *Adhesives Age* article described "Epoxy Grout Compound Bonds Pontoon Sections of Floating Bridge" (4579), and a report titled "Bridge Drains Caulked with Iron Cement" appeared in 1962 (4580). An unauthored report of 1968 discussed "Epoxy Crack Injection Repairs Highway Bridge Seat Beam" (4581). Papers were reported from England dealing with general high-strength structural bonding of bridges and related structures. McNicholas (4582), in 1969, reported his thesis on "A Critical Study of Engineering Joints in Aluminum Alloy," followed in 1970 by another McNicholas (4583) paper on "Bonded Joints in Primary Structures." Williams (4584), in the same year, also reported on "Adhesive Joints in Engineering Structures," with yet another 1971 paper by McNicholas (4585) titled "High Strength Adhesive Bonding." In 1983, Thornton (4586) reviewed his investigation of "Georgia Highway Joint Sealing," and in 1985, Cook (4587) presented his results about "Joint and Crack Sealing Needed for Bridge, Highway Upkeep."

Additional miscellaneous applications pertaining to various material bonding and sealing can be cited chronologically. Gaul and Apton (4588), in 1960, reported on "Epoxy Adhesives in Concrete," whereas Fender (4589), in 1961, offered "Procedures for Three Types of Industrial Formica Bonds." Three unauthored articles appearing in 1962 included "Bonding in Cargo Container Assembly" (4590), "Adhesive Bonding Tips and Fenules to Flexible Tubing" (4591), and "Causes of Joint Sealant Failure" (4592). Two unauthored 1965 reports discussed "Bonded Aluminum Railing Reduces Costs" (4593) and "Bonding Soundproofing Pads to Folding Steel-Door Panels" (4594). "How an Oil Company Reduces Maintenance Costs with a Resin Adhesive" (4595) appeared in a 1967 article. In 1972, two articles described "All-Glass Aquarium Assembled with Silicone Sealant" (4596) and "Adhesive Recovers Mosaic" (4597). "Tunnel Walls Sealed with Silicone Sealant" (4598) was discussed in a 1973 article. In 1977, an article described "Neoprene-Based Adhesive Bonds Shock-Absorbing Doors" (4599), and in 1978, a pertinent article reviewed how "Adhesives Stand Up to Artic Conditions on Alaskan Pipeline" (4600). In 1983, Jindal (4601) discussed his data on "Adhesives and Stress Distribution of a Plate With a Reinforced Hole."

VIII. PACKAGING APPLICATIONS

The use of adhesives for packaging applications probably started when asphaltic or natural resin or gum materials were used to seal documents in antiquity or seal the tombs of ancient monarchs to offer a kind of packaging application on a grander scale. In the movies, all of us have witnessed the use of a sealing wax to close a secret message securely, with a final press of the royal seal into the wax to make it official. The early use of aluminum foil in packaging applications to provide hermatic sealing through the use of some form of adhesive is also well known.

A variety of informative articles describing the various uses of adhesive bonding for packaging have appeared in *Adhesives Age Magazine* over a 30-yr period and can be arranged chronologically as follows. McGuire (4602), in 1961, discussed "Packaging Adhesives and the Food Law" and Lenz (4603) offered his comments on "Six Tips on Selecting Adhesives for Small Carton Sealing." An unauthored article in 1961 described "High Speed Sealing of Polyethylene Bags" (4604). In 1962, Hutchins (4605) gave conclusions about "Adhesion Problems and Food Packaging" and Fisher (4606) cited the reverse situation of "Packages for Adhesives." An unauthorized 1962 article discussed "Adhesive Bonded Food Containers" (4607). Sacharow (4608,4609) produced two technical reports in 1965 on "The Adhesive Needs of the Flexible Packager," and in 1966, he reported on "The Role of Adhesives in Packaging Laminations." Also in 1966, Debnar (4610) offered "An Introduction to Adhesives Used in Packaging and Converting Industries" together with a Bartlett (4611) article on "Adhesives for Films and Foils." A report by Meyer discussed "Equipment and Processes for Paper Bonding" (4612). In 1967, Murray (4613) added information on "Raw Materials and Adhesives for Paper Bonding" with an accompanying article on "Hot-Melt Applicator Cuts Packaging Costs" (4614).

In 1971, Egli (4615) described "How to Select Adhesives for Foil/Paper Laminations." Lambert (4616), in 1973, discussed "Recent Developments in Adhesives for Flexible Packaging" and an unauthored article reviewed how an "Automated Glue System Raises Canning Production" (4617). Estes and Kuespert (4618) projected "The Future of Adhesives Use in Packaging and Packaging Materials" in 1977.

The 1980s produced a 1981 announcement of "Packaging Converter Changes to Water-Borne Adhesives" (4619). Lawson (4620) discussed "Custom Packaging" in 1983, whereas Bentley (4621) offered his investigation of "Overcoming Substrate Problems in Waterborne Laminating" in 1986. In 1989, Fahrendorf (4622) gave his description of "Coater/Laminator Permits Material Process Flexibility," and finally, Maempel and Fricke (4623) discussed "One-Part Aqueous Dispersions for Glassy Film Laminating" in 1991.

IX. APPLIANCE APPLICATIONS

The term *appliance* is differently interpreted by different people as to the range of products involved. Because of this variance, the author has chosen to include appropriate articles that include both particular appliance-type products and processes involved in manufacturing such items using a chronological development rather than separate listing by specific product.

A significant number of *Adhesives Age* articles in 1961 introduced data and production details on a variety of manufactured products. Meyer (4624) described "Adhesive Applications at Kodak," whereas Alexander et al. (4625) reported how "Epoxy Bonding Cuts Costs of Typewriter Production." An unauthored article on typewriter manufacturing also appeared titled "Adhesive Bonding Speeds Production at Smith-Corona" (4626). Speaker production was discussed in "Room Temperature Formulation Bonds Alnico to Aluminum" (4627). Compressor repairing was the subject of "Epoxy Repairs Compressors Cracked Casing" (4628), miniaturized microphones were discussed in "Minute Quantities of Epoxy Bonds Miniaturized Microphones" (4629), and binocular manufacturing in "Bonds for Binoculars" (4630).

Carr et al. (4631), working on instrument jointures, found "Frozen Catalyzed Epoxy Solves Instrument Bonding Problem." Three unauthorized 1962 articles considered "Panel Assembly with Versatile Sprayable Adhesive'' (4632), ''Fast-Setting Adhesive Sparks Sound of Music'' (electric organ) (4633), and ''Bonded Mirror-Bright Reflectors'' (with sheet aluminum) (4634). In 1965, Madden et al. (4635) contributed a paper on ''Epoxy Resins as Bonding Agents for Filiament Wound Structures,'' whereas air conditioner production was discussed in an unauthored article titled ''Air Conditioner Quality Improved Through Use of Adhesives and Sealers'' (4636). Three unauthored presentations offered in 1966 included such widely divergent products as stereo radios in ''Stereo Radio Units Strengthened With Film Adhesives'' (4637), testing devices in ''Fast-Setting Adhesive Aids Testing Devices'' (4638), and baseboard heating units in ''Bonding Baseboard Heating Units Reduces Noise and Fatigue Problems'' (4639). In 1967, unauthorized publications considered nameplate manufacturing in ''Nameplate Manufacturer Improves Performance with a Film Adhesive'' (4640) and press fitted parts in ''Using Adhesive for Press Fitted Parts'' (4641). Boran (4642), in 1967, described ''Using Hot-Melt Adhesives in Automated Assembly of TV Coils,'' and an unauthored report discussed how ''Sealants Help Perfect Washer-Dryer Systems'' (4643).

Relevent articles seemed to diminish in the 1970s when only three articles were published. A 1974 report discussed ''Glass-to-Metal Bonding of TV Tubes'' (4644). Dietrick et al. (4645) reported their ''Optimization of Wire Adhesion by Varying Concentrations of Bonding Agents and Cure Rate'' in 1977, and Paul and Wimarski (4646) suggested ''Polyimide Adhesives for Composite Fan Blades'' in 1978.

The year 1980 began with an unauthored report on ''Acrylic System Used by Stereo Speaker/Wood Beam Makers'' (4647). In 1983, Bowles and Garrett (4648) reported ''An Appliance Case History,'' DeBarros and Beachle (4649) described ''Bonding Parabolic Reflector Flood and Spot Lamps'' in 1989, and Osterberger and Arendt (4650) discussed ''High Melt Point Additive Improves Product Assembly HM's'' in 1991.

X. MEDICAL APPLICATIONS

The use of adhesives in manufacturing medical products has some strict requirements both because of the need for a high reliability in the operational life and because the adhesive may contact human tissue and blood. Thus, they generally need to pass *U.S. Pharmacopoeia* (USP) class VI testing. Many of the plastics employed, like polyether-imide, polysulfone, polyether sulfone, polycarbonate, polyphthalate, thermoplastic polyesters, and polyphenylene ether, are among the most difficult to bond. Even polyethylene, polypropylene, ABS, and silicone rubbers, which are also widely used, require special attention to adherence to their surfaces. Francis (4651), in 1987, published an article discussing these plastics and how they might be processed.

There is a vast quantity of tubing and corresponding connectors which are usually made from PVC and polycarbonate and ABS, respectively, which are used in all hospitals. Adhesives for this market include solvent-based types, cyanoacrylates, UV-curing and RTV silicones, and UV-curing methacrylates. Possibly the highest volume market is syringe assembly, which is a high-volume use of UV-curing adhesives using an entirely automated assembly line.

Some *Adhesives Age* articles can also be cited relating to this matter. In 1961, Kemp (4652) discussed ''Pressure-Sensitive Surgical Tapes—Their History and Development.'' Page (4653), in 1966, showed how ''Tissue Adhesive Eliminates Sutures and Staples in Many Types of Surgery,'' and Krug and Marecki (4654) reported on ''Porous and Other Medical PSA's'' in 1983. In 1984, Cobian et al. (4655) summarized data on ''Epoxy Adhesive for Implantable Medical Devices.'' Finally, in 1990, Pfister (4656) described how ''Silicone PSA's Offer Flexibility for Medical, Pharmaceutical Use.''

XI. SPORTING GOODS APPLICATIONS

An increasingly wide range of adhesive-bonded products have been developed in recent years which are best classified as being related to sports activities. In most cases, these products would best be described as adhesive-bonded adaptations of already existing products, since the activities already were in existence using devices fabricated with other procedures of assembly. Probably the most influential factor which dictated the use of adhesives was the need to develop designs which were stronger structures. This, in turn, demanded stronger materials of construction of quite dissimilar chemical natures (such as combinations of metals and special plastics) which were best joined with adhesives.

A partial list of products would include skis, golf clubs, tennis rackets, windsurfer boards, snowboards, sailplanes, and archery equipment. The newer design of these kinds of products has provided relatively larger jointed areas, so that adhesive bonding provides the best way to provide maximum holding power between the dissimilar materials even in the presence of the quite high-localized stresses that can exist under typical service conditions. The use of adhesives can be shown to reduce the overall weight with increased strength and reduce the finishing costs, while offering improved durability in service.

The author was involved with the early selection of adhesive and demonstration of best composite design of materials through extensive accelerated weather durability testing for the Head ski. Unlike the earliest skis whose strength and properties were resident in a single material of construction, the modern-day ski consists of as many as 10 laminate construction layers joined together with bonded side, top, and bottom facings. This composite material construction must be carefully assembled and bonded to offer a desired strength and flexibility which incorporates outstanding resistance to impact forces, the abrasive action of contacting external surfaces, and long-term durability under all kinds of weathering service conditions. In more recent developments, the earlier employed structural adhesives used with accompanying heat curing and pressing have tended to be replaced with methacrylate adhesives using surface activators for better cost effectiveness.

The author was also involved in the early considerations of bonding aluminum tennis rackets which involved the development of special alloy extrusions and adhesive bonding. Steel-framed rackets have been similarly bonded, and opinions differ as to which metal-type racket is best or whether newer modifications of the older wooden-type rackets might still be preferable. In any case, all designs are dependent upon the strength and assembly properties of various adhesive formulations for their structural integrity. Reference can be made to a 1962 *Adhesives Age* article titled "Laminated Tennis Racket" (4657).

Similarly, the development of the adhesive-bonded aluminum golf clubs both as to shaft and head designs involved developing special alloys and extrusion components. Once again, steel and wood have been competitive materials and proponents of each can be found throughout the field of players. The refinement of design and establishment of the special qualities of each different combination of materials of construction has reached the point today where professional golf players will contend that a certain combination of materials will have optimum utility for particular kinds of golfing shots.

The very sophisticated nature of forces at work when releasing an arrow in archery has resulted in the need for computers to designate the particular shape that must be imparted to the composite plastics that are to be bonded to form the bow. Stress values of 125 MPa (18 ksi) can be attained on the wood–fiber glass laminate bonded interface.

The author has also been involved in the development work at the Alcoa Labs on aluminum baseball and softball bats, which are manufactured today in staggering numbers because of the general use in most situations outside of the professional leagues. While the major product strength in this case is not derived from adhesive joining, the use of materials to deaden the metallic sound of the impact will involve maintenance of adhesion in some form for best effect.

The substitution of both aluminum and plastic composite materials for wood in the construction of boats has been a major development of recent years. The technology used in windsurfer board and sailplane manufacturing has been reflected from those efforts. In all of these products, the use of multilayered structures to provide added stiffness and resistance to impact has introduced the use of adhesive bonding to various degrees. In addition, there is the matter of the attachment in boats of a wide variety of deck and hull hardware, which often involves the joining of dissimilar materials such as aluminum and plastics. Even the use of bouyant materials inside the hulls of both small and modest-sized ships involves the maintenance of adhesion between the two surfaces in order to provide additional strength and stiffness in the hull structure in addition to the vital function of bouyancy in case of damage to hull integrity. Three *Adhesives Age* articles relate to marine applications, starting with a 1962 article titled "Chemically Fastened Screws Resist Outboard Motor Vibration" (4658). In 1986, two unauthored reports considered "Sealant Improves Quality for Sailboat Manufacturer" (4659) and "Boat Builder Saves Time with Foam Mounting Tape" (4660).

XII. MISCELLANEOUS APPLICATIONS AND GENERAL INFORMATION REPORTS RELATIVE TO USE OF ADHESIVES IN MANUFACTURING FROM *ADHESIVES AGE MAGAZINE*

A. Different Applications Not Included Elsewhere

In 1961, Fawcett (4661) suggested "Three Epoxy Systems for Bonding Carbide Wear Parts," and an unauthored article in *Adhesives Age* titled "Adhesives Solve High-Speed Polishing Problems" (4662) appeared. In 1962, Johnson (4663) described how "Adhesive-Bonded Hose Products Withstand Torture Testing" and Middleton (4664) considered "Bonding Foundry Cores with Sodium Silicate Adhesives." An unauthored article from the same year was about "Blending Adhesives for High Speed Photographic Film" (4665). "Adhesive Coated Steel Strip and Wire Opens New Adhesive Markets" (4666) appeared in *Adhesives Age* in 1965. Two unauthored reports in 1966 considered "Silicone Rubber Bonds Solar Cells" (4667) and "How Epoxy Adhesive Simplifies Fabrication of Diamond Wheels" (4668). In 1967, unauthored articles were concerned with "Bonding of Wet Suits Reduces Repair Cost" (4669) and "Dependability of Sponge Mops Traced to Neoprene Adhesives" (4670). Finally, a 1968 article discussed "Fabricating Epoxy Resin Wind Instruments" (4671).

Three reports in the 1970s included one in 1973 on "Cyanoacrylic Adhesive Aids in Do-It-Yourself Horse Shoeing" (4672), another in 1975 about "Small Hot-Melt Systems Move into Foundries" (4673), and a 1977 article on "Polyepichlorohydrin Elastomer Solvent-Based Adhesives for Footwear" by Kozakiewicz and Penczak (4674).

Application articles of the 1980s and 1990s started in 1981 with Merrill's (4675) discussion of "Restoring Works of Art with Silicone PSA's," followed by Berger's (4676) description of how an "Art Adhesive Needs Stability, Reversibility, Compatibility" in 1985. Two unauthored articles of 1986 included "Acrylic PSA Enhances Liquid Crystal Products" (4677) and "Adhesives Improve Brewer's Product Quality and Handling" (4678). Lyons (4679), in 1990, discussed how "Ceramic Adhesive Replaces Asbestos in Sealing Furnaces," and an article titled "Adhesives to the Rescue: Bonding Lifesaving Equipment" (4680) came out in 1991.

B. General Information Articles Not Related to Specific Products

In 1961, an unauthored report described "Bonded Steel Plates Support 1250 Pound Load" (4681). Tons (4682), in 1965, discussed "Materials and Geometry in Joint Seals." In 1966, a general-interest article appeared titled "New Applications for Family of Foam Tapes" (4683). Huber and Strobel (4684), in 1967, investigated "Applications and Future Uses of Capsular Adhesives," followed in 1986 by a paper from Chao and Hoenisch (4685) on "Two-Part Acrylic Enhances Vacuum Forming Operation." Finally, in 1990, there was an unauthored article on "Urethane Hot-Melt Doubles Subassembly Production" (4686).

References

1. Sainte-Claire Deville, De l'Aluminium, ses Proprietes, sa Fabrication et ses Applications (1859 book).
2. J. D. Edwards, F. C. Frary, and Z. Jeffries, Aluminum and Its Production, McGraw-Hill, New York (1930).
3. J. D. Edwards, Combination process for alumina, Metals Technol., Tech. Publ. 1833, 12 (April 1945).
4. J. D. Edwards, The Immortal Woodshed, Dodd Mead, New York (1955).
5. J. H. Dunn and L. S. Sewell, Jr., Historical development of aluminum production and application, Aluminum (K. R. Van Horn, ed.), ASM, Metals Park, OH, Vol. 2, Chap. 1, p. 1 (1967).
6. L. M. Dunn and W. R. Tyler, Building construction, Aluminum (K. R. Van Horn, ed.), ASM, Metals Park, OH, Vol. 2, Chap. 10, p. 125 (1967).
7. R. C. Kasser, J. R. Stemler, and E. T. Wanderer, Structures for highway, electrical, petroleum and other engineering applications, Aluminum (K. R. Van Horn, ed.), ASM, Materials Park, OH, Vol. 2, Chap. 11, p. 235 (1967).
8. R. L. Horst, Jr., Structures and equipment for chemical, food, drug, beverage and atomic industries, Aluminum (K. R. Van Horn, ed.), ASM, Materials Park, OH, Vol. 1, Chap. 12, p. 259 (1967).
9. J. G. Kaufmann and E. T. Wanderer, Structures and equipment for service at cryogenic temperatures, Aluminum (K. R. Van Horn, ed.), ASM, Materials Park, OH, Vol. 2, Chap. 13, p. 297 (1967).
10. W. C. Keith, R. F. Schaffer, and J. M. Smith, Internal combustion engines, Aluminum (K. R. Van Horn, ed.), ASM, Materials Park, OH, Vol. 2, Chap. 14, p. 297 (1967).
11. W. C. Keith, J. M. Smith, and W. C. Weltman, Automotive applications outside the engine, Aluminum (K. R. Van Horn, ed.), ASM, Materials Park, OH, Vol. 2, Chap. 15, p. 343 (1967).
12. G. B. Hauser, Railroad equipment, Aluminum (K. R. Van Horn, ed.), ASM, Materials Park, OH, Vol. 2, Chap. 16, p. 367 (1967).
13. W. A. Prey, N. W. Smith, and C. L. Wood, Jr., Marine applications, Aluminum (K. R. Van Horn, ed.), ASM, Materials Park, OH, Vol. 2, Chap. 17, p. 389 (1967).
14. C. L. Burton, L. M. Mayer, and E. H. Spuhler, Aircraft and aerospace applications, Aluminum (K. R. Van Horn, ed.), ASM, Materials Park, OH, Vol. 2, Chap. 18, p. 415 (1967).
15. J. L. Faulkner and E. W. Johnson, Military vehicles and equipment, Aluminum (K. R. Van Horn, ed.), ASM, Materials Park, OH, Vol. 2, Chap. 19, p. 459 (1967).
16. P. I. Nielsen, H. H. Nuernberger, and C. G. Shiring, Bearings, tooling, instruments and other mechanical applications, Aluminum (K. R. Van Horn, ed.), ASM, Materials Park, OH, Vol. 2, Chap. 20, p. 485 (1967).
17. H. H. Caldwell, Electrical and electronic applications, Aluminum (K. R. Van Horn, ed.), ASM, Materials Park, OH, Vol. 2, Chap. 21, p. 517 (1967).

18. A. J. Haygood and R. M. Smith, Home appliances, housewares, furniture and other consumer products, Aluminum (K. R. Van Horn, ed.), ASM, Materials Park, OH, Vol. 2, Chap. 22, p. 587 (1967).
19. J. A. Lake, A. B. McKee, and R. C. Reed, Packaging applications, Aluminum (K. R. Van Horn, ed.), ASM, Materials Park, OH, Vol. 2, Chap. 23, p. 597 (1967).
20. A. G. H. Dietz, Composite Engineering Laminates, MIT Press, Cambridge, MA. (1969).
21. K. W. Allen, J. Adhes. 21:261 (1987).
22. W. Brockmann, O. D. Hennemann, and H. Kollek, Int. J. Adhes. Adhes. 2:33 (1982).
23. L. Kozma and I. Olefjord, Mater. Sci. Technol. 3:860 (1987).
24. R. D. Adams, Developments in Adhesives—2 (A. J. Kinloch, ed.). Applied Science Publishers, London, p. 45 (1981).
25. L. J. Hart-Smith, Developments in Adhesives—2 (A. J. Kinloch, ed.). Applied Science Publishers, London, p. 1 (1981).
26. A. J. Kinloch, Review-the science of adhesion, part 2: Mechanics and mechanisms of failure, J. Mater. Sci. 17:617 (1982).
27. W. C. Wake, Adhesion and the Formulation of Adhesives, 2nd ed., Applied Science Publishers, London, (1982).
28. A. J. Kinloch, L. S. Welch, and H. E. Bishop, J. Adhes. 16:165 (1984).
29. H. W. Eickner and W. E. Scholwalter, Report No. 1813, Forest Products Lab., Madison, WI (1950).
30. G. S. Kavayaski and D. J. Donnelly, Report No. D6-41517, Boeing Corp., Seattle (Feb. 1974).
31. J. D. Venables, D. K. McNamara, T. S. Sun, B. Ditchek, J. M. Chen, and R. L. Hopping, Characterization of aluminum surfaces prepared for adhesive bonding, Structural Adhesive Bonding Conf., El Segundo, CA (March 13–15, 1979).
32. P. F. A. Bijlmer, Influence of chemical pretreatments on surface morphology and bondability of aluminum, J. Adhes. 5:319 (1973).
33. J. D. Venables, Adhesion and durability of metal/polymer bonds, Adhesive Joints: Formation, Characteristics, and Testing (K. L. Mittal, ed.), Plenum Press, New York, p. 453 (1984).
34. J. D. Venables, J. Mater. Sci. 19:2431 (1984).
35. D. E. Packham, Adhesion Aspects of Polymeric Coatings (K. L. Mittal, ed.), Plenum Press, New York, p. 13 (1983).
36. J. D. Venables, D. K. McNamara, J. M. Chen, B. M. Ditchek, T. I. Morgenthaler, T. S. Sun, and R. L. Hopping, Proc. 12th Natl. SAMPE Tech. Conf., Seattle, 12:909 (Oct. 7–9, 1980).
37. G. D. Davis, J. S. Ahearn, L. J. Matinezo, and J. D. Venables, J. Mater. Sci. 20:975 (1985).
38. J. S. Ahearn and G. D. Davis, Improved durability of aluminum adhesive bonds with phosphonic acid inhibitors, J. Adhes. 28:75 (1989).
39. J. D. Minford, Comparative study of aluminum joint strength and durability with varying thickness, boehmite-type oxide surfaces, Adhesive Joints: Formation, Characteristics, and Testing (K. L. Mittal, ed.), Plenum Press, New York, p. 503 (1984).
40. N. A. DeBryne, U.S. Patent Nos. 2,499,134 (1944); 2,872,365 (1959).
41. W. E. Golf, The De Havilland Hornet, Aircraft Prod. 8:211 (1946).
42. H. Povey, The production of the Comet, Aircraft Prod. 13:134 (May); 170 (June 1951).
43. A. Hartman and J. H. Rondeel, Report M-1936, Nat. Luchtvaartlaboratorium, Amsterdam (1930).
44. H. Povey, Press bonding, Bonded Aircraft Structures, CIBA, A.R.L., Ltd., Cambridge, England, p. 37 (May 1957).
45. O. Ljungstrom, Design aspects of bonded structures 1, Bonded Aircraft Structures, CIBA, A.R.L., Ltd., Cambridge, England, p. 11 (May 1957).
46. R. J. Schliekelmann and R. Cools, Report No. R-25, Royal Netherlands Aircraft Factories Fokker, Amsterdam (1952).
47. H. F. Hardrath, et al., Tech. Note No. 3856, National Advisory Council in Aeronautics (1956).
48. E. J. van Beek, Design aspects of bonded structures, Bonded Aircraft Structures, CIBA, A.R.L., Ltd., Cambridge, England, p. 33 (May 1957).
49. Bonded Aircraft Structures, CIBA, A.R.L. Ltd., Cambridge England (A collection of papers given in 1957 at a conference in Cambridge, England), undated.
50. M. J. Bodnar (ed.), Symposium on Adhesives for Structural Applications, Interscience Publishers, New York and London (1962).
51. R. J. Schliekelmann, Adhesive bonded metal structures, Adhesion and Adhesives (R. Houwink and G. Salomon, eds.), Elsevier, Amsterdam, Vol. 2, p. 281 (1967).
52. R. J. Schliekelmann, Operational experience with adhesive bonded structures, Bonded Joints and Preparation for Bonding, AGARD Lecture Series No. 102, Air Force Materials Lab., Wright-Patterson AFB, OH, p. 1-1 (Oct. 16–17, 1979).

53. P. Albericci, Aerospace applications, Durability of Structural Adhesives (A. J. Kinloch, ed.), Applied Science Publishers, London, Chap. 8, p. 317 (1983).
54. R. J. Schliekelmann, Introduction, Adhesive Bonding of Aluminum Alloys (E. W. Thrall and R. W. Shannon, eds.), Marcel Dekker, New York, Chap. 1, p. 1 (1985).
55. R. J. Schliekelmann, REDUX for spars, Aeroplane 8:216 (1952).
56. J. H. Rondell, Comparative fatigue tests with 2024 riveted and bonded stiffened panels, Report No. 5411, Netherlands Aerospace Lab. (NLR) (1952).
57. P. F. A. Bijlmer, Fracture toughness of multiple-layer adhesive bonded aluminum alloy sheet, International Conference on Adhesion Science, Lisbon (Aug. 1978).
58. E. Preiswerk and A. von Zeerleder, Aradit, ein neues kunstharz zum verbinden von leichtmetallen, Schweiz. Arch. Angew. Wiss. Tech. 12(H4):115 (1946).
59. F. J. Plantema, Sandwich Construction, Airplane, Missle, and Spacecraft Structures, John Wiley & Sons, New York, Vol. 3 (1966). Adhesive Bonding Alcoa Aluminum, Publication of the Aluminum Company of America, Pittsburgh (1967).
60. Aircraft Structural Integrity Program, MIL-STD-1530 (Sept. 1972).
61. Airplane Damage Tolerance Requirements, MIL-A-83444.
62. Spot-weld through an adhesive? Iron Age, p. 104 (April 24, 1969).
63. Glue-spot welding joins aluminum stronger than spot, cheaper than fusion, Mater. Eng., p. 14 (Jan. 1971).
64. Weldbond flight component design/manufacturing program, Interim Tech. Report First Quarterly Contract No. F 33615-71-C-1716, Lockheed-Georgia Co. (Oct. 1971).
65. Weldbond flight component design/manufacturing program, Interim Tech. Report Sixth Quarterly Contract No. F 33615-71-C-1716, Lockheed-Georgia Co. (Jan. 1973).
66. J. Mahon, C. Vizzi, W. Sisco, and P. Nilsson, Aluminum weldbonding for 350 F service, Mater. and Proc. for the 70's: Cost Effectiveness and Reliability, SAMPE, p. 184 (1973).
67. White engineers cite truck tractor trends, Metal Prog., p. 50 (Feb. 1974).
68. J. D. Minford and E. M. Vader, Adhesive bonding of aluminum automotive body sheet, S.A.E. Int. Auto. Eng. Cong. Meetings, Paper No. 740078, Detroit, Michigan (Feb. 1974).
69. R. C. Hall, Environmental resistance of a weldbond joint, SAMPE J. (Aug.–Oct. 1974).
70. B. B. Bowen, R. E. Herfert, and K. C. Wu, Development of corrosion resistant surface treatments for aluminum alloys for spot-weld bonding, Contract No. F 33615-74-C-5-27, Northrup Aviation (March 1975).
71. K. C. Wu and B. B. Bowen, Aluminum weldbond manufacturing methods development, Interim Report IR-854-5, Contract No. F 33615-75-C-5083, Northrup Aviation (June–Nov. 1975).
72. J. D. Minford, F. R. Hoch, and E. M. Vader, Weldbonding of aluminum auto body sheet, S.A.E. Auto. Eng. Cong. Meetings, Paper No. 750462, Detroit, Michigan (Feb. 1975).
73. J. D. Minford, Comparative effect of surface contamination on the strength and performance of aluminum spot-welded or adhesive-bonded joints, ASM/SDDRG Conf. Tech. Impact of Surfaces: Relationship to Forming, Welding, and Painting, Dearborn, MI (April 1981).
74. J. D. Minford, Weldbonding aluminum in the presence of forming lubricants, S.A.E. Auto. Eng. Passenger Car Meeting, Paper No. 810816, Dearborn, MI (June 8–12, 1981).
75. Douglas Aircraft Company, Primary Adhesively Bonded Structures Technology (PABST), U.S. Air Force Contract F 33615-75-C-3016 (1975).
76. E. W. Thrall, Jr., et al., Primary adhesively bonded structure technology (PABST), Phase 1b: preliminary design, Douglas Aircraft Co., U.S. Air Force Flight Dynamics Lab. Tech. Report AFFDL-TR-76-141 (Dec. 1976).
77. R. W. Shannon and E. W. Thrall, Jr., J. Appl. Polym. Sci., Appl. Polym. Symp. 32:131 (1977).
78. E. W. Thrall, Jr., PABST program results, Adhes. Age 22(10):22 (Oct. 1979).
79. E. W. Thrall, Jr., The future of metal bonding, ASTM Symp. Adhesives and Bonding Processes for the Automobile and Aircraft Industries (Oct. 1979).
80. E. W. Thrall, Jr., Failures in adhesively bonded structures, Bonded Joints and Preparation for Bonding, AGARD Lecture Series NO. 102, Air Force Materials Lab., Wright-Patterson AFB, Ohio, p. 5–1 (1979).
81. Adhesive Bonding Alcoa Aluminum, Book Publication of the Aluminum Company of America, Pittsburgh (1967).
82. J. D. Minford, Adhesive bonding of aluminum, Midwest Engineer 20:56 (Sept. 1967).
83. J. D. Minford, L. W. Hovland, and S. A. Earnest, A guide to selection of adhesives for aluminum, Machine Design, p. 193 (Jan. 19, 1967).
84. J. D. Minford, Adhesive bonding, Aluminum (K. R. Van Horn, ed.). ASM, Materials Park, OH, Vol. 3, Chap. 15 (1967).
85. J. D. Minford, Evaluating adhesives for joining aluminum, Metals Eng. Quarterly (Nov. 1972).

86. L. H. Sharpe, Some aspects of the permanence of adhesive joints, Symp. on Structural Adhesives Bonding, Stevens Inst. Technol. Hoboken, NJ (Sept. 16, 1965).
87. W. L. Krubsack, J. D. Minford, H. F. Campbell, and E. T. Engelhart, Aluminum-faced foamed plastic sandwich panels, Adhes. Age 3(11):36 (1960).
88. J. D. Minford, Aluminum-faced sandwich panels and laminates, Int. Conf. on Aluminum Industrial Products, Pittsburgh (Oct. 30, 1973).
89. J. D. Minford and E. M. Vader, Aluminum-faced sandwich panels and laminates, Adhes. Age 18(2):30 (Feb. 1975).
90. R. L. Patrick, Introduction, Treatise on Adhesion and Adhesives (R. L. Patrick, ed.), Marcel Dekker, New York, Vol. 1, Chap. 1, pp. 3–4 (1967).
91. N. A. DeBruyne, Fundamentals of Adhesion, Bonded Aircraft Structures, CIBA, Duxford, Cambridge, England (1957).
92. M. Faraday, Faraday's select researches in electricity, Everyman's Library, Par. 355–358.
93. J. L. Leslie, Tilloch's Phil. Mag. 14:210 (1802).
94. N. A. DeBruyne and R. Houwink (eds.), Adhesion and Adhesives, Elsevier, Amsterdam (1965).
95. G. M. Kline and F. W. Reinhart, Mech. Eng. 72:717 (1950).
96. R. Bernard and J. Hertz, J. Chim. Phys. 47:95 (1950).
97. L. R. Brantley and G. Carpentier, Off. Dig. Federation Soc. Paint Technol. 324:57 (1952).
98. F. P. Bowden and D. Tabor, Structure and Properties of Solid Surfaces (R. Gomer and C. S. Smith, eds.). Univ. Chicago Press, pp. 203–239 (1953).
99. N. A. DeBruyne, Plastics Progress (P. Morgan, ed.). Lliffe and Sons, London (1951).
100. W. D. Harkins and H. N. McLaughlin, J. Am. Chem. Soc. 47:1610 (1935).
101. W. D. Harkins and E. H. Loeser, J. Chem. Phys. 18:556 (1950).
102. W. D. Harkins and H. H. King, J. Am. Chem. Soc. 41:970 (1914).
103. T. Young, Phil. Trans. Royal Soc., London, 95:65 (1805).
104. A. Dupre, Theorie Mechanique dur la Chaleur, Gauthier-Villars, Paris, p. 369 (1869).
105. W. C. Bieglow, E. Glass, and W. A. Zisman, J. Colloid Sci. 1:513 (1946).
106. I. Langmuir, J. Franklin Institute 218:143 (1934).
107. L. P. Guastalla and J. Guastalla, Mem. Serv. Chim. Etat (Paris), 34:373 (1948).
108. J. Guastalla, J. Chim. Phys. 49:250 (1952).
109. E. Jenckel and B. Rumbach, Z. Electrochem. 55:612 (1951).
110. B. V. Derjaguin, Research (London), 8:70 (1955).
111. R. Simha, H. L. Frisch, and F. R. Eirich, J. Chem. Phys. 21:365 (1953).
112. H. L. Frisch and R. Simha, J. Chem. Phys. 27:702 (1957).
113. W. A. Zisman, Ind. Eng. Chem. 55:19 (1961).
114. S. S. Voyutskii, Adhes. Age 5:30 (1962).
115. J. L. Gardon, J. Phys. Chem. 67:1935 (1963).
116. F. M. Fowkes, Ind. Eng. Chem. 56:40 (1964).
117. B. O. Baterip, Int. J. Adhes. Adhes. 9:232 (July 1981).
118. W. C. Wake, Adhesion and Formulation of Adhesives, Appl. Sci. Publ., London (1976).
119. N. A. DeBruyne and H. Houwink (eds.), Adhesion and Adhesives, Elsevier, Amsterdam (1951).
120. L. A. Girifalco and R. J. Good, J. Phys. Chem. 61:404 (1957).
121. J. M. Gardon, Encyclopedia of Polymer Science & Technology (H. R. Mark, ed.), Interscience, New York, Vol. 3 (1965).
122. D. H. Kaelble, Physical Chemistry of Adhesives, Wiley-Interscience, New York (1971).
123. K. L. Mittal, Adhesion Science and Technology (L. H. Lee, ed.), Plenum Press, New York, p. 129 (1975).
124. J. R. Huntsberger, J. Adhes. 1:289 (1969).
125. J. R. Huntsberger, Interfacial energies, contact angles, and adhesion, Treatise on Adhesion and Adhesives (R. L. Patrick, ed.), Marcel Dekker, New York, Vol. 5, Chap. 1, p. 1 (1981).
126. R. J. Good, Surf. Colloid Sci. 11:1 (1979).
127. First Report of the Adhesives Research Committee HMSO, London (1922).
128. J. W. McBain and D. G. Hopkins, J. Phys. Chem. 29:88 (1925).
129. J. W. McBain and D. G. Hopkins, Second Report of the Adhesives Research Committee HMSO, London (1926).
130. R. M. Vasenin, Adhes. Age 8(5):21 and 8(6):30 (1965).
131. K. W. Allen, Aspects of Adhesion—5 (D. J. Alner, ed.), University of London Press, London, Chap. 1, p. 23 (1969).
132. V. G. Raevskii and L. M. Pritykin, Plast Massy 2:7 (1970).
133. V. G. Raevskii, Some aspects of the theory of adhesive joints, J. Adhes. 5:203 (1973).

134. L. H. Lee, Relevance of the hard-soft acid-base (HSAB) principle to solid adhesion, 14th Annual Meet. Adhes. Soc., Clearwater, FL (Feb. 17–21, 1991).
135. V. Guttmann, The Donor-Acceptor Approach to Molecular Interaction, Plenum Press, New York (1978).
136. B. V. Derjaguin, N. A. Krotova, and V. P. Smilga, Adhesion of Solids, Eng. Trans by R. K. Johnson, Plenum Press, New York (1978).
137. W. B. Jensen, Chem. Rev. 78(1):1 (1978).
138. W. B. Jensen, Surface and Colloid Science in Computer Technology (K. L. Mittal, ed.), Plenum Press, New York, pp. 27–60 (1987).
139. W. B. Jensen, The Lewis Acid Base Concept: An Overview, Wiley-Interscience, New York (1980).
140. K. Morokuma, Acc. Chem. Res. 10:294 (1977).
141. K. Morokuma and K. Kitaura, Molecular Interactions (H. Ratajczak and W. J. Orville-Thomas, eds.), John Wiley & Sons, New York (1980).
142. F. M. Fowkes, J. Adhes. 4:155 (1972).
143. F. M. Fowkes, Recent Advances in Adhesion (L. H. Lee, ed.), Plenum Press, New York, p. 39 (1973).
144. F. M. Fowkes and M. A. Mostafa, Ind. Eng. Chem. Prod. Res. Devel. 17:3 (1978).
145. F. M. Fowkes, J. Adhes. Sci. & Technol. 1(1):7 (1987).
146. J. C. Bolger, Adhesion Aspects of Polymeric Coatings (K. L. Mittal, ed.), Plenum Press, New York, pp. 3–18 (1983).
147. J. C. Bolger and A. S. Michaels, Interface Conversion (P. Weiss and D. Cheevers, eds.), Elsevier, New York, p. 3 (1968).
148. R. G. Pearson, J. Am. Chem. Soc. 85:3533 (1963).
149. R. G. Pearson, J. Chem. Educ. 64:563 (1987).
150. R. G. Parr and W. Yang, Density-Functional Theory of Atoms and Molecules, Oxford University Press, New York (1989).
151. M. Berkowitz, S. K. Ghosh, and R. G. Parr, J. Am. Chem. Soc. 107:6811 (1985).
152. W. Yang and R. G. Parr, Proc. Natl. Acad. Sci. USA 82:6723 (1985).
153. M. Berkowitz and R. G. Parr, J. Chem. Phys. 88:2554 (1988).
154. L. H. Lee, Proc. 5th Int. Cong. Tribology (K. Holmberg and I. Nieminen, eds.), Finnish Soc. Tribology, Espoo, Finland, Vol. 3, p. 308 (1989).
155. L. H. Lee, New Trends in Physics and Physical Chemistry of Polymers (L. H. Lee, ed.), Plenum Press, New York, Chap. 12 (1990).
156. L. H. Lee, Prog. Colloid Polym. Sci. 82:1 (1990).
157. L. H. Lee, Polym. Mater. Sci. Eng. 62:881 (1990).
158. L. H. Lee, Fundamentals of Adhesion (L. H. Lee, ed.), Plenum Press, New York, Chap. 12 (1990).
159. L. H. Lee, Relevance of the density-functional theory to acid-base interactions and adhesion in solids, J. Adhes. Sci. Technol. 5(1):71 (1991).
160. K. Fukui, Science 218:747 (1982).
161. I. Fleming, Frontier Orbitals and Organic Chemical Reactions, John Wiley & Sons, London (1976).
162. R. G. Pearson, J. Org. Chem. 54:1423 (1989).
163. R. F. Hudson and G. Klopman, Tetrahedron Lett. 12:1103 (1967).
164. D. V. Keller, Jr., Status and significance of metallic adhesion research with particular reference to recent ultra high vacuum studies, Trans. Inter. Vac. Metal Conf. (E. L. Foster, ed.), American Vacuum Society, New York (1968).
165. D. H. Buckley, Influence of chemisorbed films on adhesion and friction of clean iron, NASA Rept. TND-4775 (Sept. 1968).
166. K. I. Johnson and D. V. Keller, Jr., (1) J. Appl. Phys. 38:1896 (1967); (2) J. Vac. Sci. Technol. 4:115 (1967).
167. D. V. Keller, Jr., The analysis of metallic adhesion data, Adhesion of Cold Welding of Materials in Space Environments, ASTM, Spec. Publ. No. STP-431, 181 (1968).
168. F. P. Bowden and D. Tabor, Friction and Lubrication of Solids Part II, Clarendon Press, Oxford, England (1964).
169. T. McNicholas and D. V. Keller, Jr., Utilization of contact resistance in the investigation of the metallic adhesion of iron, J. Adhes. 1:164 (1969).
170. J. A. Greenwood and J. P. Williamson, Proc. Royal Soc. A295:300 (1966).
171. G. Ehrlich, Adsorption and surface structure, Metal Surfaces, ASM, Metals Park, OH, p. 221 (1963).
172. W. P. Gilbreath, Definition of evaluation of parameters which influence the adhesion of metals, Adhesion of Cold Welding Materials in Space Environments, ASTM. Spec. Publ. No. STP 431, p. 128 (1989).
173. A. G. Akins and D. Tabor, J. Inst. Metals 94:107 (1966).
174. A. G. Atkins, A. Silverio, and D. Tabor, J. Inst. Metals, Bull. Met. Rev. 94(11):369 (1966).

175. M. Van Den Temple, Advan. Colloid and Interf. Sci. 3:137 (1972).
176. H. C. Hamaker, Physica 4:1058 (1937).
177. E. M. Lipshitz, Sov. Phys. JETP (Eng. Trans.) 2:73 (1956).
178. H. Krupp, J. Colloid Interf. Sci. 39(2):421 (1972).
179. R. S. Bradley, Phil. Mag. 13:538 (1932).
180. B. V. Derjaguin, Kolloid Zeit. 69:155 (1934).
181. B. V. Derjaguin, V. M. Muller, and Yu P. Toporov, J. Colloid Interf. Sci. 53:314 (1975).
182. K. L. Johnson, K. Kendall, and A. D. Roberts, Proc. Royal Soc. Ser. A324:301 (1971).
183. K. Kendall, J. Adhes. 7:55 (1974).
184. F. London, Z. Phyzik 63:245 (1930).
185. H. B. G. Casimir and D. Polder, Phys. Rev. 73:360 (1948).
186. K. W. Allen, Strength and structures, Aspects of Adhesion—1 (D. J. Alner, ed.), University of London Press, London, England, p. 11 (1965).
187. J. R. Huntsberger, Adhes. Age 8(11) (1965).
188. E. Orowan, J. Franklin Inst. 290:493 (1970).
189. D. Tabor, Rep. Prog. Appl. Chem. 36:621 (1951).
190. A. F. Lewis and R. J. Forrestal, Symp. Rec. ASTM Spec. Tech. Publ. No. STP-360, p. 59 (1963).
191. W. Brockmann, Nondestructive testing of adhesive properties of metal surfaces, SAMPE Conf. Aerospace adhesives and elastomers (1970).
192. F. M. Fowkes, Recent Advances in Adhesion (L. H. Lee, ed.), Gordon and Breach, London, p. 39 (1973).
193. W. H. Keesom, Phyzik. Zeit. 22:129,643 (1921).
194. P. DeBye, Phyzik. Zeit. 21:178 (1920); 22 (1921).
195. G. C. Pimental and A. L. McClellan, The Hydrogen Bond, Freeman, San Francisco (1960).
196. R. J. Good, Intermolecular and interatomic forces, Treatise on Adhesion and Adhesives (R. L. Patrick, ed.), Marcel Dekker, New York, Vol. 1, Chap. 1, p. 1 (1967).
197. N. J. DeLollis, Adhes. Age 11(12):21 (1968).
198. N. J. DeLollis, Adhes. Age 12(1):25 (1968).
199. J. E. Rutzler, Adhes. Age 2(7):28 (1954).
200. A. I. Vilenskii, E. E. Virlich, and N. A. Krotova, Soviet Plastics 5:68 (1973).
201. G. E. Koldunovich, V. G. Epshtein, and A. A. Chekanova, Soviet Rubber Technol. 29:22 (1970).
202. E. B. Trostyanskaya, G. S. Golovkin, and G. V. Komarov, Soviet Rubber Technol. 25:13 (1966).
203. R. B. Dean, Official Digest 36:664 (1964).
204. A. Ahagon and A. N. Gent, J. Polym. Sci. Polym. Phys. 13:1285 (1975).
205. E. H. Andrews and A. J. Kinloch, Proc. Royal Soc. A332:401 (1973).
206. A. N. Gent and E. C. Hsu, Macromolecules 7:933 (1974).
207. C. H. Lerchenthal, M. Brennan, and N. Yits'haq, J. Polymer Sci. Polym. Chem. 12:737 (1975).
208. C. H. Lerchenthal and M. Brennan, Polymer Eng. Sci. 16:747 (1976).
209. C. H. Lerchenthal and M. Brennan, Polym. Eng. Sci. 16:760 (1976).
210. J. L. Koenig and P. T. K. Shih, J. Colloid Interf. Sci. 36:247 (1971).
211. R. Bailey and J. Castle, J. Mater. Sci. 12:2049 (1977).
212. M. Gettings and A. J. Kinloch, J. Mater. Sci. 12:2511 (1977).
213. M. Gettings and A. J. Kinloch, Surf. Interf. Anal. 1:189 (1980).
214. B. F. Lewis, W. M. Bowser, J. L. Hurn, T. Luu, and W. W. Weinberg, J. Vac. Sci. Technol. 11:262 (1974).
215. D. E. Packham, Adhesion—7 (D. J. Alner and K. W. Allen, eds.), Transcripta Books, London, p. 51 (1973).
216. J. W. Maxwell, Trans Am. Soc. Mech. Engs. 67:104 (1945).
217. E. M. Boroff and W. C. Wake, Trans. Inst. Rubber Ind. 199,210 (1949).
218. L. E. Perrins and K. Pettett, Plastics and Polymers 39:391 (1971).
219. J. D. Venables, Adhesion—7 (K. W. Allen, ed.), Applied Science Publishers, London, Chap. 4 (1983).
220. H. E. Bair, S. Matsuoka, R. Vadinsky, and T. T. Wang, J. Adhes. 3:89 (1971).
221. T. T. Wang and H. N. Vazirani, J. Adhes. 4:353 (1972).
222. W. D. Bascom, C. O. Timmons, and R. L. Jones, J. Mater. Sci. 19:1037 (1975).
223. D. R. Mulville and R. J. Vaisnov, J. Adhes. 7:215 (1975).
224. J. R. G. Evans and D. E. Packham, J. Adhes. 11:39 (1979).
225. J. R. G. Evans and D. E. Packham, J. Adhes. 11:177 (1979).
226. J. D. Venables, D. K. McNamara, J. Chen, T. S. Sun, and R. L. Hopping, Oxide morphologies on adhesively bonded aluminum structures, Appl. Surf. Sci. 3:88 (1979).
227. J. M. Chen, T. S. Sun, and J. D. Venables, Natl. SAMPE Sump. Exhib. 22:25 (1978).

228. H. W. White, L. M. Godwin, and R. J. Elliatroglu, J. Adhes. 13:177 (1981).
229. J. D. Eick, R. J. Good, A. W. Newman, and J. R. Fromer, J. Adhes. 3:23 (1971).
230. S. S. Voyutskii and Y. L. Margolina, Rubber Chem. Technol. 30:531 (1957).
231. S. S. Voyutskii and B. V. Shtarkh, Kolloidn, Zh. 16:3 (1954).
232. S. S. Voyutskii and S. Zamazil, Kolloidn. Zh. 16:3 (1954).
233. S. S. Voyutskii, V. L. Vakula, N. I. Smeloya, and I. A. Tutorskii, Vysokomolekul Soedin 2:1671 (1960).
234. S. S. Voyutskii and V. L. Vakula, J. Appl. Polym. Sci. 7:475 (1963).
235. S. S. Voyutskii, Autohesion and Adhesion of High Polymers, Wiley-Interscience, New York, p. 127 (1963).
236. S. S. Voyutskii, Vysokomol. Soedin, Ser. A. 10:974 (1968).
237. S. S. Voyutskii, Mekh. Polim. p. 728 (1966).
238. R. M. Vasenin, RAPRA Translations 1005,1006,1010,1075, R. J. Moseley (1960–1963).
239. J. D. Skewis, Rubber Chem. Technol. 39(2):217 (1966).
240. W. G. Forbes and L. A. McLeod, Trans. Inst. Rubber Ind. 34:154 (1958).
241. J. N. Anand and H. J. Karam, Interfacial bonding in autohesion: I—Contact theory, J. Adhes. 1:16 (1960).
242. J. N. Anand and R. Z. Balwinski, Interfacial contact and bonding in autohesion: II—Interfacial forces, J. Adhes. 1:24 (1969).
243. J. N. Anand, Interfacial contact and bonding in autohesion: III—Parallel plate attraction, J. Adhes. 1:31 (1969).
244. J. N. Anand, Interfacial contact and bonding in autohesion: IV—Experimental verification in adhesion, J. Adhes. 2:16 (1970).
245. S. S. Voyutskii, J. Adhes. 3:69 (1971).
246. N. H. Sung, Polymer Eng. Sci. 19:810 (1979).
247. R. P. Campion, J. Adhes. 7:1 (1975); Adhesion—1 (K. W. Allen, ed.). Applied Science Publishers, London, Chap. 5 (1977).
248. W. V. Titow, Adhesion—2 (K. W. Allen, ed.), Applied Science Publishers, London, Chap. 12 (1978).
249. J. R. Huntsberger, The mechanisms of adhesion, Treatise on Adhesion and Adhesives (R. L. Patrick, ed.), Marcel Dekker, New York, Vol. 1, Chap. 4 (1967).
250. F. London, The general theory of molecular forces, Trans. Faraday Soc. 33:8 (1937).
251. D. Langbein, Van der Waals attraction between macroscopic bodies, J. Adhes. 1:237 (1969).
252. J. E. Lennard-Jones, Cohesion, Proc. Phys. Soc. 43:461 (1931).
253. J. N. Anand, Interfacial contact bonding in autohesion: V—Bonding of "flat" surfaces, J. Adhes. 2:23 (1970).
254. P. J. Flory, Principles of Polymer Chemistry, Cornell Press, Ithaca, NY, Chap. 13 (1953).
255. L. J. Hughes and G. E. Britt, J. Appl. Polym. Sci. 5:337 (1961).
256. A. Dobry and F. Boyer-Kawenoki, J. Polym. Sci. 2:90 (1947).
257. M. D. Ellul and A. N. Gent, The role of molecular diffusion in the adhesion of elastomers, J. Polym. Sci. Polym. Phys. Ed. 22(11):1953 (Nov. 1984).
258. B. V. Derjaguin and N. A. Krotova, Doklady Akad. Nauk SSSR, 61:849 (1948).
259. B. V. Derjaguin and N. A. Krotova, Adgeziya Poliverov, Akad. Nauk SSSR, Moscow, p. 134 (1963).
260. B. V. Derjaguin, Research 8:70 (1955).
261. B. V. Derjaguin, Research 8:365 (1955).
262. B. V. Derjaguin, N. A. Krotova, V. V. Karassey, Y. M. Kirillova, and I. N. Aleinilova, Proc. 2nd Int. Cong. Surf. Activity, Butterworths, London, Vol. III, p. 417 (1957).
263. B. V. Derjaguin and S. S. Voyutskii, Kolloid Zh. 27:524 (1965).
264. B. V. Derjaguin and V. P. Smilga, Adhesion Fundamentals and Practice, McClaren, London, p. 152 (1969).
265. B. V. Derjaguin and V. P. Smilga, Proc. 3rd Int. Cong. Surf. Activity, Cologne, Germany, II, Sec. B: 349 (1960).
266. B. V. Derjaguin and V. P. Smilga, J. Appl. Phys. 38:4609 (1967).
267. B. V. Derjaguin, Y. P. Toporov, and I. N. Aleinikova, Powder Technol. 2:154 (1968/1969).
268. B. V. Derjaguin, V. N. Muller, and Y. P. Toporov, Kolloidn. Zhurnal. 37:455,1066 (1975).
269. B. V. Derjaguin, Y. P. Toporov, V. M. Muller, and I. N. Aleinilova, Kolloidn. Zhurnal. 39(1):16 (1977).
270. B. V. Derjaguin and Y. P. Toporov, Physicochemical Aspects of Polymer Surfaces (K. L. Mittal, ed.), Plenum Press, New York, p. 605 (1983).
271. J. R. Huntsberger, J. Polym. Sci. A1:2241 (1967).
272. S. M. Skinner, R. L. Savage, and J. E. Rutzler, Jr., J. Appl. Phys. 24:438 (1953).
273. S. M. Skinner, J. Appl. Phys. 26:498 (1955).

274. S. M. Skinner, J. Gaynor, and G. W. Sohl, WADC Tech. Report, p. 56 (1956).
275. W. A. Zisman, Ind. Eng. Chem. 55:18 (1963).
276. J. J. Bikerman, The Science of Adhesive Joints, Academic Press, New York (1961).
277. D. H. Kaelble, ACS Div. Paint, Plastics, and Printing Ink Chem., Paper No. 5, Cleveland, OH (April 1960).
278. H. P. Meissner and E. W. Merrill, ASTM Bull. 151:80 (1948).
279. A. D. McClaren and C. J. Silar, J. Polym. Sci. 4:63 (1949).
280. A. J. Tobolsky, Properties and Structures of Polymers, John Wiley & Sons, New York (1960).
281. F. Bueche, Physical Properties of Polymers, Wiley-Interscience, New York (1962).
282. H. Krupp, J. Adhes. 4:83 (1972).
283. H. Krupp, J. Adhes. 5:269 (1973).
284. H. Krupp, Aspects of Adhesion—8 (K. W. Allen, ed.), Transcripta Books, London, p. 187 (1975).
285. J. Comyn, Kinetics and mechanisms of environmental attack, Durability of Structural Adhesives (A. J. Kinloch, ed.), Applied Science Publishers, London, Chap. 3, p. 120 (1983).
286. H. B. G. Casimer, Konik Ned. Akad. Wetenschap. Proc. Ser. B51:795 (1948).
287. C. Kemball, Adhesion and Adhesives (Clark, Savage, and Rutzler, eds.), John Wiley & Sons, New York, p. 70 (1954).
288. R. Simha, M. L. Frisch, and F. R. Eirich, J. Phys. Chem. 57:584 (1953).
289. E. M. Lifshitz, Zh. Eksperim i. Teor. Fiz. 21:94 (1954).
290. C. Kemball, Adhesion (D. D. Eley, ed.), Oxford University Press, London, p. 19 (1961).
291. P. H. W. DeBye, Adhesion and Cohesion (P. Weiss, ed.), Elsevier, Amsterdam, pp. 10,16 (1962).
292. A. J. Staverman, Adhesion and Adhesives (R. Houwink and G. Salmon, eds.), Elsevier, Amsterdam, Vol. 1, p. 9 (1965).
293. W. C. Wake, RIC Lecture Series No. 4, p. 1 (1966).
294. W. C. Wake, Adhesion and Formulation of Adhesives, 2nd ed., Applied Science Publishers, London, p. 9 (1982).
295. L. H. Sharpe and H. Schonhorn, Surface energetics, adhesion and adhesive joints, Advances in Chemistry Series No. 43, American Chemical Society, p. 189 (1964).
296. L. A. Girifalco and R. J. Good, J. Phys. Chem. 61:904 (1957).
297. S. Gusman, Offic. Dig. Fed. Soc. Paint Technol. 34:884 (1962).
298. J. R. Huntsberger, Advan. Chem. Ser. No. 43, Am. Chem. Soc. 43:187 (1964).
299. F. M. Fowkes, Ind. Eng. Chem. 56(12):40 (1964).
300. J. L. Gardon, Encyclopedia of Polymer Sciences & Technology, Interscience, New York, Vol. 3 (1965).
301. J. L. Gardon, Treatise on Adhesion and Adhesives (R. L. Patrick, ed.), Marcel Dekker, New York, Vol. 1, Chap. 8, p. 269 (1967).
302. J. F. Paddy and N. D. Uffindell, J. Phys. Chem. 72:1402 (1968).
303. R. J. Good, Ind. Eng. Chem. 62:54 (1970).
304. J. R. Huntsberger, Interfacial energies, contact angles, and adhesion, Treatise on Adhesion and Adhesives (R. L. Patrick, ed.), Marcel Dekker, New York, Vol. 5, Chap. 1, p. 1 (1981).
305. B. W. Cherry and M. I. Hakeen, Adhesion—10 (K. R. Allen, ed.), Elsevier, Applied Science Publishers, London, Chap. 4, p. 42 (1986).
306. N. A. DeBruyne, Nature 180:262 (1957).
307. L. E. Raraty and D. Tabor, Proc. Roy. Soc. A245:184 (1958).
308. M. J. Barbaris, Matire 215:383 (1967).
309. M. L. Mittal, Polymer Science and Technology, Plenum Press, New York, Vol. 9A, p. 129 (1975).
310. M. Levine, G. Jikka, and P. Weiss, Polym. Lett. 2:915 (1964).
311. E. Pleuddemann, J. Paint Technol. 40:1 (1968).
312. Y. Iyengar and D. E. Erickson, J. Appl. Polym. Sci. 11:2311 (1967).
313. G. A. Dykerhoff and P. J. Sell, Angew. Macromol. Chem. 21:169 (1972).
314. R. J. Good, Treatise on Adhesion and Adhesives (R. L. Patrick, ed.), Marcel Dekker, New York, Vol. 1, Chap. 2, p. 9 (1967).
315. E. H. Andrews and A. J. Kinloch, Proc. Royal Soc. A332:401 (1973).
316. D. H. Owens and R. C. Wendt, J. Appl. Polym. Sci. 13:1741 (1969).
317. H. W. Fox and W. A. Zisman, J. Colloid Sci. 5:514 (1950).
318. H. W. Fox and W. A. Zisman, J. Colloid Sci. 7:428 (1952).
319. W. A. Zisman, Advan. Chem. Ser. No. 43, Am. Chem. Soc., p. 1 (1964).
320. J. L. Gardon, Variables and interpretation of some destructive cohesion and adhesion tests, Treatise on Adhesion and Adhesives (R. L. Patrick, ed.), Marcel Dekker, New York, Vol. 1, Chap. 8, p. 314 (1967).
321. J. A. Hildebrand and R. L. Scott, The Solubility of Nonelectrolytes, Reinhold, New York, pp. 323, 402, 430 (1950).

322. J. L. Gardon, Critical review of concepts common to cohesive energy density, surface tension, tensile strength, heat of mixing, interfacial tension, and butt joint strength, J. Colloid and Interf. Sci. 59(3):582 (May 1977).
323. J. L. Gardon, J. Phys. Chem. 67:1935 (1963).
324. J. L. Gardon, J. Paint Technol. 38:43 (1966).
325. J. L. Gardon, Encyclopedia of Polymer Science and Technology (N. M. Bikeles et al., eds.), Interscience, New York, pp. 853–862 (1966).
326. J. L. Gardon, Treatise on Adhesion and Adhesives (R. L. Patrick, ed.), Marcel Dekker, New York, Vol. 1, Chap. 8, pp. 269–324 (1967).
327. R. J. Good, Surface free energy of solids and liquids: Thermodynamics, molecular forces and structure, J. Colloids and Interf. Sci. 59:398 (1977).
328. D. H. Kaelble and K. C. Vy, J. Adhes. 2:50 (1970).
329. J. H. Brooks, J. Mingins, and N. F. Owens, J. Colloid Interf. Sci. 61:215 (1977).
330. Z. Haq and J. Mingins, The contact angle of water on polymer films, Polymer Communications 25(9): 269 (Sept. 1984).
331. A. A. Vavkushevskii, V. V. Arslanov, and V. A. Ogarev, Spreading of polymer drops on smooth solid surfaces, Colloid J. USSR 46(6):940 (Nov.–Dec. 1984).
332. Y. Uyama, H. Inoue, K. Ito, A. Kishida, and Y. Ikada, Comparison of different methods for contact angle measurement, J. Colloid Interf. Sci. 141(1):275 (1991).
333. G. M. Whitesides, H. A. Biebuyck, J. P. Folkers, and K. L. Prime, Acid-base interactions in wetting, J. Adhes. Sci. Technol. 5(1):57 (1991).
334. W. Gutowski, Physico-chemical criteria for maximum adhesion. Part I: theoretical concepts and experimental evidence, J. Adhes. 19(1):29 (1985).
335. J. Gutowski, Physico-chemical criteria for maximum adhesion, Part II: A new comprehensive thermodynamic analysis, J. Adhes. 19(1):51 (1985).
336. J. Gutowski, A thermodynamic model of the adhesive bond, Proc. 10th Annual Meeting Adhesion Society (L. H. Sharpe, ed.), Gordon and Breach, New York and London, p. 295 (1988).
337. C. M. Hansen, Ind. Eng. Chem. Prod. Res. Dev. 8:2 (1969).
338. K. W. Harrison, Adhesion—3 (K. W. Allen, ed.), Applied Science Publishers, London, Chap. 9, p. 143 (1979).
339. A. M. Barton, J. Adhes. 14:33 (1982).
340. M. L. Williams, J. Appl. Polym. Sci. 13:29 (1969).
341. E. H. Andrews and A. Stevenson, Journal Material Science 13:1680 (1978).
342. J. D. Burton, W. B. Jones, and M. L. Williams, Trans. Soc. Rheol. 15:39 (1971).
343. G. P. Anderson, S. J. Bennett, and K. L. DeVries, Analysis and Testing of Adhesive Bonds, Academic Press, New York, p. 58 (1977).
344. F. M. Fowkes, Contact angle, wettability, and adhesion, Advances in Chemistry Series 43, American Chemical Society, Washington, D.C., p. 99 (1964).
345. F. M. Fowkes, J. Adhes. 4:155 (1972).
346. F. M. Fowkes, G. S. Ronay, and M. J. Schick, J. Phys. Chem. 63:1684 (1959).
347. F. M. Fowkes, J. Phys. Chem. 64:726 (1960).
348. F. M. Fowkes, Acid-base interactions in polymer adhesion, Physicochemical Aspects of Polymer Adhesion (K. L. Mittal, ed.), Plenum Press, New York, Vol. 2, p. 583 (1983).
349. Yu. S. Lipatov and L. M. Sergeeva, Adsorption of Polymers, Halsted, New York (1974).
350. P. Sorendon, J. Paint Technol. 47:31 (1975).
351. F. M. Fowkes, J. Phys. Chem. 66:382 (1962).
352. C. M. Hansen, J. Paint Technol. 39:104,505 (1967).
353. R. S. Drago, G. C. Vogel, and T. E. Needham, J. Amer. Chem. Soc. 93:6010 (1971).
354. R. S. Drago, L. B. Parr, and C. S. Chamberlain, J. Amer. Chem. Soc. 99:3203 (1977).
355. F. M. Fowkes and S. Maruchi, Organic Coatings and Plastics Chem. Prep., Amer. Chem. Soc. 37:605 (1977).
356. K. W. Allen, L. Greenwood, and W. C. Wake, J. Adhes. 16:127 (1983).
357. S. Wu, Polymer Interface and Adhesion, Marcel Dekker, New York, p. 420 (1982).
358. A. A. Berlin and V. E. Basin, Osnovy Adgezii Polimerov, Khimiya, Moscow (1969).
359. A. A. Berlin and V. E. Basin, Zh. Vses. Khim. Obschest. 14(1):9 (1969); Mekh. Polum. P. 202 (1970).
360. V. E. Basin and A. A. Berlin, Plast. Massy 2:7 (1970).
361. J. J. Bikerman, The Science of Adhesive Joints, 2nd ed., Academic Press, New York (1968).
362. J. J. Bikerman, Uspekhi Khimii 41:1431 (1972).
363. J. J. Bikerman, Vysokomol. Soedin. Ser. A. 10:978 (1968).
364. S. S. Voyutskii, Autogeziya i Adgeziya Vysokopolimerov, Khimiya, Moscow (1960).

365. V. E. Gul and L. L. Kudryasherva, Adgeziya Polimerov, Akad. Nauk SSSR, Moscow, p. 134 (1963).
366. N. I. Moskvitin, Skleivanie Polimerov, Lesnaya Prom., Moscow (1968).
367. W. Gurney, Trans I.R.I. 27:175 (1943).
368. J. W. McBain and W. B. Lee, J. Phys. Chem. 31:1676 (1927).
369. L. Pauling, The Nature of the Chemical Bond, Cornell Press, New York (1960).
370. R. J. Good, Treatise on Adhesion and Adhesives (R. L. Patrick, ed.), Marcel Dekker, New York, Vol. 1, p. 15 (1967).
371. J. D. Minford, Durability of adhesive bonded aluminum joints, Treatise on Adhesion and Adhesives (R. L. Patrick, ed.), Marcel Dekker, New York, Vol. 3, Chap. 2, pp. 101–108 (1973).
372. J. D. Minford, Adhesives, Durability of Structural Adhesives (A. J. Kinloch, ed.), Applied Science Publishers, London, pp. 176–181 (1983).
373. W. Brockmann, Adhes. Age 18(6):30 (1977).
374. R. L. M. Allen, Colour Chemistry, Nelson, London (1971).
375. E. G. Kiel and P. M. J. Heertjes, Soc. Dyers Colorists. 79:21,61 (1963).
376. C. H. Giles, H. V. Mehta, C. E. Stewart, and R. V. R. Subramanian, J. Chem. Soc., p. 4360 (1954).
377. C. H. Giles, Rev. Prog. Coloration 5:49 (1974).
378. T. Cummins, H. C. Craven, C. H. Giles, S. M. K. Rahman, J. G. Sneddon, and C. E. Stewart, J. Chem. Soc. p. 535 (1959).
379. J. P. Jeandrau, Investigation into the effect of surface treatment on the wettability and the bondability of low surface energy materials, Adhesive Joints: Formation, Characteristics, and Testing (K. L. Mittal, ed.), Plenum Press, New York, p. 121 (1984).
380. J. S. Tira, Gas plasma treatment to improve the bondability for an RTV silicone to foamed polypropylene, Dept. Energy, Contract DE-AC04-76-DP00613, BDPX-613-2224 (Rev.) (July 1979).
381. C. Mylonas, Exptl. Stress Anal. 12:129 (1955).
382. E. A. Bourcher, Nature 215:1054 (1967).
383. B. W. Cherry, Wetting kinetics and the strength of adhesive joints, J. Adhes. 2:42 (1970).
384. L. H. Lee, Relationships between surface wettability and amd glass temperatures of polymers, J. Appl. Polym. Sci. 12:719 (1968).
385. A. Mamur, Adv. Colloid and Interf. Sci. 19:75 (1983).
386. W. A. Zisman, Adhesion and Cohesion (P. Weiss, ed.), Elsevier, Amsterdam (1962).
387. W. A. Zisman, Contact Angle, Wettability, and Adhesion (R. F. Gould, ed.), American Chemical Society, p. 44 (1964).
388. W. A. Zisman, Ind. Eng. Chem. 55:18 (1963).
389. F. M. Fowkes, J. Phys. Chem. 67:2538 (1963).
390. R. J. Good and L. A. Girifalco, A theory for the estimation of surface and interfacial energies, I—Derivation and application to interfacial tension, J. Phys. Chem. 61:904 (1957).
391. R. J. Good, Contact angle wettability and adhesion, Advances in Chemistry Series No. 43, American Chemical Society, Washington, D.C., p. 74 (1964).
392. R. J. Good, A theory for the estimation of surface energies and interfacial energies, III—Estimation of surface energies from contact angle data, J. Phys. Chem. 64:561 (1960).
393. R. J. Good, Intermolecular and interatomic forces, Treatise on Adhesion and Adhesives (R. L. Patrick, ed.), Marcel Dekker, New York, Vol. 1, Chap. 1, pp. 9–68 (1967).
394. R. J. Good, The role of wetting and spreading in adhesion, Aspects of Adhesion (D. J. Alner and K. W. Allen, eds.), Transcripta Books, London, Vol. 7, p. 182 (1973).
395. R. J. Good, Contact angles and surface free energy of solids, Colloid and Surface Science (R. J. Good and R. R. Stromberg, eds.), Plenum Press, New York, Vol. 11 (1979).
396. R. J. Good and E. Eibling, Generalization of the theory for the estimation of interfacial energies: Theory of systems with high mutual solubilities and with various degrees of polarity, Ind. Eng. Chem. 62:54 (March 1970).
397. A. W. Neumann and R. J. Good, Methods of measuring contact angles, Colloid and Surface Science (R. J. Good and R. R. Stromberg, eds.), Plenum Press, New York, Vol. 11 (1979).
398. R. J. Good and F. Shoraka, The interaction of interfacial forces and viscoelastic effects in polymer adhesion and wear, Polymer Wear and Its Control (L. H. Lee, ed.), ACS Symp. Series No. 287, p. 39 (1985).
399. R. W. Johnson, Jr., and R. H. Dettre, Advan. Chem. Series No. 43, Am. Chem. Soc., p. 112 (1964).
400. R. H. Dettre and R. E. Johnson, Advan. Chem. Series No. 43, Am. Chem. Soc., p. 136 (1964).
401. M. Sheriff, J. Adhes. 7:257 (1976).
402. J. Schultz, K. Tsutsumi, and J. B. Donnet, J. Colloid Interf. Sci. 59:277 (1977).
403. R. Janardhan, P. H. Gedam, and P. S. Sampathkumaran, The effect of polymer molecular weight in the adsorption process, J. Colloid and Interf. Sci. 140(2):391 (1991).

404. M. Kawaguchi and T. Arai, Effect of surface geometry on absorption and desorption of polymer chains, Macromolecules 24(4):889 (1991).
405. N. A. DeBruyne, Aircraft Eng. 18(12):53 (1939).
406. N. A. DeBruyne, 19th Int. Conf. Pure and Appl. Chem., Symp. on Adhesion, London (July 1963).
407. J. R. Huntsberger, J. Polym. Sci. 43:58 (1960).
408. H. Schonhorn, H. L. Frisch, and T. K. Kwei, J. Appl. Phys. 37:4967 (1966).
409. T. K. Kwei, H. Schonhorn, and H. L. Frisch, J. Colloid Interf. Sci. 28:543 (1968).
410. B. W. Cherry and C. M. Holmes, Jr., J. Colloid Interf. Sci. 29:174 (1969).
411. R. H. Dettre and R. W. Johnson, Jr., The spreading of molten polymers, J. Adhes. 2:61 (1970).
412. H. Van Oene, Y. F. Chang, and S. Newman, The rheology of wetting by polymer melts, J. Adhes. 1:54 (1969).
413. C. A. Dahlquist, Adhesion, ASTM Spec. Tech. Publ. No. 360, p. 46 (1964).
414. C. A. Dahlquist, Adhesion: Fundamentals and Practice, McLaren & Son, London (1966).
415. T. Alfrey, Jr., Mechanical Behavior of Polymers, Wiley-Interscience, New York (1948).
416. J. D. Ferry, Viscoelastic Properties of Polymers, John Wiley & Sons, New York (1961).
417. J. P. Berry and A. M. Bueche, Adhesion and Cohesion (P. Weiss, ed.), Elsevier, Amsterdam (1962).
418. L. E. Nielsen, The Mechanical Properties of Polymer Solids, Reinhold, New York (1962).
419. R. F. Landel and R. F. Fedors, Fracture Processes in Polymer Solids (B. Rosen, ed.), Interscience, New York, p. 361 (1964).
420. F. Bueche and J. C. Halpin, J. Appl. Phys. 35:36 (1964).
421. M. R. Hatfield and G. B. Rathmann, J. Phys. Chem. 60:957 (1956).
422. D. H. Kaelble, Adhes. Age 3(5):37 (1960).
423. D. H. Kaelble, Adhesion and Cohesion (P. Weiss, ed.). Elsevier, Amsterdam, p. 101 (1962).
424. D. H. Kaelble, J. Colloid and Interf. Sci. 19:413 (1964).
425. D. H. Kaelble, Rheology of polymers used as adhesives, Treatise on Adhesion and Adhesives (R. L. Patrick, ed.), Marcel Dekker, New York, Vol. 1, Chap. 6, p. 195 (1967).
426. D. H. Kaelble, Peel adhesion: Influence of surface energies and adhesive rheology, J. Adhes. 1:102 (1969).
427. W. M. Bright, Adhesion and Adhesives (J. Clark, J. E. Rutzler, and R. C. Savage, eds.), John Wiley & Sons, New York, p. 130 (1954).
428. C. L. Weidner and G. J. Crocker, Rubber Chem. Technol. 33:1323 (1960).
429. H. Oberst, Acoustica 2:AB151 (1952); 4:433 (1954); Ber. Ver. Deuts. Ing. 8:100 (1956).
430. E. M. Kerwin, Jr., J. Acoust. Soc. Am. 31:952 (1959).
431. J. C. Snowden, Brit. J. Appl. Phys. 9:461 (1968).
432. C. A. Dahlquist, J. O. Hendricks, and N. W. Taylor, Ind. Eng. Chem. 43:1404 (1951).
433. C. A. Dahlquist, Adhes. Age 2(10):25 (1959).
434. H. Leaderman, Trans. Soc. Rheol. 6:361 (1962).
435. J. J. Bikerman, J. Appl. Phys. 28:1484 (1957).
436. J. J. Bikerman, Ind. Eng. Chem. 59:40 (1967).
437. J. J. Bikerman, Adhesion and Cohesion (P. Weiss, ed.), Elsevier, Amsterdam, pp. 36–45 (1962).
438. R. L. Patrick, C. M. Doede, and W. A. Vaughan, J. Phys. Chem. 61:1036 (1957).
439. D. Briggs, D. M. Brewis, and M. B. Konieczo, J. Mater. Sci. 11:1270 (1976).
440. D. T. Clark, W. J. Feast, W. K. R. Musgrave, and I. Ritchie, J. Polym. Sci. Polym. Chem. 13:857 (1975).
441. D. T. Clark, A. Dilks, and D. Shuttleworth, J. Mater. Sci. 12:2547 (1977).
442. D. Briggs, D. M. Brewis, and M. B. Konieczko, J. Mater. Sci. 12:429 (1977).
443. D. Briggs and D. M. Brewis, J. Mater. Sci. 12:2549 (1977).
444. A. Crocombe and R. Adams, Adhesion and Adhesives: Science Technology and Applications, Durham, England (1980).
445. J. J. Bikerman, J. Appl. Chem. 11:81 (1961).
446. J. J. Bikerman and D. W. Marshall, J. Appl. Polym. Sci. 7:1031 (1963).
447. D. A. Kardashov and V. L. Vakula, Zh. Vses. Khim. Otschchest 14(1):4 (1969).
448. F. H. Chung, Unified theory and guidelines on adhesion, J. Appl. Polym. Sci. 42(5):1319 (1991).
449. K. L. Mittal, Adhesion measurement: Recent progress, unsolved problems and prospects, ASTM Spec. Tech. Publ. No. 640, p. 1 (1978).
450. K. L. Mittal, J. Adhes. 7:377 (1974).
451. L. E. Muir, ASTM Spec. Tech. Publ. No. 640, p. 82 (1978).
452. K. L. Mittal, Electrocomponent Sci. Technol. 3:21 (1976).
453. K. L. Mittal, Properties of Electrodeposits: Their Measurement and Significance (R. Said, H. Leidheiser, Jr., and F. Ogburn, eds.), Electrochemical Society, Princeton, NJ, 17:272 (1975).
454. R. Jacobsson, Thin Solid Films 34:191 (1976).

455. T. T. Hitch and K. R. Bube, Basic adhesion mechanisms in thick and thin films, RCE Labs. Report., Princeton, NJ (Jan. 31, 1975).
456. L. Jacobson, Proc. IEER and EIA Electronic Components Conf., p. 474 (1971).
457. R. P. Anjard, Microelectronics and Reliability 10(4):2699 (1971).
458. T. T. Hitch, ASTM Spec. Tech. Publ. No. 640, p. 211 (1978).
459. R. W. Benoliel, Aluminum surface preparation for bonding, Document MDR 6-52-024, Boeing Company (May 1969).
460. A. W. Smith, Some basic principles of surface preparation of aluminum for adhesive bonding, Document D1-82-1003. Boeing Scientific Res. Labs, Boeing Company (Sept. 1970).
461. A. W. Smith, Surface oxide formed on aluminum etched in sulfuric acid–sodium dichromate solutions, Document D1-82-1068, Boeing Res. Labs., Boeing Company (April 1971).
462. A. W. Smith, Surface oxide on etched aluminum, J. Electrochem. Soc. Solid-State Sci. Technol. (Nov. 1973).
463. H. M. Clearfield, Surface preparation of metals, Adhesives and Sealants, Engineered Materials Handbook, ASM, Materials Park, OH, Vol. 3, pp. 258–275 (1990).
464. P. F. Kane and G. B. Larabee (eds.), Characterization of Solid Surfaces, Plenum Press, New York (1974).
465. D. H. Buckley, Surfaces, Adhesives in Manufacturing (G. L. Schneberger, ed.), Marcel Dekker, New York, Chap. 2, p. 13 (1983).
466. G. D. Davis and J. D. Venables, Surface and interfacial analysis, Durability of Structural Adhesives (A. J. Kinloch, ed.), Applied Science Publishers, London, Chap. 2, p. 43 (1983).
467. M. A. Danforth, Surface analysis, Adhesive Bonding of Aluminum Alloys (E. W. Thrall and R. W. Shannon, eds.), Marcel Dekker, New York, Chap. 5, p. 75 (1985).
468. G. D. Davis, Surface analysis techniques and applications, Adhesives and Sealants, Engineered Materials Handbook, ASM, Material Park, OH, Vol. 3, pp. 236–253 (1990).
469. J. D. Edwards, F. C. Frary, and Z. Jeffries, The Aluminum Industry: Aluminum and Its Production, McGraw-Hill, New York, p. 164 (1930).
470. K. Wefers and G. Bell, Oxides and Hydroxides of Aluminum, Alcoa Technical Paper No. 10, Alcoa Res. Labs, Alcoa Center, PA; K. Wefers and C. Misra, Oxides and Hydroxides of Aluminum, Alcoa Technical Paper No. 19, Revised, Alcoa Labs, Alcoa Center, PA (1987).
471. H. C. Stumpf, A. S. Russell, J. W. Newsome, and C. M. Tucker, Ind. Eng. Chem. 42:1398 (1950).
472. H. Ginsberg, W. Huttig, and G. Strunk-Lichtenberg, Z. Anorg. Allg. Chem. 293:33,204 (1957).
473. P. S. Moiseev, Kolloid-Zh. 9:53 (1947).
474. J. W. Teter, J. L. Ging, and C. D. Keith, U.S. Patent 2,838,375 (June 10, 1958).
475. J. Bohn, Z. Anorg. Allg. Chem. 132:1 (1924).
476. D. Papee, R. Tertian, and R. Biasis, Bull. Soc. Chim. Fr., p. 1301 (1958).
477. R. Tertian and D. Papee, J. Chim. Phys. 55:341 (1958).
478. L. D. Frederickson, Anal. Chem. 26:1883 (1954).
479. R. A. Harrington and H. R. Nelson, Am. Inst. Min. Metal Engrs., Tech. Publ. No. 1158 (1940).
480. G. Tolley, Metal Ind. 77:225 (1950).
481. J. M. Bryan, J. Soc. Chem. Ind. 69:169 (1950).
482. J. M. Bryan, Chem. Ind. 34:820 (1952).
483. D. G. Altenpohl, Use of boehmite films for corrosion protection of aluminum, Corrosion 18:143t (April 1962).
484. D. G. Altenpohl, Aluminium 29:361 (1953).
485. D. G. Altenpohl, Aluminium 31:10 (1955).
486. D. G. Altenpohl, Aluminium Metall. 9:164 (1955).
487. D. G. Altenpohl, Aluminium 38:58 (1957).
488. R. K. Hart, Trans. Faraday Soc. 50:268 (1954).
489. J. F. Murphy and H. A. Page, Preprints Div. Paint, Plastics & Printing Ink Chem., Am. Chem. Soc. 15(1):27 (1955).
490. E. Thelen, Preprint WADC TR-57-513, Wright-Patterson AFB, OH.
491. W. C. Hamilton and G. A. Lyerly, Evaluation of the adhesive bonding processes used in helicopter manufacture, Part 2—The characterization of adherend surfaces, Tech. Report No. 3185, Picatinny Arsenal, Dover, NJ (March 1971).
492. D. D. MacDonald and P. Butler, The thermodynamics of the aluminum-water system at elevated temperature, Corrision Sci. 13:259 (1973).
493. K. E. Weber and G. R. Johnston, SAMPE Quart. 6(1):16 (1974).
494. W. T. McCarvill and J. P. Bell, J. Appl. Polym. Sci. 18:335, 343, 2243 (1974).
495. W. T. McCarvill and J. P. Bell, J. Adhes. 6:185 (1974).

496. J. C. Bolger, The chemical composition of metal and oxide surfaces and how these interact with polymeric materials, 30th Ann. Tech. Conf., SPE. Chicago (1972).
497. F. Keller and J. D. Edwards, Composition and properties of the natural oxide film on aluminum, Metal Prog. 54(1):35 (July 1948).
498. F. Keller and J. D. Edwards, Composition and properties of the natural oxide film on aluminum, Metal Prog. 54(2):195 (Aug. 1948).
499. W. H. J. Vernon, Report to the atmospheric corrosion research committee, Trans. Faraday Soc. 23:150 (1927).
500. E. A. Gulbransen, A vacuum microbalance for study of chemical reactions on metals, Rev. Sci. Instr. 15:210 (1944).
501. E. A. Gulbransen and W. S. Wysong, Thin oxide films on aluminum, J. Phys. Colloid Chem. 51:1087 (1947).
502. E. A. Gulbransen, New developments in the study of surface chemistry, Metal Prog. 49:553 (1946).
503. G. Haas, Growth and structure of thin oxides, Optik 1:134 (1946).
504. W. Brockmann, Durability of adhesion between metals and polymers, J. Adhes. 29:53 (1989).
505. W. A. Zisman, Adhesion and Cohesion (P. Weiss, ed.). Elsevier, Amsterdam (1962).
506. R. L. Patrick (ed.), Treatise on Adhesion and Adhesives, Marcel Dekker, New York, Vol. 2 (1969).
507. C. A. Harper (ed.), Handbook of Plastics and Elastomers, McGraw-Hill, New York (1957).
508. J. Frados (ed.), Plastics Engineering Handbook, 4th ed., Van Nostrand-Reinhold, New York (1976).
509. I. Skeist (ed.), Handbook of Adhesives, 2nd ed., Van Nostrand Reinhold, New York (1977).
510. H. F. Mark, Future improvements in the cohesive and adhesive strength of polymers, Part I, Adhes. Age 22(7):35 (1979).
511. H. F. Mark, Future improvement in the cohesive and adhesive strength of polymers, Part II, Adhes. Age 22(9):45 (1979).
512. L. R. Whittington, Whittington's Dictionary of Plastics, 2nd ed., Technomic, Lancaster, PA (1978).
513. High Performance Adhesive Bonding, Soc. Mfg. Eng., Dearborn, MI (1983).
514. T. A. Richardson, Industrial Plastics: Theory and Application, South-Western (1983).
515. A. H. Landrock (ed.), Adhesive Technology Handbook, Noyes Publ. (1985).
516. S. R. Hartshorn (ed.), Structural Adhesives Chemistry and Technology, Plenum Press, New York (1986).
517. A. J. Kinloch, Adhesion and Adhesives Science and Technology, Chapman and Hall (1987).
518. Glossary of Terms, Engineering Plastics, Engineering Handbook, ASM, Materials Park, OH, Vol. 2 (1988).
519. R. W. Hemingway, A. H. Connor, and S. J. Branham, Adhesives from renewable sources, ACS Symposium Series No. 385, American Chemical Society, Washington, D.C. (1989).
520. L. H. Sharpe, Fundamentals of Adhesives and Sealants Technology, Glossary of Terms, Adhesives and Sealants, Engineered Materials Handbook, ASM, Materials Park, OH, Vol. 3, pp. 3–32 (1990).
521. L. H. Sharpe, Overview: Adhesive technology, Adhesives and Sealants, Engineered Materials Handbook, ASM, Materials Park, OH, Vol. 3, pp. 33–43 (1990).
522. D. J. Zalucha, Adhesives and adhesion, Adhesives and Sealants, Engineered Materials Handbook, ASM, Materials Park, OH, Vol. 3, pp. 39–43 (1990).
523. W. Brockmann, O. D. Hennemann, and H. Kollek, Adhesive Joints: Formation, Characteristics and Testing (K. L. Mittal, ed.), Plenum Press, New York, p. 469 (1984).
524. W. Brockmann, O. D. Hennemann, H. Kollek, and C. Matz, Int. J. Adhes. Adhes. 6:115 (1986).
525. G. D. Davis, T. S. Sun, J. S. Ahearn, and J. D. Venables, J. Mater. Sci. 17:1807 (1982).
526. J. S. Ahearn, G. D. Davis, T. S. Sun, and J. D. Venables, J. Mater. Sci. 18:288 (1983).
527. D. A. Hardwick, J. S. Ahearn, and J. D. Venables, J. Mater. Sci. 19:233 (1984).
528. G. D. Davis, J. S. Ahearn, and J. D. Venables, J. Vac. Sci. Technol. A2:763 (1984).
529. D. A. Harwich, J. S. Ahearn, A. Desai, and J. D. Venables, J. Mater. Sci. 21:179 (1986).
530. R. J. Davies and M. D. Ritchie, Future design concepts for the development of new pretreatments of aluminum alloy/metal matrix composites for adhesive bonding, 13th Ann. Meeting Adhes. Soc., Savannah, Georgia (Feb. 18–21, 1990).
531. J. D. Minford, Durability of adhesive bonded aluminum joints, Treatise on Adhesion and Adhesives (R. L. Patrick, ed.), Marcel Dekker, New York, Vol. 3, Chap. 2, p. 79 (1973).
532. J. D. Minford, Effect of surface preparations on adhesive bonding of aluminum, Adhes. Age 17(7):24 (1974).
533. J. D. Minford, Comparison of aluminum adhesive joint durabilities as influenced by etching and anodizing pretreatments of bonded surfaces, J. Appl. Polym. Sci., Appl. Polym. Symp. 32:91 (1977).
534. J. D. Minford, Etching and anodizing pretreatments and aluminum joint durability, SAMPE Quart. p. 18 (July 1978).

535. J. D. Minford, Aluminum adhesive bond permanence, Treatise on Adhesion and Adhesives (R. L. Patrick, ed.), Marcel Dekker, New York, Vol. 5, Chap. 3, p. 68 (1981).
536. J. D. Minford, Adhesives, Durability of Structural Adhesives (A. J. Kinloch, ed.), Applied Science Publishers, London, Chap. 4, p. 168 (1983).
537. D. M. Brewis, Aluminium adherends, Durability of Structural Adhesives (A. J. Kinloch, ed.), applied Science Publishers, London, Chap. 5 (1983).
538. A. J. Kinloch (ed.), Durability of Structural Adhesives, Applied Science Publishers, London (1983).
539. J. S. Ahearn and G. D. Davis, Proceedings of Adhesion '87, p. 292 (1987).
540. J. S. Ahearn, G. D. Davis, T. S. Sun, and J. D. Venables, Adhesion Aspects of Polymeric Coatings (K. L. Mittal, ed.), Plenum Press, New York, p. 281 (1983).
541. R. J. Davis and A. J. Kinloch, The surface characterization and adhesive bonding of aluminum, Adhesion—3 (K. W. Allen, ed.), Elsevier, Amsterdam (1989).
542. R. J. Davies and D. A. Moth, Novel techniques for oxide characterization within adhesively-bonded aluminum alloys, Proc. SAE II, University of Bristol, England (Sept. 1989).
543. A. Carre and J. Schultz, Polymer-aluminum adhesion II. Role of the adhesive and cohesive properties of the polymer, J. Adhes. 17(1):135 (1984).
544. S. R. Kurtz and R. A. Anderson, Properties of the metal-polymer inferface observed with space-charge mapping techniques, J. Appl. Phys. 60(2):681 (July 1986).
545. V. Vangani, R. Joseph, S. Devi, and A. K. Rakshit, Thermodynamics of adsorption of polybutadience on alumina and silica gel: Effect of temperature and solvent, Colloid and Polym. Sci. 269(2):242 (1991).
546. J. A. Cape and C. L. Kirby, On the meaning of surface area measurements for microporous materials, J. Colloid Interf. Sci. 138(2):515 (1991).
547. E. Papirer, J. M. Perrin, B. Siefert, and G. Philipponneau, Surface characteristics of aluminas in relation with polymer absorption, J. Colloid and Interf. Sci. 144(1):263 (1991).
548. M. A. Blesa, A. J. G. Maroto, A. E. Regazoni, Surface acidity of metal oxides immersed in water: a critical analysis of thermodynamic data, J. Colloid Interf. Sci. 140(10):287 (1990).
549. P. Berteau, M. A. Kellens, and B. Delmon, Acid-base properties of modified aluminas, J. Chem. Soc. Faraday Trans. 87(9):1425 (1991).
550. M. B. Fleisher, L. O. Golender, and M. V. Shimanskaya, Electronic charge distribution and Lewis acidity of surface aluminum atoms in gamma-alumina: A quantum chemical model, J. Chem. Soc: Faraday Trans. 87(5):745 (1991).
551. H. Kanai, G. J. Demott, and D. L. Kohlstedt, Adhesion of chromium metallization on alumina surfaces prepared by sol-gel techniques, J. Mater. Sci. 26(7):1815 (1991).
552. J. W. Richards, Aluminum, 3rd ed. Henry Carey Baird & Co., Philadelphia (1890).
553. L. Bouckere, J. Inst. Metals 71:131 (1945).
554. F. Keller and J. D. Edwards, Behavior of oxide film on aluminum, Proc. Int. Conf. Surf. Reactions, Pittsburgh (1948).
555. R. Bainbridge, P. Lewis, and J. M. Sykes, Effect of substrate preparation on the adhesion of polyethylene to aluminum, Int. J. Adhes. Adhes., p. 175 (July 1982).
556. H. A. Arbit, E. E. Griesser, and W. A. Haine, Technical Association of the Pulp and Paper Industry 40: 161 (1957).
557. P. J. Bockhoff, E. E. McConee, and J. E. Rutzler, Ind. Eng. Chem. 50:904 (1958).
558. K. Bright and B. W. Malpass, Eur. Polym. J. 4:431 (1968).
559. J. M. Sykes and T. P. Hoar, J. Polym. Sci. A-1, 7:1385 (1969).
560. D. E. Packhan, J. Bright, and B. W. Malpass, J. Appl. Polym. Sci. 18:3237 (1974).
561. R. F. Wegman, How oxides of aluminum affect bond strength, Adhes. Age 10(1):20 (1967).
562. R. F. Wegman, Effects of an inert atmosphere on the bonding of metals, Tech. Report No. TR-3206, Picatinny Arsenal, Dover, NJ (April 1966).
563. W. L. Baun, N. T. McDevitt, and J. S. Solomon, Chemistry of metal and alloy adherends by secondary ion mass spectroscopy, ion scattering spectroscopy, and Auger electron spectroscopy, Techniques for Metallurgical Applications, ASTM Spec. Publ. No. STP 596 (March 1975).
564. C. J. Powell, Surf. Anal. Technol., Appl. Surf. Anal. Workshop, University of Dayton, Ohio (June 13, 1979).
565. C. J. Powell, Appl. Surf. Sci. 4:492 (1980).
566. D. E. Hanlin and J. S. Solomon, Surf. Anal. Technol., Appl. Anal. Workshop, University of Dayton, Ohio (June 13–15, 1979).
567. A. Benninghoven, Surf. Sci. 28:541 (1971).
568. R. Schubert and J. C. Tracy, Rev. Sci. Inst. 44:487 (1973).
569. R. Schubert, J. Vac. Sci. Technol. Inst. 44:487 (1973).

570. R. Werner, Dev. Appl. Spectroscopy 7A:297 (1969).
571. N. T. McDevitt and W. L. Baun, Surface analysis of 2024 aluminum with sulfuric-chromic acid solutions, J. Electrochem. Soc., p. 1058 (July 1976).
572. N. T. McDevitt et al., Surface analysis of 6061 and 7050 alloys after conditioning chemical treatment, AFML Report TR-76-13, Air Force Materials Lab., Wright-Patterson AFB, OH (March 1976).
573. N. T. McDevitt, W. L. Baun, and J. S. Solomon, AFML Report TR-175-122. Wright-Patterson AFB, OH (1975).
574. J. A. Smith and W. E. Martinsen, Amer. Ceram. Soc. Bull. 52:855 (1973).
575. J. D. Minford, unpublished work, Aluminum Co. of America.
576. N. T. McDevitt and W. L. Baun, Some observations of the relation between chemical surface treatments and the growth of anodic barrier layer films, AFML Report TR-76-74, Air Force Materials Lab., Wright-Patterson AFB, OH (June 1976).
577. N. T. McDevitt, W. L. Baun, G. Fugate, and J. S. Solomon, Surface studies of anodic aluminum oxide layers formed in phosphoric acid solutions, AFML Report TR-77-55, Air Force Materials Lab., Wright-Patterson AFB, OH (May 1977).
578. W. L. Baun, Applications of surface analysis techniques to studies of adhesion, Appl. Surf. Sci. 4:291 (1980).
579. J. P. Wightman, Surface characterization in adhesion, Invited Lecture at Symp. Fundamentals of Adhesion: Theory, Practice, and Applications, State University New York at New Paltz (Oct. 10–12, 1984).
580. T. A. Bush, M. E. Counts, T. C. Ward, and J. P. Wightman, NASA-LARC Final Report NAS1-10646-14, Hampton, VA (Nov. 1973).
581. M. E. Counts and J. P. Wightman, NASA-LaRC Final Report, NAS1-10646-25, Hampton, VA (Nov. 1974).
582. D. W. Dwight, M. E. Counts, and J. P. Wightman, Effect of polymer properties and adherend surfaces on adhesion, Final Report, NASA Grant NSG-1124, Virginia Tech. Chem. Dept., Blacksburg, VA (Dec. 1975).
583. T. A. Bush, M. E. Counts, and J. P. Wightman, Polymer Preprints 16(1):770 (1975).
584. T. St. Clair and D. J. Progar, Polymer Preprints 16(1):583 (1985).
585. A. K. St. Clair and T. L. St. Clair, SAMPE Mater. Rev. 7:53 (1975).
586. D. J. Progar and T. L. St. Clair, Organic Coatings and Plastics Chemistry Preprints 35(2):445 (1979).
587. W. E. J. Neal, Optical examination and monitoring of surfaces, Appl. Surf. Sci. 2:445 (1979).
588. G. D. Davis, Surface analysis techniques and applications, Adhesives and Sealants, Engineered Materials Handbook, ASM, Materials Park, OH, Vol. 3, pp. 248–249 (1990).
589. G. D. Davis, Characterization of surfaces, Adhesive Bonding, (L. H. Lee, ed.), Plenum Press, New York, Chap. 6, pp. 139–170 (1991).
590. M. P. Seah, Practical Surface Analysis (D. Briggs and M. P. Seah, eds.), John Wiley & Sons, p. 181 (1983).
591. J. T. Grant, Surf. Interf. Anal. 14:271 (1989).
592. C. D. Wagner, W. M. Riggs, L. E. Davis, J. F. Moulder, and G. E. Muilenberg, Handbook of X-Ray Photoelectron Spectroscopy, Perkin-Elmer Co., Eden Praire, MN (1979).
593. C. D. Wagner, L. E. Davis, M. V. Zeller, J. A. Taylor, R. H. Raymond, and L. H. Gale, Surf. Interf. Anal. 3:211 (1981).
594. C. D. Wagner, Practical Surface Analysis (D. Briggs and M. P. Seah, ed.), John Wiley & Sons, p. 511 (1983).
595. C. D. Wagner, Anal. Chem. 49:1282 (1977).
596. C. D. Wagner, Anal. Chem. 44:1050 (1977).
597. V. I. Nefedov, N. P. Sergushin, I. M. Band, and M. B. Trzhakovskaya, Electron Spectrosc. Relat. Phenom. 2:383 (1973).
598. V. I. Nefedov, N. P. Sergushin, I. M. Band, and M. B. Trzhakovskaya, J. Electron Spectrosc. Relat. Phenom. 7:175 (1975).
599. S. Evans, R. G. Pritchard, and J. M. Thomas, J. Phys. C. Solid State Phys., p. 2483 (1977).
600. S. Evans, R. G. Pritchard, and J. M. Thomas, J. Electron Spectrosc. Relat. Phenom. 14:341 (1978).
601. L. E. Davis, N. C. McDonald, P. W. Palmberg, G. E. Riach, and R. E. Weber, Handbook of Auger Electron Spectroscopy, Perkin-Elmer Co., Norwalk, CN (1976).
602. G. E. McGuire, Auger Electron Spectroscopy Reference Manual, Plenum Press, New York (1979).
603. T. Sekine, Y. Nagasawa, M. Kudoh, Y. Sakai, A. S. Parkes, J. D. Geller, A. Mogami, and K. Hirata, Handbook of Auger Electron Spectroscopy, Japan Electron Optics Lab. Co. (1982).
604. T. A. Carlson, Photoelectron and Auger Spectroscopy, Plenum Press, New York (1974).
605. C. D. Wagner and D. M. Bickham, NIST X-ray Photoelectron Spectroscopy Database, Version 1.0. Standard Reference Database 20, Natl. Inst. of Std. & Technol. (1989).

606. C. D. Wagner, Practical Surface Analysis (D. Briggs and M. P. Seah, eds.), John Wiley & Sons, New York, p. 477 (1983).
607. C. D. Wagner, D. E. Passoja, H. F. Hillery, T. G. Kinisky, H. A. Six, W. T. Jansen, and J. A. Taylor, J. Vac. Sci. Technol. 21:933 (1982).
608. C. D. Wagner, L. H. Gale, and R. H. Raymond, Anal. Chem. 51:466 (1979).
609. T. S. Haas, J. T. Grant, and G. J. Dooley, III., J. Appl. Phys. 43:1853 (1972).
610. G. D. Davis, M. Natan, and K. A. Anderson, Appl. Surf. Sci. 15:321 (1983).
611. G. D. Davis, D. E. Savage, and M. G. Lagally, J. Electron Spectrosc. Relat. Phenom. 23:25 (1981).
612. D. E. Ramaker, Springer Series in Chemical Physics, ISISS Proc. (R. Vanselow, ed.), Springer-Verlag, Berlin, Germany (1982).
613. I. F. Ferguson, Auger Microprobe Analysis, Adam Hilger (1989).
614. S. Hofmann, Practical Surface Analysis (D. Briggs and M. P. Seah, eds.), John Wiley & Sons, p. 141 (1983).
615. D. E. Sykes, D. D. Hall, R. E. Thurstans, and J. M. Walls, Appl. Surf. Sci. 5:103 (1980).
616. A. Zalar, Surf. Interf. Anal. 9:41 (1986).
617. J. D. Geller and N. Veisfeld, Surf. Interf. Anal. 14:95 (1989).
618. R. Smith, T. P. Valkering, and J. M. Walls, Phios. Mag. A. 44:879 (1981).
619. N. Q. Lam, Surf. Interf. Anal. 12:65 (1988).
620. R. Kelly, Surf. Sci. 100:85 (1980).
621. H. M. Clearfield, G. O. Cote, K. A. Oliver, D. K. Shaffer, and J. S. Ahearn, Surf. Interf. Anal. 11:347 (1988).
622. V. Thompson, H. E. Hintermann, and L. Chollet, Surf. Technol. 8:421 (1979).
623. J. M. Walls, D. D. Hall, and D. E. Sykes, Surf. Interf. Anal. 1:204 (1979).
624. I. K. Brown, D. D. Hall, and J. M. Walls, Vacuum 31:625 (1981).
625. C. Lean and M. P. Seah, Thin Solid Films 75:67 (1981).
626. J. C. Bierlein, S. W. Gaarenstroom, R. A. Waldo, and A. C. Ottolini, J. Vac. Sci. Technol. 2:1102 (1984).
627. G. D. Davis, Surf. Interf. Anal. 9:421 (1986).
628. H. M. Clearfield, D. K. Shaffer, J. S. Ahearn, and J. D. Venables, J. Adhes. 23:83 (1987).
629. J. L. Goldstein and H. Yajowitz (eds.), Practical Scanning Electron Microscopy, Plenum Press, New York (1975).
630. G. D. Davis, H. M. Clearfield, W. C. Moshier, and G. O. Cote, Surf. Interf. Anal. 11:359 (1988).
631. J. S. Ahearn and G. D. Davis, Adhesion '87, Plastics and Rubber Institute Meetings, p. 291 (1987).
632. W. A. Coghlan and R. E. Clausing, Atomic Data 5:317 (1973).
633. H. H. Madden, J. Vac. Sci. Technol. 18:677 (1981).
634. G. K. Wehner, The aspects of sputtering in surface analysis methods, Methods of Surface Analysis (A. W. Czanderna, ed.). Elsevier, Amsterdam (1975).
635. T. S. Sun, D. K. McNamara, J. S. Ahearn, J. M. Chen, B. Ditchek, and J. D. Venables, Appl. Surf. Sci. 5:406 (1980).
636. C. C. Chang, J. Vac. Sci. Technol. 18:276 (1981).
637. A. W. Czanderna, Methods of Surface Analysis, Elsevier, Amsterdam (1975).
638. T. S. Sun, J. M. Chen, R. K. Viswanadham, and J. A. S. Green, Appl. Phys. Lett. 31:580 (1977).
639. J. S. Solomon and W. L. Baun, Surf. Sci. 51:228 (1975).
640. R. F. Wagner, Practical Surface Analysis (D. Briggs and M. P. Seah, eds.), John Wiley & Sons, p. 521 (1983).
641. D. Briggs and J. C. Riviere, Practical Surface Analysis (D. Briggs and M. P. Seah, eds.), John Wiley & Sons, p. 87 (1983).
642. Standard Practice for Elemental Identification by Auger Electron Spectroscopy, E 827, Annual Book of ASTM Stds, ASTM; Surf. Interf. Anal. 5:266 (1983).
643. W. Wei, J. Vac. Sci. Technol. 6:2576 (1988).
644. S. Ichimura, H. E. Bauer, H. Seiler, and S. Hofmann, Surf. Interf. Anal. 14:250 (1989).
645. K. Siegbahn, C. Nordling, A. Fahlman, R. Norberg, K. Hamrin, J. Hedman, G. Johansson, T. Bergmark, S. Karlsson, J. Lingren, and B. Lindberg, Nova Acta Reg. Soc. Sci. Upsalla Serv. IV, Vol. 20 (1967); Air Force Materials Lab. Tech. Report Tr-68-189, Wright-Patterson AFB, OH (Oct. 1968).
646. H. Ibach (ed.), Electron Spectroscopy for Surface Analysis, Springer-Verlag, New York (1977).
647. C. Brundle and A. D. Baker (eds.), Electron Spectroscopy-Theory, Techniques, and Applications, Academic Press, London, Vol. 4 (1981).
648. D. W. Dwight, T. J. Fabish, and H. R. Thomas (eds.), Photon, Electron, and Ion Probes of Polymer Structure and Properties, Amer. Chem. Soc. Symp. Series No. 162 (1981).
649. Applied Electron Spectroscopy for Chemical Analysis (H. Windawi and F. F. L. Ho, eds.), John Wiley & Sons, New York (1982).

650. Practical Surface Analysis by Auger and X-ray Photoelectron Spectroscopy (D. Briggs and M. P. Seah, eds.), John Wiley & Sons, New York (1983).
651. W. E. Schwartz, Jr., Anal. Chem. 45(9):789A (1973).
652. C. A. Evans, Jr., Anal. Chem. 47(9):819A (1975).
653. C. R. Brundle, J. Vac. Sci. Technol. 11(1):212 (1974).
654. W. M. Riggs and M. J. Parker, Methods of Surface Analysis (A. W. Czanderna, ed.), Elsevier, Amsterdam (1975).
655. D. T. Clark and W. J. Feist, J. Macromol. Sci. Rev. Macromol. Chem. C12(2) (1975); D. T. Clark, Polymer Surfaces (D. T. Clark and W. J. Feist, eds.), Wiley-Interscience, New York, pp. 305–385 (1978).
656. A. Dilks, Electron Spectroscopy-Theory, Techniques, and Applications (C. Brundle and A. D. Baker, eds.), Academic Press, London (1981).
657. A. Dilks, Photon, Electron, and Ion Probes of Polymer Structure and Properties (D. W. Dwight, T. J. Fabish, and H. R. Thomas, eds.). Amer. Chem. Soc. Symp. Series No. 162 (1981).
658. M. M. Millard, Industrial Applications of Surface Analysis (L. A. Caspar and C. J. Powers, eds.), ACS Symp. Series No. 199, pp. 143–202 (1982).
659. W. J. van Ooig, T. H. Visser, and M. E. F. Biemond, Surf. Interf. Anal. 6(5):197 (1984).
660. H. R. Anderson, Jr., and K. G. Sachdev, X-ray photoelectron spectroscopy in understanding adhesion phemomena, Treatise on Adhesion and Adhesives (R. L. Patrick, ed.), Marcel Dekker, New York, Vol. 6, Chap. 6, p. 213 (1989).
661. C. D. Wagner and A. Joshi, Surf. Interf. Anal. 6:215 (1984).
662. K. Yates and R. H. West, Surf. Interf. Anal. 5(5):217 (1983).
663. A. J. Kinloch and N. R. Smart, J. Adhes. 12:23 (1981).
664. M. Gettings, F. S. Baker, and A. J. Kinloch, J. Appl. Polym. Sci. 21:2375 (1977).
665. M. P. Seah, Surf. Interf. Anal. 2:222 (1980).
666. M. P. Seah, and W. A. Dench, Surf. Interf. Anal. 1:2 (1979).
667. W. J. Carter, G. K. Schweitzer, and T. A. Carlson, J. Electron Spectros. Relat. Phenom. 5:827 (1974).
668. G. D. Davis, S. Buchner, W. A. Bech, and N. E. Byer, Appl. Surf. Sci. 15:238 (1983).
669. R. S. Swingle and W. M. Riggs, CRC Crit. Rev. Anal. Chem. 5:262 (1975).
670. L. C. Feldman and J. W. Mayer, Fundamentals of Surface and Thin Film Analysis, North-Holland, Amsterdam (1986).
671. K. Siegbaun, C. Nordling, A. Fahlman, R. Norberg, K. Jamrin, J. Hedman, G. Johansson, T. Bergmark, S. E. Karlsson, J. Lindgren, and B. Lindberg, AFML Tech. Report No. TR-68-189, Wright-Patterson AFB, OH (Oct. 1968).
672. G. D. Davis, Surface Analysis Techniques and Applications, Adhesives and Sealants, Engineered Materials Handbook, ASM, Materials Park, OH, Vol. 3, p. 239 (1990).
673. K. F. J. Heinrick and D. E. Newbury, Secondary Ion Mass Spectroscopy, NBS Spec. Publ. No. 427 (1975).
674. W. Reuter, Secondary Ion Mass Spectrometry: SIMS V. (A. Benninghoven, R. J. Colton, D. S. Simons, and H. W. Werner, eds.), Springer-Verlag, Berlin, Germany, p. 94 (1986).
675. O. Ganschow, Secondary Ion Mass Spectrometry: SIMS V. (A. Benninghoven, R. J. Colton, D. S. Simons, and H. W. Werner, eds.), Springer-Verlag, Berlin, Germany, p. 79 (1986).
676. H. Oechsner, Thin Film and Depth Profile Analysis (H. Oechsner, ed.), Springer-Verlag, Berlin, Germany, p. 63 (1984).
677. H. Oechsner, Secondary Ion Mass Spectrometry: SIMS V. (A. Benninghoven, R. J. Colton, D. S. Simons, and H. W. Werner, eds.), Springer-Verlag, Berlin, Germany, p. 70 (1986).
678. C. H. Becker and K. T. Gillen, Anal. Chem. 56:1671 (1984).
679. C. H. Becker and K. T. Gillen, Secondary Ion Mass Spectrometry: SIMS V (A. Benninghoven, R. J. Colton, D. S. Simons, and H. W. Werner, eds.), Springer-Verlag, Berlin, Germany, p. 85 (1986).
680. J. B. Pallix, C. H. Becker, and K. T. Gillen, Appl. Surf. Sci. 32:1 (1988).
681. U. Schuhle, J. B. Pallix, and C. H. Becker, J. Vac. Sci. Technol. 6:936 (1988).
682. D. Briggs, Analysis of polymer surfaces by SIMS. 1-fingerprint spectra from simple polymer films, Surf. Interf. Anal. 4:151 (1982).
683. A. Benninghoven, Developments in secondary mass spectroscopy and applications to surface studies, Surf. Sci. 53:596 (1975).
684. D. Briggs, Surf. Interf. Anal. 9:391 (1986).
685. J. C. Vickerman, Methods of Surface Analysis (J. M. Walls, ed.), Cambridge University Press, London, England, p. 169 (1989).
686. S. Hofmann, Surf. Interf. Anal. 9:3 (1986).
687. D. E. Sykes, Methods of Surface Analysis (J. M. Walls, ed.), Cambridge University Press, London, England, p. 216 (1989).

688. C. W. Magee and R. E. Honig, Surf. Interf. Anal. 4:35 (1982).
689. P. K. Hansma, Tunnelling Spectroscopy: Capabilities, Applications, and New Technology, Plenum Press, New York (1981).
690. R. C. Jaklevic and J. Lambe, Phys. Rev. Lett. 17:1139 (1966).
691. J. Lambe and R. C. Jaklevic, Phys. Rev. 165:821 (1968).
692. J. Klein et al., Phys. Rev. B7:2336 (1973).
693. J. R. Kirtley and P. K. Hansma, Phys. Rev. B12:531 (1975).
694. Y. Skarlatos, R. C. Barker, G. L. Haller, and A. Yelon, Surf. Sci. 43:353 (1974).
695. A. F. Diaz, U. Hetzler, and E. Kay, J. Am. Chem. Soc. 99:6781 (1977).
696. H. T. Chu, N. K. Eib, A. N. Gent, and P. N. Henriksen, Polymer Preprint 18:113 (1977).
697. P. K. Hansma, Physics Report 30c:145 (1977).
698. H. W. White, L. M. Godwin, and T. Wolfram, A new method for the study of adhesion: Application of inelastic electron tunnelling spectroscopy, J. Adhes. 9:237 (1978).
699. H. W. White, L. M. Godwin, and T. Wolfram, Inelastic Electron Tunnelling Spectroscopy (T. Wolfram, ed.), Springer-Verlag, Heidelberg (1978).
700. H. W. White, L. M. Godwin, and R. Ellialtioglu, paper presented at 4th Ann. Meeting Adhesion Soc., Savannah, GA (Feb. 23–25, 1981).
701. P. N. Schott and B. O. Field, Surf. Interf. Anal. 1:64 (1975).
702. D. P. Smith, J. Appl. Phys. 38:340 (1967).
703. W. L. Baun, Study of adhesive bonding and bond failure surface using ISS-SIMS, Characterization of Metal and Polymer Surfaces (L. H. Lee, ed.), Academic Press, New York, Vol. 1, pp. 121–141 (1982).
704. W. L. Baun, Applications of ion beam methods to characterization of adhesive bonding materials, Industrial Applications of Surface Analysis (L. A. Casper and C. J. Powell, eds.), ACS Symp. Series No. 199, ACS, Washington, D.C., pp. 121–141 (1982).
705. H. G. Tompkins, Appl. Spectros. 28:335 (1974).
706. N. J. Harrick, Internal Reflection Spectroscopy, Interscience, New York (1967).
707. M. Delhaye, M. Dupeyrat, R. Dupyret, and Y. Levy, J. Raman Spectroscopy 8:351 (1979).
708. R. I. Iwamoto, M. Miya, K. Ohta, and S. Mima, J. Chem. Phys. 74:4780 (1981).
709. D. L. Allara, C. A. Murray, and S. Bodoff, Spectroscopy of polymer surfaces using surface enhanced Raman effect, Physicochemical Aspects of Polymer Surfaces (K. L. Mittal, ed.), Plenum Press, New York, pp. 33–43 (1983).
710. J. F. Rabolt, R. Santo, and J. D. Swalen, Appl. Spectroscopy 34:517 (1980).
711. E. Burstein, C. Y. Chen, and S. Lundquist, Light Scattering in Solids (J. L. Birman, H. Z. Cummins, and K. K. Rebane, eds.), Plenum Press, New York, p. 479 (1979).
712. T. E. Furtak and J. Reyes, Surf. Sci. 93:351 (1980).
713. Surface Enhanced Raman Scattering (R. K. Chang and T. E. Furtek, eds.), Plenum Press, New York (1982).
714. H. F. Webster, "Interaction of Thin Liquid Films with Polymer Surfaces," M.S. Thesis, Virginia Polytechnic Institute and State University, Blacksburg, VA (1984).
715. M. K. Debe and L. A. Tushaus, Reflection absorption studies of the surface diffused residues on a humidity/temperature aged urethane pressure-sensitive adhesive, J. Adhes. 15:287 (1983).
716. C. S. P. Sung, S. H. Lee, and N. H. Sung, Role of coupling agents in metal/polymer adhesion, I—The structure of the silane film at metal/polymer interface, Adhesion and Adsorption of Polymers (L. H. Lee, ed.), Plenum Press, New York, Part B, p. 757 (1980).
717. F. J. Boerio and A. C. Gosselin, Structure-property relationships of silane films on aluminum substrates, Proc. 36th Ann. SPI Reinf. Plast. Compos. Inst., Sec. 2.G, p. 1 (1981).
718. L. D'Esposito and J. L. Koenig, Fourier Transform Infrared Spectroscopy, Academic Press, New York, Vol. 1 (1978).
719. J. L. Koenig, Acc. Chem. Res. 14:171 (1981).
720. F. J. Boerio, C. A. Gosselin, R. G. Dillingham, and H. W. Liu, J. Adhes. 13:159 (1981).
721. F. J. Boerio, L. H. Schlenlein, and J. E. Greivenkamp, J. Appl. Polym. Sci. 22:203 (1978).
722. F. J. Boerio, J. W. Williams, and J. M. Burkstrand, J. Colloid Interf. Sci. 91:485 (1983).
723. F. J. Boerio and R. G. Dillingham, Adhesive Joints: Formation, Characteristics and Testing (K. L. Mittal, ed.), Plenum Press, New York, p. 541 (1984).
724. J. D. E. McIntyre and D. E. Aspnes, Surf. Sci. 24:417 (1971).
725. R. H. Muller, J. Kruger, and J. D. E. McIntyre, Advances in Electrochemistry and Electrochemical Engineering (R. H. Muller, ed.), John Wiley & Sons, Vol. 9 (1973).
726. R. Dorn and H. Luth, Phys. Rev. Lett. 33:1024 (1974).
727. H. Luth, Appl. Phys. 8:1 (1975).

728. J. D. E. McIntyre, Surface reflection spectroscopy, Optical Properties of Solids (B. O. Seraphin, ed.), North-Holland, Amsterdam (1976).
729. P. J. Fiebelman, Phys. Rev. B14:762 (1976).
730. P. Drude, Ann. Phys. (Leipzig), 32:584 (1887).
731. P. Drude, Ann. Phys. (Leipzig), 36:532 (1889).
732. P. Drude, Ann. Phys. (Leipzig), 39:481 (1890).
733. L. Tronstad, Trans. Faraday Soc. 29:502 (1933).
734. A. B. Winterbottom, Nature 140:364 (1937).
735. R. M. A. Azzam and N. M. Bashara, Ellipsometry and Polarized Light, North-Holland, Amsterdam (1977).
736. Ellipsometry, Proc. Symp. Ellipsometry in the Measurement of Surfaces and Thin Films (E. Passaglia, R. R. Stromberg, and J. J. Kruger, eds.), Bureau of Standards, Washington, D.C. (1963).
737. R. R. Stromberg and F. L. McCrackin, Ellipsometry as a tool for the characterization of surfaces, Clean Surfaces: Their Preparation, Characterization, Interfacial Study (G. Goldfinger, ed.), Marcel Dekker, New York, pp. 65–76 (1970).
738. F. L. McCracklin, E. Passaglia, R. R. Stromberg, and H. S. Steinber, J. Nat. Bur. Stds. 67A:363 (1963).
739. R. J. Archer, Opt. Soc. Amer. 52:970 (1962).
740. D. E. Aspnes, J. Vac. Sci. Technol. 18:276 (1981).
741. P. C. S. Hayfield, 1st Int. Cong. Metallic Corrosion, Butterworths, London, p. 670 (1962).
742. P. C. S. Hayfield, Soc. Chem. Ind. Monograph 28:128 (1968).
743. P. C. S. Hayfield, Surf. Sci. 56:488 (1976).
744. A. R. C. Westwood, R. K. Wiswansdham, and J. A. S. Green, Thin Solid Films 39:69 (1976).
745. J. I. Goldstein, Electron optics, Practical Scanning Electron Microscopy (J. I. Goldstein and H. Yahowitz, eds.), Plenum Press, New York (1975).
746. J. I. Goldstein, Electron beam-specimen interactions, Practical Scanning Microscopy (J. I. Goldstein and H. Yahowitz, eds.), Plenum Press, New York (1975).
747. Proc. Scanning Electron Microscopy Symp., sponsored by IITRI (1968).
748. Proc. Scanning Electron Microscopy Symp., sponsored by IITRI (1969).
749. C. W. Oatley, W. C. Nixon, and R. F. F. Pease, Electronic and Electron Physics 21:181 (1965).
750. D. E. Newberg and H. Yahowitz, Specimen preparation, special techniques, and application of scanning electron microscope, Practical Scanning Electron Microscopy (J. I. Goldstein and H. Yahowitz, eds.), Plenum Press, New York (1975).
751. E. A. Ledbury et al., Microstructural characterization of adhesive-bonded joints, Proc. 12 Nat. SAMPE Tech. Conf., Seattle, p. 935 (1980).
752. R. L. Patrick, The use of scanning electron microscopy, Treatise on Adhesion and Adhesives (R. L. Patrick, ed.), Marcel Dekker, New York, Vol. 3, Chap. 4, p. 163 (1973).
753. O. Jahari and A. V. Samudra, Scanning electron microscopy, Characterization of Solid Surfaces (P. F. Kane and G. P. Larrabee, eds.), Plenum Press, New York, pp. 107–131 (1974).
754. J. D. Venables, D. K. McNamara, J. Chen, T. S. Sun, and R. L. Hopping, Oxide morphologies on adhesively-bonded aluminum structures, Appl. Surf. Sci. 3:88 (1979).
755. H. M. Clearfield, D. K. McNamara, and G. D. Davis, Adherend surface preparation for structural adhesive bonding, Adhesive Bonding (L. H. Lee, ed.), Plenum Press, New York, Chap. 8, pp. 203–235 (1991).
756. T. H. Allen, J. Vac. Sci. Technol. 13:112 (1976).
757. T. Smith, The mechanism of adhesive failure between polymers and metallic substrates, AFML Tech. Report TR-74-73, Wright-Patterson AFB, OH (1974).
758. A. Pattniak and J. D. Meakin, Characterization of aluminum adherend surfaces, Tech. Report No. 4699, Franklin Inst. (July 1974).
759. A. Pattniak and J. D. Meakin, Characterization of aluminum adherend surfaces, Symp. on Durability of Adhesive Bonded structures, Picatinny Arsenal, Dover, NJ, p. 211 (Oct. 27–29, 1976).
760. B. B. Bowen et al., Development of corrosion resistant surface treatments on aluminum alloys for spot-weld bonding, AFML Contract No. F 33615-74-C-5027, Wright-Patterson AFB, OH (1974).
761. B. B. Bowen et al., Development of corrosion resistant surface treatments for aluminum alloys for spot welding, Interim Tech. Report NOR-79-93, Northrup Aviation (May 1974).
762. B. B. Bowen et al., Development of corrosion resistant surface treatments for aluminum alloys for spot-weld bonding, Interim Tech. Report NOR-74-231, Northrup Aviation (Aug. 1974).
763. B. B. Bowen et al., Preparation of aluminum alloy surfaces for spot-weld bonding, Natl. SAMPE Tech. Conf., Buena Park, CA (April 1974).
764. B. B. Bowen, Preparation of aluminum alloy surfaces for spot-weld bonding, SAMPE Mater. Rev. 7: 734 (1975).

765. T. S. Sun, J. D. Venables, and J. M. Chen, Effects of surface morphology and chemical composition on durability of adhesively bonded aluminum structures, Report No. MML-TR-79-37C, Martin-Marietta, Labs. (Sept. 14, 1978).
766. Std. Rec. Prac. for removal of metal surfaces for adhesive bonding, D-2651-79, ASTM, p. 785 (1979).
767. W. J. Russell, Chromate-free process for preparing aluminum for adhesive bonding, Symp. on Durability of Adhesive Bonded Structures, Picatinny Arsenal, Dover, NJ (Oct. 27–29, 1976).
768. Resistance welding of aluminum, Metals Handbook, 8th ed., Welding and Brazing, ASM, Vol. 6, p. 470 (1971).
769. W. J. Russell and W. C. Tanner, Component parts assembly with joints adhesive-mechanical, Part I, surface preparation of aluminum for weldbonding, Report No. 4610, Picatinny Arsenal, Dover, NJ (Feb. 1974).
770. P. F. A. Bijlmer, J. Adhes. 5:319 (1973).
771. J. D. Minford, Aluminum adhesive bond permanence, Treatise on Adhesion and Adhesives (R. L. Patrick, ed.), Marcel Dekker, New York, Vol. 5, Chap. 3, p. 69 (1981).
772. J. D. Minford, Aluminum joint durability in stress and condensing humidity exposure with varying surface pretreatment, Specialized Cleaning, Finishing and Coating Processes, ASM/SAMPE Conf. Proc. Los Angeles, California, Materials/Metalworking Tech. Series, ASM, Metals Park, Ohio (Feb. 5–6, 1980).
773. J. D. Minford, Effect of surface preparation on stressed aluminum joints in corrosive saltwater exposure, Adhes. Age 23(10):36 (1980).
774. J. D. Minford, Joint durability studies with abraded, etched, coated, and anodized aluminum adherends, Adhesive Joints: Formation, Characteristics, and Testing (K. L. Mittal, ed.), Plenum Press, New York, p. 485 (1984).
775. J. D. Minford, Permanence of adhesive-bonded aluminum joints, Adhesives in Manufacturing (G. L. Schneberger, ed.), Marcel Dekker, New York, Chap. 23, p. 567 (1983).
776. C. W. Jennings, J. Adhes. 4:25 (1972).
777. B. A. Wilson, Corrosion Sci. 11:527 (1971).
778. T. Smith and D. H. Kaelble, Mechanisms of adhesion failure between polymers and metallic substrates: aluminum 2024-T3 with HT-424 adhesive, Treatise on Adhesion and Adhesives (R. L. Patrick, ed.), Marcel Dekker, New York, Vol. 5, Chap. 4, p. 139 (1981).
779. P. F. A. Bijlmer and R. J. Schliekelmann, SAMPE Quart., p. 13 (Oct. 1973).
780. J. M. Chen, T. S. Sun, J. D. Venables, and R. L. Hopping, 24th Nat. SAMPE Conf. Proc., p. 1188 (1979).
781. A. J. Kinloch, H. E. Bishop, and N. R. Smart, J. Adhes. 14:105 (1982).
782. W. C. Cochran, Anodizing, Aluminum (K. R. Van Horn, ed.), ASM, Metals Park, OH, Chap. 19, p. 641 (1967).
783. A. Jenny, Electrolytic behavior of aluminum and other valve metals, The Anodic Oxidation of Aluminum and Its Alloys (English translation by W. Lewis), Chas. Griffin & Co., London, pp. 71–91 (1940).
784. J. E. Lewis and H. C. Plumb, Studies of the anodic behavior of aluminum, J. Electrochem. Soc. 105(9): 496 (1958).
785. T. P. Hoar and N. F. Mott, A mechanism for the formation of anodic films on aluminum, J. Phys. Chem. Solids 9:957 (1959).
786. F. Keller, M. S. Hunter, and D. L. Robinson, Structural features of oxide coatings on aluminum, J. Electrochem. Soc. 100(9):411 (1953).
787. C. S. Taylor, C. M. Tucker, and J. D. Edwards, Anodic coatings with crystalline structure on aluminum, Trans. Electrochem. Soc. 88:325 (1945).
788. F. Keller et al., Anodic coatings seen through the microscope, ASTM Proceedings 40:3 (1940).
789. J. S. Solomon and D. E. Hanlin, AES and SEM characterization of anodized aluminum alloy adherends for adhesive bonding applications, Appl. Surf. Sci. 4:307 (1980).
790. J. P. O'Sullivan and G. C. Wood, Proc. Royal Soc. A317:511 (1970).
791. G. C. Wood, Porous anodic films on aluminum, Oxide and Oxide Films (J. W. Diggle, ed.). Marcel Dekker, New York, Vol. 2, p. 167 (1973).
792. G. E. Thompson, R. C. Furneaux, G. C. Wood, J. A. Richardson, and J. S. Goode, Nature 272:43 (1978).
793. K. V. Heber, Electrochimica Acta 23:127 (1978).
794. D. E. Hanlin and J. S. Solomon, AES and SEM characterization of anodized aluminum alloy adherends for adhesive bonding applications, Appl. Surf. Anal. Workshop, University of Dayton, Dayton, OH (June 13–15, 1979).
795. J. W. Coburn and E. Kay, Crit. Rev. Solid State Sci. 4:561 (1974).
796. P. H. Holloway, J. Electron Spectros. Relat. Phenom. 7(3):215 (1975).
797. J. S. Solomon and W. L. Baun, J. Vac. Sci. Technol. 12:373 (1975).

798. J. Kirchner and H. W. Etzhorn, Proc. 7th Int. Vacuum Cong. and 3rd Int. Conf. Solid Surf. (R. Debrozensky, F. Rudenauer, F. Viehbock, and A. Breth, eds.), Vienna, p. 2213 (1977).
799. J. H. Manhard and F. A. Mozelewski, Plating and Surface Finishing 66:566 (1979).
800. J. S. Solomon, Appl. Spectros. 30:46 (1976).
801. J. S. Solomon and W. L. Baun, Surface characterization of contamination of adhesive bonding materials, 4th Int. Symp. Contamination Control, Washington, D.C. (Sept. 1978).
802. L. T. Drzal, Summary of the workshop held on the role of the polymer substrate interphase in structural adhesion, AFML Report No. Tr-77-129, Wright-Patterson AFB, OH (July 1977).
803. L. H. Sharpe, The interphase of adhesion, J. Adhes. 4:51 (1972).
804. S. J. Hudak, F. J. Boerio, P. J. Clark, and Y. Okamoto, XPS analysis of the interphase between an anaerobic adhesive and metal substrates, Surf. Interf. Anal. 15(2):167 (1990).
805. P. Ho, Chemistry and adhesion of metal/polymer interfaces, Appl. Surf. Sci. 41/42:559 (1989).
806. F. Nitschke, The metal/polymer interface in aluminum adhesive joints: A microanalytical study, J. Adhes. Sci. Technol. 4(1):41 (1989).
807. D. Ondrus, F. J. Boerio, and K. J. Grannen, Molecular structure of polymer/metal interphases, J. Adhes. 29(1–4):27 (1989).
808. J. S. Bishop, G. E. Thompson, and D. Jobling, The influence of the surface pretreatment and the environment on the interphasial region of adhesively-bonded aluminum joints, Adhesion '90, The Plastics and Rubber Institute Meetings, Cambridge, England (Sept. 10–12, 1990).
809. P. S. Achary and R. Ramaswamy, Modification of nitrile-phenolic bonding agent by addition of an interfacial agent: effect on the practical adhesion between nitrile rubber and metal, J. Adhes. Sci. Technol. 3(8):587 (1989).
810. T. J. Reinhart, Proc. Army Mater. Conf. 4:201 (1976).
811. P. F. A. Bijlmer, Metal Finishing, p. 30 (April 1972).
812. R. B. Mason and C. J. Slunder, Anodic reactions of aluminum and its alloys in sulfuric and oxalic acid electrolytes, Industrial Eng. Chem. 39:1602 (1947).
813. R. W. Buzzard, Anodizing of aluminum alloys in chromic acid solution of different concentrations, J. Res. Natl. Bureau Stds. 18:251 (1937).
814. R. W. Buzzard and J. H. Wilson, Deterioration of chromic acid baths used for anodic oxidation of aluminum alloys, J. Res. Natl. Bureau Stds. 18:53 (1937).
815. P. F. A. Bijlmer, Chromic acid anodize process used in Europe, Adhesive Bonding of Aluminum Alloys (E. W. Thrall and R. W. Shannon, eds.), Marcel Dekker, New York, Chap. 2, p. 21 (1985).
816. P. F. A. Bijlmer, The influence of hot water rinsing on the surface properties of Alclad 2024-T3, Fokker Report No. R-1050 (1969).
817. N. L. Rogers, Bonding to sealed chromic acid anodize, Symp. on Durability of Adhesive Bonded Structures, Picatinny Arsenal, Dover, NJ (Oct. 27–29, 1976).
818. S. C. Akers, J. Appl. Polym. Sci., Appl. Polym. Symp. 3:251 (1966).
819. N. L. Rogers, A comparative test for corrosion resistant adhesive primers, Nat. SAMPE Tech. Conf., Series 2 (1970).
820. N. L. Rogers, Sealed chromic acid anodize, Adhesive Bonding of Aluminum Alloys (E. W. Thrall and R. W. Shannon, eds.), Marcel Dekker, New York, Chap. 3, p. 41 (1985).
821. N. L. Rogers, J. Appl. Polym. Sci., Appl. Polym. Symp. 19:63 (1972).
822. N. L. Rogers, J. Appl. Polym. Sci., Appl. Polym. Symp. 32:37 (1977).
823. J. K. G. Penitz and D. J. Sharp, The effect of different aluminum alloy surface compositions on barrier anodic film formation, Electrochem Sci. Technol., p. 2227 (Oct. 1984).
824. V. Singh, P. Rama, G. J. Cocks, and D. M. R. Taplin, J. Mater. Sci. Lett. 14:745 (1979).
825. A. Csanady, I. Imre-Baan, E. Lichtenberger-Bajza, E. Szontagh, and F. Domolki, J. Mater. Sci. 15:2761 (1980).
826. H. E. Franz, Z. Werkstofftech. 14:164,290 (1983).
827. H. Kollek, Int. J. Adhes. Adhes. 5:75 (1985).
828. M. F. Abd Rabbo, J. A. Richardson, and G. C. Woods, Corrosion Sci. 16:689 (1976).
829. A. E. Yaniv, N. Fin, H. Dodiuk, and I. E. Klein, Appl. Surf. Sci. 20:538 (1985).
830. P. Poole and J. F. Watts, Int. J. Adhes. Adhes. 5:33 (1985).
831. D. J. Arrowsmith and A. W. Clifford, Int. J. Adhes. Adhes. 5:40 (1985).
832. K. W. Allen and M. G. Stevens, J. Adhes. 14:137 (1982).
833. J. C. McMillan, Bonded joints and preparation for bonding, NATO AGARD Lecture Series No. 102, Wright-Patterson AFB, OH (1979).
834. M. S. Hunter, P. F. Towner, and D. L. Robinson, Tech. Proc. Amer. Electroplating Soc. 46:220 (1959).

835. L. J. Matinezo, D. K. Shaffer, W. C. Moshier, and G. D. Davis, Environmental and adhesive durability of aluminum-polymer systems protected with organic corrosion inhibitors, J. Mater. Sci. 21(5):1601 (May 1986).
836. H. J. Lorenz and F. Mansfield, Interface and interphase corrosion inhibition, Electrochimica Acta 31(4): 467 (April 1986).
837. J. D. Venables, D. K. McNamara, J. M. Chen, T. S. Sun, and R. L. Hopping, Nat. SAMPE Tech. Conf. 10:362 (1978).
838. J. S. Noland, Adhesion Science and Technology (L. H. Lee, ed.). Plenum Press, New York, Vol. 9A, p. 413 (1975).
839. T. P. Remmel, Characterization of surfaces prior to adhesive bonding, Northrup Aviation Report AFML-TR-118 (1976).
840. W. Vedder and D. A. Vermilyea, Trans. Faraday Soc. 65:561 (1969).
841. N. T. McDevitt and J. S. Solomon, Thin anodic oxide films on aluminum alloys and their role in the durability of adhesive bonds, AFML Report No. TR-79-4216, Wright-Patterson AFB, OH (1979).
842. G. E. Thompson, R C. Furneaux, and G. S. Wood, Jr., J. Electrochem. Soc. 125(9):1480 (1978).
843. J. A. Marceau, Phosphoric acid anodize, Adhesive Bonding of Aluminum Alloys (E. W. Thrall and R. W. Shannon, eds.), Marcel Dekker, New York, Chap. 4, p. 51 (1985).
844. J. D. Venables, Adhesion and durability of metal/polymer bonds, Adhesive Joints: Formation, Characteristics, and Testing (K. L. Mittal, ed.), Plenum Press, New York, p. 453 (1984).
845. J. D. Minford, unpublished work, Aluminum Co. of America.
846. E. E. Peterson and D. B. Arnold, Relationship of interfacial compatibility to durability of adhesive-bonded joints, AFWAL Report TR-80-4176, Wright-Patterson AFB, OH (March 1981).
847. J. D. Minford, unpublished work, Aluminum Co. of America.
848. R. F. Wegman and R. Dickinson, Tech. Report No. 3300, Picatinny Arsenal, Dover, NJ (Feb. 1966).
849. J. D. Minford, unpublished work, Aluminum Co. of America.
850. T. S. Sun, J. M. Chen, J. D. Venables, and R. Hopping, Appl. Surf. Sci. 1:25 (1978).
851. J. L. Cotter, Developments in Adhesives—1 (W. C. Wake, ed.), Applied Science Publishers, London, p. 279 (1977).
852. W. T. McCarvill and J. P. Bell, J. Appl. Polym. Sci. 18:343 (1974).
853. W. Scardino and J. A. Marceau, J. Appl. Polym. Sci., Appl. Polym. Symp. 32:51 (1977).
854. J. C. Harris, Metal Cleaning Bibliographical Abstracts, ASTM Spec. Tech. Publ. No. 90 (1947); 90-A (1950); 90-D (1957).
855. S. Wernick and R. Pinner, The Surface Treatment and Finishing of Aluminum and Its Alloys, Robert Draper, England (1964).
856. D. J. George, C. J. Walton, and W. G. Zelley, Chemical pretreating and finishing, Aluminum (K. R. Van Horn, ed.), ASM, Metals Park, OH, Vol. 3, Chap. 17, p. 587 (1967).
857. N. J. Newhard, Jr., Conversion coatings for aluminum, Metal Finishing, p. 49 (July 1972).
858. S. Spring and K. Woods, Prepaint treatments-chromate coatings, Metal Finishing, p. 49 (June 1981).
859. J. D. Minford, Aluminum adhesive bond permanence, Treatise on Adhesion and Adhesives (R. L. Patrick, ed.), Marcel Dekker, New York, Vol. 5, Chap. 3, p. 55 (1981).
860. J. D. Minford, unpublished work, Aluminum Company of America.
861. G. E. Thompson, Morphology, composition, and structure of anodic films on aluminum, Adhesion '90, The Plastics and Rubber Institute Meetings, Cambridge, England (Sept. 10–12, 1990).
862. F. J. Boerio and S. G. Hong, Degradation of rubber-to-metal bonds during simulated cathodic delamination, J. Adhes. 30(1–4):119 (1989).
863. F. Watts, Mechanistic aspects of cathodic delamination of organic coatings, J. Adhes. 31(1):73 (1989).
864. J. A. Treverton and M. P. Thomas, Surface contaminants on annealed aluminum foil and their effects on adhesion, Int. J. Adhes. Adhes. 9(4):211 (1989).
865. B. R. Stromeier, Surface characterization of aluminum foil annealed in the presence of ammonium fluoroborate, Appl. Surf. Sci. 40:249 (1989).
866. J. A. Kovacich and D. Lichtman, A qualitative and quantitative study of the oxide on aluminum and silicon using AES and XPS, J. Electron Spectros. Relat. Phenom. 35(1–2):7 (Feb. 1985).
867. A. Kh. Bairemov, S. Zakipour, and C. Leygraf, An XPS investigation of dichromate and molybdate inhibitors on aluminum, Corrosion Sci. Technol. 25(10):69 (1985).
868. J. T. Grant, Methods for quantitative analysis in XPS and AES, Surf. Interf. Anal. 14(6–7):271 (1989).
869. B. R. Strohmeier, An ESCA method for determining the oxide thickness on aluminum alloys. Surf. Interf. Anal. 15(1):51 (1990).

871. J. L. Cotter and M. G. D. Hockey, Metal joining with adhesives, Rev. No. 183 of Int. Metall. Rev. 19: 103 (1974).
872. J. P. McNally and C. R. Ronan, Metal-to-metal adhesive bonding, Welding Res. Council Bull. No. 220 (Oct. 1976).
873. L. H. Sharpe, Aspects of Adhesion (D. J. Alner and K. W. Allen, eds.), Applied Science Publishers, London, England, p. 139 (1973).
874. K. L. Mittal, Polymer Eng. Sci. 17:467 (1977).
875. K. L. Mittal, Adhesion measurement of thin films, Thick Films and Bulk Coatings (K. L. Mittal, ed.), ASTM, Philadelphia, pp. 5–17 (1978).
876. K. L. Mittal, Pure and Appl. Chem. 52:1295 (1980).
877. K. L. Mittal, Adhesion Science and Technology (L. H. Lee, ed.). Plenum Press, New York, Vol. 9A, pp. 129–169 (1975).
878. B. W. Cherry, Polymer Surfaces, Cambridge University Press, London, England (1981).
879. A. J. Kinloch, J. Mater. Sci. 15:2141 (1980).
880. B. N. Chapman, J. Vac. Sci. Technol. 11:106 (1974).
881. C. Weaver, J. Vac. Sci. Technol. 12:18 (1975).
882. W. J. van Ooij, Surf. Sci. 89:165 (1979).
883. W. J. van Ooij, Interfacial interactions between polymers and other materials and their effects on bond durability, Physicochemical Aspects of Polymer Surfaces (K. L. Mittal, ed.), Plenum Press, New York, Vol. 2, pp. 1036–1091 (1983).
885. D. E. Packham, Adhesion Aspects of Polymeric Coatings (K. L. Mittal, ed.), Plenum Press, New York (1983).
886. W. V. Chang, J. Appl. Polym. Sci. 26:1759 (1981).
887. D. R. Fitchmun, S. Newman, and R. Wiggle, J. Appl. Polym. Sci. 14:2441,2457 (1970).
888. J. R. G. Evans and D. E. Packham, J. Adhes. 10:177 (1979).
889. J. R. G. Evans and D. E. Packham, J. Adhes. 9:267 (1978).
890. J. R. G. Evans and D. E. Packham, J. Adhes. 10:39 (1979).
891. F. J. Boerio, C. A. Gosselin, R. G. Dillingham, and H. W. Liu, J. Adhes. 13:159 (1981).
892. R. H. Hansen and H. Schonhorn, J. Polym. Sci. Part B, 4:203 (1966).
893. T. C. Ward, M. Sheridan, and D. L. Kotzev, Nondestructive evaluation of some bonded joints, Adhesive Joints: Formation, Characteristics, and Testing (K. L. Mittal, ed.), Plenum Press, New York, p. 347 (1984).
894. G. C. Knollmann, Int. J. Adhes. Adhes. 5:137 (1985).
895. T. K. Kwei, J. Polym. Sci. Part A, 3:3229 (1965).
896. P. S. Theocaris, J. Reinforced Plas. Compos. 3:204 (1984).
897. G. J. Howard and R. A. Shanks, J. Macromol. Sci-Phys. B19(2):167 (1981).
898. P. Peyser, Polym. Plastics Technol. Eng. 10:117 (1978).
899. P. S. Theocaris and G. D. Spatis, J. Appl. Polym. Sci. 27:3019 (1982).
900. C. U. Ko et al., Effect of surface topography on the relaxation behavior of thin polysulfone coatings on pretreated aluminum substrates, J. Adhes. 28:247 (1989).
901. R. D. Adams and W. C. Wake, Structural Adhesive Joints in Engineering, Elsevier Applied Science Publishers, London (1984).
902. H. M. Clearfield, D. K. McNamara, and G. D. Davis, Surface preparation of metals, Adhesives and Sealants, Engineered Materials Handbook, ASM, Materials Park, OH, Vol. 3, pp. 259–264 (1990).
903. P. F. A. Bijlmer, Adhesion—2 (K. W. Allen, ed.), Appl. Sci. Publ., London (1978).
904. D. J. Steel, Internal Military Vehicles and Eng. Establishment Report, Dorset, BH23 2BB, England.
905. W. A. Brockmann, The effect of surface pretreatment on the properties of metal/adhesive bonded joints, Metall. 31:245 (March 1977) (in German).
906. E. W. Thrall and R. W. Shannon, PABST Surface Treatment and Adhesive Selection, Proc. 21st Natl. SAMPE Symp. & Exhib., Azusa, CA, pp. 1004–1014 (April 6–8, 1976).
907. G. A. Carrillo, Alternative Cleaning Methods for Aluminum Honeycomb Core, Proc. 22nd Natl. SAMPE Symp. & Exhib., Azusa, CA, pp. 45–58 (April 26–28, 1977).
908. M. A. Danforth and R. J. Sunderland, Contamination of adhesive bonding surface treatment, Symp. on Durability of Adhesive Bonded Structures, Picatinny Arsenal, Dover, NJ, p. 303 (Oct. 27–29, 1976).
909. N. L. Rogers, Surface preparation of metals for adhesive bonding, Symp. Struct. Adhesive Bonding, Stevens Inst. Technol., Hoboken, NJ, p. 602 (Sept. 14–16, 1965).
910. T. T. Wang, F. W. Ryan, and H. Schonhorn, J. Appl. Polym. Sci. 16:1901 (1971).

911. R. C. Snogren, Surface treatment of joints for structural adhesive bonding, ASME Design Eng. Conf. (May 1960).
912. R. C. Snogren, Selecting surface preparation processes for adhesive bonding, ASME Design Eng. Conf. (April 1968).
913. S. Spring and L. F. Peale, On the mechanism of metal cleaning, Metal Prog. 51:102 (Jan. 1947).
914. T. C. Dumond, Metal cleaning, Materials and Methods 28(5):83 (1948).
915. D. W. Vance, Metal cleaning, Metal Finishing 49(1):55 (Jan. 1951).
916. M. A. Lesser, Metal cleaners, Soap and Sanitary Chemicals 28:42,46 (Oct., Nov., 1952).
917. P. D. Liddiard, Metal cleaning in aqueous media, Metal Industry 79:63,91,107,123,151 (July 27, Aug. 3, 10,17,24 [1951]).
918. O. M. Morgan and J. G. Lankler, Evaluation of metal cleaning compounds, I & E Chem. Anal. Ed., 14: 725 (1942).
919. J. C. Harris, R. E. Kamp, and W. H. Yanko, Application of radioactive tracer technique to metal cleaning, ASTM Bull., No. 158, p. 49 (1949).
920. J. C. Harris and R. E. Kamp, Development of metal cleaners using radioisotopic evaluation methods, Metal Finishing 48:75 (Nov. 1950).
921. J. C. Harris, R. E. Kamp, and W. H. Yanko, Improved radioactive carrier for metal cleaning studies, ASTM Bull., No. 170, p. 82 (Dec. 1950).
922. Metal Cleaning Bibliography (1842–1951 period), ASTM. Spec. Tech. Publ., Nos. 90-C, D. and E supplements for 1955, 1958, and 1960.
923. S. Spring, Metal Cleaning, Reinhold, New York (1963).
924. W. G. Zelley, Cleaning and etching of aluminum alloys, Tech. Service Report No. 155, Finishes Div., Alcoa Labs (Oct. 1965).
925. N. Chessin and V. Curran, Preparation of aluminum surfaces for bonding, J. Appl. Polym. Sci., Appl. Polym. Symp. 3:319 (1966).
926. G. L. Schneberger, Surface preparation for adhesive bonding, Presentation Amer. Soc. Mech. Eng. Design Show, Chicago (April 4, 1974).
927. L. F. Spencer, The cleaning of metals, Metal Finishing 60(4):59 (April 1962).
928. H. B. Linford, Preparation and evaluation of clean metal surfaces, Corrosion 20(7):213 (July 1964).
929. A. B. McKee, Cleaning aluminum, Modern Sanitation 7:15 (Jan. 1955).
930. W. K. Westray, Metal cleaning, Adhesives in Manufacturing (G. L. Schneberger, ed.), Marcel Dekker, New York, Chap. 17, p. 427 (1983).
931. R. C. Burrows and B. F. Crowe, Bonding contaminated surfaces, Adhes. Age 5(5):30 (May 1962).
932. C. E. Kircher, Solvent degreasing-what every user should know, ASTM Bull. No. 2129, p. 44 (Jan. 1957).
933. Handbook of Vapor Degreasing, ASTM Spec. Tech. Publ. No. 310 (1962).
934. J. D. Minford, unpublished work, Aluminum Co. of America.
935. J. D. Minford, Aluminum adhesive bond permanence, Treatise on Adhesion and Adhesives (R. L. Patrick, ed.), Marcel Dekker, New York, Vol. 5, Chap. 3, p. 82 (1981).
936. A. Pollack and P. Westphal, An Introduction to Metal Degreasing and Cleaning, Robert Draper, Teddington, England (1963).
937. Electrolytic degreasing of aluminum, Electroplating p. 337 (May 1949).
938. S. Spring, Ultrasonic cleaning and pickling, Metal Finishing 62:66 (March 1964).
939. H. B. Linford and E. B. Saubestre, A new degreasing evaluation test-the Atomizer Test, Metal Finishing 51:74 (Sept. 1953).
940. A. A. Eyles, Emulsion cleaning, Electroplating and Metal Finishing, p. 197 (June 1965).
941. C. S. Lowe, Precleaning with solvent emulsions, Metal. Ind. 64(13):301 (March 31, 1944).
942. G. B. Vos, Detergent cleaning best for thin film removal, tests show, S.A.E. J. 71(12):91 (Dec. 1963).
943. S. Spring, Metal preparation-how soil affects cleaner selection, Products Finishing, p. 34 (June 1963).
944. M. A. Lesser, Aluminum cleaners, Soap and Sanitary Chemical 22:44 (Sept. 1946).
945. G. A. Bass, Cleaning aluminum sheet prior to spot welding I and II, Metal Progress, p. 499 (April 1949); p. 668 (May 1949).
946. J. B. Mohler, Alkaline solutions-what to use for effective cleaning, Iron Age 176:59 (July 28, 1955).
947. S. Heath and G. Trolley, Dissolution of aluminum in alkaline solutions, Chem. and Ind. (Brit.), p. 367 (March 23, 1957).
948. W. J. Muller and E. Low, The dissolution of aluminum of different purities and of aluminum alloys in sodium hydroxide, Aluminium 20:257 (1938).
949. M. A. Streicher, The dissolution of aluminum in sodium hydroxide solutions—I, Trans. Electrochem. Soc. 93:285 (1948).

950. F. J. Prescott, J. K. Shaw, and J. Lilker, Sequestering agents aluminum etching, Metal Finishing 51:65 (Oct. 1953).
951. J. D. Minford, unpublished work, Aluminum Co. of America.
952. D. G. Foulke, How to clean and etch aluminum, Modern Metals 13:44 (May 1957).
953. J. Mazia, Coatings corner-chemical removal of scale and other oxides—III, Metal Finishing, p. 69 (Jan. 1980).
954. B. R. Barber, Alkaline cleaning, Boeing Process Spec. BAC-5749 Rev. E (Nov. 25, 1968).
955. J. B. Kushner, Rinsing I and II, Plating 36(8):798; 36(9):915 (1949).
956. J. B. Kushner, Rinsing technique, Metal Finishing 52:76 (Jan. 1955).
957. D. J. Trevoy and H. Johnson, Jr., The water-wettability of metal surfaces, J. Phys. Chem. 62:833 (July 1958).
958. B. D. Bhatt and V. Radhakrishnan, Some investigations on machining, etching, and sand blasting treatments of adherends with surface profile records and electron microscopy in adhesive bonding of aluminum, J. Adhes. Sci. Technol. 3:383–386 (1989).
959. V. Ramakrishnan and A. K. De, Some static characteristics of bonded metallic joints, Proc. 8th AIMTDR Conf., Bombay, India, p. 1 (1978).
960. A. C. Moloney, Surface Analysis and Pretreatment of Plastics and Metals (D. M. Bremis, ed.), Applied Science Publishers, London, p. 175 (1982).
961. N. L. Rogers, Surface preparation of metals for adhesive bonding, Appl. Polym. Symp. No. 3, Interscience, New York, p. 327 (1966).
962. A. V. Pocius and J. Claus, Replacement of sulfuric-chromic acid (FPL) etch by 3 dimensional abrasive surface conditioning for pretreatment of aerospace alloys before phosphoric acid anodization, 13th Nat. SAMPE Tech. Conf., Mt. Pocono, PA (Oct. 13–15, 1981).
963. Std. Rec. Prac. for Preparation of Metal Surfaces for Adhesive Bonding, ASTM D-2651-79 (1982).
964. H. W. Eickner, Weathering of adhesive bonded lap joints of clad aluminum alloys, Forest Prod. Lab. WADC Tech. Report 54-447, Part I (1955).
965. H. W. Eickner, Weathering of adhesive bonded lap joints of clad aluminum alloys, Forest Prod. Lab. WADC Tech. Report 54-447, Part II (1955).
966. H. W. Eickner, Weathering of adhesive bonded lap joints of clad aluminum alloys, Forest Prod. Lab. WADC Tech. Report 54-447, Part III (1958).
967. J. D. Minford, unpublished work, Aluminum Co. of America.
968. J. D. Minford, unpublished work, Aluminum Co. of America.
969. J. D. Minford and L. K. Peters, unpublished work, Aluminum Co. of America.
970. T. Smith, Mechanisms of bond endurance after surface treatment of aluminum for bonding (STAB) in nonchromate, alkaline solution, J. Adhes. 14:144 (1982).
971. T. Smith, J. Adhes. 9:313 (1977).
972. T. Smith, Surface treatment for aluminum bonding, Report to Dept. of the Army, U.S. Army Armament R&D Command, Dover, NJ (Contract No. DAAK10-78-C-0274 (Oct. 1979).
973. T. Smith, Proc. 26th Nat. SAMPE Symp., p. 664 (April 28–30, 1981).
974. H. W. Eickner and W. E. Schowalter, A study of methods for preparing clad 24S-T3 aluminum alloy surfaces for bonding, Forest Prod. Lab. Report No. 1813A (1950).
975. H. W. Eickner and W. E. Schowalter, A study of methods for preparing clad 24S-T3 aluminum alloy surfaces for bonding, Forest Prod. Lab. Report No. 1813-Information reviewed and reaffirmed (March 1956).
976. J. D. Minford, Evaluating adhesives for joining aluminum, Metals Eng. Quarterly (Nov. 1972).
977. J. D. Minford, Durability of adhesive bonded aluminum joint, Treatise on Adhesion and Adhesives (R. L. Patrick, ed.), Marcel Dekker, New York, Vol. 3, Chap. 2, p. 108 (1973).
978. J. D. Minford, Durability or permanence of adhesive joints, High-Performance Adhesive Bonding (G. DeFrayne, ed.), Soc. Mfg. Eng., Dearborn, MI, p. 73 (1983).
979. J. D. Minford, Adhesives, Durability of Structural Adhesives (A. J. Kinloch, ed.), Applied Science Publishers, London, Chap. 4, p. 168 (1983).
980. G. J. Pietsch, M. Henzler, and P. O. Hahn, Continuous roughness characterization from atomic to micron distances: Angle-resolved electron and photon scattering, Appl. Surf. Sci. 39:457 (1989).
981. M. A. Miller, Resin bonding improved by surface treatment of aluminum alloys, Adhes. Age 3(3):28 (1960).
982. M. A. McGowan, F. M. Graber, and W. M. Sutherland, Effect of adsorbed chromium on bonding surfaces upon joint strength, General Dynamics/Convair Report No. 8925-040, San Diego (Nov. 8, 1960).
983. M. A. McGowan, F. M. Graber, and W. M. Sutherland, Treatment of metal surfaces for adhesion, General Dynamics/Convair Report No. MP-59-309, San Diego (Nov. 8, 1960).

984. M. S. McGowan, N. Nalley, and W. M. Sutherland, Treatment of metal surfaces for adhesive bonding, General Dynamics/Convair Report No. MP-59-309, Add. I, San Diego (May 8, 1961).
985. T. S. Sun, J. M. Chen, J. D. Venables, and R. L. Hopping, Effects of chemical and thermal treatments on the composition of 2024 aluminum adherend surfaces, Appl. Surf. Sci. 1(2):202 (Jan. 1978).
986. J. M. Chen, T. S. Sun, and J. D. Venables, Natl. SAMPE Symp. & Exhib. 22:25 (1978).
987. J. M. Chen, T. S. Sun, and J. D. Venables, 24th Natl. SAMPE Symp. & Exhib. (May 1979).
988. L. C. Weltman, Investigation of the process of aluminum cleaning for adhesive bonding, General Dynamics/Convair Report No. FGT-1903, San Diego (March 28, 1958).
989. H. Severus-Laubenfeld, Surface treatment of aluminum in connection with the adhesion of thermoplastic and thermosetting resins, 6th Int. Light Metals, Conf., Leoben, Austria (June 1975).
990. N. Evans, Pretreatment for bonding, Eng. Mater. Design 13(2):190 (Feb. 1970).
991. R. F. Wegman, Effect of surface preparation of aluminum in bonds obtained with an epoxidized-novolac adhesive, Tech. Notes FRL-TN-69, Picatinny Arsenal, Dover, NJ (Sept. 1961).
992. M. C. Ross, R. F. Wegman, M. J. Bodnar, and W. C. Tanner, Effect of surface exposure time on bonding 2024-T3 aluminum alloys, SAMPE J. 10(1):4 (Feb. 1974).
993. M. C. Ross, R. F. Wegman, M. J. Bodnar, W. C. Tanner, SAMPE J. 10(2):4 (March/April 1974).
994. M.C. Ross, R. F. Wegman, M. J. Bodnar, and W. C. Tanner, Effect of surface exposure time on bonding 5052-H34 aluminum alloy, SAMPE J. 10:5 (May/June/July 1974).
995. R. F. Wegman and R. H. Dickerson, Effects of time elapsed between surface preparation and bonding on adhesive joint strength, Picatinny Arsenal Report No. PA-TR-3300, Dover, NJ (Feb. 1966).
996. H. R. Herrigel, Spray deoxidizing for aluminum bonding, SAMPE Eng. Ser. 15:755, SAMPE Mat. Proc. for the 1970's, Los Angeles (1969).
997. Cleaning and preparation of metal surfaces, DEF-STAN-03-2/1 Method O, chrome/sulfuric pickle aluminum alloys: HMSO Anodizing of aluminum alloys, DEF 151, HMSO.
998. J. Michaelson, Cleaning and deoxidizing aluminum alloys, Boeing Process Spec. BAC-5765 Rev. G (March 29, 1968).
999. H. M. Clearfield, D. K. McNamara, and G. D. Davis, Adhesives and Sealants, Engineered Materials Handbook, ASM, Materials Park, OH, Vol. 3, p. 260 (1990).
1000. A. W. Bethune, SAMPE J. 11:4 (1975).
1001. C. J. Almer, unpublished data, Adhesives and Sealers Div. 3M's Research Labs., St. Paul, MN.
1002. J. D. Minford, Aluminum adhesive bond permanence, Treatise on Adhesion and Adhesives (R. L. Patrick, ed.), Marcel Dekker, New York, Vol. 5, Chap. 3, p. 68 (1981).
1003. J. D. Minford, Permanence of adhesive bonded aluminum joints, Adhesives in Manufacturing (G. L. Schneberger, ed.), Marcel Dekker, New York, Chap. 23, p. 567 (1983).
1004. A. W. Smith, J. Electrochem Soc. 120:1551 (1973).
1005. W. J. Russell and E. A. Garnis, SAMPE Quart. p. 5 (April 1976).
1006. A. V. Pocius, Electrode kinetics of aluminum alloys in the sulfuric-chromic acid etch bath-relationship to surface properties and durability, 2nd Int. SAMPE Meeting, Skesa, Italy (June 1982).
1007. P. F. A. Bijlmer, Metal Finishing p. 47 (May 1975).
1008. P. F. A. Bijlmer, Metal Finishing p. 34 (Dec. 1971).
1009. N. L. Rogers, Proc. 13th Nat. SAMPE Tech. Conf., p. 640 (1981).
1010. A. Desai, J. S. Ahearn, D. K. McNamara, Cleanliness of external tank surfaces, Tech. Report MML-TR-85-65, Martin Marietta Labs, Baltimore (1985).
1011. T. J. Reinhart, Jr., Durable surface treatments for the adhesive bonding of aluminum alloys, Adhesion—2 (K. W. Allen, ed.), Applied Science Publishers, London, p. 87 (1978).
1012. J. L. Cotter, Developments in Adhesives—1 (W. C. Wake, ed.), Applied Science Publishers, London, p. 1 (1977).
1013. R. F. Wegman, W. N. Bodnar, M. J. Bodnar, and M. J. Barbarisi, Effects of deionized water immersion of prepared aluminum surface on adhesive bondability, Tech. Report No. 3495, Picatinny Arsenal, Dover, NJ (May 1967).
1014. R. F. Wegman, W. M. Bodnar, M. J. Bodnar, and M. J. Barbarisi, SAMPE J., p. 35 (Oct./Nov. 1967).
1015. J. A. Marceau and Y. Moji, Development of environmentally stable aluminum adhesive bonds, Document D6-41145, Boeing Co., Seattle, WA.
1016. Common bonding requirements for structural adhesives, Process Document BAC 5514, Boeing Commercial Airplane Co., Seattle, WA.
1017. J. A. Marceau and Y. Moji, Application of fracture mechanics testing to process control for adhesive bonding, Document D6-41141, Boeing Co., Seattle.
1018. W. L. Baun, N. T. McDevitt, and J. S. Solomon, ASTM Spec. Tech. Publ. No. 596, p. 86 (1976).
1019. N. T. McDevitt and W. L. Baun, J. Electrochem. Soc., p. 1058 (July 1976).

1020. N. T. McDevitt et al., Surface analysis of 6061 and 7050 aluminum alloys, AFML Report TR-76-13, Air Force Materials Lab., Wright-Patterson AFB, Ohio (March 1976).
1021. J. S. Solomon and W. L. Baun, Surface characterization of contamination on adhesive bonding materials, Surface Contamination: Genesis, Detection, and Control Proc., Washington, D.C. Conf., Plenum Press, New York, Vol. 2, p. 609 (1979).
1022. J. D. Minford, unpublished work, Aluminum Co. of America.
1023. J. D. Minford, unpublished work, Aluminum Co. of America.
1024. J. D. Minford, unpublished work, Aluminum Co. of America.
1025. J. D. Minford, unpublished work, Aluminum Co. of America.
1026. J. D. Minford, unpublished work, Aluminum Co. of America.
1027. J. D. Minford, unpublished work, Aluminum Co. of America.
1028. J. D. Minford, unpublished work, Aluminum Co. of America.
1029. J. D. Minford, unpublished work, Aluminum Co. of America.
1030. J. D. Minford, unpublished work, Aluminum Co. of America.
1031. R. F. Reeves and N. J. Newhard Jr., Modern Metals 24:45 (1968).
1032. R. F. Reeves, Surface preparation of aluminum painting, Aluminum Finishing Seminar, Detroit (1968).
1033. R. F. Reeves and N. J. Newhard Jr., Why you should consider conversion coating of aluminum, Metal Progress, p. 86 (1972).
1034. W. Suetake, Nippon Kinzoku Gakkaishi 32:1233 (1968).
1035. L. A. Nimon and G. K. Korpi, Chemical composition of amorphous phosphate conversion coatings on aluminum as determined by infrared spectroscopy, Plating, p. 421 (May 1972).
1036. New process for surface treatment of aluminum alloys, Japan Ind. Tech. Bull. 6(11):14 (Feb. 1979).
1037. L. K. Peters, unpublished work, Aluminum Co. of America.
1038. J. D. Minford, More economical examination of bonding durability outlined, Adhes. Age 25(4):34 (April 1982).
1039. J. D. Minford, unpublished work, Aluminum Co. of America.
1040. P. Oliva, Brush alodizing of aluminum surfaces, Boeing Process Specification MS 8.15, Rev. A (May 8, 1962).
1041. J. Michaelson, Alodizing, Boeing Process Spec. BAC-5719 Rev. E (Jan. 23, 1969).
1042. J. A. Scarlise, Energy conservation in conversion coating, S.M.E. Tech. Paper No. FC-77-613, Assoc. for Finishing Processes of Society Mfg. Engs., Dearborn, MI (1977).
1043. J. N. Tuttle, Jr., A new chemical finishing process for aluminum, Plating and Surface Finishing, p. 28 (June 1979).
1044. R. B. Mason and P. F. Fowle, J. Electrochem Soc. 101:57 (1954).
1045. Aluminum (K. R. Van Horn, ed.), ASM, Metals Park, OH, 4 vols. (1967).
1046. W. Brockmann and O. D. Hennemann, Proc. 11th Nat. SAMPE Tech. Conf., pp. 804–816 (1979).
1047. R. Exalto, Investigations in the presence of sulphur in anodic films, Fokker Report R-1759 (1974).
1048. R. Exalto, Determination of optimum chromic acid anodizing conditions for adhesive bonding, Fokker Report R-1981.
1049. W. J. Russell and C. A. L. Westerdahl, Characteristics of chromic acid anodized 2024-T3 aluminum adherends, Army Armament Res. & Devel. Command, Large Caliber Weapons Systems Lab., Dover, NJ.
1050. A. Herfert, Fundamental investigations of anodic oxide films on aluminum alloys as a surface preparation for adhesive bonding, Final Report April 1, 1975–June 30, 1976, Wright-Patterson AFB, Northrup Corp. Hawthorne, CA.
1051. N. T. McDevitt and W. L. Baun, Some observations of the relation between chemical surface treatments and the growth of anodic barrier layer films, Final AFML report June–Dec. 1975, Wright-Patterson AFB, OH (1975).
1052. N. L. Rogers, Bondable coating on aluminum and method of applying it, U.S. Patent No. 3,414,489.
1053. J. Michaelson, Chromic acid anodizing, Boeing Process Spec. BAC-5019 Rev. D (Oct. 14, 1965).
1054. R. J. Schliekelmann, Adhesive bonding in the Fokker-VFW F028 Fellowship N73-25061 AIR 39685 (Distribution by Nat. Tech. Info Service), U.S. Dept. of Commerce (1973).
1055. A. T. Elrod, Sealing, Boeing Process Spec. BAC 5000 Rev. C (July 31, 1967).
1056. MIL-A-8625, Anodic Treatment of Aluminum Alloys.
1057. Chromic Acid Anodizing, Fokker Spec. TH-6.7851. Fokker V.F.W.B.V (Aug. 1978).
1058. A. Hartman, The chromic acid and sulfuric acid anodizing, Netherlands Aerospace Lab (NRL) Report M-2166 (in Dutch) (1966).
1059. S. Wernick and R. Pinner, The Surface Treatment of Aluminum and Its Alloys, 4th ed., Robert Draper, England, p. 420 (1972).

1060. F. Keller, Anodic coatings seen through the microscope, ASTM Proc. 40:3 (1940).
1061. G. E. Best, J. G. Hecker, Jr., J. W. McGrew, and R. V. VandenBerg, Chromic acid anodizing characteristics of wrought aluminum alloys, ASTM Proc. 59:277 (1959).
1062. J. W. Diggle, T. C. Dourue, and C. W. Goulding, Anodic film oxide films on aluminum, Chem. Revs. 69:365 (1969).
1063. P. W. Diggle, Oxides and Oxide Films, Marcel Dekker, New York, Vol. 2, Chap. 3 (1973).
1064. A. Kwakernaak, The influence of rinsing and drying procedures on the surface quality of chromic/sulfuric pickled Alclad 2024-T3 material, Royal Netherlands Aircraft Factories, Fokker Report, Schiphol-Oost, Holland.
1065. P. F. A. Bijlmer, Surface treatment of aluminum for adhesive bonding, Proc. Oesterrieichischen Giesserie-Institut und Lehrkanzel fur Giesserei-kunde der Mont Hochschule, p. 320 (April 1974).
1066. Adhesive Bonding of Aluminum Alloys (E. W. Thrall and R. W. Shannon, eds.), Marcel Dekker, New York (1985).
1067. Chromic Acid Anodize, Spec. BPS FW4352 Rev. G. Bell Helicopter Co.
1068. Chromic Acid Anodize, Spec. BPS 13201, McDonnell-Douglas Corp.
1069. J. A. Marceau, Phosphoric acid anodize, Adhesive Bonding of Aluminum Alloys (E. W. Thrall and R. W. Shannon, eds.), Marcel Dekker, New York, Chap. 4, p. 51 (1985).
1070. R. R. Popley, C. A. Terry, and P. Walker, Factors affecting the adhesion of surface coatings to anodized aluminum surfaces, AWRE Report No. 0-22/67 of Atomic Weapons Res. Estab., United Kingdom Atomic Energy Authority (June 1967).
1071. G. E. Thompson, R. C. Furneaux, and G. C. Wood, The morphology of sealed anodic films formed on aluminum in phosphoric acid, Trans. Inst. Metal Finishing 53:97 (Summer 1975).
1072. A. W. Smith, Surface oxide on etched aluminum, J. Electrochem. Soc. Solid State Sci. Technol. (Nov. 1973).
1073. B. W. Malpass, D. E. Packham, and K. Bright, A study of the adhesion of polyethylene to porous alumina films using the scanning electron microscope, J. Appl. Polym. Sci. 18:3249 (1974).
1074. A. W. Bethune, Durability of bonded aluminum structure, 19th Nat. SAMPE Symp., Los Angeles (April 1974).
1075. J. A. Marceau, R. H. Firainhoc, and Y. Moji, Method for providing environmentally stable aluminum surfaces for adhesive bonding and product produced, U.S. Patent No. 4,127,451 (Nov. 28, 1978).
1076. E. F. Sutliff, BAC 5555 solution control by free phosphoric acid test, Boeing Document QCR&D Report No. 622-812 (March 20, 1974).
1077. R. W. Shannon et al., Primary adhesively bonded structure technology (PABST), U.S. Air Flight Dynamics Lab. Tech. Report No. AFFDL-TR-77-107 (Sept. 1978).
1078. J. A. Marceau and J. C. McMillan, Exploratory developments on durability of adhesive bonded joints, Air Force Materials Lab. Tech. Report TR-76-173 (Oct. 1976).
1079. M. C. Locke, W. Scardino, and H. Croop, Non-tank phosphoric acid anodize method of surface preparation of aluminum for repair bonding, SAMPE Conf., Azusa, CA, p. 488 (1975).
1080. J. C. McMillan, J. T. Quinlivan, and R. A. Davis, Phosphoric acid anodizing of aluminum for structural adhesive bonding, SAMPE Quart., p. 13 (April 1976).
1081. R. E. Horton, W. M. Scardino, and H. Croop, Adhesive Bonded Aerospace Structures Standardized Repair Handbook, U.S. Air Force Dynamics Lab., Contract AF 33615-73-C-5151, Wright-Patterson AFB, OH (1973).
1082. J. E. McCarty, R. F. Horton, M. C. Locke, M. L. Satterthwait, and B. D. Parashar, Adhesive Bonded Aerospace Structures Standardized Repair Handbook, Final Report, Oct. 1, 1975–March 31, 1976, Boeing Commercial Airplane Co., Seattle.
1083. R. F. Horton and J. E. McCarty, Adhesive Bonded Aerospace Structures Standardized Repair Handbook, Final Report, Oct. 1, 1973–Sept. 30, 1977, Boeing Commercial Airplane Co., Seattle.
1084. T. Smith, NDE method for characterizing anodized aluminum surfaces, AFML Tech. Report TR-78-146, Air Force Materials Lab., Wright-Patterson AFB, OH (Jan. 1979).
1085. O. D. Hennemann, Adhesion between metals and polymers as a three-dimensional system, J. Adhes. 22(2):87 (1987).
1086. J. D. Minford, unpublished work, Aluminum Co. of America.
1087. W. D. Bascom, Primers and coupling agents, Adhesives and Sealants, Engineered Materials Handbook, ASM, Materials Park, OH, Vol. 3, pp. 254–258 (1990).
1088. A. S. Falcone and J. E. Miller, J. Appl. Polym. Sci., Appl. Polym. Symp. 32:247 (1977).
1089. A. Herczeg, G. S. Ronay, and W. C. Simpson, Surface chemical characteristics of epoxy resins, Nat. SAMPE Tech. Conf. on Aerospace Adhesives and Elastomers, p. 221 (Oct. 6–8, 1970).

1090. J. A. Bishop, E. K. Sim, G. E. Thompson, and G. C. Wood, The adhesively bonded aluminum joint: The effect of pretreatment on durability, J. Adhes. 26:237 (1988).
1091. R. F. Foister, R. K. Gray, and P. A. Madsen, Structural adhesive bonds to primers electrodeposited on steel, J. Adhes. 24:17 (1989).
1092. R. F. Wegman, Surface Preparation Technology for Adhesive Bonding, Noyes Publications, p. 68 (1989).
1093. D. N. Shah, Rubber-modified epoxy adhesive compositions, U.S. Patent No. 4,803,232 (Feb. 1989).
1094. J. A. Marceau, An SEM analysis of adhesive primer oriented bond failures on anodized aluminum, SAMPE Quart., 9:4 (March/April 1977).
1095. N. L. Rogers, Sealed chromic acid anodize, Adhesive Bonding of Aluminum Alloys (E. W. Thrall and R. W. Shannon, eds.), Marcel Dekker, New York, Chap. 3, p. 43 (1985).
1096. R. B. Krieger, Jr., Advances in the corrosion resistance of bonded structures, SAMPE J. (Feb./March 1969).
1097. W. J. Lorenz and F. Mansfield, Interface and interphase corrosion inhibition, Electrochimica Acta 31(4):467 (April 1986).
1098. W. D. Sell, Some analytical techniques for durability testing of structural adhesives, Proceedings of 19th Nat. SAMPE Symposium (April 1974).
1099. A. J. Kinloch, J. Adhes. 10:193 (1979).
1100. G. T. Beckwith and T. Pollard, Proc. 17th SAMPE Int. Tech. Conf., p. 400 (1985).
1101. R. H. Greer, WPAFP Contract F33615-86-R-5009, Rohr Ind., P. Tydings Program Officer.
1102. A. V. Pocius and T. H. Wilson, Jr., Proc. 19th SAMPE Int. Tech. Conf. (Oct. 1987).
1103. R. A. Pike and F. P. Lamm, Preprints Amer. Chem. Soc. Div. Polymeric Mater. Sci. Eng. 56:209 (1987).
1104. R. A. Pike and F. P. Lamm, Polymer Science and Technology (L. H. Lee, ed.), Plenum Press, New York, 37:141 (1988).
1105. R. A. Pike and G. S. Golden, A water soluble hydrated metal oxide primer for adhesively bonded joints, J. Adhes. 29:111 (1989).
1106. D. M. Brewis, J. Comyn, and J. L. Tegg, Int. J. Adhes. Adhes. 1:35 (1980).
1107. J. Comyn, Kinetics and mechanisms of environmental attack, Durability of Structural Adhesives (A. J. Kinloch, ed.), Applied Science Publishers, London, Chap. 3, p. 85 (1983).
1108. R. G. Schmidt and J. P. Bell, Adv. Poly. Sci. (K. Dusek, ed.). Springer-Verlag, Berlin 75:33 (1986).
1109. E. P. Plueddemann, Adhesion through silane coupling agents, 25th SPI Reinforced Plastics/Composites Div. Conf., Washington, D.C. (Feb. 1970).
1110. E. P. Plueddemann, Silane Coupling Agents, Plenum Press, New York (1982).
1111. W. D. Bascom, Structure of silane adhesion promoter films on glass and metal surfaces, Macromolecules 5:792 (1972).
1112. W. D. Bascom, Primers and coupling agents, Engineering Materials Handbook, ASM, Materials Park, OH, Vol. 3, pp. 255–258 (1990).
1113. J. M. Park and J. P. Bell, Adhesion Aspects of Polymeric Coatings (K. L. Mittal, ed.), Plenum Press, New York, p. 205 (1982).
1114. A. J. DeNicola, Jr., and J. P. Bell, Adhesion Aspects of Polymeric Coatings (K. L. Mittal, ed.), Plenum Press, New York, p. 443 (1982).
1115. T. T. Kam and R. K. Hon, J. Coat. Technol. 55:39 (1983).
1116. S. J. Monte and G. Sugarman, Additives for Plastics (R. B. Seymour, ed.), Academic Press, New York, Vol. 1, p. 168 (1978).
1117. F. J. Boerio, Coupling agents as adhesion promoters in adhesive bonding, Treatise on Adhesion and Adhesives (R. L. Patrick, K. L. DeVries, and C. P. Anderson, eds.), Marcel Dekker, New York, Vol. 6, Chap. 7 (1989).
1118. R. G. Schmidt, J. P. Bell, and A. Garton, Chemical interactions between mercaptoester coupling agents and steel, J. Adhes. 27:127 (1989).
1119. R. G. Schmidt and J. P. Bell, Investigations of steel/epoxy adhesion durability using polymeric coupling agents III. Influence of coupling agent thickness, J. Adhes. 27:135 (1989).
1120. N. H. Sung, A. Kaul, I. Chin, and C. S. P. Sung, Polym. Eng. Sci. 22:637 (1982).
1121. E. P. Plueddemann, H. A. Clark, L. E. Nelson, and K. R. Hoffman, Modern Plastics 39:139 (Aug. 1962).
1122. E. P. Plueddemann, Modern Plastics 40:133 (June 1963).
1123. E. P. Plueddemann, 24th Ann. Tech. Conf., SPI, 19-A (1969).
1124. P. W. Erickson and E. P. Plueddemann, Interfaces in Polymer Matrix Composites (E. P. Plueddemann, ed.), Academic Press, New York (1974).

1125. H. Ishida, A review of recent progress in the studies of molecular and microstructure of coupling agents and their functions in composites, coatings, and adhesive joints, Polymer Composites 5(2):101 (1984).
1126. P. T. K. Shih and J. L. Koenig, Mater. Sci. Eng. 20:137 (1975).
1127. H. A. Clark and E. P. Plueddemann, Mod. Plastics 40:133 (1963).
1128. H. Ishida and J. L. Koenig, Appl. Spectosc. 32:462,469 (1978).
1129. H. Ishida, C. Chiang, and J. L. Koenig, Polymer 23:251 (1982).
1130. R. I. Kaas and J. L. Kardos, Polym. Eng. Sci. 11:11 (1971).
1131. G. D. Nichols, D. M. Hercules, R. C. Peek, and D. J. Vaughan, Appl. Spectrosc. 28:219 (1974).
1132. H. Ishida and J. L. Koenig, Polym. Eng. Sci. 18:128 (1978).
1133. M. Gettings and A. J. Kinloch, J. Mater. Sci. 12:2511 (1977).
1134. S. R. Culler, H. Ishida, and J. L. Koenig, J. Colloid Interf. Sci. 106(20):334 (1985).
1135. F. J. Boerio and J. W. Williams, Appl. Surf. Sci. 7:19 (1981).
1136. R. A. Gledhill and A. J. Kinloch, Weathering of plastics and rubbers, Proc. Plastics and Rubber Inst. Meetings, London, p. D12.1 (1976).
1137. M. E. Schrader and J. A. Cardamone, J. Adhes. 9:305 (1978).
1138. F. J. Boerio, R. G. Dillingham, and A. R. Snow, Org. Coat. Plast. Chem. Prepr. 50:444 (1984).
1139. A. Kaul, N. H. Sung, J. Chin, and C. S. P. Sung, Polym. Prep. 26(2):113 (1985).
1140. F. J. Boerio and D. J. Ondrus, Surface and Colloid Science in Computer Technology (K. L. Mittal, ed.), Plenum Press, New York, p. 155 (1987).
1141. N. H. Sung, S. Ni, and C. S. P. Sung, Org. Coat. Plast. Chem., p. 743 (1980).
1142. N. H. Sung and C. S. P. Sung, Proc. 35th Ann. Conf. Reinforced Plast./Compos., SPI, Section 23B (1980).
1143. R. G. Dillingham, M. S. Thesis, University of Cincinnati, Cincinnati, OH (1983).
1144. D. J. Ondrus and F. J. Boerio, J. Adhes. 22:1 (1987).
1145. F. J. Boerio and J. E. Greivenkamp, Proc. 32nd Ann. Tech. Conf. Reinforced Plastics/Compos Div. SPI, Sect. 4-A (1977).
1146. S. A. Francis and A. H. Ellison, J. Opt. Soc. Amer., 49:131 (1959).
1147. R. G. Greenler, J. Chem. Phys. 44:310 (1966).
1148. A. F. Diaz, U. Hetzler, and E. Kay, J. Am. Chem. Soc. 99:6780 (1977).
1149. H. Lee, J. Colloid Interf. Sci. 27:751 (1968).
1150. E. P. Plueddemann, J. Adhes. 2:184 (1970).
1151. H. F. Weetall and L. S. Hersh, Biochim. Biophys. Acta 206:54.
1152. P. R. Moses, L. M. Wier, J. C. Lennox, H. O. Finklea, J. R. Lenhard, and R. W. Murray, Anal. Chem. 50:576 (1978).
1153. C. H. Chiang, H. Ishida, and J. L. Koenig, J. Colloid Interf. Sci. 74:396 (1980).
1154. F. J. Boerio, S. Y. Cheng, L. Armogan, J. W. Williams, and C. Gosselin, Proc. 35th Ann. Conf. Reinforced Plastics/Composites Div. SPI, Sect. 23C (1980).
1155. W. Theidman, F. C. Tolan, P. J. Pearce, and C. E. M. Morris, J. Adhes. 22:197 (1987).
1156. J. D. Minford, unpublished work, Aluminum Co. of America.
1157. J. D. Minford, unpublished work, Aluminum Co. of America.
1158. J. D. Minford, unpublished work, Aluminum Co. of America.
1159. B. Arkles, Tailoring surfaces with silanes, CHEMTECH, p. 767 (Dec. 1977).
1160. A. J. Kinloch, W. A. Dukes, and R. A. Gledhill, Adhesion Science and Technology (L. H. Lee, ed.), Plenum Press, New York, p. 597 (June 1977).
1161. K. E. Martin and R. Rolles, Painting and other organic coating, Aluminum (K. R. Van Horn, ed.), ASM, Metals Park, OH, Chap. 21, pp. 709–742 (1967).
1162. J. E. Stillwagon and L. M. Butcher, Electrostatic treatment of aluminum foil, Modern Converter 15(7): 39 (July 1971).
1163. R. M. Sonkin, Film treatment by corona discharge-facts and myths, Publication produced by ENI Power Systems, Inc.
1164. R. H. Cram and D. V. Bibee, Proc. Paper Synth. Conf. of TAPPI, pp. 1–11 (1981).
1165. K. Rossman, J. Polym. Sci. 19:141 (1956).
1166. G. D. Cooper and M. Prober, J. Polym. Sci. 44:397 (1960).
1167. R. F. Grossman and W. A. Beasley, J. Appl. Polym. Sci. 2:163 (1959).
1168. M. M. Kadash and C. G. Seefried, Jr., Plastics Eng. 41(2):45 (1965).
1169. S. S. Voyutskii, Legkaya Prom. 1:42 (1953).
1170. V. I. Alekseenko, I. Y. Mishustin, and S. S. Voyutskii, S. S. Doki Akad. Nauk SSSR 95:93 (1954).
1171. V. I. Alekssenko, I. U. Mishustin, and S. S. Voyutskii, Kolloid Zh. 18:4 (1955).
1172. B. V. Derjaguin, S. K. Zheberkov, and A. M. Medvedeva, Kolloid Zh. 18:4 (1956).

1173. A. Ya. Korolev, Chemical modification of the surface of solids as a method of regulating their adhesive properties, Klei i Tekhnol. Skleivaniya, Sbornik Statei., pp. 35–52 (1960) (Translated from the Russian).
1174. H. P. Schreiber, C. Richard, and M. R. Wertheimer, Acid-base considerations of surface interactions in polymer systems: Control by microwave plasma treatment, Physicochemical Aspects of Polymer Surfaces (K. L. Mittal, ed.), Plenum Press, New York, Vol. 2, p. 739 (1983).
1175. R. K. Samal, H. Iwata, and Y. Ikada, Introduction of reactive groups onto polymer surfaces, Physicochemical Aspects of Polymer Surfaces (K. L. Mittal, ed.), Plenum Press, New York, Vol. 2, p. 801 (1983).
1176. T. S. Keller, A. S. Hoffman, B. D. Ratner, and B. J. McElroy, Chemical modification of Kelvar (R) surfaces for improved adhesion to epoxy resin matrices: I. Surface characterization, Physicochemical Aspects of Polymer Surfaces (K. L. Mittal, ed.), Plenum Press, New York, Vol. 2, p. 861 (1983).
1177. H. Yasuda, Adhesion of plasma polymerized films (a model study on water sensitivity of adhesion), Adhesion Aspects of Polymeric Coatings (K. L. Mittal, ed.), Plenum Press, New York, p. 193 (1981).
1178. A. Dilks and A. VanLaeken, An ESCA investigation of the plasma oxidation of poly(p-xylylene) and its chlorinated derivatives, Physicochemical Aspects of Polymer Surfaces (K. L. Mittal, ed.), Plenum Press, New York, Vol. 2, p. 749 (1983).
1179. D. Oldfield and T. E. F. Symes, Surface modification of elastomers for bonding, J. Adhes. 16:77 (1983).
1180. T. E. F. Symes and D. Oldfield, Adhesive bonding of elastomers, Treatise on Adhesion and Adhesives (J. D. Minford, ed.), Marcel Dekker, New York, Vol. 7, Chap. 2, pp. 231–331 (1991).
1181. Z. W. Kowalski, Ion bombardment modification of the surface morphology of solids, J. Mater. Sci. 25(9):3875 (1990).
1182. D. T. Clark and A. Dilks, J. Polym. Sci., Polym. Chem. Ed. 17:957 (1979).
1183. D. T. Clark and A. Dilks, J. Polym. Sci., Polym. Chem. Ed. 15:2321 (1977).
1184. A. R. Blythe, D. Briggs, C. R. Kendall, D. G. Rance, and V. J. I. Zichy, Polymer 19:1273 (1978).
1185. M. Hudis, Techniques and Applications of Plasma Chemistry (J. R. Hollahan and A. T. Bell, eds.), John Wiley & Sons, New York (1974).
1186. T. E. Nowlin and D. Foss-Smith, J. Appl. Polym. Sci. 25:1619 (1980).
1187. H. Hiraoka, J. Electrochem. Soc. 128:1065 (1981).
1188. G. N. Taylor, T. M. Wolf, and M. R. Goldrick, J. Electrochem. Soc. 128:361 (1981).
1189. D. T. Clark and A. Dilks, J. Polym. Sci. Polym. Chem. Ed. 16:911 (1978).
1190. A. Dilks and A. VanLaeken, An ESCA investigation of the plasma oxidation of poly(p-xylylene) and its chlorinated derivatives, Physicochemical Aspects of Polymer Surfaces (K. L. Mittal, ed.), Plenum Press, New York, Vol. 2, p. 749 (1983).
1191. A. T. Bell, Techniques and Applications of Plasma Chemistry (J. R. Hollahan and A. T. Bell, eds.), John Wiley & Sons, New York (1974).
1192. L. H. Sharpe, H. Schonhorn, and C. J. Lynch, Int. Sci. Technol., p. 26 (April 1964).
1193. A. Devine, W. Bodnar, E. Duda, and M. Bodnar, Effects of surface treatments on bonding to polyethylene with various type adhesives, Picatinny Arsenal Report PA-TM-1739, Dover, NJ (Oct. 1967).
1194. J. J. Bikerman, Effect of impurities on polyethylene adhesion, Appl. Chem. 11:81 (1961).
1195. H. Schonhorn and R. H. Hansen, Surface treatment of polymers for adhesive bonding, J. Appl. Polym. Sci. 11:1461 (1967).
1196. R. H. Hansen, J. V. Pascale, T. DeBenedictus, and P. M. Rentzepis, J. Polym. Sci. A3:2205 (1965).
1197. H. Schonhorn and F. W. Ryan, J. Polym. Sci. 6(A-2):231 (1968).
1198. H. Schonhorn and F. W. Ryan, Adv. Chem. Ser. 87:140 (1968).
1199. H. Schonhorn, F. W. Ryan, and R. H. Hansen, Surface treatment of polypropylene for adhesive bonding, J. Adhes. 2:93 (1970).
1200. J. A. Black, D. J. Lyman, and D. B. Parkinson, RTV silicone adhesive and potting compounds, SRI Project PRD 5046, Spec. Report No. 2, JPL Contract No. NAS 7-100 (July 1965).
1201. N. J. DeLollis and O. Montoya, Outdoor aging of adhesive sealants, Proc. Nat. SAMPE Tech. Conf., Vol. 2 (Oct. 1970).
1202. H. V. Boenig, Plasma Science and Technology, Cornell University Press, Ithaca, NY (1982).
1203. Y. Hsieh and E. Y. Chen, Ind. Eng. Chem. Prod. Res. Dev., 24:246 (1985).
1204. H. Yasuda, H. K. Sharma, and T. Yasuda, J. Polym. Sci. Phys. Ed. 19:1285 (1981).
1205. Polymer Surface Dynamics (J. D. Andrade, ed.), Plenum Press, New York (1988).
1206. M. Morra, E. Occhiello, L. Gila, and F. Garbassi, Surface dynamics vs. adhesion in oxygen plasma treated polyolefins, 13th Ann. Meeting Adhes. Soc., Savannah, GA (Feb. 18–21, 1990).
1207. V. D. McGinniss, Encyclopedia of Polymer Science and Technology, John Wiley & Sons, New York, Vol. 4, p. 418 (1986).
1208. C. Y. Kim and D. A. I. Goring, J. Appl. Polym. Sci. 15:1357 (1971).
1209. C. Y. Kim, J. Evans, and D. A. I. Goring, J. Appl. Polym. Sci. 15:1365 (1971).

1210. D. K. Owens, J. Appl. Polym. Sci. 19:265 (1975).
1211. D. Briggs and C. R. Kendall, Polymer 20:1053 (1979).
1212. D. Briggs, Adhesion—6 (K. W. Allen, ed.), Applied Science Publishers, London, pp. 111–121 (1982).
1213. A. Chew, D. M. Brewis, D. Briggs, and R. H. Dahm, Adhesion—8 (K. W. Allen, ed.), Applied Science Publishers, London, pp. 97–114 (1984).
1214. E. J. Goldman, R. A. Rosenberg, and W. E. Lee, How to use fluorocarbon plastics as a bonding agent, Adhes. Age 10:30 (Feb. 1967).
1215. N. J. DeLollis and O. Montoya, Bondability of RTV silicone rubber, J. Adhes. 3:1 (1971).
1216. Treatise on Adhesion and Adhesives (J. D. Minford, ed.), Marcel Dekker, New York, Vol. 7 (1991).
1217. B. H. River, R. H. Gillespie, and C. B. Vick, Wood as an adherend, Treatise on Adhesion and Adhesives (J. D. Minford, ed.), Marcel Dekker, New York, Vol. 7, Chap. 1, pp. 1–230 (1991).
1218. A. V. Pocius, R. D. Wald, and S. R. Hartshorn, The use of adhesives in the joining of plastics, Treatise on Adhesion and Adhesives (J. D. Minford, ed.), Marcel Dekker, New York, Vol. 7, Chap. 3, pp. 335–435 (1991).
1219. D. K. Shaffer, H. M. Clearfield, and J. S. Ahearn, Titanium as an adherend, Treatise on Adhesion and Adhesives (J. D. Minford, ed.), Marcel Dekker, New York, Vol. 7, Chap. 4, pp. 437–495 (1991).
1220. B. C. Nayer and A. W. Adamson, Adsorption and contact angle studies IV. Alcohols and water on polypropylene and polycarbonate, Physicochemical Aspects of Polymer Surfaces (K. L. Mittal, ed.), Plenum Press, New York, Vol. 2, p. 613 (1983).
1221. F. J. Holly and M. J. Owen, Adsorption on modified silicone surfaces, Physicochemical Aspects of Polymer Surfaces (K. L. Mittal, ed.), Plenum Press, New York, Vol. 2, p. 625 (1983).
1222. H. J. Jacobasch, The physicochemical surface properties of fiber forming polymers, Physicochemical Aspects of Polymer Surfaces (K. L. Mittal, ed.), Plenum Press, New York, Vol. 2, p. 637 (1983).
1223. J. Anhang and D. G. Grey, Surface characterization of polymers by inverse gas chromatography, Physicochemical Aspects of Polymer Surfaces (K. L. Mittal, ed.), Plenum Press, New York, Vol. 2, p. 659 (1983).
1224. M. Salkauskas and M. G. Klimantaviciute, Autohesion of etched polyethylene films, Physicochemical Aspects of Polymer Surfaces (K. L. Mittal, ed.), Plenum Press, New York, Vol. 2, p. 689 (1983).
1225. M. Millard, J. Burns, and H. Sachdev, Mild direct fluorination of polymers studied by XPS, Physicochemical Aspects of Polymer Surfaces (K. L. Mittal, ed.), Plenum Press, New York, Vol. 2, p. 773 (1983).
1226. T. Ohmichi, H. Tamaki, H. Kawasaki, and S. Tatsuta, Chemical characterization of surface-activated polymer films using ESCA technique, Physicochemical Aspects of Polymer Surfaces (K. L. Mittal, ed.), Plenum Press, New York, Vol. 2, p. 793 (1983).
1227. D. T. Clark, The Modification, degradation, and synthesis of polymer surfaces studied by means of ESCA, Physicochemical Aspects of Polymer Surfaces (K. L. Mittal, ed.), Plenum Press, New York, Vol. 1, p. 3 (1983).
1228. J. A. Kelber, R. R. Rye, D. R. Jennison, and J. E. Houston, Localized Auger states in polyethylene, Physicochemical Aspects of Polymer Surfaces (K. L. Mittal, ed.), Plenum Press, New York, Vol. 1, p. 83 (1983).
1229. M. A. Fortes, Wettability of polymer surfaces, Physicochemical Aspects of Polymer Surfaces (K. L. Mittal, ed.), Plenum Press, New York, Vol. 1, p. 107 (1983).
1230. F. J. Holly, Novel methods of studying polymer surfaces employing contact angle goniometry, Physicochemical Aspects of Polymer Surfaces (K. L. Mittal, ed.), Plenum Press, New York, Vol. 1, p. 141 (1983).
1231. I. C. Sanchez and C. I. Poser, Surface thermodynamics of liquid polymers: Theory, Physicochemical Aspects of Polymer Surfaces (K. L. Mittal, ed.), Plenum Press, New York, Vol. 1, p. 173 (1983).
1232. A. M. Wrobel, Surface free energy of plasma-deposited thin polymer films, Physicochemical Aspects of Polymer Surfaces (K. L. Mittal, ed.), Plenum Press, New York, Vol. 1, p. 197 (1983).
1233. W. J. van Ooij, Interfacial interactions between polymers and other materials and their effects on bond durability, Physicochemical Aspects of Polymer Surfaces (K. L. Mittal, ed.), Plenum Press, New York, Vol. 2, p. 1035 (1983).
1234. J. M. Burkstrand, Chemical interactions at polymer-metal interfaces: Studies with X-ray photoemission, Physicochemical Aspects of Polymer Surfaces (K. L. Mittal, ed.), Plenum Press, New York, Vol. 2, p. 1093 (1983).
1235. R. O. Claus and R. T. Rogers, Analysis of metal-polymer boundaries using ultrasonic interface waves, Physicochemical Aspects of Polymer Surfaces (K. L. Mittal, ed.), Plenum Press, New York, Vol. 2, p. 1101 (1983).
1236. A. J. Quant, Adhesion of encapsulants, Sandia Labs Report, Albuquerque, NM.
1237. K. E. Creed, Alumina ceramic surface cleaning for optimum bonding of resin encapsulant, General Electric Report GEPP-310A (July 1977).

1238. C. Kemball, Adhesion and Adhesives (H. T. Clark, B. Savage, and J. E. Rutzler, Jr., eds.), John Wiley & Sons, New York, p. 70 (1954).
1239. D. Taylor, Jr., and J. E. Rutzler, Jr., Ind. Eng. Chem. 50:928 (1958).
1240. F. M. Fowkes, Advan. Chem. Series No. 43, ACS, p. 99 (1964).
1241. W. A. Zisman, Handbook of Adhesives (I. Skeist, ed.), Van Nostrand Reinhold, New York (1977).
1242. L. H. Lee, Surface free energies and fracture surface energies of glassy polymers, Physicochemical Aspects of Polymer Surfaces (K. L. Mittal, ed.), Plenum Press, New York, Vol. 1, p. 523 (1983).
1243. U. T. Kreibich and H. Batzer, Angew. Makromol. Chem. 83:57 (1979).
1244. H. Batzer, F. Lohse, and R. Schmid, Angew. Makromol. Chem. 29/30:349 (1973).
1245. Y. H. Lin, Entanglement and molecular weight dependence of polymer glass transition temperature, Macromolecules 23(25):5292 (1990).
1246. M. Moreno-Villa Lobos, P. Czarnocki, and K. Piekarski, Properties of adhesives and their applications, J. Adhes. 19(2):79 (1986).
1247. R. S. Wallach, Bonding metals with adhesives, Mater. Eng. 103(4):33 (April 1986).
1248. T. Daniels, Thermal Analysis, John Wiley & Sons, New York (1973).
1249. C. W. Macosko, Adhes. Age 20:35 (1977).
1250. E. Forster, Polymere Werkstoffe (H. Batzer, ed.), Georg Thieme Verlag, Stuttgart and New York, Vol. 3, Technologie 2 (1984).
1251. U. T. Kreibich, E. Lohse, R. Schmid, and G. Wegner, Einfluss struktureller merkmale, Polymere Werkstoff (H. Batzer, ed.), Georg Thieme Verlag, Stuttgart and New York, Vol. 1, Chemie und Physik (1985).
1252. A. J. Kinloch, Epoxy Resins and Composities I (K. Dusek, ed.). Springer-Verlag, Berlin, Heidelberg, New York, Tokyo (1985).
1253. A. F. Yee and R. A. Pearson, J. Mater. Sci. 21:2462 (1986).
1254. A. F. Yee and R. A. Pearson, J. Mater. Sci. 21:2472 (1986).
1255. R. Janardhan, P. H. Gedam, P. S. Sampathkumaran, The effect of polymer molecular weight on the absorption process, J. Colloid Interf. Sci. 140(2):391 (1990).
1256. P. A. Claret, Aspects of Adhesion—3 (D. J. Alner, ed.), University of London Press, London, p. 9 (1965).
1257. E. P. McGuire, American Adhesive Index, Padric Publishing, Mountainside, NJ (1962).
1258. Adhesive Redbook, Palmerton Printing, Atlanta, GA (1982).
1259. Adhesives, 1978/1979 Book A, Cordura Publications, San Diego.
1260. Adhesives, 1978/1979 Book B, Cordura Publications, San Diego.
1261. E. P. McGuire, Adhesives Raw Materials Handbook, Padric Publishing, Mountainside, NJ (1964).
1262. C. V. Cagle, Adhesive Bonding Techniques and Applications, McGraw-Hill, New York (1973).
1263. C. V. Cagle, Handbook of Adhesive Bonding, McGraw-Hill, New York (1973).
1264. W. H. Guttman, Concise Guide to Structural Adhesives, Reinhold, New York (1961).
1265. I. Katz, Adhesive Materials, revised by C. V. Cagle, Foster Publishing, Long Beach, CA (1971).
1266. M. M. Gauthier, Types of adhesives, Adhesives and Sealants, Engineered Materials Handbook, ASM, Materials Park, OH, Vol. 3, pp. 74–93 (1990).
1267. D. T. Behm and J. Gannon, Epoxies, Adhesives and Sealants, Engineered Materials Handbook, ASM, Materials Park, OH, Vol. 3, pp. 94–102 (1990).
1268. E. F. Bushman, Phenolics, Adhesives and Sealants, Engineered Materials Handbook, ASM, Materials Park, OH, Vol. 3, pp. 103–107 (1990).
1269. C. L. Kreider, Urethanes, Adhesives and Sealants, Engineered Materials Handbook, ASM, Materials Park, OH, Vol. 3, pp. 108–112 (1990).
1270. D. P. Melody, Anaerobics, Adhesives and Sealants, Engineered Materials Handbook, ASM, Materials Park, OH, Vol. 3, pp. 113–118 (1990).
1271. D. J. Damico, Acrylics, Adhesives and Sealants, Engineered Materials Handbook, ASM, Materials Park, OH, Vol. 3, pp. 119–125 (1990).
1272. J. E. Schoenberg, Cyanoacrylates, Adhesives and Sealants, Engineered Materials Handbook, ASM, Materials Park, OH, Vol. 3, pp. 126–132 (1990).
1273. L. A. Sobrieski, Silicones, Adhesives and Sealants, Engineered Materials Handbook, ASM, Materials Park, OH, Vol. 3, pp. 133–137 (1990).
1274. E. A. Peterson, Polysulfides, Adhesives and Sealants, Engineered Materials Handbook, ASM, Materials Park, OH, Vol. 3, pp. 138–142 (1990).
1275. W. F. Harrington, Elastomeric adhesives, Adhesives and Sealants, Engineered Materials Handbook, ASM, Materials Park, OH, Vol. 3, pp. 143–150 (1990).
1276. R. D. Rossi, Polyimides, Adhesives and Sealants, Engineered Materials Handbook, ASM, Materials Park, OH, Vol. 3, pp. 151–162 (1990).

1277. S. E. Wentworth, Polyphenylquinoxalines, Adhesives and Sealants, Engineered Materials Handbook, ASM, Materials Park, OH, Vol. 3, pp. 163–168 (1990).
1278. J. W. Connell, Polybenzimidazoles, Adhesives and Sealants, Engineered Materials Handbook, ASM, Materials Park, OH, Vol. 3, pp. 169–174 (1990).
1279. W. O. Buckley and K. J. Schroeder, Adhesive modifiers, Adhesives and Sealants, Engineered Materials Handbook, ASM, Materials Park, OH, Vol. 3, pp. 175–186 (1990).
1280. L. D. Carbary, Types of sealants, Adhesives and Sealants, Engineered Materials Handbook, ASM, Materials Park, OH, Vol. 3, pp. 188–192 (1990).
1281. R. A. Giangiordano and C. Sorensson, Polysulfide (sealants), Adhesives and Sealants, Engineered Materials Handbook, ASM, Materials Park, OH, Vol. 3, pp. 193–197 (1990).
1282. J. J. Higgins and F. C. Jagisch, Butyls (sealants), Adhesives and Sealants, Engineered Materials Handbook, ASM, Materials Park, OH, Vol. 3, pp. 198–202 (1990).
1283. J. F. Regan, Urethanes (sealants), Adhesives and Sealants, Engineered Materials Handbook, ASM, Materials Park, OH, Vol. 3, pp. 203–207 (1990).
1284. J. Baghdachi, Solvent acrylics (sealants), Adhesives and Sealants, Engineered Materials Handbook, ASM, Materials Park, OH, Vol. 3, pp. 208–209 (1990).
1285. J. Johnson, Water-base acrylics and polyvinyl acetate (sealants), Adhesives and Sealants, Engineered Materials Handbook, ASM, Materials Park, OH, Vol. 3, pp. 210–214 (1990).
1286. M. D. Beers and J. M. Klosowski, Silicone (sealants), Adhesives and Sealants, Engineered Materials Handbook, ASM, Materials Park, OH, Vol. 4, pp. 215–222 (1990).
1287. A. F. Waterland III, Fluorocarbons (sealants), Adhesives and Sealants, Engineered Materials Handbook, ASM, Materials Park, OH, Vol. 3, pp. 223–227 (1990).
1288. J. Badhdachi, Polyether silicones (sealants), Adhesives and Sealants, Engineered Materials Handbook, ASM, Materials Park, OH, Vol. 3, pp. 228–233 (1990).
1289. S. Claesson, Discussions Faraday Soc. 7:321 (1949).
1290. E. Jenckel and B. Rumbach, Z. Electrochem. 55:612 (1951).
1291. H. L. Frisch and R. Simha, J. Chem. Phys. 27:702 (1957).
1292. L. E. Nielsen, The Mechanical Properties of Polymers, Reinhold, New York, p. 122 (1962).
1293. F. Bueche, Rubber Chem. Technol. 32:1269 (1959).
1294. I. Wolock, J. A. Kies, and S. B. Newman, Fracture (B. L. Averbach et al., eds.), John Wiley & Sons, New York, p. 250 (1959).
1295. F. Schwartzl and A. S. Staverman, Die Physik der Hochpolymeren (H. A. Stuart, ed.), Springer, Berlin, Vol. IV, pp. 164–213, 226–2235 (1956).
1296. E. Orowan, Rept. Progr. Phys. 12:185 (1949).
1297. R. R. Stromberg, Treatise on Adhesion and Adhesives (R. L. Patrick, ed.), Marcel Dekker, New York, Vol. 1, Chap. 3 (1967).
1298. J. C. Bolger, Structural adhesives for metal bonding, Treatise on Adhesion and Adhesives (R. L. Patrick, ed.), Marcel Dekker, New York, Vol. 3, Chap. 1, pp. 4–9 (1973).
1299. H. L. Frisch and R. Simha, J. Phys. Chem. 61:702 (1957).
1300. H. L. Frisch and R. Simha, J. Phys. Chem. 58:507 (1954).
1301. H. L. Frisch and R. Simha, J. Phys. Chem. 61:705 (1957).
1302. R. Simha, H. L. Frisch, and F. R. Eirich, J. Phys. Chem. 57:584 (1953).
1303. H. L. Frisch, R. Simha, and F. R. Eirich, J. Chem. Phys. 21:365 (1953).
1304. H. L. Frisch, J. Phys. Chem. 59:633 (1955).
1305. E. R. Gilliland and E. B. Guttoff, J. Phys. Chem. 64:407 (1960).
1306. A. J. Silberberg, J. Phys. Chem. 66:1872 (1962).
1307. A. J. Silberberg, J. Phys. Chem. 66:1884 (1962).
1308. D. H. Kaelble, Rheology of polymers used as adhesives, Treatise on Adhesion and Adhesives (R. L. Patrick, ed.), Marcel Dekker, New York, Vol. 1, Chap. 6, p. 170 (1967).
1309. D. W. Dwight, E. Sancaktor, and H. F. Brinson, Adhesion and Adsorption of Polymers (L. H. Lee, ed.), Plenum Press, New York, Part A, p. 141 (1980).
1310. G. Dolev and O. Ishai, J. Adhes. 12:283 (1981).
1311. I. Ishida, Quantitative surface FT-IR spectroscopic analysis of polymers, Rubber Chem. Technol., 80(3):497 (July–August 1987).
1312. G. U. Rakhmatkariev, T. K. Rakhmatulaeva, Sh. M. Zakhidova, and S. S. Khamaraev, IR-spectroscopic investigations of adsorption of polyacrylamide on oxides, Colloid J. U.S.S.R. 51(1):158 (1989).
1313. H. Brockmann, Chemical aspects of adhesion between metals and polymers, J. Adhes. 22(2):71 (1987).
1314. V. Kovacevic, I. Smit, and L. Kljajic-Malinovic, Properties of adhesive polymer compositions in ageing conditions, Die Angewandte Makromoleculaire Chemie 176/177:125 (1990).

1315. M. Shmorhun, A. M. Jamieson, and R. Simha, Free volume changes in epoxy adhesives during physical aging: Fluorescence spectroscopy and mechanical stress relaxation, Polymer 31(5):812 (1990).
1316. A. Zosel, Adhesive failure and deformation behavior of polymers, J. Adhes. 30(1–4):135 (1989).
1317. J. M. Tang, I. Lee, and G. S. Springer, Effects of cure pressure on resin flow, voids, and mechanical properties, J. Comp. Mater. 21:421 (1987).
1318. E. Sancaktar and H. Jozavi, Linking cure processes to adhesive bulk strength by differential thermal analysis, J. Adhes. 25:185 (1988).
1319. B. C. Ennis, Comment on "Linking Cure Processes to Adhesive Bulk Strength by DTA," J. Adhes. 30(1–4):57 (1989).
1320. E. Sancaktar and H. Jozavi, Reply to Comment by B. C. Ennis, J. Adhes. 30(1–4) (1988).
1321. H. Jozavi and E. Sancaktar, The effects of cure conditions on the relaxation behavior of thermosetting adhesives, J. Adhes. 29(1–4):233 (1989).
1322. J. Mijovic and C. H. Lee, A comparison of chemorheological models for thermoset cure, J. Appl. Polym. Sci. 38(12):2155 (1989).
1323. K. F. Pang and J. K. Giliham, Competition between cure and thermal degradation in a high glass transition temperature epoxy system—effect of time and temperature of isothermal cure on the glass transition temperature, J. Appl. Polym. Sci. 39(4):893 (1990).
1324. G. M. Bertenev, V. P. Puschenko, N. I. Shut, and M. V. Lazorenko, Determination of volume fraction of polymer bond on filler on the basis of thermophysical data, Colloid J. U.S.S.R. 48(6):932 (Nov.–Dec. 1984).
1325. S. Ponce, R. Bezinet, and H. P. Schreiber, Componet interaction and properties of pigmented epoxy systems, J. Adhes. 22(4):313 (1987).
1326. R. Kulik and R. Davies, Mechanical properties of adhesive filled with metal powders, Int. J. Adhes. Adhes. 9(4):224 (1989).
1327. R. Kulik, R. Davies, and S. M. H. Darwish, Thermal conductivity of adhesive filled with metal powders, Int. J. Adhes. Adhes. 9(4):219 (1989).
1328. W. J. Cantwell, J. W. Smith, H. H. Kausch, and T. Kaiser, Examination of the process of deformation and fracture in a silica-filled epoxy resin, J. Mater. Sci. 25(1B):663 (1990).
1329. J. N. Prassoamakis, Correlation of mechanical and acoustical properties of plasticized epoxy polymers, J. Appl. Polym. Sci. 39(10):2031 (1990).
1330. R. W. Hylands and E. H. Sidwell, J. Adhes. 11:203 (1980).
1331. L. R. F. Rose, J. Adhes. 14:93 (1982).
1332. W. D. Bascom, R. L. Cottington, and C. O. Timmons, Naval Eng. J., p. 73 (Aug. 1978).
1333. J. T. Bitner, J. L. Rushford, W. S. Rose, D. L. Hunston, and C. K. Riew, J. Adhes. 13:3 (1981).
1334. D. L. Hunston, S. S. Wang, and A. J. Kinloch, Organic Coatings Appl. Polym. Sci. Proc., 47:408 (1982).
1335. G. J. Howard and R. D. Shanks, J. Macromol. Sci.-Phys. B-19(2):167 (1981).
1336. P. Peyser, Polym. Plast. Technol. Eng. 10(2):117 (1978).
1337. Bonding Alcoa Aluminum, Aluminum Co. of America (1967).
1338. D. Hace, V. Kovacevic, D. Manojlovic, and I. Smit, The investigation of structural and morphological changes after the chlorination of rubber surfaces, Die Angewandte Makromoledulaire Chemie 177/176: 161 (1990).
1339. W. A. Pletcher and E. J. Yaroch, Water-based adhesives, Adhesives in Manufacturing (G. L. Schneberger, ed.), Marcel Dekker, New York, Chap. 16, p. 407 (1983).
1340. G. L. Schneberger (ed.), Adhesives in Manufacturing, Marcel Dekker, New York (1983).
1341. J. D. Minford, unpublished work, Aluminum Co. of America.
1342. W. B. Reynolds to W. Phillips, U.S. Patent No. 2,774,703 (1957).
1343. A. K. Doolittle and G. M. Powell, Paint, Oil, Chem. Rev. 107(7):9–11, 40–42 (1944).
1344. H. P. Brown and J. F. Anderson, Handbook of Adhesives (I. Skeist, ed.), Reinhold, New York, Chap. 19 (1962).
1345. W. H. Smarnook and S. Bonotto, Adhesion of carboxyl-containing olefins, Polym. Sci. Eng. 8(1):41 (1968).
1346. J. D. Minford, Adhesive bonded aluminum joints, Treatise on Adhesion and Adhesives (R. L. Patrick, ed.), Marcel Dekker, New York, Vol. 3, Chap. 2, pp. 83, 101–103 (1973).
1347. J. D. Minford, Adhesives, Durability of Structural Adhesives (A. J. Kinloch, ed.), Applied Science Publishers, London, Chap. 4, pp. 178–181 (1983).
1348. J. D. Minford, unpublished work, Aluminum Co. of America.
1349. J. D. Minford, unpublished work, Aluminum Co. of America.
1350. J. D. Minford, unpublished work, Aluminum Co. of America.
1351. J. V. Lindyberg, RTV silicone adhesive sealants, Adhesives in Manufacturing (G. L. Schneberger, ed.), Marcel Dekker, New York, Chap. 15, p. 387 (1983).

1352. J. M. Klosowski and G. A. Grant, The chemistry of silicone room temperature vulcanizing sealants, ACS Symp. Series No. 112, ACS, Washington, D.C. (1979).
1353. RTV silicone rubber, Technical Data Book S-35, General Electric Co., Schenectady, NY.
1354. J. D. Minford, unpublished work, Aluminum Co. of America.
1355. J. D. Minford, unpublished work, Aluminum Co. of America.
1356. J. D. Minford, Durability of aluminum bonded joints in long-term tropical exposure, Int. J. Adhes. Adhes. p. 25 (Jan. 1982).
1357. J. D. Minford, unpublished work, Aluminum Co. of America.
1358. M. D. Beers, Silicone adhesive sealants, Handbook of Adhesives (I. Skeist, ed.), Reinhold, New York (1977).
1359. R. D. Dexheimer and L. R. Vertnik, Hot melt adhesives, Adhesives in Manufacturing (G. L. Schneberger, ed.), Marcel Dekker, New York, Chap. 13, p. 325 (1983).
1360. E. F. Eastman, Ethylene copolymers-improved performance for an old molecule, Adhes. Age 26(10): 32 (1983).
1361. J. A. Filbey, L. Lavielle, and J. Schultz, Interface and bulk properties of EVA-aluminum systems, 13th Ann. Meeting Adhesion Soc., Savannah, GA (Feb. 18–21, 1990).
1362. D. Bruno (ed.), Adhesives in Modern Manufacturing. Soc. Mfg. Engs., Dearborn, MI, p. 27 (1970).
1363. R. D. Dexheimer and L. R. Vertnik, The why, how, what, when, and where of hot melts, Adhesives Ind. Conf. Proc., Tech. Conf. Assoc., El Segundo, CA, p. 63 (1974).
1364. R. D. Dexheimer and L. R. Vertnik, Handbook of Adhesives. 2nd ed. (I. Skeist, ed.), Reinhold, New York, p. 581 (1977).
1365. R. Hinterwaldner, Adhes. Age 23:18 (1980).
1366. I. Kay, Adhes. Age 21:27 (1978).
1367. W. H. Cobbs, Jr., Adhes. Age 22(12):31 (1979).
1368. D. J. St. Clair, Adhes. Age 23:30 (1980).
1369. S. R. Eddy, Adhes. Age 23:18 (1980).
1370. B. Stoops and P. Ferrier, Merging two technologies—robotics and hot-melt adhesives, Adhes. Age 26(4): 22 (1983).
1371. W. H. Cobbs, Jr., Foamability of hot-melt adhesives, Adhes. Age 22(12):31 (1979).
1372. D. Crosby, Foamed hot-melt adhesives and applicating equipment, Adhesives Ind. Conf. Proc., Tech. Conf. Assoc., El Segundo, CA, p. 131 (June 1980).
1373. F. T. Hughes, Foamed hot-melt adhesives, Adhes. Age 25(9):25 (1982).
1374. A. J. Kettleborough, Polymers, Paint Colour J. 11:62 (1978).
1375. F. Eichhorn and Th. Reiner, Elektro-Anzeiger 34(7):33 (1981).
1376. F. Annighofer, Reactive hot-melts on the basis of polyester and polyamide, J. Adhes. 22(2):109 (1987).
1377. F. Schaaf, Textilveredelung 9(1):13 (1974).
1378. R. G. Arnold, J. A. Nelson, and J. J. Verbanc, Chem. Rev. 57:47 (1957).
1379. F. Annighofer, Proc. 2nd Adhesives, Sealants, and Encapsulants Conf. (ASE), London (1986).
1380. T. E. Bugel, S. Norwalk, and R. H. Snedeker, Phenoxy resin—a new thermoplastic adhesive, 66th Annual ASTM Meeting, Atlantic City, NJ (June 23–28, 1963).
1381. J. D. Minford, unpublished work, Aluminum Co. of America.
1382. D. C. Wielinski, EnBA resins give hot melts more heat and oil resistance, Adhes. Age 32(12):30 (1989).
1383. H. P. Schreiber, Adhesion and the processing of polyolefin metls, Adhesion '90, The Plastics and Rubber Institute Meetings, Cambridge, England (Sept. 10–12, 1990).
1384. S. H. Dillman and A. Sanders, Maleated SEBS copolymers enhance polar compatibility, Adhes. Age 32(12):24 (1989).
1385. Frost and Sullivan Inc., Pressure-sensitive products and adhesives market, Report No. 614 (1978).
1386. A. D. Hamer, J. W. Hagan, and S. G. Krumenaker, Adhes. Age 23:23 (1980).
1387. J. A. Fries, Emulsion PSA's: Commercial application, Proc. Adhesive Coating Tech., PSTC Technical Seminar, Rosemont, IL (June 18–19, 1980).
1388. A. Maletsky and J. Villa, Hot-melt pressure-sensitive adhesives, The 1970's- a look back, the 1980's—a look ahead, Adhesive Coating Tech., PSTC Technical Seminar, Rosemont, IL (June 18–19, 1980).
1389. J. P. Carolan, Water-borne pressure-sensitive adhesives—a market in transition, Proc. Adhesive Coating Tech., PSTC Technical Seminar, Rosemont, IL (June 18–19, 1980).
1390. J. W. Hagan and K. C. Steuben, Pressure-sensitive adhesives, Adhesives in Manufacturing (G. L. Schneberger, ed.), Marcel Dekker, New York, Chap. 14, p. 353 (1983).
1391. M. E. Hodgson, Pressure-sensitive adhesives and their applications, Adhesion—3 (K. W. Allen, ed.), Applied Science Publishers, London, p. 207 (1979).
1392. J. D. Minford, unpublished work, Aluminum Co. of America.

1393. J. D. Minford, unpublished work, Aluminum Co. of America.
1394. J. D. Minford, unpublished work, Aluminum Co. of America.
1395. F. H. Hammond, Jr., Handbook of Pressure-Sensitive Adhesive Technology (D. Satas, ed.), Van Nostrand Reinhold, New York (1982).
1396. A. N. Magunov, Wetting of slightly nonuniform solid surfaces, Colloid J. U.S.S.R. 51(1):138 (1989).
1397. R. Bates, J. Appl. Polym. Sci. 20:2941 (1976).
1398. D. Satas (ed.), Handbook of Pressure-Sensitive Adhesive Technology, Van Nostrand Reinhold, New York (1982).
1399. C. A. Dahlquist, Handbook of Pressure-Sensitive Adhesive Technology (D. Satas, ed.), Van Nostrand Reinhold, New York (1982).
1400. J. A. Schlademan, Handbook of Pressure-Sensitive Adhesive Technology (D. Satas, ed.), Van Nostrand Reinhold, New York (1982).
1401. C. A. Dahlquist, Pressure-sensitive adhesives, Treatise on Adhesion and Adhesives (R. L. Patrick, ed.), Marcel Dekker, New York, Vol. 2, pp. 219–260 (1969).
1402. D. W. Aubrey, Developments in Adhesives—1, p. 127 (1977).
1403. J. W. Hagan, C. B. Mallon, and M. R. Rifi, Adhes. Age 22:29 (1979).
1404. A. Zosel, Physical properties and adhesion performance of PSA's, Adhes. Age 32(11):42 (1989).
1405. Y. Urahama, Effect of peel load on stringiness phenomena and peel speed of pressure-sensitive adhesive tape, J. Adhes. 31(1):47 (1989).
1406. T. Tsukatani, Y. Hatano, and H. Mizumachi, Bonding and debonding processes in tack of pressure-sensitive adhesives, J. Adhes. 31(1):59 (1989).
1407. C. Dae Han, J. Kim, and D. Man Baek, Viscoelastic behavior, thermodynamic compatibility, and phase equilibria in block copolymer-based pressure-sensitive adhesives, J. Adhes. 28(4):201 (1989).
1408. W. J. Sparks, Advances in hot melt and waterborne acrylic PSA's, Adhes. Age 25(3):38 (1982).
1409. A. D. Hickman, The design of S/B latex pressure-sensitive adhesives, Technical Association of the Pulp & Paper Industry Paper Synthetics Conf. (Sept. 1983).
1410. R. C. Oldack and R. E. Bloss, Compounding natural latex in water-based PTA's, Adhes. Age 22(4):38 (1979).
1411. B. C. Copley, Dynamic mechanical properties of silicone pressure-sensitive adhesives, Adhesive Chemistry-Developments and Trends (L. H. Lee, ed.), Plenum Press, New York (1984).
1412. J. B. Class and S. G. Chu, Viscoelastic properties of rubber-resin pressure-sensitive adhesive formulations, Adhesive Chemistry-Development and Trends (L. H. Lee, ed.), Plenum Press, New York (1984).
1413. L. H. Lee (ed.), Adhesive Chemistry-Development and Trends, Plenum Press, New York (1984).
1414. R. A. Bafford, Converting to water-based adhesives: One manufacturer's experience, Adhes. Age 22(12): 21 (1979).
1415. I. R. Houwink and G. Salomon, Adhesion and Adhesives Elsevier, New York, Vol. 1 (1965).
1416. J. M. Evans and K. E. Krajca, Waterborne adhesive tackifiers, Adhes. Age 25(3):25 (1982).
1417. M. J. Jones, Elastomer and resin modifiers for water-based adhesives, Adhesive Chemistry-Developments and Trends (L. H. Lee, ed.), Plenum Press, New York (1984).
1418. J. A. Fries, The growing uses of waterborne adhesives, Paper, Film, Foil Converter, p. 49 (Dec. 1981).
1419. J. W. Vanderhoff, E. B. Bradford, and W. K. Carrington, J. Polym. Sci., Polym. Symp. 41:155 (1973).
1420. F. M. Rosenblum, Carboxylation increased versatility of polyvinyl-acetate based adhesives, Adhes. Age 22(4):19 (1979).
1421. W. A. Lees, Bonding Method Employing a Two-Part Anaerobically Curing Adhesive Composition, U.S. Patent 3,658,624 (April 1972).
1422. M. M. Skoultchi, Rapid Curing Two-Part Adhesives: Acrylic Monomer and Copper Sacharinate: Alpha-Hydroxy Sulphone or Alpha-Amino Sulfone as Activator, U.S. Patent 4,081,308 (March 1978).
1423. A. G. Bachmann, Adhesive Composition: Quick-Setting, High Bonding Strength, U.S. Patent 4,348,503 (Sept. 1982).
1424. I. Kishi, T. Nakano, and K. Ukita, Two Liquid-Type Adhesive Compositions Comprising a Cobalt Salt Cure Accelerator in One Portion and a Hydroperoxide with an Aromatic Amine and/or a Pyridine Derivative Cure Accelerator in the Second Portion, U.S. Patent No. 4,331,795 (May 1982).
1425. W. J. Owston, Fast Curing Adhesives Butadiene Elastomer, Acrylate Methacrylic Acid Reducting Agent, U.S. Patent 3,832,274 (Aug. 1974).
1426. P. C. Briggs, Jr., Novel Adhesive Compositions: Chlorosulfonated Polyethylene or Sulfonyl Chlorides and Chlorinated Polyethylene, U.S. Patent No. 3,890,407 (June 1975).
1427. D. D. Howard and W. J. Owston, Adhesive Bonding of Polyvinyl Chloride and Other Synthetic Resins, U.S. Patent No. 3,873,640 (March 1975).

1428. L. E. Wolinski, Adhesive Compositions: Polyurethanes, Free Radical Catalysts, Acrylic and Acid Monomers, U.S. Patent No. 3,994,764 (Nov. 1976).
1429. T. H. Dawdy, E. C. Hornaman, F. H. Sexsmith, and D. J. Zalucha, Structural Adhesive Formulations: Addition Polymerizable System Containing Redox Couple Catalyst and Phosphorus Compound, U.S. Patent No. 4,223,115 (Sept. 1980).
1430. R. S. Charnock, Curable Adhesive Composition Toughened With Styrene-Butadiene Block Copolymer Rubbers, U.S. Patent No. 4,574,142 (March 1986).
1431. D. J. Zalucha, Adhes. Age 15(2):21 (Feb. 1972).
1432. J. F. Coleman, Durability studies on toughened acrylic resin adhesives, 5th Natl. SAMPE Tech. Conf., Kiamisha Lake, New York (Oct. 1973).
1433. J. F. Coleman, Modified acrylic adhesives afford durable bonding, Adhes. Age 17(2) (Feb. 1974).
1434. J. D. Minford, unpublished work, Aluminum Co. of America.
1435. P. C. Briggs, Jr., Method of Adhesive Bonding: Chlorosulfonated Polyethylene in Polymerizable Vinyl Monomer, U.S. Patent No. 4,106,971 (August 1978).
1436. P. C. Briggs, Jr., Acrylic monomers, U.S. Patent No. 4,112,013 (Sept. 1978).
1437. H. C. Gilch and G. Piestert, Adhesive Composition, Addition Polymerizable Mixture with Activator, U.S. Patent No. 4,263,419 (April 1981).
1438. J. D. Minford, unpublished work, Aluminum Co. of America.
1439. J. S. Graham, Machine Design (Oct. 7, 1976).
1440. J. S. Graham, Machine Design (Dec. 6, 1977).
1441. J. S. Graham, Modified acrylic structural adhesives, Adhesives in Manufacturing (G. L. Schneberger, ed.), Marcel Dekker, New York, Chap. 9, p. 209 (1983).
1442. F. R. Martin, Developments in Adhesives—1 (W. C. Wake, ed.), Applied Science Publishers, London, Chap. 6 (1977).
1443. A. G. Bachmann, Aerobic acrylic adhesives, Adhes. Age 25(8):19 (1982).
1444. A. G. Bachmann, Aerobic acrylic adhesives, Adhesive Chemistry-Developments and Trends (L. H. Lee, ed.), Plenum Press, New York (1984).
1445. A. G. Bachmann, Adhesive Composition: Acrylic Ester Latent Catalyst, U.S. Patent No. 4,429,088 (Jan. 1984).
1446. A. G. Bachmann, Adhesive Bonding Method, Acrylate Monomer, Elastomer Fillers, and Latent Catalyst System, U.S. Patent No. 4,432,929 (Feb. 1984).
1447. A. G. Bachmann, "Aerobic" Acrylic Adhesives (New Technology in Acrylic Adhesives, paper presented at Adhesives '84 Conf., Cleveland, OH, Soc. Mfg. Engs., Dearborn, MI (Sept. 1984).
1448. H. J. Kleiner, Process for the Preparation of Vinylphosphonic Acid; Thermal Decomposition of 2-Acetoxyenthanephosphonic Dialkyl Ester, U.S. Patent No. 4,386,036 (May 1983).
1449. D. J. Zalucha, High performance structural bonding of unprepared metals, Tech. Bull., Soc. Mfg. Engs., Dearborn, MI (Sept. 1978).
1450. D. J. Zalucha, Plastics Design Proc., p. 25 (May 1981).
1451. W. A. Lees, The science of acrylic adhesives, Brit. Polym. J. 11:64 (1979).
1452. W. A. Lees, Toughened structural adhesives and their uses, Int. J. Adhes. Adhes. 1(5) (July 1981).
1453. W. A. Lees, Advances in Adhesives, Applications, Materials, and Safety (D. M. Brewis and J. Comyn, eds.), Warwick Publishing, Birmingham, England, Chap. 6 (1982).
1454. J. D. Minford, unpublished work, Aluminum Co. of America.
1455. T. H. Dawdy, Modified Structural Adhesives; Acrylated Conjugated Polymers Reacted With Isocyanates, U.S. Patent No. 4,769,419 (Sept. 1988).
1456. T. H. Dawdy, Structural Adhesive Formulations: For Bonding Metals and Plastics; Curable Unsaturated Urethane, Butadiene Polymer, Styrene, or Acrylate Polymers or Monomer and Phosphorous Compound, U.S. Patent No. 4,452,944 (June 1984).
1457. T. H. Dawdy, Epoxy Modified Structural Adhesives Having Improved Heat Resistance; Unsaturated Organophosphorus Partial Ester, Redox Catalyst, U.S. Patent No. 4,467,071 (August 1984).
1458. D. J. Damico, Structural Adhesive Formulations: Heat Curable Addition Polymers, Polyurethanes; Bronsted Acid Catalysts, Reducible Metal Components, Sulfonyl Halides; Blends for Metal Bonding, U.S. Patent No. 4,703,089 (Oct. 1987).
1459. D. J. Damico, Bonding galvanized steel with room temperature–curing acrylics, Adhes. Age 30(10) (1987).
1460. R. M. Bennett and D. J. Damico, Structural Adhesive Formulations and Bonding Method Employing Same, U.S. Patent No. 4,855,001 (Aug. 1989).
1461. L. R. Gatechair and D. Wostratzky, Photoinitiators: An overview of mechanisms and applications, J. Radiat. Curing (July 1973).

1462. G. Pasternack, Fundamental aspects of ultraviolet light and electron beam curing, J. Radiat. Curing (July 1982).
1463. R. B. Seymour, Introduction to Polymer Chemistry. McGraw-Hill, New York (1971).
1464. R. L. Bowen and H. Argentar, A method for determining the optimum peroxide-to-amine ratio for self-curing resins, Applied Polymer Science, Vol. 17 (1973).
1465. W. H. Brendley Jr., Fundamentals of acrylics, Paint Varn. Prod. (July 1973).
1466. J. E. Yeames, A new adhesive for structural bonding of engineering materials, paper presented at Adhesives '84 Conf., Cleveland, OH, Soc. Mfg. Engs., Dearborn, MI (1984).
1467. Catalysts for the Polymerization of Acrylic Monomers, Rohm and Haas Company (June 1959).
1468. J. C. Bolger, Structural adhesives for metal bonding, Treatise on Adhesion and Adhesives (R. L. Patrick, ed.), Marcel Dekker, New York, Vol. 3, Chap. 1, pp. 22–27 (1973).
1469. J. C. Bolger, Structural adhesives: today's state-of-the-art, Adhesives in Manufacturing (G. L. Schneberger, ed.), Marcel Dekker, New York, Chap. 7, pp. 150, 155 (1983).
1470. B. C. Cope, Advances in Adhesives (D. Brewis and J. Comyn, eds.), Warwick Publishing, Birmingham, England, p. 42 (1983).
1471. D. Brewis and J. Comyn (eds.), Advances in Adhesives, Warwick Publishing, Birmingham, England (1983).
1472. A. H. Landrock, Adhesives Technology Handbook, Noyes Publications, Parkridge, NJ, Chap. 5 (1985).
1473. Adhesives Technology Handbook, Noyes Publications, Parkridge, NJ (1985).
1474. J. Shield, Adhesives Handbook, 3rd ed., Butterworth, Stoneham, MA, pp. 32–79 (1984).
1475. J. Thuen (ed.), Adhesives, 4th ed., D.A.T.A. Inc. and the International Plastics Selector Inc. (1986).
1476. Adhesive Bonding, MIL-HDBK-691B, Military Standardization Handbook, U.S. Dept. of Commerce, pp. 47–100 (1987).
1477. E. C. Millard, Epoxy adhesives, Adhesive Bonding of Aluminum Alloys (E. W. Thrall and R. W. Shannon, eds.), Marcel Dekker, New York, Chap. 7, p. 99 (1985).
1478. E. C. Millard, Adhesive selection from the user's viewpoint, Adhesive Bonding of Aluminum Alloys (E. W. Thrall and R. W. Shannon, eds.), Marcel Dekker, New York, Chap. 6, p. 89 (1985).
1479. R. H. Young and J. M. Tancrede, Phenolic adhesives and modifiers, Adhesives in Manufacturing (G. L. Scheneberger, ed.), Marcel Dekker, New York, Chap. 10, p. 237 (1983).
1480. E. J. Catchpole, Some recent European developments in the structural adhesives field, Symp. Struct. Adhesive Bonding, Stevens Inst. Technol., Vol. 1, p. 331 (Sept. 1965).
1481. J. D. Minford, Adhesive bonded aluminum joints, Treatise on Adhesion and Adhesives (R. L. Patrick, ed.), Marcel Dekker, New York, Vol. 3, Chap. 2, pp. 83–85, 101–108 (1973).
1482. J. D. Minford, Adhesives, Durability of Structural Adhesives (A. J. Kinloch, ed.), Applied Science Publishers, London, Chap. 4, pp. 176–178 (1983).
1483. M. O. Hockney, Technical Reports (a) 70081 (1970); (b) 72100 (1972); (c) 73013 (1973), Royal Aircraft Establishment, Farnborough, England.
1484. J. Koetsier, Metal-to-metal adhesive bonded aircraft structure, Proc. Natl. SAMPE Tech. Conf. 7:126 (1975).
1485. N. J. DeLollis, Durability of structural adhesives (a review), 22nd Natl. SAMPE Symp. Exhib., San Diego, p. 673 (April 1977).
1486. W. Z. Olson et al., Resistance of adhesive bonded metal lap joints to environmental exposure, Forest Prod. Lab. Report. No. WADC-TR-59-564 (Oct. 1972).
1487. G. Salomon, Adhesion and Adhesives (R. Houwink and G. Salomon, eds.), Elsevier, New York, Vol. 1, p. 325 (1965).
1488. D. C. Novelli, L-1011 adhesive bonding system, Symp. on Adhesive Bonding Processes for the Automotive and Aircraft Industries, ASTM D-14 Sponsored (Oct. 1979).
1489. H. W. Eickner and W. E. Scholwalter, A study of methods for preparing clad 24S-T3 aluminum alloy sheet surfaces for adhesive bonding, Part I, USAF-PO-(33-038) 49-469E Report No. 1813, Forest Products Laboratory (May 1950).
1490. H. W. Eickner and W. E. Scholwalter, Studies of some of the more promising cleaning methods in treatment of contaminated surfaces of clad 24S-T3 aluminum alloy sheet, Part II, USAF-PO-(33-038) 49-4696F, Report No. 1813, Forest Products Laboratory (May 1950).
1491. J. Comyn, Kinetics and mechanism of environmental attack, Durability of Structural Adhesives (A. J. Kinloch, ed.), Applied Science Publishers, London, Chap. 3, p. 119 (1983).
1492. J. D. Minford, Adhesive bonded aluminum joints, Treatise on Adhesion and Adhesives (R. L. Patrick, ed.), Marcel Dekker, New York, Vol. 3, Chap. 2, pp. 83, 84, 101–108 (1973).
1493. J. D. Minford, Aluminum adhesive bond permanence, Treatise on Adhesion and Adhesives (R. L. Patrick, ed.), Marcel Dekker, New York, Vol. 5, Chap. 3, p. 107 (1981).

1494. J. D. Minford, Adhesives, Durability of Structural Adhesives (A. J. Kinloch, ed.), Applied Science Publishers, London, Chap. 4, pp. 178–181 (1983).
1495. M. J. Bodnar and R. F. Wegman, SAMPE J., p. 51 (Aug/Sept 1969).
1496. H. F. Wolfe, C. L. Rupert, and H. S. Schwartz, Evaluation of several bonding parameters on the random bending fatigue life of adhesively bonded aluminum joints, AFWAL-TR-81-3096 (Aug. 1981).
1497. H. Lee and K. Neville, Handbook of Epoxy Resins, McGraw-Hill, New York (1967).
1498. A. F. Lewis and R. Saxon, Epoxy Resins (H. Kakwichi, ed.), Marcel Dekker, New York, Chap. 10 (1969).
1499. H. Dannenberg and C. A. May, Epoxide adhesives, Treatise on Adhesion and Adhesives (R. L. Patrick, ed.), New York, Vol. 2, pp. 3–76 (1969).
1500. P. O. Nielsen, Properties of epoxy resins, hardeners, and modifiers, Adhes. Age 25(4):42 (April 1982).
1501. R. Mulhaupt and J. H. Powell, Third generation structural adhesives, Adhesion '90, Plastics and Rubber Inst. Meetings, Cambridge, England (1990).
1502. V. P. Gupta, L. T. Drzal, and C. Y. C. Lee, The temperature-dependence of some mechanical properties of a cured epoxy resin system, Polymer Eng. Sci. 25(13):812 (Sept. 1985).
1503. C. H. Lau, K. A. Hodd, and W. W. Wright, Structure and property relationships of epoxy resins, Part I. crosslink density of cured resins, Brit. Polym. J. 17(1):18 (March 1985).
1504. G. Sharon, H. Dodiuk, and S. Kenig, Hygrothermal properties of epoxy film adhesives, J. Adhes. 30 (1–4):87 (1989).
1505. G. Sharon, H. Dodiuk, and S. Kenig, Effects of loading rate and temperature on the mechanical properties of structural adhesives containing a carrier, J. Adhes. 31(1):21 (1989).
1506. J. Kulawik, Z. Szeglowshi, T. Czapla, and J. P. Kulawik, Determination of glass transition temperature, thermal expansion and shrinkage of epoxy resins, Colloid and Polym. Sci. 267(11):970 (1989).
1507. A. Garton, G. S. Haldankar, and P. D. McLean, Modifications of free volume in epoxy adhesive formulations, J. Adhes. 29(1–4):13 (1989).
1508. V. M. Rudoi, I. A. Okulova, and V. A. Ogarev, Modification of the surface of epoxy polymers by reactive SF: Fatty acid monoepoxy esters, Colloid J. U.S.S.R. 51(2):266 (1989).
1509. G. C. Tesoro and V. Sastri, Reversible crosslinking in epoxy resins, I. feasibility studies, J. Appl. Polym. Sci. 39(7):1425 (1990).
1510. V. R. Sastri and G. C. Tesoro, Reversible crosslinking in epoxy resins, II. new approaches, J. Appl. Polym. Sci. 39(7):1439 (1990).
1511. P. I. Engelberg and G. C. Tesoro, Mechanical and thermal properties of epoxy resins with reversible crosslinks, Polymer Eng. Sci. 30(5):303 (1990).
1512. S. A. Bidstrup and C. W. Macosko, Chemorheology relationships for epoxy-amine crosslinking, J. Polym. Sci.-Chem. 28(5):691 (1990).
1513. A. Lee and G. B. McKenna, Viscoelastic response of epoxy glasses subjected to different thermal treatments, Polymer Eng. Sci. 30(7):431 (1990).
1514. Y. Nakamure, M. Yamaguchi, H. Kitayama, K. Ito, M. Okuba, and T. Matsumoto, Internal stress of epoxy resins modified with acrylic polymers produced by in situ UV radiation polymerization, J. Appl. Polym. Sci. 39(3):769 (1990).
1515. J. Massingale et al., Fundamental studies of epoxy resins for can and coil coatings, II. flexibility and adhesion of epoxy resins, J. Coating Technol. 61:31 (1990).
1516. A. J. Ryan et al., Networks by fast epoxy polymerization, Polymer Bull. 24(5):521 (1990).
1517. M. D. Glad and E. J. Kramer, Microdeformation and network structure in epoxides, J. Mater. Sci. 26(9):2273 (1991).
1518. A. C. Grillet, J. Galy, J. F. Gerard, and J. P. Pascault, Mechanical and viscoelastic properties of epoxy networks cured with aromatic diamines, Polymer 32(10):1885 (1991).
1519. W. H. Jo and K. J. Ko, The effect of physical aging on the thermal and mechanical properties of an epoxy polymer, Polym. Eng. Sci. 31(4):239 (1991).
1520. V. T. Truong and B. C. Ennis, Effect of physical aging on the fracture behavior of crosslinked epoxies, Polym. Eng. Sci. 31(8):548 (1991).
1521. J. K. Gillham, Polymer Eng. Sci. 19:676 (1979).
1522. B. A. Rozenberg, Epoxy resins and composites II, Adv. Polym. Sci. 75:113 (1986).
1523. J. Sinclair, The effect of cure temperature on epoxy resin properties, 13th Ann. Meeting Adhesion Soc., Savannah, GA (Feb. 18–21, 1990).
1524. J. R. Jones, C. Poncipe, J. M. Barton, and W. W. Wright, A radio chemical study of the kinetics of epoxide cure: Reaction on phenylglycidyl ether and other model epoxides with various imidazoles, Polymer 28(8):1358 (1987).
1525. B. Steinmann, Investigation of curing epoxides with phthalic anhydride, J. Appl. Polym. Sci. 39(9):2005 (1990).

1526. K. S. Hodd, L. Bottomley, A. Rezaifard, and C. Rezende, Novel curing agents for epoxy resin adhesives, Adhesion '90, Plastics and Rubber Institute Meetings, Cambridge, England (Sept. 10–12, 1990).
1527. S. N. Lee, Y. H. Liao, P. T. Huang, and J. L. Lee, Dielectric behavior of epoxy resin during cure: a vector voltage study, Polym. Engl. Sci. 30(4):219 (1990).
1528. D. Adolf, J. E. Martin, and J. P. Wilcoxon, Evaluation of structure and viscosity in an epoxy near the sol-gel transition, Macromolecules 23(2):527 (1990).
1529. B. Vercheve, H. Sauthereau, J. P. Pascault, C. C. Riccardi, and S. M. Moshier, Buildup of epoxy-cycloaliphatic amine networks: Kinetics, vitrification, and gelation, Macromolecules 23(3) (1990).
1530. M. L. Kaplan, A. L. Wayda, and A. M. Lyons, Lanthanide-imidazole complexes as latent curing agents for epoxy resins, J. Polym. Sci.-Chem. 28(4):731 (1990).
1531. D. J. Plazek and Z. N. Frund, Epoxy resins DEGBA, the curing and physical aging process, J. Polym. Sci.-Chem. 28(4):431 (1990).
1532. W. X. Zukas, K. J. Craven, and S. E. Wentworth, Adherend surface effects on epoxy/amine reactions, 13th Ann. Adhesion Soc. Meetings, Savannah, GA (Feb. 18–21, 1990).
1533. W. X. Zukas, K. J. Craven, and S. E. Wentworth, Model adherend surface effects on epoxy cure reactions, J. Adhes. 33(1–2):89 (1990).
1534. J. Comyn, C. Horley, D. P. Oxley, R. G. Pritchard, and J. L. Tegg, J. Adhes. 12:171 (1981).
1535. R. G. Dillingham and F. J. Boerio, J. Adhes. 23:313 (1987).
1536. R. G. Dillingham and F. J. Boerio, Analysis of the interphase region in aluminum/epoxy joints, 13th Ann. Adhesion Soc. Meeting, Savannah, GA (Feb. 18–21, 1990).
1537. F. J. Boerio and D. J. Ondrus, Molecular structure of the interphase in aluminum/epoxy and copper/epoxy joints, 13th Ann. Adhesion Soc. Meeting, Savannah, GA (Feb. 18–21, 1990).
1538. F. J. Boerio and D. J. Ondrus, Effect of substrates on the structure of polymer interphases, II. epoxy/anydride adhesives cured against aluminum and copper, J. Colloid and Inferf. Sci. 139(2):446 (1991).
1539. A. Garton, W. T. K. Stevenson, and S. P. Wang, J. Polym. Sci., Chem. Ed. 26:1377 (1988).
1540. Y. Hazony, S. J. Stadnicki, and J. K. Gillham, Low frequency thermomechanical spectrometry of polymeric materials: Computerized torsional braid experiments, I. Overview, Polym. Preprints 15(2):549 (1974).
1541. W. Alex Lee and M. J. Oliver, Study of cure of epoxy resins by tortional braid analysis, Brit. Polym. J. 15:40 (1983).
1542. R. L. Levy and D. P. Ames, Monitoring epoxy cure kinetics with a viscosity-dependent fluorescence probe, Adhesive Chemistry-Developments and Trends (L. H. Lee, ed.), Plenum Press, New York (1984).
1543. M. D. Gilbert, N. S. Schneider, and W. J. MacKnight, Mechanism of the dicyandiamide/epoxide reaction, Macromolecules 24(2):360 (1991).
1544. M. L. Kaplan, M. L. Schilling, and A. M. Mujsce, A model reaction for an epoxy resin cured with an aromatic amine, J. Polym. Sci. Polym. Chem. 24(9):599 (1991).
1545. V. Trappe, W. Burchard, and B. Steinmann, Anhydride curing of epoxy resins via chain reaction, Makromolculare Chemie Macromolecular Symp. 45:63 (1991).
1546. P. Forster, K. R. Hauschildt, and D. Wilhelm, The effect of different hardeners on the aging of cross-linked epoxy resins, Makromoleculare Chemie Macromolecular Symp. 41:165 (1991).
1547. A. V. Pocius, Adhesive Formulation and Structures, ACS Teleconference, Lecture No. 4 (May 1, 1986).
1548. F. J. McGarry and A. M. Willner, A.C.S. Div. Org. Coatings and Plastics Chem. 28:512 (1968).
1549. E. H. Rowe, A. R. Siebert, and R. S. Drake, Mod. Plastics 47:110 (1970).
1550. A. R. Siebert and C. R. Riew, ACS Preprints, Org. Coatings and Plastics Chem. 31:555 (1971).
1551. J. N. Sultan, R. C. Laible, and F. J. McGarry, J. Appl. Polym. Sci. 5:127 (1971).
1552. J. N. Sultan and F. J. McGarry, Polym. Eng. Sci. 13:29 (1973).
1553. J. N. Sultan and F. J. McGarry, Epoxy Resins (May and Tanaka, eds.), Marcel Dekker, New York (1974).
1554. A. R. Siebert and E. H. Rowe, 29th Ann. Tech. Conf., SPI (May 1974).
1555. C. K. Riew, E. H. Rowe, and A. R. Siebert, Rubber toughened thermosets, Symp. on Toughness and Brittleness of Plastics, Div. Org. Coatings and Plastics Chem., 168th ACS Meeting, Atlantic City, NJ (Sept. 1974).
1556. H. N. Vazirani, Flexible epoxy resins, Adhes. Age 23(10):31 (1980).
1557. Hycar ATBN modified epoxy adhesives, Product Bulletin AB-9, B. F. Goodrich, SP & C Div. (May 1983).
1558. Hycar reactive liquid polymer modified epoxy adhesives—diamine hardeners, Product Bulletin AB-16, B. F. Goodrich, SP & C Div. (August 1983).
1559. T. S. Chung, Cure mechanism of a modified nitrile epoxy adhesive, J. Appl. Polym. Sci. 29(12):4403 (Dec. 1984).

1560. J. A. Bishop, Novel surface and interfacial analysis techniques as aids to the development of new, high fracture toughness film adhesives, Int. J. Adhes. Adhes. 4(4):153 (Oct. 1984).
1561. R. S. Drake, Formulating epoxy structural adhesives with amine reactive polybutadiene/acrylonitrile liquid polymers, presented at Int. Adhesion Conf., Univ. Nottingham, England (Sept. 12–14, 1984).
1562. A. Takamura, B. I. Tomita, and H. Mizumachi, Dynamic mechanical properties and adhesive strength of epoxy resins modified with liquid rubber, I. Modification with ATBN, J. Appl. Polym. Sci. 30(10): 4031 (1985).
1563. Z. N. Sanjane and L. Kupchella, Dynamic mechanical analysis of rubber toughened resins, Polymer Eng. Sci. 25(18):1148 (Dec. 1985).
1564. R. S. Drake and L. L. Tolle, Formulating epoxy structural with amine reactive polybutadiene/acrylonitrile liquid polymers, presented at SPI-Epoxy Resin Formulators Meeting, Minneapolis, MN (Sept. 10–12, 1986).
1565. A. R. Siebert et al., Formulating epoxy adhesives with carboxy-terminated polybutadiene/acrylonitrile liquid polymers, presented at Adhesives and Sealants Council Seminar, Washington, D.C. (March 23–26, 1986).
1566. L. B. Kandyrin, L. G. Aleksasndrova, L. N. Borisova, and V. N. Kuleznev, Structures of epoxy-rubber films prepared from solution in a mixture of solvents, Colloid J. U.S.S.R. 48(5):957 (1986).
1567. T. Inoue et al., Structure development in rubber-modified epoxy, presented at Elastomers Conf., University of Akron, Akron, OH (1988).
1568. L. Pulliam, A. Siebert, and R. Drake, Amine reactive modifiers give epoxy formulating versatility, Adhes. Age 32(8):18 (1989).
1569. S. J. Shaw and D. A. Tod, The effect of cure conditions on rubber modified epoxy adhesive, J. Adhes. 28(4):231 (1989).
1570. S. Sankaran and M. Chanda, Chemical toughening of epoxies II. mechanical, thermal, microscopic studies of epoxies toughened with hydroxy-terminated polybutadiene/acrylonitrile, J. Appl. Polym. Sci. 39(8):1635 (1990).
1571. H. Lee and T. Kyu, Phase separation dynamics of rubber-epoxy mixtures, Macromolecules 23(2):459 (1990).
1572. D. L. Hunston and H. Mizumachi, Micromechanics of fracture in rubber-toughened epoxies, Adhesion '90, The Plastics and Rubber Institute Meetings, Cambridge, England (Sept. 10–12, 1990).
1573. T. Ricco, R. Frassine, and A. Pavan, Problems in fracture mechanics characterization of rubber-modified glassy polymers using double tortion, J. Mater. Sci. 25(3):1517 (1990).
1574. R. P. Chartoff, J. Cho, and P. S. Carlin, Morphology and mechanical properties of rubber-modified aromatic-ether bismaleimide matrix resins, Polym. Eng. Sci. 31(8):563 (1991).
1575. D. Verchere, J. P. Pascault, H. Sautereau, S. M. Moschair, C. C. Riccardi, and R. J. Williams, Rubber modified epoxies, IV. Influence of morphology on mechanical properties, J. Appl. Polym. Sci. 43(2):293 (1991).
1576. D. Verchere, J. P. Pascault, H. Sautereau, S. M. Moschair, C. C. Riccardi, and R. J. Williams, Rubber modified epoxies, II. Influence of the cure schedule and rubber concentration-on the generated morphology, J. Appl. Polym. Sci. 42(3):701 (1991).
1577. T. Iijima, M. Tomoi, T. Tochimoto, and H. Kakiuchi, Toughening of epoxy resins by modification with aromatic polyesters, J. Appl. Polym. Sci. 43(3):463 (1991).
1578. T. K. Chen and Y. H. Jan, Effect of rubber/matrix interfacial modifications on the properties of a rubber-toughened epoxy resin, Polym. Eng. Sci. 31(8):577 (1991).
1579. A. H. Gilbert and C. B. Bucknall, Epoxy resin toughening with thermoplastics, Makromolculare Chemie Macromolecular Symp. 45:289 (1991).
1580. G. Levita, S. De Petris, A. Marchetti, and A. Lazzeri, Cross-link density and fracture toughness of epoxy systems, J. Mater. Sci. 26(9):2348 (1991).
1581. H. Zhang, L. A. Berglund, and M. Ericson, Rubber-toughening of glass fiber-epoxy filament wound composites, Polymer Eng. Sci. 31(14):1057 (1991).
1582. H. J. Sue, Study of rubber-modified brittle epoxy system I. Fracture toughness measurements using the double-notch four-point-bend method, Polym. Eng. Sci. 31(4):270 (1991).
1583. G. Levita, A. Macchetti, and A. S. Lazzeri, Toughness of epoxies modified by preformed acrylic rubber particles, Makromolculare Chemie Macromolecular Symp. 41:179 (1991).
1584. C. L. Mahoney, Epoxy adhesives, 1st Natl. Adhesive Bonded Metal Structures Conf., New York University (Aug. 14–15, 1974).
1585. J. D. Minford, unpublished work, Aluminum Co. of America.
1586. J. D. Minford, unpublished work, Aluminum Co. of America.
1587. J. D. Minford, unpublished work, Aluminum Co. of America.

1588. J. D. Minford, Adhesive bonded aluminum joints, Treatise on Adhesion and Adhesives (R. L. Patrick, ed.), Marcel Dekker, New York, Vol. 3, Chap. 2, pp. 82, 91–96 (1973).
1589. J. D. Minford, Aluminum adhesive bond permanence, Treatise on Adhesion and Adhesives (R. L. Patrick, ed.), Marcel Dekker, New York, Vol. 5, Chap. 3, pp. 86, 88 (1981).
1590. J. D. Minford, Adhesives, Durability of Structural Adhesives (A. J. Kinloch, ed.), Applied Science Publishers, London, Chap. 4, pp. 182–185 (1983).
1591. J. D. Minford, unpublished work, Aluminum Co. of America.
1592. J. D. Minford, unpublished work, Aluminum Co. of America.
1593. J. D. Minford, unpublished work, Aluminum Co. of America.
1594. E. A. Blommers, E. W. Lane, S. Loshack, and B. D. Halpern, WADD Tech. Report No. 59-15 (Oct. 1959).
1595. J. Dickstein, E. A. Blommers, and W. Karo, Room temperature–curing adhesives for structures, Symp. on Adhesives for Structural Applications (M. J. Bodnar, ed.), Interscience Publishers, New York, pp. 77–81 (1962).
1596. W. C. Tanner, Symp. on Structural Adhesive Bonding, Stevens Inst. Technol., Hoboken, NJ, Vol. 1, p. 1 (Sept. 1965).
1597. W. C. Tanner, Manufacturing processes with adhesive bonding, Symp. for Processing Adhesive Bonded Structures, Stevens Inst. Technol., Hoboken, NJ (Aug. 1972).
1598. M. J. Bodnar and R. F. Wegman, Effects of outdoor aging on unstressed adhesive bonded aluminum-to-aluminum lap joints, Picatinny Arsenal Tech. Report No. 3689, Picatinny Arsenal, Dover, NJ (May 1968).
1599. A. P. Schiles, Long-term aging of epoxy bonds, Report No. ACF 412-329, ACF Industries (June 1967).
1600. J. D. Minford, unpublished work, Aluminum Co. of America.
1601. L. H. Sharpe, Some aspects of the permanence of adhesive joints, Symp. on Structural Adhesive Bonding, Stevens Inst. of Technol., Hoboken, NJ (Sept. 1965).
1602. J. D. Minford, Adhesive bonded aluminum joints, Treatise on Adhesion and Adhesives (R. L. Patrick, ed.), Marcel Dekker, New York, Vol. 3, Chap. 2, p. 92 (1973).
1603. J. D. Minford, Adhesives, Durability of Structural Adhesives (A. J. Kinloch, ed.), Applied Science Publishers, London, Chap. 4, pp. 183, 186 (1983).
1604. J. D. Minford, Aluminum adhesive bond permanence, Treatise on Adhesion and Adhesives (R. L. Patrick, ed.), Marcel Dekker, New York, Vol. 5, Chap. 3, pp. 120, 123 (1981).
1605. D. A. Wangsness, J. Appl. Polym. Sci., Appl. Polym. Symp. 32:291 (1977).
1606. E. W. Garnish, Epoxide resins as adhesives: Past and present, Brit. Polym. J. 11:72 (June 1979).
1607. J. A. Graham and J. E. O'Connor, Epoxy with low temperature cure and high temperature properties developed, Adhes. Age 21(7):20 (1978).
1608. W. Ku and J. P. Bell, Fast curing epoxy-episulfide resin for use at room temperature, Org. Coatings and Appl. Polym. Sci. Proc. 46:498 (1982).
1609. R. E. Brown and R. E. McCrea, Competition, changes for growth of epoxy adhesive markets, Adhes. Age 25(2):21 (1982).
1610. C. M. Thompson and R. Y. Ting, A study of some epoxy polymers for underwater acoustic uses, Org. Coatings and Appl. Polym. Sci. Proc. 46:661 (1982).
1611. J. D. Minford, Adhesives, Durability of Structural Adhesives (A. J. Kinloch, ed.), Applied Science Publishers, London, Chap. 4, p. 168 (1983).
1612. J. D. Minford, Adhesive bonded aluminum joints, Treatise on Adhesion and Adhesives (R. L. Patrick, ed.), Marcel Dekker, New York, Vol. 3, Chap. 2, pp. 97–100 (1973).
1613. J. D. Minford, unpublished work, Aluminum Co. of America, Alcoa Labs, Alcoa Center, PA.
1614. E. J. Hughes and J. L. Rutherford, J. Appl. Polym. Sci., Appl. Polym. Symp. 32:353 (1977).
1615. H. S. Schwartz, J. Appl. Polym. Sci., Appl. Polym. Symp. 32:65 (1977).
1616. N. J. DeLollis, Properties of silicones, polyurethanes, and epoxies, Sandia Labs. Report No. SLA-73-0365 (Oct. 1973).
1617. C. E. M. Morris, P. J. Pearce, and R. G. Davidson, Aging of structural film adhesives: changes in chemical and physical properties and the effect on joint strength, Adhesive Joints: Formation, Characteristics, and Testing (K. L. Mittal, ed.), Plenum Press, New York, p. 231 (1984).
1618. M. B. Smith and S. E. Sussman, Development of adhesives for very low temperature applications, Summary Report, Narmco Res. & Devel. (NASA Contract NAS-8-1565 (May 1963).
1619. E. C. Millard, Epoxy adhesives, Adhesive Bonding of Aluminum Alloys (E. W. Thrall and R. W. Shannon, eds.), Marcel Dekker, New York, Chap. 7, pp. 103–105 (1985).
1620. J. D. Minford, Adhesive bonded aluminum joints, Treatise on Adhesion and Adhesives (R. L. Patrick, ed.), Marcel Dekker, New York, Vol. 3, Chap. 2, pp. 112–117 (1973).

1621. J. K. Kuno, Comparison of adhesive classes for structural bonding of ultrahigh and cryogenic temperature extremes, Proceedings 7th Natl. SAMPE Symp. (May 1964).
1622. J. M. Black and R. F. Blomquist, Development of metal bonding adhesive FPL-710 with improved heat resistance properties, Forest Prod. Lab. Report NACARM 52-F19, Madison, WI (1952).
1623. J. M. Black and R. F. Blonquist, Development of metal bonding adhesives with improved heat resistance properties, Forest Prod. Lab. Report No. NACARMS 54D01, Modern Plastics 32:139 (Dec. 1954).
1624. J. M. Black and R. F. Blomquist, Industrial Engineering Chemistry 30:918 (June 1958).
1625. J. M. Black and R. F. Blomquist, Adhes. Age 5(2) and 5(3) (1962).
1626. J. D. Minford, unpublished work, Aluminum Co. of America, Alcoa Labs, Alcoa Center, PA.
1627. W. Z. Olson and R. M. Lulling, Shear strength of adhesives in stainless steel and aluminum lap joints at temperatures from −100°F (−38°C) to 1000°F (538°C), AFML Tech. Doc. Report ASD-TR-61-497, Forest Prod. Lab., Madison, WI (April 1962).
1628. J. D. Minford, Adhesive bonded aluminum joints, Treatise on Adhesion and Adhesives (R. L. Patrick, ed.), Marcel Dekker, New York, Vol. 3, Chap. 2, pp. 88, 97–100 (1973).
1629. E. Klopfenstein and H. Lee, Some thermal aspects of epoxy resins, Insulation, p. 13 (Feb. 1958).
1630. D. W. Overnall, Electron spin resonance of free radicals in epoxide resins, Polym. Letters 1:37 (1963).
1631. L. H. Lee, Thermal stability and degradation schemes of epoxy resins, Batelle Symp. Thermal Stability of Polymers, Columbus, OH (Dec. 1963).
1632. H. Lee and K. Neville, Use of dianhydrides in high-temperature epoxy resin formulations, 7th Natl. SAMPE symp., Los Angeles, p. 2-1 (May 20–22, 1964).
1633. H. P. Owen, Development of a high temperature aerospace structural adhesive, 7th Natl. SAMPE, Los Angeles, p. 10-1 (May 20–22, 1964).
1634. G. C. Grimes, Storage time effects on adhesive mechanical properties, Symp. for Processing Adhesive Bonded Structures, Stevens Inst. Technol., Hoboken, NJ (Aug. 1972).
1635. J. R. Griffith, J. G. O'Rear, and J. P. Reardon, A fluoroanhydride curing agent of heavily fluorinated epoxy resins, Adhesion Science and Technology (L. H. Lee, ed.), Plenum Press, New York, Vol. 9A, p. 429 (1975).
1636. J. R. Griffith and J. G. O'Rear, Silicone amine-cured fluoroepoxy resins, Resins for Aerospace (C. A. May, ed.), American Chemical Society, Washington, D.C. (1980).
1637. J. R. Griffith, Epoxy resins containing fluorine, CHEMTECH 12:290 (1982).
1638. A. A. Donatelli et al., Macromolecules 9(4):671 (1976).
1639. S. L. Madorsky and S. Strauss, Thermal stability of high-temperature carbon-chain polymers, Natl. Bureau of Stds. Report.
1640. R. A. Kourtides, J. A. Parker, T. W. Giants, N. Bilow, and M.-T. Hsu, Thermophysical and flammability characterization of phosphorylated epoxy adhesives, Adhesives for Industries Technol. Conf., El Segundo, CA, p. 92 (June 25, 1980).
1641. R. Jerome et al., J. Appl. Polym. Sci. 26(6):1741 (1981).
1642. K. L. Hawthorne, F. C. Henson, and R. Pinzelli, High performance tris(hydroxylphenyl) methane-based epoxy resins, Org. Coating and Appl. Polym. Sci. Prog. 46:493 (1982).
1643. Wang Zhi-Lu, Chen Dao-Yi, and Liu Xloa-Hui, A study on the properties and structure of polyester-sulfone-modified epoxy adhesives, J. Adhes. 23:67 (1987).
1644. H. Jozavi and E. Sancaktar, The effects of cure temperature and time on the stress whitening behavior of structural adhesives, Part. I. analysis of bulk tensile data, J. Adhes. 27:143 (1989).
1645. A. Garton and G. S. Haldankar, Modification of free volume in epoxy adhesive formulations, J. Adhes. 29:13 (1989).
1646. J. D. Minford, unpublished work, Aluminum Co. of America.
1647. R. F. Wegman, J. Appl. Polym. Sci., Appl. Polym. Symp. 32:1 (1977).
1648. W. F. Thomsen, Cyanoacrylate adhesive, Adhesives in Manufacturing (G. L. Schneberger, ed.), Marcel Dekker, New York, Chap. 12, p. 305 (1983).
1649. W. F. Thomsen and P. T. Von Bramer, Bonding characteristics of alkyl cyanoacrylate adhesives, 14th Natl. SAMPE symp., Cocoa Beach, FL (Nov. 1968).
1650. H. W. Coover, Cyanoacrylate adhesives, Handbook on Adhesives (I. Skeist, ed.), Van Nostrand Reinhold, New York, p. 569 (1977).
1651. H. W. Cooper, F. B. Joyner, N. H. Shearer, and T. H. Wicker, Jr., Soc. Plast. Eng. J. 15(5) (1959).
1652. J. M. Rooney, On the mechanism of oligomer formation in condensations of alkyl cyanoacrylate with formaldehyde, Polymer J. 13(10):975 (1981).
1653. G. F. Hawkins, U.S. Patent Process for Manufacture of High Purity Alpha-cyanoacrylates, Patent No. 3,465,027 (1969).
1654. T. J. Wicker and N. H. Shearer, Jr., Method of Adhesive Bonding, U.S. Patent No. 3,259,534 (1966).

1655. T. J. Wicker and N. H. Shearer, Jr., Adhesive Composition and Method of Using Alpha-cyanoacrylate Esters and Vinyl Aromatics, U.S. Patent No. 3,282,773 (1966).
1656. E. F. Donnelly, D. S. Johnston, and D. C. Pepper, Ionic and zwitter-ionic polymerization of n-alkyl 2-cyanoacrylates, J. Polym. Sci. Polym. Lett. Ed., 15:399 (1977).
1657. D. C. Pepper, Anionic and zwitterionic polymerization of alpha-cyanoacrylates, J. Polym. Sci. Symp. 62:65 (1978).
1658. H. A. R. Rasoul and H. K. Hall, Jr., Cycloaddition and polymerization reactions of methyl alpha-cyanoacrylate with electron-rich olefins, J. Org. Chem. 47:2080 (1982).
1659. J. Guthrie, M. S. Otterburn, J. M. Rooney, and C. N. Tsang, The determination of the molecular weight of poly-(ethyl-2-cyanoacrylate adhesive, Polym. Commum. 25:318 (1984).
1660. I. C. Eromosele and D. C. Pepper, The possibility of oxonium/carbonium initial cations in the zwitterionic polymerization of butyl 2-cyanoacrylate by cyclic carbonates, Macromol. Chem. Rapid Commun. 7:531 (1986).
1661. J. P. Cronin and D. C. Pepper, Zwitterionic polymerization of butyl cyanoacrylate by triphenylphosphine and pyridine, Makromol. Chem. 189:85 (1988).
1662. J. Almog and A. Gabay, A modified super glue technique—the use of polycyanoacrylate technique for fingerprint development, J. Forensic Sci. 31(1):250 (1986).
1663. Adhes. Age 32(5):32 (1989).
1664. D. L. Kotzev, T. C. Ward, and D. W. Dwight, Assessment of the adhesive bond properties of allyl 2-cyanoacrylate, J. Appl. Polym. Sci. 26:1941 (1981).
1665. D. L. Kotzev, P. C. Novakov, and V. S. Kabaivanov, Synthesis and properties of some alkenyl- and alkinyl-2-cyanoacrylates, Angew. Makromol. Chem. 92:41 (1980).
1666. K. F. Drain, J. Guthrie, C. L. Leung, F. R. Martin, and M. S. Otterburn, The effect of moisture on the strength of polycarbonate-cyanoacrylate adhesive bonds, J. Adhes. 17:81 (1984).
1667. K. F. Drain, J. Guthrie, C. L. Leung, F. R. Martin, and M. S. Otterburn, The effect of moisture on the strength of polycarbonate-cyanoacrylate adhesive bonds, Int. J. Adhes. Adhes. 5(3):133 (1985).
1668. H. W. Coover, Jr., and J. M. McIntire, Cyanoacrylate adhesives, Handbook of Adhesives (I. Skeist, ed.), Van Nostrand Reinhold, New York, pp. 569–580 (1977).
1669. E. F. Donnelley and D. C. Pepper, Solubilities, viscosities, and unperturbed dimensions of poly(ethyl cyanoacrylate)s and poly(butyl cyanoacrylate)s, Makromol. Chem. Rapid Commun. 2:439 (1981).
1670. E. Konig, Adhesive compositions containing a cyanoacrylate and itaconic anhydride, U.S. Patent No. 3,948,794 (1976).
1671. J. E. Schoenberg, 2-cyanoacrylate adhesive compositions having enhanced bond strength. U.S. Patent 4,139,693 (1978).
1672. J. E. Schoenberg and D. K. Ray-Chaudhuri, Adhesion promotor for 2-cyanoacrylate adhesive compositions, U.S. Patent No. 4,125,494 (1978).
1673. V. R. Allies and W. D. Zimmerman, Debondable cyanoacrylate adhesive compositions, Brit. Patent No. 1,529,105 (1978).
1674. E. R. Gleave, Filled cyanoacrylate adhesive compositions, U.S. Patent No. 4,102,945 (1978).
1675. A. Motegi, E. Isowa, and K. Kimura, Alpha-cyanoacrylate type adhesive composition, U.S. Patent No. 4,171,416 (1979).
1676. A. E. Litke, Thixotropic cyanoacrylate compositions, U.S. Patent No. 4,477,607 (1984).
1677. C. Petrov, B. Serafimov, and D. L. Kotzev, Strength, deformation, and relaxation of joints bonded with modified cyanoacrylate adhesives, Int. J. Adhes. Adhes. 8(4):207 (1988).
1678. C. Petrov, B. Serafimov, and D. L. Kotzev, Adhesive bond properties of ethyl-2-cyanoacrylate modified with poly(methylmethacrylate), J. Adhes. 25:245 (1988).
1679. K. G. Chorbadjiev and D. L. Kotzev, The effect of fillers upon the properties of electroconductive cyanoacrylate adhesives, Int. J. Adhes. Adhes. 8(3):143 (1988).
1680. D. J. O'Sullivan and D. Mebody, U.S. Patent 3,832,334 (1974).
1681. P. Krall, Pat. DDR No. 156,365 (1981).
1682. J. Schoenberg, Ger. Offen 2,833,842 (1978).
1683. D. L. Kotsev, Z. Z. Denchev, and V. A. Kabaivanov, Adhesive properties of ethyl 2-cyanoacrylate containing small amounts of acetic acid as adhesion promotor, Int. J. Adhes. Adhes. 7(2):93 (1987).
1684. W. Suetake, Infrared spectroscopic investigations of polymer coating-metal substrate interactions, Adhesion Aspects of Polymeric Coatings (L. H. Lee, ed.), Plenum Press, New York, p. 225 (1983).
1685. J. T. O'Connor and W. D. Zimmermann, Factors affecting adhesion of cyanoacrylate adhesives to seal bright anodized surfaces, Loctite Corp. unpublished report, Newington, CT.
1686. W. S. Garwood, Plating, p. 1323 (1966).
1687. R. F. Walton, Plating, p. 219 (1972).

1688. R. C. Spooner and W. J. Forsyth, Plating 55:36 (1968).
1689. R. C. Spooner and W. J. Forsyth, Plating 55:341 (1968).
1690. R. C. Spooner and W. J. Forsyth, Plating 55:463 (1968).
1691. L. H. Lee (ed.), Cyanoacrylate Resins-The Instant Adhesives, T. C. Publication, Div. Technol. Conf., El Segundo, Ca (1981).
1692. R. Blomquist, Cyanoacrylates-a practical guide to their chemical and physical properties, SME. Tech. Paper No. AD86-777, Soc. Mfg. Engs., Dearborn, MI (1986).
1693. R. J. Cotter, How to dispense adhesives, SME Tech. Paper NO. AD85-773, Soc. Mfg. Engs., Dearborn, MI (1985).
1694. R. Peace, Evaluating cyanoacrylates for product assembly, Adhes. Age 22(9):29 (1979).
1695. R. E. Burnett and B. W. Nordlander, U.S. Patent No. 2,628,178 (1953).
1696. B. D. Murray, M. Hauser, and J. R. Elliott, Anaerobic adhesives, Handbook of Adhesives, 2nd ed. (I. Skeist, ed.), Reinhold, New York (1977).
1697. A. P. Drafone, How to use the anaerobic adhesive, Mfg. Eng. Management, p. 46 (Jan. 1971).
1698. M. Hauser and G. S. Haviland, Anaerobic adhesives, Adhesives in Manufacturing (G. L. Schneberger, ed.), Marcel Dekker, New York, Chap. 11, p. 269 (1983).
1699. L. J. Baccei, U.S. Patent No. 4,295,909 (1981; 4,309,526 (1982).
1700. W. A. Lees, C. Ford, D. J. Bennett, J. R. Swire, and P. Harding, U.S. Patent No. 3,795,641 (1974).
1701. C. W. Boeder, Anaerobic and structural acrylic adhesives, Structural Adhesives—Chemistry and Technology (S. R. Hartshorn, ed.), Plenum Press, New York (1986).
1702. B. M. Malofsky, U.S. Patent No. 3,988,299 (1976).
1703. T. R. Baldwin, D. J. Bennett, and W. A. Lees, U.S. Patent No, 4,138,449 (1979).
1704. D. J. Stamper, Curing characteristics of anaerobic sealants and adhesives, Brit. Polym. J. 15:34 (March 1983).
1705. T. S. Fulda, Impregnation of porous metal components with anaerobic sealants, SME Tech. Paper No. FC77-532, Soc. Mfg. Engs., Dearborn, MI (1977).
1706. M. B. Pearce, Symp. on Processing for Adhesive Bonded Structures, Stevens Inst. Technol., Hoboken, NJ, p. 329 (Aug. 1972).
1707. C. Hulstein, Application of anaerobic adhesives, SME Tech. Paper No. AD76-673, Soc. Mfg. Eng., Dearborn, MI (1976).
1708. C. Hulstein, Assembling with anaerobics, Mfg. Eng., p. 37 (July 1978).
1709. C. Hulstein, Automatic application of anaerobic adhesives, Adhes. Age 22(7):25 (1979).
1710. J. D. Minford, unpublished work, Aluminum Co. of America.
1711. S. Y. Lee and K. C. Bruce, Thermal stability and aging of some new anaerobic adhesives for structural bonding, SAMPE Quart., p. 22 (July 1980).
1712. R. S. Drake and A. R. Siebert, Reactive butadiene/acrylonitrile liquid and solid elastomers: Formulating acrylic, anaerobic, and radiation-curable adhesives, Adhesives Chemistry—Developments and Trends (L. H. Lee, ed.), Plenum Press, New York (1984).
1713. B. M. Malofsky and L. J. Baccei, Anaerobic adhesives containing maleimides having improved thermal resistance, Adhesive Chemistry—Developments and Trends (L. H. Lee, ed.), Plenum Press, New York (1984).
1714. M. Hauser and J. T. Loft, Anaerobics and modified acrylics: How and what they will bond, Adhes. Age 23(13):21 (Dec. 1980).
1715. G. J. Helmstetter, The anaerobic advantage, SME Tech. Paper No. AD85-774, Soc. Mfg. Engs., Dearborn, MI (1985).
1716. G. S. Haviland, Designing with threaded fasteners, Mech. Eng. p. 17 (Oct. 1983).
1717. G. K. Junker, New criteria for self-loosening of fasteners under vibration, S.A.E. Paper No. 69005, Soc. Auto. Eng., Warrenton, PA (Jan. 13, 1969).
1718. G. S. Haviland, A logical approach to secure bolting and locking, SME Tech. Paper No. AD80-329, Soc. Mfg. Eng., Dearborn, MI (1980).
1719. R. A. Valitsky, The role of adhesives and sealants in fitting cylindrical parts, S.A.E. Paper No. 830667, SAE Cong. & Expos. Meeting, Soc. Auto. Eng., Warrendale, PA (Feb. 28–March 4, 1983).
1720. Tumbling Technology for Anaerobic Sealants, Publ. ADR-11, Loctite Corp., Newingham, CT.
1721. G. S. Haviland, Machinery Adhesives for Locking, Retaining, and Sealing, Marcel Dekker, New York (1986).
1722. W. H. O'Connor, Black and Decker's anaerobic adhesive system for assembling armatures, SME Tech. Paper No. AD79-351, Soc. Mfg. Eng., Dearborn, MI (1979).
1723. R. J. Pilarski, Robotic adhesive dispensing for bonding and sealing, Robotics Today, p. 35 (Oct. 1984).
1724. T. R. Baldwin, Overview: anaerobic adhesives, Mater. Sci. and Technol. 2(1):1 (Jan. 1986).

1725. O. Bayer, H. Rinke, W. Siefken, L. Ortner, and H. Schild, German Patent No. 728,981 to I. G. Farbenindustrie AG. (1942).
1726. O. Bayer, Angew. Chem., A59:275 (1947).
1727. J. M. DeBell, W. C. Coggin, and W. E. Gloor, German Plastics Practice (DeBell and Richardson, eds.), Cambridge, MA (1946).
1728. H. L. Hess, J. H. Saunders, M. R. Morris, B. R. Davis, and E. E. Hardy, Industrial Engineering Chemistry 46:1498 (1954).
1729. B. A. Dombrow, Polyurethanes, Reinhold, New York (1957).
1730. R. B. Godnell and S. E. Smith, Structural variations of polyurethanes: Their effect on metal adhesion at various temperatures, Adhes. Age 5(9):32 (Sept. 1962).
1731. M. Dollhausen and W. Warrach, A review of polyurethane adhesive technology, Adhesives and Sealants Council Meeting, Philadelphia (Sept. 1981).
1732. G. M. MacIver and D. P. Thompson II., Urethane structural adhesive systems, Adhesives in Manufacturing (G. L. Schneberger, ed.), Marcel Dekker, New York, Chap. 8, p. 195 (1983).
1733. B. H. Edwards, Polyurethane structural adhesives, Structural Adhesives: Chemistry and Technology (S. R. Hartshorn, ed.), Plenum Press, New York (1986).
1734. D. J. Zimmer and J. S. Murphy, The solventless approach to adhesive compliance, Technical Association of the Pulp and Paper Industry J., pp. 123–126 (Dec. 1988).
1735. M. Dollhausen, Polyurethane Adhesives Based on Baycoll, Dismocoll and Desmodur, Product Literature, Bayer AG.
1736. K. C. Frish, H. X. Xiao, and R. W. Czerwinski, Formulating polyurethane adhesives and sealants, Adhes. Age 31(10):41 (1988).
1737. D. Dietrich and J. N. Riech, Aqueous polyurethane systems and their possible uses, Adhes. Age 21(2): 24 (1978).
1738. D. Dietrich, Aqueous emulsions, dispersions, and solutions of polyurethanes: Synthesis and properties, Prog. Org. Coatings 9:281 (1981).
1739. G. B. Guise and M. B. Jackson, U.S. Patent 3,898,197 (1972).
1740. J. Pedain, N. Nachtkamp, K. Noll, and J. Grammel, DOS 2,725,589 assigned to Bayer AG (1978).
1741. N. Nachtkamp, J. Pedain, and J. Grammel, DOS 2,811,148 assigned to Bayer AG (1978).
1742. J. D. Minford, unpublished work, Aluminum Co. of America.
1743. L. J. Constanze, Improved primer for bonding polyurethane adhesives to metals, NASA Tech. Brief (Oct. 1969).
1744. J. D. Minford, unpublished report, Alcoa Labs, Alcoa Center, PA.
1745. S. B. Twiss, Structural Bonding (H. J. Bodnar, ed.), Interscience, New York, p. 455 (1966).
1746. M. B. Smith and S. E. Sussman, Development of adhesives for very low temperature application, Summary Report NASA Contract.
1747. R. W. Vaughan and C. H. Sheppard, Cryogenic temperature structural adhesives, 19th Natl. SAMPE Symp. Exhib., Vol. 19, p. 7 (1974).
1748. W. M. Frost, Strength of structural adhesives at temperatures down to −423°F (−254°C), WADC Tech. Report No. 260, Cryogenic Eng. Lab., Bureau of Standards (1959).
1749. J. Hertz, Cryogenic adhesive evaluation study, Convair Res. Program No. 111-9106, Astronics Div., Convair Corp. (Jan. 1961).
1750. J. Delmonte and E. C. Sarna, Symp. on Processing for Adhesive Bonded Structures, Stevens Inst. Technol., Hoboken, NJ, p. 320 (Aug. 1972).
1751. J. D. Minford, Aluminum adhesive bond permanence, Treatise on Adhesion and Adhesives (R. L. Patrick, ed.), Marcel Dekker, New York, Vol. 5, Chap. 3, p. 123 (1981).
1752. J. Delmonte, 19th Natl. SAMPE Symp. Exhib., Vol. 19, p. 1 (1974).
1753. A. Mondur, Multron Adhesives Formulation Chemical Bulletin, Mobay Chemical Corp., Pittsburgh.
1754. G. Eagle, Polyurethane sealants challenge plastisols in the automotive field, Plastics Eng., p. 29 (July 1980).
1755. R. O. Koch, Effects of processing variables on the bond strength of polyurethane adhesive, Report T5-6556-16, Boeing Co., Saturn Booster Branch (Oct. 4, 1967).
1756. M. E. Kimball, Polyurethane adhesives: properties and bonding procedures, Adhes. Age 24(6):21 (June 1981).
1757. M. Dollhausen and W. Warrach, Polyurethane adhesives technology, Adhes. Age 25(6):28 (June 1982).
1758. B. H. Edwards, Polyurethane structural adhesives, SME Tech. Paper No. AD85-775, Soc. Mfg. Eng., Dearborn, MI (1985).
1759. J. M. Black and R. F. Blomquist, Metal bonding adhesives for high temperature service, Natl. Advisory Comm. Aeronautics Res. Memo RM-55-126, Part 3, Forest Prod. Lab., Madison, WI (March 1956).

1760 J. M. Black and R. F. Blomquist, Development of metal-bonding adhesives with improved heat resistance, Forest Prod. Lab. Report, Madison, WI (1957).
1761. M. Naps and F. C. Hopper, Elevated temperature resistant modified epoxide resin adhesives for metals, WADC Tech. Report No. 53-126, Part 3, Forest Products Lab., Madison, WI (March 1956).
1762. R. B. Krieger, Jr., and R. E. Politi, High temperature structural adhesives, Aspects of Adhesion—3 (D. J. Alner, ed.), University of London Press, London (1967).
1763. B. B. Stewart, F. J. Riel, and S. E. Susman, Elevated temperature resistant silicone structural adhesives for metals, WADC Tech. Report No. 56-533, Part II, ASD 155-806 (Aug. 1958).
1764. L. Suffredini and J. M. McCann, Heat resistant and structural adhesives, SPE RETEC, Plastics and Adhesives in the Space Age (May 13, 1964).
1765. N. J. DeLollis, Durability of structural adhesive bonds: A review, Adhes. Age 20(9):41 (1977).
1766. A. K. St. Clair and T. L. St. Clair, A review of high temperature adhesives, NASA Tech. Memo 83141, Langley Res. Center, Hampton, VA (July 1981).
1767. L. E. Meade, Adhesives and sealants for aerospace, SAMPE Quart. (July 24, 1983).
1768. K. Nakao, Recent advances in adhesion and adhesives, Technology (II), Technocraft 18(3):24 (1985).
1769. H. H. Levine, Recent developments in high-temperature adhesives, Symp. on Adhesives for Structural Applications (M. J. Bodnar, ed.), Interscience, New York, pp. 67–76 (1962).
1770. M. S. Allison, A. C. Johnson, and R. B. Stabler, SPE Tech. Papers., Vol. VII, Washington, D.C., ANTEC, p. 26–3 (Jan. 1961).
1771. E. C. Janis, W. H. Boram, F. J. Riel, and S. E. Susman, WADC Tech. Report No. 59-11, AF33(616-5488) (1959).
1772. H. H. Levine and W. H. Boram, U.S. Navy Nord-19075 (1959).
1773. H. H. Levine, U.S. Navy NOW-61-0254c (1960).
1774. J. E. Thomas, Investigation of materials-adhesives-high temperature resistant, Report No. FGT-1898, Contract AF 33 757(7248), General Dynamics (Sept. 10, 1962).
1775. S. H. Ruetman and H. H. Levine, Research and development of high temperature structural adhesives, Prog. Report No. 5, Contract NOW-61-0254-C, Narmco Ind. (Feb. 1962).
1776. S. H. Ruetman, H. H. Levine, and W. J. Wrasidlo, Research and development of high temperature structural adhesives, Quart. Prog. Report No. 7, Narmco Contract No. NOW-61-0254-C (Aug. 1962).
1777. L. M. Roseland, Adhesives and resins for missile application, Report No. SM-44660, Douglas Aircraft Co. (Feb. 25, 1963).
1778. P. M. Hergenrother, W. J. Wrasidlo, and H. H. Levine, High temperature structural adhesives, Summary Report, Contract NOW-0421-C Narmco Ind. (April 15, 1964).
1779. P. M. Hergenrother, G. Johnston, Jr., and H. H. Levine, High temperature adhesives and resins, Quart. Prog. Report No. 1, Navy Bureau of Weapons, Narmco Contract NOW-64-05240-C (April–July, 1964).
1780. R. C. Kausen, High and low temperature adhesives—where do we stand, Proceedings 7th Natl. SAMPE SYMP. (May 1964).
1781. K. W. Humphries, High temperature adhesives, Aspects of Adhesion (D. L. Alner, ed.), University of London Press, London, p. 66 (1965).
1782. H. F. Mark, Polymers for extreme service conditions, J. Appl. Polym. Sci., Appl. Sci. Symp. 35:13 (1979).
1783. S. Oldham, High temperature polymers, SAMPE Quart., p. 1 (Jan. 1979).
1784. P. E. Cassidy, Polymers for extreme service conditions, J. Chem. Ed. 58:951 (Nov. 1981).
1785. C. L. Hendricks and S. G. Hill, Evaluation of high temperature structural adhesives for extended service, Adhesive Chemistry—Developments and Trends (L. H. Lee, ed.), Plenum Press, New York (1984).
1786. P. M. Hergenrother, Status of high-temperature adhesives, Adhesive Chemistry—Developments and Trends (L. H. Lee, ed.), Plenum Press, New York (1984).
1787. L. H. Lee, Developments in and limitations of adhesive materials for severe environments and a long service life, Adhesive Chemistry—Developments and Trends (L. H. Lee, ed.), Plenum Press, New York, p. 675 (1984).
1788. F. Riel and H. H. Levine, Structural adhesives for use above 700°F (371°C), Handbook of Adhesives, Reinhold, New York, p. 469 (1962).
1789. J. B. Whitney, High-temperature ceramic adhesives, J. Appl. Polym. Sci. 6(2):226 (March–April 1962).
1790. H. H. Levine, Research and development of high temperature structural adhesives, Quart. Prog. Report No. 1, Contract NOW-61-0254-C, Narmco Ind. (Dec. 1960).
1791. H. H. Levine, Research and developments of high-temperature stable polymers, 17th Ann. Tech. Conf. SPE., Session 14, Paper No. 1 (Jan. 1961).
1792. H. H. Levine, Research and development of high temperature structural adhesives, Summary Report Contract NORD-19075, Narmco Ind. (Sept. 29, 1961).

1793. H. H. Levine, Recent developments in high temperature adhesives, J. Appl. Polym. Sci. 6(2):184 (March–April 1962).
1794. H. H. Levine, High temperature structural adhesives, Industrial Engineering Chemistry 54(3):22 (1962).
1795. H. H. Levine, Polybenzimidazole resins for high-temperature reinforced plastics and adhesives, Conf. on Struct. Plastics, Adhesives and Filament Wound Composites (Dec. 11–13, 1962).
1796. H. H. Levine, S-Triazine polymers, Handbook of Adhesives, Reinhold, New York, p. 468 (1962).
1797. C. S. Marvel, Polyaromatics, Proc. Conf. on High-Temperature Polymer and Fluid Res. (May 8–11, 1962).
1798. C. S. Marvel, Thermally stable polymers with aromatic recurring units, SPE J. 20(3):220 (March 1964).
1799. L. H. Shenker, H. E. Edgehoffer, and C. M. Dolan, Thermal shield bonding systems for re-entry vehicles, SPE-RETEC-Plastics in the Space Age (May 13, 1964).
1800. Reliability of adhesive bonds under severe environments, Natl. Materials Advisory Board; Natl. Res. Council, USA (Dec. 1984).
1801. L. H. Lee, Adhesives and sealants for severe environments, Int. J. Adhes. Adhes. 7(2):81 (1987).
1802. A. K. St. Clair and T. L. St. Clair, The development of aerospace polyimide adhesives, Polyimides: Synthesis, Characterization, and Applications (K. L. Mittal, ed.), Plenum Press, New York, p. 1977 (1984).
1803. P. M. Hergenrother, Adhesive and composite evaluation of acetylene-terminated phenylquinoxaline resins, Polym. Sci. Eng. 21(16):1072 (1981).
1804. A. K. St. Clair, T. L. St. Clair, and S. A. Ezzell, Polyimide adhesives modified with ATBN and silicone elastomers, Adhesive Chemistry—Developments and Trends (L. H. Lee, ed.), Plenum Press, New York, p. 467 (1984).
1805. C. Arnold, Jr., Stability of high temperature polymers, J. Polym. Sci. Macromol. Rev. 14:265 (1979).
1806. W. M. Edwards, U.S. Patent No. 3,179,614 (1965).
1807. W. M. Edwards, U.S. Patent No. 3,179,364 (1965).
1808. E. Lavin, A. H. Markhart, and R. E. Kass, U.S. Patent No, 3,190,856 (1965).
1809. E. Lavin, A. H. Markhart, and R. E. Kass, U.S. Patent No. 3,347,808 (1967).
1810. L. K. English, Premium performance from polyimides, Mater. Eng. (Jan. 1986).
1811. E. C. Millard, Elevated temperature-resistant adhesives, Adhesive Bonding of Aluminum Alloys (E. W. Thrall and R. W. Shannon, eds.), Marcel Dekker, New York, Chap. 8, pp. 133–134 (1985).
1812. C. E. Sroog, J. Polym. Sci. Macromol. Rev. 11:161 (1976).
1813. I. Serlin et al., Handbook of Adhesives, 2nd ed. (I. Skeist, ed.), Van Nostrand, New York, pp. 597–618 (1977).
1814. D. A. Scola, Composites, Engineered Materials Handbook, ASM International, Materials Park, OH, Vol. 1, pp. 78–89 (1989).
1815. A. L. Landis, Handbook of Thermoset Plastics (S. H. Goodman, ed.), Noyes Publications, pp. 266–317 (1986).
1816. M. I. Bessonov et al., Polyimides: Thermally Stable Polymers, L. V. Backinowsky and M. A. Chlenov Trans., Consultants Bureau (1987).
1817. A. Scola and J. H. Vontell, CHEMTECH, pp. 112–121 (1989).
1818. F. Rodrigues, Principles of Polymer Systems, McGraw-Hill, New York, p. 17 (1970).
1819. P. M. Cotts and W. Volsen, ACS Symp. Series No. 242, ACS, pp. 227–237 (1984).
1820. M. I. Bessonov et al., Polyimides: Thermally Stable Polymers (Backinowsky and Chlenov Trans.), Consultants Bureau, p. 5 (1987).
1821. C. J. Lee, Proc. 34th Int. SAMPE Symp. 34(1):929 (1989).
1822. P. M. Hergenrother, Reactive Oligomers (F. W. Harris and H. J. Sinelli, eds.), ACS Symp. Series No. 282, ACS, pp. 1–29 (1985).
1823. H. R. Lubowitz, U.S. Patent No. 3,528,950 (1970).
1824. H. R. Lubowitz, ACS Polymer Prepr., 12(1):329 (1971).
1825. H. Stenzenberger, Brit. Polym. J. 20:383 (1988).
1826. I. K. Varma et al., Polym. News 12:294 (1987).
1827. D. Wilson, Brit. Polym. J. 20:405 (1988).
1828. K. N. Ninan, K. Krishman, and J. Mathew, J. Appl. Polym. Sci. 32:6033 (1986).
1829. S. J. Shaw and A. J. Kinloch, Int. J. Adhes. Adhes. 5(3):123 (1985).
1830. D. J. Progar, Adhesive evaluation of LARC-TPI and a water-soluble version of LARC-TPI, Int. J. Adhes. Adhes. 6(1): (1986).
1831. W. D. Roper, Spacecraft adhesives for long life and extreme environments, NASA TR-3201537 (1971).
1832. J. P. Critchley and M. A. White, J. Polym. Sci., Polym. Chem. Ed. 10:1809 (1972).
1833. P. M. Hergenrother, Polym. Prepr., 15(2):537 (1974).

1834. A. H. Landis and A. B. Naselow, Proc. Natl. SAMPE Tech. Conf., 14:236 (1982).
1835. A. K. St. Clair and T. L. St. Clair, Aluminum ion containing polyimide adhesives, NASA Tech. Brief (Fall 1980).
1836. A. K. St. Clair and T. L. St. Clair, A multi-purpose thermoplastic polyimide, SAMPE Quart., 13(1): 20 (1981).
1837. A. K. St. Clair and T. L. St. Clair, Int. J. Adhes. Adhes. 2(7):249 (1981).
1838. W. D. Burks and T. L. St. Clair, Synthesis and characterization of a melt processable polyimide, NASA Tech. Memo 84494 (1982).
1839. A. K. St. Clair and T. L. St. Clair, Addition polyimide adhesives containing various end groups, Poly. Eng. Sci. 22(1):9 (1982).
1840. A. Scola, Addition-type bismaleimides and related imide polymers, Talk at State University New York at New Paltz Polymer Series (June 1984).
1841. T. L. St. Clair, Adhesive development at NASA Langley presented to Program/Review Workshop, Virginia Tech. and State University Blacksburg, VA (May 1, 1984).
1842. T. L. St. Clair and D. A. Yamaki, Polyimides: Synthesis, Characterization, and Applications (K. L. Mittal, ed.), Plenum Press, New York, Vol. 1, p. 99 (1984).
1843. F. W. Harris et al., Polyimides: Synthesis, Characterization, and Application (K. L. Mittal, ed.), Plenum Press, New York, Vol. 1, p. 3 (1984).
1844. P. M. Sormani, R. J. Minton, and J. E. McGrath, Ring Opening Polymerization: Kinetics, Mechanisms, and Synthesis (J. E. McGrath, ed.), ACS Symp. Series No. 286, Chap. 11, p. 147 (1985).
1845. J. E. McGrath et al., Final Report for NASA Langley Res. Center, Grant No. NAG-1-343. Suppl. 10 (Oct. 1986).
1846. R. H. Bott, J. D. Summers, C. A. Arnold, L. T. Taylor, T. C. Ward, and J. E. McGrath, Synthesis and characterization of novel poly(imide siloxane) segmented copolymers, J. Adhes. 23:67 (1987).
1847. N. J. Johnston and P. M. Hergenrother, High temperature thermoplastics, Sci. Adv. Mater. Proc. Eng. Series 32 (1987): NASA TM 89104.
1848. D. J. Progar and R. A. Pike, Adhesive evaluation of water-soluble LARC-TPI, Int. J. Adhes. Adhes. 8(1):25 (Jan. 1988).
1849. D. J. Progar, T. L. St. Clair, H. D. Burks, C. R. Gautreaux, A. Yamaguchi, and M. Ohta, LARC-TPI 1590 series controlled molecular weight polyimide, 21st Int. SAMPE Tech. Conf. (Sept. 26–28, 1989).
1850. C. A. Arnold, J. D. Summers, Y. P. Chen, R. H. Bott, D. Chen, and J. E. McGrath, Polymer 30(6):986 (1989).
1851. V. M. Startsev et al., Nature of the adhesion interaction and destruction of the polyimide-aluminum foil contact during hardening, Colloid J. USSR 46(5):841 (Sept.–Oct. 1984).
1852. L. P. Buchwalter, Adhesion of polyimide to metal and ceramic surfaces: An overview, J. Adhes. Sci. Technol. 4(9):697 (1990).
1853. S. Wilkensen, Developments at NASA Lewis of high temperature thermosets, North Amer. Res. Conf. Crosslinked Polymers, Hilton Head Island, SC (Nov. 18–22, 1991).
1854. H. Ahne et al., Polyimides: Synthesis, Characterization, and Applications (K. L. Mittal, ed.), Plenum Press, New York, Vol. 2, pp. 905–918 (1984).
1855. P. G. Rickerl, J. G. Stephanie, and P. Slota, Proc. 37th Conf. Electronic Comp., Institute of Electric and Electronic Engineers, pp. 220–225 (1987).
1856. J. Pfeifer, U.S. Patent No, 4,677,186 (1987).
1857. A. A. Lin et al., Macromolecules 21:1165 (1988).
1858. M. Kojima et al., Proc. 39th Conf. Electronic Comp., IEEE, pp. 920–924 (1989).
1859. L. P. Buchwalter, J. Adhes. Sci. Technol. 1(4):341 (1987).
1860. A. D. Ulrich and W. G. Joslyn, Proc. 34th Int. SAMPE Symp. 34(1):127 (1989).
1861. R. J. Jenson and J. H. Lai, Polymers for Electronic Applications, CRC Press, Boca Raton, FL, p. 41 (1989).
1862. F. Faiycel et al., J. Appl. Phys. 65(5):1911 (1989).
1863. D. Suryanarayana and K. L. Mittal, J. Appl. Polym. Sci. 29:2039 (1984).
1864. A. V. Patsis and S. Cheng, J. Adhes. 25:145 (1988).
1865. G. Haldankar, E. Shockey, and A. Garton, Low functionality additives for polyimides and epoxy resins, 14th Ann. Meeting Adhes. Soc., Clearwater, FL (Feb. 17–20, 1991).
1866. H. W. Eickner, Strength properties of metal-bonding adhesives at temperatures from −100°F (−73°C) to 800°F (427°C), WADC Tech. Report. No. 59-152, Forest Prod. Lab., Madison, WI (July 1959).
1867. H. W. Eickner and W. Z. Olson, Statistical variations in lap-joint strength of metal-bonding adhesives at elevated temperatures, Forest Prod. Lab. Report. No. 1880, Madison, WI (Feb. 1961).

1868. Shear strengths of adhesives in stainless steel and aluminum lap joints at temperatures from −100°F (−73°C) to 1000°F (538°C), Forest Prod. Lab. Final Report Contract AF 33(606)61-06 (Feb. 1959–Dec. 1960) (April 1962).
1869. K. C. Tsou, J. Dickstein, and J. P. Halpern, Study leading to the development of high temperature resistant adhesives, Quart. Prog. Report No. 1, Contract No. DA-35-034-ORD-3501-RD, Picatinny Arsenal, Dover, NJ (July–Sept. 1961).
1870. P. M. Hergenrother, 19th Natl. SAMPE Symp. Exhib. 19:146 (1974).
1871. P. M. Hergenrother, Poly(phenyl-as-triazines) and poly(phenylquinoxalines), new cross-linked polymers, Macromolecules 7:575 (1974).
1872. P. M. Hergenrother, Highly fluorinated phenylquinoxaline polymers, Polym. Prepr. 19(2):40 (1978).
1873. P. M. Hergenrother, Poly(phenylquinoxalines) containing ethynyl groups, Macromolecules 14:891 (1981).
1874. P. M. Hergenrother, Poly(phenylquinoxalines) containing pheynlethynyl groups, Macromolecules 14:898 (1981).
1875. P. M. Hergenrother, Paper presented at 1984 Int. Chem. Conf., Pacific Basin Soc., Honolulu (Dec. 1984).
1876. J. L. Hendrick and J. W. Labadie, Synthesis of poly(arylene ether)phenylquinoxalines, Polym. Mater. Sci. Eng. 59:42 (1988).
1877. J. W. Connell and P. M. Hergenrother, Synthesis of polyphenylquinoxalines via aromatic nucleophilic displacement, Polym. Prepr., 29(1):172 (1988).
1878. P. M. Hergenrother, High temperature organic adhesives, SAMPE Quart. 3:1 (1971).
1879. P. M. Hergenrother and D. J. Progar, High temperature composite bonding with PPQ, Adhes. Age 20(11):38 (1977).
1880. C. L. Hendricks and S. G. Hill, Evaluation of high temperature structural adhesives for extended service, Adhesive Chemistry, Polymer Science & Technology, Plenum Press, New York, Vol. 29, p. 489 (1984).
1881. K. C. Brinker and I. M. Robinson, U.S. Patent 2,895,948 (1959).
1882. H. A. Vogel and C. S. Marvel, J. Polym. Sci. 50:511 (1961).
1883. H. H. Levine, Soc. Aerospace Mater. Proc. Eng. Ser., Vol. 9 (1965).
1884. J. I. Jones, Macromol. Sci. Rev. Macromol. Chem., C2(2):343 (1968).
1885. V. V. Korshak, Heat Resistant Polymers, Keter Press, p. 244 (1971) (in Hebrew).
1886. A. H. Frazer, High Temperature Resistant Polymers, Interscience, New York, p. 138 (1968).
1887. P. E. Cassidy, Thermally Stable Polymers, Marcel Dekker, New York, p. 163 (1980).
1888. E. J. Powers and G. A. Serad, High Performance Polymers: Their Origin and Development (R. B. Seymour and G. S. Kirshenbaum, eds.), p. 355 (1986).
1889. A. Buckley, D. E. Stuetz, and G. A. Serad, Encyclopedia of Polymer Science & Technology, 2nd ed., John Wiley & Sons, New York, p. 572 (1988).
1890. B. C. Ward, SAMPE J. 25(2):21 (1989).
1891. J. M. Black and R. F. Blomquist, Metal bonding adhesives for high temperature service, Modern Plastics (June 1956).
1892. R. P. Burford, Fracture toughness of polybutadiene at cryogenic temperatures, J. Mater. Sci. 18:3756 (1983).
1893. F. A. Keimel, Adhesives and Sealants Newsletter (F. A. Keimel, ed.), 9(7):5 (July 8, 1985).
1894. R. C. Kausen, Adhesives for high and low temperatures—Part I, Mater. in Design Eng. 60(2):94 (June 1964); Part II, 60(3):108 (Sept. 1964).
1895. J. Hertz, Adhesives for extreme-low-temperature applications, Electro-Technology 70(3):93 (Sept. 1962).
1896. J. Hertz, An evaluation of several structural adhesives in cryogenic applications, Adhes. Age 4(8):30 (1961).
1897. J. Hertz, Epoxy-nylon adhesives for low temperature applications, Advances in Cryogenic Engineering, Plenum Press, New York, Vol. 7 (1962).
1898. J. Hertz, Polyurethane adhesives at −423°F (253°C), Convair Astronautics Res. Prog. Report MGR-95 (Aug. 25, 1959).
1899. J. Hertz, Cryogenic adhesive evaluation study, Convair Astronautics Res. Prog. Report 111-9106 (Jan. 1961).
1900. M. J. Hiza and P. L. Barrick, Cryogenic adhesive properties of bisphenol-A epoxy resins, SPE Trans. 1:73 (April 1961).
1901. R. M. McClintock and M. J. Hiza, Epoxy resins as cryogenic structural adhesives, Modern Plastics 35(10):172,176,237 (June 1958).

1902. B. Pascuzzi and J. R. Hill, Investigations of structural adhesives for cryogenic applications, SPE-RETEC "Plastics and Adhesives in the Space Age" (May 13–14, 1964).
1903. W. M. Frost, The strength of ten structural adhesives at temperatures down to −423°F (−253°C), Advances Cryogenic Engineering, Vol. 5 (Proc. of 1959 Cryogenic Eng. Conf. at University California, Berkeley (Sept. 3–4, 1959).
1904. F. J. Claus, R. E. Mauri, E. C. Smith, and S. Drake, Evaluating the behavior of materials under space conditions, Inst. Environ. Proc., Washington, D.C. (April 5–7, 1961).
1905. S. R. Sandler, F. Berg, and G. Kitazawa, Development of improved adhesives for use at cryogenic temperatures to −423°F (−253°C), Monthly Reports to NASA Contract NAS 8-11518 (July 1964; Oct. 1964; Nov. 1964; Jan. 1965; April 1965, The Bordon Chemical Co.
1906. S. R. Sandler, F. Berg, and G. Kitazawa, Development of improved adhesives for use at cryogenic temperatures to −423°F (−253°C), Quarterly Report NASA Contract NAS 8-11518, Bordon Chemical Co. (Jan.–March 1965).
1907. L. M. Roseland, Evaluation of structural adhesives for potential cryogenic usage, 7th Natl. SAMPE Symp. (May 1964).
1908. C. S. Schollenberger, Isocyanate-based adhesives, Handbook of Adhesives (I. Skeist, ed.), Reinhold, New York, p. 333 (1962).
1909. J. H. Saunders and K. C. Frisch, High polymers, Polyurethanes: Chemistry and Technology, John Wiley & Sons, New York, Vol. XVI, Part I (1962); Part II (1964).
1910. D. H. Whetzek, R. F. Robbins, P. R. Ludtke, Y. Ohori, and R. N. Herring, Elastomeric seals and materials at cryogenic temperatures, Summary Report Natl. Bur. Stds., Contract AF 33(616)-61-04 (Nov. 1961).
1911. R. B. Gosnell, Elastomeric gasket materials development of cryogenic applications, Annual Report NASA Contract NAS 8-5053, Narmco Res. & Devel. (June 15, 1964).
1912. P. R. Ludtke, Recent developments in using elastomers for static cryogenic seals, 7th Natl. SAMPE Symp. (May 1964).
1913. F. Wilson and J. W. Feldman, Elastomeric gasket materials development of cryogenic applications, Summary Report NASA Contract NAS 8-5053, Narmco Res. & Devel. (June 15, 1964).
1914. D. E. Robinson and F. A. Schreihans, RTV silicons as sealants and adhesives for cryogenic applications, 7th Natl. SAMPE Symp. (May 1964).
1915. I. M. Zelman, R. I. Akawie, and C. J. Bahun, Development of organic sealants for applications at low temperature, Final Report NASA Contract NAS 8-2428 (1962); Summary Report (1963), Hughes Aircraft Co.
1916. R. I. Akawie, Development of seals and sealants for use at cryogenic temperatures, Summary Report NASA Contract NAS 9-2428, Hughes Aircraft Co.
1917. J. M. Klosowski, Premium sealants of the 1980's, Adhes. Age 24(11):32 (1981).
1918. J. R. Panek and J. P. Cook, Construction Sealants and Adhesives, 2nd ed., John Wiley & Sons, New York (1984).
1919. E. G. Rochow, An Introduction to the Chemistry of the Silicones, John Wiley & Sons, New York (1964).
1920. H. W. Prost, Silicones and Other Silicon Compounds, Reinhold, New York (1949).
1921. A. Damusis, Sealants, Reinhold, New York (1967).
1922. R. B. Seymour, Encyclopedia of Chemical Technology, Wiley-Interscience, New York, Vol. 15, pp. 549–558 (1982).
1923. Sealants: The Professional Guide, Sealers and Waterproofers Institute (1984).
1924. E. W. Flick, Adhesive and Sealant Compound Formulations, Noyes Publications, Park Ridge, NJ (1984).
1925. E. W. Flick, Adhesives, Sealants and Coatings for the Electronics Industry, Noyes Publications, Park ridge, NJ (1986).
1926. E. W. Flick, Construction and Structural Adhesives and Sealants—An Industrial Guide, Noyes Publications, Park Ridge, NJ (1988).
1927. K. T. Frisch, Urethanes Technology, 4(1):22 (March 1987).
1928. J. M. Klosowski, Sealants in Construction, Marcel Dekker, New York (1989).
1929. Building Seals and Sealants: Fire Standards; Building Constructions, Annual Book of ASTM Stds., Vol. 04.07 (1989).
1930. Adhesives, Sealants, and Primers, D.A.T.A. Digest, 5th ed. (Sept. 1989).
1931. R. E. Meyer, Encyclopedia of Polymer Science and Engineering, Wiley-Interscience, New York, Vol. 15, pp. 131–145 (1989).
1932. F. Keimal, The Rauch Guide to the U.S. Adhesives and Sealants Industry, Rauch Associates Inc., Bridgewater, NJ (1990).

1933. D. J. Dunn, Sealants and Sealant Technology, Adhesives and Sealants, Engineered Materials Handbook, ASM, Materials Park, OH, Vol. 3, pp. 48–55 (1990).
1934. M. D. Beer and D. M. Brassard, 3rd generation silicone gasketing technology, Paper at 90th S.A.E. Congress (Feb. 1990).
1935. M. G. Elias, R. Redman, and S. Tyler, Sealant markets and applications, Adhesives and Sealants, Engineered Materials Handbook, ASM, Materials Park, OH, Vol. 3, pp. 56–60 (1990).
1936. I. Skeist (ed.), Handbook of Adhesives, 3rd ed., Van Nostrand Reinhold, New York (1990).
1937. G. Fedor and P. Brennan, Correlation of accelerated and natural weathering of sealants, Adhes. Age 33(5):22 (May 1990).
1938. L. B. Strecker and G. A. Gettys, Insulating-glass sealants, Adhes. Age 21(11):22 (1978).
1939. J. Spetz, How fabricators can prevent insulating glass unit failures, Glass Industry, p. 10 (Feb. 1985).
1940. M. B. Berenbaum, Polysulfide polymers, Encyclopedia of Polymer Science & Technology, Interscience, New York, p. 425 (1969).
1941. J. R. Griffith, Epoxy resins containing fluoride, CHEMTECH 12:290 (1982).
1942. P. Johncock and G. F. Tudgey, Epoxy systems with improved water resistance and the non-Fickian behavior of epoxy systems during water aging, Brit. Polym. J. 15:14 (1983).
1943. H. H. Gibbs and C. V. Breder, High temperature laminating resins based on melt fusible polyimides, Amer. Chem. Soc. Polym. Preprints 15(1):775 (1974).
1944. T. L. St. Clair, Linear polyimides lecture to short course on High Temperature Polymers, State University New York at New Paltz (June 1984).
1945. S. Maudal and T. L. St. Clair, Preparation and characterization of siloxane-containing thermoplastic polyimides, Int. J. Adhes. Adhes. 4(2):87 (April 1984).
1946. H. S. Ryang, U.S. Patent 4,404,350 assigned to G. E. Co., Schenectady, NY (Sept. 13, 1983).
1947. B. L. Rathbun and P. W. Schuessler, Moisture permeation of polymer sealants and interface modifying films, Adhesive Chemistry—Developments and Trends (L. H. Lee, ed.), Plenum Press, p. 787 (1984).
1948. P. W. Schuessler, Hydrophobic resins, NASC Phase I Report (May 1981).
1949. G. F. Schmitt, Jr., Spacecraft, aircraft, missiles: Pushing the limits of adhesives, sealants, and coatings, presented at Case Western Reserve University, Cleveland, OH (Jan. 1983).
1950. R. E. Meyer, Polysulfide sealants for aerospace, Part I—Theory and background, SAMPE J., p. 6 (Nov./Dec. 1982).
1951. R. E. Meyer, Polysulfide sealants for aerospace, Part II—Application and handling, SAMPE J. (March/April 1983).
1952. W. Singh, J. W. Hutt, M. E. Williams, and V. Nuys, U.S. Patent 4,366,307, assigned to Prod. Res. & Chem. Corp., Glendale, CA (1982).
1953. J. W. Hutt and H. Singh, New high temperature and fuel resistant sealants based on Permopol P-3 polythioether polymers, Tech. Bull., Prod. Res. & Chem. Corp., Glendale, CA.
1954. H. Singh, Cyanosiloxane sealants used for space shuttle orbiter, Adhes. Age 21(4):24 (1977).
1955. H. Rosenberg and E. W. Choe, Methyl and 333-trifluoropropyl-substituted m-xylylene-siloxanylene polymers, Org. and Plastics Preprints 37(10):166 (1977).
1956. H. Rosenberg and E. W. Choe, Synthesis and properties of fluoroalkylarylene siloxanylene (FASIL) high temperature polymers, Org. Coatings and Plastic Chem. 40:793 (1979).
1957. PNF Elastomer by Firestone, Firestone Rubber and Tire Co., Akron, OH (Oct. 1979).
1958. D. F. Lohr and J. A. Bechman, PNF phosphonitrilic fluoroelastomer properties and application, Tech. Publication, Firestone Rubber and Tire Co., Akron, OH.
1959. R. J. Jones and H. N. Casey, Flexible polyimide fuel tank sealants, Adhes. Age 22(11):33 (1979).
1960. R. J. Berg, A review of seal materials for guided-missile applications, Contract DA 04-495-ORD-18, Calif. Inst. Technol. (March 1958).
1961. J. Lee, Research on high temperature-resistant seal and sealant materials for the supersonic transport, Report No. 1, Contract AF 33(657)-9609 (Oct. 1962); Report No. 2 (Jan. 1963), Republic Aviation Corp.
1962. J. W. Gooch and J. H. Daher, Elastomeric conductive sealants protect aircraft from corrosion, Adhes. Age 33(5):17 (May 1990).
1963. J. Lomax, Acrylic polymer caulks and sealants, paper at Sealant Short Course, Adhesives and Sealants Council Meeting (Jan. 1990).
1964. Standard Specification for Latex Sealing Compounds, C-834, Annual Book of ASTM Stds., ASTM, Philadelphia.
1965. Standard Practices for Use of Latex Sealing Compounds, C-790, Annual Book of ASTM Stds., ASTM, Philadelphia.
1966. Sealing Compound: Elastomeric Type, Single Component, Interim Federal Spec. TT-S-00230C, Natl. Bureau Stds., Washington, D.C.

1967. Type I Back Bedding Glazing Compounds for Use with Architectural Aluminum, AAMA Spec. 802.3, Arch. Aluminum Mfgs. Assoc.
1968. J. P. Cook, Construction Sealants and Adhesives, John Wiley & Sons, New York (1970).
1969. U.S. Patent 3,551,374, BASF Corp.
1970. Belgium Patent 689,663, BASF Corp.
1971. W. H. Brendley, Fundamentals of Acrylic Polymers, Paint, and Varnish Production, Rohm and Haas Company (July 1973).
1972. J. S. Amstock, Adhes. Age, p. 18 (Feb. 1969).
1973. W. J. Reid, Adhes. Age, p. 21 (April 1970).
1974. J. V. Fusco and P. Hous, Butyl and halobutyl rubbers, Introduction to Rubber Technology (M. Morton, ed.), Van Nostrand-Reinhold, New York (1987).
1975. E. N. Kresge, R. H. Schatz, and H. C. Wang, Isobutylene polymers, Encyclopedia of Polymer Science & Engineering, 2nd ed., John Wiley, New York, Vol. 8 (1987).
1976. C. L. Bellance, Preformed butyl tapes in glazing applications, Science & Technology of Glazing Systems (C. J. Parise, ed.), ASTM STP Publ. No. 1054 (1989).
1977. A. J. Berejka, Sealing tapes, Sealants (A. Damusis, ed.), Van Nostrand Reinhold, New York (1967).
1978. J. J. Higgins, Butyl and related solvent release sealants, Sealants (A. Damusis, ed.), Van Nostrand Reinhold, New York (1967).
1979. A. J. Berejka and J. J. Higgins, Broadened horizons for butyl sealants, Adhes. Age 16(12) (Dec. 1973).
1980. A. J. Berejka, Test methods for architectural tape sealants, Report EPL 6804-2747 (June 1968).
1981. J. E. Callan, Cross-linked butyl hot melt sealants, J. Adhes. Sealant Council 3(1) (1974).
1982. F. C. Jagisch, Polyisobutylene polymers in sealants, Adhes. Age 21(11):47 (Nov. 1978).
1983. D. Quade, New technologies in insulating glass, Glass 36 (March 1986).
1984. J. D. Minford, unpublished work, Aluminum Co. of America, Alcoa Labs, Alcoa Center, PA.
1985. J. D. Minford, unpublished work, Aluminum Co. of America, Alcoa Labs, Alcoa Center, PA.
1986. O. Bayer, Polyurethanes, Mod. Plast. 24: 149–152,250,262 (June 1947).
1987. J. H. Saunders and K. C. Frisch, Polyurethanes: Chemistry and Technology, Part I, Chemistry, Wiley-Interscience, New York (1962).
1988. J. H. Saunders and K. C. Frisch, Polyurethanes: Chemistry and Technology, Part II, Technology, Wiley-Interscience, New York (1964).
1989. B. S. Dombrown, Polyurethanes, Plastics Application Series, 2nd ed., Reinhold, New York (1965).
1990. C. E. Leyes, R. E. Weber, and R. E. Jones, Urethane, Modern Plastics Encyclopedia, Nopco Chem., p. 349 (1967).
1991. P. F. Bruins, Polyurethane Technology, Wiley-Interscience, New York (1969).
1992. D. J. David and H. B. Stanley, Polyurethanes: Analytical Chemistry of the Polyurethanes, Part III, Wiley-Interscience, New York (1969).
1993. N. Doyle, The Development and Use of Polyurethane Products, McGraw-Hill, New York (1971).
1994. I. Skeist, Polyurethanes II, Skeist Labs Inc. (Oct. 1983).
1995. R. M. Evans, Advances in Urethanes—Adhesives and Sealants, Technomic (1985).
1996. K. C. Frisch, Advances in Urethanes—Elastomers and Coatings, Technomic, Lancaster, PA (1985).
1997. K. N. Edwards, Urethane Chemistry and Applications, ACS Symp. Series No. 172, Las Vegas, NV (Aug. 1980).
1998. R. M. Evans and R. B. Greene, Urethane sealants, Building Seals and Sealants, ASTM Publ. No. STP 606, ASTM, Philadelphia, pp. 112–123 (1976).
1999. E. G. Rochow, U.S. Patent No. 2,380,995 (Aug. 1945).
2000. J. F. Hyde, U.S. Patent No. 2,490,357 (Dec. 1949).
2001. D. T. Hurd and R. C. Osthoff, U.S. Patent No. 2,737,506 (March 1956).
2002. D. T. Hurd, R. C. Osthoff, and M. L. Corrin, J. Am. Chem. Soc. 76:249 (1954).
2003. W. I. Patnode and D. F. Wilcock, J. Am. Chem. Soc. 68:358 (1946).
2004. E. L. Warrick and R. R. McGregor, U.S. Patent No. 2,431,878 (Dec. 1947).
2005. E. L. Warrick and R. R. McGregor, U.S. Patent No. 2,435,147 (Jan. 1948).
2006. J. Marsden and G. F. Roedel, U.S. Patent No. 2,469,883 (May 1949).
2007. E. C. Britton, H. C. White, and C. L. Moyle, U.S. Patent No. 2,460,805 (Feb. 1949).
2008. H. Knopf, A. Beerwald, and G. Brinkmann, West German Patent No. 957,662 (Feb. 1954).
2009. D. T. Hurd, J. Am. Chem. Soc. 77(2):988 (1955).
2010. N. Reiso and K. K. Yushi, Japanese Patent 3,738 (Jan. 1955).
2011. K. A. Andrianov, S. I. Dzhenchelskaya, and Y. K. Petrashkov, Sov. Plast. Massy 3:20 (1960).

2012. J. M. Klosowski and G. A. L. Gant, The Chemistry of Silicone Room Temperature Vulcanizing Sealants, ACS Symp. Series No. 113, Plastic Mortar, Sealants and Caulking Compounds (R. B. Seymour, ed.), American Chem. Soc., p. 1257 (1979).
2013. J. M. Klosowski, Sealants in Construction, Marcel Dekker, New York, p. 265 (1989).
2014. J. M. Klosowski, Sealants in Construction, Marcel Dekker, New York, p. 215 (1989).
2015. J. D. Minford, unpublished work, Aluminum Co. of America, Alcoa Labs, Alcoa Center, PA.
2016. M. G. Elias, R. Redman, and S. Tyler, Sealants Markets and Applications, Adhesives and Sealants, Engineered Materials Handbook, ASM, Materials Park, OH, Vol. 3, pp. 56–60 (1990).
2017. J. D. Minford, unpublished work, Aluminum Co. of America, Alcoa Labs, Alcoa Center, PA.
2018. E. Frauenglass, J. Moran, and R. Batson, New adhesives speed heat removal, Circuits Manuf. (Feb. 1984).
2019. R. E. Batson, Industrial applications for adhesives, Adhesives and Sealants, Engineered Materials Handbook, ASM, Materials Park, OH, Vol. 3, pp. 567–578 (1990).
2020. W. Hastie, Adhesives for electronics, Circuits Manuf., pp. 56–62 (Jan. 1981).
2021. J. D. Minford, unpublished work, Aluminum Co. of America, Alcoa Labs, Alcoa Center, PA.
2022. W. D. Bascom, Primers and coupling agents, Adhesives and Sealants, Engineered Materials Handbook, ASM, Materials Park, OH, Vol. 3, pp. 254–258 (1990).
2023. J. A. Bishop, E. K. Sim, G. E. Thompson, and G. C. Wood, The adhesively bonded aluminum joint: the effect of pretreatment on durability, J. Adhes. 26:237 (1988).
2024. T. J. Reinhart, Novel concepts for priming metallic adherends for structural adhesive bonding, Adhesion—2 (K. W. Allen, ed.), Applied Science Publishers, London, p. 87 (1978).
2025. A. J. Kinloch and C. M. Taig, The adhesive bonding of thermoplastic composites, J. Adhes. 21:291 (1987).
2026. E. M. Petrie, Adhesively bonded plastics: Meeting an industry challenge, Adhes. Age 32:6 (1989).
2027. R. F. Wegman, Surface Preparation Techniques for Adhesive Bonding, Noyes Publications, Parkridge, NJ, p. 68 (1989).
2028. A. R. Siebert, L. L. Tolle, and R. S. Drake, CTBN-modified epoxies work in poor bonding conditions, Adhes. Age:19 (1986).
2029. A. V. Pocius, Studies of the effect of cured primer thickness on the peel strength of metal-to-metal bonds, Int. SAMPE Meeting, Munich, Germany (Nov. 2–4, 1981).
2030. W. G. Marshall, Corrosion inhibiting primers for aluminum alloy, Aero Mater. Dept. Report No. NADC-MA-7164, Nav. Air Devel. Center, Philadelphia, PA.
2031. D. R. Croke, W. E. Robinson, Jr., and R. N. McCurdy, A test to determine the corrosion resistance of adhesive-primer systems, Mater. Res. Stds. 11(1):19 (1971).
2032. BR-127 Corrosion Inhibiting Adhesive Primer, American Cyanamid Co., Bloomingdale Div., Harve Le Grace, MD.
2033. W. D. Bascom, Structure of silane adhesion promotor films on glass and metal surfaces, Macromolecules 5:792 (1972).
2034. E. P. Plueddemann, H. A. Clark, L. E. Nelson, and K. R. Hoffman, Modern Plastics 39:139 (Aug. 1962).
2035. E. P. Plueddemann, Silanes in bonding thermoplastic polymers to mineral surfaces, Symp. Proc. for Adhesive Bonded Structures, Steven Inst. Technol., Hoboken, NJ, Vol. 1, p. 119 (Aug. 1972).
2036. Silanes, Surfaces, and Interfaces (D. E. Leyden, ed.), Gordon and Breach Science Publishers, London (1986).
2037. J. Comyn, Silane coupling agents, Structural Adhesives: Developments in Resins and Primers (A. J. Kinloch, ed.), Elsevier Applied Science Publishers, Amsterdam, Chap. 8 (1986).
2038. Structural Adhesives: Developments in Resins and Primers (A. J. Kinloch, ed.), Elsevier Applied Science Publishers, Amsterdam (1986).
2039. J. Comyn, D. P. Oxley, R. G. Pritchard, C. R. Werrett, and A. J. Kinloch, Inelastic electron tunnelling spectroscopy of some aminosilane coupling agents, J. Adhes. 28(2–3):171 (1989).
2040. A. F. Diaz, V. Hetzler, and E. Kay, J. Amer. Chem. Soc. 99:6781 (1977).
2041. T. Furukawa, N. K. Eib, K. L. Mittal, and H. R. Anderson, Jr., Surf. Interf. Anal., 4:240 (1982).
2042. T. Furukawa, N. K. Eib, K. L. Mittal, and H. R. Anderson, Jr., J. Colloid Interf. Sci. 24:204 (1983).
2043. P. N. T. Van Velzen, Surf. Sci. 140:437 (1984).
2044. D. M. Brewis et al., Surf. Interf. Anal. 6:40 (1984).
2045. F. J. Boerio, Polym. Preprints. ACS Div. Polym. Chem. 21:297 (1981).
2046. N. H. Sung, A. Kaul, I. Chin, and C. S. P. Sung, Mechanism studies of adhesion promotion by alpha-aminopropyltriethoxy silane in alpha alumina/polyethylene joint, Polym. Eng. Sci. 22:637 (1982).

2047. A. Kaul, N. H. Sing, I. Chin, and C. S. P. Sung, Durability and failure analysis of silane treated alpha alumina/polyethylene joint in wet environment, Polym. Eng. Sci. 24:493 (1984).
2048. A. Kaul, N. H. Sung, I. J. Chin, and C. S. P. Sung, Effect of bulk structure of amino silane primer on the strength and durability of aluminum/epoxy joints, Polym. Eng. Sci. 26:768 (1986).
2049. P. E. Cassidy, J. M. Johnson, and G. C. Rolls, A comprehensive study of coupling agents in urethane and epoxy adhesives, Natl. SAMPE Conf. on Aerospace Adhesives and Elastomers, p. 517 (1970).
2050. F. Liang and P. Dreyfuss, Durability of adhesive bonds between glass or metal substrates and a polybutadiene-polyurethane, J. Appl. Polym. Sci. 29:3147 (1984).
2051. J. Comyn, D. P. Oxley, R. G. Pritchard, C. R. Werrett, and A. J. Kinloch, Examination of the interaction of 3-glycidoxypropyltrimethoxy-silane with aluminum oxide by inelastic electron tunnelling spectroscopy, Int. J. Adhes. Adhes. 9(4):201 (1989).
2052. J. Comyn, D. P. Oxley, R. G. Pritchard, C. R. Werrett, and A. J. Kinloch, An examination of the interaction of silanes containing carbon-carbon double bonds with aluminum oxide by inelastic electron tunnelling spectroscopy, Int. J. Adhes. Adhes. 10(1):13 (1990).
2053. H. S. Schwartz, Effects of adherend surface treatment on stressed durability of adhesive bonded aluminum alloys, SAMPE J. 13(2) (March–April, 1977).
2054. E. C. Millard, Epoxy adhesives, Adhesive Bonding of Aluminum Alloys (E. W. Thrall and R. W. Shannon, eds.), Marcel Dekker, New York, Chap. 7, pp. 110–113 (1985).
2055. Surface Preparation and Priming of Aluminum Alloy Parts for High Durability Structural Bonding, SAE Aerospace Recommended Practice ARP 1524 (March 1978).
2056. H. Dodiuk, A. E. Yaniv, I. E. Klein, N. Fin, and L. Drori, Chemical interaction system aluminum oxide-primer-adhesive, Appl. Surf. Sci. 25(1–2):137 (1986).
2057. I. Berbezier, M. Romand, and J. M. Cuntz, Adhesion primer properties: Microstructural changes induced by aging, J. Adhes. 33(1–2):5 (1990).
2058. G. Xue, J. L. Koenig, and H. Ishida, Chemical reactions of an epoxy-functional silane in aqueous solutions, Rubber Chem. Technol. 64:152 (1991).
2059. Q. X. Yang and Q. L. Zhou, The ESCA and AES studies of the interfacial chemical bonding between aluminum and chromium (III) fumarato-coordination compound, Adhesive Chemistry—Developments and Trends (L. H. Lee, ed.), Plenum Press, New York, p. 799 (1984).
2060. R. W. Spenser, Adhesive primers for multiple state bonding and surface corrosion protection, Stevens Inst. Technol., Meetings, Hoboken, NJ (Aug. 1972).
2061. S. C. Aker, The function of adhesive primers in aircraft bonding, Stevens Inst. Technol., Meeting, Hoboken, NJ (Aug. 1972).
2062. J. D. Minford, unpublished work, Aluminum Co. of America, Alcoa Labs, Alcoa Center, PA.
2063. D. Finch, Zinc-rich primers to protect metal substrates, Mater. Eng. 103(4):41 (April 1986).
2064. R. A. Pike, F. P. Lamm, and J. P. Pinto, Factors affecting the processing of epoxy film adhesives, III. Heat-up rate, J. Adhes. 17(1):51 (1984).
2065. R. A. Pike, Proc. 17th Nat. SAMPE Tech. Conf., p. 448 (1985).
2066. R. A. Pike, Int. J. Adhes. Adhes. 5:3 (1985).
2067. R. A. Pike, Int. J. Adhes. Adhes. 6:21 (1986).
2068. R. A. Pike and F. P. Lamm, Adhesives, Sealants, and Coatings for Space and Harsh Environments (L. H. Lee, ed.), Plenum Press, New York, p. 141 (1988).
2069. W. Brockmann, 10th Int. Conf. Org. Coatings Sci. & Technol., Athens, Greece, p. 11 (1984).
2070. R. A. Pike, Inorganic adhesive primers: Effect of metal surface treatments and formation temperature, Int. J. Adhes. Adhes. 6(1):21 (1986).
2071. J. A. Filbey and J. P. Wightman, Metal alkoxide primers in titanium/epoxy bonding, J. Adhes. 28(1): 23 (1989).
2072. R. A. Pike, Water-based inorganic adhesive primers, Adhesion '90, Plastics and Rubber Inst. Meetings, Cambridge, England (Sept. 10–12, 1990).
2073. J. C. Bolger, Structural adhesives for metal bonding, Treatise on Adhesion and Adhesives (R. L. Patrick, ed.), Marcel Dekker, New York, Vol. 3, pp. 31–46 (1973).
2074. J. C. Bolger, Structural adhesives: today's state-of-the-art, Adhesives in Manufacturing (G. L. Schneberger, ed.), Marcel Dekker, New York, Chap. 7, pp. 148–166 (1983).
2075. E. C. Millard, Epoxy adhesives, Adhesive Bonding of Aluminum Alloys (E. W. Thrall and R. W. Shannon, eds.), Marcel Dekker, New York, Chap. 7, pp. 109–110 (1985).
2076. J. D. Minford, unpublished work, Aluminum Co. of America, Alcoa Labs, Alcoa Center, PA.
2077. H. Dodiuk, L. Drori, and J. Miller, The effect of moisture content on the performance of epoxy film adhesives, Part I. J. Adhes. 17:33 (1984).

2078. H. Dodiuk, L. Drori, and J. Miller, The effect of moisture content on the performance of epoxy film adhesives, Part II. J. Adhes. 19:1 (1985).
2079. H. Dodiuk, L. Drori, and J. Miller, Int. J. Adhes. Adhes. 4:169 (1984).
2080. H. Dodiuk, L. Drori, and S. Kenig, The effect of moisture content on the performance of epoxy film adhesives, Part III. J. Adhes. 24:229 (1987).
2081. H. Dodiuk, L. Drori, and N. Fin, The effect of moisture content on the performance of epoxy film adhesives, Part IV. Aluminum-filled epoxy, J. Adhes. 26:315 (1988).
2082. I. Grabovac, S. A. Mestan, and C. E. M. Morris, The effect of freeze-thaw cycling on some structural film adhesives, J. Appl. Polym. Sci. 29(12):4407 (Dec. 1984).
2083. E. J. Hughes and J. L. Rutherford, Study micromechanical properties of joints, Picatinny Arsenal Report No. 3744-AD-673745, Dover, NJ (1968).
2084. R. Seago, Natl. SAMPE Tech. Conf. 2:135 (1970).
2085. A. C. Moloney, Ph.D. Thesis, Leicester Polytechnic, Leichester, England (1979).
2086. D. M. Brewis, J. Comyn, B. C. Cope, and A. C. Moloney, Polym. Eng. Sci. 21(2):797 (1981).
2087. D. M. Brewis, J. Comyn, B. C. Cope, and A. C. Moloney, Polymer 21:344 (1980).
2088. Quarterly Prog. Report No. 2. Div. Tech. Report No. FZM-6892, General Dynamics, Fort Worth, TX.
2089. Quarterly Prog. Report No. 9, Div. Tech. Report No. FZM-7010, General Dynamics, Fort Worth, TX.
2090. J. Romanko, K. M. Liechti, and W. G. Knauss, Integrated methodology for adhesive bonded joint life predictions, AFWAL-TR-8204139, Final Report for Period July 1979 to July 1982, Wright-Patterson AFB, OH, 95433 (Nov. 1982).
2091. J. Romanko, K. M. Liechti, and W. G. Knauss, Life prediction methodology for adhesively bonded joints, Adhesive Joints: Formation, Characteristics, and Testing (K. L. Mittal, ed.), Plenum Press, New York, p. 567 (1984).
2092. L. R. Brantley and W. H. Jones, The investigation of the nature of the forces of adhesion, Tech. Report under Contract N900NR-86701, Occidental College, Los Angeles (July 1, 1951–June 30, 1952).
2093. L. R. Brantley, K. N. Bills, Jr., and R. Stabler, The investigation of the nature of the forces of adhesion, Tech. Report under Contract N90NR-86701, Occidental College, Los Angeles (July 1, 1952–June 30, 1953).
2094. L. R. Brantley, B. Stott, and J. Charnell, Tech. Report under Contract N90NR-86701, Occidental College, Los Angeles (July 1, 1953–June 30, 1954).
2095. L. R. Brantley, B. Stott, and J. Charnell, Tech. Report under Contract N00NR-86701, Occidental College, Los Angeles (July 1, 1954–June 30, 1955).
2096. D. de Jager, Economic appraisal of hot melt produced PSA's in 1976, Course Notes, Int. Hot Melt Pressure-Sensitive Adhesives, TAPPI, Amsterdam (Nov. 3–5, 1976).
2097. The Rapid Growth and Maturing Future Markets of Hot-Melt Adhesives, Sealants, Coatings 1976–1986, A confidential study by Springborn Labs., in collaboration with H. S. Hoplappa and Associates (July 1977).
2098. A. Maletsky and J. Villa, Paper, Film, and Foil Converter, p. 55 (Sept. 1976).
2099. G. Holden, Block and Graft Polymerization (R. J. Ceresa, ed.), John Wiley & Sons, New York (1973).
2100. J. T. Harland and L. A. Petershagen, Handbook of Adhesives, 2nd ed. (I. Skeist, ed.), Reinhold, New York (1977).
2101. K. C. Brinker, Raw materials for hot-melt pressure-sensitive adhesives-EVA copolymers, 1976 Course Notes, Hot-Melts, TAPPI, Boxborough, MA (May 5, 1976).
2102. R. E. Duncan and J. E. Bergerhouse, Adhes. Age p. 37 (March 1980).
2103. M. R. Rifi, Pressure-sensitive adhesive structures vs performance, in 1977 Course Notes, Hot-Melts-A Western Vista, TAPPI, San Diego (Nov. 7–10, 1977).
2104. R. R. Schmidt III and J. D. Holmes, U.S. Patent to Eastman Kodak Co., No. 4,143,858 (1979).
2105. J. D. Minford, unpublished work, Aluminum Co. of America, Alcoa Labs, Alcoa Center, PA.
2106. J. D. Minford, unpublished work, Aluminum Co. of America, Alcoa Labs, Alcoa Center, PA.
2107. J. D. Minford, unpublished work, Aluminum Co. of America, Alcoa Labs, Alcoa Center, PA.
2108. C. Bluestein, Radiant energy curable adhesives, Adhes. Age 25(12):19 (1982).
2109. Advances in Photochemistry (D. Bryce Smith, ed.), ACS, Washington, D.C., Vols. 1–12 (1968–1980).
2110. Radiation Processing, Literature, and Patents Survey, K & M Publications, Louisville, KY (1977).
2111. Bibliography 117—Radiation of Elastomers, Bibliography 127—UHF, HF, and IR Curing of Rubber and Plastics, Rubber Division Library & Information Service, University of Akron, Ohio.
2112. S. P. Pappas (ed.), UV Curing, Science, and Technology, Technology Marketing Company, Norwalk, CT (1978).
2113. R. M. Kossoff and J. W. Prane (eds.), Radiation Curable Adhesives, R. M. Kossoff & Associates, New York (1980).

2114. Radiation Curing: A Discussion of Advantages, Features, and Applications (A. H. Keough and J. Lacey, eds.), Association for Finishing Processes/Soc. Mfg. Engs., Dearborn, MI (1980).
2115. Safe Handling and Use of Ultraviolet/Electron Beam Curable Coatings, Natl. Paint and Coatings Assoc., Washington, D.C. (1980).
2116. Handbook of Radiation Measurement and Protection, Section A. General Scientific and Engineering Information; Section B. Radiation Protection Principles and Practice, CRC Press, Boca Raton, FL (1982).
2117. Reactive Cure Systems: UV-IR-EB, Captan Associates Inc. Lyndhurst, NJ (1982–1983).
2118. J. Friedman, Hand-held UV curing lamps for industrial applications, Adhes. Age 25(12):23 (1982).
2119. A. F. Readdy, Jr., Plastics fabrication by ultraviolet, infrared, induction, dielectric, and microwave radiation methods, Plastics Report R43, Plastics Tech. Evaluation Center, Picatinny Arsenal, Dover, NJ (April 1972).
2120. R. W. Harries, New developments in low-energy reactive adhesives, Adhes. Age 21(2):26 (1978).
2121. C. L. Mahoney, Structural adhesives for rapid-cure applications, Adhes. Age 22(12):26 (1979).
2122. R. Kardashian and S. V. Nablo, Electron beam curing equipment, Adhes. Age 25(12):25 (1982).
2123. Curing equipment and radiation safety, Safe handling and use of ultraviolet beam curable coatings, Natl. Paint and Coatings Association, Washington, D.C. (August 1980).
2124. OSHA safety and health standards (29CRF 1910), Occupational Safety and Health Administration, U.S. Dept. Labor.
2125. S. V. Nablo, Method of and apparatus for shielding inert zone electron irradiation of moving web materials, U.S. Patent No. 4,253,413 (Feb. 24, 1982).
2126. S. V. Nablo, Operational maintenance considerations with the electrocurtain processor (ESI TR-86), Proc. Soc. Mfg. Eng. Workshop for Radiation Curing in the Graphic Arts, Rosemont, IL (March 25–27, 1980).
2127. J. Woods, Radiation Curing of Polymers (D. R. Randell, ed.), Royal Society Chemistry, pp. 102–115 (1987).
2128. J. Woods, Radiation Curing: Science and Technology (S. P. Pappas, ed.), Plenum Press, New York (1991).
2129. S. P. Pappas, Comprehensive Polymer Science (G. Allen, J. C. Bevington, G. C. Eastmond, et al., eds.), Pergamon Press, 4(2):337–355 (1989).
2130. S. P. Pappas, Comprehensive Polymer Science 6:135 (1989).
2131. S. P. Pappas, Handbook of Organic Photochemistry (J. C. Scaiano, ed.), CRC Press, Boca Raton, FL, pp. 329–339 (1989).
2132. K. O'Hara, Radiation Curing of Polymers (D. R. Randell, ed.), Royal Society Chemistry, pp. 116–127 (1987).
2133. J. G. Kloosterboer, Advanced Polym. Sci. 84:1 (1988).
2134. S. P. Pappas, J. Radiation Curing 14(3):6 (1987).
2135. C. L. Marino, Mach. Design 56(18):50 (1984).
2136. A. H. Pincus, Proc. Polymers, Laminations, and Coatings Conf., Tech. Assoc. Pulp and Paper Ind., p. 207 (1986).
2137. A. H. Pincus, Adhes. Age 32(6):16 (1989).
2138. C. F. Lewis, Mater. Eng. 104(11):39 (1987).
2139. Radiation Curing—An Introduction to Coatings, Varnishes, Adhesives, and Inks, Ed. Comm. Radiation Curing Div. The Association for Finishing Proc. SME, Dearborn, MI, pp. 3–20 (1984).
2140. J. Woods, Tech. Paper No. FC85-414, The Association for Finishing Processes, SME, Dearborn, MI (1985).
2141. G. R. Smoluk, Mod. Plastics 65(6):250 (1988).
2142. W. J. Morris, SME Tech. Paper No. FC84-993, SME, Dearborn, MI (1984).
2143. K. C. Steuben, Adhesive Chemistry—Developments and Trends (L. H. Lee, ed.), Plenum Press, New York, p. 319 (1984).
2144. A. Barker, Adhes. Age 31(6):48 (1988).
2145. A. Barker, Adhes. Age 32(9):47 (1989).
2146. A. Schwarz, Radiation curing of PSA products, SME Tech. Paper No. FC76-497, SME, Dearborn, MI (1976).
2147. Y. Nakano, Photosensitive adhesives and sealants, Adhes. Age 16(12):28 (1973).
2148. K. C. Steuben, Radiation curing of pressure sensitive adhesives: A literature review, Adhes. Age 20(5): 16 (1977).
2149. J. H. Tsao and A. D. Ketley, U.S. Patent No. 4,156,035 (to W. R. Grace Co.) (1979).
2150. J. V. Crivello, Cationic polymerization initiated by light, CHEMTECH 10:624 (1980).

2151. J. V. Crivello and H. W. Lam, Photoinitiated cationic polymerization by dialkyl-4-hydroxyphenyl sulfonium salts, J. Ed. 18:2677,2698 (1980).
2152. W. C. Perkins, New developments in photo-induced cationic polymerizations, J. Rad. Curing 8(1):16 (Jan. 1981).
2153. A. Ledwith, Cation-radical mechanism for photopolymerization, Paper presented at Northeast Regional ACS Meeting, Rochester, NY (Oct. 1981).
2154. A. Ledwith, A. Al-Kass, D. C. Sherrinton, and P. Bonner, Ion pair dissociation equilibria for iodenium and sulphonium salts useful in photoinitiated cationic polymerization, Polymer 22:143 (1981).
2155. K. C. Steuben, Ultraviolet-cured pressure-sensitive adhesives, Adhesive Chemistry—Developments and Trends (L. H. Lee, ed.), Plenum Press, New York (1984).
2156. V. D. McGinness, Polymer and formulation design characteristics for developing adhesive bonding radiation-curable systems, Adhesive Chemistry—Developments and Trends (L. H. Lee, ed.), Plenum Press, New York (1984).
2157. J. V. Crivello, Recent progress in photoinitiated cationic polymerization, Adhesive Chemistry—Developments and Trends (L. H. Lee, ed.), Plenum Press, New York (1984).
2158. S. Torrey, Adhesive technology development since 1977, Chem. Tech. Rev. No. 95, Noyes Data, Parkridge, NJ (1977).
2159. D. J. St. Clair, Radiation curing of pressure-sensitive adhesives based on thermoplastic rubbers, 1979 Course Notes, Hot Melts—An Overview for Management, Technical Association of the Pulp and Paper Industry, SC (June 4–7, 1979).
2160. S. H. Ganslaw, Hot melt pressure-sensitive adhesives, alternatives to rubber-resin systems, 1979 Course Notes, Hot Melts—An Overview for Management, TAPPI, SC (June 4–7, 1979).
2161. T. J. Smith, The chemistry and application of UV-cured adhesives, Adhesion '90, The Plastics and Rubber Institute Meetings, Cambridge, England (Sept. 10–12, 1990).
2162. M. Fernando and A. J. Kinloch, The adhesion of photopolymers, Adhesion '90, The Plastics and Rubber Institute Meetings, Cambridge, England (Sept. 10–12, 1990).
2163. C. Decker, K. Moussa, and T. Bendaikha, Photodegradation of UV-cured coatings. II. Polyurethane-acrylate networks, J. Polymer Sci. Polymer Chem. 29(5):739 (1991).
2164. G. Berner, R. Kirchmayr, G. Rist, and W. Rutsch, J. Rad. Curing 13(4):10 (1986).
2165. S. Peeters, J. M. Loutz, and M. Philips, Polym. Paint Colour J. 179:304 (1989).
2166. S. P. Pappas and H. B. Feng, Org. Coatings 8:139 (1985).
2167. J. V. Crivello, Advanced Polym. Sci. 62:1 (1984).
2168. G. E. Green and A. C. Zahir, U.S. Patent No. 4,308,367 (to Ciba-Geigy) (1987).
2169. J. E. Gervay, European Pat. Appl. 270,945 (to Du Pont) (1987).
2170. C. R. Morgan, U.S. Patent No. 4,288,527 (to W. R. Grace) (1981).
2171. N. S. Allen, S. J. Hardy, A. Jacobine, D. M. Glaser, and F. Catalina, European Polym. J. 25:1219 (1989).
2172. S. Nakos, U.S. Patent No. 4,451,523 (to Loctite) (1987).
2173. G. Gozzelineo, A. Priola, and F. Ferrero, Makromol. Chem. Makromol. Symp. 23:393 (1989).
2174. L. A. Nativi and K. Kadziela, U.S. Patent No. 4,451,523 (to Loctite) (1984).
2175. S. Grant, Radcure Europe '87, Conf. Proc. Soc. Mfg. Eng., Dearborn, MI (July 1987).
2176. N. Siga, N. Seo, T. Harutake, and M. Shirae, U.S. Patent No, 4,778,253 (To Olympus Optical) (1988).
2177. J. M. Rooney, J. Woods, and P. Conway, U.S. Patent No. 4,525,232 (to Loctite) (1985).
2178. P. Conway, D. P. Melody, J. Woods, E. Casey, B. J. Bolger, and F. R. Martin, U.S. Patent No. 4,533,446 (to Loctite) (1985).
2179. J. Woods and P. Coakley, European Pat. Appl. 251,465 (to Loctite) (1987).
2180. R. A. A. Gonzalez, H. C. Nicolaisen, and H. P. Handwerk, U.S. Patent No. 4,882,001 (to Henkel) (1989).
2181. H. Nakahara, Complex multilayer boards vie for space in Japanese computers, Electronic Packaging Prod. 27(2):70 (Feb. 1987).
2182. R. R. Tummala and E. J. Rymaszewski (eds.), Microelectronic Packaging Handbook, Van Nostrand Reinhold, New York (1989).
2183. H. W. Markstein, Laminates support technology advances while offering alternate choices, Electronic Packaging Prod. 24(6):83 (1984).
2184. M. Schlack, Guide to high performance engineering plastics, Plastics World 45(4):30 (April 1987).
2185. M. K. Antoon, B. E. Zehner, and J. L. Koenig, Spectroscopic determination of the in-situ composition of epoxy matrices in glass fiber reinforced composites, Polym. Comp. 1(1):24 (Jan. 1980).

2186. D. A. Kourtides, J. A. Parker, T. W. Grants, N. Bilow, and M. T. Hsu, Thermophysical and flammability characterization of phosphorylated epoxy adhesives, Adhesives for Industries Technol. Conf., El Segundo, CA, p. 92 (June 25, 1980).
2187. L. H. Lee, Adhesives and sealants for severe environments, Int. J. Adhes. Adhes. 7(2):81–91 (April 1987).
2188. F. Kumar, G. M. Fohlen, and J. A. Parker, Fire-and-heat-resistant laminating resins based on maleimido-substituted aromatic cyclotriphosphazines, Macromolecules 15:1250 (1983).
2189. F. H. Sharpell, Microbial contamination and preservation of adhesives, Adhes. Age 25(4):23 (1982).
2190. P. D. Kostenbader and J. W. Flecksteiner, Biological oxidation of coke plant weak ammonia liquor, Industrial Water Eng. (Feb. 1969).
2191. F. H. Sharpell, The control of microbial contamination in waxes, Chem. Times and Trends 2(1):67079 (1978).
2192. D. L. Wedderburn, Hygiene in manufacturing plants and its effect on preservation of emulsions, J. Soc. Cosmetic Chem. 16:395 (1965).
2193. F. J. Upsher, Fungal resistance of modified epoxy structural film adhesives, Int. Biodeterioration Bull. 16(2):43 (1980).
2194. F. W. Sharpell, Industrial uses of biocides in processes and products, Developments in Industrial Microbiology, Impressions Ltd., Gaithersburg, Maryland (1980).
2195. S. J. Lederer, J. A. Jakubowski, and H. A. Birnbaum, An effective preservative for adhesives with reduced health hazards, Adhes. Age 25(4):28 (1982).
2196. A. J. Kinloch, Review-the science of adhesion, Part 2: Mechanics and mechanisms of failure, J. Mater. Sci. 17:617 (1982).
2197. D. A. Yurek, Design of adhesive bonded joints, Adhes. Age 8(12) (Dec. 1965).
2198. C. V. Cagle, Design criteria and joint design, Adhesive Bonding Techniques and Applications, McGraw-Hill, New York (1968).
2199. Design Considerations, Adhesives in Modern Manufacturing, Soc. Mfg. Engs., Dearborn, MI (1970).
2200. S. Mostovoy and E. J. Ripling, Effect of joint geometry on the toughness of epoxy adhesives, J. Appl. Polym. Sci. 15:661 (1971).
2201. G. L. Schneberger, Designing adhesive joints, Adhesives in Manufacturing (G. L. Schneberger, ed.), Marcel Dekker, New York, Chap. 4, p. 67 (1983).
2202. L. J. Hart-Smith and E. W. Thrall, Structural analysis of adhesive bonded joints, Adhesive Bonding of Aluminum Alloys (E. W. Thrall and R. W. Shannon, eds.), Marcel Dekker, New York, Chap. 13, p. 241 (1985).
2203. J. T. Quinlivan and D. Arnold, Designing with adhesives and sealants, Adhesives and Sealants, Engineered Materials Handbook, ASM, Materials Park, OH, Vol. 3, pp. 457–458 (1990).
2204. R. B. Krieger, Jr., Adhesive bonding design and analysis, Adhesives and Sealants, Engineered Materials Handbook, ASM, Materials Park, OH, Vol. 3, pp. 459–470 (1990).
2205. L. J. Hart-Smith, Rating and comparing structural adhesives: A new method, Adhesives and Sealants, Engineered Materials Handbook, ASM, Materials Park, OH, Vol. 3, pp. 471–476 (1990).
2206. F. E. Penado and R. K. Dropek, Numerical design and analysis, Adhesives and Sealants, Engineered Materials Handbook, ASM, Materials Park, OH, Vol. 3, pp. 476–500 (1990).
2207. E. Sancaktor, Fatigue and fracture mechanics, Adhesives and Sealants, Engineered Materials Handbook, ASM, Materials Park, OH, Vol. 3, pp. 501–520 (1990).
2208. F. H. Chang, Flaw detection (effects of defects), Adhesives and Sealants, Engineered Materials Handbook, ASM, Materials Park, OH, Vol. 3, pp. 521–532 (1990).
2209. R. W. Johnson, Design proof testing, Adhesives and Sealants, Engineered Materials Handbook, ASM, Materials Park, OH, Vol. 3, pp. 533–544 (1990).
2209a. DeHavilland D H 98, Mosquito, Jane's All the World's Aircraft.
2210. J. E. Thompson, Design considerations unique to sealants, Adhesives and Sealants, Engineered Materials Handbook, ASM, Materials Park, OH, Vol. 3, pp. 545–550 (1990).
2211. K. Drain and S. Chandrasekharan, Automotive applications for adhesives, Adhesives and Sealants, Engineered Materials Handbook, ASM, Materials Park, OH, Vol. 3, pp. 551–557 (1990).
2212. E. C. Clark, Aerospace applications for adhesives, Adhesives and Sealants, Engineered Materials Handbook, ASM, Materials Park, OH, Vol. 3, pp. 558–566 (1990).
2213. R. E. Batson, Industrial applications for adhesives, Adhesives and Sealants, Engineered Materials Handbook, ASM, Materials Park, OH, Vol. 3, pp. 567–578 (1990).
2214. H. K. Charles, Jr., Electronic packaging applications for adhesives, Adhesives and Sealants, Engineered Materials Handbook, ASM, Materials Park, OH, Vol. 3, pp. 579–603 (1990).

2215. K. Adams, Applications for sealants, Adhesives and Sealants, Engineered Materials Handbook, ASM, Materials Park, OH, Vol. 3, pp. 604–612 (1990).
2216. R. B. Krieger, Jr., Evaluating structural adhesives under sustained load in hostile environments, Proc. SAMPE Conf. (Oct. 1973).
2217. R. B. Krieger, Jr., Shear stress-strain properties of structural adhesives in hostile environments, American Cyanamide Co. (Oct. 1976).
2218. R. B. Krieger, Jr., Shear stress-strain properties of structural adhesives in hostile environments, J. Appl. Polym. Sci. pp. 321–339 (1977).
2219. O. Volkersen, Die Nietkraft Verteilung in Zugbeanspruchten Nietverbindungen mit Constanten Laschenquerschnitten, Luftfahrtforschung, Vol. 15, pp. 41–47 (in German) (1938).
2220. M. Goland and E. Reissner, The stresses in cemented joints, J. Appl. Mech. (Trans. ASME), 11:A17–A27 (1944).
2221. J. Sneddon, The distribution of stress in adhesive joints, Adhesives (D. Eley, ed.), Oxford University Press, London, England (1961).
2222. E. W. Kuenzi and G. H. Stevens, Determination of the mechanical properties of adhesives for use in the design of bonded joints, Report FPL-011, U.S. Forest Prod. Lab., Madison, WI (1963).
2223. D. B. Bogey, Edge bonded dissimilar orthogonal elastic wedges under normal shear loading, J. Appl. Mech. 35:460 (1968).
2224. D. B. Bogey, Two edge-bonded elastic wedges of different materials and wedge angles under surface tractions, J. Appl. Mech. 38:377 (1971).
2225. V. L. Hein and F. Erdogan, Stress singularities in a two-material wedge, Int. J. Fract. Mech. 7:317 (1971).
2226. D. B. Bogey and K. C. Wang, Stress singularities at interface corners in bonded dissimilar isotropic elastic materials, Int. J. Solid Struc. 7:993 (1971).
2227. R. D. Adams and N. A. Peppiatt, Stress analysis of adhesive-bonded joints, J. Strain Anal. 9:185 (1974).
2228. L. J. Hart-Smith, Adhesives bonded single-lap joints, NASA Report CR-112236, Douglas Aircraft Co. (Jan. 1973).
2229. T. Wah, Stress distribution in a bonded anisotropic lap joint, J. Eng. Mater. Technol. (Trans. ASME) 95:174 (1973).
2230. I. U. Ojalvo and H. L. Edinoff, Bond thickness effects on stresses in single lap joints, AIAA J. 16:204 (1978).
2231. D. Chen and S. Cheng, An analysis of adhesive-bonded single lap joints, J. Appl. Mech. 50:109 (1983).
2232. F. E. Penado and E. S. Folias, The three-dimensional stress field around a cyclindrical inclusion in a plate of arbitrary thickness, Int. J. Fract. 39:129 (1989).
2233. O. C. Zienkiewicz, The Finite Element Method, McGraw-Hill, New York (1977).
2234. K. H. Huebner and E. A. Thornton, The Finite Element Method for Engineers, John Wiley & Sons, New York (1982).
2235. K. J. Bathe, Finite Element Procedures in Engineering Analysis, Prentice-Hall, New York (1982).
2236. L. J. Segerlind, Applied Finite Element Analysis, John Wiley & Sons, New York (1984).
2237. R. K. Dropek, Numerical design and analysis of structures, Engineering Materials Handbook, ASM International, Vol. 1, pp. 464–478 (1987).
2238. F. E. Penado and R. K. Dropek, Numerical design and analysis, Adhesives and Sealants, Engineered Materials Handbook, ASM, Materials Park, OH, Vol. 3, p. 479 (1990).
2239. J. N. Reddy and S. Roy, Finite-element analysis of adhesive joints, Adhesive Bonding (L. H. Lee, ed.), Plenum Press, New York, Chap. 13, pp. 359–394 (1991).
2240. L. H. Lee (ed.), Adhesive Bonding, Plenum Press, New York (1991).
2241. J. N. Reddy and S. Roy, Finite element analysis of adhesively bonded joints, Report VPI-E-85.18, ONR, Dept. Eng. Sci. & Mech., VPI&SYU, Blacksburg, VA (Aug. 1985).
2242. S. Roy and J. N. Reddy, Nonlinear viscoelastic analysis of adhesively bonded joints, Report No. VPI-E-86.28, ONR, Dept. Eng. Sci. & Mech., VPI&SU, Blacksburg, VA (Nov. 1986).
2243. F. Erdogan and M. Ratwani, Stress distributions in bonded joints, J. Compos. Mater. 5:378 (1971).
2244. G. R. Wooley and D. R. Carver, Stress concentration factors for bonded joints, J. Aircraft 8:817 (1971).
2245. A. T. Liu, Linear elastic and elasto-plastic stress analysis for adhesive lap joints, T.A.M. Report No. 410, University of Illinois at Urbana-Champaign (July 1976).
2246. R. D. Adams and N. A. Peppiatt, Stress analysis of lap joints in fibre-reinforced composite materials, Fibre Reinf. Plast., p. 45, ICE, London (1977).
2247. E. A. Humphries and C. T. Herakovich, Nonlinear analysis of bonded joints with thermal effects, Report No. VPI-E-77.19, Dept. Eng. Sci. & Mech., Virginia Polytechnic Institute and State University, Blacksburg, VA (June 1977).

2248. U. Yucelglu and D. P. Updike, Stress analysis of bonded plates and joints, Eng. Mech. Div. ASCE 106:37 (1980).
2249. F. Delale and F. Erdogan, Viscoelastic analysis of adhesively bonded joints, J. Appl. Mech. 48:331 (1981).
2250. F. Delale and F. Erdogan, Time-temperature effect in adhesively bonded joints, J. Compos. Mater. 5: 561 (1981).
2251. S. Gali and O. Ishai, Interlaminar stress distribution within an adhesive layer in the non-linear range, J. Adhes. 9:253 (1978).
2252. Y. R. Nagaraja and R. S. Alwar, Viscoelastic analysis of an adhesive bonded plane lap joint, J. Adhes. 10:97 (1979).
2253. E. C. Francis, W. L. Hufferd, D. G. Lemini, R. E. Thompson, W. E. Briggs, and R. R. Parmeter, Time dependent fracture in adhesive bonded joints, Chem. Systems Div., Sunnyvale, CA, Interim Reports CSD 2769-1R-01/02 (May and Nov. 1982).
2254. B. Dattaguru, R. A. Everette Jr., J. D. Whitcomb, and W. S. Johnson, Geometrically nonlinear analysis of adhesively bonded joints, J. Eng. Mater. Technol. 195:59 (1984).
2255. L. R. Botha, R. M. Jones, and H. F. Brinson, Viscoelastic analysis of adhesive stresses in bonded joints, Report VPI-E-33-17, Dept. Eng. Sci. & Mech., VPI&SU, Blacksburg, VA (May 1983).
2256. M. Henriksen, Nonlinear viscoelastic stress analysis—a finite element approach, Comput. Struct. 18:133 (1984).
2257. E. B. Becker et al., Viscoelastic stress analysis including moisture diffusion for adhesively bonded joints, Report AFWAL-TR-84-4057, General Dynamics, Fort Worth, TX (Aug. 1984).
2258. S. Yadagiri and C. Papi Reddy, Viscoelastic analysis of nearly incompressibly solids, Comput. Struct. 20:817 (1985).
2259. S. Yadagiri, C. Papi Reddy, and T. Sanjeeva Reddy, Viscoelastic analysis of adhesively bonded joints, Comput. Struct. 27:445 (1987).
2260. R. A. Schapery, Further development of a thermodynamic constitutive theory: Stress formulation, A&S Report No. 69-2, Purdue University, West Lafayette, IN (Feb. 1969).
2261. R. A. Schapery, A method of viscoelastic stress analysis using elastic solutions, J. Franklin Inst. 279: 268 (1965).
2262. D. Peretz and Y. Weitsman, The non-linear thermo-viscoelastic characterization of FM-73 adhesive, J. Rheol. 26:245 (1983).
2263. W. G. Knauss and I. J. Emri, Non-linear viscoelasticity based on free volume considerations, Comput. Struct. 13:123 (1981).
2264. Y. Weitsman, An investigation of non-linear viscoelastic effects on load transfer in a symmetric double lap joint, J. Adhes. 11:279 (1981).
2265. Y. Weitsman, Stresses in adhesive joints due to moisture and temperature, J. Compos. Mater. 11:368 (1977).
2266. Y. Weitsman, Interfacial stresses in viscoelastic adhesive layers due to moisture sorption, Int. J. Solid Struct. 15:701 (1970).
2267. H. Ghoneim and Y. Chen, A viscoelastic-viscoplastic constitutive equation and its finite element implementation, Comput. Struct. 17:499 (1983).
2268. P. Czarnocki and K. Piekarski, Nonlinear numerical stress analysis of a symmetric adhesively bonded lap joint, Int. J. Adhes. Adhes. 6:156 (1986).
2269. S. Roy and J. N. Reddy, Nonlinear analysis of adhesively bonded joints, J. Non-Linear Mech. 23:97 (1988).
2270. S. Roy and J. N. Reddy, Nonlinear viscoelastic analysis of adhesively bonded joints, Tire Sci. Technol. 16:146 (1988).
2271. G. P. Anderson and K. L. DeVrie, Predicting strength of joints from test results, Int. J. Fract. 39:191 (1989).
2272. B. M. Malyshev and R. L. Salganik, The strength of adhesive joints using the theory of cracks, Int. J. Fract. Mech. 1:114 (1965).
2273. W. G. Knauss, Fracture mechanics and the time dependent strength of adhesive joints, J. Compos. Mater. 5:176 (1971).
2274. G. G. Trantina, Fracture mechanics approach to adhesive, J. Compos. Mater. 6:192 (1972).
2275. G. G. Trantina, Combined mode crack extension in adhesive, J. Compos. Mater. 6:373 (1972).
2276. W. S. Johnson and S. Mall, A fracture mechanics approach for designing adhesively bonded joints, Delamination and Debonding of Materials, ASTM Spec. Report No. 878 (W. S. Johnson, ed.), ASTM, Philadelphia, pp. 189–199 (1985).

2277. F. E. Penado and R. K. Dropek, Numerical design and analysis, Adhesives and Sealants, Engineered Materials Handbook, ASM, Materials Park, OH, Vol. 3, pp. 483–484 (1990).
2278. F. E. Penado and R. K. Dropek, Numerical design and analysis, Adhesives and Sealants, Engineered Materials Handbook, ASM, Materials Park, OH, Vol. 3, pp. 493–497 (1990).
2279. P. L. Flynn, NDE of adhesively bonded metal laminates, General Dynamics Report No. ERR-FW-1777 (Dec. 1976).
2280. L. J. Hart-Smith, Effects of flaws and porosity on strength of adhesively bonded joints, McDonnell-Douglas Report No. MDC J4699 (Nov. 1981).
2282. S. Y. Chuang, F. H. Chang, and J. R. Bell, Adhesive bond and composite strength screening system development study, Fed. Aviation Admin. Report DOT/FAA/CT-84/2 (April 1984).
2283. J. L. Caarls, R. J. Querido, and A. G. Julier, Principles and applications of the Fokker Bond Tester, Air Transport Assoc. Amer. Nondestructive Testing Forum (Sept. 1985).
2284. D. J. Hagemaier, Adhesive-bonded joints, Nondestructive Evaluation and Quality Control, Vol. 17, 9th ed., ASM Metals Handbook, Material Park, OH, pp. 610–640 (1989).
2285. Fokker F-27 and 28, Jane's All the World's Aircraft (1964).
2286. Boeing YC-14 Advanced Transport Prototype, Jane's All the World's Aircraft (1976).
2287. R. B. Aniversario, S. T. Harvey, J. E. McCarty, J. T. Parson, D. C. Peterson, L. D. Pritchett, R. D. Wilson, and E. R. Wogulis, Design, Ancillary Testing, Analysis, and Fabrication Data for the Advanced Composite Stabilizer for Boeing 737 Aircraft, Vol. II, Final Report NASA CR-166011, The Boeing Company (Dec. 1982).
2288. Bell/Boeing V-22, Jane's All the World's Aircraft (1988).
2289. Navy A-6 Intruder, Grumman/Boeing, Jane's All the World's Aircraft (1988).
2290. E. W. Thrall, Jr., et al., Primary adhesively bonded structure technology (PABST), Phase II: detail design, Douglas Aircraft Tech. Report. AFFDL-TR-77-135 (Aug. 1977).
2291. R. W. Johnson, L. W. Thomson, and R. D. Wilson, Study on utilization of advanced composites in fuselage structures of large transports, Boeing Commercial Airplane Co. Report NASA CR-172496 (Feb. 1985).
2292. R. E. Horton, et al., Damage tolerance of composites, Vol. 1, AFWAL-TR-87-3030, Air Force Aeronautical Labs., Wright-Patterson AFB, OH (July 1988).
2293. L. J. Hart-Smith, Differences between adhesive behavior in test coupons and structural joints, Paper at ASTM Committee D-14 Meeting, Phoenix, AZ (March 1981).
2294. L. J. Hart-Smith, Adhesive layer thickness and porosity criteria for bonded joints, Report AFWAL-TR-82-4172, U.S. Air Force Aeronautical Labs., Wright-Patterson AFB, OH (Dec. 1982).
2295. D. V. Chovill, S. T. Harvey, J. E. McCarty, O. E. Desper, E. S. Jamison, and H. Syder, Advanced Composite Elevator for Boeing 727 Aircraft, Vol. 1, Tech. Summary, NASA CR-3290, The Boeing Company (1981).
2296. J. E. McCarty, Full-scale testing, composites, Engineered Materials Handbook, ASM International, Materials Park, OH, Vol. 1 (1987).
2297. Building Sealants: Materials, Properties, and Performance, ASTM spec. Publ. STP No. 1069 (T. F. O'Connor, ed.), ASTM, Philadelphia, PA.
2298. J. W. Prane, Sealants and Caulks, Handbook of Adhesives (I. Skeist, ed.), Van Nostrand Reinhold, New York (1977).
2299. J. P. Cook, Construction Sealants and Adhesives, Wiley-Interscience, New York (1970).
2300. F. W. Shishler and J. M. Klosowski, Sealant stresses in tension and shear, Building Sealants: Materials, Properties, and Performance, ASTM Spec. Publ. No. 1069, ASTM, Philadelphia, PA.
2301. Joint Design for Formed-in-Place Gasketing, Publ., No. CDS-4128, General Electric Co. (1980).
2302. J. Tokarski, Formed-in-place gaskets: Concept vs. reality, Paper at Gaskets: A Symposium, SAE, Warrendale, PA (1977).
2303. Joint Sealers, One-Part Silicone Building Sealants, Rhone-Poulenc Inc.; Rhodosil 3B, Rhodosil 5C, Rodosil 6B, Rhodisil 70, Rhodosil 90, Specification Data Sheet, Construction Specifications Institute (1987–1989).
2304. Rhodosil Silicone Sealants, Data Sheet, Rhodosil 4B, Rhone-Poulenc Inc. (1989).
2305. Sealant Manual, Flat Glass Marketing Association (1983).
2306. G. P. Anderson and K. L. DeVries, Predicting bond strength, J. Adhes. 23:289 (1987).
2307. A. N. Gent and P. B. Lindley, Proc. Royal Soc. A249:195 (1958).
2308. A. N. Gent and P. B. Lindley, Proc. Inst. Mech. Eng. 173:111 (1959).
2309. E. W. Kuenzi and G. H. Stevens, U.S. Forest Prod. Lab. Report No. FPL-011, Madison, WI.
2310. R. S. Alwar and Y. R. Nagaraja, J. Adhes. 7:279 (1976).

2311. R. D. Adams, J. Coppendale, and N. A. Peppiatt, J. Strain Anal. 13:1 (1978).
2312. B. P. Holownia, J. Strain Anal. 7:236 (1972).
2313. N. L. Harrison and W. J. Harrison, J. Adhes. 3:195 (1972).
2314. G. P. Anderson and K. L. DeVries, Analysis of standard bond strength tests, Treatise on Adhesion and Adhesives (R. L. Patrick, ed.), Marcel Dekker, New York, Vol. 6, Chap. 2, pp. 55–121 (1989).
2315. C. Mylonas, Proc. Soc. Exptl. Stress Anal. 12:129 (1954).
2316. L. Greenwood, Aspects of Adhesion—5 (D. J. Alnar, ed.), University of London Press, London, p. 40 (1970).
2317. D. W. Cherry and N. L. Harrison, J. Adhes. 2:125 (1970).
2318. L. J. Hart-Smith, Report No. 6059A, Douglas Aircraft Co., Long Beach, CA (1972).
2319. L. J. Hart-Smith, Report No. 6224, Douglas Aircraft Co., Long Beach, CA (1974).
2320. R. D. Adams, S. H. Chambers, P. J. A. DelStrather, and N. A. Peppiatt, J. Strain Anal. 8:52 (1973).
2321. R. D. Adams and N. A. Peppiatt, J. Strain Anal. 8:134 (1973).
2322. W. J. Renton and J. R. Vinson, J. Adhes. 7:175 (1975).
2323. W. J. Renton and J. R. Vinson, Eng. Frac. Mech. 7:175 (1975).
2324. W. J. Renton and J. R. Vinson, J. Appl. Mech. 44:101 (1977).
2325. D. J. Allman, Mechs. Appl. Maths. 30:415 (1977).
2326. F. Thamm, J. Adhes. 7:301 (1976).
2327. O. Ishai, D. Peretz, and S. Gali, Exptl. Mech. 17:265 (1977).
2328. R. D. Adams, J. Coppendale, and N. A. Peppiatt, Adhesion—2 (K. R. Allen, ed.), Applied Science Publishers, London, p. 105 (1978).
2329. R. P. Penning, Report No. 78,042, Eng. Sci. Data Unit, London (1978).
2330. L. J. Hart-Smith, Report CR-2218, NASA, Washington, D.C. (1974).
2331. L. J. Hart-Smith, Report No. 6922, Douglas Aircraft Co., Long Beach, CA (1980).
2332. L. J. Hart-Smith, Developments in Adhesives—2 (A. J. Kinloch, ed.), Applied Science Publishers, London, p. 1 (1981).
2333. A. J. Kinloch and S. J. Shaw, J. Adhes. 12:59 (1981).
2334. R. D. Adams and J. A. Harris, Int. J. Adhes. Adhes. 4(2):65 (1984).
2335. R. D. Adams and J. A. Harris, The influence of local geometry on the strength of adhesive joints, Int. J. Adhes. Adhes. 7(2):69 (1987).
2336. E. Sancaktar, S. C. Schenck, and S. Padgilwar, Material characterization of structural adhesives in the lap shear mode I. the effect of rate, Ind. Eng. Chem. Prod. Res. Dev. 23(3):426 (1984).
2337. C. Mylonas and N. A. DeBruyne, Adhesion and Adhesives (N. A. DeBruyne and R. Houwink, eds.), Elsevier, Amsterdam, p. 91 (1951).
2338. G. J. Spies, Aircraft Eng. 30:2 (March 1953).
2339. D. H. Kaelble, Trans. Soc. Rheol. 3:161 (1959).
2340. Y. Inoue and Y. Kobatake, Appl. Sci. Res. A8:321 (1959).
2341. D. H. Kaelble, Trans. Soc. Rheol. 4:45 (1960).
2342. D. H. Kaelble, Trans. Soc. Rheol. 9:135 (1965).
2343. J. J. Bikerman, The Science of Adhesive Joints, Academic Press, New York, p. 243 (1968).
2344. A. Crocombe and R. D. Adams, Adhesion and Adhesives: Science Technology and Application, Plastics and Rubber Institute, London, p. 101 (1980).
2345. R. D. Adams and A. Crocombe, J. Adhes. 12:127 (1981).
2346. M. H. Pahdja, Stress analysis of an adhesive joint subjected to tension, shear forces, and bending moments, University of Illinois at Urbana Report (Aug. 1972).
2347. O. Ishar and S. Gali, Interlaminar stress distribution within an adhesive interlayer in the non-linear range, J. Adhes. 9:253 (1978).
2348. S. Gali, G. Dolev, and O. Ishai, Effective stress/strain concept in the mechanical characterization of structural adhesive bonding, Int. J. Adhes. Adhes. 1(3):135 (1981).
2349. R. B. Krieger, Jr., Stress analysis concepts for adhesive bonding of aircraft primary structure, Adhesively Bonded Joints: Testing, Analysis, and Design Conf., ASTM, Baltimore, MD (Sept. 10–12, 1986).
2350. Y. Gillibert, M. L. L. Klein, and A. Rigolot, Mechanical behavior assessment of epoxy adhesive in the double-lap joint, Adhesively Bonded Joints: Testing, Analysis, and Design Conference, ASTM, Baltimore, MD (Sept. 10–12, 1986).
2351. A. Alvazzadeh, M. Bechara, A. Ghazal, and G. Verchery, Special mixed finite elements for interfacial stress analysis of adhesively bonded joints, Adhesively Bonded Joints: Testing, Analysis, and Design Conference, ASTM, Baltimore, MD (Sept. 10–12, 1986).
2352. K. M. Liechti and T. Hayashi, On the uniformity of stresses in some adhesive deformation specimens, J. Adhes. 29(1–4):25 (1989).

2353. W. C. Carpenter and R. Barsoum, Two finite elements for modeling adhesive in bonded configurations, J. Adhes. 30(1–4):25 (1989).
2354. D. A. Bigwood and A. D. Crocombe, Elastic analysis and engineering design formulae for bonded joints, Int. J. Adhes. Adhes. 9(4):229 (1989).
2355. A. N. Gent and C. W. Lin, Comparison of peel and lap shear bond strengths for elastic joints with and without residual stresses, J. Adhes. 30(1–4):1 (1989).
2356. A. Amijima and T. Fujii, Extension of a one-dimensional finite element model program for analyzing elastic-plastic stresses and progressive failure of adhesive bonded joints, Int. J. Adhes. Adhes. 9(4):242 (1989).
2357. T. Sawa, K. Temma, and H. Ishikawa, Three-dimensional stress analysis of adhesive butt joints of solid cyclinders subjected to external stresses, J. Adhes. 31(1):33 (1989).
2358. H. L. Groth, Calculation of stresses in bonded joints using the substructuring technique, Int. J. Adhes. Adhes. 6(1):31 (1986).
2359. J. E. Ritter, Adhesion of thin polymer coatings under contact stresses, Adhesion '90, The Plastics and Rubber Institute Meetings, Cambridge, England (Sept. 10–12, 1990).
2360. Y. Q. Hu, P. N. Li, and D. Y. Ju, Boundary element method for stress analysis of adhesive bonded structures, Adhesion '90, The Plastics and Rubber Inst. Meetings, Cambridge, England (Sept. 10–12, 1990).
2361. X. Zhao and R. D. Adams, A new approach to determining the bending moment factor in single lap joints, Adhesion '90, The Plastics and Rubber Institute Meetings, Cambridge, England (Sept. 10–12, 1990).
2362. D. A. Bigwood and A. D. Crocombe, Non-linear adhesive bonded joint design analysis, Int. J. Adhes. Adhes. 10(1):31 (1990).
2363. R. Davies and A. A. Khalil, Design and analysis of bonded double containment corner joints, Int. J. Adhes. Adhes. 10(1):31 (1990).
2364. D. Chen and S. Cheng, Stress distribution in plane scarf and butt joints, J. Appl. Mech. 57(1):78 (1990).
2365. G. W. Wycherley, S. A. Mestan, and I. Grabovac, A method of uniform shear stress-strain analysis of adhesives, J. Test. Eval. 18(3):208 (1990).
2366. S. Suzuki, Stress analysis of cemented orthotropic lap joints, J. Strain Anal. for Eng. Design 25(1):12 (1990).
2367. J. J. Dike and G. C. Johnson, Residual stress determination using acoustoelasticity, J. Appl. Mech. 57(1):12 (1990).
2368. W. Dannenberg, SPE J. 21(7):669 (1965).
2369. W. T. Chen and C. W. Nelson, IBM J. Res. Devel. 23(2):179 (1979).
2370. N. A. Debruyne and H. Houwink, Adhesion and Cohesion, Elsevier, London, p. 98 (1951).
2371. R. Rolf, J. E. Jombock, and L. K. Peters, Adhes. Age 14(7) (1971).
2372. G. P. Anderson, K. L. DeVries, and G. Sharon, Evaluation of adhesive test methods, Adhesive Joints: Evaluation, Characteristics and Testing (K. L. Mittal, ed.), Plenum Press, New York, p. 269 (1984).
2373. L. J. Hart-Smith, Report 6707, Douglas Aircraft Co., Long Beach, CA (1978).
2374. O. Volkersen, Const. Met. 4:3 (1965).
2375. L. J. Hart-Smith, Report CR-112235, Douglas Aicraft Co., Long Beach, CA (Jan. 1973).
2376. O. Volkersen, The Distribution of the Forces on the Rivets in Stretched Joints With Constant Strap Cross Sections, J Helledoren, Trans., Durand Reprinting Committee, California Institute of Technology.
2377. N. A. DeBruyne, The strength of glued joints, Aircraft Eng. 16:115–118, 140 (1940).
2378. L. J. Hart-Smith, Design and analysis of bonded repairs for metal aircraft structures, Bonded Repair of Aircraft Structures (A. A. Baker and R. Jones, eds.), Martinus Nijhoff, pp. 31–47 (1988).
2379. C. Mylonas and N. A. DeBruyne, Adhesion and Adhesives (N. A. DeBruyne and R. Houwink, eds.), Elsevier, Amsterdam, p. 91 (1951).
2380. F. Erdogan and M. Ratnani, J. Adhes. 5:378 (1971).
2381. F. E. Penado and R. K. Dropek, Numerical design and analysis, Adhesives and Sealants, Engineered Materials Handbook, ASM, Materials Park, OH, Vol. 3, p. 487 (1990).
2382. L. J. Hart-Smith and E. W. Thrall, Structural analysis of adhesive bonded joints, Adhesive Bonding of Aluminum Alloys (E. W. Thrall and R. W. Shannon, eds.), Marcel Dekker, New York, Chap. 13, pp. 247–250 (1985).
2383. L. J. Hart-Smith and E. W. Thrall, Structural analysis of adhesive bonded joints, Adhesive Bonding of Aluminum Alloys (E. W. Thrall and R. W. Shannon, eds.), Marcel Dekker, New York, Chap. 13, pp. 241–244 (1985).
2384. M. D. Wright, Composites 9:259 (1978).
2385. G. N. Sage, Adhesion—3 (K. W. Allen, ed.), Applied Science Publishers, London, p. 123 (1979).

2386. J. M. Miller, J. L. Hammill, and K. E. Luyk, Understanding effects of adhesive ductility and bondline geometry on tube-and-socket joint performance, ASTM Int. Symp. on Adhesively Bonded Joints: Testing, Analysis, and Design, Baltimore, MD (Sept. 10–12, 1986).
2387. J. M. Miller, Evaluation of adhesive-bonded extrusion to simulated casting joints, unpublished report, Alcoa Labs, Alcoa Center, PA.
2388. M. Bresko, The behavior of tube-and-socket adhesively bonded joints including mechanical fasteners, unpublished report, Alcoa Labs, Alcoa Center, PA.
2389. Production Design Guide for Adhesive Bonding Sheet Metal, Amer. Iron & Steel Inst., Washington, D.C., no date.
2390. R. L. Lubkin and E. Reissner, Trans. ASME 78:1213 (1956).
2391. R. D. Adams and N. A. Peppiatt, J. Adhes. 9:1 (1977).
2392. J. D. Minford, unpublished work, Aluminum Co. of America.
2393. A. Beevers and A. C. P. Kho, The performance of adhesive bonded thin-gage sheet metal structures with particular reference to box-section beams, Adhesive Joints: Formation, Characteristics, and Testing (K. L. Mittal, ed.), Plenum Press, New York, p. 627 (1984).
2394. H. H. G. Jellinek, Surfaces and polymer stability, Physicochemical Aspects of Polymer Surfaces (K. L. Mittal, ed.), Plenum Press, New York, pp. 255–281 (1983).
2395. H. A. Willis and V. J. I. Zichy, Polymer Surfaces (D. T. Clark and W. J. Feast, eds.), John Wiley & Sons, New York, Chap. 15 (1978).
2396. D. M. Brewis and D. Briggs, Polymer 22:7 (1981).
2397. H. Schonhorn, Polymer Surfaces (D. T. Clark and W. J. Feast, eds.), John Wiley & Sons, New York, Chap. 10 (1978).
2398. D. Briggs, D. M. Brewis, and M. B. Konieczko, J. Mater. Sci. 14:1344 (1979).
2399. A. Dilk, Anal. Chem. 53:802A (1981).
2400. D. T. Clark and W. J. Feast (eds.), Polymer Surfaces, John Wiley & Sons, New York (1978).
2401. D. W. Dwight and H. R. Thomas, Polym. Prepr. 22(1):302 (1981).
2402. A. D. Briggs, Elastic and viscoelastic adhesion, Physicochemical Aspects of Polymer Surfaces (K. L. Mittal, ed.), Plenum Press, New York, pp. 669–687 (1983).
2403. H. Hertz, J. Reine Angew. Math. 92:156 (1881).
2404. K. L. Johnson, K. Kendall, and A. D. Roberts, Proc. Royal Soc. A324:301 (1971).
2405. J. A. Greenwood and K. L. Johnson, Phil. Mag. A43:697 (1981).
2406. J. F. Archard, Tribology Int. 7:213 (1974).
2407. D. J. Whitehous and J. F. Archard, Proc. Royal Soc. A316:97 (1970).
2408. K. N. G. Fuller and D. Tabor, Proc. Royal Soc. A345:327 (1975).
2409. G. A. D. Briggs and B. J. Briscoe, Nature 260:313 (1976).
2410. G. A. D. Briggs and B. J. Briscoe, J. Phys. D: Appl. Phys. 10:2453 (1977).
2411. D. Maugis and M. Barquins, J. Phys. D: Appl. Phys. 11:1989 (1978).
2412. D. Maugis, Adherence and Fracture Mechanics, Adhesive Bonding (L. H. Lee, ed.), Plenum Press, New York, Chap. 11, pp. 303–335 (1991).
2413. R. S. Rivlin and A. G. Thomas, J. Polym. Sci. 10:291 (1973).
2414. G. I. Barenblatt, The mathematical theory of equilibrium cracks in brittle fracture, Adv. Appl. Mech. 7:55 (1962).
2415. P. C. Paris and G. C. Sih, Stress analysis of cracks, Fracture Toughness Testing and Its Applications, ASTM Spec. Publ., STP No. 381, pp. 30–83, Philadelphia (1965).
2416. C. Gurner and J. Hunt, Quasi-static crack propagation, Proc. Roy. Soc., London, Ser: A-299:508 (1967).
2417. J. R. Rice, J. Appl. Mech. 35:379 (1968).
2418. E. H. Andrews and B. J. Walker, Proc. Roy. Soc., London, Ser: A-324:57 (1971).
2419. M. L. Williams, J. Adhes. 4:307 (1972).
2420. J. D. Landes and J. A. Begley, ASTM, Spec. Publ., STP No. 514, pp. 1–23 (1972).
2421. J. G. Williams, Stress Analysis of Polymers, Longman, London (1973).
2422. J. F. Knott, Fundamentals of Fracture Mechanics, Butterworths, London, p. 94 (1973).
2423. E. H. Andrews, J. Mater. Sci. 9:887 (1974).
2424. M. L. Williams and G. P. Anderson, "Fracture 1977", 4th. Int. Cong. Fracture 1:643 (1977), University of Waterloo Press, Waterloo, Ontario, Canada (1978).
2425. S. J. Burns, J. C. Pollet, and C. Lun Chow, Nonlinear fracture mechanics, Int. J. Fract. 14:311–326 (1978).
2426. J. W. Hutchison and P. C. Paris, Stability analysis of J controlled crack growth, Elastic-Plastic Fracture, ASTM, Spec. Tech. Publ., STP No. 668, Philadelphia (1979).
2427. J. G. Williams, Stress Analysis of Polymers, Ellis Horwood, Chichester, England, p. 291 (1980).

2428. D. Maugis and M. Barquins, Fracture mechanics and adherence of viscoelastic solids, Adhesion and Adsorption of Polymers (L. H. Lee, ed.), Plenum Press, New York, Part A, pp. 203–277 (1980).
2429. D. Maugis, Subcritical crack growth, surface energy and fracture toughness of brittle materials, Fracture Mechanics of Ceramics (R. C. Bradt, A. G. Evans, D. P. H. Hasselman, and F. F. Lange, eds.), Plenum Press, New York, Vol. 8, pp. 255–272 (1986).
2430. E. H. Andrews, Fracture Mechanics Approach, Adhesive Bonding (L. H. Lee, ed.), Plenum Press, New York, Chap. 12, pp. 337–358 (1991).
2431. T. Smith and D. H. Kaelble, Mechanisms of adhesive failure between polymers and metallic substrates: Aluminum 2024-T-3 with HT-424 adhesives, Treatise on Adhesion and Adhesives (R. L. Patrick, ed.), Marcel Dekker, New York, Vol. 5, Chap. 4, pp. 139–233 (1981).
2432. O. Kubaschewski and B. E. Hopkins, Oxidation of Metals and Alloys, 2nd ed., Butterworths, London (1962).
2433. F. Keller, M. S. Hunter, and D. L. Robinson, J. Electrochem. Soc. 101:355 (1954).
2434. F. Liechti and W. D. Treadwell, Helv. Chim. Acta 30:1204 (1947).
2435. J. D. Minford, Adhesives, Durability of Structural Adhesives (A. J. Kinloch, ed.), Applied Science Publishers, London, Chap. 4, p. 168 (1983).
2436. R. A. Harrington and H. R. Nelson, Trans. AIME 137:62 (1940).
2437. H. W. L. Phillips, Inst. Met. Monograph and Report Series No. 13:237 (1952).
2438. J. C. Grosskreutz, Mechanical properties of metal oxide films, J. Electrochem. Soc. Solid State Science 116(9) (Sept. 1969).
2439. J. C. Grosskreutz, The effect of oxide films on dislocation-surface interactions in aluminum, Surf. Sci. 8:173 (1967).
2440. D. K. Benson and J. C. Grosskreutz, Suppression of fatigue cracking through control of surface conditions, AFML-TR-67-343, Air Force Mater. Lab., Wright-Patterson AFB, OH (Sept. 1968).
2441. W. D. Bascom, Macromolecules 5:792 (1972).
2442. R. F. Wegman, A study of adhesive joint failures by evaporative rate analysis, Picatinny Arsenal Report No. PA-TR-3788, Picatinny Arsenal, Dover, NJ (Jan. 1969).
2443. H. F. Brinson, The viscoelasticity of adhesives and composites including failure and fracture concepts, Lecture at Symp. on Fundamentals of Adhesion Theory, Practice, and Applications, New York University at New Paltz (Oct. 10–12, 1984).
2444. L. H. Lee, Significant Advances and Developments in Adhesion and Adhesives (L. H. Lee, ed.), Plenum Press, New York, Vol. A, p. 1 (1975).
2445. L. H. Lee, Adhesion and Adsorption of Polymers, Plenum Press, New York (1979).
2446. A. J. Kinloch, The science of adhesion—Part 1, surface and interfacial aspects, J. Mater. Sci. 15:2141 (1980).
2447. W. S. Sampath and P. J. Wilbur, A new rapid technique for characterizing microstructures of films produced by ion beams in the transmission microscope, J. Mater. Sci. Eng. 90:367 (June 1987).
2448. H. Ishida, Quantitative surface FT-IR spectroscopic analysis of polymers, Rubber Chem. Technol. 60(3):497 (July–Aug. 1987).
2449. J. A. Gilbert, Performance capabilities of intermediate film build electrocoat versus high film build electrocoat, J. Coat. Technol. 62(782):29 (1990).
2450. C. G. Zimba, V. M. Hallmark, S. Turell, and J. D. Swalen, Application of Fourier transform raman spectroscopy to studies of thin polymer films, J. Phys. Chem. 94(2):939 (1990).
2451. J. W. Ager, D. K. Veirs, and G. M. Rosenblatt, Raman intensities and interference effect for thin films adsorbed on metals, J. Chem. Phys. 92(3):2067 (1990).
2452. W. Russell, Fracture analysis of adhesive-bonded joints, SAMPE Quart. p. 8 (July 1978).
2453. M. L. Williams, R. F. Landerl, and J. D. Ferry, J. Am. Chem. Soc. 77:3701 (1955).
2454. A. N. Gent and C. T. R. Pulford, Micromechanics of fracture in elastomers, J. Mater. Sci. 19(11):3612 (Nov. 1984).
2455. H. F. Enderle, H. G. Kilian, and Th. Vilgis, Irreversible deformation of macromolecular networks, Colloid and Polym. Sci. 262(9):696 (Sept. 1984).
2456. A. Carpinteri, Size effects in material strength due to crack growth and material non-linearity, Theoret. Appl. Fract. Mech. 2(1):39 (1984).
2457. K. Ravi-Chander and W. G. Knauss, An experimental investigation into dynamic fracture II. microstructural aspects, Int. J. Fract. 26(1):65 (Sept. 1984).
2458. L. Relland and L. J. Broutman, Surface embrittlement of ductile polymers: A fracture mechanics analysis, Polymer Eng. Sci. 25(4):207 (March 1985).
2459. G. B. Sinclair, Some inherently unreliable practices in present day fracture mechanics, Int. J. Fract. 28(1):3 (May 1985).

2460. C. L. Chow and J. Xu, Mixed mode ductive fracture using the strain energy density criterion, Int. J. Fract. 28(1):17 (May 1985).
2461. G. Karami and R. T. Fenner, Analysis of mixed mode fracture and crack closure using the boundary integral equation method, Int. J. Fract. 30(1):13 (Jan. 1986).
2462. S. Melin, When does a crack grow under Mode II conditions, Int. J. Fract. 30(2):103 (Feb. 1986).
2463. A. N. Gent and C. Wang, Fracture mechanics and cavitation in rubber-like solids, J. Mater. Sci. 26(12): 3392 (1991).
2464. Y. Termonia, Multiaxial deformation of polymer networks, Macromolecules 24(5):1128 (1991).
2465. D. Jangblad, P. Gradin, and T. Stenstrom, Determination and verification of elastic parameters for adhesives, Adhesively Bonded Joints: Testing, Analysis, and Design Conf., ASTM, Baltimore, MD (Sept. 10–12, 1986).
2466. J. H. Crews, K. N. Shivakumar, I. S. Raju, Factors influencing elastic stresses in double cantilever beam specimens, Adhesively Bonded Joints: Testing, Analysis, and Design Conf., ASTM, Baltimore, MD (Sept. 10–12, 1986).
2467. L. G. Stringer, Comparison of the shear stress-strain behavior of some structural adhesives, J. Adhes. 18(3):185 (1985).
2468. A. G. Atkins and Y. W. Mai, Residual strain energy in elastoplastic adhesive and cohesive fracture, Int. J. Fract. 30(3):203 (March 1986).
2469. D. Post, R. Czarnek, J. D. Woods, and D. Joh, Deformation and strains in a thick adherend lap joint, Adhesively Bonded Joints: Testing, Analysis, and Design Conf., ASTM, Baltimore, MD (Sept. 10–12, 1986).
2470. K. Temma, T. Sawa, and A. Iwatas, Two-dimensional stress analysis of adhesive butt joints subjected to cleavage loads, Int. J. Adhes. Adhes. 10(4):285 (1990).
2471. K. Temma, T. Sawa, H. Uchida, and Y. Nakano, A two-dimensional stress analysis of butt adhesive joints having a circular hole defect in the adhesive subjected to external bending moments, J. Adhes. 33(1–2):133 (1990).
2472. K. Temma, T. Sawa, and Y. Tsunoda, Three-dimensional stress analysis of adhesive butt joints with disbond areas and spew fillets, Int. J. Adhes. Adhes. 10(4):294 (1990).
2473. T. Corson, Y. H. Dai, and D. A. Dillard, Peel stress distribution between adherends with varying curvature mismatch, J. Adhes. 33(1–2):107 (1990).
2474. T. S. Chao, Viscoelastic properties of glass forming polymers, J. Mater. Sci. 25(2A):957 (1990).
2475. J. P. Korb, B. Sapoval, C. Chachaty, and A. M. Tischenko, Nuclear relaxation and fractal structure of a cross-linked polymer, J. Phys. Chem. 94(2):953 (1990).
2476. M. Parvin and W. Knauss, Damage induced constitutive response of a thermoplastic related to composites and adhesive bonding, Int. J. Fract. 42(1) (1990).
2477. W. J. Cantwell, J. W. Smith, H. H. Kausch, and T. Kaiser, Examination of the process of deformation and fracture in a silica-filled epoxy resin, J. Mater. Sci. 25(1B):633 (1990).
2478. E. A. A. van Hartingsveldt and J. J. van Aartsen, Strain-rate dependence of interfacial adhesion in particle-reinforced polymers, Polymer 32(8):1482 (1991).
2479. A. J. Kinloch and R. J. Young, Fracture Behavior of Polymers, Applied Science Publishers, London (1983).
2480. D. L. Hunston, J. L. Bitner, J. L. Rushford, and J. Oroshnik, J. Elast. Plastics 12:133 (1980).
2481. A. J. Kinloch, S. J. Shaw, and D. L. Hunston, Proc. Int. Conf. on Yield, Deformation and Fracture, Plastics and Rubber Inst., Cambridge, England (1982).
2482. A. K. El-Senussi and J. P. H. Webber, On the double cantilever beam technique for studying crack propagation, J. Appl. Phys. 56(4):885 (Aug. 1984).
2483. S. N. Zhurkov, V. S. Kurksenko, and V. A. Petrov, Principles of the kinetic approach of fracture prediction, Theoret. Appl. Fract. Mech. 1(3):271 (1984).
2484. K. F. Fischer, Review of brittle fracture criteria in case of static and cyclic mixed mode loading, Theoret. Appl. Fract. Mech. 1(2):117 (1984).
2485. H. Jozavi and E. Sancaktar, The effects of cure temperature and time on the bulk fracture properties of a structural adhesive, J. Adhes. 18(1):25 (1985).
2486. Y. B. Zeng, L. Z. Zhang, W. Z. Peng, and Q. Yu, Microstructure, mechanical properties, and fracture behavior of liquid rubber toughened thermosets, J. Appl. Polym. Sci. 42(7):1905 (1991).
2487. D. L. Hunston, W. D. Bascom, E. E. Wells, J. D. Fahey, and J. L. Bitner, Adhesion and Adsorption of Polymers (L. H. Lee, ed.), Plenum Press, New York, Part A, p. 321 (1980).
2488. G. P. Anderson, S. J. Bennett, and K. L. DeVries, Analysis and Testing of Adhesive Bonds, Academic Press, New York (1977).
2489. M. L. Williams, J. Appl. Polym. Sci. 14:1121 (1970).

2490. I. N. Sneddon, The distribution of stress in the neighborhood of a crack in an elastic solid, Proc. Royal Soc. A-187:229 (1946).
2491. M. L. Williams, On the stress distribution at the base of a stationary crack. Trans. ASME, J. Appl. Mech. 24:109–114 (1957).
2492. E. J. Ripling, S. Mostovoy, and R. L. Patrick, Materials Res. Lab. Studies 4:129 (1960).
2493. G. R. Irwin, Fracture mechanics applied to adhesives systems, Treatise on Adhesion and Adhesives (R. L. Patrick, ed.), Marcel Dekker, New York, Vol. 1, p. 233 (1967).
2494. G. P. Anderson, V. L. Ruggles, and G. S. Stibor, Int. J. Fract. Mech. 7:63 (1971).
2495. G. P. Anderson, K. L. DeVries, and M. L. Williams, Finite element in adhesion analyses, Int. J. Fract. Mech. 9(4) (Dec. 1973).
2496. S. Y. Elliott, Techniques for evaluation of adhesives, Handbook for Adhesive Bonding (C. V. Cagle, ed.), McGraw-Hill, New York (1973).
2497. M. L. Chang, K. L. DeVries, and M. L. Williams, Exptl. Mech. 14:89 (1974).
2498. G. P. Anderson, K. L. DeVries, and M. L. Williams, Mixed mode stress field effect in adhesive fracture, Int. J. Fract. Mech. 10(4) (Dec. 1975).
2499. W. A. Jemian and M. B. Ventrice, The fracture toughness of adhesive bonded joints, J. Adhes. 1:190 (July 1969).
2500. W. A. Jemain and R. C. Wilcox, Study of the onset of permanent deformation in structural bonded joints, Contract DAAH01-70-C-1425 (Jan. 1972).
2501. E. J. Ripling, S. Mostovoy, and R. L. Patrick, Application of fracture mechanics to adhesive joints, Final Report Contract NONR-3544(00)(X), Materials Res. Lab., Glenwood, IL (1963).
2502. E. J. Ripling, S. Mostovoy, and R. L. Patrick, Mater. Res. Stds. 4:129 (1964).
2503. E. J. Ripling, S. Mostovoy, and R. L. Patrick, Application of fracture mechanics to adhesive joints, ASTM Spec. Publ. No. STP 360, p. 5 (1964).
2504. S. Mostovoy, P. B. Crosley, and E. J. Ripling, J. Mater. 2:881 (1965).
2505. S. Mostovoy and E. J. Ripling, J. Appl. Polym. Sci. 10:1351 (1966).
2506. S. Mostovoy, P. B. Crosley, and E. J. Ripling, Use of crackline loaded specimens for measuring plane-strain fracture toughness, J. Mater. 2(3) (1967).
2507. S. Mostovoy and E. J. Ripling, J. Appl. Polym. Sci. 13:1083 (1969).
2508. S. Mostovoy, E. J. Ripling, and C. F. Bersch, Fracture toughness of adhesive joints, J. Adhes. 3:125 (1970).
2509. E. J. Ripling, S. Mostovoy, and C. F. Bersch, J. Adhes. 3:145 (1971).
2510. E. J. Ripling, S. Mostovoy, and H. T. Corten, Fracture mechanics: A tool for evaluating structural adhesives, J. Adhes. 3:107 (1971).
2511. S. Mostovoy and E. J. Ripling, J. Appl. Polym. Sci. 15:641 (1971).
2512. S. Mostovoy, H. R. Smith, R. G. Lingwall, and E. J. Ripling, Eng. Fract. Mech. 3:291 (1971).
2513. S. Mostovoy and E. J. Ripling, J. Appl. Polym. Sci., Appl. Polym. Symp. 19:395 (1972).
2514. S. Mostovoy and E. J. Ripling, Fracturing characteristics of adhesive joints, Final Report Contract N00019-71-C-0329, Mater. Res. Lab., Glenwood, IL (1972).
2515. S. Mostovoy and E. J. Ripling, Final Report Contract N00019-73-C-0113, Mater. Res. Lab., Glenwood, IL (Dec. 31, 1972).
2516. S. Mostovoy and E. J. Ripling, Final Report Contract N00019-73-C-0113, Mater. Res. Lab., Glenwood, IL (1974).
2517. S. Mostovoy and E. J. Ripling, Adhesion Science and Technology (L. H. Lee, ed.), Plenum Press, Part B, p. 513 (1975).
2518. S. Mostovoy and E. J. Ripling, Fracturing characteristics of adhesive joints, Final Report Contract No. N000010-74-C-0274, Mater. Res. Lab., Glenwood, IL (Jan. 1974–Jan. 1975).
2519. S. Mostovoy and E. J. Ripling, Fracturing characteristics of adhesive joints, Final Report Contract No. N00019-76-C.0323, Mater. Res. Lab., Glenwood, IL (Jan. 1975–Jan. 1976).
2520. S. Mostovoy and E. J. Ripling, Fracturing characteristics of adhesive joints, Final Report Contract No. N00019-76-C-0323, Mater. Res. Lab., Glenwood, IL (Jan. 1976–Jan. 1977).
2521. S. Mostovoy and E. J. Ripling, J. Appl. Polym. Sci. 13:1083 (1979).
2522. W. D. Bascom and R. L. Cottingham, J. Adhes. 4:193 (1972).
2523. K. M. Liechti, D. Ginsburg, and E. C. Henson, A comparison between measurements and finite element predictions of crack opening displacements near the front of an interface crack, J. Adhes. 23(2): 123 (1987).
2524. S. Barsoum, Asymptotic fields in adhesive fracture, J. Adhes. 29(1–4):149 (1989).
2525. A. Needleman, An analysis of decohesion along an imperfect interface, Int. J. Fract. 42(1) (1990).

2526. R. R. Reynolds, K. Kokini, and G. Chen, The mechanics of the interface crack using the finite element method, J. Eng. Mater. & Technol. 112(1):38 (1990).
2527. U. Schom-Rode and T. Berger-Bocker, Detection of microcrack accumulation in tensile loaded specimens, Eng. Fract. Mech. 35(4/5):771 (1990).
2528. P. G. Charalambides, H. C. Cao, J. Lund, and A. G. Evans, Development of a test method for measuring the mixed mode fracture resistance of bimaterial interfaces, Mech. of Mater. 8(4):269 (1990).
2529. M. Ortiz and J. A. Blume, Effect of decohesion and sliding on bimaterial crack-tip fields, Int. J. Fract. 42(2) (1990).
2530. P. B. Crosley and E. J. Ripling, A thick adherend instrumented double-cantilever-beam specimen for measuring debonding of adhesive joints, J. Test. Eval. 19(1):24 (1991).
2531. M. L. Williams, Review of continuum mechanics factors in adhesive fracture, Adhesive Joints: Formation, Characteristics, and Testing (K. L. Mittal, ed.), Plenum Press, New York, p. 703 (1984).
2532. G. P. Anderson, Applied Adhesive Fracture Mechanics, Ph.D. Dissertation, University of Utah, Salt Lake City (1973).
2533. M. L. Williams and F. N. Kelly, Rubber Chem. Technol. 42:1175 (1969).
2534. A. N. Gent, Rubber Chem. Technol. 47:202 (1974).
2535. A. J. Kinloch and S. J. Shaw, Developments in Adhesives—2 (A. J. Kinloch, ed.), Applied Science Publishers, London, p. 82 (1981).
2536. W. B. Jones and D. H. Kaelble, Investigation of fatigue and crack propagation behavior of adhesives, AFML Tech. Report. Air Force Materials Lab, Wright-Patterson AFB, Ohio.
2537. G. P. Anderson and K. L. DeVries, Predicting bond strength, Proc. 10th Ann. Meet. Adhes. Soc., Williamsburg, VA, Gordon and Breach, p. 751 (1988).
2538. M. L. Williams, Bull. Seismological Soc. Amer. 49(2):199 (1959).
2539. E. H. Dill, A. L. Deak, and W. F. Schmidt, Handbook of Engineering Structural Analysis Solid Propellents, CPIA Publ. No. 214, Chap. 5 (1971).
2540. G. P. Anderson and K. L. DeVries, Predicting bond strength, Proc. 10th Ann. Meeting Adhes. Soc., Williamsburg, VA, Gordon and Breach, London, p. 753 (1988).
2541. G. P. Anderson, K. L. DeVries, and M. L. Williams, J. Colloid and Interf. Sci. 47:600 (1972).
2542. W. S. Johnson and P. D. Mangalgiri, NASA Tech. Memo NO. 87571 (July 1985).
2543. E. H. Andrews, Generalized fracture mechanics approach to adhesion, Adhesive Bonding (L. H. Lee, ed.), Plenum Press, Chap. 12, p. 341 (1991).
2544. W. D. Bascom, R. L. Cottington, R. L. Jones, and P. Peyser, J. Appl. Polym. Sci. 19:2545 (1975).
2545. W. D. Bascom and R. L. Cottington, J. Adhes. 7:333 (1976).
2546. W. D. Bascom et al., J. Appl. Polym. Sci. 19:2545 (1975).
2547. S. S. Wang, J. F. Mandell, and F. J. McGarry, Int. J. Fract. 14:39 (1978).
2548. R. A. Gledhill and A. J. Kinloch, Polym. Eng. Sci. 19:82 (1979).
2549. R. Y. Ting and R. L. Cottington, J. Adhes. 12:243 (1981).
2550. A. Muscati, and D. J. Lee, Elastic-plastic finite element analysis of thermally loaded cracked structures, Int. J. Fract. 25(4):227 (Aug. 1984).
2551. D. G. O'Connor and H. F. Brinson, Tech. Report VPI-E-79-31, VPI&SU, Blacksburg, VA, p. 65 (Sept. 1979).
2552. A. A. Khalil and M. R. Bayoumi, Effect of loading rate on fracture toughness of bonded joints, Int. J. Adhes. Adhes. 11(1):25 (1991).
2553. C. W. Woo and C. L. Chow, Fatigue crack propagation in aluminum and PMMA, Int. J. Fract. 26(2): 37 (Oct. 1984).
2554. A. S. Kobayashi, M. Ramuly, M. S. Dadkheh, K. H. Yang, and B. S. J. Kang, Dynamic fracture toughness, Int. J. Fract. 30(4):275 (April 1986).
2555. R. D. Thomson and J. W. Hancock, Ductile failure by void nucleation, growth, and coalescence, Int. J. Fract. 26(2):99 (Oct. 1984).
2556. L. Lorenzo and H. T. Hehn, Effect of ductility on the fatigue behavior of epoxy resins, Polym. Eng. Sci. 25(4):274 (Feb. 1986).
2557. A. J. Kinloch, D. G. Gilbert, and S. J. Shaw, A mechanism for ductile crack growth in epoxy polymers, J. Mater. Sci. 21(3):1051 (March 1986).
2558. M. Belgacen and T. Brethear, A local study of the mechanical interaction between a rigid inclusion and the surrounding plasticized matrix, Scripta Metallurgica 19(3):285 (March 1985).
2559. E. J. Ripling, P. B. Crosley, and W. S. Johnson, A comparison of pure Mode I and mixed Mode I-III cracking of an adhesive containing an open knit cloth carrier, Adhesively Bonded Joints: Testing, Analysis, and Design Conf., ASTM, Baltimore, MD (Sept. 10–12, 1986).
2560. A. A. Griffith, Phil. Trans. Roy. Soc. A221:163 (1920).

2561. A. A. Griffith, Proc. 2nd. Int. Cong. Appl. Math., Delft, Holland, p. 54 (1924).
2562. E. Orowan, Rep. Progr. Phys. 12:185 (1948).
2563. J. Rice, Fracture—An Advanced Treatise (H. Liebowitz, ed.), Academic Press, New York, Vol. 2, p. 191 (1968).
2564. A. N. Gent, A. Ahagon, N. J. Kim, and Y. Kumagi, Rubber Chem. Technol. 48:896 (1975).
2565. E. H. Andrews, Developments in Polymer Fracture—1 (E. H. Andrews, ed.), Applied Science Publishers, London, p. 1 (1979).
2566. L. H. Lee, Physicochemical Aspects of Polymer Surfaces (K. L. Mittal, ed.), Plenum Press, New York, Vol. 1, p. 523 (1983).
2567. L. H. Lee, Adhesive Chemistry—Developments and Trends (L. H. Lee, ed.), Plenum Press, New York, Vol. 1 (1984).
2568. D. H. Kaelble, Proc. 23rd Int. Cong. on Pure and Appl. Chem., Butterworths, London, Vol. 8, p. 265 (1971).
2569. D. H. Kaelble, A relationship between the fracture mechanics and surface energetics failure criteria, J. Appl. Polym. Sci. 18:1869 (1974).
2570. Shaw Ming Lee, An in-situ failure model for adhesive joints, J. Adhes. 18:1 (1985).
2571. R. B. Krieger, Jr., Stiffness characteristics of structural adhesives for stress analysis in hostile environments, American Cyanamid Company, Harve de Grace, MD (Oct. 1975).
2572. E. J. Hughes, W. Althof, and R. B. Krieger, Jr., Mechanical properties of adhesives, Adhesive Bonding of Aluminum Alloys (E. W. Thrall and R. W. Shannon, eds.), Marcel Dekker, New York, Chap. 9, pp. 167–175 (1985).
2573. D. H. Kaelble, J. Adhes. 2:66 (1970).
2574. H. Dannenberg, J. Appl. Polym. Sci. 5:125 (1961).
2575. B. M. Malyshev and R. L. Solganik, Int. J. Fract. Mechs. 1:114 (1965).
2576. T. Hata, M. Gamo, and Y. Doi, Dobienshi Kagaku. 22:152 (1965).
2577. A. N. Gent and R. P. Petrich, Proc. Roy. Soc., London A310:433 (1969).
2578. P. B. Lindley, J. Inst. Rubber Ind. 5:243 (1971).
2579. H. W. Westergaard, J. Appl. Mech. A6:46 (1939).
2580. G. R. Irwin, Fracture, Handuch der Physik, Springer-Verlag, Berlin, Vol. 6, p. 551 (1958).
2581. G. R. Irwin, Fracture mechanics, Structural Mechanics (Goodier and Hoff, eds.), Pergamon Press, New York, p. 557 (1960).
2582. G. R. Irwin, Appl. Mats. Res. 3:65 (1964).
2583. F. Erdogan, J. Appl. Mech. 87:232 (1963).
2584. F. Erdogan, J. Appl. Mech. 89:829 (1965).
2585. J. R. Rice, J. Appl. Mech. 87:418 (1963).
2586. F. Erdogan and G. Gupta, Int. J. Solid Struct. 7:39 (1971).
2587. F. Erdogan and G. Gupta, Int. J. Solid Struct. 7:1089 (1971).
2588. F. Erdogan and K. Arlin, Int. J. Eng. Sci. 10:115 (1972).
2589. P. S. Theocaris, Acta. Mech. 24:99 (1976).
2590. D. R. Mulville and P. W. Mast, Eng. Fract. Mech. 8:555 (1976).
2591. R. Calhoun and M. Lowengrub, Int. J. Eng. Sci. 16:423 (1978).
2592. A. Piva and E. Viola, Eng. Fract. Mech. 13:143 (1980).
2593. K. Ikemami and K. Kamiya, Effect of flaws in the adhering interface on the strength of adhesive-bonded butt joints, J. Adhes. 14:1 (1982).
2594. G. P. Anderson and K. R. DeVries, Evaluation of adhesive testing methods, Adhesive Joints: Formation, Characteristics, and Testing (K. L. Mittal, ed.), Plenum Press, New York, p. 277 (1984).
2595. G. P. Anderson and K. L. DeVries, Analysis of standard bond strength tests, Treatise on Adhesion and Adhesives (R. L. Patrick, ed.), Marcel Dekker, New York, Vol. 6, p. 78 (1989).
2596. E. H. Rowe and C. K. Riew, What failure mechanisms tell about toughened epoxy adhesives, Plast. Eng. (March 1975).
2597. K. M. Liechti and T. Freda, On the use of laminated beams for the determination of pure and mixed-mode fracture properties of structural adhesives, J. Adhes. 28(2–3):145 (1989).
2598. D. V. Grilitskii, E. V. Guina, O. I. Dumanskii, and Yu. V. Sorokatyi, Experimental determination of the stress intensity factors in structures with cracks, Soviet Mater. Sci. 20(1):23 (Jan.-Feb. 1984).
2599. Y. C. Lem and J. F. Williams, The effect of contact stress intensity factors on fatigue crack propagation, Theoret. Appl. Fract. Mech. 1(2):193 (1984).
2600. T. Wah, Stress intensity factors determined by use of Westergaard's stress functions, Eng. Fract. Mech. 20(1):65 (1984).

2601. R. M. L. Foote and V. T. Buchwald, An exact solution for the stress intensity factor for a double cantilever beam, Int. J. Fract. 29(3):125 (Nov. 1985).
2602. C. Ruiz and J. Epstein, On the variation of the stress intensity factor along the front of a surface, Int. J. Fract. 38(4):231 (Aug. 1986).
2603. Tetelman and McEvily, Fracture of Structural Materials, John Wiley & Sons, New York (1967).
2604. Yokobori, An Interdisciplinary Approach to Fracture and Strength of Solids, Noordhoff (1968).
2605. Liebowitz, Fracture, Academic Press, New York, 7 vols. (1968).
2606. E. H. Andrews, Fracture in Polymers, Elsevier, Amsterdam (1968).
2607. Jenkins, Polymer Science, A Materials Science Handbook, North-Holland, Amsterdam (1972).
2608. H. H. Kausch, Hassell, and Jaffee, Deformation and Fracture of High Polymers, Plenum Press, New York (1972).
2609. A. J. Kinloch and R. J. Young, Fracture Behavior of Polymers, Applied Science Publishers, London (1983).
2610. J. G. Williams, Fracture Mechanics of Polymers, Ellis Horwood Series in Eng. Sci. (1984).
2611. M. F. Kanninen and Popelar, Advanced Fracture Mechanics, Oxford Scientific Publications, London, England (1985).
2612. R. A. Schapery, Correspondence principles and a generalized J-integral for large deformation and fracture analysis of viscoelastic media, Int. J. Fract. 25(3):195 (1984).
2613. H. J. Schindler, On the relationship between J-integral and crack opening displacement, Eng. Fract. Mech. 20(2):281 (1984).
2614. R. H. Dodde and D. T. Read, Experimental and analytical estimates of the J-integral for tensile panels containing short center cracks, Int. J. Fract. 28(2):39 (June 1985).
2615. A. N. Gent and J. Jeong, Contribution of bending energy losses to the apparent tear energy, Int. J. Fract. 29(3):157 (Nov. 1985).
2616. R. J. Ferguson, G. P. Marshall, and J. G. Williams, Polymer 14:451 (1973).
2617. J. D. G. Sumpter and C. E. Turner, Cracks and Fracture, ASTM Spec. Publ. STP 601, p. 3, ASTM (1976).
2618. A. Saleemi and J. A. Narin, The plane strain essential work of fracture as a measure of the fracture toughness of ductile polymers, Polym. Eng. Sci. 30(4):211 (1990).
2619. A. Lee and G. B. McKenna, Viscoelastic response of epoxy glass subjected to different thermal treatments, Polym. Eng. Sci. 30(7):431 (1990).
2620. W. S. Johnson and S. Mall, A fracture mechanics approach for designing adhesively bonded joints, NASA Tech. Memo 85694, Natl. Tech. Infor. Service (1983).
2621. S. Mall, W. S. Johnson, and R. A. Everett, Jr., Cyclic debonding of adhesively bonded composites, Adhesive Joints: Formation, Characteristics, and Testing (K. L. Mittal, ed.), Plenum Press, New York, pp. 639–658 (1984).
2622. S. Mall and W. S. Johnson, Characterization of Mode I and Mixed Mode failure of adhesive bonds between composite adherends, ASTM Spec. Publ. STP 893, pp. 322–334 (1986).
2623. A. J. Kinloch and S. J. Shaw, Developments in Adhesion—2 (A. J. Kinloch, ed.), Applied Science Publishers, London, pp. 83–124 (1981).
2624. E. Sancaktar and J. Tang, Mixed-mode fracture in adhesively bonded joints under dynamic loading, Proc. of Adhesion '87, Plastics and Rubber Inst. Meetings, pp. 18–1, 18–7 (1987).
2625. E. Sancaktar, Elastoplastic fracture behavior of structural adhesives, Proc. 5th Int. Joint Govt-Industry Symp. on Struct. Adhesive Bonding, Amer. Def. Preparedness Assoc., pp. 120–139 (1987).
2626. E. Sancaktar, Elastoplastic fracture behavior of structural adhesives, Mechanical Behavior of Adhesive Joints (G. Verchery and A. H. Cardon, eds.), Editions Pluralis, pp. 351–362 (1987).
2627. W. S. Johnson and P. D. Mangalgiri, Influence of resin on interlaminar mixed mode failure, NASA Tech. Memo. 87571, Natl. Tech. Info. Service (1985).
2628. H. Jozavi and E. Sancaktar, The effects of cure temperature and time on the bulk fracture properties of a structural adhesive, J. Adhes. 18:25 (1985).
2629. J. Benbow, Proc. Phys. Soc. 78:970 (1961).
2630. J. P. Berry, Fracture Processes in Polymeric Solids (Rosen, ed.), Interscience, New York, p. 157 (1964).
2631. A. T. DiBenedetto and K. L. Trachte, J. Appl. Polym. Sci. 14:2249 (1970).
2632. L. J. Broutman and S. Sahu, Mater. Sci. Eng. 8:98 (1971).
2633. J. S. Foot and I. M. Ward, J. Mater. Sci. 7:367 (1972).
2634. I. Wolock and S. B. Newman, Fracture Processes in Polymeric Solids (Rosen, ed.), Interscience, New York, p. 235 (1964).
2635. F. A. Johnson and J. C. Radon, Eng. Fract. Mech. 4:555 (1972).

2636. J. Springarn, Chevron-notched specimens for measuring fracture toughness, Adhesively Bonded Joints: Testing, Analysis, and Design Conf., ASTM, Baltimore, MD (Sept. 10–12, 1986).
2637. R. Herrera, H. D. Metias, E. A. Stocchi, and L. A. de Verdie, A simple method for fatigue precracking specimens for fracture mechanics tests, J. Test. and Eval. 15(5):281 (Sept. 1987).
2638. J. A. Hinkley and S. L. Mung, Fracture toughness of polyimide films, Polymer 31(1):75 (1990).
2639. G. C. Adams, R. G. Bender, B. A. Crouch, and J. G. Williams, Impact fracture toughness tests on polymers, Polym. Eng. Sci. 30(4):241 (1990).
2640. A. N. Gent and A. J. Kinloch, J. Polym. Sci. A-2, 9:283 (1971).
2641. E. Orowan, Proceedings of the Symposium on Fatigue and Fracture of Metals, John Wiley & Sons, New York, p. 139 (1950).
2642. H. W. Greensmith, L. Mullins, and A. G. Thomas, Chemistry and Physics of Rubberlike Substances (L. Bateman, ed.), Maclaren, London, Chap. 10 (1963).
2643. A. N. Gent and A. W. Henry, Proceedings of the International Rubber Conference, Brighton, England (1967), Maclaren, London, p. 193 (1968).
2644. A. N. Gent and A. J. Kinloch, Adhesion of viscoelastic materials to rigid substrates, III. Energy criterion for failure, J. Polym. Sci. A-2, 9:659 (1971).
2645. E. H. Andrews and A. J. Kinloch, Proc. Roy. Soc., A-332:401 (1973).
2646. E. H. Andrews and A. J. Kinloch, J. Polym. Sci., Symp. 46:1 (1974).
2647. E. H. Andrews and N. E. King, J. Mater. Sci. 11:2004 (1976).
2648. S. J. Bennett, K. L. DeVries, and M. L. Williams, Int. J. Fract. 10:33 (1974).
2649. G. P. Anderson, K. L. DeVries, and M. L. Williams, J. Colloid Interf. Sci. 47:600 (1974).
2650. M. L. Williams, Proc. 5th U.S. Cong. Appl. Mechs., p. 451 (1966).
2651. W. B. Jones and M. L. Williams, The measurement of adhesive energy in fracture investigations, UTEC DO-68-019, University of Utah, Salt Lake City (Feb. 1968).
2652. J. D. Burton, W. B. Jones, and M. L. Williams, Trans Soc. Rheol. 15:39 (1971).
2653. M. L. Williams, Proceedings of the 9th Conference on Adhesion and Adhesives, City University of London (April 1971).
2654. M. L. Williams, Review of Continuum Mechanics Factors in Adhesive Fracture, Adhesive Joints: Formation, Characteristics, and Testing (K. L. Mittal, ed.), Plenum Press, New York, pp. 708–709 (1984).
2655. M. L. Williams, J. Adhes. 5:81 (1973).
2656. M. L. Williams, K. L. DeVries, and P. R. Despain, J. Dent. Res. 52:517 (1973).
2657. P. R. Despain, K. L. DeVries, R. D. Luntz, and M. L. Williams, J. Dent. Res. 52:674 (1973).
2658. R. K. Fletcher, S. M. Breitling, J. K. Adamich, and M. L. Williams, Vitro 13:857 (1977).
2659. D. Maugis, Adherence of elastomers: fracture mechanics aspects, J. Adhes. 23(1):61 (1987).
2660. T. Tsukatani, Y. Hatano, and H. Mizumachi, Bonding and debonding processes in tack of pressure sensitive adhesives, J. Adhes. 31(1):59 (1989).
2661. Y. Urahama, Effect of peel load on stringiness phenomena and peel speed of pressure-sensitive adhesive tape, J. Adhes. 31(1):47 (1989).
2662. V. Weissburg and M. Arcan, Mechanical evaluation of structural adhesives: Stiff versus flexible adherents, Adhesion '90, The Plastics and Rubber Institute meetings, Cambridge, England (Sept. 10–12, 1990).
2663. D. R. Mulville, D. L. Hunston, and P. W. Mast, J. Eng. Mater. Technol. 100:25 (1978).
2664. Y. W. Mai and A. S. Vipond, J. Mater. Sci. 13:2280 (1978).
2665. A. D. Jonath, Adhesion and Adsorption of Polymers (L. H. Lee, ed.), Plenum Press, New York, Part A, p. 175 (1980).
2666. W. D. Bascom and J. Orashnik, J. Mater. Sci. 13:1411 (1978).
2667. R. A. Gledhill, A. J. Kinloch, S. Yamini, and R. J. Young, Polymer 19:574 (1978).
2668. D. L. Hunston and W. D. Bascom, Rubber-modified thermoset resins, Adv. Chem. Ser. 208 (C. K. Riew and J. K. Gillham, eds.), Amer. Chem. Soc., Chap. 7 (1983).
2669. A. J. Kinloch, S. J. Shaw, and D. L. Hunston, Proc. Int. Conf. on Yield, Deformation, and Fracture, The Plastics and Rubber Institute Meetings, Cambridge, England (1982).
2670. D. L. Hunston, A. J. Kinloch, S. J. Shaw, and S. S. Wang, Fracture behavior of adhesive joints, Adhesive Joints: Formation, Characteristics, and Testing (K. L. Mittal, ed.), Plenum Press, New York, p. 789 (1984).
2671. D. L. Hunston, W. D. Bascom, E. E. Wells, J. D. Fahey, and J. L. Bittner, Adhesion and Adsorption of Polymers (L. H. Lee, ed.), Plenum Press, New York, Part A, p. 321 (1980).
2672. S. I. Rokhlin, Adhesive joint characterization by ultrasonic surface and interface waves, Adhesive Joints: Formation, Characteristics, and Testing (K. L. Mittal, ed.), Plenum Press, New York, pp. 307–345 (1984).
2673. S. I. Rokhlin, M. Hefets, and M. Rosen, J. Appl. Phys. 51:3579 (1980).

2674. W. D. Bascom, R. L. Cottington, and C. O. Timmons, Nav. Eng. J. 88:73 (1976).
2675. J. N. Sultan and F. J. McGarry, Microstructural characteristics of toughened thermoset polymers, Research Report R69-59, School of Engineering, Massachusetts Institute of Technology, Cambridge (1969).
2676. F. J. McGarry and A. M. Willner, Toughening of an epoxy resin by an elastomeric second phase, Research Report R68-8, School of Engineering, Massachusetts Institute of Technology, Cambridge (1969).
2677. R. Drake and A. Siebert, SAMPE Quart. 6(4):11 (1975).
2678. J. C. Bolger, Structural adhesives for metal bonding, Treatise on Adhesion and Adhesives (R. L. Patrick, ed.), Marcel Dekker, New York, Vol. 3, Chap. 1, pp. 17–20 (1973).
2679. J. C. Bolger, Structural adhesives: State-of-the-art, Adhesives in Manufacturing (G. L. Schneberger, ed.), Marcel Dekker, New York, Chap. 7, p. 145 (1983).
2680. W. D. Bascom, R. L. Cottington, and C. O. Timmons, J. Appl. Polym. Sci., Appl. Polym. Symp. 32:165 (1977).
2681. R. Y. Ting and R. L. Cottington, Adhes. Age 24(6):35 (1981).
2682. G. P. Marshall, L. H. Coutts, and J. G. Williams, J. Mater. Sci. 9:1409 (1974).
2683. M. Marvin and J. G. Williams, J. Mater. Sci. 10:1883 (1975).
2684. R. A. Gledhill, A. J. Kinloch, and S. J. Shaw, J. Mater. Sci. 14:1769 (1979).
2685. M. Takashi, K. Ogawa, R. Katita, and T. Kuno, Proc. Japan Congr. Mater. Res. 16:211 (1973).
2686. A. D. Roberts and A. G. Thomas, Adhesion—1 (K. W. Allen, ed.), Applied Science Publishers, London, p. 141 (1977).
2687. E. H. Andrews and A. Stevenson, Adhesion—3 (K. W. Allen, ed.), Applied Science Publishers, London, p. 81 (1979).
2688. H. P. Meissner and G. H. Baldauf, Trans. ASME 73:697 (1951).
2689. S. W. Losoki and G. Kruas, J. Polym. Sci. 18:359 (1955).
2690. N. Bredzs and H. Schwartzbart, Welding J. (Welding Res. Suppl.) p. 610 (1956).
2691. A. G. H. Dietz, ASTM Spec. Tech. Publ., STP 194, p. 19 (1957).
2692. J. J. Bikerman and C. R. Huang, Trans. Soc. Rheol. 3:5 (1959).
2693. W. C. Wake, Adhesion (D. O. Eley, ed.), Oxford University Press, London, p. 191 (1961).
2694. R. W. Bryant and W. A. Dukes, Aeronautic and Space Eng. Mfg. Meeting, Los Angeles, Paper No. 670855 (Oct. 1967); S.A.E. Meeting, New York (1968).
2695. W. A. Dukes and R. W. Bryant, J. Adhes. 1:48 (1969).
2696. R. L. Patrick, J. A. Brown, N. M. Cameron, and W. G. Geraman, J. Appl. Polym. Sci., Appl. Polym. Symp. 16:87 (1971).
2697. B. W. Cherry and K. W. Thomson, Fracture Mechanics and Technology, Sijthoff and Nordoff, Holland, p. 723 (1971).
2698. R. C. Wilcox and W. A. Jemain, Poly. Eng. Sci. 13:40 (1973).
2699. P. D. Hilton and G. P. Gupta, Design and Eng. Conf., Paper No. 73-3e-21, ASME, Philadelphia (April 1973).
2700. R. L. Patrick, The use of scanning electron microscopy, Treatise on Adhesion and Adhesives (R. L. Patrick, ed.), Marcel Dekker, New York, Vol. 3, Chap. 4, p. 163 (1973).
2701. A. N. Gent, J. Polym. Sci. A2:283 (1974).
2702. M. Takashi, K. Ogawa, and T. Kumo, Proc. Japan Cong. Mater. Res. 17:192 (1974).
2703. N. W. Mai, J. Adhes. 7:141 (1975).
2704. S. Mostovoy and E. J. Ripling, J. Appl. Polym. Sci. 13:1083 (1979).
2705. R. A. Gledhill, A. J. Kinloch, and S. J. Shaw, J. Adhes. 11:3 (1980).
2706. E. H. Andrews and A. Stevenson, J. Adhes. 11:17 (1980).
2707. J. C. McMillan, Developments in Adhesives—2 (A. J. Kinloch, ed.), Applied Science Publishers, London, p. 243 (1981).
2708. S. Mostovoy and E. J. Ripling, Influence of water on stress corrosion cracking of epoxy bonds, J. Appl. Polym. Sci. 13(6):1083 (1969).
2709. S. Mostovoy and E. J. Ripling, Effect of joint geometry on the toughness of epoxy adhesives, J. Appl. Polym. Sci. 15:661 (1971).
2710. S. Mostovoy and E. J. Ripling, J. Appl. Polym. Sci., Appl. Polym. Symp. 19:395 (1972).
2711. W. J. Renton, Exptl. Mech. 16(11):409 (1976).
2712. E. Sancaktar, The Viscoelastic Shear Behavior of a Structural Adhesive, Ph.D. Thesis, Virginia Polytechnic Institute and State University, Blacksburg, VA (1979).
2713. W. J. Renton, Structural properties of adhesives, Quart. Prog. Report No's 1-7, Vought Corp. ATC Report No. B-94400/8CR-26 (1977-1978).
2714. F. Delale and F. Erdogan, Viscoelastic analysis of adhesively bonded joints, Report for NASA-NGR 39-007-001, Lehigh University, Bethlehem, PA (Feb. 1981).

2715. F. Delate and F. Erdogan, Time-temperature effect in adhesively bonded joints, Report for NASA-NGR 39-007-001, Lehigh University, Bethlehem, PA (Feb. 1981).
2716. M. Chmura and E. McAlbee, Correlation of mechanical properties of resins obtained in an adhesive joint and bulk form, Picatinny Arsenal Report No. TR-3300, Picatinny Arsenal, Dover, NJ (April 1966).
2717. H. R. Brinson, C. T. Herkovich, and M. P. Renieri, Rate and time dependent failure of structural adhesives, Fracture Mechanics of Composites, ASTM Bull. STP No. 617, ASTM, Philadelphia, p. 177 (1975).
2718. E. Sancaktar and H. F. Brinson, The viscoelastic shear behavior of a structural adhesive, Adhesion and Adsorption of Polymers (L. H. Lee, ed.), Plenum Press, New York, Part A, p. 279 (1980).
2719. V. Niranjan, Bonded joints—a photoelastic study, University of Toronto, UTIAS T. N. No. 138 (July 1970).
2720. R. A. Everett, Jr., The Significance of Peel Stresses in Cyclic Debonding, M.S. Thesis, Old Dominion University, Richmond, VA (May 1980).
2721. S. Padgilwar and E. Sancakatar, The rate and time dependent material characterization of LARC-3 structural adhesive, Clarkson College Report No. MIE-069 (April 1981).
2722. D. W. Dwight, M. E. Counts, and J. P. Wightman, Surface analysis and adhesive bonding, II—Polyimides, Colloid and Interfacial Science (M. Kerker, ed.), Academic Press, New York, p. 143 (1976).
2723. D. W. Dwight, J. G. McGrath, and J. P. Wightman, ESCA analysis of polymer structure and bonding, J. Appl. Polym. Sci., Appl. Polym. Symp. 34:35 (1978).
2724. J. S. Sugama, L. E. Kukelka, and N. Carcielle, Nature of interfacial interaction mechanisms between polyacrylic acid macromolecules and oxide metal surfaces, J. Mater. Sci. 19(12):4045 (Dec. 1984).
2725. G. C. Knollman, Variation of shear modulus through the interfacial bond zone of an adhesive, Int. J. Adhes. Adhes. 5(3):137 (July 1985).
2726. G. C. Knollman and J. J. Hartog, Experimental determination of the variation in shear modulus through the interfacial zone of the adhesive, J. Adhes. 17(4):251 (1985).
2727. R. B. Clough, F. S. Biancaniello, H. N. G. Wadley, and U. R. Kattner, Fiber and interface fracture in single-crystal aluminum/silicon carbide fiber composites, Metallurgical Trans. 21(10):2747 (1990).
2728. M. J. Napolitane and A. Moet, Dissipative processes in interfacial failure, J. Adhes. 33(1–2):149 (1990).
2729. K. L. DeVries, M. L. Williams, and C. D. Chang, Adhesive fracture of a lap-shear joint, Exptl. Mech. p. 89 (March 1974).
2730. J. S. Cartner and H. F. Brinson, The non-linear viscoelastic behavior of adhesives and chopped fiber composites, Virginia Tech. Report No. VPI-E-78-21, Virginia Polytechnic Institute and State University, Blacksburg, VA (Aug. 1978).
2731. K. M. Liecht and W. G. Knauss, Crack propagation at material interfaces: II. Experiments on mode interaction, GA1-CIT Report No. SM 80-21, California Institute of Technology, Pasadena (1981) (Preliminary Version).
2732. H. F. Brinson, D. H. Morris, and Y. T. Yeow, A new experimental method for the accelerated characterization and prediction of the failure of polymer-based composite laminates, 6th Int. Conf. Exptl. Stress Anal., Munich, West Germany (Sept. 1978).
2733. W. I. Griffith, D. H. Morris, and H. F. Brinson, The accelerated characterization of viscoelastic composite materials, Virginia Tech. Report VPI-E-80-15 (April 1980).
2734. T. Tanaka and K. Tanimana, 19th Japan Cong. Mater. Res., Tokyo, Japan, p. 179 (1975).
2735. H. Nakayama, 22nd Japan Conf. Mater. Res. Kyoto, Japan, p. 162 (1978).
2736. J. D. Minford, unpublished work, Aluminum Co. of America.
2737. R. D. Adams and J. Coppendale, J. Adhes. 10:49 (1979).
2738. J. J. Bikerman, Science of Adhesive Joints, Academic Press, New York, p. 273 (1965).
2739. N. A. DeBruyne, Aircraft Eng. 16:115 (1944).
2740. N. A. DeBruyne, Aircraft Eng. 16:140 (1944).
2741. K. Wellinger and U. Rembold, VDI Zeitscrift 199:41 (1958).
2742. R. A. Kline, Stress analysis of adhesively bonded joints, Adhesive Joints: Formation, Characteristics, Testing (K. L. Mittal, ed.), Plenum Press, New York, p. 588 (1984).
2743. K. M. Liechti and Y-M. Liang, The effect of adhesive layer thickness on interfacial crack initiations, 13th Ann. Meeting Adhes. Soc., Savannah, GA (Feb. 18–21, 1990).
2744. M. Chang, J. Adhes. 4:221 (1972).
2745. Y. Wertsman, J. Adhes. 11:279 (1981).
2746. W. Carpenter, AIAA J. 18:350 (1980).
2747. R. W. Bryant and W. A. Dukes, Brit. J. Appl. Phys. 16:101 (1965).
2748. R. W. Bryant and W. A. Dukes, J. Appl. Polym. Sci., Appl. Sci. Symp. 3:81 (1966).

2749. S. Bandyopadhyay, Review of the microscopic and macroscopic aspects of fracture of unmodified and modified epoxy resins, Mater. Sci. Eng. A125:157 (1990).
2750. I. M. Low, Effects of residual stresses on the failure micromechanics in toughened epoxy systems, J. Mater. Sci. 25(4):2144 (1990).
2751. V. K. Srivastava and P. S. Shembekar, Tensile and fracture properties of epoxy resin filled with flyash particles, J. Mater. Sci. 25(8):3513 (1990).
2752. W. L. Wu, T. Wu, and D. L. Hunston, Structural heterogeneity in epoxies, Polym. Eng. Sci. 30(16): 835 (1990).
2753. S. Wu, Chain structure, phase morphology and toughness relationships in polymers and blends, Polymer Eng. Sci. 30(13):753 (1990).
2754. S. Y. Lee, Thermomechanical properties of polymeric materials and related stresses, SAMPE Quart., p. 48 (April 1990).
2755. D. H. Kaelble and C. L. Ho, Trans Soc. Rheol. 18:219 (1974).
2756. J. L. Gardon, J. Appl. Polym. Sci. 7:625 (1963).
2757. J. L. Gardon, J. Appl. Polym. Sci. 7:643 (1963).
2758. C. Jouwersma, J. Polym. Sci. 45:253 (1960).
2759. C. Mylonas, Proceedings of the 4th International Congress on Rheology, Wiley-Interscience, New York, Part 2, p. 423 (1965).
2760. W. C. Wake, Adhesion and Formulation of Adhesives, Applied Science Publishers, London, p. 121 (1976).
2761. A. N. Gent and G. R. Hamid, Polymer Eng. Sci. 17:462 (1977).
2762. J. Johnston, Adhes. Age 11(4):20 (1968).
2763. D. W. Aubrey, G. N. Welding, and T. Wong, J. Appl. Polym. Sci. 13:2193 (1969).
2764. T. Igarashi, J. Appl. Polym. Sci. 19:2129 (1975).
2765. F. Yamato, S. Yamakawa, and S. Tsuru, J. Polym. Sci., Polymer Phys. Ed. 18:1847 (1980).
2766. D. Satas and F. Egan, Adhes. Age 9(8):22 (1966).
2767. D. W. Aubrey, T. A. Jackson, and J. D. Smith, J. Inst. Rubber Ind. 3:265 (1969).
2768. K. Kendall, J. Phys. D. Appl. Phys. 11:1519 (1978).
2769. G. Lake and A. Stevenson, J. Adhes. 12:13 (1981).
2770. W. C. Wake, Trans. I.R.I. 35:145 (1959).
2771. C. Mylonas and N. A. DeBruyne, Adhesion and Adhesives (N. A. DeBruyne and R. Houwink, eds.), Elsevier, Amsterdam, p. 91 (1951).
2772. P. Czarnocki and K. Piekarski, Yielding of adhesives, J. Mater. Sci. 21(12):4296 (1986).
2773. P. Czarnocki and K. Piekarski, Thin membranes loaded with adhesive joints, Int. J. Adhes. Adhes. 7(2):103 (1987).
2774. D. Y. Wang, The effect of stress distribution on the fatigue behavior of adhesive bonded joints, Final Report ASD-TAR-63-93, Air Force Mater. Lab., Wright-Patterson AFB, OH (July 1963).
2775. D. Y. Wang, Influences of stress distribution on fatigue strength of adhesive bonded joints, Exptl. Mech. 4:173 (1964).
2776. E. Sancaktar, Fatigue and fracture mechanics, Adhesives and Sealants, Engineered Materials Handbook, ASM, Materials Park, OH, Vol. 3, pp. 500–520 (1990).
2777. N. N. S. Chen, P. I. F. Niem, and R. C. Lee, Fatigue behavior of adhesive joints, J. Adhes. 21:115 (1987).
2778. R. D. Adams and W. C. Wake, Structural Adhesive Joints in Engineering, Elsevier, Amsterdam (1984).
2779. E. Sancaktar and P. Lawry, A photoelastic study of stress distribution in adhesively bonded joints with prebent adherends, J. Adhes. 11:233 (1980).
2780. W. S. Johnson, Overview of NASA programs directed toward adhesive joint service life prediction, DOD/NASA Workshop on Adhesive Joint Service Life Prediction, NASA (1981).
2781. W. S. Johnson and S. Mall, A fracture mechanics approach for designing adhesively bonded joints, NASA Tech. Memo No. 85694, NASA (1983).
2782. R. D. Adams, R. W. Atkins, J. A. Harris, and A. J. Kinloch, Stress analysis and failure properties of carbon-fibre-reinforced-plastic/steel double-lap joints, J. Adhes. 20:29 (1986).
2783. D. A. Jablonski, Fatigue crack growth in structural adhesives, J. Adhes. 11:125 (1980).
2784. K. Matsui, Size-effects on average ultimate shear stresses of adhesive-bonded rectangular or tubular lap joints under tension-shear, Int. J. Adhes. Adhes. 10(2):81 (1990).
2785. K. Matsui, Effect of size on nominal ultimate tensile stresses of adhesive-bonded circular or rectangular joints under bending or peeling load, Int. J. Adhes. Adhes. 10(2):90 (1990).
2786. H. L. Groth, Viscoelastic and viscoplastic stress analysis of adhesive joints, Int. J. Adhes. Adhes. 10(3): 207 (1990).

2787. S. A. Hashim, M. J. Cowling, and I. E. Winkle, Design and assessment methodologies for adhesively bonded structural connections, Int. J. Adhes. Adhes. 10(3):139 (1990).
2788. A. D. Crocombe, D. A. Bigwood, and G. Richardson, Analyzing structural adhesive joints for failure, Int. J. Adhes. Adhes. 10(3):167 (1990).
2789. P. A. Fay and G. D. Suthurst, Redesign of adhesively bonded box beam sections for improved impact performance, Int. J. Adhes. Adhes. 10(3):128 (1990).
2790. D. H. Kaelble, J. Colloid Sci. 19:102 (1969).
2791. D. H. Kaelble and R. S. Reylek, J. Adhes. 1:124 (1969).
2792. D. W. Aubrey and M. Sherriff, J. Polym. Sci., Polym. Chem. Ed. 18:2597 (1980).
2793. D. W. Aubrey, Adhesion—3 (K. W. Allen, ed.), Applied Science Publishers, London, p. 191 (1979).
2794. J. L. Cotter, Rev. High Temp. Mater. 3(4):277 (1973).
2795. R. E. Yeager, J. Appl. Polym. Sci., Appl. Polym. Symp. 3:169 (1969).
2796. S. R. Sandler and F. R. Berg, J. Appl. Polym. Sci. 9:3809 (1965).
2797. J. P. Thomas, J. Appl. Polym. Sci., Appl. Polym. Symp. 3:109 (1960).
2798. A. F. Lewis, R. A. Kinsmouth, and R. P. Kreahling, J. Adhes. 3:249 (1972).
2799. A. F. Lewis, Adhes. Age 15(6):38 (1972).
2800. W. Spath, Adhąsion 17(10):348 (1973).
2801. K. W. Allen and M. E. R. Shanahan, J. Adhes. 7:161 (1975).
2802. K. W. Allen and M. E. R. Shanahan, J. Adhes. 8:43 (1976).
2803. V. Rayatskas and V. Pekarskas, J. Appl. Polym. Sci. 20:1941 (1976).
2804. R. A. Gledhill and A. J. Kinloch, Polymer 17:727 (1976).
2805. W. C. Wake, K. W. Allen, and S. M. Dean, Elastomer: Criteria for Engineering Data (C. Hepburn and R. J. W. Reynolds, eds.), Applied Science Publishers, London, p. 113 (1979).
2806. W. Spath, Adhasion 16:416 (1972).
2807. J. Peterka, Adhasion 17:288 (1973).
2808. J. R. Beatty and E. C. Dalguish, Rubber Chem. Technol. 47:188 (1974).
2809. R. J. Kuhbander and T. J. Aponyi, Adhes. Age 19(9):27 (1976).
2810. J. A. Marceau, Adhes. Age 21(4):37 (1978).
2811. W. D. Bascom and S. Mostovoy, Amer. Chem. Soc. Org. Coatings Plastics Chem., Prepr. 38:152 (1978).
2812. J. R. Romanko and W. G. Knauss, Developments in Adhesives—2 (A. J. Kinloch, ed.), Applied Science Publishers, London, p. 173 (1981).
2813. A. J. Kinloch, Metal Sci. 14:305 (1980).
2814. A. J. Kinloch, and J. G. Williams, J. Mater. Sci. 15:987 (1980).
2815. S. Mostovoy and E. J. Ripling, Adhesion Science and Technology (L. H. Lee, ed.), Plenum Press, New York, Part B, 513 (1975).
2816. P. C. Paris and F. Erdogan, Trans ASME 85:528 (1963).
2817. P. C. Paris, 10th Sagamore Army Mater. Res. Conf., p. 107 (1964).
2818. J. D. Minford, Aluminum adhesive bond permanence, Treatise on Adhesion and Adhesives (R. L. Patrick, ed.), Marcel Dekker, New York, Vol. 5, Chap. 3, pp. 47–52 (1981).
2819. N. A. DeBruyne, The extent of contact between glue and adherend, Bull. No. 168, Tech. Service Dept., Aerospace Research Ltd., Duxford, Cambridge, England (Dec. 1956).
2820. W. D. Bascom, The origin and removal of microvoids in filiment wound composites, NRL Report 6268 (May 24, 1965).
2821. A. Mahoon, Contributions of corrosion to adhesive bond failure in aluminum alloys, Adhesion (England) 4:71 (1980).
2822. R. H. Greer, 2nd Natl. SAMPE Conf. (1970).
2823. N. L. Rogers, Processing for Adhesive Bonded Structures (M. Bodnar, ed.), Interscience, New York, NY (1972).
2824. W. M. Scardino and J. A. Marceau, Symp. on Durability of Adhesive Bonded Structure, ARADCOM (Oct. 1976).
2825. R. W. Benoliel, 10th Natl. SAMPE Conf. (Oct. 1978).
2826. A. Stevenson, On the durability of rubber/metal bonds in seawater, Int. J. Adhes. Adhes. 5(2):81 (1985).
2827. A. Stevenson, Electrochemical and corrosion effects on adhesive joints, Adhesives and Sealants, Engineered Materials Handbook, ASM, Materials Park, OH, Vol. 3, pp. 629–636 (1990).
2828. H. Leidheiser, W. Wang, and L. Igetofft, The mechanism for the delamination of organic coatings from a metal surface, Prog. Org. Coatings, Vol. 2, p. 19 (1983).
2829. W. Wang and H. Leidheiser, A method for the quantitative interpretation of cathodic delamination, Pourbaix Symp. of the Electrochemical Soc., New Orleans (Oct. 1984).

2830. F. Sedillot and A. Stevenson, Laminated rubber articulation for the deep water gravity tower, J. Energy Resour. Technol. Trans. ASME 105:480 (1983).

2831. J. Bullock, R. Thomas, J. Thornton, and R. Rushing, Durability of rubber/metal bonds in marine environments, Conf. of the Amer. Chem. Soc., Rubber Div., Dallas (April 1988).

2832. W. McIntosh and A. Stevenson, Operating experience with elastomer composites on the Hutton Tension Leg Platform, Proc. 9th Int. Conf. on Offshore Mechanics and Artic Engineering, Vol. III, Part A, pp. 127–138 (1990).

2833. R. Baboian and G. Haynes, Corrosion in Natural Environments, ASTM Spec. Tech. Publ. STP 558, pp. 171–184 (1974).

2834. S. E. Castle and J. F. Watts, Corrosion Control by Organic Coatings, Natl. Assoc. Corr. Eng. (NACE), p. 78 (1981).

2835. A. Stevenson, The effect of electrochemical potentials on the durability of rubber/metal bonds in sea water, J. Adhes. 21:313 (1987).

2836. F. J. Boerio, S. J. Hudak, M. A. Miller, and S. G. Hong, Cathodic disbondment of neoprene from steel, Proc. 10th An. Adhes. Soc. Meetings (L. H. Sharpe, ed.), Gordon and Breach, London, p. 567 (1988).

2837. M. Pourbaix, Atlas of Electrochemical Equilibria, NACS (1974).

2838. E. Mattson, British Corrosion J. 13(1):5 (1978).

2839. J. A. Richardson and G. C. Wood, J. Electrochem. Soc. 102(2):193 (1973).

2840. J. L. Cotter and M. D. Hockney, Metallurgical Rev. 19:103 (1974).

2841. T. Smith, 7th Natl. SAMPE Conf. (Oct. 1975).

2842. W. E. Bascom, S. Gadomski, and R. L. Jones, 9th Natl. SAMPE Tech. Conf. (Oct. 1977).

2843. R. L. Coble and N. M. Parikh, Fracture: An Advanced Treatise (H. Liebowitz, ed.), Academic Press, New York, New York (1972).

2844. R. J. Riel, SAMPE J. 7(5) (Aug.–Sept., 1971).

2845. R. J. Riel, Corrosion of adhesive bonded aluminum alloy panels, Rohr Corp. Report 24-2047 (Aug. 1968).

2846. R. H. Greer, Corrosion resistant adhesive bonding, Aerospace Adhesives and Elastomers, Natl. SAMPE Tech. Conf. Series, Vol. 2, p. 561 (1970).

2847. N. L. Rogers, Corrosion of adhesive bonded clad aluminum, S.A.E. Paper No. 720344, S.A.E. Nat. Bus. Aircraft Meeting, Wichita, KS (1972).

2848. N. L. Rogers, A comparative test for bondline corrosion: Clad versus bare aluminum alloys, Proc. 5th Natl. SAMPE Tech. Conf., Kiemesha Lake, NY, p. 160 (1973).

2849. W. T. McFarlen, Environmental testing of adhesive bonded metallic structures, Composite Materials in Engineering Design (B. R. Noton, ed.), ASM, Materials Park, OH, p. 678 (1973).

2850. T. R. Walton and J. E. Cowling, Report No. 7077, Naval Res. Lab., Washington, D.C. (June 1970).

2851. R. A. Gledhill and A. J. Kinloch, J. Adhes. 6:315 (1974).

2852. A. J. Kinloch, NATO AGARD Lecture Series 102, Air Force Materials Lab. Meeting, Wright-Patterson AFB, OH (March 1979).

2853. J. D. Minford, Comparative aluminum joint evaluations in varying seawater exposure conditions, J. Adhes. 18:19 (1985).

2854. J. S. Hammond, J. W. Holubka, and R. A. Dickie, Preprints Div. Org. Coat. Plast. Chem. 39:505 (1978).

2855. J. S. Hammond, J. W. Holubka, and R. A. Dickie, J. Coating Technol. 51:45 (1979).

2856. R. A. Dickie, J. W. Holubka, and J. S. Hammond, Preprints Div. Org. Coat. Plast. Chem. 41:499 (1979).

2857. J. W. Holubka, J. S. Hammond, J. E. DeVries, and R. A. Dickie, J. Coatings Technol. 52:63 (1980).

2858. J. S. Hammond, J. W. Holubka, J. E. DeVries, and R. A. Dickie, Corrosion Sci. 21:239 (1981).

2859. W. J. van Ooij, Kautschuk und Gummi. Kunststoffe 30:739,833 (1977).

2860. W. J. van Ooij and A. Kleinhesselink, Appl. Surf. Sci. 4:325 (1980).

2861. W. J. van Ooij, A. Kleinhesselink, and S. R. Leijenaar, Surf. Sci. 89:165 (1979).

2862. T. Smith, Natl. SAMPE Tech. Conf. 7:349 (1975).

2863. A. S. Tetelman and A. J. McEvily, Fracture of Structural Materials, John Wiley & Sons, New York, p. 50 (1954).

2864. W. Althof, The influence of moisture on adhesive bonded joints, Adhesion—5 (K. W. Allen, ed.), Applied Science Publishers, London, p. 15 (1980).

2865. W. C. Wake, Effect of environment on joint performance, Adhesion and Formulation of Adhesives, 2nd ed. (W. C. Wake, ed.), Applied Science Publishers, London, p. 170 (1982).

2866. J. Comyn, Kinetics and mechanism of environmental attack, Durability of Structural Adhesives (A. J. Kinloch, ed.), Applied Science Publishers, London, Chap. 3, p. 89 (1983).

2867. D. J. DeLollis and O. Montoya, Mode of failure in structural adhesive bonds, J. Appl. Polym. Sci. 11: 983 (1967).
2868. J. C. Bolger, Structural adhesives for metal bonding, Treatise on Adhesion and Adhesives (R. L. Patrick, ed.), Marcel Dekker, New York, Vol. 3, Chap. 1, pp. 22–27, 34–42 (1973).
2869. B. I. Buck and M. G. D. Hockney, Effect of outdoor exposure on adhesive bonded joints, Aspects of Adhesion—7 (K. W. Allen, ed.), Transcripta Books, London, p. 242 (1973).
2870. A. J. Kinloch, Interfacial fracture mechanical aspects of bonded joints, AGARD Lecture Series No. 102, Wright-Patterson AFB, OH, pp. 2-2 to 2-15 (Oct. 16–17, 1979).
2871. D. W. Levi, R. F. Wegman, M. C. Ross, and E. A. Garnis, Use of hot water aging for estimating lifetime of adhesive bonds to aluminum, SAMPE Quart. 7(3):1 (1971).
2872. D. J. Falconer, N. C. MacDonald, and P. Walker, Chemical Industries, p. 1230 (1964).
2873. M. Gettings, F. S. Baker, and A. J. Kinloch, Use of AES and XPS to study the locus of failure of structural adhesive joints, J. Appl. Polym. Sci. 21:2371 (1977).
2874. C. Kerr, N. C. MacDonald, and S. Orman, J. Appl. Chem. 17:62 (1967).
2875. R. I. Butt and J. L. Cotter, J. Adhes. 8:11 (1976).
2876. C. Kerr, N. C. MacDonald, and S. Orman, Brit. Polym. J. 2:67 (1970).
2877. A. J. Kinloch, Interfacial fracture mechanical aspects of adhesive bonded joints, AGARD Lecture Series No. 102, Wright-Patterson AFB, OH, p. 2-3 (Oct. 16–17, 1979).
2878. J. Bardeleben, Notes on the diffusion of water through epoxy resins, Kunstoffe 53:162 (1963).
2879. J. A. Manson and E. H. Chiu, Penetration of water in a filled epoxy resin, J. Polym. Sci., C-41:95 (1973).
2880. J. Crank, The Mathematics of Diffusion, Oxford University Press, London (1958).
2881. J. H. Wilson, T. S. Sudarshan, and H. H. Mabie, Humidity effects on the fracture mode transition in anodized 2024-T351 aluminum, J. Mater. Sci. 3(9):773 (Sept. 1984).
2882. M. S. Majerus, D. S. Soong, and J. M. Parusnitz, Experimental measurements and Monte-Carlo simulation of water diffusion into epoxy matrices, J. Appl. Polym. Sci. 29(8):2453 (August 1984).
2883. P. Johncock and G. F. Tudgey, Some effects of structure, composition, and cure on the water absorption and glass transition temperature of amine-cured epoxies, Brit. Polymer J. 18(5):292 (1986).
2884. H. Dodiuk, G. Sharon, and S. Kenig, Hygrothermal properties of adhesively bonded joints and their correlation with bulk adhesive properties, J. Adhes. 33(1-2):45 (1990).
2885. G. Whitaker, M. I. Darby, G. H. Westenholm, B. Yates, M. H. Collins, A. R. Lyle, and B. Brown, Influence of temperature and hydrostatic pressure on moisture absorption in polymer resins, J. Mater. Sci. 26(1):49 (1991).
2886. K. Nakamura and T. Maruno, unpublished data.
2887. K. Nakamura, T. Maruno, and S. Sasaki, Theory for the decay of the wet shear strength of adhesion and its application to metal/epoxy/metal joints, Int. J. Adhes. Adhes. 7(2):97 (1987).
2888. K. Nakamura and M. Kondo, Japan J. Appl. Phys. 12:994 (1973).
2889. K. Nakamura, J. Metal Finish Soc., Japan 25:403 (1974).
2890. W. D. Bascom and R. L. Patrick, The surface chemistry of bonding metals with polymer adhesives, Adhes. Age 17(10):25 (1974).
2891. W. Bascom, J. Adhes. 2:161 (1970).
2892. L. H. Little, Infrared Spectra of Absorbed Species, Academic Press, New York (1966).
2893. M. L. Hair, Infrared Spectroscopy in Surface Chemistry, Marcel Dekker, New York (1967).
2894. J. R. Dacey, Chemistry and Physics of Interfaces (S. Ross, ed.), American Chemical Society, Washington, D.C., p. 151 (1965).
2895. H. A. Resing, Adv. in Molec. Relaxation Processes 1:109 (1967).
2896. F. P. Bowden and W. R. Throssell, Proc. Royal Soc. p. 209 (1951).
2897. J. C. Henniker, Rev. Mod. Phys. 21:322 (1949).
2898. H. T. Corten, Fracture, An Advanced Treatise (H. Liebowitz, ed.), Academic Press, New York, Vol. 7, p. 676 (1972).
2899. A. J. Kinloch, Interfacial fracture mechanical aspects of adhesive bonded joints, AGARD Lecture Series No. 102, Wright-Patterson AFB, OH, p. 2-4 (Oct. 16–17, 1979).
2900. J. D. Minford, Durability of aluminum bonded joints, Treatise on Adhesion and Adhesives (R. L. Patrick, ed.), Marcel Dekker, New York, Vol. 3, Chap. 2, pp. 111–114 (1973).
2901. Y. Weitsman, Stresses in adhesive joints due to moisture and temperature, J. Comp. Mater. 11:378 (1977).
2902. A. J. Kinloch, Interfacial fracture mechanical aspects of adhesive bonded joints, AGARD Lecture Series No. 102, Wright-Patterson AFB, OH, p. 2-5 (Oct. 16–17, 1979).
2903. K. M. Takahashi, A study of interfacial moisture ingression by A.C. impedence measurements, 11th Ann. Adhes. Soc. Meeting, Charleston, SC (Feb. 21–24, 1987).

2904. D. R. Lefebvre, S. Roy, D. A. Dillard, H. F. Brinson, and J. N. Reddy, The diffusion of gases and vapors in polymers: Effect of strain and boundaries with applications to the durability of adhesive joints, 11th Ann. Adhes. Soc. Meeting, Charleston, SC (Feb. 21–24, 1987).

2905. D. R. Lefebvre, D. A. Dillard, and H. F. Brinson, The durability of adhesive joints: an engineering study, Final Report, Office Naval Research Project CASS/ESM-88-7, Virginia Tech. Report No. VPI-E-88-19, Virginia Polytecnic Institute and State University, Blacksbury, VA.

2906. D. M. Brewis, J. Comyn, A. K. Raval, and A. J. Kinloch, Is there a critical relative humidity for the weakening of joints in wet air? Adhesion '90, The Plastics and Rubber Inst. Meetings, Cambridge, England (Sept. 10–12, 1990).

2907. D. R. Lefebvre, K. M. Takahashi, and V. R. Raju, Degradation of epoxy coatings in humid environments: The critical relative humidity, 13th Ann. Adhes. Soc. Meeting, Savannah, GA (Feb. 18–21, 1990).

2908. M. R. Pigott, Acta Metallurgica 12:803 (1964).

2909. J. D. Minford, Aluminum adhesive bond permanence, Treatise on Adhesion and Adhesives (R. L. Patrick, ed.), Marcel Dekker, New York, Vol. 5, Chap. 3, p. 123 (1981).

2910. J. Comyn, Kinetics and mechanisms of environmental attack, Durability of Structural Adhesives (A. J. Kinloch, ed.), Applied Science Publishers, London, Chap. 3, pp. 85–131 (1983).

2911. J. S. Carslaw and J. C. Jaeger, Conduction of Heat in Solids, 2nd ed., Oxford University Press, London (1959).

2912. H. Fujita, Adv. Polym. Sci. 3:1 (1961).

2913. D. M. Brewis, J. Comyn, R. J. A. Shalash, and T. L. Tegg, Polymer 21:357 (1980).

2914. D. H. Kaelble and P. J. Dynes, J. Mater. Eval. 35:103 (1977).

2915. E. L. McKague, J. D. Reynold, and J. E. Halkias, J. Appl. Polym. Sci. 22:1643 (1978).

2916. C. D. Shirrell and J. Halpin, ASTM Spec. Tech. Publ., STP 617, p. 514 (1977).

2917. E. L. McKague, J. D. Reynolds, and J. E. Halkias, Trans. ASME 98:92 (1976).

2918. T. S. Cook, D. E. Walrath, and P. H. Francis, Natl. SAMPE Symp. Exhib., 22:339 (1977).

2919. E. R. Long, NASA Tech. Paper No. 1474 (1979).

2920. N. D. Daniely and E. R. Long, J. Polym. Sci. Chem. Ed. 19:2443 (1981).

2921. G. Marom and L. J. Broutman, J. Appl. Polym. Sci. 26:1493 (1981).

2922. M. K. Antoon and J. L. Koenig, J. Macromol. Sci-Rev. Macromol. Chem. C-19:135 (1980).

2923. D. M. Brewis, J. Comyn, and J. L. Tegg, Polymer 21:134 (1980).

2924. J. M. Augl and A. E. Berger, Natl. SAMPE Tech. Conf. 8:383 (1976).

2925. D. Y. Perera and P. M. Heertzes, J.O.C.C.A. 54:395 (1971).

2926. M. Mozisek, Jąd. Energ. 22:448 (1976).

2927. W. Althof, Proc. 11th Natl. SAMPE Tech. Conf., p. 309 (1979).

2928. T. C. Wong and L. J. Broutman, Moisture diffusion in epoxy resins, Part I. non-Fickian sorption processes, Polym. Sci. and Eng. 25(9):521 (1985).

2929. T. C. Wong and L. J. Broutman, Water to epoxy resins, Part II. diffusion mechanism, Polym. Sci. Eng. 25(9):529 (1985).

2930. P. J. Pearce, R. G. Davidson, and C. E. M. Morris, Hydrolytic stability of some uncured epoxy resins, J. Appl. Polym. Sci. 26:2363 (1981).

2931. S. G. Abbott and N. Brumpton, The effect of moisture on polyurethane adhesives, J. Adhes. 13:44 (1981).

2932. K. F. Droin, J. Guthrie, C. I. Leung, F. R. Martin, and M. S. Otterburn, The effect of moisture on the strength of the steel/steel cyanoacrylate bond, J. Adhes. 17:71 (1984).

2933. G. L. Schneberger, Polymer structure and adhesive behavior, Adhes. Age 17:17 (April 1974).

2934. Moisture effects, life assurance of composite structures, AFML Report No. TR-75-51, General Dynamics, Fort Worth Div., Vol. 1 (May 1975).

2935. Fatigue behavior of adhesively bonded joints, Report FZM-6697, Quarterly Prog. Report No. 5, General Dynamics, Fort Worth Div. (Oct. 1977).

2936. S. N. Zhurkov and E. E. Tomashevsky, Physical Basis of Yield and Fracture, Institute of Physics, London, p. 200 (1966).

2937. S. N. Zhurkov and V. E. Korsukov, J. Polym. Sci., Polym. Phys. 12:385 (1974).

2938. M. R. Bowditch, The relationship between hydrolytic stability of adhesive joints with equilibrium water content, Adhesion '90, Plastics and Rubber Inst. Meetings, Cambridge, England (Sept. 10–12, 1990).

2939. G. D. Small and P. A. Fay, Creep of adhesive lap-joints in dry and high humidity environments, Adhesion '90, Plastics and Rubber Inst. Meetings, Cambridge, England (Sept. 10–12, 1990).

2940. N. Bianchi, G. Garbassi, R. Pucciariello, and G. Romano, Study of water sorption effect on toughened acrylic adhesives, Int. J. Adhes. Adhes. 10(1):19 (1990).

2941. L. ElSa'ad, M. I. Darby, and B. Yates, Moisture absorption by epoxy resins: The reverse thermal effect, J. Mater. Sci. 25(8):3577 (1990).
2942. W. Althof, DFVLR-Forschungsbericht 79-06 (1979).
2943. W. Althof, Aluminium 55:600 (1979).
2944. D. M. Brewis, J. Comyn, A. C. Moloney, and J. L. Tegg, European Polym. J. 17:127 (1981).
2945. J. A. Barrie, Water in polymers, Diffusion in Polymers (J. Crank and G. S. Park, eds.), Academic Press, London, p. 259 (1968).
2946. C. Kerr and S. Orman, The effect of water on aluminum/epoxide bonds, 7th Conf. on Adhesion, City University of London, England (1969).
2947. M. K. Antoon, J. L. Koenig, and T. Serafini, J. Polym. Sci., Phys. Ed. 19:1567 (1981).
2948. M. K. Antoon and J. L. Koenig, J. Polym. Sci., Phys. Ed. 19:197 (1981).
2949. R. L. Levy, D. L. Fanter, and C. J. Summers, J. Appl. Polym. Sci. 24:1643 (1974).
2950. C. E. Browning, Polym. Eng. Sci. 18:16 (1978).
2951. C. E. Browning, Nat. SAMPE Symp. Exhib. 22:365 (1977).
2952. A. Mazor, L. J. Broutman, and B. H. Eckstein, Nat. Tech. Conf., SPE (Prepr.), p. 77 (1976).
2953. N. R. Farrar and K. H. G. Ashbee, J. Phys. D. 11:1009 (1978).
2954. J. Nicholas and K. H. G. Ashbee, J. Phys. D. 11:1015 (1978).
2955. A. Apicella, L. Nicolais, G. Astarita, and E. Drioli, Polymer 20:1143 (1979).
2956. A. Apicella, L. Nicolais, G. Astarita, and E. Drioli, Polym. Eng. Sci. 21:17 (1981).
2957. A. Apicella and L. Nicolais, Industrial Engineering Chemistry Product Research & Development 20:138 (1981).
2958. K. Mizutani and T. Iwatsu, J. Appl. Polym. Sci. 26:3447 (1981).
2959. T. G. Fox, Bull. Amer. Phys. Soc. 1:123 (1956).
2960. F. N. Kelley and F. Beuche, J. Polym. Sci. 50:549 (1961).
2961. J. A. McMillan and S. C. Los, Nature 206:806 (1961).
2962. M. Suisake, H. Suga, and S. Seki, Bull. Chem. Soc. Japan 41:2591 (1968).
2963. D. H. Rasmussen and A. P. MacKenzie, J. Phys. Chem. 75:967 (1971).
2964. R. Frank, Water, Plenum Press, New York (1972).
2965. R. J. Morgan and E. T. Mones, 11th Natl. SAMPE Tech. Conf., p. 218 (1979).
2966. C. E. Browning, G. E. Husman, and J. M. Whitney, ASTM Spec. Tech. Publ., STP 617, p. 481 (1977).
2967. R. J. A. Shalash, Ph.D. Thesis, Leicester Polytechnic, Leicester, England (1980).
2968. S. Gazit, J. Appl. Polym. Sci. 22:3547 (1978).
2969. Y. J. Weitman, Compos. Mater. 11:378 (1977).
2970. J. P. Sargent and K. H. G. Ashbee, J. Adhes. 11:175 (1980).
2971. A. E. H. Love, Treatise on the Mathematical Theory of Elasticity, Cambridge University Press, Cambridge, England, Chap. XXII, p. 455 (1959).
2972. D. E. Jesson and J. P. Sargent, The deformation of the adherends in an adhesive joint undergoing water uptake, J. Adhes. 14:119 (1982).
2973. N. J. DeLollis, 22nd Nat. SAMPE Symp. Exhib., p. 673 (April 1977).
2974. J. Boutilier, E. J. Hughes, and J. L. Rutherford, Proc. 26th Natl. SAMPE Symp., Los Angeles, p. 781 (1981).
2975. Y. J. Weitsman, Residual stresses in adhesive joints, Final Report, AFWAL-TR-81-4121, Wright-Patterson AFB, OH (1981).
2976. E. J. Hughes and J. L. Rutherford, Singer Contract D-AA21-67-C-0500, Symp. on Durability, Picatinny Arsenal, Dover, NJ (Aug, 1965).
2977. E. J. Hughes, J. Boutilier, and J. L. Rutherford, The effect moisture on the dimensional stability of adhesively-bonded joints, Kearfott Div. Singer Co., Little Falls, NJ 07424.
2978. G. Good, Combined temperature-moisture-mechanical stress effects on adhesive joints, Adhesives and Sealants, Engineered Materials Handbook, ASM, Materials Park, OH, Vol. 3, pp. 651–655 (1990).
2979. D. Fox, M. M. Labes, and A. Weissberger, Physics and Chemistry of the Organic Solid State, John Wiley & Sons (1965).
2980. M. Gordon, C. S. Hope, L. D. Loan, and R. J. Roe, Proc. Royal Soc. A 258:215 (1960).
2981. J. Crank and G. J. Park (eds.), Diffusion in Polymers, Academic Press, New York (1968).
2982. J. B. Enns and J. K. Gillham, Effect of the extent of cure on the modulus, glass transition, water absorption and density of an amine-cured epoxy, J. Appl. Polym. Sci. 28:2831 (1983).
2983. W. Brockmann, Durability of polymer bonds, Adhesion Aspects of Polymeric Coatings (K. L. Mittal, ed.), Plenum Press, New York, p. 265 (1982).
2984. C. Kerr, N. C. MacDonald, and S. Orman, Effect of hostile environments on adhesive joints, J. Appl. Chem. 17:62 (1967).

2985. R. I. Butt and J. L. Cotter, J. Adhes. 8:11 (1976).
2986. H. S. Katz and J. V. Milewski (eds.), Handbook of Fillers and Reinforcements for Plastics, Van Nostrand, New York (1978).
2987. J. R. Griffiths and J. D. Bultman, Industrial Engineering Chemistry, Prod. Res. Dev. 17(1):8 (1978).
2988. J. S. Ahearn, G. D. Davis, T. S. Sun, and J. D. Venales, Correlation of surface chemistry and durability of aluminum/polymer bonds, Adhesion Aspects of Polymeric Coatings (K. L. Mittal, ed.), Plenum Press, New York, p. 288 (1983).
2989. Composite Materials (Vol. 6), Interfaces in Polymer Matrix Composites (E. P. Plueddemann, ed.), Academic Press, New York (1974).
2990. A. Ahagon, A. N. Gent, and E. C. Hsu, Adhesion Science and Technology (L. H. Lee, ed.), Plenum Press, New York, Vol. 9A, p. 281 (1975).
2991. M. Gettings, F. J. Baker, and A. J. Kinloch, Use of AES and XPS to study the locus of failure of structural adhesive joints, J. Appl. Polym. Sci. 21:2375 (1977).
2992. P. D. Calvert, R. R. Lalanandham, and D. R. M. Watson, Adhesion Aspects of Polymeric Coatings (K. L. Mittal, ed.), Plenum Press, New York, p. 457 (1983).
2993. A. N. Gent and J. Schultz, J. Adhes. 3:281 (1972).
2994. S. Orman and C. Kerr, Aspects of Adhesion (D. J. Alner, ed.), University of London Press, London, p. 64 (1971).
2995. J. Comyn, D. M. Brewis, R. J. A. Shalash, and J. L. Tegg, Adhesion—3, Applied Science Publishers, London, p. 13 (1979).
2996. H. Ishii and Y. Yamaguchi, Kogakuin, Daigaku Kenkyu Kokoku 44:652 (1978).
2997. W. V. Chang and J. S. Wang, J. Appl. Polym. Sci. 25:1759 (1981).
2998. A. J. Kinloch, Review: the science of adhesion, Part 2. mechanics and mechanism of failure, J. Mater. Sci. 17:644,645 (1982).
2999. M. Gettings and A. J. Kinloch, J. Mater. Sci. 12:2049 (1977).
3000. J. Comyn, Kinetics and mechanism of environmental attack, Durability of Structural Adhesives (A. J. Kinloch, ed.), Applied Science Publishers, London, pp. 119–121 (1983).
3001. K. K. Knoch and M. C. Locke, Proc. SAMPE Tech. Conf., Mt. Pocono, PA, p. 445 (1981).
3002. J. D. Venables, 20th Conference on Adhesion and Adhesives, City University of London, England (April 1982).
3003. J. D. Venables, Adhesion and durability of metal/polymer bonds, Adhesive Joints: Formation, Characteristics, and Testing (K. L. Mittal, ed.), Plenum Press, New York, pp. 460–461 (1984).
3004. J. D. Minford, Aluminum adhesive bond permanence, Treatise on Adhesion and Adhesives (R. L. Patrick, ed.), Marcel Dekker, New York, Vol. 5, Chap. 3, p. 82 (1981).
3005. R. G. Donnelly and R. E. Cohen, Enchancement of the stability of common polymeric materials against undersea degradation, Report No. MTSG 81-14, Massachusetts Institute of Technology, Cambridge (Oct. 1981).
3006. A. A. Fahny and J. C. Hart, Stress dependence of water diffusion in epoxy resins, Polymer Composites 1:77 (1980).
3007. O. Gilat and L. J. Broutman, ASTM Spec. Tech. Report No. STP 685, p. 61.
3008. C. D. Bailey, Failure analysis of a bonded honeycomb structure—a case study employing SEM, ASTM Spec. Tech. Publ., STP 493 (Symp. Annual Meeting, Toronto, Ontario), p. 55 (July 1971).
3009. H. Dodiuk, A. E. Naniv, and N. Fin, Characterization of oxidized aluminum-1100 exposed to hygrothermic environments, Appl. Surf. Sci. 29(10):67 (1987).
3010. R. A. Just, Environmental effects on fracture of adhesively bonded joints, Adhesively Bonded Joints: Testing, Analysis, and Design Conf., ASTM, Baltimore (Sept. 10–12, 1986).
3011. D. M. Brewis, J. Comyn, A. K. Ravel, and A. J. Kinloch, The effect of humidity on the durability of aluminium-epoxide joints, Int. J. Adhes. Adhes. 10(4):247 (1990).
3012. W. Brockmann and H. Kollek, Natl. SAMPE Exhib. 23:1119 (1978).
3013. O. D. Hennemann and W. Brockmann, J. Adhes. 12:297 (1981).
3014. P. F. A. Bylmer, Metal Finishing, pp. 22–26 (Aug. 1977).
3015. J. S. Ahearn, T. S. Sun, Fraebe, J. D. Venables, and R. L. Hopping, SAMPE Quart., pp. 39–45 (Oct. 1980).
3016. American Society for Testing Materials, Philadelphia.
3017. Adhesives, Volume 15.06 Annual Book of ASTM Standards, American Society for Testing Materials, Philadelphia.
3018. Building Seals and Sealants: Fire Standards: Building Construction, Volume 04.07 Annual Book of ASTM Standards, American Society for Testing Materials, Philadelphia.
3019. Society of Automotive Engineers, Warrendale, PA.

3020. Technical Association for the Pulp and Paper Industry (TAPPI), Atlanta, GA.
3021. J. Nardone, Specifications and Standards for Adhesives and Sealants, Adhesives and Sealants, Engineered Materials Handbook, ASM, Materials Park, OH, Vol. 3, pp. 61–64 (1990).
3022. A. H. Landrock, Guide to general information, Adhesives and Sealants, Engineered Materials Handbook, ASM, Materials Park, OH, Vol. 3, pp. 65–72 (1990).
3023. D. Meinhold, Mathematical modeling and statistical inference, Adhesives in Manufacturing (G. L. Schneberger, ed.), Marcel Dekker, New York, Chap. 24, p. 603 (1983).
3024. H. Koski and G. L. Schneberger, Testing adhesives, Adhesives in Manufacturing (G. L. Schneberger, ed.), Marcel Dekker, New York, Chap. 22, p. 551 (1983).
3025. H. Koski and G. L. Schneberger, Testing adhesives, Adhes. Age 27(5):8 (May 1984).
3026. K. L. DeVries and G. P. Anderson, Analysis and design of adhesive bonded joints, AGARD-NATO Lecture Series No. 102, Wright-Patterson AFB, OH, p. 3-1 (March 1979).
3027. G. P. Anderson, K. L. DeVries, and G. Sharon, Evaluation of adhesive test methods, Adhesive Joints: Formation, Characteristics, and Testing (K. L. Mittal, ed.), Plenum Press, New York, p. 269 (1984).
3028. G. P. Anderson and K. L. DeVries, Analysis of testing procedures, Treatise on Adhesion and Adhesives (R. L. Patrick, ed.), Marcel Dekker, New York, Vol. 6, pp. 55–121 (1989).
3029. D. A. Dillard and D. LeFebvre, Testing and analysis, Adhesives and Sealants, Engineered Materials Handbook, ASM, Materials Park, OH, Vol. 3, pp. 313–314 (1990).
3030. T. A. Osswald and J. Rietveld, Measuring constitutive properties, Adhesives and Sealants, Engineered Materials Handbook, ASM, Materials Park, OH, Vol. 3, pp. 315–324 (1990).
3031. R. D. Adams, Failure strength tests and their limitations, Adhesives and Sealants, Engineered Materials Handbook, ASM, Materials Park, OH, Vol. 3, pp. 325–334 (1990).
3032. K. M. Liechti, Fracture testing and failure analysis, Adhesives and Sealants, Engineered Materials Handbook, ASM, Materials Park, OH, Vol. 3, pp. 335–348 (1990).
3033. E. Sancaktar, Static and dynamic fatigue testing, Adhesives and Sealants, Engineered Materials Handbook, ASM, Materials Park, OH, Vol. 3, pp. 349–372 (1990).
3034. M. G. Allen, Special tests for membranes and miniature components, Adhesives and Sealants, Engineered Materials Handbook, ASM, Materials Park, OH, Vol. 3, pp. 373–381 (1990).
3035. D. R. Lefebvre, Special tests for sealants and elastomeric/foam materials, Adhesives and Sealants, Engineered Materials Handbook, ASM, Materials Park, OH, Vol. 3, pp. 382–390 (1990).
3036. L. T. Drzal and P. J. Herrera-Franco, Composite fiber-matrix bond tests, Adhesives and Sealants, Engineered Materials Handbook, ASM, Materials Park, OH, Vol. 3, pp. 391–405 (1990).
3037. D. E. Packham, Microstructural analysis, Adhesives and Sealants, Engineered Materials Handbook, ASM, Materials Park, OH, Vol. 3, pp. 406–419 (1990).
3038. D. Roylance, Thermal properties and temperature effects, Adhesives and Sealants, Engineered Materials Handbook, ASM, Materials Park, OH, Vol. 3, pp. 420–427 (1990).
3039. K. M. Takahashi, Electrical properties, Adhesives and Sealants, Engineered Materials Handbook, ASM, Materials Park, OH, Vol. 3, pp. 428–440 (1990).
3040. D. W. Schmueser, Evaluating test geometries, Adhesives and Sealants, Engineered Materials Handbook, ASM, Materials Park, OH, Vol. 3, pp. 441–455 (1990).
3041. M. Romand, F. Gillard, M. Charbonnier, and A. Roche, Recent developments in the spectroscopic characterization of modified surfaces for bonding or painting, J. Adhes. 23(1):1 (1987).
3042. H. Tokutake, K. Nishimori, J. Matsura, and V. Pinet, Scanning Auger electron microscope resolution determined by a quantitative AES method, Surf. Sci. 186(3):339 (1987).
3043. J. P. Petrakian and P. Renucci, Theoretical analysis of AES depth profiling in multilayers, application to C/W multilayers, Surf. Sci. 186(3):447 (1987).
3044. H. E. Bishop, D. P. Moon, P. Marriott, and P. R. Chalker, Application of a high spatial resolution combined AES/SIMS instrument, Vacuum 30(10):929 (1989).
3045. A. J. Pertsin and Yu. M. Pashunin, Differential charging in XPS studies of polymer/metal interfaces, Appl. Surf. Sci. 33(3):171 (1990).
3046. L. J. Atanasoska, S. G. Anderson, H. M. Meyer, and J. H. Weaver, XPS study of chemical bonding at polyimide interfaces with metal and semiconductor overlays, Vacuum 40(1–2):85 (1990).
3047. N. H. Turner, Surface analysis: X-ray photoelectron spectroscopy and Auger electron spectroscopy, Anal. Chem. 62:113R (1990).
3048. J. F. Watts and E. M. Gibson, The determination of acid-base properties of inorganic surfaces by XPS, Adhesion '90, Plastics and Rubber Inst. Meetings, Cambridge, England (Sept. 10–12, 1990).
3049. S. Akhter, K. C. Cannon, and J. M. White, Distribution of compositional defects in solvent cast polyvinyl alcohol films: Angular dependent XPS and reactivity study, Appl. Surf. Sci. 44(1):49 (1990).

3050. K. Wittmaack, Charge compensation in SIMS analysis of polymer foils using negative secondary ions, 10(6):311 (1987).
3051. G. Save, P. Hakansson, and B. J. R. Sundquist, Spin-deposited sub-micrometer films of organic molecules for secondary-ion mass spectroscopy studies, Anal. Chem. 59(17):2059 (Sept. 1987).
3052. K. J. Hook, T. J. Hook, J. H. Wandass, and J. A. Gardella, Secondary-ion formation from functional polymer systems in static secondary-ion mass spectrometry. Appl. Surf. Sci. 44(1):29 (1990).
3053. M. Kodema, K. Kuramoto, and I. Karino, ESCA and FT-IR studies on boundary-phase structure between blend polymers and polyimide substrates, J. Appl. Polym. Sci. 34(5):1889 (Sept. 1987).
3054. Gi. Xue, Shi-Ying Liu, Yi Jin, and Shan Geng Jiang, In-situ studies of coatings on metal wires by FT-IR external resistance spectroscopy, Appl. Spectros. 41(7):1172 (1987).
3055. M. W. Urban and E. M. Salazar-Rojaz, Probing organic-inorganic interactions and curing processes by photocoustic Fourier transform infrared spectroscopy, J. Polym. Sci.-Chem. 28(4):1593 (1990).
3056. T. Cosgrove, C. A. Prestidge, and B. Vincent, Chemisorption of linear and cyclic polymethylsiloxane on alumina studied by Fourier transform infrared spectroscopy, J. Chem. Soc.-Faraday Transactions 9:1377 (1990).
3057. J. Comyn, D. P. Oxley, R. G. Pritchard, and C. R. Werrett, Inelastic electron tunnelling spectroscopy (IETS) of silane coupling agents on aluminum oxide, 11th Ann. Adhes. Soc. Meeting, Charleston, SC (Feb. 21–24, 1988).
3058. G. K. Wertheim, New method for bulk quantitative analysis by ESCA, J. Electron Spectros. 50(Nos. 1 & 2):31 (1989).
3059. J. L. Koenig, The application of solid state NMR spectroscopy to the study of interpenetrating polymer networks in the interface of glass reinforced epoxy composites, 11th Ann. Adhes. Soc. Meeting, Charleston, SC (Feb. 21–24, 1988).
3060. H. N. Cheng, Using analytical techniques to characterize adhesives, Adhes. Age 31(13):34 (1988).
3061. A. O. K. Nieminen and J. L. Koenig, NMR imaging of the interfaces of epoxy joints, J. Adhes. 30(1–4):47 (1989).
3062. F. S. Blum, Magnetic resonance of polymers at surfaces, Colloids and Surfaces 45:361 (1990).
3063. R. A. Kinsey, Solid state NMR of elastomers, Rubber Chem. Technol. 63:167 (1990).
3064. W. Kuhn and U. Eichoff, NMR spectroscopy: Limitations and applications in materials sciences, Die Makromoleculare Chemie Macromolecular Symposia 34:287 (1990).
3065. K. Siegbaun, From x-ray to electron spectroscopy and new trends, J. Electron Spectros. 51:11 (1990).
3066. D. D. McDonald, Some advantages and pitfalls of electrochemical impedance spectroscopy, J. Sci. Eng. Corrosion 46(3):229 (1990).
3067. J. Titz, G. H. Wagner, H. Spahn, M. Ebert, K. Juttner, and W. J. Lorenz, Characterization of organic coatings on metal substrates by electrochemical impedance spectroscopy, J. Sci. and Eng. Corrosion 46(3):221 (1990).
3068. J. S. Marceau and R. H. Firminhac, Basic studies on bonding development—a study of surface treatments and environmental effects using radioactive tracers and electrical measurements, Boeing Document T6-5523 (May 1969).
3069. J. S. Marceau and R. H. Firminhac, Basic studies on bonding developments—2nd report: A study of surface treatments and environmental effects listing radioactive tracers, Boeing Document T6-5612 (1970).
3070. C. A. L. Westerdahl and J. R. Hall, The location of adhesive bond failures by using radioactive tracers, Picatinny Arsenal Report No. TR-4498 (March 1973).
3071. J. R. Jones, C. Poncipe, J. M. Berton, and W. W. Wright, A radiochemical study of the kinetics of epoxide cure: reaction of glycidyl ether and other model epoxides with various imidazoles, Polymer 28(8):1358 (1987).
3072. I. Kelson, Delayed desorption- a proposed new utilization of radioactive atomic nuclei for surface studies, Appl. Phys. 20(8):1049 (Aug. 1987).
3073. D. F. Adams and D. E. Walrath, Current status of the Iosipescu shear test methods, J. Comp. Mater. 21:494 (June 1987).
3074. J. A. Barnes, M. Kumosa, and D. Hull, Theoretical and experimental evaluations of the Iosipescu shear test, Compos.Sci. Technol. 28(4):252 (1987).
3075. M. J. P. Pindera, P. Ifju, and D. Post, Iosipescu shear characterization of polymeric and metal matrix composites, Exptl. Mech. 30(1):101 (1990).
3076. M. G. Allen, R. T. Howe, and S. D. Senturia, Microfabricated structures for the in-situ measurement of residual stress, Young's modulus and ultimate strain of thin films, Appl. Phys. Lettr. 51(4):241 (June 1987).

3077. W. S. Sampath and P. J. Wilbur, A new rapid technique for characterizing microstructures of films produced by ion beams in the transmission electron microscope, Mater. Sci. Eng. 90:143 (June 1987).

3078. K. M. Liechti and T. Freda, On the use of laminated beams for determining pure and mixed-mode fracture toughness of adhesives, 11th Ann. Adhes. Soc. Meeting, Charleston, SC (Feb. 21–24, 1988).

3079. R. Herrera, H. D. Mejias, E. A. Stocchi, and L. A. de Vedia, A simple method for fatigue precracking specimens for fracture mechanics tests, J. Test. Eval. 15(5):291 (Sept. 1987).

3080. E. F. Crawley and M. C. van Schoor, Material damping in aluminum and metal matrix composites, J. Comp. Mater. 21:553 (1987).

3081. C. J. Budziak and A. W. Newmann, Automation of the capillary rise technique for measuring contact angles, Colloids and Surf. 43(2/4):279 (1989).

3082. J. K. Spelt, Solid surface tension: the use of thermodynamic models to verify its determination from contact angles, Colloids and Surf. 43(2/4):389 (1989).

3083. J. M. Ager, D. K. Viers, and G. M. Rosenblatt, Raman intensities and interference effect for thin films adsorbed on metals, J. Chem. Phys. 93(3):2067 (1990).

3084. V. M. Sura and R. Rhinehart, Identation test for adhesion measurement of polyimide films, J. Adhes. Sci. Technol. 4(2):161 (1990).

3085. J. E. Ritter, Adhesion of thin polymer coatings under contact stresses, Adhesion '90, Plastics and Rubber Inst. Meetings, Cambridge, England (Sept. 10–12, 1990).

3086. S. Klotz, J. von Seggers, M. Kung, and H. J. Cantow, Imaging polymer interfaces by element specific electron microscopy and electron-energy loss spectroscopy, Polymer Communications 31(8):332 (1990); Polymer Communications 31(8):332 (1990).

3087. M. Manoharan, J. P. Hirth, and A. R. Rosenfield, A suggested procedure for combined Mode I/Mode III fracture toughness testing, J. Test. Eval. 18(2):106 (1990).

3088. J. Johnston, Adhes. Age 26:34 (1983).

3089. F. Urushizaki, H. Yamaguchi, and H. Mizumachi, J. Adhes. Soc. Japan 20:295 (1984).

3090. H. Mizumachi and T. Saito, J. Adhes. 20:83 (1986).

3091. H. Mizumachi, Materials Technol. (Japan) 2:6 (1984).

3092. H. Mizumachi, J. Adhes. Soc. Japan 20:522 (1984).

3093. H. Mizumachi, J. Adhes. Soc. Japan 30:2675 (1985).

3094. H. Mizumachi and Y. Hatano, Proceedings of 10th Annual Meeting of the Adhesion Soc. (L. H. Sharpe, ed.), Gordon and Breach, London, pp. 75–84 (1988).

3095. F. H. Hammond, Jr., ASTM Spec. Publ. No. 360, ASTM, Philadelphia (1963).

3096. J. D. Minford, unpublished work, Aluminum Co. of America.

3097. J. D. Minford, unpublished work, Aluminum Co. of America.

3098. H. T. Corten, J. Adhes. 3:103 (1971).

3099. W. D. Bascom and C. O. Timmons, Adhesion Science and Technology (L. H. Lee, ed.), Plenum Press, New York, Vol. 9B, p. 501 (1975).

3100. W. D. Bascom, R. Y. Ting, R. J. Moulton, C. K. Riew, and A. R. Siebert, J. Mater. Sci. 16:2657 (1981).

3101. W. D. Bascom, D. L. Hunston, and C. O. Timmons, Org. Coat. Plast. Prepr. 39:179 (1978).

3102. W. D. Bascom, R. J. Moulton, E. J. Rowe, and A. R. Siebert, Org. Coat. Plast. Prepr. 39:1964 (1978).

3103. W. D. Bascom and D. L. Hunston, Adhesion—6 (K. W. Allen, ed.), Applied Science Publishers, London, Chap. 14 (1982).

3104. D. L. Hunston and W. D. Bascom, Comp. Tech. Rev. 5:118 (1983).

3105. W. D. Bascom, G. W. Bullman, D. L. Hunston, and R. M. Jensen, Proc. SAMPE Symp. Reno, NV (April 1984).

3106. D. L. Hunston, Comp. Tech. Rev. 4(Winter):176 (1984).

3107. D. L. Hunston and W. D. Bascom, Rubber-Modified Thermoset (C. K. Riew and J. L. Gillham, eds.), ACS Symp. Ser. 208, American Chemical Society, Washington, D.C., Chap. 7 (1984).

3108. D. L. Hunston and G. W. Bullman, Int. J. Adhes. Adhes. 5(2):69 (1985).

3109. A. J. Kinloch, S. J. Shaw, D. A. Tod, and D. L. Hunston, Polymer 24:1341, 1355 (1983).

3110. D. L. Hunston, A. J. Kinloch, S. J. Shaw, and S. S. Wang, Characterization of the fracture beahvior of adhesive joints, Adhesive Joints: Formation, Characterization, and Testing (K. L. Mittal, ed.), Plenum Press, New York, pp. 789–807 (1984).

3111. W. A. Jemian, Structural precursors to fracture in adhesive joints, Adhesive Joints—Formation, Characteristics, and Testing (K. L. Mittal, ed.), Plenum Press, New York, pp. 809–827 (1984).

3112. A. H. Cottrell, The Mechanical Properties of Matter, John Wiley & Sons, New York (1964).

3113. A. C. Wilcox and W. A. Jemian, Metall. Rev. 1:13 (1972).

3114. A. C. Wilcox and W. A. Jemian, Polym. Engl. Sci. 13:40 (1973).

3115. K. Masubuchi and R. E. Keith, Final Report on "Fundamentals of Deformation Characteristics of Adhesive-Bonded Joints and Metal/Adhesive Interfaces, Contract DA-01-021-AMC-14693(z), Batelle Inst., Columbus, OH (1967).
3116. G. C. Grimes, Aero Adhes. Elast., pp. 249–258 (1970).
3117. W. J. Renton and J. R. Vinson, The Analyses and Design of Composite Material Bonded Joints under Static and Fatigue Loadings, Contract AFOSR-1760-72, University of Delaware, Newark (1973).
3118. A. N. Gent, J. Mater. Sci. 5:925 (1970).
3119. T. Alfrey, Treatise on Adhesion and Adhesives (R. L. Patrick, ed.), Marcel Dekker, New York, Vol. 1, pp. 151–169 (1967).
3120. C. B. Buchnall, Toughened Plastics, Applied Science Publishers, London (1977).
3121. R. P. Kambour, J. Polym. Sci. D7:1 (1973).
3122. S. Rabinowitz and P. Beardmore, CRC Crit. Rev. Macromol. Sci. 1:1 (1972).
3123. A. S. Argon, J. G. Hannoosh, and M. M. Salama, Fract. 1:445 (1977).
3124. J. L. Gardon, Treatise on Adhesion and Adhesives (R. L. Patrick, ed.), Marcel Dekker, New York, Vol. 1, p. 291 (1967).
3125. A. W. Bryant and W. A. Dukes, Structural Adhesives Bonding (M. J. Bodnar, ed.), Interscience, New York, p. 81 (1966).
3126. W. C. Wake, Adhesion (D. D. Eley, ed.), Oxford University Press, London, p. 191 (1961).
3127. W. Diem, G. Elssner, T. Suga, and G. Petzow, The influence of layer thickness and internal stresses on the bond strength of metal-to-ceramic joints, Adhesive Joints: Formation, Characteristics, and Testing (K. L. Mittal, ed.), Plenum Press, New York, pp. 855–869 (1984).
3128. J. Friedel, Dislocations, Pergamon Press, London (1964).
3129. A. A. Donetelli, C. T. Mooney, and J. C. Bolger, Fracture toughness of elastomer-modified epoxy adhesives, Adhesive Joints: Formation, Characteristics, and Testing (K. L. Mittal, ed.), Plenum Press, New York, pp. 829–878 (1984).
3130. E. J. Ripling, J. S. Santner, and P. B. Crosley, Fracture of composite-adhesive-composite systems, Adhesive Joints: Formation, Characteristics, and Testing (K. L. Mittal, ed.), Plenum Press, New York, pp. 755–787 (1984).
3131. E. J. Hughes and J. L. Rutherford, Study of micromechanical properties of adhesive bonded joints, Contract DAAA21-67-C-0500, Picatinny Arsenal, Dover, NJ (Aug. 1968).
3132. E. J. Hughes, W. Althof, and R. B. Krieger, Jr., Mechanical properties of adhesives, Adhesive Bonding of Aluminum Alloys (E. W. Thrall and R. W. Shannon, eds.), Marcel Dekker, New York, Chap. 9, p. 141 (1985).
3133. E. J. Hughes, W. Althof, and R. B. Krieger, Jr., Mechanical properties of adhesives, Adhesive Bonding of Aluminum Alloys (E. W. Thrall and R. W. Shannon, eds.), Marcel Dekker, New York, Chap. 9, pp. 167–175 (1985).
3134. ASTM Designation D-2236-70, Standard Method of Test for Dynamic Mechanical Properties of Plastics, ASTM, Philadelphia.
3135. M. L. Williams, Int. J. Fract. Mech. 1(4):202 (1966).
3136. G. P. Anderson and K. L. DeVries, Analysis of standard bond strength tests, Treatise on Adhesion and Adhesives (R. L. Patrick, ed.), Marcel Dekker, New York, Vol. 6, Chap. 3, pp. 84–109 (1989).
3137. A. M. Messner, Stress distribution in poker chip tensile specimens, Bull. Working Group on Mechanical Behavior, CIPA Publ. No. 27 (1963).
3138. G. H. Lindsay, Triaxial fracture studies, J. Appl. Phys. 38:4843 (1967).
3139. F. J. Plantema, De Schreifspannung in eme lijmnaad, Report M1181, Nat. Luchtvaartlaboratorium, Amsterdam (1949).
3140. G. R. Wooley and D. R. Carver, J. Aircraft 8(19):817 (1971).
3141. T. R. Guess, R. E. Allred, and F. P. Gerstle, Comparison of lap-shear test specimen, J. Test. Eval. 5(2): 84 (1977).
3142. B. H. River, A method for measuring adhesive shear properties, Adhes. Age 24(12):30 (1981).
3143. A. N. Gent, Strength of adhesive bonds-plastic and viscoelastic effects, Proc. 46th Nat. Colloid Symp., Amherst, MA (1972).
3144. K. L. DeVries, M. L. Williams, and M. D. Chang, A fracture mechanics analysis of adhesive failure on a single lap-shear joint, Preprint. Soc. Exptl. Stress Anal., Westport, CT, Cleveland Meeting (1972).
3145. S. S. Wang and J. F. Yau, Interface cracks in adhesively bonded lap-shear joints. Int. J. Fract. 19:295 (1982).
3146. G. P. Anderson and K. L. DeVries, Analysis of standard bond strength tests, Treatise on Adhesion and Adhesives (R. L. Patrick, ed.), Marcel Dekker, New York, Vol. 6, Chap. 3, pp. 70–84 (1989).

3147. D. A. Yurek, Private communication, 3M's Test Method No. C-294, 3M's Adhesives, Coatings, and Sealers Div. (June 1972).
3148. D. M. Brewis, J. Comyn, B. C. Cope, and A. C. Moloney, Lab Practice 28(7):743 (1979).
3149. J. D. Minford, unpublished work, Aluminum Co. of America.
3150. T. B. Frazier, A computer assisted thick adherend test to characterize the mechanical properties of adhesives, SAMPE Tech. Conf. (1970).
3151. R. A. Seago, A practical application of the thick adherend shear test, Symp. on Durability of Structural Adhesives, Picatinny Arsenal, Dover, NJ, p. 499 (Oct. 27–29, 1976).
3152. S. Amijima, A. Yoshida, and T. Fujii, Proc. 2nd Int. Conf. Composite Materials, Toronto, Canada, Met. Soc. AIME, New York (1978).
3153. J. K. Sen and R. M. Jones, Stresses in double lap joints bonded with a viscoelastic adhesive: Part I. Theoretical and experimental corroboration, AIAA J. 18(10):1237 (1980).
3154. J. K. Sen and R. M. Jones, Stresses in double lap joints bonded with a viscoelastic adhesive: Part II. Parametric study and joint design, AIAA J. 18(11):1376 (1980).
3155. W. N. Sharpe and T. J. Muha, Comparison of theoretical and experimental shear stress in the adhesive layer of a lap joint model, Proc. Army Symp. Solid Mechanics, Army Mater. & Mech. Res. Center, Watertown, MA, AMMRC-MS-8, p. 23 (1974).
3156. G. R. Irwin, Trans. ASME J. Appl. Mech. 24:361 (1957).
3157. F. Erdogan and G. C. Sin, Trans. ASME J. Basic Eng. 85D:519 (1963).
3158. G. G. Trantina, Combined mode crack extension in adhesive joints, T. & AM Report No. 352, University of Illinois, Urbana (1971).
3159. W. D. Bascom and D. L. Hunston, The fracture of epoxy and elastomer modified epoxy polymers, Treatise on Adhesion and Adhesives (R. L. Patrick, ed.), Marcel Dekker, New York, Vol. 6, pp. 33, 134, 171, 172 (1989).
3160. W. D. Bascom and C. O. Timmons, unpublished results, Naval Res. Lab.
3161. T. R. Brussat, S. T. Chiu, and S. Mostovoy, Fracture mechanics of structural adhesive bonds, Report Lr 27614-5, Air Force Mater. Lab., Wright-Patteron AFB, OH (1976).
3162. W. D. Bascom, D. L. Hunston, and C. O. Timmons, Org. Coat. Plast. Prepr. 39:179 (1978).
3163. W. S. Johnson and S. Mall, ASTM Spec. Publ. No. 876, p. 185 (1985).
3164. N. T. McDevitt and W. L. Baun, The three-point bend test for adhesive joints, Adhesive Joints: Formation, Characteristics, and Testing (K. L. Mittal, ed.), Plenum Press, New York, p. 369 (1984).
3165. N. L. Rogers, Natl. SAMPE Tech. Conf., Vol. 2, p. 571 (1970).
3166. N. L. Rogers, Proc. Symp. on Durability of Adhesive Bonded Structures, Picatinny Arsenal, Dover, NJ, p. 57 (1976).
3167. V. S. Srinath, Proc. Structural Adhesives and Bonding, Tech. Conf. Assoc., El Segundo, CA, p. 24 (March 1979).
3168. W. D. Bascom, Adhes. Age 24(4):28 (1979).
3169. W. L. Baun, Air Force Mater. Lab. Report TR-79-4126, Wright-Patterson AFB, OH (Sept. 1979).
3170. W. L. Baun, The role of surface and bulk characterization, Adhesive Joints: Formation, Characteristics, and Testings (K. L. Mittal, ed.), Plenum Press, New York, p. 3 (1984).
3171. W. Althof and W. Brockmann, Bicentennial of Materials Progress, SAMPE Symp. Proc., 21:581 (April 1976).
3172. W. Althof, New Horizons—Materials and Processes for the 80's, SAMPE NCTS 11:309 (Nov. 1979).
3173. W. Althof, Effects of low cycle loading on shear stressed adhesive bondlines, Adhesive Joints: Formation, Characteristics, and Testing (K. L. Mittal, ed.), Plenum Press, New York, p. 659 (1984).
3174. Y. Gilibert and G. Verchery, Influence of surface roughness on mechanical properties of joints, Adhesive Joints: Formation, Characteristics, and Testing (K. L. Mittal, ed.), Plenum Press, New York, p. 69 (1984).
3175. E. J. Hughes, W. Althof, and R. B. Krieger, Mechanical properties of adhesives, Adhesive Bonding of Aluminum Alloys (E. W. Thrall and R. W. Shannon, eds.), Marcel Dekker, New York, Chap. 9, pp. 167–175 (1985).
3176. J. A. Marceau and E. W. Thrall, Environmental-durability testing, Adhesive Bonding of Aluminum Alloys (E. W. Thrall and R. W. Shannon, eds.), Marcel Dekker, New York, Chap. 10, p. 187 (1985).
3177. L. J. Hart-Smith and E. W. Thrall, Structural analysis of adhesive bonded joints, Adhesive Bonding of Aluminum Alloys (E. W. Thrall and R. W. Shannon, eds.), Marcel Dekker, New York, Chap. 13, p. 241 (1985).
3178. Development and evaluation of the climbing peel method for testing adhesive bonds in sandwich and metal-to-metal construction, WADC Tech. Report No. 56-386 (Sept. 1956).

3179. C. F. Carter, Durability of adhesive peel joints while stressed in water, ASTM Spec. Publ. STP No. 401, p. 28 (Aug. 1966).
3180. M. C. Locke, unpublished data, Boeing Commercial Airplane Co. (1980).
3181. A. J. Kinloch, Review-the science of adhesion, Part 2—Mechanics and mechanisms of failure, J. Mater. Sci. 17:631 (1982).
3182. J. A. Marceau, Y. Moji, and J. C. McMillan, A wedge test for evaluating adhesive bonded surface durability, 21st Natl. SAMPE Symp., Vol. 21 (April 6–8, 1976).
J. A. Marceau, Y. Moji, and J. C. McMillan, A wedge test for evaluating adhesive-bonded surface durability, Adhes. Age 20(10):28 (1977).
3183. J. D. Minford, unpublished work, Aluminum Co. of America.
3184. J. Cognard, Quantitative measurements of the energy of fracture of an adhesive joint using the wedge test, J. Adhes. 22:87 (1987).
3185. J. Cognard, J. Adhes. 20:1–13 (1986).
3186. J. Cognard, Int. J. Adhes. Adhes. 6:215 (1986).
3187. A. J. Kinloch and S. J. Shaw, J. Adhes. 12:15 (1981).
3188. H. W. Eickner, Environmental exposure of adhesive bonded metal joints, Forest Products Lab. Report WADC 59-567, Part I, Madison, WI (Feb. 1960).
3189. H. W. Eickner, Environmental exposure of adhesive bonded metal joints, Forest Products Lab. Report WADC 59-564, Part I, Madison, WI (1962).
3190. H. W. Eickner, W. Z. Olsen, and R. M. Lulling, Resistance of adhesive bonded metal lap joints to environmental exposure, Forest Products Lab. Report WADC Tr-59-654, Part II, Madison, WI (Oct. 1962).
3191. G. F. Carter, Outdoor durability of adhesive joints under stress, Adhes. Age 10(10):32 (Oct. 1967).
3192. J. A. Harris and R. D. Adams, The impact strength of adhesive lap joints, Adhesive Joints: Formation, Characteristics, and Testing (K. L. Mittal, ed.), Plenum Press, New York, p. 611 (1984).
3193. J. A. Harris and R. D. Adams, unpublished data (1981).
3194. A. J. Kinloch, The impact performance of structural adhesive joints, 10th Ann. Adhes. Soc. Meetings, Charleston, SC (Feb. 21–24, 1987).
3195. G.L. Roderick, R. A. Everett, Jr., and J. H. Crews, Jr., Fatigue of composite materials, ASTM Publ. STP 569, pp. 295–306, ASTM, Philadelphia (1975).
3196. T. R. Brussat, S. T. Chiu, and S. Mostovoy, Fracture mechanisms for structural adhesive bonds, AFML-TR-163, Air Force Materials Lab, Wright-Patterson AFB, OH (1977).
3197. B. Dattaguru, R. A. Everett, Jr., J. D. Whitcomb, and W. S. Johnson, Geometrically-nonlinear analysis of adhesive bonded joints, 23 AIAA/ASME/ASCE/AHS Structures, Structural Dynamics and Materials Conf., New Orleans, LA (May 1982).
3198. J. Romanko and W. G. Knauss, Fatigue behavior of adhesively bonded joints, AFWAL-TR-80-4037, Air Force Materials Lab., Wright-Patterson AFB, OH, Vol. I (1980).
3199. M. J. Jacobsen, Sonic fatigue of advanced composite panels in thermal environments, Report AIAA-81-1699, American Institute of Aeronautics and Astronautics, p. 1–9 (1981).
3200. R. A. Everett Jr., The role of peel stresses in cyclic debonding, NASA TM-84504, Natl. Aero. & Space Admin., Washington, D.C.
3201. S. Mall, W. S. Johnson, and R. A. Everett, Jr., Cyclic debonding of adhesively bonded composites, Adhesive Joints: Formation, Characteristics, and Testing (K. L. Mittal, ed.), Plenum Press, New York, pp. 639–658 (1984).
3202. T. B. Frazier and A. D. Lajoie, Durability of adhesive bonded joints, Tech. Report AFML-TR-74-26, Bell Helicopter Company (March 1974).
3203. J. Romanko, Behavior of adhesively bonded joints under cyclic loading, AGARD Lecture Series No. 102, pp. 4–1 to 4–42, Wright-Patterson AFB, OH (Oct. 16 and 17, 1979).
3204. H. C. Schjelderup, A Compilation of Mechanical Property Data FM-73, Task 610 and part of 609, Douglas Aircraft Co., McDonnell-Douglas Corp. Data Report MDC-J6074 (Sept. 28, 1977).
3205. W. J. Renton, Structural Properties of Adhesives, Final Tech. Report, AFML-TR-78-127, Vol. 1, Wright-Patterson AFB, OH (Dec. 1978).
3206. Fatigue Behavior of Adhesively Bonded Joints, Contract F33615-76-C-5220, sponsored by USAF AFML/MBC and conducted by General Dynamics, Fort Worth Div. (July 1976–March 1980).
3207. Structural Properties of Adhesives, Contract F33615-76-C-5224, sponsored by USAF AFML/MBC and conducted by Vought Corporation, Advanced Technol. Center, Dallas (May 1976–May 1978).
3208. Primary Adhesively Bonded Structure Technology (PABST), Contract F33615-75-C-3016, sponsored by USAF AFFDL/Wright-Patterson AFB, OH and conducted by Douglas Aircraft Co., McDonnell-Douglas Corp., Long Beach, CA (March 1975–Dec. 1978).

3209. A Matting and G. Ulmer, Kautschuk und Gummi-Kunststoffe 16:334 (1963).
3210. Kuhbander and Aponyi, Proc. 21st Natl. SAMPE Symp., pp. 589–605 (1975).
3211. Klapprott, Mahoney, and Aponyi, Proc. 20th Natl. SAMPE Symp. Exhib., p. 546 (1975).
3212. May and Adsit, Proc. Natl. SAMPE Symp. Exhib., p. 546 (1975).
3213. S. Hashemi and J. G. Williams, Size and loading mode effects in fracture toughness testing of polymers, J. Mater. Sci. 19(11):3746 (Nov. 1984).
3214. Y. Sewaki, N. Aoyama, and T. Kawasaki, Determination of fatigue fracture toughness by conventional rotary bending specimens, Int. J. Fract. 26(10):43 (Sept. 1984).
3215. M. E. Chaplya and N. M. Gvozdyuk, A new method of investigating fatigue crack growth in lamellar composite materials, Soviet Mater. Sci. 20(1):73 (Jan.–Feb. 1984).
3216. K. R. Wentz and H. F. Wolfe, Development of random fatigue data for adhesively bonded and weldbonded structures subjected to dynamic excitation, ASME Trans. J. Eng. Mater. Technol. 100:70 (Jan. 1978).
3217. W. S. Johnson, W. C. Rister, and T. Spamer, Spectrum growth on adhesively bonded structure, ASME Trans. J. Eng. Mater. Technol. 100:57 (Jan. 1978).
3218. J. M. Anderson, C. S. Chu, and W. M. McGee, Growth characteristics of a fatigue crack approaching and growing beneath an adhesively bonded doubler, ASME Trans. J. Eng. Mater. Technol. 100:52 (Jan. 1978).
3219. M. M. Ratwani, A parametric study of fatigue crack growth behavior in adhesively bonded metallic structures, ASME Trans. J. Eng. Mater. Technol. 100:46 (Jan. 1978).
3220. T. R. Brussat and S. T. Chiu, Fatigue crack growth of bondline cracks in structural joints, ASME Trans. J. Eng. Mater. Technol. 100:39 (Jan. 1978).
3221. N. J. Pfeiffer and J. A. Alic, Fatigue crack propagation in 8- and 22-layer 7075-T6 aluminum alloy laminates, ASME Trans. J. Eng. Mater. Technol. 100:32 (1978).
3222. D. R. Mulville, D. L. Hunston, and P. W. Mast, Developing failure criteria for adhesive joints under complex loading, ASME Trans. J. Eng. Mater. Technol. 100:25 (Jan. 1978).
3223. N. Parida, S. K. Das, P. C. Cooper, and O. N. Mohanty, Probability, confidence, and sample size in fatigue testing, J. Test. Eval. 18(6):385 (1990).
3224. E. J. Hughes, J. Boutiller, and J. L. Rutherford, The effect of moisture on the dimensional stability of adhesively bonded joints, Adhesive Joints: Formation, Characteristics, and Testing (K. L. Mittal, ed.), Plenum Press, New York, p. 137 (1984).
3225. E. J. Hughes, W. Althof, and R. B. Krieger, Jr., Mechanical properties of adhesives, Adhesive Bonding of Aluminum Alloys (E. W. Thrall and R. W. Shannon, eds.), Marcel Dekker, New York, pp. 142–159 (1985).
3226. R. B. Krieger, Jr., Stress analysis of metal-to-metal bonds in a hostile environment, Amer. Cyanamid Co., Harve-de-Grace, MD, Adhes. Age 21(6):(1978).
3227. R. B. Krieger, Jr., Analyzing joint stresses using an extensometer, Adhes. Age 28(11):26 (1985).
3228. F. C. Bossler, M.C. Franzblau, and J. L. Rutherford, J. Phys. E. Ser. 2, 1:820 (1968).
3229. E. J. Hughes, J. L. Rutherford, and F. C. Bossler, Rev. Sci. Instrum. 39(5):666 (1968).
3230. H. K. Kim, F. A. Mohamed, and J. C. Earthman, A novel specimen geometry for double spear creep experiments, J. Test. Eval. 199(2):93 (1991).
3231. L. C. McDonald and K. T. Hartwig, Cryogenic creep testing, J. Test. Eval. 19(2):107 (1991).
3232. C. Elster and J. Honerkamp, Modified maximum entrophy method and its application to creep data, Macromolecules 21(1):310 (1991).
3233. E. J. Hughes, W. Althof, and R. B. Krieger, Jr., Mechanical properties of adhesives, Adhesive Bonding of Aluminum (E. W. Thrall and R. W. Shannon, eds.), Marcel Dekker, New York, Chap. 9, pp. 159–167 (1985).
3234. E. J. Hughes, J. Boutilier, and J. L. Rutherford, The effect of moisture on the dimensional stability of adhesively bonded joints, Adhesive Joints: Formation, Characteristics, and Testing (K. L. Mittal, ed.), Plenum Press, New York, pp. 137–150 (1984).
3235. H. Dannenberg, J. Appl. Polym. Sci. 14:125–134 (1961).
3236. A. N. Gent, New and improved tests for adhesion, 10th Ann. Meet. Adhes. Soc., Williamsburg, VA (Feb. 21–24, 1987).
3237. D. A. Dillard, The constrained blister test—a novel approach to measuring adhesive fracture toughness, 11th Ann. Meet. Adhes. Soc., Charleston, SC (Feb. 21–24, 1988).
3238. M. G. Allen and S. D. Senturia, Peeling of thin films using island blister, 11th Ann. Meet. Adhes. Soc., Charleston, SC (Feb. 21–24, 1988).
3239. M. G. Allen and S. D. Senturia, Application of the island blister test for thin film adhesion of thermosetting adhesives, J. Adhes. 29(1–4):219 (1989).

3240. A. Moet, Thermodynamics of the blister test, 11th Ann. Meet. Adhes. Soc., Charleston, SC (Feb. 21–24, 1988).
3241. D. Dillard, Developments in blister tests for adhesives, Adhesion '90, Plastics and Rubber Inst. Meet., Cambridge, England (Sept. 10–12, 1990).
3242. Y. H. Lai and D. A. Dillard, An elementary plate theory prediction strain energy release rate of the contrained blister test, J. Adhes. 31(2–4):177 (1990).
3243. M. Fernando and A. J. Kinloch, Use of the "inverted blister" test to study the adhesion of photopolymers, Int. J. Adhes. Adhes. 10(2):69 (1990).
3244. R. J. Schliekelmann, Non-destructive testing of adhesive bonded metal-to-metal joints, Non-Destructive Testing (April 1972).
3245. R. J. Schliekelmann, Non-destructive testing of bonded joints—recent developments in testing systems, Non-Destructive Testing (April 1975).
3246. A. Matting and K. Ulmer, Kautschuk und Gummi. Kunststoffe 16:280,387 (1963).
3247. W. Thomson (Lord Kelvin), Phil. Mag. 5:82 (1898).
3248. P. F. A. Bijlmer, Surface Contamination: Genesis, Detection, and Control (K. L. Mittal, ed.), Plenum Press, New York, Vol. 2, pp. 723–748 (1979).
3249. A. Kwakernaak, R. Exalto, and H. A. van Hoof, Surface characteristics of anodic oxides on aluminum alloys by means of surface potential difference, surface impedance, and surface morphology, Adhesive Joints: Formation, Characteristics, and Testing (K. L. Mittal, ed.), Plenum Press, New York, p. 103 (1984).
3250. C. E. Michelson, J. Electrochem. Soc. 115:213 (1969).
3251. A. T. Fromhold, Jr., and J. Kruger, J. Electrochem. Soc. 120:722 (1973).
3252. T. Smith and G. Lindberg, Surf. Technol. 9:1 (1979).
3253. E. T. Englehart and D. J. George, Mater. Protection 3:24 (Nov. 1964).
3254. H. Birtel and W. Leute, Aluminium 43:93 (1967).
3255. A. Ott, Aluminium 52:491 (1976).
3256. R. J. Schliekelmann, Non-Destructive Testing 5:79 (1972).
3257. J. R. Rienka, The resonance/impedence and the Volta potential methods for the nondestructive testing of bonded joints, 8th World Conf. NDT, Cannes, France (1976).
3258. R. D. Adams, P. Cawley, and C. C. H. Guyott, Nondestructive inspection of adhesively-bonded joints, Proc. 10 Ann. Adhes. Soc. Meet. (L. H. Sharpe, ed.), Gordon and Breach, London, pp. 103–114 (1988).
3259. F. J. Meyer and G. B. Chapman, Adhes. Age 23(4) (April 1980).
3260. R. J. Schliekelmann, Nondestructive testing of adhesive bonded metal structures, Part I—Theoretical considerations, Adhes. Age 10(8):35 (1967).
3261. Nondestructive testing of adhesive bonds in aircraft, Adhes. Age 10(8):35 (Aug. 1967).
3262. R. T. Lennon and C. R. Wang, Nondestructive inspection of adhesive bonded metal helicopter rotor blades, Defense Conf. on Nondestructive Testing, Boston (Oct. 4–6, 1966).
3263. P. E. J. Vogel, Infrared multiple-scan bond inspection system, Defense Conf. Nondestructive Testing, Boston (Oct. 4–6, 1966).
3264. R. L. Jarvis, Inspection of bonded assemblies, Amendment 1, American Aviation Corp., Cleveland, OH (July 5, 1968).
3265. A. O. K. Nieminen and J. L. Koenig, Macroscopic and modern microscopic NDE methods for adhesive bonded structures, Int. J. Adhes. Adhes. 11(1):5 (1991).
3266. J. R. Kraska and H. W. Kamm, Evaluation of sonic methods for inspecting adhesive bonded honeycomb structures, AFML Tech. Report TR-69-283, Wright-Patterson AFB, OH (1970).
3267. R. Botsco and R. Roopenian, Nondestructive determination of bond integrity of metal insert fasteners in nonmetallic honeycomb structures, Mater. Eval. 26:90 (May 1968).
3268. R. D. Adams, A. M. Allen, and P. Cawley, Proc. 11th Western Conf. N.D.T., Las Vegas (1985).
3269. Yu. V. Lange, and I. I. Teumin, Soviet J. NDT. 7:157 (1971).
3270. R. Schroeer et al., The acoustic impact technique—a versatile tool for nondestructive evaluation of aerospace structures and components, Mater. Eval. 28(2) (1970).
3271. P. Cawley, NDT Int. 17:59 (1984).
3272. R. Botsco, The eddy-sonic test method, Mater. Eval. 26:21 (1968).
3273. G. J. Curtis, Bonded Structure Testing, U.K. Atomic Energy Authority, Report AERE R-6098 (1967).
3274. M. T. Clark, Definition and nondestructive detection of critical bondline flaws, AFML Tech. Lab. Report TR-78-108, Wright-Patterson AFB, OH (1978).
3275. D. J. Hagemaier and R. Fassbender, Nondestructive testing of adhesive bonded structures, SAMPE Quart. 9(4) (1978).

3276. D. J. Hagemaier, Nondestructive inspection, Adhesive Bonding of Aluminum Alloys (E. W. Thrall and R. W. Shannon, eds.), Marcel Dekker, New York, Chap. 15, pp. 337–342, 365–380 (1985).
3277. H. E. Babb, Ultrasonic inspection of adhesively-bonded composites, J. Appl. Polym. Sci., Appl. Polym. Symp. 3:405 (1966).
3278. W. D. Lanier, Pulse-echo ultrasonic methods for nondestructive testing of C5 adhesive bonded aluminum composites, Spring Conf. Amer. Soc. Nondestructive Testing, Los Angeles (March 10–13, 1969).
3279. W. Althof, Nondestructive testing of metal/adhesive bonds with ultrasonic pulse-echo instruments, Report Til/T-5680 (Aug. 1966) (English translation N67-18274 by Ministry of Aviation, London).
3280. J. B. Beal, Ultrasonic emission detector evaluation of the strength of bonded materials, Res. Achievements Rev. Vol. II, Report No. 5, NASA-Marshall Space Flight Center (Sept. 29, 1966).
3281. J. B. Beal, Ultrasonic emission detector evaluation of strength of bonded materials, 2nd Tech. Status and Trends Symp., Nondestructive Testing: Trends and Techniques, Marshall Space Flight Center (Oct. 26–27, 1966).
3282. W. N. Clotfelter, The nondestructive evaluation of adhesive bonded composite materials, 2nd Tech. Status and Trends Symp., Nondestructive Testing: Trends and Techniques, Marshall Space Flight Center (Oct. 26–27, 1966).
3283. J. S. Arnold, Development of nondestructive tests for structural adhesive bonds, WADC Tech. Report No. TR-54-231 (1957).
3284. J. S. Arnold and C. T. Vincent, Development of nondestructive tests for structural adhesive bonds, Std. Res. Inst. (Dec. 1959).
3285. G. Schmitz and L. Frank, Nondestructive testing for evaluation of strength of bonded material, NASA Report CR-67983, Contract NAS8-11456, N66-11761 (Sept. 1965).
3286. C. C. Kammerer, Project Apollo automated ultrasonic bond quality inspection, SAMPE J. 2(30):16 (April/May 1966).
3287. J. P. Reese and V. H. Boruff, Evaluation of ultrasonic test devices for inspection of adhesive bonds, Martin Co., Summary Report, Contract Noa(s) 59-6266-C (July 1961).
3288. N. B. Miller and V. H. Boruff, Evaluation of ultrasonic test devices for inspection of adhesive bonds, Martin Co., Final Report No. 12, Contract Nos(s) 59-6266-C (1962).
3289. S. A. LoPilato and S. W. Carter, Unbond detection using ultrasonic phase analysis, Mater. Eval 24:690 (Dec. 1966).
3290. J. Schijve, Investigations on the ultrasonic testing of glued metal joints, Aircraft Eng. 30:269,355 (1958).
3291. H. M. Gonzales and C. V. Cagle, Nondestructive testing of adhesive bonded joints, ASTM Spec. Tech. Publ. No. 360 (1964).
3292. D. F. Smith and C. V. Cagle, Ultrasonic testing of adhesive bonds using the Fokker Bond Tester, Mater. Eval. 24 (July 1966).
3293. J. L. Rose and P. A. Meyer, J. Appl. Polym. Sci., Appl. Polym. Symp. 32:389 (1977).
3294. J. L. Rose and P. A. Meyer, Mater. Eval. 31(6):109 (1973).
3295. H. G. Tattersall, The ultrasonic pulse-echo technique as applied to adhesion testing, J. Appl. Phys. 6 (1973).
3296. W. Lechodziejewski, Int. J. NDT (Dec. 1972).
3297. T. C. Adameder, Elastomer-to-metal adhesive bonds—adhesive quality determination by ultrasonic techniques, Naval Underwater Systems Center Report (April 1972).
3298. P. A. Meyer and J. L. Rose, Ultrasonic attenuation effects associated with the physical modeling of adhesive bonds, J. Appl. Phys. 48:3705 (1977).
3299. P. A. Meyer and J. L. Rose, Ultrasonic procedure for the determination of bond strength, Materials Evaluation 42:109 (1973).
3300. J. L. Rose and P. A. Meyer, Ultrasonic procedures for the determination of bond strength, AFOSR Interim Scientific Report AFOSR-73-2480A (April 1975).
3301. H. M. Gonzales and R. P. Marschel, Ultrasonic inspection of Saturn II tank insulation, Mater. Conf. of N.D.T., Los Angeles (1964).
3302. C. C. Kammerer, L. Furon, and N. Jeffras, Project Apollo automated ultrasonic bond quality inspection, SAMPE J. 2(3):16 (April/May 1966).
3303. D. Hagemaier and R. Fassbender, Nondestructive inspection-primary adhesively bonded structure technology, Industrial Review, Book 2, p. 233 (Sept. 14–16, 1977).
3304. F. H. Chang, P. L. Flynn, D. E. Gordon, and J. R. Bell, Determination of adhesive bondline quality by ultrasonic techniques, General Dynamics, Fort Worth Div. Report ERR-FW-1664 (Dec. 1975).
3305. F. H. Chang, J. C. Couchman, and B. G. W. Yee, Transmission frequency spectra of ultrasonic waves through multi-layer media, Ultrasonics Symp. Proc., IEEE Group on Sonics and Ultrasonics (Nov. 1973).

3306. F. H. Chang, J. C. Couchman, J. R. Bell, and D. E. Gordon, Correlation of NDE parameters with adhesive bond strength in multi-layered structures, Proc. ASNT 10th Symp. on NDT (April 1975).
3307. D. O. Thompson, Interdisciplinary program for quantitative flaw detection, Sci. Center, Rockwell Int. Corp., Report No. SC 595.14SA (Jan. 1976).
3308. W. A. Dukes and A. J. Kinloch, Nondestructive testing of bonded joints, an adhesion science viewpoint, Non-Destructive Testing Res. Practice (1974).
3309. M. X. Li, W. Z. Ding, and J. M. Chen, NDT Int. 15:137 (1982).
3310. R. M. Gonzales and R. P. Merschell, Ultrasonic inspection of Saturn S-II tank insulation bonds, Mater. Eval. 24 (July 1966).
3311. R. Newschafer, Assuring Saturn quality through nondestructive testing, Mater. Eval. 27(7) (1969).
3312. D. Hagemaier and R. Fassbender, Non-destructive testing of adhesive bonded structures, SAMPE Quart. 9(4) (1978).
3313. K. J. Rienks, The resonance impedance and the Volta potential methods for the non-destructive testing of joints, 8th World Conf. on Non-Destructive Testing, Cannes, France (1976).
3314. D. J. Hagemaier, End product nondestructive evaluation of adhesive bonded metal joints, Adhesives and Sealants, Engineered Materials Handbook, ASM, Materials Park, OH, Vol. 3, pp. 753–758 (1990).
3315. F. H. Chang, P. L. Flynn, D. E. Gordon, and J. R. Bell, Principles and applications of ultrasonic spectroscopy in NDE of adhesive bonds, IEEE Trans. in Sonics and Ultrasonics, Su-23:334–338 (Sept. 1977).
3316. D. A. Hagemaier, Ultrasonic maintenance inspection of aircraft structures, Mater. Eval. 34:9 (Jan. 1976).
3317. D. A. Hagemaier, Ultrasonic applications in the aerospace industry, SAE, Natl. Aerospace Eng. & Mfg. Meeting, San Diego, p. 31 (Oct. 1–3, 1974).
3318. D. O. Thompson, Requirements for and advances in nondestructive evaluation, Proceedings of the New York Institute of Electrical and Electronic Engineering Inc., pp. 642–652 (1974).
3319. C. J. Palmer, The effect of water path variation in ultrasonic through-transmission sensitivity-for adhesively-bonded aircraft honeycomb structures, Mater. Eval. 33:128–132, 134 (June 1945).
3320. G. A. Alers, P. L. Flynn, and M. J. Buckley, Ultrasonic techniques for measuring the strength of adhesive bonds, Mater. Eval. 35:77 (April 1977).
3321. P. L. Flynn, Cohesive bond strength predictions for adhesive joints, J. Test. Eval. 7(3) (1979).
3322. B. G. Martin et al., Interference effects in using the ultrasonic pulse-echo technique on adhesive bonded metal parts, Mater. Eval. 37(5) (1979).
3323. O. R. Gericke and B. L. Monagle, Determination of delaminations by ultrasonic spectroscopy, IEEE Trans. Sonics and Ultrasonics, Su-23:399–445 (Sept. 1976).
3324. T. Smith, NDT techniques for the prediction of adhesive failure loci prior to bonding, Mater. Eval. 33: 101 (May 1975).
3325. J. L. Rose, M. J. Avioli, and R. Bilgram, A feasibility study on the nondestructive evaluation of an adhesively bonded metal-to-metal bond: an ultrasonic pulse-echo approach, J. Nondestructive Testing 25(2) (March 1983).
3326. G. A. Matzkanin, C. C. Gardner, and G. I. Burkhart, Improved inspection for bonded C5A structure, Nondestructive Testing Research Practice (1974).
3327. H. T. Clark, Definition and nondestructive detection of critical adhesive bondline flaws, AFML Final Report, Wright-Patterson AFB, OH (July 1978).
3328. M. H. Loew, J. M. Fitzgerald, and A. N. Mucciardi, AFML Final Report NO. TR-78-206, Wright-Patterson AFB, OH (Dec. 1978).
3329. F. Fassbender, NDT-210 bondtest inspection, Douglas Aircraft Co. Report DPS 4.733-2 (June 1977).
3330. A. Hartman and J. B. de Jonge, Nondestructive and destructive tests on Redux bonded single and double joints with various glueline thicknesses, Netherlands Aircraft Board Report NLR-TNM 2099 (March 1962).
3331. J. B. de Jonge, The behavior of the Fokker Bond Tester on double-lap joints, Netherlands Aircraft Board Report No. NLR-TMN 2098 (March 1962).
3332. H. C. Schjederup and R. W. Shannon, Primary adhesively bonded structure technology (PABST)—a technology analogous to composites, American Institute of Aeronautics and Astronautics Inc., pp. 450–457 (1977).
3333. H. M. Gonzales and C. V. Cagle, Nondestructive testing of adhesive bonded joints, 4th Pac. Area Natl. Meeting, ASTM Paper No. 79, Los Angeles, Oct. 1–5 (1962).
3334. E. J. Regalado, Nondestructive testing at Norair, Part II, evaluation of Fokker Bond Tester System for nondestructive testing of FM-47 adhesive bonded honeycomb and metal-to-metal structures, 11th Ann. Aircraft and Missile Conf., Los Angeles (Nov. 1961).

3335. J. Hertz, Investigation of bond deterioration by use of the Fokker Bonds Tester, Report GD/A-ERR-AN-682, General Dynamics (Dec. 31, 1964).
3336. C. S. Phelan, Appl. data Fokker Bond Tester System covering recent developments and permanent records, 6th Symp. Nondestructive Evaluation of Aerospace and Weapons Systems Comp. and Materials, San Antonio (April 17–19, 1967).
3337. MIL-STD-860 (USAF), Fokker Ultrasonic Adhesive Bond Test (1978).
3338. R. E. Clemens, Application of NDT on adhesive bonded structures for Northrup aircraft F5 and T-38 aircraft, SAMPE Quart. p. 170 (1970).
3339. G. J. Curtis, Ultrasonic Testing, Non-Conventional Testing Techniques, John Wiley & Sons, Chichester, England (1982).
3340. F. Fassbender, Fokker Bond Test Inspection, Douglas Aircraft Co., Report NO. DPS 4.733-2.
3341. Technical Handbook Th 14.7150E, Fokker VFW, The Netherlands.
3342. Fokker F-28 Nondestructive Testing Manual—Part 6, Fokker VFW, The Netherlands.
3343. E. P. Lloyd and A. F. Brown, Adhesion—2 (K. W. Allen, ed.), Applied Science Publishers, London, p. 133 (1978).
3344. A. Pilarski, Ultrasonic evaluation of the adhesion degree in layered joints, Mater. Eval. 43(8):765 (May 1986).
3345. D. K. Rehbein, R. B. Thompson, and O. Buck, Fatigue crack characterization by ultrasonic inspection, J. Test. Eval. 18(6):421 (1990).
3346. R. J. Patton, Incorporation of additives in adhesives for radiographic inspection of adhesive-bonded honeycomb structures, North Amer. Aviation, Tech. Summary Report No. SID-64-2087 (Jan. 29, 1965).
3347. D. J. Hagemaier, End product evaluation of adhesive bonded metal joints, Adhesives and Sealants, Engineered Materials Handbook, ASM, Materials Park, OH, Vol. 3, p. 759 (1990).
3348. W. E. Dance and D. H. Peterson, J. Appl. Polym. Sci., Appl. Polym. Symp. 32:399 (1977).
3349. G. J. Curtis, Non-Destructive Testing, p. 249 (Oct. 1975).
3350. W. Brockmann, Acoustic emission analyzer as a method of testing adhesive metal joints, Proc. 22nd Natl. SAMPE Symp. & Exhib., San Diego, California, pp. 277–300 (April 26–28, 1977).
3351. R. A. Hill, The use of acoustic emission for characterizing adhesive joint failures, NDT Int. 10:63 (April 1977).
3352. A. A. Pollock, Stress wave emission in NDT, Non-Destructive Testing 2:178 (1969).
3353. J. Rodgers and S. Moore, Applications of acoustic emission to sandwich structures, Acoustic Emission Technol. Corp., Sacramento (1980).
3354. J. Rodgers and S. Moore, The use of acoustic emission for detection of active corrosion and degraded adhesive bonding in aircraft structures, Sacramento Air Logistics Center (SM/ALC/MMET), McClellan AFB, CA.
3355. The sign of a good panel is silence, Aviation Eng. Maint. 3(4) (April 1979).
3356. W. N. Clotfelter, Acoustic techniques for the nondestructive evaluation of adhesively-bonded composite materials, NASA Report TMX-53219 (March 17, 1965).
3357. F. H. Matusik, Nondestructive techniques for the nondestructive evaluation of adhesion, Adhes. Age 11(12):32 (Dec. 1968).
3358. A. E. Holt, Nondestructive evaluation by Schlieren method of adhesively-bonded structures, Conf. on Plastic/Composite Structures, Air Force Materials Lab, Wright-Patterson AFB, OH.
3359. D. Leroy and J. M. Claeys, Acousto-optic method for nondestructive testing, J. Nondestructive Eval. 4(10):43 (March 1984).
3360. R. Padmanabhan, N. Suriyayothin, and W. E. Wood, Analysis of acoustic emissions generated during stress corrosion cracking, J. Test. Eval. 12(5):280 (Sept. 1984).
3361. A. Arora, Acoustic emission characterization of corrosion reactions in aluminum alloys, Corrosion 40(8):459 (Sept. 1984).
3362. C. B. Scruby, G. R. Baldwin, and K. A. Stacey, Characterization of fatigue crack extension by quantitative acoustic emission, Int. J. Fract. 28(4):201 (Aug. 1985).
3363. D. Wells, NDT of sandwich structures by holographic interometry, Mater. Eval. 27(11) (1969).
3364. D. J. Hagemaier, Nondestructive inspection of bonded aluminum alloy, Adhesive Bonding of Aluminum Alloys (E. W. Thrall and R. W. Shannon, eds.), Marcel Dekker, New York, Chap. 15, pp. 385–388, 390–392 (1985).
3365. R. J. Schliekelmann, Non-destructive testing of bonded joints-recent developments in testing systems, Non-Destructive Testing (April 1975).
3366. D. H. Collins, Acoustical holography scanning techniques for imaging flaws in reactor pressure vessels, Proc. 9th Symp. NDE, San Antonio (April 25–27, 1973).

3367. A Sample of Acoustical Holographic Imaging Tests, Holosonics Inc., Richmond, WA.
3368. C. R. Pond, M. T. Holden, and R. I. Gellert, The application of holographic interferometry to nondestructive testing, Symp. NDT of Welds and Materials Joining, Los Angeles (March 11–13, 1968).
3369. R. F. Erf, R. M. Gagosz, J. P. Waters, K. A. Stetson, and H. G. Aas, Nondestructive holographic techniques for structure inspection, Final Report United Aircraft, East Hartford, CT (July 1, 1971–April 30, 1974).
3370. M. J. Barbarisi and B. R. Chisolm, Initial feasibility study employing holographic vibrational analysis to locate nonbonds, Picatinny Arsenal Report, Dover, NJ.
3371. R. J. Schliekelmann, Interference holography: A means for non-destructive testing, Royal Netherlands Aircraft Factories Fokker, Schiphol-Oost, Holland.
3372. R. J. Quirido, Holographic non-destructive testing of advanced composite materials in aerospace constructions, Proc. 8th World Conf. on Non-Destructive Testing, Cannes, France (Paper 3A-6) (1976).
3373. M. F. Vallat, P. Smigielski, P. Martz, and J. Schultz, The application of coherent optics to the study of adhesive joints II. holographic interferometry, J. Appl. Polym. Sci. 30(10):3953 (1985).
3374. L. O. Heflinger, R. F. Wuerker, and H. Spetzler, Thermal expansion coefficient measurement of diffusely reflecting samples by holographic interferometry, Rev. Sci. Instr. 44:629 (1973).
3375. R. Jones, Strain distribution and elastic-constant measurement using holographic and speckle-pattern inferferometry, J. Strain Anal. 9:4 (1974).
3376. R. Jones and D. Biji, A holographic interferometric study of the effects associated with the four-point bending technique for measuring Poisson's ratio, J. Phys. E. Sci. Instr. 7:357 (1974).
3377. B. Bruder, How to distinguish surfaces and sub-surface cracks using electromagnetic NDT methods, NDT Inter. 17(4):221 (Aug. 1984).
3378. L. Brateman, Amer. J. Roentgenl. 146:971 (1986).
3379. P. T. Callahan and C. D. Eccles, J. Magnetic Resonance 71:426 (1987).
3380. J. L. Koenig, The application of solid state NMR spectroscopy to the study of interpenetrating polymer networks in the interface of glass reinforced epoxy composites, 11th Ann. Adhes. Soc. Meeting, Charleston, SC (Feb. 21–24, 1988).
3381. A. O. K. Nieminen and J. L. Koenig, J. Adhes. Sci. Technol. 2:407 (1988).
3382. A. O. K. Nieminen and J. L. Koenig, Bruker NMR Users Conf., Lowell, MA (1988).
3383. A. O. K. Nieminen and J. L. Koenig, NMR imaging of the interfaces of epoxy joints, J. Adhes. 30(104): 47 (1989).
3384. A. O. K. Nieminen and J. L. Koenig, 12th Ann. Adhes. Soc. Meeting, Hilton Head Island, SC (1989).
3385. A. O. K. Nieminen and J. L. Koenig, Using NMRI to evaluate adhesives and bonded joints, Adhes. Age 32(12):17 (1989).
3386. A. O. K. Nieminen, J. Liu, and J. L. Koenig, J. Adhes. Sci. & Technol. 3:455 (1989).
3387. J. Liu, A. O. K. Nieminen, and J. L. Koenig, Appl. Spectros. 43:1260 (1989).
3388. F. S. Blum, Magnetic resonance of polymers at surfaces, Colloids and Surfaces 45:361 (1990).
3389. R. A. Kinsey, Solid state NMR of elastomers, Rubber Chem. and Technol. 63:167 (1990).
3390. W. Kuhn and U. Eichoff, NMR spectroscopy: limitations and applications in material sciences, Die Makromoleculare Chemie Macromolectular Symposia 34:287 (1990).
3391. A. Udawaga, Y. Yamamoto, and R. Chujo, Two-dimensional NMR characterization of cycloaliphatic epoxy resins, Polymer 31(12):2425 (1990).
3392. B. Blumich, P. Blumler, E. Gunther, G. Schauss, and H. W. Speiss, Nondestructive evaluation of polymer materials by solid state NMR imaging, Makromolculare Chemie Macromolecular Symp. 44:37 (1991).
3393. D. J. Hagemaier, Bonded joints and non-destructive testing: Bonded honeycomb structures, Nondestructive Testing, Part I (Dec. 1971), Part II (Feb. 1972).
3394. P. R. Vettito, A thermal I.R. inspection technique for bond flaw inspection, J. Appl. Polym. Sci., Appl. Polym. Symp. No. 3 (1966).
3395. D. J. Hagemaier, Nondestructive inspection, Adhesive Bonding of Aluminum Alloys (E. W. Thrall and R. W. Shannon, eds.), Marcel Dekker, New York, Chap. 15, pp. 388–390 (1985).
3396. R. J. Schliekelmann, Non-destructive testing of adhesive bonded joints, AGARD-NATO Lecture Series No. 102, Wright-Patterson AFB, OH, pp. 8–4,8–5 (1979).
3397. D. J. Hagemaier, End product nondestructive evaluation of adhesive bonded joints, Adhesives and Sealants, Engineered Materials Handbook, ASM, Materials Park, OH, Vol. 3, pp. 762–763 (1990).
3398. H. A. Johnson and C. L. Etheridge, Nondestructive test development for advanced airplanes, J. Mater. 1:823 (Dec. 1966).
3399. F. E. Azofon and W. A. Rohr, Detection of flaws in adhesive bonded metallic honeycomb by infrared nondestructive testing, Soc. Nondestructive Testing, Spring Convention, Los Angeles (March 7–10, 1966).

3400. E. Kutzscher, K. H. Zimmermann, and J. L. Botkin, Thermal and infrared methods for nondestructive testing of adhesive-bonded structures, Mater. Eval. 26:143 (July 1968).
3401. E. W. Kutzscher and K. H. Zimmermann, A scanning infrared inspection system to nondestructive testing of bonded aerospace structures, Appl. Optics 7:1715 (Sept. 1969).
3402. C. Searles, Thermal image inspection of adhesive bonded structures, Proc. Symp. NDT of Welds and Material Joining, Los Angeles (1968).
3403. W. E. Woodmansee, Cholesteric liquid crystals and their application to thermal N.D.T. Mater. Eval. p. 564 (1966).
3404. W. Woodmansee and H. Southworth, Detection of materials discontinuities with liquid crystals, Mater. Eval. 26(8) (1968).
3405. W. Woodmansee and H. Southworth, Thermal nondestructive testing with cholesteric acid liquid crystals, Canadian Council for Nondestructive Technol., 5th Int. Conf. Nondestructive Testing, Montreal (May 21–26, 1967).
3406. S. Brown, Cholesteric crystals for nondestructive testing, Mater. Eval. 26:163 (1968).
3407. T. A. Simcox, Liquid crystals-their application to nondestructive testing, 7th Ann. Region III Convention, IEEE, Cocoa Beach, FL (Nov. 18–20, 1968).
3408. S. E. Cohen, Nondestructive testing for evaluation strength of bonded materials, Bibliographic Listing NASA CR-61503, Lockheed-Georgia Corp. (Sept. 1966).
3409. M. Treca, NDI of bonded structures, AGARD Non-Destructive Inspection Practices, pp. 529–577.
3410. R. J. Schliekelmann, Non-destructive testing of adhesive bonded joints, NATO-AGARD Lecture Series No. 102, Wright-Patterson AFB, OH, pp. 8-1 to 8-37 (1979).
3411. D. J. Hagemaier, Nondestructive inspection, Adhesive Bonding of Aluminum Alloys (E. W. Thrall and R. W. Shannon, eds.), Marcel Dekker, New York, Chap. 15, p. 337–423 (1985).
3412. D. J. Hagemaier, End product nondestructive evaluations of adhesive bonded metal products, Adhesives and Sealants, Engineered Materials Handbook, ASM, Materials Park, OH, Vol. 3, pp. 743–776 (1990).
3413. R. D. Adams and H. P. Cawley, NDT Int. 21:208 (1988).
3414. C. C. Guyott, H. P. Cawley, and R. D. Adams, The nondestructive testing of adhesively bonded structures: A review, J. Adhes. 20(2):129 (1986).
3415. Y. Bar-Cohen and A. K. Mal, End product nondestructive evaluation of adhesive bonded composite joints, Adhesives and Sealants, Engineered Materials Handbook, ASM, Materials Park, OH, Vol. 3, pp. 777–784 (1990).
3416. A. K. Mal, C. C. Yin, and Y. Bar-Cohen, The influence of material dissipation and imperfect bonding on acoustic wave reflection from layered solids, Review of Progress in Quantitative Nondestructive Evaluation (D. O. Thompson and D. E. Chimenti, eds.), Plenum Press, New York, Vol. 7A, pp. 927–934 (1988).
3417. D. J. Hagemaier, NDT of adhesive bonded structure, Paper 6652 presented to Air Trans. Assoc. of Amer., Nondestructive testing Forum, Douglas Aircraft Co. (Sept. 13–15, 1977).
3418. Y. Bar-Cohen, V. Arnon, and M. Meron, Defect detection and characterization in composite sandwich structures by ultrasonics, SAMPE J. 14(1):4 (1978).
3419. A. K. Mal, Wave propagation in layer composite laminates under periodic surface loads, Wave Motion 10:257 (1988).
3420. D. E. Chimenti and A. H. Nayfeh, Experimental ultrasonic reflection and guided waves propagation in fibrous composite laminates, Proc. Joint ASME and SES Appl. Mech. & Eng. Sci. Conf. (A. K. Mal and T. C. T. Ting, eds.), AMD 90:29–38 (June 20–22, 1988).
3421. W. Arnold and H. Reiter, Adhesion—9 (K. W. Allen, ed.), Elsevier Applied Science Publishers, London, p. 85 (1984).
3422. D. M. Thaker and N. J. Burton, Adhesion—11 (K. W. Allen, ed.), Applied Science Publishers, London, p. 56 (1987).
3423. S. Persson and E. Oestman, Appl. Optics 24:4095 (1985).
3424. S. Persson, Polymer 29:802 (1988).
3425. Adhesive Bonded Aerospace Structures Standardized Repair Handbook, Section 10.0 on Nondestructive Inspection, AFML Tech. Report No. TR-77-206, Final Report (Dec. 1977).
3426. Specification P.S. 21233.1, Acceptance Criteria for Nondestructive Testing of F-15 Bonded Assemblies, McDonnell-Douglas Corp., St. Louis, MO.
3427. DPS 4.732, Eddy-Sonic Inspection of Adhesive Bonded Metal Assemblies, Douglas Aircraft Co. (July 1971).
3428. D. J. Hagemaier, End product nondestructive evaluation of adhesive bonded metal joints, Adhesives and Sealants, Engineered Materials Handbook, ASM, Materials Park, OH, Vol. 3, p. 772 (1990).

3429. J. D. Minford, Durability evaluation of adhesive bonded structures, Adhesive Bonding (L. H. Lee, ed.), Plenum Press, New York, Chap. 9, pp. 239–290 (1991).
3430. L. K. Peters, unpublished work, Aluminum Co. of America.
3431. J. D. Minford, Aluminum adhesive bond permanence, Treatise on Adhesion and Adhesives (R. L. Patrick, ed.), Marcel Dekker, New York, Vol. 5, Chap. 3, pp. 116–126 (1981).
3432. J. D. Minford, Durability of Adhesive Bonded Aluminum Joints, Treatise on Adhesion and Adhesives (R. L. Patrick, ed.), Marcel Dekker, New York, Vol. 3, Chap. 2, pp. 111–110 (1973).
3433. J. D. Minford, Adhesives, Durability of Structural Adhesives (A. J. Kinloch, ed.), Applied Science Publishers, London, Chap. 4, pp. 186–192 (1983).
3434. J. D. Minford, Permanence of Adhesive Bonded Aluminum Joints (G. L. Schneberger, ed.), Marcel Dekker, New York, Chap. 23, pp. 580, 588–594 (1983).
3435. D. M. Brewis, Aluminium adherends, Durability of Structural Adhesives (A. J. Kinloch, ed.), Applied Science Publishers, London, Chap. 5, pp. 219–222 (1983).
3436. J. A. Marceau and W. Scardino, Durability of adhesive-bonded joints, AFML Report No. Tr-75-3, Wright-Patterson AFB, OH (Feb. 1975).
3437. E. McAbee and D. W. Levi, J. Appl. Polym. Sci. 11:2067 (1967).
3438. E. McAbee and D. W. Levi, Use of reaction rate method to predict failure times of adhesive joints at constant stress, Picatinny Arsenal Tech. Report No. 4105, Mater. Eng. Lab., Picatinny Arsenal, Dover, NJ (Dec. 1970).
3439. E. McAbee and D. W. Levi, Prediction of failure times.
3440. W. C. Jones, E. McAbee, and D. W. Levi, Use of multiple regression analysis to develop predictive models for failure times of adhesive bonds at constant stress, II—aluminum adherends, Picatinny Arsenal Tech. Report No. TM-2066 (March 1973).
3441. E. M. Vader, unpublished work, Aluminum Co. of America.
3442. B. R. Appleman, Survey of accelerated test methods for anticorrosive coatings performance, J. Coatings Technol. 62:57 (1990).
3443. T.C. Simpson, P. J. Moran, H. Hampel, G. D. Davis, B. S. Shaw, C. O. Arah, T. L. Fritz, and K. Zankel, Electrochemical monitoring of organic coating degradation during atmospheric or vapor phase exposure, J. Sci. Eng. Corr. 46(4):331 (1990).
3444. S. Nakahara, Coatings, Adhesive Bonding of Aluminum Alloys (E. W. Thrall and R. W. Shannon, eds.), Marcel Dekker, New York, Chap. 12, pp. 227–240 (1985).
3445. A. Stevenson, Electrochemical and corrosion effects on adhesive joints, Adhesives and Sealants, Engineered Materials Handbook, ASM, Materials Park, OH, Vol. 3, pp. 628–629 (1990).
3446. T. J. Reinhart, Chemical effects on adhesive joints, Adhesives and Sealants, Engineered Materials Handbook, ASM, Materials Park, Ohio, Vol. 3, pp. 637–643 (1990).
3447. K. T. Kern, S. A. T. Long, W. G. Witte, and W. L. Harries, Radiation and vacuum effects on adhesive materials in bonded joints, Adhesives and Sealants, Engineered Materials Handbook, ASM, Materials Park, Ohio, Vol. 3, pp. 644–650 (1990).
3448. J. D. Minford, unpublished work, Aluminum Co. of America.
3449. W. Brockmann, Adhasion, pp. 335–342 (1969).
3450. W. Brockmann, Adhasion, pp. 448–460 (1969).
3451. W. Brockmann, Adhasion, pp. 52–66 (1970).
3452. W. Brockmann, Adhasion, pp. 250–252 (1970).
3453. W. Brockmann, Adhasion, pp. 72–83 (1973).
3454. W. Brockmann, Chemical aspects of adhesion between metals and Polymers, Proc. 10th Ann. Meeting Adhesion Soc. (L. H. Sharpe, ed.), Gordon and Breach, London (1988).
3455. R. F. Blomquist, Importance of evaluation of adhesives for all environmental conditions and permanence studies, Symp. on Adhesives for Structural Applications (M. J. Bodnar, ed.), Interscience, New York, pp. 43–48 (1962).
3456. N. J. DeLollis and O. Montoya, Mode of failure in structural adhesive bonds, J. Appl. Polym. Sci. 11: 983 (1964).
3457. M. J. Bodnar, Journal of Applied Polymer Science, Applied Polymer Symposia No. 3, Wiley-Interscience, New York (1966).
3458. Durability of Adhesive Joints, Spec. Tech. Publ., (STP) 401 (1966).
3459. I. G. Zwei, F. Fleshner, H. Dodiuk, and L. Drori, Research report: durability of structural adhesive joints, Int. J. Adhes. Adhes. 4(3):137 (July 1984).
3460. A. Carre and J. Schultz, Polymer-aluminum adhesion III. effect of a liquid environment, J. Adhes. 18(3):171 (1985).

3461. M. K. Omar, A. G. Atkins, and J. K. Lancaster, The adhesive-failure wear of metals, Wear 107(3):279 (Feb. 1986).

3462. J. A. Harris, The damage tolerance of adhesive joints, Adhesion '90, Plastics and Rubber Inst. Meetings, Cambridge, England (Sept. 1990).

3463. H. M. Hand, C. O. Arah, D. K. McNamara, and M. F. Macklenberg, Effects of environmental exposure on adhesively bonded joints, Int. J. Adhes. Adhes. 11(1):15 (1991).

3464. J. D. Minford, Durability evaluation of adhesive bonded structures, Adhesive Bonding (L. H. Lee, ed.), Plenum Press, New York, Chap. 9, pp. 239–284 (1991).

3465. A. F. Lewis and R. T. Natarajan Gounder, Permanence and endurance of structural adhesives joints, Treatise on Adhesion and Adhesives (R. L. Patrick, ed.), Marcel Dekker, New York, Vol. 5, Chap. 7, p. 313 (1981).

3466. H. Liebowitz, Fracture of non-metals and composites, Fracture: An Advanced Treatise in 7 volumes (H. Liebowitz, ed.), Academic Press, New York (1969–1972).

3467. F. N. Kelley and M. L. Williams, Rubber Chem. and Technol., 42(4):1174 (1969).

3468. J. A. Manson and R. H. Hertzberg, Crit. Rev. Macromol. Sci. 1(4):433 (1973).

3469. C. E. Inglis, Proc. Inst. Naval Architects (March 14, 1913).

3470. J. P. Berry, J. Appl. Phys. 34:62 (1963).

3471. G. R. Irwin, Trans. ASME, Ser. D82(2):417 (1960).

3472. T. Hata, Recent Advances in Adhesion (L. H. Lee, ed.), Gordon and Breach, London, p. 269 (1973).

3473. E. M. Prot, Rev. Metallurgy 45(12) (1948); English trans. E. J. Ward, WADC Tech. Report, p. 52 (Sept. 1952).

3474. H. S. Loveless, C. W. Deeley, and D. L. Swanson, SPE Trans. 2(2) (April 1962).

3475. H. S. Loveless and D. E. Williams, Polym. Eng. Sci. 10(3) (April 1962).

3476. A. F. Lewis and R. T. Natarajan Gounder, Permanence and endurance of structural joints, Treatise on Adhesion and Adhesives (R. L. Patrick, ed.), Marcel Dekker, New York, Vol. 5, Chap. 7, pp. 322–334 (1981).

3477. J. R. McDowell, unpublished report, Lord Corp. Hughson Chem. Div., Erie, PA (June 1974).

3478. J. D. Minford, Adhesive bonded aluminum joints, Treatise on Adhesion and Adhesives (R. L. Patrick, ed.), Marcel Dekker, New York, Vol. 5, Chap. 3, pp. 121–126 (1981).

3479. K. W. Allen, S. M. Smith, W. C. Wake, and A. D. van Raalte, The concept of an endurance limit for adhesive joints, Int. J. Adhes. Adhes. 5(1):23 (Jan. 1985).

3480. R. S. Lenk, Plastics Rheology, John Wiley & Sons, New York, p. 125 (1968).

3481. W. F. Watson, High Polymers (E. M. Fattes, ed.), John Wiley & Sons, Vol. 19, New York (1964).

3482. G. G. Winspear, Vanderbilt Rubber Handbook (G. G. Winspear, ed.), R. T. Vanderbilt Co., New York, p. 316 (refer to ASTM Test D-623) (1968).

3483. P. C. Paris and F. Erdogan, Trans. ASME 85:528 (1963).

3484. J. M. Barsom, Fatigue-crack propagation in steels of various yield strength, 1st Natl. Cong. Pressure Vessels and Piping (May 10, 1971).

3485. A. H. Lepie and A. Adicoff, J. Appl. Polym. Sci. 18:2165 (1974).

3486. R. F. Boyer, Rubber Chem. Tech. 36:1303 (1963).

3487. M. B. Bever and M. Shen, Mater. Sci. Eng. 15:145 (1974).

3488. A. F. Lewis and R. Saxon, Epoxy Resins (C. May and T. Tanaka, eds.), Marcel Dekker, New York (1973).

3489. M. B. Neiman, B. M. Kovarskaya, L. I. Golubenkova, A. S. Strinznkova, I. I. Laventovskaya, and M. S. Akutin, J. Polym. Sci. 56:383 (1962).

3490. M. B. Neiman, Aging and Stabilization of Polymers (M. B. Neiman, ed.), Consultants Bureau, New York (1965).

3491. R. B. Krieger and R. W. Politi, Aspects of Adhesion (D. J. Alner, ed.), Chemical Rubber Co., Cleveland, OH, Vol. 3, p. 46 (1969).

3492. J. D. Minford, Adhesive bonded aluminum joints, Treatise on Adhesion and Adhesives (R. L. Patrick, ed.), Marcel Dekker, New York, Vol. 3, Chap. 2, p. 79 (1973).

3493. J. D. Minford, Aluminum adhesive bond permanence, Treatise on Adhesion and Adhesives (R. L. Patrick, ed.), Marcel Dekker, New York, Vol. 5, Chap. 3, p. 45 (1981).

3494. J. D. Minford, Adhesives, Durability of Structural Adhesives (A. J. Kinloch, ed.), Applied Science Publishers, London, Chap. 4, p. 135 (1983).

3495. J. D. Minford, Adhesive bonded structure durability, Adhesive Bonding (L. H. Lee, ed.), Plenum Press, New York, Chap. 9, pp. 270–275 (1991).

3496. W. Brockmann, Steel adherends, Durability of Structural Adhesives (A. J. Kinloch, ed.), Applied Science Publishers, London, Chap. 7, p. 281 (1983).

3497. D. R. Askins, Environmental durability testing of structural adhesives, Part I, Air Force Materials Lab Contract No. TR-78-35, Wright-Patterson AFB, OH (Dec. 1978).
3498. D. R. Askins, Environmental durability testing of structural adhesives, Part II, Air Force Materials Lab Contract No. Tr-78-35, Wright-Patterson AFB, OH (Dec. 1978).
3499. D. K. Owens, J. Appl. Polym. Sci. 14:1725 (1970).
3500. H. Schonhorn, 12th An. Conf. Adhes. & Adhes., City University of London, England (April 1974).
3501. H. Schonhorn and H. L. Frisch, J. Polym. Sci., Polym. Phys. 11:1005 (1973).
3502. J. C. Halpin, J. R. Korf, and W. Goldberg, J. Comp. Mater. 4:970).
3503. J. C. Halpin, M. E. Waddoups, and T. A. Johnson, Int. J. Fract. Mech. 8:465 (1972).
3504. J. C. Halpin, J. Comp. Mate. 6:208 (1972).
3505. D. H. Kaelble, P. J. Dynes, L. H. Crane, and L. Hans, ASTM Composite Reliability Conf., Las Vegas (April 1974).
3506. D. H. Kaelble, P. J. Dynes, and L. H. Crane, J. Adhes. 7:25 (1975).
3507. A. W. Davis and T. A. Johnson, ASTM Composite Reliability Conf. Las Vegas (April 1974).
3508. M. E. Waddoups, R. V. Wolfe, and D. J. Wilkins, Reliability of Complex Large Scale Composite Structures: Proof of Concept, AFML Report No. TR-73-160 (July 1973).
3509. D. H. Kaelble, P. J. Dynes, and E. H. Cirlin, J. Adhes. 6:23 (1974).
3510. A. F. Lewis and L. J. Forrestal, Symp. Rec. Dev. Adhes. Sci., ASTM Spec. Tech. Publ. No. 360, p. 59 (1963).
3511. A. F. Lewis and R. T. Natarajan Gounder, Preprint Booklet, Org. Coatings and Plastics Div., Spring meet., Philadelphia (April 1975).
3512. D. Bulgin, G. D. Hubbard, and M. H. Walters, Proc. 4th Rubber Tech. Conf., London, p. 173 (1962).
3513. D. F. Moore and W. Geyer, Wear 22:113 (1972).
3514. A. F. Lewis and R. T. Natarajan Gounder, Permanence and endurance of structural adhesive joints, Treatise on Adhesion and Adhesives (R. L. Patrick, ed.), Marcel Dekker, New York, Vol. 5, Chap. 7, pp. 351–358 (1981).
3515. M. Avrami, J. Chem. Phys. 7:1103 (1939).
3516. M. Avrami, J. Chem. Phys. 8:212 (1940).
3517. M. Avrami, J. Chem. Phys. 9:177 (1941).
3518. W. A. Johnson and F. R. Mehl, Trans. AIME 134:416 (1939).
3519. R. T. Natarajan Gounder and T. Davidson, J. Polym. Sci., Phys. Ed. 10:2209 (1972).
3520. W. Brockmann, SAMPE Nat. Tech. Conf., Buena Park, CA, p. 198 (April 1974).
3521. S. M. Wiederhorn and L. H. Polz, J. Amer. Ceram. Soc. 53:543 (1970).
3522. S. M. Wiederhorn, J. Amer. Ceram. Soc. 50:407 (1967).
3523. A. F. Lewis and R. T. Natarajan Gounder, Permanence and endurance of structural adhesive joints, Treatise on Adhesion and Adhesives (R. L. Patrick, ed.), Marcel Dekker, New York, Vol. 5, Chap. 7, pp. 344–359 (1981).
3524. J. D. Minford, Durability of adhesive bonded aluminum joints, Treatise on Adhesion and Adhesives (R. L. Patrick, ed.), Marcel Dekker, New York, Vol. 3, Chap. 2, pp. 111–119 (1973).
3525. W. B. Jones, Natl. SAMPE Tech. Conf., Buena Park, CA, p. 75 (April 1974).
3526. A. J. Kinloch, Environmental failure of structural adhesive joints: A literature survey, AD784890 or ERDE TN95 (Aug. 1973).
3527. A. J. Kinloch, W. A. Dukes, and R. A. Gledhill, ACS-Coatings and Plastic Preprints 35(1):546 (1975).
3528. S. Matsuoka, Polym. Eng. Sci. 3:162 (1974).
3529. Gulf Oil Corp., Effects of moisture on physical properties of Gulf Nylon, Technical Bulletin.
3530. R. D. Deanin, Polymer Structure, Properties, and Applications, Cahners Books, Boston, p. 75ff (1972).
3531. D. H. Kaelble, Physical Chemistry and Adhesion, Wiley-Interscience, New York, pp. 295, 360 (1971).
3532. P. Hittmar and R. Ullman, J. Appl. Polym. Sci. 6:191 (1962).
3533. L. J. Broutman, SPE J., p. 283 (March 1965).
3534. A. R. C. Westwood, J. Mater. Sci. 9:1871 (1974).
3535. W. J. Renton and J. R. Vinson, Analysis and design of anisotropic bonded joints, AFOSR Tr-75-0125 (August 1975).
3536. W. J. Renton and J. R. Vinson, J. Aircraft 12(5):442 (1975).
3537. Army Materials and Mechanics Res. Center, The role of mechanics in design, Structural Joints, Proc. Army Symp. Solid Mech., AMMRC-MS-74-8 (Sept. 1974).
3538. F. N. Kelley and J. L. Trout, AIAA Paper No. 72-1085 AIAA/SAE 8th Joint Propulsion Specialist Conf., New Orleans, Nov. 29–Dec. 1 (1972).
3539. Testing for prediction of material performance in components and structures, NMAB NO. 288, Nat. Res. Council, Washington, D.C. (May 1972).

3540. G. W. Walter, A critical review of the protection of metals by paints, Corr. Sci. 26(1):27 (1986).
3541. L. J. Matinezo, D. K. Shaffer, W. C. Moshier, and G. D. Davis, Environmental and adhesive durability of aluminum-polymer systems protected with organic corrosion inhibitors, J. Mater. Sci. 21(5):1601 (May 1986).
3542. W. J. Lorenz and F. Mansfeld, Interface and interphase corrosion inhibition, Electrochimica Acta 31(4): 467 (April 1986).
3543. J. E. O. Mayne, Protection by epoxy films, Brit. Corr. J. 25(3):189 (1990).
3544. J. D. Minford, unpublished work, Aluminum Co. of America.
3545. J. D. Minford, unpublished work, Aluminum Co. of America.
3546. J. D. Minford, unpublished work, Aluminum Co. of America.
3547. N. A. Debruyne, The extent of contact between glue and adherend Bull. 168, Tech. Service Dept., Aero Research Ltd., Duxford, Cambridge, England (Dec. 1956).
3548. W. D. Bascom, The origin and removal of microvoids in filiment wound composites, NRL Report No. 6268 (May 24, 1965).
3549. J. D. Minford, unpublished report, Alcoa Labs, Alcoa Center, PA.
3550. J. D. Bikerman, ASTM Spec. Tech. Publ., No. 640, pp. 30–40 (1978).
3551. J. J. Bikerman, Adhesion and Cohesion (P. Weiss, ed.), Elsevier, Amsterdam, pp. 36–45 (1962).
3552. J. J. Bikerman, Industrial Engineering Chemistry 59(9):40 (1967).
3553. J. J. Bikerman, J. Paint Technol. 43:98ff (1971).
3554. R. J. Good, Recent Advances in Adhesion (L. H. Lee, ed.), Gordon and Breach, New York (1973).
3555. R. J. Good, J. Adhes. 4:133 (1972).
3556. R. J. Good, ASTM Spec. Tech. Publ. No. 640, pp. 18–29 (1978).
3557. W. L. Baun, J. Adhes. 7:261 (1976).
3558. D. M. Wyatt, R. C. Gray, J. C. Carver, D. M. Hercules, and L. W. Masters, Appl. Spect. 28(5):439 (1974).
3559. J. A. Laird and F. W. Nelson, The effect of glass surface chemistry on glass/epoxy systems, SPE Trans. (April 1964).
3560. P. W. Erickson, A. Volpe, and F. R. Cooper, Effects of glass surfaces on laminating resins, Modern Plastics (Aug. 1969).
3561. J. Outwater and D. Kellogg, A simple experiment to show the origin of water debonding of resin against glass, Report for Contract V-3219 (01) (X) (Sept. 15, 1961).
3562. W. A. Zisman, Surface chemistry of glass reinforced plastics, Symp. Glass-Resin Interface of the Soc. Plastics Ind., 19th Ann. Exhib. and Conf., Chicago (Feb. 6, 1964).
3563. S. Sterman and J. B. Toogood, How to promote adhesion with silicones and silanes, Adhes. Age (July 1965).
3564. J. A. Laird, Glass surface chemistry for glass reinforced plastics, Final Report, Navy Contract W-0679-C (FBM) (June 1963).
3565. J. D. Minford, Adhesive bonded aluminum joints, Treatise on Adhesion and Adhesives (R. L. Patrick, ed.), Marcel Dekker, New York, Vol. 3, Chap. 2, p. 114 (1973).
3566. J. D. Minford, Aluminum adhesive bond permanences, Treatise on Adhesion and Adhesives (R. L. Patrick, ed.), Marcel Dekker, New York, Vol. 5, Chap. 3, p. 123 (1981).
3567. A. Mazor, L. J. Broutman, and B. H. Eckstein, Natl. Tech. Conf. SPE (Prepr.) p. 77 (1976).
3568. J. R. Huntsberger, J. Polym. Sci. A1, p. 1339 (1963).
3569. W. M. Bright, Adhesion and Adhesives, John Wiley & Sons, New York, p. 130 (1954).
3570. C. L. Brett, J. Appl. Sci. 18:315 (1974).
3571. D. M. Brewis, Mat. Sci. Technol. 2:761 (1986).
3572. W. L. Baun, ASTM STP No. 640, ASTM, Philadelphia, p. 41 (1978).
3573. J. S. Crompton, An examination of interfacial failure in adhesively bonded aluminum, J. Adhes. 28: 135–143 (1989).
3574. W. M. Baun, N. T. McDevitt, and J. S. Solomon, Chemistry of metal and alloy adherends by SIMS, ISS, and AES, Surface Analysis Techniques for Metallurgical Applications, ASTM, STP Publ. No. 596, p. 86 (March 1975).
3575. N. T. McDevitt and W. M. Baun, Surface Analysis of 6061 and 7050 Alloys, AFML Report No. TR-76-13 (March 1976).
3576. J. D. Minford, unpublished work, Aluminum Co. of America.
3577. J. D. Minford, unpublished work, Aluminum Co. of America.
3578. J. D. Minford, unpublished work, Aluminum Co. of America.
3579. J. D. Minford, unpublished work, Aluminum Co. of America.
3580. J. D. Minford, unpublished work, Aluminum Co. of America.

3581. J. D. Minford, Aluminum adhesive bond permanence, Treatise on Adhesion and Adhesives (R. L. Patrick, ed.), Marcel Dekker, New York and Basel, Vol. 5, Chap. 3, p. 91 (1981).
3582. J. D. Minford, unpublished work, Aluminum Co. of America.
3583. B. H. River, C. B. Vick, and R. H. Gillespie, Wood as an adherend, Treatise on Adhesion and Adhesives (J. D. Minford, ed.), Marcel Dekker, New York and Basel, Vol. 7, Chap. 1, pp. 131–133 (1991).
3584. J. D. Minford, unpublished work, Aluminum Co. of America.
3585. J. D. Minford, unpublished work, Aluminum Co. of America.
3586. J. D. Minford, Adhesive bonded aluminum joints, Treatise on Adhesion and Adhesives (R. L. Patrick, ed.), Marcel Dekker, New York, Vol. 3, Chap. 2, p. 94 (1973).
3587. J. D. Minford, Adhesive bonded aluminum joints, Treatise on Adhesion and Adhesives (R. L. Patrick, ed.), Marcel Dekker, New York and Basel, Vol. 3, Chap. 2, p. 95 (1973).
3588. G. D. Spathis, E. P. Sideridis, and P. S. Theocaris, Int. J. Adhes. Adhes., p. 195 (April 1981).
3589. J. D. Minford, unpublished work, Aluminum Co. of America.
J. L. Hranica, unpublished work, Aluminum Co. of America.
3590. J. D. Minford, unpublished work, Aluminum Co. of America.
3591. W. C. Wake, Adhesion (D. D. Eley, ed.), Oxford University Press, Oxford, England.
3592. H. Foulkes, J. Shields, and W. C. Wake, J. Adhes. 2:254 (1970).
3593. J. D. Minford, Aluminum adhesive bond permanence, Treatise on Adhesion and Adhesives (R. L. Patrick, ed.), Marcel Dekker, New York, Vol. 5, Chap. 3, pp. 81–83 (1981).
3594. J. D. Minford, Permanence of adhesive bonded aluminum joints, Adhesives in Manufacturing (G. L. Schneberger, ed.), Marcel Dekker, New York, Chap. 23, p. 581 (1983).
3595. J. D. Minford, unpublished work, Aluminum Co. of America.
3596. D. M. Brewis, Aluminium adherends, Durability of Structural Adhesives (A. J. Kinloch, ed.), Applied Science Publishers, London, Chap. 5, p. 232 (1983).
3597. P. Poole and J. F. Watts, Effect of alloy composition and surface pretreatment on the durability of adhesive-bonded aluminum alloy joints, J. Adhes. Adhes. 5(1):33 (Jan. 1985).
3598. A. Price, T. Baldwin, and P. J. Gregson, Effect of simple pretreatments on the durability of adhesive-bonded Al-Li-Cu-Mg alloy, Int. J. Adhes. Adhes. 11(1):11 (1991).
3599. J. D. Minford, unpublished work, Aluminum Co. of America.
3600. J. D. Minford, unpublished work, Aluminum Co. of America.
3601. P. M. Stifel, Durability testing of adhesive bonded joints, 19th Natl. SAMPE Symp. & Exhib. (April 23–25, 1974).
3602. A. J. Kinloch, Surface analysis and bonding of aluminum-magnesium alloys, J. Adhes. 14:105 (1982).
3603. A. J. Kinloch, Adhesion—6 (K. W. Allen, ed.), Applied Science Publishers, London (1982).
3604. G. D. Davis, Surface Analysis technology and applications, Adhesives and Sealants, Engineered Materials Handbook, ASM, Materials Park, OH, Vol. 3, pp. 237–238 (1990).
3605. G. D. Davis, Characterization of surfaces, Adhesive Bonding (L. H. Lee, ed.), Plenum Press, New York, Chap. 6, pp. 141–144 (1991).
3606. C. D. Wagner, Anal. Chem. 44:6 (1972).
3607. J. D. McMillan, Surface preparation-the key to bondment durability, AGARD Lecture Series No. 102, Wright-Patterson AFB, OH, p. 7–4 (1979).
3608. J. D. Minford, unpublished work, Aluminum Co. of America.
3609. J. D. Minford, unpublished work, Aluminum Co. of America.
3610. J. D. Minford, Durability of adhesive bonded aluminum joints, Treatise on Adhesion and Adhesives (R. L. Patrick, ed.), Marcel Dekker, New York, Vol. 3, Chap. 2, pp. 82–85, 91–108 (1973).
3611. J. D. Minford, Adhesives, Durability of Structural Adhesives (A. J. Kinloch, ed.), Applied Science Publishers, London, pp. 182–190 (1983).
3612. J. D. Minford, Durability of adhesive bonded aluminum joints, Treatise on Adhesion and Adhesives (R. L. Patrick, ed.), Marcel Dekker, New York, Vol. 3, Chap. 2, pp. 83–85 (1973).
3613. J. D. Minford, Adhesives, Durability of Structural Adhesives (A. J. Kinloch, ed.), Applied Science Publishers, London, pp. 176–181 (1983).
3614. J. D. Minford, Aluminum adhesive bond permanence, Treatise on Adhesion and Adhesives (R. L. Patrick, ed.), Marcel Dekker, New York, Vol. 5, Chap. 3, pp. 86–91 (1981).
3615. J. C. Bolger, Structural adhesives for metal bonding, Treatise on Adhesion and Adhesives (R. L. Patrick, ed.), Marcel Dekker, New York, Vol. 3, Chap. 1, pp. 22–27, 34–42 (1973).
3616. J. C. Bolger, Structural adhesives: State of the art, Adhesives in Manufacturing (G. L. Schneberger, ed.), Marcel Dekker, New York, Chap. 7, pp. 150–159 (1983).
3617. R. H. Young and J. M. Tancrede, Phenolic adhesives and modifiers, Adhesives in Manufacturing (G. L. Schneberger, ed.), Marcel Dekker, New York, Chap. 10, pp. 263–268 (1983).

3618. R. D. Ebewele, B. H. River, and J. A. Koutsky, Relationship between phenolic adhesive chemistry and adhesive joint performance, J. Appl. Polym. Sci. 31(7):2275 (1986).
3619. B. I. Buck and M. G. D. Hockney, Effect of outdoor exposure on adhesive bonded joints, Aspects of Adhesion—7 (R. W. Allen and D. J. Alner, eds.), Applied Science Publishers, London.
3620. G. Salomon, Adhesion and Adhesives (I. R. Houwink and G. Salomon, eds.), Elsevier, New York, p. 325 (1965).
3621. J. Comyn, D. M. Brewis, and S. T. Treadwell, Bonding of aluminum alloys with some phenolic adhesives and a modified epoxide adhesive, and strength changes on exposure to moist air at 50°C, J. Adhes. 21:59 (1987).
3622. P. Albericci, Aerospace applications, Durability of Structural Adhesives (A. J. Kinloch, ed.), Applied Science Publishers, London, Chap. 8, p. 347 (1983).
3623. J. D. Minford, Durability of adhesive bonded aluminum joints, Treatise on Adhesion and Adhesives (R. L. Patrick, ed.), Marcel Dekker, New York, Vol. 3, Chap. 2, pp. 91–92 (1973).
3624. D. W. Levi et al., SAMPE Quart., p. 1 (April 1976).
3625. D. W. Levi, J. Appl. Polym. Sci., Appl. Polym. Symp. 32:189 (1977).
3626. J. C. Bolger, Structural adhesives for metal bonding, Treatise on Adhesion and Adhesives (R. L. Patrick, ed.), Marcel Dekker, New York, Vol. 3, Chap. 1, pp. 17–20, 46, 47 (1973).
3627. W. D. Bascom and D. L. Hunston, The fracture of epoxy and elastomer-modified epoxy polymers, Treatise on Adhesion and Adhesives (R. L. Patrick, ed.), Marcel Dekker, New York, Vol. 6, Chap. 4, pp. 146–159 (1989).
3628. W. A. Lees, Advances in Adhesive Application Materials and Safety Book, Warwick Publishing, Birmingham, England, Chap. 6 (1982).
3629. J. D. Minford, Durability of adhesive bonded aluminum joints, Treatise on Adhesion and Adhesives (R. L. Patrick, ed.), Marcel Dekker, New York, Vol. 3, Chap. 2, pp. 97–100 (1973).
3630. M. J. Bodnar and R. G. Wegman, Outdoor aging of aluminum joints, Report No. 3689, Picatinny Arsenal, Dover, NJ (May 1968).
3631. G. C. Grimes, J. Appl. Polym. Sci., Appl. Polym. Symp. 32:247 (1977).
3632. J. Luckyam and A. E. Vardy, Fatigue performance of two structural adhesives, J. Adhes. 26:273 (1988).
3633. R. R. N. Jones, J. Swamy, J. Bloxham, and A. Bouderbalah, Int. J. Cement Composites 2:91 (1976).
3634. M. D. MacDonald and A. J. J. Carlder, Int. J. Adhes. Adhes. 2:119 (1982).
3635. P. Albrecht, J. Struct. Eng. ASCE p. 113 (1987).
3636. D. M. Martin, Ph.D. Thesis, University of Dundee, Dundee, Scotland (1985).
3637. R. J. Lark and G. C. Mays, WBRU Interim Report IR37, University of Dundee, Dundee, Scotland (May 1982).
3638. R. J. Lark, WBRU Interim Report IR34, Rev. A., University of Dundee, Dundee, Scotland (Nov. 1983).
3639. J. D. Minford, Adhesive bonded aluminum joints, Treatise on Adhesion and Adhesives (R. L. Patrick, ed.), Marcel Dekker, New York, Vol. 3, Chap. 2, p. 118 (1973).
3640. J. Peterka, Adhasion, p. 200 (1975).
3641. D. R. Askins and H. S. Schwartz, 9th Natl. SAMPE Tech. Conf., Atlanta, GA, 9:329 (Oct. 1977).
3642. D. R. Askins and H. S. Schwartz, J. Appl. Polym. Sci., Appl. Polym. Symp. 32:217 (1977).
3643. R. W. Shannon and E. W. Thrall Jr., J. Appl. Polym. Sci., Appl. Polym. Symp. 32:217 (1977).
3644. R. W. Shannon et al., General material property data, U.S. Air Force Flight Dynamics Lab. Tech. Report TR-77-107 (Sept. 1978).
3645. J. D. Minford, unpublished work, Aluminum Co. of America.
3646. J. D. Minford, unpublished work, Aluminum Co. of America.
3647. J. D. Minford, unpublished work, Aluminum Co. of America.
3648. J. D. Minford, unpublished work, Aluminum Co. of America.
3649. J. D. Minford, unpublished work, Aluminum Co. of America.
3650. J. D. Minford, unpublished work, Aluminum Co. of America.
3651. J. D. Minford, Adhesive joining aluminum to engineering plastics: I. Polyester fiberglass composite, Physicochemical Aspects of Polymer Surfaces (K. L. Mittal, ed.), Plenum Press, New York, Vol. 2, pp. 1139–1160 (1983).
3652. J. D. Minford, Adhesive joining aluminum to engineering plastics: II. Engineering grade styrene and cross-linked styrene, Physicochemical Aspects of Polymer Surfaces (K. L. Mittal, ed.), Plenum Press, New York, pp. 1164, 1166, 1168, 1171 (1983).
3653. J. D. Minford, unpublished work, Aluminum Co. of America.
3654. J. D. Minford, unpublished work, Aluminum Co. of America.
3655. M. Angelovivi, H. Dodiuk, and S. Kenig, Toughened acrylics short cure aids in underwater bonding, Adhes. Age 33(3):45 (1990).

3656. R. W. Drisko, J. B. Crilly, and R. M. Staples, Adhesives for use underwater, NTIS Report No. TN-1367, AD-A004 938 (1947).
3657. J. D. Clark, J. Sharp, and M. Bowditch, An underwater adhesive-based repair method for offshore structures, I—Mech. E. Conf. Publ. No. 6, Int. Cong. Struct. Adhes. Eng., pp. 113–121 (1986).
3658. R. C. Allen, J. Bird, and J. Clarke, The use of adhesives in the repair of cracks in ship structures, I—Mech. E. Conf. Publ. No. 6, Int. Conf. Struct. Adhes. Eng., pp. 169–178 (1986).
3659. J. V. Sharp, Adhesives for underwater repairs, Eng. Digest. 33(7):34 (1987).
3660. R. W. Drisko, Underwater marine applications of coatings and adhesives, Am. Chem. Soc. Meet., Div. Org. Coat. & Plastics 31(1):584 (1974).
3661. C. R. Hegedus, Displacement of water from a steel surface, J. Coat. Technol. 56:711 (1984).
3662. R. A. Oliver, Adhesives for structural bonding underwater or in wet conditions, European Adhesives and Sealants, pp. 15–18 (1988).
3663. K. W. Allen, T. Hatzinilolaou, and K. B. Armstrong, Research report: a comparison of acrylic adhesives for bonding aluminum alloys after using various surface preparation methods, Int. J. Adhes. Adhes. 4(3):133 (July 1984).
3664. K. W. Allen, L. Greenwood, and K. B. Armstrong, A comparison of different grades of an acrylic adhesive for bonding an aluminum alloy, Int. J. Adhes. Adhes. 5(3):149 (July 1985).
3665. D. L. Kotzev, T. C. Wood, and D. W. Dwight, Assessment of the adhesive bond properties of allyl 2-cyanoacrylate, J. Appl. Polym. Sci. 26:1941 (1981).
3666. D. L. Kotzev, P. C. Novakov, and J. S. Kabaivanov, Synthesis and properties of some alkenyl- and alkinyl 2-cyanoacrylates, Angew. Makromol. Chem. 92:41 (1980).
3667. K. F. Drain, J. Guthrie, C. L. Leung, F. R. Martin, and M. S. Otterburn, The effect of moisture on the strength of steel-steel cyanoacrylate adhesive bonds, J. Adhes. 17:81 (1984).
3668. J. D. Minford, unpublished work, Aluminum Co. of America.
3669. H. K. Lauer and M. R. Boyaner, J. Appl. Polym. Sci., Appl. Polym. Symp. 32:301 (1977).
3670. J. D. Minford, unpublished work, Aluminum Co. of America.
3671. B. D. Murray, J. Appl. Polym. Sci., Appl. Polym. Symp. 32:411 (1977).
3672. W. Wrasidlo, 19th Natl. SAMPE Symp. Exhib. 19:120 (1974).
3673. T. J. Aponyi and C. B. Delano, 19th Natl. SAMPE Symp. Exhib. 19:178 (1974).
3674. R. W. Vaughan and R. L. Jones, NASA CR-112003 (Dec. 1971).
3675. R. A. Kirchoff, U.S. Patent No. 4,540,763 (1985).
3676. R. A. Kirchoff et al., Proceedings 18th Int. SAMPE Conf., pp. 476–489 (1986).
3677. C. Chih-Ching, Polym. Sci. Technol. 37:227 (1988).
3678. L. H. Lee, Int. J. Adhes. Adhes. 81 (1988).
3679. F. Mercer, Int. Patent Application 688,794 (1985).
3680. S. Choe et al., Polym. Mater. Sci. Eng. 56:827 (1987).
3681. E. C. Chenevey, European Patent Application 87301367.6 (1987).
3682. W. MacKnight, European Patent Application 87301367.6 (1987).
3683. G. Guerra et al., Macromolecules 21:231 (1988).
3684. P. M. Hergenrother, Quinoxaline/phenylquinoxaline copolymers, J. Appl. Polym. Sci. 21:2157 (1977).
3685. P. M. Hergenrother, Polyphenylquinoxalines containing pendant phenylethynyl groups: Preliminary mechanical properties, J. Appl. Polym. Sci. 28:355 (1983).
3686. C. L. Hendricks, S. G. Hill, J. N. Hale, and W. G. Dumars, Evaluation of high temperature structural adhesives for extended service, Phase V. NASA Cr 178176 CA, NASA, 1987; CA Vol. 110, 155627.
3687. S. J. Havens, F. W. Harris, and P. M. Hergenrother, Polyphenylquinoxalines containing alkylenedioxy groups, J. Appl. Polym. Sci. 32:5957 (1986).
3688. J. D. Allen and C. I. Yates, Preparation and characterization of high temperature syntactic foams, 19th Natl. SAMPE Symp. Exhib. 19:42 (1974).
3689. C. L. Segal, 19th Natl. SAMPE Symp. Exhib. 19:51 (1974).
3690. B. G. Kimmel, 19th Natl. SAMPE Symp. Exhib. 19:55 (1974).
3691. A. M. Usmani, Environmental considerations unique to sealants, Adhesives and Sealants, Engineered Materials Handbook, ASM, Materials Park, OH, Vol. 3, pp. 673–679 (1990).
3692. V. R. Foster, Polymers in caulk and sealant materials, J. Chem. Ed. 64(10):861 (Oct. 1987).
3693. J. D. Minford, unpublished work, Aluminum Co. of America.
3694. J. D. Minford, unpublished work, Aluminum Co. of America.
3695. J. D. Minford, unpublished work, Aluminum Co. of America.
3696. J. D. Minford, unpublished work, Aluminum Co. of America.
3697. J. D. Minford, unpublished work, Aluminum Co. of America.
3698. J. D. Minford, unpublished work, Aluminum Co. of America.

3699. J. D. Minford, unpublished work, Aluminum Co. of America.
3700. J. D. Minford, unpublished work, Aluminum Co. of America.
3701. R. H. Brown and L. A. Willey, Constitution of alloys, Aluminum (K. R. Van Horn, ed.), ASM, Metals Park, OH, Vol. 1, Chap. 2, pp. 31–54 (1967).
3702. W. A. Dean, Effects of alloying elements and impurities on properties, Aluminum (K. R. Van Horn, ed.), Metals Park, OH, Vol. 1, pp. 163–208 (1967).
3703. W. W. Binger, E. H. Hollingsworth, and D. O. Sprowls, Resistance to corrosion and stress corrosion, Aluminum (K. R. Van Horn, ed.), ASM, Metals Park, OH, Vol. 1, Chap. 7, pp. 209–276 (1967).
3704. W. E. Sicha, Properties of commercial casting alloys, Aluminum (K. R. Van Horn, ed.), ASM, Metals Park, OH, Vol. 1, Chap. 8, pp. 277–302 (1967).
3705. J. A. Nock, Jr., Properties of commercial wrought alloys, Aluminum (K. R. Van Horn, ed.), ASM, Metals Park, OH, Vol. 1, Chap. 9, pp. 303–336 (1967).
3706. J. P. Lyle, Jr., Properties of commercial wrought alloys, Aluminum (K. R. Van Horn, ed.), Metals Park, OH, Vol. 1, Chap. 10, pp. 337–358 (1967).
3707. J. D. Minford, unpublished work, Aluminum Co. of America.
3708. J. D. Minford, unpublished work, Aluminum Co. of America.
3709. J. D. Minford, unpublished work, Aluminum Co. of America.
3710. J. D. Minford, unpublished work, Aluminum Co. of America.
3711. J. D. Minford, unpublished work, Aluminum Co. of America.
3712. J. D. Minford, Durability of adhesive bonded aluminum joints, Treatise on Adhesion and Adhesives (R. L. Patrick, ed.), Marcel Dekker, New York, Vol. 3, Chap. 2, pp. 109–110 (1973).
3713. L. K. Peters and J. D. Minford, unpublished work, Aluminum Co. of America.
3714. J. D. Minford, unpublished work, Aluminum Co. of America.
3715. J. D. Minford, unpublished work, Aluminum Co. of America.
3716. J. D. Minford, unpublished work, Aluminum Co. of America.
3717. J. D. Minford, unpublished work, Aluminum Co. of America.
3718. J. D. Minford, unpublished work, Aluminum Co. of America.
3719. R. F. Wegman, W. M. Bodnar, E. S. Duda, and M. J. Bodnar, Adhes. Age (10):22 (Oct. 1967).
3720. J. D. Minford, unpublished work, Aluminum Co. of America.
3721. K. W. Allen and M. Smith, Paste forms of a cold chromate etch treatment for aluminum, J. Adhes. 18(3):167 (1985).
3722. H. M. Clearfield, D. K. McNamara, and G. D. Davis, Adherend surface preparation for structural adhesive bonding, Adhesive Bonding (L. H. Lee, ed.), Plenum Press, New York, Chap. 8, pp. 203–2327 (1991).
3723. W. Brockmann, The nature of adhesion mechanisms and the influence of surface treatment on the behavior of bonded joints, NATO AGARD Lecture Series No. 102, Air Force Materials Lab, Wright-Patterson AFB, OH (1979).
3724. S. Noviroz, J. L. Koenig, and H. Ishida, J. Adhes. 18:93 (1985).
3725. P. Walker, J. Coat Technol. 52:29 (1980).
3726. D. K. Shaffer, H. M. Clearfield, and J. S. Ahearn, Martin-Marietta Tech. Lab Report TR-86-76c (unpublished).
3727. J. D. Minford, unpublished work, Aluminum Co. of America.
3728. J. D. Minford, unpublished work, Aluminum Co. of America.
3729. P. A. Bijlmer, Chromic acid anodize process as used in Europe, Adhesive Bonding of Aluminum Alloys (E. W. Thrall and R. W. Shannon, eds.), Marcel Dekker, New York, Chap. 2, pp. 24–25 (1985).
3730. J. D. Minford, unpublished work, Aluminum Co. of America.
3731. D. J. Arrowsmith and A. W. Clifford, Int. J. Adhes. Adhes. 5:40 (1985).
3732. C. J. Amore and J. F. Murphy, Metal Finishing 63:50 (1965).
3733. A. J. Kinloch, Adhesion—6 (K. W. Allen, ed.), Applied Science Publishers, London, p. 95 (1982).
3734. J. S. Solomon, D. Hanlin, and N. T. McDevitt, Adhesion and Adsorption of Polymers (L. H. Lee, ed.), Plenum Press, New York, p. 103 (1980).
3735. D. M. Brewis, Aluminium adherends, Durability of Structural Adhesives (A. J. Kinloch, ed.), Applied Science Publishers, London, Chap. 5, p. 241 (1983).
3736. J. D. Minford and L. K. Peters, unpublished work, Aluminum Co. of America.
3737. J. D. Minford, unpublished work, Aluminum Co. of America.
3738. J. D. Minford, unpublished work, Aluminum Co. of America.
3739. The Tropical Durability of Metal Adhesives, RAE Tech. Note. CHEM 1349 (Feb. 1959).
3740. R. F. Wegman, W. M. Bodnar, and M. J. Bodnar, Effects of outdoor aging on unstressed, adhesive-bonded aluminum-to-aluminum lap shear joints, Part I—Report PA-TR-3304; Part II—Report PA-TR-3603; Part III—Report PA-TR-3699, Picatinny Arsenal, Dover, NJ (May 1968).

3741. C. I. Hause, W. C. Pagel, and A. G. McKown, The effect of south Florida weather aging on structural adhesive joints, 3M's Company Tech. Report (Oct. 1965).
3742. Effect of Outdoor Exposure on Stresses and Unstressed Bonded Joints, RAE Tech. Report 70081 (May 1970).
3743. G. Fedor and P. Brennan, Correlation of accelerated and natural weathering of sealants, Adhes. Age 33(5):22 (1990).
3744. S. L. Lehmann, Primers for structural adhesives, High Performance Adhesive Bonding (G. DeFrayne, ed.), Soc. Mfg. Eng., Dearborn, MI, pp. 36–54 (1983).
3745. G. DeFrayne (ed.), High Performance Adhesive Bonding, Soc. Mfg. Eng., Dearborn, MI (1983).
3746. S. L. Lehmann, C. L. Mahoney, and T. F. Mika, New Approaches to Corrosion Inhibition, SAMPE, 23:1078 (1978).
3747. C. J. Pedersen, J. Am. Chem. Soc. 89:7071 (1967).
3748. C. J. Pedersen, J. Am. Chem. Soc. 92:386 (1970).
3749. D. Bright and M. R. Truder, J. Chem. Soc. B:1545 (1970).
3750. F. Vogtle and E. Weber, Agnew. Chem. Int. Edt. 13(2):149 (1974).
3751. D. St. C. Black, Commun., p. 1004 (1968).
3752. E. M. Laboratories, Inc., Kryptofix 211, 221, and 222, E. Merck, Darmstadt, Germany.
3753. T. J. Reinhart, Novel concepts for priming metallic adherends for structural adhesive bonding, Adhesion—2 (K. W. Allen, ed.), Applied Science Publishers, London, p. 87 (1978).
3754. G. F. L. Ehlers, Correlation between structure and thermal stability of epoxy resins, Polymer 1:304 (1960).
3755. D. K. Klapprott and D. L. Paradis, U.S. Patent No. 3,678,130.
3756. K. K. Rice, C. L. Mahoney, D. K. Klapprott, S. L. Lehmann, Correlations of adhesive primer performance with chemical structure and bulk properties, SAMPE, 10:35 (1978).
3757. H. Van Hoorn, A dynamic mechanical study of the effect of chemical variations on the internal mobiltiy of linear epoxy resins (polyhydroxyethers), J. Appl. Polym. Sci. 12:871 (1968).
3758. J. D. Minford, unpublished work, Aluminum Co. of America.
3759. Nak-Ho. Sung, Moisture effects on adhesive joints, Adhesives and Sealants, Engineered Materials Handbook, ASM, Materials Park, OH, Vol. 3, pp. 622–627 (1990).
3760. J. D. Minford, unpublished work, Aluminum Co. of America.
3761. J. D. Minford, unpublished work, Aluminum Co. of America.
3762. S. Sterman and J. G. Marsden, Silane Coupling Agents, Industrial Engineering Chemistry 58:33 (1966).
3763. R. L. Patrick, The use of scanning electron microscopy, Treatise on Adhesion and Adhesives (R. L. Patrick, ed.), Marcel Dekker, New York, Vol. 3, Chap. 4, p. 218 (1973).
3764. F. J. Boerio, Coupling agents as adhesion promotors in adhesive bonding, Treatise on Adhesion and Adhesives (R. L. Patrick, ed.), Marcel Dekker, New York, Vol. 6, Chap. 7, pp. 262–267 (1989).
3765. A. J. Kinloch, Predicting and increasing durability of structural adhesive joints, Adhesion—3 (K. W. Allen, ed.), Applied Science Publishers, London, p. 1 (1978).
3766. E. J. Ripling, P. B. Crosley, W. D. Bascom, and E. Munley, The use of adhesives to replace welded connections in bridges, Proc. 20th SAMPE Int. Tech. Conf., Session C2 (Sept. 1988).
3767. E. J. Ripling, P. B. Crosley, and W. D. Bascom, Use of adhesives to replace welded connections in bridges, Contract DTFH 61-87-C-00029, U.S. Federal Highway Administration (1989).
3768. J. D. Minford, unpublished work, Aluminum Co. of America.
3769. G. F. Carter, Low-cost durability testing of stressed lap-shear adhesive joints, Adhes. Age 12(6):21 (June 1969).
3770. L. H. Sharpe, Some aspects of the permanence of adhesive joints, Appl. Polym. Symposia 3:353 (1966).
3771. H. D. Steffens and W. Brockmann, Aging resistance of light metal bonded joints using new surface treatment methods, Royal Aircraft Establishment Library Translation No. 1662 (1972).
3772. W. Althof and W. Brockmann, Adhes. Age 20(9):28 (1977).
3773. R. F. Wegman, Durability of adhesive bonded aluminum joints, Picatinny Arsenal Report No. 4169, Dover, NJ (June 1971).
3774. J. D. Minford, Durability of Adhesive Bonded Aluminum Joints, Treatise on Adhesion and Adhesives (R. L. Patrick, ed.), Marcel Dekker, New York, Vol. 3, Chap. 2, pp. 111–119 (1973).
3775. W. D. Bascom, Adhes. Age 22(4):28 (1979).
3776. J. A. Marceau and W. Scardino, Stevens Inst. Technol. Conf. (Aug. 1972).
3777. P. E. Erickson, A. Volpe, and F. R. Cooper, Modern Plastics (Aug. 1969).
3778. J. D. Minford, unpublished work, Aluminum Co. of America.
3779. J. D. Minford, unpublished work, Aluminum Co. of America.
3780. G. E. Nordmark, Fatigue performance of aluminum alloy joints for autobody applications, SAE Cong. Exhib. Meetings, Detroit, SAE Paper No. 780397 (Feb. 27–March 3, 1978).

3781. E. M. Vader, F. R. Hoch, and G. E. Nordmark, unpublished work, Aluminum Co. of America.
3782. J. D. Minford, unpublished work, Aluminum Co. of America.
3783. W. F. Marwick and P. G. Sheasby, Evaluation of adhesives for aluminum structured vehicles, S.A.E. Tech. Paper No. 870151, Int. Cong. & Expos. Meetings, Detroit, MI (Feb. 23–27, 1987).
3784. P. G. Selwood, F. J. Law, P. G. Sheasby, and M. J. Wheeler, The evaluation of an adhesively bonded aluminum structure in an Austin-Rover Metro Vehicle, S.A.E. Tech. Paper No. 870149, Int. Cong. & Expos. Meetings, Detroit, MI (Feb. 23–27, 1987).
3785. J. D. Minford and E. M. Vader, unpublished work, Aluminum Co. of America.
3786. J. D. Minford, unpublished work, Aluminum Co. of America.
3787. Adhesive Bonding of Aluminum Body Sheet Alloys, T-14 Report of The Aluminum Association (1st Ed. Nov. 1975).
3788. J. D. Minford, unpublished work, Aluminum Co. of America.
3789. J. D. Minford, Comparative effect of surface contamination on the strength and performance of aluminum spot-welded or adhesive-bonded joints, Technological Impact of Surfaces: Relationship to Forming, Welding, and Painting, ASM Conf. Proc. sponsored by ASM and ADDRG, Dearborn, MI (April 14–15, 1981); pp. 207–227 in ASM book copywrighted in 1982.
3790. J. D. Minford, unpublished work, Aluminum Co. of America.
3791. E. C. Kron, Adhesive bonding of die casting, 15th Ann. Natl. Lead Anal. & Phys. Testing Symp., Highstown, NJ (Oct. 15, 1965). (Reprinted from an Engineering Report by Doehler-Jarvis Div. National Lead, Toledo, OH).
3792. M. D. Anderson, Adhesive bonding of die-cast aluminum 380 alloy to 4140 steel, Picatinny Arsenal Report PA-TM-1502, Dover, NJ (Jan. 1965).
3793. R. W. Wieland, A comparison of chemical versus mechanical cleaning of aluminum sand castings, McDonnell Aircraft Report 32A-345 (Feb. 2, 1961).
3794. B. Manor, Adhesive bonding of die castings-a case history, Tech. Paper No. 104, 5th Natl. Die Casting Cong. Meeting, Detroit, MI (Nov. 4–7, 1968).
3795. I. Skeist, Handbook of Adhesives, 2nd ed., Van Nostrand Reinhold, New York (1977).
3796. J. D. Minford, unpublished work, Aluminum Co. of America.
3797. J. D. Minford, unpublished work, Aluminum Co. of America.
3798. J. D. Minford, unpublished work, Aluminum Co. of America.
3799. B. H. River, C. B. Vick, and R. H. Gillespie, Wood as an adherend, Treatise on Adhesion and Adhesives (J. D. Minford, ed.), Marcel Dekker, New York, Vol. 7, Chap. 1, pp. 6–19 (1991).
3800. H. Mizumachi, Activation energy of the curing reaction of urea resin in the presence of wood, Wood Sci. 6(1):14 (1973).
3801. D. Narayanamurti, R. C. Gupta, and G. M. Verna, Influence of extractives in the setting of adhesives, Holzforschung und Holzverwertung 14(5/6):85 (1962).
3802. J. D. Minford, unpublished work, Aluminum Co. of America.
3803. H. Baumann, Glues and Contact Adhesives, Springer-Verlag, Berlin (1967).
3804. G. Giereng, Adhesive bonding of thin sheet-metal structures with PVC plastisols, notice of Deutsche Forschungsgesellschroft fur Blechverarbertung und Oberflochenbehandlung EV., 19(7):95 (1968).
3805. Natural Joining with Adhesives, Publication of 3M's Co., St. Paul, MN, USA.
3806. K. Ruhsland and B. Winkler, Process for making high-grade adhesive bends without surface treatment of bending parts, GDR Patent WP 112 658; C 09 J, 5/02.
3807. S. Semerdfrev and P. Pavov, On adhesive bonding of metal parts on unclean surfaces, Paper at meeting for experts for adhesive and casting resin technology, Halle, Germany (1969).
3808. K. Ruhsland, Possibilities of integrating surface treatment of bonding parts in the adhesive bonding process (K. L. Mittal, ed.), Plenum Press, New York, pp. 257–266 (1984).
3809. R. A. Rosty, R. G. Wegman, K. M. Adelson, E. A. Garvis, and D. W. Levi, U.S. Army Armament Res. & Devel. Center, Tech. Report ARSCE-TR-83011, Dover, NJ (1983).
3810. M. Debski, M. E. R. Shanahan, and J. Schultz, Int. J. Adhes. Adhes. 6(3):145 (1985).
3811. D. O. Bowen and C. L. Volkmann, 17th Natl. SAMPE Tech. Conf., p. 532 (1985).
3812. R. Puttocher and K. Starkow, East German Patent DD-222615.
3813. Kokar Tokkyo Koho, Japanese Patent No. JP60-106827 (1985).
3814. P. Canmeicon and J. P. Wightman, J. Adhes. 22:13 (1987).
3815. T. V. Natale, Automotive Eng. 96(7):23 (1988).
3816. K. Ruhsland and B. Winkler, Adhesive bonding of metals without surface treatment of bonding parts, Adhasion 21(1):6 (1977).
3817. K. Ruhsland, FV adhesive bonding technique and its possibilities of application, Fertigungstechnik und Betrab 28(5):296 (1978).

3818. K. Ruhsland, Results of application-technical testing of Epasol FV/Z15 939, Z15 Notice 21(10):1070 (1979).
3819. K. Ruhsland, Comparison of properties of oily metal adhesive bonds, Z15 Notice 22(10):1172 (1980).
3820. K. Ruhsland, Vibrational adhesive bonding, Adhasion 23(6):184 (1979).
3821. Coated valves fight corrosion, Adhes. Age 3(11):30 (1960).
3822. A. W. Niconchuk, Testing adhesives for the vulcanized-in-sole process, Adhes. Age 4(3):32 (1961).
3823. S. Yurenka, How to test structural adhesives, Adhes. Age 4(11):34 (1961).
3824. R. C. Burrows and B. F. Crowe, Bonding contaminated surfaces, Adhes. Age 5(5):30 (1962).
3825. R. B. Gosnell, M. B. Smith, and S. E. Susman, Structural variations of polyurethanes: Their effect on metal adhesion at various testing temperatures, Adhes. Age 5(9):32 (1962).
3826. P-S label withstands acid washings, Adhes. Age 5(10):29 (1962).
3827. R. F. Wegman and W. C. Tanner, Effects upon the strength of epoxy adhesives when stressed to failure in miliseconds, Adhes. Age 8(9):32 (1965).
3828. J. Bjorksten, Preventing new product failure, Adhes. Age 8(12):33 (1965).
3829. C. F. Donovan, How to test the strength of bonded joints, Adhes. Age 9(3):24 (1966).
3830. Resin-based adhesive helps coat rivets with anti-corrosion liquid, Adhes. Age 9(7):34 (1966).
3831. J. A. Carlson and L. P. Sapetta, Stresses in assemblies bonded with thermosetting adhesives, Adhes. Age. 10(12):26 (1967).
3832. R. F. Wegman, W. M. Bodnar, E. S. Duda, and M. J. Bodnar, How weathering and aging affect bonded aluminum, Adhes. Age 10(10):22 (1967).
3833. Epoxy resurfacing eliminates corrosion, Adhes. Age 10(10):37 (1967).
3834. C. F. Carter, Low cost durability testing of stressed lap-shear adhesive joints, Adhes. Age 12(6):21 (1969).
3835. Developing testing standards for building joint sealants: The job of ASTM Committee C-24, Adhes. Age 16(11):47 (1973).
3836. W. Althof and W. Brockmann, New test method for the prediction of environmental resistance of adhesive bonded joints, Adhes. Age 20(9):27 (1977).
3837. J. A. Marceau, Y. Moji, and J. C. McMillan, A wedge test for evaluating adhesive-bonded surface durability, Adhes. Age 20(10):28 (1977).
3838. J. D. Minford, Durability of structural adhesive bonded aluminum joints, Adhes. Age 21(3):17 (1978).
3839. J. A. Marceau, J. C. McMillan, and W. M. Scardino, Cyclic stress testing of adhesive bonds, Adhes. Age 21(4):37 (1978).
3840. R. F. Wegman, M. J. Bodnar, and M. C. Ross, A new technique for assessing durability of structural adhesives, Adhes. Age 21(7):38 (1978).
3841. L. C. Jackson, Improving adhesion by gas-plasma contaminant removal, Adhes. Age 21(9):34 (1978).
3842. W. D. Bascom, Stress corrosion of structural adhesive bonds, Adhes. Age 22(4):28 (1979).
3843. J. L. Rutherford and E. J. Hughes, Creep in adhesive-bonded metal joints, Adhes. Age 22(11):55 (1979).
3844. T. E. Baker, G. L. Fix, P. Granas, and J. S. Judge, Characterization of a hot-melt adhesive system for severe environments, Adhes. Age 23(2):25 (1980).
3845. D. F. Clark, Modifying pressure-sensitive test for weather stripping, Adhes. Age 25(3):47 (1982).
3846. S. K. Joneja and G. M. Newaz, Evaluating SMC bonds using a wedge test, Adhes. Age 28(11):18 (1985).
3847. L. E. Rantz, Proper surface preparation: bonding's critical first step, Adhes. Age 30(7):10 (1987).
3848. G. S. Haviland, Challenges and opportunities in adhesive engineering, Adhes. Age 30(7):10 (1987).
3849. R. Lovald, Influencing adhesive evolution: developments on substrates, Adhes. Age 30(7):18 (1987).
3850. J. P. Wightman, Surface analysis examines fundamental adhesion questions, Adhes. Age 30(9):30 (1987).
3851. G. Fedor and P. Brennan, Correlation of accelerated and natural weathering of sealants, Adhes. Age 33(5):22 (1990).
3852. T. J. Silva and S. Spindel, Testing elastomeric sealants to predict performance, Adhes. Age 34(6):12 (1991).
3853. Adhesives and Sealants, Engineered Materials Handbook, ASM, Materials Park, OH, Vol. 3 (1990).
3854. L. H. Sharpe, Glossary of terms, Adhesives and Sealants, Engineered Materials Handbook, ASM, Materials Park, OH, Vol. 3, pp. 3–32 (1990).
3855. J. Hagquist, F. K. Meyer, and S. K. M. Swanson, Adhesives markets and applications, Adhesives and Sealants, Engineered Materials Handbook, ASM, Materials Park, OH, Vol. 3, pp. 44–55 (1990).
3856. D. A. Khardashov (ed.), Bonding Agents and the Technology of Adhesive Bonding, Wright-Patterson AFB, OH, Report FDT-TT-62-1617/1 translation (June 6, 1963).
3857. H. R. Merriman, Selecting adhesives for structural metal bonding, Adhes. Age 7(8):22 (1964).
3858. J. D. Minford, Adhesive bonding, Aluminum (K. R. Van Horn, ed.), ASM, Metals Park, OH, Vol. 3, Chap. 15, pp. 549–574 (1967).

3859. R.T. McIntyre, F. L. Duer, and E. A. Gallagher, Effect of varying process parameters in the fabrication of adhesive bonded structures, Part VI: Production methods, Picatinny Arsenal Report No. 4162, Dover, NJ (Contrast No. DAAA-21-70C-0220).
3860. Design guide—assembling with adhesives, Machine Design, p. 1 (Aug. 18, 1966).
3861. F. E. Hall, Bonding aluminum parts, Prod. Eng. (Feb. 28, 1966).
3862. D. J. Conway, Metal-to-metal adhesives for structural applications at elevated temperature, Adhes. Age 11(5):30 (1968).
3863. N. J. DeLollis, Adhesives for Metals—Theory and Technology, Industrial Press, New York (1970).
3864. D. K. Rider, Principles and applications of adhesives—a prototype module, Educational Modules for mater. Sci. & Eng. (Aug. 1978). Reprinted from Mater. Technol., Vol. II of a set of texts on Physical Design for Electronic Systems, Copyright 1970 by Bell Laboratories.
3865. P. L. Layman, Chem. Eng. News, p. 8 (Nov. 1, 1982).
3866. H. Ellerhorst, Jr., Industrial adhesive markets—a profile for the 1080's, Adhes. Age 25(1):42 (1982).
3867. J. D. Bittence, Adhesives update, Mater. Eng., p. 34 (Aug. 1983).
3868. A. Brief and I. Skeist, The adhesives industry—25 years of growth, Adhes. Age 26(6):12 (1983).
3869. M. Giudice, Adhesive and sealant sales to hit $26.5 billion by 1998, Adhes. Age 33(7):28 (1990).
3870. G. S. Haviland, Manufacturing aspects of six adhesive cure systems, SME Tech. Paper AD 76-666, Soc. Mfg. Eng., Dearborn, MI (1976).
3871. J. D. Minford, unpublished work, Aluminum Co. of America.
3872. J. A. Marshall et al., Research on the application of electrical curing to heat resistant structural adhesives, Report AFML TR-67-202, Brunswick Corp., Marion, VA (Aug. 1967).
3873. J. D. Minford, unpublished work, Aluminum Co. of America.
3874. W. Althof, Strength of heat-resistant metal-to-metal bonded joints by combining two adhesives, Report DLB-FB-73-22, Inst. for Aircraft Construction, Braunschweig, Germany (1973).
3875. W. Althof, Increase in strength of heat-resistant adhesive bonded metal joints by combining two adhesives, Tech. Translation ESRO TT-45, Inst. for Aircraft Construction, Braunschweig, Germany (1974).
3876. A. Chapiro, Radiation Chemistry of Polymeric Systems, Interscience, New York (1961).
3877. C. B. Rybny, J. C. Trevellas, and J. A. Vona, New materials for radiation curable coatings, Paper FC 75-304, SME 2nd Int. Rad. Curing Conf., Cincinnati, OH (May 6–8, 1975).
3878. R. B. Mesrobian, Year of progress for radiation curing, Paper FC 75-301, SME 2nd Int. Rad. Curing Conf., Cincinnati, OH (May 6–8, 1975).
3879. V. D. McGinniss, Photoinitiation of acrylate systems for UV curing, Paper FC 76-486, SME 3rd Int. Rad. Conf., Rosemont, IL (Sept. 1976).
3880. S. V. Nablo et al., Electron beam processor technology, Nonpolluting Coatings and Coating Procedures (J. L. Gardon and J. W. Prane, eds.), Plenum Press, New York (1973).
3881. C. R. Hoffman, Shielding and safety requirements for electron accelerator systems, Radiation Phys. Chem. 9:131 (1977).
3882. Occupational Safety and Health Standard, Federal Register 37, No. 202, Part II, 21158 (Oct. 1972).
3883. S. V. Nablo and E. P. Tripp III, Electron curing of adhesives and coatings, Adhes. Age 22(3):24 (1979).
3884. S. V. Nablo and E. P. Tripp III, Electron curing of adhesives and coating, High Performance Adhesive Bonding (G. DeFrayne, ed.), Society Manufacturing Engineers, Dearborn, Michigan (1983).
3885. R. Dowlenko et al., Radiation polymerization for pressure-sensitive adhesives, CHEMTECH p. 539 (Sept. 1974).
3886. H. Chu, W. Oraby, E. Bettencourt, and W. K. Walsh, Radiation curable binders for non-wovens: structure property relationships, Proc. Tech. Symp. Non-Woven Tech., INDA, New York, pp. 107–123 (1977).
3887. C. Bluestein and M. S. Cohen, Visible light curables offer dispensing and exposure pluses, Adhes. Age 29(9):42 (1986).
3888. J. D. Delmonte and R. Fullerton, Influence of application variables on properties of an epoxy adhesive, J. Appl. Polym. Sci., Appl. Polym. Symp. 3:397 (1966).
3889. P. M. Cocanour and G. C. Anderson, Non-woven webs as thermal-bonding adhesives, Adhes. Age 23(10):25 (1980).
3890. G. S. Haviland, Automating adhesive joining, Loctite Corp., Newington, CT.
3891. W. Devlin, Metering and mixing equipment, Adhesives and Sealants, Engineered Materials Handbook, ASM, Materials Park, OH, Vol. 3, pp. 687–692 (1990).
3892. M. Petterborg, Barrier/injection kits reduce costs and production time, Adhes. Age (July 1988).
3893. J. Drake, Impact of encapsulation compounds on the selection of dispensing equipment, Electr. Manuf. (Nov. 1989).
3894. J. E. DeVries, Dispensing and application equipment for adhesives and sealants, Adhesives and Sealants, Engineered Materials Handbook, ASM, Materials Park, OH, Vol. 3, pp. 693–702 (1990).

3895. H. J. Borstell and V. Wheeler, Adhesive bonding preparation, application, and tooling, Adhesives and Sealants, Engineered Materials Handbook, ASM, Materials Park, OH, Vol. 3, pp. 703–708 (1990).
3896. T. Taricco and M. Moulding, Adhesives and sealant curing and cure controls, Adhesives and Sealants, Engineered Materials Handbook, ASM, Materials Park, OH, Vol. 3, pp. 709–715 (1990).
3897. H. Turner, Automation and robotics for adhesives and sealants, Adhesives and Sealants, Engineered Materials Handbook, ASM, Materials Park, OH, Vol. 3, pp. 716–725 (1990).
3898. B. L. Davies, H. Rozban, and A. K. Forrest, Use of robots and adaptive control in automated dispensing of adhesives, Adhesion '90, Plastics and Rubber Inst. Meetings, Cambridge, England (1990).
3899. N. A. DeBruyne and R. Houwink, Adhesion and Adhesives, Elsevier, New York, p. 413 (1951).
3900. Chlorination of Elastomeric Products, Report BL-305, E. I. DuPont de Nemours & Co.
3901. J. H. Gerstenmaier, Rubber Age 73:495 (1953).
3902. D. M. Alstadt, Rubber World, 133(2):221 (1955).
3903. S. L. Brams, Rubber World 137:888 (1958).
3904. S. Buchan, Rubber-to-Metal Bonding, Palmerton, New York (1959).
3905. G. W. Painter, Adhes. Age 3(2):36 (1960).
3906. W. M. DeGrease, Rubber Age 87:1013 (1960).
3907. J. W. Gallagher, Rubber World 144:75 (1961).
3908. J. D. Hutchison, Elastomer bonding: a guide to material selection, processing, and troubleshooting, Elastomerics p. 35 (April 1978).
3909. J. W. Gallagher, Rubber-to-metal bonding, Adhes. Age 11(1):28 (1968).
3910. E. J. Spearman, Evaluation of rubber-to-metal bonding agents, Report RIA-68-1558, Army Weapons Command Sci. & Technol. Lab., Rock Island, IL (June 1968).
3911. K. Bond, Rubber-to-metal waterborne eliminate emissions and odors, Adhes. Age 33(2):22 (1990).
3912. R. Y. Ting, A study on elastomer/metal bonds applicable in underwater sonar systems, Adhesive Joints: Formation, Characteristics, and Testing (K. L. Mittal, ed.), Plenum Press, New York, pp. 555–564 (1984).
3913. E. Cutts, Developments in Adhesives—2 (A. J. Kinloch, ed.), Applied Science Publishers, London, pp. 367–404 (1981).
3914. C. H. Peterson, Adhes. Age 7:30 (1964).
3915. C. H. Peterson, Adhesives for bonding rubber to various metals, High Performance Adhesive Bonding (G. DeFrayne, ed.), SME, Dearborn, MI (1983).
3916. H. B. Reed, High frequency curing of adhesives, Adhes. Age 4(3):38 (1961).
3917. E. W. Jones, Some tips on production line handling of contact adhesives, Adhes. Age 5(8):24 (1962).
3918. Improving bonding accuracy with thermistor controllers, Adhes. Age 5(9):30 (1962).
3919. A. B. Klinetob, Proper care of spreader reduces gluing problems, Adhes. Age 9(1):25 (1966).
3920. Methods of bonding curved-form parts, Adhes. Age 8(11):35 (1965).
3921. How to prevent solvent hazards in adhesive plants, Adhes. Age 10(7):34 (1967).
3922. F. V. Sebbard, Mishandling of adhesives reduces machineability, Adhes. Age 11(8):35 (1968).
3923. Infrared oven solves laminating problems, Adhes. Age 11(1):32 (1968).
3924. Application equipment for anaerobic adhesives, Adhes. Age 15(5):44 (1972).
3925. Application equipment for adhesives and sealants-a survey of the manufacturers, Adhes. Age 16(5):22 (1973).
3926. B. P. Spearman and J. D. Hutchison, Post vulcanizing bonding concepts, Adhes. Age 17(4):24 (1974).
3927. A. B. Reighard, Hot-melt labeling equipment-present and future, Adhes. Age 17(5):31 (1974).
3928. C. W. Jennings, Two techniques for characterizing surfaces for adhesive bonding, Adhes. Age 20(2):29 (1977).
3929. N. R. Roobol, D. C. Hutchison, J. W. Baldwin, and J. L. Mand, Determining airless sprayability of adhesives, Adhes. Age 21(6):39 (1978).
3930. C. Hulstein, Automatic application of anaerobic adhesives, Adhes. Age 22(7):25 (1979).
3931. L. A. Mushel, Laboratory and pilot coater/laminators, Adhes. Age 22(10):41 (1979).
3932. Whats new in machinery and equipment, Adhes. Age 23(5):18 (1980).
3933. F. C. Herot, Engineered dispensing stations for adhesives, Adhes. Age 23(5):23 (1980).
3934. Whats new in machinery and equipment, Adhes. Age 24(5):24 (1981).
3935. Whats new in machinery and equipment, Adhes. Age 25(5):23 (1982).
3936. R. Mosher, The rectilinear system of robot dispensing, Adhes. Age 25(5):30 (1982).
3937. K. A. Jacobs, Adhesive application equipment for two-part reactive materials, Adhes. Age 25(5):35 (1982).
3938. An approach to process control of adhesive bonding, Adhes. Age 26(2):17 (1983).
3939. Whats new in machinery and equipment, Adhes. Age 26(3):13 (1983).

3940. N. Dueweke, Robotics and adhesives—an overview, Adhes. Age 26(4):11 (1983).
3941. D. Morat, Industrial bonding and automation in France, Adhes. Age 27(4):18 (1984).
3942. Whats new in machinery and equipment, Adhes. Age 28(3):19 (1985).
3943. Whats new in machinery and equipment, Adhes. Age 29(3):27 (1986).
3944. Whats new in machinery and equipment, Adhes. Age 30(3):20 (1987).
3945. Company halves materials costs using foamed HM technology, Adhes. Age 29(12):30 (1986).
3946. Dispenser uses less adhesive to produce improved product (lipstick case application), Adhes. Age 30(9):28 (1987).
3947. Visible light curing provides low shrinkage and good depth, Adhes. Age 31(4):20 (1988).
3948. H. N. Cheng, Using analytical techniques to characterize adhesives, Adhes. Age 31(12):37 (1988).
3949. Whats new in machinery and equipment, Adhes. Age 32(3):20 (1989).
3950. S. Kingsley, Dispenser eliminates solvents and improves profitability, Adhes. Age 32(11):36 (1989).
3951. C. Elliott, T. Millard, and P. Wing, Hybrid technology fulfills adhesive application needs, Adhes. Age 33(8):22 (1990).
3952. Versatile coater/laminators aid in two-ply production, Adhes. Age 33(11):34 (1990).
3953. Automated dispenser increases production and reduces rework, Adhes. Age 33(11):32 (1990).
3954. Whats new in machinery and equipment, Adhes. Age 34(3):24 (1991).
3955. K. McGinnis and G. Benham, Hand-held dispensing equipment reduces waste and saves cost, Adhes. Age 34(11):18 (1991).
3956. K. W. Carroll, Selecting a high production adhesive system, Adhes. Age 9(7):20 (1966).
3957. F. T. Wooten, Adhesives for coupling large strains, Adhes. Age 12(3):30 (1969).
3958. B. Gould, Guide to adhesives selection with a pictoral checklist, Adhes. Age 16(10):19 (1973).
3959. D. M. Weggemans, Adhesives selection charts, Adhes. Age 16(10):36 (1973).
3960. Checklist for adhesive selection, Adhes. Age 16(10):27 (1973).
3961. R. L. Downey, Sprayable adhesive systems, Adhes. Age 17(3):35 (1974).
3962. Adhesives replace fasteners in shipping containers, Adhes. Age 17(2):32 (1974).
3963. G. M. Green, Effectiveness of cell edge adhesives, Adhes. Age 17(3):28 (1974).
3964. C. L. Mahoney, Structural adhesives for rapid-cure applications—Part I, Adhes. Age 22(10):22 (1979).
3965. C. L. Mahoney, Structural adhesives for rapid-cure applications—Part II, Adhes. Age 22(12):26 (1979).
3966. W. A. Lees, Selecting adhesives by computer, Adhes. Age 26(3):20 (1983).
3967. The adhesives industry-25 years into the future, Adhes. Age 26(6):19 (1983).
3968. W. A. Lees, Designing and producing toughened structural adhesives, Adhes. Age 27(11):26 (1984).
3969. R. D. Hermansen and S. A. Tunick, Formulating custom-tailored thermal transfer adhesives, Adhes. Age 32(11):38 (1989).
3970. B. D. Ludbrook, High performance adhesives meet many industry needs, Adhes. Age 33(12):25 (1990).
3971. G. M. LeFave, F. Y. Hayashi, and R. Gamero, One-part polysulfide sealants, Adhes. Age 5(10):34 (1962).
3972. J. Zakim and M. Shihadeh, A comparative guide to sealants and caulking compounds, Adhes. Age 8(8):19 (1965).
3973. C. S. Peterson, The sealant jungle, Adhes. Age 10(6):30 (1967).
3974. Silicone building sealant handles cold weather, Adhes. Age 17(11):18 (1974).
3975. A. T. Devine, Sealants: a comparative evaluation of performance in two typical joint configurations, Adhes. Age 17(11):37 (1974).
3976. S. A. Brady, Silicone sealants for solar energy systems, Adhes. Age 20(11):36 (1977).
3977. S. A. Brady, Silicone sealants, types available, and typical uses, Adhes. Age 21(11):33 (1978).
3978. J. Brower, Silicone sealants, general properties and application details, Adhes. Age 21(11):36 (1978).
3979. Sealants: New product and application profiles, Adhes. Age 22(11):47 (1979).
3980. C. E. Oxley, Elastomers in sealants, Adhes. Age 23(2):33 (1980).
3981. K. Kishita, Liquid sealants as gasket materials, Adhes. Age 23(11):53 (1980).
3982. Extruded PVC foam sealant yields six-figure savings, Adhes. Age 32(5):26 (1989).
3983. F. J. Modic, Silicone pressure-sensitive adhesives: their properties and applications, Adhes. Age 5(12):36 (1962).
3984. C. Wangman, Pressure sensitive tapes: types and applications, Adhes. Age 20(9):23 (1977).
3985. J. Miron and I. Skeist, Trends in pressure-sensitive and heat-seal materials, Adhes. Age 21(1):35 (1978).
3986. J. T. Hayes, Water-based acrylic PSA's, Adhes. Age 22(3):34 (1979).
3987. D. F. Merrill, Silicone PSA's: types, properties, and uses, Adhes. Age 22(3):39 (1979).
3988. L. M. Kegley, Future trends for PSA's, Adhes. Age 23(3):35 (1980).
3989. Solvent acrylic PSA's resist high temperature and humidity, Adhes. Age 30(7):17 (1987).

3990. L. A. Sobrieski and T. J. Tangney, Silcone PSA's perform well at high and low temperatures, Adhes. Age 31(13):23 (1988).
3991. A. Zosel, Physical properties and adhesion performance of PSA's, Adhes. Age 32(11):42 (1989).
3992. H. F. Carlson, Bonding with hot-melts, Adhes. Age 5(11):32 (1962).
3993. R. Bucksot, Hot-melt adhesives lower costs of pallet stabilization, Adhes. Age 20(8):34 (1977).
3994. R. E. Duncan and J. E. Bergerhouse, EVA and VAE copolymers for hot-melt PSA's, Adhes. Age 23(3): 37 (1980).
3995. P. Borg and J. Boutillier, Hot-melt copolymers exhibit high cohesion and compatibility, Adhes. Age 29(8):31 (1986).
3996. E. McBride, Ethylene terpolymer expands hot-melt technology to PVC, Adhes. Age 34(8):20 (1991).
3997. T. R. McClellan and K. W. Rausch, Polyurethane sealants: Some improved system, Adhes. Age 9(11): 23 (1966).
3998. S. Trippe and J. Burke, PU adheres many substrates offers production flexibility, Adhes. Age 33(7):18 (1990).
3999. D. J. Kelly, Neoprene contact adhesives: some special advantages and production application, Adhes. Age 5(2):22 (1062).
4000. R. A. Martin, Neoprene-phenolic adhesives, Adhes. Age 9(5):28 (1966).
4001. S. Kiriyama, Metal-containing epoxy resins as adhesives, Adhes. Age 12(7):18 (1969).
4002. M. Plonchak, Applications grow for anaerobic adhesives and sealants, Adhes. Age 15(11):45 (1972).
4003. Epoxy-based materials-an aid to OSHA compliance, Adhes. Age 16(2):28 (1973).
4004. H. N. Vazirani, Flexible epoxy resins, Adhes. Age 23(10):31 (1980).
4005. P. Briggs, G. Rose, K. Rzetclny, and G. T. Sivy, Structural methacrylates, yield process, performance benefits, Adhes. Age 34(7):17 (1991).
4006. Leather to aluminum bonding, Adhes. Age 4(3):42 (1961).
4007. Plastic letters bonded to metal station front, Adhes. Age 4(12):26 (1961).
4008. G. Epstein and S. Litvak, The influence of some selected fabrication variables on structural metal-to-metal bonding, Adhes. Age 4(6):36 (1961).
4009. J. P. Bush and J. A. Scott, Epoxy adhesive bonds extruded sections of aluminum light pole, Adhes. Age 4(8):23 (1961).
4010. Bonded metal mural, Adhes. Age 5(5):34 (1962).
4011. Epoxy adhesive speeds production of bonded glass-aluminum assemblies, Adhes. Age 5(7):34 (1962).
4012. Aluminum and fiberglass faced plywood panels, Adhes. Age 5(12):30 (1962).
4013. L. C. Jackson, Bonding plastic laminates onto metals, Adhes. Age 5(1):30 (1962).
4014. J. H. Valentine, How to assure successful metal and glass joints, Adhes. Age 5(12):26 (1962).
4015. Diffusion process offers new ways to bond metals (adhering aluminum parts), Adhes. Age 9(5):37 (1966).
4016. D. M. Preiss, Bonding polyurethane to metal, Adhes. Age 9(4):22 (1966).
4017. S. J. Dastin, Bonded beryllium structures, Adhes. Age 9(5):24 (1966).
4018. A. F. Lewis, Post-formable adhesively bonded metal laminates, Adhes. Age 10(5):20 (1967).
4019. S. Y. Yoshino, M. A. Nadler, and D. H. Richter, Fabrication methods for PBI adhesive beryllium sandwich structures, Adhes. Age 10(8):26 (1967).
4020. Laminating method for metal panels allows post-forming, Adhes. Age 11(6):35 (1968).
4021. S. Litvak, Polybenimidazole adhesives for bonding stainless steel, beryllium, and titanium alloys, Adhes. Age 11(1):17 (1968).
4022. D. J. Conway, Metal-to-metal adhesives for structural applications at elevated temperatures, Adhes. Age 11(5):30 (1968).
4023. H. E. Rowe and S. F. Radtke, How to join lead with adhesives, Adhes. Age 11(9):22 (1968).
4024. T. F. Duncan, Adhesives and methods for bonding metal and plastics to porous substrates, Adhes. Age 12(4):24 (1969).
4025. Vacuum impregnation for sealing porosity of powdered metal parts, Adhes. Age 16(3):33 (1973).
4026. D. Fields, Summary of the weldbonding process, Adhes. Age 16(9):41 (1973).
4027. F. P. Darmory, Extreme high temperature polyimide adhesive for bonding titanium and stainless steel, Adhes. Age 17(3):22 (1974).
4028. R. D. Dexheimer and L. R. Vertnik, Metal bonding with polyamide hot-melt adhesives, Adhes. Age 17(8):31 (1974).
4029. W. D. Bascom, The surface chemistry of bonding metals with polymer adhesives, Adhes. Age 17(10):25 (1974).
4030. J. J. Bethke and S. J. Ketcharn, Polysulfide sealants for corrosion protection of spot-welded aluminum alloys, Adhes. Age 17(11):29 (1974).

4031. R. D. Paul and J. McGivern, Electrochemical characterization and control of titanium surfaces for adhesive bonding, Adhes. Age 17(12):41 (1974).
4032. W. Brockmann, Interface reactions and their influence on the long-term properties of metal bonds, Adhes. Age 20(5):30 (1977).
4033. R. W. Vaughan, C. H. Sheppard, and R. Baucom, Polyimide adhesives for weld-bonding titanium, Adhes. Age 20(7):19 (1977).
4034. R. V. Wolff, Effects of moisture on mean strength of bonded composite-to-metal joints, Adhes. Age 21(6):33 (1978).
4035. O. Ishai and T. Girshegorn, Strength of bonded aluminum-CFRP single lap joints, Adhes. Age 21(7):25 (1978).
4036. Relaxation time, bond strength of epoxy resins to aluminum, Adhes. Age 26(2):25 (1983).
4037. A. R. Bunk, M. A. Roe, and M. Luttinger, Epoxies for residential copper plumbing, Adhes. Age 27(13): 21 (1984).
4038. Tape reduces production costs and aluminum skin marring, Adhes. Age 30(13):21 (1984).
4039. H. Muller, Strong flexible polyester HM's bond metal-to-metal systems, Adhes. Age 33(12):32 (1990).
4040. J. Osterndorf and R. B. Bonk, Evaluating adhesives for aluminum-to-copper bonding, Adhes. Age 34(7):24 (1991).
4041. Synthetic rubber compounds seal plastic skylights, Adhes. Age 4(10):37 (1961).
4042. Bonded vinyl patches, Adhes. Age 4(12):25 (1961).
4043. V. Vohralik, Vacuum covering with adhesive coated PVC, Adhes. Age 5(3):30 (1962).
4044. H. B. Russell, Bonding Teflon FEP film with conventional adhesives, Adhes. Age 5(4):29 (1962).
4045. M. St. Cyr, Methods of bonding fluorocarbon plastics to structural materials, Adhes. Age 5(8):31 (1962).
4046. Bonding Tedlar to steel with acrylic adhesives, Adhes. Age 5(10):38 (1962).
4047. Better adhesion to plastic bottles with resin adhesive films, Adhes. Age 9(3):32 (1966).
4048. How to safely bond facing panels of ABS plastic to sensitive styrene foam, Adhes. Age 9(8):26 (1966).
4049. D. Debnar, How to bond decorative laminates to wood, Adhes. Age 10(9):35 (1967).
4050. J. Delmonte, Bonding thermoplastics with improved urethanes, Adhes. Age 11(4):27 (1968).
4051. A. T. Devine and M. J. Bodnar, Effect of various surface treatments on adhesive bonding of polyethylene, Adhes. Age 12(5):35 (1969).
4052. H. W. Rauhut, Pretreating polyethylene for optimum structural adhesive joints, Adhes. Age 12(12):28 (1969).
4053. R. M. Lerner, Plasma treatment of Delrin for improved adhesive bonding, Adhes. Age 12(12):35 (1969).
4054. R. L. Bersin, How to obtain strong bonds via plasma treatment, Adhes. Age 15(3):37 (1972).
4055. R. F. Karle, Adhesive bonding PVC in water demineralizing systems, Adhes. Age 15(12):34 (1972).
4056. J. Panek, Polysulfide sealants for plastics, Adhes. Age 16(11):32 (1973).
4057. R. A. Bragole, Adhesive bonding polyolcfins, Adhes. Age 17(4):24 (1974).
4058. J. C. Smith, Silicone adhesives for joining plastics, Adhes. Age 17(6):27 (1974).
4059. E. M. Petrie, Joining the "Engineering Plastics" how adhesives compare with other bonding methods, Adhes. Age 23(8):14 (1980).
4060. R. A. Lombardi, Reactive acrylic polymer gives better plastic film adhesion, Adhes. Age 30(2):18 (1987).
4061. E. W. Kuenzi and W. S. Erickson, Shear stability of flat panels or sandwich construction, Forest Products Lab, Madison, WI, Report No. 1560 (1955).
4062. M. E. Reville, Deflection and stresses in a uniformly loaded, simply supported, rectangular sandwich plate, Forest Products Lab., Madison, WI, Report No. 1847 (Dec. 1955).
4063. C. B. Norris, Compressive buckling design curves for sandwich panels with isotropic facings and orthotropic cores, Forest Products Lab., Madison, WI, Report No. 1854 (Feb. 1956).
4064. G. Epstein, Adhesive Bonding of Metals, Reinhold, New York (1954).
4065. H. F. Hardroth et al., Fatigue crack propagation in aluminum alloy box beams, Tech. Note No. 3856, Natl. Advisory Council in Aeronautics (1956).
4066. F. H. Blair, Space age adhesives, Adhes. Age 2:20 (1959).
4067. P. M. Belcher, A. L. Eshleman, and J. D. Van Dyke, Development of aircraft structures to withstand acoustic loads, Aerospace Eng. 18(6) (1959).
4068. W. D. Myers, Adhesives for honeycomb cores, Adhes. Age 3(11):22 (1960).
4069. G. Epstein, Adhesive bonds for sandwich construction, Adhes. Age 6(8):30 (1963).
4070. G. Epstein, The influence of some selected fabrication variables in structural metal-to-metal bonding, Adhes. Age 4(6):36 (1961).
4071. N. W. Evans, The development of autoclave bonding of metal subassemblies in the British aircraft industry, Adhes. Age 4(1):20 (1961).

4072. L. H. Lunsford, Design of bonded joints, Symposium on Adhesives for Structural Applications (M. J. Bodnar, ed.), Interscience, New York, pp. 9–14 (1962).
4073. E. J. Gunter and M. I. Sanger, Low temperature curing adhesives for critical aerospace applications, 7th Natl. SAMPE Symp., Los Angeles, p. 1–4 (1964).
4074. R. J. Hilton, Use of room temperature-curing adhesive film on the 727 airplane, J. Appl. Polym. Sci., Appl. Polym. Symp. No. 3, Wiley-Interscience, New York, p. 191 (1966).
4075. E. B. McMullon and E. W. Garnish, Structural adhesive bonding of metals, Metals and Materials, p. 398 (Dec. 1967).
4076. J. T. Jensen, Application of metal bonding in glider structures, Aero-Revue 43:431, 441, 442 (1968).
4077. L. Keemar and R. Strong, BOAC, Shell Aviation News No. 366 (1968).
4078. R. B. Krieger, Jr., Advances in the corrosion resistance of bonded structures, SAMPE J. 5(2):25 (Feb./March 1969).
4079. J. Hong, Advanced bonding for large aircraft, SAE Tech. Paper No. 700863, Soc. Auto. Eng. Meeting, Detroit, MI (Oct. 1970).
4080. R. C. Snogren, Space-age bonding techniques, Mech. Eng. p. 23 (May 1970).
4081. M. A. Melcon and V. D. Moss, Controlled structural fatigue through adhesive bonding, ASME Publ. 71-DE-27 (1971).
4082. Primary adhesively bonded structure technology (PABST), Tech. Bulletin No. 3, USAF Contract F 33615-75-C-3016, Douglas Aircraft Co., Long Beach, CA (July 28, 1975).
4083. Idem, Tech. Bulletin No. 4, Idem, (Sept. 28, 1975).
4084. Idem, Alert Tech. Bulletin, Idem, (March 30, 1976).
4085. Idem, Tech. Bulletin No. 8, Idem, (March 30, 1976).
4086. Idem, Tech. Bulletin No. 9, Idem, (July 28, 1976).
4087. Idem, Tech. Bulletin No. 10, Idem, (Nov. 28, 1976).
4088. Idem, Tech. Bulletin No. 13, Idem, (May 28, 1977).
4089. Idem, Tech. Bulletin No. 14, Idem, (July 28, 1977).
4090. Idem, Tech. Bulletin No. 20, Idem, (Nov. 5, 1980).
4091. E. W. Thrall, Jr., PABST- a technology demonstrator, AIAA Aircraft Systems Technology Meeting, Los Angeles, CA (Aug. 4–7, 1975).
4092. Industrial production-savings in gluing aircraft together, Business Week, p. 88 (Feb. 21, 1977).
4093. E. W. Thrall, Jr. and R. W. Shannon, PABST surface treatment and adhesive selection, Adhes. Age 20(7):37 (1977).
4094. R. W. Shannon, Bus. Aircraft Meeting, Topeka, KS (1979).
4095. E. W. Thrall, Jr., The future of metal bonding, ASTM Symp. on Adhesives and Bonding Processes for the Automobile and Aircraft Industries, Monteray, CA (Oct. 10, 1979) (Douglas Aircraft Paper 6816).
4096. E. W. Thrall, Jr. and R. W. Shannon (eds.), Adhesive Bonding of Aluminum Alloys, Marcel Dekker, New York (1985).
4097. Adhesive bonding of large fuselage panels (third article in a series), Aircraft Eng. (May 1972).
4098. R. J. Dauksys, Research and development of aerospace adhesive bonded systems and concepts, SAMPE Quart., p. 1 (Oct. 1973).
4099. B. L. Reynolds, Proc. Army Mater. Conf. 4:605 (1976).
4100. C. Arnold, Bonding development of improved adhesives for acoustic structures, Boeing Aircraft Co. Report.
4101. R. F. Wegman, M. C. Ross, S. A. Slota, and E. S. Duda, Evaluation of the adhesive bonding processes used in helicopter manufacture, Part I, Durability of adhesive bonds obtained as a result of processes used in durability of adhesives bonds in UH-1 helicopter, Picatinny Arsenal Report TR-4186 (Sept. 1971).
4102. R. V. Wolff and G. H. Lemon, Reliability predictions for adhesive bonds, AFML-TR-72-121, AFML Lab., Wright-Patterson AFB, OH (March 1972).
4103. T. J. Reinhart, Jr., New adhesive needs and concepts for the aircraft industries, SAE Tech. Report No. 720118, Soc. Auto. Eng. Meetings, Detroit, MI (Jan. 1972).
4104. Spoilers, inboard and outboard, open position ultimate load static test, Airplane (A3) test results, F111A/B program, Report 12A6169, Addendum 1, General Dynamics Corp. (June 16, 1969).
4105. Structural repair instructions, FB-111A aircraft, Tech. Manual T.O IF-111 (b) A-3 (May 9, 1969).
4106. F-111A 100% flight loads program results, Report FZS-12-1039, Addendum 1, General Dynamics Corp. (Dec. 1, 1972).
4107. Report FZS-12-8165, General Dynamics Corp., Vol. 7 (Feb. 1972).
4108. Static testing of F-111 outboard spoiler, Test Report T6-5918 No. 1, Boeing Co. (April 26, 1976).

4109. F. F. Rudder and H. E. Plumbee, Jr., Sonic fatigue design guide for military aircraft, AFFDL Report TR-74-122 (AD-Boof-600L), Air Force Flight Dynamics Lab., Wright-Patterson AFB, OH (May 1975).
4110. A. Zimmerman, Gluing airplanes together, American Machinist, p. 124 (July 1977).
4111. K. R. Wentz and H. F. Wolfe, Development of random fatigue data for adhesive bonded and weld-bonded structures subjected to dynamic excitation, J. Eng. Mater. Technol., p. 100 (Jan. 1978).
4112. D. L. Potter et al., Primary adhesively bonded structure technology (PABST) design handbook for adhesive bonding, AFFDL Tech. Final Report TR-79-3129, Douglas Aircraft Co., Long Beach, CA (Nov. 1979).
4113. Contract F33615-76-C-5178, Laminated Wing Structures, sponsored by USAF AFML/FIBC, Wright-Patterson AFB, Ohio 45433; conducted by General Dynamics, Fort Worth Division, Fort Worth, TX (June 1975–June 1980).
4114. Contract F33615-76-C-3138, Advanced Technology Wing Structure, sponsored by USAF AFML/FIBC, Wright-Patterson AFB, Ohio 45433; conducted by General Dynamics, Fort Worth Division, Fort Worth, TX (Nov. 1976–March 1980).
4115. J. Romanko and W. G. Knauss, Development in Adhesives—2 (A. J. Kinloch, ed.), Applied Science Publishers, London, p. 173 (1980).
4116. K. M. Liechti and W. G. Knauss, Advances in Aerospace Structures and Materials, AD-01, p. 51 (1981).
4117. J. Romanko, K. M. Liechti, and W. G. Knauss, Integrated Methodology for Adhesive Bonded Joint Life Predictions, AFWAL-TR-82-4139, Final Report for Period July 1979 to July 1982, Wright-Patterson AFB, OH (Nov. 1982).
4118. W. D. Sanders, Bonding aircraft wing panels improves weight and durability, Adhes. Age 29(9):38 (1986).
4119. A. C. Fehrle and R. L. McDougal, Bonding and durability, S.A.E. Paper No. 790561, Bus. Aircraft Meeting, Witchita, KS (April 3–6, 1979).
4120. Structural Adhesives with Emphasis on Aerospace Applications, A Report of the ad hoc Committee on Structural Adhesives for Aerospace Use, National Materials Advisory Board, National Research Council, Treatise on Adhesion and Adhesives (R. L. Patrick, ed.), Marcel Dekker, New York, Vol. 4 (1976).
4121. J. E. McCarty, R. E. Horton, et al., Adhesive Bonded Aerospace Structures Standardized Repair Handbook, AFML Tech. Report TR-75-158, Boeing Commercial Airplane Co. (Aug. 1975).
4122. J. E. McCarty, R. E. Horton, et al., Adhesive Bonded Aerospace Structure Standardized Repair Handbook, AFML Tech. Report TR-75-201, Boeing Commercial Airplane Co. (Dec. 1976).
4123. R. E. Horton, Demonstration of an improved method for repair of bonded aircraft structure, SAMPE Quart., p. 20 (July 1979).
4124. W. D. Doetsch, R. W. Nickel, and W. F. Routh, Low Cost Aircraft Structural Repair and Maintenance Study, Air Force Flight Dynamics Lab Tech. Report TR-76-63, Rockwell International, Los Angeles (March 1977).
4125. M. H. Kuperman and R. E. Horton, Adhesive bonded aluminum structural repair, Adhesive Bonding of Aluminum Alloys (E. W. Thrall and R. W. Shannon, eds.), Marcel Dekker, New York, Chap. 16, pp. 425–493 (1985).
4126. M. H. Kuperman and R. E. Horton, Repair of aluminum aircraft structures, Adhesives and Sealants, Engineered Materials Handbook, ASM, Materials Park, OH, Vol. 3, pp. 801–820 (1990).
4127. Metallic Materials and Elements for Aerospace Vehicles Structures, MIL-HDBK-5, Military Standardization Handbook, U.S. Depart. Defense (May 1989).
4128. Aluminum and Aluminum Alloy Plate and Sheet, General Specification for QQ-A-250.
4129. Cloth, Glass, Finished, for Resin Laminates, MIL-C-9084, Military Specification.
4130. Polymer Matrix Composites, Vol. 1, MIL-HDBK-17, Military Standardization Handbook, U.S. Dept. Defense (June 1989).
4131. Core Material, Aluminum, for Sandwich Construction, MIL-C-7438, Military Specification.
4132. Core Materials, Plastic Honeycomb, Laminated Glass Fabric Base, for Aircraft Structural and Electronic Applications, MIL-C-8073, Military Specification.
4133. Core, Honeycomb, Fibrous Aramid Base, Phenolic Coated, AMS 3711, Aerospace Materials Specification.
4134. Adhesive, Film Form, Metallic Structural Sandwich Construction, MIL-A-25463, Military Specification.
4135. Adhesives, Heat Resistant, Airframe Structural, Metal-to-Metal, MMM-A-132, Military Specification.
4136. Sealing Compound, Temperature Resistant, Integral Fuel Tanks and Fuel Cell Cavities, High Adhesion, MIL-S-8802, Military Specification.
4137. Sealing and Coating Compound, Corrosion Inhibitive, MIL-S-81733, Military Specification.

4138. Sealing Compound, Topcoat, Fuel Tank, Buna-N, MIL-S-4383, Military Specification.
4139. S. Y. T. Chan and K. B. Armstrong, Cold-setting adhesives for repair processes using various surface preparation methods, Int. Conf. Defense Applications of Advanced Repair Technology for Metal and Composite Structures, Naval Res. Lab., Washington, D.C. (July 1981).
4140. K. G. Kibler, Characterization of composition variables in a structural adhesive, SAMPE Quart., p. 39 (April 1982).
4141. R. Hinrichs and J. Thuen, Natl. SAMPE Symp. 24:404 (1979).
4142. J. F. Carpenter, Natl. SAMPE Symp. 24:446 (1979).
4143. A. A. Wickham, D. D. Rice, and R. J. Dubois, Natl. SAMPE Symp. 24:506 (1979).
4144. G. L. Hagnauer, Polylmer Composites 1:81 (1980).
4145. T. A. Sewell, Chemical Characterization and Quality Control for Adhesives, Contract No. N00019-79-C-0064 (Feb. 1980).
4146. R. A. Britton, Acceptance criteria for aerospace structural adhesives, AFML Tech. Report TR-78-53, McDonnell Aircraft Company, St. Louis, MO (May 1978).
4147. C. J. Alner and A. V. Pocius, Physical properties of aerospace structural adhesives displaying resistance to open time in high humidity, 12th Natl. SAMPE Tech. Conf. (Oct. 7–9, 1980).
4148. G. C. Grimes, The adhesive-honeycomb relationship, Appl. Polym. Symposia 3:157 (1966).
4149. C. L. Mahoney, The influence of adhesive components on the corrosion of aluminum honeycomb, Non-metallic materials selection, application, and environmental effects, Natl. SAMPE Tech. Conf., Azusa, CA, pp. 411–424 (1972).
4150. M. C. Locke, Adhesive bonded square edge aluminum honeycomb durability, 9th Natl. SAMPE Tech. Conf., Atlanta, GA (Oct. 4–6, 1977).
4151. C. A. May et al., Cure monitoring techniques for adhesive bonding techniques, Air Force Wright Aeronautical Labs (AFWAL) Tech. Report TR-80-4171, Lockheed Missile and Space Company, Sunnyvale, CA (Nov. 1980).
4152. W. D. Brown and G. K. McKeegan, Raw Materials Quality Control, Adhesives and Sealants, Engineered Materials Handbook, ASM, Materials Park, OH, Vol. 3, pp. 729–734 (1990).
4153. R. W. Roberts, Processing quality control, Adhesives and Sealants, Engineered Materials Handbook, ASM, Materials Park, OH, Vol. 3, pp. 735–742 (1990).
4154. A. Hartman, Fatigue test on single-lap joints in clad 2024-T3 aluminum alloy manufactured by a combination of riveting and adhesive bonding, Report NLR-TN-M.2170, Nat. Lucht-en Ruimte-vaartlaboratorium, Amsterdam (Dec. 1966).
4155. J. Fasold, Polysulfide helps prevent icing on vanes of B-1B aircraft, Adhes. Age 29(11):20 (1986).
4156. R. E. Horton, R. S. Whitehead et al., Damage Tolerance of composites, Vol. 1, AFWAL-TR-87-3030, Wright-Patterson AFB, OH (July 1988).
4157. L. J. Hart-Smith, Differences between adhesive behavior in test coupons and structural joints, Paper at ASTM Comm. D-14 Meeting, Phoenix (March 1981).
4158. L. J. Hart-Smith, Adhesive layer thickness and porosity criteria for bonded joints, AFWAL Report TR-82-4172, Wright-Patterson AFB, OH (Dec. 1982).
4159. H. C. Tsai, Approximate solution for skin-stiffener separation including effect of interfacial shear stiffness, NADC Report No. 8313-66, Naval Air Development Center (Nov. 1983).
4160. R. W. Johnson, Design proof testing, Adhesives and Sealants, Engineered Materials Handbook, ASM, Materials Park, OH, Vol. 3, pp. 533–544 (1990).
4161. J. Hertz, Cryogenic adhesive evaluation study, Mat. Res. Report ERR-AN-032, Conviar Astronautics Div. (Jan. 25, 1961).
4162. C. E. Eppinger and W. J. Love, Bonding plastic-to-metal for high strength at low temperatures, Advances in Cryogenic Engineering, Plenum Press, New York (1960).
4163. J. J. Whelton, Interim report on room temperature-curing adhesives for SATURN S-II, Part I—Screening program, Report SDL-324, North American Aviation Inc. (March 5, 1962).
4164. E. F. Hess, Curing structural adhesives, Adhes. Age 7(7):19 (1964).
4165. J. R. Hill, Adhesive bonding of space structures, Report D2-24154-1, Boeing Co., Seattle (June 6, 1966).
4166. C. C. Kammerer, L. Furon, and N. Jeffras, Project Apollo automated ultrasonic bond quality inspection, SAMPE J. 2(3):16 (April/May 1966).
4167. I. K. Spiker and R. Barnes, Adhesives in space—Apollo and future manned spacecraft, Aerospace Adhesives and Elastomers, Proc. Natl. SAMPE Tech. Conf., Anaheim, CA, Vol. 2 (1970).
4168. B. J. Mulrey and D. M. Mazenko, Structural adhesives for space systems, Proc. SAMPE Technol. Conf., El Segundo, CA, p. 340 (March 1979).
4169. G. Epstein and J. J. Smith, Adhesive bonding requirements for space environments, Air Force Mater. Symp., Miami Beach, FL (May 1970).

4170. G. Epstein, Advances for space systems, Part I—Requirements and applications, Adhes. Age 13(6):18 (1972).
4171. G. Epstein, Adhesives for space systems, Part II—Launch and re-entry systems, Adhesives Age 13(7):27 (1972).
4172. W. S. Stiles, Secondary bonding, NASA Report N66-34135, Marshall Space Flight Center (March 1966).
4173. B. Pascuzzi, Adhesives role in the development of Mariner Venus/Mercury 1973 Spacecraft, Adhes. Age 16(9):20 (Sept. 1973).
4174. J. Srinivas, Analysis of bonded joints, NASA Report TN D-7855 (April 1975).
4175. B. D. Murray, Machine Design 53:67 (1981).
4176. W. A. Lees, Production Engineer 61:66 (1982).
4177. H. L. Novak and D. A. Comer, Modified cyanoacrylate ester bonds solid rocket booster, Adhes. Age 33(2):28 (1990).
4178. R. A. Pike and R. S. Williams, Ultrasonic assessment of cure rate effects in bonded honeycomb structures, Adhesive Joints (K. L. Mittal, ed.), Plenum Press, New York, p. 369 (1984).
4179. R. S. Williams and P. E. Zwieke, Mater. Eval. 40:312 (1982).
4180. C. L. Mahoney, S.A.E. Cong. & Expos., Detroit, MI, Paper No. 790151, Detroit, Michigan (Feb. 26, 1979).
4181. J. Thuen and R. Hinricks, Proc. 25th Nat. SAMPE Symp., p. 126 (May 6, 1980).
4182. E. Sancaktar, H. Jozavi, and R. M. Klein, 5th An. Adhesion Soc. Meeting, Mobile, AL (Feb. 22, 1982).
4183. J. F. Carpenter and T. T. Bartels, SAMPE Quart. 7(1):1 (1976).
4184. C. A. May, D. K. Hadad, and C. E. Browning, Polym. Eng. Sci. 19:545 (1979).
4185. J. E. Twichell, J. O. Walker, and J. B. Maynard, J. Chromatog. Sci. 17:259 (1979).
4186. D. L. Paradis, Structural Adhesives and Bonding, Tech. Assoc. Conf., El Segundo, CA, p. 110 (1979).
4187. G. W. Lawless, Proc. 24th Nat. SAMPE Symp., 2:979 (1979).
4188. R. A. Pike, F. P. Lamm, and J. P. Pinto, J. Adhes. 12:143 (1981).
4189. C. E. M. Morris, A. G. Moritz, and R. G. Davidson, Adhesion and Adsorption of Polymers (L. H. Lee, ed.), Plenum Press, New York, p. 313 (1980).
4190. J. J. Pitts, P. J. Pearce, T. W. Rosewarne, R. G. Davidson, B. C. Ennis, and C. E. M. Morris, J. Macromol. Sci., Chem. A17:227 (1982).
4191. P. J. Pearce, R. G. Davidson, and C. E. M. Morris, J. Appl. Polym. Sci. 26:2363 (1981).
4192. P. J. Pearce, R. G. Davidson, and C. E. M. Morris, J. Appl. Polym. Sci. 27:4501 (1982).
4193. P. J. Pearce, R. G. Davidson, and C. E. M. Morris, J. Appl. Polym. Sci. 28:283 (1983).
4194. L. R. Sanders, R. S. Baxter, and R. J. Jurgens, Titanium Science and Technology (R. I. Jaffe and H. M. Burte, eds.), Plenum Press, New York, Vol. 1, p. 105 (1973).
4195. M. A. Steinberg, Sci. Amer. 255:66 (1986).
4196. M. J. Donachie, Jr. (ed.), Titanium and Titanium Alloys, Source Book, ASM, Metals Park, OH, p. 3 (1983).
4197. A. Mahoon, Titanium adherends, Durability of Structural Adhesives (A. J. Kinloch, ed.), Applied Science Publishers, London, p. 255 (1983).
4198. K. W. Allen, H. S. Alasalim, and W. C. Wake, J. Adhes. 6:153 (1974).
4199. Y. Moji and J. A. Marceau, Method of Anodizing Titanium to Promote Adhesion, U.S. Patent No. 32,959,091, The Boeing Company (1976).
4200. A. Mahoon and J. L. Cotter, SAMPE Tech. Conf. Ser. 10:425 (1978).
4201. M. E. Schrader and J. A. Cardamone, J. Adhes. 9:305 (1978).
4202. B. M. Ditchek, T. I. Morgenthaler, T. S. Sun, and J. D. Venables, Proc. 12th SAMPE Tech. Conf., p. 882 (1980).
4203. J. L. Cotter and A. Mahoon, Int. J. Adhes. Adhes. 2:47 (1982).
4204. R. F. Wegman and D. W. Levi, Proc. 27th Natl. SAMPE Symp., p. 440 (1982).
4205. A. Mahoon, Proc. 27th Natl. SAMPE Symp., p. 150 (1982).
4206. P. D. Peters, E. A. Ledbury, C. L. Hendricks, and A. G. Miller, Proc. 27th Natl. SAMPE Symp., p. 940 (1982).
4207. S. R. Brown, Proc. 27th Natl. SAMPE Symp., p. 363 (1982).
4208. M. Natan, J. D. Venables, and K. R. Breen, Proc. 27th Natl. SAMPE Symp., p. 178 (1982).
4209. J. A. Filbey, J. P. Wightman, and D. J. Progar, J. Adhes. 20:283 (1987).
4210. D. J. Progar, J. Adhes. Sci. Technol. 1:135 (1987).
4211. J. A. Skiles and J. P. Wightman, J. Adhes. 26:301 (1988).
4212. C. Matz, Int. J. Adhes. Adhes. 8:17 (1988).
4213. J. A. Filbey and J. P. Wightman, J. Adhes. 28:1 (1989).
4214. A. C. Kennedy, A. Kohler, and P. Poole, Int. J. Adhes. Adhes. 3:133 (1983).

4215. E. P. Plueddemann, Int. J. Adhes. Adhes. 1:305 (1981).
4216. S. J. Monte and G. Sugarman, Elastomers 115:30 (1983).
4217. S. J. Monte and G. Sugarman, Polym. Eng. Sci. 24:1369 (1985).
4218. J. Comyn, Structural Adhesives: Development in Resins and Primers, Applied Science Publishers, London, p. 269 (1982).
4219. P. D. Calvert, R. P. Lolanandham, and D. R. M. Walton, Adhesion Aspects of Polymeric Coatings (K. L. Mittal, ed.), Plenum Press, New York, p. 457 (1983).
4220. S. J. Monte and G. Sugarman, Adhesion Aspects of Polymeric Coatings (K. L. Mittal, ed.), Plenum Press, New York, p. 421 (1983).
4221. M. H. Stone, J. Adhes. 26:101 (1988).
4222. J. A. Filbey and J. P. Wightman, J. Adhes. 28:23 (1989).
4223. V. C. Bell, B. L. Stump, and H. Gazern, J. Polym. Sci., Polym. Chem. Ed. 14:2275 (1976).
4224. A. K. St. Clair, W. S. Slemp, and T. L. St. Clair, Sci. Adv. Mater. Proc. Eng. Series 23:113 (1978).
4225. J. P. Wightman and S. Diaz, The application of thermodynamics and spectroscopic techniques to adhesion in the polyimide/Ti 6-4 and polyphenylquinoxaline/Ti 9-4 systems, Adhesive Chemistry-Developments and Trends (L. H. Lee, ed.), Plenum Press, New York (1984).
4226. Sealing the X-15, Adhes. Age 3(11):34 (1960).
4227. K. F. Hahn, Lap-shear and creep performance in metal-to-metal bonds (aircraft and missiles), Adhes. Age 4(12):34 (1961).
4228. Critical area of X-15 sealed with silicone rubbers, Adhes. Age 5(1):25 (1962).
4229. Film bonding adhesive and the Convair 990 jet, Adhes. Age 5(12):22 (1962).
4230. Bonded weatherstripping solves Boeing problems, Adhes. Age 5(2):29 (1962).
4231. Bonding de-icer boats to helicopter blades, Adhes. Age 8(7):29 (1965).
4232. Aircraft struts and pontoons-adhesive assembly permits weight and cost reductions, Adhes. Age 8(10):33 (1965).
4233. Pre-fitted bonded parts for aircraft construction, Adhes. Age 9(1):26 (1966).
4234. Pressure-sensitive aluminum tape shields aircraft, Adhes. Age 9(6):35 (1966).
4235. Using adhesives in a floating airfield, Adhes. Age 9(12):38 (1966).
4236. Bonded helicopter blade meets tough competition in Vietnam, Adhes. Age 10(10):31 (1967).
4237. Sealant kits speed airplane overhauls, Adhes. Age 10(11):33 (1967).
4238. J. F. Callahan, Building supersonic wings and stabilizers, Adhes. Age 11(8):19 (1968).
4239. Adhesive bonding jumbo jet airframes, Adhes. Age 11(11):34 (1968).
4240. Sealants play major role in construction of military helicopters, Adhes. Age 12(4):36 (1969).
4241. Bonded landing pads provide safer stops for British crafts, Adhes. Age 12(4):36 (1969).
4242. M. S. Howeth and B. C. Cornelius, Advanced composite production of wing trailing edge panels, Adhes Age. 14(9):22 (1971).
4243. D. A. Petrino, L-1011 TriStar places new demands on aircraft sealants and adhesives, Adhes. Age 15(2):15 (1972).
4244. Helicopter research group verifies integrity of industrial adhesives, Adhes. Age 15(9):36 (1972).
4245. C. N. Powis, Structural adhesive bonding in Hovercraft, Adhes. Age 16(9):44 (1973).
4246. L. J. Hart-Smith, Designing adhesive bonds, Adhes. Age 21(10):32 (1978).
4247. R. J. Jones and H. N. Cassey, Flexible polyimide fuel tank sealants, Adhes. Age 22(11):33 (1979).
4248. T. J. Reinhart, Use of structural adhesives by the U.S. Air Force, Adhes. Age 24(8):20 (1981).
4249. Fuel tank bonding project cuts Air Force repair costs, Adhes. Age 25(6):42 (1982).
4250. B. Silverman and A. Norrbom, Polysulfide sealants in aircraft composite structures, Adhes. Age 26(7):28 (1983).
4251. H. L. Novak, Wet epoxy primer improves bondline shear strength (aerospace), Adhes. Age 29(5):33 (1986).
4252. Pressure-sensitive tape helps catch USAF space capsules, Adhes. Age 4(3):31 (1961).
4253. RTV silicone rubbers seal space capsule, Adhes. Age 4(4):40 (1961).
4254. L. E. Gatzek and C. R. Feller, Space-age adhesive materials, Adhes. Age 5(5):26 (1962).
4255. Adhesives in the space race, Adhes. Age 8(10):29 (1965).
4256. L. M. Roseland, Developing structural adhesives for space applications, Adhes. Age 9(4):32 (1966).
4257. Spaceman to bond self to vehicle with adhesive capsules, Adhes. Age 11(12):26 (1968).
4258. A. C. Robertson and E. L. Brown, A bonded bulkhead for Saturn, Adhes. Age 11(12):26 (1968).
4259. Adhesive plays vital role in Apollo heat-shield structure, Adhes. Age 12(10):29 (1969).
4260. Using weldbonding to assemble spacecraft, Adhes. Age 14(9):28 (1971).
4261. Epoxy adhesive used in spacecraft, Adhes. Age 17(7):29 (1974).
4262. H. Singh, Cyanosiloxane sealants used in construction of space shuttle orbiter, Adhes. Age 21(4):24 (1978).

4263. N. A. DeBruyne, The physics of adhesion, J. Sci. Instrum. 24(2):29 (Feb. 1947).
4264. E. W. Kuenzi, Sandwich studies, Forest Products Lab. and University of Dayton Symp. (June 1957).
4265. L. W. Hovland and E. R. McLaughlin, Aluminum foamed plastic panels, Soc. Plast. Eng. J. 15:12 (Dec. 1959).
4266. R. T. Schwartz and D. V. Rosato, Structural sandwich construction, Composite Engineering Laminates (A. G. H. Dietz, ed.), Massachusetts Institute of Technology Press, Cambridge, MA (1969).
4267. W. D. Myers, Adhesives for honeycomb cores, Adhes. Age 3(11):22 (1960).
4268. W. M. Leeser, Bonded honeycomb-core steel doors, Adhes. Age 4(11):24 (1961).
4269. How adhesives are used in honeycomb, Adhes. Age 4(4):37 (1961).
4270. Seven layer bonded sandwich provides skiis with permanent camber, Adhes. Age 5(3):33 (1962).
4271. J. P. Tipton, Haystack antenna system features honeycomb sandwich construction, Adhes. Age 8(7): 30 (1965).
4272. M. S. Hochberg, Evaluating adhesive bonded temperature-resistant sandwich construction, Adhes. Age 9(3):19 (1966).
4273. R. Kantner and S. Litvak, Adhesives for bonding large high-temperature sandwich structures, Adhes. Age 12(11):24 (1969).
4274. A. F. Lewis and G. B. Elder, Selecting adhesives for vibration damping metal laminate applications, Adhes. Age 12(10):31 (1969).
4275. C. S. Phelan, The forgotten corner of bonding, how sandwich panels are attached, Adhes. Age 17(9):23 (1974).
4276. K. R. Berg and J. E. Ramsey, Viscoelastic adhesive effects on sandwich structure, Adhes. Age 17(12):33 (1974).
4277. Waterborne cuts cleanup and eliminates compliance costs (stress skin and structural panel fabrication), Adhes. Age 30(2):26 (1987).
4278. K. Rose, Sandwich materials, Materials and Methods Manual No. 103 (March 1954).
4279. R. K. Humke, Selection guide for sandwich panel core, Prod. Eng. (Jan. 20, 1958).
4280. C. B. Hemming, Molded laminates and sandwich materials, Composite Eng. Laminates (A. G. H. Dietz, ed.), Massachusetts Institute of Technology Press, Cambridge, MA (1969).
4282. High speed infrared panel cure, Adhes. Age 5(3):38 (1962).
4283. Applying facings with roll-coating techniques, Adhes. Age 9(2):31 (1966).
4284. Hot spray adhesives speed production of sandwich panels, Adhes. Age 10(5):30 (1967).
4285. Latex adhesive effectively bonds insulation to metal panels, Adhes. Age 11(3):31 (1968).
4286. D. Debnar, Selecting suitable adhesives for expanded polystyrene, Adhes. Age 15(3):41 (1972).
4287. J. D. Minford, unpublished work, Aluminum Co. of America.
4288. J. D. Minford, unpublished work, Aluminum Co. of America.
4289. J. D. Minford, unpublished work, Aluminum Co. of America.
4290. J. D. Minford, unpublished work, Aluminum Co. of America.
4291. J. D. Minford, unpublished work, Aluminum Co. of America.
4292. J. D. Minford, unpublished work, Aluminum Co. of America.
4293. Sprayable adhesives improves urethane foam bonding, Adhes. Age 11(4):20 (1968).
4294. C. Strobech, Bonding lightweight insulated sandwich elements with polyurethane adhesives, Adhes. Age 20(5):23 (1977).
4295. S. L. DeGisi and C. H. Smith, Improved bondability of molded rigid urethane foam, Adhes. Age 21(7): 35 (1978).
4296. J. D. Minford, unpublished work, Aluminum Co. of America.
4297. J. D. Minford, unpublished work, Aluminum Co. of America.
4298. J. D. Minford, unpublished work, Aluminum Co. of America.
4299. Dow Ceramic Foam, Chem. Eng., p. 44 (Jan. 11, 1971).
4300. J. D. Minford, unpublished work, Aluminum Co. of America.
4301. J. D. Minford and E. M. Vader, unpublished work, Aluminum Co. of America.
4302. J. D. Minford, unpublished work, Aluminum Co. of America.
4303. W. E. Willis (ed.), Wood Extractives, Academic Press, New York (1962).
4304. R. C. Pettersen, The chemical composition of wood, The Chemistry of Solid Wood (R. M. Rawell, ed.), Advances in Chemistry Series No. 207, American Chemical Society, Washington, D.C. (1984).
4305. D. Fengal and G. Wegener, Wood: Chemistry, Ultrastructure, Reactions, Walter de Gruyere, New York (1984).
4306. D. Narayanamurti, R. C. Gupta, and G. M. Verna, Influence of extractives on the setting of adhesives, Holzfarshung und Holzverwertung 14(5/6):85 (1962).
4307. J. D. Minford, unpublished work, Aluminum Co. of America.

4308. J. D. Minford, unpublished work, Aluminum Co. of America.
4309. U. T. Krebich and A. F. Mercantonio, New developments in structural adhesives for the automobile industry, J. Adhes. 22(2):153 (1987).
4310. J. W. Paris, Bonding with anaerobic adhesives in the automotive industry, S.A.E. Paper No. 770089, Int. Auto. Eng. Cong. & Exhib., Detroit, MI (Feb. 28–March 4, 1977).
4311. P. E. Wright, Incentives for using automotive bonding methods, Adhes. Age 15(9):30 (1972).
4312. G. Gialanella, Application systems for structural adhesives in the automotive and construction industries, Adhes. Age 15(5):37 (1972).
4313. W. A. Lees, Use of adhesives in constructing vehicles, Adhes. Age 24(2):23 (1981).
4314. D. Schilkey, An automotive case history, Adhes. Age 26(4):18 (1983).
4315. M. Wall, Automaker halves man hours with robotics HM dispenser, Adhes. Age 30(3):34 (1987).
4316. G. L. Schneberger, Using adhesives effectively, S.A.E. Paper No. 730134, S.A.E. Cong. & Exhib. Meetings, Detroit, MI (Feb. 1973).
4317. G. L. Schneberger, Low-energy curing adhesives—their uses, advantages, and limitations, S.A.E. Paper No. 760086, S.A.E. Cong. & Exhib. Meetings, Detroit, MI (Feb. 1976).
4318. G. L. Schneberger, Fundamentals of adhesive selection, S.M.E. Tech. Paper No. AD75-791 (1975).
4319. G. L. Schneberger, Automotive adhesives-why, what, where, and how, S.A.E. Paper No. 780191, S.A.E. Cong. & Exhib. Meetings, Detroit, MI (Feb. 27–March 3, 1978).
4320. G. L. Schneberger, Automotive adhesives-past, present, and future, S.A.E. Paper No. 800210, S.A.E. Cong. & Exhib. Meetings, Detroit, MI (Feb. 25–29, 1980).
4321. Transactions of the S.A.E.: 1911–1979, Soc. Auto. Eng., Warrendale, PA.
4322. Engineering Index: 1970–1979, Engineering Index Inc., New York.
4323. Y. F. Chang, M. F. Milewski, and K. P. Tremonti, Adhesives Technol. for Auto. Eng. Appl: Proc. SME Conf., Paper No. AD87-527, Soc. Mfg. Eng., Dearborn, MI (Oct. 1987).
4324. Friction materials have longer life than bonded, Adhes. Age 8(9):28 (1965).
4325. CIBA-GEIGY Adhesives for Brake and Clutch Bonding, Instruction Manual RTB1F, Ciba-Geigy Corp. (1983).
4326. Auto maker disperses with trunk welding job, Adhes. Age 28(3):28 (1985).
4327. W. Chun, T. Kish, and K. P. Tremonti, Bonding automotive hoods to eliminate spot welds, Adhes. Age 29(2):27 (1986).
4328. J. Schibley, Automated hood assembly controls cost and quality, Adhes. Age 33(7):24 (1990).
4329. Anaerobic compound increases engine core plug strength, Adhes. Age 16(6):29 (1973).
4330. Car makers decrease costs with HM sealant applicator, Adhes. Age 29(12):22 (1986).
4331. W. E. Leuchten, Automotive fixed glass sealing, Adhes. Age 9(7):27 (1966).
4332. C. E. Dietrich and R. J. Glatki, Bonding processes advance design of automobile lights, Adhes. Age 31(2):35 (1988).
4333. R. Grunder and H. W. Koehler, Automatic light source potting reduces cost and lead time, Adhes. Age 32(11):26 (1989).
4334. Tape replaces metal fasteners on parts of cars, Adhes. Age 11(9):31 (1968).
4335. T. M. Richards and R. D. Weltzen, Adhesive strengthens riveted assembly, Adhes. Age 8(12):35 (1965).
4336. F. J. Meyer and R. J. Zienert, Improving FRP bonding efficiency, Adhes. Age 24(4):31 (1981).
4337. R. Kurtz, Dispenser assures quality of SMC panel repair, Adhes. Age 33(13):18 (1990).
4338. Automotive Engineering 96(2):6 (1988).
4339. P. G. Gheyara, Paper presented at Adhesives Technology for Automotive Engineering Conf., Soc. Mfg. Eng., Dearborn, MI (Nov. 1989).
4340. Automation catches up with formed-in-place gaskets, Adhes. Age 15(5):48 (1972).
4341. Automobile tug-or-war tests cyanoacrylate adhesive bond, Adhes. Age 15(9):40 (1972).
4342. T. L. Wilkinson, Acrylic adhesives: new way to bond aluminum auto parts, Adhes. Age 21(6):20 (1978).
4343. J. G. Bolger, Epoxies for manufacturing cars, buses, and trucks, Adhes. Age 23(12):14 (1980).
4344. T. L. Wilkinson and D. P. Tyler, Acrylics improve for bonding automotive aluminum alloys, Adhes. Age 24(12):34 (1981).
4345. D. J. Damico, Bonding galvanized steel with R.T. curing acrylics, Adhes. Age 30(11):25 (1987).
4346. D. H. Shah, Epoxy's rapid cure reduces cycle time for automaker.
4347. Two-part acrylic used to bond sensors to fuel tanks bottoms, Adhes. Age 32(5):29 (1989).
4348. P. G. Phillips and W. E. Broxterman, Adhesives, sealants in autos to grow at 250% of GNP, Adhes. Age. 32(9):40 (1989).
4349. A. M. Chasser and J. M. Maklouf, Epoxies reinforce metals for automotive industry, Adhes. Age 34(3): 38 (1991).

4350. Adhesive weights balance wheels, Adhes. Age 10(1):20 (1967).
4351. N. Bartrug and R. Koleg, Improving the adhesion of fiber glass cord to rubber, Adhes. Age 11(6):27 (1968).
4352. Epoxy formulation solves battery-production problems, Adhes. Age 19(3):30 (1976).
4353. G. W. Beaver, Replacing clamp with adhesive saves money for auto maker, Adhes. Age 32(5):18 (1989).
4354. Battery drain tube adhesive reduces capital expenditures, Adhes. Age 33(11):40 (1990).
4355. K. J. Schroeder and K. F. Drain, Paper presented at Adhesives Technology for Automotive Eng. Appl. Conf., SME, Dearborn, MI (Nov. 1989).
4356. J. M. Margolis, Automotive development in the 1980's—composites versus metals, SAMPE J. 16(4):7 (July/August 1980).
4357. J. F. Norton, Structural adhesives in the motor car industry, Sheet Metal Ind., p. 803 (Nov. 1981).
4358. B. Scheidle, CIBA-GEIGY Resin Aspects 21:8 (1989).
4359. J. D. Minford, unpublished work, Aluminum Co. of America.
4360. Y. F. Chang, T. G. Kish, K. P. Tremonti, and W. Chun, Induction curing adhesives for automotive hoods to eliminate spot welds, SME Adhesives Conf., Atlanta (Sept. 10–12, 1985).
4361. Y. F. Chang, T. G. Kish, K. P. Tremonti, and W. Chun, Induction curing adhesives for automotive hoods, Adhes. Age 29(2):27 (1986).
4362. G. H. Clamer, The development of the coreless induction furnace, Metals and Alloys 6:119 (1935).
4363. F. W. Curtis, High-Frequency Induction Heating, McGraw-Hill, New York (1944).
4364. J. Davies and P. Simpson, Induction Heating Handbook, McGraw-Hill, London (1979).
4365. I. Namiki, J. Appl. Polym. Sci. 20:1833 (1976).
4366. R. C. Rains, Mater. Res. Symp. Proc. 124:323 (1988).
4367. K. Wongkamolsesh, Proc. 43rd An. SPI Conf., Soc. Plast. Ind., p. 24E (1988).
4368. D. H. Shah, Proc. SME Autocom 88' Conf. Soc. Mfg. Eng., Dearborn, MI, Paper No. AD88-232 (May 1988).
4369. I. Spinu, J. G. Dillard, J. W. Grant, Durability of aluminum composite bonds, 13th Ann. Meeting Adhes. Soc., Savannah, GA (Feb. 18–21, 1990).
4370. E. W. Petrie, Joining the engineering plastics, Adhes. Age 23(8) (1980).
4371. E. W. Petrie, Joining the engineering plastics, High Performance Adhesive Bonding (G. DeFrayne, ed.), Society Manufacturing Engineers, Dearborn, MI, pp. 193–202 (1983).
4372. C. A. Harper (ed.), Handbook of Plastics and Elastomers, McGraw-Hill, New York, Chap. 10 (1975).
4373. I. Skeist (ed.), Handbook of Adhesives, 2nd ed., Van Nostrand Reinhold, New York, Chap. 41 (1978).
4374. Bonding and joining, parts 1 and 2, Plastics Technology (Aug. 1970).
4375. H. G. Henry, Teflon joined by fusion bonding, Mater. and Methods 38:114 (1953).
4376. J. D. Minford, E. M. Vader, and M. McGinnis, unpublished work, Aluminum Co. of America.
4377. E. M. Vader, unpublished work, Aluminum Company of America.
4378. L. A. Sobrieski and T. J. Tangney, Silicone PSA's perform well at high and low temperatures, Adhes. Age 31(3):23 (1988).
4379. A. M. Chasser and J. M. Makhlouf, A body reinforcing adhesives, Adhesive Technol. Automotive Eng. Applications, Society Manufacturing Engineers Clinic.
4380. V. Wigotsky, Plast. Eng., p. 23 (Sept. 1989).
4381. M. T. Cocozzoli, Sealants and Structural Adhesives, Fisher Body Division, General Motors Corporation (March 1982).
4382. M. A. Sermokins, Paper presented at Adhesives Technology for Automotive Eng. Applications Conf., Soc. Mfg. Eng., Dearborn, MI (Oct. 1987).
4383. M. D. Kirby, S.A.E. Tech. Paper No. 870308, S.A.E. Cong. & Exhib. Meetings, Detroit, MI (1987).
4384. L. Zabel and M. Pfeiffer, Automotive glass bonding technology, Adhesive Technol. Automotive Eng. Appl., SME Clinic, Dearborn, MI (Nov. 1–2, 1990).
4385. I. N. Moody, P. A. Fay, and G. D. Suthurst, Adhesion II (K. W. Allen, ed.), Elsevier, Amsterdam, Chap. 7, p. 97 (1987).
4386. W. F. Warwick and J. H. Powell, Proc. "88" SME Conf., Soc. Mfg. Eng., Dearborn, MI (Oct. 1988).
4387. A. J. Kinloch and G. A. Kiodikian, J. Adhes. 24:109 (1987).
4388. M. Fischer and M. Pasquier, Constr. Build. Mater. 3(1):31 (1989).
4389. S. B. Twiss, Adhesives in the automobile industry, Handbook of Adhesives (I. Skeist, ed.), Reinhold, New York, Chap. 57, pp. 637–646 (1962).
4390. S. B. Twiss, How adhesives are used in the modern automobile, Adhes. Age 6(13):16 (1963).
4391. S. B. Twiss, Adhesive requirements in the automotive industry, Symp. on Recent Developments in Adhesion Science, ASTM Tech. Publ. No. 360, p. 96 (1963).

4392. S. B. Twiss, Adhesives of the future, Appl. Polym. Symp. No. 3, p. 455 (1966).

4393. S. B. Twiss, Where are automotive adhesives going? Automotive Ind. 141(11):42; 144(12):29; 145(2):31 (1971).

4394. G. F. DeFrayne and S. B. Twiss, Bonding processes in automotive manufacture, Adhesive Bonded Structures Appl. Polym. Symp. No. 19, Interscience, New York, p. 291 (1972).

4395. G. F. DeFrayne and S. B. Twiss, Bonding processes in automotive manufacture, S.A.E. Tech. Paper No. 730136, Int. Auto. Eng. Cong. & Exhib. Meetings, Detroit, MI (Jan. 8–12, 1973).

4396. R. H. Beck and D. A. Yurek, Structural adhesive bonding of the automotive body, Body Eng. J. (Oct. 1975).

4397. J. C. Bolger, Epoxy bonded school buses, S.A.E. Tech. Paper No. 780192, S.A.E. Cong. & Exhib. Meetings, Detroit, MI (Feb. 1978).

4398. J. C. Bolger, Epoxy adhesives in automotive assembly operations, S.A.E. Tech. Paper, S.A.E. cong. & Exhib. Meetings, Detroit, MI (Feb. 25–29, 1980).

4399. R. D. Murray, New approach to selecting structural adhesives, Machine Design 53(12):67 (1981).

4400. W. A. Lees, The use of adhesives in the manufacturing industries, Prod. Engineer 61(3):66 (1982).

4401. M. Hauser and G. S. Haviland, Anaerobic adhesives, Adhesives in Manufacturing (G. L. Schneberger, ed.), Marcel Dekker, New York, Chap. 11, pp. 290–295 (1983).

4402. R. J. Slingerland, Past, present, and future of elastomeric body sealers in automotive applications, S.A.E. Tech. Paper No. 800269, Soc. Auto. Eng. Cong. & Exhib. Meetings, Detroit, MI (Feb. 25–29, 1980).

4403. K. Adams, Applications for sealants, Adhesives and Sealants, Engineered Materials Handbook, ASM, Materials Park, OH, Vol. 3, pp. 608,609 (1990).

4404. R. Grunder and H. W. Koehler, Automatic light source potting reduces costs and lead time, Adhes. Age 32(11):26 (1989).

4405. T. DeBarros and M. Beschle, Bonding parabolic reflector flood and spot lamps, Adhes. Age 32(3):42 (1989).

4406. J. Nowicki, Liquid polyurethanes as foam-in-place gaskets, Adhesive Technol. for Automotive Eng. Applications, Society Manufacturing Engineers Clinic, Dearborn, MI (Nov. 1–2, 1990).

4407. A. G. Bachmann, Enhancing quality and processes through aerobic adhesives and coatings, Adhesive Technol. for Automotive Eng. Applications, Society Manufacturing Engineers Clinic, Dearborn, MI (Nov. 1–2, 1990).

4408. D. F. Accettura, More and less: An engine oil pan gasket dispensing solution, Adhesive Technol. for Automotive Eng. Applications, Society Manufacturing Engineers Clinic, Dearborn, MI (Nov. 1–2, 1990).

4409. P. T. Leung and C. T. Chmiel, Waterborne adhesive for thermoforming vinyl to rigid surfaces, Adhesive Technol. for Automotive Eng. Applications, Society Manufacturing Engineers, Dearborn, MI (Nov. 1–2, 1990).

4410. B. L. Powell, Silicone elastomeric adhesives for plastic automotive bumper systems, Adhesive Technol. for Automotive Eng. Applications, Society Manufacturing Engineers Clinic, Dearborn, MI (Nov. 1–2, 1990).

4411. R. L. Hauser, Hot-melt adhesives for automotive headlights, Adhesive Technol. for Automotive Eng. Applications, Society Manufacturing Engineers Clinic, Dearborn, MI (Nov. 1–2, 1990).

4412. R. A. Martin, One-part water-based adhesives for vacuum-forming automotive interior trim, Adhesive Technol. for Automotive Eng. Applications, Society Manufacturing Engineers Clinic, Dearborn, MI (Nov. 1–2, 1990).

4413. Electronics for Automotive Applications, Publication 600, Delco Electronics Division, General Motors Corporation.

4414. K. Adams, Applications for sealants, Adhesives and Sealants, Engineered Materials Handbook, ASM, Materials Park, OH, Vol. 3, pp. 610–612 (1990).

4415. Application of noise control materials in the automotive industry, Paper presented at SAE Noise Vibration Conf., Traverse City, MI, Society Automotive Engineers Inc. (May 1989).

4416. D. Klempner, Recent Developments in Polyurethane (K. C. Frisch, ed.), I.P.N.'s Technomic, Lancaster, PA, p. 50 (1988).

4417. J. J. Baron, Aluminium 30(5) (1954).

4418. C. S. King, Proc. Inst. Mech. Eng. 197(64) (1983).

4419. D. Kewley, S.A.E. Tech. Paper No. 850103, Cong. & Exhib. Meetings, Detroit, MI (1985).

4420. P. G. Sheasby, M. J. Wheeler, and D. Kewley, Aluminum Technology, Book 391, Institute of Metals (1986).

4421. D. Kewley, The BL technology ECVs energy conservation vehicle, S.A.E. Tech. Paper No. 850103, Cong. & Expos. Meetings, Detroit, MI (Feb. 25–March 1, 1985).

4422. D. Keeley, I. G. Campbell, and J. E. Wheatley, Manufacturing feasibility of adhesively bonded aluminum for volume car production, S.A.E. Tech. Paper No. 870150, Cong. & Expos. Meetings, Detroit, MI (Feb. 23–27, 1987).
4423. P. G. Selwood, F. J. Law, P. G. Sheasby, and M. J. Wheeler, Evaluation of an adhesively bonded aluminum structure in an Austin Rover Metro Vehicle, S.A.E. Tech. Paper No. 870149, Cong. & Expos. Meetings, Detroit, MI (1987).
4424. M. J. Wheeler, P. G. Sheasby, and D. Kewley, Aluminum structured vehicle technology- a comprehensive approach to vehicle design and manufacturing in aluminum, S.A.E. Tech. Paper No. 870146, Cong. & Expos. Meetings, Detroit, MI (Feb. 23–27, 1987).
4425. T. G. Johannisson, ASEA Flex Forming Process, I.D.D.R.G., 14th Biennial Congress, p. 219 (1986).
4426. Amino Fluid Forming Process, Die Manufacturing & Die Maintenance Systems, Amino Engineering Co. (March 1986).
4427. W. F. Marwick and P. G. Sheasby, S.A.E. Tech. Paper No. 870151, Cong. & Expos. Meetings, Detroit, MI (1987).
4428. G. L. Leone and B. Altshuller, S.A.E. Tech. Paper No. 840292, Cong. & Expos. Meetings, Detroit, MI (1984).
4429. E. P. Patrick, J. R. Autil, and T. S. Sun, S.A.E. Technical Paper No. 840291, Cong. & Expos. Meetings, Detroit, MI (1984).
4430. A. Seeds, D. Nardini, and F. Cassese, The development of a center cell structure in bonded aluminum for the Ferrari 408 research vehicle, S.A.E. Tech. Paper No. 890717, Cong. & Expos. Meetings, Detroit, MI (Feb. 27–March 3, 1989).
4431. W. F. Marwick and P. G. Sheasby, Evaluation of adhesives for aluminum structured vehicles, S.A.E. Tech. Paper No. 870151, Cong. & Expos. Meetings, Detroit, MI (Feb. 23–27, 1987).
4432. D. Nardini and A. Seeds, Structural design considerations for bonded aluminum structured vehicles, S.A.E. Tech. Paper No. 890716, Cong. & Expos. Meetings, Detroit, MI (Feb. 27–March 3, 1989).
4433. M. J. Bull, A. R. Carr, P. Payne, and C. Suffell, The application of a knowledge based CAD system to press tool design, S.A.E. Tech. Paper No. 900797, Cong. & Expos. Meetings, Detroit, MI (Feb. 26–March 2, 1990).
4434. I. J. McGregor, A. D. Seeds, and D. Nardini, The design of impact absorbing members for aluminum structured vehicles, S.A.E. Tech. Paper No. 900796, Cong. & Expos. Meetings, Detroit, MI (Feb. 26–March 2, 1990).
4435. D. Nardini, I. J. McGregor, and A. D. Seeds, Analysis and testing of adhesively bonded aluminum structural components, S.A.E. Tech. Paper No. 900795, Cong. & Expos. Meetings, Detroit, MI (Feb. 26–March 2, 1990).
4436. E. M. Vader, unpublished work, Aluminum Co. of America.
4437. L. H. Linebarrier, Adhesives and fasteners join in rivet bonded assembly, Mater. Eng. (Sept. 1975).
4438. J. D. Minford, unpublished work, Aluminum Co. of America.
4439. J. D. Minford, unpublished work, Aluminum Co. of America.
4440. Epoxy repairs casting (refrigerator car maintenance), Adhes. Age 5(5):25 (1962).
4441. Floor covering bonded to tracked vehicles, Adhes. Age 9(4):31 (1966).
4442. Versatile adhesive automates production of mobile homes, Adhes. Age 9(7):33 (1966).
4443. Truck bodies quickly repaired with metal-backed pressure-sensitive tape, Adhes. Age 9(12):39 (1966).
4444. Apache Buffalo camping trailer utilizes high-resistant adhesive, Adhes. Age 10(1):31 (1967).
4445. Metal foil tape fixes trucks, Adhes. Age 12(11):33 (1969).
4446. C. Sullivan, PVC foam tape new sealant for recreational vehicles, Adhes. Age 16(4):17 (1973).
4447. Adhesives for recreational vehicles kept uniform with portable mixers, Adhes. Age 17(7):30 (1974).
4448. Adhesive/sealant cuts riveting and prevents leaks for hauler, Adhes. Age 30(13):26 (1987).
4449. C. S. Adderley, Mater. Des. 9(5):289 (1988).
4450. Auto Body Repair Manual, Publication 28326e, Ciby-Geigy Corp.
4451. L. K. Peters, unpublished work, Aluminum Co. of America.
4452. Repair (epoxy) instructions for aluminum automotive radiators, Service Bulletin HES-62-31, Harrison Radiator Division, General Motors, Lockport, NY (April 1962).
4453. J. D. Minford, unpublished work, Aluminum Co. of America.
4454. L. S. Buchoff, Adhesives in the electronic/electrical industry, what we need? Adhes. Age 16(9):38 (1973).
4455. R. E. Batson, Industrial applications for adhesives, Adhesives and Sealants, Engineered Materials Handbook, ASM, Materials Park, OH, Vol. 3, pp. 574,575 (1990).
4456. N. J. DeLollis, The use of adhesives and sealants in electronics, IEE Trans. on Parts, Materials, and Packaging, PMP-1:4 (Dec. 1965).

4457. H. K. Charles, Jr., Electronic packaging applications for adhesives and sealants, Adhesives and Sealants, Engineered Materials Handbook, ASM, Materials Park, OH, Vol. 3, pp. 579–603 (1990).
4458. G. Moore, VLSI, what does the future hold? Electron. Aust. 42:14 (1980).
4459. S. M. Sze, VLSI Technology, McGraw-Hill, New York (1983).
4460. C. P. Wong, Integrated circuit device encapsulants, Polymers for Electronic Applications (J. H. Lai, ed.), CRC Press, Boca Raton, FL, p. 63 (1989).
4461. C. P. Wong, Integrated circuit encapsulants, Polymers in Electronics, John Wiley & Sons, New York, p. 638 (1986).
4462. J. Czochralski, Z. Phys. Chem. 92:219 (1918).
4463. E. P. A. Metz, R. C. Miller, and R. Mazelsky, J. Appl. Phys. 33:2016 (1962).
4464. H. K. Charles, Jr. and G. V. Clatterbaugh, Thin-film hybrids, Packaging Electronic Materials Handbook, ASM, Materials Park, OH, Vol. 1, p. 313 (1989).
4465. A. B. Glasser and G. E. Subak-Sharpe, Diffusion, Integrated Circuit Engineering, Design, Fabrication, and Applications, Addison-Wesley, Reading, MA, p. 197 (1979).
4466. L. Holland, Vacuum Deposition of Thin Films, Chapman and Hall (1966).
4467. B. Chapman, Glow Discharge Processes, John Wiley & Sons, New York (1980).
4468. C. S. Pai, E. Cabreros, S. S. Lau, T. E. Seidel, and I. Suni, Appl. Phys. Lett. 47(7):652 (1985).
4469. J. M. Poate, K. N. Tu, and J. W. Meyer (eds.), Thin Films-Interdiffusion and Reactions, John Wiley & Sons, New York (1987).
4470. J. J. Steppan, J. A. Roth, L. C. Hall, D. A. Jeannotte, and S. P. Carbone, A review of corrosion failure mechanisms during accelerated tests: Electrolytic metal migration, J. Electrochem. Soc. Solid-State Sci. Technol., p. 175 (Jan. 1987).
4471. N. G. Einspruch and G. B. Larrabee (eds.), VLSI Electronics: Microstructure Science, Academic Press, New York, Vol. 6 (1983).
4472. J. H. Pugh and R. S. Williams, J. Mater. Res. 1(2):343 (1986).
4473. R. Beyers, K. B. Kim, and R. Sinclair, J. Appl. Phys. 61(6):2195 (1987).
4474. E. Wolfgang, Future trends in electron beam testing, Microelectron Eng. 7:434 (1987).
4475. H. K. Charles, Jr., Electrical interconnection, Packaging, Electronic Materials Handbook, ASM, Materials Park, OH, Vol. 1, p. 224 (1989).
4476. C. J. Leedecke, P. C. Baird, and K. D. Orphanides, Glass-to-metal seals, Packaging, Electronic Materials Handbook, ASM, Materials Park, OH, Vol. 1, p. 455 (1989).
4477. O. M. Uy and R. C. Benson, Package sealing and passivation coatings, Packaging, Electronic Materials Handbook, ASM, Materials Park, OH, Vol. 1, p. 237 (1989).
4478. S. A. Gee, W. F. VandenBogert, V. R. Akylas, and R. T. Shelton, Strain gauge mapping of die surface stresses, Proc. 39th Electronic Components Conf., IEEE, p. 343 (1989).
4479. H. K. Charles, Jr. and G. V. Clatterbaugh, Solder joint reliability-design implications from finite element modeling and experimental testing, ASME Annual Winter Meeting, Paper 89-WA/EP-22 (Dec. 1988).
4480. D. J. Bodendorf, K. T. Olsen, J. F. Trenko, and J. R. Ninnard, Active silicon chip carrier, IBM Tech. Disclosure Bull. 15(2):656 (July 1972).
4481. C. H. Mitchell and H. Berg, Thermal studies of a plastic dual-in-line package, IEEE Trans. Compon. Hybrids Manuf. Technol. CHMT-2 (4) (1979).
4482. C. V. Clatterbaugh and H. K. Charles, Jr., Design optimization and reliability testing of surface mount solder joints, Int. J. Hybrid Microelectronics 8:31 (1985).
4483. C. V. Clatterbaugh and H. K. Charles, Jr., Thermochemical behavior of soldered interconnects for surface mountings—a comparison of theory and experiment, Proc. 39th Electronic Components Conf., IEEE, p. 60 (May 1985).
4484. W. Borland, Thick film hybrids, Packaging, Electronic Materials Handbook, ASM, Materials Park, OH, Vol. 1, p. 332 (1989).
4485. H. Nakahara, Complex multilayer boards vie for space in Japanese computers, Electron. Packag. Prod. 27(2):70 (Feb. 1987).
4486. R. R. Tummala and R. J. Rymaszewski (eds.), Microelectronic Packaging Handbook, Van Nostrand Reinhold, New York (1989).
4487. H. W. Markstein, Laminates support technology advances while offering alternate choices, Electron. Packag. Prod. 24(6):83 (1984).
4488. M. Schlack, Guide to high performance engineering plastics, Plastics World 45(4):30 (1987).
4489. M. K. Antoon, B. E. Zehner, and J. L. Koenig, Spectroscopic determination of in-situ composition of epoxy matrices in glass-fiber-reinforced components, Polym Compos. 1(1):24 (1980).

4490. C. P. Wong, Integrated circuit device encapsulants, Polymers for Electronic Applications (J. H. Lai, ed.), CRC Press, Boca Raton, FL, p. 69 (1989).
4491. S. Yamanaka, T. Ibara, T. Maeda, T. Takikawa, and H. Yoshino, Applications of ceramic thin film technology in hybrid microelectronics, Proc. Int. Microelectronics Symp., Int. Soc. for Hybrid Microelectronics, p. 439 (Oct. 1989).
4492. R. W. Berry, P. M. Hall, and M. T. Harris, Thin Film Technology, Van Nostrand, New York, p. 271 (1986).
4493. L. A. Maissel and R. Glang, Handbook of Thin Films, McGraw-Hill, New York, Chaps. 3, 4 (1970).
4494. A. C. Adams, Dielectric and polysilicone film deposition, VLSI Technology (S. M. Sze, ed.), McGraw-Hill, New York, p. 73 (1983).
4495. W. Kern and V. S. Ban, Chemical vapor deposition of inorganic thin films, Thin Film Processes (J. L. Vossen and W. Kern, eds.), Academic Press, New York, p. 257 (1978).
4496. W. Kern and G. L. Schnable, Low pressure chemical vapor deposition for very large scale intergration processing—a review, IEEE Trans. Electron. Dev., Vol. ED-26, p. 647 (1979).
4497. J. M. Meyer and S. S. Lau, Electronic Materials Sciences: For Integrated Circuits in Silicon and GaAs, Macmillan, New York, p. 283 (1990).
4498. A. C. Adams and C. D. Capio, The deposition of silicon dioxide films at reduced pressure, J. Electrochem. Soc. 126:1042 (1979).
4499. W. A. Pliskin, Comparison of properties of dielectric films deposited by various methods, J. Vac. Sci. Technol., 14:1064 (1977).
4500. H. K. Charles, Electronic packaging applications for adhesives and sealants, Sealants and Adhesives, Engineered Materials Handbook, ASM, Materials Park, OH, Vol. 3, pp. 594–596 (1990).
4501. Idem., pp. 596–597.
4502. Idem., pp. 597–599.
4503. Idem., pp. 599–600.
4504. Idem., pp. 600–601.
4505. Idem., pp. 601–602.
4506. Epoxy-sealed marine radios, Adhes. Age 3(11):28 (1960).
4507. Bonding computer components, Adhes. Age 3(5):42 (1960).
4508. Epoxy for electrical applications, Adhes. Age 3(5):30 (1960).
4509. D. K. Rider, Adhesive clips facilitate telephone wire installation, Adhes. Age 4(4):20 (1961).
4510. Adhesive-coated copper simplifies production of die-stamped circuits, Adhes. Age 4(12):20 (1961).
4511. R. C. Snogren, Adhesives bond heat sinks to printed circuit boards, Adhes. Age 11(6):21 (1968).
4512. J. E. Hess and L. C. Jackson, Designing adhesives and interfaces to optimum electrical conductivity, Adhes. Age 12(11):41 (1969).
4513. C. W. Armstrong, New concepts in adhesive bonding (laminated structural materials, paper, foil, fabric lamination, and motor wire bonding), Adhes. Age 15(2):33 (1972).
4514. H. C. Selya, Encapsulation of electrical units with thermosets, Adhes. Age 16(9):46 (1973).
4515. D. L. Holland, Adhesives for flexible printed circuits, Adhes. Age 16(4):17 (1973).
4516. New silver epoxy speeds electronic production, Adhes. Age 17(6):35 (1974).
4517. E. J. Leech, Adhesive coated laminates for additional circuits, Adhes. Age 18(6):31 (1975).
4518. W. J. O'Malley, Silicone pressure-sensitive adhesives for flexible printed circuits, Adhes. Age 18(6):17 (1975).
4519. H. Heuring, Bonding RF absorbent material to microwave enclosure covers, Adhes. Age 22(10):44 (1979).
4520. UV adhesive lowers magnet assembly cost, Adhes. Age 25(5):40 (1982).
4521. Simplified wiring in auto printed circuit boards, Adhes. Age 26(1):37 (1983).
4522. N. Schultz, Potting and encapsulating electronic circuits, Adhes. Age 26(8):19 (1983).
4523. S. M. Pandiri, The behavior of silver flakes in conductive epoxy adhesives, Adhes. Age 30(11):31 (1987).
4524. M. Moreau, UV curing cuts time and costs for solenoid manufacturer, Adhes. Age 31(4):18 (1988).
4525. D. Leone and A. Noyes, Automated potting reduces urethane waste and rework (transducers), Adhes. Age 33(11):26 (1990).
4526. M. Clark, Tapes and structural bonding reduce radar unit cost, Adhes. Age 34(7):28 (1991).
4527. Turn-key potting benefits OEM electronics producer, Adhes. Age 34(3):18 (1991).
4528. H. Kraus, Adhesives for Microelectronics, Abelstik Labs.
4529. F. Kulesza, Conductive epoxy solves surface mount problem, Electronic Prod., p. 83 (March 1984).
4530. Conductive epoxy is tested for SMT solder replacement, Electronics Packaging Prod. (Feb. 1985).
4531. Thermally conductive acrylic bonds solid state relays, Adhes. Age 29(11):28 (1986).

4532. J. Coleman, Sticky issues in electronic assembly, Assembly Eng., p. 29 (1989).
4533. R. D. Hermansen and S. A. Tunick, Formulating custom-tailored thermal transfer adhesives, Adhes. Age 32(1):38 (1989).
4534. J. C. Jagt, Adhesive concepts for the fixation of electronic components, Adhesion '90, Plastics and Rubber Inst. Meetings, Cambridge, England (1990).
4535. R. L. McKenzie, Metal curtain wall sealants, Adhes. Age 3(5):28 (1960).
4536. L. Sheppard, Jr., Use of adhesives in preparing prefinished surfaces for the housing industry, Adhes. Age 4(10):26 (1961).
4537. Computer room's floating floor bonded in place, Adhes. Age 4(6):35 (1961).
4538. Bonded panels for a city under ice, Adhes. Age 4(8):38 (1961).
4539. Ceramic-to-glass bonds in 60 seconds, Adhes. Age 4(11):33 (1961).
4540. Epoxy helps put up a good front (redecorating store fronts), Ahdes. Age 5(10):35 (1962).
4541. Unusual roof design requires new coating technique, Adhes. Age 5(3):25 (1962).
4542. Bonding prefab panels of mosaic tiles, Adhes. Age 5(4):28 (1962).
4543. Caulking and sealing Chicago's new airport, Adhes. Age 5(6):28 (1962).
4544. Reinforced bonded handles for multiwalls, Adhes. Age 5(12):35 (1962).
4545. Sealing stained glass, Adhes. Age 5(6):27 (1962).
4546. W. P. Ellis, Adhesives for thermal insulation, Adhes. Age 5(4):22 (1962).
4547. J. V. Fitzgerald, Adhesives for ceramic tile, Adhes. Age 5(11):34 (1962).
4548. Building panels strengthened with wide-range adhesive, Adhes. Age 8(5):26 (1965).
4549. Adhesives for wallboard, Adhes. Age 8(5):30 (1965).
4550. Pedestal floor constructed with contact adhesive, Adhes. Age 8(10):34 (1965).
4551. Chemical bonding of stained glass windows, Adhes. Age 8(6):35 (1965).
4552. Pavillion constructed with bonded sandwich panels proves successful, Adhes. Age 9(2):28 (1966).
4553. Window wall systems utilize a variety of sealers, Adhes. Age 9(4):35 (1966).
4554. Sealants play a vital part in building the largest structure, Adhes. Age 10(2):34 (1967).
4555. Commercial and private buildings utilize a one-part epoxy system, Adhes. Age 10(4):35 (1967).
4556. Aggregate epoxy protects facade, Adhes. Age 10(7):37 (1967).
4557. Experimental arch house features adhesive bonded stressed skin roof, Adhes. Age 10(9):25 (1967).
4558. J. H. Valentine, Glass and glazing joints, Adhes. Age 12(4):19 (1969).
4559. New butyl rubber offers advantages in window and construction sealant, Adhes. Age 12(9):30 (1969).
4560. A. F. Santaniello, Polysulfide sealant meets the needs of the World Trade Center, Adhes. Age 15(11):32 (1972).
4561. Roll-out roofing meets need of modular home construction, Adhes. Age 15(2):40 (1972).
4562. Construction adhesive saves time in two building applications, Adhes. Age 15(4):34 (1972).
4563. Adhesive dispensing system cuts prefab housing costs, Adhes. Age 15(10):36 (1972).
4564. Silicone sealant waterproofs spires of museum towers, Adhes. Age 15(12):27 (1972).
4565. M. Meinertzhagen, Using adhesives in thermal insulation, Adhes. Age 16(6):31 (1973).
4566. T. G. Chase, Multipurpose construction adhesive increases structural soundness and minimizes costs, Adhes. Age 16(6):42 (1973).
4567. G. B. Lowe, Structural adhesives and compounds for use in the construction industry, Adhes. Age 16(12): 41 (1973).
4568. Crosslinked butyl solves roofing problem, Adhes. Age 16(2):34 (1973).
4569. Primer/conditioner seals concrete floor in space conversion, Adhes. Age 16(4):35 (1973).
4570. Manufacturing dual-seal insulating glass units, Adhes. Age 21(11):30 (1978).
4571. C. C. Booth, A guide to building construction adhesives, Adhes. Age 22(2):31 (1979).
4572. J. Bax, Formulating solvent-free, water-resistant adhesives for ceramic tiles, Adhes. Age 24(10):22 (1981).
4573. Sealant enhances building design, Adhes. Age 25(7):36 (1982).
4574. F. A. Slautterback, Monterey's $32 million aquarium, bonding fiberglass with hot-melt, Adhes. Age 26(13):15 (1983).
4575. G. J. Bouchey, Task group examines glazing for insulating glass units, Adhes. Age 27(8):27 (1984).
4576. L. B. Sandberg and J. M, Klosowski, Structural glazing: Behavior details of double-bead installations, Adhes. Age 29(5):26 (1986).
4577. J. G. Halbmaier, Fast cure hot-melt increases Parquet flooring assembly, Adhes. Age 33(12):17 (1990).
4578. Waterborne solves builder's production and EPA problems, Adhes. Age 34(7):36 (1991).
4579. Epoxy grout compound bonds pontoon sections of floating bridge, Adhes. Age 4(9):40 (1961).
4580. Bridge drains caulked with iron cement, Adhes. Age 5(8):30 (1962).
4581. Epoxy crack injection repairs highway bridge seat beam, Adhes. Age 11(9):34 (1968).

4582. J. B. McNicholas, A Critical Study of Engineering Joints in Aluminum Alloy, Ph.D. Thesis, University of Salford (1969).
4583. J. B. McNicholas, Bonded joints in primary structures, Institute of Structural Engineers and Sheffield University Conference, Sheffield, England (July 1970).
4584. B. H. Williams, Adhesive joints in engineering structures, Institute of Structural Engineers and Sheffield University Conference, Sheffield, England (July 1970).
4585. J. B. McNicholas, High strength adhesive bonding, Welding and Metal Fabrication, p. 103 (March 1971).
4586. J. B. Thornton, Georgia highway joint sealing, Adhes. Age 26(9):24 (1983).
4587. J. P. Cook, Joint and crack sealing needed for bridge, highway upkeep, Adhes. Age 28(5):29 (1985).
4588. R. W. Gaul and A. J. Apton, Epoxy adhesives in concrete, Adhes. Age 3(5):24 (1960).
4589. E. E. Fender, Procedures for three types of industrial Formica bonds, Adhes. Age 4(10):30 (1961).
4590. Bonding in cargo container assembly, Adhes. Age 5(3):39 (1962).
4591. Adhesive bonding tips and ferrules to flexible tubing, Adhes. Age 5(4):37 (1962).
4592. J. Panek, Causes of joint sealant failure, Adhes. Age 5(7):23 (1962).
4593. Bonded aluminum railing reduces costs, Adhes. Age 8(5):31 (1965).
4594. Bonding soundproofing pads to folding steel-door panels, Adhes. Age 8(11):29 (1965).
4595. How an oil company reduces maintenance costs with a resin adhesive, Adhes. Age 10(4):28 (1967).
4596. All-glass aquarium assembled with silicone sealant, Adhes. Age 15(1):36 (1972).
4597. Adhesive recovers mosaic, Adhes. Age 15(10):26 (1972).
4598. Tunnel walls sealed with silicone sealant, Adhes. Age 16(1):40 (1973).
4599. Neoprene-based adhesive bonds shock-absorbing doors, Adhes. Age 20(7):34 (1977).
4600. Adhesives stand up to artic conditions on Alaskan pipeline, Adhes. Age 21(1):32 (1978).
4601. U. C. Jindal, Adhesives and stress distribution of a plate with reinforced hole, Adhes. Age 26(8):25 (1983).
4602. E. P. McGuire, Packaging adhesives and the food law, Adhes. Age 4(11):30 (1961).
4603. E. C. Lenz, Six tips on selecting adhesives for small carton sealing, Adhes. Age 4(6):34 (1961).
4604. High speed sealing of polyethylene bags, Adhes. Age 4(11):28 (1961).
4605. W. T. Hutchins, Adhesion problems and food packaging, Adhes. Age 5(10):23 (1962).
4606. R. G. Fisher, Packages for adhesives, Adhes. Age 5(11):28 (1962).
4607. Adhesive bonded food containers, Adhes. Age 5(8):27 (1962).
4608. S. Sacharow, The adhesive needs of the flexible packager, Adhes. Age 8(10):26 (1965).
4609. S. Sacharow, The role of adhesives in packaging laminations, Adhes. Age 9(7):30 (1966).
4610. D. Debnar, An introduction to adhesives used in the packaging and converting industries, Adhes. Age 9(2):30 (1966).
4611. F. P. Bartlett, Adhesives for films and foils, Adhes. Age 9(2):34 (1966).
4612. W. Meyer, Equipment and processes for paper bonding, Adhes. Age 9(9):22 (1966).
4613. E. Murray, Raw materials and adhesives for paper bonding, Adhes. Age 10(9):22 (1967).
4614. Hot-melt applicator cuts packaging costs, Adhes. Age 10(7):29 (1967).
4615. G. A. Egli, How to select adhesives for foil/paper laminations, Adhes. Age 14(9):38 (1971).
4616. P. Lambert, Recent developments in adhesives for flexible packaging, Adhes. Age 16(7):22 (1973).
4617. Automated glue system raises canning production, Adhes. Age 16(4):38 (1973).
4618. G. M. Estes and D. Kuespert, The future of adhesives use in packaging and packaging materials, Adhes. Age 20(1):35 (1977).
4619. Packaging converter changes to water-borne adhesives, Adhes. Age 24(6):41 (1981).
4620. B. Lawson, Custom packaging, Adhes. Age 26(3):31 (1983).
4621. D. J. Bentley, Jr., Overcoming substrate problems in water-borne adhesives, Adhes. Age 29(2):22 (1986).
4622. P. Fahrendorf, Coater/laminator permits material process flexibility, Adhes. Age 32(11):34 (1989).
4623. L. Maempel and D. I. Fricke, One-part aqueous dispersion for glassy film laminating, Adhes. Age 34(2): 17 (1991).
4624. G. Meyer, Adhesive applications at Kodak, Adhes. Age 4(8):20 (1961).
4625. A. P. Alexander, C. F. Langenhagen, Jr., and M. Nierenberg, Epoxy bonding cuts costs of typewriter production, Adhes. Age 4(11):22 (1961).
4626. Adhesive bonding speeds production at Smith-Corona, Adhes. Age 4(4):25 (1961).
4627. Room temperature formulation bonds Alnico-to-Aluminum, Adhes. Age 4(4):31 (1961).
4628. Epoxy repairs compressors cracked casing, Adhes. Age 4(8):29 (1961).
4629. Minute quantities of epoxy bond miniaturized microphones, Adhes. Age 4(6):28 (1961).
4630. Bonds for binoculars, Adhes. Age 4(10):23 (1961).

4631. W. A. Carr, R. Dunaltz, and C. Shiners, Frozen catalyzed epoxy solves instrument bonding problem, Adhes. Age 5(3):22 (1962).
4632. Panel assembly with versatile sprayable adhesive formulation, Adhes. Age 5(4):26 (1962).
4633. Fast-setting adhesive sparks sound of music (electric organ), Adhes. Age 5(5):34 (1962).
4634. Bonded mirror-bright reflectors (sheet aluminum), Adhes. Age 5(7):32 (1962).
4635. J. J. Madden, A. S. Burhans, and C. F. Pitt, Epoxy resins as bonding agents for filament wound structures, Adhes. Age 8(6):21 (1965).
4636. Air conditioner quality improved through use of adhesives, Adhes. Age 8(9):38 (1965).
4637. Stereo radio units strengthened with film adhesives, Adhes. Age 9(5):30 (1966).
4638. Fast-setting adhesive aids testing devices, Adhes. Age 9(7):26 (1966).
4639. Bonding baseboard heating units reduces noise and fatigue problems, Adhes. Age 9(11):27 (1966).
4640. Nameplate manufacturer improves performance with film adhesive, Adhes. Age 10(6):34 (1967).
4641. Using adhesive for press fitted parts, Adhes. Age 10(7):26 (1967).
4642. E. A. Boran, Using hot melt adhesives in automated assembly of TV coils, Adhes. Age 12(9):39 (1969).
4643. Sealants help perfect washer-dryer systems, Adhes. Age 11(4):34 (1968).
4644. Glass-to-metal bonding of TV tubes, Adhes. Age 17(2):22 (1974).
4645. M. J. Dietrick, D. E. Emnert, and D. A. Lederer, Optimization of wire adhesion by varying concentration of bonding agents and cure rate, Adhes. Age 20(1):27 (1977).
4646. R. D. Paul and H. G. Wimarski, Polyimide adhesive for composite fan blades, Adhes. Age 21(4):31 (1978).
4647. Acrylic system used by stereo speaker/wood beam makers, Adhes. Age 23(4):36 (1980).
4648. P. J. Bowles and L. W. Garrett, An appliance case history, Adhes. Age 26(4):26 (1983).
4649. T. DeBarros and M. Beachle, Bonding parabolic reflector flood and spot lights, Adhes. Age 32(3):42 (1989).
4650. L. D. Osterberger and W. D. Arendt, High melt point additive improves product assembly HM's, Adhes. Age 34(12):22 (1991).
4651. D. Francis, Plastic process and materials selection, MDDI, pp. 45–49 (Nov. 1987).
4652. M. H. Kemp, Pressure sensitive surgical tapes—their history and development, Adhes. Age 4(12):22 (1961).
4653. R. C. Page, Tissue adhesive eliminates sutures and staples in many types of surgery, Adhes. Age 9(12):27 (1966).
4654. K. Krug and N. M. Marecki, Porous and other medical PSA's, Adhes. Age 26(12):19 (1983).
4655. K. E. Cobian, A. H. Jevne, J. C. Schrantz, A. J. Coury, P. T. Callahan, and J. M. Kendall, Epoxy adhesives for implantable medical devices, Adhes. Age 27(13):17 (1984).
4656. W. R. Pfister, Silicone PSA's offer flexibility for medical, pharmaceutical use, Adhes. Age 33(13):20 (1990).
4657. Laminated tennis racket, Adhes. Age 5(11):37 (1962).
4658. Chemically fastened screws resist outboard motor vibration, Adhes. Age 5(7):40 (1962).
4659. Sealant improves quality for sailboat manufacturer, Adhes. Age 29(5):30 (1986).
4660. Boat builder saves time with foam mounting tape, Adhes. Age 29(10):32 (1986).
4661. W. W. Fawcett, Three epoxy systems for bonding carbide wear parts, Adhes. Age 4(3):22 (1961).
4662. Adhesives solve high-speed polishing problems, Adhes. Age 4(3):34 (1961).
4663. H. S. Johnson, Adhesive-bonded hose products withstand "torture testing," Adhes. Age 5(9):22 (1962).
4664. A. B. Middleton, Bonding foundry cores with sodium silicate adhesives, Adhes. Age 5(12):24 (1962).
4665. Blending adhesives for high-speed photographic film, Adhes. Age 5(2):28 (1962).
4666. Adhesive coated steel strip and wire opens new adhesive markets (rubber-to-metal applications), Adhes. Age 8(8):28 (1965).
4667. Silicone rubber bonds solar cells, Adhes. Age 9(1):21 (1966).
4668. How epoxy adhesive simplifies fabrication of diamond wheels, Adhes. Age 9(11):33 (1966).
4669. Bonding of wet suits reduces repair cost, Adhes. Age 10(2):38 (1967).
4670. Durability of sponge mops traced to neoprene adhesives, Adhes. Age 10(8):41 (1967).
4671. Fabricating epoxy resin wind instruments, Adhes. Age 11(5):29 (1968).
4672. Cyanoacrylate adhesive aids in do-it-yourself horse shoeing, Adhes. Age 16(2) (1973).
4673. J. C. Hunt, Small hot-melt systems move into foundaries, Adhes. Age 18(5):34 (1975).
4674. J. Kozaliewicz and P. Penczak, Polyepichlorohydrin elastomer solvent-based adhesives for the footwear industry, Adhes. Age 20(7):29 (1977).
4675. D. F. Merrill, Restoring works of art with silicone PSA's, Adhes. Age 24(2):39 (1981).
4676. G. A. Berger, Art adhesive needs stability, reversibility, compatibility, Adhes. Age 28(3):30 (1985).
4677. Acrylic PSA enhances liquid crystal products, Adhes. Age 29(10):30 (1986).

4678. Adhesives improve brewer's product quality and handling, Adhes. Age 29(12):34 (1986).
4679. B. T. Lyons, Ceramic adhesive replaces asbestos in sealing furnaces, Adhes. Age 33(7):22 (1990).
4680. Adhesives to the rescue: bonding lifesaving equipment, Adhes. Age 34(3):44 (1991).
4681. Bonded steel plates support 1250 pound load, Adhes. Age 4(11):29 (1961).
4682. F. Tons, Materials and geometry in joint seals, Adhes. Age 8(9):22 (1965).
4683. New applications for family of foam tapes, Adhes. Age 9(8):37 (1966).
4684. N. F. Huber and M. G. Strobel, Applications and future uses of capsular adhesives, Adhes. Age 10(11): 28 (1967).
4685. Y. H. Chao and F. P. Hoenisch, Two-part acrylic enhances vacuum forming operation, Adhes. Age 29(2):18 (1986).
4686. Urethane hot-melt doubles subassembly production, Adhes. Age 33(9):48 (1990).

Note: Some references were edited out of the text and removed from the reference list. They are reference numbers 870, 884, 2281, and 4281.

Author Index

Key: Author, reference number(page on which reference is cited).

Subject Index